Svalbard (Norway)
Arctic Circle
See inset below
RUSSIA
ASIA
North Sea
EUROPE
KAZAKHSTAN
MONGOLIA
Kuril Is. (Russia)
GEORGIA
ARMENIA
UZBEKISTAN
KYRGYZSTAN
TURKMENISTAN
TAJIKISTAN
TURKEY
NORTH KOREA
JAPAN
SOUTH KOREA
TUNISIA
LEBANON
SYRIA
AZERBAIJAN
CHINA
ISRAEL
IRAQ
IRAN
AFGHANISTAN
JORDAN
PACIFIC OCEAN
ALGERIA
LIBYA
EGYPT
KUWAIT
BAHRAIN
PAKISTAN
NEPAL
BHUTAN
QATAR
UNITED ARAB EMIRATES
SAUDI ARABIA
INDIA
TAIWAN
Tropic of Cancer
MYANMAR (BURMA)
OMAN
BANGLADESH
LAOS
MALI
NIGER
CHAD
SUDAN
ERITREA
YEMEN
Wake Island (U.S.)
THAILAND
VIETNAM
Northern Mariana Is. (U.S.)
MARSHALL ISLANDS
AFRICA
DJIBOUTI
NIGERIA
BENIN
PHILIPPINES
Guam (U.S.)
CENTRAL AFRICAN REP.
ETHIOPIA
SRI LANKA
FEDERATED STATES OF MICRONESIA
CAMBODIA
PALAU
TOGO
CAMEROON
BRUNEI
SOMALIA
MALAYSIA
UGANDA
KENYA
MALDIVES
KIRIBATI
GABON
RWANDA
SINGAPORE
Equator
CONGO, REP.
CONGO, DEM. REP.
BURUNDI
NAURU
INDONESIA
PAPUA NEW GUINEA
SOLOMON ISLANDS
TANZANIA
SEYCHELLES
INDIAN OCEAN
EAST TIMOR
ANGOLA
MALAWI
COMOROS
TUVALU
ZAMBIA
MOZAMBIQUE
VANUATU
FIJI
NAMIBIA
ZIMBABWE
MADAGASCAR
MAURITIUS
BOTSWANA
Réunion (France)
New Caledonia (France)
AUSTRALIA
SOUTH AFRICA
SWAZILAND
LESOTHO
NEW ZEALAND
Prime Meridian
Kerguelen Is. (France)
Antarctic Circle
ANTARCTICA
Europe
FINLAND
NORWAY
SWEDEN
ESTONIA
IRELAND
UNITED KINGDOM
North Sea
DENMARK
LATVIA
LITHUANIA
RUSSIA
NETHERLANDS
BELARUS
ATLANTIC OCEAN
BELGIUM
GERMANY
POLAND
LUXEMBOURG
CZECH REPUBLIC
FRANCE
LIECHTENSTEIN
SLOVAKIA
UKRAINE
SWITZERLAND
AUSTRIA
HUNGARY
MOLDOVA
SLOVENIA
CROATIA
ROMANIA
MONACO
SAN MARINO
BOSNIA AND HERZEGOVINA
PORTUGAL
ANDORRA
SPAIN
Corsica (France)
YUGOSLAVIA
BULGARIA
Black Sea
GEORGIA
Balearic Is. (Spain)
ITALY
ALBANIA
MACEDONIA
Gibraltar (U.K.)
Mediterranean Sea
Sardinia (Italy)
GREECE
TURKEY
ASIA
0 250 500 mi
0 250 500 km
Sicily (Italy)
MALTA
Crete (Gr.)
CYPRUS
SYRIA
LEBANON

The WORLD ECONOMY

For Josie

The WORLD ECONOMY

RESOURCES,

LOCATION, TRADE,

AND DEVELOPMENT

Fifth Edition

FREDRICK P. STUTZ
San Diego State University

AND

BARNEY WARF
Florida State University

Upper Saddle River, NJ 07458

Library of Congress Cataloging-in-Publication Data

Stutz, Fredrick P.
The world economy : resources, location, trade and development/ Fredrick P. Stutz and Barney Warf. —5th ed.
p. cm.
ISBN 0-13-243689-2
1. Economic geography. 2. Economic history I. Warf, Barney II. Title
HC59.S8635 2007
330.9—dc22

2006035255

Publisher, Geosciences: *Dan Kaveney*
Associate Editor: *Amanda Brown*
Executive Managing Editor: *Kathleen Schiaparelli*
Assistant Managing Editor: *Beth Sweeten*
Production Editor: *Pine Tree Composition/Patty Donovan*
Editorial Assistant: *Jessica Neumann*
Manufacturing Buyer: *Alan Fischer*
Senior Managing Editor, AV Production & Management: *Patricia Burns*
Manager, Production Technologies: *Mathew Haas*
Managing Editor, Art Management: *Abigail Bass*
Art Production Editor: *Rhonda Aversa*
Illustrations: *Argosy Publishing*
Cartography: *The GeoNova Group*
Director of Creative Services: *Paul Belfanti*
Art Director: *Jayne Conte*
AV Editor: *Rhonda Aversa*
Director, Image Resource Center: *Melinda Patelli*
Manager, Rights and Permissions: *Zina Arabia*
Manager, Visual Research: *Beth Brenzel*
Image Permission Coordinator: *Debbie Latronica*
Photo Researcher: *Yvonne Gerin*
Manufacturing Manager: *Alexis Heydt-Long*
Manufacturing Buyer: *Alan Fischer*
Cover Design: *Bruce Kenselaar*
Cover Photo: Aerial view of cargo containers Michael S. Yamashita/CORBIS

Pearson Prentice Hall
Pearson Education, Inc.
Upper Saddle River, New Jersey 07458

Printed in the United States of America

10 9 8 7 6 5 4 3 2

ISBN 0-13-243689-2

Pearson Education Ltd., *London*
Pearson Education Australia Pty. Ltd., *Sydney*
Pearson Education Singapore, Pte. Ltd.
Pearson Education North Asia Ltd., *Hong Kong*
Pearson Education, Canada, Ltd., *Toronto*
Pearson Educación de Mexico, S.A. de C.V.
Pearson Education–Japan, *Tokyo*
Pearson Education Malaysia, Pte. Ltd.

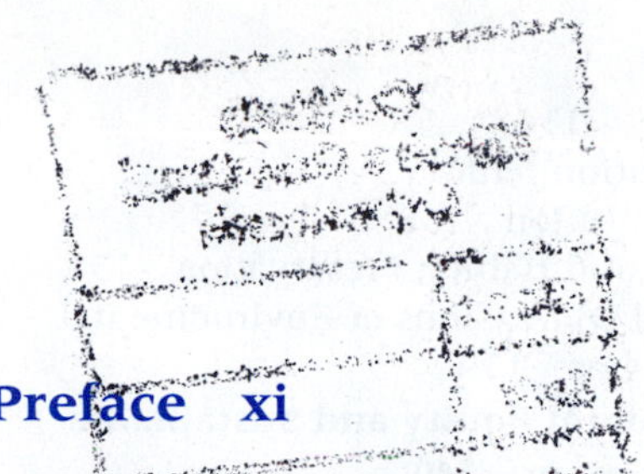

CONTENTS

PREFACE

The World Economy: Resources, Location, Trade, and Development, now in its fifth edition, offers an overview of the field of economic geography and its linkages to related issues of development and underdevelopment, international business, and the global economy. In an age of increasing globalization, an understanding of these issues is central to both liberal arts and professional educations, for the concerned voter to the engaged business practitioner.

This work is designed as a comprehensive introduction to the ways in which economic activity is stretched over the space of the earth's surface. Economists too rarely take the spatial dimension seriously, a perspective that implies all economic activity occurs on the head of a pin. Geographers, in contrast, are interested in the manner in which social relations and activities occur unevenly over space, the ways in which local places and the global economy are intertwined, and the difference that location makes to how economic activity is organized. No social process occurs in exactly the same way in different places; thus, where and when economic activity occurs has a profound influence on *how* it occurs. Space, then, can no longer be relegated to the sidelines. As globalization has made small differences among places increasingly important, space has become more, not less, important.

This new edition differs from the previous one in several respects. It has updated empirical data found throughout. Some traditional material has been trimmed to keep the book up to date with changing ideas and approaches. In keeping with the discipline's growing concern for political and cultural issues, which recognizes that the economy cannot be treated separately from other domains of social activity, this volume offers more emphasis on the historical context and political economy of capitalism, including class and gender relations. A new chapter has been added on consumption, which has long been marginalized. Throughout, it synthesizes diverse perspectives—ranging from mainstream location theory to post-structuralism—to reveal capitalism as a profoundly complex, important, and fascinating set of social and spatial relations. It explores issues ranging from the locational determinants of firms to the role of the state in shaping market economies. Additionally, it offers much more discussion about services, including the multiple reasons for the growth of the service economy, its labor market impacts, and the fundamental role played by telecommunications in the global services economy. Finally, it approaches international development in an intellectually critical manner, emphasizing multiple theoretical views concerned with the origins and operations of the global economy.

Some students wrongly assume that economic geography is dominated by dry, dusty collections of facts and maps devoid of interpretation. This volume aims to show them wrong. Others are intimidated by economics, equating it with abstract and difficult mathematical equations. While this book uses both maps and some diagrams to make various points, it does not presume that the student has an extensive background in economics. There are several forms of economics, including neoclassical views and political economy. The volume at hand uses both of these and other perspectives as well, in an attempt to raise the readers understanding to a level above that of the lay public but not to the degree of sophistication expected of an expert. In doing so, the book hopes to show that economic geography offers insights that make the world more meaningful and interesting; it is simultaneously an academic exercise, in the sense that it sheds light on how and why the world is structured in some ways and not others, and a very practical one, i.e., as a useful narrative for those studying business, trade, finance, planning, and other applied fields. Each chapter includes a summary, key terms, study questions, suggested readings, and useful websites for those curious enough, brave enough, and energetic enough to explore further.

We are grateful to many people who helped us in this endeavor. Numerous colleagues in the discipline of geography, within our departments and throughout North America and Europe, have inspired us in many ways, often without knowing it! Dan Kaveney of Prentice Hall brought the authors together, cajoled and guided them, and oversaw the review process. Patty Donovan meticulously edited every chapter, clarifying points and polishing the writing. James

Rubenstein, author of *The Cultural Landscape An Introduction to Human Geography,* graciously allowed us to purloin much of his wonderful artwork, and we owe him a real debt of gratitude. Finally, we thank our friends and families, who put up with us in our grouchier moments as we struggled with the text.

For Santa and Derek.

Frederick Stutz
Dept. of Geography
San Diego State University
San Diego, California
http://rohan.sdsu.edu/faculty/fstutz

Barney Warf
Dept. of Geography
Florida State University
Tallahassee, Florida
http://www.fsu.edu/~geog

The World Economy

CHAPTER

1

ECONOMIC GEOGRAPHY: AN INTRODUCTION

OBJECTIVES

- To acquaint you with the discipline of geography, its major paradigms, and the subfield of economic geography
- To introduce the major problems, constraints, and disparities in economic development
- To indicate how geography can help to resolve these problems
- To understand the principal political economies of the world
- To state the four major questions important to understanding the world economy

"This aerial view of the San Francisco Bay region includes Silicon Valley, one of the world's leading centers of computer technology. It reveals the uneven geographies produced by networks of production and consumption as they are stretched over space, reflecting the multitude of factors that lead different kinds of firms and workers to cluster in some places and not others.

Geographic Perspectives

Anything that happens on the earth's surface is geographic. All social processes, events, problems, and issues, from the most local—your body—to the most global, are inherently geographic; that is, they take place in space, and where they are located influences their origins, nature, and trajectories over time. Everything that is social is also spatial; that is, it happens someplace. Where you are sitting now, how you got there, where you live and work, the patterns of buildings and land uses in your school or city, transport routes, and the ways people move through them all are different facets of geography. So are the distributions of the world's cultures, the patterns of wealth and poverty, the flows of people, goods, disease, and information.

Geography is the study of space, of how the earth's surface is used, of how societies produce places, how human activities are stretched among different locations. In many respects, geography is the study of space in much the same way that history is the study of people in time. This conception is very different from simplistic popular stereotypes that portray geographers as a boring bunch concerned only with drawing boundaries and obsessed with memorizing the names of obscure capital cities. Essentially, the *discipline of geography examines why things are located where they are.* Geographers are interested in explaining the processes that give rise to spatial distributions, not simply mapping those patterns. Thus, they examine not only where people and places are located, but also how people understand those places, give them meaning, change them, and are in turn changed by them. Because this set of topics involves both social and environmental topics, geography is the study of both the distribution of human and natural phenomena and lies at the intersection of the social and physical sciences.

All social processes and problems are simultaneously spatial processes and problems, for everything social occurs somewhere, and, more important, *where* something occurs shapes *how* it occurs. Place is not some background against which we study social issues, but it is part of the nature and understanding of those issues. Geographers ask questions related to location: Why are there skyscrapers downtown? Why are there famines in Africa? How does the sugar industry affect the Everglades? Why is Scandinavia the world's leader in cell phones? How is NAFTA (North American Free Trade Agreement) reshaping the American and Mexican economies? How has the microelectronics revolution changed productivity and competitiveness and the global locational dynamics of this sector? What can be done about inner-city poverty? Much more interesting than simply finding patterns on the earth's surface, as any atlas can demonstrate, is the *explanation* linking the spatial outcomes to the social and environmental processes that give rise to them.

To view the world geographically is to see space as socially produced, a product of social relations, a set of patterns and distributions that change over time. This means that geographic landscapes are social creations, in the same way that your shirt, your computer, your school, and your family are also social creations. Geographers maintain that the production of space involves different spatial scales, ranging from the smallest and most intimate—the body—to progressively larger areas, including neighborhoods, regions, nations, and the least intimate of all, the global economy.

Because places and spaces are populated—inhabited by people, shaped by them, and given meaning by them—geographers argued that all social processes are embodied. The body is the most personal of spaces, the "geography closest in." Individuals create a geography in their daily life as they move through time and space in their ordinary routines. Societies are formed by the movements of people through space and time in everyday life. In local communities, neighborhoods, and cities—the next larger scale—these movements form regular patterns that reflect a society's organization, its division of labor, cultural preferences and traditions, and political opportunities and constraints. Geographies thus reflect the class, gender, ethnic, age, and other categories into which people sort themselves. At larger scales, spatial patterns reflect the organization of resources, the technologies of production and transportation, and legal and regulatory systems. These are found, for example, in cities and nation-states. Finally, the global economy itself—an intertwined complex of markets and countries—involves planetwide patterns of production, transportation, and consumption, with vast implications for the standard of living and life chances of people in different areas.

A common measure of national economic activity is the gross national product (GNP) of countries. Because there are such wide variations in the size and productivity of national economies around the world, a map of the sizes of national economies (Figure 1.1) bears little resemblance to the global distribution of people. Some of the largest countries, such as India, have relatively small GNPs, testimony to their small economies and low standards of living. Dividing each country's GNP by its population yields per capita GNP, a frequently used yardstick of quality of life (Figure 1.2). It is important to remember that maps and tables of abstract numbers

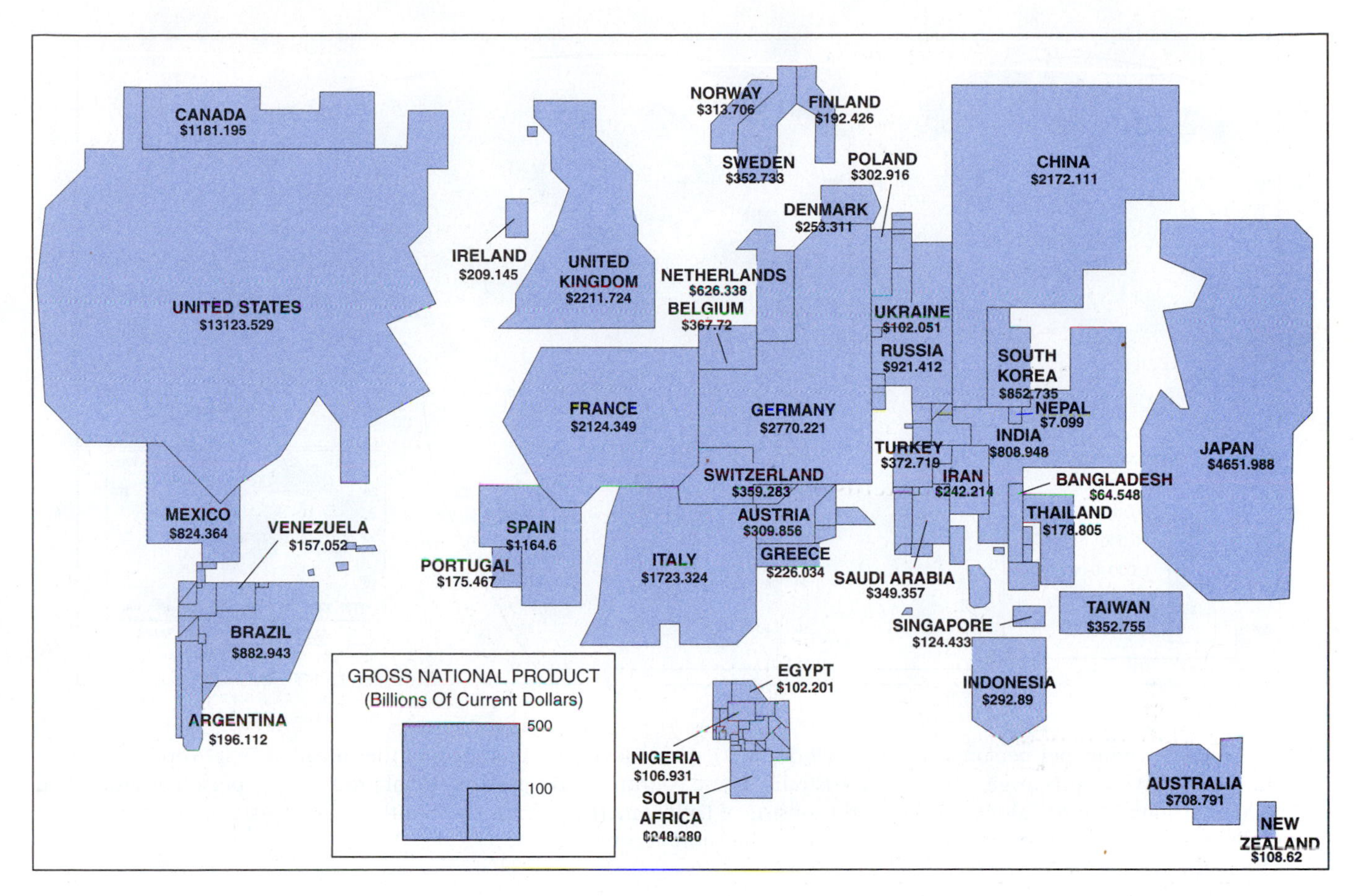

FIGURE 1.1
Relative sizes of national economies, 2005. This map shows the size of each country's gross national product, not its actual land area. On this cartogram, the whole African continent is smaller than Canada even though it has over 20 times the population of Canada. South America, China, and Russia also have small national economies relative to Japan, Europe, and the United States.

reflect real-world conditions in which people live, work, suffer, and die.

Geographers study how societies and their landscapes are intertwined. To appreciate this, we recognize that social processes and spatial structures shape each other in many ways. Societies involve complex networks that tie together economic relations of wealth and poverty, political relations of power, cultural relations of meanings, and environmental processes as well. Geographers examine how societies and places produce one another, including not only the ways in which people organize themselves, but also how they view their worlds, how they represent space, and how they give meaning to it. Divorcing one dimension, say the economic, from another, such as the political, is ultimately fruitless, but to make the world intelligible, we must approach it in manageable chunks. This text centers upon only one aspect of this set of phenomena, economic landscapes.

Economic Geography

Economic geography is a subdiscipline concerned with the spatial organization and distribution of economic activity (production, transportation, communication, and consumption), the use of the world's resources, and the geographic structure and expansion of the world economy. Economic geographers address a wide range of topics at different spatial scales using different theories and methodologies. Different generations of economic geographers have sought to explain local and global economic landscapes in different ways at different moments in time. In short, economic geography is an evolving discipline whose ideas are in constant flux. There is no "one" economic geography; there is a large array of different economic geographies from which to choose.

In the 1960s, the introduction of computers and statistical techniques provided a framework for analyzing and interpreting location decisions of firms and

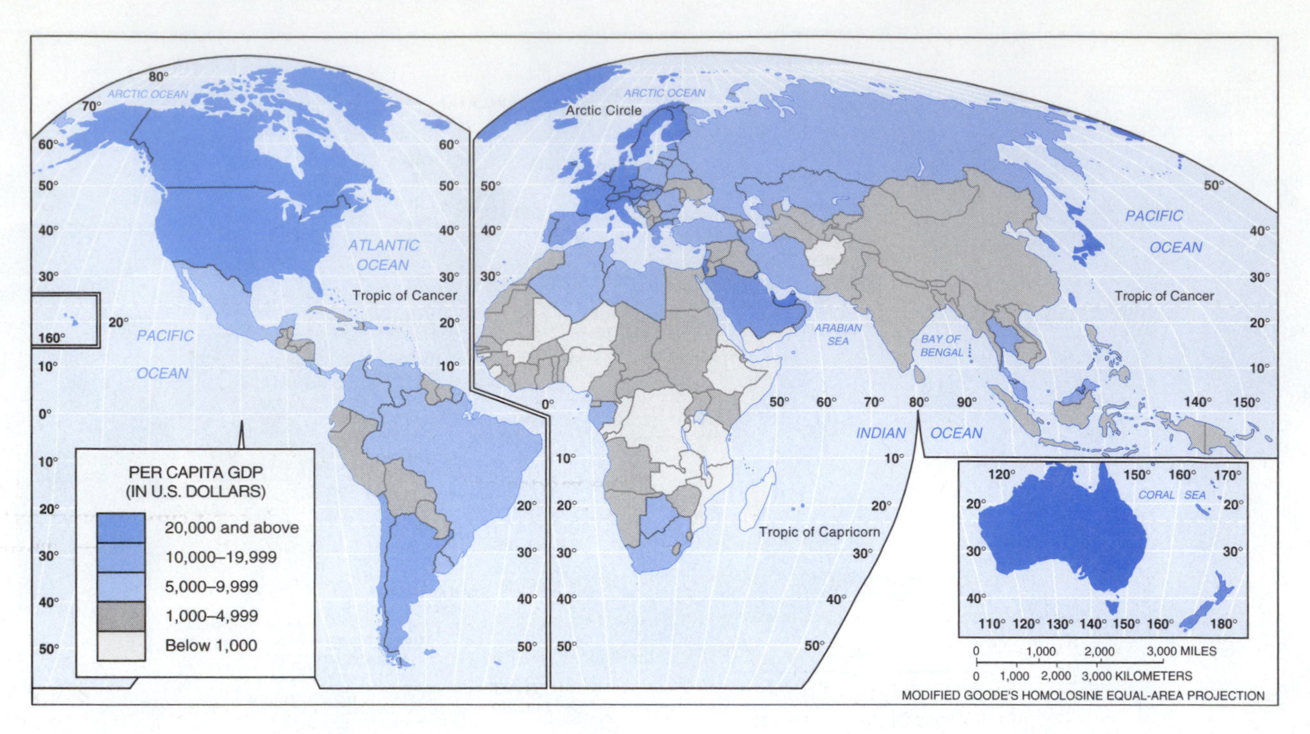

FIGURE 1.2
Gross domestic product per person around the world, 2005. More developed countries of the so-called First World, including North America, Western Europe, Japan, and Australia. These countries have gross national products per person at more than $20,000 per capita. The remainder of the world consists of Latin America, Africa, and South and East Asia. Here, incomes are generally less than $2,500 per capita. (See color insert for more illustrative map.)

individuals (e.g., market versus raw material location, accessibility versus transportation costs) and spatial structures (e.g., land-use patterns, industrial location, settlement). This approach aimed is called *logical positivism,* which emphasizes the scientific method in the analysis of economic landscapes, including the formulation of hypotheses, data collection, and predictive models.

An important part of this perspective, *location theory,* attempts to explain and predict geographic decisions that result from aggregates of individual decision making. The main aim of location theorists is to integrate the spatial dimension into classical economic theory. Location theorists made great contributions to the understanding of economic space. Many modeled *spatial integration and spatial interaction,* the linking of points through the construction of transport networks and the corresponding flows of people, goods, and information, including commuting and migration fields and shopping networks. Others sought to uncover *spatial structures,* the location of the elements of distribution with respect to each other, such as the hierarchy of cities. Spatial structures limit, channel, or control spatial processes; because they are the result of huge amounts of cumulative investment over years and even centuries, large alterations to the spatial structures of towns, regions, or countries are difficult to make. Spatial structure and social process are circularly causal: Structure is a determinant of process, and process is a determinant of structure. For example, the existing distribution of regional shopping centers in a city will influence the success of any new regional shopping center in the area.

Location theorists emphasized the building and use of models derived from neoclassical economics and analyzing business and industry in a world of pure competition that assumed that entrepreneurs are completely rational and attempt to maximize profits with perfect knowledge of the cost characteristics of all locations. This image of an entrepreneur became known as *homo economicus* ("economic person"), an omniscient, rational individual who is driven by a single goal—to maximize utility (or happiness, for consumers) or profits (for producers). Obviously this view is an oversimplification, but it offers one useful starting point.

Critics of spatial analysis noted its tendency to emphasize form at the expense of process, how it portrayed geographies as frozen and unchanging. The positivist approach is silent about historical context and politics, class, gender, ethnicity, struggle, power, and conflict. It tends to represent people as simply points on a map, abstracting them from their social worlds, as if they did not think and feel about their surroundings. Critics questioned the relevance of overly abstract mathematical models based on questionable assumptions that failed to capture the richness of political and social life. As the spatial analysis tradition was challenged, several alternative approaches rose to take its place. Economic geographers today thus have a suite of different theoretical approaches from which to choose.

Behavioral geographers challenged the simplistic view of behavior offered by *homo economicus* and pointed to the complex ways in which spatial information is acquired perceptually and interpreted cognitively in a world of imperfect information. This view was useful in introducing notions such as imperfect information, uncertainty, and suboptimal behavior to decision making, including probabilistic models, and made the field more realistic about how people understand space and move through it.

Humanistic geographers, drawing from the philosophical tradition of phenomenology, emphasized that human life occurs only through the world of subjective experience and symbolic meanings. For example, resources have no existence apart from human wants, which are specific to different kinds of societies and cultural priorities. This view is important in focusing on the intentions that motivate people, their ideologies and worldviews, the symbolic systems used to make sense of reality. It emphasizes that economic landscapes are made by actors, not impersonal social forces devoid of human content, and serves to remind us that issues of human socialization and social reproduction are critical to the analysis of how space is produced and that economies and geographies are produced by human actors, not abstract forces devoid of people.

Structuralist geographers, many of whom were influenced by marxism, charged that traditional theories of spatial organization obscure more than they reveal. In their view, location theories are narrowly conceived and blind to historical process—thus they are designed primarily to serve the goals of those who wield power. These geographers believe that a focus on the political organization of society and space—the ways in which power is organized and exerted to control resources—is fundamental to understanding space. Power is a fundamental part of how any social system is organized, and power and wealth are always closely linked. Power is always unequally distributed among and within societies. Any understanding of economic geography, of who is relatively rich and powerful and who is poor and holds less power, must therefore invoke some notion of economic class, as well as gender, ethnicity, and other types of social relations. Structuralists argued that the positivist, behavioral, and humanist views of human behavior are undersocialized; that is, they ignore the social contexts in which people live and which deeply shape what and how they think. This perspective maintains that social relations cannot be reduced to individual behavior, that societies are more than the sum of their parts, and that to understand economic landscapes we must understand their historical development, the class structure of a society, its gender relations and ethnicity, and how these are tied to culture and ideology. For structuralists, economic landscapes are the products of changing social relations of power and wealth that organize space in historically distinctive forms.

More recently, *poststructuralists* in geography and other disciplines have initiated yet another wave of change in how we view the economy and economic landscapes. This perspective includes a wide diversity of views, but essentially they maintain that the dynamics of capitalism cannot be understood independently of the modes of thought used to conceive, represent, and understand them. Capitalism doesn't simply exist outside of people's minds, but also inside of them. Thus, capitalism is as much "cultural" as it is "economic" and "political," and these distinctions are arbitrary. Poststructuralists initiated a "cultural turn" in geography that holds that the economy must be embedded within culture (i.e., that economic relations are always ones among people, emphasizing the role of signs and language in the production process). Poststructuralists tend to put consumption as an economic and social phenomenon on a par in importance with production. In this view, there is no single, objective view of the world, only multiple, partial perspectives, each of which is tied to different power interests. The dominant views that naturalize the world thus tend to be those of the powerful, although there is always room to challenge them.

Economic geography has recently been characterized by major changes in thinking, and today several schools of thought are evident simultaneously. While the discipline retains its long-standing interest in location theory and quantitative modeling, it has also come to reduce the boundaries between analyses of the economic and the political, between economy and culture, between society and nature. Such bifurcations often distort more than they clarify, and economic geography today borrows freely from many points of view. Students of economic geography can learn from all of these perspectives and combine them in creative ways.

Because the reality of the world is inevitably understood from and through a particular worldview, it is essential that we are aware of different theoretical systems, their assumptions, strengths, limitations, and conclusions. For this reason, this text uses a comparative approach in which different perspectives are explained and contrasted. Looking at the world through different ideological lenses better enables us to meet the challenge of world development problems. The way in which a society answers the central questions of economic geography depends on its historical context, class and gender relations, role of the state, position in the world system, and cultures and ideologies.

ECONOMIC GEOGRAPHY OF THE WORLD ECONOMY

The focus of this book is the world economy, the networks, processes, and institutions that shape the planetary system of resource distribution, create wealth and poverty in different parts of the globe, and contribute to the rise and fall of different national powers. The scale of the global is only one way in which economic geography can be studied, but given the massive processes of globalization that have been at work, particularly over the last 30 years, it is highly appropriate for understanding what goes on the world around you. The world economy links far-flung people and places so that what happens in one place shapes what happens in another through networks of interdependence. For example, the probability is that the clothes you are wearing now were made in a developing country; that the gas you put in your car came from a foreign source; that financial decisions being made in London or New York shape your credit availability and interest rates you pay. Every trip to the grocery store is a window to the global economy. Seen in this light, the global economy and everyday life are two sides of the same coin.

The world economy is constantly transformed by a combination of technological and geopolitical forces. This combination is creating a globalization of culture, a globalization of the economy, and a globalization of environmental issues. Technological changes—improvements in transportation and communications—are reducing the friction of distance and barriers to worldwide exchange. The principal instruments of the globalization of culture are worldwide television, music, and consumption patterns. The principal instruments of globalization of the economy are multinational corporations, which through their activities are producing new efficiencies in production, distribution, and the use of the world's resources (Table 1.1). The collapse of communism around the world in the 1990s and the reduction of government roles facilitated a new round of globalization by removing institutional and trade barriers to investment and trade. This increased pace of globalization helps explain the nature of competitive advantage of countries, states, and regions.

Changes in the world economy are simultaneously cultural, technological, political, and environmental. Transportation cost reductions have improved exchanges of people and goods. Advancements in telecommunications, including fiber-optic networks and satellite technology, have rapidly increasing ease, speed, quantity, and quality of information transactions. Worldwide political changes, ranging from the collapse of communism to widespread deregulation, have diminished the role of the state and increased the power of corporations. Rising populations in the developing world, and stagnant demographic growth in the developed, have altered the global supply, demand, and cost of labor, shaping migration patterns. Globalization has accelerated international economic, political, and cultural ties, ranging from corporate investment to trade to tourism to terrorism. And cultural changes, including the commodification and westernization of the world's many cultures, simultaneous secularization and growing religious fundamentalism in different places, and mounting awareness of international issues, have played a role in reshaping local and global social movements, consumption, and civil society.

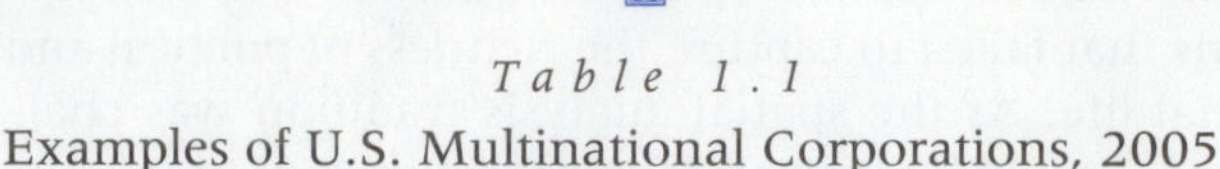

Table 1.1

Examples of U.S. Multinational Corporations, 2005

Company	*Total Revenues ($ billions)*	*Foreign as % of Total Revenues*
Exxon Mobile	298.0	70
ConocoPhillips	136.9	30
Dow Chemical	40.2	63
United Technologies	37.4	51
Caterpillar	30.2	53
Halliburton	20.5	78
Manpower	14.9	86
Computer Sciences Corp.	14.8	33
U.S. Steel	14.1	20
Texas Instruments	12.6	84
Occidental Petroleum	11.5	30
Carnival Corp.	9.7	40
Apple Computer	8.3	41
Jabil Circuit	6.2	84
Monsanto	5.5	46
Qualcomm	4.9	79

Source: Information Week www.informationweek.com/global50/rank/.

Because transportation and communication costs have fallen rapidly, many services and goods that were provided locally are becoming internationally mobile. World communication systems now allow for companies to subcontract their production and financial operations across continents, wherever price is cheapest and quality is the best. Information has become more mobile than ever before, and a new global economy exists in information transmission, corporate consultancy, cable television news, Internet information services, and software systems design, programming, and application. International finance has also become both global and computerized, and capital markets are now highly mobile for all forms of marketable equities and securities, stocks, bonds, and currency transactions. The globalization of finance has been aided by financial deregulation—the removal of state controls over interest rates, tariffs, barriers into banking, and other financial services.

The world is full of problems—debt, unemployment, poverty, inadequate access to health care, food shortages, and environmental degradation—that are

The grocery store exemplifies how everyday life and the global economy are shot through with each other. Access to a wide diversity of high-quality products at low prices that most, but not all, Americans enjoy reflects the highly advantaged position of their society within the contemporary global order. Learning to link your observations about your personal world to social trends and historical contexts is at the heart of social science.

rooted in the structure and development of the world economic system. Understanding the reasons for such problems begins by recognizing the long domination of the world by developed countries and the existence of an international economic order established as a framework for an international economic system.

The term *world economy* refers to the capitalist world economy, a multistate economic system that was created in the late fifteenth and early sixteenth centuries. As this system expanded, it took on the configuration of a core of dominant countries with a periphery of dominated countries. A division of countries is into First, Second, and Third Worlds, a categorization that is a product of the politics of the Cold War. The *First World* includes the economically developed countries of Europe, North America, Australia, New Zealand, and Japan (Figure 1.3)

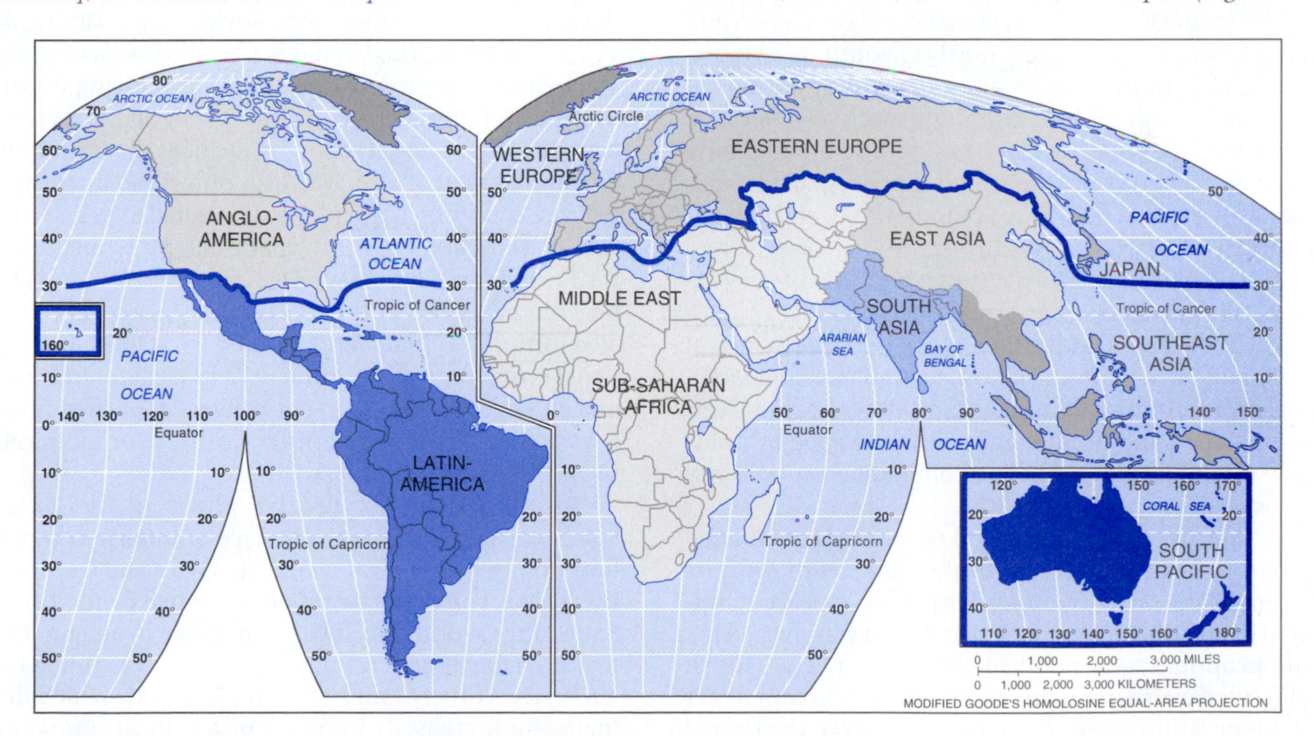

FIGURE 1.3
Major world regions can be divided into the First and Third Worlds, also called the global North and South. The former Second World, which consisted of the Soviet Union and its client states, expired with the collapse of communism in the 1990s and has been divided between the two. The First World, or North, includes the developed states in Europe, Russia, Japan, and North America, as well as Australia and New Zealand. The Third World, or global South, includes everyone else, the relatively less developed countries in Latin America, Africa, the Muslim world, South Asia, East Asia, and Southeast Asia.

(i.e., countries within the political orbit of the United States when the world system was defined by U.S. opposition to the Soviet Union in the cold war, 1945–1991). The defining feature of these countries, which comprise about one-quarter of humanity, is their relatively high standard of living, characterized by a large middle class. The *Second World* was represented by the Soviet Union and Eastern Europe, a designation that has lost its meaning in the post–cold war era (the 1990s and since), when the Second World disappeared, to be divided between the First and Third.

The dominated countries, in the underdeveloped *Third World* in the global South, are sometimes also called developing or, a bit more accurately, less developed countries (LDCs). The Third World consists of Latin America, Africa, the Middle East, and Asia, a broad set of diverse societies with a great range of cultures, historical backgrounds, and standards of living. Generally speaking, the Third World consists of relatively low-income countries, although it must be remembered that these range from impoverished societies such as Mali to quite prosperous countries such as Singapore, which resembles Europe in its quality of life. Whether Russia should be considered a First or Third World country is open to debate; its GNP per capita, after all, is lower than that of Mexico. Some observers even identify a Fourth World as a subset of the Third World, the poorest countries on earth (located mostly in Africa) with incomes of less than $300 per capita per year.

Egyptian farmer tilling the soil. This field is being prepared for growing cotton, to meet a worldwide demand for cotton clothing. In the future, the poorer countries of the world will have to rely on agriculture to raise their standards of living and to supply the capital they need to create industries. Agricultural production, therefore, must be increased. Some developing countries, such as Egypt, grow a disproportionate amount of nonfood crops for the export revenue it generates. *(Photo: United Nations)*

The strong geographic regionalism of developed and underdeveloped countries, north to south, is striking, leading to synonyms such as North and South to describe the First and Third Worlds, respectively. Because terms like *First World* or *Third World* are so broad, many observers have taken recently to more regionally specific labels, such as East Asia, Southeast Asia, and Central America.

The *international economic system* or *world economy* includes the institutions and relations of global capitalism such as the globalization of capital, international trade, and flows of information, technology, and labor. Because international markets and flows of resources, capital, labor, and products are always shaped by politically sovereign states, the international economic system is also a political system.

At any given time, the world economy is dominated by one or more core states. In the nineteenth century, the era of the Pax Britannica, or period of peace dominated by the British Empire, Great Britain was the world's only economic and political superpower. The British navy ruled every ocean in the world, and the sun never set on the British empire. By 1900, however, the United States overtook Britain as the world's largest national economy, and after World War II the United States displaced Britain as the world's leading superpower. In the post–World War II period preceding the 1970s, the United States was the principal power of the West, although the world political system was divided between the superpowers, including the Soviet Union. During the 1950s and 1960s, the United States was unquestionably the largest economy in the world, producing a large share of its manufactured goods and agricultural products. The international economy includes the institutions established after World War II to reflect the interests of the core countries, particularly the United States, such as the World Bank, the International Monetary Fund (IMF), and the World Trade Organization (WTO). As the hegemonic power in the world, the United States created institutions that were required to establish international economic order in tune with its ideals of free trade and investment (although critics allege that "free trade" is a smokescreen for powerful countries to economically invade less powerful ones).

By the 1970s, the relative power of the United States began to decline in the face of intense competition from rival core states such as Japan and Germany. By the late 1970s, the world order created by the United States after World War II began to come to an end. One major factor in generating this change was the petroleum crises of 1973 and 1979 (Figure 1.4), which dramatically increased the price of a critical input into industrialized economies. This action dealt a significant blow to the world economy, driving up heating and transportation costs, plunging most industrialized countries into recession, exacerbating unemployment,

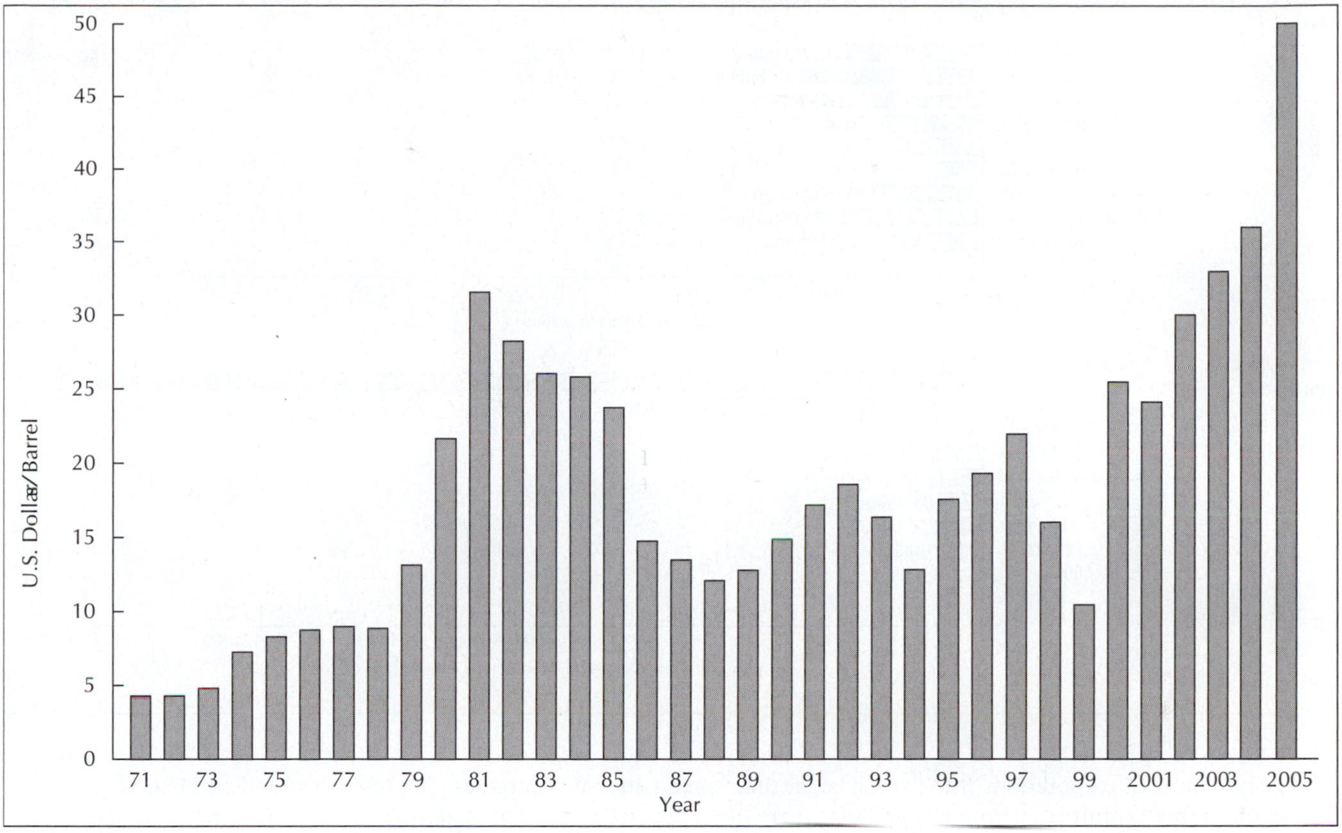

FIGURE 1.4
Average price per barrel of oil, 1971–2005 (in $U.S.). The nominal price peaked in 1981 with the OPEC oil embargo at $33/barrel. It dropped throughout the 1980s, then rose slightly in the 1990s. Growing global demand, including China, and gradually diminishing reserves drove the price up to new highs in 2005.

accelerating deindustrialization, and curtailing many people's standards of living. A major reason for the breakdown of the postwar world was a decline in the rate of profit of many firms in the industrial West. Faced with intense global competition, firms had to restructure themselves, including organizational and technological changes as well as relocating parts of their operations to the developing world. Some firms went out of business, but others responded to the challenge to automate and to "go international." They became more international partly due to the rising speed of travel and the new technologies of information handling.

Out of the old order came the birth of a new one in the 1980s and 1990s. This new world system, in which the Soviet bloc disappeared, left the United States as the world's only superpower. However, U.S. economic hegemony has been increasingly challenged by the rise of the newly industrializing countries (NICs), particularly those in East Asia and especially China. This global order is characterized by highly developed international markets dominated by multinational corporations, many of which are larger in their gross output than some countries (Figure 1.5). States and national governments also played a role in giving rise to the new system, by reducing trade barriers (e.g., through the World Trade Organization, the European Union, and the North American Free Trade Agreement) and by enormous international flows of capital and information. The American economy has become progressively more globalized, partly as a result of the influx of foreign investment from a variety of nations, mostly in Europe, Canada, and Japan (Figure 1.6). Simultaneously, the microelectronics revolution unleashed an enormous wave of change that dramatically affected all domains of production and consumption, particularly in telecommunications and finance, accelerating the globalization of services as well as manufacturing. We will explore the economic and spatial architecture of this world in great detail later.

GLOBALIZATION

Globalization refers to worldwide processes that make the world, its economic system, and its society more uniform, more integrated, and more interdependent.

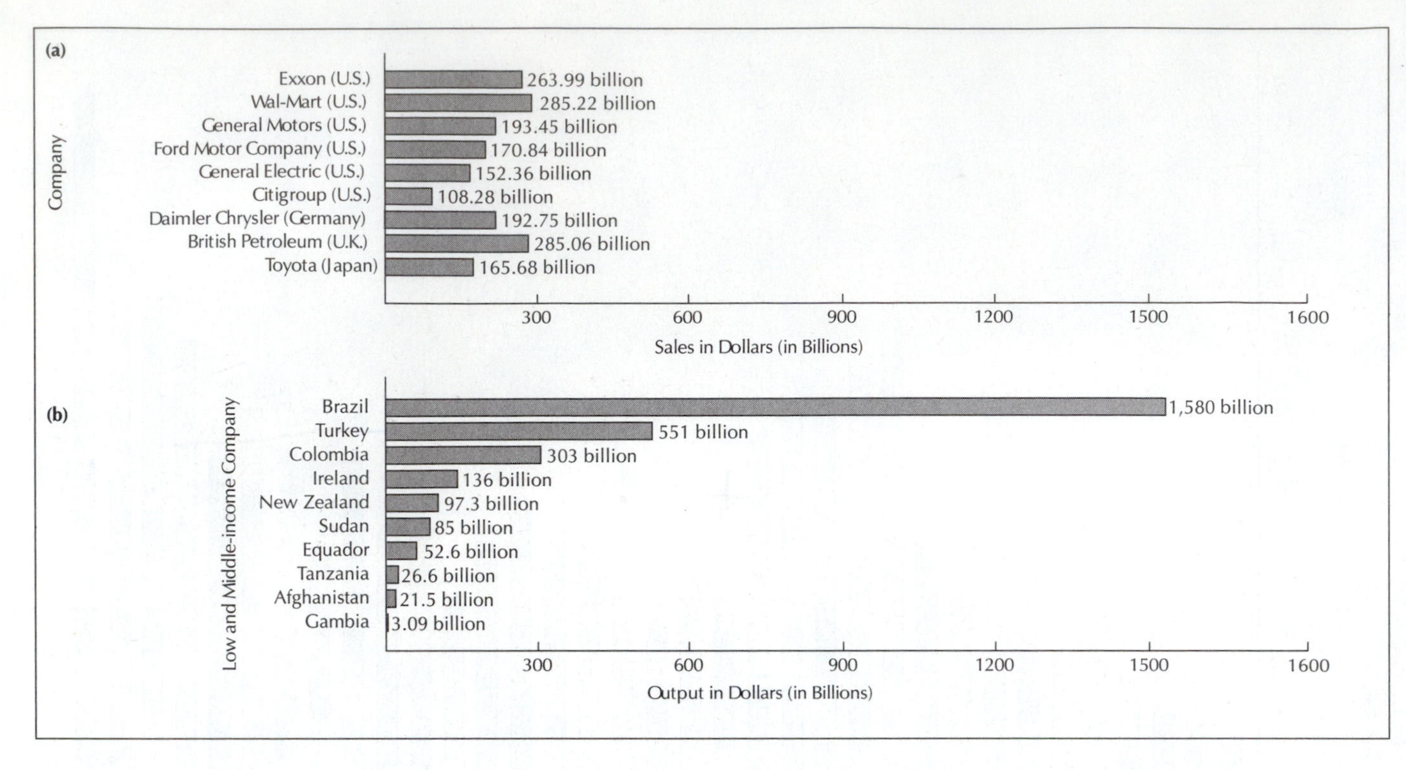

FIGURE 1.5
Many multinational corporations (MNCs) are larger than some national economies. The relative size of an MNC is important to small countries whose economies can are often drastically affected by decisions a global corporation makes.

Globalization is the process of the economy becoming worldwide in scope (i.e., an expansion in the scope, scale, and velocity of international transactions). The globalization process is a useful way to explain the movement of people, goods, and ideas within and among various regions of the world and their cultural, political, and environmental systems. Globalization is a process that shrinks the world by reducing the transport and communication times and costs places. This process has exposed different people in the world to an increasingly homogeneous global culture (largely American in origin), a global economy, and global environmental changes on a scale never been seen before. In some regions, social and political problems, as well as economic problems, result from a tension between the processes promoting global culture, economy, and environment, on the one hand, and a practice and preservation of local economic isolation, cultural tradition, and the localization of environmental problems, on the other hand. We now take a brief look at some of the most important dimensions of globalization that are occurring at an ever-increasing rate in the world today: globalization of culture, consumption, telecommunications, and economic activity, including transnational corporations, foreign investment, work, services, and information technology.

Globalization of Culture

Culture is the total learned way of life of a society. Culture can be defined as that body of beliefs, customs, traditions, social forms, and material traits constituting a distinct social tradition of a group of people. Cultural practices include religion, and attitudes toward family size as well as language, which is the transmission of ideas through symbols, signs, and dialects. Material traits involve food, clothing, and shelter. All of the world's peoples consume food, wear clothing, and construct shelter, but various cultural groups incorporate and produce these necessities in a variety of ways, including:

1. The globalization of culture is based on an increasing level of shared beliefs, social forms, and material traits worldwide.
2. Societies display fewer cultural differences than in the past.
3. Globalization of cultural and consumer preferences is being created by telecommunication systems worldwide and a globalized media.
4. The penetration of global culture in different regions across the earth is taking place at different rates. All peoples do not share access to globalization to the same degree.

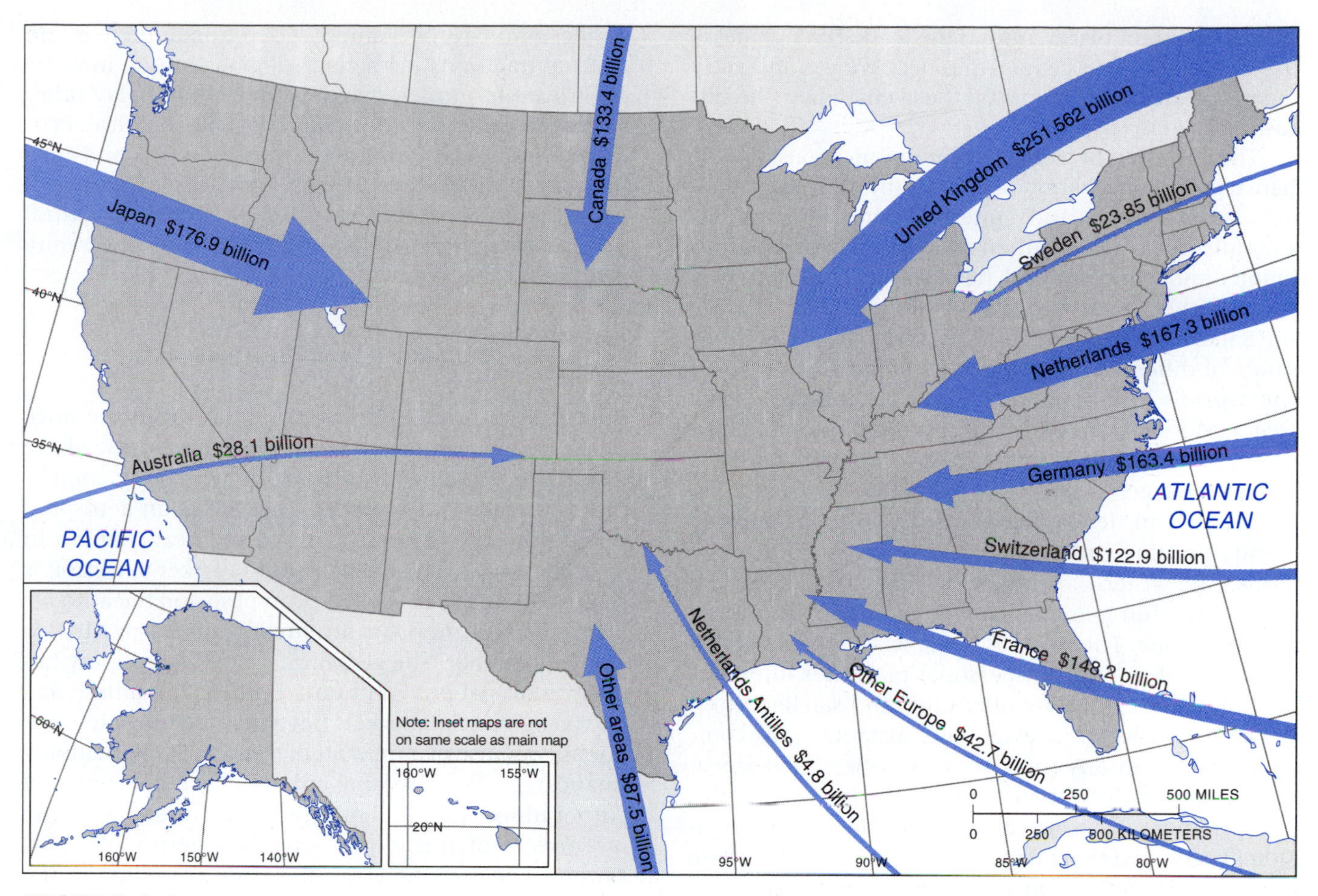

FIGURE 1.6
Foreign direct investment (FDI) in the United States comes primarily from Europe and Japan. Developed countries of the world are the primary investors in other developed countries. This diagram demonstrates the globalization of finance and the globalization of the world economy.

Globalization of Consumption

The survival of a culture's distinctive customary beliefs, social forms, and material traits is being threatened by the global diffusion of consumption preferences. For example, large numbers of the world's young people enjoy wearing blue jeans and Nike shoes, consuming Coca-Cola and Pepsi, smoking Marlboro cigarettes, eating McDonald's hamburgers, listening to Madonna, or watching American action movies. Consumption preferences in food, clothing, shelter, and leisure activities are displaying a globalization of culture today. The globalization of culture is based primarily on diffusion of lifestyles and products from more developed countries, especially from the United States, to less developed ones. The reverse flow is minuscule. Regardless of cultural tradition, people around the world are inspired to drive an automobile, watch television, and own a house with a stove and indoor plumbing.

Students of globalization observe an increasingly, homogenized global landscape in office towers, stores, restaurants, and service stations. Retail chains create recognizable logos and visual appearances that do not vary from one region to another. Customers recognize these logos and building designs in magazines in whatever landscape or part of the world they may find themselves in.

Telecommunications

Cultural groups in different regions share similar beliefs, social forms, and material traits, thanks to enhanced telecommunications. Because of cable television and international news services, we know a great deal about political and economic events happening worldwide within a few hours. Faraway places are less remote and more accessible now than they were just 10 years ago. Through television programming and the Internet, we can reach into the countries and cities of the world's people that are far away.

We can learn of places far away via television and the Internet. We can receive pictures and messages worldwide at the touch of a button or the click of a mouse. We can communicate almost instantly with people in distant places around the world through desktop computers and cellular phones. We can instantly see people in distant places on the evening news broadcast on television.

This diffusion of global telecommunications has enhanced the globalization of culture through shared beliefs, social forms, and material traits. Africans, for example, have often shifted away from traditional religions and animism and have adopted Christianity south of the Sahara and Islam in the countries sharing the Sahara Desert. The world's peoples still speak thousands of different languages, but English has become the world's primary international language—*lingua franca*. More international communications and global business activities are conducted in English than any other language.

Citizens in developed countries take for granted these telecommunication innovations, such as MTV (Music Television), communications across oceans by telephones and fax, and satellite communications via the cell phone. The world is being wired up into global networks of millions of personal machines interconnected to each other by fiber-optic and satellite links. These networks allow essentially instantaneous communications to anyone else on the Net. That interchange can include mail, documents, books, pictures and photographs, voice and music, video and television images, and programs and film. The largest of these networks, the *Internet*, includes over 10,000 universities and colleges with databases of computer programs and other information.

What these changes mean is that we are quickly moving toward the time when anyone can get any information to almost anyone else at almost any time throughout the entire world. This globalized network is transmitting data and will be the prime conduit for the capital commodity of the coming age: information.

This global information network is far more refined in the United States and Canada than any other region and therefore allows North America to project its culture to the rest of the world. The information network is the perfect tool that will allow the sophisticated knowledge worker to mine the databases and knowledge bases of a huge number of sources. This information age is exacerbating one of the major problems of the coming decades, the increasing disparity between have and have not nations and have and have not social groups within countries.

The world contains significant handfuls of people here and there who are isolated and sheltered, who have never seen television, used a phone, or ridden in a motor vehicle. Access to communications and transportation of the information age is restricted by an uneven division of wealth worldwide. Even within a region, access may be restricted because of uneven distribution of wealth or because of discrimination against a tribe, ethnicity, or women. A determination to retain cultural traditions in the face of globalization in some cases leads to intolerance of people who display other beliefs, social forms, and material traits. Political, economic, and social disputes and unrest, wars, and conflict erupt in places such as Southeastern Europe, Africa, and the Middle East, where different cultural groups have been unable to share the same geography peacefully.

Globalization of the Economy

Companies, societies, and individuals that were once unaffected by events and economic activity elsewhere now share a single economic world with other companies, societies, and workers. The fate of an aerospace worker in Los Angeles is tied to political changes in Eastern Europe. The job of an auto worker in Detroit is related to the fall of the Mexican peso and the auto industry's investment in production plants in Mexico along the border. The globalization of the economy has meant that national and state borders and differences between financial markets have become much less important because of a number of trends: (1) international finance; (2) the increasing importance of transnational corporations; (3) foreign direct investment from the *core regions* of the world—North America, Western Europe, and East Asia; (4) global specialization in the location of production; (5) globalization of the tertiary sector of the economy; (6) the globalization of office functions; and (7) global tourism.

Globalization involves international financial flows. In the deregulated, hypermobile electronic world of international banking today, telecommunications allowed a single global capital market. Computers allow traders to monitor and trade in national currencies, stocks, bonds, and annuities listed anywhere in the world instantaneously. Banks, financial houses, and corporations can operate worldwide partly because of the decision centers that control the global economy. Consequently, banks and corporations can react immediately to changes in the value of commodities or gold on the world market and the rate of exchange between the dollar, the euro, the yen, and other currencies.

Transnational Corporations

The globalization of the economy has been spearheaded by transnational corporations (TNCs), sometimes referred to as multinational corporations (MNCs), or multinational enterprises (MNEs). A transnational corporation may conduct research, operate industries, and sell products in many countries, not just where its headquarters are based. Most transnational corporations maintain their headquarters in one of the three regions of the core countries—North America (United States and Canada), Western Europe (especially Germany, France, the United Kingdom, and the Netherlands), and Japan. Most of the factories and firm locations of transnational corporations are within these three regions, as well as their headquarters. In 2000, transnational corporations employed 100 million in the core regions, with 20 million elsewhere.

In 1970, the world's 15 richest nations were headquarters to 7500 MNCs. However, by 2000 these same countries hosted 25,000 MNCs. Today, there are some 53,000[1] MNCs in the world controlling about 40 percent of all private sector assets and accounting for a third of the goods produced for the world's market economies. They employ 100 million people directly, which is 4 percent of the employment in developed regions and 12 percent in developing regions. Some countries, such as Canada, have extremely high proportions of total production by MNCs, especially in mining, manufacturing, and petroleum sectors of their economy. MNCs also play a disproportionately dominant role in other developed countries, such as Belgium, France, the Netherlands, Italy, Britain, and Japan. Very large MNCs have sales of goods and services exceeding $100 billion annually. The sales of the largest MNCs are larger than most countries' total economies. This is important to a small country's economy that can be affected by a decision of a global corporation. Today, multinational corporations control greater than half of total international trade simply via intracorporation transfers of components, services, investments, profits, and managerial talent among their scattered plants and offices in various countries. Most of this intrafirm trade is not finished products and services, but components, subassemblies, parts, and semifinished products.

Today, MNCs, not countries, are the primary agents of international trade, largely between and within their organizations. In so doing, MNCs change countries' reserves of resources in effect by moving human and physical capital and technology from one part of the world to the other, creating a new asset base, and allowing production and manufacturing to occur in outsourced locations where they may not have happened otherwise. An MNC will produce in a country where a set of characteristics taken together is more attractive: location, resource endowments, size and nature of market, and political environment. Further, the MNC is able to use transfer pricing, the practice of price setting for goods and services provided by subsidiaries, so as to transfer taxable profits to the lowest tax country possible and minimize tax overhead—to shift the MNC's profits to countries that have the lowest tax rates.

[1]United Nations World Investment Report, 1998 Trends and Determinants, p. 1.

Multinational corporations are able to compete on a world scale because they can operate with greater information efficiency and share that with their subsidiaries and branches throughout the world via the Internet and satellite and fiber-optic communication systems. This transnational communications ability is a tremendous advantage to the MNC in that it can stay aware of markets, products, labor, and business opportunities. Other advantages of the MNC are its large store of capital, technology, managerial skill, and scale economies.

Foreign direct investment (FDI) indicates investment by foreigners in factories that are operated by the foreign owners of a multinational corporation. U.S. transnational corporations are most likely to invest in Europe, Canada, and Latin America. Western European transnationals are most likely to invest in Eastern European, Russian, and African markets, as well as in North America. Japanese transnationals are most likely to invest in Asia and in North America.

Since the 1980s, governments in the three regions where transnational corporations are based, North America, Europe, and East Asia, have changed their forms of governance to accommodate international corporate capital, altering tax codes and regulations that formerly hindered transnational operations. Other countries where transnational corporations wish to invest, especially in developing countries, have also modified their laws and regulations to encourage transnational operations within their borders. These changes, often labeled "neoliberalism," have changed the relationships between countries and corporations, favoring the latter over the former.

Globalization of Investment

The globalization of the world economy is centered in the core regions—North America, Western Europe, and Japan, as well as the Tigers of the Pacific Rim. From the three major world cities, or command centers, in New York, London, and Tokyo, orders are sent instantaneously to factory shops and research centers around the world. Manufacturing enterprises have located their production and assembly lines and lower-cost offices outside the high-cost core countries. For example, most U.S. sportswear companies, centered in New York and Los Angeles, have moved their production to Asian countries. Latin America, Africa, and Asia contain three-fourths of the world population and almost all of its population growth. Countries in these realms find themselves on the periphery of the world economy or outer edge of global investment decisions made by transnational corporations, the result of centuries of colonialism and a world order in which the rules often work against them.

Three trends demarcate FDI in developing countries. First, the proportion of FDI that core countries are allocating to periphery countries is declining. Core countries increasingly invest in one another. Second, FDI is becoming more geographically selective. Countries that attract the greatest FDI from the core countries are those that have chosen the export-led strategy of economic growth. Export-led industrialization is characterized by countries welcoming foreign investment to build factories that will manufacture goods for international markets and employ local labor. Export-led policies rely on global capital markets to facilitate international investment and global marketing networks to distribute the products. The countries that have grown the fastest in recent decades have generally followed the export-led approach as opposed to the alternate approach, import substitution.

Global Specialization of Work

Every location in the global economy plays a distinctive role based on its particular combinations of assets and weaknesses as they are produced historically. Transnational corporations assess the economic and locational assets of each place in the world economy. Specialization in the location of global production shows a declining role of population and resources, the original factors of production in global development. Today, brain power has largely replaced muscle power as the primary source of wealth in the world and *transmaterialization* (substitutability among inputs) has changed the nature of resources. Input factors and components move intrafirm, final goods are fabricated close to the point of consumption, and national boundaries count much less than they did in the past global economy.

In the new global economy, transnational corporations maintain a competitive edge by correctly identifying geographic factors and the optimum location of each of the activities, including engineering systems, raw material extraction, production, storage, office functions, marketing, and management. Suitable places for each activity may be clustered in one country or may be disbursed in countries around the world. The resulting globalization of the economy has increased economic differences among places in a have and have-not world. Factories are closed in some locations and reopened in other countries. Some countries become centers of technical research, whereas others become centers for low-skilled manual tasks. Changes in the geography of production have created a spatial division of labor in which regions specialize in particular functions. Transnationals decide where to locate in response to characteristics of local labor force, skill level, prevailing wage, and attitudes toward unions, tariffs, and transportation rates. A transnational may close factories in regions with high wage rates and strong labor unions.

Globalization of Services

The globalization of services and consumption is also important. For example, U.S. service exports generate one-third of the nation's foreign revenues and dwarf auto exports. Business services provide essential inputs to transnational corporations as the transnational corporations expand into the world arena. The international tertiary sector includes the business services of legal counsel, business consulting, accounting, marketing, sales, advertising, billing, and computer services. Many professionals, such as architects, software designers, and business consultants, market their skills throughout the world. Globalization of services includes international tourism, visiting foreign students, entertainment, TV, music recordings on CDs, and world distribution of U.S. movies. By 2000, 60 percent of the gross revenues of the five largest U.S. motion picture studios derived from outside the United States. Globalization of services displays the same trends as the globalization of manufacturing. The globalization of services operates in a world of a declining role of the nation-state, but the continuing reinforcement of cultural differences at both national and regional scales.

The mega service firms of North America are not really national and will respond in nonnational ways. Large corporations are not international or transnational; they are nonnational and will increasingly see economics and politics without a particular national flavor.

As the global transmission of television becomes ubiquitous and intercontinental information networks allow international subscribers access to a huge amount of American culture, the United States' most powerful export, the United States will grow in influence. This will have both positive and negative effects on the culture and the disposition of the world's peoples. It is likely to broaden the links of commonality among the younger generations of the developed world, especially those who are Internet and World Wide Web savvy. At the same time, it threatens to alienate the United States from more conservative elements in those cultures, many of whom have turned to religious fundamentalism.

Globalization of Tourism

Tourism is one the world's fastest growing industries, employing 200 million people in 2000, and contributing to about 12 percent of the world's gross domestic product. The tourism industry is already one of the leading export sectors and is expected to grow at an annual rate of 3 percent worldwide. The highest rate of growth will take place in some developing countries, especially tropical regions and areas with picturesque scenery.

The countries best endowed for global tourism provide both natural and cultural attractions to their visitors, along with pleasant climates, good beaches, and attractive social and political milieus. Political stability is critical to this industry.

Information Technology and Globalization

Communication improvements mean that a globalization of the world economy is moving forward rapidly to a point where many people in any location can receive and send information to others in any other location almost at any time. The world economy increasingly depends on moving information instead of people. The global network, which will export and import not products but information, will allow innovations to sweep the world at rapid rates. This trend will exacerbate and increase the disparity between the have and have-not nations. "Fast" and "slow" societies in the future will refer to the effect that information technology will have on the tempo of human affairs. This information-based economy means that the relative success of individuals or groups is based on access to information, more than money or products, more than natural resources, labor pools, and other traditional metrics of power and wealth. A global information network of the information age is a pool that will allow a *knowledge worker* of the global economy to mine the databases and knowledge bases of important information bases of the world. The world will be interdependent more than ever before, which will facilitate the interchange of information among researchers as never before. Information systems will be the facilitators of the knowledge-based world economy.

Real-time information systems mean that information becomes available as it happens, or at least as soon as possible after software programs process it and make it available. Real-time implies immediate accessibility so that everyone on the Net can seek critical information by accessing a computer screen. Real-time is one of the essential differences between the world economy in the future and the world economy in the past. With real-time information systems, more people will be making more decisions in a customized world economy, as the number of people who interface with customers becomes part of a self-managing business unit. Companies, individuals, and MNCs need feedback on their decisions as soon as it becomes available so that they can learn faster and make constant adaptations to meet customers' needs better than anyone else. The globalization of the world economy in the future will demand real-time information systems so that world business decisions will be made with the minimum of bureaucracy.

The communications and information technology (IT) revolution is based on the networking of individual computers. The engine of change for the twenty-first century is based on the communications revolution of computers being linked to global networks of information databases and personal computers linked to one another by fiber-optic and satellite communications. These communication networks, such as the Internet, allow instantaneous communication to anyone else on the network. Communication linkages can include photographs, voice and music, videos, television images and programs, films, documents, books, pictures, mail, and spreadsheets. With large commercial networks such as America Online or Compuserve, one can teleshop at 1000 stores, make reservations at hotels in Europe, buy airplane tickets, monitor the weather and stock market, pay bills, and download files from newspapers, magazines, and encyclopedias.

Globalization versus Local Diversity

Globalization has affected different regions in different ways, and generally does not completely destroy its unique local diversity. Many current political, social, and economic problems result from the tension between forces promoting globalization of the culture and economy versus the preservation of local cultural traditions and economic self-sufficiency. The desire of some people to retain their traditional economies and cultural preferences against the rising tide of increasing globalization has led to political conflict, social chaos, and market fragmentation in more traditional regions of the world.

Globalization and local diversity will coexist and shape each other, a theme some geographers call "glocalization." In this regard, the hypothesis of uneven fragmentation—the view that the world economy produces different results in different places—allows for continuing antagonism between globalizing and localizing tendencies that, even if unevenly, will accommodate each other. For this accommodation to take place, individuals have to appreciate that they can advance both local and global values without either penalizing the other. It is important to comprehend that the development of multiple loyalties to different local, national, and transnational affiliations need not be mutually exclusive. People can be loyal to their family, community, country, and the worlds' people simultaneously. In a globalized world, more and more people become aware of the extent to which their well-being is dependent on events and trends elsewhere in the world.

World Development Problems

The world economy is highly uneven in its behavior and impacts over time and space. Temporally, world economic growth rates have waxed and waned, dropping during the global recession of the early 1990s and rising slightly during the first years of the new millennium (Figure 1.7). Frequently, the average growth rate of developing countries exceeded that of the developed

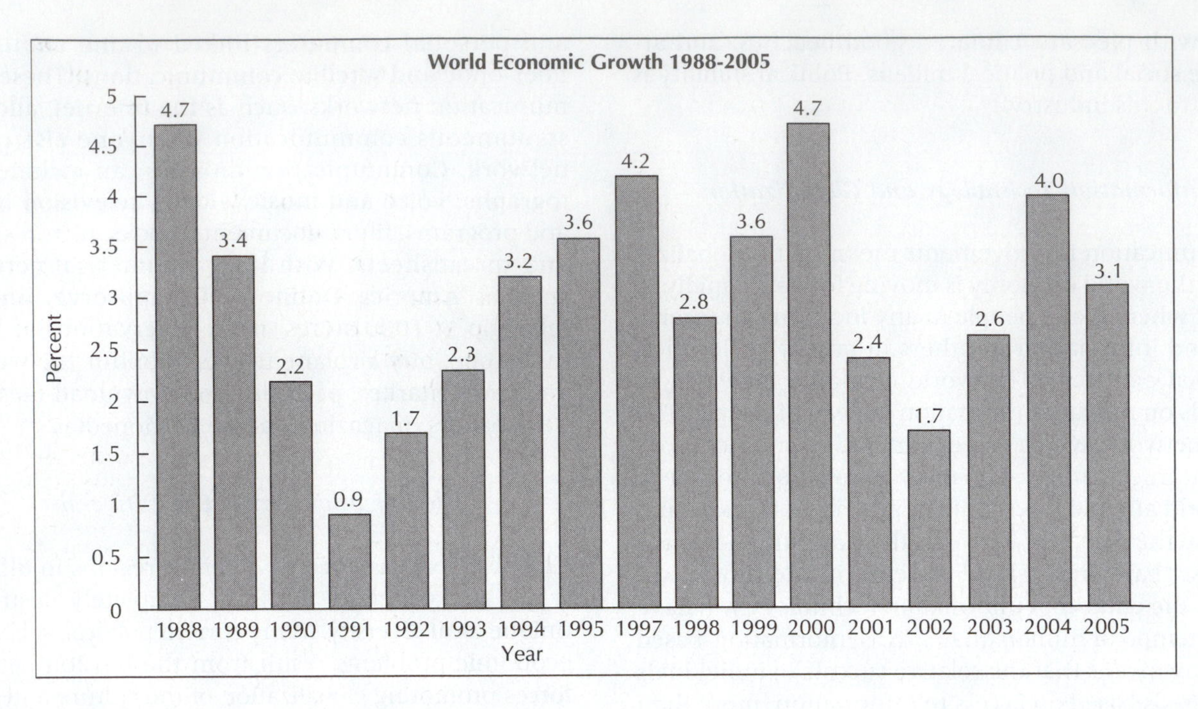

FIGURE 1.7
World average economic growth rates, 1988–2005. Fluctuations in average growth reflects international recessions, the price of oil, productivity growth (including the impacts of new technologies), catastrophic events such as the Indian Ocean tsunami, changes in government policies, and political turmoil or the lack thereof.

ones, widening the gap between the world's haves and have-nots. Economic growth rates are very uneven among different world regions (Figure 1.8), and while they have been comparatively high in East Asia, they have been much lower in many other areas such as Latin America. Low economic growth rates mean that people's standards of living increase slowly, or not at all, depending on labor markets, rates of inflation, unemployment, and population growth.

Despite the economic progress that has been made in many parts of the world, there are still vast areas of the planet in which billions of people remain mired in deep poverty. Much of the world has not benefitted from globalization. Economic development, and the lack of it, are thus important questions for economic geographers.

Development is a concept full of hope, even though the consequences of the jolting and dislocating process

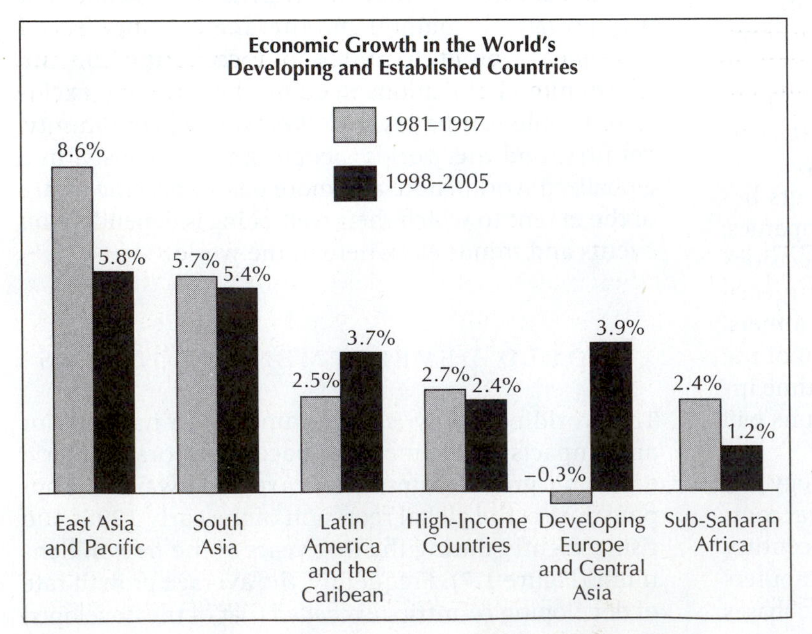

FIGURE 1.8
Economic growth in the world's major geographic regions, 1981–2005. Worldwide averages in growth rates mask large variations among its major regions. For the last 40 years, East Asia has been the world's most rapidly growing set of economies, followed by South Asia.

can be horrendous for people when long-standing traditions and relationships break down. The purpose of development is to improve the quality of people's lives—that is, to provide secure jobs, housing, adequate nutrition and health services, clean water and air, affordable transportation, and education. Whether development takes place depends on the extent to which social and economic changes and a restructuring of geographic space help or hinder in meeting the basic needs of the majority of people (see Chapter 13).

Problems associated with the development process occur at all scales, ranging from a Somalian villager's access to food and a health clinic to the international scale of trade relations between rich and poor countries. Our attempts to understand development problems at the local, regional, and international levels must consider the principles of resource use, as well as the principles surrounding the exchange and movement of goods, people, and ideas.

Two critical issues require immediate attention. One is the challenge to economic expansion posed by the environmental constraints of energy supplies, resources, and pollution (see Chapter 4). The other element is the enormous and explosive issue of disparities in the distribution of wealth between rich and poor countries, city and rural areas, wealthy and poor people, dominant and subordinate ethnic groups, and men and women (see Chapter 13).

Environmental Constraints

The world environment—the complex and interconnected links among the natural systems of air, water, and living things—is caught in a tightening vise. The environment is being stressed by the massive overconsumption and a wasteful consumer culture of the developed world. On the other hand, the environment is being squeezed by the poor people in developing countries who must often destroy their resource base in order to stay alive. The constraints of diminishing energy supplies, resource limitations, and environmental degradation are three obstacles that threaten the possibility of future economic growth.

There is a significant energy problem in much of the developing world. Oil is an unaffordable luxury for much of the world's population, who cook and heat with fuelwood, charcoal, animal wastes, and crop residues. In countries such as India, Haiti, Indonesia, Malaysia, Tanzania, and Brazil, fuelwood collection is a major cause of deforestation—one of the most severe environmental problems in the underdeveloped world.

The fragility of the environment poses a formidable obstacle to economic growth. Are there limits to growth? Is the world overpopulated? Some of our present activities, in the absence of controls, may lead to a world that will be uninhabitable for future generations. Topsoil, an irreplaceable resource, is being lost because of overcultivation, improper irrigation, plowed grassland, and deforestation. Water tables are falling, including in the United States, where, for example, the Ogallala water basin under the Great Plains is in increasing danger of being rapidly depleted. Forests are being torn down by lumber and paper companies and by farmers in need of agricultural land and wood to keep warm or cook their food. Water is being poisoned by domestic sewage, toxic chemicals, and industrial wastes. The waste products of industrial regions are threatening to change the world's climate. Accumulated pollutants in the atmosphere—carbon dioxide, methane, nitrous oxide, sulfur dioxide, and chlorofluorocarbons—are said to be enhancing a natural *greenhouse effect* that may cause world temperatures to rise. El Niño events, or periodic warming of the Pacific Ocean just 0.25 degrees, caused violent weather disruptions worldwide, with billions of dollars worth of damage from floods, mudslides, and loss of life. Chlorofluorocarbons, which are used as aerosol propellants and coolants and in a variety of manufacturing processes, are blamed for damaging the earth's ozone layer, which protects life from ultraviolet radiation from the sun. Yet another hazard to the environment is the fallout from nuclear bomb tests that took place in the 1950s and 1960s and from nuclear power reactor accidents such as those at Three Mile Island, Pennsylvania, and Chernobyl, Ukraine.

Disparities in Wealth and Well-being

The world economy generates great variations in economic structures, standards of living, and quality of life around the globe. The world's richest and poorest nations have enormous differences in wealth and standard of living as measured by economic statistics such as GNP per capita are paralleled by social and demographic ones such as life expectancy, infant mortality rates, literacy, and caloric consumption. In short, maps of economic measures are simultaneously maps of other dimensions of people's lives, including how long they live, the chances that their babies grow into adults, their ability to read and write, and the quality of the food they eat. These numbers point to the multifaceted nature of poverty and development, which are not just economic but also social and political.

Poverty afflicts relatively few people in economically developed countries, although there are nonetheless disturbingly large numbers of poor in wealthy societies such as the United States, including hunger and malnutrition among families in Appalachia or Native American reservations, bankrupt farmers on the Minnesota prairie, unemployed factory workers in Detroit, and single mothers on welfare in New York. Deeply entrenched, institutionalized poverty confines billions of people to lives of inadequate food, shelter, health care, transportation, education, and access to other resources. Mass poverty is the single most important world development problem of our time. You cannot doubt this assertion when you see maimed people on the streets of Bombay, begging children in Mexico City, desperate farm laborers in Brazil, emaciated babies in Mali, or women and children carrying firewood on their backs in the countryside north of Nairobi. Mass poverty is ethically intolerable and a critical issue that we must try to overcome.

Who are the world's poor? They are the 15 million children in Africa, Asia, and Latin America who die of hunger every year. They are the 1.5 billion people, or 24 percent of the world's population, who do not have access to safe drinking water. They are the 1.4 billion without sanitary waste disposal facilities. They are the 3 billion people—50 percent of the world's population—who live in countries in which the per capita income was less than $400 in 2005. Half of the world (largely in Africa) earns one dollar a day or less. These numbers are characteristic of impoverished countries in which much economic activity takes place outside of the market. These people are caught in a vicious cycle of poverty (Figure 1.9), often with few ways out, and lead lives of quiet desperation and hopelessness. Their life expectancies tend to be short, infant mortality levels high, and access to energy, medical care, transportation, and education often minimal. The economic geography of the world is at its core concerned with these social and spatial discrepancies among and within countries.

The poor of the world overwhelmingly live in developing countries, most of them former European colonies, which failed for one reason or another to keep up with the economic levels of the West over the last 500 years. During the worldwide economic boom that occurred in the three decades following World War II (1945–1975), the GNP of the developed countries more than doubled. Although per capita real income in developing countries also rose, incomes in developed countries rose much more quickly. Developed countries enjoyed 66 percent of the world's increase, whereas half of the world's population in underdeveloped countries (excluding China) made do with one-eighth of the world's income. By 1982, the national income of the United States (then 235 million people) was about equal to the total income of the Third World (more than 3 billion people). In short, over the last half-century the rich have become much richer, and the poor have gained only slightly.

The developing world is far from a homogeneous entity; that is, there are enormous differences among and within developing countries in terms of their historical background, cultures, economies and standards of living, and when and how they were incorporated into the world system. So great are the variations among countries, and often within, that it is simplistic to speak of a single developing world without immediately acknowledging its differences. To lump, say, South Korea, with a standard of living similar to southern Europe, together with Mozambique, one of the world's poorest states, is to fail to understand the profound differences that separate them.

With the debt crisis of the 1980s, the United States finally discovered it had a real stake in the prosperity of the developing world. The inability of some countries to make payments on their debt placed the financial

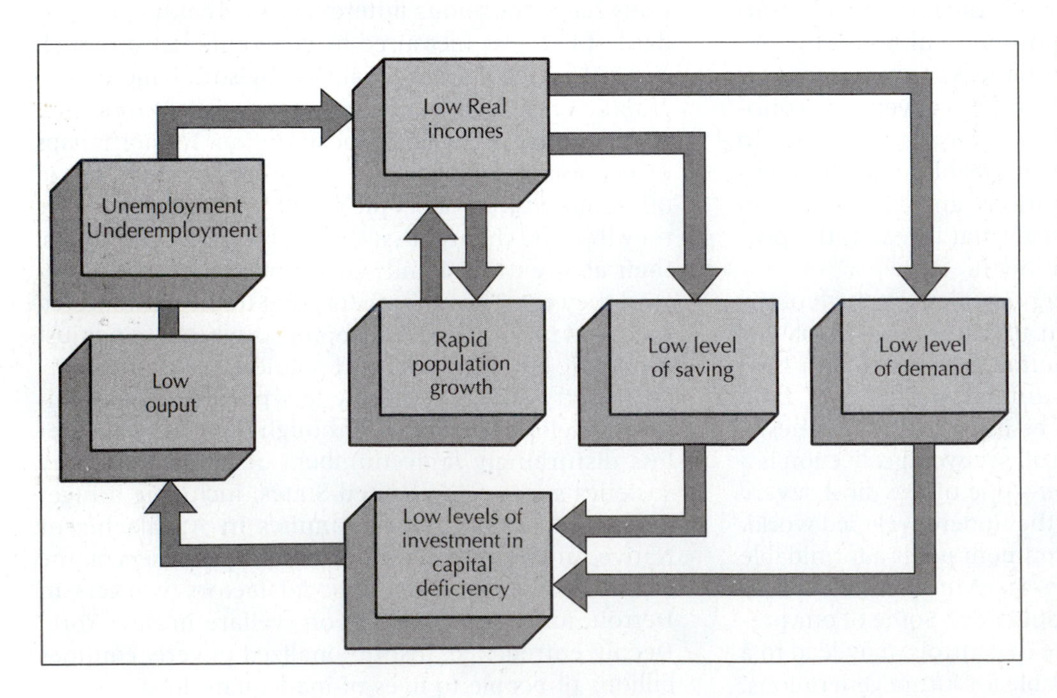

FIGURE 1.9
The cycle of poverty in Third World countries. Most Third World nations have low per capita income, which leads to a low level of saving and a low level of demand for consumer goods. This makes it very difficult for these nations to invest and save. Low levels of investment in physical and human capital result in low productivity for the country as a whole, which leads to underemployment and low per capita income. In addition, many of these countries are faced with rapid population growth, which contributes to low per capita incomes by increasing demand without increasing supply or output.

structures of the United States and some European nations in jeopardy. Many U.S. banks, including some of the largest, would technically be insolvent if their loans to developing nations were declared in default. This led to enormous pressure to resolve the immediate problems of the debt crisis, many of which were directly related to the poor economic performance of the economies of the debtor nations.

Unfortunately, for many debtors, the solution often proved to be more painful than the problem itself. Under strict rules imposed by the International Monetary Fund and other international agencies, which believed in market fundamentalism (the narrow notion that only free markets can alleviate social problems), stringent limits were placed on the economic policies of debtors, with the result that a majority of citizens in these nations often found themselves worse off. The goals of *IMF conditionality,* as it came to be called, were to restore growth, reduce central government involvement in the economy, and expand the exports of goods and services while reducing imports so that the debtor would have sufficient earnings of foreign revenue to make payment on the interest and principal of their debt.

There is little evidence that these policies helped to restore economic growth, and they even lowered many people's quality of life, as the former chief economist of the World Bank, George Stiglitz (2002), noted, by forcing drastic cutbacks in necessary government services and forcing currency devaluations that drive up the cost of imports. However, such changes did result in export surpluses that made debt servicing easier. As a consequence, the 1990s saw a remarkable reversal in the flow of financial resources—from the flow from rich nations to poor nations to assist in the development effort, to a flow from poor to rich. Debt repayments have become a serious obstacle to further economic development in poor countries where capital and financial resources are scarce and every dollar lost has repercussions through the economy.

FOUR MAJOR QUESTIONS OF THE WORLD ECONOMY

A central problem of economic geography deals with *scarcity* in space and the methods to overcome it so that a fair distribution of resources can occur. The problem of scarcity does not strike people in the United States or Canada in the same way as it affects the developing world. Most inhabitants of the First World are used to seeing grocery stores and shopping malls loaded with food and consumer goods, but it is quite the opposite situation in much of the rest of the world. Scarcity in space has long been a problem for humankind, and much of this text aims at describing the processes whereby scarcity has originated and how the world can overcome it.

Four basic questions of economic geography and the world economy arise for each country and each economic system in an effort to overcome every society's scarce resources (Figure 1.10). The answers to the questions will vary by the type of political economy—capitalist, *command economy,* traditional economy, or some hybrid form.

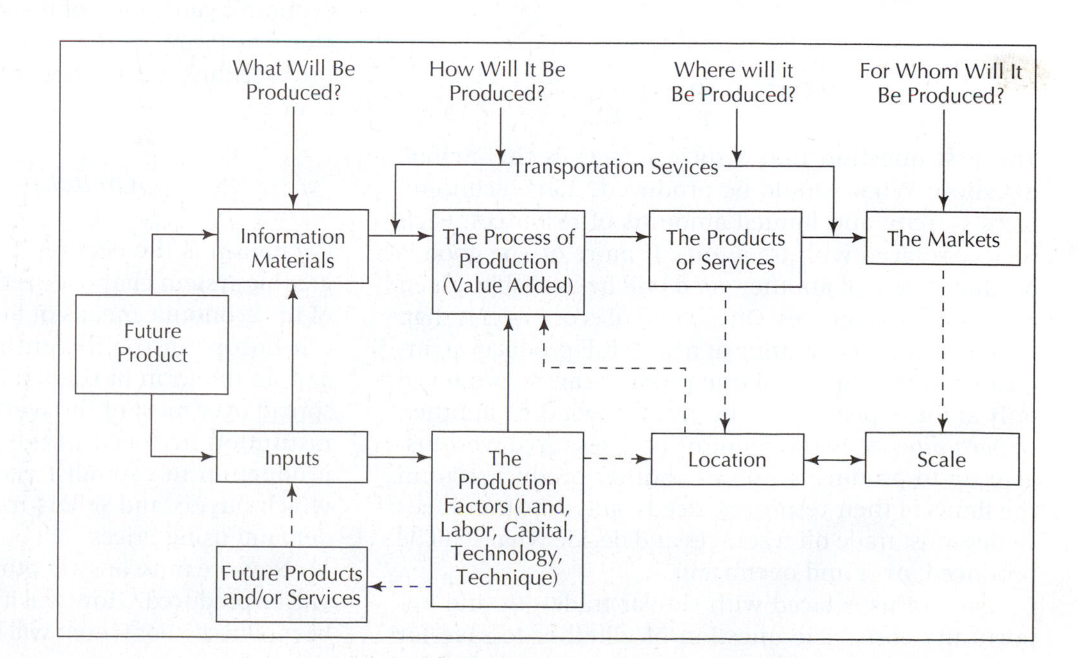

FIGURE 1.10
The four questions of the world economy.

Question 1: With limited resources available, exactly *what* should be produced and at what level or scale of production? What is the mix of outputs that a society produces with a given mix of inputs?

Question 2: *How,* or with what combination of techniques, labor and capital inputs, and other resources, should the output be produced? What dictates the investment levels in an economy, its productivity growth, and how many workers will be employed?

Question 3: *Where* should the output be produced? A related question is, Why is production located where it is? Why are some types of economic activity clustered in space whereas others are dispersed over vast areas?

Question 4: Which groups should receive what share of the goods and services? How should a society's output be divided by class, gender, age, ethnicity, and among regions? That is, *who* will receive the goods and services it generates?

There are no simple, easy answers to these issues. They involve complex mixtures of elements that include various facets of a society's historical background; its culturally constructed notions of fairness, efficiency, and equality; the particular position it has in the world system; levels of productivity growth; and the impacts of the policies of its government. How these questions are addressed determines to a large extent a society's overall well-being and the life chances of its inhabitants. No two societies answer these issues in precisely the same way, which means that the types and levels of inputs, the ways in which they are combined to produce outputs, and the social and spatial distribution of those outputs must be understood geographically.

What Should Be Produced?

The first question that a society faces is the "what" question: What should be produced? Each economy faces scarcity and limited amounts of resources; each society is faced with trade-offs. If more of one good is produced, less of another good will be available, given a fixed set of resources. One axiom of economics is that, for an economy operating at nearly full production, increases in the output of one product can be produced only at the expense of, or by producing less of, another. *Opportunity cost* is the amount of a resource we must give up to produce a unit of another product. Within the limits of their resources, needs, and time frame, societies must trade off resources and decide what shall be produced, over and over again.

Each of us is faced with similar trade-offs and opportunity costs. The question of whether to save for graduate school or buy a new car may be a difficult choice. We may choose to live near our downtown office on expensive land but have a short commute, saving time. Or we may choose to live in a distant suburb on a cheaper piece of land (per unit area), consume more of it than the downtown condo would require, but have a farther (and more costly from an opportunity-cost perspective) commute.

Answering the "How" Question

The goal of firms under capitalism is to maximize profits, which generally means increasing levels of production and consumption. This involves techniques to measure increased benefits from higher production and answering the question "for whom," so that a society can measure the benefit of distributing goods to social groups. These factors, used to determine how production takes place and who receives the goods, depend on the type of political economy—capitalism, command economy, mixed systems, or a traditional system, as previously described.

Political Economies

Briefly, we discuss the political economic systems of the world: *capitalism;* the *command economy* of the communist states; and the *traditional economy* of some developing nations. The term *political economy* is often applied to these types of production because they are more than simply economic systems but involve political relations of power, including class, that shape the allocation of outputs among their inhabitants. Of these, capitalism is by far the most important and widespread; indeed, the economic geography of the globe can largely be understood in terms of the dynamics of capitalism and its intersections with other forms of political economic systems.

Capitalist Economies

Capitalism is the economic, social, political, and geographic system characterized by the private ownership of the economic means of production. It arose in western Europe in the fifteenth and sixteenth centuries, and, in the form of colonialism, ultimately came to be spread over most of the world today. The fundamental institution involved in the organization of factors of production in capitalist economies is the market by which buyers and sellers interact through supply and demand using prices.

Now we must answer our four basic questions: What will be produced? How will it be produced? Where will it be produced? For whom will it be produced? The guiding imperative in capitalist economies is profit, the difference between revenues a firm receives and its production costs. Only profitable products will be produced, based on market demand and price. Prices reflect the utility and value of the good, based on consumers maximizing their own interests, although demand is often constructed through advertising. How and where the goods are produced is based on labor and technology efficiency and the spatial distribution of production costs. In competitive market economies, the most efficient producers are the survivors; their production processes and locations will dictate how and where goods will be produced.

Capitalism features two major groups of decision makers—private households (and individuals) and businesses or corporations. The mechanisms that operate to bring households and businesses together are the *resource market* and the *product market*, which refer to the supply and demand for the inputs and outputs of the production process, respectively (Figure 1.11). Thus, resource markets organize capital, land, and labor together to produce goods and services; product markets consist of buyers and sellers of those outputs. These markets are tied together through flows of capital (between businesses and resource markets), labor and wages (between households and resource markets), consumption expenditures for goods and services (between product markets and households), and sales revenues and profits (between product markets and businesses).

Despite widespread popular opinion, capitalism consists of more than just markets. A commonly held view of capitalism is that is synonymous with the free market with minimal governmental intervention, a system sometimes called laissez-faire. However, historically truly free markets (with zero government rules) have never really existed. Governments have always been an integral part of such systems, including the provision of the infrastructure, protecting property rights, providing public services such as education, and protecting producers from foreign competition, including immigrant labor. Indeed, the argument can be made that markets could not exist without some role of the state. This means that the various forms of capitalism are mixed systems in which both markets and governments are important decision makers, including vital domains such as transportation, education, and health care. The balance between markets and governments will vary widely among countries and over time, ranging from those with high levels of government intervention, such as in Scandinavia, to those with relatively little, such as the United States.

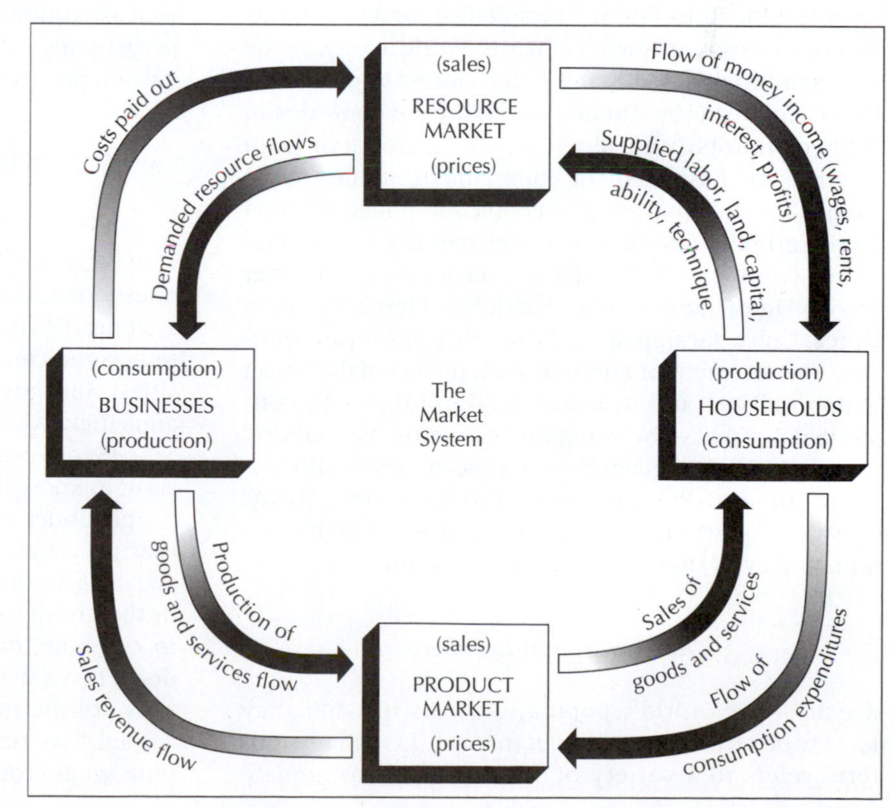

FIGURE 1.11
Circular flows in the capitalist economy. The circular flow in the capitalist economy involves a resource market where households supply resources to businesses and where businesses provide money income to households. It also consists of the product market where businesses manufacture and produce goods and services for households, while households provide money revenue from their wages and income to consume such goods and services. In the resource market, shown in the upper half of the diagram, households are on the supply side and businesses are on the demand side. The bottom half of the diagram shows the product market; households are on the demand side and businesses are on the supply side.

The Command Economy

The system of political economy that stands in greatest contrast to capitalism is the *command economy*, sometimes known as communism. It is a form of totalitarianism, in which the means of production and property and resources are owned and controlled by a central government that serves as the economic planning agency. Prices, levels of production, plant location and size, and any other major decision regarding the use of the means of production and resources, as well as the distribution of output and price levels, are set by the central bureaus of the command economy. Often government ownership occurs through publicly owned corporations, as was the case in the former Soviet Union and its allies. Manufacturing plants and agricultural facilities are mostly owned by the government (although in some former communist states, such as Poland, private ownership in some sectors, such as agriculture, remained significant). Production quotas and output targets are established not on the basis of supply and demand but according to government quota directives. Consumer goods frequently are produced secondarily to capital goods in such economies, or an emphasis is placed on heavy industry, military might, and collectivized farming on huge state-run farms.

How are the four critical questions of economic geography answered in the command economy? The answers to the "how," "what," "where," and "for whom" questions posed are determined on the basis of the state bureaucratic plans, not on market factors.

It is difficult to find pure capitalism or a pure command economy anywhere in the world today. Hong Kong and the United States come closest to pure capitalism, followed by Canada, many of the countries of Western Europe, and Japan, as well as some in Latin America and East Asia. The command economics of the former Soviet Union and the People's Republic of China were certainly the largest experiments of this kind. Other countries imitated the policies of the former Soviet Union and China, including Vietnam, North Korea, Cuba, Nicaragua, and a number of African countries, to one extent or another. With the fall of the Soviet Union in 1991 and its disintegration into its 15 constituent republics, the command economy, as a political economic form of geography, has become virtually extinct. China has become socialist in name only. Today, only Cuba and North Korea practice versions of this system, and even there its future is uncertain.

The Traditional Economy

Much of the world's population lives in what may loosely be termed a *traditional economy*. Essentially, this term refers to a variety of social forms that predate capitalism, often involving elements of feudalism, but occasionally involving aspects of nomadism, tribal societies, and even slavery. These forms of political economy existed for thousands of years and still remain in a few areas of the world, but most have been annihilated by the onslaught of global capitalism. Most such societies are rural in nature and very poor. Generally, ownership of land is a key variable in structuring social class. Most people are peasants and farmers (except in hunting and gathering or nomadic societies), often living in relatively remote rural areas and depending on the land, or similar extractive activities, for their livelihoods. Elements of traditional economies can be found in New Guinea, parts of Africa, segments of the Arab world, and in regions of India, Burma, and South America.

In traditional economies, the "what," "how," "where," and "for whom" questions are answered with the observance of culture, habit, and custom. Price, exchange, distribution, and income are regulated by social convention, kinship, and custom rather than by impersonal market forces. In southern Asia, much of Latin America, parts of the Muslim world, and most of Africa, economic class and caste and heredity define the social and economic roles of individuals and their upward mobility or lack thereof. In deeply traditional societies, there is often a strong fatalism, that is, an ideology that maintains that economic conditions are locked and will not improve and that one's individual position in society cannot change. Religious and cultural values frequently dictate the societal norms as well as economic aspirations, and the status quo is self-perpetuated. Often such systems involve aspects of capitalism, particularly in urban areas.

GEOGRAPHICAL INFORMATION SYSTEMS

One increasingly common way of studying geographic questions and issues consists of geographic information systems (GISs). The use of GISs has grown dramatically to become commonplace in many businesses, universities, and governments. GISs are now used for an amazingly wide range of applications (Figure 1.12).

GISs are used to link different data sets. Suppose we want to know the death rate due to cancer among those people under 10 years old in each county. Suppose also that (as usual) we have the numbers of people of this age in each county in one file and the numbers of death in the group for each county in another file. We need to combine, or link, the two data files, which a GIS does. If we wish to answer questions such as, "What soils are the most productive as far as wheat is concerned?" we need to overlay the two data sets and compute what crop productivity exists on each and every

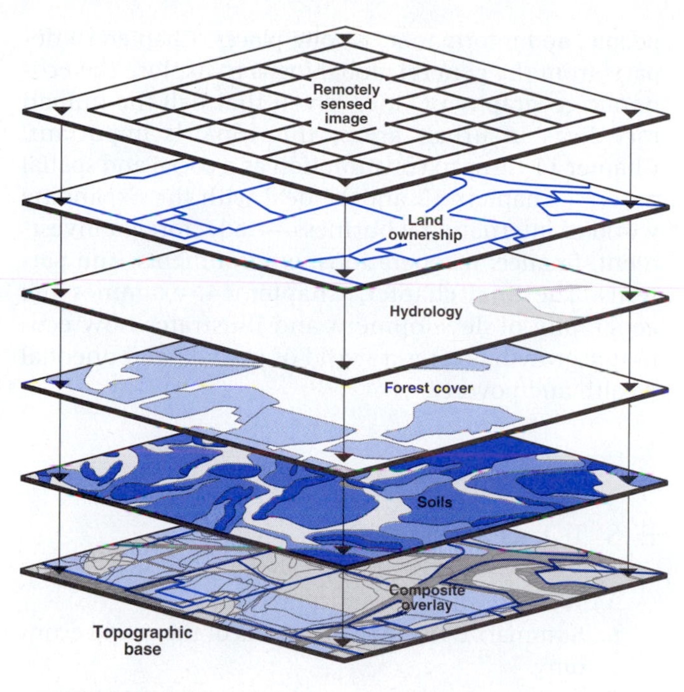

FIGURE 1.12
A geographic information system (GIS) organizes spatial information into overlapping layers, each of which contains a type of human or environmental information.

type of soil. In principle, this is like laying a map on tracing paper over another and noting the combinations of soil and crop productivity. The important point is that a GIS can do these operations because it uses geography or space as the common key between the data sets—information is linked together if it relates to the same space as does another set of information. The largest collection of geographic data yet assembled is the amazing volumes of satellite imagery collected from space. Unlike much other (vector) geographic data, these come in raster (or grid) form—small square areas of ground are each represented by one or more numbers that describe the properties of the ground area.

The applications of GIS are varied. In Europe, the main effort went into the building of land registration systems and environmental databases. Land-use suitability mapping is a GIS technique that can help find the best location for a variety of land-use developmental actions, given a set of goals and other criteria. The mapping technique is based on environmental and human processes, and it analyzes the interactions among three sets of factors: location, land-use development actions, and environmental effects. The technique can yield maps that show (1) what land use will cause the least change in environmental processes; (2) qualitative predictions of environmental impacts of proposed land-use developments, given certain land-use developmental actions to be carried out and specific environmental actions to be controlled; and (3) the most and least suitable locations for those land-use actions.

At the heart of environmentally sensitive, systematic land planning and in the pivotal position between analyses and the definition of alternatives are suitability maps, which assess the ability of each increment of land under study to support a given use. As the technique is now developing, the assessment of suitability is commonly based on predictions of the results likely to come about if a certain development is placed on a particular piece of land. Thus, in suitability mapping, environmental impact analysis and land planning can be effectively merged into a spatial decision support system. The suitability mapping process, then, begins with a collection of information forming coherent descriptions to these three component sets of factors and a means of defining the connections among them. These three sets of information—describing locations, developmental actions, and environmental effects—form the three legs of the tripod on which the suitability mapping process is built.

Generally, at the most basic level, these include the horizontal layers of the biosphere, starting with bedrock composition and other geological factors and proceeding upward through soil types, hydrology, plant and animal communities, and microclimates. To this we often add human contributions to the environment, such as existing land uses, transportation routes, accessibility, and political and social boundaries. Soil types can be grouped by various characteristics, such as bearing capacity, expansion potential, or porosity. Hydrology might include groundwater basins and their capacity and recharge areas. Often, streams and tributaries are mapped to second, third, or fourth orders. Sometimes rates of flow are included. Human-made factors can likewise vary in their level and type of detail.

SUMMARY AND PLAN

This book explores the economic geography of capitalism, especially at a global scale. Although it is important to understand the local and national levels of economic activity, the rapid growth of the world economy has increasingly focused attention on processes, problems, and policies at an international scale. In this chapter, we present the geographer's perspective. We provide a definition of the field and introduce the main concepts geographers use to interpret and explain world development problems at a variety of scales, ranging from small areas and regions to big chunks of the world.

The following chapters of this text, which progress in logical sequence, are organized around the themes of distribution and economic growth. Chapter 2 provides a historical overview of the development of capitalism. Chapters 3 and 4 deal with population and resources, respectively, issues of major significance in economic geography. Chapter 5 summarizes many of the concepts and theories that inform the analysis of economic landscapes. Chapters 6 through 8 apply these ideas to the primary, secondary, and tertiary economic sectors, respectively, or agriculture, manufacturing, and services, examining the unique dynamics of industries in each sector and how they change over time and space. Chapter 9 dwells on transportation and communications, fundamental industries in the movement of goods, people, and information among places. Chapter 10 departs from the general global focus to explore the economic geography of cities; given that half the human race lives in urban areas, this topic is important. Chapter 11 turns to consumption as a social and spatial process. Chapters 12 and 13 deal with the expanding world of international business—trade, foreign investment, finance, its operations, environments, and patterns. The final chapter, Chapter 14, examines the geography of development and illustrates how economic growth creates a world of uneven and unequal wealth and poverty.

STUDY QUESTIONS

1. What defines the geographic perspective?
2. Define the term *globalization* and list reasons why it has occurred.
3. Is a world culture emerging? Why or why not?
4. How do MNCs contribute to globalization?
5. Who are the poor in the world?
6. Summarize the four questions of the world economy.
7. Constrast capitalistic and command economies.

KEY TERMS

abstract space
accessibility
capital
circular flowing market system
command economy
comparative advantage
demand
development
distance decay effect
economic geography
entrepreneurial skill
environmental determinism
factors of production
Four Tigers
friction of distance
gross domestic product (GDP)
gross national product (GNP)
greenhouse effect
hegemonic power
interest
international economic systems
international economic order
internationalization of capital
labor
land
law of demand
law of supply
location theory
logical positivism
mixed economic systems
opportunity cost
phenomenological approach
political economy
product market
profit
purchasing power parity
raw materials
relative location
rent
spatial integration
spatial interaction
spatial organization
spatial structure
structuralist geographers
Third World debt crisis
traditional market
transnational corporation
world economy

SUGGESTED READINGS

Dicken, P. 2004. *Global Shift: The Internationalization of Economic Activity*, 4th ed. New York: Guilford Press.
Friedman, T. 2005. *The World Is Flat: A Brief History of the Twenty-first Century.* New York: Farrar, Straus and Giroux.
Knox, P., Agnew, J., and L. McCarthy. 2003. *The Geography of the World Economy,* 4th ed. London: Edward Arnold.
Rubenstein, J. 2001. *The Cultural Landscape: An Introduction to Human Geography,* 7th ed. New York: Macmillan.
Stiglitz, J. 2002. *Globalization and Its Discontents.* New York: W.W. Norton.

WORLD WIDE WEB SITES

USGS: NATIONAL MAPPING PROGRAM

http://www.usgs.gov

This site provides accurate and up-to-date cardiographic data and information for the United States. These data products and information provide a framework of spatial information needed by federal, state, and local government agencies, as well as the private sector, to deal with such problems as conserving our natural resources, identifying and mitigating hazards, defining and studying ecosystems, and supporting economic development.

USGS: THE GEOGRAPHIC NAMES INFORMATION SYSTEM

http://www.usgs.gov

The GNIS, developed by the USGS in cooperation with the U.S. Board on Geographic Names, contains information about almost 2 million physical and cultural geographic features in the United States. The GNIS is the official repository of domestic geographic names in the United States.

WORLD TRAVEL GUIDE

http://www.wtgonline.com/country/

Maps, climate graphs, and country information.

YAHOO

http://www.yahoo.com

Regional, country, or U.S. states; extensive information about all geographic aspects of places; Society and Culturelinks to cultural information about countries; links to many other sites.

YAHOO ALTA VISTA WEB PAGES

http://av.yahoo.com/bin/

Many links, such as to the CIA world fact book, with diverse information.

ESRI—ENVIRONMENTAL SYSTEMS RESEARCH INSTITUTE

homepage: http://www.esri.com

jumpstation:
http://www.esri.com/services/jumpstation/jumpstation.html

Links to government sites, commercial sites, and educational sites with geographic information.

CIA PUBLICATIONS AND HANDBOOKS

http://www.odci.gov/cia/publications/pubs.html

World factbook with diverse geographical data about countries.

USGS

http://www.usgs.gov/

Topographic maps, other maps, physical geographic data.

U.S. BUREAU OF THE CENSUS

http://www.census.gov

For a good introduction to desktop mapping and the latest news about TIGER and other data sources, visit **http://www.wessex.com**

INTERNATIONAL GOVERNMENT'S META-PAGES—NORTHWESTERN UNIVERSITY

http://www.library.nwuedu/search

Mike McCaffrey-Noviss of the Northwestern University Library Government Publications department provides no-nonsense, utilitarian pointer pages to over 60 country government pages and nearly 100 international government organizations (IGOs). Sites included are the official government or organization sites; pages are updated biweekly and aim for breadth of coverage. Depth of coverage of both government and IGO sites varies by country. International regional and topical pages are forthcoming.

"GLOBAL LEGAL STUDIES JOURNAL" INDIANA UNIVERSITY SCHOOL OF LAW

http://www.law.indiana.edu/glsj/glsj.html

Indiana University's *Global Legal Studies Journal,* a biannual "peer-reviewed interdisciplinary journal focusing on the intersections of global and domestic legal regimes, markets, politics, technologies, and cultures," is highlighted by a symposium issue each fall. The latest symposium issue concerns feminism and globalization of market forces.

CHAPTER

2

THE HISTORICAL DEVELOPMENT OF CAPITALISM

OBJECTIVES

- To explore the historical context of capitalism
- To provide an overview of the characteristics of capitalist economies
- To document the importance of the Industrial Revolution and its impacts
- To shed light on the relations between colonialism and global capitalism

"A factory during the Industrial Revolution reflects the exploitative labor relations that accompanied and underpinned the growth of modern capitalism, including the frequent use of child labor."

Geographies are not created overnight. The spatial distribution of people and economic activities reflects the imprint of processes that often take years, even centuries, to unfold. For this reason, a historical understanding of economic landscapes is necessary. Because the present is produced out of the past, and shaped by it in countless ways, any serious understanding of economic geography must include an appreciation of how the contemporary world came to be. A historical appreciation reminds us that the construction of the modern world took a long time to occur and that the landscapes of the present are constantly changing.

This chapter provides a historical appreciation of capitalism in several ways. First, it delves into the context in which capitalist economies and societies were born and developed, particularly feudalism. Second, it explores the characteristics of capitalism, the features that make it unique. *Capitalism*—the dominant form of production and consumption around the world—is not the only way in which human beings have organized themselves but came into being in the sixteenth and seventeenth centuries, mostly in western Europe. Third, this chapter turns to the Industrial Revolution, which began in the eighteenth century and marked an exponential increase in the scale and speed of capitalist activities. Finally, it addresses the relations between capitalism and colonialism, the process by which capitalism "went global," spilling out of Europe and effectively conquering the rest of the globe.

Feudalism and the Birth of Capitalism

Human beings have developed many ways of organizing resources and production systems and providing for themselves over time. For the vast bulk (90% or more) of human existence, we were hunters and gatherers, food collectors depending on nature for food and other necessities. The agricultural revolution that began roughly 10,000 years ago (Chapter 6) saw a major transition in the ways in which people worked and lived, including settlements and the first class-based societies. Many types of societies were based on slavery, which in Europe culminated in the Roman Empire that ended in the fifth century A.D.

Prior to capitalism, the prevailing form of economic and social relation was *feudalism*, which lasted for more than a millennium (approximately from the fifth to the fifteenth centuries). Sometimes called the Middle Ages or Dark Ages, feudalism was a deeply entrenched society that was stable for a long period. Feudalism was not unique to Europe, as other places had similar types of society, including Japan and to some extent India. Politically, this type of system was manifested in Europe as a changing set of empires, including the Frankish kingdoms, the Normans, the Holy Roman Empire, tsarist Russia, and the Austro-Hungarian Empire, which ended only in World War I. Indeed, one of the major differences between Europe and the United States is the impact that feudalism had in Europe: In North America, capitalism emerged on a landscape that had not been shaped by more than a millennium of feudalism, including its land use and property system, cities, and class and gender relations.

Characteristics of Feudalism

Feudalism was marked by a distinct set of interlocking characteristics that made it qualitatively different from capitalism. This aspect is important to remind us of the uniqueness of the economy and social system in which we live and work.

Compared with the dynamic, ever-changing world in which we exist today, feudalism was a remarkably stable and conservative world that exhibited very little change. To an observer of feudal France in the eighth century and Poland in the eighteenth century, there would appear to be relatively few differences. Most people's lifestyles—working the land in a cycle of endless drudgery—would be unchanged from one generation to another. Almost everyone lived like their fathers and mothers before them and their grandparents before them. Tradition was the dominant shape of human experience and gave everyone a sense of where they fit into the world. In this sense, feudalism actively discouraged experimentation and change. Under capitalism, in contrast, novelty is the norm, for firms use it to sell goods and services in the market.

However, it is erroneous to think of the Middle Ages as completely static. Indeed, this view arose during the Renaissance, when historians sought to contrast the changes of their day with the alleged stasis of the past. In fact, during the later Middle Ages there were significant changes. Universities were established, new types of farming introduced, wetlands drained, forests were cleared, plagues and diseases swept through the land, new technologies were introduced, and political conflicts caused enduring changes. The introduction of the longbow in the fourteenth century, for example, made knights essentially obsolete.

In Europe, the church (i.e., the Catholic Church until the Protestant Reformation of the sixteenth century) was the predominant political/ideological institution. Most people were extremely religious, and their belief in God informed every aspect of their behavior and everyday life. The population fatalistically accepted its lot in life, and the idea of progress, of change for the better, was largely unthinkable. In most towns, the church or cathedral was the largest and most impressive building. Local priests, who were often the only ones who could read and write (and even many of them were illiterate), were important actors in the community's spiritual and intellectual life. Education and schooling emphasized the Bible. In Rome, the pope exerted great power over kings and nobles throughout the continent, often appointing leaders and threatening to excommunicate those who did not obey. Popes were masters of politics, wealthier than anyone else, and often corrupt. The church owned farmland and hunting estates, raised taxes, and even had its own armies.

The ruling class of feudalism was an aristocratic nobility whose power lay in the ownership of land. There were many tiers within this ruling class, including a variety of lords, dukes, earls, and others. In an overwhelmingly rural society, in which the productivity of agriculture was comparatively low, the vast majority of people were peasants and farmers. Ownership of land was the basis of wealth and political power, and unlike capitalism, there was no effective division between public and private property. The aristocracy controlled the reins of government, including the military and penal system.

Farming under feudalism was based on animate sources of energy, that is, living human and animal muscle power. Peasants and draft animals worked the fields, collected firewood, drew water, and performed the innumerable tasks necessary to keep their society working. Child labor on the farms was the norm, and birthrates were high. The population lived in small hamlets and villages, some of which were self-sufficient, producing their own food, clothing, and other necessities. Peasants would be almost completely illiterate, ignorant even of events a few miles away, unaware even of what century they lived in.

In the manorial system that characterized feudalism, the extraction of surplus value occurred through the payments of rent. Tenant farmers paid tribute to their local lords, who owned the land. Often they paid one-half or more of their output as rent, in exchange for protection from invasions and robbers. Thus, rent payments were the primary form of wealth transfer from the poor to the wealthy, which occurred through the state rather than a market.

Markets existed under feudalism but typically were small and poorly developed, and only the wealthy had the income required to buy luxury goods. Typically, markets consisted of seasonal fairs where itinerant merchants sold metal goods, silks, or jewelry. Thus, feudalism was not a type of society in which markets were the central institution that governed the allocation of resources; rather, this function lay in the state.

A substantial share of the rural population, but not everyone, consisted of *serfs* (Figure 2.1). Serfdom was a uniquely feudal institution that differed both from slavery and from capitalism. A serf was not a slave, that is, he or she was not owned by a master. Rather, serfs were bound to the land by feudal law and custom. Serfs and other farmers (including pools of "freemen") lived a monotonous life in which each day was identical to the day before, doing the same chores, eating the same food, and seeing the same people.

Standards of living under feudalism were universally low, except for a small group of the aristocratic elite. Most people lived very simply. Diets were typically inadequate, and malnutrition was common. Famines broke out every few years. Life expectancy in feudal Europe was typically under 50 years. Many women died in childbirth, and infant mortality rates were high. Water supplies were often infected by bacteria, and diseases such as cholera, plague, and tuberculosis took an enormous toll in human lives and suffering.

Agricultural work was organized around the rhythms of the seasons, with different tasks for the spring, summer, fall, and winter. Most people lived in extended families, in which several generations may co-inhabit one dwelling; grandparents, parents, nieces and nephews, cousins, and infants lived in crowded

In feudal cities, the cathedrals were usually the largest and most ornate buildings, testifying to the power and wealth of the church, as well as its hold over the local population.

FIGURE 2.1
Serfs were the mainstay of the feudal labor force, producing the agricultural surplus that supported the nobility and the state. Serfs lived monotonous lives, paying a large share of their output as "rent" to their local lord. They exemplify a noncapitalist form of labor organization (i.e., without labor markets).

rooms, often without paved floors. The vast majority of people were illiterate; even many kings, queens, aristocrats, and priests were unable to read and write. Peasants and farmers passed lessons to their children through rhymes and stories, and the populace was generally very superstitious.

Although feudal society was predominantly rural and agricultural, there were a few cities and towns. Urban areas under feudalism were very different from those of today. Because agricultural productivity rates were low, and the ability of farmers to support urbanites correspondingly limited, cities were small. Most farming hamlets did not exceed 200 or 300 people at most, and cities over 10,000 people were rare. Of course, there were a few metropolitan areas, such as Constantinople or London, but these were few and far between. Feudal cities were densely populated, with the inhabitants crowded together, often in vary unsanitary conditions. There was no running water or sewer system, and the streets were often covered with mud and animal waste. The centers of feudal cities often consisted of a walled fortress, often with a small palace located within where the local lord lived. As the town grew, new walls would be constructed, leading to concentric rings. Because land was not a commodity to be bought and sold, but allocated on the basis of power, there was little differentiation among land uses. Commercial and residential land uses were mixed together, and there was no effective distinction between home and work.

Within the cities, feudal *guilds*, or associations of craft workers and artisans, produced a variety of goods. Guilds consisted of skilled workers with years of experience, and were organized by the type of good they produced. There were, for example, blacksmiths' guilds, weavers' guilds, goldsmiths' guilds, and guilds for bakers, paper making, glass workers, and shoemakers. Young men who were chosen to work in the guild spent years as apprentices learning the trade before becoming craftsmen in their own right.

The End of Feudalism

The so-called Dark Ages in Europe changed relatively little from the sixth to the fourteenth centuries. Innovation and change were discouraged, and feudal society was remarkably stable. However, the late medieval period, starting around the eleventh century, saw a gradual agricultural revolution based on the introduction of the heavy plow, waterwheels, the horseshoe, stirrup, the three-field system of farming, and several other innovations, which were introduced from other, more advanced societies. Other imports included cotton, the compass, sugar, rice, silk, paper, printing, the needle, the zero, and the windmill. Indeed, feudal Europe was the western terminus of a much larger world system that stretched across the Mediterranean, the Middle East, the Indian Ocean, and into Eastern Asia (Figure 2.2).

The introduction of new innovations, which constituted a commercial revolution of sorts, improved agricultural productivity. The supply of agricultural land expanded as peasants and farmers in the late Middle Ages cut down forests and drained swamps to make room for new farmland. This set of circumstances led to a gradual increase in the urban population. By the fifteenth century, much of western Europe was carpeted with a growing network of cities, called "newtowns" in Britain and villanovas in Spain. The increase in productivity underscored a round of cathedral construction.

It must be emphasized, however, that compared with much of the rest of the world, feudal Europe was relatively primitive. Standards of living and rates of innovation in Europe from the fifth to the fifteenth centuries were much lower than the wealthier, more powerful, and more sophisticated societies of the Arab world, India under the Mughals, or China under the Sung and Ming Dynasties. Indeed, many of the luxury goods and innovations that Europeans imported came from these wealthier societies.

Carcassone, France, reflects the geography of the feudal city, with defensive walls used for military purposes.

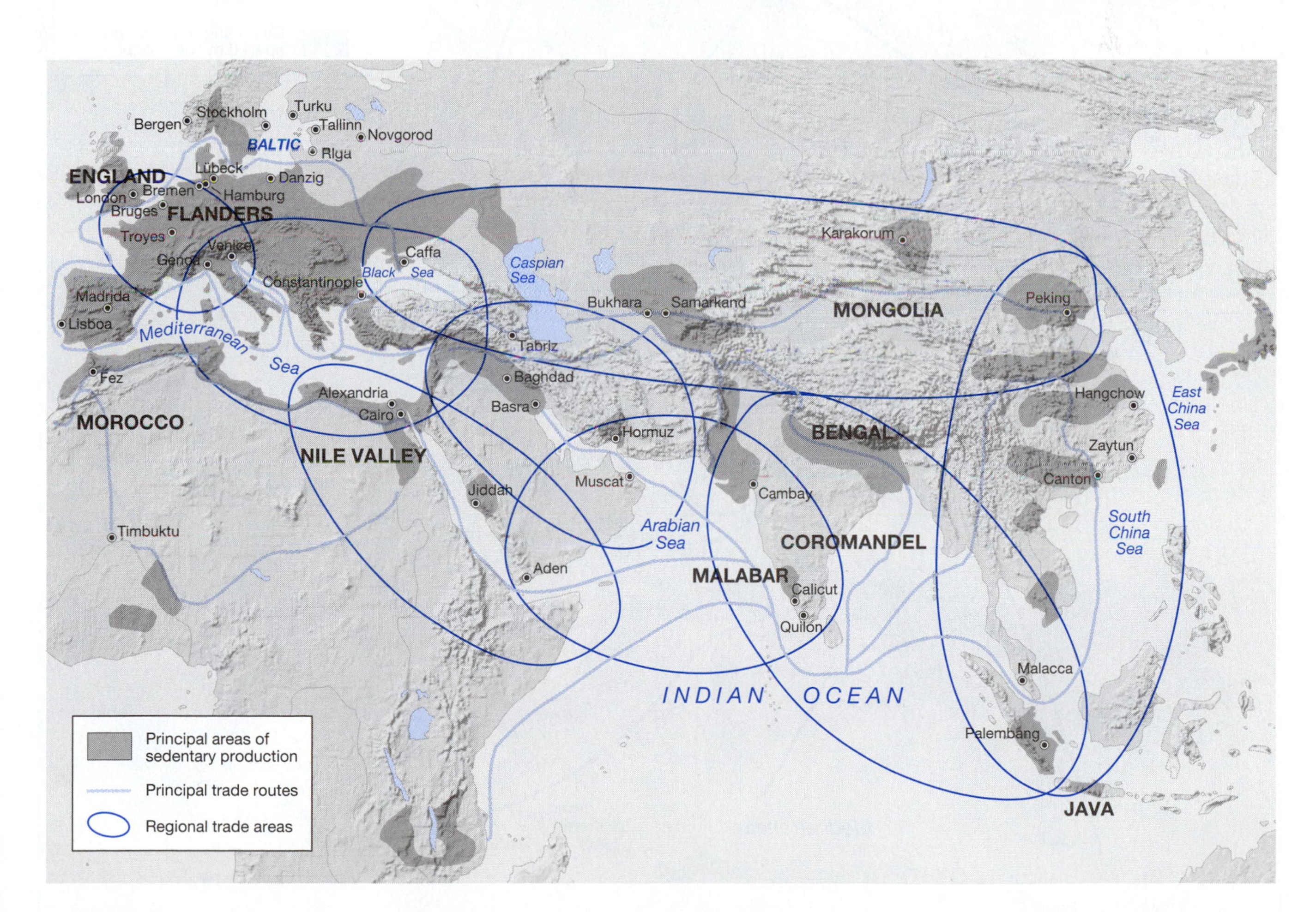

FIGURE 2.2
The precapitalist world-system of the fourteenth century. A diverse array of societies stretching from feudal Europe to the Middle East formed one network embedded within a much broader trade system. Across the Indian Ocean, southern India, Indonesia, and China under the Sung and Ming Dynasties constituted another set of networks. Overland trade along the Silk Road routes formed yet a third circuit. Unlike the capitalist world system that arose later, in the sixteenth century, this one lacked a distinct core.

Among the other things introduced to Europe was a bacterium, *yersinia pestis,* which causes a deadly disease commonly known as bubonic plague. In Asia, this disease was common in rodent populations in the steppes, or grasslands. Among humans, *bubonic plague* is highly contagious and was called the Black Death for the dark, swollen lymph glands it produced. In 1347, a ship carrying plague landed in Genoa, Italy, which was part of the expanding trade network between Europe, the Middle East, and Asia. Within four years, one-quarter of Europe's population—more than 25 million people—was dead. The plague raced through the crowded, unsanitary cities, in places annihilating the majority of inhabitants. From southern Europe, it spread north, to Germany, Britain, Scandinavia, and Russia (Figure 2.3).

Several historians have speculated that the plague played a major role in knocking feudal Europe off of its base, destabilizing it and opening the door to a new type of society. Within a few years, much of the continent experienced labor shortages; Europe went from being a land-poor, people-rich set of societies to a people-poor, land-rich group. The disintegration of legal systems allowed serfs to run away without fear of being caught and returned. Others have argued that feudalism was suffering from numerous problems anyway and would have collapsed without the plague. In some

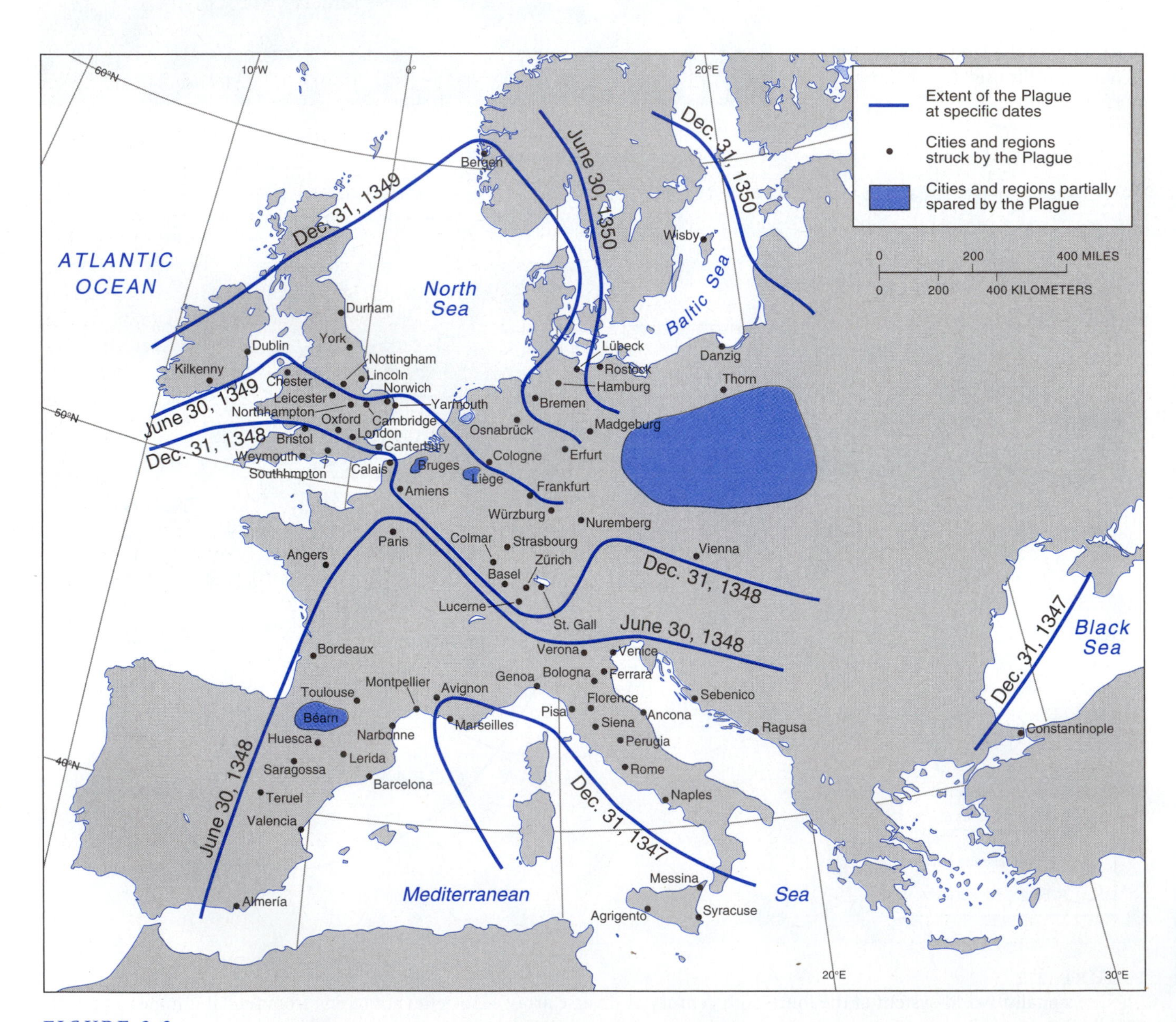

FIGURE 2.3
Diffusion of the Bubonic Plague through fourteenth-century Europe. The plague's devastation knocked feudal Europe off its equilibrium, creating labor shortages and destabilizing the social structure. Some historians argue that the plague facilitated the way for capitalism to emerge.

cities, such as Florence, Italy, capitalist social institutions such as banks were already emerging before the fourteenth century. In any case, combined with other changes, including a mini–Ice Age, the Hundred Years War between France and England, and the Crusades, feudalism in Europe began to crumble, and in its ashes a new economic, political, and social system emerged: capitalism.

The Emergence and Nature of Capitalism

Over several hundred years, from the fifteenth to the nineteenth centuries, feudalism in Europe was gradually replaced by a new kind of society, capitalism. Thus, feudalism served as a womb that incubated what would ultimately become the most powerful type of economic and political system in the world.

If capitalism can be said to have a birthplace, it would most likely be in northern Italy. The city-states of this peninsula, such as Florence, Venice, Pisa, and Genoa, played a key role in fomenting the new kind of society. They had large groups of wealthy merchants with active commercial ties to the Middle East (Figure 2.4), including the famous Medici family of Florence, with vast holdings in silver mines, silk production, and banking. In northern Europe, a network of cities traded in the Hanseatic League, which stretched from Russia and Scandinavia across northern Germany and into the North Sea (Figure 2.5). In these centers, the rising groups of *burghers*, or merchants, accumulated wealth and power that would make them the dominant figures in the new social formation.

Like feudalism, capitalism possesses a distinct set of characteristics that define it and give it its unique form. The major features of capitalism include those discussed in the following subsections.

Markets

Unlike feudalism and slavery, in which political power through the state is the principal way in which resources are allocated, under capitalism the most important institution is the *market*. Markets consist of buyers and sellers of *commodities*, which are goods and services bought and sold for a price. Thus, the expansion of market societies saw the steady commodification of different goods and services, including food, housing, clothing, transportation, education, medical care, and other domains. Not everything under capitalism is a commodity (e.g., air); only scarce goods (i.e., those that command a price and can generate a profit for producers) can be classified as such.

There is a huge variety of markets based on the type of commodity being produced and consumed, as well as the amount and nature of competition. Markets range from being free wheeling and highly competitive, with many small producers, to large ones dominated by a few major producers, or oligopolists. In market-based societies, private property, and the right to own it, are key requirements to production. Markets thus do not exist or function well without appropriate legal guarantees to protect property rights. The incentive of producers to sell goods and services is *profit*, the difference between gross revenues and production costs. Thus, markets involve production for exchange, rather than subsistence or use. In this sense, capitalism involved the triumph of the private sphere.

Because markets involve competition among different producers, there is a strong incentive to produce goods and services cheaply, efficiently, and to please consumers. Thus, markets contain a powerful incentive to innovate, which is largely responsible for making capitalism a dynamic society. In this sense, capitalism differs considerably from noncapitalist societies in that it rewards innovation, change, and risk taking.

Markets are not unique to capitalism. There were markets in slave-based societies such as Rome. In feudal Europe there were occasional markets in the forms of annual fairs and festivals, which brought traveling merchants and local populations together. However, markets are unique in their *importance* to capitalism: Only under capitalism are markets the major way in which resources are allocated. It is worth noting that markets are not the *only* way in which resources are organized, for even in ostensibly free market societies the state, or government, plays a key role (Chapter 5).

Class Relations

Although capitalist societies are sometimes depicted as being devoid of class, market societies in fact do have a system of classes of different kinds. In contrast to feudalism, the class system of capitalism reflected a broad-based shift from a hierarchy based on tradition to one based on money, from born rank to earned status. Historically, this process saw the ascendancy of the merchant class, or what is sometimes called the bourgeoisie (the middle class of feudalism, the ruling class of capitalism).

As the merchants and burghers of Europe gained wealth, power, and prestige in the fifteenth and sixteenth centuries, they came to control increasingly large domains of their societies and to enjoy mounting political power (Figure 2.6). This process was resisted by the aristocracy, which correctly perceived the newcomers to be a lethal threat to their centuries of rule. The demise of the feudal aristocracy came gradually in some places, such as England (e.g., during the civil wars of the seventeenth century), and suddenly in others, such as France, where the aristocracy collapsed in the Revolution of 1789.

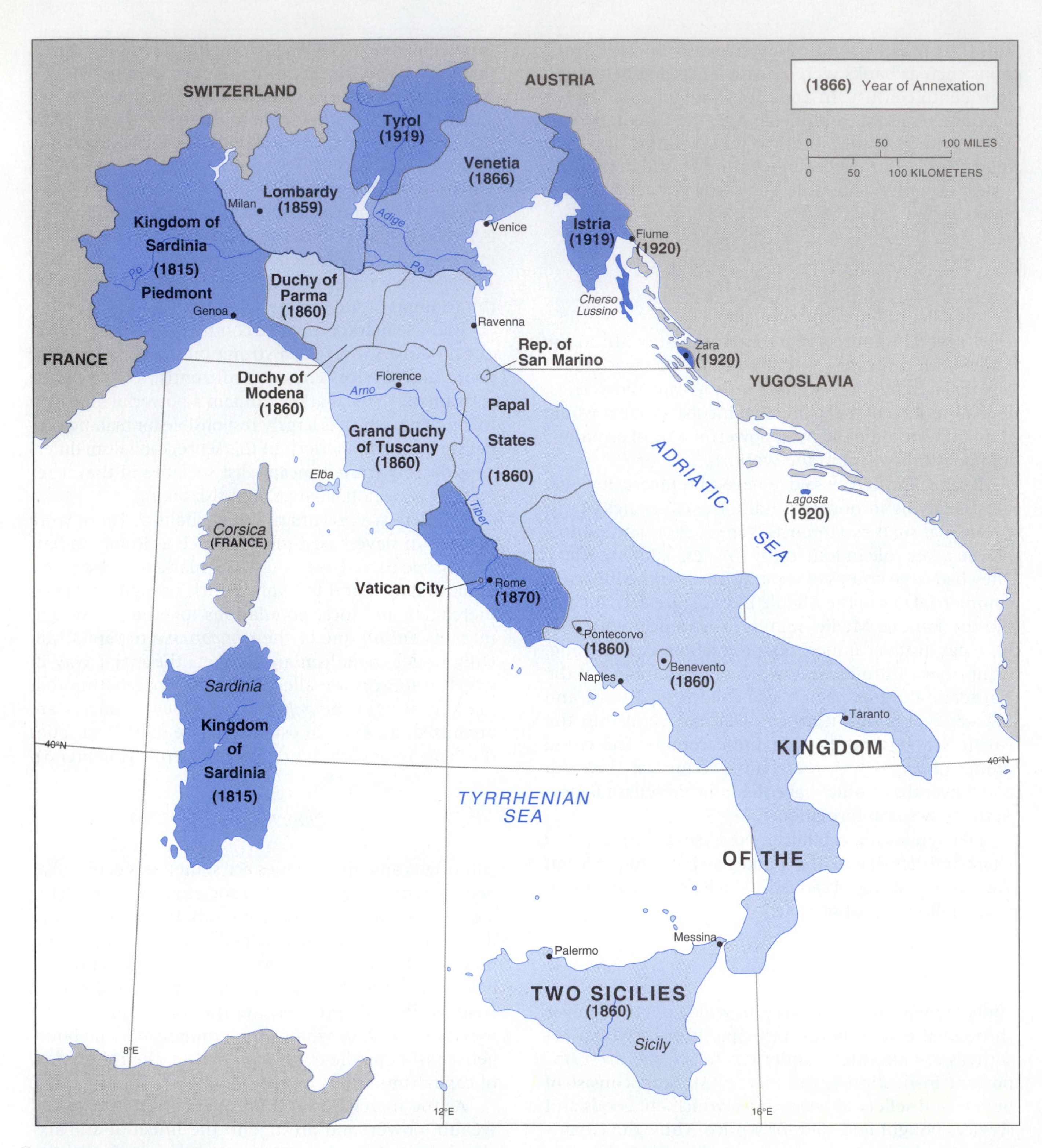

FIGURE 2.4
Italian city-states in the nineteenth century, prior to unification in 1861. Capitalism began in northern Italy, and for much of its history its political geography was characterized by city-states, not nation-states. Italian states such as Florence, Genoa, and Venice, which had extensive trade relations with the rest of the Mediterranean world, also saw wealthy merchant families rise to power.

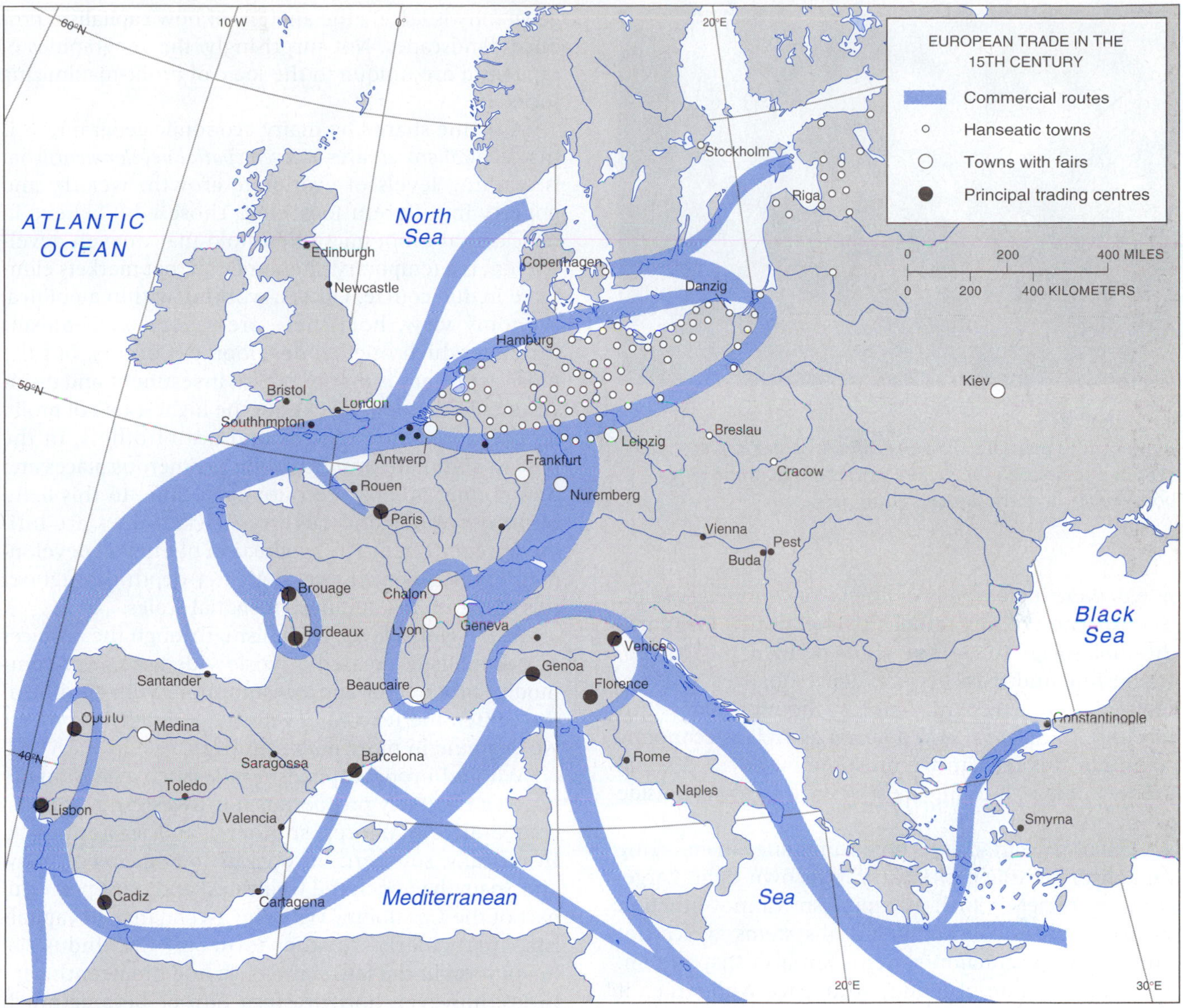

FIGURE 2.5
The Hanseatic League of fifteenth-century Europe was a loose connection of cities and principalities that stretched across the Baltic and North Seas. The emerging burgher or merchant class in those places would be at the forefront of class relations that accompanied the emergence of capitalism.

In addition to the ruling classes, capitalism changed the role and nature of the population of workers. In particular, labor itself became a commodity, bought and sold for a price (wages) in labor markets. Over several centuries, the peasants and serfs of Europe were gradually forged into a working class. In the context of industrial Europe, this population became known as a proletariat. Unlike feudal Europe, workers under capitalism must sell their labor power to survive. Thus, the process of commodification extended to include the capacity to labor.

Finance

The growth of capitalism saw a deep and fundamental change in the role of money, which increasingly became the measure of all worth. In noncapitalist societies, *barter* plays a major role in how economic relations are organized. Goods may be traded for one another, or labor traded for goods. This approach has the advantage of providing transparency in the exchange process. In the context of barter-based economies, money is relegated to a relatively small role. Obviously money existed before capitalism—the Romans, for example, had vast quantities of coins—but under market-based societies money assumed a new level of importance. As the cash system replaced barter, money became standardized and ubiquitous as a measure of value. The commodification of time, space, and labor were measured in monetary terms, so that even human life came to have a financial value. Wealth and

FIGURE 2.6
Rembrandt's painting *Sampling Officials of the Drapers' Guild* (1662) exemplified the wealth and prestige of the new bourgeoisie in early capitalist Europe.

power were increasingly defined along monetary lines, that is, economically rather than politically. Many traditional social roles that were defined by kinship, friendship, and trust became depersonalized and formalized as they were mediated by money. Money thus has important social and political as well as economic roles. For this reason, we must see money as a social product, that is, it cannot exist or have meaning outside of society.

The organization and control of money in emerging capitalism became an industry in its own right. Large, complex societies such as capitalism cannot function without well-established financial systems, which not only reflect production systems but also shape them. The turnover rate of money—the pace with which it changes hands—is important to the process of capital accumulation. With origins in the goldsmiths of feudal Europe, modern banking arose to become a huge industry, linking savers and borrowers of different types and with different needs. By the seventeenth century, commercial credit became widespread, and with it, different types of banks and insurance firms. Joint stock companies spread the risks of large investments over many small producers. Accounting became an important profession. By the nineteenth century, financial systems were increasingly regulated by the state through central banks, which sought to control money supplies and thus interest, inflation, and exchange rates (Chapter 8).

Territorial and Geographic Changes

If capitalism fundamentally changed the rules of societies, it also reshaped how they were organized geographically. Because capitalism is overwhelmingly the most significant economic and political system worldwide, economic geography is largely the analysis of how capitalism produces landscapes. Not surprisingly, the geographies of capitalism are unique to the logic of profit-maximizing societies.

A theme shared by many economic geographers is that capitalism creates *uneven spatial development*, that is, varying levels of economic growth, wealth, and poverty in different locations. Those who believe in neoclassical economics often hold that uneven development is a temporary phenomenon that markets eliminate in due course. Others, working within a political economy view, hold there are several mechanisms through which uneven development occurs, but the most significant is that of capital investment and disinvestments. As capital seeks out the highest rate of profit, it flows into some regions and out of others, in the process simultaneously creating prosperous places and abandoning others to economic decline. In this light, wealthy regions and poverty-stricken ones are intimately connected. The production of uneven development, which we explore at greater depth throughout this book, occurs at different spatial scales.

At the global level, capitalism, through the auspices of colonialism, created a worldwide system of commodity production and consumption, with Europe at the center and its colonies in the periphery, a process we examine in more detail shortly.

Within Europe, capitalism unleashed a division between a relatively prosperous northwestern part and a comparatively poorer southern and eastern part. Historically, southern Europe, including Greece, Italy, and Spain, had been the wealthiest and most powerful part of the Continent. With the ascendancy of capitalism, particularly in the form of the Industrial Revolution in the late eighteenth and nineteenth centuries, however, northwestern Europe became much more advanced economically, particularly following the defeat of Napoleon in 1815 (Figure 2.7). To this day, the societies of western Europe remain wealthier than those elsewhere in that continent.

Within the individual countries of Europe, as well as other nation-states dominated by capitalism, there emerged a steady division between cities and the countryside; as rural areas were reshaped by waves of enclosures and the commodification of agriculture in the form of cash crops, large numbers of people migrated to the urban areas. Under feudalism, in contrast, there were relatively few differences in standards of living between urban and rural areas.

Finally, within cities, urban land, like labor, became a commodity, organized through land markets (see Chapter 10). In the process, profit became the mechanism for separating different land uses, including a division between home and work, areas of production and social reproduction. Gradually, as cities grew larger, particularly under the Industrial Revolution in the nineteenth century, the distances between home

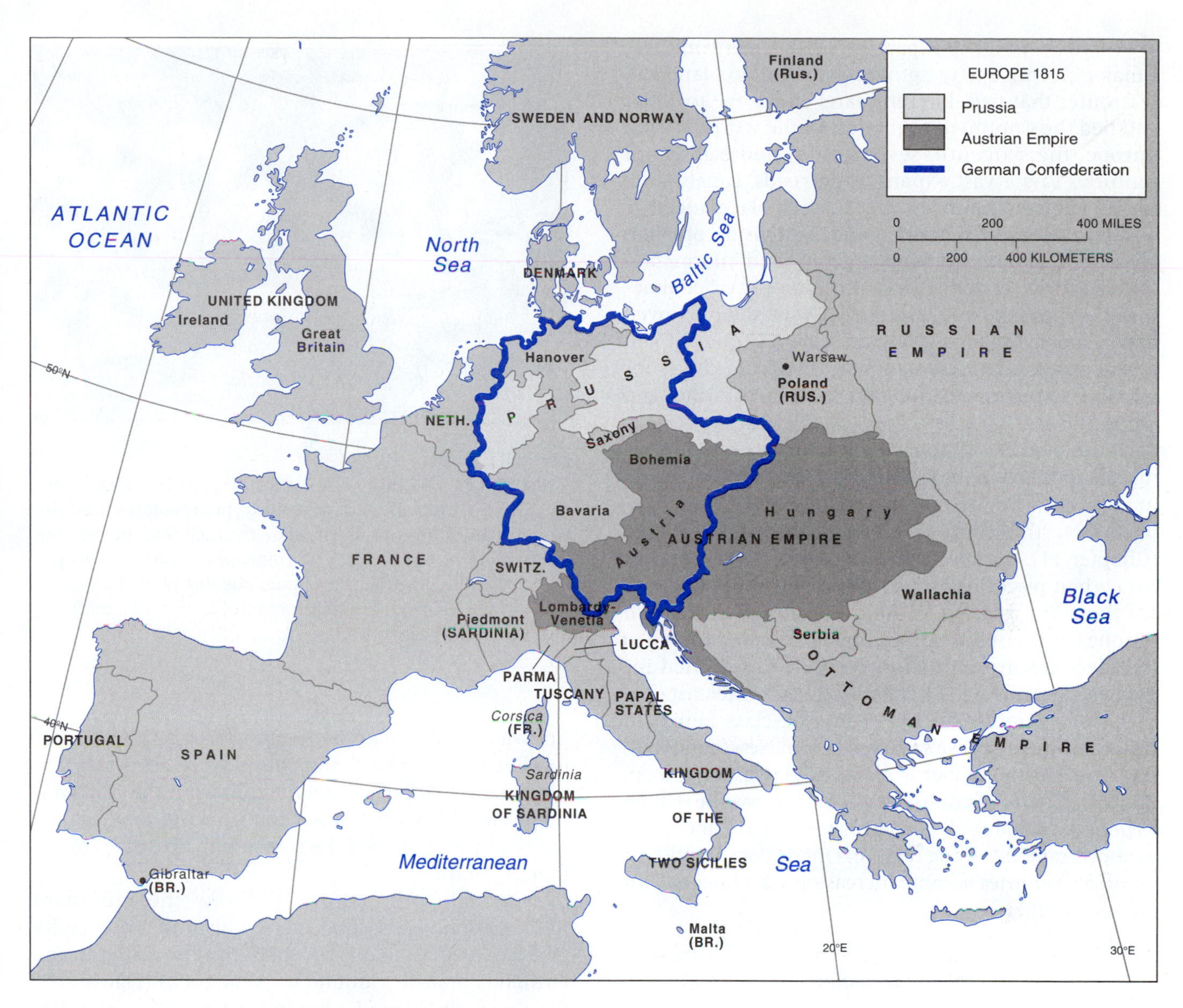

FIGURE 2.7
Europe in 1815, at the end of the Napoleonic Wars. The turbulence in Europe after the French Revolution of 1789 saw the nation-state emerge as the primary political entity in the continent.

and work were stretched to where workers engaged in mass commuting.

The key for the economic geographer is to see all of these scales as different versions of the same process (i.e., uneven development manifested at the global, continental, national, and local scales).

Long-Distance Trade

The ability to buy and sell goods over long distances is a fundamental part of capitalist societies. Trade reflects the geographic organization of exchange, linking producers and consumers who may never see one another. As capitalism took hold and became entrenched across the European continent, trade networks proliferated in diversity and extent. Of course, there was trade prior to capitalism. In feudal Europe, trade with the Muslim world, and through the Silk Road, with Asia, allowed the influx of many goods. Indeed, Europe was the western terminus of a much larger fourteenth-century trading system that stretched across the Middle East, India, and into Southeast Asia and China (see Figure 2.2). Yet prior to capitalism, trade was largely confined to precious goods, such as spices, silks, porcelain, and precious metals. The consumers were only aristocrats, who had the means to purchase such luxuries.

If long-distance trade was peripheral to feudalism—those societies could have survived without such goods—it is integral to capitalism. In market-based societies, trade occurs in all sorts of goods, from luxuries to everyday ones. The expansion of trade networks was a major incentive to the growing networks of land and sea routes that tied different parts of Europe together and tied the Continent to the rest of the world. Within Europe, the sixteenth, seventeenth, and eighteenth centuries saw a vast expansion in roads, canals, and, somewhat later, railroads, that sutured places together into an increasingly interdependent division of labor. The speed with which people, goods, and information circulated accelerated, a process commonly called time-space compression (Chapter 9). New ships allowed Europeans to sail long distances relatively quickly, and in the process they created new maps and charts for nautical navigation, learning the behavior of winds and currents.

If trade reflects differences among places in the nature of production, it also helps to shape those places. In economics and economic geography, this idea is reflected in the concept of comparative advantage (Chapter 11), the specialization of production that occurs when places begin to trade extensively with one another. The growth of long-distance trade within Europe, and between Europe and its colonies, helped to fuel declines in production costs and associated increases in standards of living. By the seventeenth century, for example, an upper-middle-class family in Britain could purchase salted cod from Newfoundland, furs from Russia, timber from Scandinavia, wines from France, blown glass from what is now the Czech Republic, and olive oil and citrus from Spain or Greece. In short, capitalist trade relations made consumers better off as countries became increasingly interdependent on one another.

FIGURE 2.8
Guttenburg's invention of the printing press in 1450, using the Chinese innovation of movable type, revolutionized the way in which ideas could circulate through late medieval or early capitalist Europe. It is no coincidence that the cheap books that the printing press made possible played a big role in the European Enlightenment of the seventeenth and eighteenth centuries.

New Ideologies

Capitalism, it should be understood, is not simply an economic, political, or geographic set of relations, it is all of these simultaneously. Just as feudalism consisted of multiple dimensions, including an ideological one dominated by religion, so too does capitalism exist both in the economy and in other domains. If market-based systems existed in the ways in which people were organized and interrelated, they also were to be found in how people perceived the world. The emergence of capitalism brought with it a vast panoply of ideological changes that revolutionized the ideas, science, and culture of the modern world.

As early as 1450, the introduction of the printing press made the production of books easier, faster, and cheaper (Figure 2.8). This innovation had a huge impact on the societies of Europe. Ideas of many sorts began to circulate around the continent, and larger numbers of people learned to read and write. In the sixteenth and seventeenth centuries, starting in Italy, Europe witnessed the explosion of artistic and scientific knowledge known as the Renaissance. Although most intellectuals of the Renaissance were religious, the movement marked a dramatic shift from the God-centered view prevalent under feudalism to one that increasingly emphasized the role of human beings in the making of the world.

The sixteenth century also saw the Protestant Reformation, the second great schism in Christianity (the first being the division between the Catholic and Orthodox branches around the year 1000) (Figure 2.9). Starting with Martin Luther in Germany, Protestantism offered a different view of life and God than that of Catholicism. In particular, Protestantism emphasized the role of the individual and his/her direct relation to God, one facet of a much broader growth of individualism in different social spheres. Bypassing priests as intermediaries between individuals and God, Protestantism spread with the growth of literacy in the aftermath of the printing press. The famous sociologist Max Weber studied the relations between Protestantism and the development of industrial capitalism in northwest Europe, arguing that the "Protestant ethic," which stressed delayed gratification, savings, and material success as a sign of God's grace and potential entry into heaven, was instrumental to the development of markets. His view held that Protestantism elevated work into a moral obligation, paving the way for capital accumulation. This perspective has been challenged by those who maintain that Protestantism followed in the wake of market relations rather than causing them, or that at least Protestantism and capitalism coevolved.

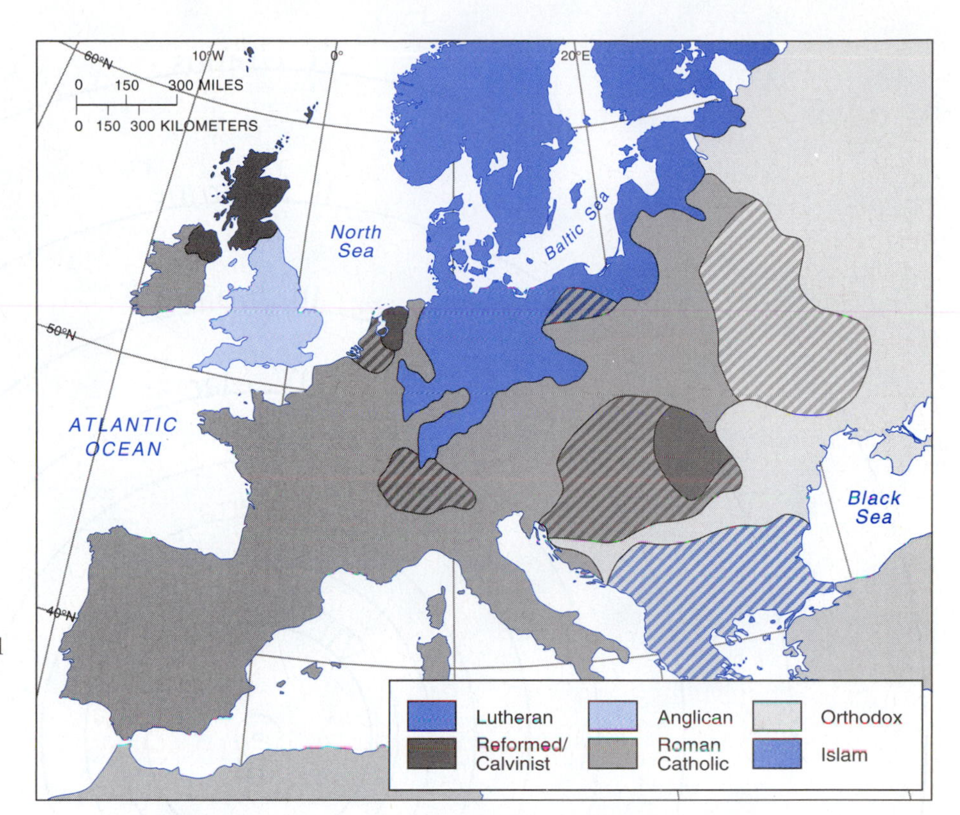

FIGURE 2.9
Europe in the seventeenth century was gripped by severe religious wars following the Protestant Reformation. The rise of Protestantism in northern and northwestern Europe—the second great schism in Christianity—stressed individualism, one of several lines of thought that comprised the new ideological universe of emerging capitalism.

Science in the sixteenth and seventeenth centuries played a critical role in restructuring how people viewed the world and their place in it. The Copernican revolution, augmented by Galileo, led to a heliocentric view of the universe rather than the older Aristotelian one, which placed the earth in the center (Figure 2.10), a shift that accentuated the gradually emerging secularism of the time. In the seventeenth century, modern science was born as scientists such as Newton, Boyle, Bacon, Pascal, and Lavoisier made enormous strides in the understanding of physics and chemistry and their applications to gravity, optics, and other fields.

By the seventeenth, eighteenth, and nineteenth centuries, western societies underwent the *Enlightenment*, an important explosion of science and secular political thought. In Britain, France, Germany, and elsewhere, advances were made in geology, chemistry, physics, and biology, including the discovery of atoms, electromagnetism, and bacteria, leading to the germ theory of disease. The Darwinian revolution of the nineteenth century revolutionized our understanding of evolution and ecosystems. These discoveries demystified nature, subjecting it to scientific law, a process that was accelerated by the proliferation of universities and institutionalized learning through academic societies. A gradual, widespread secularization of culture was underway. In political thought, thinkers such as John Locke, David Hume, and Adam Smith developed a worldview that stressed secularism, individualism, rationality, progress, and democracy.

The Nation-State

Capitalism is not simply an economic system for organizing the production and consumption of resources, it is also a political one. Thus, the emergence of capitalism witnessed a series of political changes across the landscape that accompanied the rise of market-based societies. One of the most important of these was the rise to prominence of the *nation-state*.

A nation is a group of people who share a common culture, language, history, and territory, often manifested in a common identity. Feudal empires had many nations within their borders; the Holy Roman Empire (Figure 2.11) and the Austro-Hungarian Empire, for example, contained dozens of different ethnicities. Individual and collective identity was largely defined in religious terms (e.g., "Christiandom") or as subjects of a particular king. By the eighteenth century, however, as these empires began to disintegrate, the nation had increasingly come to be the primary source of identity of many peoples. Nation and state came to overlap in the forms of nationalism, self-determination, and sovereignty. However, even in the most classic textbook examples of the nation-state, such as France, ethnicity was hardly homogeneous; there were always different ethnic, linguistic, and religious minorities, including, for example, French Basques, Bretons, and Corsicans.

As national and ethnic identities displaced older feudal ones, the political geography of Europe was redefined along ethnic lines. This process played out in key

FIGURE 2.10
The heliocentric Copernican universe, as distinct from the older, Aristotelian geocentric one, formed an important component in the increasingly secularized view of nature that accompanied the scientific and political revolutions of early modernity. Note that the shift from a geocentric to heliocentric perspective on the universe caused an enormous intellectual and political uproar.

events such as the Treaty of Westphalia in 1648, which effectively put an end to the Holy Roman Empire and legitimated the nation-state as the primary unit of international law and relations (Figure 2.12). The French Revolution of 1789 was a major turning point in the breakdown of the old feudal social order and the rise of the new modern one, creating new forms of political identity such as citizens. The Napoleonic Wars that ended in 1815 likewise laid much of the geographic basis for the nation-states of the continent. In western Europe, the centralized monarchies of feudalism were gradually replaced by constitutional republics, something that did not occur in eastern Europe until after World War I. By the twentieth century, the nation-state

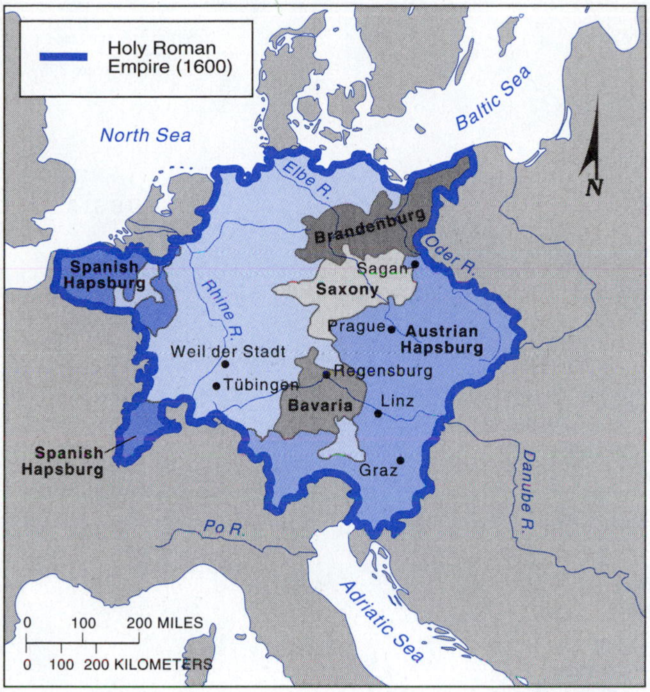

FIGURE 2.11
The Holy Roman Empire in 1600 was characteristic of the ethnically diverse states of medieval Europe. Although it was centered in Germany, at various times it included parts of what are today France, Italy, Austria, Poland, and other places. This social formation exemplifies how ethnicity was not a primary unit of political identity until the emergence of the nation-state.

had become securely entrenched as the dominant form of political organization throughout the world, including many former European colonies.

The emergence of market societies facilitated the growth of nation-states in several ways. Rising wealth, the diffusion of mass literacy, growing cadres of the intelligentsia, and political parties demanding a role in the newly democratic societies were all part of this process. Other institutions were also important, such as the creation of national banks and currency, a military draft, the media, and national rail systems, which tied together the diverse parts of the emerging nation-states. Thus, capitalism is not simply a society that produces markets, it makes *both* states and markets. Conversely, the new nation-states facilitated markets in several ways, including, for example, the construction of an infrastructure, the provision of public services (e.g., schools, transportation), and the protection of domestic producers from foreign competition. These relations lead some social scientists to discard the conventional view that markets were born free of the shackles of state in favor of a perspective that maintains markets *required* the state to survive.

It is important to remember that capitalism long preceded the nation-state and that there is no necessary correspondence between the two, an observation with important implications in the current age of globalization. Capitalism began in the city-states, not nation-states, of northern Italy in the fifteenth and sixteenth centuries, in contrast to the nation-state, which is largely a product of the industrialization of the eighteenth and nineteenth centuries. The ascendancy of the nation-state marked a new scale at which capitalist social relations could be contained and managed, including trade and production networks. Capitalism thus produced the scale of the national at the same time as it produced the international.

THE INDUSTRIAL REVOLUTION

Capitalism has been a dynamic, ever-changing society since its inception. The growth and development of market-based societies was, by historical standards, very rapid. This pace of change, which would ultimately become normalized in the formation of the modern world, accelerated greatly in the eighteenth and nineteenth centuries during the Industrial Revolution. It is important to note that the Industrial Revolution occurred long after capitalism began; indeed, for most of capitalism's history, it involved preindustrial forms of production, including labor-intensive artisanal and household types of production. However, starting in the mid-1800s, an explosive increase in the speed and productivity of capitalist production in Europe, North America, and Japan occurred that transformed the worlds of work, everyday life, and the global economy. Somewhat later, in the twentieth century, the industrialization spread to eastern Europe and the Soviet Union, and since the 1970s has spread to the newly industrializing countries in selected parts of the developing world (Chapter 12).

Industrialization is a complex process that involves multiple transformations in inputs, outputs, and technologies. Three dimensions are particularly important here.

Inanimate Energy

If preindustrial societies relied upon *animate sources of energy* (i.e., human and animal muscle power) to get things done, industrialization can be defined loosely as the harnessing of *inanimate sources of energy*. This process marked a major milestone in human economic evolution. There are several types of inanimate energy that have been tapped historically. The first of these involves running water, or water in a particular stage of the hydraulic cycle when it is moving overland from higher to lower elevations. This source had been used in the late Middle Ages to grind corn and flour. By the early eighteenth century, some producers of textiles began to use water-powered mills to run their machines. Indeed, running water was a major source of energy in the earliest stages of the Industrial Revolution, but it constrained firms to locating near streams and rivers. Many

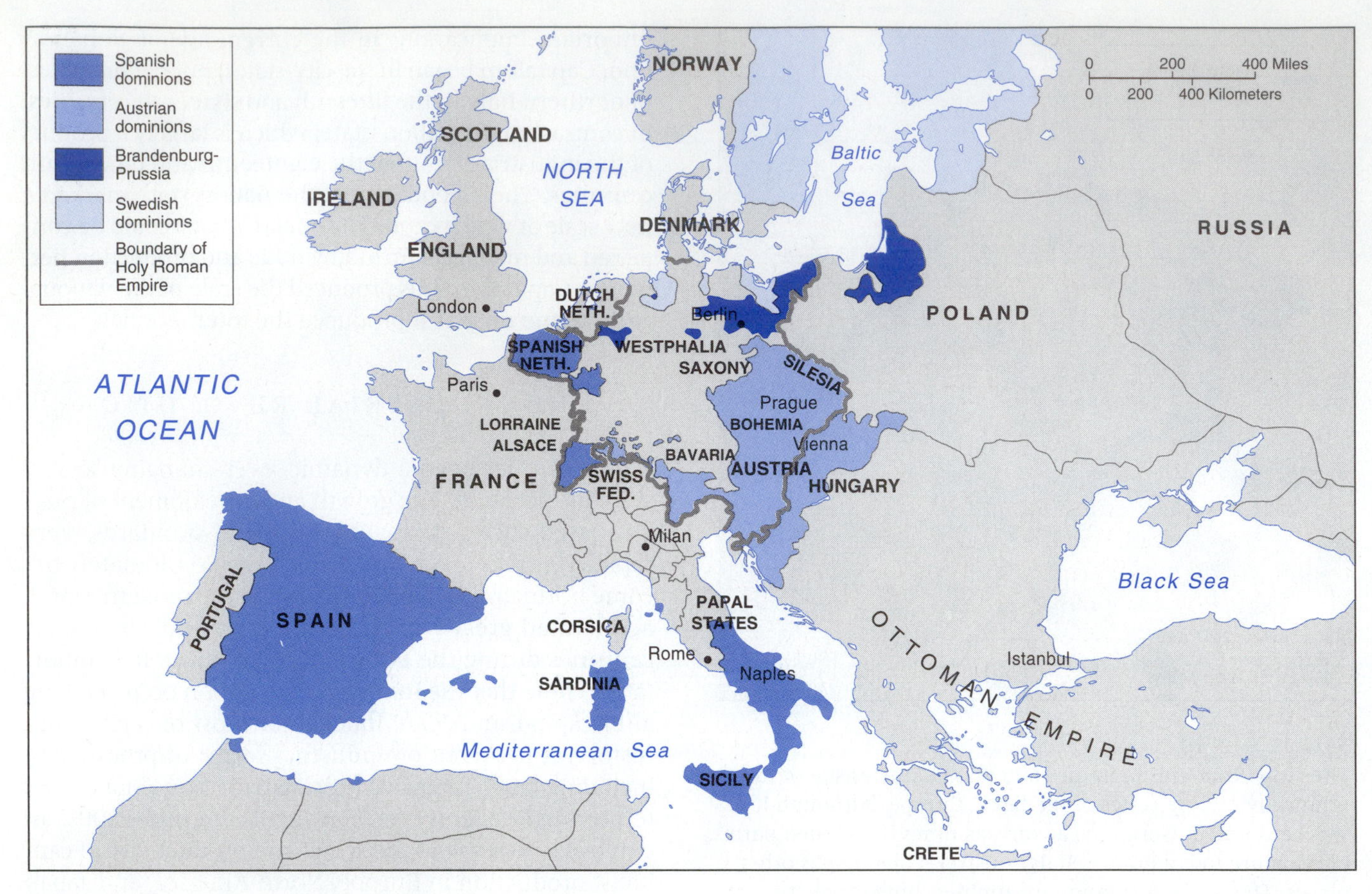

FIGURE 2.12
The treaty of Westphalia signed in 1648 not only set the boundaries among the nation-states of early modern Europe, but it also legitimized the nation-state, and the principle of noninterference, in international politics. The emerging geography of capitalism was as much political as economic in nature.

textile plants in southern New England, for example, used this strategy. There were severe drawbacks to this approach. Many streams, for example, are annual, meaning they may dry up in the summers if there has been insufficient rainfall in their catchment area. And locating on a stream may put the producer inconveniently far away from the market.

A more efficient source of inanimate energy involves coal and the steam engine, the designs for which were laid by Newcomen in 1712 (Figure 2.13). The first operating steam engine was built by the Scottish engineer James Watt in 1769 and marked a turning point in the process of industrialization. The steam engine, originally designed to pump seawater out of mine shafts that penetrated under the ocean, used relatively cheap and abundant fuels and could do the work of dozens of men far more efficiently. Wood provided the first major source of fuel for this invention, which required heating water into steam in order to drive the engine's pistons. As producers began to cut down forests in England in large numbers, deforesting much of the

FIGURE 2.13
The steam engine, designed by Thomas Newcomen in 1712 but first built by James Watt in 1769, was the key invention of the Industrial Revolution. It was the first device to harness inanimate energy on a mass basis and revolutionized both production and transportation for the two centuries.

country, wood supplies began to dwindle, and the rising cost eroded profits.

As wood became scarce, producers switched to coal, which could be mined in large quantities. Thus, as England industrialized, several areas became major coal producing centers, including Wales and Newcastle. As the Industrial Revolution spread across the face of Europe, the large coal deposits of the northern European lowlands became increasingly important. This relation exemplifies how nature helps to shape the formation of geographies. In the United States as well, coal deposits in Appalachia played a key role in the nation's industrialization.

In the nineteenth century, coal was joined by other fossil fuels, particularly petroleum and to a lesser extent natural gas (Chapter 4). The abundance of cheap energy was the lifeblood of industrialization, and production became increasingly energy intensive as a result. This substitution of inanimate for animate energy both freed tens of millions of people from drudgery and allowed for large numbers to live relatively comfortable lives in the expanding middle class that industrialization produced.

Technological Innovation

As we have seen, capitalism is a very dynamic economic system. Firms, under the lure of profits and threat of ruin, engage in frequent innovation as a way to reduce costs and increase revenues. While there were certainly new technologies that emerged prior to industrialization, the Industrial Revolution witnessed an explosive jump in the number, diversity, and applications of new technologies (Table 2.1). A *technology* is a means of converting inputs to outputs. These can range from extremely simple to sophisticated. An otter using a rock to open an oyster is employing a technology, as you are when you use a pen or computer. As industrialization produced an increasingly sophisticated division of labor, opportunities for new inventions rose rapidly. These were employed in agriculture, in manufacturing, in transportation and communications, and in services. Figure 2.14 illustrates a variety of technologies that accompanied the long-term increases in sophistication in transportation and production.

Table 2.1

Some Major Innovations of the Industrial Revolution

1708	mechanical seed sower
1712	steam engine
1758	threshing machine
1765	spinning jenny
1787	power loom
1793	cotton gin
1807	steamboat
1828	railroad
1831	electric generator
1834	reaper
1839	photography, vulcanized rubber
1844	telegraph
1846	pneumatic tire
1849	reinforced concrete
1850	refined gasoline
1851	refrigeration; sewing machine
1857	pasteurization
1859	gasoline engine
1866	open hearth furnace
1867	dynamite
1873	typewriter
1876	telephone
1877	phonograph
1878	microphone
1879	electric lightbulb
1884	rayon
1886	hydroelectric power plant
1888	camera; radio waves
1892	diesel engine
1895	X-rays
1896	wireless telegraphy
1899	aspirin
1900	zeppelin
1903	airplane
1906	vacuum tube
1925	television

In the Industrial Revolution, a major reorganization in the nature of work occurred with the development of the *factory* system. Prior to this era, industrial work was organized on a small scale basis, including home-based work. The early textile industry, for example, used the "putting out" system of independent workers and contractors. By the late eighteenth century, however, firms in different industries were grouping large numbers of workers together under one roof. Some of the largest factories held thousands of workers. This process effectively created the industrial working class. Never before in human history had so many workers been concentrated on a permanent basis, a feature that changed how they lived and viewed each other, and themselves. Inside factories, workers used vast amounts of capital (i.e., many types of machines). The introduction of interchangeable parts, invented by American gun maker Eli Whitney, made machines more reliable. By the early twentieth century, Henry Ford introduced the moving conveyor belt, which further accelerated the tempo of work and the ability of workers to produce.

FIGURE 2.14
Throughout human history, increasing technological sophistication has been tied to the development of energy resources.

Productivity Increases

As a consequence of the massive technological changes of the Industrial Revolution, productivity levels surged. *Productivity* refers to the level of output generated by a given volume of inputs; productivity increases refer to higher levels of efficiency (i.e., greater levels of output per unit of input [e.g., labor hour or unit of land]) or, conversely, fewer inputs per unit of output.

As Figure 2.15 indicates, productivity levels rose exponentially in the nineteenth century. There were several important repercussions of this phenomenon. As the cost of producing goods declined, standards of living rose. Now, most workers labored long hours and endured standards of living still quite low to those we enjoy today. But nonetheless, over several decades, industrialization saw many kinds of goods become increasingly affordable. Because wage rates have been linked historically to the marginal productivity of labor, the working class became better off. Clothing, for example, which was scarce before the Industrial Revolution, became relatively cheap and ceased to be as accurate an indicator of class status as it had been previously.

Most important in this regard concerns the industrialization of agriculture. As machine after machine was introduced into farming, food became progressively cheaper, and diets improved as more people ate more and better food than ever before. With the notorious exception of the Irish Potato Famine of the 1840s, hunger and malnutrition gradually declined throughout Europe, although they appeared in the aftermath of wars.

The Geography of the Industrial Revolution

Like all major social processes, the Industrial Revolution unfolded very unevenly over time and space. Whereas capitalism had its origins in Italy, industrialization was very much a product of northwestern Europe. Some observers put the first textile factories in Belgium, in cities such as Liege and Flanders. However, without question it was Britain that became the world's first industrialized nation. By the end of the eighteenth century, Britain stood virtually alone as the world's only industrial economy, a fact that gave it an enormous advantage over its rivals. Britain's industrial base, for example, allowed it to triumph over France in their eighteenth-century rivalry for global hegemony and to flood its competitors' markets with cheap textiles. Cities in the Midlands of Britain, such as Leeds and Manchester, were known as the workhouses of the world for their high concentrations of workers, capital, and output, becoming centers of the British textile and metal working industries (Figure 2.16). Others, such as London, Glasgow, and Liverpool, became centers for shipbuilding, which, in a maritime-based world economy, was a major industry in its own right. In many cities, networks of producers in guns, watches, shoes, metals, and light industry formed dense industrial districts of small firms with intricate linkages of inputs and outputs.

Why Britain? There are no simple answers to this question. Britain had already enjoyed a network of long-distance trade relations with its colonies in North America and elsewhere. Agriculture in Britain was well advanced into the process of commodification compared with the European continent. And Britain enjoyed large deposits of coal and was the locale where the steam engine was invented.

A half century after it began in Britain, the Industrial Revolution diffused to the European Continent (Figure 2.17). In Europe, this saw the formation of industrial complexes in the lower Seine River and Paris. In Italy, the Po River valley became a major producer of textiles and shoes. In Scandinavia, cities such as Stockholm became major ship producers. In Germany, which was late to industrialize (following unification in 1871), the Ruhr region became a global center of steel, automobile, and petrochemicals firms. By the early nineteenth century, the Revolution began to spread worldwide (Figure 2.18), including leapfrogging

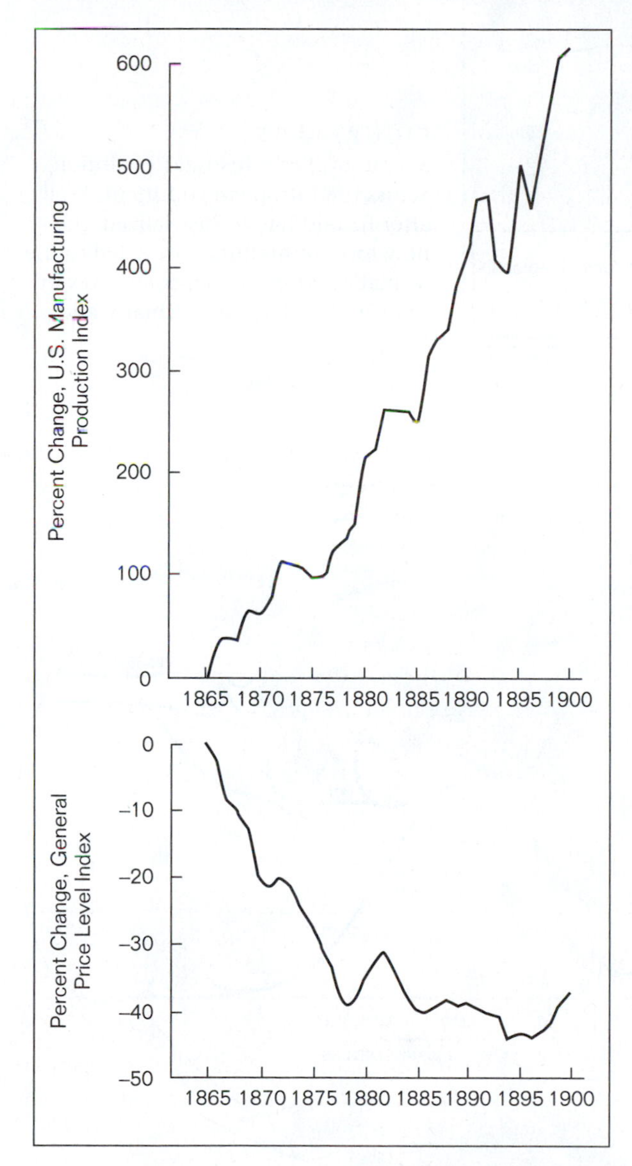

FIGURE 2.15
Manufacturing productivity in the United States rose exponentially in the latter nineteenth century. After the Civil War, the Industrial Revolution began in earnest. As productivity increased, the prices of goods dropped accordingly, and standards of living rose. Geographically, this period saw the emergence of the Manufacturing Belt along the southern shores of the Great Lakes.

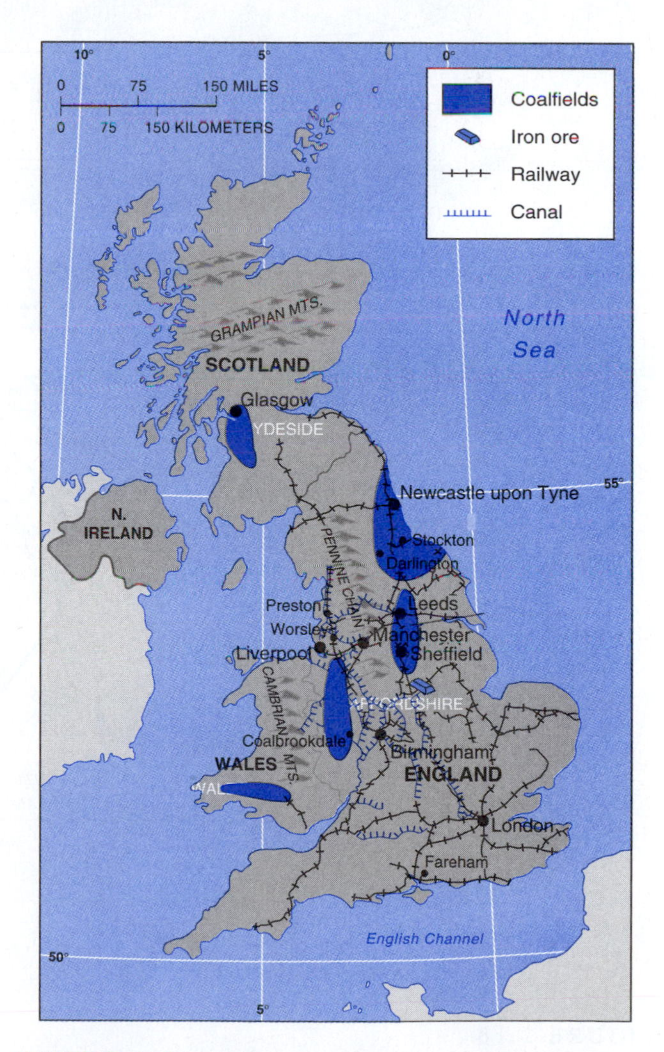

FIGURE 2.16
Britain's industrial areas, the sources of the Industrial Revolution. Coal from Wales and Newcastle fueled the development of the factory system, particularly in the Midlands cities such as Manchester, Sheffield, Birmingham, Leeds, and Liverpool.

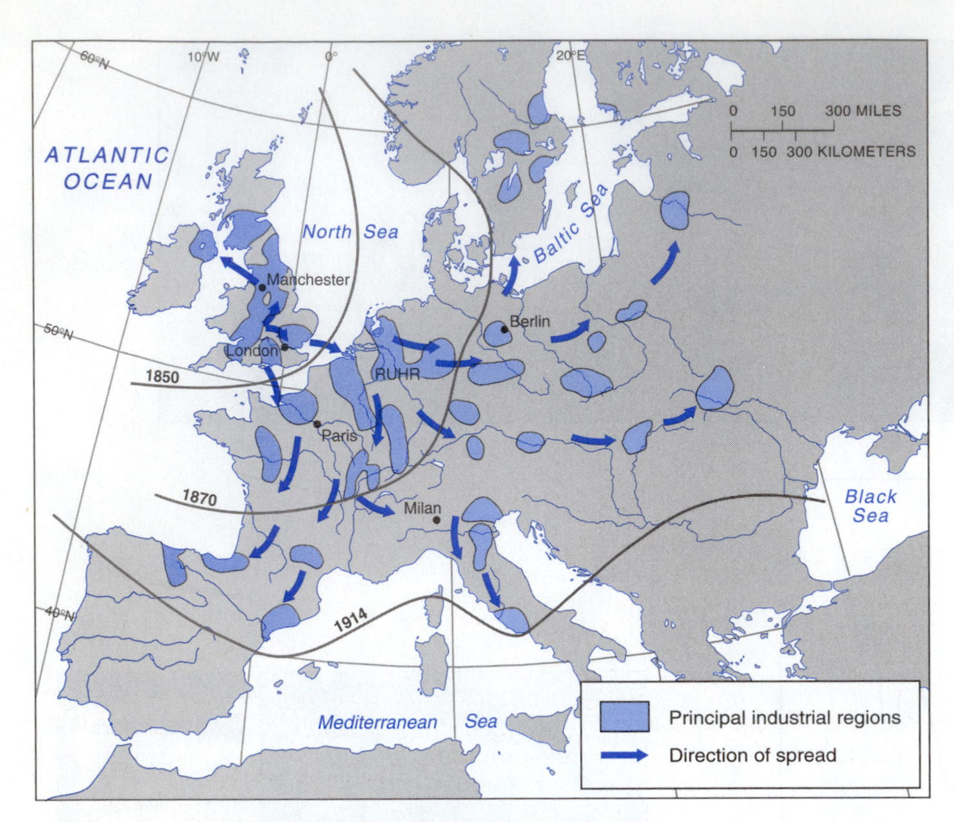

FIGURE 2.17
Spread of the Industrial Revolution across the European continent. Well after Britain had industrialized, the new form of manufacturing led to the formation of industrial complexes in France, then later in Germany and Italy.

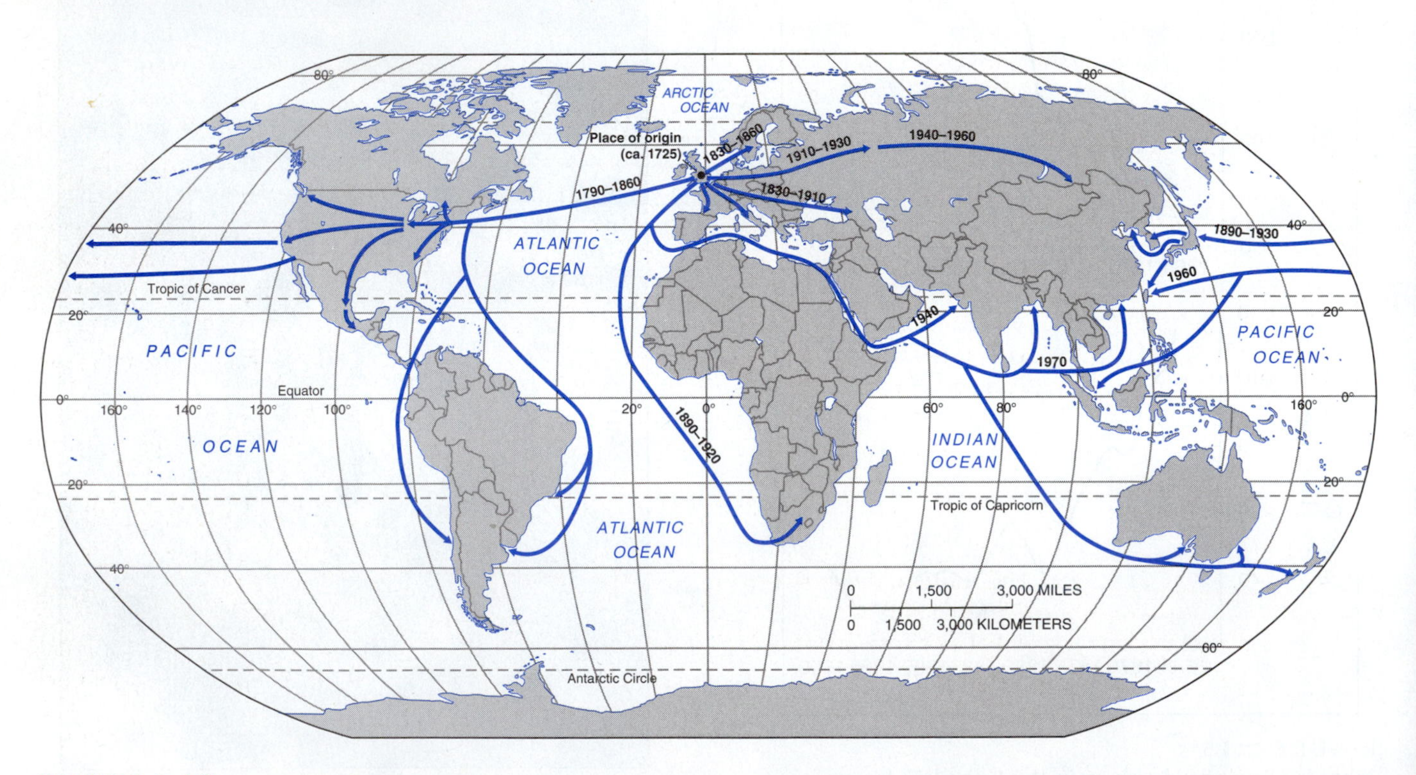

FIGURE 2.18
The global diffusion of the Industrial Revolution. By the early nineteenth century, the process had become entrenched in New England and later took root across the rest of North America. Japan emerged from a long period of hibernation in 1868 and became the first non-Western industrial power shortly thereafter. Eastern Europe lagged well behind the western part, and Russia only industrialized in the 1920s under the Stalinist government in the Soviet Union. The East Asian newly industrialized countries (NICs) started to industrialize in the 1960s. The process continues today in selected parts of the developing world.

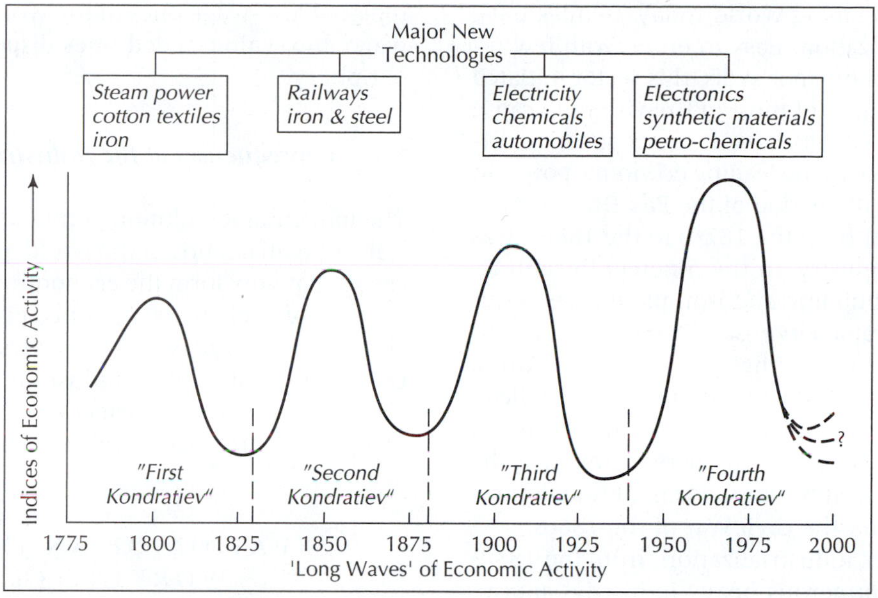

Major New Technologies

Steam power
cotton textiles
iron

Railways
iron & steel

Electricity
chemicals
automobiles

Electronics
synthetic materials
petro-chemicals

Indices of Economic Activity

"First Kondratiev"
"Second Kondratiev"
"Third Kondratiev"
"Fourth Kondratiev"

1775 1800 1825 1850 1875 1900 1925 1950 1975 2000

'Long Waves' of Economic Activity

FIGURE 2.19
Long waves of economic activity, named after their discoverer, Kondratiev. The historical development of capitalism is often organized into four such waves, each of which was centered around a different technology. The textile industry dominated the first, from the late eighteenth to the early nineteenth centuries. The second wave, lasting roughly from 1820 to 1880, saw the widespread application of steam power, including the railroad and steamship. The third, from about 1880 to 1930, witnessed the growth of heavy industry, particularly steel making and the automobile, and ended in the Great Depression of the 1930s. The fourth, which began in earnest after World War II, saw the rapid growth of petrochemicals and aerospace. Many argue that we are living in a fifth wave that began in the 1980s, propelled by electronics and business services.

across the Atlantic, igniting the industrialization of southern New England. Starting in the 1870s, Japan became the first non-Western country to join the club of industrialized nations as the old feudal order there collapsed following the Meiji Restoration of 1868. Russia, which flirted with industrialization under Peter the Great, did not become fully industrialized until the 1920s, when under Stalin the Soviet Union leaped to become the world's second largest economy in the span of a decade. In the twentieth century, the process of industrialization diffused to many developing countries, particularly in East Asia, where it has had profound consequences for millions of people. In a sense then, the industrialization of the developing world, which is partial and incomplete, is a continuation of a long-standing historical process. Although this process changed over the years, its broad outlines remained the same. The industrial complexes formed by the diffusion of the Industrial Revolution remain highly important to the global economy today and are discussed in more detail in Chapter 7.

Cycles of Industrialization

Just as the process of industrialization occurs in different places at different times, so too did the nature and form of industrialization vary in successive historical periods. As we shall see in more detail in Chapter 5, capitalism is prone to long-term cyclical shifts in its industries, products, labor markets, and geographies, a concept often referred to as Kondratiev waves of roughly 50 to 75 years' duration (Figure 2.19). For our purposes, this means that industrialization saw the rise of different industries at different times. Industrialization was thus not one process but a series of them that varied over time and space.

The first wave of the Industrial Revolution (1770s–1820s) centered on the textile industry (Figure 2.20). In Britain, as in the rest of Europe, North America,

FIGURE 2.20
Power looms in an English textile mill, 1820. Textiles formed the first great wave of the Industrial Revolution, as they are small in scale, with few barriers to entry, and use unskilled labor. The power loom obviated the need for heavy manual labor and paved the way for the widespread use of female and child labor in this industry.

Japan, and the developing world today, textiles have *always* led industrialization. Easy to enter, with few requirements of capital or labor skills, this sector initiated the industrial landscapes of most of the world. Because this wave was centered in Britain, it saw that nation catapulted to prominence as the leading economic power in the world, initiating the period of the Pax Britannica.

The second wave, from the 1820s to the 1880s, was a period of heavy industry. In the nineteenth century, sectors such as shipbuilding and iron plants were critical. Large scale and capital intensive, these types of firms differed markedly from the light industry of textiles. They required massive capital investments, were difficult to enter, and moved toward an oligopoly rather than a competitive market. This was the period in which the U.S. Manufacturing Belt began to form, although most of its growth was after the Civil War of the 1860s.

The third wave of industrialization, from the 1880s to the 1930s, saw numerous heavy industries appear, including steel, rubber, glass, and automobiles. This was a period of massive technological change, including capital intensification and automation of work, as well as economic changes. As local markets gave way to national markets, most sectors experienced as steady oligopolization, or concentration of output and ownership in the hands of few large firms. Many companies became multiestablishment corporations. Not surprisingly, this wave saw the rise to power of robber barons such as Carnegie (steel), Rockefeller (oil), Duke (tobacco), Dupont (chemicals), J. P. Morgan (banking), and John Deere (agricultural machinery).

In the fourth wave of industrialization, which started during or immediately after the Depression of the 1930s and lasted until the oil shocks of the 1970s, the primary growth sectors were petrochemicals (including plastics) and automobiles. With a relatively stable global economy, this era saw the domination of the world system by the United States, which produced a huge share of the planet's industrial output.

The fifth wave of industrialization, often held to begin after the oil shocks of the 1970s, has been led by the electronics industry, which was powered by the microelectronics revolution and by the explosive growth of producer services (Chapter 8). This epoch saw rapid productivity growth in household electronics and information-processing technologies.

It is important to note that during each era, the major propulsive industry was commonly featured as the "high-tech" sector of its day. Thus, just as electronics is often celebrated at this historical moment for its innovativeness and ability to sustain national competitiveness, so too were the textile industry in the eighteenth century and steel industry in the nineteenth century associated with high levels of productivity and wages. What was a leading industry at one moment would become a lagging one at the next historical epoch, as high-wage, high-value-added sectors replaced low-wage ones in the world's core and as low-wage, low-value-added ones dispersed to the world's periphery.

Consequences of the Industrial Revolution

The Industrial Revolution permanently changed the social and spatial fabric of the world, particularly in the societies that now form the economically developed world. No part of their social systems, economy, technology, culture, or everyday life was left untouched. Within a century of its inception, industrialization changed a series of rural, poverty-stricken societies into relatively prosperous, urbanized, and cosmopolitan ones.

Creation of an Industrial Working Class

As we noted earlier, a significant part of industrialization was the reorganization of work along the lines of the factory system. For the first time in human history, large numbers of workers labored together using machines. These conditions were quite different from those facing agricultural workers, who were dispersed over large spaces and relied on animate sources of energy. Industrialization gave rise to organized labor markets in which workers were paid by the hour, day, or week.

In short, as firms created a new form of labor, they created a new form of laborer, a proletariat or industrial working class, which became socialized into the new

Child laborers in Britain working in the textile mills.

conditions of work. This process was not easy, given how brutally exploitative working conditions were during this time. Workers typically labored for 10, 12, even 14 hours per day, six days per week, for relatively low wages. (The eight-hour day and Saturdays free from work were products of workers' movements in the 1930s.) Often the work was unsanitary and dangerous, even lethal, as workers were subjected to accidents, poor lighting, and poor air quality. Child labor was also common, subjecting those as young as four or five to horrendous and exploitative conditions such as are now found in the developing world.

As a result of this process, time—like space, and so much else—became a commodity, something bought and sold. The transition from agricultural time to industrial time was important. Prior to the Industrial Revolution, people experienced time seasonally, and rarely felt the need to be conscious of it. Time was simply lived, without worry about the precise beginnings and endings of events. With industrialization, however, time was measured and divided into discrete units, as signaled by the factory whistle, bell, and stop watch. This change marked the commodification of time through the labor market.

If industrialization produced a working class, it also produced labor unions. The first resistance to employers included the British Luddites in the eighteenth century, named after their leader, "General" Ludd. Luddites blamed their miserable working conditions on the machines they used, and often destroyed them in attempts to halt their exploitation. In France, workers wearing large wooden shoes, or "sabots," jammed them into the machinery in an act of sabotage. By the late nineteenth century, organized labor had created a number of unions, which in the United States included the Knights of Labor, the American Federation of Labor, and in the twentieth century, the Industrial Workers of the World and the Congress of Industrial Organizations. Industrialization was thus often a period of considerable class conflict.

Urbanization

Geographically, the Industrial Revolution was closely associated with the growth of cities. Almost everywhere, industrialization and *urbanization* have been virtually simultaneous processes. Manufacturing firms concentrated in cities, especially large ones, such as the British Midlands (e.g., Leeds, Manchester) or throughout the U.S. Manufacturing Belt (e.g., Pittsburgh, Cleveland, Detroit, Chicago, and Milwaukee).

The reasons why firms concentrated in cities are important. Often, there is a tendency to attribute this phenomenon to the presence of workers in urban areas. Which came first, firms or workers? Cities were clearly centers of capital as much as they were centers of labor. This poses something of a "chicken or egg" problem. Yet cities were very small when the Industrial Revolution began, and through agricultural mechanization (which reduced rural job opportunities) and rural-to-urban migration, the urban labor supply was created.

As Chapter 5 documents more fully, there are powerful reasons for firms to concentrate, or agglomerate, in cities. Most firms benefit by having close proximity to other firms, including suppliers of parts and ancillary services. Concentration allows firms to share a specialized infrastructure, information, and labor force. The cities of the nineteenth century were composed of dense webs of industrial firms, with intricate input and output relations tying them together.

Industrialization changed societies from predominantly rural to predominantly urban in character. In Europe, North America, and Japan, for the first time in history, the majority of people lived in cities. The growth of cities in industrial societies is frequently depicted using an urbanization curve (Figure 2.21), which illustrates the percentage of people living in urban areas over time. In the United States, for example, the first

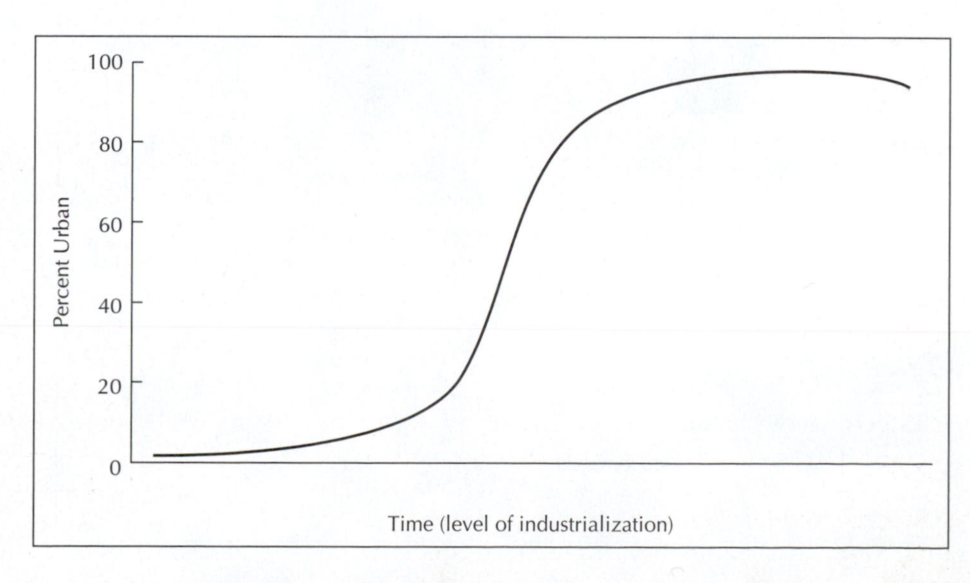

FIGURE 2.21
An urbanization curve expresses the proportion of a country's population that lives in cities at different stages in industrialization. Preindustrial societies are agricultural, rural ones; because manufacturing is concentrated in urban areas, industrialization causes cities to grow more quickly than the countryside.

national census of 1790 showed that 95 percent of Americans lived in rural areas. This proportion increased throughout the nineteenth century, and by 1920, 50 percent of the nation's population lived in cities. Today, it is roughly 85 percent. This pattern is similar in virtually every other country that has industrialized.

Population Effects

Industrialization changed more than simply the geographic distribution of people (i.e., in cities), it also shaped the growth rates and demographic composition. We explore how populations change in more detail in Chapter 3, but it should be noted that the Industrial Revolution unleashed drastic changes in both the rate of growth and the health of the population.

At the eve of the Industrial Revolution, the famous theorist Thomas Malthus predicted that rapid population growth would create widespread famine. Yet Malthus was soon proved to be wrong, at least in the short run. The industrialization of agriculture generated productivity increases greater than the rate of population growth, and the creation of a stable and better food supply improved most people's diets. As a result, life expectancy rose. Industrialization also lowered death rates, particularly as malnutrition declined and infant mortality rates dropped. Eventually, public health measures and cleaner water helped to control the spread of most infectious diseases. As death rates dropped, the populations of industrializing countries increased dramatically (Figure 2.22). This change was also accompanied by a shift from the extended to the nuclear family. Eventually, as Chapter 3 explains, industrialization also led to a decline in the birthrate, families had fewer children, and growth rates declined.

Growth of Global Markets and International Trade

Yet another impact of the Industrial Revolution concerned the global economy. Capitalism had formed a loose network of international trade well before the eighteenth century. Indeed, as we saw earlier, there were extensive linkages across Eurasia and the Indian Ocean in the early fourteenth century. The harnessing of inanimate energy for transportation, however, dramatically accelerated the speed of both land and water transportation, forming a significant round of time-space compression (Chapter 9). Whereas both sailing ships and horse-drawn transport traveled at roughly 10 mph, for example, steamships could reach speeds of 40 mph and railroads more than 65 mph. Moreover, the new, industrialized forms of transportation were not only faster but also cheaper, resulting in cost-space convergence as well.

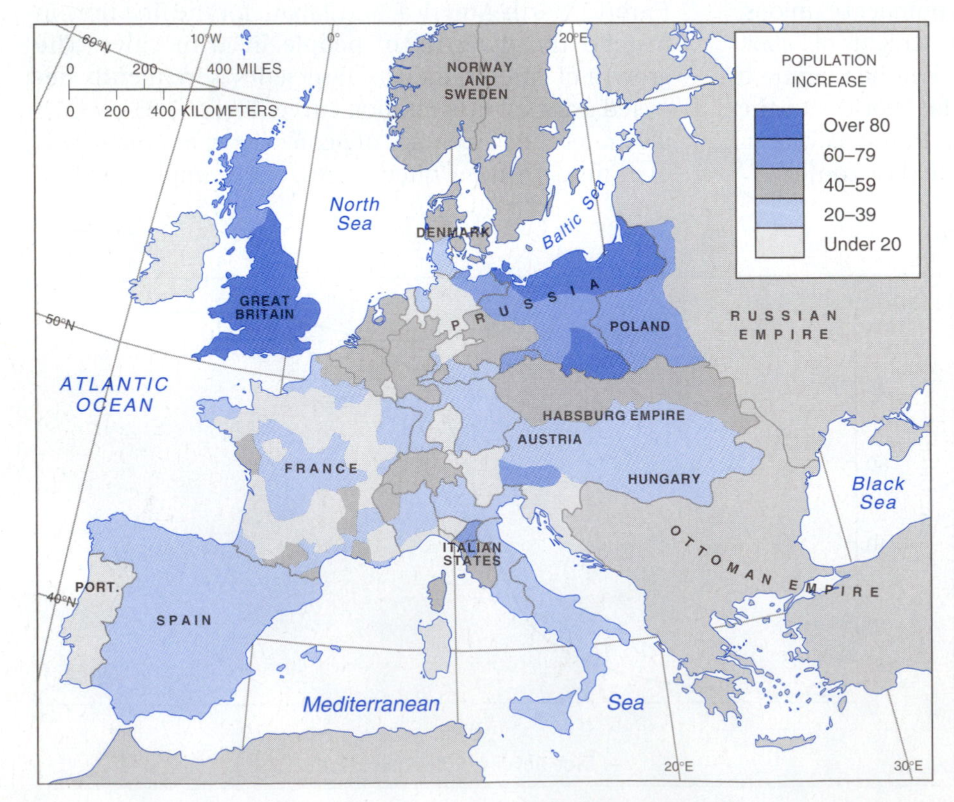

FIGURE 2.22
Population growth in Europe from 1800 to 1850. The capacity of industrialized societies to support large, dense nucleations of people led to higher rates of population growth and larger numbers of people in northern Europe than in the south, which industrialized later and less completely.

These changes dramatically lowered the barriers to trade, and the volume of imports and exports internationally began to soar. Europe, starting with Britain, could import unprocessed raw materials, including cotton, sugar, and metal ores, and export high-value-added finished goods, a process that generated large numbers of jobs in Europe and contributed to a steady rise in the standard of living. It is worth noting that classical political economists such as Adam Smith and David Ricardo began to demolish the philosophy of mercantilism, which preached state protection against imports, at precisely this historical moment.

Finally, the industrial world economy saw an explosion of international finance. British banks, largely concentrated in London, for example, began to extend their activities on an international basis, lending to clients and investing in markets overseas. Much of the capital that financed the American railroad network came from Britain. The globalization of production was thus accompanied by the steady globalization of money and credit.

The timing of industrialization was significant to individual nations. There existed an important difference between early and late industrializers in this regard. Early industrializers (e.g., Britain, the United States) faced little competition internationally. Thus, their light industries (i.e., textiles) associated with the first wave of industrialization could develop relatively slowly, with minimal government intervention. Firms in sectors such as textiles, with few barriers to entry and low infrastructural demands and which were quite competitive internationally, were important in shaping national political climates characterized by laissez faire politics and minimal government intervention.

In contrast, relatively late industrializers faced a significantly different international climate, one dominated by early industrializers. Countries such as Germany and Japan, which did not begin industrializing until the late nineteenth century—long after Britain and the United States—faced significant competition in industries such as textiles. Consequently, these nations tended to experience relatively short periods in which their economies were dominated by light industry and moved rapidly into heavier sectors. Germany, for example, developed a comparative advantage in steel, armaments, and automobiles. In countries in which heavy industry dominates and places significantly higher demands on the state for labor training, infrastructure, and trade protection, national political cultures that look favorably upon state intervention are more likely to develop. This is true of newly industrializing countries today. Thus, the internal political culture within countries was strongly affected by the timing of their entry into the international division of labor.

COLONIALISM: CAPITALISM ON A WORLD SCALE

Intimately associated with the development of capitalism in Europe was Europe's conquest of the rest of the globe. This process, euphemistically called the "Age of Exploration," can be viewed as the expansion of capitalism on a global scale. If the geographies of capitalism are typified by uneven spatial development, as noted earlier, then colonialism involved the construction of uneven development on a global scale, with Europe at the center and its colonies on the world periphery. This theme is found in many theories of world development such as world-systems theory (Chapter 13).

Colonialism was simultaneously an economic, political, and cultural project. It was also an act of conquest, by which a small group of European powers came to dominate a very large group of non-European countries. Culturally, as Said (1978) points out, this process involved the emergence of the distinction between the "West" and the "Rest." In conquering the "Orient," Europeans came to discover themselves as Westerners, often in contrast to other people whom they represented in highly erroneous terms.

Everywhere, colonized people fought back against colonial rule. Examples include the Inca rebellions against the Spanish, Zulu attacks on the Dutch Boers, the great Indian Sepoy uprising of 1857, and the Boxer Rebellion in China in 1899–1901. Yet Western powers, armed with guns, ships, and cannon, effectively dominated the entire planet. While a few countries were nominally independent, such as Thailand, the only one to escape colonialism substantively was Japan, which, under the Tokugawa Shogunate, closed itself off from the world until 1867.

Colonialism had profound implications for both the colonizers and the colonized, which is why a sophisticated understanding of economic geography must include some understanding of this process. Globally, colonialism produced the division between the world's developed and less developed countries, a theme explored in detail in Chapter 13. Colonialism changed European states too, deepening the formation of capitalist social relations and markets as well as the nation-states in western Europe. Prior to colonialism, Europe was a relatively poor and powerless part of the world, compared with the Muslim world, India, or China; afterward, Europe became the most powerful collection of societies on the planet.

The Unevenness of Colonialism

It is important to note that colonialism did not occur in the same way in different historical moments and different geographic places. Temporally, there were two major waves of colonialism (Figure 2.23), one associated

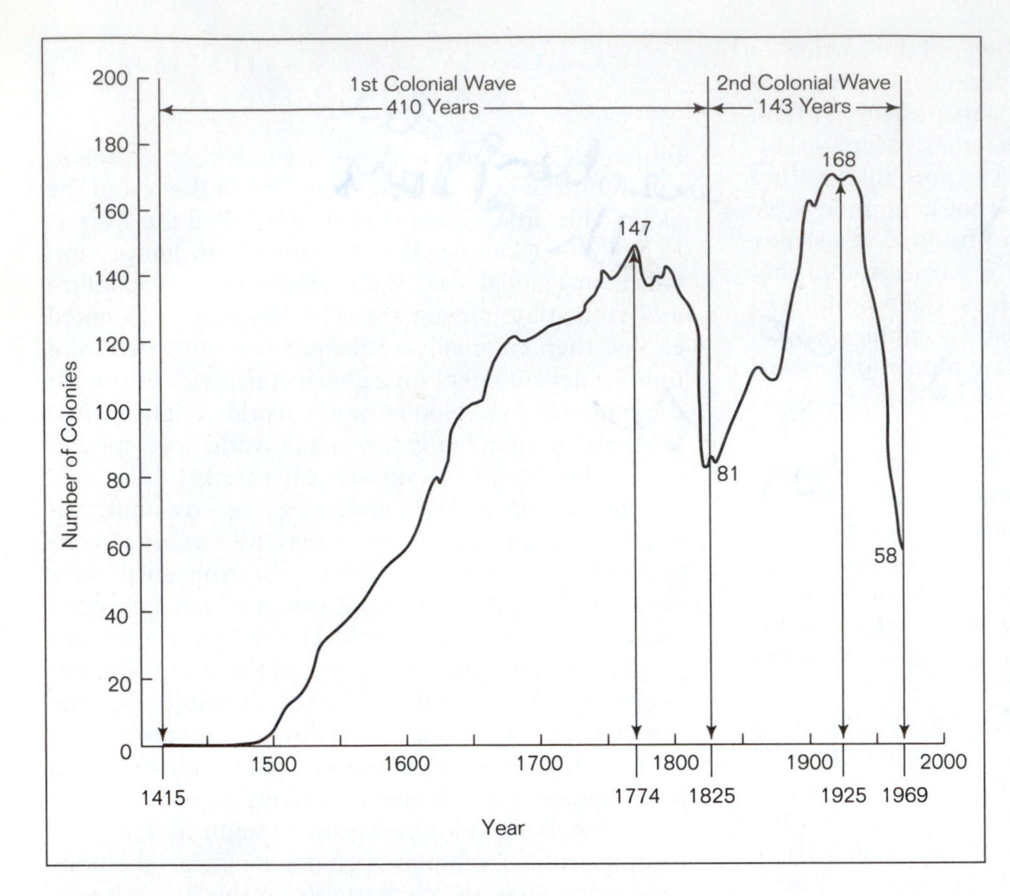

FIGURE 2.23
Waves of European colonialism. The first major wave, lasting from the sixteenth century until 1815, was dominated by the Spanish and Portuguese conquest of the New World. The Napoleonic Wars, however, weakened Spain and saw the independence of its colonies in the Americas, leading to a decline in the total. The second wave, from 1815 until the 1960s, witnessed the European conquest of Africa, Asia, and the Middle East, particularly by the British and French. After World War II, decolonization saw the breakup of the colonial empires and the rise in the number of nominally independent countries in the world.

with the preindustrial mercantile era and the other with the Industrial Revolution. From the sixteenth century, when colonialism began, until the early nineteenth century, Western economic thought was characterized by mercantilism, in which state protection of private interests was justified as necessary for the national well-being. During this period, the primary colonial powers were Spain and Portugal, and their primary colonies were in the New World and parts of Africa.

Following the Napoleonic wars, which ended in the Treaty of Vienna in 1815, European powers were relatively weak. This opening provided an opportunity for nationalists in Latin America, led by Simon Bolivar, to break away and become independent countries. Thus, the number of colonies declined sharply in the early nineteenth century. However, during the subsequent phase of industrialization, characterized economically by the ideology of free trade, the number of colonies grew again. This time Britain emerged as the world's premier power, and along with France, colonized large parts of Africa and Asia. Finally, as Figure 2.23 illustrates, the number of colonies declined rapidly after World War II amid the era of decolonization.

Spatially, colonialism was also uneven. Different colonial empires had widely varying geographies, as illustrated by the distribution of empires at their peak in 1914, the eve of World War I (Figure 2.24). The British Empire, which encompassed one-quarter of the world's land surface, stretched across every continent of the globe, including large parts of western Africa, the Indian subcontinent, and parts of Southeast Asia. The French ruled in parts of western Africa and in Indochina. The Portuguese had Brazil, chunks of Africa, Goa in India, and parts of Indonesia and Chinese Macau. The Belgians possessed Congo, the private holdings of Leopold II. Portugal retained control over vast swaths of Africa. Italy was in Libya and Ethiopia. Even Germany, late to unify and to industrialize, controlled parts of Africa (Togo, Namibia, Tanganyika) and New Guinea.

How Did the West Do It?

What allowed Europe to conquer the rest of the planet? The answers to this question are not simple. Obviously they do not lie in any innate superiority of Europeans. Indeed, for much of history, Europe was relatively weaker and poorer than the countries it conquered. By the sixteenth century, however, Europe did come to possess several technological and military advantages over its rivals.

Jared Diamond (1999), in his Pulitzer Prize–winning book *Guns, Germs, and Steel,* maintains that Western societies enjoyed a long series of advantages by virtue of geographic accident. Agriculture in the West, centered

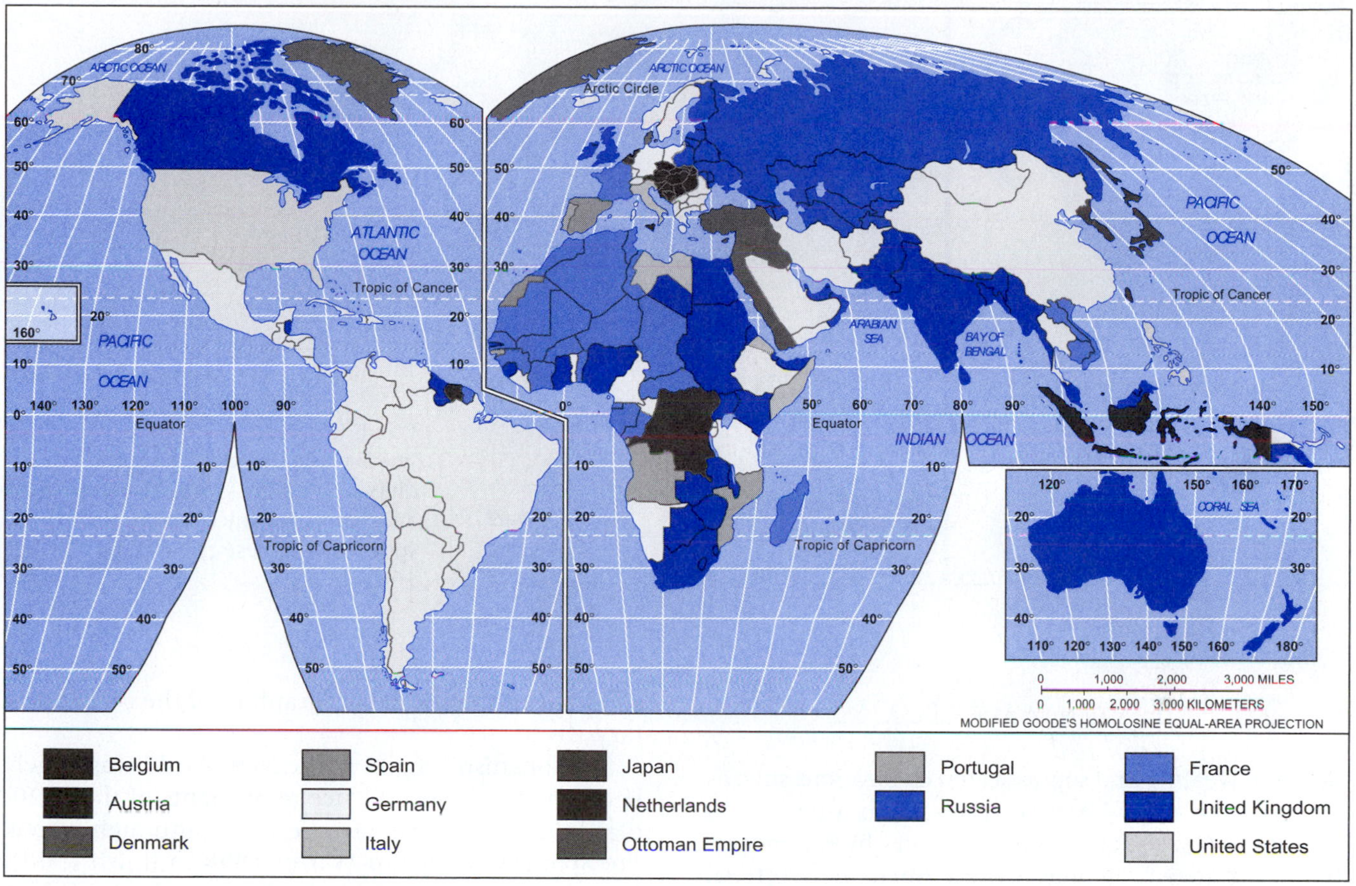

FIGURE 2.24
The geography of colonial empires in 1914, the eve of World War I and the peak period of European influence. The British Empire encompassed vast areas in Africa, South Asia, and Australia. France ruled over most of western Africa and Indochina. Portugal retained its hold over Angola and Mozambique. Belgium controlled the Congo. Indonesia belonged to the Dutch. Even late-developing Italy and Germany controlled parts of Africa. Japan was becoming a new colonial power in Asia, and the United States had become both a colonial power in the Philippines and Puerto Rico, as well as an emerging neocolonial one.

on wheat, was productive and could sustain large, dense populations. Unlike the New World, Old World societies were often stretched across vast East-West axes, or regions with common growing seasons, which facilitated the diffusion of crops. There was a long Western history of metalworking, which not only increased economic productivity but also allowed for weapons such as guns and cannons. By the Renaissance, Europeans had become highly skilled at building ships and navigating the oceans. And by the late eighteenth century, the West had discovered inanimate energy, which offered numerous economic and military advantages. In addition, the Europeans unleashed diseases (if unintentionally), particularly smallpox and measles, on the New World, which provided them with an unintended advantage but also led to labor shortages.

Others maintain that the West's advantages were not simply technological, but political. The Western "rational" legal and economic system stressed secular laws and the importance of property rights. Yet others point out that unlike the Arabs, Mughal India, or China, Europe was never united politically. Indeed, every time one European power attempted to conquer the others, it was defeated, as exemplified by France in the early nineteenth century and Germany twice in the twentieth. The lack of centralized political authority created a climate in which dissent and critical scholarship was tolerated. For example, the French Huguenots, Protestants in a predominantly Catholic country, could flee persecution by moving to Switzerland, where they started the Jura district watch industry. Similarly, when Columbus failed to obtain financing for his voyages from the Italians, he could switch to Spain, whose king ultimately consented.

A Historiography of Conquest

To appreciate colonialism, it is necessary briefly to delve into its specifics in different times and places. The following review is intended to demonstrate that colonialism meant quite different things under different contexts (i.e., it was historically and geographically specific).

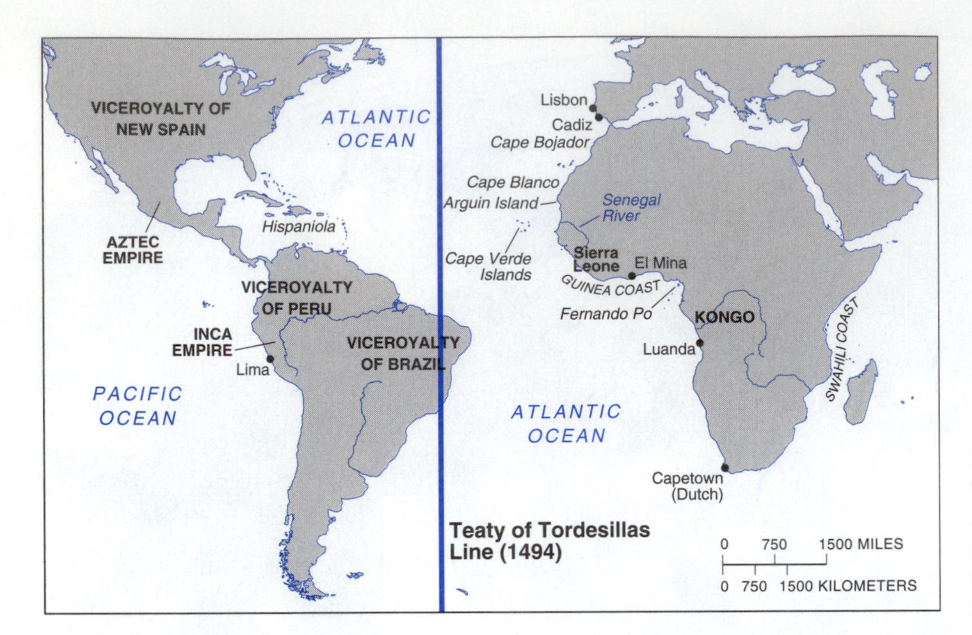

FIGURE 2.25
The Treaty of Tordesillas in 1494 divided the New World between Spain and Portugal, which is why Brazil speaks Portuguese rather than Spanish.

Latin America

Home to wealthy and sophisticated civilizations such as the Inca, Mayan, and Aztec cultures, Latin America was one of the first regions to be taken over by Europeans. Two years after Columbus arrived, Spain and Portugal struggled over who owned the New World, a contest settled by the pope at the Treaty of Tordesillas in 1494 (Figure 2.25). The conquistadors who spread out over Mexico and Peru annihilated the Aztecs and Incan states, respectively. In large part, this was accomplished through the introduction of smallpox, which killed 50 to 80 million people within a century of Columbus's arrival, the greatest act of genocide in human history.

Under the philosophy of mercantilism, which stressed bullion, or precious metals, as the key to national wealth, the Spanish took home large quantities of silver from the New World. The silver mines in central Mexico were among the largest in the world, and in the Potosi mines in Bolivia, 2 million Aymara Indians perished digging silver. Argentina takes its name from the Latin word for silver, and is home to the Rio Plata, or silver river, down which the metal flowed from Bolivia. Most of this metal was taken back to Spain in galleons and provided an enormous base of capital that financed economic activities throughout Europe.

Spain also introduced the *encomienda* land grant system into the New World, giving large tracts of lands to potential rivals to the Spanish throne to lure them from Iberia. This was an extension of the latifundia system practiced in Iberia. As a small landed aristocracy consolidated its hold, the distribution of farmland became highly uneven, with a few wealthy landowners and large numbers of landless *campesinos.* This pattern continues in the present, indicating how colonialism shapes the geographies of the contemporary world.

The Spanish Empire in the New World was largely ended by the independence movements of the 1820s that followed the Napoleonic Wars (although it took the Spanish-American War of 1898 to finish it off). Upon independence, the Spanish empire broke up into a series of independent countries stretching from Mexico to Argentina. The Portuguese Empire in Brazil did not fragment in the same way, leaving that nation as the giant of Latin America.

North America

The colonialization of North America proceeded along somewhat different lines. Here, the Spanish were active in Florida and in the Southwest (Figure 2.26). A century before the Pilgrims arrived at Plymouth Rock in 1620, the Spanish had control of what is now Texas and California. The French took over Quebec and the St. Lawrence River valley, only to lose it to Britain in the eighteenth century, and the Mississippi River valley, with the key port of New Orleans, only to sell it to the United States in the Louisiana Purchase of 1803. The Russians crossed the Bering Straits and seized Alaska but sold it to the United States in 1867. The Dutch established colonies in New Amsterdam, including Haarlem and Brueklyn, but the British captured the area in 1664 and renamed it New York. Britain emerged as the dominant power in North America, controlling New England and the Piedmont states along the eastern seaboard. Canada was colonized largely through the famous Hudson Bay company, then the world's largest, which controlled much of the fur and fish trade.

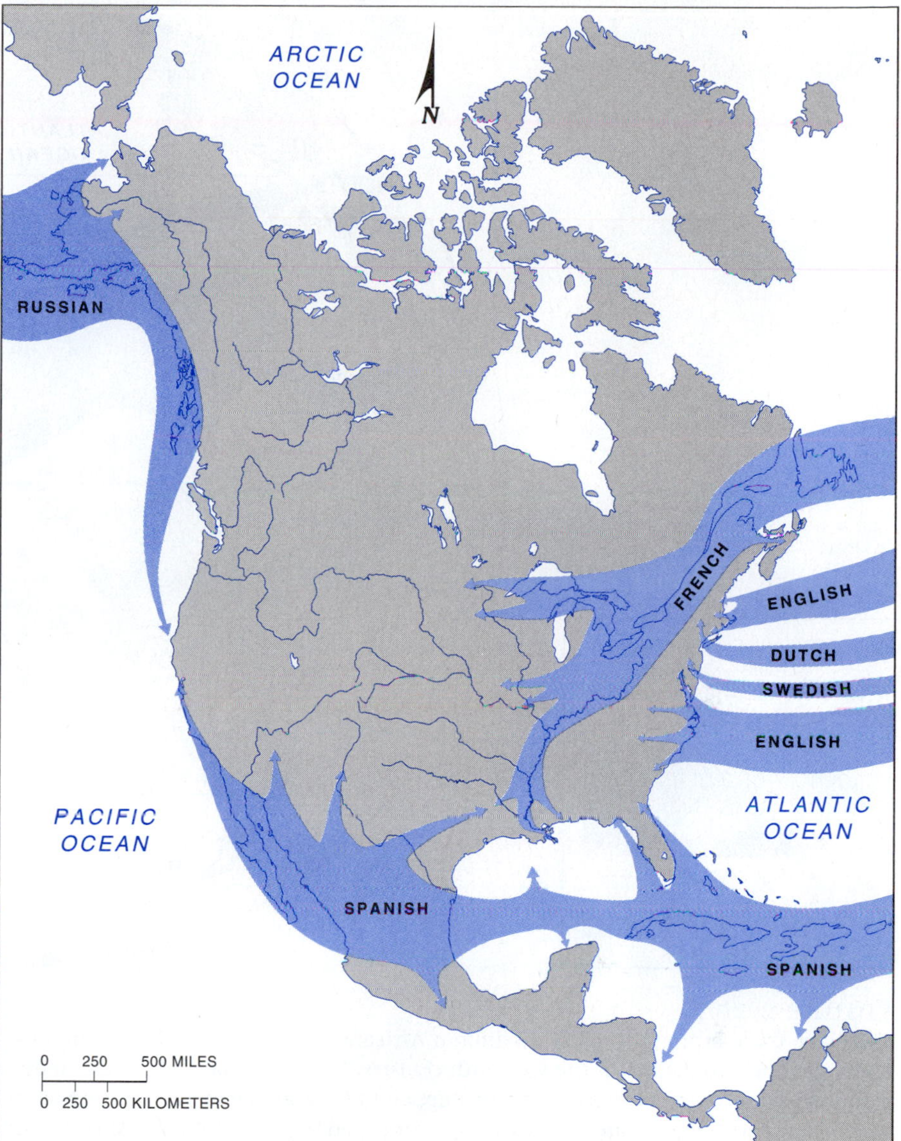

FIGURE 2.26
The colonial conquest of North America saw a variety of European powers. The Spanish were in Florida and the southwestern United States long before any other western power. The French occupied the St. Lawrence River valley and the Mississippi, before selling the Louisiana Purchase to the United States in 1803. Russia occupied Alaska, sending explorers as far south as California. Britain had colonies in New England and the Piedmont, and ruled Canada through the Hudson Bay Company.

British colonialism in North America began with a series of port cities on the east, typically at the mouths of rivers (e.g., Boston, New York, Philadelphia). After the independence of these colonies in the Revolutionary War, settlers moved west, across the Appalachians in the early nineteenth century and across the Great Plains and Rocky Mountains somewhat later. Railroads opened up this region to the East. As in Latin America, this process involved the wholesale eradication of Native American peoples, theft of their land, and commodification of territory.

Africa

Colonialism in Africa was unique, as it was everywhere. This process included slavery, the kidnapping of roughly 20 million people and exporting them to the New World (Figure 2.27), where they were used to compensate for the labor shortages brought on by the decimation of Native Americans. The capture of slaves robbed these societies of young adults in their prime working years and sometimes occurred with the assistance of local kings who sought to profit from the trade. Slaves were generally taken from western Africa, and the largest numbers were brought to work the sugar plantations in Brazil and the Caribbean. Others were brought to work the cotton and tobacco plantations of the American South. In eastern Africa, there long existed a smaller system of slavery operated by the Arabs.

Africa is rich in minerals, and colonialists were quick to seize upon that fact. In the nineteenth century, when European powers penetrated from the coastal areas into the interiors, mines for copper, gold, and diamonds soon opened, often using slave labor. These remain the basis of many African economies today.

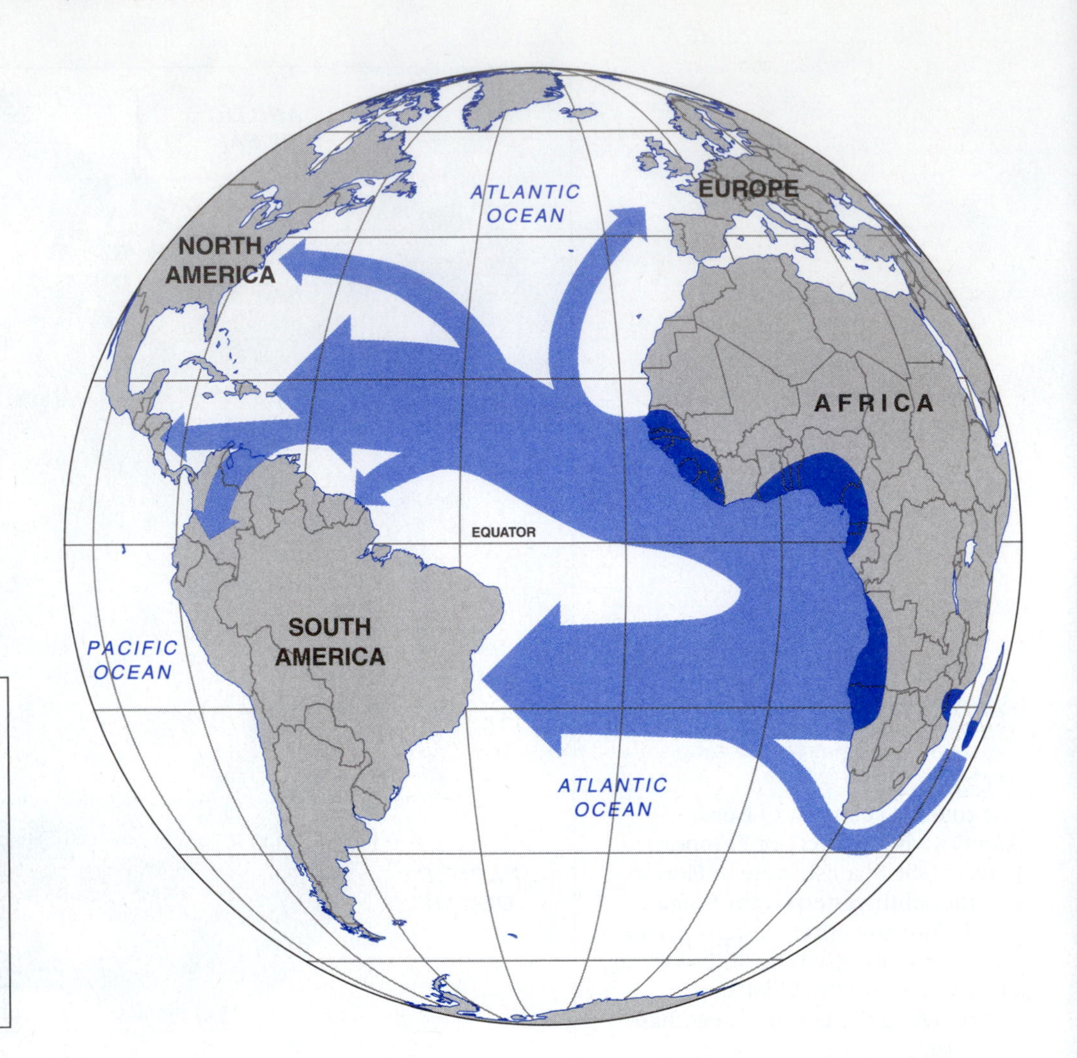

AFRICANS EXPORTED AS SLAVES (estimated) 1500–1870

3 –5 million

Fewer than 1/2 million

FIGURE 2.27
The slave trade brought roughly 20 million Africans to the New World to compensate for the labor shortages induced by the great smallpox epidemics of the sixteenth century. Slavery robbed African societies of many of their young people in the prime working years. Most slaves were shipped to the Caribbean and South America to work on sugar and fruit plantations, while in the southern United States slaves were used to grow and harvest cotton and tobacco.

Perhaps the most important colonial impact on Africa was the political geography that Europe constructed. Following the famous Berlin Conference of 1884, when European powers drew maps demarcating their respective areas of influence, the colonies of Africa bore no resemblance whatsoever to the distribution of indigenous peoples. Roughly 1000 tribes were collapsed into about 50 states (Figure 2.28). Some groups were separated by colonial boundaries; many others, with widely different cultures and economic bases, were lumped together. Not surprisingly, upon independence in the 1950s and 1960s, African states have been wracked by numerous civil wars and tribal conflicts in which millions have perished, including Angola, Congo, Rwanda, Liberia, and Sudan.

The Arab World

The Arab world, which was one of the most powerful and sophisticated centers of world culture from the seventh through the fifteenth centuries, had been colonized long before the Europeans arrived. The expansion of the Ottoman Empire over several centuries saw one group of Muslims, Turks, who are not Arabs, dominated another group of Muslims, the Arab peoples of the Middle East and North Africa (Figure 2.29). Gradually, starting in the nineteenth century, European powers encroached on this vast domain. In 1798, the French seized Egypt from the Ottomans, only to lose it to Britain three years later. The British used Egypt as a source of cotton, growing large plantations along the Nile River and building the Suez Canal in 1869. France also seized Northwest Africa.

After World War I, the Ottoman Empire collapsed, and the French and British seized its Arab colonies. The French took over Syria and Lebanon. The British assumed control over Palestine, much of which became Israel in 1948, as well as Iraq and the sheikdoms of the Persian Gulf. Many Arabs, who initially welcomed the Europeans as liberators, quickly learned that the new boss was very similar to the old one in its suppression of indigenous liberties.

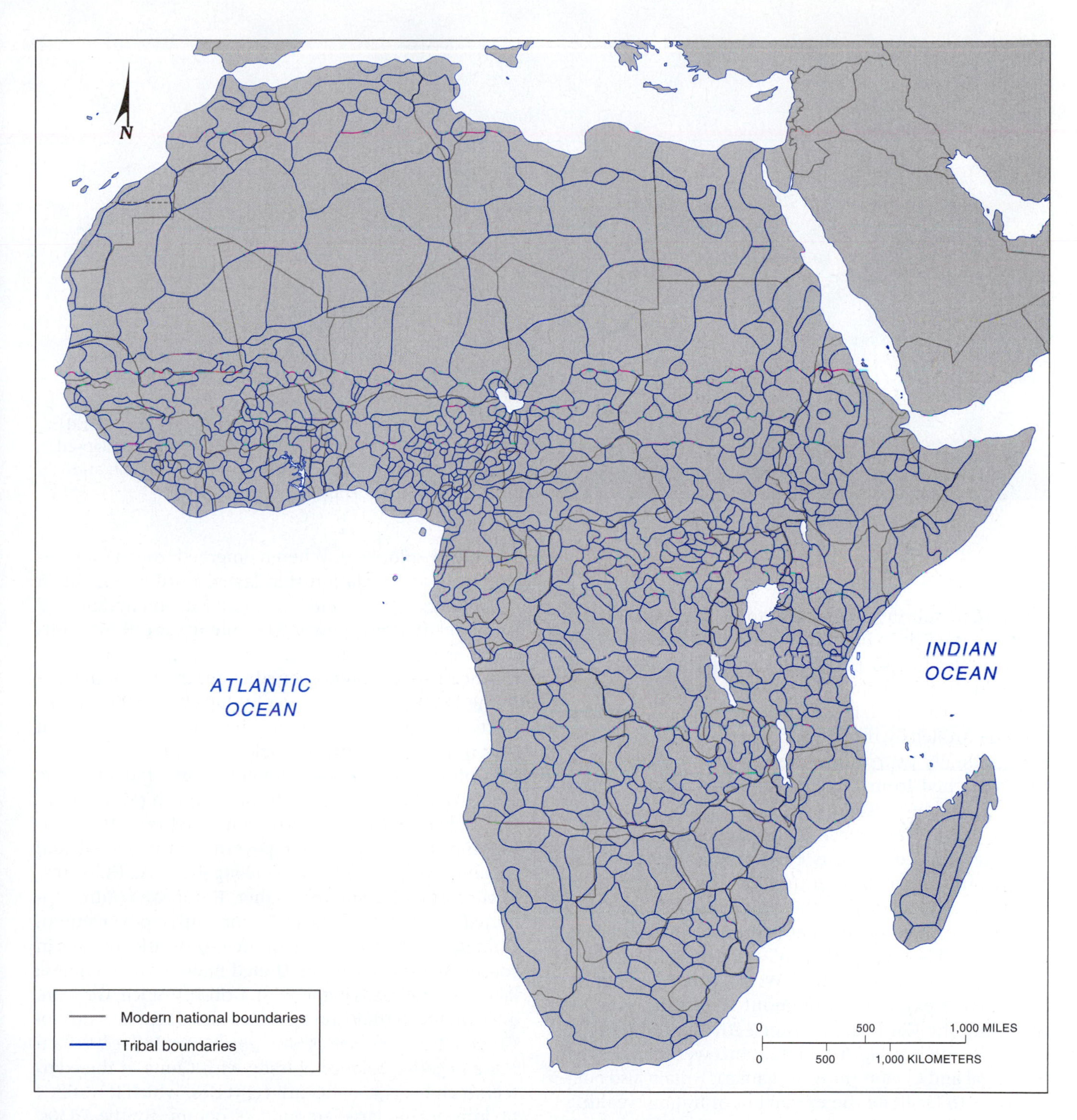

FIGURE 2.28
The Berlin Conference of 1884 redrew the map of Africa, collapsing roughly 1000 tribes into 50 different states. Because the boundaries of what would become Africa's states bore no resemblance to the geography of the people who lived there, many tribes were split in two, while others, without any cultural or economic similarities, were lumped together. The political geography of Africa has played out disastrously in the form of numerous tribal conflicts and civil wars today that have claimed the lives of tens of million of people.

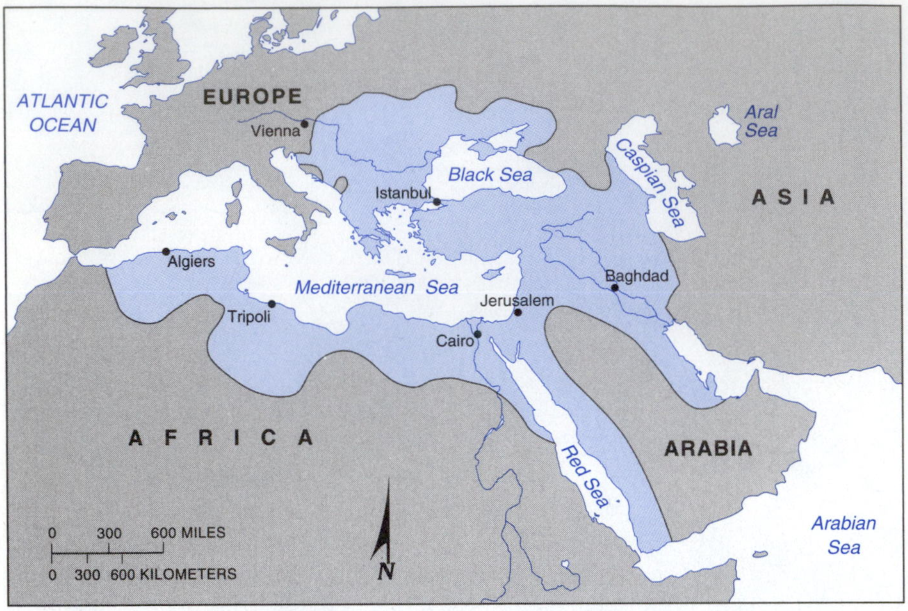

FIGURE 2.29
Turkey's Ottoman Empire stretched across North Africa, the Middle East, and southeastern Europe. Although Turks are not Arabs, this was the dominant power in the Arab world until World War I, when it collapsed, paving the way for the British and French colonialists.

SOUTH ASIA

The Indian subcontinent came to be the jewel in the crown of the British Empire. Starting in the seventeenth century, the British East Indies company established footholds in this domain, founding the city of Calcutta in 1603. India, which is predominantly Hindu, had long been controlled by the Muslim Mughals, whose power was gradually usurped by the foreigners. A vast land that stretched from Muslim Afghanistan in the east through Hindu India to Buddhist Burma, the Indian colony was the largest colonial possession on earth.

Britain had enormous economic impacts on this land. In the nineteenth century, British textile imports flooded India, destroying the Mughal textile industry, a classic example of colonial deindustrialization. This event later became symbolically important in the independence movement after World War II. Indian laborers were exported throughout the British empire, including the Caribbean, eastern Africa, and parts of the Pacific such as Fiji. Tea plantations were established in Bengal and Ceylon (now Sri Lanka). Britain also built railroads to facilitate the extraction of Indian wealth.

In 1857, a mass uprising against the British took place known as the Sepoy Rebellion. Encouraged by the local raj, or Mughal ruler of Calcutta, the revolt claimed the lives of tens of thousands of Indians when it was crushed. Although it failed, the action forced the British crown to assume direct control over this land rather than administer it through the East Indies Company.

EAST ASIA

East Asia, comprising China, Korea, and Japan, also had a unique colonial trajectory. Japan, as noted earlier, was never colonized. When it emerged from a long period of feudal isolation that lasted until 1868, Japan rapidly westernized and industrialized, and became the only non-Western power to challenge the West on its own terms, gradually expanding its power through northeast Asia. Korea, opened under the threat of force in the 1870s, was taken over by Japan in 1895 and annexed in 1905, and Japan held it until the end of World War II. A similar situation held in Taiwan.

China, however, was a different story. Under the rule of the Manchus, foreigners from the north (Manchuria) who ruled the Chi'n dynasty from 1644 until the revolution of 1911, the Chinese government was weak and corrupt. Except for a few cities along the coast, China was never formally colonized; rather, European control operated through a pliant and cooperative government. Chinese coolie labor was exported to British colonies in southeast Asia and to the United States, where chinese labor built railroads in the West. British, French, German, and American trade interests purchased vast amounts of Chinese tea, silks, spices, and porcelains. In fact, Britain ran a negative balance of trade with China in the eighteenth and early nineteenth centuries, which it rectified by introducing large amounts of opium. By the 1830s, opium addiction was widespread in China. Opium is highly addictive, and the introduction of mass quantities created severe social disruptions in Chinese society. Nonetheless, profits were more important, and the British trade balance was restored (Figures 2.30a and 2.30b).

In a rare moment of defiance, the Manchu government resisted opium imports, and Britain and China fought two short, nasty conflicts, the Opium Wars of the 1840s. The British won easily. As compensation, the British seized "treaty ports," coastal cities such as Hong Kong and Shanghai, where Western, not Chinese, law applied. Britain held Hong Kong until 1997.

British Sales of Opium to China

Number of Chests (Thousands)

0 5 10 15 20 25

1729 1790 1810 1823 1832

Years

(a)

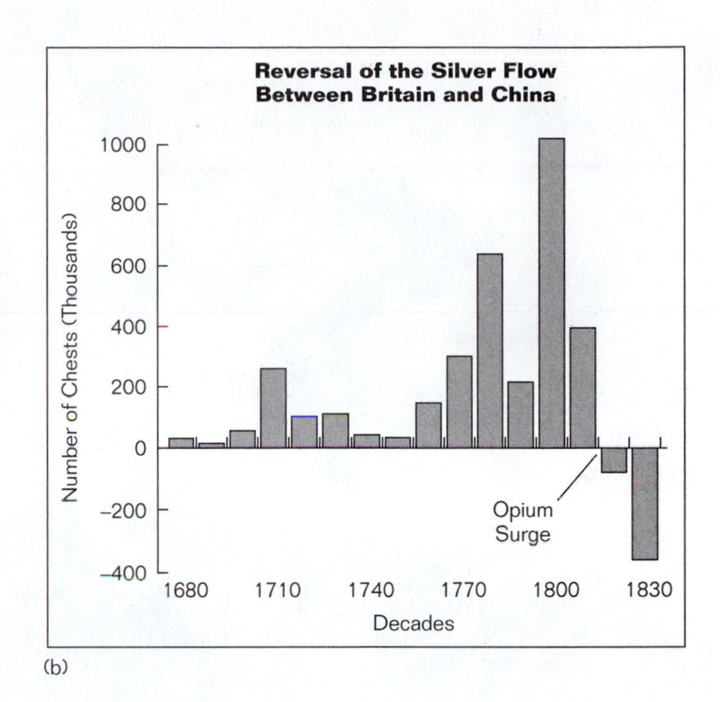

(b)

FIGURE 2.30
British sales of opium to China in the nineteenth century represented a response to Britain's trade deficit with that nation. Economically, opium had exactly the intended effect, reversing the balance of payments as indicated by silver flows. Socially, it was catastrophic: Opium is highly addictive, and as much as 12 percent of China's adult population became hooked on opium as a result of cheap British imports. The Manchu regime's opposition prompted the Opium Wars of the 1840s, which Britain won handily, opening China to foreign treaty ports.

Chinese resentment against the Manchus culminated in the Taiping Rebellion, a huge uprising in the southern part of the country led by Chinese Christians. Lasting from 1851 to 1864, this rebellion led to the deaths of more than 20 million people but was ultimately crushed by the Chinese government with Western backing. The shorter Boxer Rebellion of 1899–1901 was more explicitly anti-Western. These revolts set the stage for the successful nationalist revolution of 1911, which ended Manchu rule and initiated the Republic of China.

SOUTHEAST ASIA

The peninsula of Indochina and the islands of Southeast Asia, long home to a rich and diverse series of peoples and civilizations, were conquered by a variety of different European powers (Figure 2.31). The Philippines was Spain's only Asian colony and served as the western terminus of the trans-Pacific galleon trade. Administered as part of Mexico, the Philippines was heavily shaped by Spanish rule, which affected land-use patterns (including sugar plantations) and the language and made it the only predominantly Catholic country in Asia. Spanish rule ended in 1898, when the United States took over, and it became independent after World War II.

The French controlled much of Indochina, including Vietnam, Laos, and Cambodia. French rule shaped the design and architecture of cities such as Saigon, and large numbers of Vietnamese converted to Catholicism. French domination did not end until 1954, with the defeat at Dienbienphu, an event that laid the foundations for American military involvement there.

Britain was also a major power in southeast Asia, controlling Burma and, only informally, the economy of Thailand. These were made into rice exporters for other parts of the British Empire. The British controlled the colony of Malaya (later Malaysia), including the strategically critical Malacca Straits. They founded the city of Singapore as a naval station and commercial center to exercise control over this region. Malaya, like other colonies in the area, became a major producer of rubber products as well as timber and tin.

Indonesia, now the fourth most populous country in the world, was dominated by the Dutch for several hundred years. Dutch rule, starting with the founding of the Batavia colony on Java in the eighteenth century, gradually expanded to include the other islands. The primary institution involved was the Dutch East Indies Company, a chartered crown monopoly similar to the British East Indies and British West Indies Companies. Indonesia became a significant source of spices, tropical hardwoods, rubber, cotton, and palm oils. In all of southeast Asia, Chinese immigrants came to play major roles in the economy as bankers and shopkeepers.

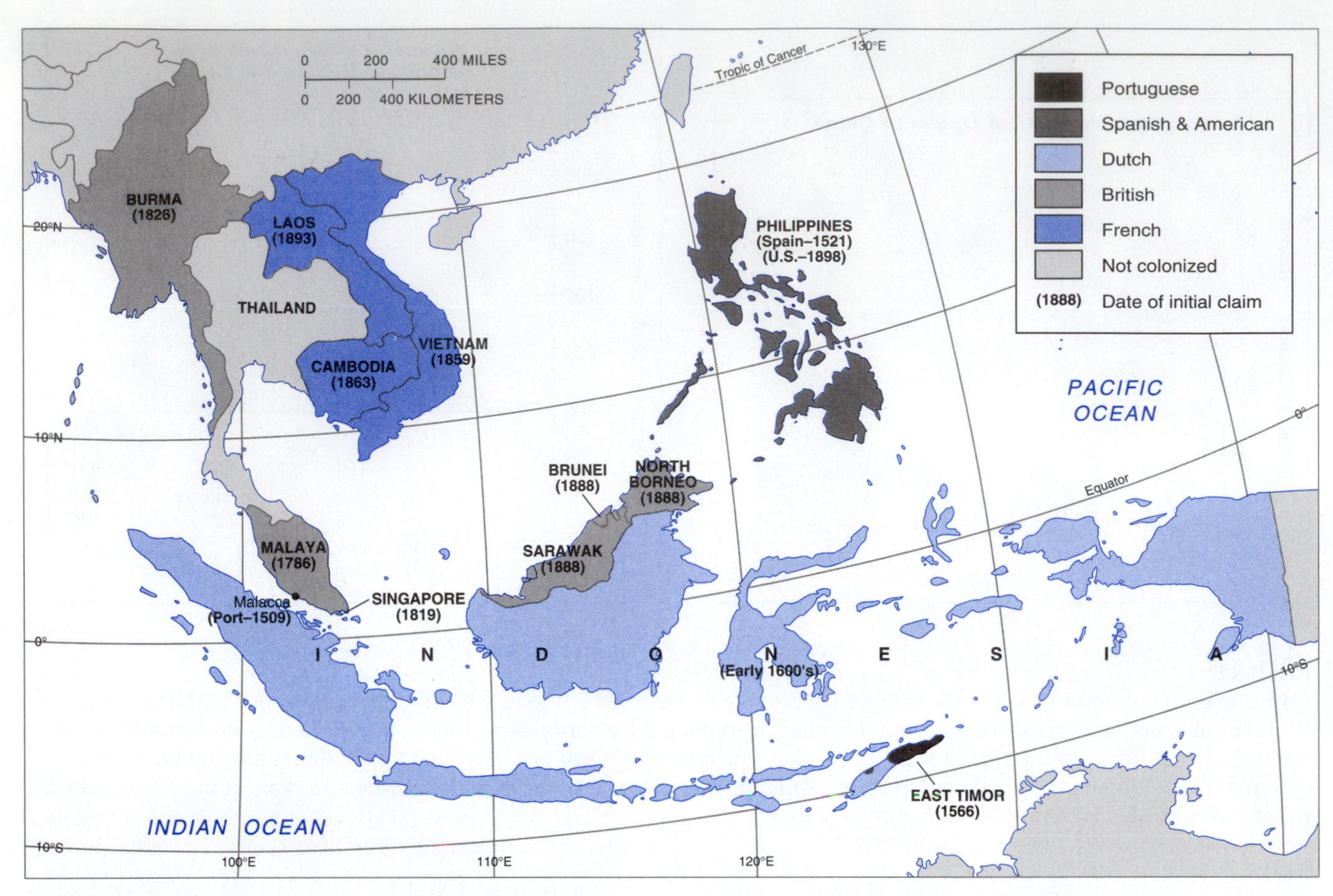

FIGURE 2.31
Colonialism in Southeast Asia saw a variety of different powers. The French occupied Vietnam, Laos, and Cambodia. Britain ruled Burma, Malaya, and Singapore and indirectly controlled Thailand. The Spanish held the Philippines until 1898, when the United States took it away. The Dutch ruled Indonesia until 1947, when it finally achieved independence.

Oceania

Australia, New Zealand, and the Pacific Ocean islands, which are commonly grouped as the region of Oceania, formed yet another domain of colonialism. Australia was inhabited for thousands of years by indigenous aborigines, most of whom were eradicated by the British as they exerted control. The native Tasmanian populated was completely exterminated. The continent served originally as a penal colony for criminals. In the nineteenth century, it became a significant exporter of wheat and beef.

New Zealand, also a British colony, had a much larger native population, the Maori, who continue to form a significant presence there. The introduction of refrigerated shipping in the late nineteenth century turned this island nation into a major producer of lamb and dairy products.

Finally, the countless islands of the Pacific Ocean, home to varying groups of Polynesians, Micronesians, and Melanesians, were conquered by the British and French. Captain Cook sailed through in the seventeenth century, paving the way for followers. Fishing and whaling interests used these islands to refuel in the nineteenth century. Following World War II, when the United States drove Japan out of the Pacific, America became the leading political force in the area.

The Effects of Colonialism

By now, it is abundantly evident that colonialism had enormous consequences on colonized places. It is worth recapitulating these in more systematic form.

Annihilation of Indigenous Peoples

Often, colonialism involved traumatic consequences for the people who were conquered. At times, this involved open genocide, such as in Australia, in which more than 90 percent of the aboriginal population was exterminated. In the New World, disease led to the deaths of tens of millions. The African slave trade devastated tribal societies on that continent. While this is the bluntest expression of colonial control, it serves as a reminder that the European conquest of the world was often violent.

Restructuring Around Primary Economic Sector

The incorporation of colonies into a worldwide division of labor led above all to the development of a primary economic sector in each of them. Primary economic activities are those concerned with the extraction of raw materials from the earth, including logging, fishing, mining, and agriculture. In Latin America, Africa, and Southeast Asia, cash crops such as sugar, cotton, fruits, rubber, and tobacco were grown for sale abroad with the plantation system (Figure 2.32). Silver, tin, gold, and other metallic ores were mined using slave labor or peasants working in slavelike conditions. Mercantilist trade policies worked to suppress industrial growth in the colonies. This process is largely responsible for the fact that many developing countries today export low-valued goods and must import high-valued ones.

Formation of Dual Society

Colonialism brought with it great inequality in colonized societies. Often, colonial powers utilized a small native elite to assist them in governing the colonies, typically drawn from an ethnic minority. For example, the French utilized the Alawites in Syria, a sect of Islam neither Sunni nor Shiite. The Germans and Belgians favored lighter skinned Tutsi over the darker skinned Hutu in Rwanda and Burundi. The British relied on the Muslim Mughal rulers to govern Hindu India.

FIGURE 2.32
A sugar plantation in the Caribbean Antilles. Plantations used cheap labor, often slaves, working under deplorable conditions. They represented the first wave in the commercialization of many crops worldwide.

For the bulk of the population, colonialism entailed declining economic opportunities, a theme central to dependency theory (Chapter 13). Traditional patterns of agriculture were disrupted, often with disastrous effects. Land-use patterns favored colonialists, while indigenous peasants had to pay for the costs of their own exploitation with taxes. People living in dry climates in western Africa, for example, coped well with drought until the British forced them into a system of cash cropping. Huge famines shook India, Egypt, and China in the nineteenth century.

Polarized Geographies

As colonial societies became polarized, so too did the spaces they comprised. Ports, which were central to European maritime trade and control, became important centers of commerce, often to the detriment of traditional capitals further inland. For example, in Peru, Lima displaced the Incan city of Cuzco; in western Africa, the famous trade city of Timbuktu declined as new maritime routes flourished; in India, coastal cities such as Calcutta, Bombay (now Mumbai), and Madras displaced the Mughal capital of Delhi; in Burma (now Myanmar), Mandalay fell far behind coastal Rangoon; in Vietnam, the imperial capital Hue declined in the face of Saigon; and in Indonesia, the traditional center of Jogjakarta was marginalized by the Dutch port city of Batavia, later Jakarta. As cities grew and offered more opportunities, millions of people left the poorer rural areas in waves of rural-to-urban migration.

From the coasts, railroads extended colonial control into the interior (Figure 2.33), often reaching into mineral-rich regions or plantations. A long-term consequence of this design is that the road and rail networks of developing countries often bear little resemblance to the distribution of the population that lives there and their needs; rather, they are constructed to facilitate the export of raw materials to the colonizing country, and today, to the global economy.

Transplantation of the Nation-State

The nation-state, as we observed earlier, was fundamentally a European creation. Nonetheless, it was widely dispersed around the world as colonies were made into states. In Africa, where this process was the most notorious, it led to the formation of unstable states with highly artificial borders. Similarly,

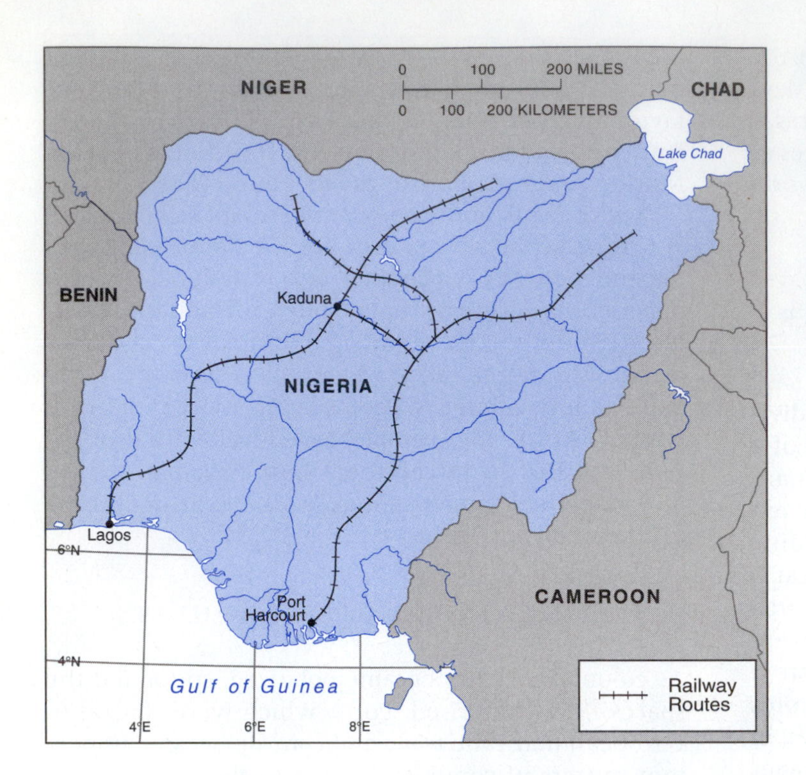

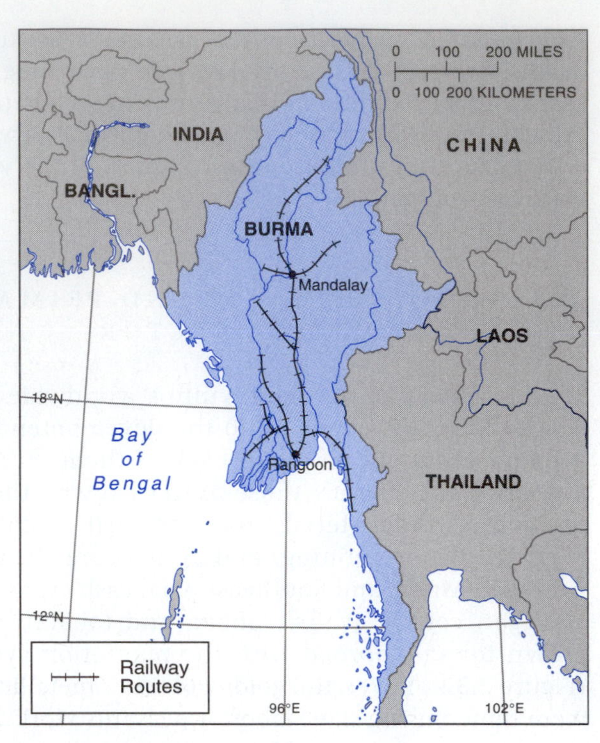

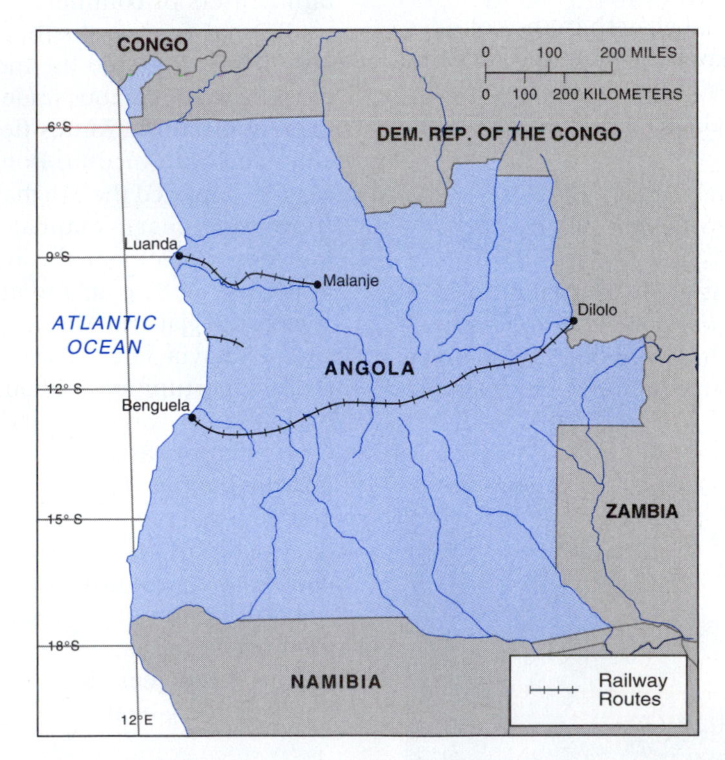

FIGURE 2.33
Railroads in Nigeria, Burma, and Angola reveal the colonial propensity to establish a transportation infrastructure that linked coastal port cities with the resources of the interior, which were typically mines or plantations.

Burma and Afghanistan were creations of the British. Even India, which is more culturally diverse than all of Europe, was stitched into one country; surprisingly, it divided only into three (including Pakistan and Bangladesh). Unlike Europe, where states were centered on some degree of ethnic similarity, the states of much of the developing world were too diverse to be understood in the same terms as in Europe. In such societies, where local religious, ethnic, and tribal loyalties supersede nationalism, political conflicts can impair economic growth and development. Of course, the United States is also very diverse ethnically and culturally but emerged under very different historical circumstances than did the former European colonies in the developing world. Above all, the United States was primarily a colonizer rather than a colony, able to exert military and economic power over other countries and ultimately rising to become the globe's premier superpower, and that makes all the difference.

CULTURAL WESTERNIZATION

As noted previously, colonialism is not simply an economic or political process, but also a cultural and ideological one. Western economic and political control was accompanied by the imposition of Western culture. Missionaries, for example, spread Christianity throughout the colonial world, sometimes successfully (e.g., Latin America, the Philippines) and sometimes not (e.g., China). School systems in colonies, generally set up to benefit the ruling elite, offered extensive instruction in the history and culture of the colonial country, but little about the society in which the students lived. More broadly, colonialism may be seen as one chapter in the broader process by which global capitalism homogenizes lifestyles, values, and role models around the world, turning disparate sorts of peoples into ready consumers.

THE END OF COLONIALISM

The European empires were long lived, lasting almost half a millennium. Yet ultimately, they collapsed. In Latin America, this process began relatively early, following the Napoleonic wars, which weakened the core sufficiently to allow the periphery to break away. In Africa and Asia, the end of colonialism came much later, following World Wars I and II, which, similar to the global geopolitical situation of the early nineteenth century, saw the European powers self-destruct.

The international environment following World War II provided an ideal opening for various nationalist and independence movements in the Arab world, Africa, India, and Southeast Asia. Sometimes communists were involved. The Japanese occupation of Indochina, the Philippines, and Indonesia destroyed the myth of European invincibility. Often, independence movements were led by intellectuals educated in the West, such as Ghana's Kwame Nkrumah, Vietnam's Ho Chi Minh, and India's Mohandas Gandhi. Moreover, the cold war rivalry between the United States and the Soviet Union allowed political leaders in the developing world to play the superpowers off against one another.

Independence movements succeeded, sometimes peacefully, as in India, and often violently, as exemplified by the Vietnamese and the Algerian defeat of the French in their respective countries. As a result of this process of decolonization, the number of independent states multiplied rapidly in the 1950s, 1960s, and 1970s. Today, there are very few official colonies remaining. Whether colonialism is truly dead, however, is another matter; in Chapter 13 we take up the notion of neocolonialism, colonialism in practice but not in name.

SUMMARY

This chapter has introduced you to the historical foundations of the world economy. To understand the economic geography of the world today, it is necessary to appreciate how it originated and came to be. This entails knowing something about capitalism, the type of society that emerged in the sixteenth and seventeenth centuries and now dominates virtually the entire globe. To appreciate what capitalism is, and how it constructs geographies, is to understand that it is one of many possible ways in which human beings have organized themselves historically and geographically.

First, the chapter described feudal society, both to outline in some detail what a noncapitalist society looks like and to sketch the historical context in which capitalism emerged. Then we proceeded to describe the fundamental features of capitalism, which, uniquely, is dominated by markets as the primary way in which resources are organized. Markets are not unique to capitalism, but their importance is. However, this does not mean that markets are the only way in which economic activity is shaped, for the state plays a role even in the "freest" of market societies. Capitalism also entails a specific set of class relations, including the commodification of labor and a working class. Capitalism also creates landscapes of uneven spatial development.

Next, we delved into the Industrial Revolution, the period in the eighteenth and nineteenth centuries that saw a fantastic transformation in capitalist social relations. The ability to harness inanimate energy led to a wave of technological innovations and productivity increases. Industrialization created the factory system, and with it, the working class. In commodifying time and space the Industrial Revolution produced radically new geographies. Industrialization, which began in Britain, catapulted that country to become the most powerful in the world. It also unleashed a new international economy characterized by significantly higher levels of trade and investment, which were brought on in part by the application of the steam engine to land and ocean transportation. The rising speed of transportation and communication unleashed great waves of time-space compression. Yet industrialization was a complex process that varied over time, with the rise of different industries, products, and technologies.

The latter part of this chapter explored the multiple dimensions of colonialism, the expansion of capitalism on a global scale. We traced some of the advantages that Europeans enjoyed that allowed them to conquer many societies much larger than themselves. The chapter examined major world regions conquered by different European powers, noting that colonialism entailed different impacts in different parts of the globe. Finally, we summed up the impacts of colonialism, which produced the division between the world's economically developed and underdeveloped nations.

STUDY QUESTIONS

1. Why is historical context important to the analysis of contemporary economic geography?
2. What is feudalism?
3. How is capitalism a unique type of economy and society?
4. Are there classes under capitalism?
5. Is capitalism the only type of society to have markets? Why or why not?
6. When and where did capitalism begin?
7. What is industrialization?
8. When and where did the Industrial Revolution begin?
9. What are some major economic, social, and geographic impacts of the Industrial Revolution?
10. How did Europe manage to colonize the rest of the world?
11. How did colonialism differ among Latin America, Africa, and Asia?
12. How did colonialism affect the societies and geographies of the colonies?
13. When did colonialism come to an end, and why?

KEY TERMS

Bubonic plague
burghers
capitalism
commodities
feudalism
guilds
market
serfs

SUGGESTED READINGS

Abu-Lughod, J. 1989. *Before European Hegemony: The World System A.D. 1250–1350.* New York: Oxford University Press.

Chirot, D. 1985. "The Rise of the West." *American Sociological Review* 50:181–195.

Diamond, J. 1999. *Guns, Germs, and Steel.* New York: W. W. Norton.

Hindess, B. and P. Hirst. 1975. *Pre-Capitalist Modes of Production.* London: Routledge and Kegan Paul.

Landes, D. 1969. *The Unbound Prometheus: Technological Change and Industrial Development in Western Europe from 1750 to the Present.* Cambridge: Cambridge University Press.

Mann, M. 1986. *The Sources of Social Power.* Cambridge: Cambridge University Press.

Said, E. 1978. *Orientalism.* London: Routledge.

Thompson, E. P. 1966. *The Making of the English Working Class.* New York: Vintage Press.

Wallerstein, I. 1979. *The Capitalist World-Economy.* Cambridge: Cambridge University Press.

WORLD WIDE WEB SITES

WORLD FACTBOOK ON COUNTRIES
http://www.odci.gov/cia/publications/pubs.html

THE WORLD BANK
http://www.worldbank.org/
A leading source for country studies, research, and statistics covering all aspects of economic development and world trade. Its home page provides access to the contents of its publications, to its research areas, and to related Web sites.

WEBEC: WORLD WIDE WEB RESOURCES
http://netec.wustl.edu/WebEc/WebEc.html
An extensive set of site listings concerning the economy.

CHAPTER

3 POPULATION

OBJECTIVES

- To describe and account for the world distribution of human populations
- To examine the economic causes and consequences of population change
- To describe the Malthusian argument, its extensions, and weaknesses
- To describe the major demographic and economic characteristics of a population
- To outline the Demographic Transition
- To discuss the growth and impacts of the baby boom
- To describe and explain economic migrations, past and present

Crowds of people gather at the Syracuse University Carrier Dome sports complex.

Human beings are the most important element in the world economy. People are not only the key productive factor, but their welfare is also the primary objective of economic growth and analysis. People are the producers as well as the consumers of goods and services. As the world's population continues to grow, we face the critical question of whether there is an imbalance between producers and consumers. Does population growth prevent the sustainability of development? Does it lead to poverty, unemployment, and political instability?

To help answer these questions, this chapter examines the determinants and consequences of population change for developed and developing countries. It analyzes population distributions, characteristics, and trends. It also reviews competing theories on the causes and consequences of population growth.

GLOBAL POPULATION DISTRIBUTION

While there is a widespread belief that there are "too many" people in the world, in fact the presence of large numbers of human beings is a relatively recent phenomenon (Figure 3.1). For the vast bulk of human existence—upward of 95 percent—people existed as hunters and gatherers, collecting food from various wild plants and stalking animals. This mode of production yields relatively few calories per unit area and supports only low population densities. Further, it is demographically very stable over time: Because births equaled deaths, there was virtually no increase in the number of people in the world. Following the Agricultural Revolution of roughly 8000 B.C., the capacity of many societies to support more people increased somewhat. However, the really big gains in population did not occur until the onset of the Industrial Revolution of the eighteenth and nineteenth centuries, when the transformation of agriculture freed millions from lives of rural toil and allowed for large, dense urban settlements. Thus, exponential population growth is largely a product of modernity and the modern world.

At the end of 2005, there were approximately 6.5 billion people in the world. The diverse populations that inhabit the world were very unevenly distributed geographically. Most people are concentrated in but few parts of the world (Figure 3.2), particularly along coastal areas and the floodplains of major river systems. Four major areas of dense settlement are (1) East Asia, (2) South Asia, (3) Europe, and (4) the eastern United States and Canada. In addition, there are minor clusters in Southeast Asia, Africa, Latin America, and along the U.S. Pacific coast. Figure 3.3 is a cartogram, a map that deliberately distorts areas in proportion to a variable, in this case, population size, revealing the large masses of humanity in Asia. Asia's population is the largest, as it has been for several centuries. In 2005, Asia contained 3.66 billion people or 56 percent of the world's population. Six of the top 10 countries in population size—China, India, Indonesia, Japan, Bangladesh, and Pakistan—are in Asia. Europe (664 million) and Russia (143 million) combined were home to about 12 percent, Africa (915 million) to 14 percent, Latin America (554 million) to 8.5 percent, North America (331 million) to 5 percent, and Oceania (34 million) to less than 1 percent. The populations of the developing world—Africa, Asia (excluding Japan and Russia), and Latin America—accounted for three out of every four humans.

National population figures show even more variability. Ten out of the world's nearly 200 countries

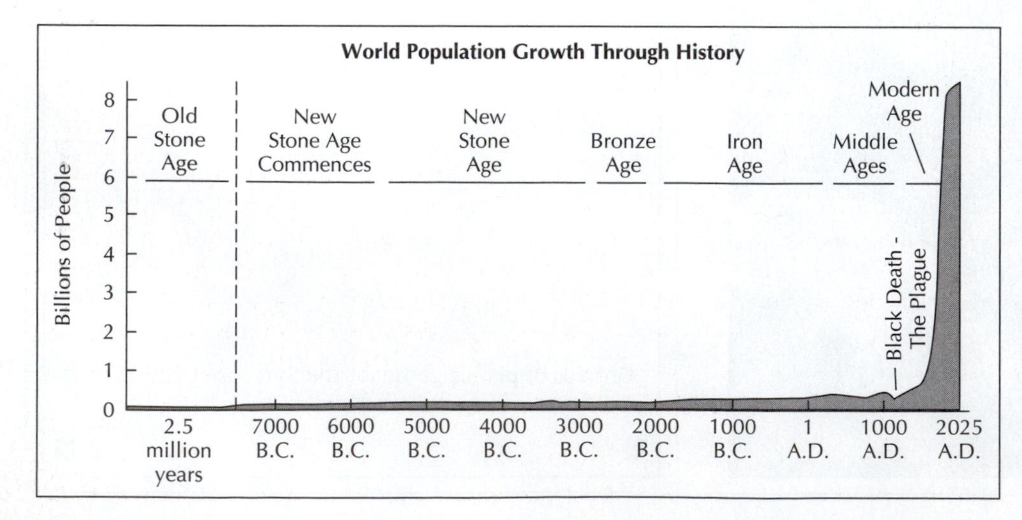

FIGURE 3.1
World population growth throughout history. For most of human existence, population levels were low and growth rates were zero. Only with the Industrial Revolution that created the modern age did growth rates begin to rise, leading to an exponential increase in the numbers of people.

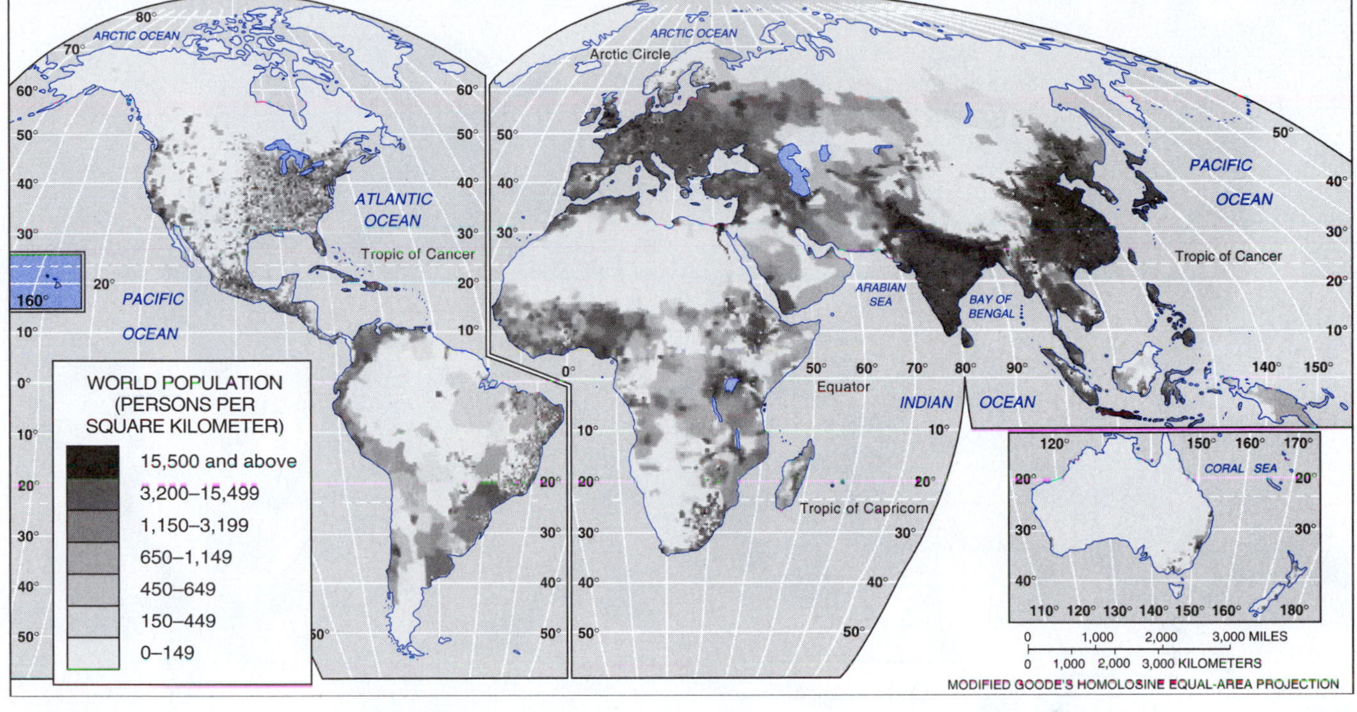

FIGURE 3.2
Population dot map of the world. This map shows population clusters within each country. Population density is shown on this map as a concentration of dots, with one dot representing 10,000 people. Population density is defined as the number of phenomena occurring within a real unit. Population density worldwide is normally expressed as the number of people per square mile. Population density in East Asia, notably China and Japan, as well as South Asia, including Bangladesh, India, and Pakistan, is extremely high. Population density in Northern Asia, Africa, and South and North America is quite low, comparatively speaking. Three major and two minor areas of world population concentration occur. These are (1) East Asia, (2) South Asia, (3) Europe, (4) Northeastern United States and Southeastern Canada, and (5) Southeast Asia, especially the country of Indonesia and the island of Java.

account for two-thirds of the world's people (Table 3.1). Five countries—China, India, the United States, Indonesia, and Brazil—contain half of the world's population. With 1.3 billion people, China is the world's most populous country. India, with 1.1 billion, is second, but is growing more quickly, and in roughly 50 years will surpass China in population. The United States, with 296 million in 2005, is third, and is by far the most populous of the developed nations. Indonesia, the world's largest Muslim nation, is fourth, with 222 million. Other nations with large populations include Brazil (184 million), Pakistan (162 million), Bangladesh (144 million), Russia (143 million, but declining), Nigeria (132 million), Japan (127 million, but declining), and Mexico (105 million). Only two of the 10 most populous nations are considered to be economically developed (the United States and Japan).

Population Density

Because countries vary so greatly in size, national population totals tell us nothing about crowding. Consequently, population is often related to land area. This ratio is called population density—the average number of people per unit area, usually per square mile or square kilometer. Several countries with the largest populations have relatively low population densities. For example, the United States is the fourth most populous country, but in 2005 it had a population density of only 84 people per square mile. Although they have significant and dense metropolitan areas, the United States and Canada form one of the more sparsely populated areas of the world.

Excluding countries with a very small area (such as Singapore), Bangladesh is the world's most crowded nation, where more than 144 million people are crowded into an area the size of Iowa. Three of the 10 most densely populated countries—the Netherlands, Japan, and Belgium—are economically developed, whereas another three—South Korea, Taiwan, and Israel—are newly industrializing countries (NICs). The remainder are clearly less developed nations, reminding us that there is no clear relationship between population density and economic development.

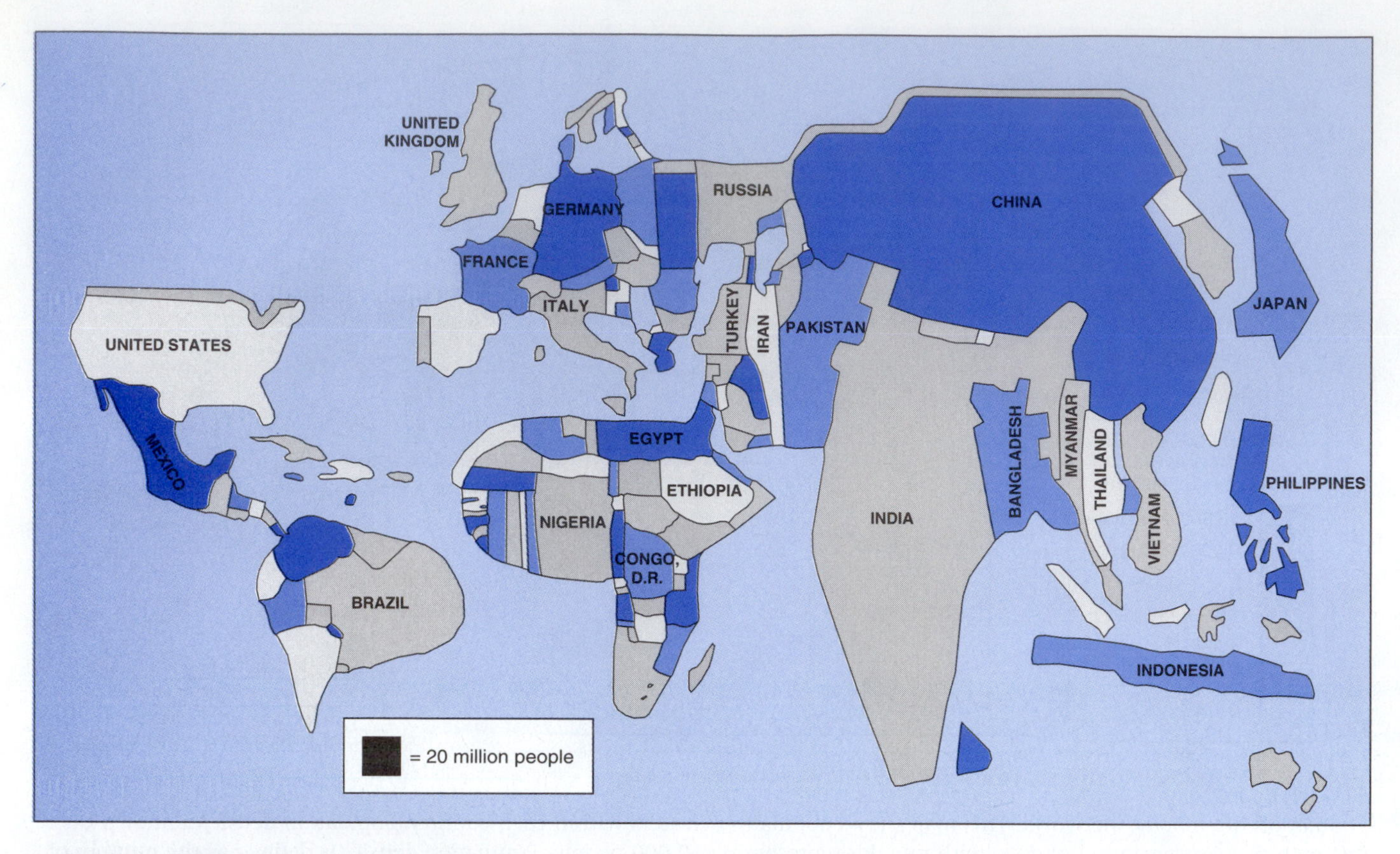

FIGURE 3.3
Cartogram of world population. This map shows the area of each country in proportion to its population. Geographic space has been transformed into population space. Asia dominates the map, especially China, with 1.3 billion people, and India, with 1.1 billion people. Europe is much larger than on "normal" map of territories. Both South America and Africa show up much smaller than their territorial size because their populations are relatively small.

Contrary to popular opinion, not all crowded countries are poor. In fact, in the Sahel states of Africa, population densities are very low. But what explains the fact that many people in the Netherlands or Singapore live well on so little land? Part of the explanation lies in their historical development and position within the colonial and contemporary world systems. Part lies in their industrious people and their ability to adapt to change. Part

Table 3.1
The World's Ten Most Populous Countries, 2005

Country	*Population 2005 (millions)*	*% Annual Growth Rate*	*Estimated Population 2050 (millions)*
China	1,304	0.6	1,437
India	1,104	1.4	1,628
United States	296	0.6	420
Indonesia	222	1.4	308
Brazil	184	1.0	260
Pakistan	162	2.0	295
Bangladesh	144	2.1	231
Russia	143	20.4	119
Nigeria	132	2.4	258
Japan	127	0.0	101

Source: World Population Data Sheet.

lies in the policies of their governments, which encourage economic growth. And part of the explanation lies in their history of trade or their relative locations. Singapore is on one of the great ocean crossroads of the world. But being on a crossroads has worked no similar miracle for Panama. In 2005, Singapore had a per capita income ($29,700) that was four times that of Panama ($7,300).

National population densities are abstractions that conceal much variation within countries, as well as among them. Egypt had a reasonably low figure of 71 people per square kilometer in 2005, but 96 percent of the population lives on irrigated, cultivated land along the Nile Valley where densities are extremely high. In China, the vast majority of people live in the eastern third of the country, near the Pacific Coast, where most of the large cities are concentrated (Figure 3.4). Similarly, in the United States there are very densely populated and sparsely populated areas (Figure 3.5). Large areas to the west of the Mississippi are essentially devoid of people, whereas the Northeast is densely settled. The island of Manhattan, for example, has a density that is roughly the same as Hong Kong.

Factors Influencing Population Distribution

What explains the uneven distribution of the world's people? One factor among many is the physical environment. Most of the world's people tend to be concentrated along the edges of continents, river valleys, at low elevations, and in humid midlatitude and subtropical climates. Lands deficient in moisture, and hence inhospitable to agriculture (at least without widespread

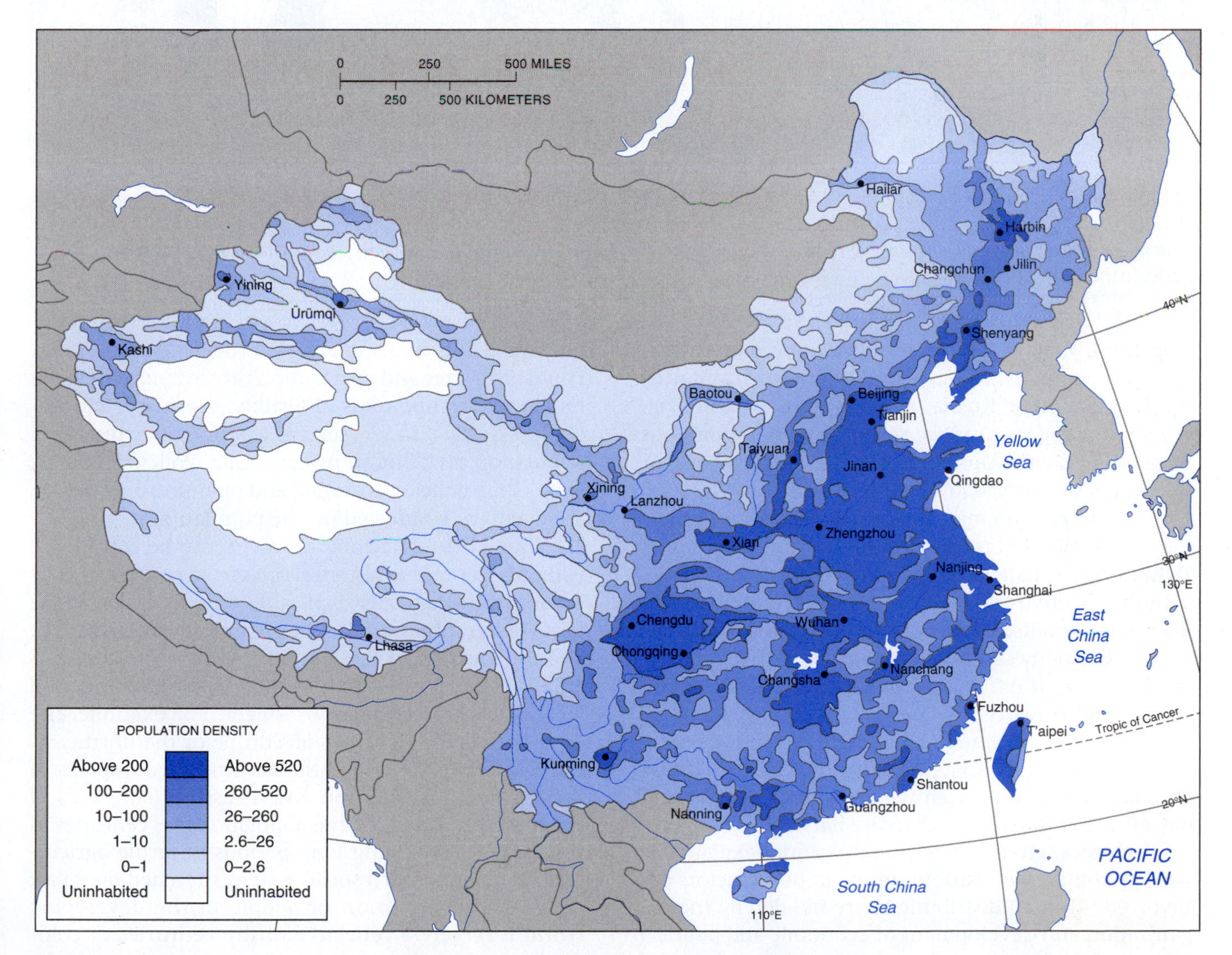

FIGURE 3.4
The distribution of China's population. With 1.3 billion people, most of China's population is clustered in the eastern half of the country, where moister climates allow for an agricultural base. The western half, in contrast, is mountainous or very dry.

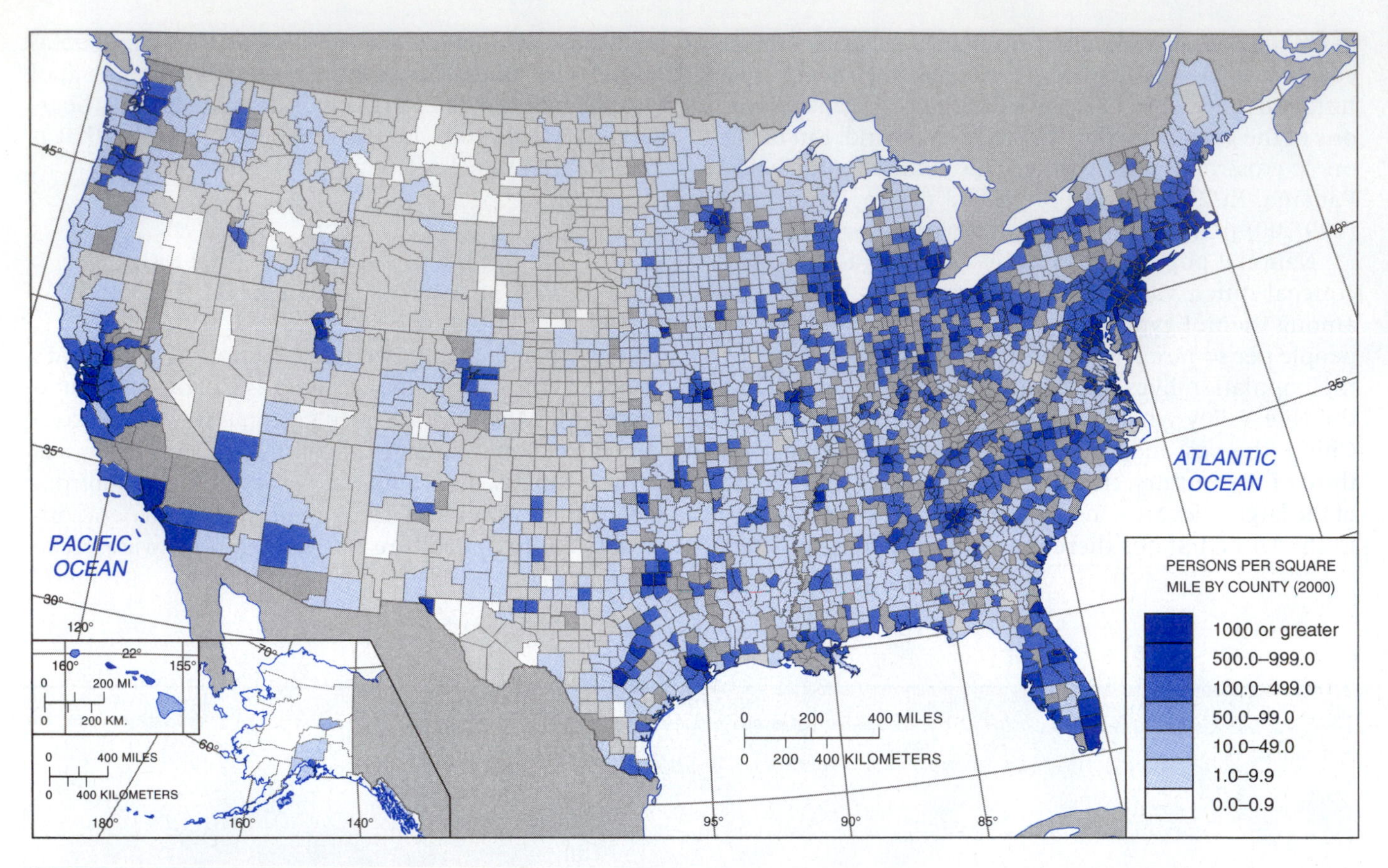

FIGURE 3.5
Distribution of the U.S. population, 2000. The densest regions of the United States include the metropolitan areas in the Northeast, Midwest, and California. Much of the Midwest and intermontane West, in contrast, consists of low-density agricultural and ranching areas.

irrigation), such as the Sahara Desert, are sparsely settled. Few people live in very cold regions, such as northern Canada, arctic Russia, and northern Scandinavia, where growing seasons are short. Many mountainous areas—whether because of climate, thin stony soils, or steep slopes—are also low-density habitats.

Extreme caution must be exercised in ascribing population distribution to the natural environment alone. To hold that climate or resources control population distribution is environmental determinism, a view long discredited because it is simplistic and often factually incorrect. Certainly climatic extremes, such as insufficient rainfall, present difficulties for human habitation and cultivation. However, given the forces of technology, the deficiencies of nature increasingly can be overcome. Air conditioning, heating, water storage, and irrigation are examples of the extensive measures that technology offers to residents of otherwise harsh environments.

If physical environments alone cannot explain the world's population distribution, what other factors are involved? Human distributions are molded by the organization and development of economic and political systems. It is the geography of economic activity—the labor markets, job opportunities, and infrastructures of urban areas—that generate large, dense populations in developed and, increasingly, underdeveloped countries. Population sizes and distributions are influenced by demographic components of fertility, mortality, and migration. Social disasters such as war or famine may alter population distribution on any scale. Policy decisions, such as tax policies or zoning and planning ordinances, are eventually reflected on the population map.

None of these factors, however, can be considered without reference to historical circumstance. Present population distribution is explicable only in terms of the past. Geographies are never created instantaneously, and the location of the world's people is the accumulation of forces operating at the global, national, and local scales for centuries or longer. For example, the high-population densities of Europe or the northeastern United States reflect the accumulated impacts of the Industrial Revolution and its associated waves of urbanization. China's large population was centuries in the making, reflecting long periods of fertile agriculture, irrigation, and a social system stretched over vast areas. The distribution of people in the developing world is largely a reflective of the centuries of colonialism, which focused growth on coastal areas, the locations of the port cities that were the centers of maritime trade in the colonial world economy.

POPULATION GROWTH OVER TIME AND SPACE

The geography of the world's population is never static but is in constant change. The world's population is increasing, albeit at a decreasing rate. Each year an additional 80 million people inhabit the earth, roughly equivalent to the size of Mexico, which means the planet adds 270,000 people daily, about three per second. Between 2000 and 2005, many countries in Europe, including Russia as well as Japan, lost population as their deaths exceed births. The major locus of world population growth is in the developing countries, in which more than three-fourths of humankind dwell. With 6.5 billion people already and another billion expected by the year 2015, how will the developing world manage? How will the vast population increase affect efforts to improve living standards? Will the developing world become a permanent underclass in the world economy? Or will the reaction to an imbalance between population and resources be waves of immigration and other spillovers to the developed countries?

Population Change

The current rapid growth rate of the world population is a recent phenomenon. It took from the emergence of humankind until 1850 for the world population to reach one billion. The second billion was added in 80 years (1850–1930), the third billion was added in 30 years (1931–1960), the fourth in 16 years (1960–1976), the fifth in only 11 years (1977–1987), and the sixth in 12 (1987–1990).

Like all living things (and some that are inanimate, such as your savings account), human populations have the capacity to grow rapidly. Thus, the historical pattern of population growth shown in Figure 3.1 looks like an explosion. We can express this idea using the notion of *doubling times*—the number of years that it takes a population to double in size, given a particular rate of growth. In general, the doubling time for a population can be determined by using the *rule of seventy,* which means that you divide 70 by the average annual rate of growth. For example, at 1.3 percent, the rate of world increase, the doubling time is 54 years, which means that in the year 2060 the world may have 13 billion people *if* current rates of growth continue unchanged. At an annual increase of 0.9 percent per year, the doubling time for the U.S. population is 80 years (meaning, if the current rate continues unchanged, the U.S. will have 592 million people in 2085). As growth rates increase, doubling times decrease accordingly.

The vast bulk of the world's population growth is occurring in the developing world. Of the continents, Africa has the fastest rate of growth. In 2005, the population of Africa was growing by 2.5 percent per year. Kenya, with a population growth rate of 2.1 percent per year, will see its population of 33 million double in just 33 years.

Rapidly declining death rates and continued high birthrates are the cause of this explosion. Death rates have been falling to fewer than 10 deaths per 1000 people each year in Asia and Latin America, and to about 13 per 1000 in Africa. Crude birthrates are changing less spectacularly. They are highest in Africa (38 births per 1000 people annually), Latin America (22 per 1000), and Asia (20 per 1000). These latter figures compare with crude birthrates of 10 per 1000 in Europe and 14 per 1000 in North America.

The world population was 6.5 billion in 2005, and the United Nations projects it will reach 8.5 billion in 2025. Almost all of this increase will occur in the developing countries. The largest absolute increase is projected for Asia, reflecting its huge population base. Future population growth will further accentuate the uneven distribution of the world's population. In 2005, 80 percent of the world's population lived in the developing world, but by the year 2025, the proportion will increase to 86 percent.

The rate of natural increase for a country or a region is measured as the difference between the birthrate and the death rate. Births and deaths represent two of the three basic population change processes; the third is migration. Every population combines these three processes to generate its pattern of growth. We can express the relationship among them using the equation

$$\text{Population change} = \text{Births} - \text{Deaths} + \text{In-migration} - \text{Out-migration, or,}$$

$$\Delta P = BR - DR + I - O,$$

where ΔP represents the rate of population change, BR is the crude birthrate, DR is the crude death rate, I is the total in-migration rate, and O is the total out-migration (or emigration, if internationally) rate. Natural growth (NGR)—the most important component of population change in most societies—is defined as the difference between the birth and death rates:

$$NGR = BR - DR,$$

while net migration rates (NMR) are the difference between in-migration and out-migration rates:

$$NMR = I - O.$$

Thus,

$$\Delta P = NGR + NMR,$$

where *NGR* is the natural growth rate and *NMR* is the net migration rate.

For the world as a whole, net migration is obviously zero. However, for any scale smaller than the globe, both natural growth and net migration must be included. Natural increase accounts for the greatest population growth in most societies, especially in the short run. However, in the long run, migration contributes far more than the number of people moving into an area because the children of immigrants add to the population base.

Fertility and Mortality

The immediate cause for the surge in the growth of the world population is the difference between the crude birthrate and the crude death rate. The crude *birthrate* is the number of babies born per 1000 people per year, and the crude *death rate* is the number of deaths per 1000 per year. For example, the U.S. birthrate in 2005 was 14, and the death rate was 8. During that year, the growth rate was 14 minus 8, or 6 per 1000, which is a *natural growth rate* of 0.6 percent.

Birthrates fluctuate over time, in response to changing economic and political circumstances and cultural values about having children. In the United States, crude birthrates rose sharply after World War II, producing the baby boom. The children of the baby boom, Generation X, were born largely in the 1980s and 1990s.

FIGURE 3.6
Thomas Robert Malthus was an influential theorist who started the idea of overpopulation. His pessimistic views of food and demographic growth influenced many early political economists, earning the discipline the nickname "the dismal discipline."

MALTHUSIAN THEORY

One of the first social scientists to tackle the matter of population growth and its consequences was the British Reverend Thomas Robert Malthus (Figure 3.6). Malthus's ideas, contrived in the early days of the Industrial Revolution in the late eighteenth century, had an enormous impact on the subsequent understanding of this topic. He offered his most concise explanation in his 1798 book, *Essay on the Principle of Population Growth.* Malthus was concerned with the growing poverty evident in British cities at the time, and his explanation was largely centered on the high rates of population growth that he observed, which are common to early industrializing societies. Thus, it is with Malthus that the theory of overpopulation originates. His pessimistic worldview earned economics the label of the "dismal science" and stood in sharp contrast to the utopian socialism emanating from France in the aftermath of the French Revolution of 1789.

The essence of Malthus's line of thought is that human populations, like those of most animal species, grow exponentially (or in the parlance of his times, geometrically). A geometric series of numbers increases at an increasing rate of time. For example, in the sequence 1, 2, 4, 8, 16, 32, and so on, the number doubles at each time period, and the increase rises from 1 to 2 to 4 to 8 and so forth. Exponential population growth, in the absence of significant constraints, is widely observed in bacteria and rodents, to take but a few examples from zoology. Note that there is an important assumption regarding fertility embedded in Malthus's analysis here: He portrayed fertility as a biological inevitability, not a social construction. This argument was in keeping with the large size of British families at the time and the excess of fertility over mortality. In short, in Malthus's view, humans, like animals, always reproduced at the biological maximum; they were, and are, portrayed as prisoners of their genetic urges to reproduce. It is worth noting that Malthus's argument carried with it a strong moral dimension: It was not just anyone who reproduced rapidly, he observed, but most particularly the poor.

Second, Malthus maintained that food supplies, or resources more generally, grew at a much slower rate than did the population. Specifically, he held that the food supply grew linearly (or arithmetically, in his terminology). An arithmetic sequence of numbers, in contrast to an exponential one, grows at a constant rate over time. For example, in the sequence 1, 2, 3, 4, 5, and so on, the difference from one number to the next is always the same. Malthus's view that agricultural outputs increased linearly over time reflected the preindustrial farming systems that characterized his world. In such circumstances, without economies of scale, an increase in outputs is accomplished only with a proportional increase in inputs such as labor, reflective of what economists call a linear production function. However, this view of agricultural output is actually rather optimistic by Malthus's reckoning. He argued that in the face of limited inputs of land and capital, agricultural output was likely to suffer from *diminishing marginal returns*. For example, as farmers moved into areas that were only marginally hospitable for crops, perhaps because they are too dry, too wet, too cold, or too steep, they would need increases in inputs that are proportionately much larger than the increases in output. Diminishing returns, he held, would actually lead to increases in agricultural output that were smaller than a linear production function (with no economies of scale, see Chapter 5) would generate (Figure 3.7).

When one plots the exponential growth of population against the linear growth of food supplies (Figure 3.8), it is clear that sooner or later, the former must exceed the latter. Thus, in the Malthusian reading, populations always and inevitably outstrip their resource bases, and people are condemned to suffering and misery as a result. Malthus blamed much of the world's problems on rapid population growth, and subsequent generations of theorists influenced by his thoughts have invoked overpopulation to explain everything from famine to crime rates to deviant social behavior. Malthus himself entered into a famous debate with his friend David Ricardo over whether the British government should subsidize food for the poor, Malthus maintaining that such subsidies only encouraged the poor to have more children and thus exacerbated poverty in the long run.

Malthus refined his argument to include checks to population growth. Given that natural population growth is the difference between fertility and mortality, "positive checks" are factors that reduce the fertility rate. Contraceptives are an obvious case in point, although Malthus objected to their use on religious grounds, advocating instead moral restraint or abstinence. Other positive checks include delayed marriage and prolonged lactation, which inhibits pregnancy. Should positive checks fail, as he predicted they would, population growth would ultimately be curbed by "negative checks" that increased the mortality rate, particularly the familiar horsemen of the Apocalypse: death, disease, famine, and war. Malthusianism thus attributes to rapid population growth a variety of social ills, including poverty, hunger, and disease.

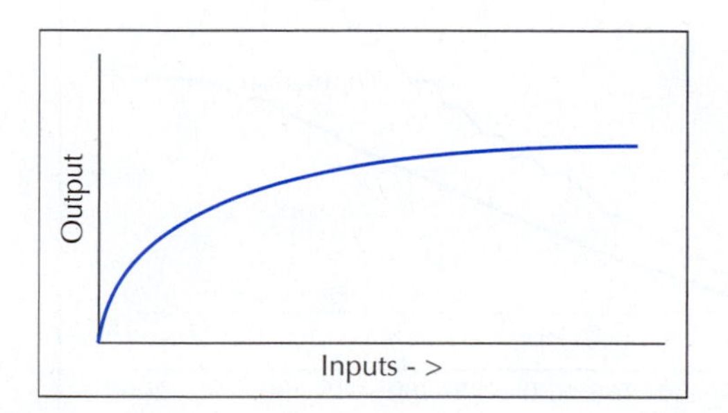

FIGURE 3.7
Diminishing marginal returns, proposed by Malthus, set in when increases in inputs (e.g., land, labor) fail to generate equal increases in outputs. This process leads to declining productivity growth. To some extent, diminishing returns can be offset by technological change.

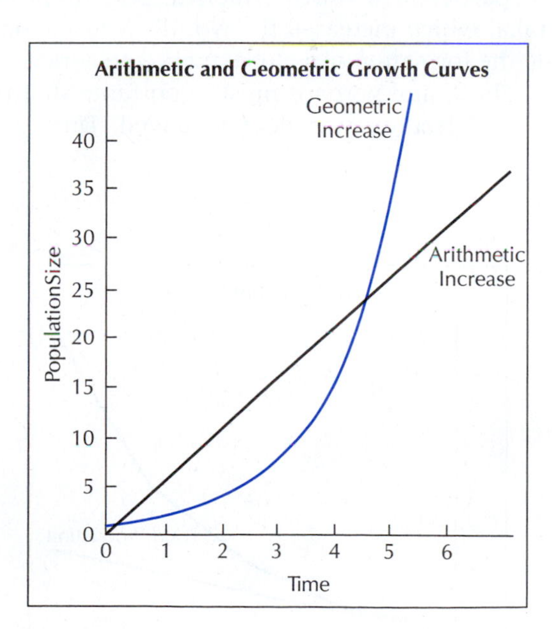

FIGURE 3.8
Malthus's argument contrasted the geometric (exponential) increases in population with arithmetic (linear) increases in the food supply. Eventually, the demand for food, or resources more generally, must exceed the supply, leading to famine. However, critics noted that Malthus's conceptions were flawed by their simplistic treatment of production, as well as the determinants of fertility.

Malthus's ideas became widely popular in the late nineteenth century, particularly as they were incorporated into the prevailing social Darwinism of the time, which represented social change in biological terms, often naturalizing competition as a result. However, to many observers it became increasingly apparent that his predictions of widespread famine were wrong. The nineteenth century saw the food supply improve, prices decline, and famine and malnutrition virtually disappear from Europe (except for the Irish potato famine of the 1840s). By the early twentieth century, Malthusianism was in ill repute.

Critics noted that Malthus made three major errors. First, he did not foresee, and probably could not have foreseen, the impacts of the Industrial Revolution on agriculture; the mechanization of food production simply rendered the assumption of a linear increase untenable (Figure 3.9). Indeed, the world's supply of food has consistently outpaced population growth, meaning that productivity growth in agriculture has been higher than the rate of increase in the number of people. This observation implies that there is plenty of food to feed everyone in the world and that hunger is not simply caused by overpopulation but by a variety of other factors, including politics.

Second, Malthus did not foresee the impacts of the opening up of midlatitude grasslands in much of the world, particularly North America, Argentina, and Australia, which increased the world's wheat supplies during the formation of a global market in agricultural goods. Third, and perhaps most important, Malthus's analysis of fertility was deeply flawed. During the Industrial Revolution, fertility rates declined and family sizes decreased. Thus, contrary to his expectation, humans are not mere prisoners of their genes, and the birthrate is a socially constructed phenomenon, not a biological destiny.

In the 1960s, when the world experienced population growth rates in excess of 2.6 percent annually, Malthusianism underwent a revival in the form of *neo-Malthusianism*. Neo-Malthusians acknowledged the errors that Malthus made but maintained that while he may have been wrong in the short run, much of his argument was correct in the long run. In keeping with the growing environmental movement of the times, neo-Malthusians also added an ecological twist to Malthus's original argument. The most famous expression of neo-Malthusian thought was the Club of Rome, an international organization of policy makers, business executives, scholars, and others concerned with the fate of the planet. The Club of Rome funded a famous study of the planet's future, published as *The Limits to Growth* (1972), which modeled the earth's population growth, economic expansion and resource consumption, and energy and environmental impacts (Figure 3.10). It concluded that the rapid population and economic growth rates of the post–World War II boom could not be sustained indefinitely, and that ultimately there would be profound worldwide economic, environmental, and demographic crises. Much of this argument was framed in terms of the exhaustion of nonrenewable resources and ecological catastrophe. Unlike Malthus, neo-Malthusians advocated sharply curtailing population growth through the use of birth control and had

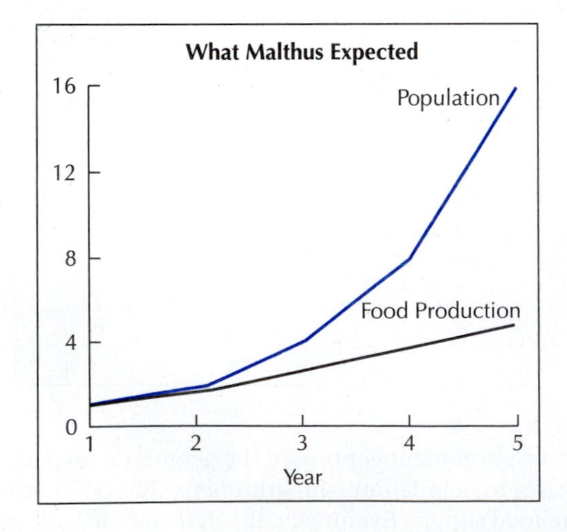

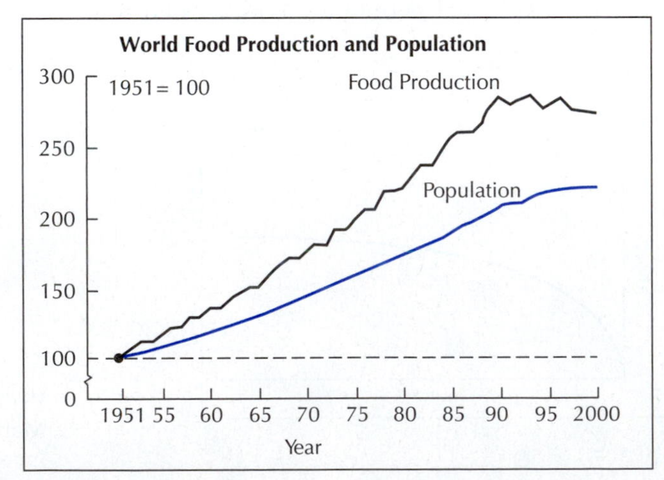

FIGURE 3.9
Malthus's predictions of catastrophe were belied by the productivity gains and declining fertility unleashed by the Industrial Revolution. Since World War II, the world's food supply has increased more quickly than its population, indicating that the causes of hunger are not simply reducible to population growth but involve complex political questions about colonial legacies, uneven development, corrupt governments, and war and conflict.

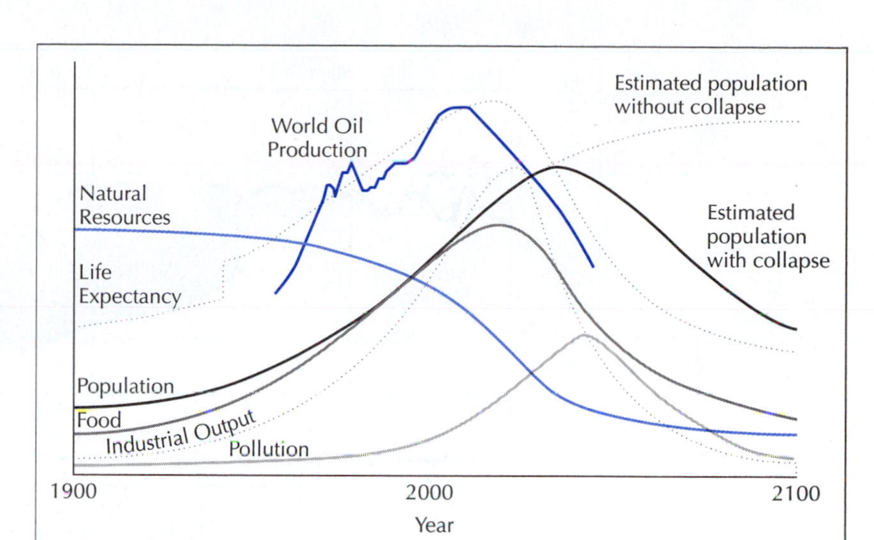

FIGURE 3.10
Neo-Malthusians such as the Club of Rome revived Malthus's arguments in the 1960s by using computer models of the world economy, population growth, and resource usage. They included ecosystems in their analysis and predicted that in the long run, his predictions still had merit. Unlike Malthus, however, neo-Malthusians advocate the use of birth control.

an important impact of international programs promoting contraceptives and family planning, such as the Peace Corps and Agency for International Development.

While neo-Malthusianism retains a credibility that the original Malthusian doctrine does not, it too suffered from a simplistic understanding of how resources are produced (when the price of oil rises, corporations find more oil). In addition, family planning programs in the developing world have often failed to live up to expectations, often for the simple reason that advocating contraception to curb population growth ignores the reasons why people in impoverished countries have large families and many children. In short, whatever its merits, neo-Malthusianism must be viewed in light of other models of population growth that originate from different premises and often arrive at different conclusions.

The trains are crowded in Bangladesh. Here, in one of the world's most fertile lands, the combination of high-population density and the concentration of land into fewer and fewer hands contributes to continuing poverty and hunger.

Demographic Transition Theory

Developed by several demographers in the 1920s, the demographic transition theory stands as important alternative to Malthusian notions of population growth. Essentially, this is a model of a society's fertility, mortality, and natural population growth rates over time. Because this approach is explicitly based on the historical experience of western Europe and North America as they went through the Industrial Revolution, time in this conception is a proxy for industrialization and all of its economic, social, and geographic characteristics. In short, the demographic transition examines how birth, death, and growth rates change, and, more important, why they change, as a society moves from a rural, impoverished, and traditional context into a progressively wealthier, urbanized, and modern one. This approach can be demonstrated with a graph of birth, death, and natural growth rates over time that divides societies into four major stages (Figure 3.11). Each stage is discussed here in detail.

Stage 1: Preindustrial Society

In the first stage, a traditional, rural, preindustrial society and economy, fertility rates are high; families are large, and most women are pregnant much of the time (Figure 3.12). Thus, impoverished countries such as Rwanda have exceptionally large families (8.5 children per mother, in contrast to wealthy ones in Europe, North America, and some of the newly industrializing countries (NICs) in Asia, where women have on average less than two children apiece).

Traditionally, fertility rates in preindustrial societies have been very high for a variety of reasons. In agrarian economies, often short of labor, children are

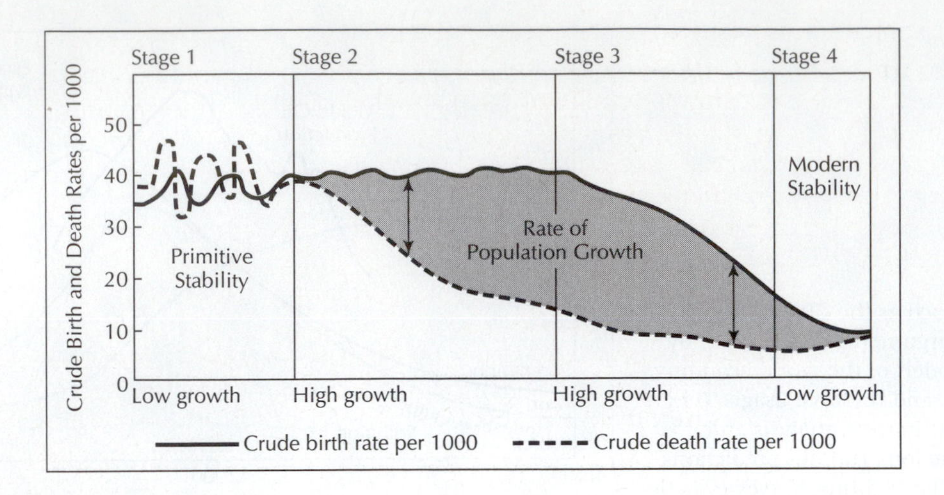

FIGURE 3.11
The demographic transition. Four stages of demographic change are experienced by countries as they develop from preindustrial to modern. Stage 1: Both high birthrates and crude death rates occur; the resulting natural increase is quite low. Stage 2: Declining death rate, but a continually high birthrate, results in an increasing population growth rate. Stage 3: Death rates remain low, while birthrates start to decline rapidly, resulting in a rapid but declining growth rate. Stage 4: Both death rates and birthrates are low, resulting in a low natural increase. The important issue here is *why* these rates decline (i.e., the social forces that change birthrates and death rates).

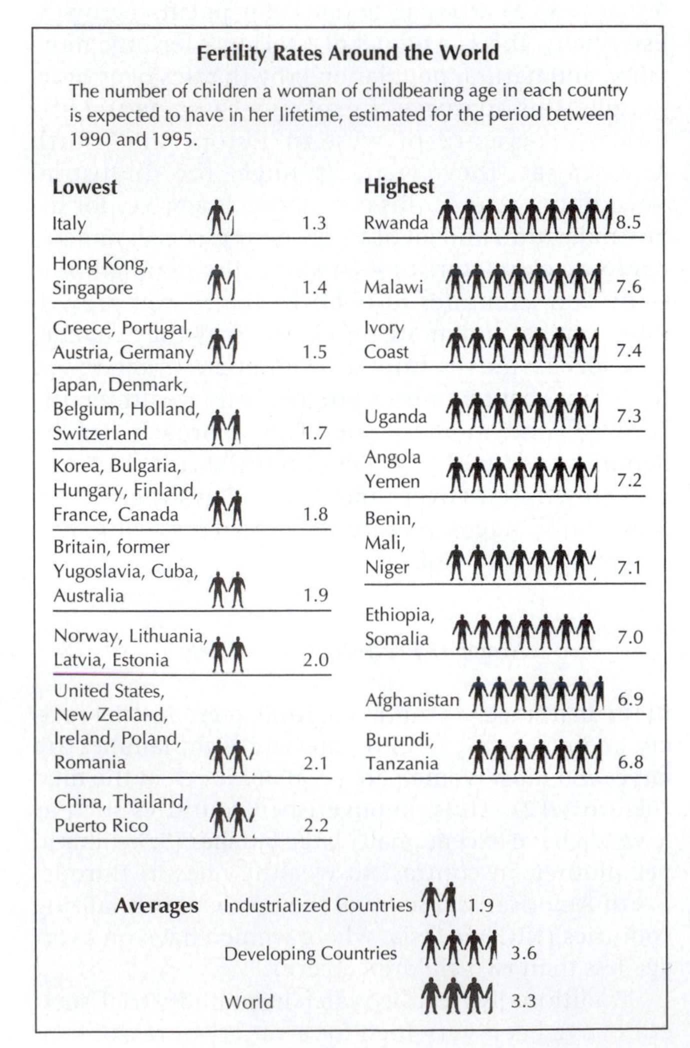

FIGURE 3.12
Fertility rates around the world vary widely. The average number of children each woman may expect to have ranges from as little as 1.3 in Italy to as high as 8.5 in impoverished nations such as Rwanda. These patterns reflect the dynamics of the demographic transition and the social forces that lead people in different societies to have large or small families.

a vital source of farm labor, helping to plant and sow crops, tending to farm animals, performing chores, carrying water and messages, and helping with younger siblings. Econometric studies reveal that even children as young as four can generate more income than they consume. Even in North America, the creation of summer vacations was made necessary so that school kids could return to the farms to help their families.

Families in this context are typically extended, with several generations living together. In addition, children are important resources to take care of their elderly parents; in the absence of institutionalized social programs such as Social Security, the aged depend on their offspring for assistance. Finally, in such societies with high infant mortality rates, having many children is a form of insurance that some proportion will survive until adulthood. In short, there are very clear reasons why poor societies have high crude birthrates. In contrast to Malthusianism, which explains this observation by appeal to human genetics, the demographic transition portrays high total fertility rates as a rational strategic response to poverty.

Thus, a map of crude birthrates around the world (Figure 3.13) reveals that the poorest societies have the highest rates in the world, particularly in Africa and most of the Middle East. In contrast, for reasons we shall soon see, crude birthrates in North America, Europe and Russia, Japan, Australia, and New Zealand are relatively low. The world's lowest birthrates are found in Spain and Italy. In societies with high birthrates, the age distribution of populations tends to be young. Thus, the proportion of the population aged less than 18 (the median age in many developing countries) is a reflection of high fertility rates (Figure 3.14).

However, in preindustrial societies, mortality rates are also typically quite high, which means that average life expectancy is relatively low. The primary causes of death in poor, rural contexts are the result of inadequate diets, particularly protein, which weaken the immune system, as well as unsanitary drinking water and bacterial diseases. The most common diseases in this context are diarrheal ones, which lead to dehydration, including cholera, as well as others such as dengue fever, schistosomiasis, bilharzia, malaria, tuberculosis, plague, and measles, although historically smallpox was

Child labor is a mainstay of the workforce in developing countries. In agrarian economies, children perform a variety of important tasks essential to household survival and take care of their parents in their old age. Thus, under these conditions, high fertility levels are a rational demographic response to poverty.

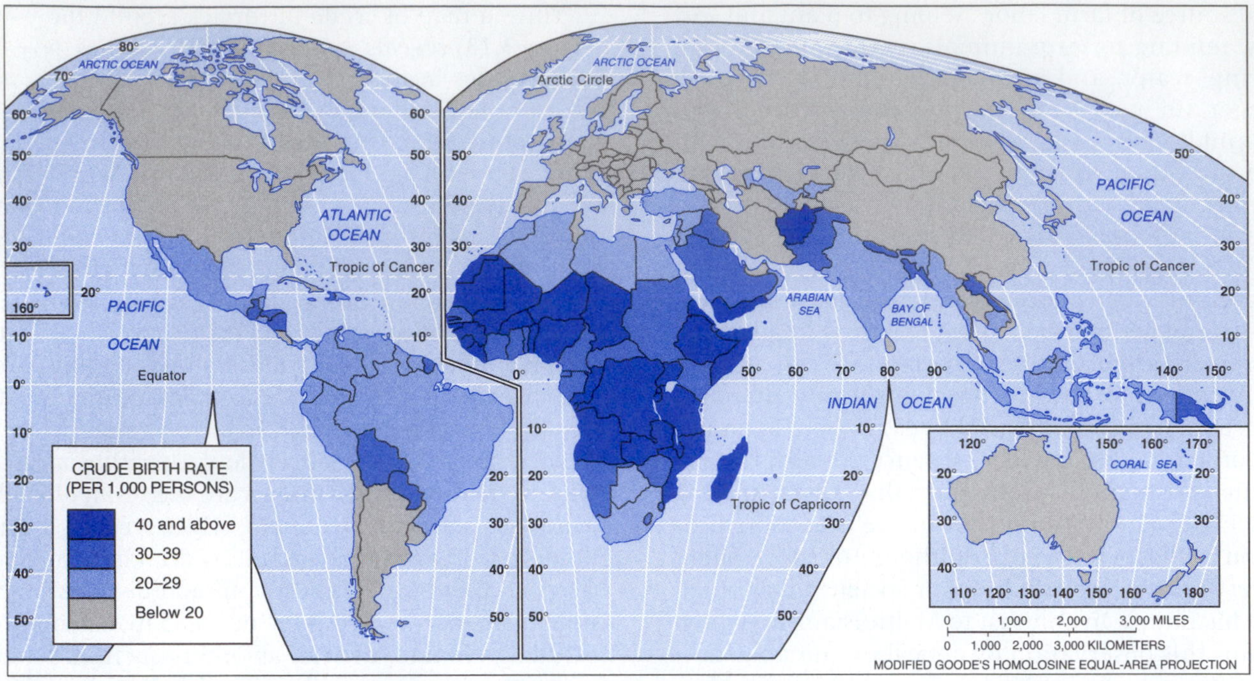

FIGURE 3.13

The geography of crude birthrates around the world in 2000 closely reflects the level of economic development. Generally, the poorest societies have the largest families. Africa and much of the Muslim world tend to have the highest fertility levels. In these circumstances, the need for child labor, particularly on farms, is the predominant motive. In economically advanced societies, this need is mitigated, and the opportunity costs of raising children rises, so families are smaller. Thus, birth rates in Europe, Japan, the United States and Canada, and Australia are low.

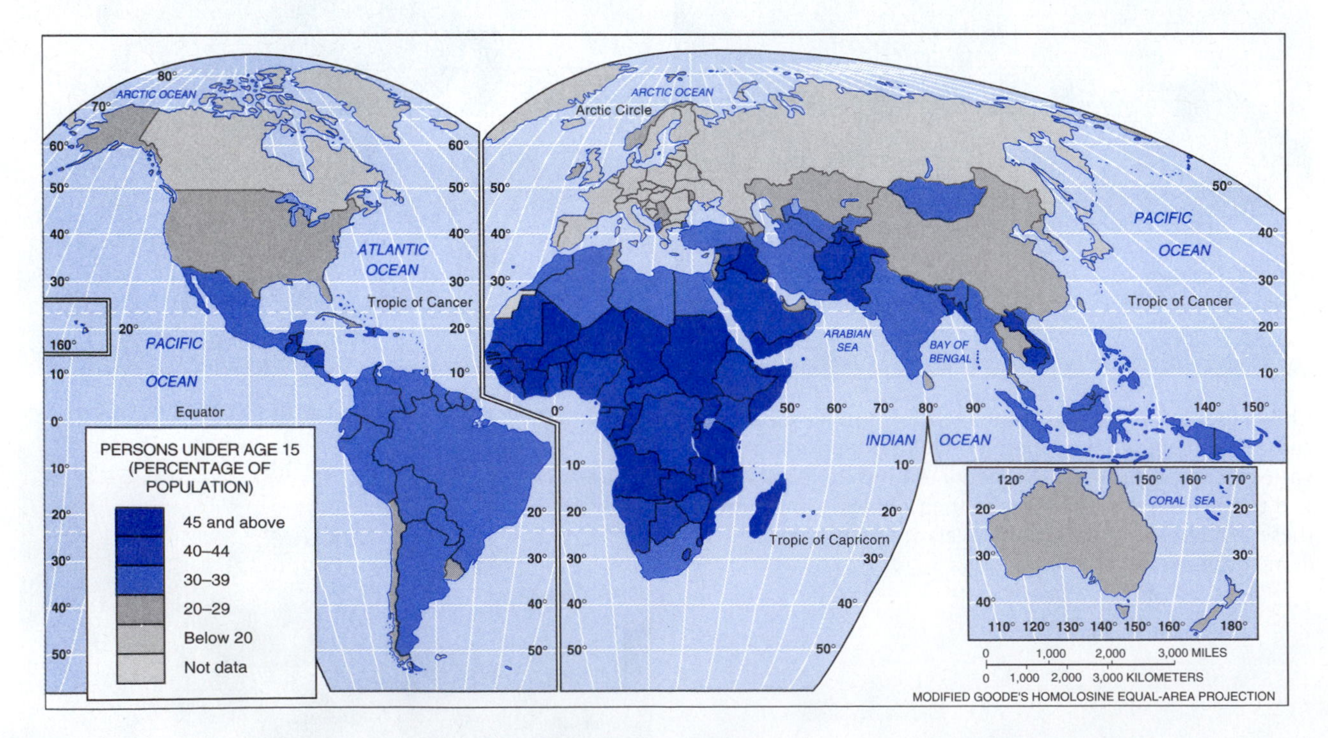

FIGURE 3.14

The proportion of each country's population below the age of 15 is closely correlated with fertility rates. In poor, rural societies, where birthrates are high and families large, a large share of the people (often more than half) are younger than 15, particularly in Africa and parts of the Arab world. In contrast, in more developed states with lower birthrates and smaller families, the median age rises; in the First World, the fastest growing age groups are the middle aged and elderly.

Table 3.2

World's Most Dangerous Infectious Diseases, 2004
(millions of victims annually)

Respiratory infections	3.9
AIDS	2.8
Diarrheal diseases	1.8
Tuberculosis	1.6
Malaria	1.3
Measles	0.6

Source: World Health Organization.

also important. Table 3.2 lists the most dangerous infectious diseases in the world in 2003, including respiratory infections brought on by pneumonia and influenza (which kill 3.5 million annually); AIDS, which takes 2.3 million lives annually (especially in Africa); diarrheal diseases, which deplete the body's nervous system of electrolytes; and tuberculosis, which kills 1.5 million. Because disease and malnutrition are ever-present threats to people in poor societies, including particularly the most vulnerable, the very young, infant mortality rates are also high, and a significant proportion of babies do not live to see their first birthday.

The world geography of death rates (Figure 3.15) thus closely reflects the wealth or poverty of societies (including their historical development and role in the global economy), which in turn is manifested in a variety of issues that shape national mortality rates: the amount, quality, and consistency of adequate food; access to health care; the public health infrastructure; care for expectant mothers, babies, and young children; smoking rates; and several other factors. Countries with the highest death rates—and thus lowest life expectancies—are found primarily in sub-Saharan Africa, although Afghanistan, Pakistan, Russia, and central Asian states are also relatively high. Conversely, the developed world, as well as Latin America, China, and India, has relatively low death rates.

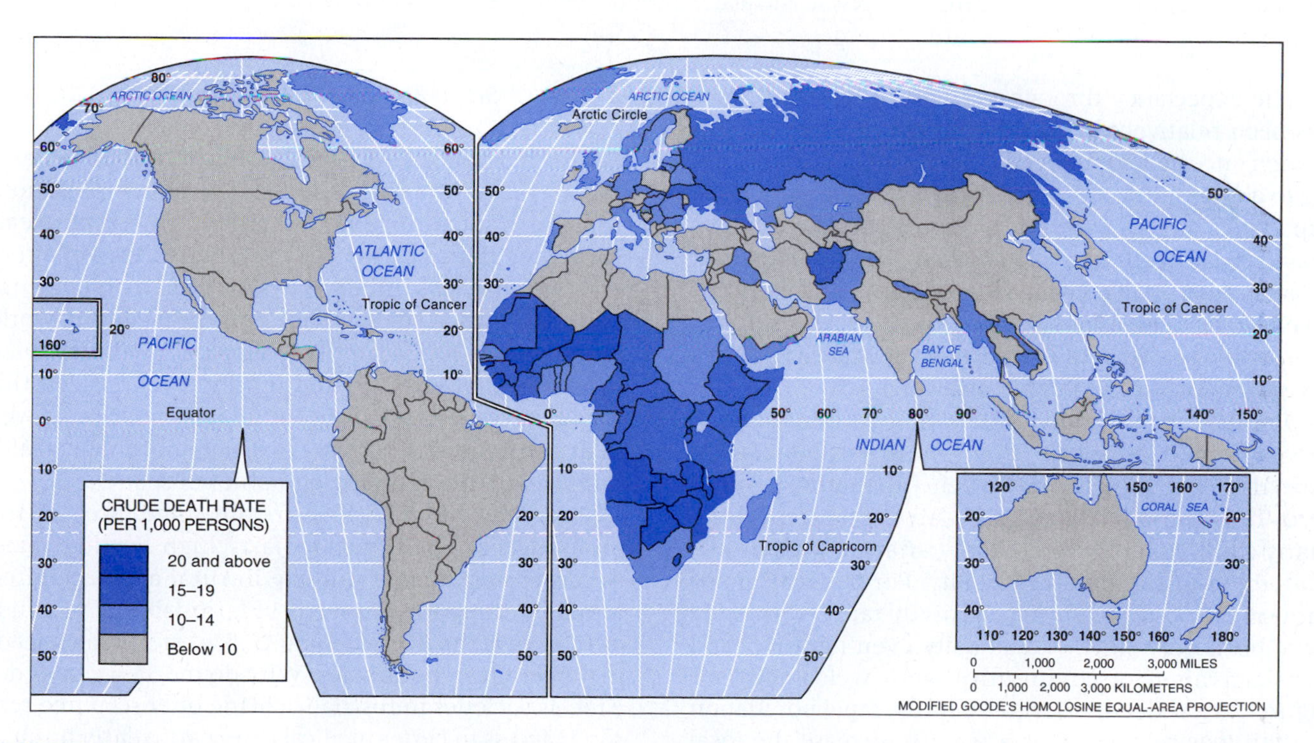

FIGURE 3.15
Like birthrates, the geography of crude death rates in 2000 also closely mirrors the level of economic development. The poorest countries, especially in sub-Saharan Africa, have high death rates, largely due to inadequate diets and infectious diseases such as malaria and contaminated water supplies. Russia, which is relatively more developed, has a remarkably high death rate due to the economic collapse it suffered in the 1990s and the spread of diseases such as tuberculosis. Much of the rest of the world, in contrast, including even poor countries such as China and India as well as the First World, has brought down death rates and increased life expectancy by providing adequate food supplies and public health infrastructures.

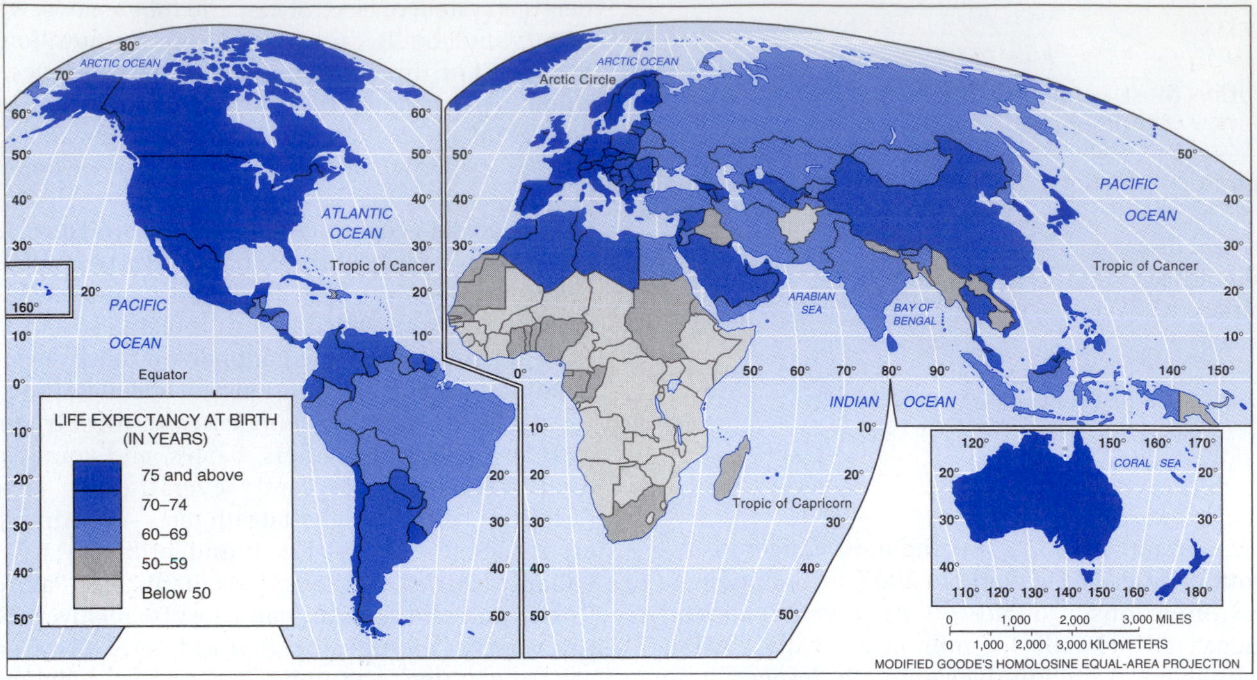

FIGURE 3.16
A world map of life expectancy at birth in 2000 reveals that people in wealthier countries live longer than those in poorer ones. Whereas the inhabitants of impoverished societies such as Mali or Chad cannot expect to live beyond age 50, the average age at death for those in Europe, Japan, Australia, and the United States and Canada is over 75. Life expectancies are thus closely correlated with people's access to adequate nutrition, public health, and health care.

Life expectancy throughout most of human history has been relatively low, often only in the twenties, although once people survived infancy their chances of living to old age improved considerably. The geography of life expectancy around the world (Figure 3.16) closely reflects that of crude death rates but is also shaped by differences in standards of living. Living for a long time is a luxury enjoyed by the populations of economically developed societies, while people in most of Africa, the Middle East, and Russia tend to die before they reach age 70.

Because both fertility and mortality rates are high, the *difference* between them—natural population growth—is relatively low, often fluctuating around zero. Thus, although families are large and parents have many children, growth rates are curtailed by malnutrition, disease, and infant mortality. For this reason, for thousands of years, human growth rates worldwide have been very slow, occasionally even negative, and new arrivals to a community were welcomed (see Figure 3.1). Indeed, prior to Malthus, rapid population growth was celebrated as a way to increase the local labor force, diversify the division of labor, and raise the standards of living. While relatively few societies in the world live in the circumstances described here—that is, few people today live isolated from the world economy and its demographic aftermath—Stage 1 may be held to describe certain tribes in parts of central Africa, Brazil, or Papua New Guinea.

Stage 2: Early Industrial Society

The second stage of the demographic transition pertains to societies in the earliest phases of industrialization, when manufacturing jobs are growing in urban areas. Such conditions pertain, for example, to the Britain of Malthus's day, the United States in the early nineteenth century, or selected countries in the developing world today, such as Mexico or the Philippines (although these countries have experienced fertility declines). In this context, economic changes in the labor market as well as in consumption, particularly diet and public health care, have important demographic consequences.

Early industrial societies retain some facets of the preindustrial world, particularly high fertility rates. Because most people still live in rural areas, children remain an important source of farm labor. The major difference is the decline in mortality rates, which leads to longer life expectancies. Why do mortality rates decline as societies industrialize? One often-claimed reason is access to better medical care, particularly hospital and vaccinations from diseases, an assertion advocated by those in the health care occupations. However, the historical evidence does not sustain this view; early hospitals were often filthy, and patients may have been more likely to die in them than if they stayed home! Moreover, the introduction of vaccinations often came *after*, not before, declines in deaths due to the many diseases that occurred. For example, in the United States, the vaccines for measles, scarlet fever, typhoid, diphtheria, tuberculosis, and pneumonia, which were invented in the 1920s, 1930s, and 1940s, all occurred after the majority of the declines in the death rate from each disease in the first two decades of the twentieth century (Figure 3.17). Indeed, rather than private health care providers, it was public health measures, particularly clean drinking water and sewers, that played a significant role in lowering diseases. Better housing was also important. Finally, the industrialization of agriculture and cheaper food, which led to better diets, was vital in improving immune systems and raised life expectancies, including lowering *infant mortality rates* (deaths of babies under age one).

Death rates for different demographic groups do not drop evenly as an economy develops over time. Infant mortality rates tend to drop earliest and most quickly. It was not uncommon in premodern societies to find an infant mortality rate in excess of 300 infant deaths per 1000 live births—20 percent or more of all babies died before reaching their first birthday. Nowhere in the world today are rates that severe, but the highest rates are found in sub-Saharan Africa (Figure 3.18), where poverty, lack of adequate diets, disease, inadequate drinking water, and insufficient

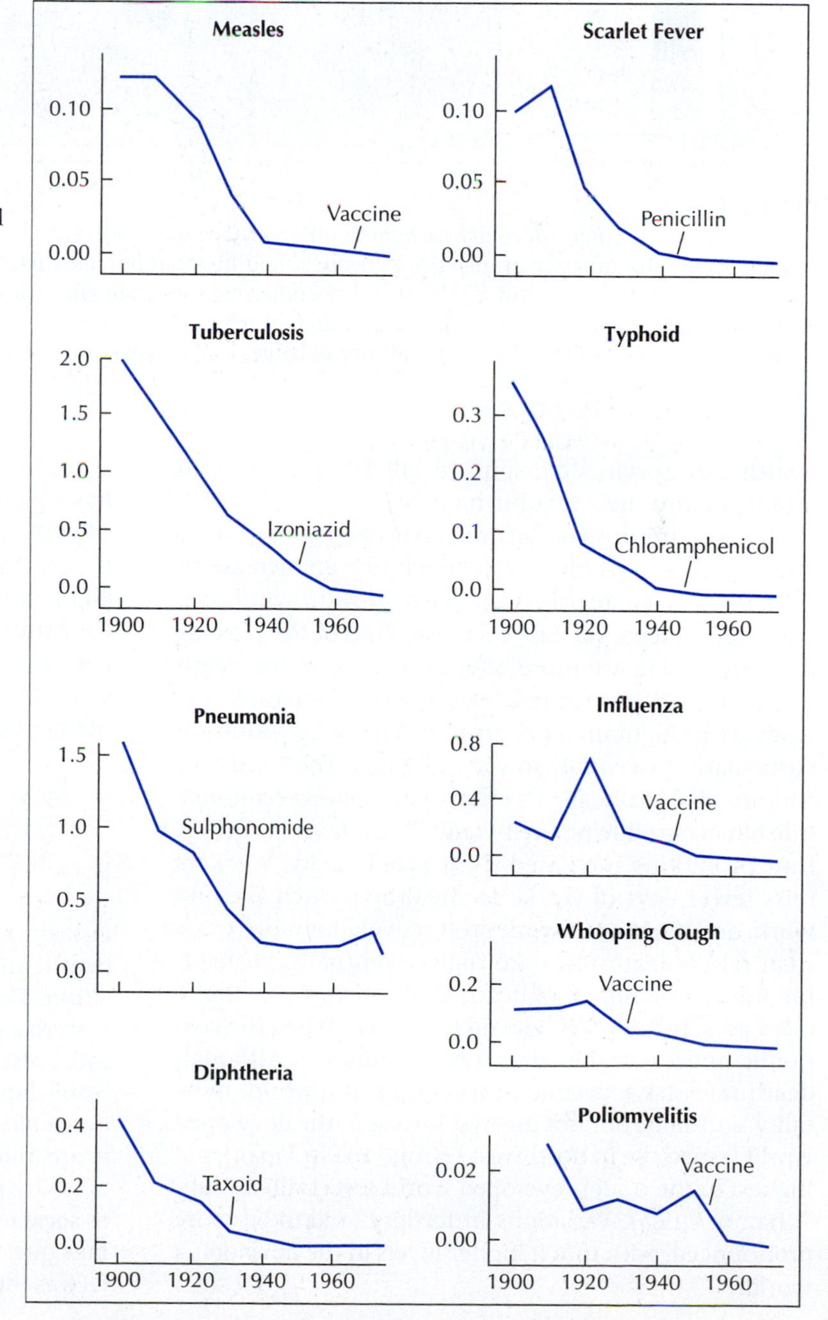

FIGURE 3.17
Death rates due to nine infectious diseases in the United States from 1900 to 1973 reveal that the bulk of the declines occurred prior to the introduction of inoculations or medical cures. Thus, the major reasons for lower mortality rates as societies develop economically are not linked to physicians and hospitals, but to better diets and public health measures such as clean water and garbage removal.

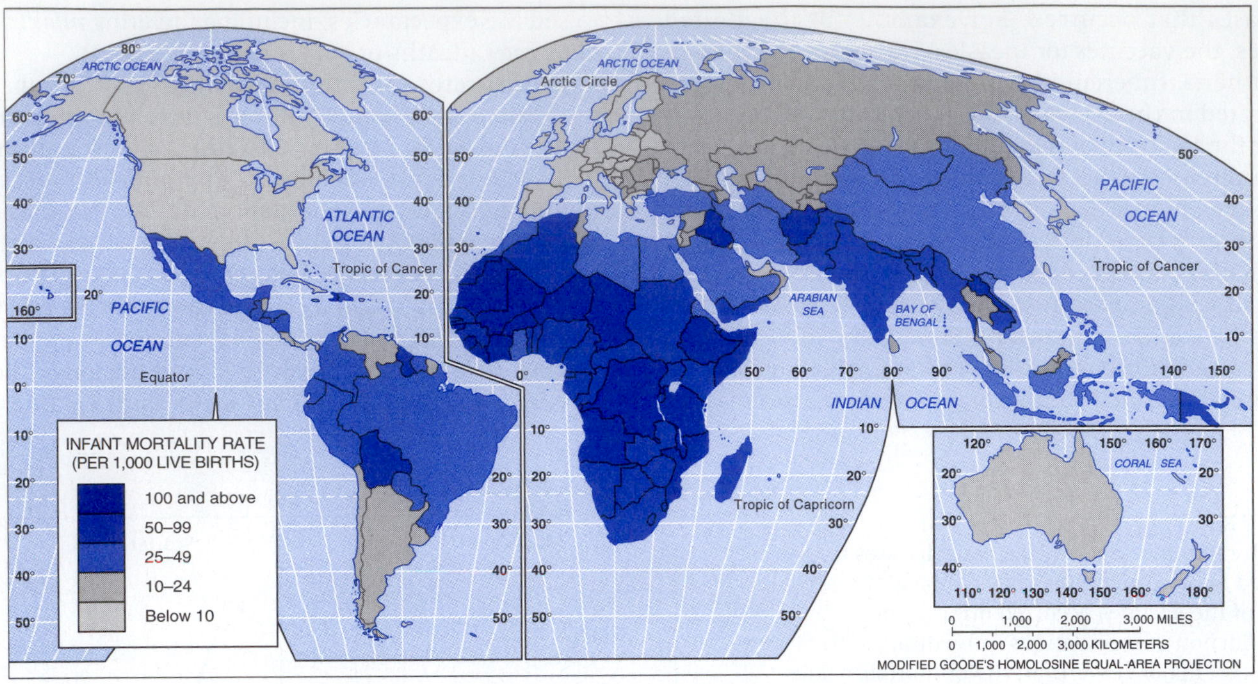

FIGURE 3.18
The geography of infant mortality rates in 2000 is perhaps the most sensitive and optimal measure of economic development. The capacity of societies to protect their most vulnerable members—babies—reflects their parents' access to nutrition and clean water and exposure to infectious diseases. Poor countries suffer from high rates of infant deaths, often exceeding 100 deaths per 1000 births (e.g., much of Africa, Afghanistan, Iraq); people in the economically developed world, in contrast, can expect that the vast majority of babies will live past their first birthday.

health care services conspire to kill 10 percent of all infants before their first birthday.

Because the drop in the death rate disproportionately affects the very young, it acts much like an increase in the birthrate—more babies survive to grow to adulthood. Life expectancies likewise increase. One of the reasons the very young are more affected is that, as the death rate drops, it does so initially because communicable diseases are brought under control, and the very young are particularly susceptible to such diseases. The control of communicable disease has the serendipitous economic side effect of reducing the overall illness level in society, thus promoting increased labor productivity. Workers miss fewer days of work, are healthier when they do work, and are able to work productively for more years than when death rates are high. Eventually, as death rates drop, the timing of death shifts increasingly to the older ages, to the years beyond retirement when the economic impact on the labor force is minimal. Although death rates have declined throughout the world, mortality is usually, but not always, lowest in the developed world (especially in northwest Europe and in Japan) and highest in the underdeveloped world (especially in sub-Saharan Africa). Variations in fertility tend to be more pronounced, with much higher levels in the developing world.

In early industrial societies, because the death rate has dropped but the birthrate has not, the natural growth rate grows explosively. This situation characterized the world Malthus observed at the end of the eighteenth century and is evident in a wide number of countries in the developing world today. In short, poor countries have rapid increases in population not because they have more children than before but because fewer people die earlier.

Stage 3: Late Industrial Society

Societies in the throes of rapid industrialization, in which a substantial share, if not the majority, lives in cities, exhibit a markedly different pattern of birth, death, and growth rates than those earlier in the transition. Death rates remain relatively low, for the reasons discussed earlier. What is different is that fertility rates also exhibit a steady decline. It is important to note that declines in crude birthrates almost always occur *after* declines in the death rate; societies are much more amenable to death control than birth control.

Why do crude birthrates fall and families get smaller as societies become wealthier? The answer to this important question lies in the changing incentives that people face as their worlds shift from primarily rural to primarily urban, with a corresponding increase in the size and complexity of labor markets. For many people, the decision to have, or not to have, children is the most important question they will ever face, with profound consequences not only for their personal well-being but also for society at large. Essentially, urbanization and industrialization lead to smaller families because the benefit/cost ratio of children changes over time. This assertion is not meant to reduce children to simple economic commodities. However, in societies in which a large number of women enter the paid labor force—become commodified labor outside of the home, rather than unpaid workers inside of it—the constraints to child rearing become formidable. For women, who typically have primary responsibility for child care, working outside of the home and taking care of young children in urbanized societies pose extraordinarily difficult obstacles. Many women understandably drop out of the labor market, if only temporarily, to take care of their kids. As a result, they do not earn an income while staying at home, relying on their husbands for support. Economically, this process generates an opportunity cost to having children: The more children a couple has, or the longer a mother refrains from working outside of the home, the greater the opportunity cost she faces or they face as a family (Figure 3.19). As women's incomes rise, either over time or comparatively within a society, the opportunity cost of children rises accordingly, leading to lower fertility rates.

In distinct contrast to neo-Malthusian family planning, which tends to ignore the social circumstances pertaining to fertility, in this model there is a clear link between labor markets and fertility behavior. Getting women to work outside of the home in commodified labor markets is the surest form of birth control. As fertility rates decline, so too does the natural growth rate. In short, relatively prosperous societies tend to have smaller families, and there is frequently a corresponding shift from extended to nuclear families in the process.

Historically, fertility levels fell first in Western Europe, followed quickly by North America, and more recently by Japan, and then the remainder of Europe. In all of those areas of the world, reproductive levels are below the level of generational replacement; the United States is the only major country still above that level, and just barely. Elsewhere in the world, however, crude birthrates remain at much higher levels, although in China and Southeast Asia birthrates are dropping very quickly. There has been a modest decline in South Asia, the Middle East, much of Latin America, and parts of sub-Saharan Africa. Overall, then, almost all the world's nations are experiencing more births than deaths each year, with the biggest gap being found in the less developed nations and the narrowest difference existing in the more developed nations. In addition to these patterns of natural increase, many areas of the world are also impacted by migration.

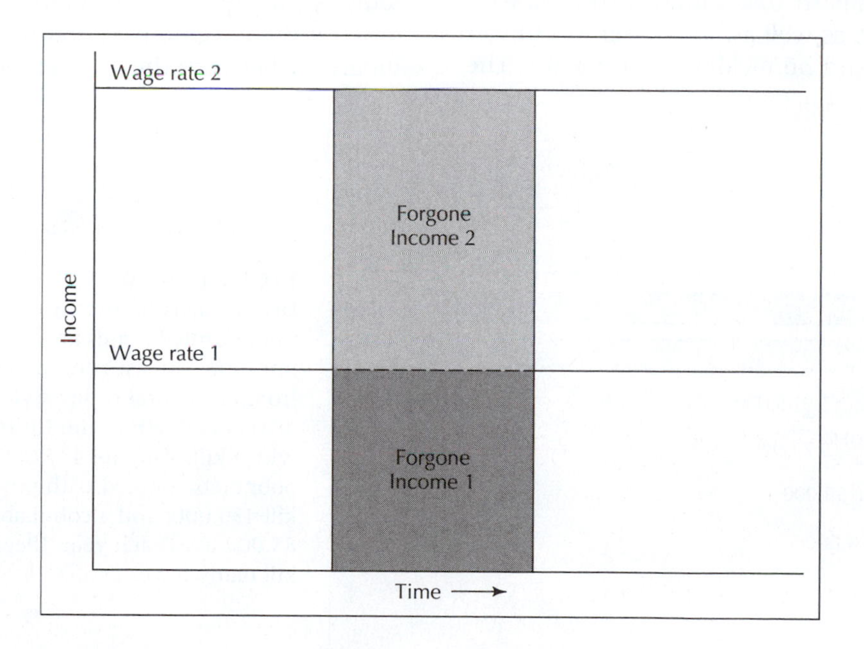

FIGURE 3.19
The opportunity costs of having children rise with income as mothers drop out of the paid labor force to take care of young children. The amount a family gives up to raise a young child for a given period of time is the wage rate multiplied by the period of time the mother is absent from paid work. (For example, a woman earning $20,000 per year who drops out for five years to raise a child sacrifices $100,000 in forgone income.) Among low-income families, including those in poor countries or the economically disadvantaged in the First World, the amount forgone is relatively low (hatched area) because incomes are low. As societies develop and family incomes rise, however, the amount forgone rises (shaded area). Thus, the economic cost of raising a child rises with the level of economic development; for this reason, poor families tend to be large families, both among and within countries, and wealthier families tend to be smaller ones.

Increasing life expectancy in a region or country is an important indicator of social progress. Between 1980 and 2002, the world's average life expectancy at birth increased from 61 to 68 years. In developed regions, average life expectancy is 71 years for males and 79 years for females; in developing regions, 62 years for males and 65 years for females; in least developed countries, the values are 51 and 54, respectively.

Stage 4: Postindustrial Society

The fourth and final stage of the demographic transition, postindustrial society, depicts wealthy, highly urbanized worlds with their own configurations of birth, death, and growth rates. In this context, indicative of Europe, Japan, and North America today, death rates are very low, and life expectancies correspondingly high. Do death rates ever reach zero? No, for that would mean life expectancies become infinite! Even the declines in the death rates will not exhibit much improvement, and they face diminishing marginal returns: The easy causes of death have been largely eliminated, and overcoming the remaining ones will be much harder.

As societies industrialize and become progressively wealthier, the causes of death change from infectious diseases, the bane of the preindustrial world, to lifestyle-related ones, particularly those associated with smoking and obesity, as well as, to a lesser extent, car accidents, suicides, and homicides (Figure 3.20). The leading causes of death in the United States today, for example, are heart disease, cancers of all forms, and strokes (Figure 3.21).

Birthrates too, continue to fall in such contexts, as families grow smaller and many couples elect to go childless or have only one. Around the world, national income and population growth rates are inversely related (Figure 3.22). When crude birthrates drop to the level of crude death rates, a society reaches zero population growth (ZPG). When birthrates drop below death rates, as they have in virtually all of Europe, the society experiences negative population growth. Japan, the oldest major society in the world (older demographically than Florida), will see its population decline by 30 percent or more in the next 50 years. Such situations are characterized by large numbers of the elderly, a high median age, and a relatively small number of children, all of which have dramatic implications for public services. Often, in such contexts, governments take steps to increase the birthrate with ample rewards for childbirth (e.g., long paid parental leaves) and publicity campaigns. In societies with extremely low or negative population growth, the major cause of demographic growth often changes to immigration.

Globally, uneven economic development—the legacy of colonialism, different national policies, position in the global system, and national cultural systems—generates uneven patterns of natural population growth; the geography of natural growth rates is the difference between the geographies of birth and death

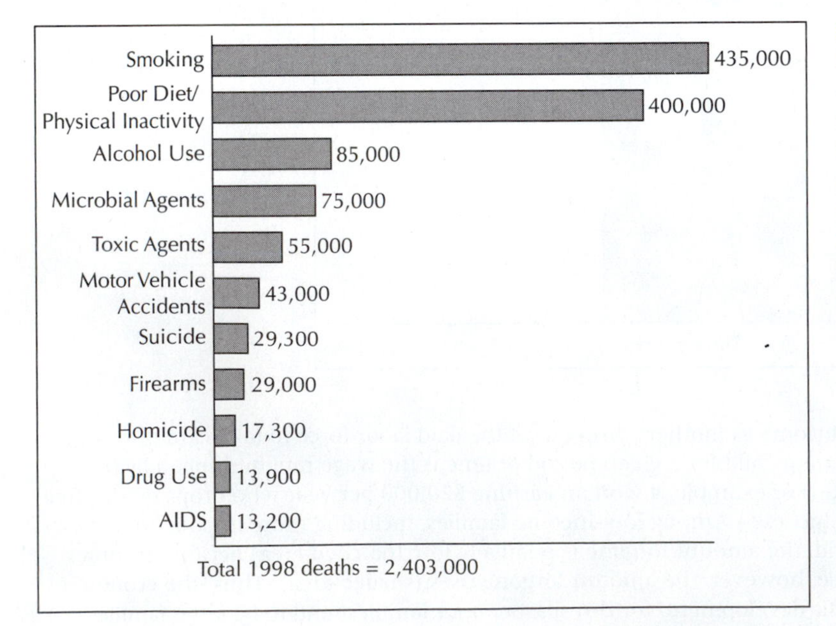

FIGURE 3.20
Deaths in economically advanced societies such as the United States are less due to hunger and infectious diseases and much more likely to result from behavioral or life style choices. The primary drivers of death in the United States are smoking, which kills roughly 435,000 Americans annually; poor diets coupled with physical inactivity, which kill 430,000; and alcohol abuse, which takes 85,000 lives each year. Illegal drugs, in contrast, kill many fewer people (less than 20,000).

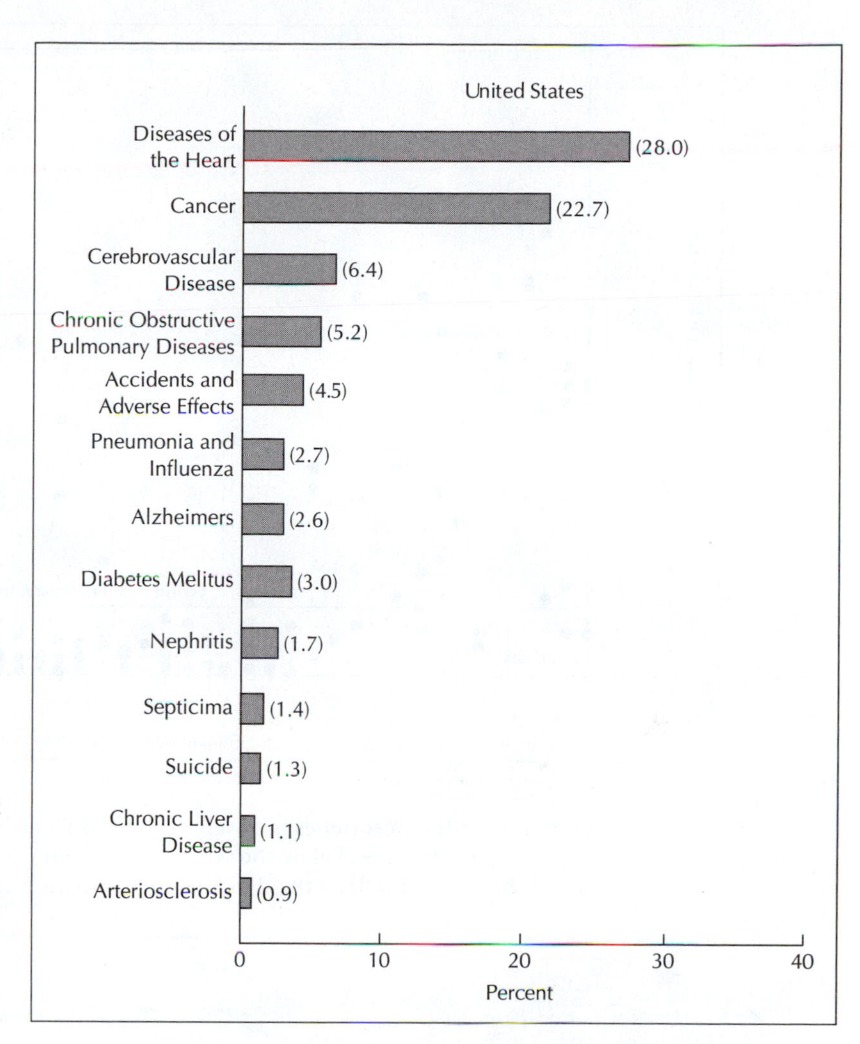

FIGURE 3.21
The drivers of death are manifested in different biological outcomes. Heart disease, which is closely related to smoking, alcohol abuse, and obesity, is the largest single killer in the United States, responsible for more than 28 percent of all deaths, followed by cancers of all forms combined (22.7%). Cerebrovascular disease (strokes) is third.

rates (Figure 3.23). The most rapid rates of increase are found throughout the poorer parts of the developing world (i.e., in Africa and the Arab and Muslim worlds). The economically developed nations, including North America, Japan, Europe, Australia, and New Zealand, in contrast, have low rates of population growth, often hovering around zero or even lower.

These patterns have significant implications for the nature and future of the world's people. Although the world's average natural growth rate has been slowly declining, it still adds approximately 100 million people per year (Figure 3.24), roughly the population of Mexico. Projecting into the future, declining fertility levels are believed to lead to slower rates of demographic growth throughout the twenty-first century (Figure 3.25). However, because there are so many people of child-bearing age in the developing world, the total population of the planet is projected to rise to roughly 10 billion by the year 2100 (Figure 3.26). The vast bulk of these additions will be in the Third World. However, these projections are based on different assumptions about the future of fertility (Figure 3.27). Should fertility rates decline more rapidly than expected, the increase in the world's people may not be as dramatic as some believe.

However, the population explosion in the developing world will have enormous impacts. Not all of the world's problems are reducible to population growth, but many are not independent of it either. Rapid population growth will accelerate, among other things, agricultural overcultivation and soil depletion, overfishing, poaching, deforestation, depletion of water resources, loss of biodiversity, and rural-to-urban migration (Figure 3.28). Further, by generating an infinite pool of poor people, it keeps wages low, not only in the developing world, but also in the developed, as globalization pits the labor forces of countries against one another. Thus, it is important to keep the dynamics of population growth in mind in relation to environmental degradation, economic development, and international issues.

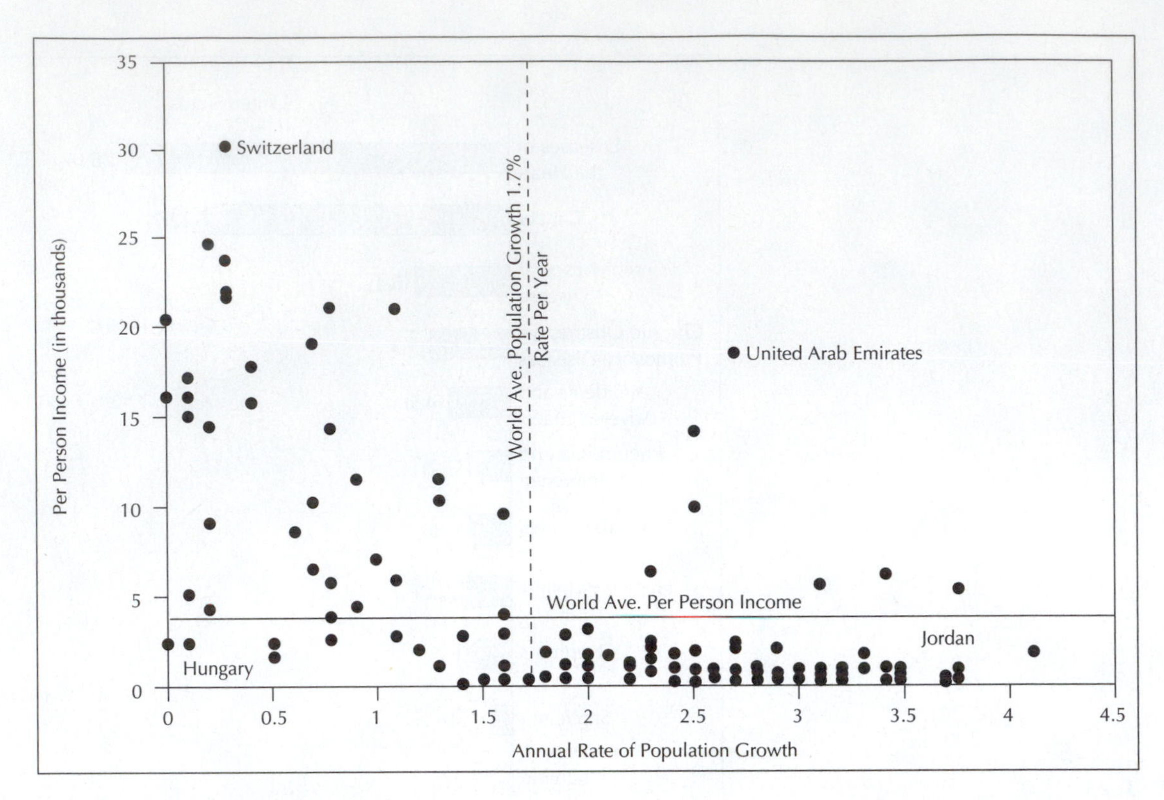

FIGURE 3.22
The rate of population growth in different societies is closely associated with their average income. Countries with high rates of demographic growth tend to have incomes below the world average; conversely, those with high per capita incomes tend to have low rates of growth. This pattern reflects the dynamics of the demographic transition discussed earlier.

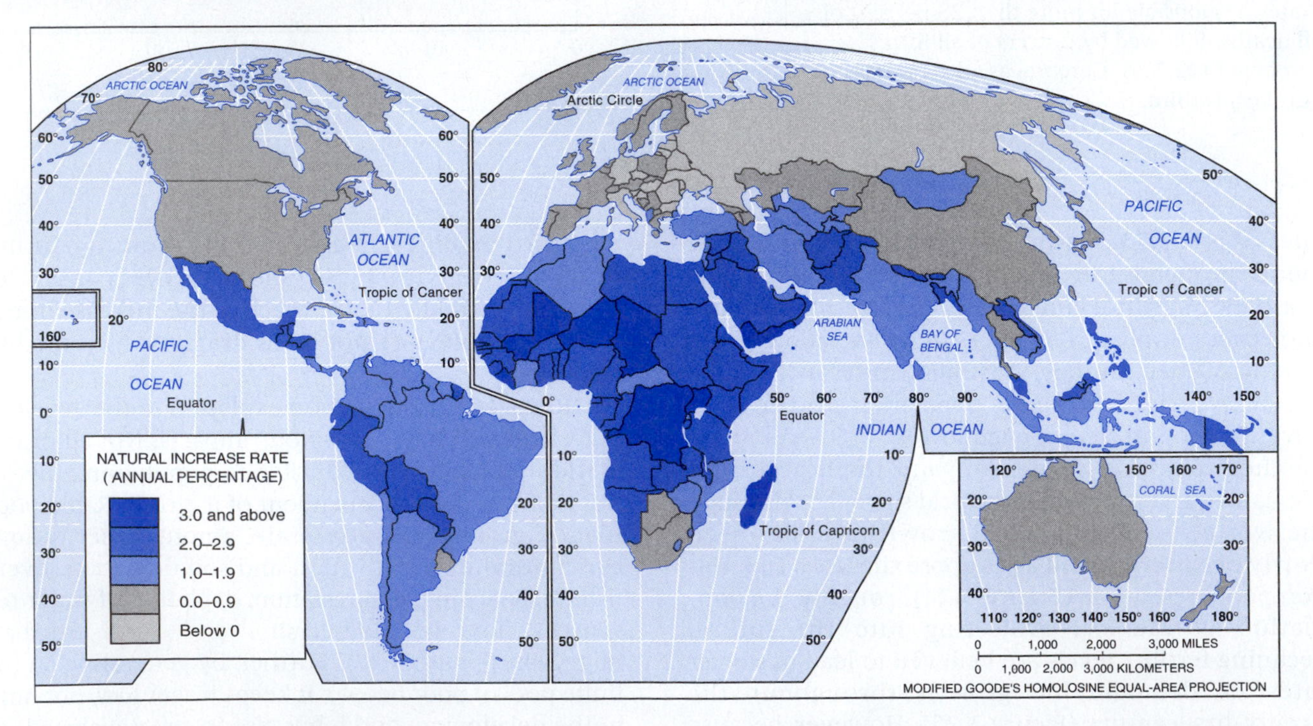

FIGURE 3.23
Natural annual increase in world population, 2000. The fastest growing areas of the world include Africa, Central America, and Southwest Asia. Here growth rates exceed 2.5 percent, with a number of countries in Central America and Africa actually exceeding 3 percent per year. Three percent growth per year does not seem like a high growth rate; however, it indicates total population doubling time for a country of only 23 years. With a 2 percent growth rate, a country would double in 35 years. For a 1 percent growth rate, a country may double in 70 years. Natural increase in Europe as a whole is only 0.2 percent, and several countries have negative rates of natural growth.

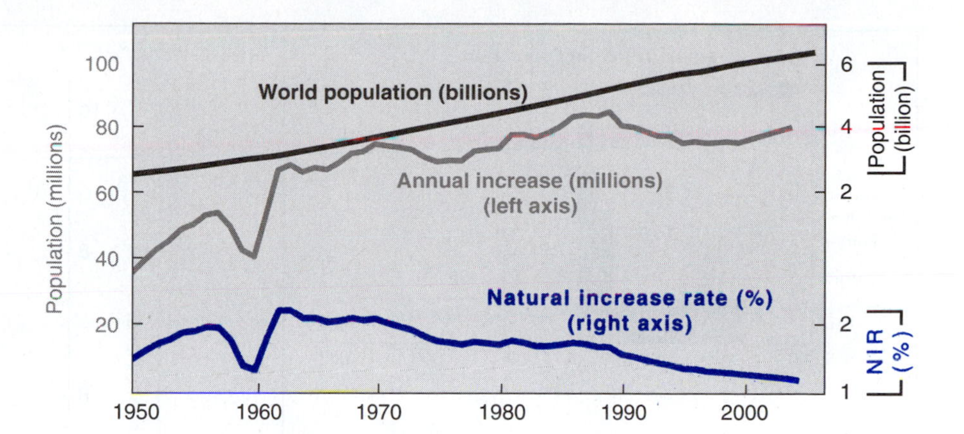

FIGURE 3.24
World population growth rates have slowly declined as fertility levels gradually dropped. Despite lower rates of increase, the world still adds roughly 100 million people each year.

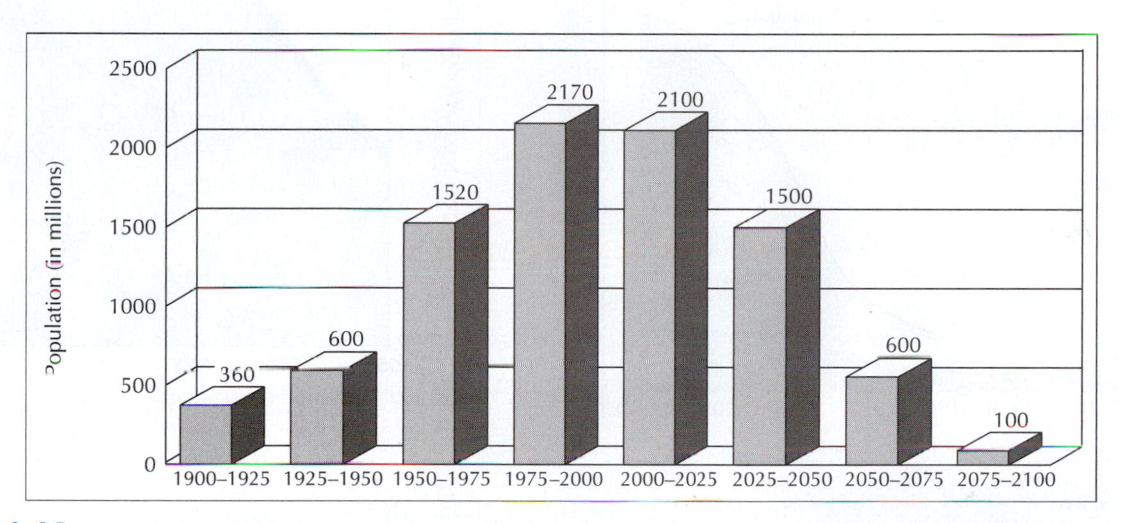

FIGURE 3.25
Net additions to the world's population between 1900 and 2100. The world's population has increased exponentially in the twentieth century. The period from 1975 to 2025 shows the largest quarter decade of increase, with more than 4 billion people being added to the world's population. From there, the world's population should slow down in rate of increase as fertility levels decline everywhere.

Contrasting the Demographic Transition and Malthusianism

By now you may have realized that the assumptions, analyses, and conclusions offered by Malthusian theory and the demographic transition are markedly different. Whereas Malthusianism tends to take fertility for granted—arguing that people are prisoners of biological imperatives to reproduce uncontrollably—the demographic transition reveals fertility is socially constructed (i.e., families have children, or don't have them, as the case may be, depending on the costs and benefits that children offer). Moreover, whereas Malthusian scenarios inevitably depict the population as growing uncontrollably, to the point of resource exhaustion, the demographic transition predicts steadily declining levels of world population growth as crude birthrates converge upon death rates. The evidence supports this view. After accelerating for two centuries, the overall rate of world population growth is slowing down. In 1992, population was growing at 1.7 percent a year, down from a peak of 2 percent in the late 1960s. The rate of growth declined to about 1.3 percent in 2005. However, the absolute size of population will continue to increase because the size of the base population to which the growth rate applies is so large.

Criticisms of Demographic Transition Theory

Although the demographic transition has wide appeal because it links fertility and mortality to changing socioeconomic circumstances, it has also been criticized on several grounds. Some critics point out that it is a model derived from the experience of the West and then applied to many non-Western societies, as if they are bound to repeat the exact sequence of fertility and

Regional Trends in Population

Region	2005 Population	2025 Projected Population
North America	331	369
Latin America & Caribbean	554	709
Europe	807	718
Africa	915	1,495
Asia	3,668	4,960

(millions)

Total World Population
Underdeveloped Regions
Developed Regions
Population (billions)
Year

FIGURE 3.26
The vast bulk of additions to the world's population have occurred, and will continue to occur, in the world's economically underdeveloped regions. Natural growth rates in the First World are low, often negative, whereas they tend to be relatively high in poorer countries.

mortality stages that occurred in Europe, Japan, and North America. There is no inevitable logic that assures the developing world must meekly follow in the footsteps of the West. Some have pointed out that the developing world is in many ways qualitatively different from the West, in no small part because of the long history of colonialism. Further, demographic changes in the developing world have been much more rapid than in the West. For example, whereas it took decades, or even centuries, for mortality rates in Europe to decline to their modern levels, in some developing countries the mortality rate has plunged in only one or two generations. Because mortality rates do not vary geographically as much as fertility rates, most of the spatial differences in natural growth around the world are due to differences in fertility. These caveats are useful in cautioning us to examine the historical context in which all theories and explanations emerge and to be wary of blindly importing models developed in one social and historical context into radically different ones for which they were not originally intended.

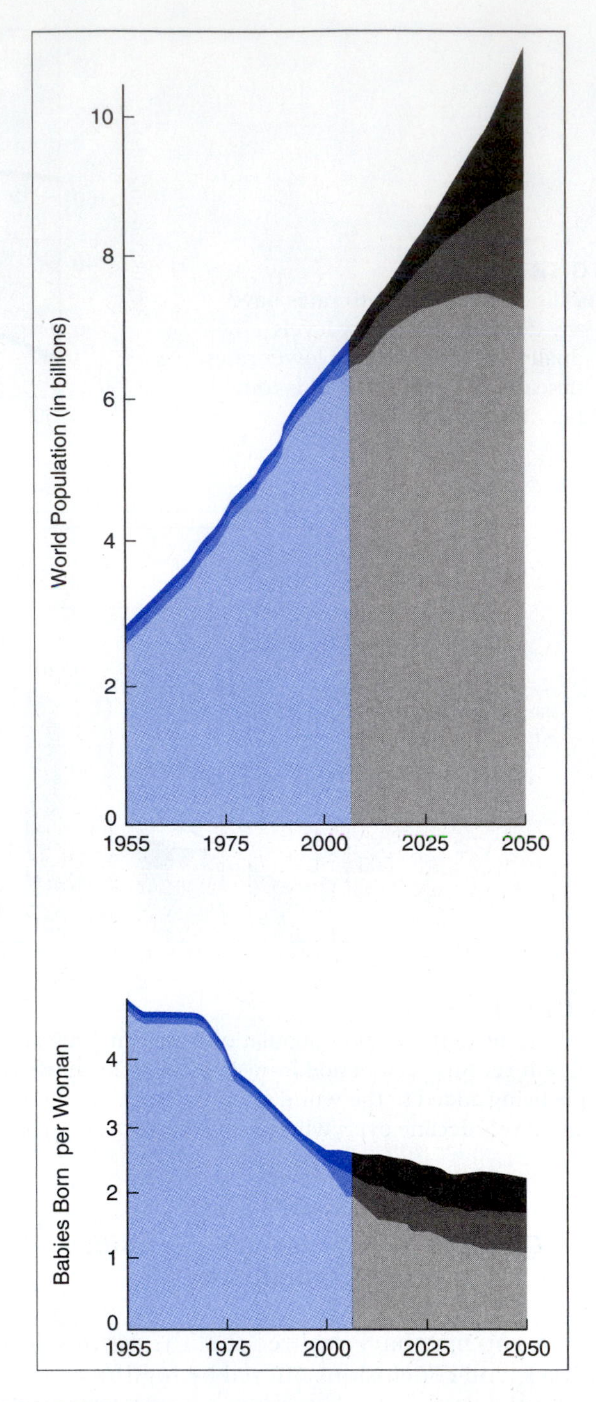

FIGURE 3.27
The future of the world's population size depends on how fertility levels change. If current fertility levels are maintained, the number of people in the world could exceed 10 billion by 2050. However, if fertility levels drop, as they have been doing, the rate of increase will slow down; at a minimum, the world can expect to hold more than 7 billion people in 50 years.

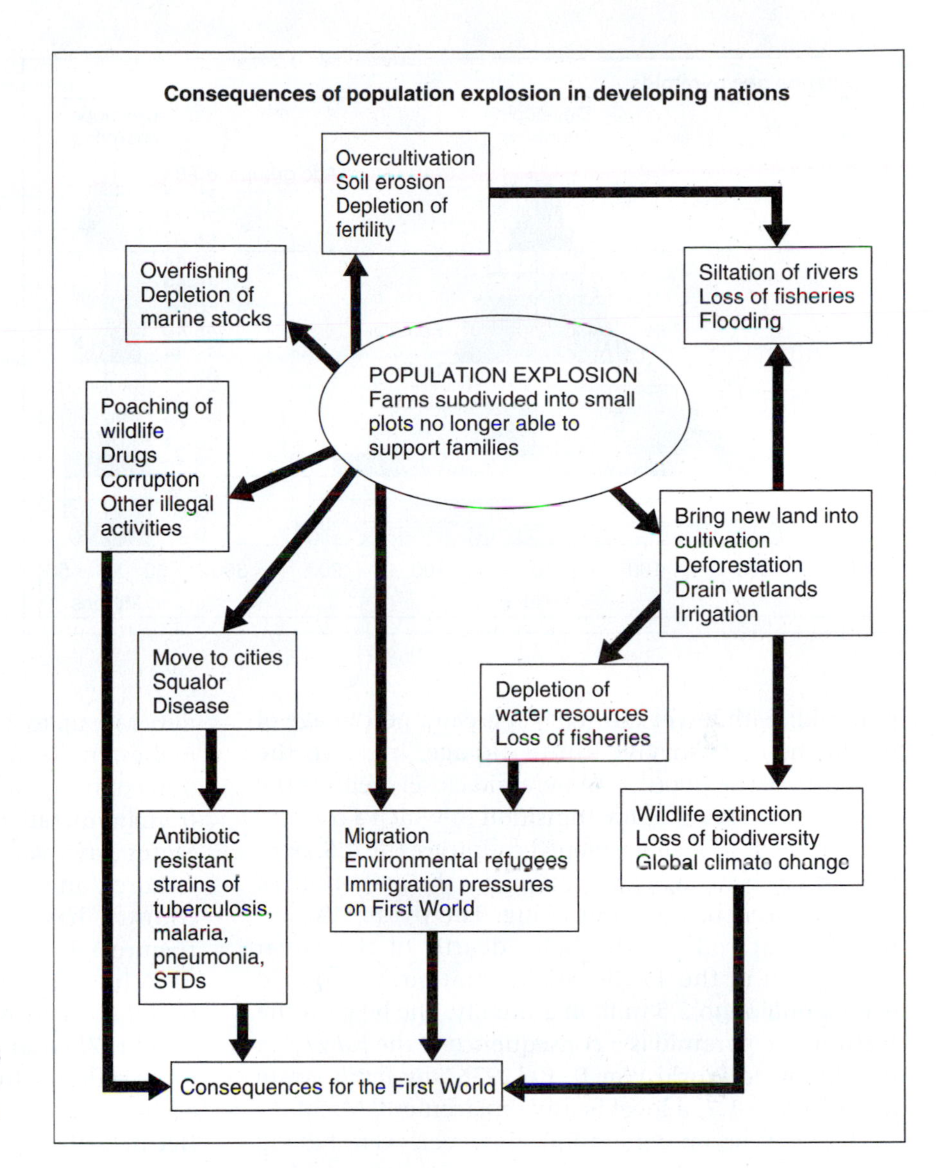

FIGURE 3.28
The population boom in developing countries will have numerous repercussions. Not all of the world's problems can be blamed on population growth, but rapid rates of increase are likely to exacerbate environmental degradation, such as soil erosion and wildlife loss, and generate large numbers of people seeking better opportunities in cities and in the economically developed world.

POPULATION STRUCTURE

Except for total size, the most important demographic feature of a population is its age-sex structure. The age-sex structure affects the needs of a population as well as the supply of labor; therefore, it has significant policy implications. A rapidly growing population implies a high proportion of young people under the working age. A youthful population also puts a burden on the education system. When this cohort enters the working ages, a rapid increase in jobs is needed to accommodate it. By contrast, countries with a large proportion of older people must develop retirement systems and medical facilities to serve them. Therefore, as a population ages, its needs change from schools to jobs to medical care.

The age-sex structure of a country is typically summarized or described through the use of *population pyramids*. They are divided into five-year age groups, the base representing the youngest group, the apex the oldest. Population pyramids show the distribution of males and females of different age groups as percentages of the total population. The shape of a pyramid reflects long-term trends in fertility and mortality and short-term effects of baby booms, migrations, wars, and epidemics. It also reflects the potential for future population growth or decline.

Two basic, representative types of pyramid may be distinguished (Figure 3.29). One is the squat, triangular profile. It has a broad base, concave sides, and a narrow tip. It is characteristic of developing countries having high crude birthrates, with a young average age, and relatively few elderly. Natural growth rates in such societies tend to be high.

In contrast, the pyramid for economically developed countries, including the United States, describes a slowly growing population. Its shape is the result of a history of declining fertility and mortality rates, augmented by substantial immigration. With lower fertility, fewer people have entered the base of the

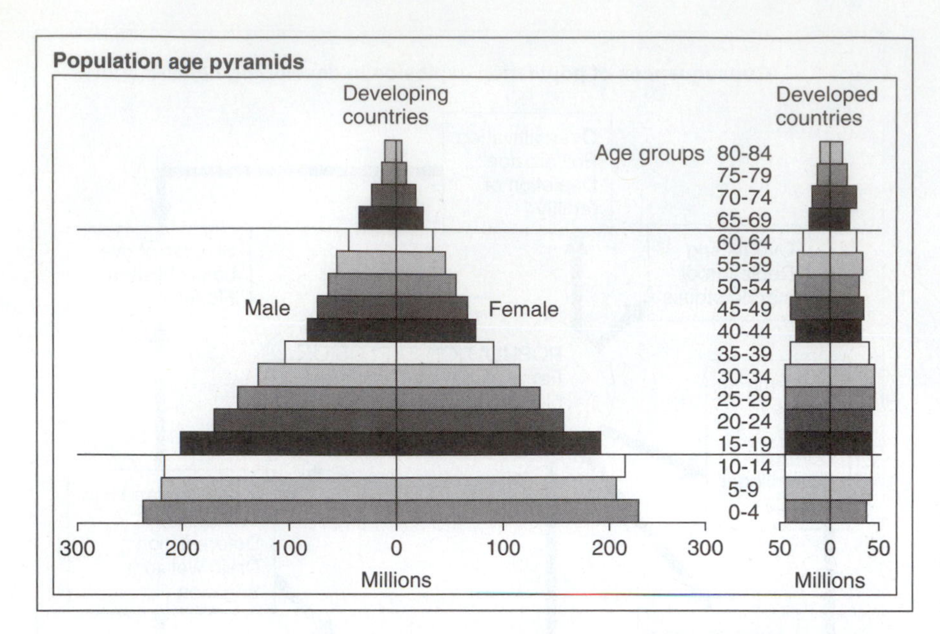

FIGURE 3.29
Population pyramids reflect the age and sex structure of a society. Poor countries, with high birthrates and a large share of the population at early ages, have pyramids with a broad base and narrow top. Economically wealthier societies, with low rates of natural growth, have older populations; their pyramids have proportionately smaller bases and a larger share of people in the middle aged and elderly groups.

pyramid; with lower mortality, a greater percentage of the births have survived until old age. In short, the structure of the population pyramid closely reflects the stage of the demographic transition in which a country is positioned. Like all developed countries, the U.S. population has been aging, meaning that the proportion of older persons has been growing. The pyramid's flattened chest reflects the baby dearth of the Great Depression in the 1930s, when total births dropped from 3 million to 2.5 million annually. The bulge at the waist of the pyramid is a consequence of the *baby boom* that followed World War II. By 1976, the fertility rate had fallen to 1.7, a level below replacement. Members of the baby-boom generation, however, were having children in the 1980s and 1990s, driving the total fertility rate up to 2.0, almost at replacement level and the highest in the developed world. Thus, the U.S. population continues to grow from natural increase as well as from immigration. Because different parts of the United States have very different socioeconomic conditions, cultures, and migration patterns, various places in the country have very different population pyramids (Figure 3.30).

A few developed countries have very low rates of population growth—in some cases *zero population growth* (ZPG) or *negative population growth.* They have low crude birthrates, low death rates, and, in some cases, net out-migration. France is an example. Because of very low fertility, the country is experiencing negative population growth, and although there is

Chinese children in Shanghai welcoming visitors from the United States. By 2002, China's population reached 1.3 billion, or more than one-fifth of the world population. Considering its Third World status and its huge size, it has a remarkably low 1.2 percent average natural increase. One child per family is the norm, and it is enforced by the government with a series of economic sanctions.

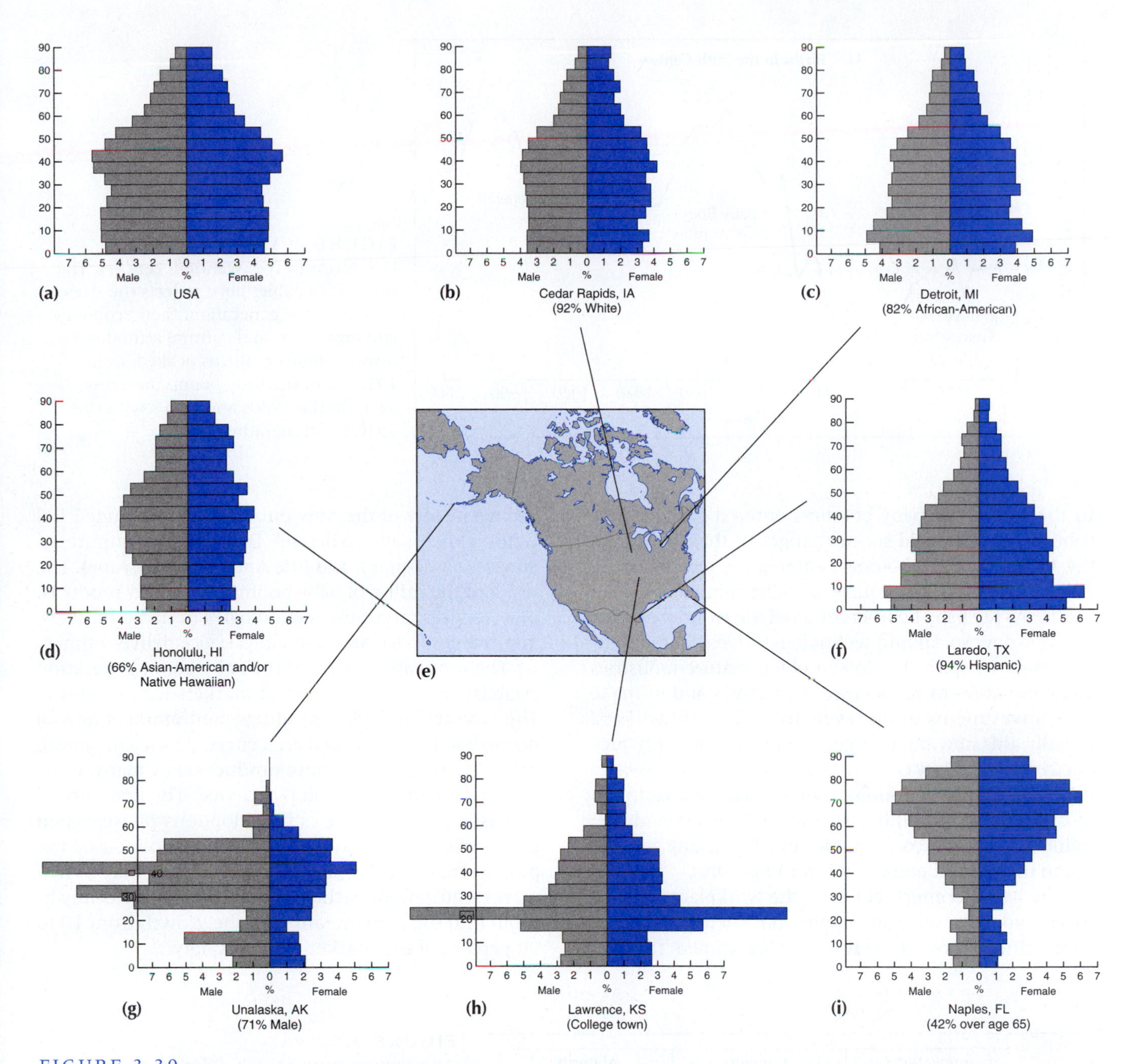

FIGURE 3.30
Population pyramids for selected communities in the United States reflect a diverse suite of demographic conditions. In college towns, a disproportionately large number of young adults is evident. In retirement communities, such as Naples, Florida, there are far more elderly than young people. Note the imbalance between males and females in Unalaska, a reflection of the labor market there.

a steady stream of foreigners (especially Arabs) into the country, France tries to limit immigration. Population decline is an economic concern to many European countries, as well as Japan, the world's oldest nation demographically. Who will fill the future labor force? Is the solution the immigration of guest workers from developing countries? In these respects, demographic changes have profound influence on immigration policies and the size and nature of the labor force.

THE BABY BOOM AND ITS IMPACTS

The so-called baby boom—everyone born between 1946 and 1964—is the largest generation in world history, 90 million strong (Figure 3.31). Baby boomers were the children of the "greatest generation," those who lived through the Great Depression and fought World War II. After the war, with the American economy booming and standards of living rising, birthrates increased dramatically, giving rise to a flood of children in the schools.

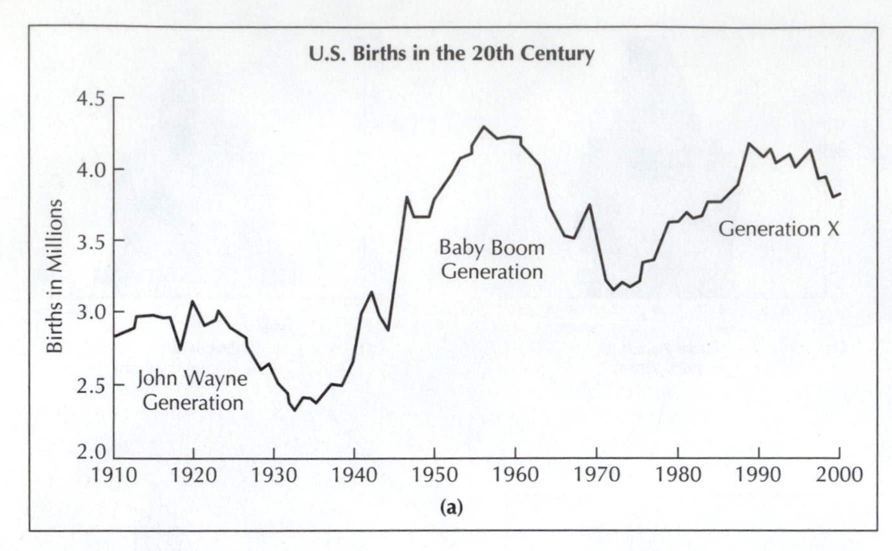

FIGURE 3.31
U.S. births in the twentieth century. The number of babies born reflects the size of their parents' generation, their economic circumstances, and cultural attitudes toward children. Births peaked in the 1950s with the baby boom, then rose again in the 1980s and 1990s with the arrival of Generation X.

In the 1960s, the baby boomers entered college, contributing to the rapid social changes of that decade. In the 1970s, the baby boomers entered the workforce, 90 million strong, including more working women than ever before. As the baby boomers entered the workforce, they brought new social and technological ideas and an entrepreneurial spirit that forced older, mature industries and companies to restructure themselves and to make large investments at all levels to train new workers. Initially, this new boomer generation was not very productive and had low earnings and savings rates. This large swell entering the workforce required a huge investment in capital stock and infrastructure—office space, desks, training programs, computer terminals, parking garages, not to mention cafeterias and clothing stores.

The baby boomers redefined the workplace, causing social and technological change, but their conformist, civic-minded bosses were not accustomed to such change. A few of the new entrepreneurs included Bill Gates (Microsoft), Michael Dell (Dell Computers), Steven Jobs (Apple), and Eric Anderson (Netscape). The result of the influx of baby boomers was new products, new services, and new technologies in niche markets, improving service and reducing service delivery times.

The innovation swell of the baby boom generation established the industries and markets for the future. The new technologies, products, and markets moved through a life cycle called an S curve, or a *logistic growth curve* (Figure 3.32). A new product takes many years and a trial period before it is accepted. The first third of a logistic growth curve is the adoptions phase, when the market may be suspicious of the product, or the product may be overpriced and not user friendly. When market saturation reaches 10 percent, adoptions mushroom into the mainstream, and the growth from 10 to 90 percent of the market occurs rapidly.

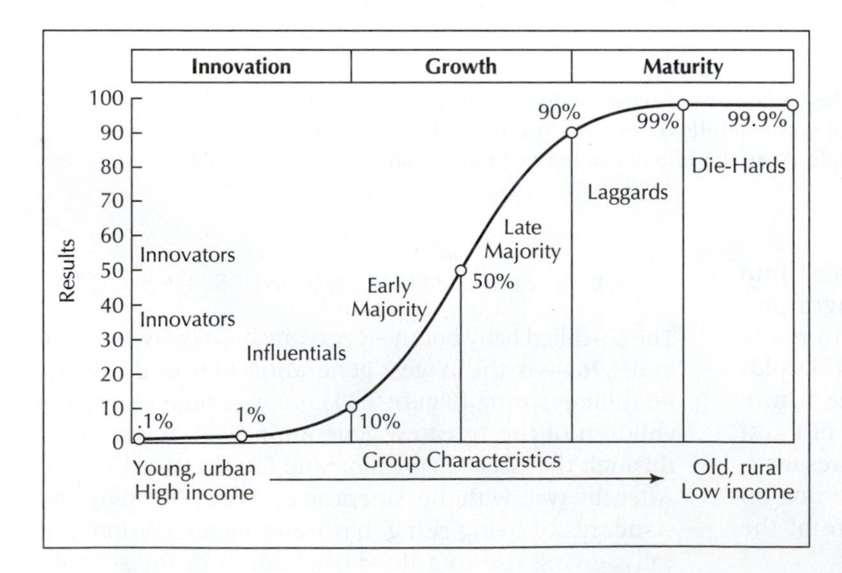

FIGURE 3.32
An adoption curve reveals different groups of people who are likely to accept an innovation or purchase a new product, such as cell phones or the Internet, at different moments in time, and thus is widely used in marketing as well as in economic geography.

Large retooling and training investments that were needed to absorb the baby boomers into the job market required consumers to pay a high inflation tax to finance them. It took a large investment to launch new technologies and industries. Technological innovations are expensive to start up, and there is a correlation between inflation and such revolutions. It was no accident that America saw the highest inflation in U.S. history in the late 1970s as the largest generation in history, the baby boomers, went through their early job years. During this time, the most powerful innovations and technologies in history started to emerge—the digital revolution from personal computers to fax machines to cellular phones.

Now that the baby boom is preparing to retire, labor force growth is low. What this means for business demographics is that companies will not have to train masses of new workers and charge them the inflation tax. However, the present workforce must be redeployed so that everyone becomes more productive to keep up with the high growth rates of the industry and market demand. This will be possible because of the microelectronics revolution, which allows workers to leverage information technologies and become more efficient in their work patterns.

The baby boomers are just moving into their most important spending years. Family spending follows a predictable life cycle that results in maximum spending between ages 46 and 50. The great number of baby boomers moving as consumers in and out of their peak spending years causes booms and busts in the economy. The baby boom spending wave began in the 1980s. North American baby boomers will move into their peak years of total spending for durable goods around the year 2010, creating the biggest business boom in the history of the world.

Migration

Migration is a movement involving a change of permanent residence. It is a complex phenomenon that raises a lot of questions. Why do people move? What factors influence the intensity of a migratory flow? What are the effects of migration? What are the main patterns of migration?

Causes of Migration

Most people move for economic reasons. They relocate to take better-paying jobs or to search for jobs in new areas. They also move to escape poverty or low living standards, to find a better life for their children, to escape adverse political conditions, or to fulfill personal dreams.

The causes of migration can be divided into push-and-pull factors. Push factors include high unemployment rates, low wages, poverty, shortages of land, famine, or war. In the late 1970s and early 1980s, various communist purges in Vietnam, Kampuchea, and Laos pushed approximately 1 million refugees, who resettled in the United States, Canada, Australia, China, Hong Kong, and elsewhere. Pull factors include job and educational opportunities, relatively high wages, the hope for agricultural land, or the "bright lights" of a large city. The rich oil-exporting countries in the Middle East act as a pull factor for millions of immigrants seeking employment. In Kuwait and the United Arab Emirates, nearly 80 percent of the workforce was composed of foreigners in 2001. Spatial differences in economic opportunities, therefore, go far toward explaining why young people often leave rural areas, the influx of Mexicans into the southwestern United States, or the immigration of non-Westerners into Europe, including Indians and Bangladeshis into Britain, Turks into Germany, and Arabs into France.

Migrations can be voluntary or involuntary and reflect the historically specific matrix of cultural, economic, and political circumstances in which migrants live. Most movements are voluntary, such as the westward migration of pioneer farmers in the United States and Canada. Involuntary movements may be forced or impelled. In forced migration, people have no choice; their transfer is compulsory. Examples include the African slave trade, in which 20 million people were stolen from their homelands and shipped to the New World, and the deportation of British convicts to the United States in the eighteenth century and to Australia in the nineteenth. In impelled migration, people choose to move under duress. In the nineteenth century, many Jewish victims of the Russian pogroms elected to move to the United States and the United Kingdom. Civil wars in Central America in the 1980s led hundreds of thousands of Salvadorans and Nicaraguans to emigrate to the United States. In Africa, multiple civil wars have displaced millions of refugees into neighboring nations.

The Economics of Migration

This subsection examines migration of a voluntary nature due to economic incentives, which comprises the largest single category of human migrations throughout history. Unfettered migration is an expression of a free market for labor, although markets are never as free for labor as they are for capital. Unrestricted migration is generally only feasible within the borders of the same nation-state. Among states, restricting the flow of immigrants is one of many ways states as well as markets shape economic landscapes.

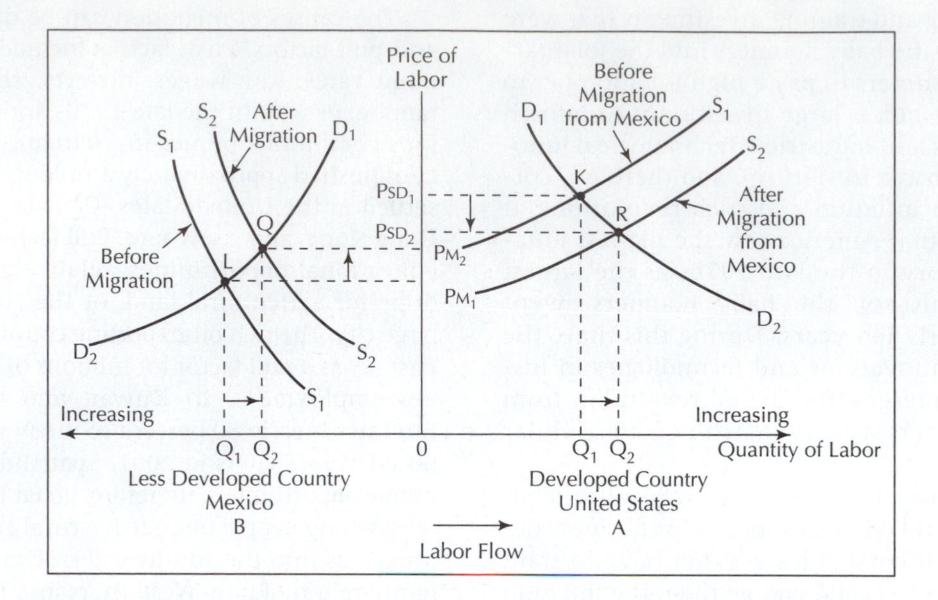

FIGURE 3.33
Migration and wage differentials. The quantity of labor is measured on the horizontal axis, and the price of labor is measured on the vertical axis. As the price increases on the vertical axis, the demand for labor decreases in both the developed country and the less developed country. The supply curves of both developed and undeveloped countries slope upward, suggesting that a greater supply of labor is available at higher prices. Because the equilibrium price of labor is higher in the developed country than in the undeveloped country, labor migrations occur from less developed to developed, or from B to A, to take advantage of higher wages. The greater the wage differential, both the greater the flow and the longer the distance of flow. The wage rate differential between San Diego and Tijuana, for example, is 8:1 for comparable occupations.

Consider two regions as shown in Figure 3.33. Region A on the right, is a highly industrialized country (e.g., the United States or Germany). Region B is a less developed country. A labor market exists within each nation. The quantity of labor supplied and demand is measured on the horizontal axis and the price of labor (wages) is measured on the vertical axis. As the price of labor increases, the demand decreases (i.e., employers face a demand curve), and as the price decreases, the demand increases. The supply curve S_1 increases upward to the right, suggesting that a greater supply of labor is available at higher prices. As the price for labor increases, individuals who would not care to work at lower wages now come into the market. They substitute work at higher prices for staying home and taking care of children, going to school, or being in retirement.

In order to facilitate the analysis for the less developed country, such information is shown on the left side of Figure 3.33. Instead of measuring the quantity of labor from zero to the right for the less developed country, we now measure it from zero to the left. Price remains on the vertical axis. The supply curve in the less developed country slopes upward to the left and the demand curve slopes downward to the left. In this manner, we get a back-to-back set of supply and demand curves for a less developed country and a developed country.

Observe the equilibrium position of the supply and demand for labor in the developed country before migration occurs, which occurs at point K. Also note the equilibrium position for the supply and demand of labor in the less developed country. This equilibrium position occurs at L. In classical *labor migration theory,* there is an assumption that information about job availability and wage differentials is widely available and held by workers (i.e., there is little uncertainty). Labor in the less developed country finds out about jobs available in the developed country at higher wage rates. Because the equilibrium price in Region A is higher than that in Region B, labor migrates from Region B to Region A to take advantage of higher wages (assuming there are no barriers to migration). The greater the differential in wages, the greater the flow of labor will be. In Region A, extra labor is now coming into the region, which is used to working for lower wages. Because the extra labor is supplied over and above the indigenous supply and the labor is used to working at lower wages, the supply curve moves downward toward the right to the new equilibrium R. In addition, because the labor pool has left Region B, the supply of labor is reduced. The supply curve moves upward and to the right in Region B, thus raising the equilibrium price to Q. The new equilibrium price in Region A, at R, is at a lower level than it was prior to labor migration. Thus, migration will continue as long

FIGURE 3.34
Flows of millions of illegal immigrants from Mexico to the United States take place at a variety of places along the border. Often migrants undertaking such journeys are exposed to very dangerous conditions, and may die in transit .

as there is a difference between the wages of Region A and Region B, which exceed a cost associated with migration. In the case of flows from Mexico to the United States, most categories of employment are paid two to five times the rate in Mexico. Consequently, the flows both on a daily basis and on a longer term basis continue to occur at high levels, including many illegal or undocumented workers (Figures 3.34, 3.35).

In classical migration theory, transportation costs and other costs associated with moving an individual or family are included, such as selling a property and purchasing one in the new region. The costs of labor in the different regions will not be exactly equalized. But classical trade and migration theory tell us that the long-run price of labor in the two regions should come into close harmony with one another. Relocation and similar migration costs should be split up over the period of work remaining in the life of the mover.

However, when we observe real world labor movements and price differentials, we find that wage rates do not seem to be converging among regions. Major discrepancies occur in the wages paid in and among various regions of the United States and countries of Europe, as well as in South America and India. If neoclassical theory held, there would be less difference in national averages of per capita incomes. One reason that labor differential rates exist is because of the imperfections in the availability of knowledge about opportunities. Many workers in the less developed countries do not know that jobs they may be qualified for in more highly developed countries even exist. As an individual contemplates changing locations and even countries, he or she is beset by a series of social factors, including lack of friends and knowledge and the feeling of uneasiness in the new setting. Consequently, the largest numbers of international labor migrants are young males who do not have families to relocate.

Cultural differences are also important to understanding migration. The cultural shock of living in a new environment, especially when one does not have the resources to live adequately or does not speak the native language, presents social problems. Institutional barriers also exist, such as the status of immigration or the length of time allowed in the host country. African Americans, Puerto Ricans, Chinese immigrants, Mexican Americans, and women have encountered

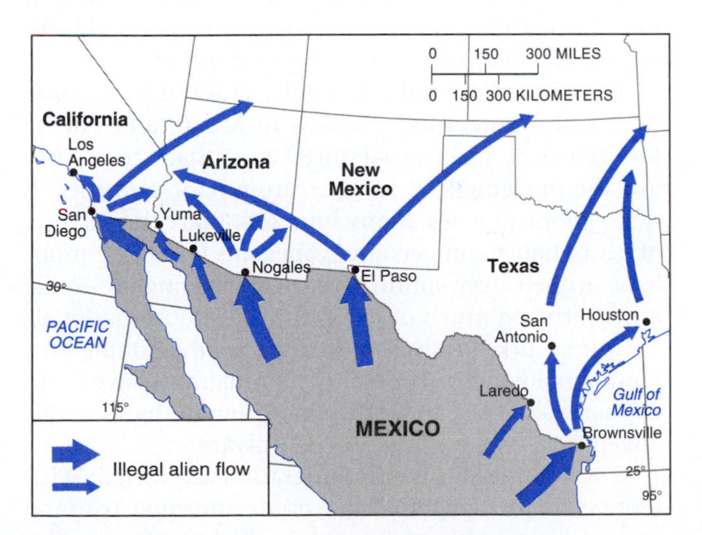

FIGURE 3.35
Illegal or undocumented migration flows from Mexico to the United States.

such resistance in the past in the United States in their search for improved working conditions and wages, often in the face of xenophobic racism and ostracism.

Consequently, we cannot expect that economic forces alone will lead to a total eradication of wage inequities throughout a country or throughout the world. Barriers to migration, including legal obstacles and immigration restrictions, imperfect information, lack of skills, inability to afford transportation, and the powerful bonds that hold people in place work to prevent a free flow of labor often within countries. At best, only a small portion of the population in a low-wage region has the ability to gain access to higher pay in developed nations. Therefore, there will continue to be a discrepancy in per capita earnings between less developed countries and developed countries and between depressed regions and economically healthy regions within countries. The demand curves for labor also shift down and to the right as the more highly qualified, productive labor leaves, and up and to the right as new workers arrive. Migration is thus intertwined with local and national labor markets in complex ways that shape average incomes and unemployment rates.

The availability of work and wage rates account for major labor flows throughout the world, from countries lacking in jobs and high rates to countries with jobs available at relatively higher wage rates. Major labor flows occur (1) from Mexico and the Caribbean to the United States and Canada; (2) from South American countries to Argentina, Venezuela, and Peru; (3) from North Africa and southern European nations to northwestern Europe; (4) from Africa and Asia to Saudi Arabia; and (5) from Indonesia to Malaysia, Singapore, and Australia. Migrants vary by age. Young adults are most likely to be migrants because of their desire for an improved life and greater ability to travel and overcome hardships.

Barriers to Migration

All countries regulate the flow of immigration. The United States limits legal immigration to approximately 600,000 people annually, although a total of roughly 1.1 million enter the United States legally or illegally every year. Altogether, about 33.1 million immigrants live in the United States, comprising 11 percent of the population. Of this group, an estimated total of 5 million people live in the country illegally, often under constant threat of being caught and deported. Billions of dollars are spent annually to police the borders of the United States, much of which is used to try to keep Mexicans out.

Characteristics of Migrants

Some countries have higher rates of migration than do others—both into and within the country. In general, the countries that have long histories of migration, such as the United States, Canada, and Australia, have higher migration rates in the modern world than do other countries, such as China, where migration is far less common (although recently China has witnessed a surge of people leaving rural areas for the more prosperous coastal cities). When people do move, they are far more likely to be young adults than they are any other group. Young adults have the longest working life ahead of them and thus stand to gain more than the elderly by the accumulated benefits accrued to them by moving to a relatively more high-wage region. In the developed world, migration rates tend to drop significantly by the time people have established families and purchased homes, typically in their thirties.

Consequences of Migration

Migration has demographic, social, and economic effects, due especially to the fact that migrants tend to be young adults and are often the more ambitious and well-educated members of a society. Obviously, the movement of people from Region A to Region B causes the population of A to decrease and B to increase. Because of migratory selection, the effects are more complicated. If the migrants are young adults, their departure increases the average age, raises the death rate, and lowers the birthrate in Region A. For the region of immigration, B, the opposite is true (i.e., their arrival tends to lower the average age and the death rate but increase the fertility rate). If migrants to Region B are retirees, their effect is to increase the average age, raise the death rate, and lower the birthrate. Arizona and Florida, for example, have attracted a large number of retirees, resulting in a higher-than-average death rate.

Social conflict is a fairly frequent social consequence of migration. It often follows the mass movement of people from poor countries to rich. There were tensions in Boston and New York after the Irish arrived in the 1840s; fleeing the Potato Famine, they were the first Catholics to arrive in large numbers. Similar tensions have come with recent migrants—Cubans to Miami and Puerto Ricans to New York. Social unrest and instability also follow the movement of refugees from poor countries to other poor countries. Many immigrants are the subjects of xenophobia and become scapegoats for all the problems in their new country, especially during economic downturns. In much of Europe, for example, nationalists blame Turks, Arabs, Pakistanis, and other immigrants for unemployment. Generally, poor migrants have more difficulty adjusting to a new environment than the relatively well educated and socially aware.

The economic effects of migration are varied. With few exceptions, migrants contribute enormously to the economic well-being of places to which they come. For example, guest workers were indispensable to the economy of West Germany prior to reunification in 1991. Without them, assembly lines would have closed down, and patients in hospitals and nursing homes would have been unwashed and unfed. Without Mexican migrants, fruits and vegetables in Texas and California would go unharvested and service in restaurants and hotels would be much more expensive. Migrants to the United States also pay income and sales taxes, but illegal ones do not reap the benefits of programs such as Social Security.

In the short run, the massive influx of people to a region can cause problems. The U.S. Sunbelt states have benefited from new business and industry but are hard pressed to provide the physical infrastructure and services required by economic growth. In Mexico, migrants to Mexico City accelerate the competition for scarce food, clothing, and shelter. Despite massive relief aid, growing numbers of refugees in the developing world impoverish the economies of host countries.

Emigration can relieve problems of poverty by reducing the supply of labor. External migration relaxed the problem of poverty in Jamaica and Puerto Rico in the 1950s and 1960s. However, emigration can also be costly. Some of the most skilled and educated members of the population of Third World countries migrate to developed countries. Each year, the income transferred through the "brain drain" to the United States amounts to significant quantities of funds, although billions of dollars are also sent back home in the form of remittances to family members who stayed behind. Indeed, remittances are often a major source of income for impoverished villages in the developing world.

Patterns of Migration

To examine patterns of migration, it is helpful to consider migration internationally or within a country. It is also convenient to subdivide external migration into intercontinental and intracontinental, and internal migration into interregional, rural-urban, and intrametropolitan. International migrations are greatly exceeded by internal population movements, especially to and from cities.

The great transoceanic exodus of Europeans and the Atlantic slave trade to the New World are spectacular examples of intercontinental migration. In the five centuries before the economic depression of the 1930s, these population movements contributed greatly to a redistribution of the world's population. It has been estimated that between 9 and 10 million slaves, mostly from Africa, were hauled by Europeans into the sparsely inhabited Americas. The importance of the "triangular trade" of Europe, Africa, and the Americas can hardly be exaggerated, especially for British economic development. Africans were purchased with British manufactured goods. They were transported to work on plantations where they undertook production of sugar, cotton, indigo, molasses, and other tropical products. The processing of these products created new British industries. Plantation owners and slaves provided a new market for British manufacturers whose profits helped to finance Britain's Industrial Revolution.

The Atlantic slave trade, however, was dwarfed by the voluntary intercontinental migration of Europeans. Mass emigration began slowly in the 1820s and peaked on the eve of World War I, when the annual flow reached 1.5 million (Figure 3.36). At first, migrants came from densely populated northwestern Europe. Later they came from poor and oppressed parts of southern and eastern Europe. Between 1840 and 1930, at least 50 million Europeans emigrated. Their main destination was North America, but the wave of migration spilled over into Australia and New Zealand, Latin America (especially Argentina), and southern Africa. These new lands were important for Europe's economic development. They offered outlets for population pressure and provided new sources of foodstuffs and raw materials, markets for manufactured goods, and openings for capital investment. Another large-scale intercontinental migration was the Chinese diaspora of the nineteenth and early twentieth centuries, especially into Southeast Asia.

Since World War II, the pattern of intercontinental migration has changed. Instead of heavy migratory flow from Europe to the New World, the tide of migrants is overwhelmingly from developing to developed countries (Figure 3.37). Migration into industrial Europe and to North America has been spurred partly by widening economic inequality and by rapid rates of population increase in the developing world. Immigrants thus form significant populations in many countries around the world (Figure 3.38), particularly in the United States, western Europe, Australia, and South Asia. Some of them are refugees, while others are unskilled workers seeking jobs outside of their native lands.

The era of heavy intercontinental migration is over. Mass external migrations still occur, but at the intracontinental scale. In Europe, forced and impelled movements of people in the aftermath of World War II have been succeeded by a system of migrant labor. Thousands of people from Latin America, particularly Mexicans, many of whom are illegal aliens, find their way to the United States each year to work (Figure 3.39). Similarly, the most prosperous industrial countries of Europe attract workers from the agrarian periphery (Figure 3.40). France and Germany are the main receiving countries of labor migrants to Europe. France attracts workers from North Africa. West Germany drew workers from Italy, Greece, and Turkey. Migrant workers from southern Europe usually have low skills and perform jobs unacceptable to indigenous workers.

The system of extraterritorial migrant labor also exists in the developing world. In Africa, laborers move great distances to work in mines and on plantations. In West Africa, the direction of labor migration is from the

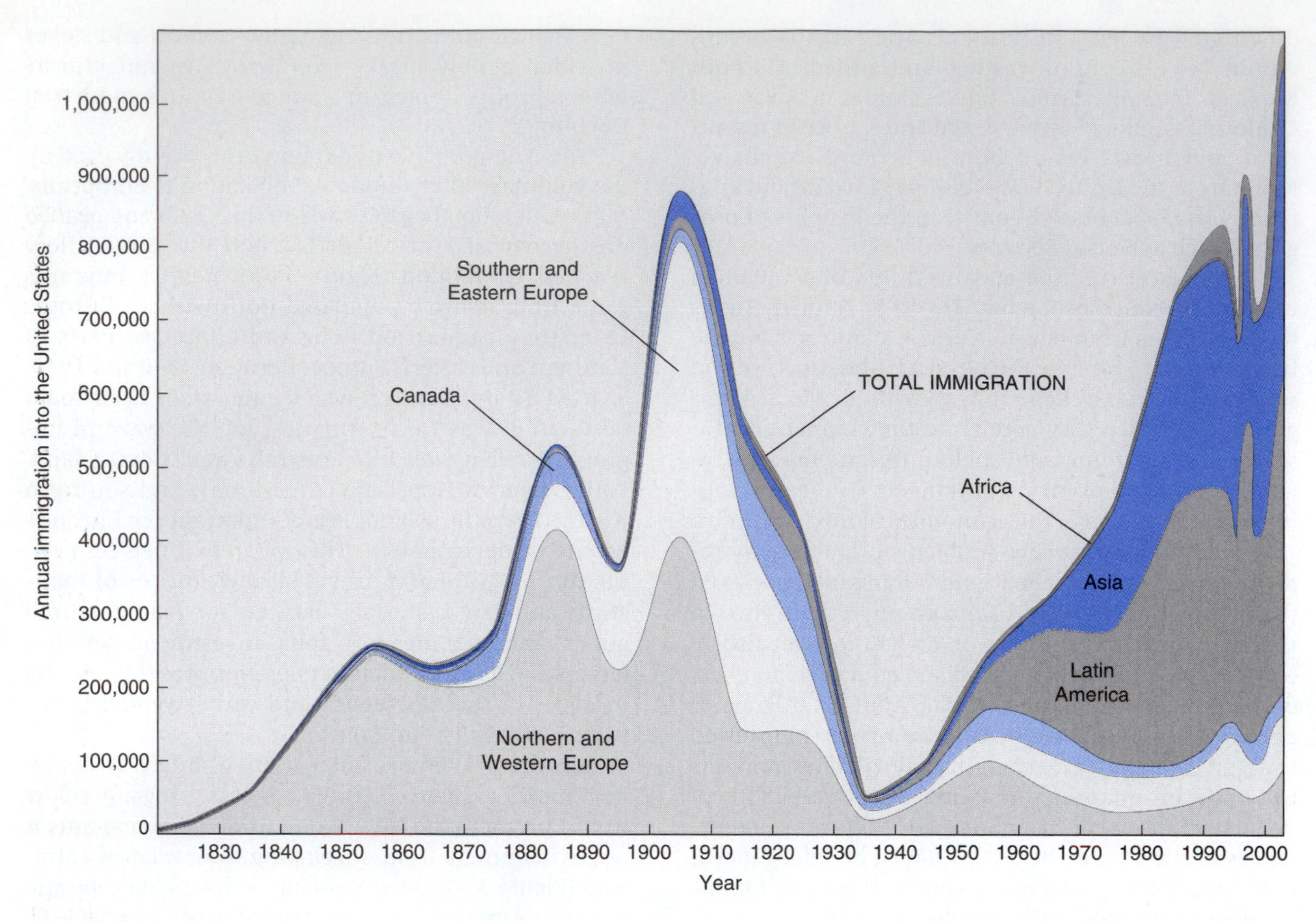

FIGURE 3.36
Immigration to the United States by region of the world. European countries provided more than 90 percent of all immigrants to the United States during the 1800s. Up until 1960, Europeans continued to provide more than 80 percent of the total migration to the United States. But since the 1960s, Latin America and Asia have supplanted Europe as the most important source of immigrants to the United States. During the Great Depression and World War II years, immigration was at an all-time low. Add the taxes that an immigrant and his or her descendants are likely to generate over their lifetimes; then subtract the cost of the government services they are likely to consume. The result is that each new immigrant yields a net gain to the government of $80,000. States and cities lose $25,000, while the impact on the federal treasury is $105,000.

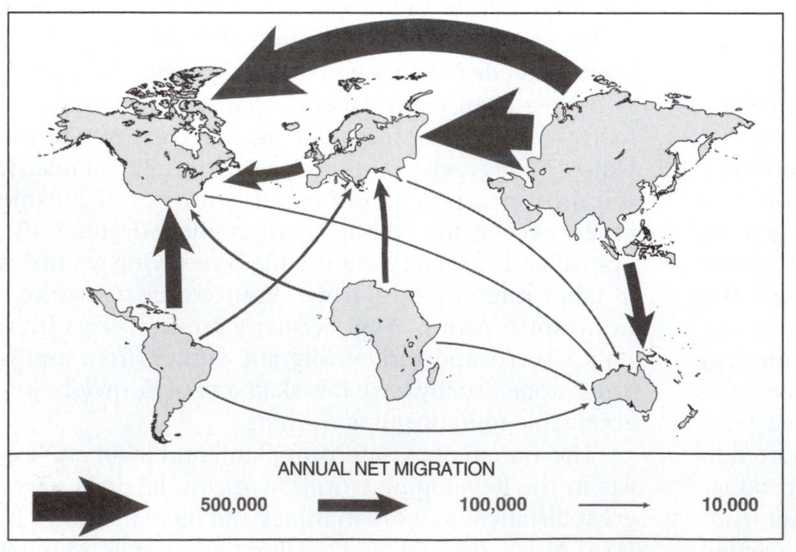

FIGURE 3.37
Intercontinental migration flows are dominated by people leaving the developing world in search of opportunities in the developed world, including immigrants from Asia and Latin America to the United States and Canada and Africans and Asians to Europe.

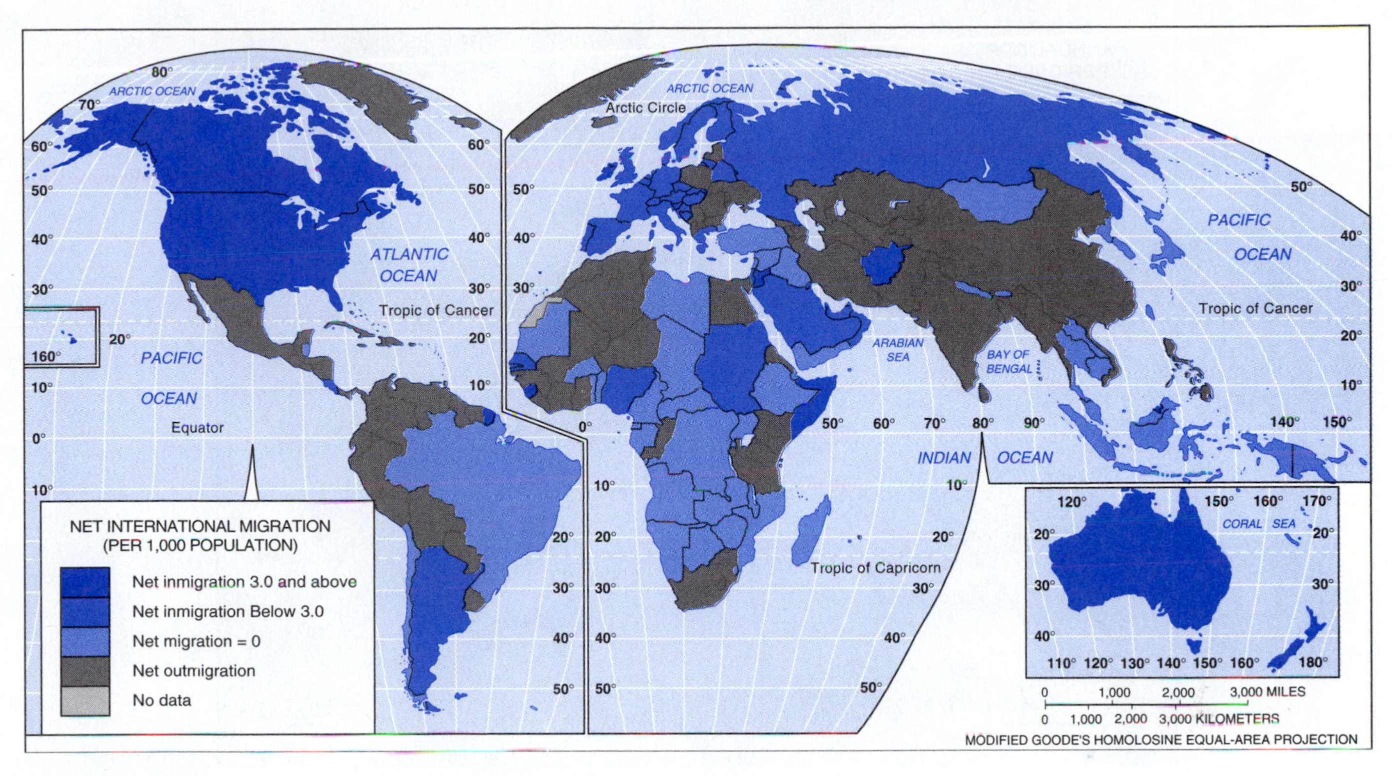

FIGURE 3.38
Net international migration rates around the world. The United States has the largest group, although other populations are found in France, Britain, Germany, and Australia, as well as India, Pakistan, Iran, and Saudi Arabia. The latter includes many unskilled workers from Asia.

interior to coastal cities and export agricultural areas. In East Africa, agricultural estates attract extraterritorial labor, typically refugees (Figure 3.41). In southern Africa, migrants focus on the mining-urban-industrial zone that extends from southern Zaire in the north, through Zambia's Copper Belt and Zimbabwe's Great Dyke, to South Africa's Witwatersrand in the south.

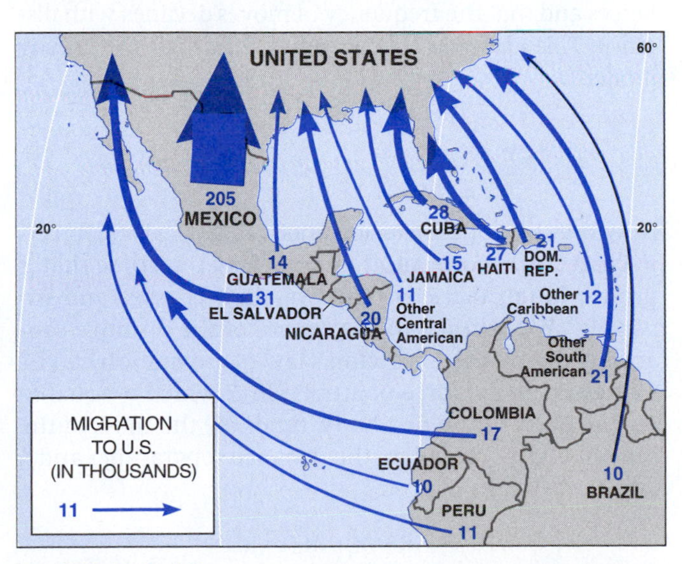

FIGURE 3.39
Migration flows to the United States from Latin America. The largest source of emigrants is Mexico, with smaller streams from various parts of the Caribbean, Central America, and northern South America.

Today, refugee generating and receiving countries are concentrated in Africa (6 million people), Southeast Asia (4 million), and Latin America (2 million). The causes of refugee movement typically include wars (e.g., Vietnam, World War II, Afghanistan); racial and ethnic persecution (e.g., South Africa, Bosnia-Herzogovina); economic insufficiency increased by political turmoil (e.g., Sudan); and natural and human-caused disasters (e.g., Central American hurricanes).

Colonizing migration and population drift are two types of interregional migration. Examples of colonizing migration include the nineteenth-century spontaneous trek westward in the United States and the planned eastward movement in the USSR beginning in 1925. General drifts of population occur in almost every country, and they accentuate the unevenness of population distribution. Between the two world wars, there was a drift of African Americans from the rural South to the cities of the nation's industrial heartland in the Northeast and Midwest. Since the 1950s, there has been net out-migration from the center of the United States to both coasts and a shift of population from the Rustbelt states to the Sunbelt (Figure 3.42). Today, the majority of Americans live in the South and West, as opposed to the North and Midwest, although the vast expanses of land in the

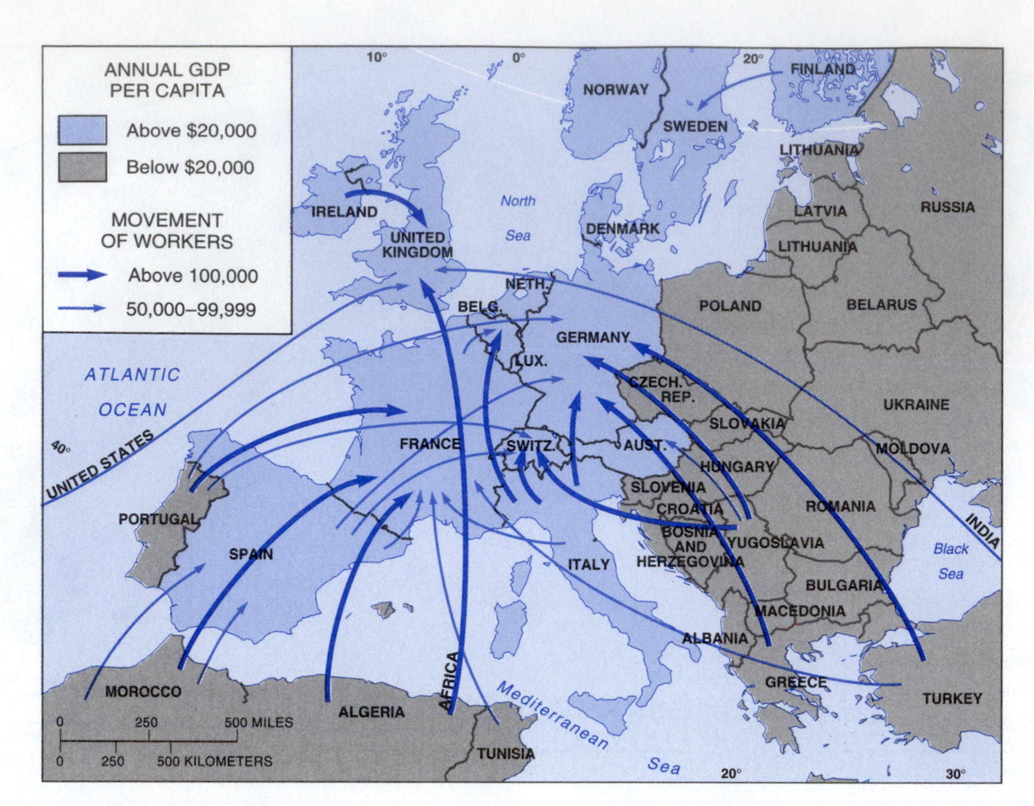

FIGURE 3.40
Migrant flows to and within Europe are generally from poorer countries with a labor surplus to wealthier ones with better employment opportunities. For example, Turks have long served as "guest workers" in Germany, and France hosts a growing population of Arabs from Algeria and Morocco. There are also net migration streams out of Spain, Portugal, Greece, and Italy to France, Switzerland, Germany, and other states.

Sunbelt states generate lower population densities than in the Northeast. The steady growth of the Sunbelt has shifted the geographic center of gravity of the U.S. population steadily westward over time (Figure 3.43).

The most important type of internal migration is rural-urban migration, which is usually for economic motives. The relocation of farm workers to industrial urban centers was prevalent in developed countries during the nineteenth century. Since World War II, migration to large urban centers has been a striking phenomenon in nearly all developing countries. Burgeoning capital cities, in particular, have functioned as magnets attracting migrants in search of "the good life" and employment.

In highly urbanized countries, intermetropolitan migration is increasingly important. Although many migrants to cities come from rural areas and small towns, they form a decreasing proportion. Job mobility is a major determinant of intercity migration. So too is ease of transportation, especially air transportation. For intermetropolitan migrants from New York, the two most popular destinations are Miami and Los Angeles.

Modeling Migration

Because migration is important, there is a long history of trying to model it analytically. The intensity of a migration flow is affected by migrant characteristics, political restrictions, and distance. In the nineteenth century, the British sociologist Ravenstein studied migration in England and concluded that most people move short distances and that the frequency of moves declines with distance. This idea was further refined through the gravity model.

The Gravity Model

Newton's law of gravitation states that any two objects in space attract one another according to a force that is proportional to the product of their masses and inversely proportional to the square of the distance separating them. Thus, Newton's law of gravitation can be expressed as the force of attraction F, which is equal to M_i, the mass of the first body, times M_j, the mass of the second body, divided by the distance separating i and j, which is given as:

$$F = K \cdot M_i \cdot M_j / d^2,\ d_{ij}^2$$

where K is the gravitational constant. Force equals mass of i times mass of j divided by distance between i and j.

The *gravity model* helps to understand the volume of migration flows between two places, i and j. The flow,

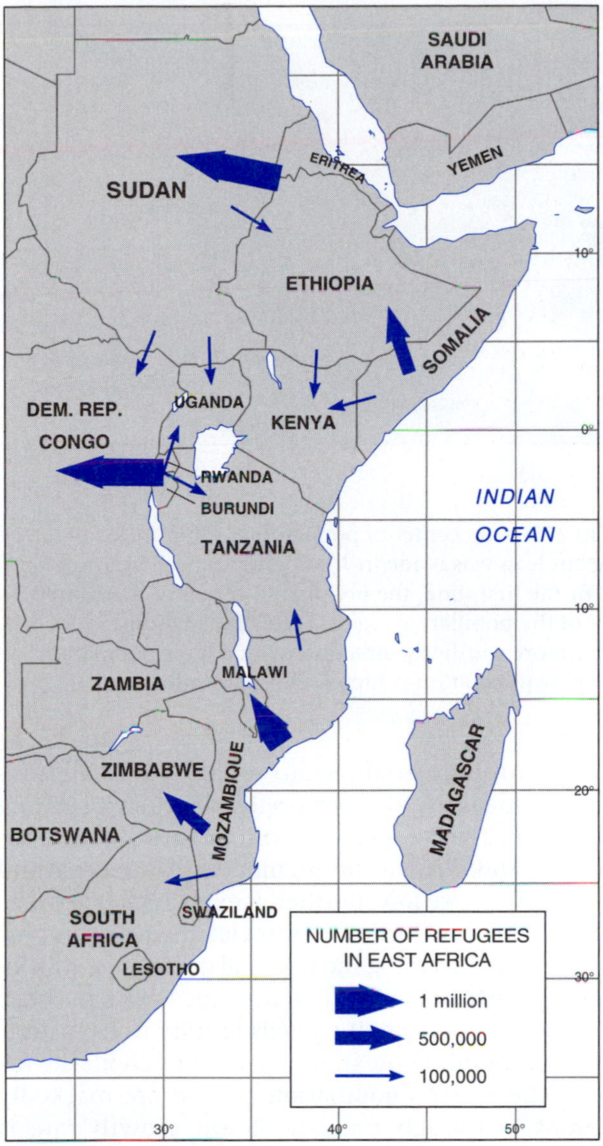

FIGURE 3.41
Arrows represent involuntary migration in East Africa in the late twentieth century. Eritreans, Ethiopians, Somalis, and Sudanese have been forced to move because of raging civil wars. To escape civil war, Hutus have been forced to migrate from Rwanda, and residents of Mozambique have fled to Malawi and other neighboring states.

or interaction, I_{ij}, is equal to the mass or population, P_i, of the first place, times the mass or population, P_j, of the second place, divided by the distance between the two, raised to a power given by

$$I_{ij} = P_i \cdot P_j \text{ divided by } d_{ij} \text{ squared.}$$

The exponent of the distance variable (which can be measured in time or cost rather than simply distance) reflects the friction of distance that migrants encounter when they move; higher exponents indicate that movements are relatively short over space. In this way the model captures the important roles played by city size and distance of movement.

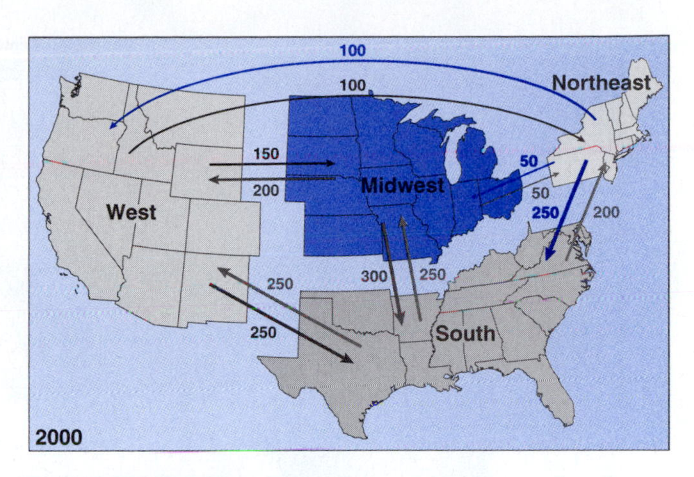

FIGURE 3.42
Migration flows among the four major census regions of the United States. While all regions exchange people with each other, the largest flows are to and from the South. These flows are one factor in the growth of the Sunbelt (South and West), which is now more populous than the Snowbelt or Rustbelt (Northeast and Midwest).

The gravity model can be summarized by two principles: (1) Larger places have a greater drawing power for flows of commodities, individuals, and information than smaller ones; and (2) places that are more distant have a weaker attraction for one another than closer places. The gravity model is limited to flows between places, taking two places at a time, and while it may be useful in more refined versions to predict migration flows (say in traffic engineering), it is really more of a statistical description than a causal explanation.

Summary

Because people are the single most important element in the world economy, it is essential to learn about population distribution, qualities, and dynamics. In this chapter, we began by examining the uneven distribution of people over the earth's surface. The vast majority of the 6.5 billion people alive today live in the developing world, particularly in Asia, where more than half reside. We emphasized that the distribution of people is a reflection of centuries, or even millennia, of uneven economic development, particularly following the Industrial Revolution and the formation of global colonial empires in the eighteenth and nineteenth centuries, which dramatically concentrated populations in cities and along the coasts. We noted population density is not an adequate variable to account for economic well-being: Some of the world's poorest countries are sparsely inhabited, and some of the wealthiest, such as the Netherlands, are very dense.

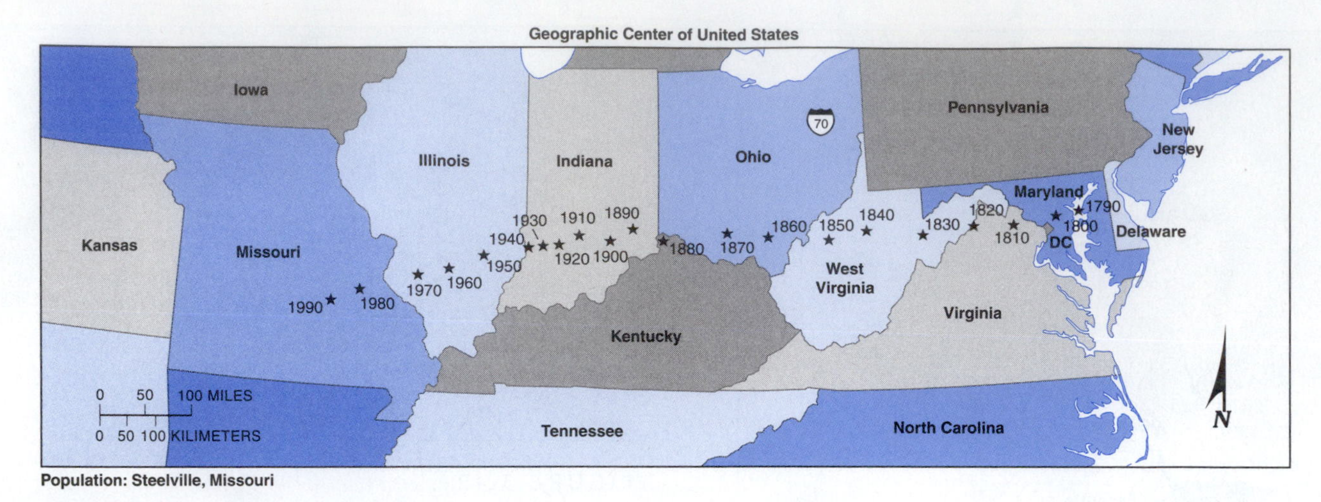

FIGURE 3.43
The population center of the United States. Since the first U.S. census in 1790, the center of population of the United States has moved steadily west and, at a somewhat slower pace, south. The giant leap was made in 1960 when Hawaii and Alaska were added to the United States as the forty-ninth and fiftieth states. For the first time, the population center of the United States moved west of the Mississippi River in 1980. In 2000, the center of the population was at Steelville, Missouri, 100 miles southwest of St. Louis. The new pattern seems to have shifted in a more southerly direction, with a large population attraction to the southwestern United States, especially California. The new population center is drifting southwest.

We also examined the processes of population change. The two major components of population change are migration and natural increase. The principal force affecting world population distribution used to be migration; now it is natural increase, the difference between fertility and mortality rates. This chapter explored the nature of population structures, particularly the age-sex distribution portrayed by population pyramids. This device is useful in contrasting the distribution of people by age and sex among different countries or the same country over time, particularly to reveal how changing economic circumstances, by changing fertility and mortality rates, create larger or smaller pools of young and elderly. This line of thought is useful in forecasting the future population status of regions or countries, including, for example, the changing size and composition of labor forces.

Malthusian conceptions of population change held that the growth of people must inevitably outstrip the resource base of the planet, or parts thereof. While this position is useful in noting that uncontained population growth cannot go on unchecked indefinitely, it also falls victim to simplistic views of why people have children and was derailed by the productivity growth of the Industrial Revolution. Indeed, by focusing on population growth as the source of the world's complex problems, Malthusianism tends to blame the victim (i.e., the poor) and ignore other, more important political and economic forces.

The demographic transition offers a superior model of population growth in that it links fertility and mortality rates, and thus natural growth, to the economic dynamics of industrialization, urbanization, and expanding capitalism. This perspective offers a compelling explanation as to why crude birthrates are high in poor countries and why in the premodern context natural growth rates were low. Further, it explains the declines in death rates and birthrates associated with economic development, including the essential question as to why couples have fewer children as their incomes rise. Thus, unlike Malthusianism, it embeds fertility in its historical and economic context. Its conclusions about the future of the world's population growth are markedly different. Although the population growth rate is falling, the world's population is projected to increase for decades to come, due to the large momentum built into the vast and youthful population of the developing world.

In addition to fertility and mortality, migration is a major force in shaping the geography of population. Excluding involuntary migration such as slavery or impelled migration (typically refugees from wars), we noted that spatial discrepancies in economic opportunities are the major forces driving migration. The causes include both push and pull factors, but typically center upon unemployment rates and average incomes. The great transcontinental migration streams of the nineteenth century have given way to flows from the developing to the developed world. Young people, particularly males, constitute the largest group of migrants. Migration has important impacts on local labor markets, affecting the supply of labor, and thus wage rates. Thus, economic and demographic forces are fused unevenly over the geographic landscape.

STUDY QUESTIONS

1. Summarize the spatial distribution of the world's population.
2. How is urbanization related to level of economic development?
3. Where will most of the world's population growth occur in the twenty-first century? Why?
4. Define *fertility rates* and *mortality rates.*
5. Summarize the major causes of international migration.
6. What were the two largest periods of immigration in the United States?
7. What is a population pyramid?
8. How and why does educational level affect demographic behavior?
9. Describe the changing composition of the U.S. labor force over time in terms of primary, secondary, and tertiary sector.
10. How does population size affect economic development?
11. How does Generation X compare in size to the baby boom?
12. What is the gravity model?

KEY TERMS

achieved characteristics
ascribed characteristics
baby boom
baby bust
birthrate (crude)
capital
carrying capacity
death rate (crude)
demographic transition
doubling time
fertility rate (total)
geographic information system (GIS)
human capital
human suffering index
infant mortality rate
labor force
labor migration theory
law of diminishing returns
limits to growth
migration
natural increase
negative population growth
neo-Malthusian
optimum population size
physiologic density
population composition
population pyramid
primary economic activity
primate city
push-and-pull factors
rate of natural increase
regional growth forest
rule of seventy
secondary economic activity
tertiary economic activity
unemployment
urbanization
zero population growth (ZPG)

SUGGESTED READINGS

Kirk, D. 1996. "Demographic Transition Theory." *Population Studies* 50: 361–88.

Newbold, K. 2006. *Six Billion Plus: Population Issues in the Twenty-first Century.* Boulder, CO: Rowman and Littlefield.

Peters, G. and R. Larkin. 2005. *Population Geography: Problems, Concepts and Prospects.* New York: Kendall/Hunt.

World Resources Institute. 2004. *World Resources.* New York: Oxford University Press.

WORLD WIDE WEB SITES

DEMOGRAPHY AND POPULATION STUDIES
http://lcweb.loc.gov/homepage/lchp.html
Library of Congress; links libraries and online catalogs.
http://www.worldbank.org
World Bank.

UPDATED U.S. POPULATION ESTIMATES [.ZIP]
http://www.census.gov/population/www/projections/popproj.html
The U.S. Census Bureau, in association with the Federal-State Cooperative Program for Population Estimates (FSCPE), has recently released updated population estimates on the national, state, and county level. All files are available as ASCII text, with some of the larger ones compressed in .zip format. Documentation and layout of files is available.

U.S. CENSUS INTERNATIONAL DATABASE
http://www.census.gov/ipc/www/idbnew.html

THE CENSUS BUREAU
http://www.census.gov
The Census Bureau Web site was designed to enable "intuitive" use and is intended to be visually appealing, concise, and quick loading. It was designed so users can effectively locate and utilize the resources the site has to offer, such as the "Population Clock" and its small search engine.

THE INTER-UNIVERSITY CONSORTIUM FOR POLITICAL AND SOCIAL RESEARCH
http://www.icpsr.umich.edu
ICPSR provides a wealth of data on national, state, and local elections; Census information; political behavior attitudes; poll and survey data.

LIBRARY OF CONGRESS
http://www.loc/.gov
Astounding variety of resources and exhibits.

MAPQUEST
http://www.mapquest.com
Maps of cities and regions in the United States and the world.

POPULATION REFERENCE BUREAU
http://www.prb.org//Content/NavigationMenu/PRB/Journalists/World_Population_Clock/2005_World_Population_Clock.htm
World Population Clock.

CHAPTER

4

Resources and Environment

OBJECTIVES

- To describe the nature, distribution, and limits of the world's resources
- To examine the nature of world food problems and to make you aware of the difficulties of solving them
- To describe the distribution of strategic minerals and the time spans for their depletion
- To consider the causes and consequences of the energy crisis and to examine present and alternative energy options
- To examine the nature and causes of environmental degradation
- To compare and contrast growth-oriented and balance-oriented lifestyles

"Petroleum storage facilities reflect its critical importance to the world economy."

Economic growth and prosperity depend partly on the availability of natural resources and the quality of the environment. There is growing concern that the consumption of inputs and goods in developed countries, and increasingly in developing countries, is depleting the world's stock of resources and irreparably degrading the natural environment. What can be done to effectively manage resources and protect the environment?

Optimists believe that economic growth in a market economy can continue indefinitely; they see relatively few limits in raw materials and great gains in technological productivity. In contrast, pessimists assert that there are inherent limits to growth imposed by the finiteness of the earth—by the fact that air, water, minerals, space, and usable energy sources can be exhausted or ecosystems overloaded. They believe these limits are near and, as evidence, point to existing food, mineral, and energy shortages and to areas now beset by deforestation and erosion.

Some scholars think that the long running debate on resources and the environment, which waxes and wanes over time, is counterproductive, evading practical issues that demand our immediate attention. We need to keep our purposes in mind and try to understand how to achieve our ends. If our purpose is to create a habitable and sustainable world for generations to follow, how can we redirect present and future output to serve that end? One solution is to transform our present a *growth-oriented lifestyle,* which is based on a goal of ever-increasing production and consumption, to a *balance-oriented lifestyle* designed for minimal environmental impact. A balance-oriented lifestyle would include an equitable and modest use of resources, a production system compatible with the environment, and appropriate technology. The aim of a balance-oriented world economy is maximum human well-being with a minimum of material consumption. Growth occurs, but only growth that truly benefits all people, not just the elite few. However, what societies, rich or poor, are willing to dismantle their existing systems of production to accept a lifestyle that seeks satisfaction more in quality and equality than in quantity and inequality? Are people programmed for maximum consumption by a value system constantly reinforced through advertising willing to change their ways of thinking and behaving?

This chapter, which discusses growth-oriented versus balance-oriented philosophies of resource use, deals with the complex components of the population-resources issue. Have population and economic growth rates been outstripping food, minerals, and energy? What is likely to happen to the rate of demand for resources in the future? Could a stable population of 10 billion be sustained indefinitely at a reasonable standard of living utilizing currently known technology? These are the salient, critical questions with which this chapter is concerned.

Resources and Population

Popular opinion in the industrial West generally appreciates the need to reduce population growth but overlooks the need to limit economic growth that exploits resources. Most people in the economically developed world suffer from a view that holds resources are limitless and do not appreciate that our rapid consumption of them ultimately threatens our affluent way of life. The First World is, in short, liquidating the resources on which our way of life was built. The growth of some developing countries is aggravating the situation. Their growing populations put increasing pressure on resources and the environment, and many aspire to affluence through Western-style urban industrialization that depends on the intensive use of resources. Poor countries generally do not have the means for running the high-energy production and transportation systems manifest in the industrial West. Even by conservative estimates, the production of a middle-class basket of goods requires six times as much in resources as a basket of essential or basic goods. The expansion of GNP through the production of middle-class baskets means that only a minority of people in poor countries would enjoy the fruits of economic growth. Resource constraints prevent the large-scale production of consumer goods for the growing populations of the developing countries.

However, numerous measures of material well-being (e.g., per capita incomes, calories consumed, life expectancy) show that people in most, but not all, countries are better off today than their parents were. But there are problems with this optimistic assessment. These improvements are based on averages; they say nothing about the distribution of material well-being. Another difficulty is that the world may be achieving improvements in material well-being at the expense of future generations. This would be the case if economic growth were using up the world's resource base or environmental carrying capacity faster than new discoveries and technology could expand them.

Carrying Capacity and Overpopulation

The population-resources problem is much debated, particularly during periods of economic shortages and rising prices. Neo-Malthusian pessimists believe that the world will eventually enter a stationary state at *carrying capacity,* which is the maximum population that can be supported by available resources. They point to recurring food crises and famines in Africa as a result of overpopulation. However, carrying capacity, an idea borrowed from ecology, is simplistic in that it ignores the historical, political, and technological context in which the production and consumption of goods occurs. Human beings are not mindless products of an unchanging nature, and are capable of modifying their environment and altering the constraints and opportunities it presents. On the other hand, optimists believe in the saving grace of modern technology. Technological advances in the last 200 years have raised the world's carrying capacity, and future technical innovations as well as the substitution of new raw materials for old hold the promise of raising carrying capacity still further.

The answer to the population-resources problem also depends on the standard of living deemed acceptable. To give people a minimal quality of life instead of one resembling the American middle class would require vast quantities of additional resources. The establishment of an economy that provides for the basics of life—sufficient food, housing, education, transportation, and health care—depends on our capacity to develop alternatives to the high-energy, material-intensive production technologies characteristic of the industrial West. Already, there are outlines of a theory of resource use suited to the needs of a basic goods economy. Some of the main ideas are (1) the adoption of a sun-based organic agriculture; (2) the use of renewable sources of energy; (3) the use of appropriate technology, labor-intensive methods of production, and local raw materials; and (4) the decentralization of production in order to increase local self-reliance and minimize the transport of materials. These productive forces would minimize the disruption of ecosystems and engage the unemployed in useful, productive work. Typically, economies that produce essential goods for human consumption face neither excessive unemployment nor overpopulation. Moreover, secure supplies of basic goods provide a strong motivation for reducing population size, as families no longer require many children to ensure economic prosperity.

Some of these African mothers can do little to save their children's lives. Children are suffering the complications of undernutrition and malnutrition. Asia has managed to increase production fast enough to stay well ahead of population growth, but Latin America and Southwest Asia, the Middle East, and North Africa have barely managed to stay even with the population. The worst report is from Africa, where population growth is outstripping food production year by year. Per capita food production in Africa has been falling steadily for the last 25 years. Many factors contribute to these fluctuations, including drought, changing world prices, and civil and ethnic unrest.

Types of Resources and Their Limits

All economic development comes about through the use of human resources (e.g., labor power, skills, and intelligence). In order to produce the goods and services people demand in today's global economy, we need to obtain natural resources. What are natural resources and what are their limits?

Resources and Reserves

Natural resources have meaning only in terms of historically specific technical and cultural appraisals of nature and are defined in relation to a particular level of development. *Resources,* designated by the larger box in Figure 4.1, include all substances of the biological and physical environment that may someday be used under specified technological and socioeconomic conditions. Because these conditions are always subject to change, we can expect our determination of what is useful to also change. For example, uranium, once a waste product of radium mining of the 1930s, is now a valuable ore. Taconite ores became worthwhile in Minnesota only after production from high-grade, nonmagnetic iron ores declined in the 1960s.

At the other end are reserves, designated by the upper-left-hand box in Figure 4.1. *Reserves* are quantities of resources that are known and available for economic exploitation with current technologies at current prices. When current reserves begin to be depleted, the search for additional reserves is intensified. Estimates of reserves are also affected by changes in prices and technology. *Projected reserves* represent estimates of the quantities likely to be added to reserves because of discoveries and changes in prices and technologies projected to occur within a specified period, for example, 50 years.

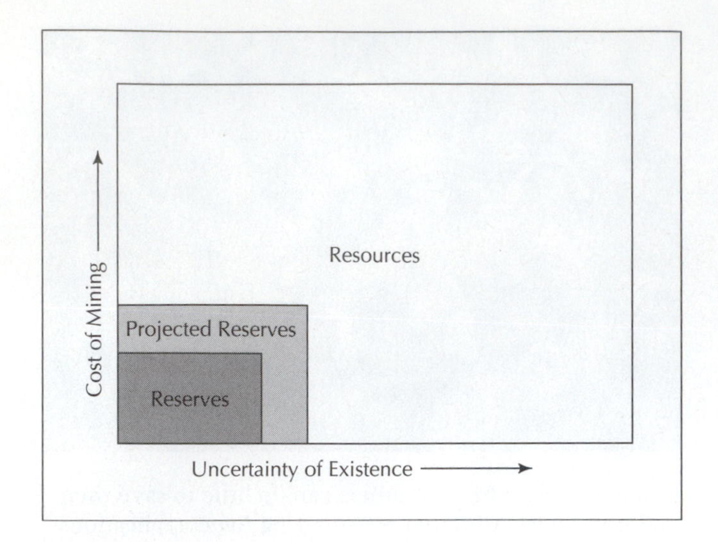

FIGURE 4.1
Classification of resources. Resources include all materials of the environment that may someday be used under future technological and socioeconomic conditions. Reserves are resources that are known and available with current technologies and at current prices. Projected reserves are reserves based on expected future price trends and technologies available.

Renewable and Nonrenewable Resources

A major distinction is between nonrenewable and renewable resources. *Nonrenewable resources* consist of finite masses of material, such as fossil fuels and metals, which cannot be used without depletion. They are, for all practical purposes, fixed in amount, or in some cases, such as soils, because they form slowly over time. Consequently, their rate of use is important. Large populations with high per capita consumption of goods deplete these resources fastest.

Many nonrenewable resources are completely altered or destroyed by use; petroleum is an example. Other resources, such as iron, are available for recycling. Recycling expands the limits to the sustainable use of a nonrenewable resource. At present, these limits are low in relation to current mineral extraction.

Renewable resources are those resources capable of yielding output indefinitely without impairing their productivity. They include *flow resources* such as water and sunlight, and *stock resources* such as soil, vegetation, fish, and animals. Renewal is not automatic, however; resources can be depleted and permanently reduced by misuse. Productive fishing grounds can be destroyed by overfishing. Fertile topsoil, destroyed by erosion, can be difficult to restore and impossible to replace. The future of agricultural land is guaranteed only when production does not exceed its maximum sustainable yield. The term *maximum sustainable yield* means maximum production consistent with maintaining future productivity of a renewable resource.

In our global environment, the misuse of a resource in one place affects the well-being of people in other places. The misuse of resources is often described in terms of the *tragedy of the commons,* a term coined by biologist Garrett Hardin in 1968. This metaphor refers to the way public resources are ruined by the isolated actions of individuals, which occurs when the costs of actions are not captured in market prices. Originally it referred to the tendency of shepherds to use common grazing land; as each one sought as much of the commons as possible, it became overgrazed. Similarly, people who fish are likely to try to catch as many fish as they can, reasoning that if they don't, others will. Thus, it exemplifies a market failure, a problem generated by individual actors who behave "rationally" but collectively create an irrational and self-destructive outcome. Similarly, dumping waste and pollutants on public waters

The Kenyan rangelands on which these herders cattle graze are in jeopardy. With growing grazing pressures, more than 60 percent of the world's rangelands and at least 80 percent of African, Asian, and Middle Eastern rangelands are now moderately to severely desertified. About 65 million hectares of once productive land in Africa have become desert during the last 50 years.

and land or into the air is the cheapest way to dispose of worthless products. Firms are generally unwilling to dispose of these materials by more expensive means unless mandated by law.

Sometimes resources are unavailable, not because they are depleted but because of politics. Resources are under the control of sovereign nation-states. Many wars in the twentieth century have been resource wars. For example, Japan invaded Korea and Taiwan in the 1890s largely in order to obtain arable land and coal. The Iraqi invastion of Kuwait in 1990 and the U.S. invasion of Iraq in 2003 were largely motivated by concerns over the region's oil supplies. In the Middle East, fierce national rivalries make water a potential source of conflict: While some parts are blessed with adequate water supplies, most of the region is insufficiently supplied. Some observers predict that political tension over the use of international rivers, lakes, and aquifers in the Middle East may escalate to war in the next few years.

Food Resources

Thanks to scientific advances in farming, world food production has been increasing faster than population (Figure 4.2). While there is sufficient food to feed everyone in the world, there are huge geographic variations in people's access to a sufficient number and quality of calories (Figure 4.3). The populations of the industrialized world are generally well fed; indeed, in the United States, the major dietary problem is an overabundance of calories and an epidemic of obesity. In the developing world, in contrast, hundreds of millions of people worldwide still go hungry daily. With demand for food expected to grow at 4 percent per year over the next 20 to 30 years, the task of meeting that need will be more difficult than ever before. A record explosion in the world's population coupled with the problem of poverty threaten the natural resources on which agriculture depends, such as topsoil. To make matters worse, environmental degradation perpetuates poverty, as degraded ecosystems diminish agricultural returns to poor people.

The gulf between the well fed and the hungry is vast. Average daily calorie consumption is 3300 in developed countries and 2650 in developing countries (Table 4.1). But these are average figures. There are people in the developed world who go hungry and some people in Africa with plenty to eat. Averages mask the extremes of *undernutrition*—a lack of calories—and overconsumption. Even with a high-calorie satisfaction, people may suffer from *chronic malnutrition*—a lack of enough protein, vitamins, and essential nutrients. The most important measure in assessing nutritional standards is the daily per capita availability of calories, protein, fat, calcium, and other nutrients. In the world today, the sharpest nutritional differences are not from country to country or from one region to another within countries. They are between rich and poor people. The poor of the earth are the hungry, and those with the least political power often suffer in terms of an insufficient food supply.

Hunger among the poor of the world is often attributed to deforestation, soil erosion, water-table depletion,

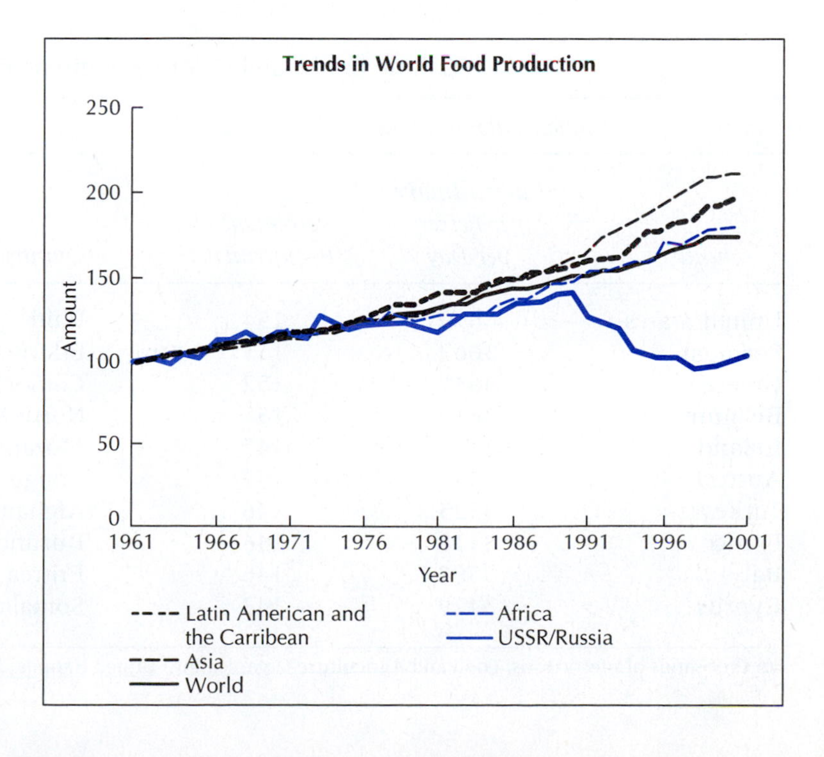

FIGURE 4.2
Trends in world food production, 1961–2001. Note the precipitous drop in food production in the former Soviet Union.

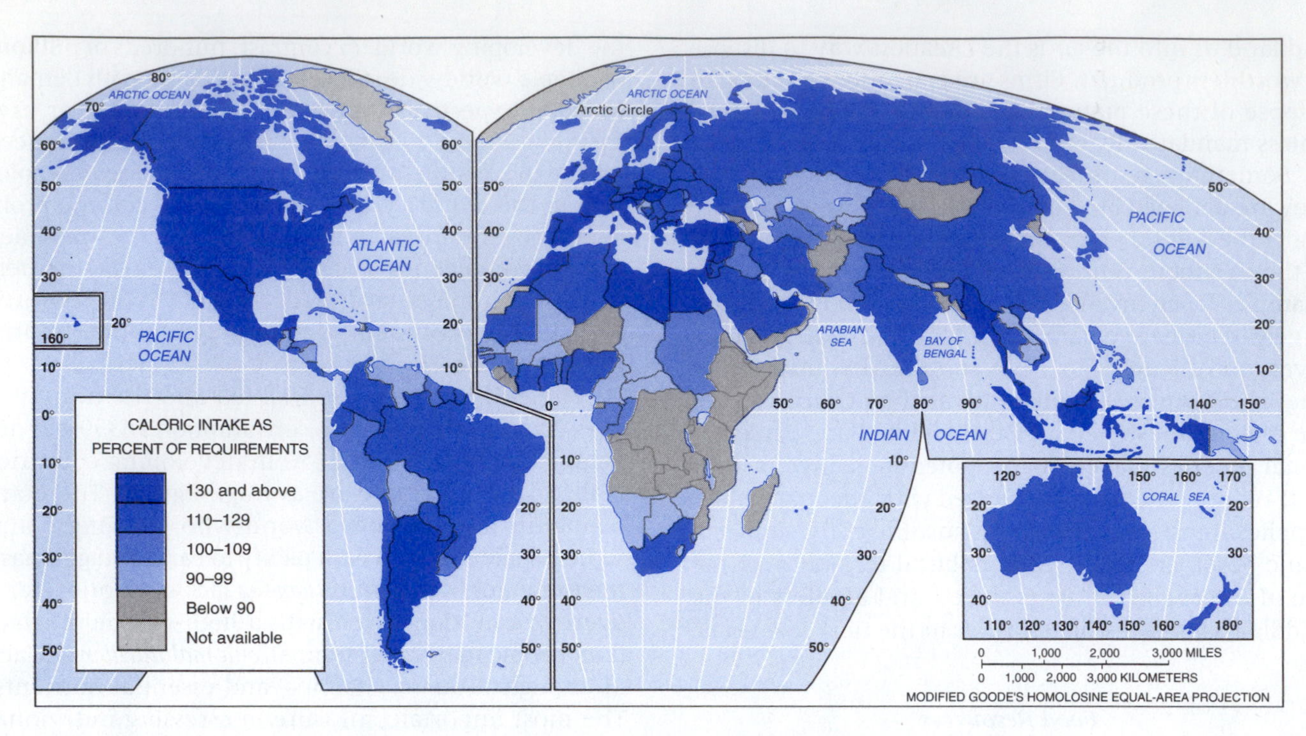

FIGURE 4.3
Caloric intake as a percentage of adult daily requirements. Highly developed regions of the world receive, on the average, 130 percent of the daily caloric requirements (2400 calories per day) set by the United Nations Food and Agricultural Organization (FAO). Some countries in South America, South Asia, and many countries in Africa receive less than 90 percent of the daily caloric requirements needed to sustain body and life. Averages must be adjusted according to age, gender, and body size of the person and by regions of the world. Although it appears from this map that the great majority of world populations are in relatively good shape with regard to calories per capita (food supply), remember that averages overshadow destitute groups in each country that receive less than their fair share. Again, the situation is most severe in the Sahel or center belt of Africa.

Table 4.1
Calorie Intake and Calorie Requirement Satisfaction, 2004.

Highest Calorie Intake			*Lowest Calorie Intake*			
Country	*Calorie Intake per Person per Day*	*Percentage of Requirements*	*Country*	*Calorie Intake per Person per Day*	*Percentage of Requirements*	*Food Aid**
United States	3699	154	Haiti	1869	79	51
Portugal	3667	153	Ethiopia	1858	77	23
Greece	3647	152	Comoros	1858	77	44
Belgium	3619	151	North Korea	1837	76	0
Ireland	3537	147	Mozambique	1832	76	33
Austria	3536	147	Congo	1755	73	486
Turkey	3525	146	Afghanistan	1745	72	51
France	3518	146	Burundi	1685	70	136
Italy	3507	146	Eritrea	1622	67	1304
Cyprus	3429	143	Somalia	1566	65	32

*In thousands of metric tons. Food and Agriculture Organization, United Nations, 2005.

the frequency and severity of droughts, and the impact of storms such as hurricanes. Although the environment does have a bearing on the food problem, it has limited significance compared with the role of social conditions such as war and a world economy whose rules are tilted against the impoverished. Subsidized agricultural exports from the United States, for example, have bankrupted millions of farmers in the developing world, reducing those countries' ability to feed themselves.

Population Growth

Population growth is one of many causes of the food problem, and Malthusian views often influence the public's opinion of this issue. However, presently, at the global level, there is no food shortage. In fact, world food production grew steadily from 1961 to 2005. Even over the next several decades, production increases, assuming continuing high investments in agricultural research, are likely to be sufficient to meet effective demand and rising world population. However, some are more pessimistic about future world food production. They argue that food production will be constrained by the limits to the biological productivities of fisheries and rangelands, the fragility of tropical and subtropical environments, massive overfishing of the world's oceans, the increasing scarcity of fresh water, the declining effectiveness of additional fertilizer applications, and social disintegration in many developing countries.

The success of global agriculture has not been shared equally. In Africa, per capita food production has not been able to keep up with population growth. By contrast, Asia and to a lesser extent Latin America have experienced tremendous success in per capita terms. The reasons for this are complex, and have to do with the relative equality in patterns of land ownership, government policies toward farmers (e.g., price ceilings on agricultural crops), the respective ability of countries to build infrastructures and extend credit to small farmers, and the role of different states in the world economy.

The food and hunger problem is most severe in Africa. Fifteen countries are experiencing exceptional food emergencies. Of the 28 countries with food-security problems, 23 are in sub-Saharan Africa (Figure 4.4). Indeed, famine, the most extreme expression of poverty, is now mainly restricted to Africa. The fact that famine has been held at bay for decades in Latin America and Asia suggests that famine can be eliminated. But how? Certainly, bringing an end to Africa's multiple civil wars would go a long way toward eradicating famine. Africa has witnessed countless brutal conflicts that have killed tens of millions of people. Such conflicts divert resources from civilian use, interrupt the production of crops, terrorize populations, destroy the

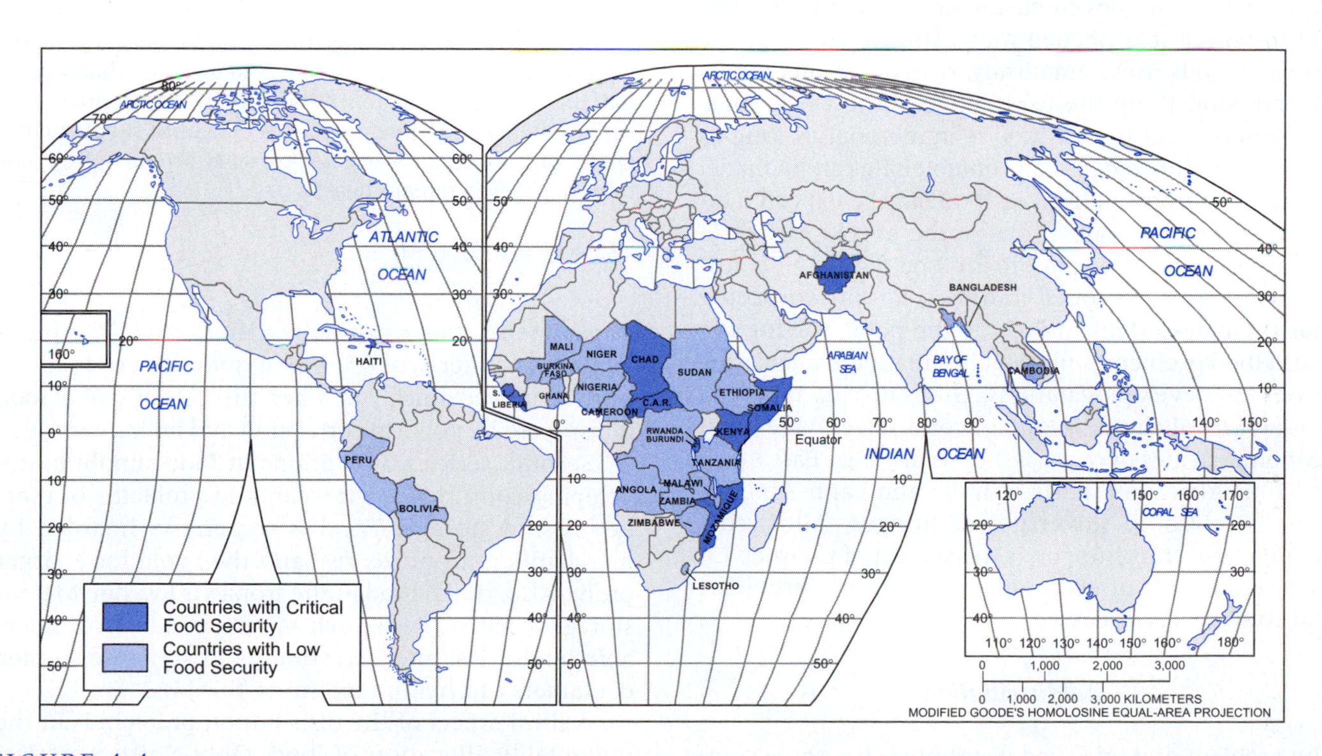

FIGURE 4.4
Developing countries with low or critical food security indexes, 2005. Africa remains the continent most seriously challenged by food shortages. Fifteen countries in the region are facing critical food emergencies. Of the 28 countries with household food security problems, 23 are in sub-Saharan Africa.

infrastructure, destabilize markets, and complicate the stability of the governments, creating famine and prohibiting the flow of developmental aid. But peace is not in itself a sufficient condition for removing acute hunger. Appropriate policies and investments are needed to stimulate rural economic growth that underpins food security and to provide safety-net protection for the absolute poor. Rural infrastructure development, credit to farmers, and land redistribution are also necessary steps in this regard. Price controls on food crops create disincentives to produce, and heavily subsidized food imports from the developed world bankrupt farmers. Often elites in the developing world care more about their foreign bank accounts than the well-being of their own populations.

The pace of urbanization in the developing countries has also contributed to the food problem. In recent decades, hundreds of millions of people who previously lived in rural areas and produced some food have relocated to urban areas, where they must buy food. As a result of urbanization, there is a higher demand for food in the face of lower supply.

Poverty

The inequitable allocation of food is directly related to poverty, the single greatest cause of the hunger problem. Hungry people are inevitably poor people who lack the purchasing power to feed themselves. Under capitalism, food goes to customers who can afford it, not to where it is needed most. During famines, the prices of foods rise dramatically, with disastrous results for the poor. From the perspective of the world market, where food is produced is immaterial as long as costs are minimized and a profitable sale can be made. Thus, in the midst of hunger, food may be exported for profit. Since the populations of the developed world can afford to pay much more for food than their counterparts in less developed countries, it is not surprising that the market fails to include the poor. Solving the world food problem is ultimately a matter of alleviating poverty in developing nations. This is no easy task, and while parts of the developing world have made great economic strides over the last 40 years (e.g., East Asia), much of Africa and parts of India and Latin America remain mired in poverty and hunger. Alleviating poverty, and thus hunger, is the subject of Chapter 13, in which a host of economic development problems and strategies is discussed.

Maldistribution

The problem of world food distribution has three components. First, there is the problem of transporting food from one place to another. Although transport systems in developing countries lack the speed and efficiency of those in developed countries, they are not serious impediments under normal circumstances. The problem arises either when massive quantities of food aid must be moved quickly or when the distribution of food is disrupted by political corruption and military conflict.

Third World farmers, such as these in Indonesia, depend on high rice yields. Rice is the staple food for more than one-half the world's population. While rice and other grains supply energy and some protein, people must supplement grains with fruits, nuts, vegetables, dairy products, fish, and meat in order to remain healthy.

Second, serious disruptions in food supply in developing countries are traceable to problems of marketing and storage. Food is sometimes hoarded by merchants until prices rise and then sold for a larger profit. Also, much food in the tropics is lost due to poor storage facilities. Pests such as rats consume considerable quantities, and investments in concrete storage containers can help to minimize this loss.

A third aspect of the distribution problem is in the inequitable allocation of food. Only North America, Australia, and Western Europe have large grain surpluses. But food grain is not always given when it is most needed. Food aid shipments and grain prices are inversely related. Thus, U.S. food aid was low around

1973, a time of major famine in the Sahel region of Africa, because cereal prices were at a peak.

Closely associated with poverty as a cause of hunger in developing countries is the structure of agriculture, including land ownership. Land is frequently concentrated in a few hands. In Bangladesh, less than 10 percent of rural households own more than 50 percent of the country's cultivable land; 60 percent of Bangladesh's rural families own less than 2 percent. A similar situation applies in Latin America (Chapter 14). Many rural residents own no land at all. They are landless laborers who depend on extremely low wages for their livelihoods. But without land, there is often no food.

Civil Unrest and War

Political conflict is an important cause of hunger and poverty. Occasionally governments withhold food to punish rebellious populations. Devastating examples of depriving food to secessionist areas include the government in Nigeria starving the Biafrans in the 1970s and the government in Ethiopia starving the Eritreans into submission, with 6 million people dying in the process. In Sudan, the Arab government's war against the African population in Darfur has led millions to starve. In Zimbabwe, the government of Robert Mugabe denied food to his opponents in order to quash domestic opposition. Civil wars, which are frequent in developing countries whose political geographies were shaped by colonialism and which have unstable governments, devastate agricultural production. Without a stable political environment, the social mechanisms necessary to produce and distribute food to the hungry cannot operate.

Environmental Decline

As population pressure increases on a given land area, the need for food pushes agricultural use to the limits, and marginal lands, which are subject now to *desertification* and *deforestation* (Figure 4.5), are brought into production. Removal of trees allows a desert to advance, because the wind break is now absent. The cutting of trees also lowers the capacity of the land to absorb moisture, which diminishes agricultural productivity and increases the chances of drought. Desertification and deforestation are symptoms as well as causes of the food problem in developing countries. Natural resources are mined by the poor to meet the food needs of today; the lower productivity resulting from such practices is a concern to be put off until tomorrow.

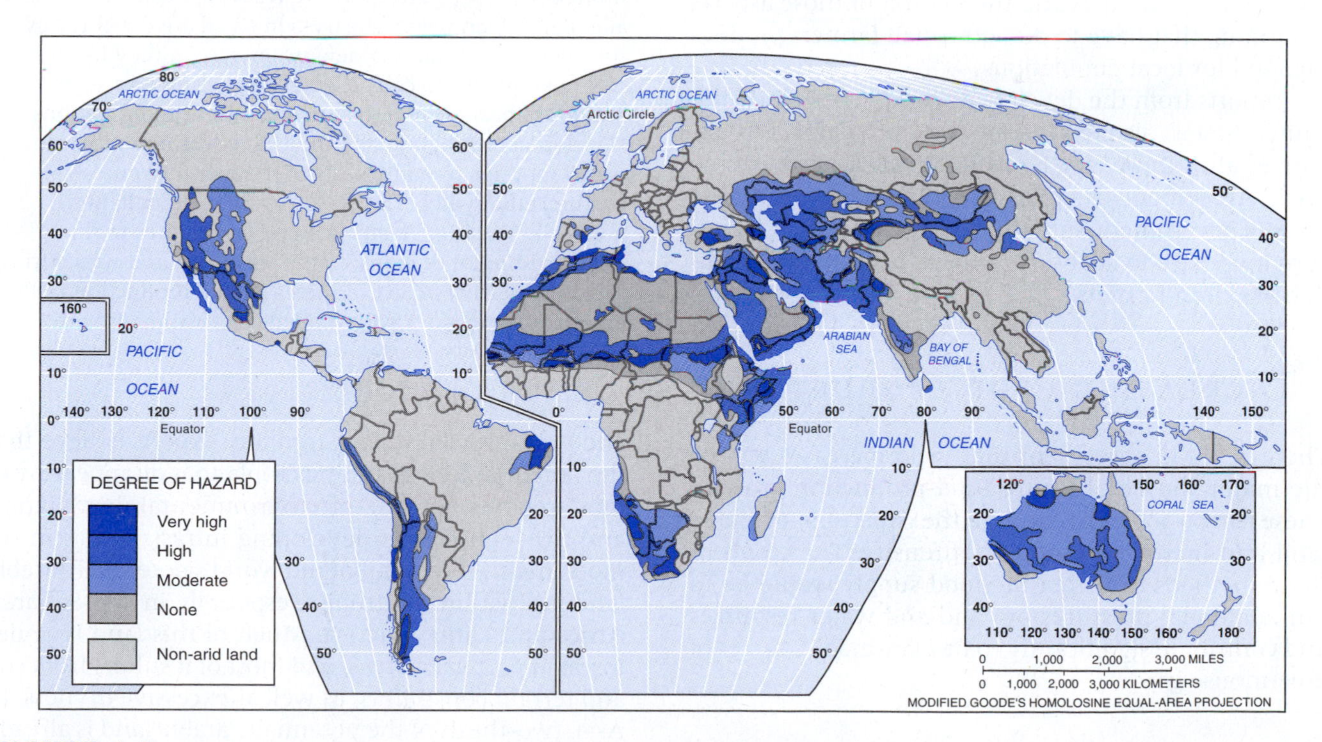

FIGURE 4.5
World desertification. The main problem is overuse by farmers and herdsmen. Approximately 10 percent of the earth's surface has lost its topsoil due to overuse of lands by humans, creating desertification. An additional 25 percent of the earth's surface is now threatened. Topsoil is being lost at a rate of approximately 30 billion metric tons per year. Approximately 20 million acres of agricultural land are wasted every year to desertification by agricultural overuse. When plants are uprooted by overplowing or by animals, the plants that stabilized shifting soil are removed. When the rains come, water erosion can wash away the remaining topsoil.

Government Policy and Debt

In many developing countries, government policies have emphasized investment in their militaries and cities at the expense of increasing agricultural production. In addition, some governments in Africa have provided food at artificially low prices in order to make food affordable in cities. While this practice keeps labor affordable for multinational corporations and placates the middle class, it robs farmers of the incentive to farm. Farmers cannot make a living from artificially low commodity prices.

The average debt of many developing countries runs into the billions. In 2002, the aggregate debt of African countries stood at $220 billion. Simply put, African countries have no surplus capital to invest in their infrastructure or food production systems. Instead, they have to enforce austerity, reducing levels of government services in support of economic growth, particularly agricultural growth. Debt repayments subsume a large share of foreign revenues, decreasing funds available for investment.

In recent decades, agriculture in developing countries has expanded. This expansion is in the export sector, not in the domestic food-producing sector, and it is often the result of deliberate policy. Governments and private elites have opted for modernization through the promotion of export-oriented agriculture. The result is the growth of an agricultural economy based on profitable export products and the neglect of those aspects of farming that have to do with small farmers producing food for local populations.

Imports from the developed world, particularly the United States, also exacerbate food problems. For example, after the passage of the North American Free Trade Agreement (NAFTA) in 1994, massive U.S. exports of government-subsidized corn caused the price of corn in Mexico to fall by 70 percent, bankrupting 1 million Mexican farmers.

Water for Chad. Water is an important ingredient to sustain human life. Fifty percent of the world's people do not have adequate, clean water. Villagers in Chad are delighted as the water pours out of a new water system they have worked together to construct. The system is part of an antidesertification project funded by the United Nations Development Program and the U.S. government. Acute water shortage in many parts of the world requires solutions that will be costly, technically difficult, and politically sensitive. Water scarcity contributes to the impoverishment of many countries in east and west Africa, threatening their ability to increase food production fast enough to keep pace with modern population growth.

Increasing Food Production

There is broad agreement that yield increases will be the major source of future food production growth. These can be achieved through the expansion of arable land and increased cropping intensity. The result of these methods of increasing food supply would be to put additional pressures on land and water resources and contribute significantly to human-made sources of greenhouse gases.

Expanding Cultivated Areas

The world's potentially farmable land is estimated to be about twice the present cultivated area. Vast reserves are theoretically available in Africa, South America, and Australia, and smaller reserves in North America, Russia, and Central Asia. However, many experts believe that the potential for expanding cropland is disappearing in most regions because of environmental degradation and the high cost of developing infrastructure in remote areas. About half of the world's potentially arable land lies within the tropics, especially in sub-Saharan Africa and Latin America. Much of this land is under forest in protected areas, and most of it suffers from soil and terrain constraints, as well as excessive dryness. In Asia, two-thirds of the potentially arable land is already under cultivation; the main exceptions are Indonesia and Myanmar. South Asia's agricultural land is almost totally developed.

The expansion of tropical agriculture into forest and desert environments contributes to deforestation and desertification. Since World War II, roughly half of the world's rain forests in Africa, Asia, and Latin America have disappeared. Conversion of this land to agriculture has entailed high costs, including the loss of livelihoods for the people displaced, the loss of biodiversity, increased carbon dioxide emissions, and decreased carbon storage capacity. *Desertification*—the growth of deserts due to humanly caused factors, typically on the periphery of natural deserts—threatens about one-third of the world's land surface and the livelihood of nearly a billion people. Many of the world's major rangelands are at risk. The main factor responsible for desertification is overgrazing, but deforestation (particularly the cutting of fuel wood), overcultivation of marginal soils, and salinization caused by poorly managed irrigation systems are also important influences. Deforestation and desertification are destroying the land resources on which the development of the developing countries depends.

Raising the Productivity of Existing Cropland

The quickest way to increase food supply is to raise the productivity of land under cultivation. Remarkable increases in agricultural yields have been achieved in developed countries through the widespread adoption of new technologies. Corn yields in the United States are a good example. Yields expanded rapidly with the introduction of hybrid varieties, herbicides, and fertilizers. Much of the increase in yields came through successive improvements in hybrids.

The approach for increasing yields in developed countries has been adopted in developing countries. This approach is known as the *Green Revolution,* in which new high-yielding varieties of wheat, rice, and corn are developed through plant genetics, including crops that grow more quickly, perhaps yielding several harvests per year, are more pest and drought resistant, and have higher protein content. The Green Revolution has had enormous impacts in Asia and Mexico, increasing the food supply, but it is not a panacea. It depends on machinery, for which the poor lack sufficient capital to buy. It depends on new seeds, which poor farmers cannot afford. It depends on chemical fertilizers, pesticides, and herbicides, which have contaminated underground water supplies, as well as streams and lakes. It depends on large-scale, one-crop farming, which is ecologically unstable because of its susceptibility to pestilence. It depends on controlled water supplies, which have increased the incidence of malaria, schistosomiasis, and other diseases. It is confined largely to a group of 18 heavily populated countries, extending across the tropics and subtropics from South Korea to Mexico (Figure 4.6). It is also benefiting countries that include half of the world's population. This approach

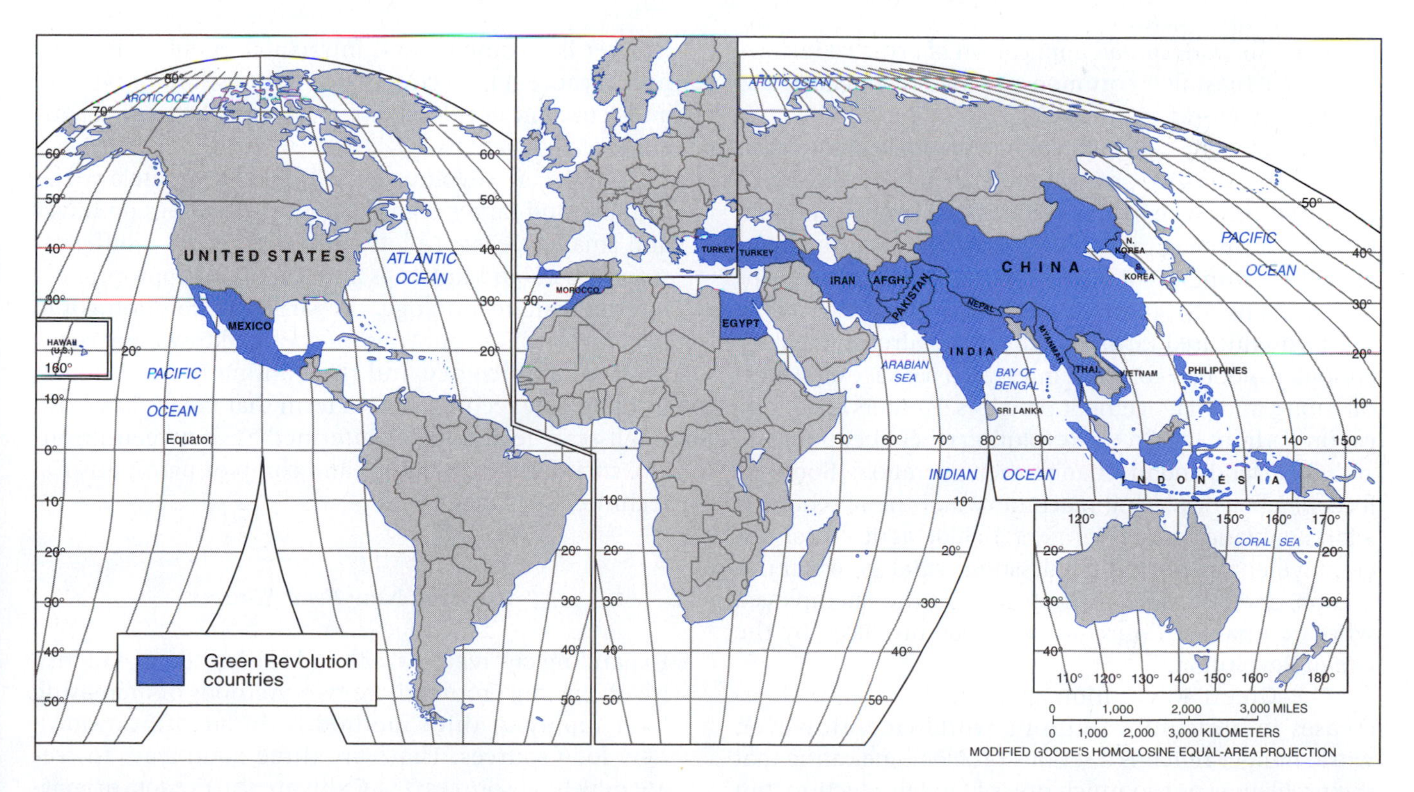

FIGURE 4.6
The chief benefiting countries of the Green Revolution. The Green Revolution was the result of plant scientists genetically developing high yielding varieties of staple food crops such as rice in East Asia, wheat in the Middle East, and corn in Middle America. By crossing "super strains" that produced high yields with more genetically diverse plants, both high yield and pest resistance were introduced.

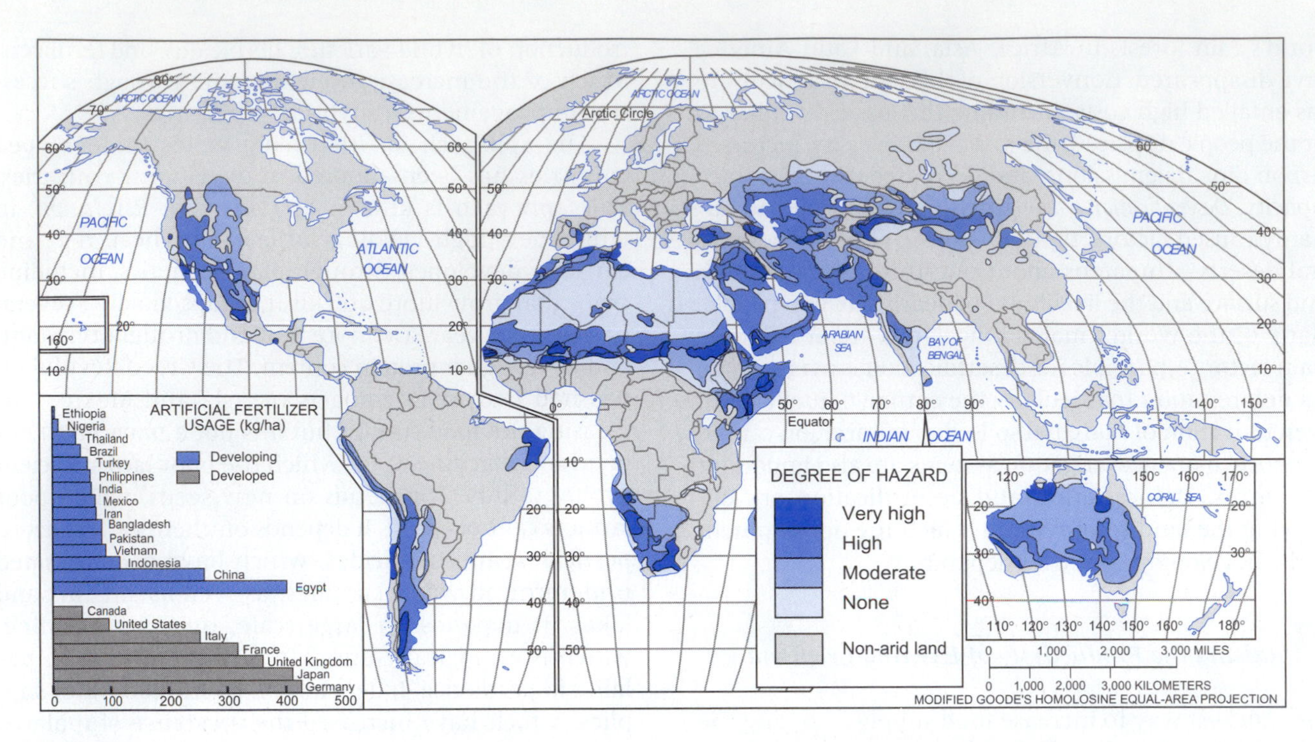

FIGURE 4.7
Artificial fertilizer usage. The application of artificial fertilizers, as opposed to natural ones obtained from people and animals, may enhance agricultural productivity but also makes economies more dependent on petroleum.

involves the widespread application of artificial fertilizers, an increasingly common practice throughout the developing world (Figure 4.7).

Politically, the Green Revolution promises more than it can deliver. Its sociopolitical application has been largely unsatisfactory. Even in areas where the Green Revolution has been technologically successful, it has not always benefited large numbers of hungry people without the means to buy the newly produced food. It has benefited mainly Western-educated farmers, who were already wealthy enough to adopt a complex integrated package of technical inputs and management practices. Farmers make bigger profits from the Green Revolution when they purchase additional land and mechanize their operations. Some effects of labor-displacing machinery and the purchase of additional land by rich farmers include agricultural unemployment, increased landlessness, rural-to-urban migration, and increased malnutrition for the unemployed who are unable to purchase the food produced by the Green Revolution.

The Green Revolution generated substantial increases in agricultural output worldwide. However, world hunger remains a serious problem, indicating that the problem is not so much one of food production, but of food demand in the economic sense (i.e., purchasing power). Unfortunately, the Green Revolution does nothing to increase the ability of the poor to buy food. Hunger is a complex and intractable problem in large part because it is so closely tied to questions of poverty and economic development, not simply increasing agricultural productivity.

The Green Revolution has helped to create a world of more and larger commercial farms alongside fewer and smaller peasant plots. However, given a different structure of land holdings and the use of appropriately intermediate technology, the Green Revolution could help developing countries on the road toward agricultural self-sufficiency and the elimination of hunger. Intermediate technology is a term that means low-cost, small-scale technologies intermediate between primitive stick-farming methods and complex agroindustrial technical packages.

Creating New Food Sources

Expanding cultivated areas and raising the productivity of existing cropland are two methods of increasing food supply. A third method is the identification of new food sources. There are three main ways to create new food sources: (1) Cultivate the oceans, or mariculture; (2) develop high-protein cereal crops; and (3) increase the acceptability and palatability of inefficiently used present foods.

Cultivating the Oceans

Fishing and the cultivation of fish and shellfish from the oceans is not a new idea. At first glance, the world seems well supplied with fisheries because oceans cover three-fourths of the earth. However, fish provide a very small proportion—about 1 percent—of the world's food supply.

World fish consumption in the late twentieth century increased more rapidly than did the population, and even exceeded beef as a source of animal protein in some countries. However, since 1987, fish caught by commercial fishing fleets leveled off and declined as a result of overfishing (Figure 4.8). Overfishing has been particularly acute in the North Atlantic and Pacific Oceans. Countries such as Iceland and Peru, whose economies rely heavily on fishing, are sensitive to the overfishing problem. Peru's catch of its principal fish, the anchovy, has declined by over 75 percent because of overfishing. The Peruvian experience demonstrates that the ocean is not a limitless fish resource, as did the quest for whales a century earlier.

Commercial fishing fleets employ sophisticated techniques but catch what nature has provided, much like hunters and gatherers. An alternative approach is to follow the example of animal husbandry by devising methods for commercial fish farming. *Mariculture*, or fish farming, is now expanding rapidly and accounts for 5 percent of the world's fish caught yearly. The cultivation of food fish such as catfish, trout, and salmon is big business in the United States, Norway, Japan, and other fishing countries.

High-Protein Cereals

Another source of future food production rests in higher protein cereal crops. Agricultural scientists seek to develop high-yield, high-protein cereal crops in the hope that development of hybrid seeds will be able to help the protein deficiency of people in developing countries who do not have available meats from which to gain their protein needs, as do people in developed countries.

Fortification of present rice, wheat, barley, and other cereals with minerals, vitamins, and protein-carrying amino acids is an approach that also deserves attention. This approach is based on the fortified food production in developed countries and stands a greater chance of cultural acceptance because individual food habits do not necessarily need to be altered. But developing countries rely on unprocessed, unfortified foods for 95 percent of their food intake. Large-scale fortification and processing would require major technological innovation and scale economies to produce enough food to have an impact on impoverished societies.

More Efficient Use of Foods

In many developing countries, foods to satisfy consumer preferences as well as religious taboos and cultural values are becoming limited. The selection of foods based on social customs should be supplemented with information concerning more efficient use of foods presently available. An effort should be made to increase the palatability of existing foods that are plentiful.

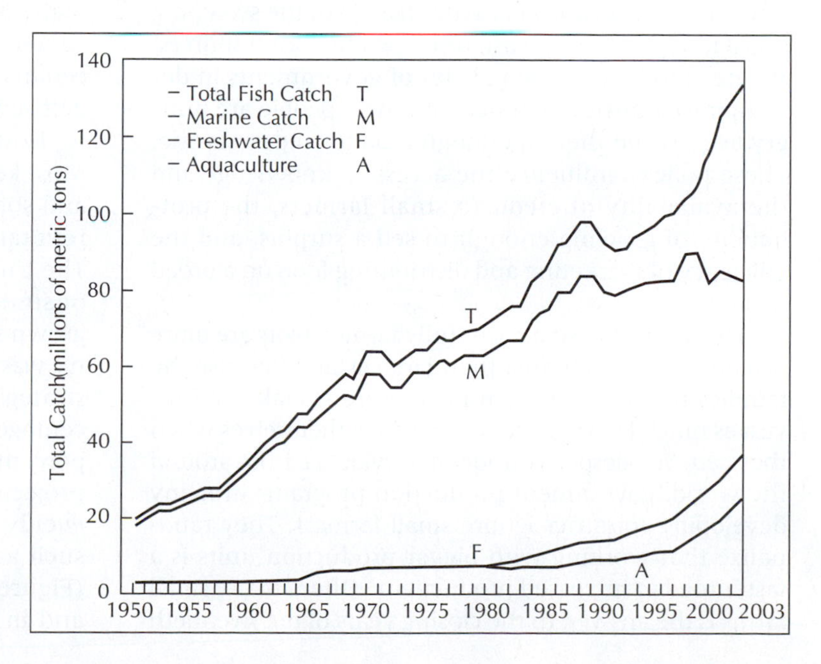

FIGURE 4.8
Global fish catch, 1950–2003. Rising demand and increasingly efficient industrial fishing methods not only have yielded dramatically higher catches, but also have increasingly depleted the world's oceans of many species. See Richard Ellis, *The Empty Ocean*.

Fish meal is a good example. Presently, one-third of the world's fish intake is turned into fodder for animals and fertilizer. Fish meal is rich in the Omega 3 fatty acids and amino acids necessary for biological development. However, in many places, the fish meal is not used because of its taste and texture.

Another underused food resource is the soybean, a legume rich in both protein and amino acids. Most of the world's soybeans wind up being processed into animal feed or fertilizer and into production of nondigestible industrial materials. World demand for tofu and other recognizable soybean derivatives is not large. By contrast, hamburgers, hot dogs, soft drinks, and cooking oils made partially from soybeans are more acceptable.

A Solution to the World Food Supply Situation

As we have emphasized, there is a widely shared belief that people are hungry because of insufficient food production. But the fact is that food production is increasing faster than population, and still there are more hungry people than ever before. Why should this be so? It could be that the production focus is correct, but soaring numbers of people simply overrun these production gains. Or it could be that the diagnosis is incorrect—scarcity is not the cause of hunger, and production increase, no matter how great, can never solve the problem.

The simple facts of world grain production make it clear that the overpopulation/scarcity diagnosis is incorrect. Present world grain production can more than adequately feed every person on earth. Ironically, the focus on increased production has compounded the problem of hunger by transforming agricultural progress into a narrow technical pursuit instead of the sweeping social task of releasing vast, untapped human resources. We need to look to the policies of governments in developing countries to understand why people are hungry even when there is enough food to feed everyone. These policies influence the access to knowledge and the availability of credit to small farmers, the profitability of growing enough to sell a surplus, and the efficiency of marketing and distributing food on a broad scale.

The fact is that small, carefully farmed plots are more productive per unit area than large estates because the families that tend to them have more at stake and invest as much labor as necessary to feed themselves when they can. Yet, despite considerable evidence from around the world, government production programs in many developing countries ignore small farmers. They rationalize that working with bigger production units is a faster road to increased productivity. Often, many small farms is the answer. In the closing years of the twentieth century, many agricultural researchers, having gained respect for traditional farming systems, agree with this conclusion.

NONRENEWABLE MINERAL RESOURCES

Although we can increase world food output, we cannot increase the supply of minerals. A mineral deposit, once used, is gone forever. A *mineral* refers to a naturally occurring inorganic substance in the earth's crust. Thus, silicon is a mineral whereas petroleum is not, since the latter is of organic origin. Although minerals abound in nature, many of them are insufficiently concentrated to be economically recoverable. Moreover, the richest deposits are unevenly distributed and are being depleted.

Except for iron, nonmetallic elements are consumed at much greater rates than metallic ones. Industrial societies do not worry about the supply of most nonmetallic minerals, which are plentiful and often widespread, including nitrogen, phosphorus, potash, or sulfur for chemical fertilizer, or sand, gravel, or clay for building purposes. Those commodities the industrial and industrializing countries do worry about are the metals.

Location and Projected Reserves of Key Minerals

Only five countries—Australia, Canada, South Africa, the United States, and Russia—are significant producers of at least six strategic minerals vital to defense and modern technology (Figure 4.9). Of the major mineral-producing countries, only a few—notably the United States and Russia—are also major processors and consumers. The other major processing and consuming centers—Japan and western European countries—are deficient in strategic minerals.

How large is the world supply of strategic minerals? Most key minerals will be exhausted within 100 years, and some will be depleted within a few years at current rates of consumption, assuming no new reserves. The United States is running short of domestic sources of strategic minerals. Its dependence on imports has grown steadily since 1950; prior to that year, the country was dependent on imports for only four designated strategic minerals. When measured in terms of percentage imported, U.S. dependency increased from 50 percent in 1960 to over 82 percent in 2003. Minerals projected as future needs by the United States are unevenly distributed around the world. Many of them, such as manganese, nickel, bauxite, copper, and tin (Figure 4.9), are concentrated in Russia and Canada and in developing countries. Whether these critical

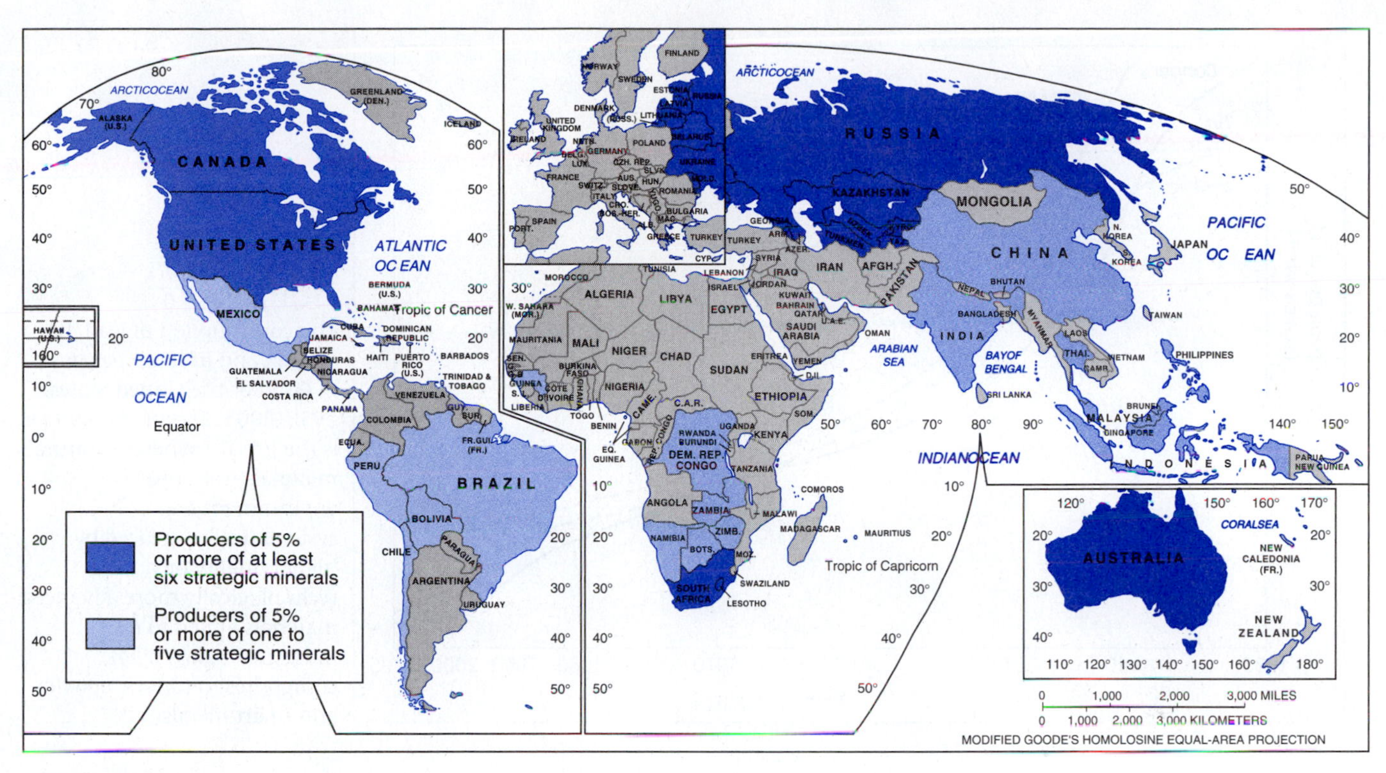

FIGURE 4.9
Major producers of strategic minerals.

substances will be available for U.S. consumption may depend less on economic scarcity and more on international tensions and foreign policy objectives.

Solutions to the Mineral Supply Problem

Affluent countries are unlikely to be easily defeated by mineral supply problems. Human beings, the ultimate resource, have developed solutions to the problem in the past. Will they in the future? Although past experience is never a reliable guide to the future, there is no need to be unduly pessimistic about the exhaustion of minerals as long as we develop alternatives.

If abundant supplies of cheap electricity ever became available, it might become possible to extract and process minerals from unorthodox sources such as the ocean. The oceans, which cover 71 percent of the earth, contain large quantities of dissolved minerals. Salt, magnesium, sulphur, calcium, and potassium are the most abundant of these minerals and amount to over 99 percent of the dissolved minerals. More valuable minerals that are also present include copper, zinc, tin, and silver. Improved efficiency of production has reduced the demand for various minerals per unit of national output (Figure 4.10). Some minerals such as bromine and magnesium are being obtained electrolytically from the oceans at the present time.

Much more feasible than mining the oceans is devoting increased attention to improving mining technology, especially to reducing waste in the extraction and processing of minerals. Equally feasible is to utilize technologies that allow minerals to be used more efficiently in manufacturing. Also, if social attitudes were to change, encouraging lower per capita levels of resource use, more durable products could be manufactured, saving not only large amounts of energy but also large quantities of minerals.

Reusing minerals is still another option to our mineral problems. Every year in the United States and other affluent countries, huge quantities of household and industrial waste are disposed of at sanitary landfills and open dumps. These materials are sometimes called "urban ores" because they can be recovered and used again. For years, developed countries have been recycling scarce and valuable metals such as iron, lead, copper, silver, gold, and platinum, but large amounts of scrap metals are still being wasted. Although we could recover a much greater proportion of scrap, this is unlikely when prices are low or when virgin materials are cheaper than recycled ones.

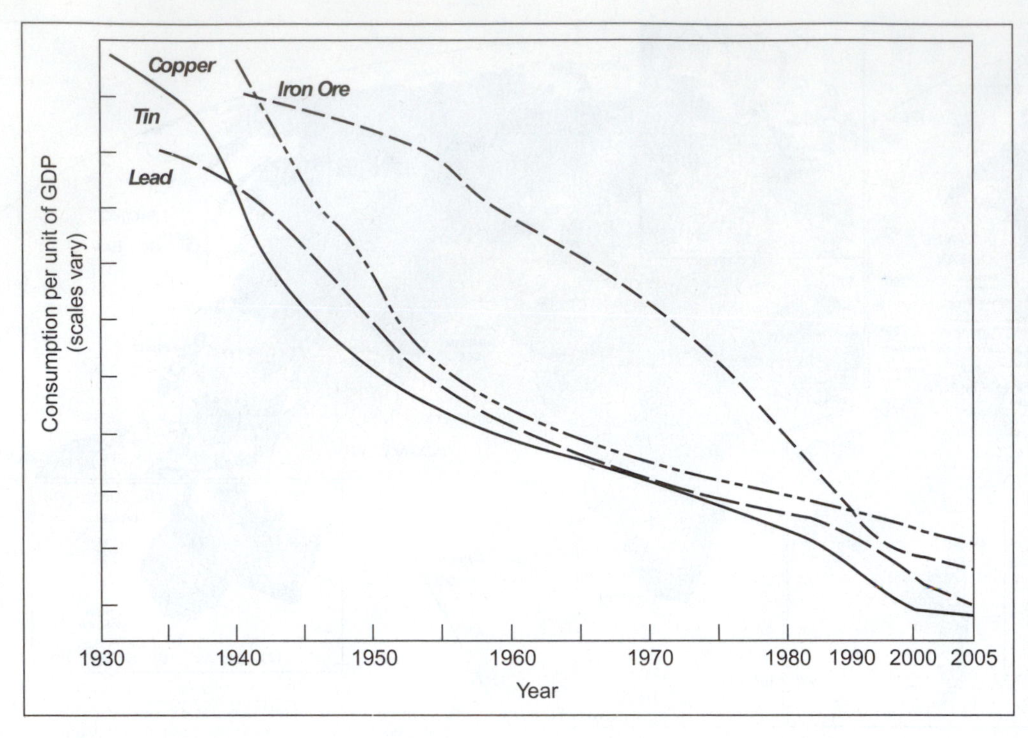

FIGURE 4.10
The consumption of lead, tin, copper, and iron ore per unit of GDP for the United States, 1930–2005. *Transmaterialization* is the process whereby natural materials from the environment are systematically replaced by higher quality or technologically more advanced materials linked to new industries—glass fibers, composites, ceramics, epoxies, and smart metals.

Environmental Impact of Mineral Extraction

Mineral extraction has a varied impact on the environment, depending on mining procedures, local hydrological conditions, and the size of the operation. Environmental impact also depends on the stage of development of the mineral—exploration activities usually have less of an impact than mining and processing mineral resources.

Minimizing the environmental impacts of mineral extraction is in everyone's best interest, but the task is difficult because demand for minerals continues to grow and ever-poorer grades of ore are mined. For example, in 1900 the average grade of copper ore mined was 4 percent copper; by 1973, ores containing as little as 0.5 percent copper were mined. Each year more and more rock has to be excavated, crushed, and processed to extract copper. The immense copper mining pits in Montana, Utah, and Arizona are no longer in use because foreign sources, mostly in the developing countries, are less expensive. As long as the demand for minerals increases, ever lower quality minerals will have to be used and, even with good engineering, environmental degradation will extend far beyond excavation and surface plant areas.

ENERGY

The development of energy sources is crucial for economic development. Today, commercial energy is the lifeblood of modern economies. Indeed, it is the biggest single item in international trade. Oil alone accounts for about one-quarter of the volume (but not value) of world trade. The U.S. economy consumes vast amounts of energy, overwhelmingly consisting of fossil fuels (Figure 4.11). These form the inputs that, along with labor and capital, run the economic machine that feeds, houses, and moves the population. As Figure 4.12 indicates, the primary uses of petroleum are transportation and industrial purposes, whereas the major uses of coal are for electrical power generation.

Until the energy shocks of the 1970s, commercial energy demands were widely thought to be unproblematic, that is, always there to generate rising affluence. Suddenly, higher prices brought energy demands in the industrial countries to a virtual standstill, generating inflation, unemployment, and accelerating deindustrialization (Figure 4.13). Thousands of factories were shut down, and more than 3 million workers were laid off. They learned firsthand that when energy fails, everything fails in an urban-industrial economy. During the 1980s and 1990s, oil prices decreased from $30 in 1981 to $14 in 1999. OPEC, once considered an invincible cartel, saw its share of world oil output drop as non-OPEC countries expanded production. Many developing countries, strapped by heavy energy debts, were relieved to see prices falling. Oil-exporting developing countries, such as Mexico, Venezuela, and Nigeria, which came to depend on oil revenues for an important source of income, were hurt the worst. By 2006, however, world spot oil prices had risen again to a nominal price of $60 per barrel.

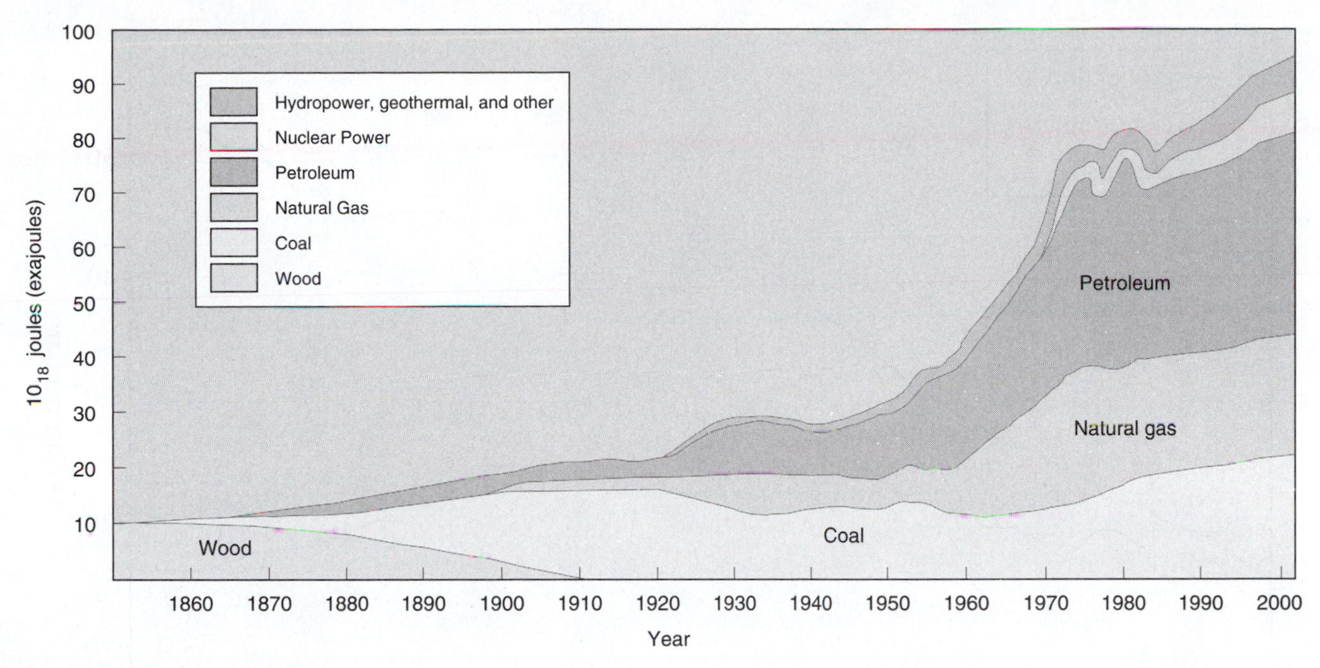

FIGURE 4.11
U.S. energy consumption in joules, 1850–2000. The U.S. economy contains 5 percent of the world's people but uses one-third of its energy. The three principal sources of fossil fuels are coal, natural gas, and petroleum. In 1850, the burning of wood provided the nation's energy supply. By 1900, wood had been supplanted by coal. After World War II, petroleum and natural gas surpasses coal as the chief source of energy in the United States. Hydro and nuclear have also increased recently.

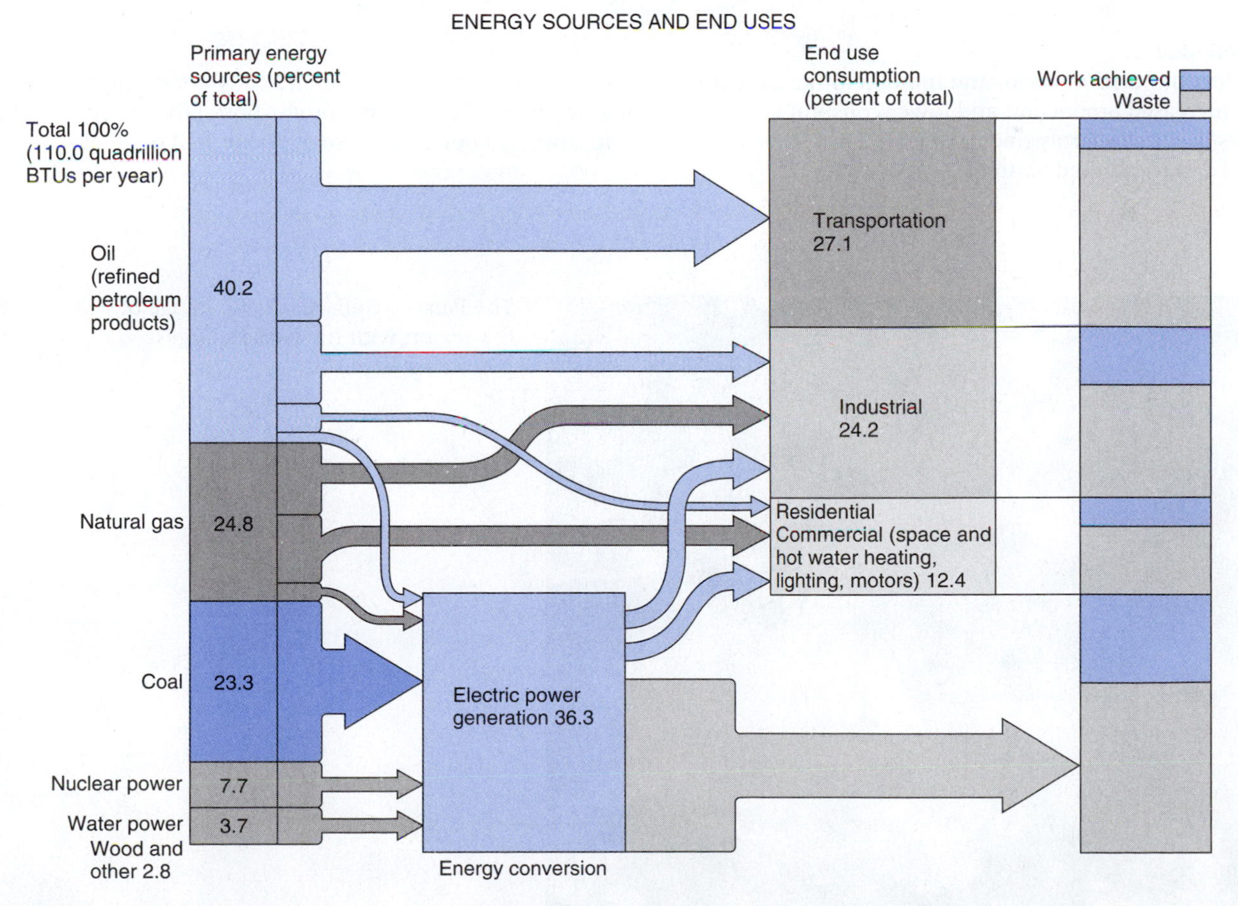

FIGURE 4.12
U.S. energy sources and end uses. Different energy inputs are applied to different uses. While coal is still widely used for electrical generation, petroleum is the most common energy source for transportation and industrial production.

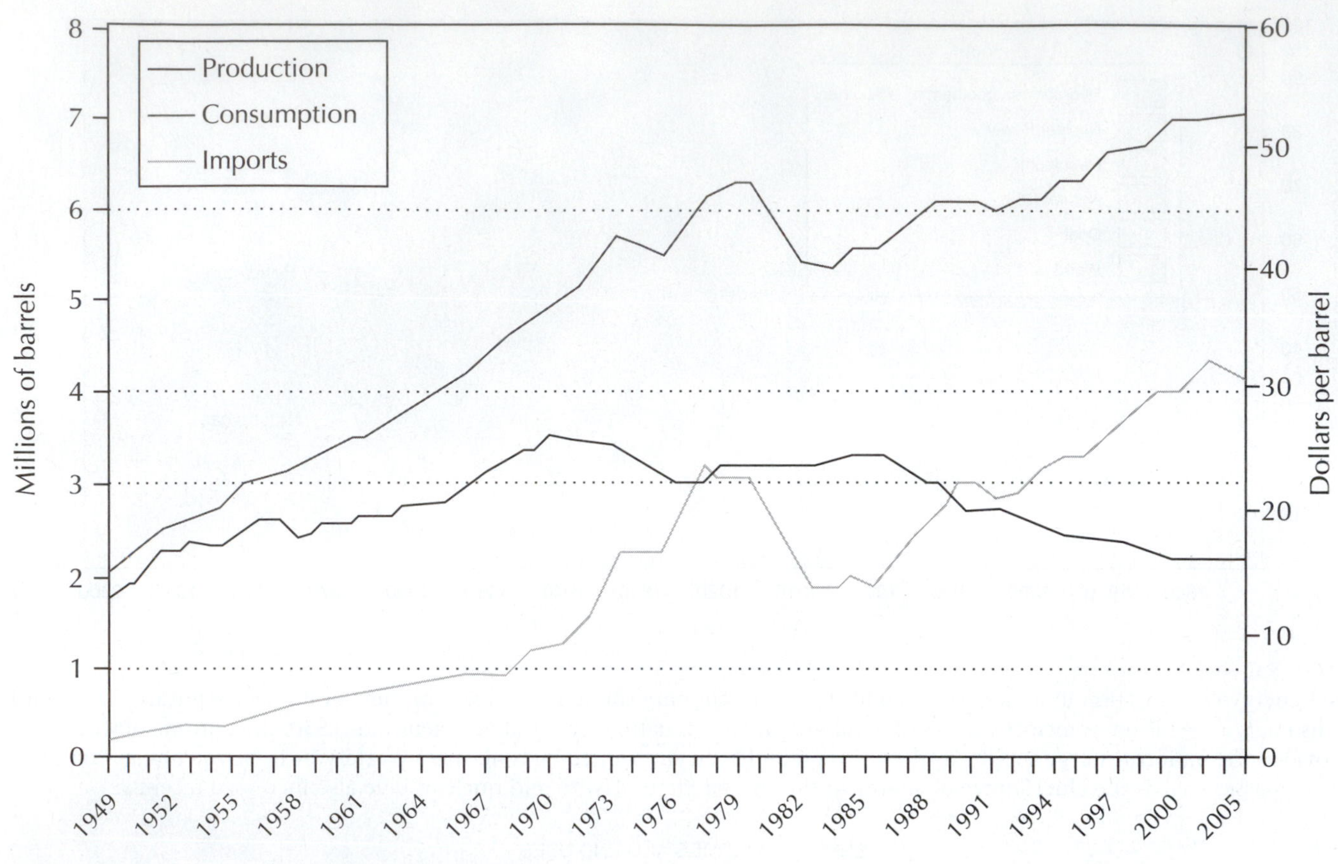

FIGURE 4.13
Oil production, consumption, and imports in the United States, 1950–2003. In the late 1970s and early 1980s, high prices created increased production and lower consumption; also, the Alaskan oil fields came into production. From 1980 onward, oil prices declined sharply due to the decline in OPEC's oligopolistic power. Imports comprise over one-half of all oil consumed in the United States.

The Persian Gulf. A satellite image of the region with the world's largest petroleum deposits.

Energy Production and Consumption

Most commercial energy produced is from nonrenewable resources. Most renewable energy sources, particularly wood and charcoal, are used directly by producers, mainly poor people in the developing countries. Although there is increasing interest in renewable energy development, commercial energy is the core of energy use at the present time. Only a handful of countries produce much more commercial energy than they consume. If we take petroleum consumption and production as an example, the main energy surplus countries include Saudi Arabia, Iraq, Mexico, Iran, Venezuela, Indonesia, Algeria, Kuwait, Libya, Qatar, Nigeria, and the United Arab Emirates. Saudi Arabia is by far the largest exporter of petroleum and has the largest proven reserves. Nearly one-half of all African countries are energy paupers. Several of the world's leading industrial powers—most notably Japan, many western European countries, and the United States—consume much more energy than they produce, making them heavily reliant on imported oil, largely from the Middle East. This fact profoundly shapes the foreign policies of countries such as the United States.

The United States leads the world in total energy use, but leaders in per capita terms also include Canada, Norway, Sweden, Japan, Australia, and New Zealand (Figure 4.14). With 5 percent of the world's population, the United States consumes roughly one-quarter of the world's energy, largely for transportation, which consumes 40 percent of American energy

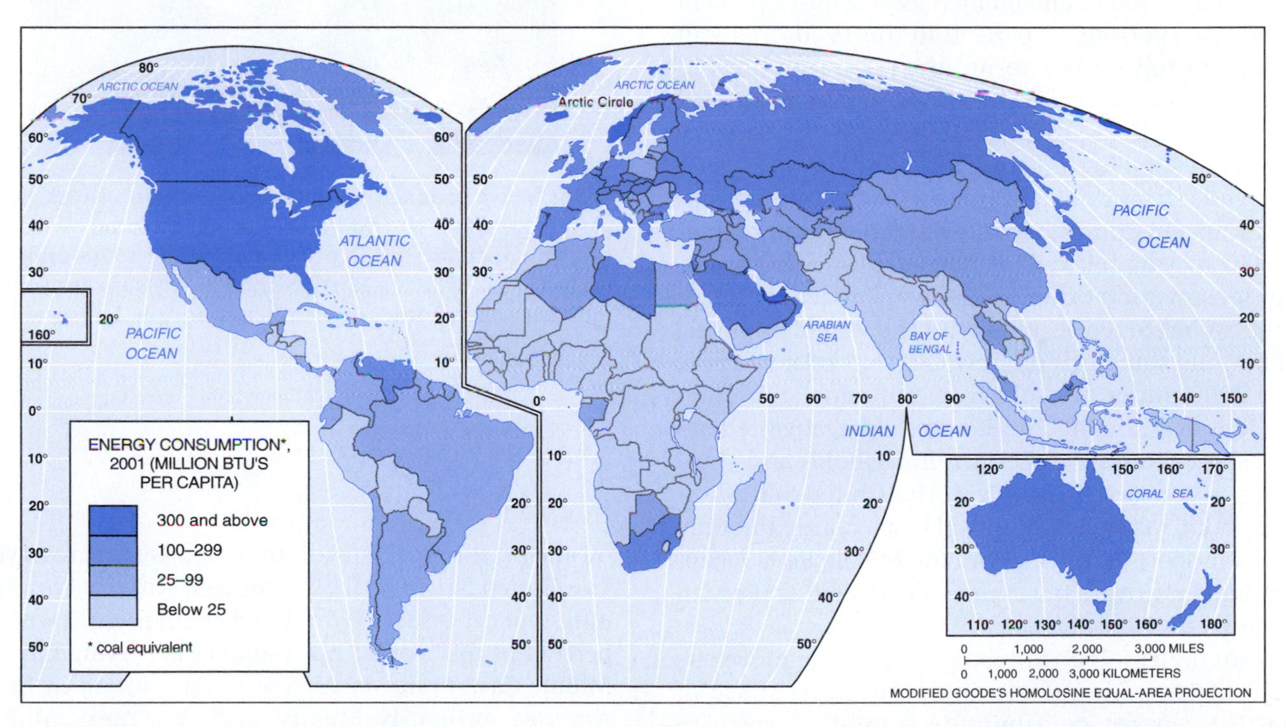

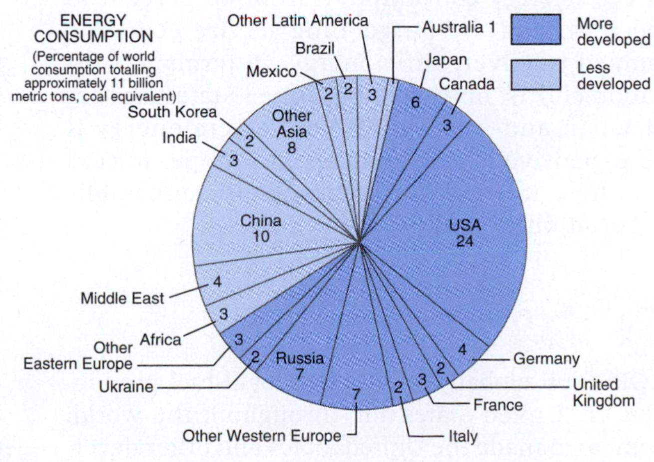

FIGURE 4.14
World per capita electricity consumption, 2001. The United States, Canada, and the Scandinavian countries consume more electricity per capita than any other countries. When the electricity usage of the United States, Canada, Europe, and Russia is combined, 75 percent of electricity usage in the world is accounted for, but only 20 percent of the people. By comparing this map to the map of crude petroleum proven reserves, the deficit areas of the world such as Europe and Japan, which have high energy needs but low fossil fuel resources, can be seen. In addition, there are areas in the world with high fossil fuel resources but low energy needs, such as the Middle East countries, Mexico, Venezuela, Indonesia, Argentina, Algeria, Nigeria, and China.

inputs. The automobile, for all the convenience it offers, is a highly energy-inefficient way to move people. In contrast, developing countries consume about 30 percent of the world's energy but contain about 80 percent of the population. Thus there exists a striking relationship between per capita energy consumption and level of development. Most developing countries consume meager portions of energy, well below levels required with even moderate levels of economic development. Commercial energy consumption in developed countries has been at consistently high levels, whereas in developing countries it has been at low but increasing levels.

Oil Dependency

Americans were seriously affected by the 1973 and 1979 Arab oil embargos, because imported oil as a proportion of total demand increased from 11 percent in the late 1960s to 50 percent in the 1970s to about 58 percent today. As a result, the administrations of the United States have repeatedly called for a national policy of oil self-sufficiency to reduce U.S. dependency on foreign supplies of petroleum, without much success. Under heavy political pressure from corporations and campaign donors, air and water pollution regulations have been relaxed, and tax credits for home energy conservation expenditures were ended. The U.S. Congress has toyed at times with the idea of imposing stricter fuel standards on new cars, but relaxed these after 2000 under pressure from automobile producers. These conflicting policies worked against federal efforts to encourage American households and companies to conserve fossil fuels. The United States imports about 58 percent of the oil it consumes, but only a small proportion comes from the Middle East. Japan, Italy, and France are much more dependent on Persian Gulf oil.

Nonetheless, U.S. industry did become more energy efficient. Manufacturing reduced its share of total U.S. energy consumption from 40 percent to 36 percent, and the burgeoning service economy consumed relatively little energy. In terms of conservation efforts, however, the United States lags behind Japan and western Europe, where energy is more expensive. Gasoline taxes in Europe, for example, help to fund more energy-efficient public transportation.

A Shell/Esso production platform in Britain's North Sea gas field. British oil exploration was stimulated by a dramatic increase in the price of oil in the 1970s and early 1980s. Britain's North Sea oil and gas investment may keep the country self-sufficient for the foreseeable future.

Production of Fossil Fuels

The OPEC oil embargo stimulated fossil fuel production in the United States and throughout the world. The embargo made the United States and other developed countries aware of their dependency on imported oil and on the world distribution of fossil fuel reserves. The United States is richly endowed with coal but has only modest reserves of oil and natural gas. Over 65 percent of the world's oil resources are located in the Middle East. Other large reserves are found in Latin America, primarily Mexico and Venezuela, and in Russia and Nigeria. Natural gas, often a substitute for oil, is also unevenly distributed, with nearly 40 percent in Russia and central Asia and 34 percent in the Middle East (Figure 4.15).

The unevenness of the world's supply and demand for petroleum creates a distinct pattern of trade flows of petroleum (Figure 4.16), the most heavily traded (by volume) commodity in the world. Primarily, these flows represent exports from the vast reserves of the Middle East to Europe, East Asia, and North America, although the United States also imports considerable quantities from South America and Nigeria. The differences between crude oil production and consumption for each major world region are sketched in more detail in Figure 4.17.

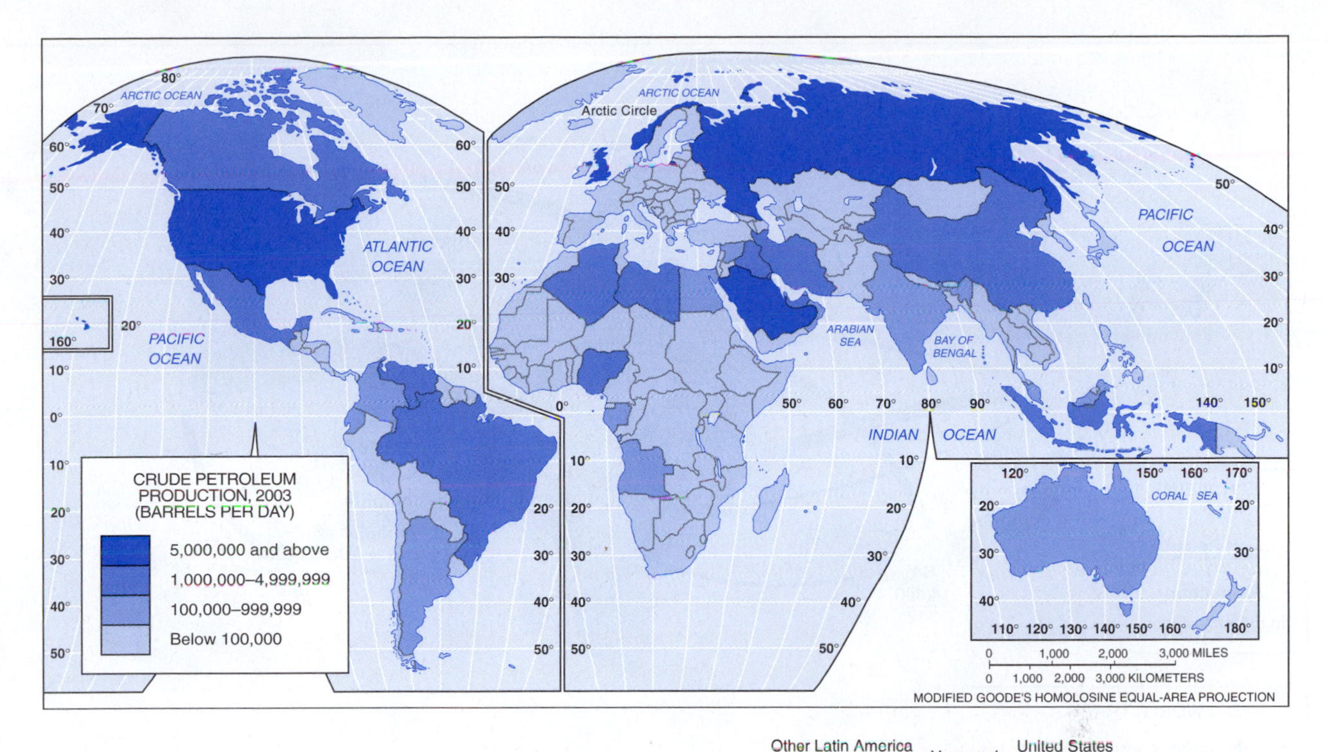

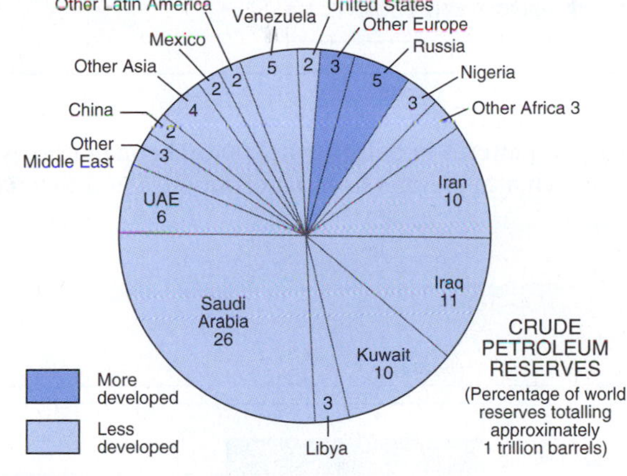

FIGURE 4.15
World crude petroleum production and reserves, 2003. Saudi Arabia is the largest producer and has the largest proven reserves; the United States is a major producer but with limited reserves.

Adequacy of Fossil Fuels

In the next few decades, energy consumption is expected to rise significantly, especially because of the growing industrialization of developing countries. Most of the future energy production to meet increasing demand will come from fossil energy resources—oil, natural gas, and coal. How long can fossil fuel reserves last, given our increasing energy requirements? Estimates of energy reserves have increased substantially in the last 20 years, and therefore there is little short-term concern over supplies; consequently, energy prices are relatively low. If energy consumption were to remain more or less at current levels, which is unlikely, proved reserves would supply world petroleum needs for 40 years, natural gas needs for 60 years, and coal needs for at least 300 years. Although the size of the world's total fossil fuel resources is unknown, they are finite, and production will eventually peak and then decline.

Oil: Black Gold

Most of the world's petroleum reserves are heavily concentrated in a few countries, mostly in politically unstable regions. Although reserves increased by 170 percent between 1978 and 2003, most of this increase is attributed to new discoveries in the Middle East. Regionally, however, reserves have been declining in important consuming countries. For example, reserves in Russia declined by 9 percent between 1991 and 2004. They also declined by 9 percent in the United States during the same period. Despite new discoveries, Europe's reserves are likely to be depleted by 2050. Moreover, exports of oil from Africa and Latin America will probably cease by 2050. The Middle East will then be the only major exporter of oil, but political turmoil (e.g., wars, revolutions) there could cause interruptions in oil supplies, creating problems for the import-dependent regions of North America, western Europe, Japan, and the East Asian newly industrializing countries (NICs).

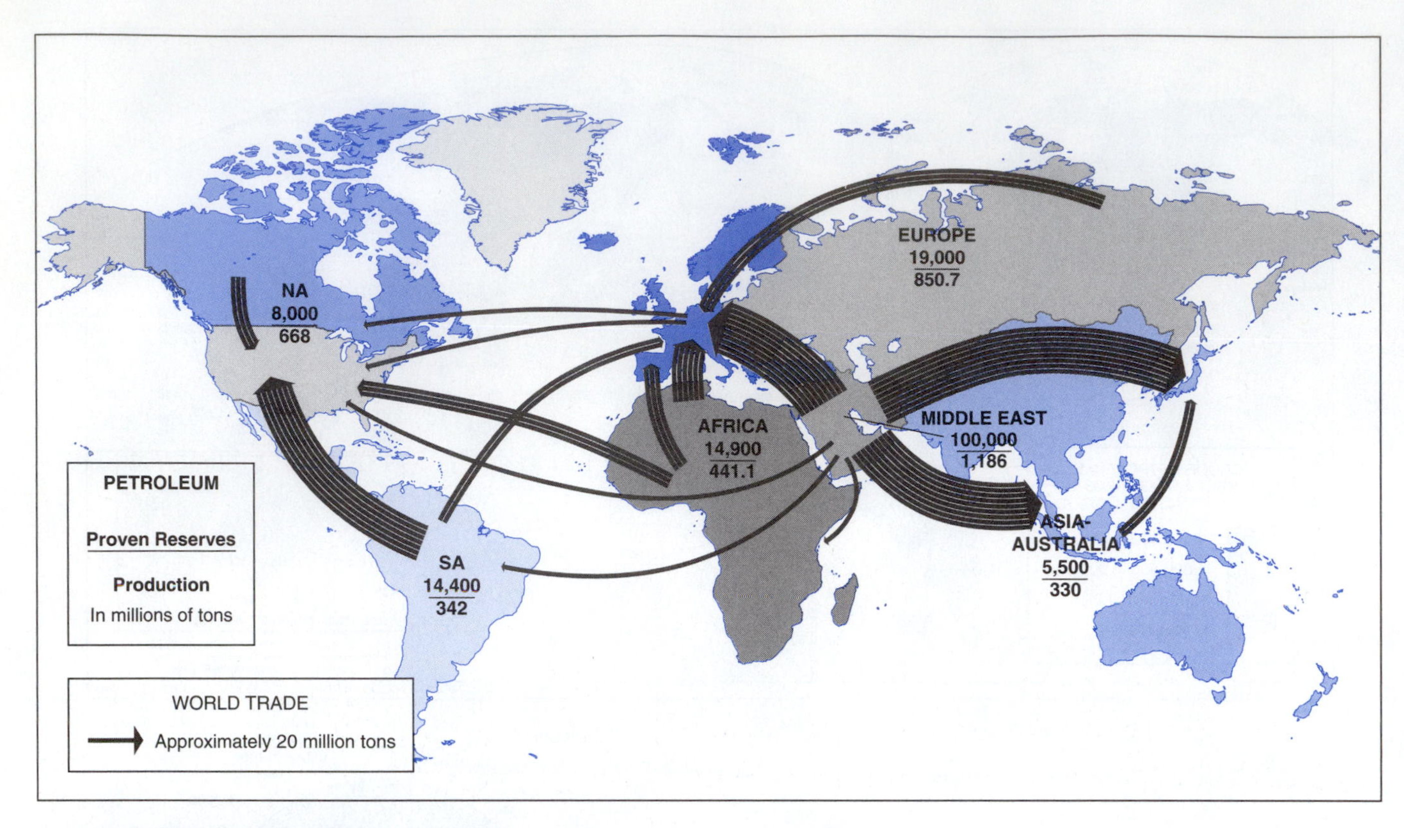

FIGURE 4.16
World trade patterns in petroleum, 2004. The major flows are from the Middle East to Europe, East Asia, and the United States, which also imports from Latin America and Nigeria.

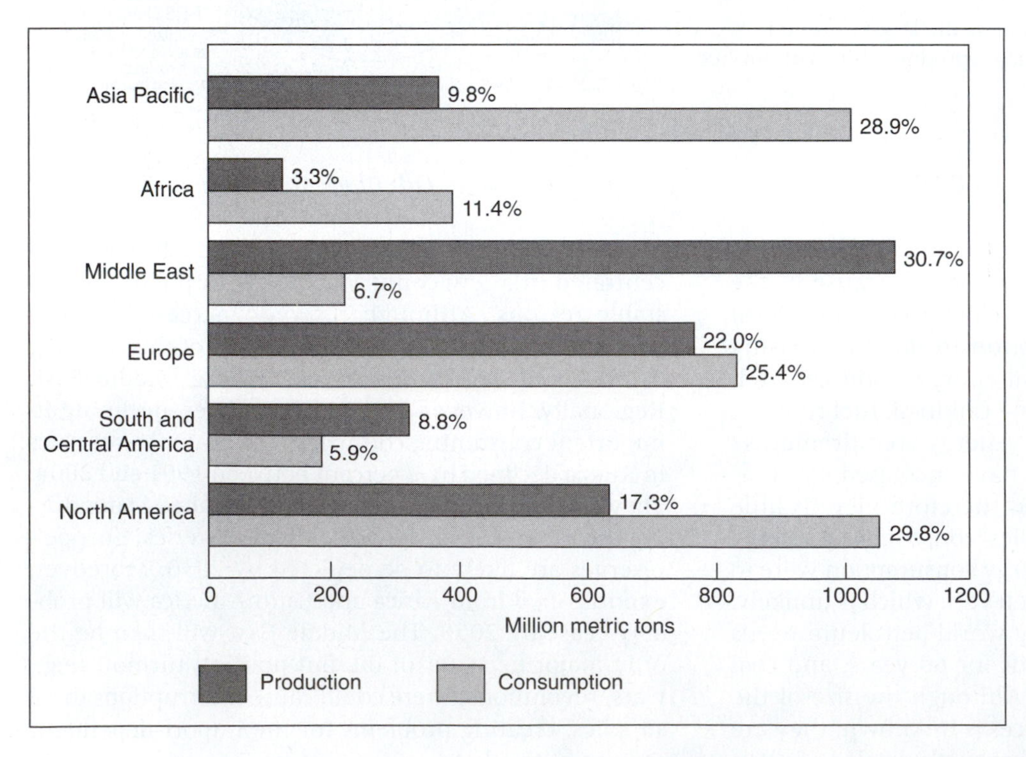

FIGURE 4.17
The production and consumption of crude oil by major world regions in millions of metric tons, 2004. The developed market economies of Europe, North America, and Asia, especially Japan, consume a far greater proportion of energy resources than they produce. Conversely, the Middle East, especially the Persian Gulf region, produces the most crude oil of any world region but consumes only one-sixth of its production. Latin America, Africa, and the former Soviet Union consume less crude oil than they produce.

Natural Gas

The political volatility of the world's oil supply has increased the attractiveness of natural gas, the fossil fuel experiencing the fastest growth in consumption. Natural gas production is increasing rapidly, and so too are estimates of proven gas reserves. Estimates of global gas reserves have increased during the last decade, primarily due to major finds in Russia, particularly in Siberia, and large discoveries in China, South Africa, and Australia. Reserves have also been increasing in western Europe, Latin America, and in North America. Gas production will eventually peak, probably in the first two or three decades of the twenty-first century. As a result, gas supplies will probably last a bit longer than oil supplies.

The distribution of natural gas differs from that of oil. It is more abundant than oil in the former Soviet Union, western Europe, and North America, and less abundant than oil in the Middle East, Latin America, and Africa. A comparison of Figures 4.18 and 4.19 shows that natural gas also differs in its pattern of production and consumption. Because of the high cost of transporting natural gas by sea, the pattern of production is similar to that of consumption.

Coal

Coal is the most abundant fossil fuel, and most of it is consumed in the country in which it is produced

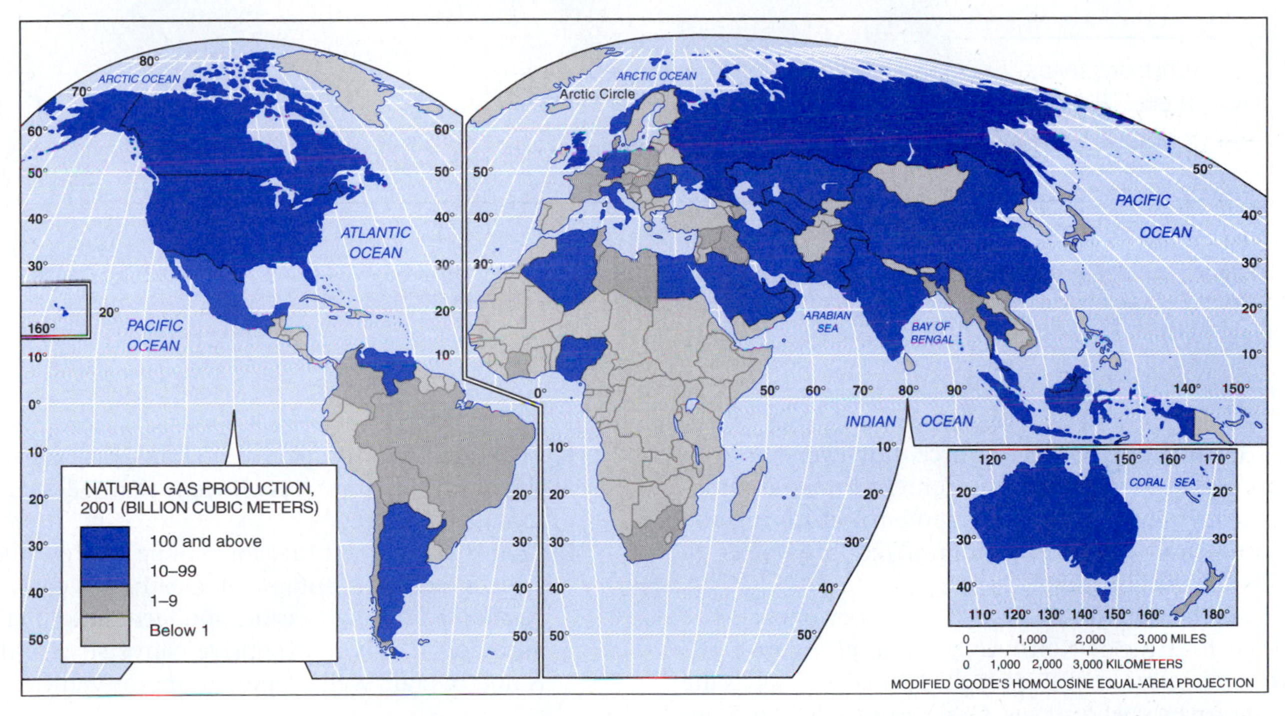

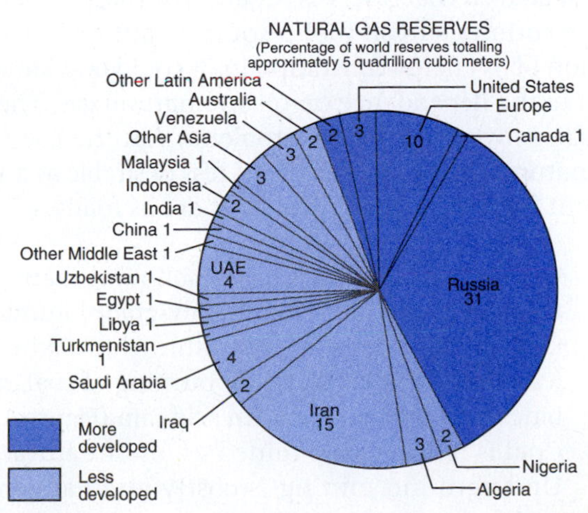

FIGURE 4.18
Production and consumption of natural gas by major world regions, 2001.

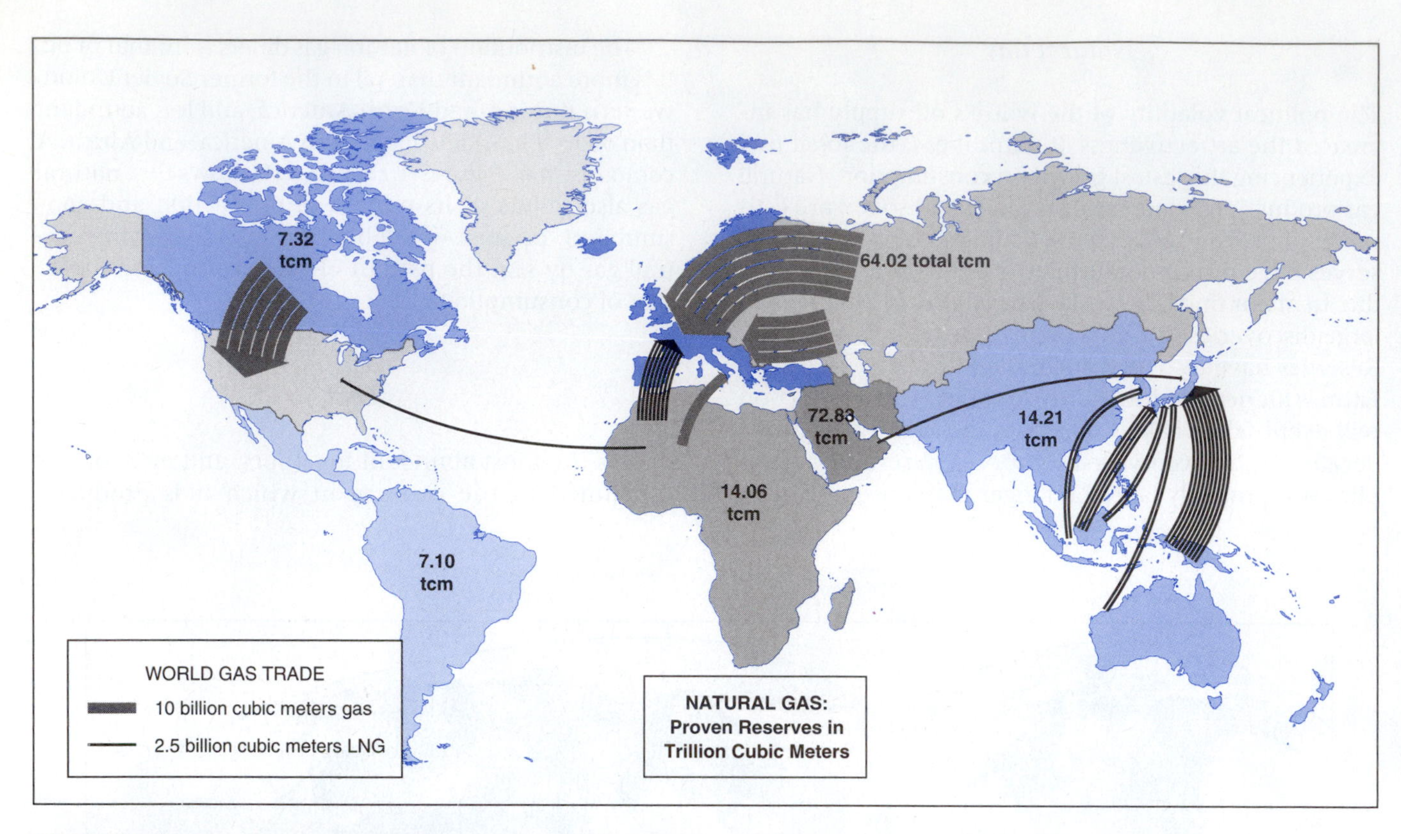

FIGURE 4.19
World trade patterns in natural gas, 2004. Russia is the world's largest exporter, primarily to Europe. Japan receives most of its gas from Southeast Asia, whereas the United States imports most from Canada.

(Figure 4.20). Use of this resource, however, has been hampered by inefficient management by the international coal industry, the inconvenience of storing and shipping, and the environmental consequences of large-scale coal burning.

China is the world's largest consumer of coal. The principal fossil fuel in North America is coal (Figure 4.21). With the exception of Russia, the United States has the largest proven coal reserves. Coal constitutes 67 percent of the country's fossil fuel resources, but only a small fraction of its energy consumption. It could provide some relief to the dependence on oil and natural gas. The use of coal, however, presents problems that the use of oil and natural gas do not, making it less desirable as an important fossil fuel. These problems are as follows:

1. Coal burning releases more pollution than other fossil fuels, especially sulfur. Low-grade bituminous coal has large amounts of sulfur, which, when released into the air from the burning of coal, combines with moisture to form acid rain (Figure 4.22).
2. Coal is not as easily mined as oil or natural gas. Underground mining is costly and dangerous, and open-pit mining leaves scars difficult to rehabilitate to environmental premining standards.
3. Coal is bulky and expensive to transport, and coal slurry pipelines are less efficient than oil or natural gas pipelines.
4. Coal is not a good fuel for mobile energy units such as trains and automobiles. Although coal can be adapted through gasification techniques to the automobile, it is an expensive conversion, and it is not, overall, well adapted to motor vehicles.

ENERGY OPTIONS

The age of cheap fossil fuels will eventually come to an end. As societies prepare for that eventuality, they must conserve energy and find alternatives to fossil fuels, especially alternatives that do not destroy the environment. How viable are the options?

Conservation

One way to reduce the gap between domestic production and consumption in the short run is for consumers to restrict consumption. Energy conservation stretches finite fuel resources and reduces environmental stress. Conservation can substitute for expensive, less environmentally desirable supply options and help to buy

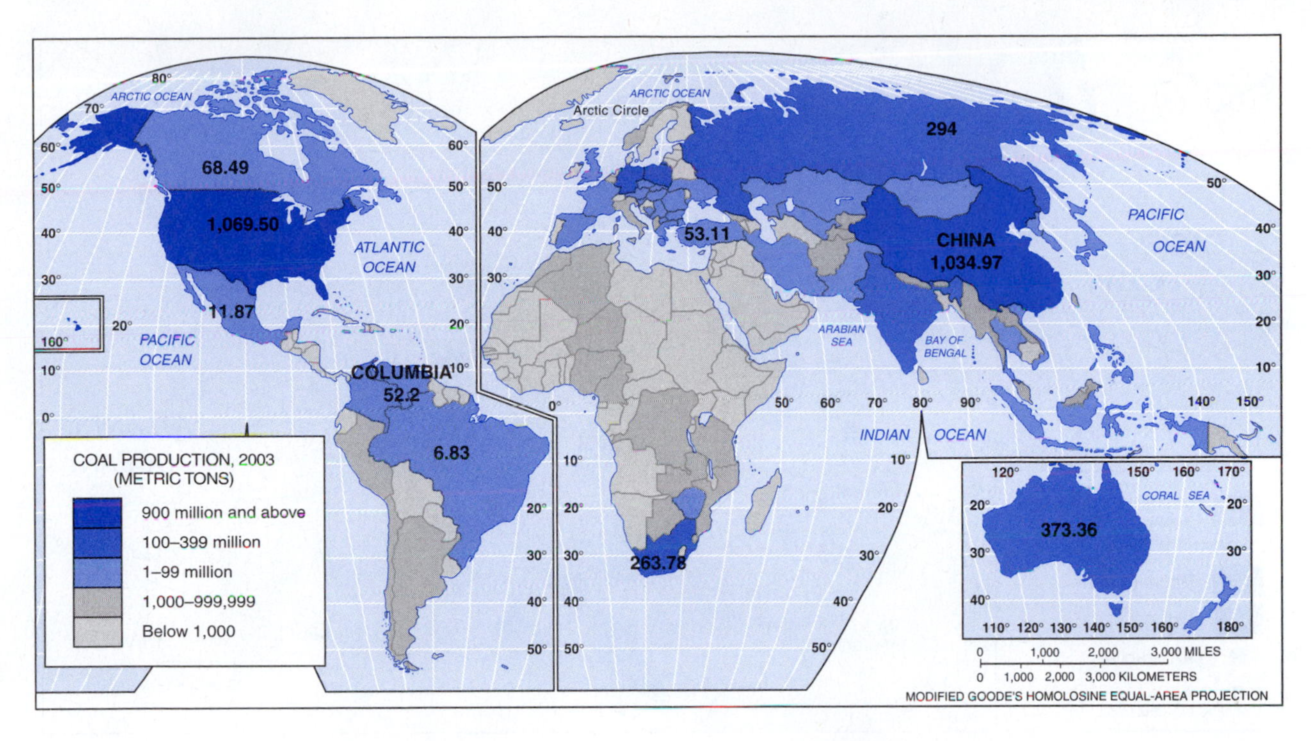

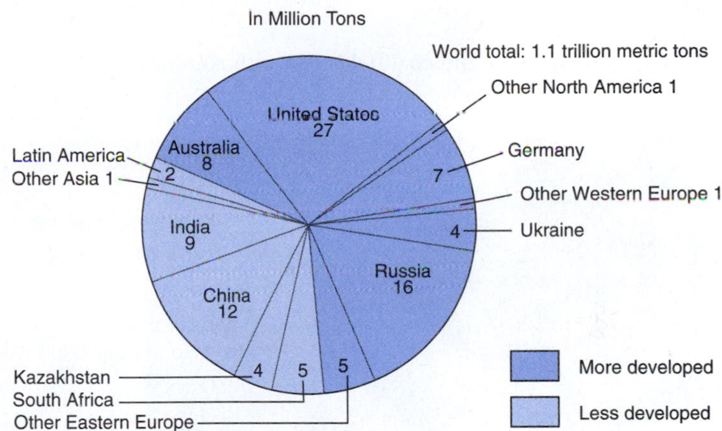

FIGURE 4.20
Coal production and reserves, 2003. The United States, China, and Russia lead the world in major coal deposits. Australia, Canada, and Europe also have major deposits.

time for the development of other more acceptable sources of energy.

Many people believe that energy conservation means a slow-growth economy; however, energy growth and economic growth are not inextricably linked. In the United States, from the early 1870s to the late 1940s, GNP per capita increased sixfold, whereas energy use per capita only slightly more than doubled. Energy efficiency, the ratio of useful energy output to total energy input, increased steadily throughout this century, partly as a result of industries installing better equipment. Even greater improvement can be expected in the new century.

Nuclear Energy

The form of nuclear energy currently in use commercially—*nuclear fission*—involves splitting large uranium atoms to release the energy within them. But nuclear fission causes many frightening issues, which became alarmingly clear after the nuclear accidents at the Three Mile Island plant in Pennsylvania in 1979 and at Chernobyl in the Ukraine in 1986. Concerns over nuclear energy range from environmental concerns caused by radiation to problems of radioactive waste disposal. Early radioactive wastes were dumped in the ocean in drums that soon began leaking. Likewise, many sites throughout the United States have contaminated groundwater supplies and leak radioactive wastes. One hotly discussed strategy currently underway is to store much of the nation's nuclear waste at Yucca Mountain, Nevada, miles away from major towns and cities. Another problem associated with the use of nuclear energy is the danger of terrorists stealing small amounts of nuclear fuel to construct weapons, which, if detonated, would wreak world havoc. Yet another is its high costs: Each plant costs billions of dollars to build and needs elaborate engineering and backup systems, as well as precautionary measures.

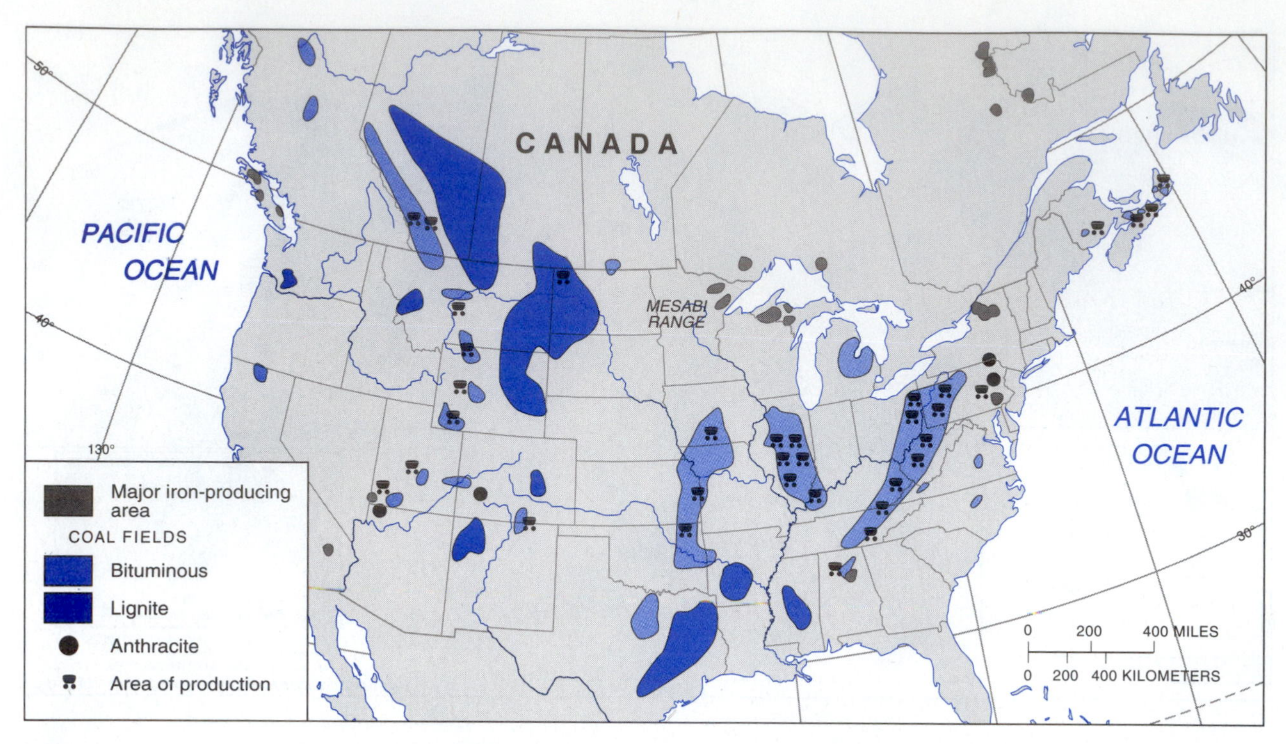

FIGURE 4.21
Major iron producing areas and coal fields of North America.

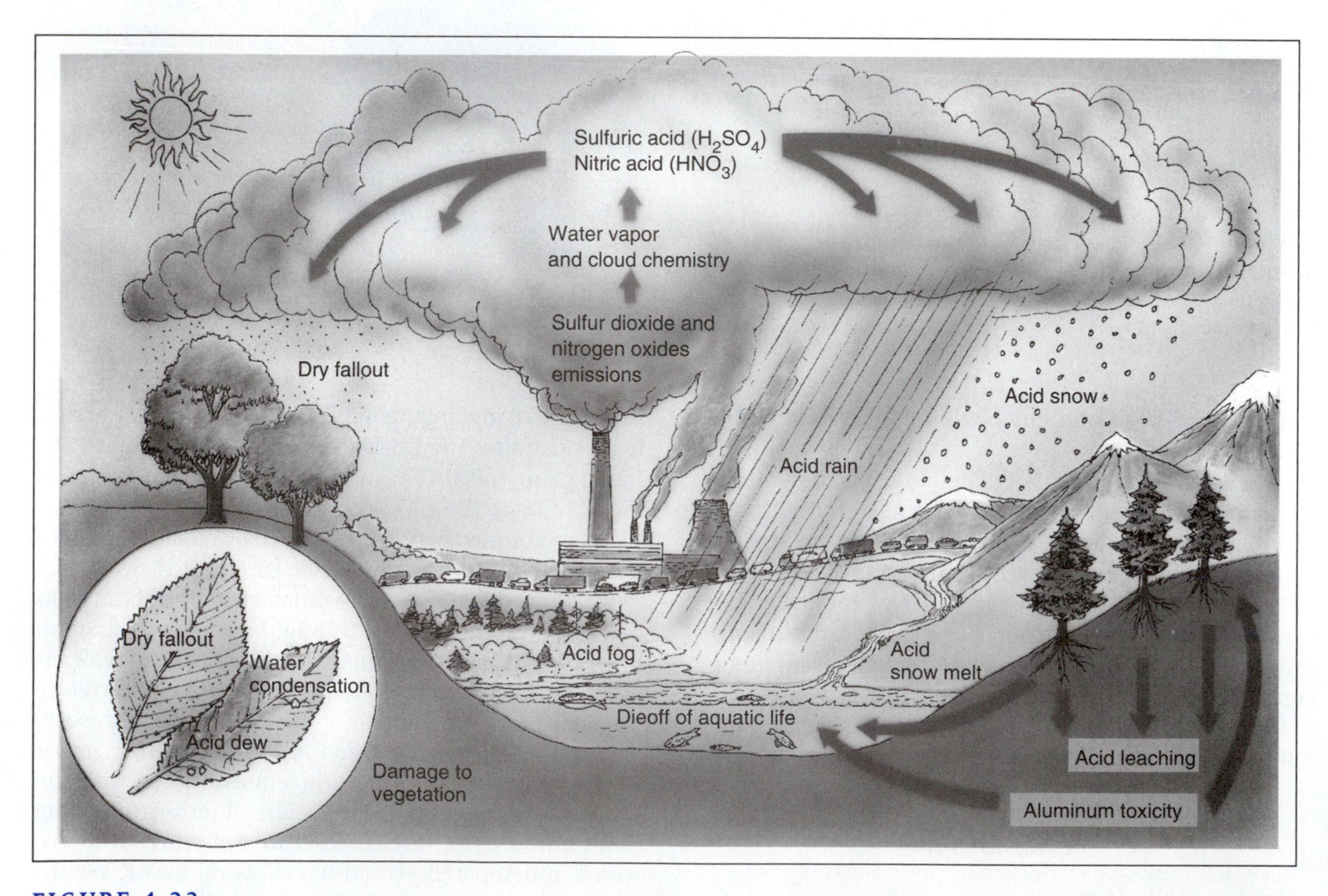

FIGURE 4.22
Acid rain creation. When sulfur is released into the atmosphere from the burning of coal and oil, it combines with moisture to create acid rain.

Nuclear power is less acceptable in North America than in some western European countries and Japan (Figure 4.23). In France, one-half of its energy comes from nuclear power; in Japan, 25 percent. Belgium, France, Hungary, and Sweden produce more than one-half of their energy from nuclear power plants, while Finland, Germany, Spain, Switzerland, South Korea, and Taiwan are also major producers and users. In the United States and Canada, countries that are less dependent on nuclear energy, the eastern portions rely on nuclear power plants more than do the western portions. For example, New England draws most of its electricity from nuclear power plants. Interestingly, some countries have decided to throw in the towel on nuclear power

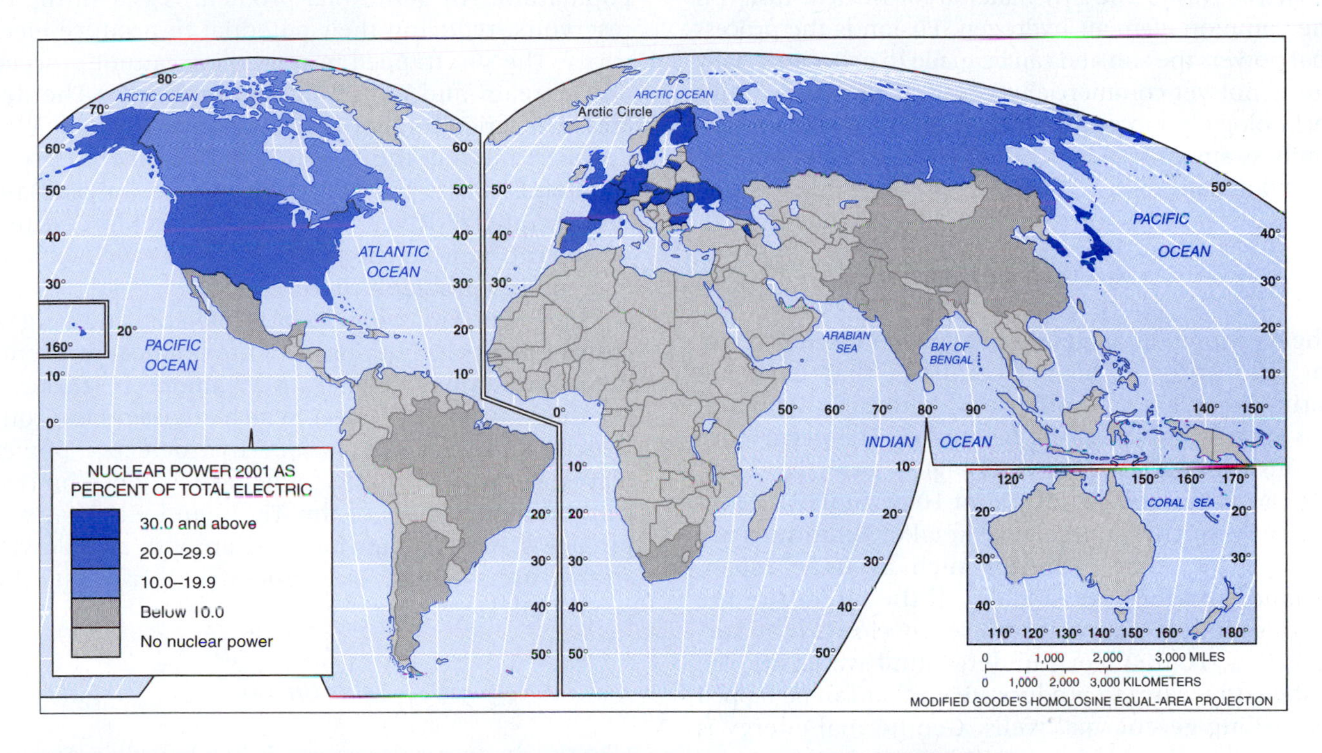

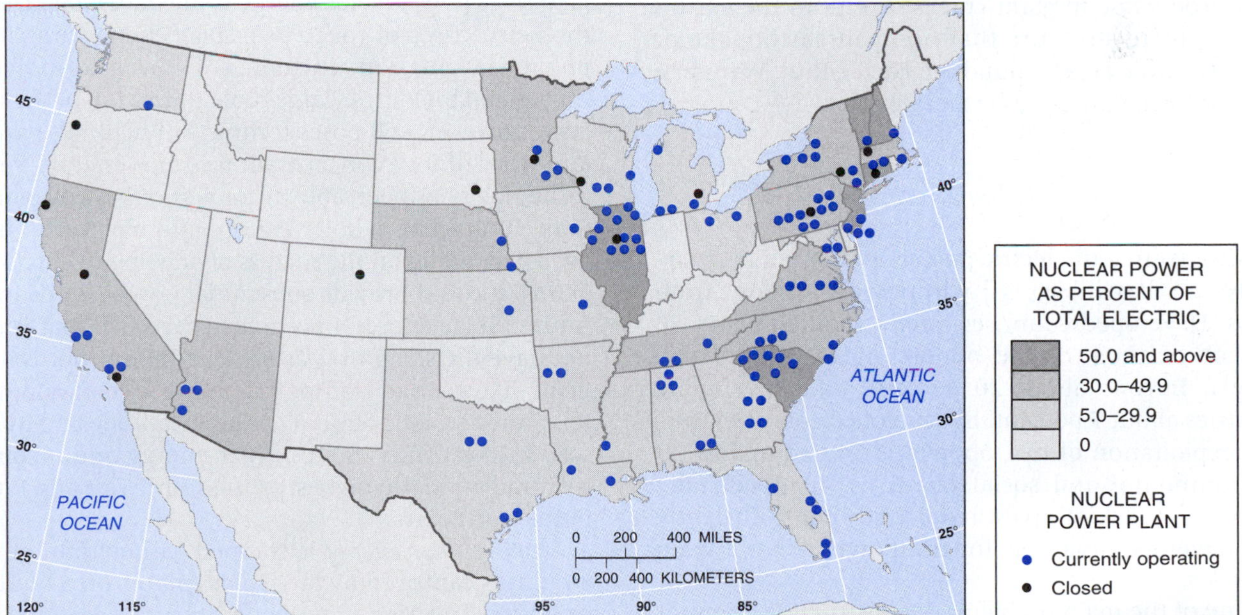

FIGURE 4.23
Nuclear power as a percentage of total energy use, 2001. The most important areas of nuclear energy production in the world include Western Europe and Japan. These are areas that have a relatively small amount of fossil fuels to satisfy local demand for energy. In Europe, for example, France, Germany, Sweden, and Finland produce more than 50 percent of their electrical energy from nuclear sources. Nuclear power is much less prevalent in the developing nations of the world because of extremely high-scale economies, or start-up costs, and the need for expensive uranium fuels.

generation because of high risk and high cost. Sweden began phasing out its nuclear plants in 1995 and is expected to be completed by 2010.

Nuclear fusion, the combination of smaller atoms to release their energy (the process that fuels the sun), has the potential to be a solution to the environmental concerns of nuclear fission because it does not release radioactive waste. The raw material for nuclear fusion is the common element hydrogen. Fusion is the process that powers the sun and can be made to occur artificially but is not yet commercially viable. If research on this technology is successful, nuclear fusion would provide limitless amounts of very cheap electricity and pose no radiation dangers.

Geothermal Power

The development of geothermal power holds promise for the future in several countries that have hot springs, geysers, and other underground supplies of hot water that can easily be tapped. The occurrence of this renewable resource is highly localized, however. New Zealand obtains about 10 percent of its electricity from this source, and smaller quantities are utilized by other countries such as Italy, Japan, Iceland, and the United States. If the interior of the earth's molten magma is sufficiently close to the surface (i.e., 10,000 feet), underground water may be sufficiently warm to produce steam that can be tapped by drilling geothermal wells. Geothermal energy is most producible in giant cracks or rifts in the earth's tectonic plate structure that occur in earthquake or active volcanic areas around the Pacific Rim. Wyoming and California are noted examples.

Hydropower

Another source of electric power, and one that is virtually inexhaustible, is hydropower—energy from rivers. Developed countries have exploited about 50 percent of their usable opportunities, Russia and Eastern Europe about 20 percent, and developing countries about 7 percent. In developed countries, further exploitation of hydropower is limited mainly by environmental and social concerns. In developing countries, a lack of investment funds and sufficiently well developed markets for the power are the main obstacles.

One of the main problems of constructing dams for hydropower is the disruption to the natural order of a watercourse. Behind the dam, water floods a large area, creating a reservoir; below the dam, the river may be reduced to a trickle. Both behind and below the dam, the countryside is transformed, plant and animal habitats are destroyed, farms are ruined, and people are displaced. Moreover, the creation of reservoirs increases the rate of evaporation and the salinity of the remaining water. In tropical areas, reservoirs harbor parasitic diseases, such as schistosomiasis. For example, since the construction of the Aswan High Dam in the 1970s, schistosomiasis has become endemic in lower Egypt, infecting up to one-half of the population. An additional problem is the silting of reservoirs, reducing their potential to produce electricity. The silt, trapped in reservoirs, cannot proceed downstream and enrich agricultural land. The decrease in agricultural productivity from irrigated fields downstream from the Aswan High Dam has been substantial. China is completing the Three Gorges Dam on the Yangtze River, which will be the largest hydroelectric dam on the planet, and may well experience similar problems there.

The long-term hydrological, ecological, and human costs of dams easily transform into political problems on international rivers. An example is Turkey's Southeast Anatolia Project, which envisages the construction of 22 dams and 19 hydroelectric power plants. Because the project is being developed on two transboundary rivers—the Tigris and Euphrates—problems and disputes have arisen with two downstream users—Syria and Iraq—whose interests the project affects.

Solar Energy

Like river power, solar energy is inexhaustible. During the petro crises of the 1970s, solar energy caught the public imagination, including a few solar-powered homes and buildings. Large-scale utilization of solar energy, however, still poses technical difficulties, particularly that of low concentration of the energy. So far, technology has been able to convert only slightly more than 30 percent of solar energy into electricity; however, depending on the success of ongoing research programs, it could provide substantial power needs in the future. Solar energy's positive aspects are that it does not have the same risks as nuclear energy, nor is it difficult, like coal, to transport, and it is free of pollution. It is almost ubiquitous but varies by latitude and by season. In the United States, solar energy and incoming solar radiation are highest in the southwest and lowest in the northeast.

Passive solar energy is trapped rather than generated. It is captured by large glass plates on a building or house. The greenhouse effect is produced by shortwave radiation from the sun. Once the rays penetrate the glass, they are converted to long-wave radiation and are trapped within the glass panel, thus heating the interior of the structure or water storage tank. The other way of harnessing solar energy is through *solar energy,* including photovoltaic cells

The large hyperbolic cooling tower and reactor containment dome of the Trojan nuclear power plant in Rainier, Oregon. Safety issues surrounding the use of nuclear energy are fraught with turmoil. Most OECD countries expanded their nuclear energy production during the last 20 years, with France and Japan in the lead. Expansion of nuclear capacity had slowed by 1998 because of cost concerns and the chilling effects of the accidents at Three Mile Island in Pennsylvania and at Chernobyl in the Ukraine. New energy sources, such as geothermal, solar, biomass, and wind energy, have increased and now provide up to 5 percent of total primary energy requirements in Australia, Austria, Canada, Denmark, Sweden, and Switzerland.

made from silicon. A bank of cells can be wired together and mounted on the roof with mechanical devices that maximize the direct sun's rays by moving at an angle proportional to the light received. Another type of active solar energy system is a wood or aluminum box filled with copper pipes and covered with a glass plate, which collects solar radiation and converts it into hot water for homes and swimming pools.

Cost is the main problem with solar energy. High costs are associated with the capturing of energy in cloudy areas and high latitudes. But unlike fossil fuels, solar energy is difficult to store for long periods without large banks of cells or batteries. Currently, solar energy production is more expensive than other sources of fuel. To promote the development of innovative energy supplies when the Arab oil embargo hit in the 1970s, the U.S. government offered tax incentives, including tax deductions for solar units mounted on housetops. Although this tax deduction offset the high costs of constructing solar energy systems, maintenance and reliability soon become a problem. Families that move lose their investment, because most systems installed are rarely recoverable in the sale price of homes.

Wind Power

The power of the wind provides an energy hope for a few areas of the world where there are constant surface winds of 15 mph or more. The greatest majority of *wind farms* in the United States are in California. However, wind machines are an expensive investment, and the initial cost plus the unsightliness of the wind machine has ended most wind farm projects. Wind farm potential in California has never matched expectations, and wind farming stagnated; however, it is currently experiencing a revival.

Biomass

Still another form of renewable energy is biomass—wood and organic wastes. Today, biomass accounts for about 14 percent of global energy use. For Nepal, Ethiopia, and Tanzania, more than 90 percent of total energy comes from biomass. The use of wood for cooking—the largest use of biomass fuel—presents enormous environmental and social problems because it is being consumed faster than it is being replenished. Fuelwood scarcities—the poor world's energy crisis—affect 1.5 billion people and could affect 3 billion in the future unless corrective actions are taken.

With good management practices, biomass is a resource that can be produced renewably. It can be converted to alcohol and efficient, clean-burning fuel for cooking and transportation. Its production and conversion are labor intensive, an attractive feature for developing countries that face underemployment and unemployment problems. But the low efficiency of photosynthesis requires huge land areas for energy crops if significant quantities of biomass fuels are to be produced.

ENVIRONMENTAL DEGRADATION

On some days in Los Angeles, pollution levels reached what is called locally a *level three alert,* although conditions there have improved recently. People are advised to stay indoors, cars are ordered off the highways, and strenuous exercise is discouraged. In Times Beach, Missouri, which is some 50 miles south of St. Louis, dioxin levels from a contaminated plant became so high that the Environmental Protection Agency required the town to be closed and the residents to be relocated. Around Rocky Flats, nuclear wastes of plutonium have degraded the soil so that radioactivity levels are five times higher than normal. In New England, acid rain has become so bad that it has killed vegetation and fish in rivers and lakes.

Environmental problems, caused mainly by economic activity, may be divided into three overlapping categories: (1) pollution, (2) wildlife and habitat preservation, and (3) environmental equity.

Pollution

Pollution is a discharge of waste gases and chemicals into the air, land, and water. Such discharge can reach levels sufficient to create health hazards to plants, animals, and humans, as well as to reduce and degrade the environment. The natural environment has the capacity to regenerate and cleanse itself on a normal basis; however, when great amounts of gases and solids are released into it from industrial economic activity, recycling and purification needs are sometimes overwhelmed. From that point on, the quality of the environment is reduced as pollutants create health hazards for humans, defoliate forests, inundate land surfaces, reduce fisheries, and burden wildlife habitats.

Air Pollution

Air pollutants, the main sources of which are illustrated in Figure 4.24, are normally carried high into

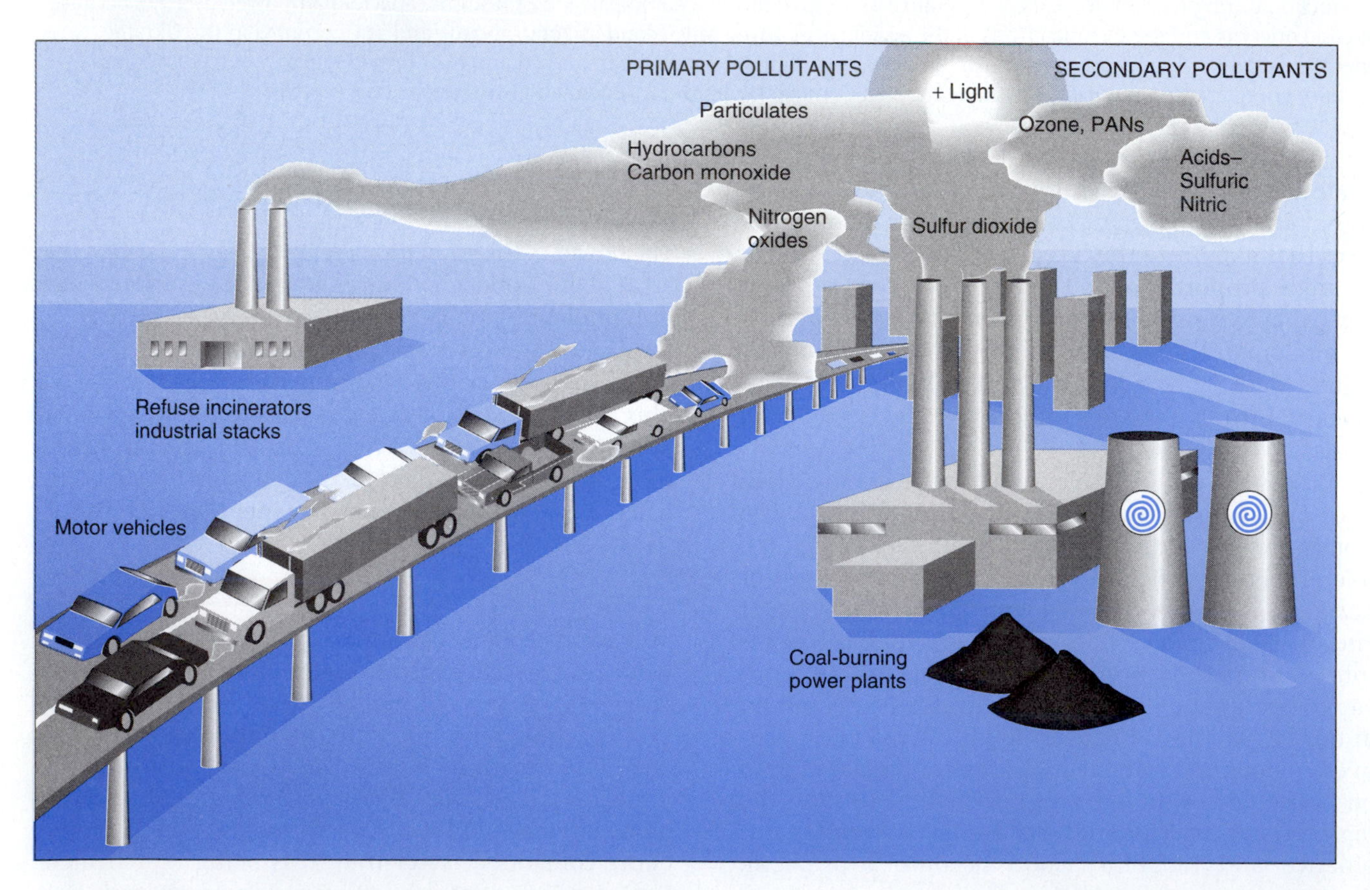

FIGURE 4.24
The primary sources of major air pollutants. Industrial processes are the major sources of particulates; transportation and fuel combustion cause the lion's share of the other pollutants.

the atmosphere, but occasionally, and in some places more than others, a temperature inversion prevents this from occurring. Inversions are caps of warm air that prevent the escape of pollutants to higher levels. Under these conditions, the inhabitants of a place are under an even greater risk. These conditions promote the formation of smog that blocks out sunshine, causes respiratory problems, stings the eyes, and creates a haze over large, congested cities everywhere.

Air pollution gives rise to different concerns at different scales. Air pollution at the local scale is a major concern in cities because of the release of carbon monoxide, hydrocarbons, and particulates. Air pollution at the regional scale is exemplified by the problem of acid precipitation in eastern North America and Eastern Europe (Figure 4.25).

At the global scale, air pollution may damage the atmosphere's *ozone layer* and contribute to the threat of *global warming*. The earth's protective ozone layer is thought to be threatened by pollutants called *CFCs (chlorofluorocarbons)*. When CFCs such as freon leak from appliances such as air conditioners and refrigerators, they are carried into the stratosphere, where they contribute to ozone depletion. As a result of the 1987 Montreal Protocol, developed countries stopped using CFCs by the year 2000 and developing countries must stop by 2010. Scientists hope that this international agreement will effectively reduce ozone depletion.

Concern about global warming centers around the burning of fossil fuels in ever greater quantities, which increases the amount of carbon dioxide in the air, which in turn makes the atmosphere more opaque, reducing thermal emission to space. Heat-trapping gases, such as carbon dioxide, warm the atmosphere, enhancing a natural *greenhouse effect* (Figure 4.26). The vast majority of these are produced by industrialized economies. Since the 1890s, the average temperature of the earth's surface has increased by 2 degrees Fahrenheit. This increase in temperature may or may not be humanly induced, however. There are divergent views on the issue. Nonetheless, even if the observed global warming is consistent with natural variability of the climate system, most scientists

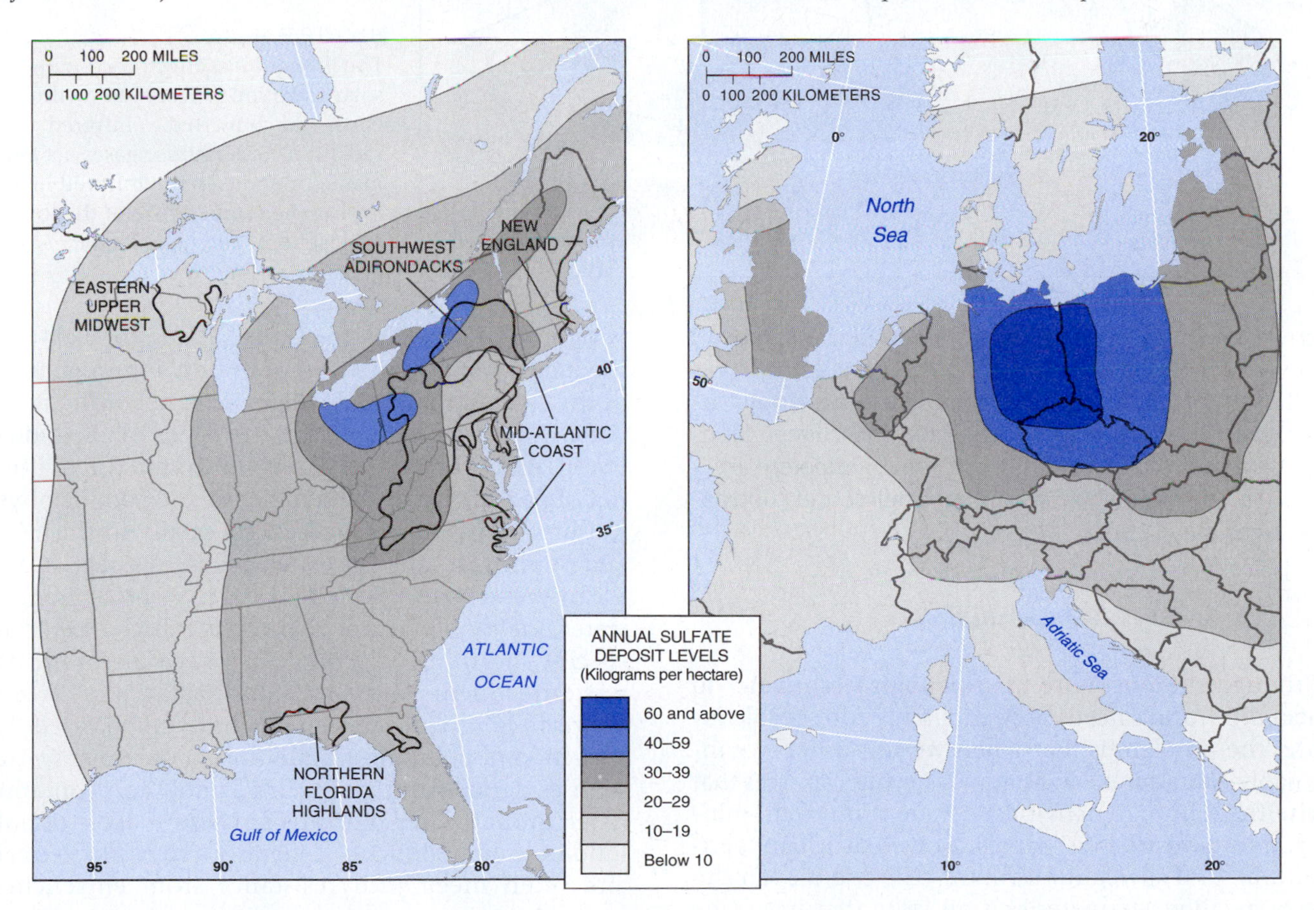

FIGURE 4.25
The worst inflicted areas of acid rain in North America occur downwind from the principal polluting regions of the industrialized Midwest. Ohio, western Pennsylvania, and northern New York State are the areas most heavily inflicted with acid rain deposits. Acid rain and snow deposits are also well documented in Europe, which is in a belt of prevailing wind coming from the west, as is the United States. Sulfur, released into the atmosphere from the burning of coal, combines with water vapor to produce sulfuric acid. Such acid creates substantial air pollution and etches away at limestone buildings, monuments, and markers on the earth. Acid precipitation can also kill plant and animal life, especially aquatic life. Literally thousands of lakes in Sweden and Norway no longer support the fish they once did.

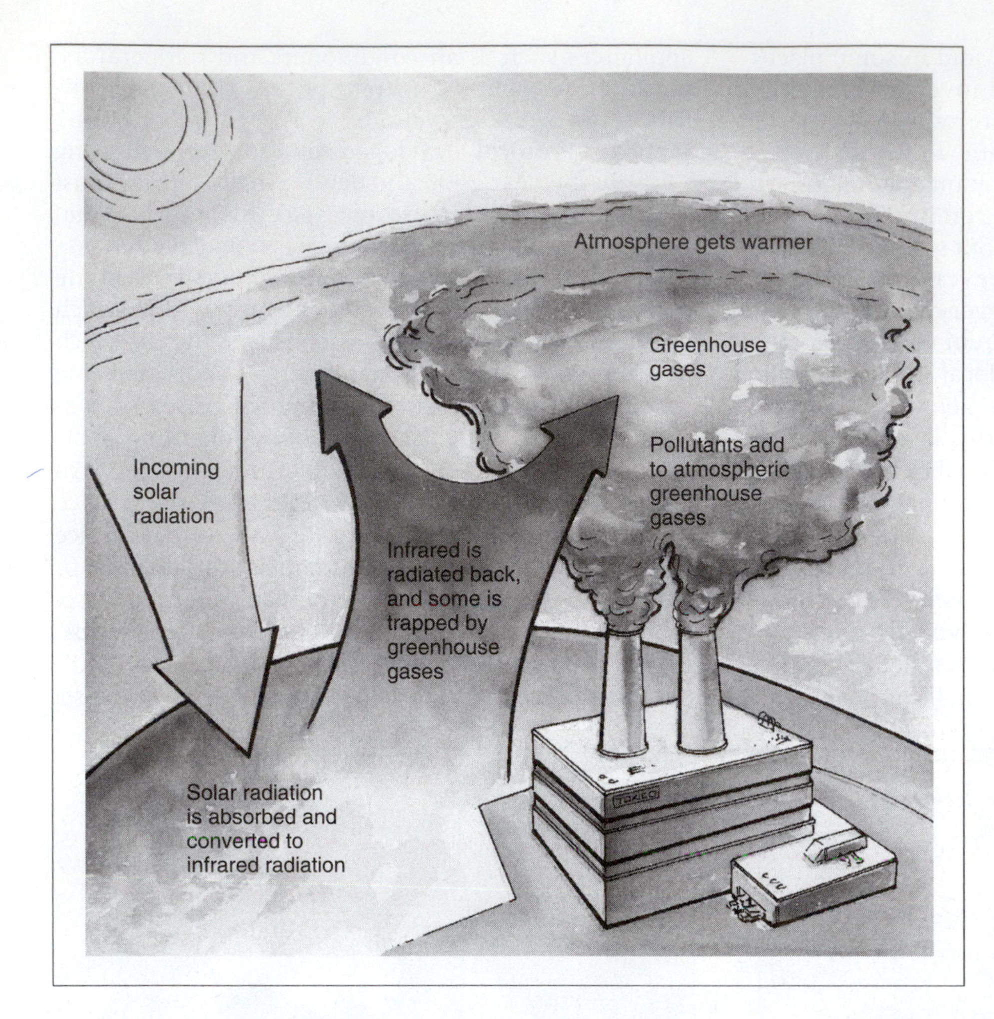

FIGURE 4.26
The greenhouse effect. Incoming solar radiation is absorbed by the earth and converted to infrared radiation. Greenhouse gases act as an insulator and trap the infrared heat, raising the temperature of the lower atmosphere. The overall effect is global warming.

agree that it is socially irresponsible to delay actions to slow down the rate of anthropogenic greenhouse gas buildup. For example, continued warming would increase sea levels, disrupt ecosystems, and change land-use patterns. While agriculture in some temperate areas may benefit from global warming, tropical and subtropical areas may suffer.

Water Pollution

Although there is more than enough fresh water to meet the world's needs now and in the foreseeable future, the problem is that when we use water, we invariably contaminate it. Major wastewater sources that arise from human activities include municipal, mining, and industrial discharges, as well as urban, agricultural, and silvicultural runoff. The use of water to carry away waste material is an issue that has come into prominence, because water is being used more heavily than ever before. As populations and standards of living rise, problems of water utilization and management increase. These problems are most acute in developing countries, where some 1 billion people already find it difficult or impossible to obtain acceptable drinking water. But water is also an issue in developed countries. In the United States, for example, the major water management problem through most of the twentieth century focused on acquiring additional water supplies to meet the needs of expanding populations and associated economic activities. Recently, water management has focused on the physical limits of water resources, especially in the West and Southwest, and on water quality. Passage of the Clean Water Act in 1972 resulted in improvement in water quality of streams that receive discharges from specific locations or *point sources* such as municipal waste-treatment plants and industrial facilities. Recent efforts to improve water quality have emphasized the reduction of pollution from diffuse or *nonpoint sources* such as agricultural and urban runoff and contaminated groundwater discharges. These sources of pollution are often difficult to identify and costly to treat, and often meet with resistance from entrenched agribusiness interests.

Wildlife and Habitat Preservation

Wildlife and habitat preservation for plants and animals called *renewable natural resources* are in danger throughout the world. These natural environments are critical reserves for many species of plants and animals. Wildlife, forest lands, and wetlands, including lakes, rivers and streams, and coastal marshes, are subject to acid rain, toxic waste, pesticide discharge, and urban pollution. They are also endangered by encroachment of land development and transportation facilities worldwide. The demand for tropical hardwoods, such as teak and mahogany, has already stimulated waves of destruction in tropical rainforests. In the United States alone, expanding economic activity has consumed forests and wetlands, depleted topsoil, and polluted local ecosystems at a rapid rate. Many species of plants and animals have been reduced, including the grizzly bear, American bison, prairie dog, gray wolf, brown pelican, Florida panther, American alligator and crocodile, and a variety of waterfowl and tropical birds. All in all, the exponential growth of human beings has been closely associated with a corresponding dramatic decline in the number of species in the world (Figure 4.27), and future economic growth may threaten yet another catastrophic round of loss in the planet's biodiversity.

The problem of wildlife and habitat preservation is exacerbated by the need for economic gain and corporate profit. A variety of questions beset wildlife managers and environmental farmers. Should farmers be permitted to drain swamps in Louisiana to farm the land, thus destroying the habitat for alligators? Should forest fires started by lightning be allowed to burn themselves out, as has been the practice on western U.S. forests and rangelands? The trade-off of residential lands versus wetlands, wildlife migration versus forest management, highway safety versus habitat preservation, and conservation versus real economic development and growth of the U.S. economy are difficult issues. It is difficult to select the best alternative, and policies may reflect the power of entrenched economic interests as much as the public welfare.

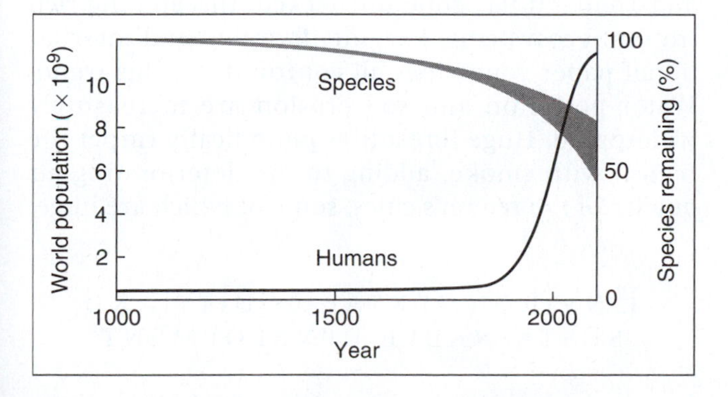

FIGURE 4.27
An inverse relationship exists between human population size and the survival of species worldwide. Uncertainty about the extent of decreasing biodiversity is reflected in the width of the species curve.

Regional Dimensions of Environmental Problems

Because economic structures and change are not uniform around the world but exhibit great variation among the world's continents (as well as within them), and because population growth, cultural patterns, standards of living, and state policies regulating problems such as pollution vary widely, the environmental problems unleashed by capitalism and demographic change worldwide differ considerably.

In North America, the major environmental problems include acid rain downwind from industrial source areas (Figure 4.28). In addition, water pollution and withdrawal of groundwater are serious issues. Because the U.S. economy is so huge and energy intensive, the pollutants generated there are major sources of global problems such as the depletion of the ozone layer and global warming. Across the face of North America, other issues such as wetlands destruction, saltwater intrusion, and urban air quality are serious predicaments.

In Latin America, the primary environmental issue is deforestation, particularly in the Amazon River basin (Figure 4.29), where farmers and logging companies are stripping away the forest cover of some of the world's richest ecosystems. Further, degradation of ground cover leads to mud slides that can be lethal to thousands. Overgrazing and soil erosion are other important consequences, and in many Latin American cities, such as Mexico City, air quality is poor.

In Europe, these issues range from pollution of the Mediterranean Sea, whose coasts are populated by dense clusters of cities, as well as acid rain in Germany and Poland (Figure 4.30). Rising sea levels pose a threat to the Netherlands, much of which lies below sea level and uses dikes to hold the ocean back. Air and water pollution are constant problems requiring government intervention.

Environmental problems in Russia and its neighbors are tangled up in decades of mismanagement by the Soviet regime and an economy that collapsed in the 1990s (Figure 4.31). The Chernobyl disaster in 1985 left parts of the Ukraine polluted with nuclear waste. The extraction of water from the Aral Sea has left it nearly dead. And air pollution and acid rain take a toll on the region's enormous forests.

The predominantly Muslim world of North Africa and the Middle East has its own environmental problems (Figure 4.32). Soil erosion and overgrazing have contributed to desertification, particularly in the Sahara. Egypt's Aswan High Dam has been a mixed blessing, reducing floods on the Nile but also reducing the siltation that keeps farmlands fertile. Heavy use of river water contributes to soil salinization. All this takes place in a region with one of the world's highest rates of population growth.

Sub-Saharan Africa's environmental problems, framed in the context of extreme poverty and rapid population

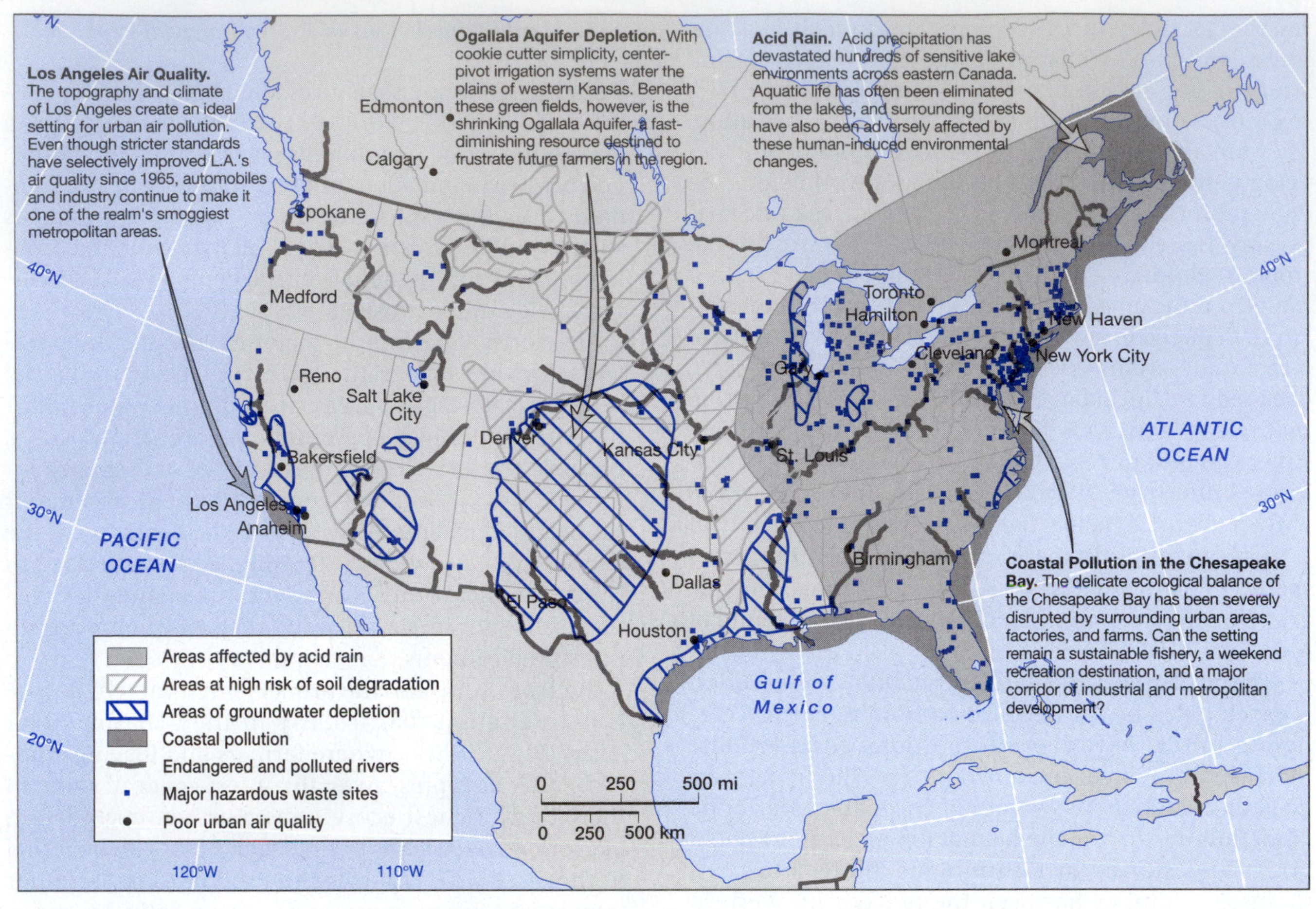

FIGURE 4.28
Environmental concerns and issues of North America.

growth, include widespread deforestation in western Africa (Figure 4.33). A continent in which islands of people were surrounded by oceans of wildlife has become one of islands of wildlife surrounded by oceans of people. Overgrazing and soil erosion threaten agricultural land as well as biodiversity, and the region is rocked by wars, drought, famines, and diseases such as AIDS and malaria.

In South Asia, the Indian subcontinent, home to more than 1.5 billion people, has seen widespread soil salinization and water and air pollution, which conspire to reduce the supply of agricultural land (Figure 4.34). Green Revolution farming accentuates these issues. Deforestation in the Himalaya mountains has increased flooding downstream in Bangladesh.

East Asia, home to 1.3 billion Chinese as well as hundreds of millions more in Korea and Japan, exhibits similar problems (Figure 4.35). The encroachment of farmers into forests has reduced the habitats of many species. Trying to stop periodic floods on the Yangtze River, the Chinese government is finishing the Three Gorges Dam, which will modify the basin and ecosystems of Asia's largest river. Japan supports 127 million people in a country with little arable land, and its dense cities exhibit severe air pollution levels, although the rapid growth of Chinese cities has rendered the air quality there considerably worse yet.

Southeast Asia, one of three major regions in the world that sustain rainforests, has enjoyed rapid economic growth over the last two decades. Its environmental problems have grown proportionately (Figure 4.36). Deforestation in Indonesia, Malaysia, and Thailand has gone unchecked, threatening rich tropical ecosystems. Logging, farming, and international paper companies all contribute to this trend. Water pollution and soil erosion are increasingly widespread. Huge forest fires periodically carpet the region with smoke, adding to the deteriorating air quality of the region's cities, some of which are huge.

Environmental Equity and Sustainable Development

Economic development policies and projects all too often carry as many costs as benefits. Build a dam to bring hydroelectric power or irrigation water, and fertile river bottomlands are drowned, farmers are displaced, waterborne diseases may fester in the still

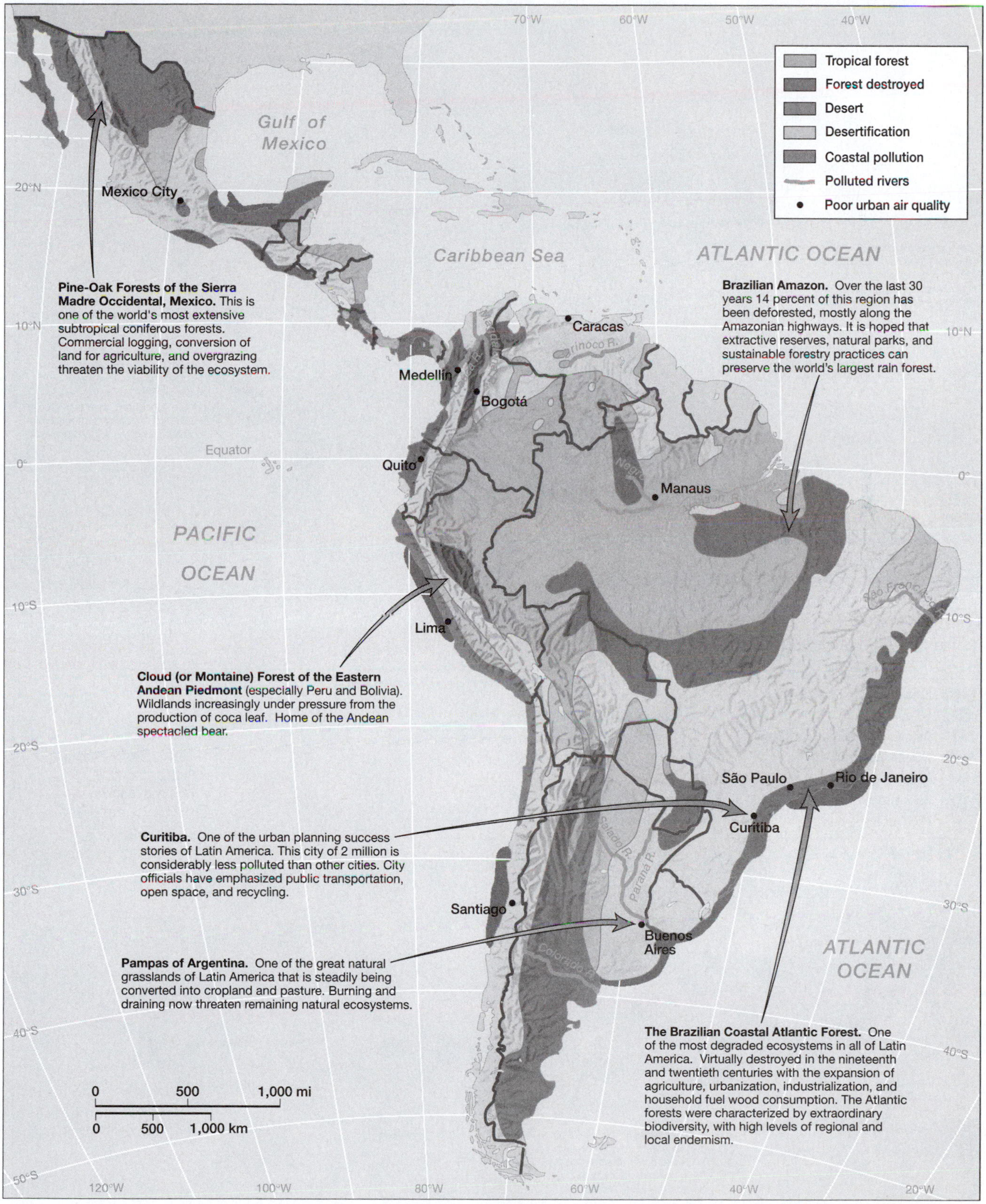
Tropical forest
Forest destroyed
Desert
Desertification
Coastal pollution
Polluted rivers
Poor urban air quality
70°W
60°W
50°W
40°W
20°N
10°N
0°
10°S
20°S
30°S
40°S
50°S
120°W
100°W
80°W
60°W
40°W
20°W
Gulf of Mexico
Caribbean Sea
ATLANTIC OCEAN
PACIFIC OCEAN
ATLANTIC OCEAN
Equator
Mexico City
Caracas
Medellín
Bogotá
Quito
Manaus
Lima
São Paulo
Rio de Janeiro
Curitiba
Santiago
Buenos Aires
Pine-Oak Forests of the Sierra Madre Occidental, Mexico. This is one of the world's most extensive subtropical coniferous forests. Commercial logging, conversion of land for agriculture, and overgrazing threaten the viability of the ecosystem.
Brazilian Amazon. Over the last 30 years 14 percent of this region has been deforested, mostly along the Amazonian highways. It is hoped that extractive reserves, natural parks, and sustainable forestry practices can preserve the world's largest rain forest.
Cloud (or Montaine) Forest of the Eastern Andean Piedmont (especially Peru and Bolivia). Wildlands increasingly under pressure from the production of coca leaf. Home of the Andean spectacled bear.
Curitiba. One of the urban planning success stories of Latin America. This city of 2 million is considerably less polluted than other cities. City officials have emphasized public transportation, open space, and recycling.
Pampas of Argentina. One of the great natural grasslands of Latin America that is steadily being converted into cropland and pasture. Burning and draining now threaten remaining natural ecosystems.
The Brazilian Coastal Atlantic Forest. One of the most degraded ecosystems in all of Latin America. Virtually destroyed in the nineteenth and twentieth centuries with the expansion of agriculture, urbanization, industrialization, and household fuel wood consumption. The Atlantic forests were characterized by extraordinary biodiversity, with high levels of regional and local endemism.
0
500
1,000 mi
0
500
1,000 km

FIGURE 4.29

Environmental concerns and issues of Latin America.

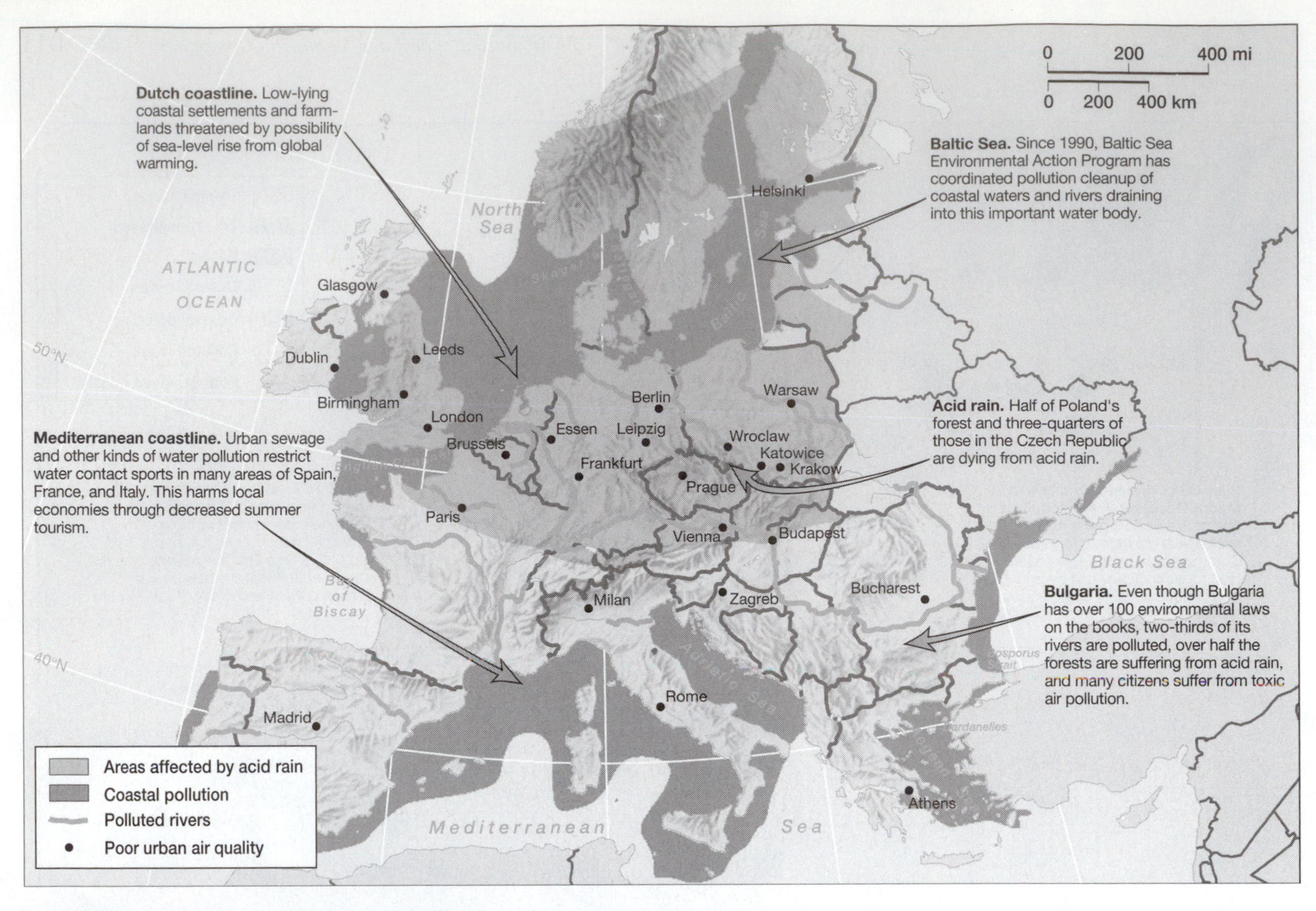

FIGURE 4.30
Environmental concerns and issues of Europe.

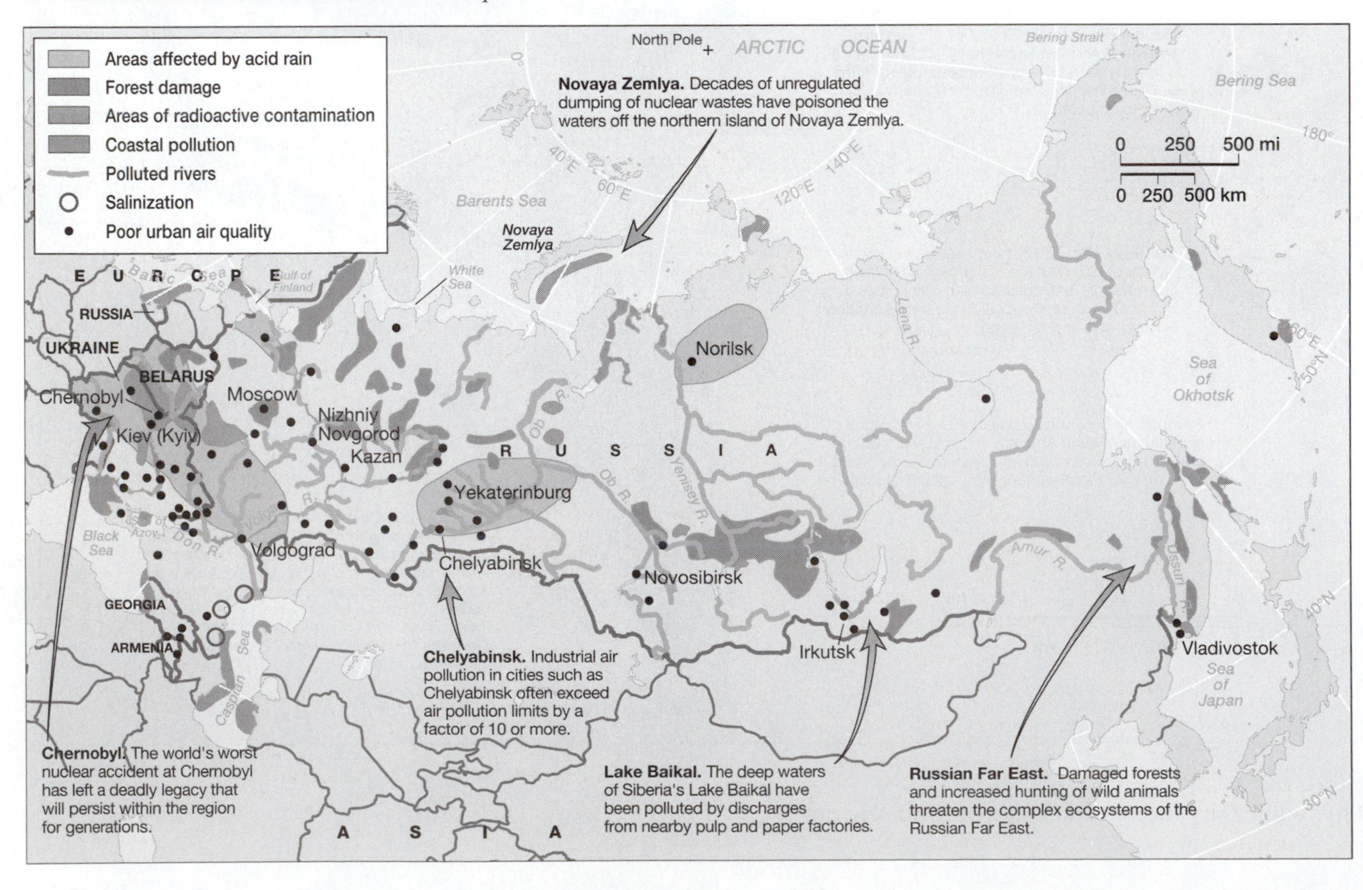

FIGURE 4.31
Environmental concerns and issues of Russia and neighbors.

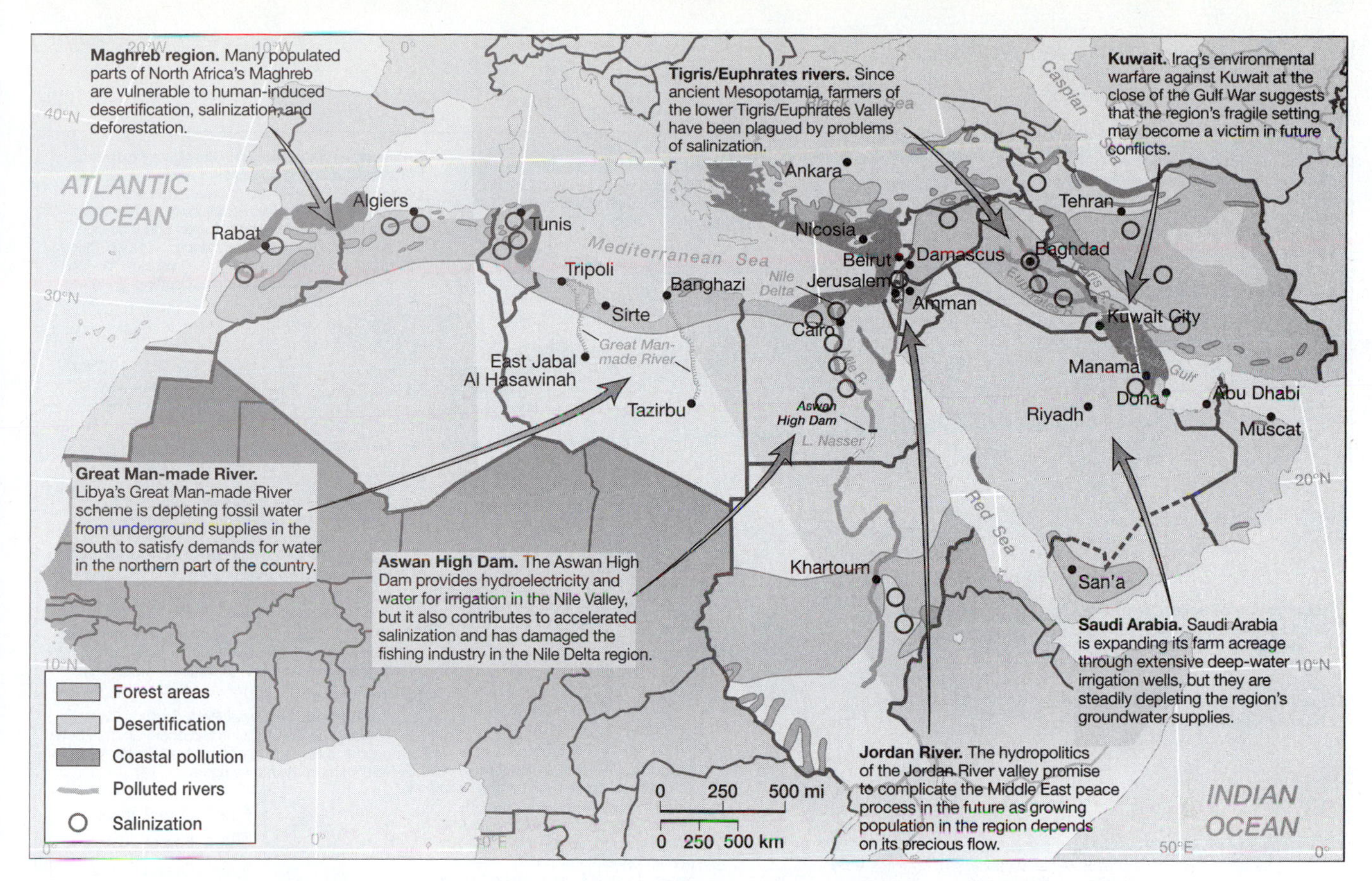

FIGURE 4.32
Environmental concerns and issues of Southwestern Asia and Northern Africa.

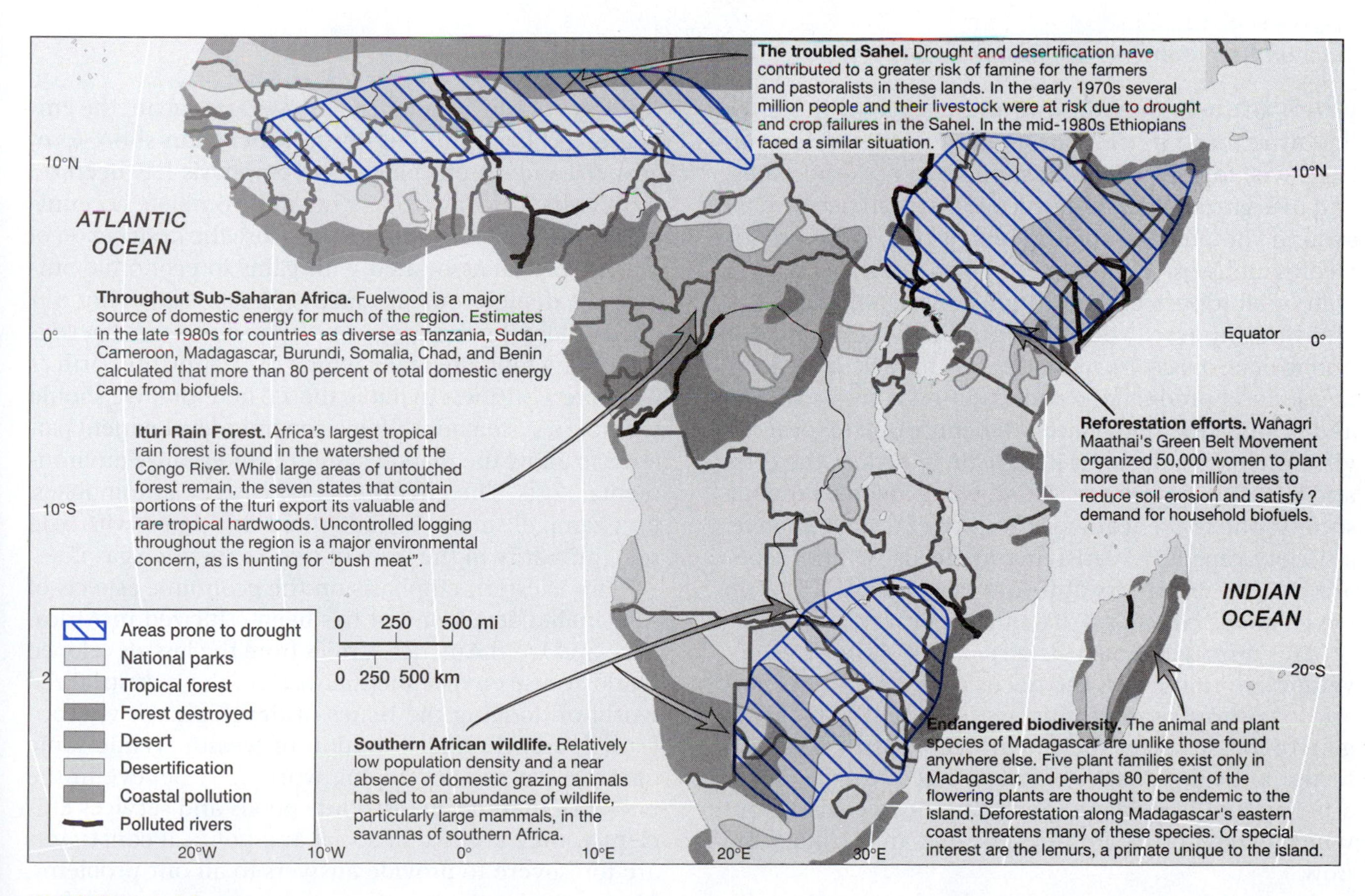

FIGURE 4.33
Environmental concerns and issues of Subsaharan Africa.

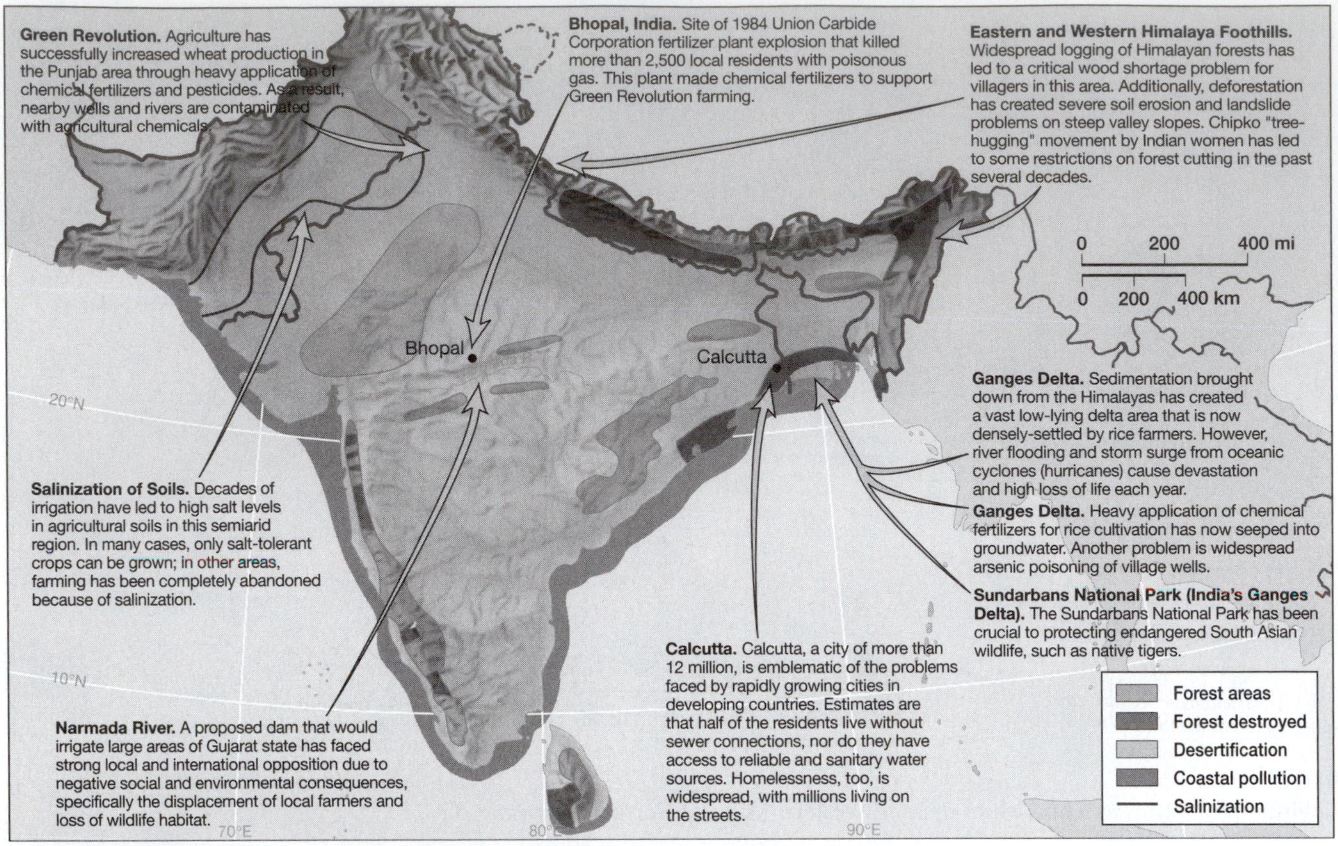

FIGURE 4.34
Environmental concerns and issues of South Asia.

waters of the reservoir, and the course of the river downstream to its delta or estuary is altered forever. Dig a well to improve water supplies in dry rangelands, and overgrazing and desertification spread outward all around the points of permanent water. Mine ore for wealth and jobs, and leave despoiled lands and air and water pollution. Build new industries, shopping malls, and housing tracts, and lose productive farmland or public open space. Introduce a new miracle crop to increase food production, and traditional crop varieties and farming methods closely synchronized to local environments disappear. Is it possible to reckon the costs and benefits of "progress"? Can we develop a humane society, one that encourages both equity and initiative, a society capable of satisfying its needs without jeopardizing the prospects of future generations? How do we go about creating a sustainable society?

The term *sustainable development* is defined as development that meets the needs of the present generation without compromising the ability of future generations to meet their own needs. Most accept its focus on the importance of long-range planning, but as a policy tool it is vague, providing no specifics about which needs and desires must be met and fulfilled and how.

In the debate on sustainable development, two different emphases have emerged. In the industrialized countries of North America, Europe, and Japan, the emphasis has been on long-term rather than short-term growth, and on efficiency. The emphasis has been on economics: If today we rely on an incomplete accounting system, one that does not measure the destruction of natural capital associated with gains in economic output, we deplete our productive assets, satisfying our needs today at the expense of our children. There is something fundamentally wrong in treating the earth as if it were a business in liquidation. Therefore, we should promote a systematic shift in economic development patterns to allow the market system to internalize environmental costs. The environmental costs of automobiles, for example, should include those associated with acid rain, primarily in the form of more expensive gasoline.

This Western emphasis on the economic aspects of sustainable development has been criticized in Africa, Asia, and Latin America. Critics from the less-developed world accuse environmentalists from the industrialized world of dodging the issues of development without growth and the redistribution of wealth. While some observers in the developing world may believe in the power of markets to distribute goods and services efficiently, they argue that social and political constraints are too severe to provide answers to all our problems. Many criticize the West's excessive consumption of resources. Many advocates in the developing world put

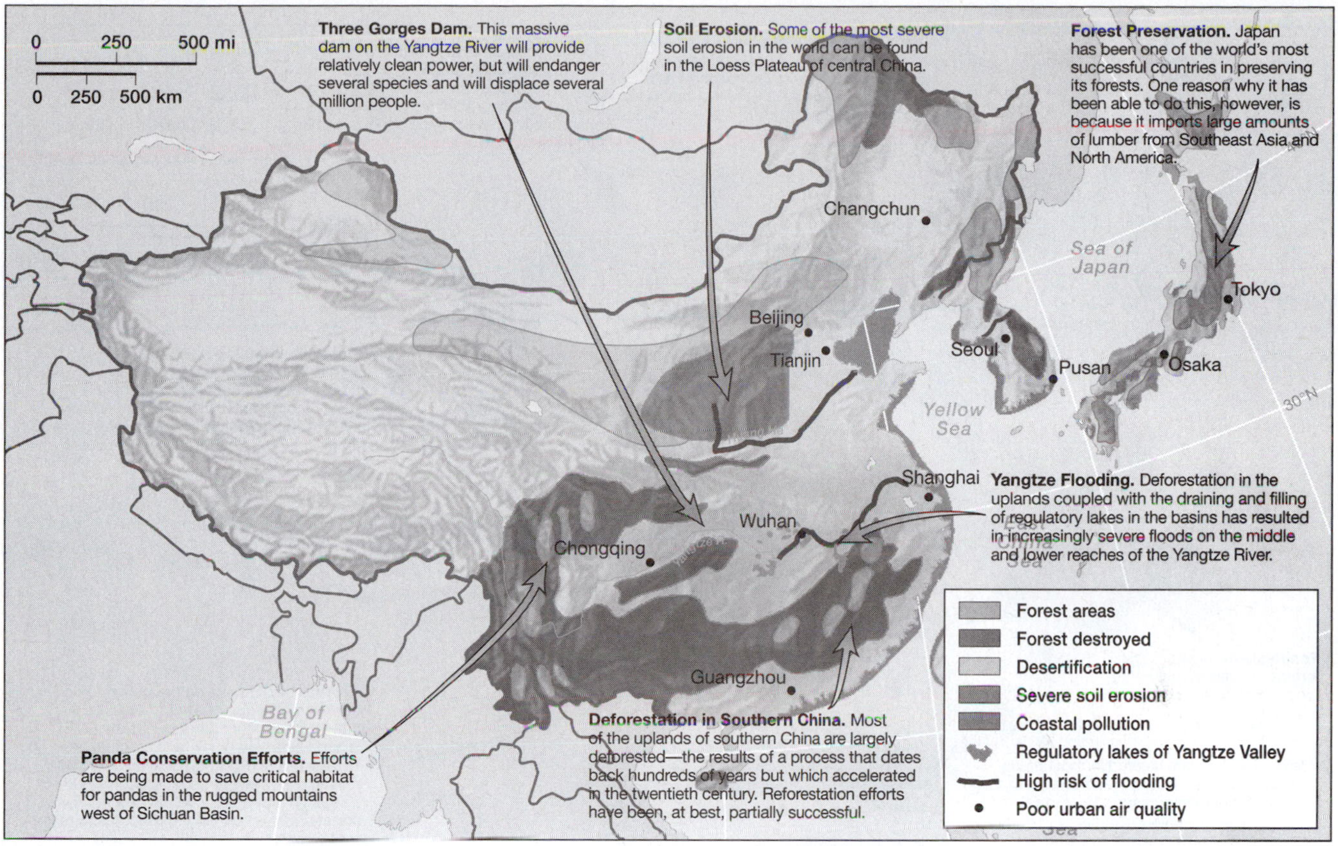

FIGURE 4.35
Environmental concerns and issues of East Asia.

basic human needs ahead of environmental concerns. Let us work toward a sustainable future, but let us do so by ensuring food, shelter, clean water, health care, security of person and property, education, and participation in governance for all. An extension of this sentiment is a desire to protect basic values as well—to respect nature rather than dominate it, and to use the wisdom of indigenous groups to reexamine current, mostly Western, structures of government, and the relationships that people have with the environment.

Is the consumerist West ready to listen to those with different values and priorities? Surely there are many paths to a sustainable future, each determined by individualized priorities of what is desired and therefore worthy of sustaining. Surely too, in following those paths, we must recognize that future growth will be constrained by resources that are finite or whose availability is difficult to determine. Finally, we must realize that no region can achieve sustainability in isolation. A desirable and sustainable future will be the result of many social and policy changes, some small and at the local level, others international and far reaching. If we accept that the futures of rich and poor are inextricably linked, perhaps we will achieve the humility necessary for compromise. A world that only rewards the rich, however, at the expense of the poor is doomed to social inequality and environmental destruction.

From a Growth-Oriented to a Balance-Oriented Lifestyle

Given the dynamics of the market system, it is unlikely that energy availability will place a limit on economic growth on the earth; however, ultimately, drastic changes in the rate and nature of the use of energy resources are certain. The ultimate limits to the use of energy will be determined by the ability of the world's ecosystems to dissipate the heat and waste produced as more and more energy flows through the system.

In countless ways, energy improves the quality of our lives, but it also pollutes. As the rate of energy consumption increases, so too does water and air contamination. Sources of water pollution are numerous: industrial wastes, sewage, and detergents; fertilizers, herbicides, and pesticides from agriculture; and coastal oil spills from tankers. Air pollution reduces visibility; damages buildings, clothes, and crops; and endangers human health. It is especially serious in urban-industrial areas, but it occurs wherever waste gases and solid particulates are released into the atmosphere.

Pollution is the price paid by an economic system emphasizing ever-increasing growth as a primary goal. Despite attempts to do something about pollution problems, the growth-oriented lifestyle characteristic of Western urban-industrial society continues to widen

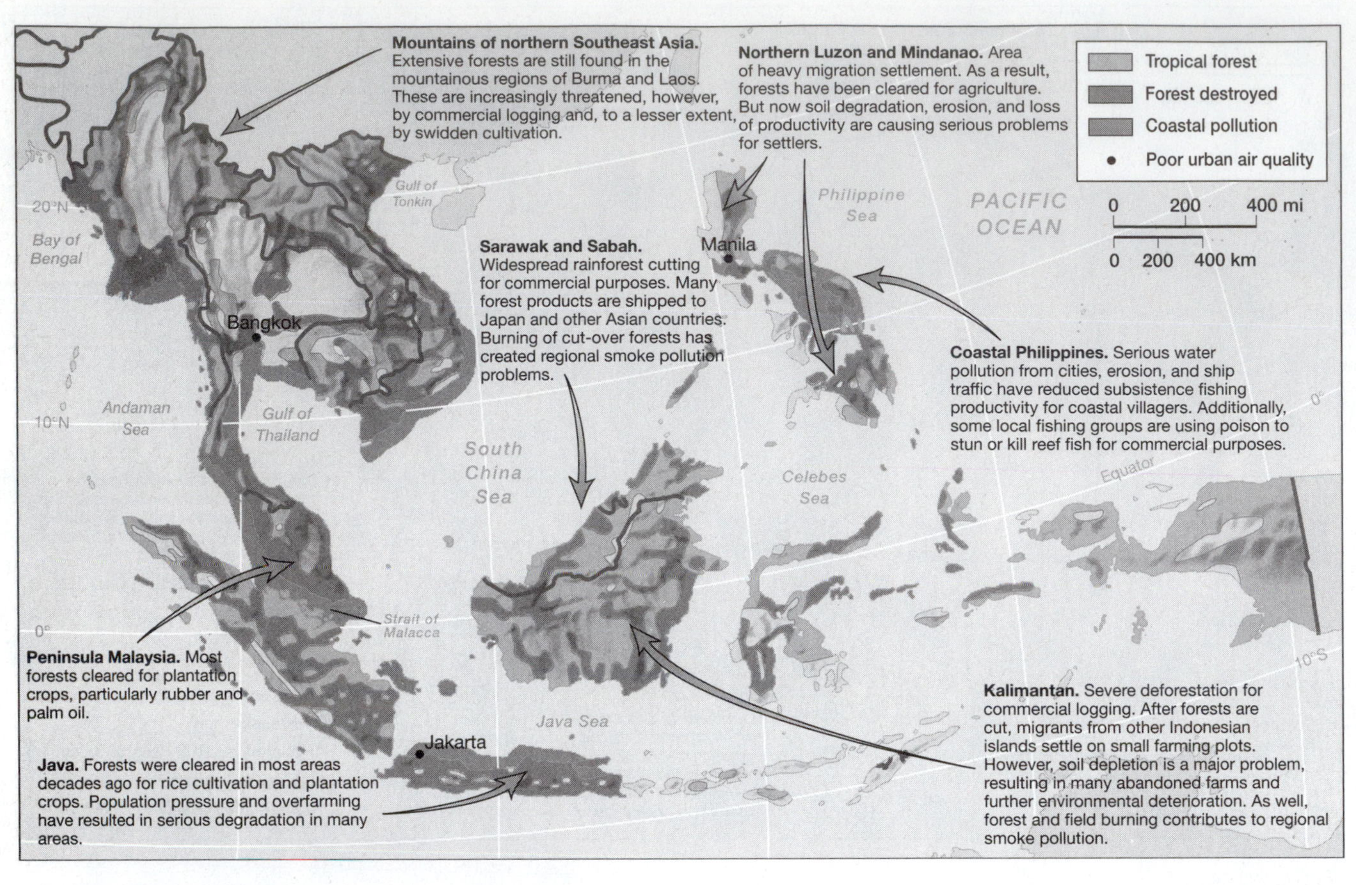

FIGURE 4.36
Environmental concerns and issues of Southeast Asia.

the gap between people and nature. "Growthmania" is a road to nowhere. Many argue that we must transform our present linear or growth-oriented economic system into a balance-oriented system. A balance-oriented economy explicitly recognizes natural systems. It recognizes that resources are exhaustible, that they must be recycled, and that input rates must be reduced to levels that do not permanently damage the environment. A balance-oriented economy does not mean an end to growth but a new social system in which only desirable low-energy growth is encouraged. It requires a deemphasis on the materialistic values we have come to hold in such high esteem. If current resource and environmental constraints lead us to place a higher premium on saving and conserving than on spending and discarding, then they may be viewed as blessings in disguise.

SUMMARY

We conclude this chapter by restating the resources-population problem. It is possible to solve resource problems by (1) changing societal goals, (2) changing consumption patterns, (3) changing technology, and (4) altering population numbers. In the Western world, much of the emphasis is on technological advancement and population control.

Following a review of renewable and nonrenewable resources, we explored the question of food resources. The food "crisis" is essentially a consequence of social relations, including war and disruptions of agricultural systems. Food production is increasing faster than population growth, yet more people are hungry than ever before. In the course of transforming agriculture into a profit base for the wealthy in the developed and in the less developed worlds, the Third World poor are being forced into increasingly inhospitable living conditions. Famine, like poverty, is a social construction, not a natural event, and its origins, like its solutions, must be found in the uneven distribution of resources among and within countries.

Unlike food, which is replenishable, nonrenewable minerals and fossil fuels, once used, are gone forever. We discuss some of the alternatives to fossil fuels and point to energy conservation as a potent alternative with potential that remains to be fully exploited. Finally, the comparison between growth-oriented and balance-oriented lifestyles underscores the importance of social concerns as they relate to economic growth. Growth and inequality are inextricably linked parts of social change and environmental protection.

STUDY QUESTIONS

1. What is meant by carrying capacity?
2. Differentiate renewable from nonrenewable resources.
3. What are the major causes of Third World hunger?
4. What are three methods of expanding world food production?
5. What was the Green Revolution? Where was it largely located?
6. Summarize major flows of oil on the world market.
7. Where are the major world coal deposits located?
8. What are some alternative energy options to fossil fuels?
9. What are some environmental consequences of high energy use? Be specific.

KEY TERMS

acid rain
balance-oriented lifestyle
biomass
California Environmental Quality Act (CEQA)
carrying capacity
conservation
deforestation
depletion curves
desertification
energy
fossil fuels
geothermal energy
Green Revolution
growth-oriented lifestyle
intermediate technology
limits to growth
marine fisheries
maximum sustainable yield
minerals
mine tailings
National Environmental Policy Act (NEPA)
NIMBY and LULU effects
nonrenewable resources
normal lapse rate
Organization of Petroleum Exporting Countries (OPEC)
overpopulation
price ceiling
recycling
renewable resources
reserve
reserve deficiency minerals
resource
second law of thermodynamics
solar energy
stationary state
strategic minerals
tragedy of the commons
transmaterialization
triage
undernutrition
wind farm

SUGGESTED READING

Castree, N. and B. Braun, eds. 2001. *Social Nature: Theory, Practice, and Politics.* London: Blackwell.

Ellis, R. 2003. *The Empty Ocean: Plundering the World's Marine Life.* New York: Shearwear Books.

Falola, T. and A. Genova. 2005. *The Politics of the Global Oil Industry.* New York: Praeger.

Klare, M. 2002. *Resource Wars: The New Landscape of Global Conflict.* New York: Owl Books.

Zimmerer, K. 2006. *Globalization and New Geographies of Conservation.* Chicago: University of Chicago Press.

WORLD WIDE WEB SITES

CONSERVATION DATABASES—WCMC

http://www.unep-wcmc.org/index.html?
http://www.unep-wcmc.org/cis/~main

The World Conservation Monitoring Centre, whose purpose is the "location and management of information on the conservation and sustainable use of the world's living resources," provides five searchable databases. Users can search by country for threatened animals and plants (plants are available for Europe only), protected areas of the world, forest statistics and maps, marine statistics and maps, and national biodiversity profiles (twelve countries only at present). Information is drawn from several sources, and database documentation varies from resource to resource.

STATE OF THE WORLD'S FORESTS—FAO
http://www.fao.org/forestry/index.jsp

The United Nations Food and Agriculture Organization presents information on the current status of the world's forests, major developments over the reporting period, and recent trends and future directions in the forestry sector. SOFO provides information on global forest cover, including estimates for 1995, change from 1990, and revised estimates for forest cover change.

"MAPPING THE WORLD BY HEART"
http://www.mapping.com

Has links to worldwide geographic and educational resources, such as time zone maps and other related global information.

ENVIRONMENTAL PROTECTION AGENCY
http://www.epa.gov

This site provides everything you ever wanted to know about environment and material resources.

ATLAPEDIA
http://www.atlapedia.com/index.html

Atlapedia Online contains key information on every country of the world. Each country profile provides facts and data on geography, climate, people, religion, language, history, and economy, making it ideal for students of all ages.

AG-ECON
http://agecon.lib.umn.edu

AgEcon contains information for agricultural and applied economics.

CHAPTER

5

THEORETICAL CONSIDERATIONS

OBJECTIVES

- To present the basic factors underlying the location decisions of firms
- To summarize the Weberian model of transport costs
- To show how production technique, scale, and location are interrelated
- To illustrate how and why firms grow and change over time and space
- To reveal the geographic organization of corporations
- To emphasize the role of social contexts and relations in the behavior of firms
- To depict the role of business cycles, particularly Kondratiev longwaves
- To point out the role of the state in shaping economic landscapes

"Electronic assembly plant."

Theory is a way of looking at the world, an explanation, a way of making sense of the relationships among variables. Theory is what separates description from explanation. A theory allows us to establish causality, to test hypotheses, to justify arguments, and to make claims to truth. Theories are simplifications about the world that allow us to gain understanding. Theories are thus indispensable to knowing how the world works, whether in the formal intellectual world or in our everyday lives at home and at work. Understanding theory is not a choice, because theory is inescapable. We all use theories every day, whether we're aware of them or not. Economic geographers use theory as well, ranging from the simple to the very sophisticated, to understand the order and chaos of economic landscapes.

There are a variety of theoretical frameworks for interpreting economic landscapes, including traditional industrial location theory, the behavioral approach, and the political economy, or structural, perspective. Industrial location theory derives from and shares the conservative ideology of neoclassical economics, using abstract models to search for best, or optimal, locations. The behavioral approach focuses on the psychology of the decision-making process: Rather than considering how decisions *should* be made in firms, it examines how decisions *are* made. This approach recognizes the possibility of suboptimal behavior. The political economy perspective challenges the ideology of the normative and behavioral industrial theories, which approach the question of location from a strictly managerial perspective, and instead emphasizes the social and historical context within which industrial activity takes place.

This chapter offers an overview of some of the major conceptual approaches and topics in economic geography. It begins with a discussion of the factors of location that influence industrial locations of single-plant firms. *Industrial* in this context should not be interpreted to be the same as manufacturing; every major sector of the economy is an industry, including many services, such as finance, legal services, and accounting. These factors include labor, land, capital, and management. Second, this chapter delves into the first, and still highly influential, model of industrial location, introduced by Alfred Weber a century ago. Third, it explores the interrelations among production technique, scale of output, and geographic location. Fourth, it discusses how and why firms grow over time and space. Fifth, it turns to the geographic organization of firms, moving from simple single-establishment ones to large, multiestablishment, multiproduct corporations with well-developed internal divisions of labor. Sixth, it emphasizes the social forces and contexts that firms simultaneously produce and are produced by, including conflicts between capital and labor. Seventh, it turns to the cyclicality of capitalism, its tendency toward boom and bust cycles, which in turn shape uneven spatial development. Finally, the chapter concludes by reciting some of the many ways the state, or government, plays a critical role in building, changing, and reproducing the economic geographies of capitalism.

Factors of Location

There are numerous variables that influence the location of firms and industries, which are aggregations of firms. The locational decision of a firm is thus quite complex, and many companies spend considerable time and effort in choosing the optimal location. After all, investments in inappropriate locations can be disastrous. Thus, firm decision making is a rational process, if an imperfect one, and is subject to the iron laws of market competition. Although personal considerations such as climate or the owner's preferences may figure in from time to time on the margins, firms cannot choose locations arbitrarily or they will be forced out of business by their more rational competitors. The major factors that shape firms' locations include labor, land, capital, and managerial and technical skills. All these are necessary for production, and all exhibit spatial variations in both quantity and quality.

Labor

For most industries (except in the primary sector), labor is the most important determinant of location, especially at the regional, national, and global scales. When firms make location decisions, they often begin by examining the geography of labor availability, productivity, and skills. The degree to which firms rely on labor, however, varies considerably among different sectors of the economy, and even among different firms, which may adopt different production techniques.

Labor is required for all forms of economic production, but the relative contribution of labor to the cost structure and value added varies considerably among industries. For example, the contribution of labor costs is high in the automobile industry but very low in the petroleum industry. The demand for labor depends on how labor intensive or capital intensive a given production process is. In highly capital-intensive industries, labor costs may be irrelevant. Thus it is a mistake, but a common one, to assume that all industries seek out low cost labor. Over time, most industries have become increasingly capital intensive, that is, they have substituted capital for labor, particularly when production in large quantities justifies the investments involved. Much of the productivity growth of capitalism historically results from capital intensification.

The supply of labor in a given region also greatly affects its cost. In countries with high birthrates, the supply tends to be relatively high, and labor costs are low.

Labor is essential to the production process, although the amount, skill, and productivity vary widely from sector to sector, as well as across space and time.

In economically advanced countries, late in the demographic transition, the birthrate is low, and labor is relatively expensive. These trends shape the willingness of firms to become more capital intensive: when labor is cheap, there is little incentive to mechanize.

There are great variations among industries in terms of their preferences for different types of workers. Because some firms demand particular types of workers in terms of their age or sex, the demographic structure of a region also shapes the supply of certain types of employees. The fast-food industry, for example, prefers young workers who will work for minimum wage, and the supply of this demographic segment is more limited in certain regions and not others. Finally, since labor is mobile over space (but not perfectly so), migration (or internationally, immigration) also shapes the local supply of labor. In regions that can attract labor easily, wage rates will tend to be low, all else constant. When the supply is limited by, say, immigration restrictions, wage rates tend to go up. For example, U.S. congressional limitations on immigration after World War I contributed to the rise in wages in the 1920s. At the local level, housing costs can also constrain the supply of labor if they are so high that workers cannot find affordable places to live, an increasingly common problem in the coasts of the United States.

In theory, because labor is relatively mobile, regional differences in the supply and demand for labor should move toward equilibrium over time. A high demand for labor in one place, and thus high wages, and an excess supply in another, and thus low wages, should be brought into equilibrium by labor migration. Examples are the nineteenth-century migration of people from rural areas to the city in countries such as Britain and the United States, and the twentieth-century movement of labor from the Third World to the wealthy countries of northern Europe: from the Caribbean and South Asia to Britain, from North Africa to France, and from Turkey to Germany (Chapter 3).

As in the case of all production factors, the response of labor to changing market conditions is not instantaneous. Labor has a relatively high degree of inertia, especially in the short run. People are generally reluctant to leave familiar places, even if jobs are plentiful in other areas. They will sit out short periods of unemployment or accept a smaller net income than could be earned elsewhere. Public policies, such as unemployment payments and workers' compensation, have reduced the plight of the unemployed and underemployed in advanced industrial countries in this century. The lack of instantaneous adjustment in labor demand and supply has resulted in variations in the cost of labor within and among countries. In the United States, wages are often higher in the more industrial states and in highly unionized environments.

It is simply not true that firms always want the cheapest labor possible. Cost is only one of many factors. Equally important is productivity. Productivity is largely a function of the skills present in the local labor force, or *human capital*, which in turn are derived from formal and informal educational systems, on-the-job training, and years of experience. Firms will pay relatively high wages for skilled, productive labor. To illustrate this point, consider that if labor costs were central to the location of all firms, then very low-wage countries, say Mozambique, should attract vast quantities of capital, which they don't, and high wage countries, such as Germany or the United States, should see a rapid exodus of jobs. The reality of the geography of labor is much more complex, of course, and involves national and local labor markets in which jobs are constantly created and destroyed; skills are produced, reproduced, and changed; new technologies come into play; and other cultural, economic, and social forces interact. What firms seek as much as cheap labor is productive labor.

The skill level of a given occupation greatly affects the size of its labor market. Generally, skilled labor markets tend to be geographically larger than unskilled ones. Workers may migrate hundreds or thousands of miles for well-paying positions, and the market for many skilled jobs is global in reach, including university professors. Unskilled positions, in contrast, typically draw from a relatively small labor shed; few people would travel cross-country, for example, to take a job as a janitor or cashier.

It is worth emphasizing that labor is also important because the labor process is saturated with politics. Labor is the only factor input that is sentient, and it alone is able to resist the conditions of exploitation, to go on strike, to engage in slowdowns, or to unionize

(Figure 5.1). Unionization rates vary widely across the United States (Figure 5.2), adding to differentials in the cost of labor. The South is generally much less unionized than the North. Thus, in addition to the cost of labor, firms must consider working conditions, health and safety standards, pensions and health benefits, vacations and holidays, demands for worker training, subsidized housing, and the role of labor unions, all of which shape local wage rates and productivity levels.

At the world scale, developed countries have higher wage rates than newly industrializing countries. One factor responsible for differential wage rates is the level of worker organization. Higher rates of unionization are associated with higher wages. Unionization is generally more prevalent in the older, established industrial countries than in the newly industrializing countries. Thus, considerable advantages can be gained by companies that relocate to, or purchase from, newly industrializing countries, especially if these countries are characterized by low levels of capital-labor conflict. The capital-labor conflict, manifested in industrial disputes, is a powerful force propelling the drift of industrial production outward from the center to the periphery of the world system.

FIGURE 5.1
Labor, unlike other inputs into production, alone is sentient and capable of changing the conditions of work, as with strikes or slowdowns.

Land

At the local scale (i.e., within a particular metropolitan commuting area, in which labor costs are relatively constant), land availability and cost are the single most important locational factors affecting firms' location decisions. The cost of land reflects the highly localized supply and demand, and different types of firms require different quantities in the production process. Generally, larger firms, particularly in manufacturing, require more land and are thus more sensitive to the costs, although in some sectors, such as producer services, firms pay very high costs (in rent or by purchasing a site). Firms often engage in intensive examination of several selected possible sites before settling on an optimum location.

The primary determinant of the cost of land is its accessibility. Transport costs (the measure of accessibility) determine the location rent of parcels at different distances form the city. Thus, because land downtown is the most accessible, it is by far the most expensive; in most cities, land costs decline exponentially away from the city center (Figure 5.3).

However, not all firms necessarily seek out low-cost land. The imperative to do so depends on the trade-off between land and transportation costs that firms make to maximize their profits. Firms that must have accessible land—generally labor-intensive firms that must maximize their accessibility to labor, to each other, and to urban services—will pay very high rents in order to locate near the city center. Firms that do not require access to clients, suppliers, and services, on the other hand—such as large manufacturing firms in suburban industrial parks—make a different trade-off, choosing to locate on the urban periphery where land costs are low but transport costs are higher. The demand for land and the need to agglomerate are thus inversely related.

Within metropolitan areas, there has been a centrifugal drift of manufacturing to suburban properties since the 1970s. Factories historically needed a large amount of land, preferably in a one-story building. Large parcels of industrialized land are more likely to be available in the suburbs than in central-city locations, where accessibility makes land relatively expensive. However, computerization and just-in-time inventory systems have diminished the need for land in some cases.

More reasons why industrial properties have expanded into the suburbs include locations that are easily accessible to motor freight by interstate highway and beltway, and access to suburban services and infrastructure,

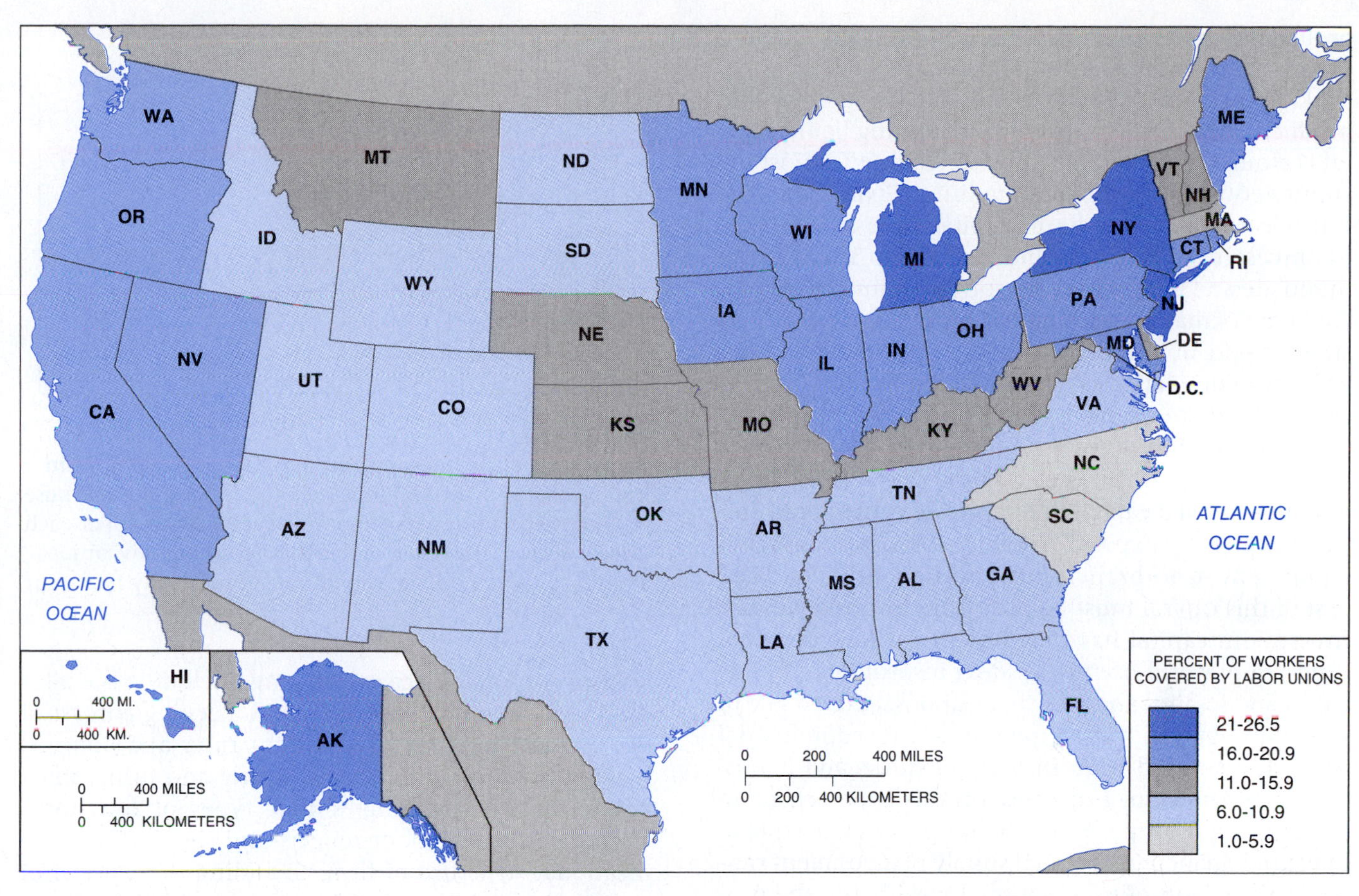

FIGURE 5.2
Percent of U.S. workers covered by labor unions, 2000. Unlike other inputs into the production process, labor alone can struggle to improve the conditions of its work. Unionization has declined sharply in the United States to 15 percent of the labor force from a high of 45 percent in 1950, largely due to deindustrialization. Unions are much less well represented in services than manufacturing and less prevalent in "right to work" states, particularly in the South, where they are not legally entitled to require workers at an establishment to join.

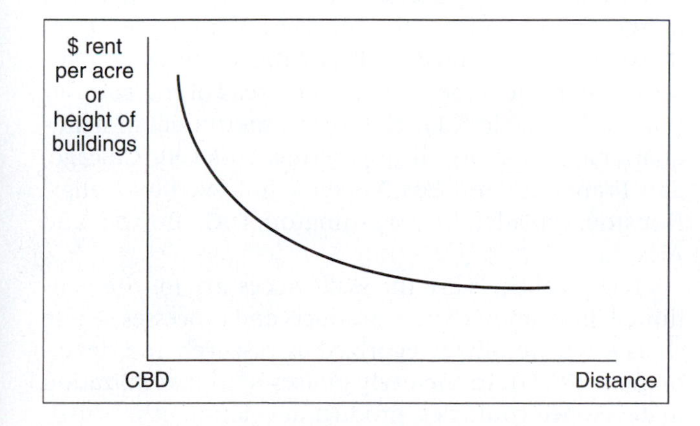

FIGURE 5.3
Land costs decline with distance from the central business district in most cities because accessibility to parcels in the periphery declines, and transport costs rise. Firms and households all make trade-offs between rents and transport costs depending on their specific locational needs.

including ample sewer, water, parking, and electricity. Industries may also be attracted to the suburbs because of nearness to amenities and residential neighborhoods. Suburban locations minimize labor's journey to work, which helps to hold wages down.

Capital

Under capitalism, capital plays a dominant role in structuring the production process. Capital takes one of two major forms: *fixed capital* and *liquid,* or *variable, capital.* Fixed capital includes machinery, equipment, and plant buildings. Besides the installation and construction costs, firms must budget for maintenance and repair and depreciation. The age of the capital stock of a region greatly affects its overall productivity levels: Places with newer fixed investments are more productive. Liquid capital includes intangible revenues, including corporate profits, savings, loans, stocks, bonds, and other financial instruments. The rate of capital formation reflects variables such as corporate profitability (including market prices, production costs), savings rates, interest rates, and taxation levels.

Liquid capital or finance, is by far the most mobile production factor. The cost of transporting liquid capital is almost zero, and it can be transmitted almost instantaneously over fiber-optic lines. Fixed capital is much less mobile than liquid capital; for example, capital invested in buildings and equipment is obviously immobile and is a primary reason for industrial inertia. Any type of manufacturing that is profitable has an assured supply of liquid capital from corporate revenues or borrowing (depending on its credit rating). Interest rates—the cost of capital—hardly vary within individual countries, but do vary among them. Most types of manufacturing, however, initially require large amounts of fixed capital to establish the operation—or, periodically, to expand, retool, or replace outdated equipment or to branch out into new products. The cost of this capital must be paid from future revenues. Investment capital has a variety of sources: personal funds; family and friends; lending institutions, such as banks and savings and loan associations; and the sale of stocks and bonds. Most capital in advanced industrial countries is raised from the sale of stocks and bonds, although American firms rely on this approach more than firms in Europe, where banks play a larger role in industrial financing. The total supply of investment capital is a function of total national wealth and the proportion of total income that is saved. Savings become the investment capital for future expansion.

Whether a particular type of industry, or a given firm, can secure an adequate amount of capital depends on several factors. One factor is the demand for capital, which varies from place to place and from time to time. Of course, capital can always be obtained if users are willing to pay high enough interest rates. Beyond supply-and-demand considerations, investor confidence is the prime determinant of whether capital can be obtained at an acceptable rate. Investor confidence in a particular industry may exist in one area but be lacking in another.

Capital is important as well because firms can substitute capital for labor in a process of capital intensification (Figure 5.4). The history of capitalism is largely one of steady capital intensification in different industries, particularly in agriculture, in which only a very small fragment of the labor force in industrialized countries now works. Capital intensification can increase productivity, but it may also displace workers. Only if the cost of goods drops sufficiently to increase real incomes and worker expenditures can it generate job growth in the long run. Historically this has generally been the case.

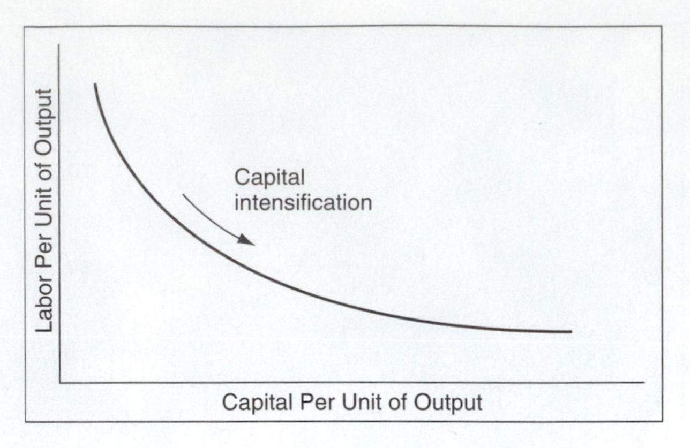

FIGURE 5.4
The process of capital substitution for labor is a long-term trend under capitalism. Firms seek to reduce costs and raise productivity by using machines instead of workers per each unit of output. However, when labor costs are low, or jobs are difficult to automate, capital substitution may not occur.

Managerial and Technical Skills

Managerial and technical skills are also required for any type of production. Management involves the nuts and bolts of corporate decision making, including the allocation of resources, raising capital, keeping abreast of the competition and government rules and policies, making investment decisions, hiring and firing, marketing, and performing similar types of functions. Corporate management reflects and shapes the organizational structure of firm, including the pattern of ownership and how decisions are made. Firm management forms may range from sole proprietorships to partnerships and be either publicly or privately owned (the former sell stocks and are owned by shareholders, the latter are typically family owned).

Within firms, management forms an important part of the corporate division of labor (i.e., the headquarters as compared with branch plants). Corporate headquarters decide a firm's overall competitive strategy, what markets and products to focus on, labor policies, mergers and acquisitions, and types of financing. Thus these tend to be skilled, well-paying, white-collar jobs. Most are in the largest urbanized areas of the country (Figure 5.5; Table 5.1). The top 10 metropolitan headquarter areas for large firms are New York City, Chicago, San Francisco, and Los Angeles, followed by Dallas, Houston, Philadelphia, Washington, D.C., Boston, and Atlanta.

Technical skills are the skills necessary for the continued innovation of new products and processes. These skills are generally categorized as research and development (R&D). In the early phases of industrialization in developed countries, product development was usually carried out in tandem with production by small firms, many of which, together with their innovations, failed to survive. Today, the R&D required for new products is typically a large and expensive process, involving long lead times between invention and production, a process that is often beyond the scope of

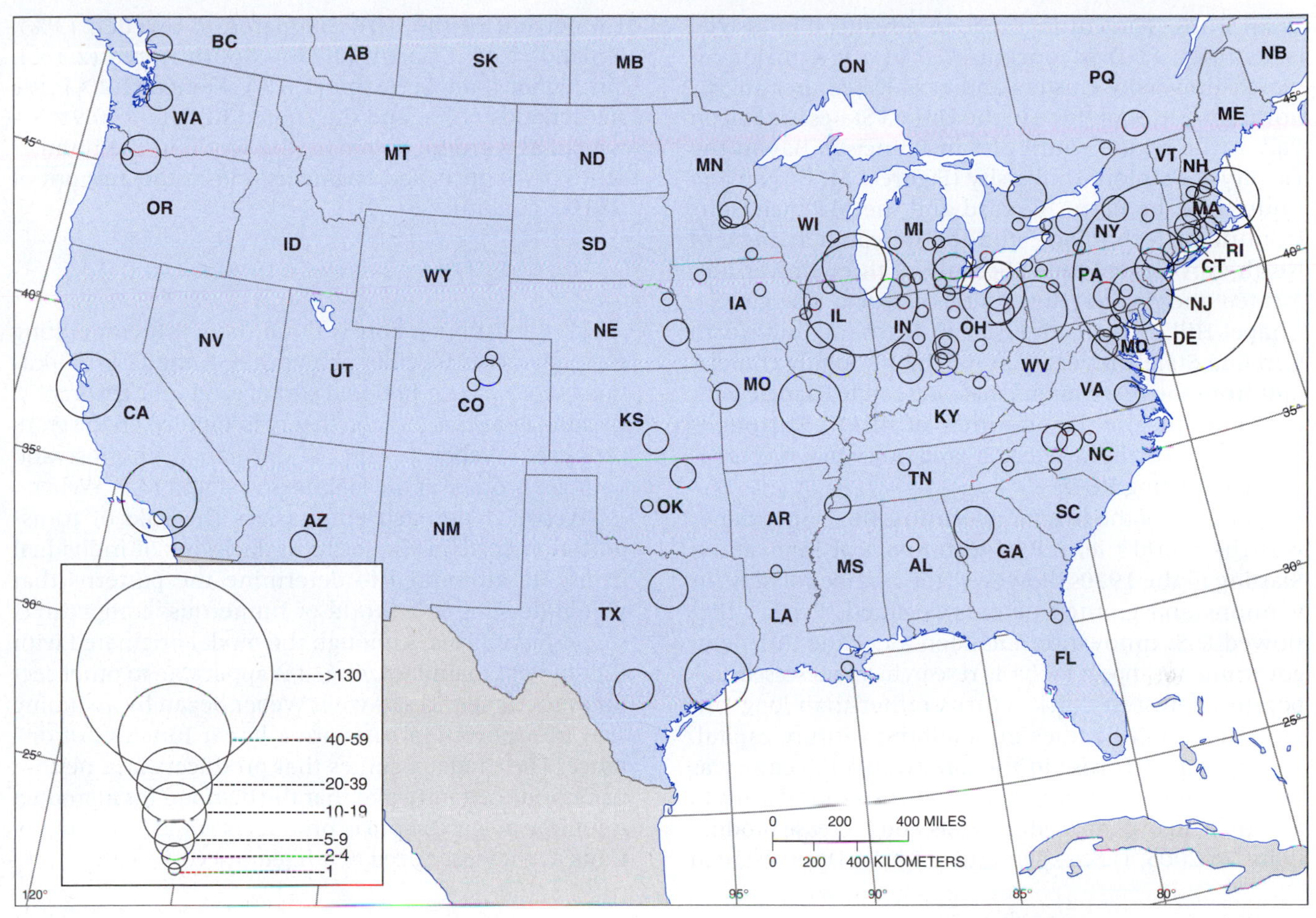

FIGURE 5.5
Corporate headquarters in the United States, 2004. Headquarters are the site for the producer services and decision-making functions required to make all companies operate effectively and profitably, including management and administration, public relations, advertising, legal matters, and research and development. Because they have numerous linkages to other firms, particularly other high value-added services, headquarters generally cluster in large cities. Cities in the northeastern and midwestern United States, with skilled labor pools and dense agglomerations of ancillary services, have been the most attractive. Historically, New York City has been the nation's headquarters, although there has been a gradual decentralization to other cities lower in the urban hierarchy.

Table 5.1

Location of Major U.S. Corporate Headquarters, 2004

City	Number	City	Number
New York	239	St. Louis	39
Chicago	109	Cleveland	35
San Francisco	91	Detroit	34
Los Angeles	85	Miami	31
Dallas–Ft. Worth	76	Denver	27
Houston	70	Milwaukee	26
Philadelphia	70	Nashville	25
Washington, D.C.	66	Phoenix	23
Boston	66	Tampa	20
Atlanta	53	Seattle	19
Minneapolis	50	San Diego	18

Source: Klier, T. and W. Testa. 2002. "Location Trends of Large Company Headquarters during the 1990s." *Economic Perspectives.* Chicago: Federal Reserve Bank of Chicago, pp. 12–26.

small firms. The cutting edge of advanced industrial economies, R&D is concentrated in a few major research-university clusters and established areas of innovation. Three of these in the United States are Silicon Valley, the region south of San Francisco Bay in the vicinity of Stanford University (Figure 5.6); Boston and Route 128, home to Harvard and the Massachusetts Institute of Technology; and the Research Triangle of North Carolina, so called because of three universities located there—the University of North Carolina at Chapel Hill, Duke University in Durham, and North Carolina State University at Raleigh. Roughly equidistant from the three main cities, Research Triangle Park is home to the laboratories of IBM, Burroughs Wellcome, Northern Telecom, and other major companies conducting R&D.

For most of the twentieth century, the United States was the world's leader in technological innovation. Starting in the 1970s, however, the number of new inventions and granted patents declined. Factors that slowed U.S. innovation included a decline in federal government support for basic research, an increased corporate interest in quick returns rather than long-run growth, and difficulties in obtaining venture capital. Outlays for R&D rose in the late twentieth century as foreign competition and high labor costs forced firms to automate in order to reduce costs and increase productivity. In 2005, U.S. expenditures for R&D represented 1.6 percent of the GDP, compared to Sweden (3%), Finland (2.4%), Japan (2.2%), South Korea (2.1%), but higher than Germany (1.6%), France (1.2%), the Netherlands (1%), and the United Kingdom (0.9%)—which, as a group, dominate the world in the number of R&D scientists and engineers and in the amount of R&D expenditures.

The Weberian Model

A list of location factors is not a theory. Incorporating these elements together analytically is the task of location theory, a time-honored part of economic geography. Classical industrial location theory is founded on the work of Alfred Weber (1929), a German economist and younger brother of the famous sociologist Max Weber.

Weber's approach emphasizes the role of transportation costs in the location decisions of individual firms. He attempted to determine the patterns that would develop in a world of numerous, competitive, single-plant firms. Although the model originated with the study of manufacturing, it is applicable to other sectors such as services as well. Weber began by assuming that transportation costs are a linear function of distance. The model assumes that producers face neither risk nor uncertainty and that the demand for a product is infinite at a given price; producers could sell as many units as they produced at a fixed price.

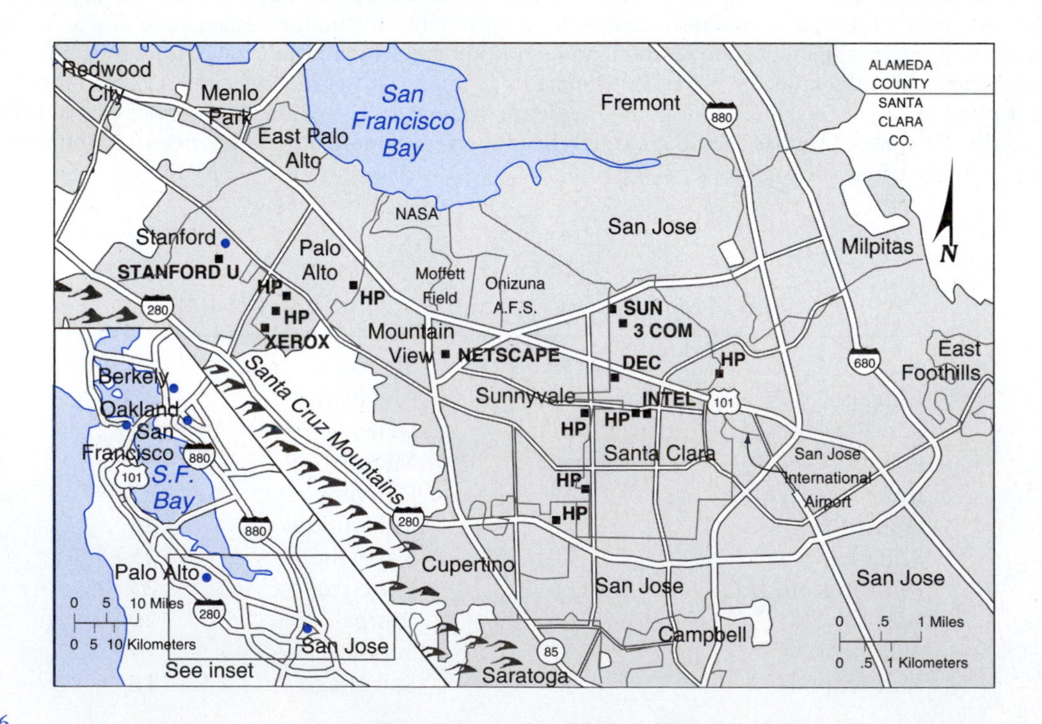

FIGURE 5.6
Silicon Valley, just south of San Francisco, California, is the world's largest region to produce computer software. It exemplifies the tendency of skilled, high-value-added economic activity to cluster in distinct areas, much like Italy's Emilia-Romagna or Germany's Badden-Wurtenburg.

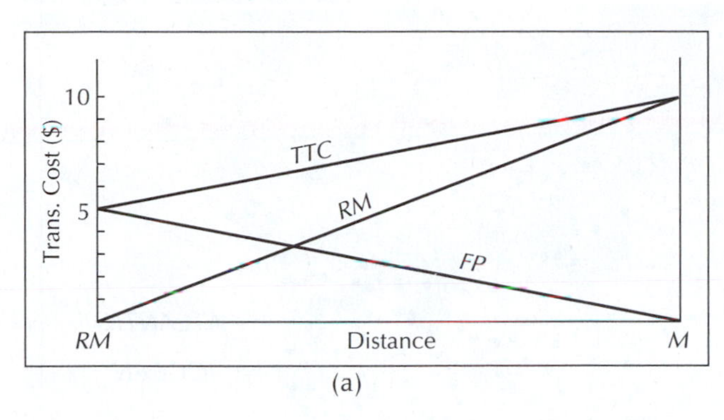

(a)

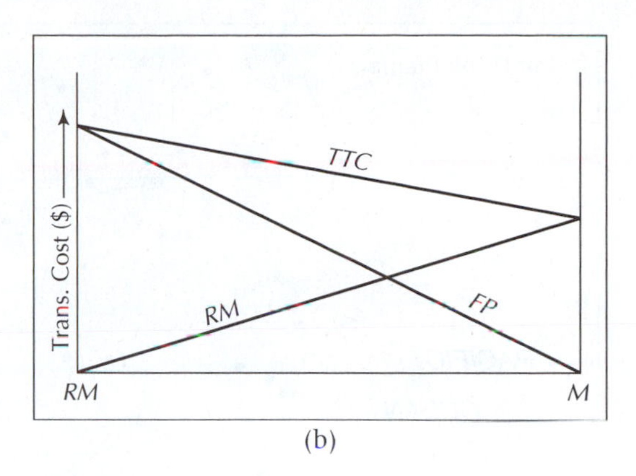

(b)

FIGURE 5.7
Weber's model: one localized raw material, located at *RM*. The line *RM* represents assembly costs, while the line *FP* represents finished product distribution costs. In Figure 5.7a, total transport costs are lowest at the raw material, as in primary sector activities. In Figure 5.7b, total costs (TTC) are lowest at the market. In this example of a weight-losing industry, total transportation costs are minimized by locating the plant at the site of the market.

Weber taught geographers to think about the distinction between material- and market-oriented industries. The first cost faced in the production process is that of assembling raw materials. Raw materials, such as coal, are found only at specific locations; their transportation costs are a function of the distance that they must be moved. For each case we must consider (1) the costs of assembling raw materials (*RM*), (2) the costs of distributing the finished product (*FP*), and (3) the total transportation costs (*TTC*) (Figure 5.7). The best location for a manufacturing plant is the point at which total transportation costs are minimized.

FIGURE 5.8
Gravel mining operations reflect an industry in which transport costs for the inputs are high, which leads firms to locate near the raw material.

Localized Pure Raw Materials

In Figure 5.7a, the assembly costs for the localized raw material (*RM*) are minimized at point *RM*. Finished-product distribution costs are minimized at *M*. Thus, in cases where the cost of shipping raw materials outweighs the costs of shipping the final product, the firm will locate near the raw materials, such as an iron ore mining company (Figure 5.8). In Figure 5.7b, raw materials, once processed, add to the weight of the finished product, so the total transportation costs (*TTC*) are minimized at the market (*M*).

Bottled and canned soft-drink manufacturing exemplifies the use of one pure raw material (syrup concentrate). If the plant locates at the market, the ubiquitous raw material (water), which makes the largest contribution to the weight of the finished product, does not need to be moved (Figure 5.9).

Communications media, especially the printed medium, are an example of a perishable industry where location near the market is essential. Newspapers must be printed near the location where they are consumed to minimize the distance that these bulky items must be transported. Consequently, producing a nationwide newspaper is a difficult logistic problem. The *New York Times*, the *Wall Street Journal*, and *U.S.A. Today* are not printed and published in New York City or Washington, D.C., and distributed to the entire United States. Instead, these companies transmit news via satellite to scattered locations such as Atlanta, Chicago, and Los Angeles, where the papers are printed and then distributed to the nation.

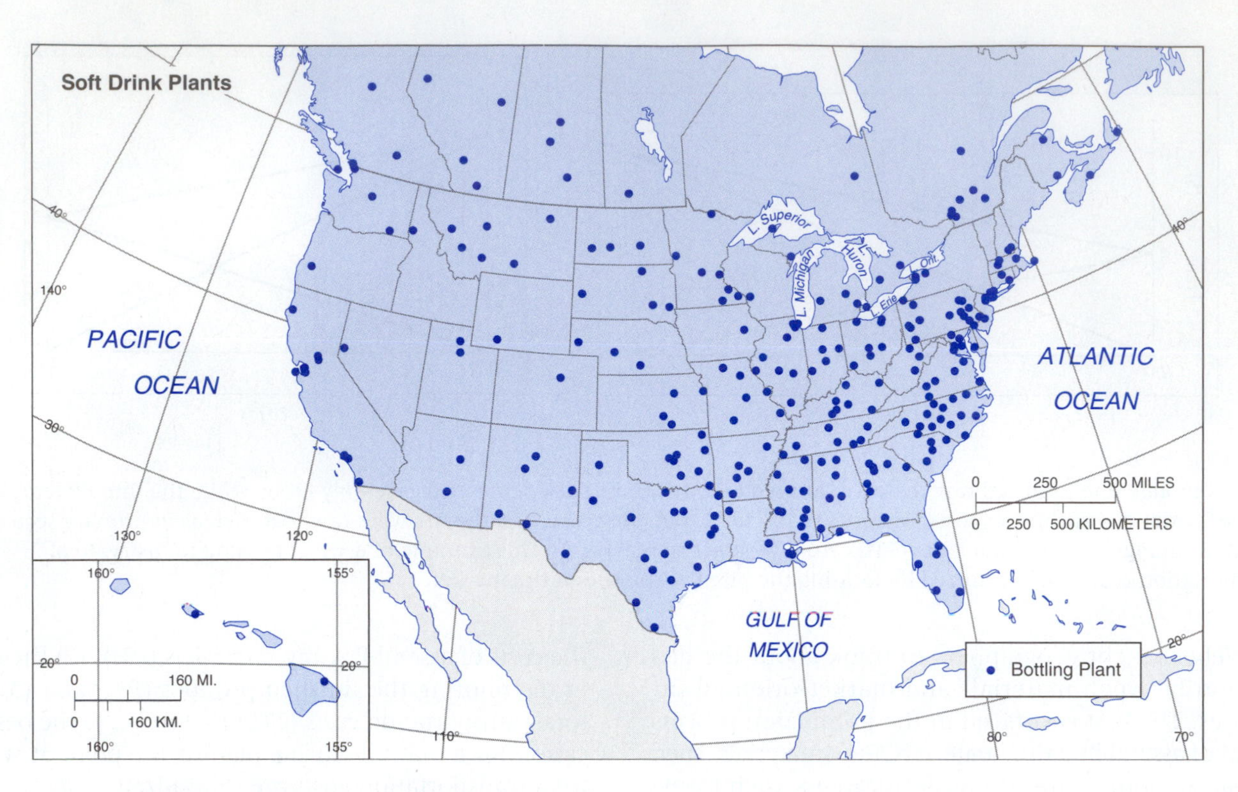

FIGURE 5.9
Bottled and soft drink manufacturing exemplifies a form of production that is market oriented. Most concentrate on the peripheries of metropolitan areas.

Other specialized manufacturers, such as designer-clothing manufacturers, are attracted to large markets as well. Being near accessible locations where buyers can view the merchandise is important. High-fashion clothing distribution is also a perishable industry because of the speed with which clothing styles change. As a result of the decisions by national and international buyers, merchandising, production, and sales occur at the market. New York City's garment and manufacturing district is a good example of this type of specialized manufacturer. In addition, cloth manufacturers and suppliers of shoulder pads, clasps, pins, zippers, elastic, and thread cluster near the principal garment manufacturers in New York City.

Weber in Today's World

Weber's model was originally developed for the analysis of manufacturing, although it can be applied to other sectors such as services as well. It is useful to explore the influence of transportation costs in a rigorous way, and many economic geographers still work in this tradition. Despite its broad appeal, there are several developments that limit its applicability. First, not all firms need to minimize transport costs; as we saw with the rent-transport trade-off that all firms must make, some firms will accept higher transport costs and locate on the urban periphery. Second, the production process is much more complex than it was in the early twentieth century, when Weber developed his model. Many plants begin with semifinished items and components rather than with raw materials. Producers' goods seldom lose large amounts of weight; therefore, there is not much tendency toward material orientation.

Weber's model has also been criticized for its unrealistic view of transportation costs as a linear function of distance. Because of fixed costs, especially terminal costs, long hauls cost less per unit of weight than short hauls do. Plants tend to locate at material or market points rather than at intermediate points, unless there is an enforced change in the transportation mode, such as at a port. However, with the expansion of the modern trucking industry and its flexibility in short hauls, the disadvantages of intermediate locations have been reduced.

Three other developments have a bearing on how industrial location choices have changed:

1. **Transportation costs have been declining in the long run.** This decline increases the importance of other locational factors, particularly labor costs and productivity. This is most obvious in firms producing high-value and high-tech products. For these firms, transportation costs are relatively unimportant. Yet for firms that distribute consumer goods (e.g., soft drinks) to dispersed markets, transportation costs remain a significant factor (Table 5.2).

Table 5.2

Transport Costs as a Percentage of Product Prices

Stone, clay, and glass	27
Petroleum	24
Lumber and wood	18
Food	13
Furniture	12
Paper products	11
Primary metals	9
Textiles	8
Fabricated metals	8
Transport equipment	8
Rubber and plastics	7
Tobacco	5
Machinery	5
Instruments	4
Apparel	4
Printing and publishing	4
Electronics	4
Leather products	3

Source: Compiled by authors from U.S. Department of Commerce national input-output tables.

2. **Brainpower has been steadily displacing muscle and machine power and transforming natural resources.** Natural resources are no longer as important in the growth of economies as they were historically. Instead, there has occurred a widespread transmaterialization of resources as smaller, lighter products are made from resources to which high technology and brainpower have been added.

3. **Finally, real-world patterns are evolutionary;** they are not the result of decisions made by optimizers. Most real-world decisions do not result in best (most profitable) locations. Locational decisions, once made, often lead to *industrial inertia,* the tendency to continue investing in a nonoptimal site even if more optimal locations exist. Tensions develop between ideal spatial patterns and the patterns produced by localized resources. As technology (especially transportation) improves, ideal spatial patterns (from the entrepreneur's viewpoint) become more feasible, but the inertia resulting from past actions exerts a constant deterrent on actualizing these patterns.

Technique and Scale Considerations

The establishment of any manufacturing plant in a market economy involves the interdependent decision-making criteria of *scale*—the size of the total output—and *technique*—the particular combination of inputs that are used to produce an output.

Technique, or the particular combination of inputs used to produce a given product, has an important effect on a firm's locational decision. A certain amount of land (resources), labor, and capital is needed to produce any finished product, but, within economic and technological limits, firms can substitute among inputs to minimize their costs: Capital may be substituted for labor. The greater the differences are among inputs in terms of their prices and productivity levels, the more the incentive that firms have to substitute among them.

The most evident trend has been substitution of capital in the form of machinery for labor. This trend is most evident in agriculture, which has progressively changed from very labor intensive to capital intensive over time. More and more manufacturing systems, which apply sophisticated technology to improve the quality and efficiency of production, are replacing the use of labor, including robotics. In services, capital intensification is also evident in the computerization of the office. Whether substitution between production factors occurs depends on the relative costs and productivity of the two inputs and the scale and locational decisions already made by the firm. If, for example, labor costs rise at a given location, the firm may choose to substitute capital for labor at that location, or it may opt to change locations to take advantage of lower labor costs and thus maintain the same labor-to-capital ratio.

The limits to substitution among inputs vary considerably from industry to industry and are fixed for given periods of time by technological constraints and the prices of inputs. Firms must choose from an available suite of options as to how they want to produce, and the costs of labor, land, and capital will greatly affect these choices. Some industries lend themselves more to capital intensification than do others. Petroleum refining, for example, can be readily automated, whereas garment manufacture cannot. The garment industry, therefore, is much more sensitive to changes in labor costs than is petroleum. In the late nineteenth and early twentieth centuries, the U.S. textile industry shifted from old multistoried New England mills to new mills in the Southern Piedmont as labor costs rose in the Northeast. This is an example of the influence that options in technique exert to determine the locational decision. The increased labor costs in the Northeast outweighed the costs of moving the industry to the Southeast. Of course, the wage advantage of the South did not persist indefinitely; as new industry moved south, wages there rose. Eventually, textile firms migrated farther afield—to Mexico, Brazil, South Korea, and Singapore. If capital substitution had been a viable option, the textile industry might not have moved. Many times a firm may want to change its scale to increase output and to earn extra profits. A change in scale may also require a change in location and/or technique.

Scale Considerations

Economic scale, or the level of output, along with location and technique, is also very important. But scale is also important because producers are concerned with the unit cost of production, and adjustments in scale can produce considerable variations in unit cost. Scale is the means by which production is "tuned" to meet demand. In some economies, this tuning may be done by the state; in others, by private entrepreneurs.

Principles of Scale Economies

Along with standardization of parts, the *division of labor* is a primary feature of mass production. Workers who perform one operation in the production process over and over are much more efficient than those who are responsible for all phases. The division of labor not only increases the efficiency of production but also facilitates the use of relatively unskilled labor. A worker can learn one simple task in a short time, whereas the skills required to master the entire operation might take years to learn. This process was instrumental, for example, in the early growth of the U.S. automobile industry in the system pioneered by Henry Ford. Division of labor, however, requires a relatively large scale because a large pool of workers is necessary. A common way to measure the size of a firm is by the number of employees. Capital, once invested in machinery and buildings, becomes fixed capital and produces income only when in operation. A three-shift firm makes much more efficient use of its fixed capital than a single-shift firm does. The three-shift firm is three times larger in scale, measured by employment, yet its fixed capital investment may be no more than that of the single-shift firm.

Large firms generally pay much less for material inputs than small firms do. For example, Ford Motor Company can obtain tires for a much lower unit price than an individual dealing with the same tire company can, because Ford buys millions of tires a year. Increasing scale, in other words, generally lowers the unit cost of inputs.

Economists portray scale economies as a curve of long-run average costs (LRACs), which graphs the unit costs as a function of scale. Several possible LRAC curves are indicated in Figure 5.10. Notice that unit costs decrease, reach an optimum point, and then began to increase. The rise in the curve is termed *diseconomies of scale* (diminishing marginal returns to scale) and occurs when a firm becomes too large to manage and operate efficiently. The optimum scale

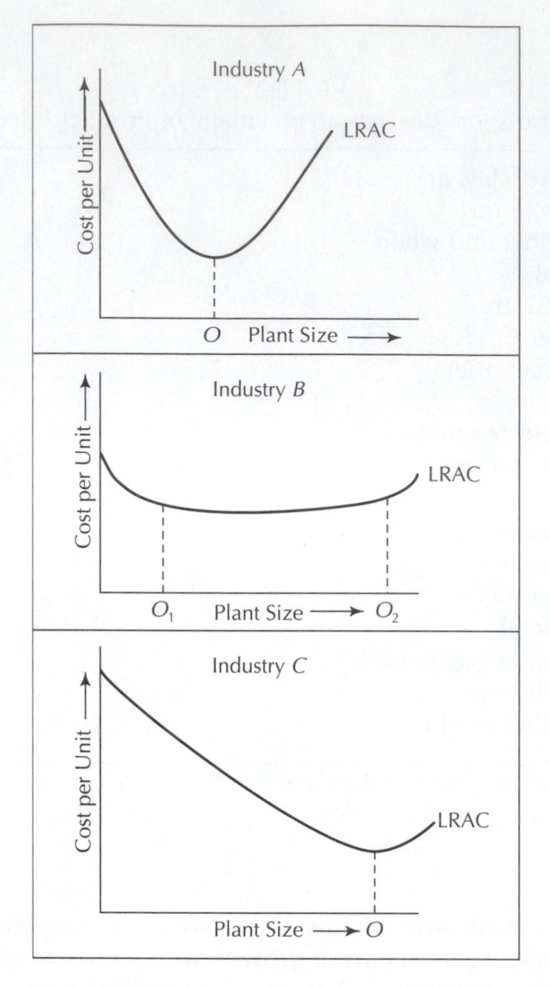

FIGURE 5.10
Variations in long-run average cost (LRACs) curves. Organizational and technical constraints make the optimal level of output variable on the characteristics of the industry and firm as it tries to achieve economies of scale and avoid diseconomies of scale.

of operation is very small in Industry *A*, very large in industry *C*, and fairly wide ranging in industry *B*. Firms in Industry *A* should be small, they should be large in Industry *C*, and they can range from small to large in Industry *B*.

Possible scale economies also indicate how firms in an industry can expand production. A firm in Industry *A*, for example, can build a branch plant; increasing the size of operations on the original site would produce diseconomies of scale. Firms in Industry *B*, however, can increase production either by expanding existing plants or by building new ones.

Vertical and Horizontal Integration and Diversification

Besides simply increasing plant size, two other means are commonly employed for effecting scale changes.

Some firms purchase raw material sources or distribution facilities. This is called *vertical integration* (or vertical merger) in that the firm controls more steps "up and down" in the total production process. Some early large automobile firms, for example, owned their own iron and coal mines and produced their own steel; over time, however, auto producers have become more vertically disintegrated, that is, the production of different parts has been taken over by specialized firms. Large oil companies are often vertically integrated; they control exploration, drilling, refining, and retailing. In contrast, *horizontal integration* (or horizontal merger) occurs when a firm gains an increasing market share of a given niche of a particular industry. This is typical when markets become oligopolistic, that is, controlled by a handful of large firms.

Another trend among corporations in the United States, Japan, and Western Europe has been a strategy of *diversification.* Many large corporations, through conglomerate mergers and acquisitions, control the production and marketing of diverse products. A company may produce many unrelated products, each with elements of horizontal and vertical integration. Diversification spreads risk and increases profits. Diminishing demand for the products of one division may be offset by rising sales in another.

Most industrial location theory is based on a small, single-plant operation producing a single product. Large corporations are much more complex, but they deal with all the variables of location theory and must still make locational decisions. Although large enterprises may seem to be more concerned with technique and scale decisions, each locational decision has an effect on scale and technique. We should consider two points: First, large firms may be able to operate in less than optimal locations and still have a significant effect on the market through the control that they exert over government policies and the prices and sources of raw materials. Conversely, large firms may be able to make optimal locational choices through their employment of the scientists and technical personnel who help top management make more profitable decisions.

Interfirm Scale Economies: Agglomeration

So far, we have been concerned with intrafirm scale economies, that is, within one company. However, scale economies also apply to clusters of firms in the same or related industries—for example, the computer firms localized in California's Silicon Valley. By clustering, unit costs can be lowered for all firms, in ways that could not occur if they were spatially separate. These economies, called *agglomeration economies,* take several forms. *Production linkages* accrue to firms locating near other producers that manufacture their basic raw materials. By clustering, distribution and assembly costs are reduced.

Service linkages occur when enough small firms locate in one area to support specialized services. The advertising industry in New York offers an example (Figure 5.11). Advertising agencies must cluster within a short distance of Madison Avenue in order to avail themselves of the dense networks of interfirm linkages to be found there, including information and gossip on the latest trends, markets, clients, hires, and products. For a firm to be successful, it must be near this complex; otherwise, it might as well be on the other side of the moon.

In addition to production and service linkages, there are *marketing linkages,* which occur when a cluster is large enough to attract specialized services. The small firms of the garment industry in New York City have collectively attracted advertising agencies, showrooms,

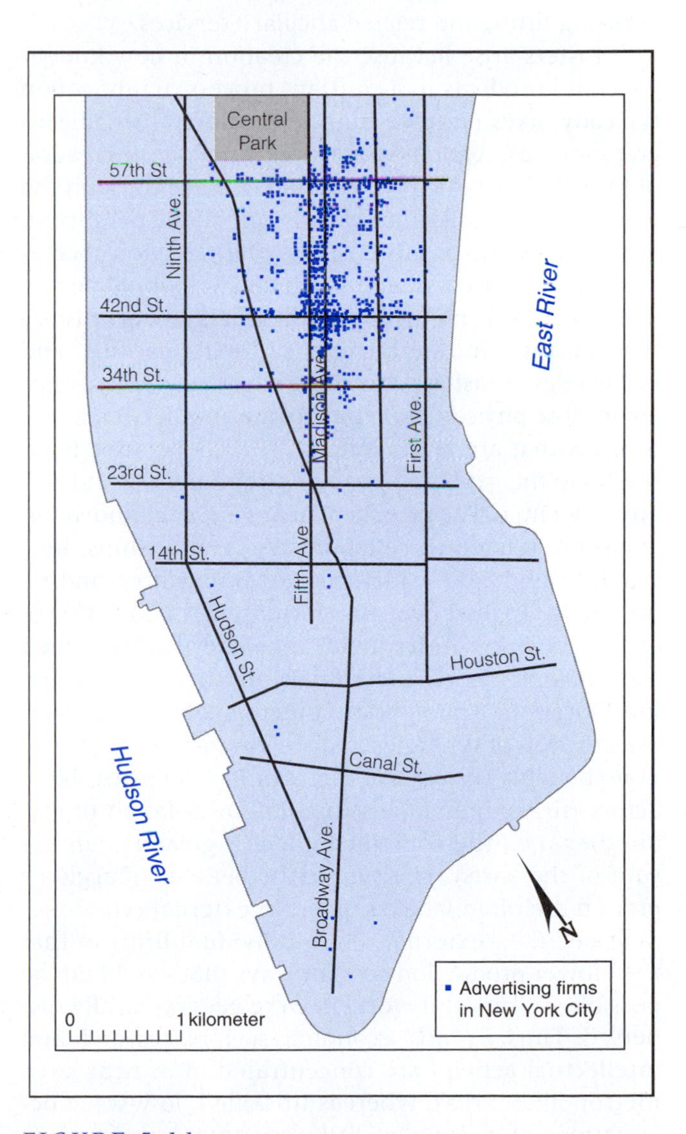

FIGURE 5.11
Advertising agencies in Manhattan must cluster near Madison Avenue to achieve the agglomeration economies so essential to labor-intensive, information-intensive, white-collar, high-value-added functions.

buyer listings, and other aspects of finished-product distribution that deal exclusively with the garment trade. Firms within the cluster have a cost advantage over isolated firms that must provide these specialized services for themselves or that must deal with New York firms at a considerable distance and cost.

Cities provide markets, specialized labor forces and services, utilities, and transportation connections, as well as access to specialized information. Clusters of production thus arise from the numerous interconnections of people, goods, services, and information that suture individual entrepreneurs, firms, their supplier networks, ancillary services, and related institutions together. In addition to firms, other entities in clusters may include government agencies and offices, public/private partnerships, trade associations, universities, legal services and patent attorneys, accounting firms, specialized advertising firms, and related ancillary services.

Clusters arise because the creation of new knowledge and products in a constant process of innovation typically takes place within the confines of small geographic areas. Agglomeration is essential to the creation of knowledge, which typically takes place through the interactions of skilled individuals, and to the production of synergies, the positive aspects of interaction such as new ideas, processes, and products that would be unlikely to arise from firms or individuals acting in isolation of one another. Organizational learning and knowledge transfer occur most effectively when firms are in close physical proximity to one another. Proximity of firms that are both rivalrous and cooperative is essential to the generation of an entrepreneurial and creative dynamic. The process of discovery and innovation is closely related to collaborative relationships, networking, and tight spatial linkages among firms and individuals. In essence, the formation of knowledge spillovers relies on frequent, repeated, and sustained interactions among individuals and firms in a given local location. An important outcome of this process is the creation of synergies, that is, interactions that generate benefits (i.e., ideas) that would not be possible if actors (firms, individuals) operate in isolation of one another (i.e., the combined effect is greater than the sum of the parts). The synergistic benefits of agglomeration are often labeled "positive external economies of scale" (i.e., external to an individual firm) in that they lower production costs in ways that would not be possible if firms and workers were geographically dispersed. Thus, in most economic sectors, research and intellectual activity are concentrated in or near large metropolitan areas, whereas unskilled, low-wage occupations that involve little creative activity (e.g., branch plants, back offices) tend to disperse to smaller towns. *Urbanization economies,* therefore, are a combination of production, service, and marketing linkages concentrated at a particular location.

Evaluation of Industrial Location Theory

Are industrial location patterns rational? Do firms search for optimal locations? Here, we distinguished between two types of decision making leading to locational patterns: *adoptive* and *adaptive behavior.*

From the perspective of behavioral geography, the main defect of normative industrial location theory is that it fails to say what decision makers actually do. How do enterprises actually select profitable sites for branch plants? Economic geographers often find that corporate executives confirm the validity of the framework of locational search, learning, and choice evaluation. The first phase of the decision-making process is the recognition that a growth problem exists with respect to demand. The possible responses are in situ expansion, relocation, acquisition, or construction of a new plant. A new plant involves a three-stage search procedure, the outcome of which leads to a decision and, finally, to the allocation of resources. It also generates feedback into learning behavior and into the decision-making environment. It is important to make a distinction between short- and long-term responses of the organization. Classical industrial location theory tends to be static and ignore the time horizon within which firms operate. The behavioral approach is much more realistic in recognizing that the environment in which the enterprise operates is in a constant state of flux.

Location theory may be evaluated by comparing optimal patterns against real world ones. The final location chosen by an industry is not always determined by transportation costs, as was Weber's principal question, nor by production costs at the site, including land, labor, capital, and managerial and technical skills. There are several factors that complicate locational decisions:

1. A firm may have more than one critical site or situation factor, each of which suggests a different location.
2. Even if a firm clearly identifies its critical factor, more than one critical location may emerge.
3. A firm cannot always precisely calculate costs of situation and site factors within the company or at various locations because of unknown information.
4. A firm may select its location on the basis of inertia and company history. Once a firm is in a particular community, expansion in the same place is likely to be cheaper than moving to a new location.
5. The calculations of an optimal location can be altered by a government grant, loans, or tax breaks.
6. Noneconomic factors play important roles in *footloose industries* that have gravitated to coastal areas in the Sunbelt of the United States because of recreational opportunities and availability of amenities.

Industrial location theory has also been criticized by structuralists, primarily because it focuses on individual firms as abstract entities, without embedding and contextualizing them within the rest of the economy. For structuralists, locational analysis begins at the top, with the world's capitalist system, not at the bottom, with individual firms. The actual behavior of the individual firm takes on its meaning in the broader economic context of the class and historical dynamics of capitalism as an integrated system. Working up from the bottom can explain neither the individual elements nor the system as a whole. According to structuralists, then, industrial location theory is unrealistic because it focuses on only a small part of reality. This criticism extends to the new approaches in which the simple conception of the single-plant firm has been replaced with a model of the firm as a complex organizational structure.

How and Why Firms Grow

Most large companies operate at the maximum scale possible on the LRAC curve. In fact, evidence of increasing returns to scale has led to a reappraisal of the theory of the firm. The tendency toward increasing scales of operation is therefore based on the motivating force of growth. Firms expand for two reasons: survival and growth. Both goals are promoted by horizontal and vertical expansion and by diversification.

The view that corporate growth is part of a natural progression is deterministic, however, and it flies in the face of reality. The majority of firms in an economy remain small and peripheral. Only some firms, especially those that manufacture capital goods, have the potential to develop into large corporations. Financial barriers prevent most firms from making successive transitions from a small regional base to larger national organizations and then to multinational operations. Access to finance—banking capital, venture capital, and international bond and currency markets—has become increasingly uneven, favoring some firms and not others. Because these finance gaps have become wider, a small firm has a much less chance of evolving into a corporate giant today than it did a hundred years ago.

How a firm grows depends on the *strategy* that it follows and the *methods* that it selects to implement its strategy. As we discussed earlier in the chapter, growth strategies are *integration* or *diversification*. In the United States, horizontal integration predominated from the 1890s to the early 1900s, vertical integration came to the fore in the 1920s, and diversification has been the principal goal since the 1950s. This three-stage sequence provides a framework for understanding the interrelationship of the various strategies. The early growth of large enterprises involves the removal of competition by absorption, leading to oligopoly. This is followed by a period in which the oligopoly protects its sources of supply and markets by vertical integration, buying firms "upstream" and "downstream" in the production process. Once a dominant position is achieved, rapid corporate growth can proceed only with diversification.

Methods for achieving growth are internal or external to the firm. Growth can be financed internally by the retention of funds or new share issues. Or it can be generated externally by acquiring the assets of other firms through mergers. Most large firms employ both means, but external growth is particularly important for the largest- and fastest-growing enterprises.

Whatever strategy and method are adopted, corporate growth typically involves the addition of new factories and, thus, a change in geography. Initially, much of the employment and productive capacity of a firm

The classical principles of industrial location theory are evidenced in the river valley and railroad site of Bethlehem Steel Corporation's plant at Bethlehem, Pennsylvania. This huge plant, which extends for nearly eight kilometers along the south bank of the Lehigh River, converts raw materials—Appalachian coking coal and Minnesota iron ore—into structural shapes, large open-die and closed-die forgings, forged steel rolls, cast steel and iron rolls, ingot molds, and steel, iron, and brass castings. The main market for the steel products is the American Manufacturing Belt with its abundance of metal-using industries.

concentrates in the area in which it was founded. As enterprises grow, they become more widely dispersed multiplant operations, which is sometimes accompanied by decreasing dominance of the home region. Exceptions tend to be companies confined to one broad product area and based in a region where there is a historical specialization within that product area.

The choice of growth strategy affects corporate geography. Horizontal integration frequently involves setting up plants over a wider and wider area. The geographic consequences of vertical integration vary according to whether the move is backward ("down" in the production process) or forward ("up" in the production process). *Backward integration,* in which a firm takes over operations previously the responsibility of its suppliers, can lead a firm into resource-frontier areas. An example is the development of iron-ore deposits by U.S. and Japanese companies in Venezuela and Australia. Conversely, *forward integration,* in which a firm begins to control the outlets for its products, can lead a resource-based organization to set up plants in market locations. Diversification does not have such predictable consequences for the geography of large enterprises.

The method of growth also affects the geography of multiplant firms. When growth is achieved internally, enterprises can carefully plan the location of new branch plants. When growth is achieved externally, enterprises inherit facilities from acquired firms; hence, there is less control over their locations. Moreover, the attractiveness of new facilities often lies in their economic, financial, and technical characteristics. Nonetheless, geography does play a role in the decision process. Firms typically confront the uncertainty and risk of expansion by investing first in geographically adjacent or culturally similar environments.

Geographers have developed models of how firms grow. Most of these models postulate a single development path beginning with a small, single-plant operation and culminating with the multinational enterprise. This kind of evolution along a path from a local to a national and then to an international company is exceptional. Unequal access to finance makes it difficult, if not impossible, for many firms to expand beyond the subnational scale. In the late twentieth century, the size distribution of firms resembles a broad-based pyramid in which fewer and fewer firms can move from one level to another. Rather than the single development sequence that may have existed in the nineteenth century, today multinationals follow a distinctive path through a series of discrete development sequences.

GEOGRAPHIC ORGANIZATION OF CORPORATIONS

Multiestablishment, multiproduct corporations, which include headquarters, manufacturing plants, research laboratories, education centers, offices, warehouses, and distribution terminals, have their own distinctive geographies. To appreciate the internal geography of these systems, two issues must be considered: (1) the ways in which corporations are organized to maximize efficiency, and (2) the influence of hierarchical management structures on the location of employment.

Organizational Structure

Companies organize themselves hierarchically in a variety of ways to administer and coordinate their activities. The basic formats are (1) functional orientation, (2) product orientation, (3) geographic orientation, and (4) customer orientation. A fifth format, which is a combination of at least two of the basic formats, is called a *matrix structure.* Different companies may select different formats, but all formats are always subject to review and modification.

The organizational format that is based on various corporate functions—manufacturing, marketing, finance, and research and development—is illustrated in Figure 5.12. With this framework, all the company's functional operations are concentrated in one sector of the enterprise. An example of a company with this type of organizational structure is Ford Motor Company. This form of organization works well for companies with relatively narrow ranges of products.

Figure 5.12b illustrates the product-orientation organizational structure. Product groups can be cars, trucks, buses, and farm equipment for a major motor vehicle manufacturer. Although a corporate central staff is needed to provide companywide expertise and to provide some degree of assistance to each product group, each group also has its own functional staff. Thus, a fairly high degree of managerial decentralization is required. The product-orientation format works well for companies with diverse product lines. Westinghouse is an example of a company organized in accordance with this format.

A third organizational format is based on geographic orientation—either the geographic location of customers or of the company's productive facilities. The company is organized around regions rather than functions or products. Under this form of organization, most or all the corporation's activities relating to any good or service that is bought, sold, or produced within a region are under the control of the regional group head. Each geographic region is under a separate profit center. This organizational format is best suited for companies with a narrow range of products, markets, and distribution channels. It is popular among oil companies and major money-center banks.

Some companies organize according to the types of customers that they want to serve rather than the locations of customers. For example, commercial banks

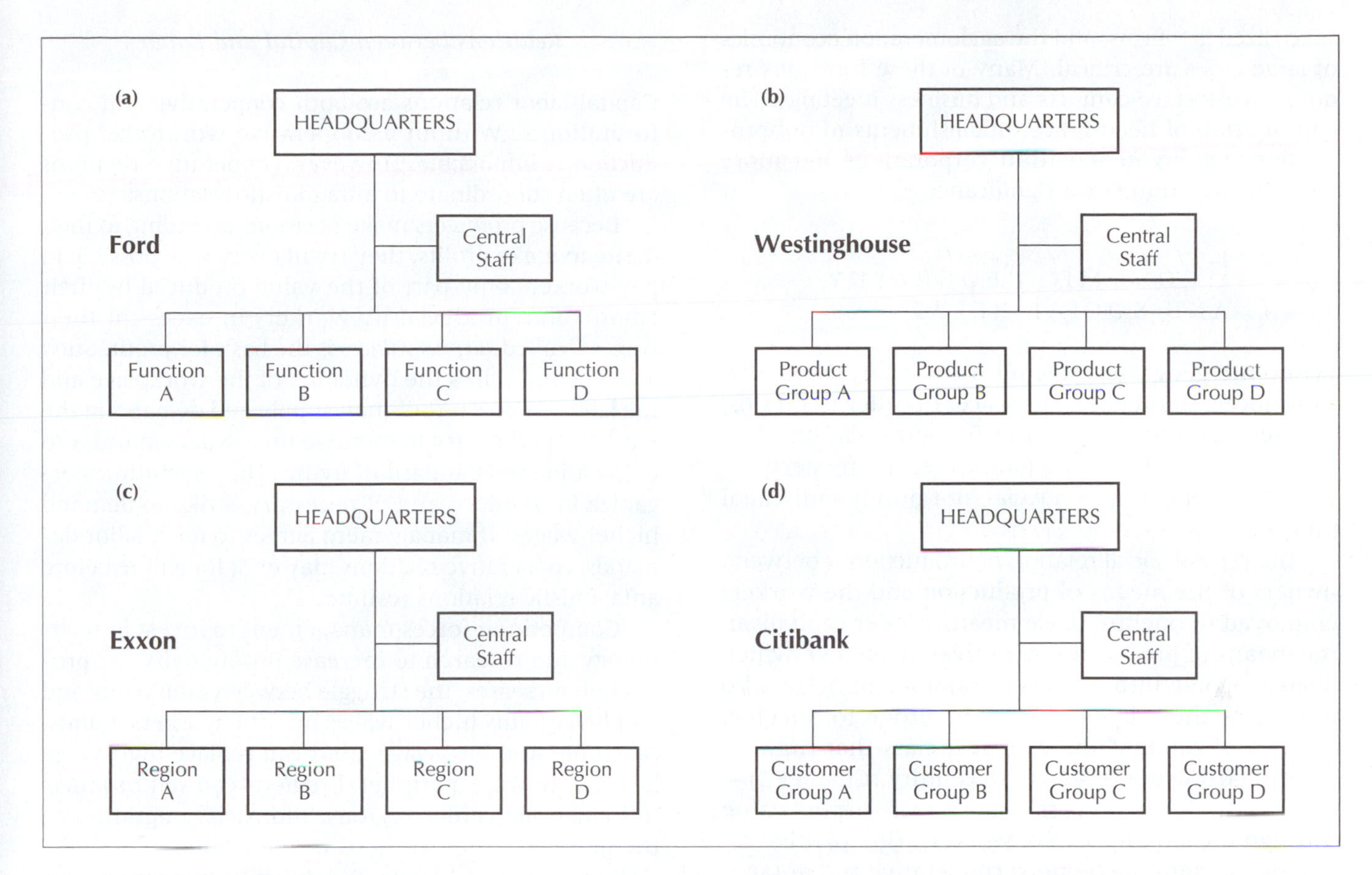

FIGURE 5.12
Organizational structures: (a) functional organization, such as the Ford Motor Company; (b) product orientation, such as Westinghouse; (c) geographic orientation, such as Exxon; (d) customer orientation, such as Citibank.

are commonly organized around groups such as the personal, corporate, mortgage, and trust departments. Alternatively, manufacturing corporations might be structured around industrial, commercial, and governmental divisions according to the prevalent type of customer for each group.

The various organizational structures all have advantages, but none is ideal for all companies. Indeed, it is safe to say that these formats have drawbacks for most or all the companies that have adopted them. Nonetheless, a company usually chooses one basic format as the most satisfactory structure for its needs at a particular time in its evolution—or it creates a combination of two or more types.

For the small, single-plant firm, strategy and production functions are not geographically separated; hence, there is no need for an intermediate tier of coordinating activities. As firms grow to become multilocational companies, more complex functional and spatial divisions of labor develop. One of the best-known forms of spatial organizations draws on the characteristics of large electronics companies. Strategic functions such as research and marketing are performed at the headquarters. Coordinating functions are dispersed to regional offices that control interdependent production facilities. This organizational structure represents a clear-cut distinction between the functions of conception and execution, a division that mirrors the distinction between nonproduction and production employment.

Administrative Hierarchies

A large proportion of the employees of large corporations, even those primarily in manufacturing, are involved in nonproduction activities. The proportion is increasing because of the substitution of capital for labor and because of the growth of various activities associated with administration, management, research, advertising, finance, legal services, and so forth. The ratio of nonproduction to production employment is less important from a geographic perspective than is the relative location of these activities.

Strategic head-office functions tend to cluster in a relatively few large metropolitan areas, especially in the case of huge firms with a financial orientation rather than a production orientation. Headquarter functions rely heavily on access to other firms, including clients, suppliers, advertisers, repair and maintenance, and specialized law firms, and the agglomeration economies of large cities are critical. Many of these functions require face-to-face contacts and business meetings. The contribution of head-office establishments to nonproduction employment within corporations has more strategic than numerical significance.

Economic Geography and Social Relations

Economic geography includes the analysis of how firms locate and behave in space. In order to do this, it is necessary to embed firms in their social and historical context. From the political economy perspective, the production is a *social*, not purely individual process.

The crucial social relation of production is between owners of the means of production and the workers employed to operate these means. Under capitalism, the means of production are privately owned, which divides people into owners of capital and those who must sell their labor power in order to survive. Owners of capital—the capitalist class that came to power starting in the sixteenth century (Chapter 2)—control the labor process and extract surplus value through the exploitation of workers. Competitive relations exist among owners; cooperative and antagonistic relations emerge between owners (capital) and workers (labor).

Relations among Owners

Capitalists make independent production decisions under competitive conditions. A raw competitive struggle for survival is fundamental to an appreciation of capitalist development. Competition requires producers to apply a minimum of resources to achieve the highest output. It forces companies to minimize costs, which means extreme labor specialization and subordination of workers to machine automation. It demands large-scale production to lower costs and to control a segment of the market. It also entails the acquisition of linked or competing companies and the investment of capital in new technology and in research and development (R&D).

Competition is the source of capitalism's immense success as a mode of production. But the tensions between opposing elements cannot be solved without fundamental change. Consider, for example, the critical environmental issues generated by the contradiction between capitalism and the natural environment. For productive forces to continue to expand without a reduction in living standards, new values must be built into the production system. These values are already evidenced by the use of renewable energy sources and the imposition of pollution controls.

Relations between Capital and Labor

Capital-labor relations are both cooperative and confrontational. Without a cooperative workforce, production is impossible. However, cooperative relations are often subordinate to antagonistic relations.

Because producers make decisions according to their desire to make profits, they try in every way possible to pay workers only part of the value produced by their labor. Value produced by workers in excess of their wages—called *surplus value*—is the basis for profit. Such a view emphasizes the dynamics of the workplace and the labor process rather than supply and demand in the market. Workers try to increase their wages in order to enjoy a higher standard of living. They sometimes organize into unions and, if necessary, strike to demand higher wages. If management agrees to meet labor demands, cooperative relations may exist for a time before antagonistic relations resume.

Competition forces management to invest in technology and research to increase productivity. As production increases, the struggle between employers and employees puts higher wages into the workers' hands. Machines and low-wage labor can replace high-wage labor. Low-wage peripheral regions can sell products to high-wage center regions. Industrial migration to the periphery removes jobs in the center, which disciplines organized labor. Pressures to increase wages slacken, and mass demand decreases. A problem of underconsumption develops. Thus, in capitalism, the solution to one problem may be the breeding ground for new problems.

Competition and Survival in Space

Relations among owners and between capital and labor are sources of change in the geography of production. Competition among owners may cause a company to relocate all or parts of its operation to a place where it can secure low-wage labor. From the company's perspective, this strategy is mandatory for survival; if other companies lower their costs and it does not, it will inevitably lose in the competitive struggle. Capitalists must expand to survive, and the struggle for existence leads to the survival of the biggest. In their search for profits, giant corporations have extended their reach so that few places in the world remain untouched.

The incessant struggle of companies to compete successfully is especially evident in the entrepreneurial response to differential levels of capital-labor conflict. Old industrial regions of the core—Europe, North America—have high-conflict levels. In contrast, peripheral regions have various combinations of lower conflict levels and lower wages. Organized labor in the old industrial areas induced the owners of capital to switch production and investment to countries that were not yet industrialized or to newly industrializing countries. The reason that mobile capital could avoid the demands of organized labor was the development of productive forces—an increased ability to traverse space and conquer the technical problems of production—and the emergence of a huge alternative labor force in the developing world following the colonial revolution in Asia, Africa, and the Caribbean.

These dramatic changes in the 1970s and 1980s ended the original international division of labor that was formalized in the late nineteenth century. Under the old imperial system, the advanced powers were the industrialists, and the colonies were the agriculturalists and producers of raw materials. After decolonization, light industry and even some heavy industry began to emerge in the former colonies. The advanced economies assisted this process. The increasing globalization of production was accompanied by a new international division of labor. The world became a "global factory," in which the developed countries produced the sophisticated technology, and the developing countries were left with the bulk of the low-skill manufacturing jobs. The emergence of this new international division of labor, mainly a consequence of the activities of the footloose multinational corporations, resulted in deindustrialization in the old industrial regions of advanced economies and a precarious export-led industrial revolution in parts of the developing world.

Business Cycles and Regional Landscapes

Capitalism is a society and economy notorious for its instability. The history of capitalist economies is replete with boom and bust periods, epochs of rapid growth, high profits, and low unemployment followed by periods of crisis, economic depression, bankruptcies, and high unemployment. Why does this instability exist? How does this instability affect national, regional, and local economies?

The most famous depiction of this process is that of *Kondratiev cycles,* named after the Soviet economist Nikolai Kondratiev, who first identified them in the 1920s. Examining historical data on changes in output, wages, prices, and profits, Kondratiev hypothesized that industrial countries of the world experienced successive waves of growth and decline since the beginning of the Industrial Revolution. Based on the emergence of key technologies, these long cycles have a periodicity of roughly 50 to 75 years' duration (Figure 5.13). The reasons that underlie this duration

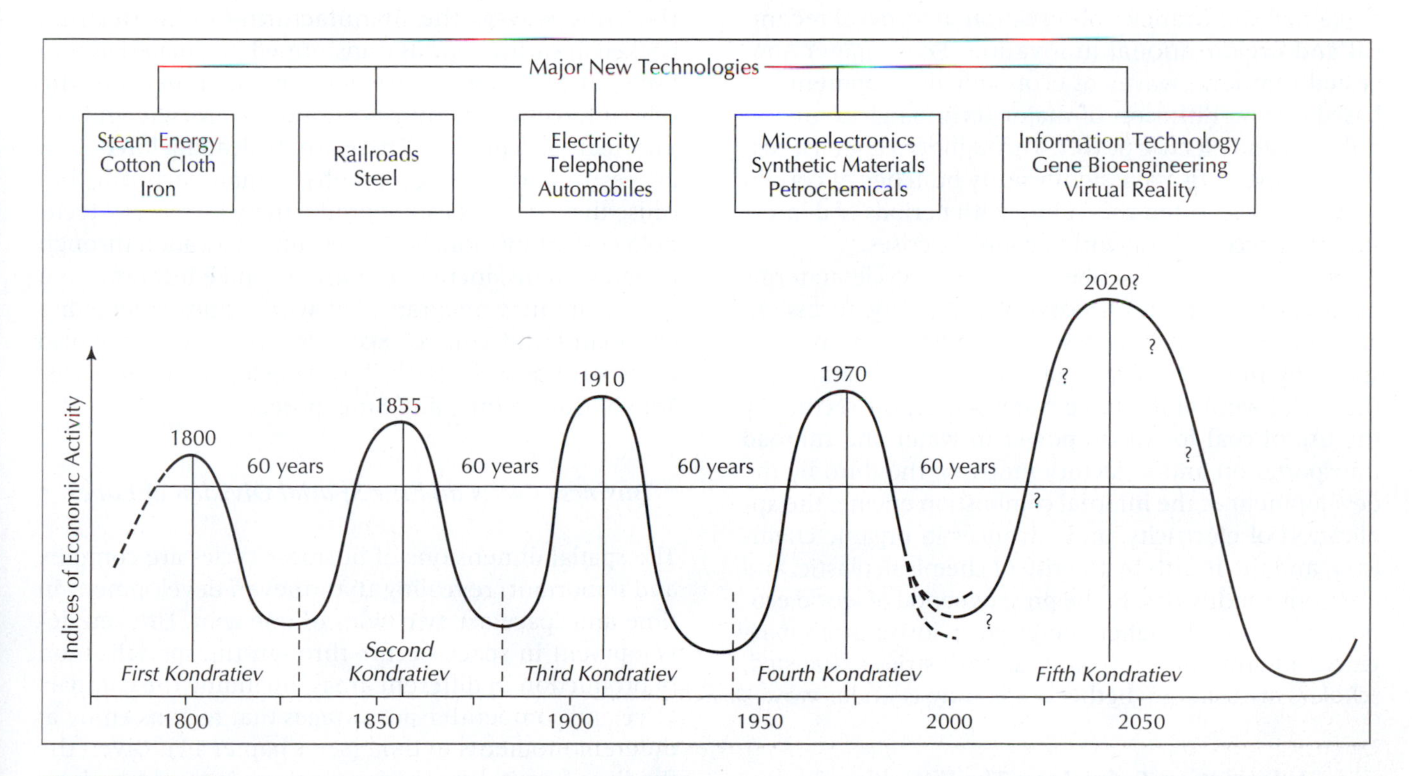

FIGURE 5.13
Kondratiev longwaves of economic activity. Kondratiev waves last approximately 50 years each and have four phases of activity, including boom, recession, depression, and recovery. Each period of economic activity has its associated major technological breakthroughs that power economic growth and employment. On the horizon, the world appears to be a new boom cycle based on information technology, biotechnology, space technology, energy technology, and materials technology.

of these waves reflect long-term trends in the rate of capital formation and depreciation; as fixed capital investments reach the end of their useful economic life, the drag on productivity they create generates incentives to search for new technologies.

The first Kondratiev wave, at the dawn of the Industrial Revolution, was centered on the textile industry and lasted from roughly 1770 to the 1820s, when the West was swept by a series of recessions, bankruptcies, and bank failures. The second wave, focused on railroads and the iron industry, originated in the 1820s, peaked in the 1850s, and ended in the great round of consolidation in the 1880s and 1890s, particularly the depression of 1893, the second worst in world history. The third wave, associated with Fordist industries such as automobiles but also including electricity and chemicals, arose at the end of the nineteenth century, peaked around World War I, and collapsed suddenly during the Great Depression of the 1930s. The fourth Kondratiev, which was propelled by World War II, peaked in the 1960s, corresponded to the postwar wave of growth, and included major propulsive industries such as petrochemicals and aerospace; it ended with the petroshocks of the 1970s. Many believe that we live in the midst of a fifth wave, starting in the 1980s and centered around services.

Joseph Schumpeter, a famous German economist, explained Kondratiev's observation in terms of technical and organizational innovation. Schumpeter suggested that long waves of economic development are based on the diffusion of major technologies, such as railways and electric power. Throughout capitalist history, innovations have significantly bunched at certain points in time, often coinciding with periods of depression that accompany world economic crises.

Simon Kuznets described Kondratiev cycles in terms of successive periods of recovery, prosperity, recession, and depression. The upswing of the first cycle was inspired by the technologies of water transportation and the use of wind and captive water power; the second by the use of coal for steam power in water and railroad transportation and in factory industry; the third by the development of the internal combustion engine, the application of electricity, and advances in organic chemistry; and the fourth by the rise of chemical, plastic, and electronics industries. In the present period of world economic crisis, with higher energy costs, lower profit margins, and growth of the old basic industries exhausted, scholars are asking whether a *fifth wave* is under way.

Information Technology: The Fifth Wave?

Some scholars argue that a fifth Kondratiev cycle began in the 1980s, and is associated with *information technology*. Information technology production is based on microelectronic technologies, including microprocessors, computers, robotics, satellites, fiber-optic cables, and information handling and production equipment, including office machinery and fax machines. Information technology, production, and use has a strong production pattern in Japan, in East Asian newly industrializing countries, in the United States and Canada, and in Western Europe, notably Germany, Sweden, and France. The importance of information technology results from the convergence of communications technology and computer technology. Communications technology involves the transmission of data and information, whereas computer technology is concerned primarily with the processing, analysis, and reporting of information. A new technoeconomic paradigm does seem to be emerging based on the extraordinarily low costs of storing, processing, and communicating information. In this perspective, the 1980s and 1990s were a prolonged period of social adaptation to the growth of this new technological system, which is affecting virtually every part of the economy, not only in terms of its present and future employment and skill requirements but also in terms of its future market prospects.

At the front of the contemporary wave are business services such as advertising, purchasing, auditing, inventory control, and financing (*producer services*). The defining characteristic of these new services is that they create and manipulate knowledge products in almost the same way as the manufacturing industries that peaked in earlier rounds transformed raw materials into physical products. These services have become the salient forces of the new postindustrial society and the information economy. These are the leading forces that are restructuring the geography of manufacturing, because they are the basis of productivity increases—technological innovation, better resource allocation through expert systems, increased training, and education. As a result, the new geography of world cities emerged; a command and control economy centered on world cities such as New York, Los Angeles, London, Paris, Tokyo, Hong Kong, and Singapore.

Business Cycles and the Spatial Division of Labor

The spatial dimensions of business cycles are complex and important, revealing that uneven development in time and space are two sides of one coin. Uneven development in space occurs through the specialization of production in different areas, including the comparative and competitive advantages that regions enjoy at different moments in time (see Chapter 11). Given the fluidity of capitalism, however, there is no reason for a region or country to enjoy its advantages in production indefinitely. Capital, labor, and information move across space, changing the conditions of profitability in different places. Put differently, uneven development in time is manifested when a region's or nation's comparative advantage is created and lost as capital creates and destroys regions over successive business cycles. The loss of comparative advantage makes a region attractive to firms: It offers pools of unemployed, and hence cheap labor, an infrastructure, and often other advantages as well. In short, regions abandoned by capital may be ready to be recycled for a new use.

Over different business cycles, several industries may locate in one region, each leaving its own imprint on the local landscape. Each industry constructs a labor force, invests in buildings, and shapes the infrastructure in ways that suit its needs (and profits). From the perspective of each region, therefore, business cycles resemble waves of investment and disinvestment. Each wave, or Kondratiev cycle, deeply shapes the local economy, landscape, and social structure and leaves a lasting imprint on a region that is not easily erased.

For example, the textile industry in New England created its industrial landscape in the nineteenth century, ranging from small mills located on streams to the large factories in Lowell, Massachusetts, or Manchester, New Hampshire. These landscapes, including the people who inhabit them, persisted long after the industry abandoned New England for the South in the early twentieth century. For many years New England was a relatively poor part of the country, with high unemployment rates. By the 1980s, however, a new wave of production had centered on the region, the electronics industry. Firms producing computer hardware and software found the human resources of the Boston metropolitan region attractive, including the famous Route 128 corridor, and the local geography of this industry was shaped to no small extent by the residues of earlier ones.

In short, each set of investment/location decisions in a region is prestructured by earlier sets of decisions. Thus, as their comparative advantage changes, regions accumulate a series of different imprints: Each wave is shaped by and transforms the vestiges of past waves. Thus, regions are unique combinations of layers of investment and disinvestment over time. Such an approach explains the unique characteristics of places through their economic histories. Because capitalism constantly reproduces spatial inequality by diverting capital from low-profit to higher-profit regions, individual places are perpetually open to the lure of new forms of investment and vulnerable to the risk of being abandoned by capital. This view allows us to integrate the specifics of regions with broader understandings of capitalist processes.

THE STATE AND ECONOMIC GEOGRAPHY

Contrary to much popular opinion, the economic landscapes of capitalism are not simply the products of "free markets," but also involve the role of the state (government in all its forms and functions). In noncapitalist societies, particularly feudalism, the state was the major means by which resources were allocated; there was, effectively, no division between public and private power. Under capitalism, markets are the *primary* means through which decisions are made to hire people, use land, or determine how to utilize capital, but they are not the *only* means. The state does what no individual or firm can do, tackling problems too big for private firms and providing necessary but unprofitable services. Given the long history of the state—often hidden by notions that celebrate a mythical "free market" that has never existed in fact—it is dubious that capitalism could survive without the state. Even the most unfettered of markets, for example, such as a garage sale, presupposes the existence of money, property rights, and an infrastructure such as a road system.

The degree of state intervention varies historically and geographically—it waxes and wanes depending on

Large, capital-intensive infrastructural projects are almost always built, owned, and maintained by the state, which greatly affects the economic landscapes of capitalist societies.

the political forces and economic imperatives at work—but it is never zero. In the nineteenth century, particularly in the United States, state intervention was considerably less than today: There were few public services, although the federal government did subsidize railroads and erect tariffs against imports. Starting with the Great Depression of the 1930s, when the market generated extreme suffering for millions of people, the role of the state changed, including the so-called "welfare state," which offered numerous social protections and safeguards (e.g., Social Security, the minimum wage). Since the late twentieth century, under the pressure of globalization, the welfare state has retreated around much of the world, particularly in the United States, and has been replaced with a deregulated, "neoliberal" state based on the premise that the market is the optimal way of allocating public as well as private resources. Any account of how economic landscapes are produced, therefore, must include some understanding of the role of the state. Several dimensions are sketched here.

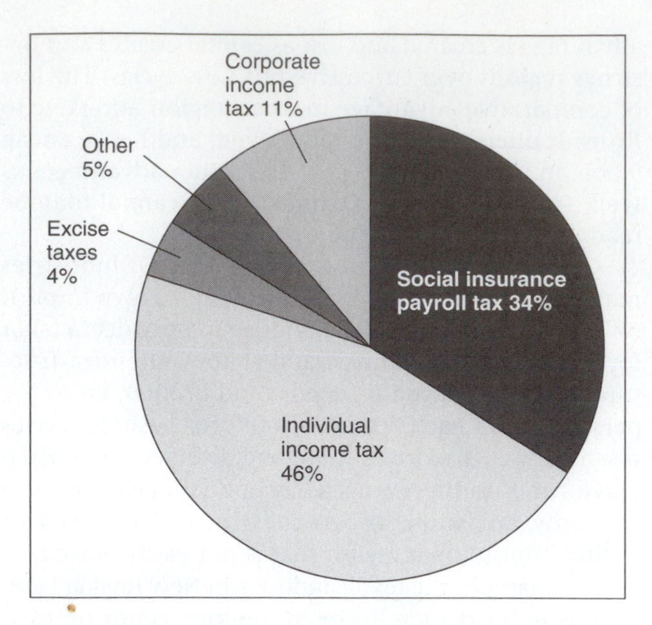

FIGURE 5.14
Federal government taxes, 2002. Government receipts come in a variety of forms, primarily from individual income taxes. Generous tax cuts for corporations have reduced their share to 11 percent. Although taxes are not the most important locational consideration for firms, they constitute a flow of resources among classes and regions, finance public sector activity, and represent one of many ways in which the state shapes economic landscapes.

One way that states shape economies is through the creation and enforcement of a legal system. Laws, which carry the moral authority of right and wrong, also act economically to protect property rights: Without the right to buy and sell, to have assets secured against forcible appropriation, markets would simply not exist. The development of capitalism thus entailed a secularized legal system. Laws and regulations encompass a vast array of activities that both constrain behavior and protect some parties from the actions of others, including, to take but a handful, health and safety regulations (e.g., environmental protection, workplace safety rules, restaurant inspections), antidiscrimination ordinances, and antitrust laws. The state enforces laws through the police, judiciary, and the military and maintains a monopoly over the legal use of violence.

The state is also heavily involved in setting fiscal and monetary policy. Fiscal policy—which determines how governments spend their money—has enormous impacts on localities throughout a nation state. In the United States, for example, the federal government's budget is larger than $2 trillion annually. Governments collect revenues through a variety of taxes and fees, including, most important, through individual income taxes, but also including to a lesser extent the social insurance payroll tax and corporate income taxes, which generate only 11 percent of the federal government's revenues (Figure 5.14).

Conversely, government expenditures, which are uneven across the landscape, have huge impacts on local areas, generating jobs, subcontracts, revenues, and profits. In some cases, such as Washington, D.C., entire metropolitan regions exist due to these expenditures. Public outlays include transfer payments such as Social Security, veterans' benefits, and other entitlement programs. The government also subsidizes producers of many goods, particularly agriculture and dairy products, spending far more on corporate welfare than aid to poor people.

In the same vein, the government controls the money supply, which in turn affects inflation, interest, and exchange rates, all of which have geographically uneven consequences in highly complex ways. Almost all nations have a national bank that attempts to manage their money supply; in the United States, it is the Federal Reserve (Chapter 8).

The state has a huge impact on economic landscapes through the construction of an infrastructure, including transportation and communication networks (roads, highways, bridges, airports, ports), water and electrical supply systems, hydroelectric dams, sewers, and the like. Without the infrastructure, the circulation of people, goods, and ideas that are fundamental to markets would be impossible. Thus, the Federal Interstate Highway System (Figure 5.15), the largest project the federal government has undertaken, plays an enormous role in facilitating the movement of people and goods among the nation's cities. Similarly, public services include public education, health care, fire and police departments, libraries and swimming pools, trash and snow removal, public transportation and housing, parks, and so on. In the United States, most public

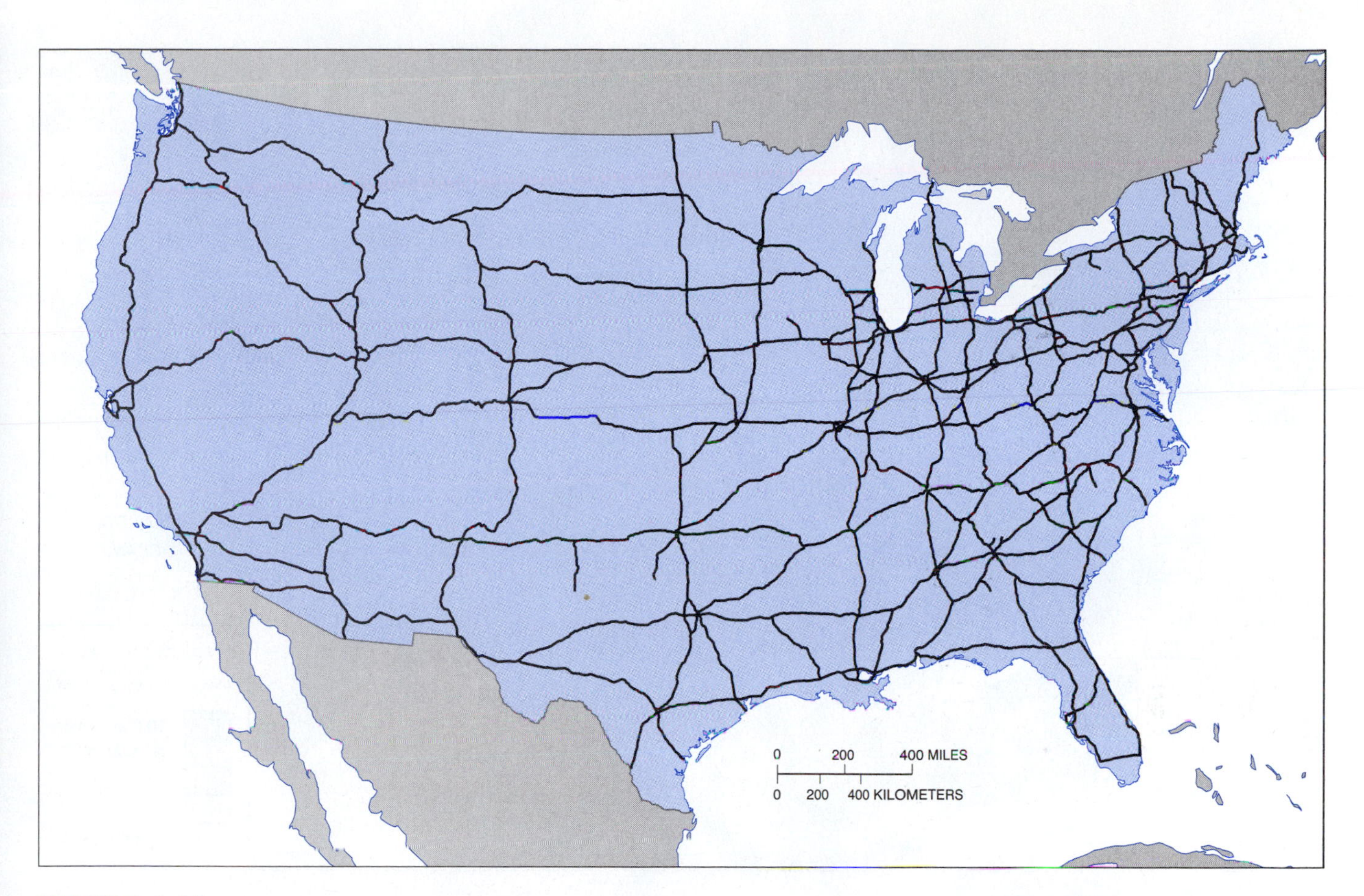

FIGURE 5.15
The Federal Interstate Highway System. The infrastructure is a hugely important form of state intervention in capitalist societies. The Interstate Highway System, the largest project the federal government ever undertook, has facilitated a dramatic round of time-space compression among American cities. Like most such projects, its costs are born publicly while the benefits remain appropriated privately.

services are provided at the local municipal or county level; federal government services (as opposed to transfer payments) are primarily confined to the post office and defense.

The state shapes the labor markets of capitalist societies in many ways, both directly and indirectly. In the United States, the federal government is the nation's largest single employer, with more than 2 million people in 2003 out of a national labor force of roughly 130 million. State and local government employment is even higher, amounting to approximately 20 million in 2003. Additionally, the state shapes labor markets through interventions such as the minimum wage, health and safety codes, antidiscrimination rules, and regulations concerning benefits, overtime, and vacations. The state may offer training grants and seek to generate human capital through the education system, which in turn affects private employers.

Housing markets are another area that the state shapes. Private housing markets are hugely affected by government-influenced interest and inflation rates. Public housing, which in the United States amounts to only 2 percent of the total housing stock, is an important source for low-income people in many cities. Zoning codes on population density, minimum lot size, and architectural details affect the supply and demand for housing. In some cities, the municipal government imposes some forms of rent control, which typically create a market for housing at below-market price prices. More broadly, the federal government owns 40 percent of the area of the United States (Figure 5.16), primarily in the western United States, for national parks and military facilities.

Finally, the state acts as an agent in international issues. Governments shape trade in many ways, a topic we explore in more depth later, with tariffs, quotas, and nontariff barriers, as well as subsidies to exporters. Many governments attempt to manipulate exchange rates to make their exports competitive or to improve their trade balance. The state also controls the international movement of labor, putting restrictions on immigration, thus affecting the supply of workers domestically. These policies work unevenly across the

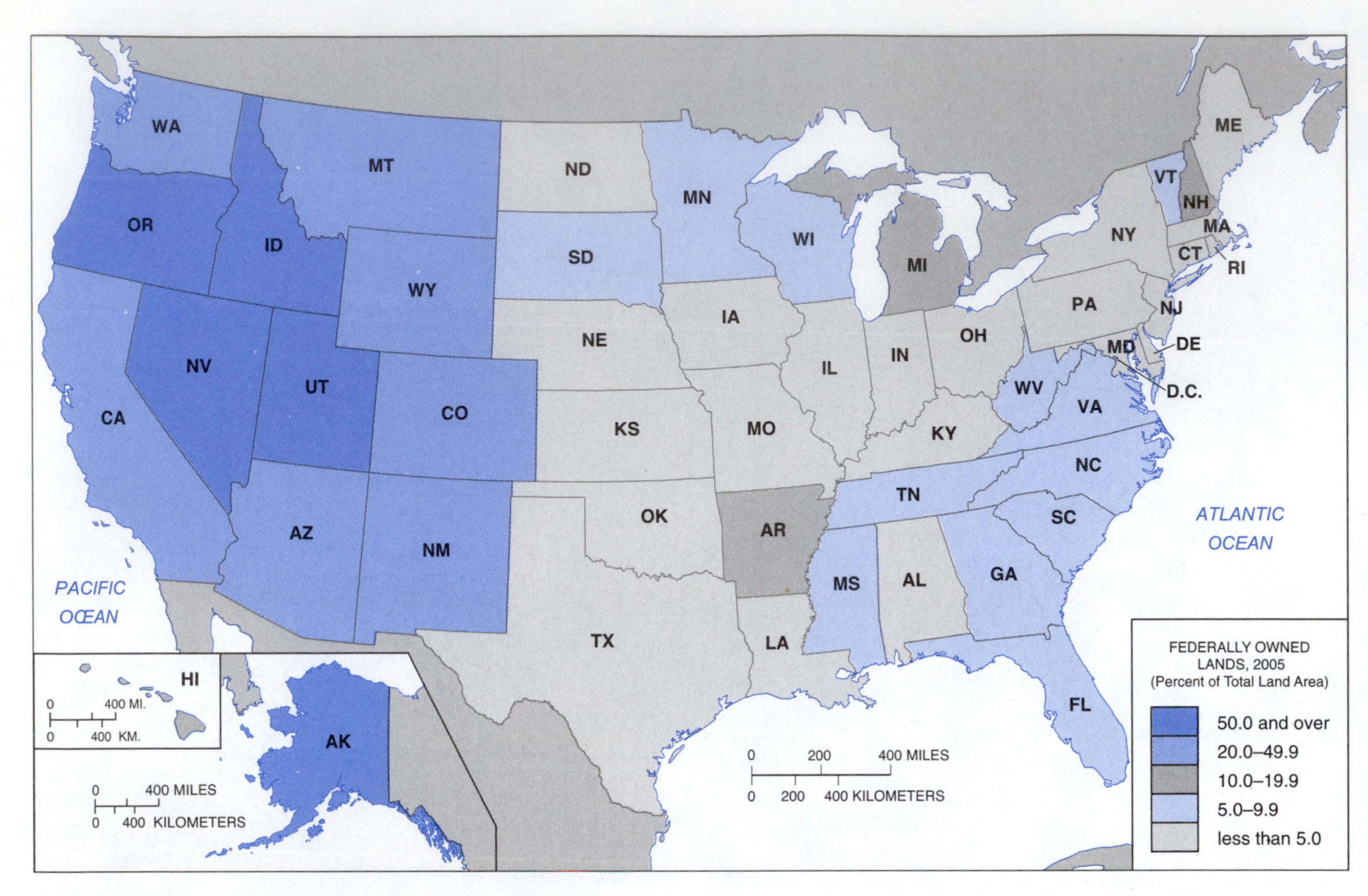

FIGURE 5.16
Federally owned lands, 2002. The federal government owns roughly 40 percent of the nation's territory, especially in the West, in many forms, including national parks, highways, forests, and military facilities.

national landscape, as some regions are more connected to the international economy than are others. All these examples serve to demolish the myth of the "free market," demonstrating that capitalism is a system in which both markets and state are in operation, although not to an equal extent.

Summary

In this chapter, we examined various theoretical and conceptual dimensions pertaining to how economic landscapes are constructed. It opened by listing the major factors of location that influence the locations of firms, including labor, land, capital, and management skills. We noted labor productivity was as important as cost. The uneven distribution of these phenomena leads to geographically uneven patterns in their use by firms.

The chapter explicated in some depth the famous Weberian theory of location, which centers on transport-cost minimization. Classical location theory stresses that manufacturing patterns are caused by geographic characteristics—*locational factors*—rather than by underlying social relations. Assembly costs are incurred because the raw materials required for a particular kind of manufacturing are distributed in different places. Production costs vary because of the areal differences in the costs of labor and land; while the costs of capital investments may also vary geographically, the costs of financial capital are much more uniform due to its much greater mobility. Finished-product distribution costs are incurred when producers must sell to dispersed or widely scattered markets. All these costs are collectively called *locational costs.* Classical location theory provides a rationale to help find the points of production at which locational costs are minimized.

The chapter discussed the behavior of firms in time and space. Most geographers now question the usefulness of traditional location theory in light of the multiproduct, multiplant, multinational operations characteristic of the global structure of production and consumption. Accordingly, we devoted a portion of the chapter to the spatial behavior of large industrial enterprises. Attention was given to trends in industrial organization, the relationship of large firms to small ones, the reasons for corporate growth, and the internal geography of corporate systems.

The chapter illustrated how firms face a choice between their choice of production technique, which reflects the costs of inputs, and the scale of output, which generates economies of scale. Few issues are as important as economies of scale in understanding where firms locate, the nature of the market they are in, and how they change over time. The chapter also explored the growth of firms, ranging from their strategies to the roles of vertical integration and disintegration. It also delved into the division of labor within firms, including corporate administrative hierarchies and the separation of headquarters and production functions.

Finally, it embedded firms in their social context, pointing out that firms are always part of a broader nexus of capital and labor relations, which may include divisions among the owners of capital and between capital and labor. Firms are thus not isolated decision makers floating in a world without constraints but are part of the process of commodity production, transportation, and consumption.

Because capitalism is a tremendously dynamic society beset with constant, often wrenching changes, the advantages of regions over others are temporary in nature, rarely surviving more than one round of long-term growth, or Kondratiev wave. The periodicity of capitalism is reflected in the rise of different industries at different moments in time—textiles, steel, automobiles, electronics, producer services—each of which is the high-tech sector of its day. The Kondratiev model reminds us that the present world economic system may be in the midst of a fifth upswing, this one based on a cluster of microelectronics and information technologies. We sutured business cycles to the spatial division of labor to understand how local landscapes are created through the successive imposition of different layers of investment over time.

Finally, we ended the chapter by noting that capitalism is not synonymous with the "free market" because markets are always and everywhere shaped by the state to one degree or another. The state makes and enforces laws and regulations that enforce property rights, collects and spends money, affects interest and inflation rates, generates jobs and subcontracts and regulates working conditions, regulates land use, builds the infrastructure and provides public services, shapes housing markets, and intervenes internationally in trade, exchange rates, and immigration. In all these ways, and more, the state is an actor as important to the construction of the economic geographies of capitalism as are market forces.

STUDY QUESTIONS

1. How is labor different from other inputs firms use?
2. What determines an industry's demand for labor?
3. Do firms always pursue the cheapest labor? Why not?
4. How does land figure into the locational calculus of firms?
5. What is capital? Differentiate fixed and liquid capital.
6. How are technical skills related to innovation?
7. What is the Weberian model of firm location?
8. What are scale and agglomeration economies?
9. What is the behavioral approach to industrial location?
10. What are forward and backward integration?
11. What are multiplant and multiproduct enterprises?
12. What are Kondratiev cycles?
13. How do business cycles shape local landscapes?
14. What are six ways the state affects economic landscapes?

KEY TERMS

adaptive behavior
adoptive
agglomeration economies
backward integration
diseconomies of scale
vertical integration
diversification
division of labor
fifth wave
fixed capital
footloose industries
forces of production
forward integration
gross raw materials
horizontal integration
human capital
industrial-complex economies
industrial inertia

information technology
integration
Kondratiev cycles
liquid, or variable, capital
localized raw materials
locational costs
locational factors
material index
marketing linkages
matrix structure
methods
mode of production
producer services
production linkages
pure raw materials
scale
service linkages
social relations of production
spatial margins to profitability.
strategy
surplus value
technique
ubiquitous raw materials
urbanization

SUGGESTED READINGS

Cox, K., ed. 1997. *Spaces of Globalization: Reasserting the Power of the Local.* New York: Guilford.

Dicken, P. 2004. *Global Shift: Industrial Change in a Turbulent World,* 4th ed. London: Harper & Row.

Herod, A. 1998. *Organizing the Landscape: Geographical Perspectives on Labor Unionism.* Minneapolis: University of Minnesota Press.

Herod, A., G. Ó Tuathail, and S. Roberts, eds. 1997. *An Unruly World? Globalization, Governance, and Geography.* London: Routledge.

Knox, P., Agnew, J., and L. McCarthy. 2003. *The Geography of the World Economy,* 4th ed. London: Edward Arnold.

Krugman, P. 1991. *Geography and Trade.* Cambridge, Mass.: University Press.

Martin, R., and P. Sunley. 1998. "Slow Convergence? The New Endogenous Growth Theory and Regional Development." *Economic Geography* 74:201–27.

Massey, D. 1984. *Spatial Divisions of Labor: Social Structures and the Geography of Production.* New York: Methuen.

Porter, M. 2003. "The Economic Performance of Regions." *Regional Studies* 37:549–78.

Porter, M. E. 1990. *The Competitive Advantage of Nations.* New York: Free Press.

Storper, M. 1997. *The Regional World: Territorial Development in a Global Economy.* New York: Guilford.

Thrift, N., and K. Olds. 1996. "Reconfiguring the Economic in Economic Geography." *Progress in Human Geography* 20:311–37.

WORLD WIDE WEB SITES

WORLD FACTBOOK ON COUNTRIES
http://www.odci.gov/cia/publications/pubs.html

THE WORLD TRADE ORGANIZATION
http://www.wto.org/
The principal agency of the world's multilateral trading system. Its home page includes access to documents discussing international conferences and agreements, reviewing its publications, and summarizing the current state of world trade.

THE WORLD BANK
http://www.worldbank.org/
A leading source for country studies, research, and statistics covering all aspects of economic development and world trade. Its home page provides access to the contents of its publications, to its research areas, and to related Web sites.

U.S. DEPARTMENT OF COMMERCE
http://www.doc.gov/
Charged with promoting American business, manufacturing, and trade. Its home page connects with the Web sites of its constituent agencies.

BUREAU OF LABOR STATISTICS WEB SITE
http://stats.bls.gov/
Contains economic data, including unemployment rates, worker productivity, employment surveys, and statistical summaries.

ECONOLINK
http://www.progress.org/econolink
Econolink has "the best Web sites that have anything to do with economics." Its listings are selective with descriptions of side content; many referenced sites themselves have Web links.

U.S. INTERNATIONAL TRADE IN GOODS AND SERVICES HIGHLIGHTS
http://www.census.gov/indicator/www/ustrade.html

THE WORLD TRADE ORGANIZATION
http://www.wto.org/
The principal agency of the world's multilateral trading system. Its home page includes access to documents discussing international conferences and agreements, reviewing its publications, and summarizing the current state of world trade.

THE WORLD BANK
http://www.worldbank.org/
A leading source for country studies, research, and statistics covering all aspects of economic development and world trade. Its home page provides access to the contents of its publications, to its research areas, and to related Web sites.

U.S. DEPARTMENT OF COMMERCE
http://www.doc.gov/

Charged with promoting American business, manufacturing, and trade. Its home page connects with the Web sites of its constituent agencies.

BUREAU OF LABOR STATISTICS WEB SITE
http://stats.bls.gov/
Contains economic data, including unemployment rates, worker productivity, employment surveys, and statistical summaries.

WEBEC: WORLD WIDE WEB RESOURCES
http://netec.wustl.edu/WebEc/WebEc.html
A more extensive set of site listings though more purely "economic" than "economic geographic."

CHAPTER

6

AGRICULTURE

OBJECTIVES

- To discuss the origin and diffusion of agriculture
- To help you appreciate the effects of agricultural practices on the land
- To describe world subsistence agricultural forms and regions
- To acquaint you with commercial agricultural practices and world regions
- To describe the agricultural policies of the United States and their shortcomings
- To summarize the von Thünen model of agricultural production

Peasants plowing rice fields prior to planting in central Luzon, Philippines.

Agriculture, the world's most space-consuming activity and one of humanity's leading occupations, is the science and art of cultivating crops and rearing livestock in order to produce food (and fiber) for sustenance or for economic gain.

The historical geography of agriculture has been critical to the survival and success of the human species. Throughout the vast bulk of our existence (98% or more), we were hunters and gatherers, not agriculturalists; people were food collectors, not producers. The Neolithic Revolution, roughly 10,000 years ago, which saw the invention or discovery of agriculture, made possible a nonnomadic existence; it paved the way for a social surplus and the rise of cities, denser social forms and more refined divisions of labor, and fostered the development of new technologies such as metalworking.

Until the nineteenth century, however, agriculture produced little food per worker, so most of the population worked full time or part time on the land. The small surplus released few people for other pursuits. Not until the industrialization of agriculture that started in Europe during the last 200 years did large-scale employment in manufacturing and service activities become possible.

The shift of labor from the agricultural sector to other sectors constitutes one of the most remarkable changes in the world economy in modern times. In the United States and the United Kingdom, less than 2 percent of the economically active population now works directly in agriculture. In contrast, about 70 percent of the population in a number of African and Asian countries are engaged in the agricultural sector.

Economic geographers are concerned with problems of agricultural development and change, as well as with patterns of rural land use. Where was agriculture discovered? How did it diffuse? Why do farmers so often fail to prevent environmental problems? What are the characteristics of the main agricultural systems around the world? What is the effect of industrialized agriculture on farmers and the rural countryside? What principles can help us understand the spatial organization of rural land use? What are the consequences of government policy on farming? In this chapter, we seek answers to these questions.

Of critical importance to many of the issues addressed in this chapter is the decision-making environment of land users and land managers. Frequently, individual farmers make direct land-use decisions, but they often must choose from a predetermined range of options. Farmers may be denied access to common property resources, such as water or grazing land. Landlords, multinational corporations, the state, and market demand may force them into economic marginalization. They may be faced with fluctuations in prices for export commodities. To appreciate the response of land managers to changes in their circumstances, we must recognize the significance of different scales. Patterns of production and land use are the outcome of a series of forces operating at a series of spatial scales.

THE RISE OF AGRICULTURE

The rise of agriculture was one of the most important milestones in the development of human beings. Agriculture was the first instance of human land use that significantly altered the natural environment and is still one of the major means by which humans transform the natural environment. Through a series of accidents and deliberate experiments over thousands of years, several prehistoric peoples learned how to produce food and fiber plants and how to herd animals and to control animal breeding. Domestication of plants and animals probably emerged as an extension of food-gathering activities of preagricultural hunters and gatherers and as a response to a slow, sustained increase in population pressure; even primitive agriculture generates more calories per unit area than does hunting and gathering and so supports denser populations. Agriculture transformed people from food collectors to food producers and made possible the emergence of permanent settled communities during the Neolithic Revolution.

The first agricultural revolution began in the *Fertile Crescent* of the Middle East nearly 10,000 years ago (Figure 6.1). Although this is a dry climate, the

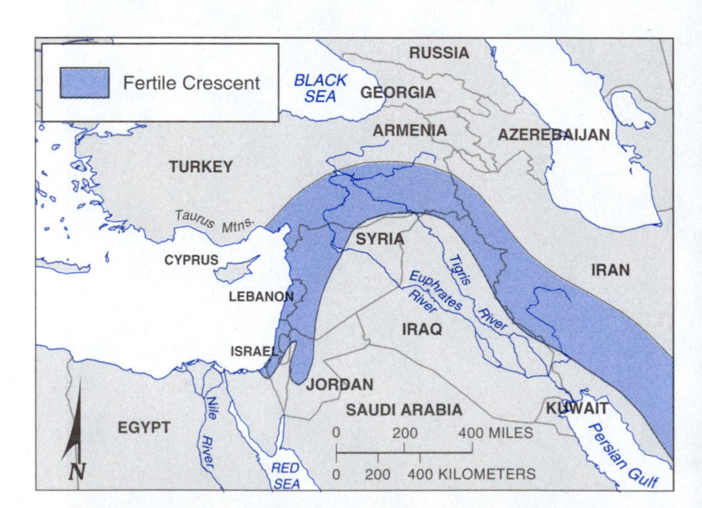

FIGURE 6.1
The Fertile Crescent. This area stretches from the Persian Gulf in the southeast, north to the southern border of Turkey, and to the Mediterranean Sea in the southwest. The Nile Valley of Egypt, a rich agricultural region, and its Delta are sometimes included in the Fertile Crescent. The territory was one of the most important in terms of successions of great empires throughout world history, including the Assyrians, Babylonians, Medes, Persians, and Turks. Many important early developments in the domestication of agricultural plants originated in the region between the Tigris and Euphrates rivers, which comprise present-day Syria, Iraq, Turkey, and Iran.

floodplains of the Tigris and Euphrates Rivers were a well-watered area, extending from the highlands of the eastern Mediterranean through the foothills of the Taurus and Zagros mountains. Archaeological finds also indicate that domestication began early in parts of Central and South America, northern China, the Indus River, and Southeast Asia, as well as several different places in Africa. The spread of agriculture from these centers was slow (Figure 6.2). For example, archaeologists have calculated that it took from 6000 B.C. to 3000 B.C. for a form of shifting cultivation to spread along the Danube and Rhine corridors. Another 1000 years elapsed before agriculture reached southern England.

A reliable food supply liberated some people from food gathering, leading to a more complex division of labor and higher productivity levels. Increased security and leisure, resulting from the new way of life, allowed time for arts and crafts. Communities became involved in spinning, weaving, and dyeing cloth and in manufacturing pottery and containers. Adequate food supplies also allowed for the exchange of specialized goods in markets. In addition, plant cultivation weakened the forces that scattered populations and strengthened the forces that concentrated them. The new way of life allowed people to live in villages and towns, which, although small by our standards, reached population densities far higher than those of preagricultural communities.

Farming practices changed very slowly until the creation of a feudal hierarchy in medieval Europe. In this hierarchy, secular or religious overlords protected the serfs, who farmed fields and paid taxes in *kind* (goods or commodities) or money according to the custom of the manor. English religious manors extracted the *tithe*—one-tenth of a farmer's annual production. Bishops and abbots put little into the farming business and took none of the risks but harassed farmers at harvest time. Similar feudal systems arose in India and China, where peasants provided the surplus value that sustained merchants, craft workers, the nobility, and the military.

The most important innovations associated with farming in medieval Europe were the heavy plow, the replacement of oxen with horses for plowing, and the development of the open-field system, consisting of two or three large fields on each side of a village. These advances increased agricultural production, intensified human concentration in villages and towns, accelerated commerce, and changed patterns of environmental exploitation. For example, the forested lowlands of northern Europe were gradually cleared when the

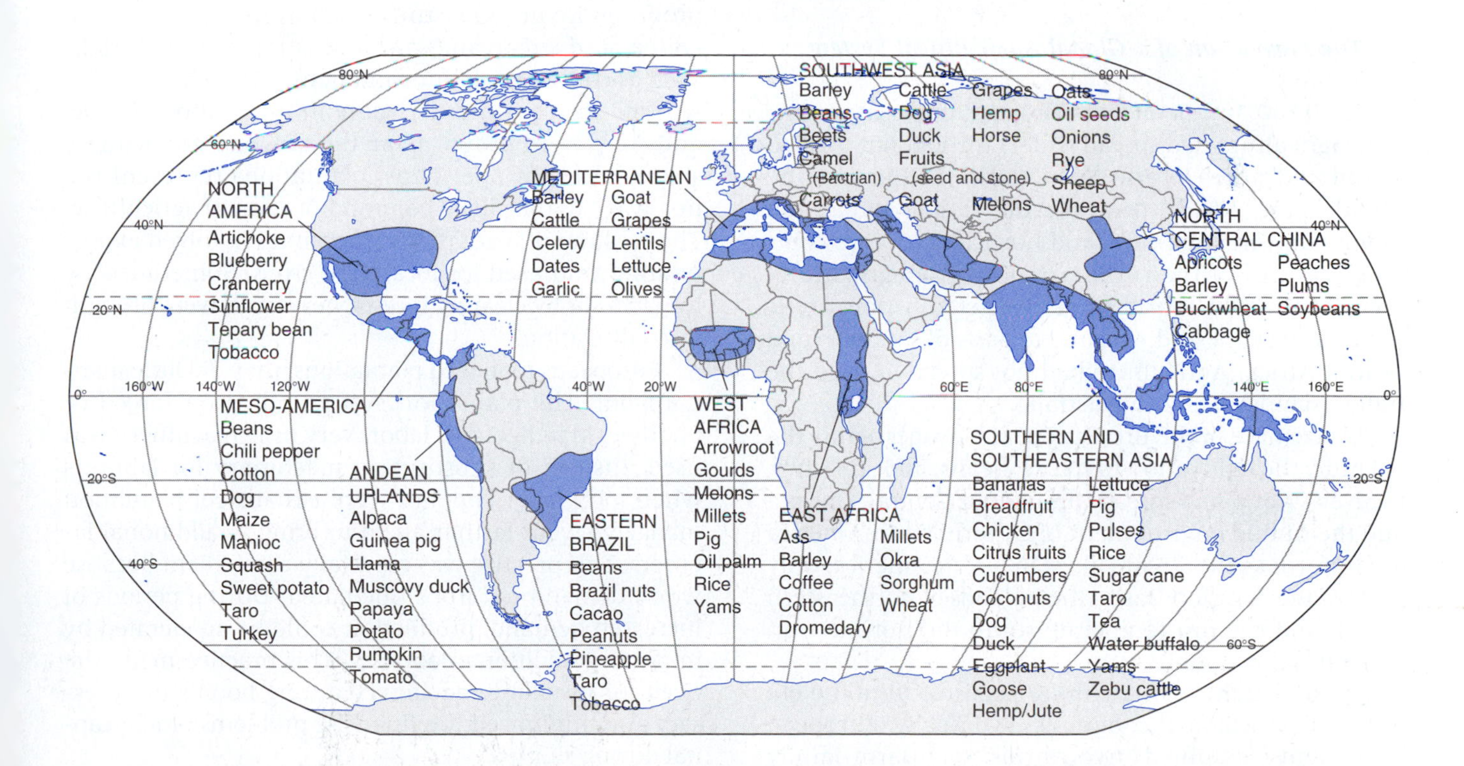

FIGURE 6.2

Origins of plant and animal domestication. Animals were probably first domesticated as household pets by prehistoric peoples. The domestication of plants came from two sources. The first was *vegetated planting,* which was the planting of pieces of existing plants, such as stems and roots. Plants first found growing wild that were useful to the household were cut up and transplanted. The second source was *seed agriculture,* which involved the planting of seeds and was the direct result of natural, annual fertilization of plants.

heavy plow was invented. Clay lowland soils could not be cultivated with the old Mediterranean scratch plow, which was suitable only for light limestone soils.

Medieval farming methods prevailed in western Europe until capitalism invaded the rural manor. The rise of market forces in agriculture transformed food into a commodity, something to be bought and sold for a profit. Land uses changed to crops that generated the highest rate of profit. The demise of the manorial system replaced subsistence agriculture with market-oriented agriculture. Open fields were enclosed by fences, hedges, and walls. Crop rotation replaced the medieval practice of fallowing fields. Seeds and breeding stock improved. New agricultural areas opened up in the Americas. Farm machines replaced or supplemented human or animal power. The family farm came to represent the core model of commercial agriculture. This transformation resulted from a vast population increase in the new trading cities that depended on the countryside for food and raw materials. Another force that brought the market into the countryside was the alienation of the manorial holdings. Lords, who needed cash to exchange for manufactured goods and luxuries, began to rent their lands to peasants rather than having them farmed directly through labor-service obligations. Thus, they became landlords in the modern sense of the term.

The Formation of a Global Agricultural System

By A.D. 1500, on the eve of European overseas expansion, agriculture had spread widely throughout the Old World and much of the New World. In Europe, the Middle East, Africa, central Asia, China, India, and Indonesia, cereal farming and horticulture were common features of the rural economy. Nonagricultural areas of the Old World were restricted to the Arctic fringes of Europe and Asia and to parts of southern and central Africa. Agriculture had not spread beyond the Indonesian islands into Australia.

By the time of the first European voyages across the Atlantic, the cultivation of maize, beans, and squash in the New World had spread throughout Central America and the humid environment of eastern North America as far north as the Great Lakes. In South America, only parts of the Amazon Basin, the uplands of northeastern Brazil, and the dry temperate south did not have an agricultural economy.

These patterns of agriculture persisted until the era of European colonial conquests. Eventually, European settlements assumed two forms: (1) farm-family colonies in the middle latitudes of North America, Australia, New Zealand, and South Africa; and (2) plantation colonies in the tropical regions of Africa, Asia, and Latin America. These two types of agricultural settlements differed considerably.

For example, farm colonization in North America depended on a large influx of European settlers whose agricultural products were initially for a local market rather than an export market. Europeans introduced the farm techniques, field patterns, and types of housing characteristic of their homelands, yet they often modified their customs to meet the challenge of organizing the new territory. For example, the checkerboard pattern of farms and fields that characterizes much of the country west of the Ohio River resulted from a federal system of land allocation (the *Township and Range System*). It involved surveying a baseline and a principal meridian, the intersection of which served as a point of origin for dividing the land into six-by-six-mile townships, then into square-mile sections, and still further into quarter sections a half-mile long. This orderly system of land allocation prevented many boundary disputes as settlement moved into the interior of the United States.

In tropical areas, Europeans, and later Americans and Japanese, imposed a plantation agricultural system that did not require substantial settlement by expatriates. *Plantations* are large-scale agricultural enterprises devoted to the specialized production of one tropical product raised for the market (Chapter 2). They were first developed in the 1400s by the Portuguese on islands off the tropical West African coast. Plantations produced luxury foodstuffs, such as spices, tea, cocoa, coffee, and sugarcane, and industrial raw materials, such as cotton, sisal, jute, and hemp. These crops were selected for their market value in international trade, and they were grown near the seacoast to facilitate shipment to Europe. Thus, plantations represent the first wave in the global commodification of agriculture. The creation of plantations sometimes involved expropriating land used for local food crops. Sometimes, by irrigation or by clearing forests, new lands were brought into cultivation.

Europeans managed plantations; they did little manual labor. The plantation system relied on forced or poorly paid indigenous labor. Very little machinery was used. Instead of substituting machinery for laborers when local labor supplies were exhausted, plantation managers went farther afield to bring in additional laborers. This practice was especially convenient because world demand for crops fluctuated. During periods of increased demand, production could be accelerated by importing additional laborers. This practice made the need for installing machinery during booms unnecessary and minimized the financial problems of idle capital during slumps.

The effect of centuries of European overseas expansion was to reorganize agricultural land use worldwide. Commercial agricultural systems have become a feature of much of the habitable world. Hunting and gathering, the oldest means of survival, has virtually

On the world's largest rubber plantation, at Harbel, Liberia, more than 36,000 hectares, or 30 percent of the total land area of Liberia, are cultivated by Firestone. The company has also established plantations in Brazil, Ghana, Guatemala, and the Philippines. How do plantations benefit host societies?

disappeared. Pastoralists, such as the Masai of Kenya and Tanzania who drive cattle in a never-ending search for pasture and water, have declined in numbers. Subsistence farming still exists, but only in areas where impoverished farmers, especially in developing countries, barely make a living from tiny plots of land. Few completely self-sufficient farms exist; most farmers, even in remote areas of Africa and Asia, trade with their neighbors at local markets.

The Industrialization of Agriculture

In the nineteenth and twentieth centuries, a third agricultural revolution took place. This revolution points to the resolution of the distinction between family and corporate models of agriculture. In other words, it signifies the elimination of distinct agrarian economies and communities. Industrial agriculture has become the dominant form in most developed countries and is being applied to export enclaves of developing countries. Key elements of industrial agriculture are extreme capital intensity, high-energy use, concentration of economic power, and a quest for lower unit costs of production. Although industrial agriculture has increased output per unit of input, it has also depleted water and soil resources, polluted the environment, and destroyed a way of life for millions of farm families.

The industrialization of agriculture drastically reduced the number of farmers in North America. In the United States, the number of farmers declined from 7 million in 1935 to around 1.9 million in 2004. In Canada, 600,000 farm operators existed in 1951, but less than half that number were still in operation in 2002. Europe witnessed similar trends. In Britain, for example, an annual 1.5 percent decline in the number of farm workers has occurred for decades. Today, the percentage of a labor force engaged in primary economic activities (agriculture, logging, fishing, mining) is a useful measure of economic development around the world (Figure 6.3). Economically advanced countries have a small share of their workers in agriculture, such as in North America, Europe, Japan, and Australia, whereas in much of the developing world significant proportions of workers are engaged in farming. In most of Africa and East Asia, for example, more than 60 percent of the employed population works as peasant farmers.

Human Impacts on the Land

The emergence of agriculture and its subsequent spread throughout the world has meant that little if any land still can be considered natural or untouched. Almost everywhere, nature has been modified extensively by human beings, making it difficult to speak of nature as a phenomenon separate from human impacts. Vegetation has been changed most noticeably. Virtually all vegetation zones show signs of extensive clearing, burning, and the browsing of domestic animals. The impoverishment of vegetation has led to the creation of successful agricultural and pastoral landscapes, but it has also led to *land degradation* or a reduction of land capability.

Hunters and gatherers hardly disturb vegetation, but farmers must displace vegetation to grow their crops and

A tapper on the Firestone plantation in Liberia makes an incision in a rubber tree. The latex will flow down the incision through a spout and into a cup attached to the tree. Some of the latex is carried in pails by women to collecting stations. What would this tapper do if he were not working for Firestone?

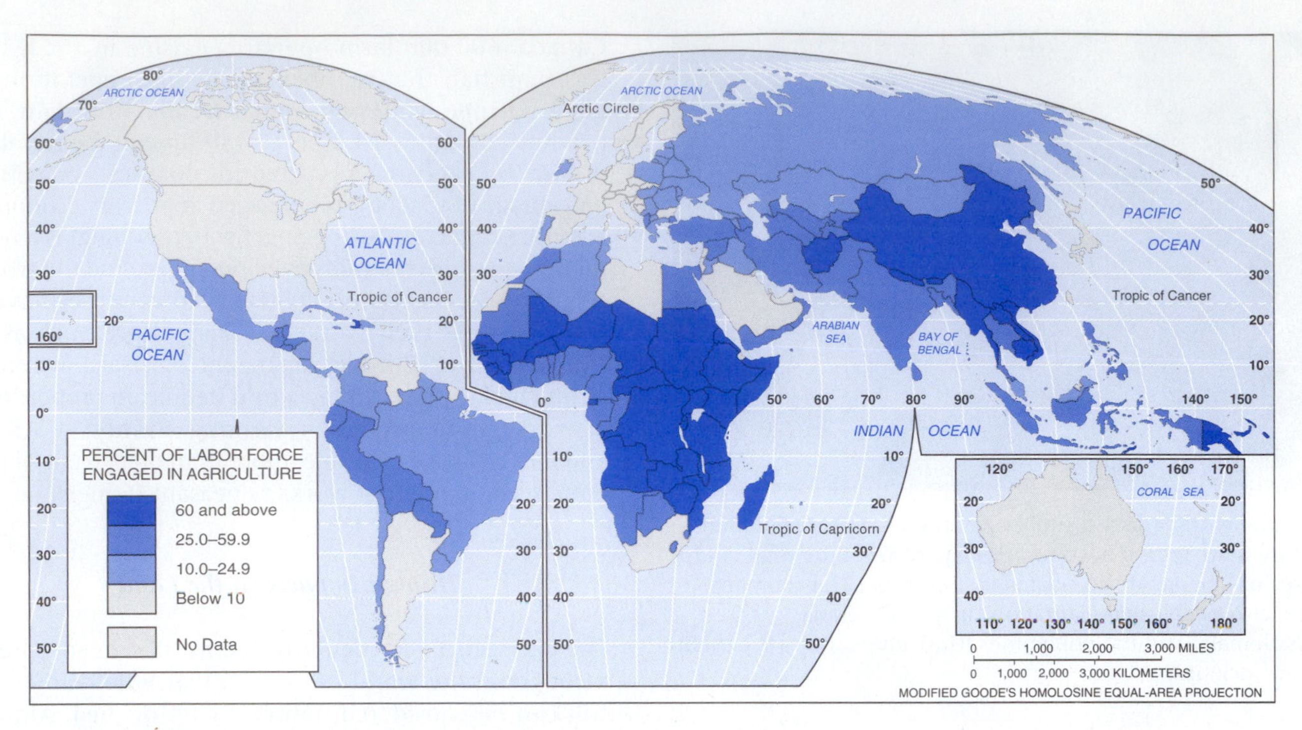

FIGURE 6.3
The proportion of the labor force engaged in agriculture is an important measure of economic development. In economically advanced countries, very few people are needed to generate a food supply for everyone, whereas in developing countries a much larger share of workers are engaged in farming. (See color insert for more illustrative map.)

to tend their livestock. Farmers are land managers; they upset an equilibrium established by nature and substitute their own. If they apply their agrotechnology with care, the agricultural system may last indefinitely and remain productive. On the other hand, if they apply their agrotechnology carelessly, the environment may deteriorate rapidly. How farmers actually manage land depends not only on their knowledge and perception of the environment but also on their relations with groups in the wider society—in the state and the world economy.

As agriculture intensifies, environmental alteration increases. Anthropologist Ester Boserup proposed a simple but famous model of agricultural systems based on five stages. Stage 1, forest-fallow cultivation, involves cultivation for 1 to 3 years followed by 20 to 25 years of fallow. In Stage 2, bush-fallow cultivation, the land is cultivated for 2 to 8 years, followed by 6 to 10 years of fallow. In Stage 3, short-fallow cultivation, the land is fallow for only 1 to 2 years. In Stages 4 and 5, annual cropping and multicropping, fallow periods are

The development of the center pivot irrigation system in the 1950s enabled large-farm operators to transform huge tracts of land in sandy or dry regions of the United States into profitable cropland. Here, alfalfa is being irrigated in Montana.

either very short—a few months—or nonexistent. Boserup noted that the transition from one form of agriculture to another was accompanied by an increasing population density, improved tools, increasing integration of livestock, improved transportation, a more complex social infrastructure, more permanent settlement and land tenure, and more labor specialization.

In contrast, *permanent agriculture* (annual cropping and multicropping) usually occurs in areas of high-potential environmental productivity and of high-population pressure. Under permanent cultivation, the land becomes totally transformed. Yet the fertility of the land may not be impaired. For example, soils of the Paris Basin have been cultivated intensively for hundreds of years, and still they remain highly productive. In many parts of East Asia, carefully terraced hillsides have maintained the productivity of valuable soil resources after thousands of years.

In general, industrialized farming practices pose the main danger to the environment. Clean tillage on large fields, monoculture (the cultivation or growth of a single crop), and the breaking down of soil structure by huge machines are a few factors that may destroy the topsoil. Droughts and dust storms during the 1930s, 1970s, 1980s, and 1990s in the Great Plains of the United States gave testimony of how nature and industrial agriculture can combine to destroy the health of a steppe landscape.

Agriculture threatens ecological balances when people begin to believe that they have freed themselves from dependence on land resources. In developed countries, there is a tendency to exploit the land as a result of pressure to maximize profits. Corporate producers want to make land use more efficient and productive; thus, farming is often viewed as just another industry. However, we must remember that land is more than a means to an end; it is finite, spatially fixed, and ecologically fragile.

Factors Affecting Rural Land Use

Rural land-use patterns, which are arrangements of fields and larger land-use areas at the farm, regional, or global level, are difficult to understand. Worldwide, hundreds of farm types exist. The most interesting aspect of the world's agricultural land-use areas or regions is not their extent but the uniformity of land-use decisions farmers make within them. Given any farming region, why do farmers make similar land-use decisions? Several variables determine land use, including site characteristics and cultural preferences and perception, which are discussed next.

Site Characteristics

Variations in rural land use depend partly on site characteristics, such as soil type and fertility, slope, drainage, exposure to sun and wind, and the amount of rainfall and average annual temperature. As an example, consider the climate in which crops grow. Plants require particular combinations of temperature and moisture. Absolute physical limits of the crop are "too wet," "too dry," "too cold," and "too hot." Absolute climatic limits are wide for some crops, such as maize and wheat, but narrow for others, such as pineapples, cocoa, bananas, and certain wine grapes.

Cultural Preferences and Perceptions

Food preferences, often with religious origins, are one variable affecting the type of agricultural activity at a given site. For example, many Africans avoid protein-rich chickens and their eggs. Hindus abstain from eating beef; Jews and Muslims do not eat pork (Figure 6.4). Most people in East and Southeast Asia abstain from drinking milk or eating milk products. In the United States, a consumer preference for meat leads American farmers to put a greater proportion of their land in forage crops for animal feed than European farmers, who grow more food crops.

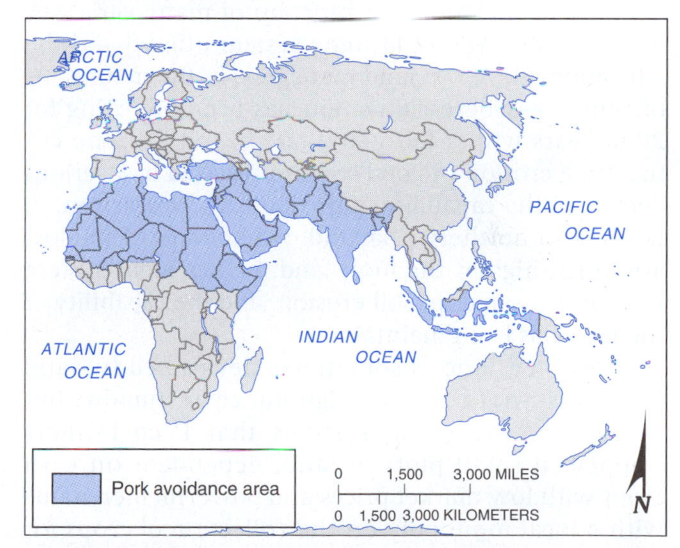

FIGURE 6.4
Pork avoidance areas of the world. Cultural preferences and perceptions against certain foods by a particular culture are called *food taboos*. The pork avoidance, or pork taboo, for North Africa and the Middle East arose in biblical times among Jews. Pigs were considered unclean because of their environmental setting. Today, the Muslim areas of the world, which subscribe to some Old Testament biblical laws, are the largest areas of pork avoidance. The Muslim faith spread throughout the Middle East to India and Southeast Asia, areas that are not nomadic herding regions; nonetheless, they have adopted the religious pork taboo. India is only 15 percent Muslim and therefore does not have a taboo against pork by the majority of the population; however, a beef taboo is practiced in this world region by the Hindu majority.

People interpret the environment through different cultural lenses. Their agricultural experiences in one area influence their perceptions of environmental conditions in other areas. Consider the settlement of North America. The first European settlers were Anglo-Saxons accustomed to moist conditions and a tree-covered landscape. They equated trees with fertility. If land was to be suitable for farming, it should, in its natural state, have a cover of trees. Thus, the settlers of New England and the East Coast realized their expectations of a fertile farming region. When Anglo-Saxons edged onto the prairies and high plains west of the Mississippi River, they encountered a treeless, grass-covered area. They underestimated the richness of the prairie soils, in particular, and the area became known as the "great American desert." In the late nineteenth century, a new wave of migrants from the steppe grasslands of eastern Europe appraised the fertility of the grass-covered area more accurately than the Anglo-Saxons who preceded them did. The settlers from eastern Europe, together with technological inventions such as barbed-wire fencing and the moldboard plow, helped to change the perception of the prairies from the great American desert to the great American breadbasket.

Land degradation is a function of many variables, including the type of farming system utilized and the educational levels of land managers. In the mountains of Ethiopia, where cultivation has been occurring for 2000 years with a fairly low rate of soil loss, the cumulative erosion of good soil has resulted in a serious decline in the capability of the land. In comparison, in the hills of northern Thailand, where rates of soil loss are much higher, the local land management system has compensated for soil erosion, and the capability of the land has been maintained.

However, land is sometimes devastated by land managers—not because of ignorance or stupidity but because of the social systems that keep farmers trapped in small plots of land, dependent on cash crops with low market prices and powerful merchants with a local monopoly on the collection of crops for sale to the market. Local farmers may well be aware of the causes of land degradation and attempt to combat it with fertilizers, mulching, and terracing. However, without real land reform, many peasants around the developing world are trapped in desperate circumstances reflecting the dynamics of the global economy and oppressive national and local social structures.

Systems of Agricultural Production

Systems of agricultural production set their imprint on rural land use. Like manufacturing, the agricultural endeavor is carried out according to two basic systems of production, peasant or precapitalist systems and capitalist, commodified agriculture. The major distinction between these is the labor commitment of the enterprise. In the peasant system, production comes from small units worked entirely, or almost entirely, by family labor. In the capitalist system, family farming is still widespread, but labor is a commodity to be hired and dismissed by the enterprise according to changes in the scale of organization, the degree of mechanization, and the level of market demand for products.

In any geographic region, one system of production dominates the others. For example, capitalist agriculture dominates parts of South America, whereas peasant agriculture dominates other parts. Capitalist agriculture finds expression in a vast cattle ranching zone extending southwest from northeastern Brazil to Patagonia; in Argentina's wheat-raising Pampa, which is similar to the U.S. Great Plains; in a mixed livestock and crop zone in Uruguay, southern Brazil, and south central Chile, which is comparable to the U.S. Corn Belt; in a Mediterranean agriculture zone in middle Chile; and in a number of seaboard tropical plantations in Brazil, the Guianas, Venezuela, Colombia, and Peru. Peasant agriculture dominates the rest of the continent. There is shifting cultivation in the Amazon Basin rainforest, rudimental sedentary cultivation in the Andean plateau country from Colombia in the north to the Bolivian Altiplano in the south, and a wide strip of crop and livestock farming in eastern Brazil between the coastal plantations and livestock ranching zones.

Subsistence, or Peasant Mode, of Production

Subsistence agriculture, also called *peasant agriculture,* occurs entirely in developing countries where market relations have not fully encompassed all domains of economic life. It is relatively labor intensive, involving endless hours of backbreaking toil. In such societies, the bulk of the population lives in rural areas. Farmers are small-scale producers who invest little in mechanical equipment or chemicals. They are interested mainly in using what they produce rather than in exchanging it to buy things that they need. Food and fiber are exchanged, particularly through interaction with capitalist agriculture at global, national, and local scales, but farm families consume much of what they produce. To obtain the outputs required to be self-supporting, peasant farmers are frequently willing to raise inputs of labor to very high levels, especially in crowded areas where land is rarely available. Highly intensive peasant agriculture occurs in the extensive rice fields of South, East, and Southeast Asia. Most of the paddies are prepared by ox-drawn plow, and the rice is planted and harvested by hand—millions of hands.

Another example of the peasant mode of production exists in the semiarid zone of East Africa. This zone includes the interior of Tanzania, northeast Uganda, and the area surrounding the moist high-potential heartland of Kenya. As in most parts of the developing world, peasant agriculture in this region has been complicated by the colonial and postcolonial experience. People in this area earn a living by combining several activities. They eat their crops and livestock and sell or exchange agricultural surpluses at markets. They grow cash, or export, crops such as cotton. They maintain beehives in the bush and sell part of the honey and wax. They brew and sell beer. They hunt, fish, and collect wild fruits. They earn income by cutting firewood, making charcoal, delivering water, and carrying sand for use in construction. Some of them have small shops or are tailors. Most important, people sell their labor, both short term and long term, nearby and far away.

To farm and herd successfully in the semiarid zone, land managers must meet certain requirements set by the environment and by the nature of crops and animals. Livestock require water, graze, salt, and protection from disease and predators. To meet these needs day after day, year after year, land managers must have considerable skill and knowledge. They must know a great deal not only about the ability of animals to withstand physiological stress but also about environmental management—which grass to save for late grazing and where and when to establish dry-season wells to enable the stock to withstand the rigor of the daily journey between water and graze. With respect to crops, land managers must know about plant-moisture and nutrient needs. They must also be sensitive to the variability of rainfall.

Most of the time, this system of agriculture in East Africa provides peasants with an adequate and varied food supply. In bad times, there are mechanisms for sharing hardship and loss so that those farmers who are hardest hit can usually rebuild their livelihoods after bad times end. However, the peasant mode of production has been forced to adjust to pressures from governments and the world economy during colonial and postcolonial periods, including competing with subsidized grain imports from countries such as the United States.

Subsistence Agriculture: Crops and Regions

Most of the world's farmers, including the people of Latin America, Africa, and Asia, practice subsistence agriculture. These regions have several characteristics in common:

1. The majority of workers are engaged in agriculture instead of manufacturing or services.
2. Agricultural methods and practices are technologically primitive. Farms and plots are small in comparison with those of the developed world, labor is used intensively, and mechanization and fertilization are used only infrequently.
3. Agricultural produce that is harvested on the farms is used primarily for direct consumption. The family, or the extended family, subsists on the agricultural products from the farm. Although in certain years surpluses may be produced, this is rarely the case.

There are several major categories of subsistence agriculture: shifting cultivation in the tropics, pastoral nomadism in North Africa and the Middle East, and intensive subsistence agriculture in South and East Asia, where rice is grown (Figure 6.5).

Shifting Cultivation

Shifting cultivation or *swidden agriculture* is practiced in three main tropical rainforest areas of the world: (1) the South American Amazon region, (2) central Africa, and (3) Southeast Asia, Indonesia, and New Guinea (Figure 6.5). Rainfall is heavy in these regions, vegetation is thick, and soils are relatively poor in quality.

When shifting cultivation is practiced, the people of a permanent village clear a field adjacent to their settlement by slashing vegetation. After the field is cleared with axes, knives, and machetes, the remaining stumps are burned. Daily rain returns the ash and nutrients to the soil, temporarily fertilizing it. (Because of this clearing technique, shifting cultivation is sometimes called *slash-and-burn* agriculture.) The field is used for several years. At the end of this time, the soil is depleted, and the village turns elsewhere to clear another field. Eventually, the forest vegetation again takes over, and the area is refoliated. The soil is thus allowed to replenish itself. Swidden agriculture survives in areas of the humid tropics that have low-potential environmental productivity and low population pressure. Under ideal conditions, this form of agriculture leaves much of the original vegetation intact. Farmers make small, discontinuous clearings in forests. They cut down some trees, burn the debris, and prepare the soil for a variety of crops—groundnuts, rice, taro, sweet potato. Because no fertilizer is used, soil nutrients are quickly depleted. Using hoes or knives, farmers plant the fields by hand with tubers or seeds. An indentation is made in the soil. A stem of a plant is submerged or a seed is dropped into a hole, and soil is pushed over the opening by hand. Mechanization and animals are not used for plowing or for harvesting. The most productive farming occurs in the second or third year after burning. Following this, surrounding vegetation rapidly regenerates, weeds grow, and soil productivity dwindles. The plot, sometimes called a *swidden* or *milpa,* is abandoned. Then a new site is selected nearby. Usually, the

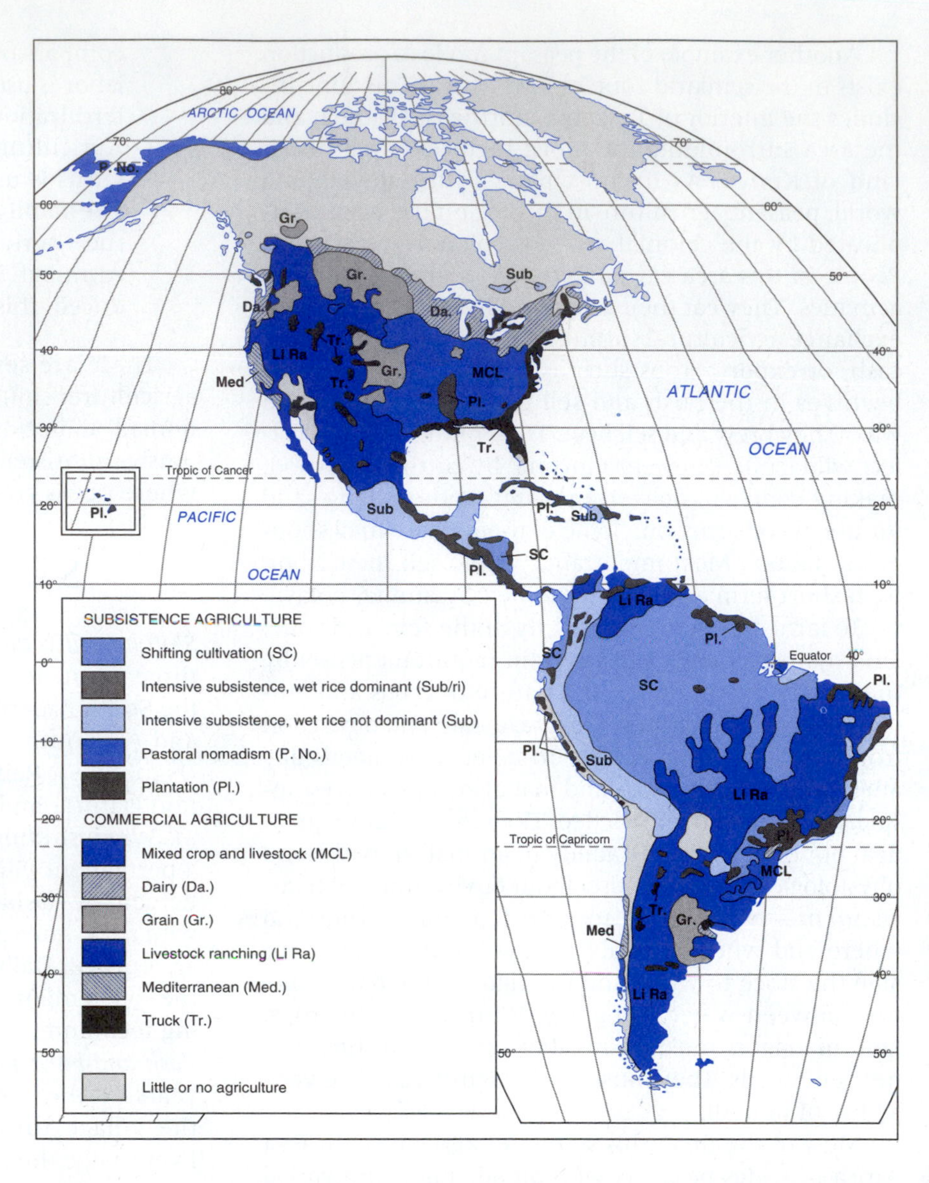

FIGURE 6.5
World agricultural regions. Africa, Southeast Asia, and the Amazon basin are the principal regions of shifting agriculture. Intensive subsistence agriculture is found in East Asia, South Asia, and Southeast Asia. Plantation agriculture is primarily in tropical and subtropical regions of Latin America, Asia, and Africa. Commercial regions include mixed crop and livestock farming, which exists primarily in the northern Unites States, southern Canada, and central Europe. Grain cultivation exists in Argentina, the Great Plains of the United States, the central Asian Plains of the former Soviet Union, and Australia.

village does not permanently relocate. The villagers commonly return to the abandoned field after 6 to 12 years, by which time the soil has regained enough nutrients to grow crops again. Except on steep slopes, where soil erosion can be a serious problem, shifting cultivation can be a sustainable system of agriculture. It allows previous plots to regenerate natural growth. However, shifting cultivation can lead to degradation when an increasing population demands too much of the land, reducing the fallow period.

The predominant crops grown in shifting agricultural areas include corn and manioc (cassava) in South America, rice in Southeast Asia, and sorghum and millet in Africa. In some regions, yams, sugarcane, and other vegetables are also grown. The patchwork of a swidden is quite complex and seemingly chaotic. On one swidden, a variety of crops can be grown, including those just mentioned, as well as potatoes, rice, corn, yams, mangoes, cotton, beans, bananas, pineapples, and others, each in a clump or small area within the swidden.

Only 5% of the world's population engages in shifting cultivation today. This low percentage is not surprising because tropical rainforests are not highly populated areas. However, shifting cultivation occupies approximately 25% of the world's land surface and therefore is an important type of agriculture. The amount of land devoted to this type of agriculture is decreasing because governments in these regions deem

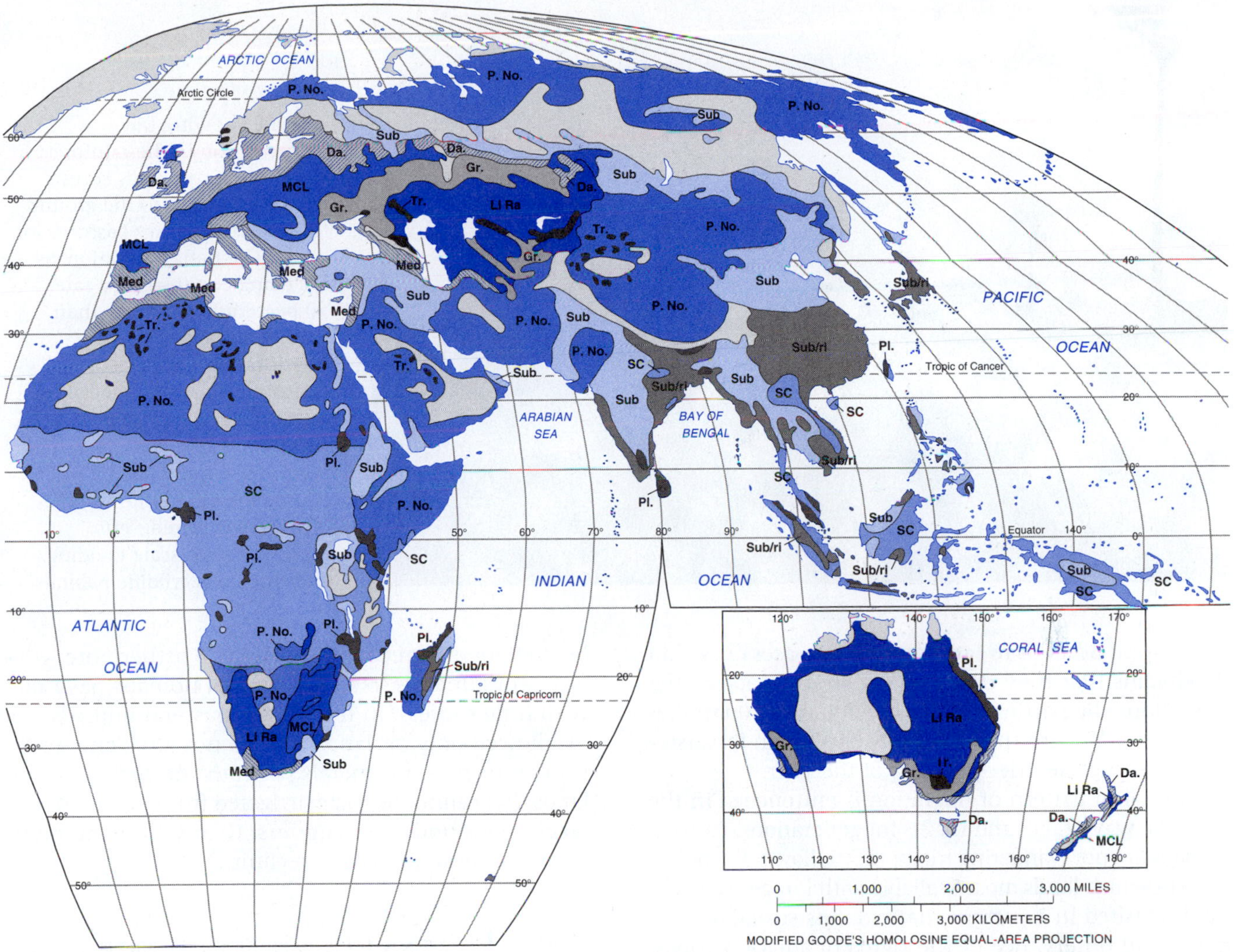

Livestock ranching includes areas too dry for plant cultivation, including western Canada and the United States, southeastern South America, central Asia, and large portions of Australia. Mediterranean agriculture specializes in horticulture and includes areas surrounding the Mediterranean Sea, regions of the southwestern United States, central Chile, and the southern tip of Africa. Finally, truck farming—commercial gardening and fruit farming—is found in the southeastern United States and in southeastern Australia.

shifting cultivation to be economically unimportant. Consequently, governments in developing countries are selling and leasing land to commercial interests that destroy the tropical hardwoods and rainforests.

Pastrol Nomadism

Shifting cultivation and *pastoral nomadism* can be classified as extensive subsistence agriculture. Areas in which pastoral nomadism is practiced include North Africa and the Middle East, the eastern plateau areas of China and Central Asia, and eastern Africa's Kenya and Tanzania (Figure 6.5). Only 15 million people are pastoral nomads, but they occupy 20 percent of the earth's land area. The pastoral nomads occupy areas that are climatically opposite those of shifting cultivators. The lands occupied by pastoral nomads are dry, usually less than 10 inches of rain accumulate per year, and typical agriculture is normally impossible, except in oases areas.

Instead of depending on crops as most other farmers do, nomads depend on animal herds for their sustenance. Everything that they need and use is carried with them from one forage area to another. Tents are constructed of goats' hair, and milk, clothing, shoes, and implements are produced from the animals. Pastoral nomads consume mostly meat and grain. Sometimes, in exchange for the meat, other needed goods are obtained from sedentary farmers in marginal lands near the nomads' herding regions. It is common

A scene showing the deforestation in the rainforest in Acro, Western Brazil. Forests have been burnt to the ground to create temporary pastures for cattle. The nation's rainforests are being cut down at a rate 50 percent faster today than they were 10 years ago. Rainforest loss creates or contributes to a number of intractable problems: It contributes to greenhouse warming, eliminates the cleansing of the atmosphere, creates new semideserts, increases large-scale flooding, and threatens wildlife habitats.

for pastoral nomads to farm areas near oases or within floodplains that they occupy for a short period of the year. Nomadic parties usually include 6 to 10 families who travel in a group, carrying bags of grain for sustenance during the drier portions of the year.

A cyclical pattern of migration is entrenched in the nomadic way of life, and it lasts for generations. Pastoral nomads are not wandering tribes; they follow a 12-month cycle in which lands most available with forage are cyclically revisited in a pattern that exhibits strong territoriality and observance of the rights of adjacent tribes. The exact migration pattern of today's pastoral nomads have developed from a precise geographic knowledge of the region's physical landforms and environmental provision.

Nomads must select animals for their herds that can withstand drought and provide the basic necessities of the herdsmen. The camel is the quintessential animal of the nomad because it is strong, can travel for weeks without water, and can move rapidly while carrying a large load. The goat is the favorite small animal because it requires little water, is tough, and can forage off the least green plants. Sheep are slow moving and require more water, but they provide other necessities: wool and mutton. Small tribes need between 25 and 60 goats and sheep and between 10 and 25 camels to sustain themselves.

Before the railroad and telegraph, pastoral nomads were the communication agents of the desert regions, carrying with them innovations and information. This is no longer the case as nomadic societies have fallen before the territorial imperatives of the nation-state and its fixed boundaries. However, nomadic herding remains because these vast dry areas of the world cannot be used for other economic activity. Furthermore, government attempts to settle pastoral nomads have met with little success. In the future, pastoral nomads will be allowed only on lands that do not have energy resources or precious metals beneath the surface, or on lands that cannot be easily irrigated from nearby rivers, lakes, or groundwater aquifers. In any case, the number of pastoral nomads is declining.

INTENSIVE SUBSISTENCE AGRICULTURE

Intensive subsistence agriculture is practiced by large populations living in East, South, and Southeast Asia, Central America, and South America (Figure 6.5). Whereas shifting cultivation and pastoral nomadism are extensive low-density, marginal operations, intensive subsistence agriculture, as the term implies, is a higher intensity type of agriculture in the majority of the densely populated developing areas of the world. Rice is the predominant crop because of its high levels of carbohydrates and protein. Most farmers using intensive subsistence rice agriculture use every available piece of land, however fragmented, around their villages. Most often, a farm encompasses only a few acres.

Intensive subsistence agriculture is characterized by several features:

1. Most of the work is done by hand, with all family members involved. Occasionally farm animals are used, such as water buffalo or oxen. Almost no mechanization is used because of lack of capital

to purchase such equipment and because plots are tiny.
2. Plots of land are extremely small by Western standards. Almost no piece of land is wasted. Even roads through agricultural regions of intensive subsistence are made narrow so that all cultivatable areas can be used.
3. The physiological density (i.e., the number of people that each acre of land can support) is very high.
4. Principal regions that are cultivated are river valleys and irrigated fields in low-lying, moist regions in the middle latitudes.

Because rice is a crop that has a high yield per acre and is rich in nutrients, it is a favorite in intensive subsistence agricultural regions (Figure 6.6). First, the field is plowed with a sharpened wooden pole that is pulled by oxen. Next, the field is flooded with water and planted with rice seedlings by hand. Another method is to spread dry seeds over a large area by hand. When the rice is mature, having developed for three-fourths of its life underwater, it is harvested from the rice paddy. To separate the husks from the rice itself, the farmers thrash the rice by beating it on a hard surface or by trampling it underfoot. Sometimes it is even poured on heavily traveled roads. The chaff is thus removed from the seeds, and sometimes the wind blows the lightweight material far from the pile of rice itself, a process known as *winnowing*.

Some year-round, tropical, moist areas of the world allow *double cropping*. This means that more than one crop can be produced from the same plot throughout the year. Occasionally in wet regions, two rice crops are grown, but more frequently a rice crop and a different crop, which requires less water, are produced. The field crop is produced in the drier season on nonirrigated land. In the higher latitudes of East Asia, rice is mixed with other crops and may not be the dominant crop. In western India and the northern China plain, wheat and barley are the dominant crops, with oats, millet, corn, sorghum, soybeans, cotton, flax, hemp, and tobacco also produced.

Problems Faced by Subsistence Agriculturalists

Subsistence agriculture is subjected to variations in soil quality, availability of rain from year to year, and, in general, environmental conditions that can harm crop-production levels and endanger life. In addition, subsistence agriculturalists lack tools, implements, hybrid seeds, fertilizer, and mechanization that developed na-

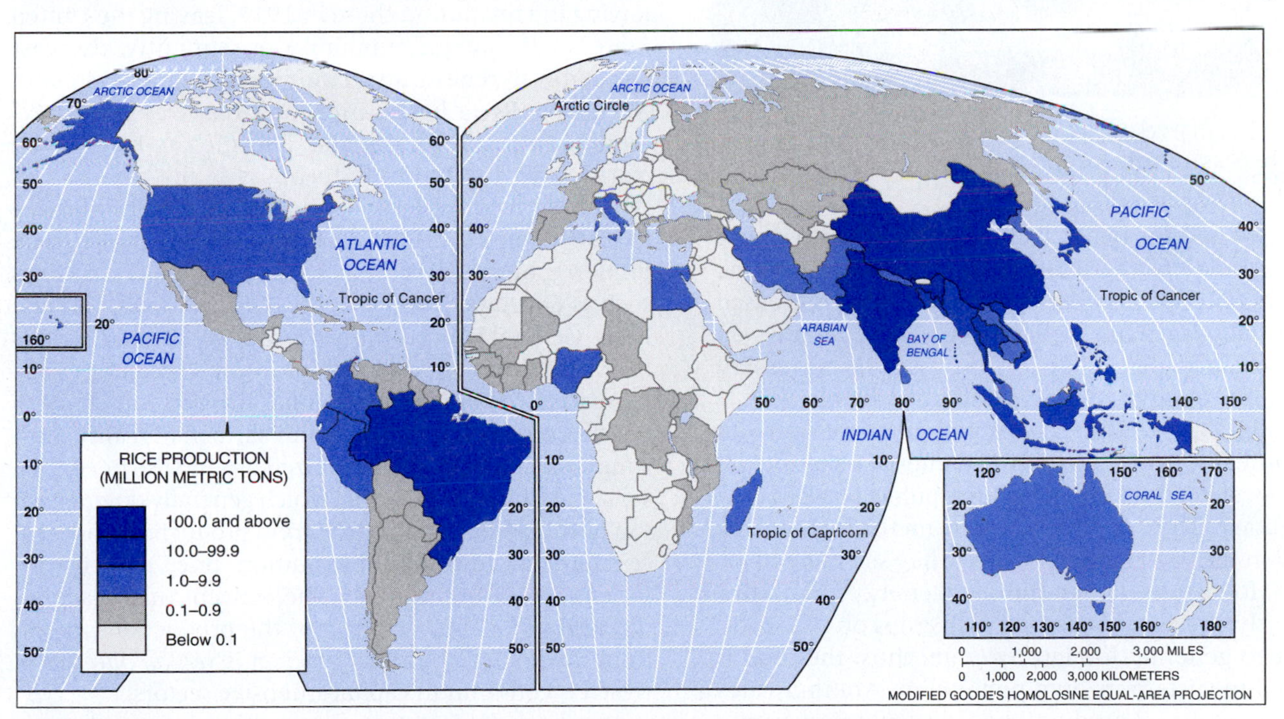

FIGURE 6.6
World rice production. Anthropologists tell us that rice was domesticated in East Asia more than 7000 years ago. Unlike corn in the United States, rice is almost exclusively used for human consumption. About 2 billion people worldwide are fed chiefly by rice. It is the most important crop cultivated in the most densely settled areas of the world, including China in East Asia, India in South Asia, and Southeast Asia. These areas produce more than 90 percent of the world's rice. Populations in these regions use rice for between one-third and one-half their total food intake. Rice can be grown in only very humid and tropical regions because it depends on water to develop. It requires a large amount of labor and tedious work, but the returns are bountiful. Rice produces more food per unit of land than any other crop; thus, it is suitable for the most densely populated areas of the world, such as China and India. (See Color Insert for more illustrative map.)

In Indonesia, harvesting rice is an example of labor-intensive peasant culture.

tions have had for nearly 100 years. With such drawbacks, subsistence agriculturalists can barely provide for their families, and net yields have not increased substantially for many generations. These families do not have enough capital to purchase the necessary equipment to improve their standard of living.

Finally, all too often, developing countries turn to their limited sources of export revenues to generate the cash flow needed for infrastructure, public services, and the military. They must produce something that they can sell in the world market. Often, they sell mineral resources, foodstuffs, and nonmineral energy fuels. Most frequently, these countries sell cash crops on the world market to generate foreign revenue; thus, the food is not used to sustain its own population. Another category of agricultural products that can generate revenue is nonfood or not-nourishing crops, such as sugar, hemp, jute, rubber, tea, tobacco, coffee, and a growing harvest of cotton to satisfy the world's need for fabric and denim. All these commodities, however, command very low prices on the global market. How can impoverished nations feed themselves when a large proportion of their agricultural productivity and acreage is devoted to nonfood crops? This is the plight of many African, South American, Central American, and Asian countries today. As a result, sometimes alternative sources of income are inviting, even if they are illegal, such as cocoa or opium.

Commercial Agriculture

Commercial agricultural areas dominated by capitalist social relations include the United States and Canada, Argentina and portions of Brazil, Chile, Europe, Russia and Central Asia, South Africa, Australia, New Zealand, and portions of China.

Agriculture in the United States epitomizes the contemporary capitalist system of food production. The American agricultural system developed in the nineteenth century as part of the unfolding of the European "frontier" across North America. Railroads and steamships dramatically lowered transport costs to the markets along the East Coast, and agricultural trading centers and ports, such as Chicago, St. Louis, and New Orleans grew rapidly. By the turn of the century, the United States had become a major supplier of wheat and other commodities to Europe. The only other major producer, Russia, effectively withdrew from world markets following the revolution there in 1917, leaving the United States as the major supplier. Consequently, the vast agricultural region stretching across the Ohio and Mississippi River basins into the Great Plains, and extending into central Canada, became the core of the North American food-producing system.

U.S. agriculture today is a huge and very productive industry dominated by a handful of large *agribusiness* firms. Agribusiness is dominated by giant food companies such as ConAgra, Dole, Nabisco, Ralston Purina, General Mills, General Foods, Hunt-Wesson, and United Brands to control the whole food chain from "seedling to supermarket." Whereas the popular imagination clings to the stereotype of the small family farmer, in reality most American agriculture is organized around the needs of a small handful of large firms, which generally do not own the farmland but control the food production and processing (e.g., canning), distribution, price and cost information, and marketing. The concept that describes the food companies' control of the production process from raw material to final product is *vertical integration,* which is common in capital-intensive sectors.

Agribusiness is extremely *capital intensive* and *energy intensive.* Farmers rely on copious quantities of chemicals, tractors, harvesters, airplanes, and other equipment—most of it very sophisticated, computer controlled, and expensive—to keep labor inputs low and productivity levels high. Only 2 percent of the U.S. labor force is employed in agriculture, and it not only feeds the other 98 percent of the populace but exports vast amounts of food as well. The very high per capita

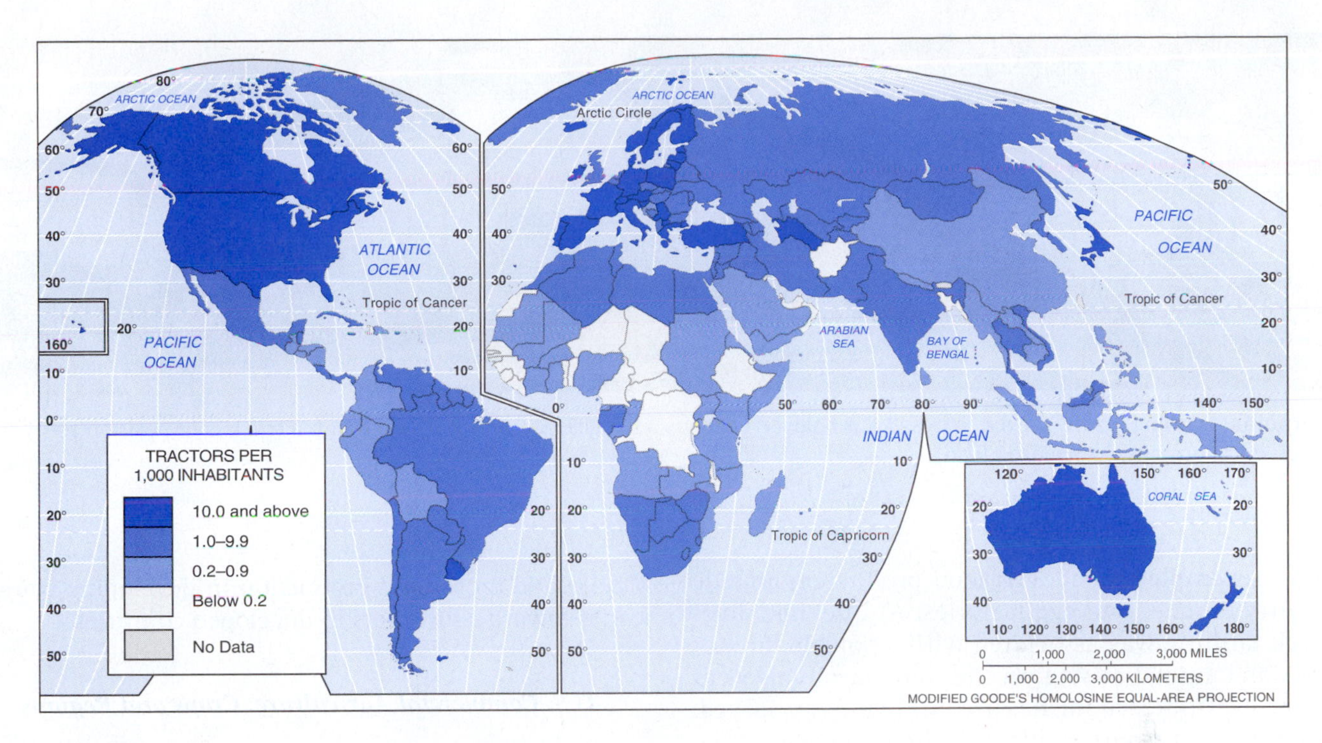

FIGURE 6.7
Tractors per 1000 hectares of farmland. Tractor usage is an indicator of the degree to which agriculture is mechanized. Farming systems in economically developed countries rely heavily on tractors and similar equipment—as well as fossil fuels—whereas low tractor usage is associated with labor-intensive farming in less developed countries.

productivity has resulted in long-term rural depopulation. For example, the use of tractors worldwide (Figure 6.7) is a measure of the capital-intensity of agricultural production: Countries with highly commodified agricultural systems rely extensively on this technology, freeing people from the farm, whereas in developing nations with a peasant-based system of agriculture, tractors are relatively uncommon.

The importance of corporate farming is growing in *market gardening,* which is sometimes called *truck farming.* Modern food-production truck farms specialize in intensively cultivated fruits, vegetables, and vines, and they depend on migratory seasonal farm laborers to harvest their crops. In the United States, California is the epicenter of fruit and vegetable farms, although they are widespread in Florida as well.

Agribusiness has also extended livestock farming immensely, including the mechanized raising and slaughter of cattle, often under inhuman conditions.

Rice is the mainstay of the diet in East Asia, and terraced hillsides, which create cultivable land out of hillsides, have been formed over millennia.

Corporate farming in the United States. An employee watches a television monitor to see when a truck is in position to receive its computer-calculated load.

Other examples of modern food production include poultry ranches and egg factories. At one time, livestock farming was associated with a combination of crop and animal raising on the same farm. In recent years, livestock farming has become highly specialized. An important aspect of this specialization has been the growth of factorylike feedlots, which raise thousands of cattle and hogs on purchased feed, generating huge quantities of animal waste. Feedlots are common in the western and southern states, in part because winters there are mild. These feedlots raise more than 60 percent of the beef cattle in the United States.

Thus, corporate agriculture is an industry similar to the production of other goods such as cars. Agriculture's major backward linkages—its purchases from other sectors—include petroleum and machinery; notably, labor is only a small part of the production costs, given how capital intensive the sector is. The forward linkages of this sector—its sales to other parts of the economy—include the food processing sector and meat production; a large share of cereals, especially corn, are grown for animal feed.

Modern American farming is quick to respond to new developments, such as new production techniques. Consequently, farmers with sizable investments of money, materials, and energy can create drastic changes in land-use patterns. For example, farmers in the low-rainfall areas of the western United States have converted large areas of grazing land to forage and grass production with the use of center-pivot irrigation systems. Other farmers grow sugar beets and potatoes in western oases through federally subsidized water projects.

American corporate farming is also extending overseas to become a worldwide food-system model. Family farming is still dominant in western Europe, but beef feedlots are found in parts of Italy. Poultry-raising operations in Argentina, Pakistan, Thailand, and Taiwan are increasingly similar to those in Alabama or Maryland. Enterprises such as United Brands, Del Monte, Unilever, and Brooke Bond Oxo are diverting more and more food production in developing countries toward consumers in developed countries.

U.S. Commercial Agriculture: Crops and Regions

The main characteristic of *commercial agriculture* is that it is produced for sale off the farm, at the market. Following are some of the characteristics of commercial agriculture:

1. Populations fed by commercial agriculture are urban populations engaged in other types of economic activity, such as manufacturing, the services, and information processing.
2. Only a small proportion of the population is engaged in agriculture.
3. Machinery, fertilizers, and high-yielding seeds are used extensively, with high-energy inputs.
4. Farms are extremely large, and the trend is toward even larger farms.
5. Agricultural produce from commercial agriculture is integrated with other agribusiness, and a vertical integration exists that stretches from the farm to the table.

Commercial Agriculture and the Number of Farmers

The percentage of laborers in developed countries working in commercial agriculture is less than 5 percent overall. In contrast, in some portions of the developing world where intensive subsistence agriculture is practiced, 90 percent of the population is directly engaged in farming, and the average is 60 percent overall. Today, U.S. farmers on average produce enough food for themselves and 70 other families.

In 2005, the United States had approximately 2.1 million farms, compared with 5.7 million in 1950 (Figure 6.8). This reduction in the number of farm families as a percentage of the population is a result of push factors and pull factors. *Push factors* are economic factors

Harvesting wheat by combine in the United States exemplifies capital-intensive and energy-intensive agriculture. Grain, such as the wheat shown here, is often a major crop on most farms. Commercial grain agriculture is different from mixed crop and livestock farming in that the grain is grown primarily for consumption by humans rather than by livestock. In developing countries, the grain is directly consumed by the farm family or village, whereas in commercial grain farming, output is sold to manufacturers of food products.

(such as the high cost of equipment or high interest rates) that drive families off the farms. Farming is difficult, often dangerous work, and low crop prices can be ruinous. *Pull factors* include the advantages of urban life in the United States, Europe, and other developed countries. The opportunity for college education and occupations in the cities have long lured farm children off the land. One serious problem, and a push factor, is the encroachment of the metropolitan area onto the best farmland, directly adjacent to the urban area through the expansion of housing subdivisions and shopping centers. Suburban sprawl, brought by interstate highways that reduce the commute and penetrate into the countryside, has usurped viable topsoil and farmland around many metropolitan areas in the United States, as well as around Amsterdam, Paris, and Buenos Aires.

Millions
Farms
7
6
5
4
3
2
1
0
1950 1960 1970 1980 1990 1995 2000

FIGURE 6.8
The number of U.S. farms, 1950–2000. The number of U.S. farms has decreased by almost two-thirds, to approximately 2.1 million. At the same time, the average farm size has increased steadily.

Machinery and Other Resources in Farming

The second aspect unique to commercial agriculture, besides the small percentage of farmers in the population, is the heavy reliance on expensive machinery, tractors, combines, trucks, diesel pumps, and heavy farm equipment, all amply fueled by petroleum and gasoline resources, to produce the large output of farm products. To this has been added miracle seeds that are hardier than their predecessors and that produce more impressive tonnages. Commercial agriculture is also fertilizer intense.

Improvements in transportation to the market have resulted in less spoilage. Products arrive at the canning and food-processing centers more rapidly than they did earlier. By 1850, many American farms were well connected to cities by rail transportation. More recently, the motor truck has supplanted rail transportation, and the advent of the refrigerator car and the refrigerator truck meant that freshness is preserved. Cattle also arrive at packing houses by motor truck as fat as when they left the farm, unlike the mid-nineteenth century, when long cattle drives were the order of the day, connecting cattle-fattening areas in Texas, Oklahoma, and Colorado with the Union-Pacific rail line stretching from St. Louis to Kansas City to Denver.

Agricultural experiment stations are now part of every state and are usually affiliated with land-grant universities. These stations have made great strides and improvements in agricultural techniques, not only in improved fertilizers and hybrid plant seeds but also in hardier animal breeds and new and better insecticides

Feed lots for beef cattle in California. According to the "Code of the West," cattle ranchers owned little land, only cattle, and grazed open land wherever they pleased. New cattle breeds introduced from Europe, such as the Hereford, offered superior meat but were not adapted to the old ranching system of surviving the winter by open grazing. In moist areas, crop growing supplanted ranching because it generated a higher income per acre. Some cattle are still raised on ranches, but most are sent for fattening to feedlots along major interstate highways or railroad routes. Many feedlots are owned by agribusiness and meat processing companies rather than by individuals.

and herbicides, which have reduced pestilence. In addition, local and state government farm advisors can provide information about the latest techniques, innovations, and prices so that the farmer can make wise decisions concerning what should be produced, when it should be produced, and how much should be produced.

Types of Commercial Agriculture

We can divide commercial agriculture into six main categories: mixed crop and livestock farming, dairy farming, grain farming, cattle ranching, Mediterranean cropping, and horticulture and fruit farming (see Figure 6.5).

Mixed Crop and Livestock Farming

Mixed crop and livestock farming is the principal type of commercial agriculture, and it is found in Europe, Russia, Ukraine, North America, South Africa, Argentina, Australia, and New Zealand. The primary characteristic of mixed crop and livestock farming is that the main source of revenue is livestock, especially beef cattle and hogs. In addition, income is produced from milk, eggs, veal, and poultry. Although the majority of farmlands are devoted to the production of crops such as corn, most of the crops are fed to the cattle. Cattle fattening is a way of intensifying the value of agricultural products and reducing bulk. Because of the developed world's preference for meat as a major food source, mixed crop and livestock farmers have fared well during the last 100 years.

In developed nations, the livestock farmer maintains soil fertility by using a system of crop rotation in which different crops are planted in successive years. Each type of crop adds different nutrients to the soil. The fields become more efficient and naturally replenish themselves with these nutrients. Farmers today use the *four-field rotation system,* wherein one field grows a cereal, the second field grows a root crop, the third field grows clover as forage for animals, and the fourth field is fallow, more or less resting the soil for that year.

Most cropping systems in the United States rely on corn (Figure 6.9) because it is the most efficient for fattening cattle. Some corn is consumed by the general population in the form of corn on the cob, corn oil, or margarine, but most is fed to cattle or hogs. The second

Corporate cattle farming in the United States.

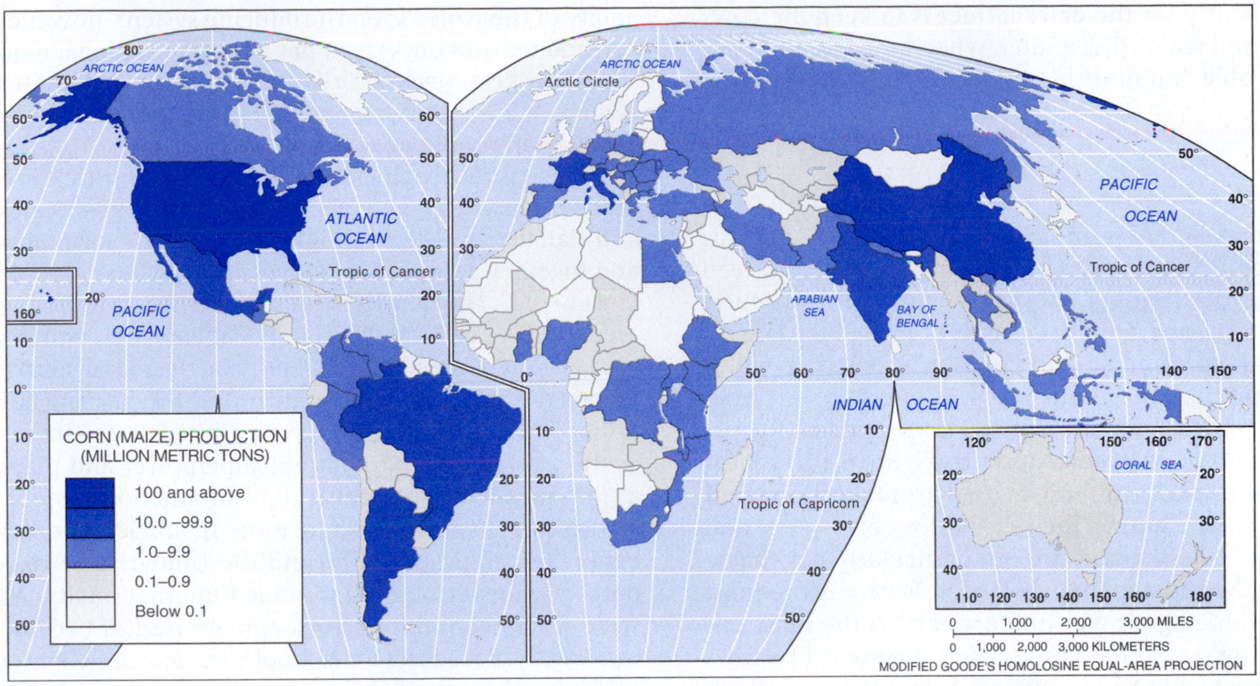

FIGURE 6.9
World corn (maize) production. Corn was domesticated in Central America more than 5000 years ago and exported to Europe in the fifteenth century. The United States accounts for more than 30 percent of the world's corn production, and 90 percent of this corn crop is fed to cattle and livestock for fattening and meat production. Outside the United States, corn is called maize. China is the second leading producer of corn, but the majority of the crop is consumed by humans. Because meat produces a greater market value per pound than selling corn does, U.S. farmers convert corn into meat by feeding it to livestock on farms and feedlots. In the United States, the Corn Belt is also the livestock region of North America, and it is located in the western Midwest and the eastern Great Plains. Argentina and Brazil, as well as Europe, also have sizable corn-production areas. (See Color Insert for more illustrative map.)

most important crop in mixed crop and livestock farming regions of central North America and the eastern Great Plains is the soybean. The soybean has more than 100 uses, but it is used mainly for animal feed. In China and Japan, tofu is made from soybean milk and is used as a major food source high in protein and low in fat.

Dairy Farming

Dairy farming accounts for the most farm acreage in the northeastern United States and northwestern Europe and accounts for 20 percent of the total output by value of commercial agriculture. Ninety percent of the world's milk supply is produced in these few areas of the world. Most milk is consumed locally because of its weight and perishability. Dairy farming is an intensive land-use activity. Because milk is heavy and highly perishable, dairy farms are often located near cities, and the milk is trucked into the cities daily.

Some dairy farms produce butter and cheese as well as milk. In general, the farther the farm is from an urban area, the more expensive the transportation of fluid milk, and the greater proportion of production in more *high-value-added* commodities, such as cheese and butter. For example, the Swiss discovered ways of transforming their milk products into high-value-added chocolates, cheeses, and spreads that are distributed worldwide. These processed products are not only lighter but also less perishable. On the other hand, in the United States, the proximity of farms to Boston, New York, Philadelphia, Baltimore, and Washington, D.C., on the East Coast, and to Chicago and Los Angeles in the Midwest and West, means that these farms primarily produce liquid milk. Farms throughout the remaining areas of the United States primarily produce butter and cheese. Worldwide, remote locations such as New Zealand, for example, devote three-fourths of their dairy farms to cheese and butter production, whereas three-fourths of the farms in Britain, with a much higher population density at close proximities, produce fluid milk.

Dairy farms are relatively labor intensive because cows must be milked twice a day. Most of this milking is done with automatic milking machines. However, the cows still must be herded into the barn and washed, the milking machines must be attached and disassembled, and the cows must be herded back out and fed.

The difficulty for the dairy farmer is to keep the cows milked and fed during winter, when forage is not readily available and must be stored.

Grain Farming

Commercial grain farms are usually in drier territories that are not conducive to dairy farming or mixed crop and livestock farming. Most grain, unlike the products of *mixed crop and livestock* farms, is produced for sale directly to consumers. Only a few places in the world can support large grain-farming operations. These areas include China, the United States, Canada, Russia, Ukraine, Argentina, and Australia (Figure 6.10). Wheat is the primary crop and is used to make flour and bread. Other grains include barley, oats, rye, and sorghum. These grains are not particularly perishable and can be shipped long distances. Wheat is the most highly valued grain per unit area and is the most important for world food production. Figure 6.11 shows that grain yield and production increased markedly in developing countries between 1970 and 2003, while cropland area has increased only slightly. However, production per capita is much more disappointing in developing countries, particularly Africa. However, the ability of the world's food-producing system, however constrained and imperfect, has allowed the global food supply to keep pace with world population increases (Figure 6.12), denying, or at least forestalling, the Malthusian prophecy (Chapter 3).

Wheat is the leading international agricultural commodity transported between nations. The United States and Canada are the leading export nations for grains and together account for 50 percent of wheat exports worldwide. The North American wheat-producing areas have been appropriately labeled the world's breadbasket because they still provide the major source of food to many deficit areas, including the starving nations of Africa (Figure 6.13). The United States stands as the world's only agricultural superpower and plays a unique role in the global food production system. Agricultural exports generate more than $50 billion in export revenues annually, and the United States exports 20 percent of all food traded internationally. As with other economic sectors, agriculture has become thoroughly globalized; for example, farmers in Nebraska and Kansas are well aware that next year's revenues will be shaped by weather and political events in markets in Europe and Asia.

In North America, the Spring *Wheat Belt* is west of the mixed crop and livestock farming area of the

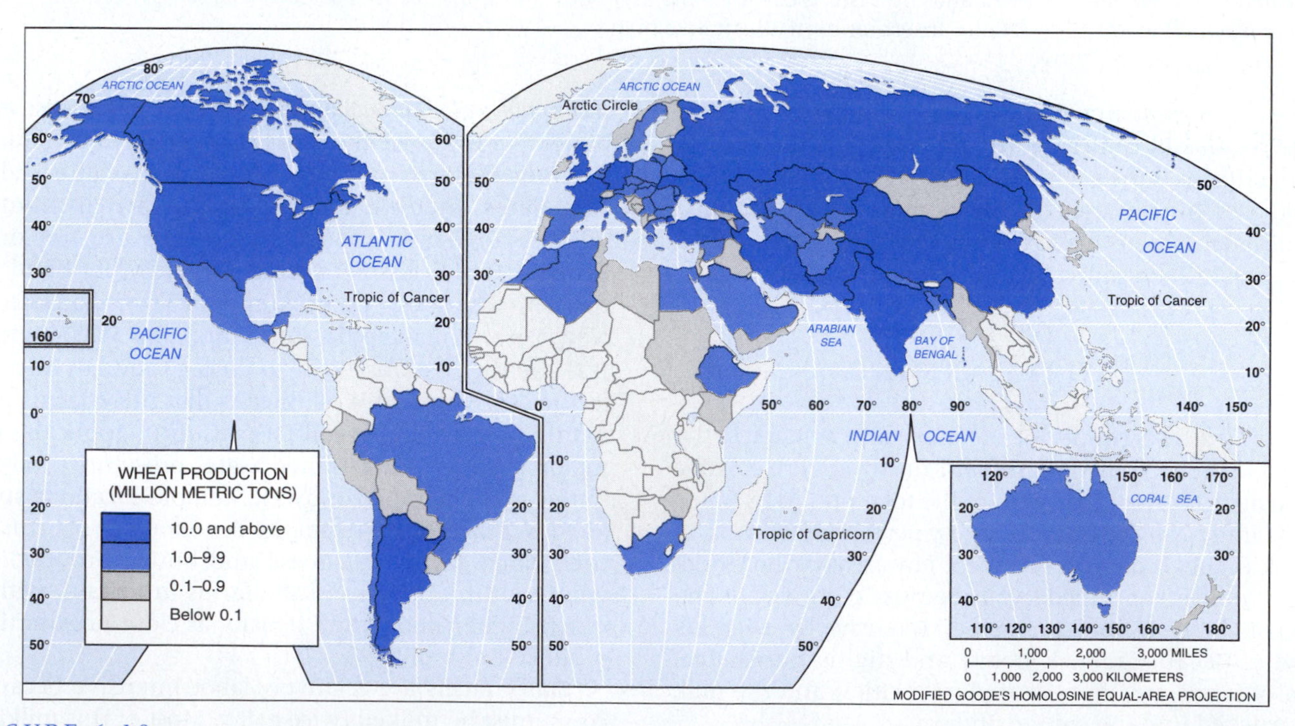

FIGURE 6.10
World wheat production. China is the world's leading wheat producer, followed by the United States and Russia. The United States, Canada, Australia, and Argentina are the primary wheat exporters, whereas Russia, Kazakhstan, India, and China import the most wheat. Wheat can be stored in grain elevators. Therefore, current wheat prices worldwide reflect not only growing conditions for that year, but also supplies from commercial and subsistence operations that have been stored throughout the world. (See Color Insert for a more illustrative map.)

FIGURE 6.11
The world's output of major food crops has increased dramatically during the last 30 years. The most dramatic increase has been in cereals, from 1.3 billion metric tons in 1970 to over 2.4 billion metric tons in 2003. Milk, meat, fish, fruit, and vegetables have also made gains in production worldwide. In the last 30 years, every region of the world has increased production. The figure shows that the index of grain production has increased substantially during the last 20 years primarily as a result of an increase in yields, instead of an ever increasing cropland.

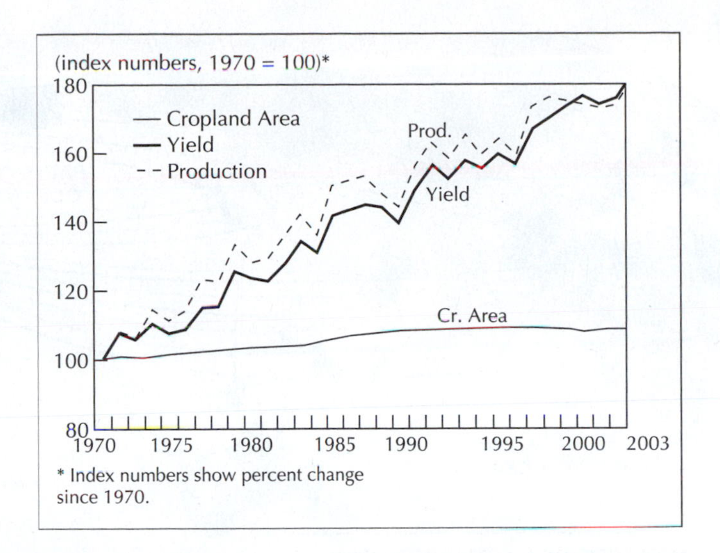

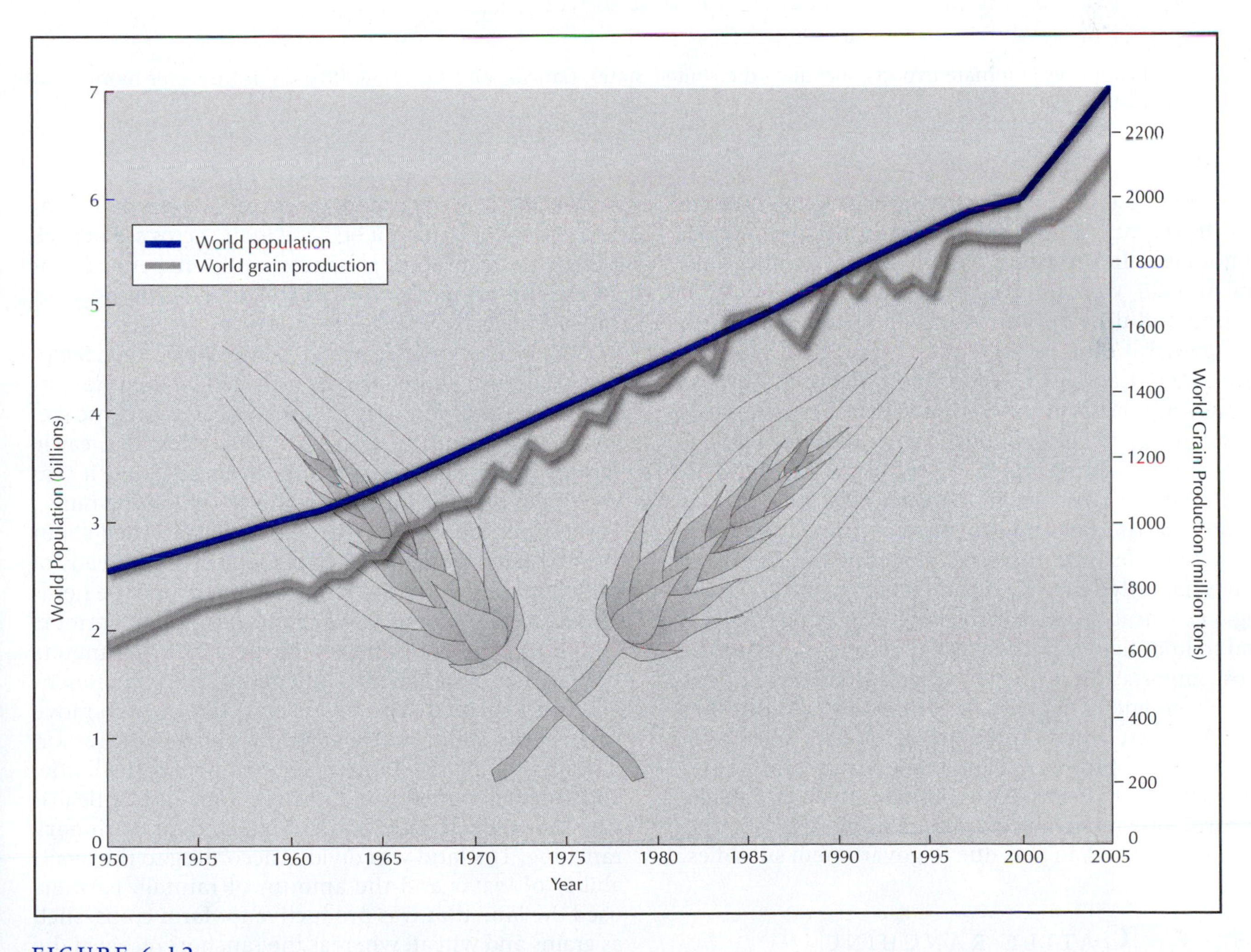

FIGURE 6.12
World population growth and grain production, 1950–2005. Despite persistent concerns about diminishing returns and environmental degradation, the world's food supply system has either refuted or postponed Malthusian predictions of massive famine.

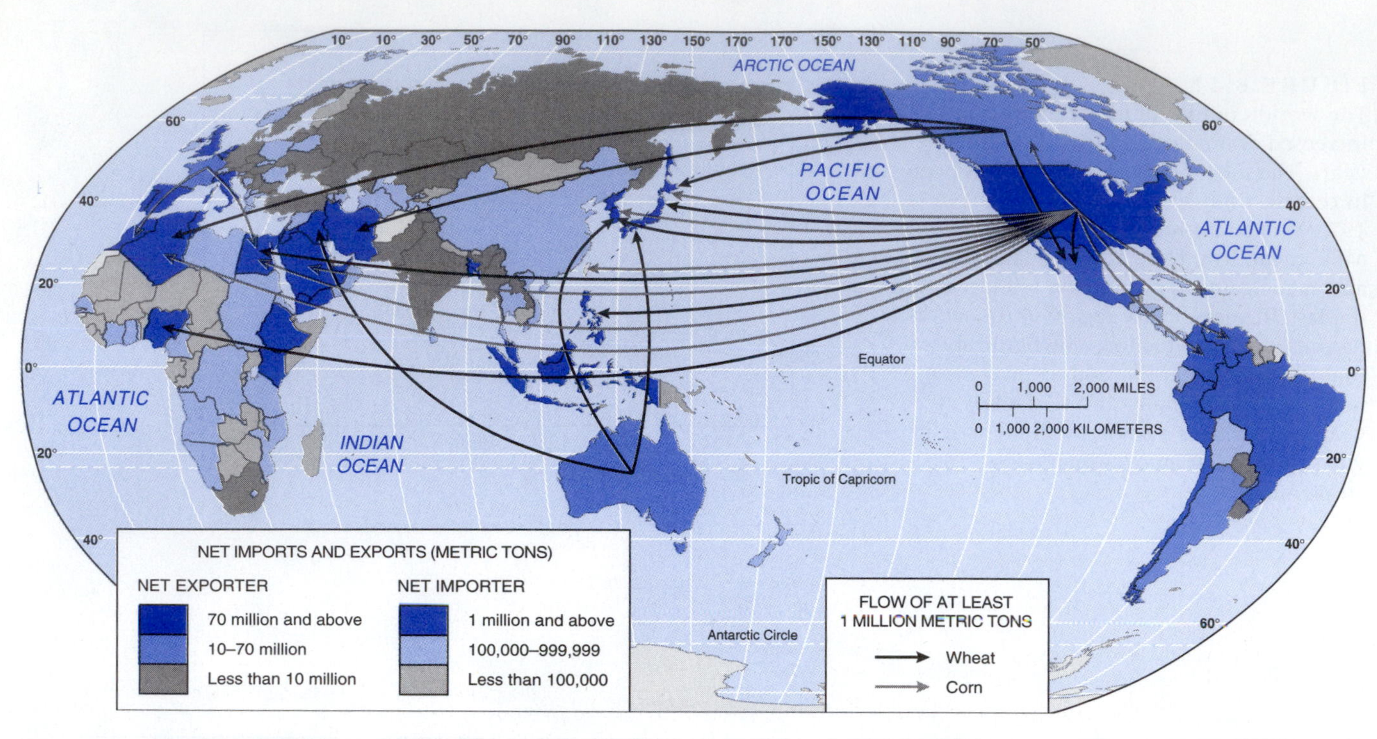

FIGURE 6.13
A handful of countries dominate exports, including the United States, Canada, and Australia. (See Color Insert for more illustrative map.)

Midwest and is centered in Minnesota, North Dakota, South Dakota, and Saskatchewan (Figure 6.14). Major cities in the Great Plains, such as Minneapolis and St. Paul, were often established as centers of flour milling and distribution. Another region, just south of the Wheat Belt, is the Winter Wheat Belt, which is centered in Kansas, Colorado, and Oklahoma. Because winters are harsh in the Spring Wheat Belt, the seeds would freeze in the ground, so instead *spring wheat* is planted in the spring and harvested in the fall; the fields are fallow in the winter. *Winter wheat,* however, is planted in the fall and moisture accumulation from snow helps fertilize the seed. It sprouts in the spring and is harvested in early summer. Like corn-producing regions, wheat-producing areas are heavily mechanized and require high inputs of energy resources. Today the most important machine in wheat-producing regions is the combine, which not only reaps but also threshes and cleans the wheat. Large storage devices called *grain elevators* are a prominent landscape feature as one traverses the Great Plains of the United States and Canada. In part, these reflect the enormous surpluses that farmers have accrued, in part due to government subsidies.

Cattle Ranching

Cattle ranching is practiced in developed areas of the world where crop farming is inappropriate because of aridity and lack of rainfall. Cattle ranching is an extensive agricultural pursuit because many acres are needed to raise cattle. In some instances, cattle are penned near cities, and forage is trucked to cattle-fattening pens called *feedlots.*

Major cities grew up across the western United States partly because of the services provided by their slaughterhouses and stockyards. Denver, Dallas, Chicago, Kansas City, and St. Louis are examples. If a cattle farmer could get a steer to one of these cities, it was worth ten times as much as it was worth on the range. Early American ranchers were not as concerned about owning territory as they were about owning heads of cattle. Consequently, the range was open, and the herds grazed as they went toward market. Later, farmers bought up the land and established their perimeters with barbed-wire fences. Until about 1887, the ranchers cut the barbed-wire fences and continued to move their herds about wherever they pleased. However, after that point, the farmers seemed to win the battle, and ranchers were forced to switch from long cattle drives and wide territories of rangeland to stationary ranching. The land was divided according to the availability of water and the amount of rainfall. Farmers used the land that was productive for farm crops, such as grains and wheat, whereas the ranchers received the land that was too dry for farming. Ironically, given the frequent hostility of cattle ranchers to the state, today 60 percent of cattle grazing occurs on land leased from

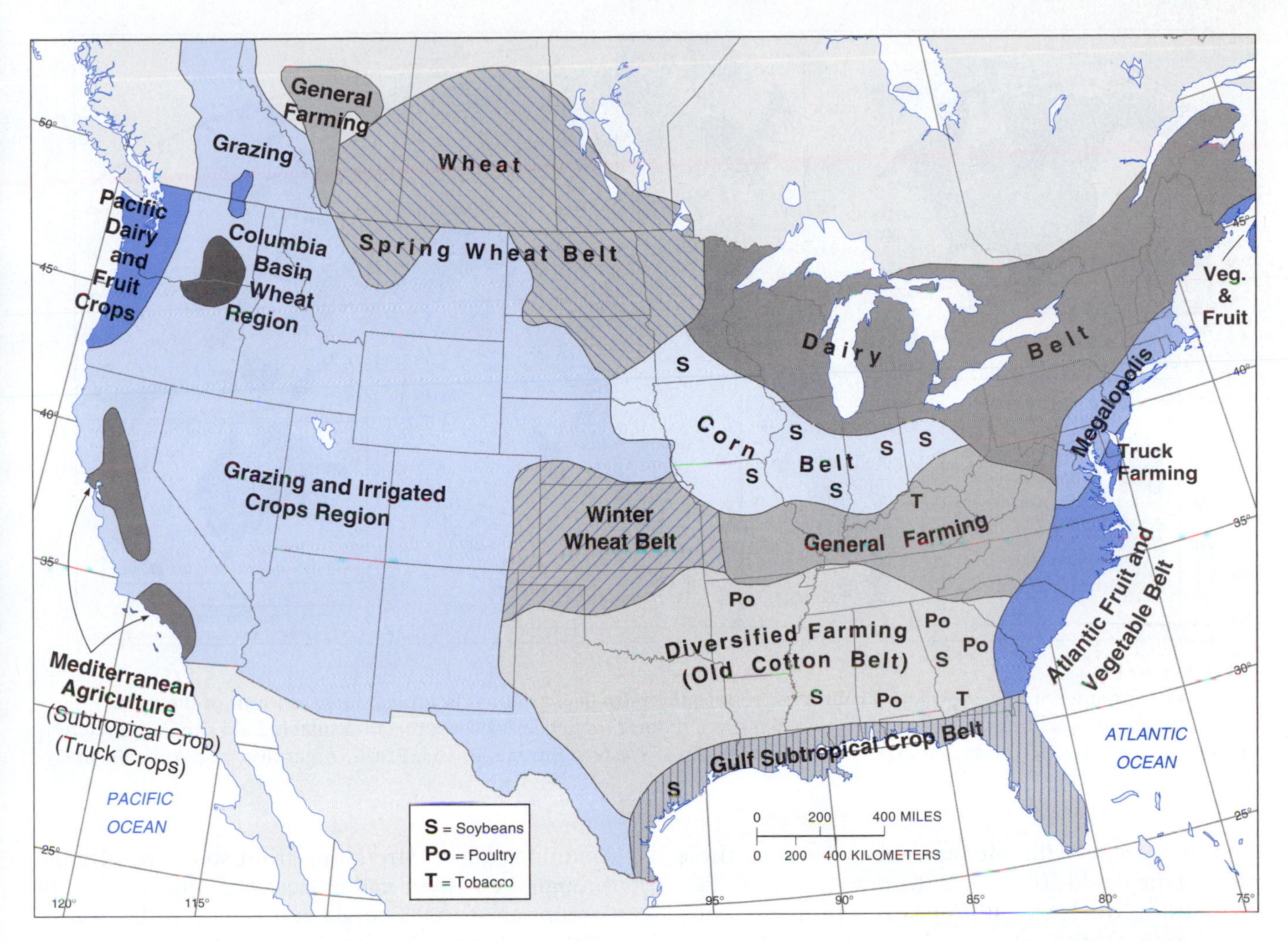

FIGURE 6.14
Major agricultural regions of the United States and Canada.

the U.S. government, with ranchers paying fees well below market rates. Today, ranches in Texas and the West cover thousands of acres because the semiarid conditions mean that several acres alone are required to raise one head of cattle. Some extremely large ranches are owned by meat-packing companies that can fatten the cattle, slaughter them, and package the meat all on the same ranch.

South America, Argentina, Uruguay, and Brazil all have significant cattle-ranching industries (Figure 6.15). These regions, as well as Australia and New Zealand, followed a similar pattern of cattle-ranching development. First, cattle were grazed on large, open, government tracks with little regard for ranch boundaries. Later, when a conflict with farming interests occurred, cattle ranches moved to drier areas. When irrigation first began to be used in the 1930s and 1940s, farms expanded their territory, and ranchers moved to even drier areas and centered much of their herds on feedlots near railroads or highways directed toward the markets. Today, ranching worldwide has become part of a vertically integrated agribusiness meat-processing industry.

Mediterranean Cropping

Mediterranean regions of the world grow specialized crops, depending on soil and moisture conditions. These regions include the lands around the Mediterranean Sea, southern California, central Chile, South Africa, and southern Australia. In these regions, summers are dry and hot, and winters are mild and wet. The Mediterranean Sea countries produce olives and grapes. Two-thirds of the world's wine is produced in Mediterranean Europe, especially Spain, France, and Italy. In addition, these countries and Greece produce the world's largest supply of olive oil. In California, the crop mix is slightly different because of consumer demand and preferences. Most of the land devoted to *Mediterranean agriculture* is taken up by citrus crops, principally oranges, lemons, and grapefruit.

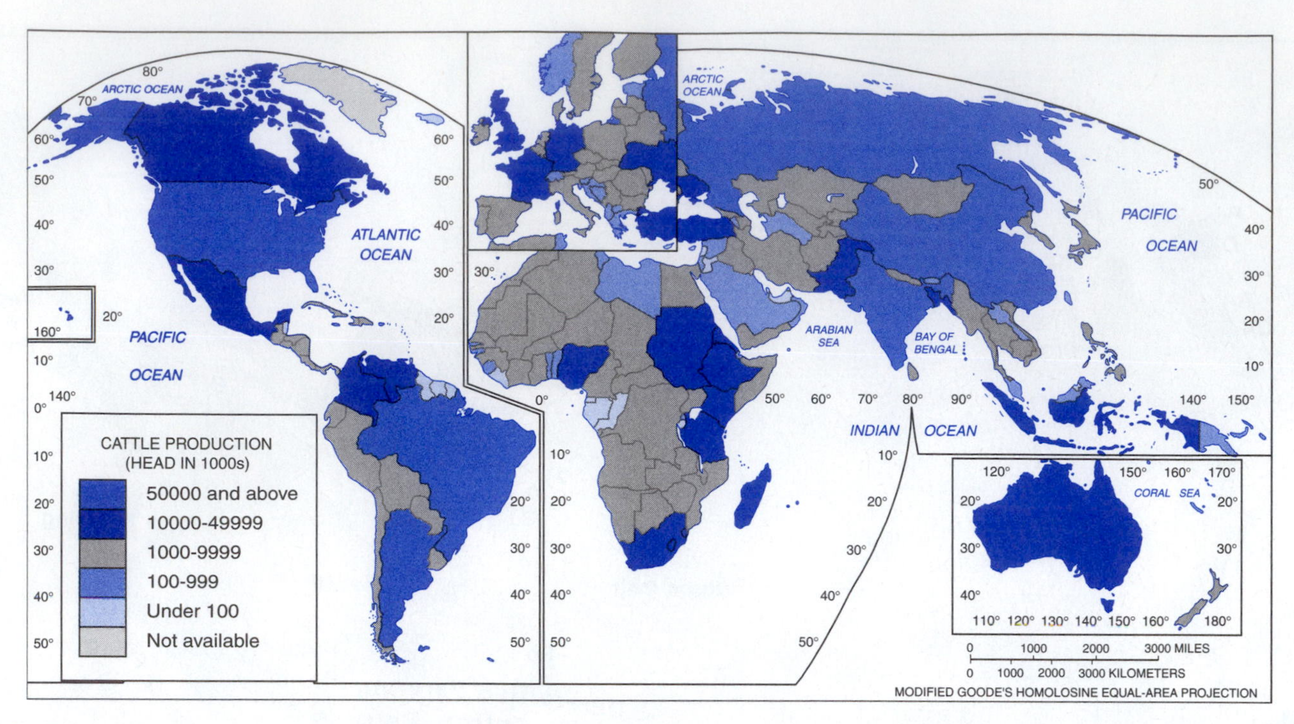

FIGURE 6.15
World cattle production. The developed countries produce the most beef products because a large amount of the grain crops, particularly corn, can be fed to cattle to fatten them. Poorer nations must consume all available food supplies directly or use them as revenue-producing exports. The United States, Western Europe, Russia, Brazil, Argentina, and Australia are the leading producers worldwide.

Unfortunately for Mediterranean farmers, these areas of the world are some of the most prized for their climates. Northern Europeans turned many Mediterranean Sea areas into tourist rivieras. No discussion is necessary to describe the land use changes associated with the burgeoning growth of southern California. Condominium projects, time shares, and burgeoning suburban developments for major cities, especially Los Angeles, are rapidly dwindling our Mediterranean agricultural lands. Ninety percent of the Chilean population lives in the Mediterranean lands in the central one-third of the country, centered on Santiago.

Horticulture and Fruit Farming

Because of consumer preferences, purchasing power, and a severe winter season, there is a tremendous demand in U.S. East Coast cities for fruits and vegetables not grown locally. Shoppers in Philadelphia, New York, Washington, D.C., Baltimore, and Boston pay dearly for *truck farm* fruits and vegetables, such as apples, asparagus, cabbage, cherries, lettuce, mushrooms, peppers, and tomatoes. Consequently, a horticulture and fruit-farming industry exists as close as possible to this portion of the United States as temperature and soil conditions allow. Stretching from southern Virginia through the eastern half of North Carolina and South Carolina to coastal Georgia and Florida is the *Atlantic Fruit and Vegetable Belt* (see Figure 6.14). This is an intensively developed agricultural region with a high value per acre. The products are shipped daily to the northeastern cities for direct consumption or for fruit and vegetable packing and freezing.

As in the case of Mediterranean agriculture and subtropical cropping in southern California, the Atlantic fruit and vegetable horticulture industry relies heavily on inexpensive labor. In California, the laborers are primarily from Mexico and Central America and often enter the United States illegally. On the Atlantic Coast, the laborers are primarily from the Caribbean, and their immigration status is also often questionable. Farm workers often work under brutally exploitative conditions for extremely low wages, typically well below minimum wage. Inexpensive labor is the major way that specialized agriculturalists maintain profits in areas under pressure for urban growth and expansion. Often illiterate and politically powerless, they suffer at the hands of unscrupulous employers. Farm workers may be cheated out of their pay, exposed to dangerous pesticides, and not be able to afford a place to live. In part, our cheap fruits and vegetables come at the expense of human misery (Figure 6.16).

FIGURE 6.16
Farm workers in the United States, many of whom are illegal immigrants, work long hours at very low rates of pay. Agriculture is exempted from minimum wage laws, and many of these impoverished workers, who make possible fresh fruits and vegetables at low prices, are highly vulnerable to exploitative employers.

U.S. AGRICULTURAL POLICY

In the historical development of America, farms were family owned, small, and served local markets. In those days, farm prices were relatively stable and predictable, although persistent overproduction in the late nineteenth century gradually depressed prices and bankrupted many farmers. By the early twentieth century, farms had become much larger, more highly mechanized, and technological improvements revolutionized agriculture. An individual farm family could manage as many as a thousand acres with new mechanized equipment. With improved transportation to the markets and between countries, the U.S. farmer now served a much wider market area. The early twentieth century was a prosperous time for U.S. farmers, especially during World War I, when they provided a large amount of food for Allied troops. However, many farmers lost their fortunes during the 1920s and 1930s with the twin economic and environmental catastrophes of the Great Depression, and many farms ceased to operate. World War II created another upswing for agricultural pursuits as farmers once again provided food for a much larger army, the Western allies, and a hungry nation.

After the war, the farmers' fortunes dwindled in the 1950s and 1960s until the U.S. government agreed to major grain trade agreements with the Soviet Union in the 1970s, which increased American exports. Since then, world markets for U.S. grain have dwindled as many foreign countries have become better able to produce more of their own food. For example, as a result of Green Revolution technology, India, formerly a net food importer, is now a net food exporter. Relatively high-interest rates made the cost of borrowing fertilizer, seeds, and equipment frequently prohibitive, and periodic increases in the value of the U.S. dollar relative to other currencies made American exports uncompetitive. At the same time, the prices of farm operation—including machinery, fertilizer, land, and transportation—have increased drastically. These less profitable times for farmers, in which costs have outrun income, have continued. Compared with other sectors of the U.S. economy, the farm sector has faced higher operating costs and lower revenues, although it also enjoys subsidies.

The Farm Problem in North America

One reason that agricultural markets are currently in such desperate straits is that the demand for farm products is price-inelastic. Consumers do not demand more food when farm prices are low, so the reduction in price does not lead to a substantial increase in the quantity demanded. This phenomenon is coupled with the fact that the yield from agriculture has increased manyfold during the last 100 years. Technological and mechanical improvements and hybrid seeds have increased yields so much that U.S. farm productivity is the highest in the world. The quantity of farm products has increased much more rapidly than has demand. These three factors have pushed down prices, as shown in Figure 6.17.

In Figure 6.17, we see the three tendencies of American farming during the last 100 years: drastically increased supply, moderately increased demand, and falling prices. The result has been relatively low returns to farm families and has spelled disaster for many farmers. With lower prices and increased quantities, more and more farmers cannot afford the rapidly rising costs for machinery, fertilizer, transportation, and labor. World farm prices have likewise fallen (Figure 6.18). The result to U.S. farmers has been a continuing reduction in return for their investment. Many farmers have sought to now move

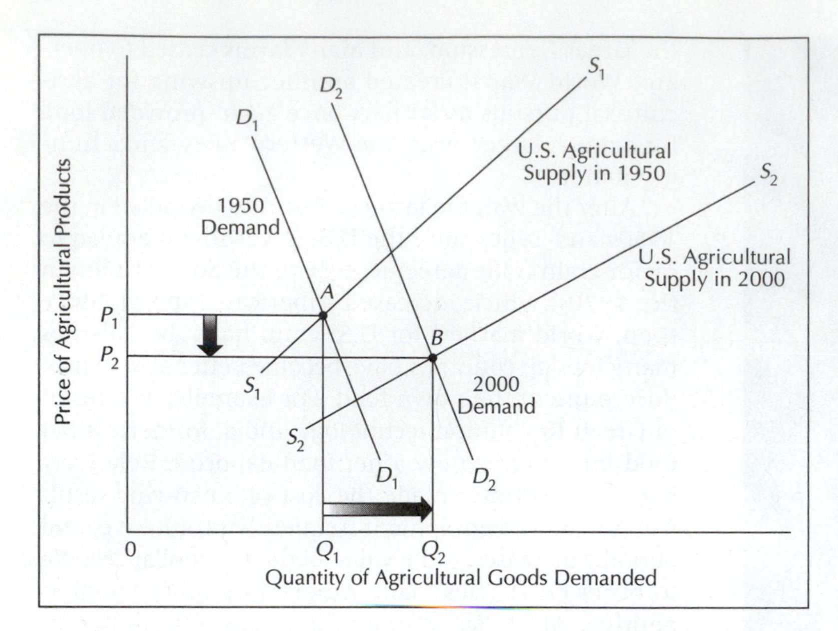

FIGURE 6.17
Dynamics of the U.S. agricultural sector. During the last 100 years, U.S. farm production has burgeoned remarkably because of increased productivity, increased mechanization, and improved fertilizers, pesticides, and hybrid seeds. The U.S. farm supply in 2000 was substantially more than it was in 1950, despite the Soil Bank program and other methods used to keep land out of production. At the same time, because food is an income-inelastic commodity, demand has increased, but not as much as supply. The result has been increased outputs and reduced prices.

their productivity to other, more profitable industries. However, unlike a store that can change hands and change function or a high-tech manufacturing plant that can change products, a farm is difficult to adapt to a new economic use.

As a result of persistent overproduction and low prices, there has been a large movement of farm families and farm labor away from the farm. In 1910, 35 percent of the U.S. population lived and worked on farms. By 2005, this figure dropped to less than 1 percent of the total population. However, considering present production, prices, and consumption, we still have too many farmers in America today. In a normal market situation, resources would have shifted away from agriculture into other economic activities. However, because of U.S. government price supports for farms, this has not been the case.

The U.S. Farm Subsidy Program

In 1933, with farming in deep crisis, Congress passed the Agricultural Adjustment act to aid American farmers. This act was designed to help a large proportion of the population (up to 33%) who lived in rural areas. The act artifically raised farm prices so that farmers could enjoy a "fair price," or *parity price*, for their products. A parity price was defined as "equality between the price farmers could sell their products for, and the price they would spend on goods

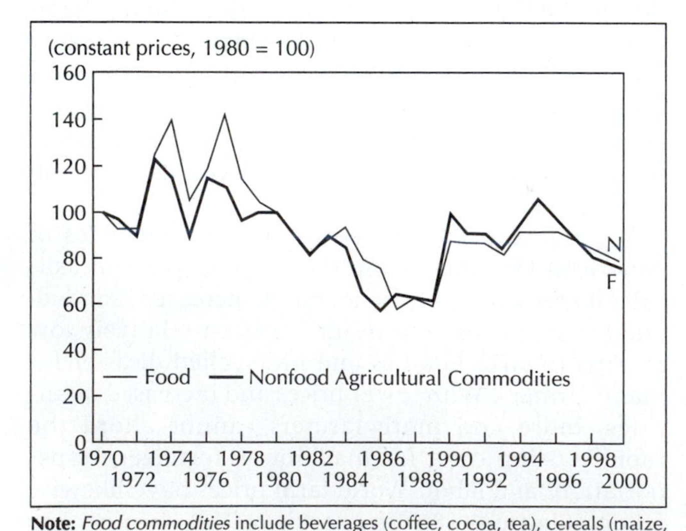

Note: *Food commodities* include beverages (coffee, cocoa, tea), cereals (maize, rice, wheat, grain sorghum), fats and oils (palm oil, coconut oil, groundnut oil, soybeans, copra, groundnut meal, soybean meal), and other food (sugar, beef, bananas, oranges). *Nonfood agricultural commodities* include cotton, jute, rubber, and tobacco.

FIGURE 6.18
Trends in world agricultural commodity prices, with 1980 as a base year. The trends show declines worldwide since 1970. Most farm prices have dropped to 60 percent of their 1970 levels. The decline in agricultural commodity prices has negatively affected many developing countries that base their economies on farm exports. This worldwide decline in farm commodity prices is a result of the inelasticity of food demand in developed nations, increased food supplies from Green Revolution techniques, increases in food productivity, subsidies to producers in industrialized economies, and low-income growth of farmers in developing nations. In the United States, for example, export subsidies designed to use farm overproduction abroad has kept world prices down by increasing the supply.

and services to run the farm." The period selected to determine parity prices was from 1910 to 1914, when farm prices were relatively high in comparison with other products.

Since 1933, however, the ratio between farm prices and all consumer goods declined until 2000, when it was approximately 30 percent of the original 1914 parity established in 1933. In other words, without parity, farmers could sell products and purchase only 30 percent of what they could in the earlier period. Admitting that markets had created widespread irrationality in agriculture, the federal government stepped in to establish a subsidy program, or a price floor, for key agricultural commodities, a guaranteed price above the market price. These supports were minimum prices that the government could assure farmers. For example, the government bought all corn and wheat from farmers and sold it at what the market would bear. It stored many of these commodities in its own storage facilities; thus public funds are used to encourage farmers to grow more than the market can consume and to store the surplus. In 1994, the U.S. government offered farmers *target pricing*, which is similar to the price supports of the 1950s through the 1980s. With target pricing, the government pays directly to the farmer the difference between the market selling price and the target price that the government has set; the government no longer takes control of the product.

Figure 6.19 shows the effects of price supports on agricultural products:

1. The market cannot arrive at an equilibrium price through its normal market mechanism.
2. Farmers produce a larger amount of surplus goods than consumers are willing to buy.
3. Buyers pay more than they would if market conditions prevailed.
4. Farmers' incomes are artificially raised by government subsidies, and consumers' incomes are artificially lowered.

As shown in Figure 6.19, with the parity price artificially high, farmers will supply the intersection of the price line and the supply curve at K for a total quantity of q_1. However, with higher prices, the consumers will demand q_2. The difference between q_1 and q_2 is the surplus that the government would purchase under the old price-support plan. Regardless of whether the subsidization is in the form of price supports or target pricing, the result is an extra cost to the taxpayer.

These price-support and target-price programs created artificially high agricultural prices to U.S. agriculture for the last 60 years. The hope, of course, was that market prices would rise to parity, and they did during World War II and during the Soviet grain trade agreements of the 1970s. However, most of the time, the price of farm products was much less than the parity price. The government also attempted to reduce production with the Soil Bank program, which paid farmers to keep acreage out of production. Initially, this approach worked, but the per-acre yields increased amazingly and completely overshadowed the lost acreage in terms of total yield.

The small American farmer as a cultural and economic institution is an endangered species. For years, the government price-support programs kept inefficient farmers in business. Government subsidies, however, favor large farmers over small ones. Because subsidies are a function of a farm's output, the subsidy program benefits the largest farmers, the corporate agribusinesses, not the small family farms

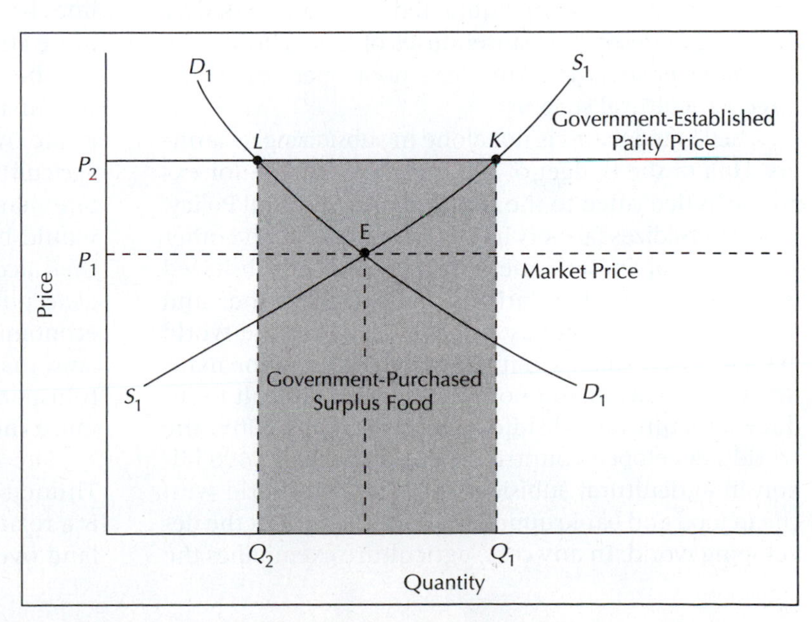

FIGURE 6.19
U.S. agricultural price supports. The U.S. government has supported farmers by establishing a parity price above the equilibrium price, according to supply and demand factors. For the last 60 years, the effect of the price support has been to establish prices higher than they would normally be, thus producing a surplus that the government was required to purchase with tax dollars. The market cannot obtain an equilibrium at E because surplus goods are produced and too often wasted. The farmers' incomes are artificially raised, but buyers in the marketplace must pay more than the goods would warrant under normal conditions. Unfortunately, resources are artificially allocated and therefore misallocated as price shifts from P_1 to P_2, demand drops back to point *L*, while supply moves up to point *K*.

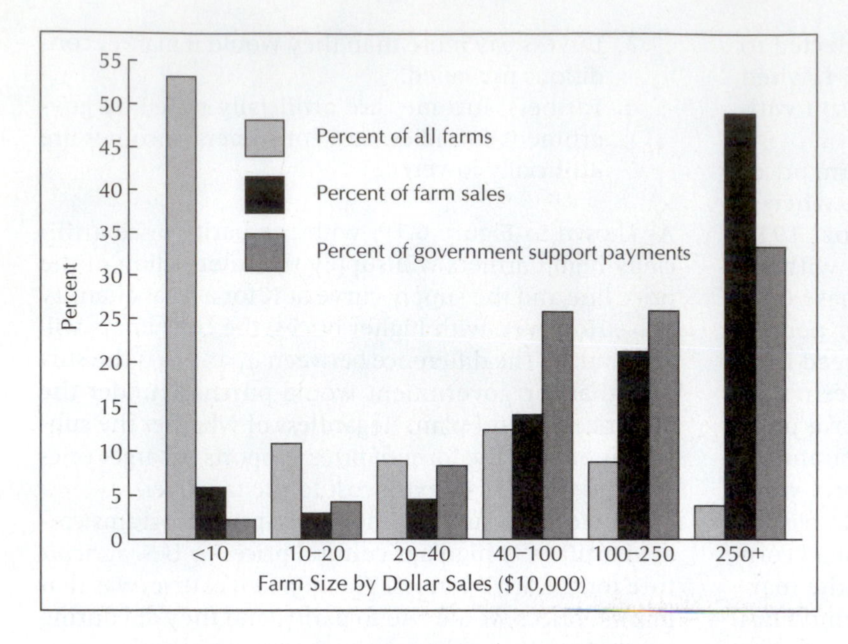

FIGURE 6.20
The proportion of farms, the proportion of farm sales, and the percentage of government support payments by farm size and dollar sales. Most U.S. farms are small and produce small proportions of total farm sales. However, most government support payments go to large farms. With regard to farming and farm policy, the rich appear to be getting richer, whereas the poor or small farms, who were the original focus of price supports, are getting relatively poorer and scarcer.

as originally intended (Figure 6.20). The U.S. corn sector is the largest recipient of government payments, including $10 billion in 2003. In essence, the large corporate agribusiness farms have become richer and, with the lion's share of U.S. government subsidy, forced food prices even lower. This has, in effect, continued to force the small farmers off the land. Due to the political power of farmers, the government has found it politically difficult to trim agricultural subsidies.

The obvious solution to America's farm problems is to design new uses for farm products that are not currently demanded in the United States. One example is an attempt to generate gasohol from corn and other agricultural products to fuel automobiles. A second use of the food surplus is *Food-for-Peace programs,* which allow agricultural surpluses to be distributed to starving nations instead of being liquidated or exterminated by dumping or destroying storehouses of food. Locally, the *food stamp program* for America's poor operates off the large agricultural surplus.

The United States is not alone in subsidizing its farmers. Half of the budget of the European Union, for example, is dedicated to the Common Agricultural Policy, which subsidizes farmers in France, Germany, and other countries. Japanese farmers are very heavily subsidized in a country with relatively little arable land, and Japanese consumers pay prices well above the world average as a result. Farming is often draped in the mantle of nationalism, and politicians find it difficult to reduce agricultural subsidies everywhere. In 2005, the world's developed countries spent more than $600 billion in agricultural subisides, flooding the world with cheap food and bankrupting farmers throughout the developing world. In any case, agriculture exemplifies the powerful role of the state in almost all market-dominated societies; there is certainly no "free market" in farming.

THE VON THÜNEN MODEL

An important factor that shapes individual farmers' land-use decisions is the relative location or situation of a place in terms of its access to other places. At one time, before commercial agriculture, a farmer's site relations—links with soil, sun, rain, and crops—were overwhelmingly important considerations in earning a living. Given the time-space compression of worldwide markets, the importance of situational components in agriculture increased as market exchange economies grew to be much more important, including transport lines between farms and the market, linking them ever more strongly to a wider spatial economy.

The importance of relative location in rural land use was first discussed by Johann von Thünen, a German estate owner interested in economic theory and local agricultural conditions. From his experiences as an estate manager, he observed that identical plots of land would be used for different purposes depending on their accessibility to the market. His book *The Isolated State,* published in 1826, was one of the first models in economic geography. Von Thünen's aim was to uncover laws that govern the interaction of agricultural prices, transport costs, and land uses as landlords seek to maximize their income.

The concept of economic rent is central to von Thünen's model of agricultural land use. *Economic rent* is a relative measure of the advantage of one parcel of land over another. More precisely, it is the difference in net profits between two units of land. Differential rents may result from variances in productivity of different parcels of land and/or variances in the distance from the market. Von Thünen demonstrated that rent is the price of accessibility to the market. In other words, rents decline with the distance from a market center. Geographers often use the term *location rent* as opposed to *economic rent* to express the concept of decline in rents with an increase in distance from the market.

The Isolated State

To explain agricultural land use, von Thünen described an idealized agricultural region about which he made certain assumptions. He envisioned an isolated state with a large city serving as the only market. A uniform plain surrounded the city. There were no extraneous disturbances in this idealized landscape; social classes and government intervention were absent. Transport to the central town increased at a rate proportional to distance.

What pattern of cultivation will take shape in these conditions? Near the town will be grown those products that are heavy or bulky in relation to their value and that are consequently so expensive to transport that the remoter districts are unable to supply them. Here also we find the highly perishable products, which must be used very quickly. With increasing distance from the town, the land will progressively be given over to products cheap to transport in relation to their value. For this reason alone, concentric rings or belts will form around the town, each with its own particular staple product. From ring to ring, the staple product will change. Thus, von Thünen suggested that in a landscape free from all complicating factors, locational differences were sufficient to produce a varied pattern of land use.

To illustrate von Thünen's concept of differential rent, assume an isolated state producing one commodity (say, wheat) grown at a single intensity on yield per acre per year. Further assume that the market price of wheat is $100 per hectare per year, that it costs every farmer in the state $40 to produce a hectare of wheat, and that transport costs are $5 for each hectare of wheat (Figure 6.21). Farmers adjacent to the market pay no transport costs; therefore, their net profits would simply be market price ($100) minus production costs ($40)—or $60. Farmers located 1 kilometer from the market pay $5 in transport charges; thus, their net profits would be $55. At 6 kilometers from the market, farmers would earn a net profit of $30, and at 12 kilometers, net profits would be zero. Beyond 12 kilometers, it would be unprofitable to grow a crop for the market.

Farmers near the central market pay lower transport costs than farmers at the margin of production. Clearly, the net profits of the closer farmers are greater, and the difference is known as *economic rent*. Farmers recognize this condition, and they know that it is in their best interest to bid up the amount that they will pay for agricultural land closer to the market. Bidding

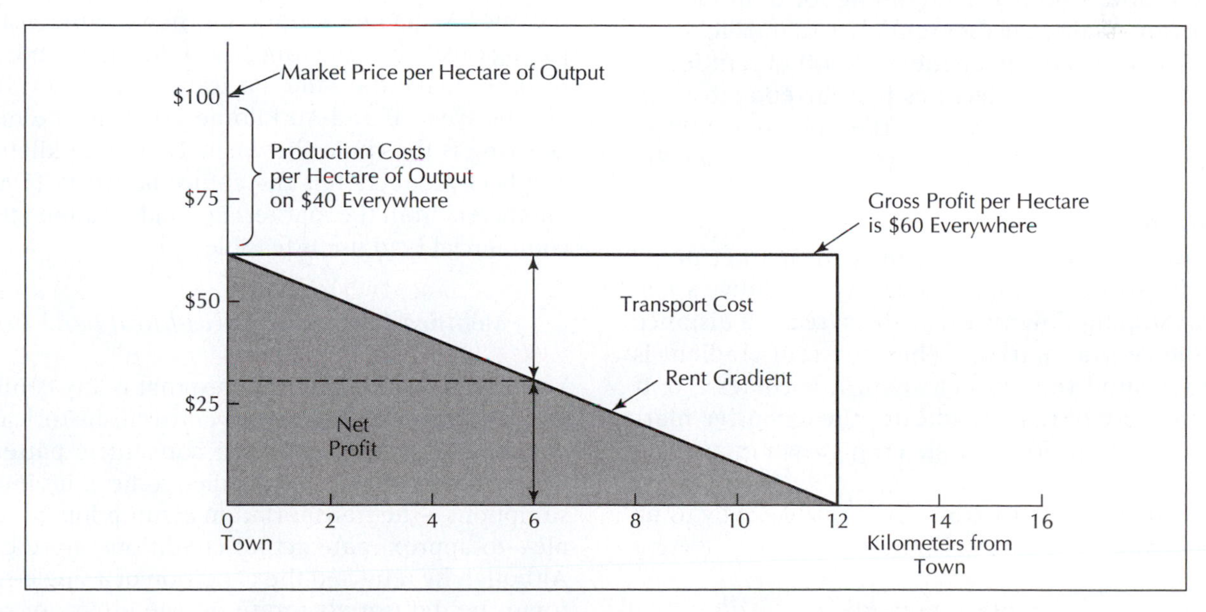

FIGURE 6.21
Net income from wheat production. Von Thünen suggested that the net income would decline with increasing distance from the town where the product had to be sold. Transportation costs would eat into the net income as one moved farther away from the town. In this example, production costs are $40 per hectare at every location, and the margin of profitability is 12 kilometers.

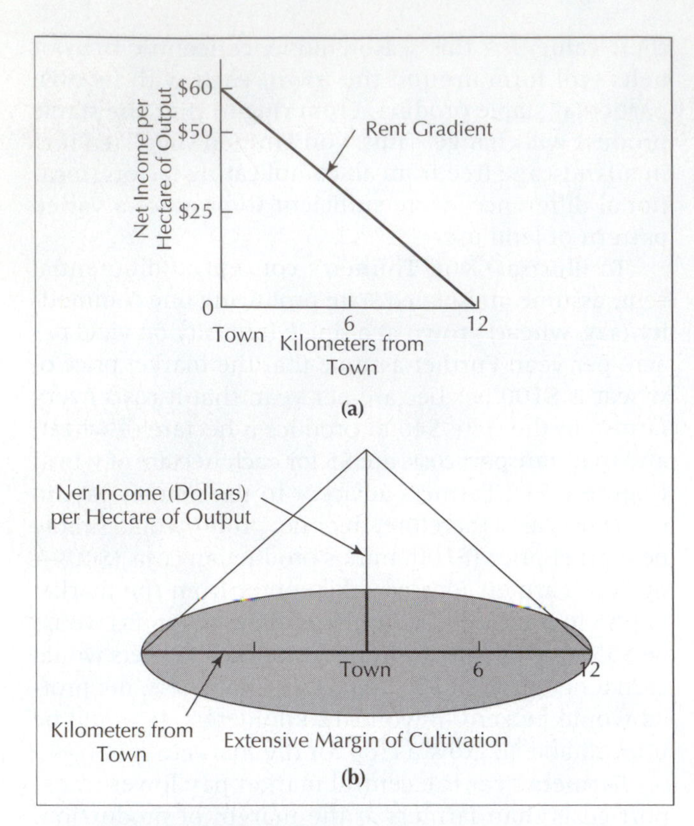

FIGURE 6.22
From a rent gradient to a rent cone.

continues until bid rent equals location rent. At that price, farmers recover production and transport costs, and landowners receive location rents as payments for their land. Competitive bidding for desirable locations cancels income differentials attributable to accessibility. The *bid rent,* or the trade-off of rent levels with transport costs, declines just far enough from the market to cover additional transport costs; hence, farmers are indifferent to their distances from the market.

We can simplify Figure 6.21 by including production costs in a single expression with market price. This is illustrated in Figure 6.22, which shows a *rent gradient* sloping downward with increasing distance from the central market. When the rent gradient is located around the market town, it becomes a rent cone, the base of which indicates the extensive margin of cultivation for a single crop grown at a single intensity.

Location rent for any crop can be calculated by using the following formula:

$$R = E(p - a) - Efk,$$

where
R = location rent per unit of land
E = output per unit of land
k = distance to the market
p = market price per unit of output
a = production cost cost per unit of product (including labor)
f = transport rate per unit of distance per unit of output

Thus, if we assume that a wheat farmer 20 kilometers from the market obtains a yield of 1000 metric tons/km^2, has production expenses of \$50/ton, spends \$1/ton/km to transport grain to the market, and receives a market price of \$100/ton at the central market, the location rent accruing to 1 square kilometer of the farmer's land can be calculated as follows:

$$R = 1000\ (\$100 - \$50) - 1000\ (\$1 \cdot 20) = \$50{,}000 - \$20{,}000 = \$30{,}000.$$

At 50 kilometers from the market, the location rent per square kilometer of land in wheat is \$0. Obviously, no rational farmer in a competitive market economy would grow wheat beyond 50 kilometers from the market.

Location-Rent Gradients for Competing Crops

In von Thünen's analysis, patterns of agricultural land use form according to the principles of highest and best use as measured by the location rent at each distance from the market. Consider location-rent gradients for an isolated state in which farmers have three land-use choices: vegetable production, dairying, and beef production (Figure 6.23). A farmer close to the market could profitably carry on any one of the three activities, but which activity would maximize the farmer's income? Vegetable production has the highest rent-paying capability. All farmers seeking to maximize their incomes make the same decision: They grow vegetables between 0 and 10 kilometers from the market. Dairying is the choice between 10 and 25 kilometers, and beef production is the choice between 25 and 50 kilometers from the market. Beyond 50 kilometers, no commercial land use is feasible.

Modified Patterns of Agricultural Land Use

Von Thünen was acutely aware that many conflicting factors—physical, technical, cultural, historical, and political—would modify the concentric patterns of agricultural land use. He modified some of his initial assumptions—the transportation assumption, for example—to approximate actual conditions more closely. Although he retracted the condition of a single-market town, he did not elaborate on the effects of several competing markets and a system of radiating highways. We can presume, however, that the tributary areas of competing markets would have a variety of crop zones enveloped by those of the principal market town and

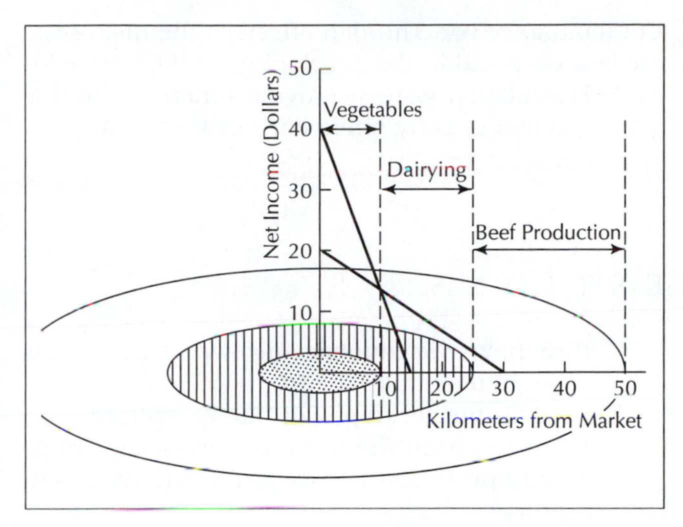

FIGURE 6.23
Location rent gradients for competing crops. Because of the different sloping bid rent curves based on market prices and transportation costs for various crops, the highest and best use and the highest level of net income will result from a combination of crops being grown at different distances from the market. Vegetables have the highest rent-paying ability because of their relatively high yield per hectare. Due to their weight, however, the bid rent curve drops off dramatically, as transportation costs eat into profits. Vegetables are the best choice up to 10 kilometers from the market, beyond which dairying becomes a better choice. At approximately 24 kilometers from the market, dairying and its associated relatively heavy and perishable products become more expensive to ship than the less expensive beef production. From this point out to the margin of production at 50 kilometers, beef is the wisest choice. Beyond that point, no commercial land use is feasible.

that a radiating highway system would have produced a "starfish" pattern.

SUMMARY

The invention or discovery of agriculture, roughly around 10,000 B.C., in several parts of the world marked a momentous change in how human beings live, enabling settled communities, class-based societies, the state, and a more specialized division of labor than hunting and gathering, which comprised 95 percent or more of human existence.

All people depend on agriculture for their wellbeing, but in our agricultural pursuits, we necessarily modify the land. The production of agricultural landscapes thus involves the large scale transformation of nature in many ways. When geographers speak of agricultural regions, they refer to the artificial division of the world among a variety of socially produced farming types. The most intriguing aspect of farming regions is not their number or extent but the similarity of land-use decisions farmers make within them. After reviewing agricultural origins and dispersals, we identified basic factors that influence agricultural land-use patterns: site characteristics, cultural preferences and perceptions, systems of production, and relative location.

Agriculture in developing countries is primarily of the subsistence variety. Subsistence agriculture includes shifting cultivation, intensive farming, and pastoral nomadism. Developing countries often have more than 50 percent of their labor force engaged in agriculture and use relatively little mechanized equipment.

By contrast, agriculture in developed countries employs a small percentage of the labor force and uses large-scale mechanized equipment and large inputs of energy and fertilizers. While outputs per hectare are comparable to intensive farming and even somewhat less, outputs per worker are as much as 50 times greater. The industrialization of agriculture—a process that is still ongoing in much of the world today—freed millions from the boring and arduous toil on the farms. Farmers who raise crops and livestock on huge farms are part of a vast agricultural system that includes machinery manufacturers; fertilizer, pesticide, and energy supplies; grain mills and slaughterhouses; food processing; and wholesale and retail distribution. Although mixed crops and livestock farms are the most common in developed countries, other types of commercial agriculture also exist, including dairy farming; commercial grain farming, usually centered on wheat; cattle ranching; Mediterranean horticulture; and tree farms.

Like many sectors under capitalism, agriculture exhibits the powerful role of the state in shaping markets and landscapes. Far from being a free market, agriculture in most societies, including the United States, is heavily subsidized by the government in programs that reflect the political power of the farm lobby. Price floors and subsidies contribute to the chronic overproduction and low prices that farmers typically face, and subsidies go disporportionately to large corporate agribusinesses that dominate the industry rather than to small family farms.

One theory that helps us to understand the local organization of agriculture was formalized by Johann Heinrich von Thünen in the early nineteenth century. We described the model that von Thünen developed to explain patterns of local land use. We then presented some of the conclusions that can be drawn from this model: (1) There is an inverse relationship between location rent and transport costs; (2) there is a limit to commercial farming on a homogeneous plain with an isolated market town at its center; (3) land values and intensity of land use increase toward the market; and (4) crop types compete with one another and are ordered according to the principle of the highest economic rent. We saw how the basic Thünian principles can be applied to agricultural land-use patterns at scales ranging from the village to the world. Contemporary von Thünian effects at the microscale are best observed in the developing world, where localized circulation systems provide a transport setting similar to that of early nineteenth-century Europe.

STUDY QUESTIONS

1. What proportion of the U.S. population works in agriculture? Why is it so low?
2. When and where did the world's first agricultural revolution occur?
3. Describe the impacts of European colonialism on global patterns of agricultures.
4. What are plantations?
5. What is meant, exactly, by Boserop's agricultural intensification?
6. Summarize the peasant mode of production.
7. What is the main characteristic of commercial agriculture?
8. How have the number and size of U.S. farms changed over time?
9. Describe the geography of U.S. agriculture.
10. What has been the trend in agricultural commodity prices worldwide in the late twentieth century?
11. What is von Thünen's model of agriculture? What is economic rent?

KEY TERMS

agribusiness
Atlantic Fruit and Vegetable Belt
capital intensive
commercial agriculture
environmental perception
four-field rotation system
highest and best use
high value added
inelastic supply
intensive margin of cultivation
intensive subsidence agriculture
labor intensive
location rent
marginal product
margin of cultivation
Mediterranean cropping
mixed crop and livestock farming
parity price
pastoral nomadism
peasant agriculture
price floor
rent
shifting cultivation
slash-and-burn agriculture
stages of production
subsistence agriculture
target pricing
use value
vertical integration
virgin and idle lands program
von Thünen model
wheat belts

SUGGESTED READINGS

Hart, J. 2003. *The Changing Scale of American Agriculture.* Charlottesville: University of Virginia Press.

Rubenstein, J. 2001. *The Cultural Landscape: An Introduction to Human Geography,* 7th ed. New York: Macmillan.

Watts, M. 1996. "Development III: The Global Agrofood System and Late Twentieth-Century Development." *Progress in Human Geography* 20:230–45.

WORLD WIDE WEB SITES

AG AGENT HANDBOOK
http://edis.ifas.ufl.edu/index.html
This online handbook from the University of Florida has links to information on soils, field and forage crops, turfgrass, vegetable crops, sustainable production practices, forestry, dairy cattle, beef cattle, poultry, rabbit production, swine, control of small mammals and birds, beekeeping, agricultural engineering, farm and resource economics, and pest control.

LIFE ON THE FARM
http://web2.airmail.net/bealke
What is it like to live on a farm? Visit Chuck Bealke's virtual farm and get his biweekly remembrance or comment on farm life—much of it based on farming (soybeans, wheat, corn, hay, and polled herefords) west of St. Louis. Links to many agricultural online resources.

NATIONAL AGRICULTURAL LIBRARY (NAL)
http://www.nalusda.gov
The NAL is part of the Agricultural Research Service of the U.S. Department of Agriculture and is one of four national libraries in the United States. NAL is a major international source for agriculture and related information.

UNITED STATES DEPARTMENT OF AGRICULTURE HOME PAGE
www.nass.usda.gov/QuickStats/

CHAPTER 7

MANUFACTURING

OBJECTIVES

- To explore the fundamental nature of the manufacturing process
- To acquaint you with the major manufacturing regions of the world
- To describe recent global shifts in the globalization of world manufacturing
- To summarize the geography of U.S. manufacturing, major industries and their changes over time
- To show the trend toward flexible manufacture, flexible labor, and the flexible economy

A maquiladora textile worker in Ciudad Juarez, Mexico.

To manufacture is to make things—to transform raw materials into goods that satisfy needs and wants. Manufacturing is crucial because it produces goods that sustain human life, provides employment, and generates economic growth. It has played this role since the Industrial Revolution in England in the late eighteenth century, when manufacturing generated the working classes of Europe, North America, and Japan.

Geographers who study manufacturing emphasize the locational behavior of firms and the structures of the places they create. First, this chapter examines the nature of manufacturing, including the basic steps involved in the transformation of raw materials into final products. Second, it turns to the major regions in the world that produce goods—North America, Europe, and Japan. Manufacturing is highly unevenly located around the world. How did these clusters come to exist? Third, it explores three crucial industries in more depth: textiles, automobiles, and electronics. Fourth, it summarizes the geography of U.S. manufacturing and the changes wrought by globalization. Fifth, it introduces the notions of flexibility, post-Fordist production, and just-in-time systems, which have revolutionized the manufacturing production and delivery process.

The Nature of Manufacturing

Manufacturing involves deciding what is to be produced, gathering together the raw materials and semifinished inputs at a plant, reworking and combining the inputs to produce a finished product, and marketing the finished product. These phases are called *selection, assembly, production,* and *distribution.* The assembly and distribution phases require transportation of raw materials and finished products, respectively. Industrial location theory attempts to identify the locations that will minimize these transportation costs. The production phase—changing the form of a raw material—involves combining land, labor, capital, and management, factors that vary widely in cost from place to place. Thus, each of the steps of the manufacturing process has a spatial or a locational dimension.

Changing the form of a raw material increases its use or value. Flour milled from wheat is more valuable than raw grain. Bread, in turn, is worth more than flour. This process is termed *value added by manufacturing.* The value added by manufacturing is quite low in an industry engaged in the initial processing of a raw material. For example, turning sugar beets into sugar yields an added value of about 30 percent. In contrast, changing a few ounces of steel and glass into a watch yields a high added value—more than 60 percent. The cost and productivity of labor, and the availability of skills, plays an important role in high value-added manufacturing.

Concentrations of World Manufacturing

Manufacturing capacity and employment are highly unevenly distributed around the world and go far to explain the uneven spatial development that typifies the world economy. Three major areas account for approximately 80 percent of the world's manufacturing (Figure 7.1): North America, Europe (including Russia and Ukraine), and Japan.

North America

North American manufacturing is largely centered in the northeastern United States and southeastern Canada (Figure 7.2), the North American Manufacturing Belt, which accounts for one-third of the North American population and nearly two-thirds of North American manufacturing employment. This area was settled by Europeans in the seventeenth and eighteenth centuries, and grew rapidly in the nineteenth century. It was tied to the European markets and possessed the raw materials, iron ore, coal, and limestone necessary to produce the heavy machinery and manufactured items on which the industrialization of America was based. In addition, this region had many markets and a large labor pool.

The transportation system included the St. Lawrence River and the Great Lakes, which were connected to the East Coast and the Atlantic Ocean by the Mohawk and Hudson Rivers. This transportation system allowed the easy movement of bulky and heavy materials. Later, canals and railroads supplemented the river and lake system.

Within the North American Manufacturing Belt, there are several districts. The oldest is southern New England, centered on the greater Boston metropolitan area. Historically, this area was the textile and clothing manufacturing center of the early nineteenth century. Cotton was brought from the Southern states to be manufactured into garments, many of which were consumed locally and some of which were exported to Europe. As the low-wage European immigrant laborers settled and unionized, wages became higher, and the textile industry moved to the South in the early twentieth century. Today, New England manufacturing centers on electrical machinery, fabricated metals, and electronic products. The region is noted for highly skilled labor, with nearby universities—including Boston College, Boston University, Massachusetts Institute of Technology, and Harvard University—providing the chief supply (Figure 7.2).

The Middle Atlantic district includes the metropolitan areas of New York, Baltimore, Philadelphia, and Wilmington, Delaware. The Great Lakes industrial traffic terminates in New York City via the

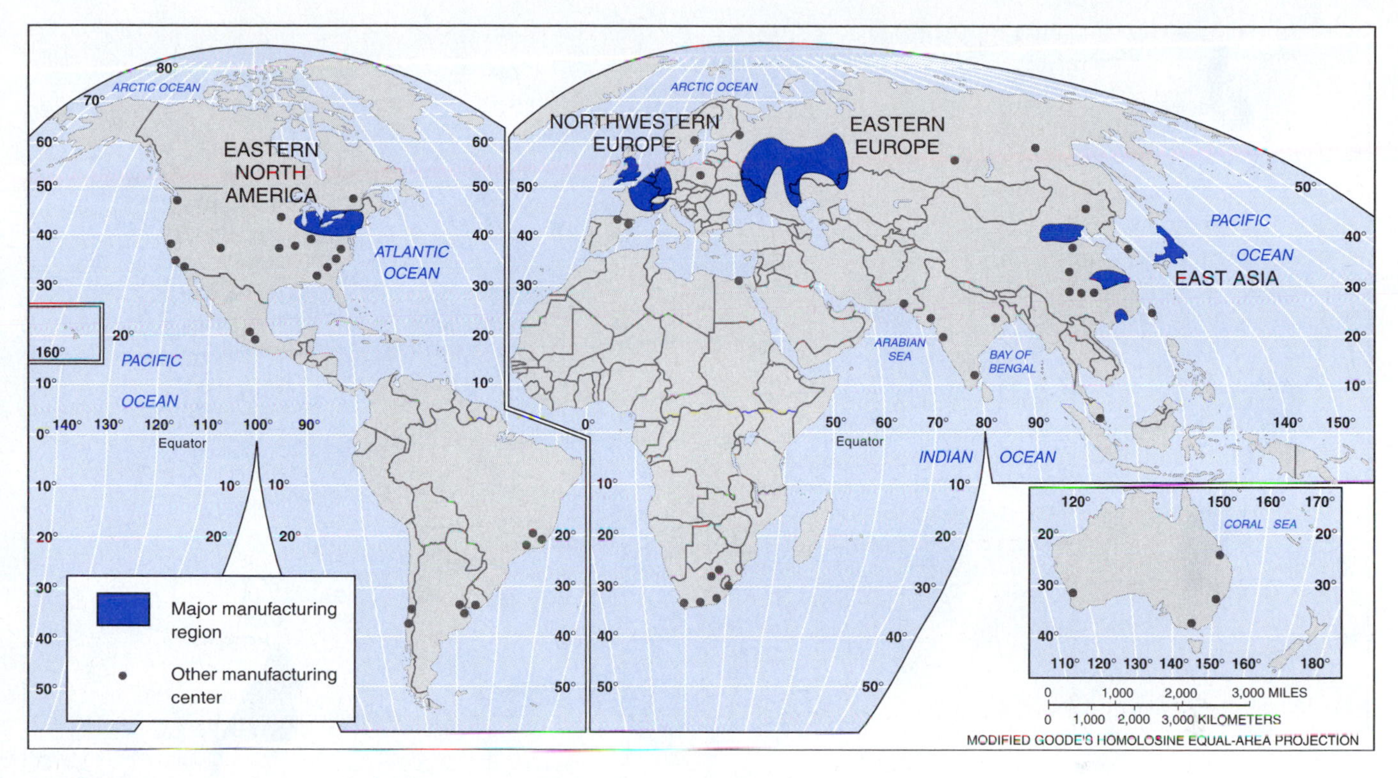

FIGURE 7.1
Worldwide distribution of manufacturing. The four main manufacturing regions include the northeastern United States and southern Great Lakes region, the northwestern European region, the eastern Soviet Union and Ukraine region, and the Japan/South Korea region. See the text for a detailed elaboration of districts within each of these regions, as well as other manufacturing regions shown as dots.

Mohawk and Hudson rivers. From New York City, foreign markets and sources of raw materials can be reached. New York City is the largest market and has the largest labor pool. Because of the enormous agglomeration economies it offers, many of the *Fortune* 500 firm headquarters are in this district. The New York district is in proximity not only to trade with the rest of the world but also to the population centers and manufacturing hubs of America (Figure 7.2). It is also near financial, communications, and news and media industries, which are important for advertising and distribution. This region produces apparel, iron and steel, chemicals, machinery, fabricated metals, and a variety of processed foods. In addition, it is the headquarters of the North American publishing industry. Many major book and magazine publishing companies are found in this region.

The central New York and Mohawk River valley district produces electrical machinery, chemicals, optical machinery, and iron and steel. These industries agglomerate along the Erie Canal and the Hudson River, the only waterway connecting the Great Lakes to the U.S. East Coast. Abundant electrical power produced by the kinetic energy of Niagara Falls provides inexpensive electricity to this district and explains the attraction of the aluminum industry, which requires large amounts of electricity. The New York industrial cities of Buffalo, Rochester, Syracuse, Utica, Schenectady, and Albany are situated in this district.

The Pittsburgh–Cleveland–Lake Erie district, centered in western Pennsylvania and eastern Ohio, is the oldest steel-producing region in North America. Pittsburgh was the original steel-producing center because of the iron ore and coal available in the nearby Appalachian Mountains. When the iron ore became depleted, new supplies were discovered in northern Minnesota and transported in via the Great Lakes system. Besides iron and steel, electrical equipment, machinery, rubber, and machine tools are produced in this region.

The western Great Lakes industrial region is centered on Detroit in the east and Chicago in the west (Figure 7.2). In addition, Toledo, Ohio, in the east and Milwaukee in the west were long known for the production of transportation equipment, iron and steel, automobiles, fabricated metals, and machinery. Detroit and surrounding cities, of course, have the preeminent position of automobile manufacture, and Chicago has produced more railroad cars, farm tractors and implements, and food products than any other city in the United States. The convergence of railroad and highway transportation routes in this area makes it readily accessible to the rest of the country and a good distribution point to a national market.

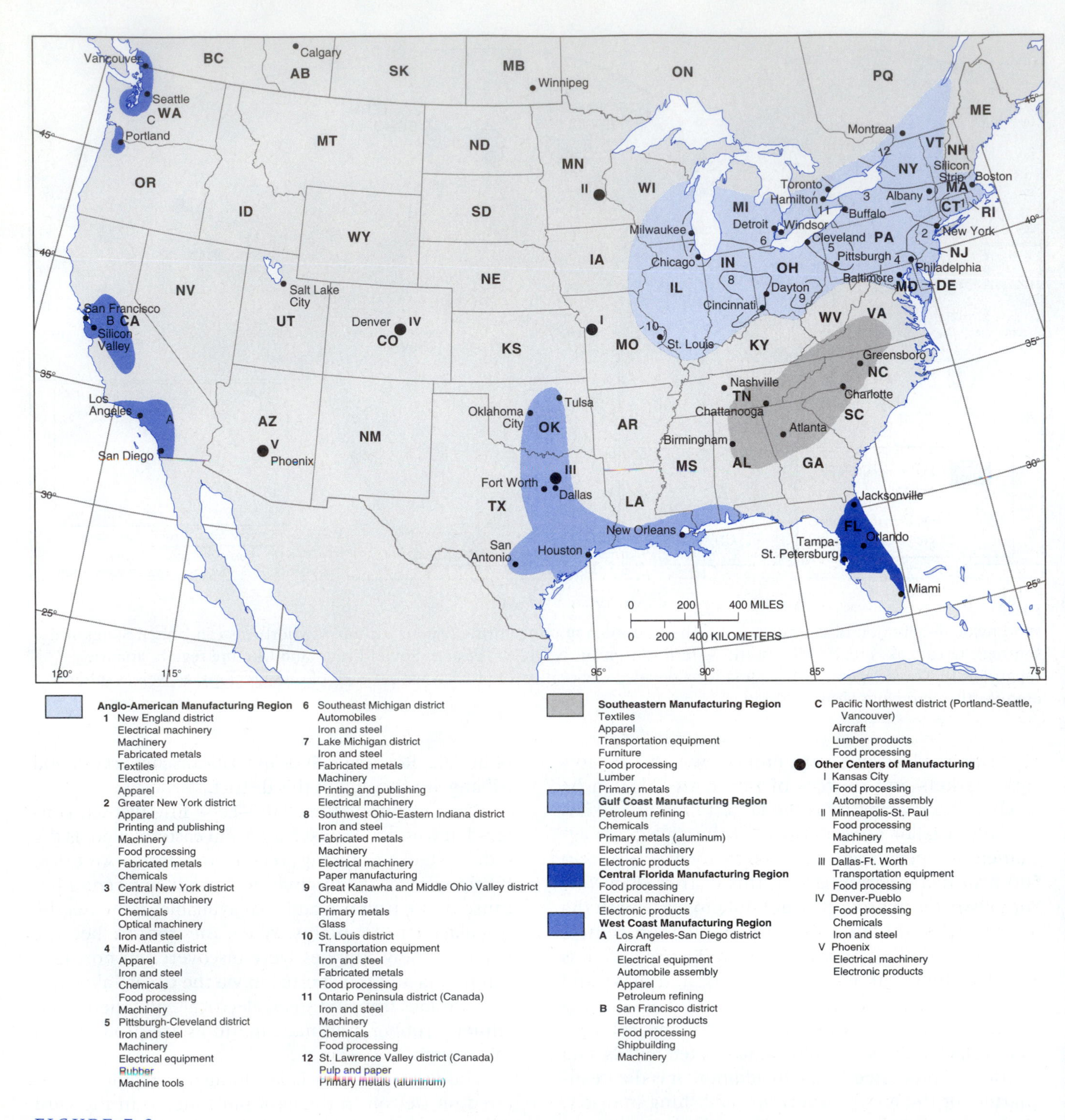

FIGURE 7.2
Manufacturing regions and districts throughout the United States and Canada.

Canada's most important industrial region stretches along the St. Lawrence River Valley, on the north shore of the eastern Great Lakes (Figure 7.2). This area has access to the St. Lawrence River–Great Lakes transportation system, is near the largest Canadian markets, has skilled and plentiful labor, and is supplied with inexpensive electricity from Niagara Falls. Iron and steel, machinery, chemicals, processed foods, pulp and paper, and primary metals, especially aluminum, are produced in this district. For example, Toronto is a leading automobile-assembly location in Canada, whereas Hamilton is Canada's leading iron and steel producer.

The southeastern manufacturing region of the United States stretches south from central Virginia through North Carolina, western South Carolina, northern Georgia, northeastern Alabama, and northeastern

New England became the first and foremost textile manufacturing region in the United States in the nineteenth century. The mills pictured here are in Lawrence, Massachusetts. By the 1940s, the textile region of southern New England had been in decline for more than 20 years. Firms left the region in search of more profitable operating conditions, and workers were forced to seek other employment. The region has experienced a revival as new industries, notably electrical engineering, have replaced the older, declining ones.

Tennessee (Figure 7.2). It wraps around the southern flank of the Appalachian Mountains because of poor transportation connections across the mountains. Textiles are the main product, the industry having moved from the Northeast to the South to take advantage of less expensive, nonunion labor. Transportation equipment, furniture, processed foods, and lumber are also produced. Aluminum manufacturers moved to this region because of the inexpensive electricity produced by the more than 20 dams built by the Tennessee Valley Authority, and Birmingham was long the iron and steel center of the southeastern United States because of the plentiful iron-ore and coal supplies nearby. Figure 7.3 shows the distribution of American textile employment today. The primary concentration is in the Southeast, particularly in North and South Carolina. Secondary concentrations are found in New England, a remnant of its heyday in the early twentieth century, as well as southern California, where sweatshops using immigrant labor are common.

The Gulf Coast manufacturing region stretches from southeastern Texas through southern Louisiana, Mississippi, and Alabama, to the tip of the Florida panhandle, including cities such as Houston, Texas; Baton Rouge, Louisiana; Mobile, Alabama; and Pensacola, Florida. Because of nearby oil and gas fields, petroleum refining and chemical production are important. The region also produces primary metals, including aluminum, and electrical machinery and electronic products.

The Los Angeles and San Diego district in southern California specializes in aircraft and aerospace manufacture and electrical equipment. In the 1930s, the airline industry chose this location because favorable weather much of the year meant unimpeded test flights and savings on heating and cooling the large aircraft plants. Federal government subsidies during World War II reinforced this advantageous location. Because of the myriad electronic parts and equipment and the associated high-tech sensing and navigational devices required in aircraft manufacture, the electronics industry was also attracted to this region and was anchored there 30 years later (Figure 7.4). Today, aircraft, apparel manufacture, and petroleum refining are important in Los Angeles, whereas San Diego also specializes in pharmaceutical production and in military and transport equipment industries.

The San Francisco Bay Area is another important West Coast manufacturing region. Electronic products, processed foods, ships, and machinery are produced in the district. *Silicon Valley,* the world's largest manufacturing area for semiconductors, microprocessors, software, and computer equipment, is located just south of San Francisco in the Santa Clara Valley.

The Pacific Northwest district includes the cities of Seattle, Washington, and Portland, Oregon. Boeing Aircraft is the single largest employer, followed closely by the paper, lumber, and food-processing industries.

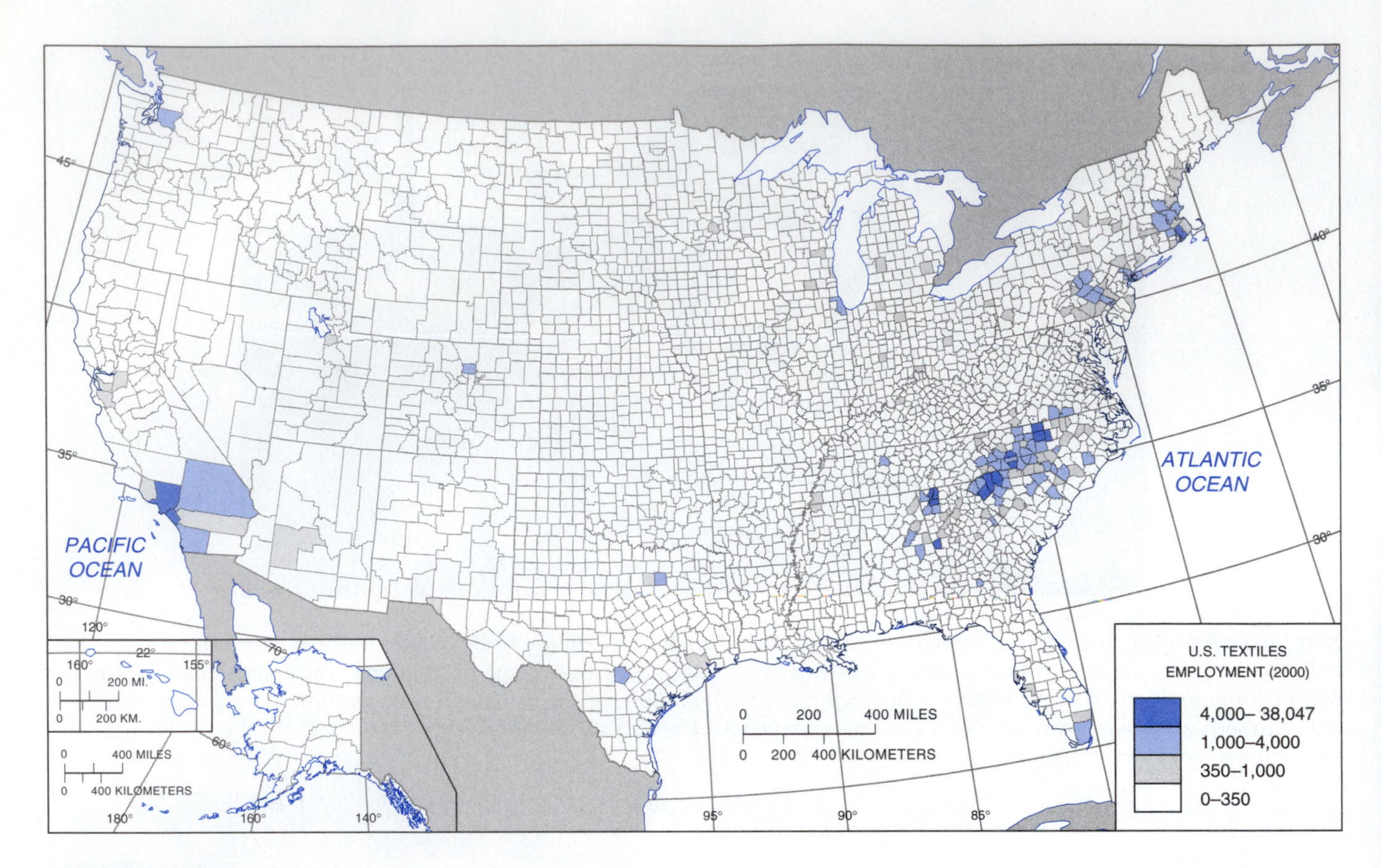

FIGURE 7.3
The distribution of U.S. textiles employment, 2000.

Europe

Europe has some of the world's most important industrial regions (Figure 7.5). They are located in a north-south linear pattern, starting from Scotland and extending through southern England, continuing from the mouth of the Rhine River valley in the Netherlands, through Germany and France, to northern Italy. Good supplies of iron ore and coal provide fuel to the countries in these industrial regions.

The Industrial Revolution started in the United Kingdom in the mid-eighteenth century. It had its basis in textile and woolen manufacture. Because many dependent nations have since learned to produce their own iron, steel, and textiles, the world currently has an oversupply of these items, and the market for British goods has decreased substantially. Britain's outmoded factories, high labor costs, slow productivity growth, and deteriorating infrastructures reduced its overall global competitiveness for products. In contrast, Germany and

San Jose, California. A $1 billion downtown renaissance has helped establish San Jose, capital of Silicon Valley, as California's third and the United States' eleventh largst city.

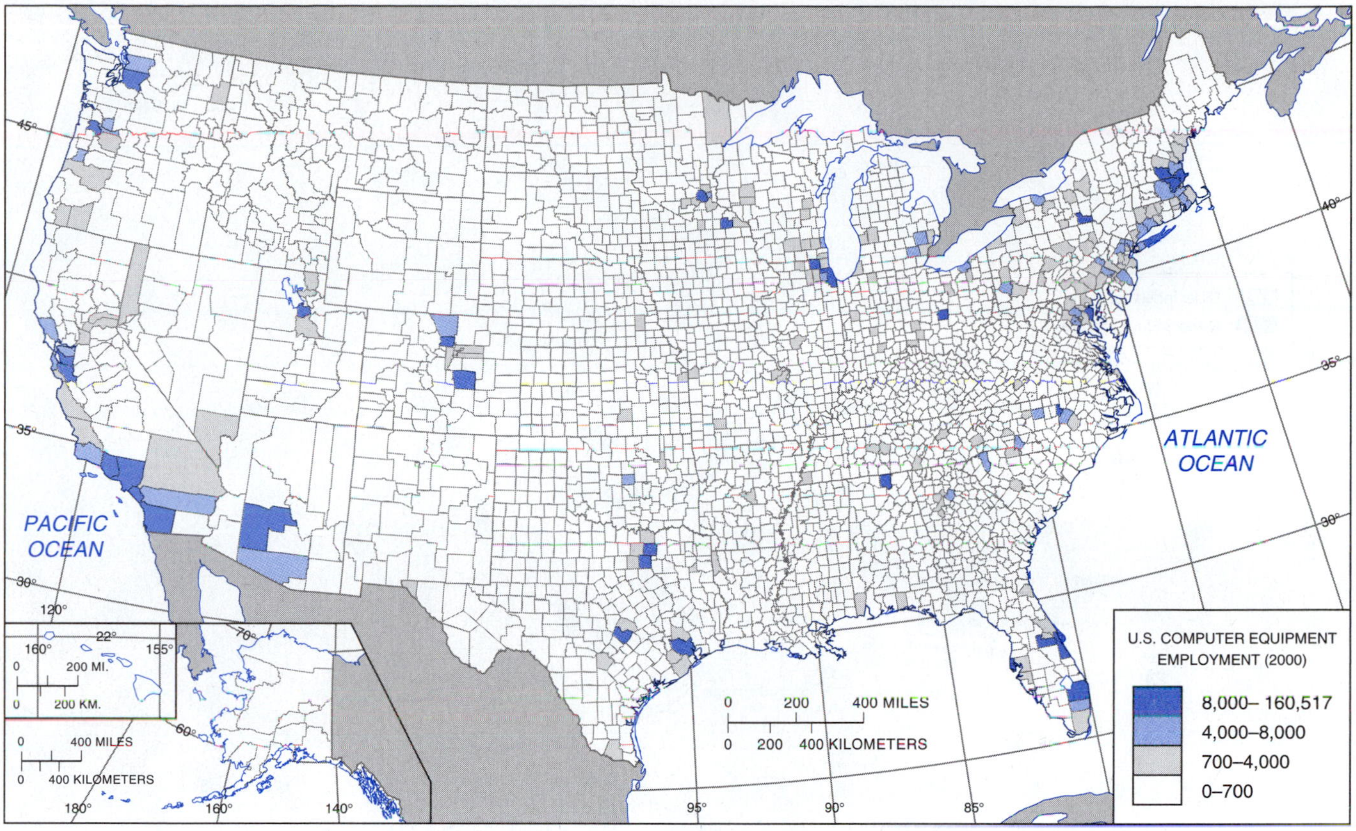

FIGURE 7.4

The distribution of U.S. computer equipment employment, 2000. California has the largest number of firms, followed by a cluster in the Northeast. Electronics manufacturing plants gravitate to highly skilled labor. The finished products are more valuable than are items of clothing, and higher wages can be paid. There is a concentration of employment in California because early electronics production was linked to the aircraft manufacturing industry centered on the West Coast; the microelectronics revolution and the rise of Silicon Valley furthered this growth.

Japan, with U.S. assistance, rebuilt after World War II, modernizing their plants and industrial processes at the same time. As a result, Germany and Japan are industrial successes in the world today, whereas Britain, more so than any other modern industrialized country, has suffered persistent industrial decline.

The largest European manufacturing region today is in the northern European lowland countries of Belgium and the Netherlands, northwestern Germany, and northeastern France. In this region, the Rhine and Ruhr Rivers meet. This region's backbone has been the iron and steel industry because of its proximity to coal and iron-ore fields. Production of transportation equipment, machinery, and chemicals helped lead Western Europe into the industrial age long before the rest of the world. The Rhine River, which is the main waterway of European commerce, empties into the North Sea in the Dutch city of Rotterdam. Consequently, because of its excellent location, Rotterdam has become the world's largest port, although it has been surpassed by those in East Asia. While exports of iron and steel are down from what they were 30 years ago, the region has been better able to avoid the depression of the United Kingdom because of its greater internal conversion of steel into high-quality finished products, which are in demand worldwide.

The Upper Rhine–Alsace-Lorraine Region (called Mid-Rhine in Figure 7.5) is in southwestern Germany and eastern France. It is the second most important European industrial district, after the Rhine–Ruhr River valley. Because of its central location, this area is well situated for distribution to population centers throughout western Europe. The main cities on the German side include Frankfurt, Stuttgart, and Mannheim. Frankfurt became the financial and commercial center of Germany's railway, air, and road networks. Stuttgart, on the other hand, is a center for precision goods and high-value, volume, manufactured goods, including the Mercedes Benz, Porsche, and Audi automobiles. Mannheim, located along the Rhine River, is noted for its chemicals, pharmaceuticals, and inland port facilities. The western side of this district, in France, is known as Alsace-Lorraine and produces a large portion of the district's iron and steel.

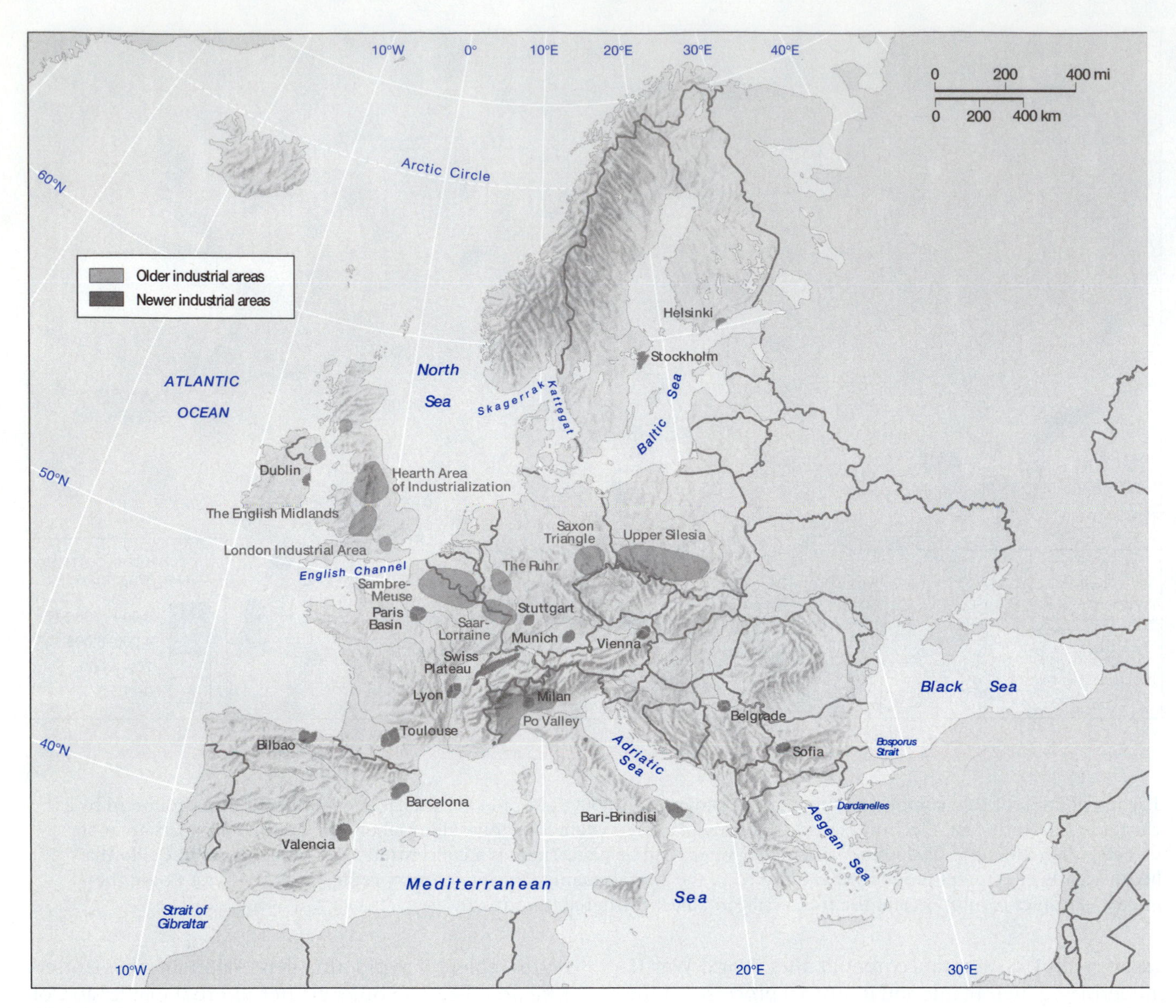

FIGURE 7.5

European manufacturing regions. Much European manufacturing exists in a linear belt from Scotland to the Midlands of Britain, to the South, including the London area. This belt continues onward from the low countries of Belgium, Luxembourg, and the Netherlands, south along the Rhine River, including portions of France and Germany, and into northern Italy. These areas became major manufacturing regions not only because of the concentration of skilled laborers but also due to the availability of raw materials, principally coal and iron ore. In addition, good river transportation was available, as well as large consuming markets for finished products.

The Po River valley in northern Italy includes Turin, Milan, and Genoa, including only one-fourth of Italy's land but more than 70 percent of its industries and 50 percent of its population. This region specializes in iron and steel, transportation equipment (especially high-value automobiles), textile manufacture, and food processing. The Emilia-Romagha region is Italy's high technology center. The Alps, a barrier to the German and British industrial regions, give the Italian district a large share of the southern European markets. The mountains also provide Italian industries with cheap hydroelectricity and therefore reduced operating costs. Compared with workers in the American Manufacturing Belt, Italian laborers are willing to work for lower wages, and thus this region attracts labor-intensive industries, such as textiles, from northern Europe.

The Ukraine industrial region relies on the rich coalfield deposits of the Donets Basin. The iron and steel industry base is the city of Krivoy Rog, with nearby Odessa as the principal Black Sea all-weather port (Figure 7.6).

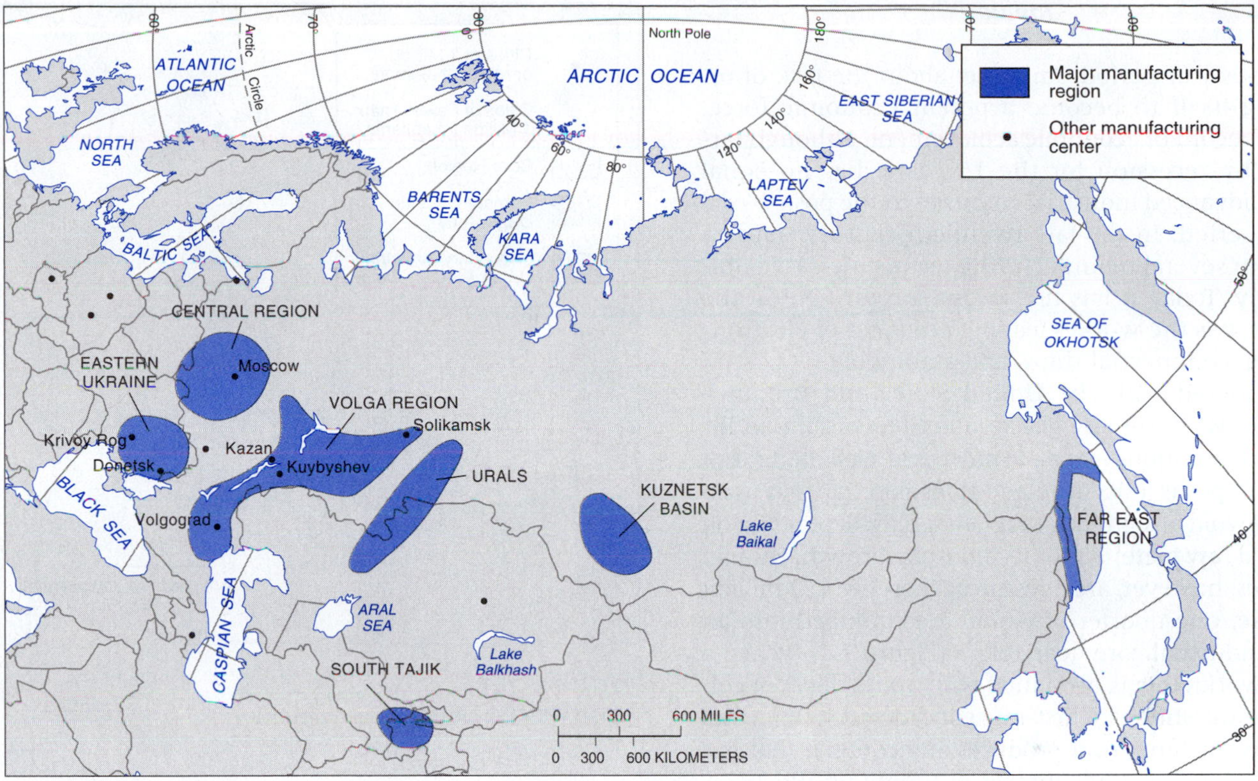

FIGURE 7.6
Manufacturing regions of Russia, Ukraine, and Central Asia.

The area is collectively known as Donbass. Like the German Ruhr area, Ukraine's industrial district is near iron-ore and coal mines, a dense population, and a large agricultural region and is served by good transportation facilities.

In Russia, the Moscow industrial region is near the population center of Russia and takes advantage of a large, skilled labor pool as well as a large market, even though natural resources are not plentiful. The largest single item produced is textiles: linen, cotton, wool, and silk fabrics. East of the Moscow industrial region is the linear Volga region, extending northward from Volgograd (formerly Stalingrad) and astride the Volga River. The Volga River, a chief waterway of Russia, has been linked via canal to the Don River and thereby to the Black Sea. This industrial region developed during and after World War II because it was just out of reach of the invading Nazi army that occupied the Ukraine. It is the principal location of substantial oil and gas production and refining. Recently, a larger oil and gas field was discovered in West Siberia. Nonetheless, the Volga district is Russia's chief supplier of oil and gas, chemicals, and related products. Recently, one of the largest automobile plants in the world opened in Toglaitti, producing Fiat automobiles.

Just east of the Volga region are the low-lying Ural Mountains that separate European Russia from Asian Russia (Figure 7.6). The Ural Mountains have the largest deposits of industrial minerals found anywhere in the former Soviet Union. Mineral types include iron, copper, potassium, magnesium, salt, tungsten, and bauxite. The central-lying Ural district was important during World War II because it also was beyond the reach of the German army. Although coal must be imported from the nearby Volga district, the Urals district provides Russia with iron and steel, chemicals, machinery, and fabricated metal.

The Kuznetsk Basin—also called *Kuzbass* and centered on the towns of Novosibirsk, along the trans-Siberian railroad, and Novokuznetsk—is the chief industrial region of Russia east of the Urals. Again, as in the case of the Ukraine and the Urals districts, the Kuznetsk industrial district relies on an abundant supply of iron ore and the largest supply of coal in the country. The Kuznetsk Basin is a result of the grand design of former Soviet city planners. These planners poured heavy investments into this region, hoping that it would become self-perpetuating and eventually the industrial supply region for Soviet Central Asia and Siberia.

Japan

In the postwar period, Japan set about the task of rebuilding itself to become a potent economic force. Japan's record of economic achievement, although tarnished by recession for the 15 years, has no equal among advanced industrial countries in the post–World War II period. In the late twentieth century, Japan's output in several manufacturing sectors increased dramatically. Today it has the second largest GNP in the world and is the world's leading producer of electronics, steel, commercial ships, and automobiles.

Compared with the United States and Britain—countries with the physical resources to sustain an industrial revolution—Japan is much less well off. Except for coal deposits in Kyushu and Hokkaido, Japan is practically devoid of significant raw materials, depending on imported raw materials for its industrial growth. Human resources, however, are not scarce. There are 127 million Japanese, one-quarter of whom are crammed into an urban-industrial core near Tokyo (Figure 7.7). When a strong work ethic is combined with a high level of collective commitment, a first-rate educational system, and government support, it produces an economy that has enjoyed significant advantages in the world economy, although Japan has suffered with the crash of the "bubble economy" in the 1990s.

Although permanent workers in large firms are well paid, especially when viewed in relation to part-time workers in small- and medium-sized firms, the relative cost of labor is lower in Japan many other industrial countries. Savings and profits are directed by the state toward whatever goals are set forth by a unique collaborative partnership between MITI—Japan's Ministry of International Trade and Industry—and private enterprise. Under MITI's guidance, Japan has relocated industries such as steel making and shipbuilding "offshore" in the newly industrializing countries (NICs), where labor costs are lower. Now, countries such as South Korea and Singapore have developed their own higher valued-added industries (e.g., consumer electronics).

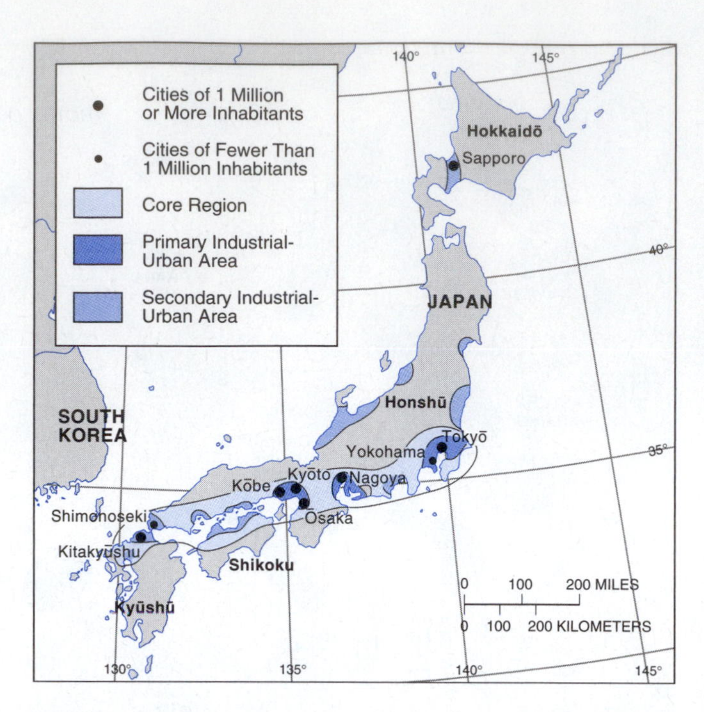

FIGURE 7.7
Japan's core region and selected cities.

THE GLOBALIZATION OF MANUFACTURING

As capitalism underwent one of its period restructurings, the new *international division of labor* asserted itself in the 1970s and 1980s, the rate of world economic growth declined, a process that coincided with the end of the long boom period after World War II. The world economic crisis started with a deep recession in 1974–1975 following the first oil shock in 1973. One of its most visible effects in the advanced economies was deindustrialization—reflected by the loss of manufacturing jobs. As firms restructured or went out of business in a climate of intense international competition, workers were laid off.

Dramatic changes in the geography of world manufacturing accelerated in the crisis of the 1970s and 1980s. Although the advanced countries maintained a huge share in world manufacturing output, their output grew less quickly than that of the less developed countries. The manufacturing output growth of the advanced economies slowed dramatically in the 1970s, and the number of workers in manufacturing in the advanced countries declined. Britain lost 28 percent of its manufacturing workers between 1974 and 1983; West Germany lost 16 percent; France, 14 percent; and the United States, 8 percent. The highest rate of manufacturing job loss in the United States was in the Midwest—more than 11 percent between 1975 and 1982. However, manufacturing employment increased in new industrial areas of the Sunbelt and areas successful in restructuring their industrial bases, such as New England.

Since 1960, manufacturing output has increased sharply in lower-wage industrializing periphery countries. Between 1974 and 2000, the advanced industrial countries lost 24 million jobs, whereas the newly industrializing countries gained 19 million jobs. Jobs lost in the advanced industrial countries paid from $9 to $31 an hour, but those gained in the newly industrializing countries typically paid only $8 per hour or less. The gains from expansion in the newly industrializing countries were more than offset by the losses in the advanced industrial countries. Indeed, the shift

In this textile factory in Fortaleza, Brazil, women constitute the largest part of the workforce. Brazil is a major exporter of textiles to advanced industrial countries. Despite high rates of growth in many Third World countries, only a handful of economies in East Asia actually managed to narrow the gap with the northern, industrialized countries. Moreover, increased polarization between the richer and poorer countries has been accompanied by a rising trend in income inequality within countries such as in Brazil. The income share of the richest 20 percent has grown almost everywhere, while those in the lower ranks have experienced no rise in incomes to speak of. Even the middle classes have experienced little improvement.

led to lower global wage shares that may contribute to stagnation.

The most rapid growth of manufacturing output occurred in East and South Asia, the world region with the fastest rate of economic growth since World War II. Japan, with high wages but relatively low labor militancy, set the model for much of this region, a phenomenon sometimes called the "flying geese formation." This notion compares industrialization to a flock of birds flying in a V formation, with Japan at the head, the NICs closely behind, and other countries lagging somewhat behind. Several newly industrializing countries equaled or exceeded Japan's annual growth rate in manufacturing output, including South Korea, Taiwan, Hong Kong, and Singapore.

The traditional view is that deindustrialization in some places and industrialization in others are mirror images of each other. Industrial growth and decline are offsetting tendencies, representing a zero-sum, or even a positive, global game. The shift of production processes from the industrial heartland to the periphery releases a skilled labor force for more sophisticated forms of production in developed countries and allows labor in the developing countries to move from relatively unproductive employment to more highly productive employment in industry. The shift may lead to some transitional unemployment, but job losses in the industrial heartland are of little significance compared with the enormous rewards attached to a global reallocation of production.

Nomadic capital, although it may serve individual company interests, can be socially inefficient. Those who hire labor control the work process, and distribution is always in favor of those who control the production location. Corporate allocation of production and investment is guided primarily by profitability concerns, where profitability is determined by the price of labor and the amount of work that can be extracted at this price. Nomadic capital can also be socially inefficient because giant corporations are rarely faced with the full social costs of their locational decisions. Shifting production from country to country means that the advanced industrial countries must absorb not only most of the social costs of communities that are now abandoned because they can no longer

Manufacturing Reeboks at the Lotus Plant, Philippines.

be industrially competitive but also the costs of the social infrastructure required by the newly industrializing countries.

Globalization of Major Manufacturing Sectors

Globalization occurs differently in different industries. Because various forms of manufacturing have their own specific technical, labor, and locational requirements, and because nation-states have approached different industries from a variety of regulatory perspectives, the formation of worldwide production changes is unique to each industry. Four industries—textiles and garments, steel, automobiles, and electronics—dramatize the similarities and differences that exist among manufacturing sectors as they adapt to the realities of a global market.

Textiles and Garments

The textile and garment industries dramatically reveal the globalization of manufacturing. Textile manufacture is the creation of cloth and fabric, whereas clothing manufacture uses textiles to produce wearing apparel. Textiles and garments comprise the classic low-tech industry: labor intensive, with relatively little technological sophistication, small firms, and few economies of scale. Textiles were the leading sector of the Industrial Revolution, not only in Britain but also in the remainder of Europe, the United States, and Japan. By continually looking for new sources of cheap, easily exploitable labor, this industry has exhibited a very fluid geography over time, a pattern that continues with its globalization and entry into the developing world in the late twentieth century.

Textile and garment production in the economically developed world declined steadily from the 1970s onward as the industry was confronted with cheap imports from abroad. Conversely, in the developing nations, especially in Asia, production and employment increased significantly. Today, major cotton fabric producers include China, where the garment industry has grown explosively, and India. In Asia, the industry is notorious for its exploitation of young, predominantly female labor, many of whom work long hours for very low wages. Other significant producers in the world are the United States, Russia, Brazil, and Italy (Figure 7.8).

Steel

The steel industry has played an enormously important role in the development of industrialized societies. Iron and steel production generate a wide variety of outputs essential to many other sectors as well, including, for example, parts for automobiles, ships, and aircraft; steel girders for buildings, dams, and other large projects; industrial and agricultural machinery; pipes, tubing, wire, and tools; furniture; and many others uses. The inputs into steel making are relatively simple, including iron ore, which is purified into pig iron; large amounts of energy, generally in the form of coal; and limestone, which is used in the purification process. Steel production is a very capital-intensive process, requiring huge sums of investment, and because the barriers to entry are high, it has generally been very oligopolistic. Transport costs have traditionally been high in this sector, which has made it an ideal candidate for Weberian locational analysis (Chapter 5).

The historical geography of steel production includes the important Midlands cities of Britain, such as Sheffield and Birmingham, which were critical in the early stages of the Industrial Revolution. A similar complex of steel production arose in the Ruhr region

Steel exemplifies forms of manufacturing that are highly capital intensive, relying on huge economies of scale to produce their output profitably.

COTTON WOVEN FABRIC (IN MILLION SQUARE METERS)

10,000 and above
1,000–9,999
100–999
1–99
Below 1

MODIFIED GOODE'S HOMOLOSINE EQUAL-AREA PROJECTION

FIGURE 7.8
Working conditions in many contemporary textile plants resemble the conditions Charles Dickens described in 19th century London.

in western Germany on the banks of the Rhine River. In the northeastern United States, the earliest iron and steel producers were very small and localized, using wood and charcoal as fuel and serving local markets. A number of changes in the third Kondratiev wave of the late nineteenth century, however, dramatically reshaped the industry. The Bessemer open hearth furnace made the production of steel relatively cheap, using huge amounts of energy in plants that were open 24 hours per day. Geographically, the industry came to center on the steel towns of the Manufacturing Belt, including, above all, Pittsburgh, but also Hamilton, Ontario; Buffalo, New York; Youngstown and Cleveland, Ohio; Gary, Indiana; and Chicago (Figure 7.9). These locations allowed easy access to coal from Appalachia, iron ore from Minnesota and upper Michigan, and cheap transportation via the Great Lakes and the railroads. The rise of the U.S. Steel Company under Andrew Carnegie saw 30 percent of the nation's steel output in the hands of one firm in 1900.

The United States dominated the world's steel industry in the early and mid-twentieth century, producing as much as 63 percent of the world's total output after World War II. However, the rise of new competitors, first in the rebuilt factories of Europe (particularly Germany, France, and Spain) and Japan, and later in some developing countries (e.g., Brazil, South Korea) (Figure 7.10), saw a gradual decline in the share of world steel produced in the United States to 8.3 percent of the world total in 2005 (Figure 7.11).

The decline of the U.S. steel industry generated enormous problems for families living in the former steel towns. Waves of plant closures in the 1970s and

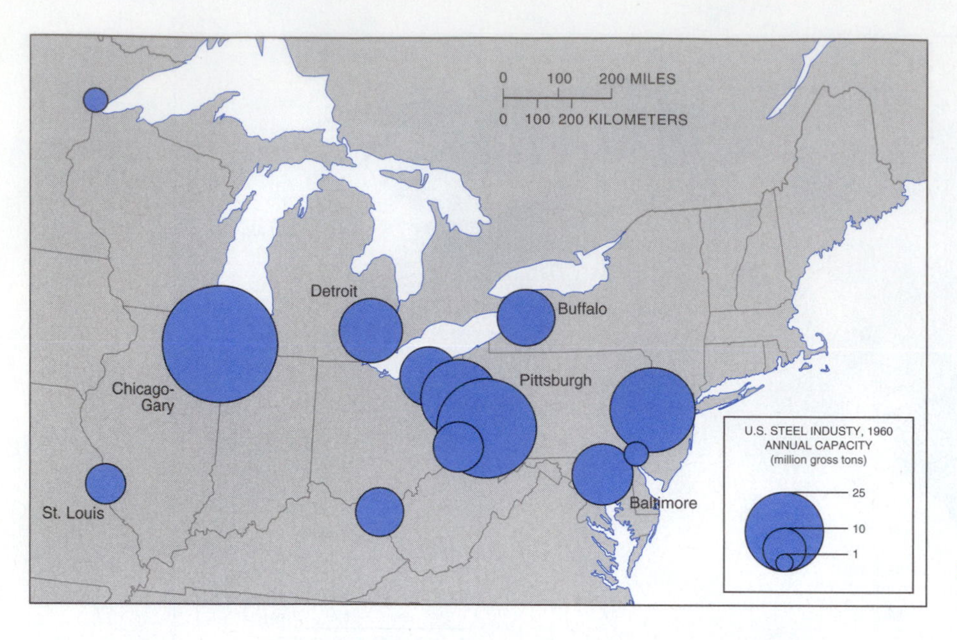

FIGURE 7.9
U.S. steel employment cities, 1950.

1980s generated high unemployment, rising poverty, depressed property values, and out-migration. Today, former giants of steel production in the Manufacturing Belt produce very little steel; Pittsburgh produces none at all. The industry's response to crisis, other than plant closures, was both to call for protectionism from imports and to restructure. The introduction of computerized technology led to widespread changes in steel production, including the emergence of highly automated "minimills" that use scrap metal as inputs (Figure 7.12), are not generally unionized, and produced specialized outputs for niche markets, all of which are symptomatic of the growth of post-Fordist production to be discussed later in this chapter.

Automobiles

From its inception at the dawn of the twentieth century, the automobile industry has unleashed major effects on cities and everyday life throughout the world. Originally Henry Ford standardized a European invention, auto production, when the industry centered upon the Detroit region, adding another layer of investment to the Manufacturing Belt. Ford introduced the moving assembly line and a highly detailed division of labor to make automobiles affordable to the middle class. He also raised wages to reduce turnover rates for his workers.

The automobile industry today comprises giant transnational corporations. Highly capital intensive, with high start-up costs, has a long history of oligopolization. Today, in no other industry do so few companies dominate the world scene. For example, the world's 10 leading automobile manufacturers produce over 70 percent of the world's automobiles. Each of these, from General Motors and Ford to the smallest automobile producer, has foreign assembly plants in other countries. Many have full-blown vertically integrated manufacturing operations, where all parts in the final assembly are foreign supplied.

Worldwide automobile production has increased rapidly. Between 1960 and 2004, there was a worldwide increase of 400 percent. Figure 7.13 shows the 2004 world distribution of automobile production and assembly. Three major nodes of automobile production exist—Japan, the United States, and Europe. In 2004, Europe accounted for 27 percent of the world's automobiles; Japan, 20 percent; and the United States,

Japan utilizes one-half of all industrial robots in the world, an example of the increasing capital intensity of the production process, particularly in manufacturing.

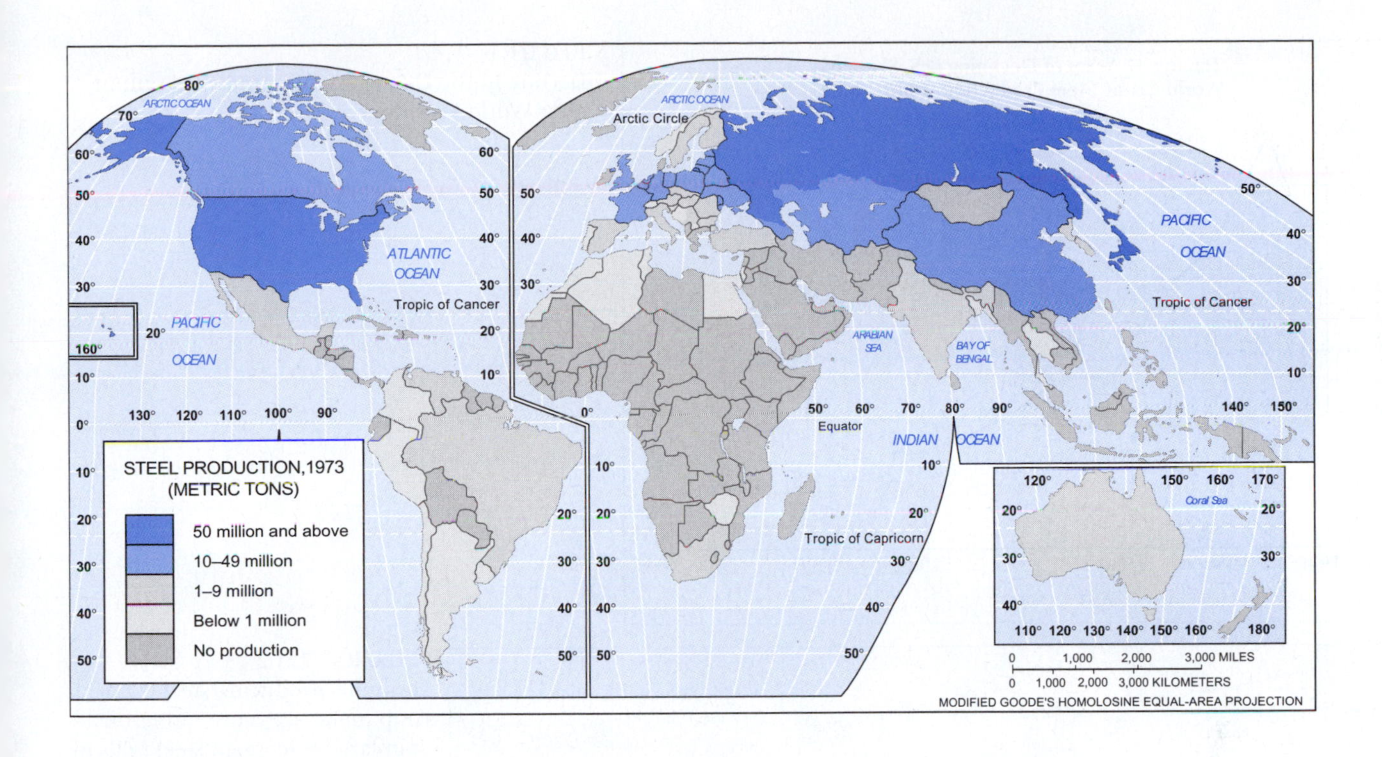

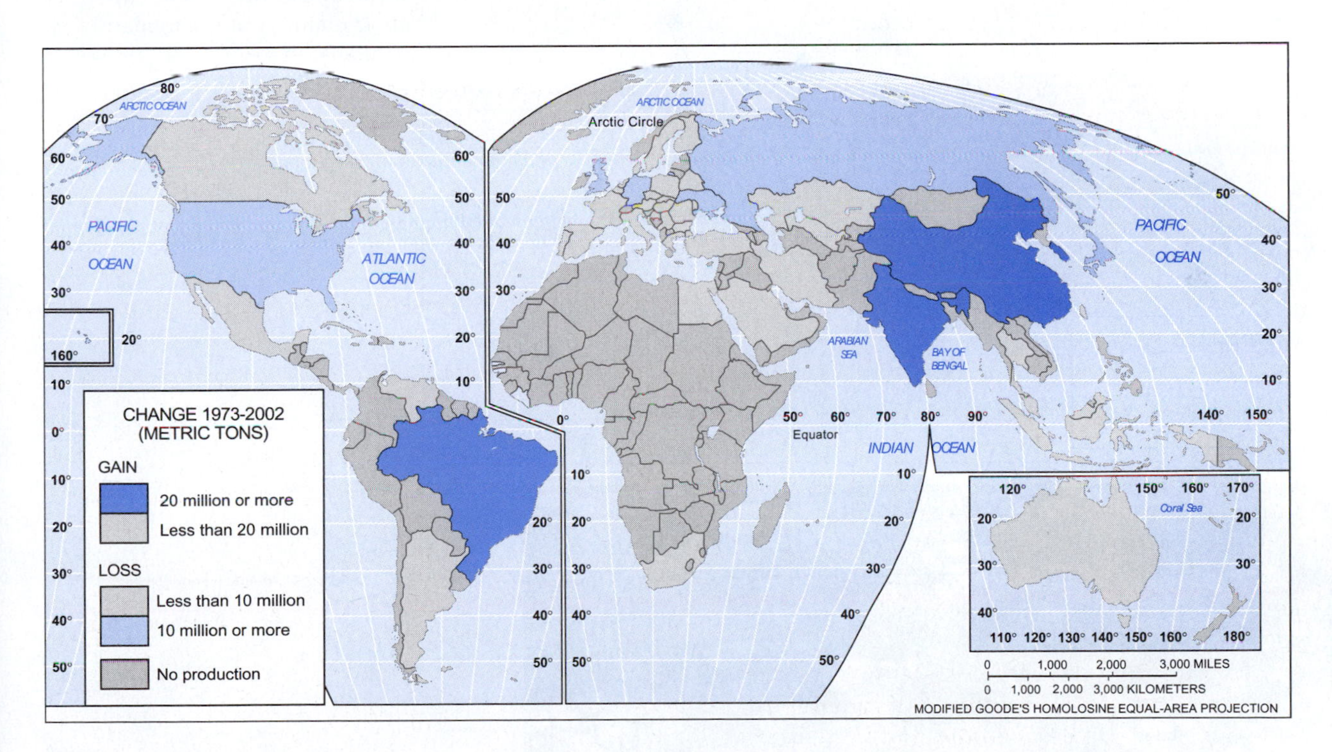

FIGURE 7.10
Worldwide steel production shifts between 1973 and 2002. In 1973, North America and Europe accounted for 90 percent of the world's steel output. Production by 2002 had shifted dramatically to developing countries in Latin American, South Asia, and East Asia. Global steel production remained approximately the same during this period.

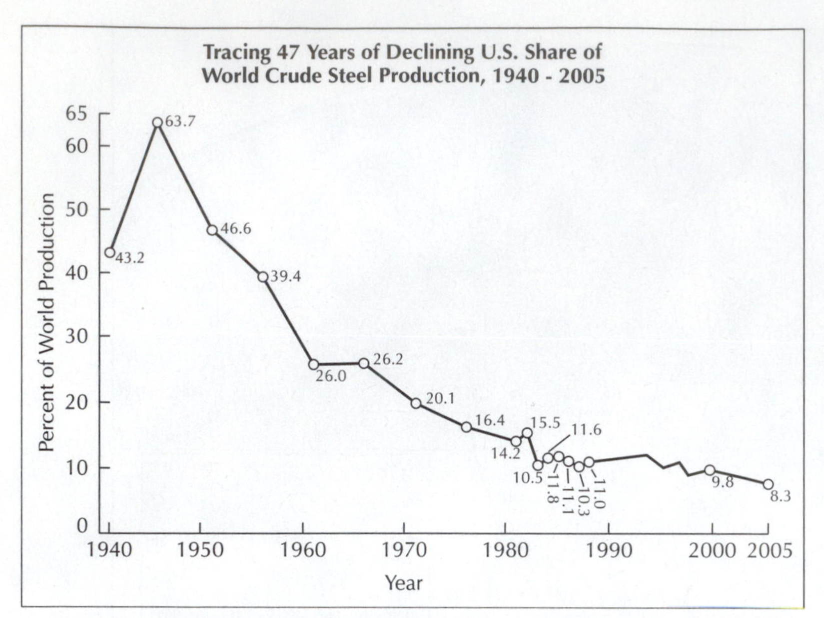

FIGURE 7.11
Decline in the U.S. share of world steel output after World War II.

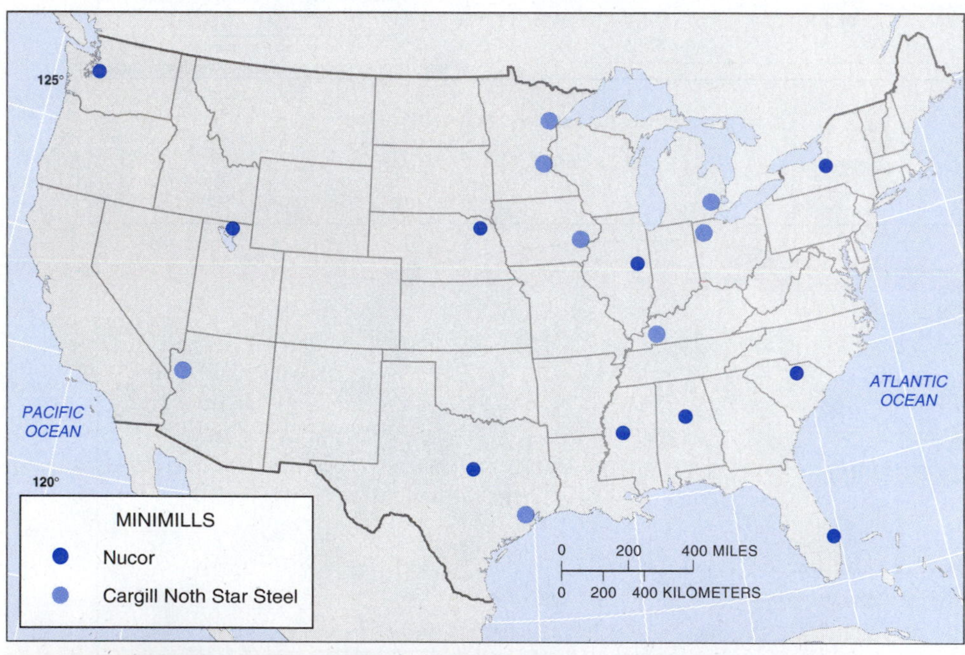

FIGURE 7.12
Minimills producing steel from scrap metal are more numerous than large integrated steel mills in the United States. They are located near markets because their main input is abundantly available there.

Firmly linked to the global economy by its reliance upon international trade, Japan's economy is heavily export-oriented as seen by these cars waiting on loading docks.

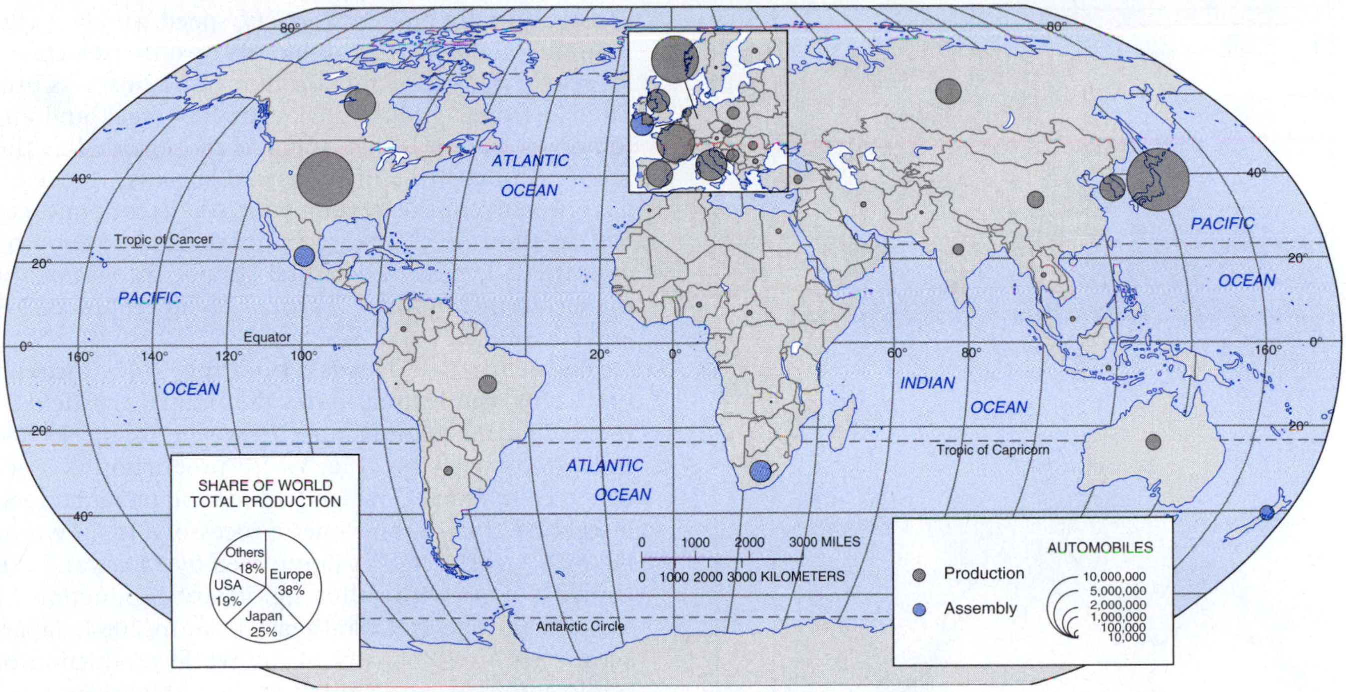

FIGURE 7.13
World distribution of automobile production and assembly. In 2002, Japan produced 10.3 million automobiles, which was 19 percent of the world's total output. In 1960, the United States produced half of the world's total automobiles, but by 2004, that proportion had dropped to 16 percent. In Europe, Germany, France, and Italy are the largest producers.

21 percent. Smaller production centers exist in Brazil, Russia, and Australia. The three developed regions of the world, East Asia, North America, and Europe, accounted for 72 percent of the automobiles produced. Unlike the trend in the textile and clothing industries, the developed countries clearly cornered the market in automobile manufacture. Only a few developing economies have shown a significant increase in automobile assembly but not in their full-scale production.

The most dramatic shift in the automobile industry was the tremendous increase in Japan's productivity between 1960 and 2000, from 165,000 to over 10 million cars annually, almost one-quarter of the world's total output, and almost surpassing U.S. output. In 1960, the United States produced more than half the world's automobiles, but by 2005, only 21 percent. Japanese firms also invested heavily in the United States in the late twentieth century, setting up factories in much of the Midwest (Figure 7.14) that allowed them access to the American market, low transport costs, and freedom from fluctuating exchange rates and threats of U.S. protectionism.

Electronics

Although its roots extend into the nineteenth century, the electronics industry underwent enormous changes with the microelectronics revolution of the late twentieth century. Microelectronic technology is the dominant technology of the present historical moment, transforming all branches of the economy and many aspects of society.

The radio was invented and produced as early as 1901, but the modern electronics industry was not born until Bell Telephone Laboratories built the transistor in the United States in 1948. The transistor supplanted the vacuum tube, which had been used in most radios, televisions, and other electronic instruments. The microelectronic transistor was a solid-state device made from silicon and acted as a semiconductor of electric current. By 1960, the *integrated circuit* was produced, which was a quantum improvement because transistors could be connected on a single small silicon chip. By the early 1970s, a computer so tiny that it could fit on a silicon chip the size of a fingernail came into production. Thus, the *microprocessor,* which could do the work of a roomful of vacuum tubes, was born. With these changes, information changed from analogue to digital format in binary code, which made it much easier for computers to use. The microprocessor made possible the microcomputer, which in turn revolutionized the collection and analysis of information, particularly office work (Figure 7.15).

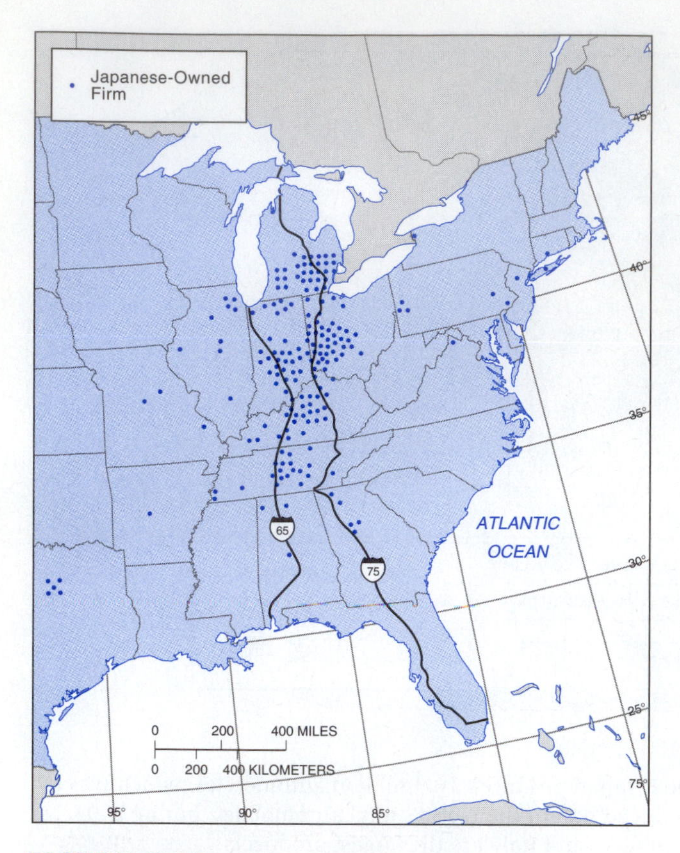

FIGURE 7.14
Distribution of Japanese-owned automobile parts manufacturing plants in the United States, 2000.

Increasing power and miniaturization progressed, and at the same time, new applications for the electronics industry were discovered, including calculators, electronic typewriters, computers, industrial robots, aircraft-guidance systems, and combat systems. New discoveries were applied to automobile construction for guidance, safety, speed, and fuel regulation. An entire new range of consumer electronics also became available for home and business use. The electronics industry, like textiles, steel, and automobiles before it, has come to be regarded as the modern touchstone of industrial success. Hence all governments in the developed market economies, as well as those in the more industrialized developing countries, operate substantial support programs for the electronics industry, particularly microprocessors and computers.

For nearly two decades, from the 1960s through the 1970s, the United States dominated the field of semiconductor manufacture. However, by the 1990s, Japan took over this role. World production of electronic components, which includes semiconductors, integrated circuits, and microprocessors, is shown in Figure 7.16. The field is dominated by Japan and the United States, with other significant production in western Europe and Southeast Asia. In 2004, Japan accounted for 40 percent of the world production of semiconductors; the United States, 21 percent; and Europe, 11 percent. In Southeast Asia, South Korea, Malaysia, Taiwan, Thailand, and Hong Kong were significant manufacturers.

The most important component of the semiconductor industry is computer memory, *RAM* (random access memory). Although the United States produced 100 percent of the world's total output in 1974, by 2004 it produced only approximately 15 percent, and Japan had claimed 70 percent. Similar to the shift in the automobile industry, there has been a tremendous global shift in the direction of East Asia, primarily to Japan, in RAM production.

The world manufacture of consumer electronics is much more widely spread than that of the semiconductor industry. Although in the semiconductor industry, the United States, Japan, and Europe account

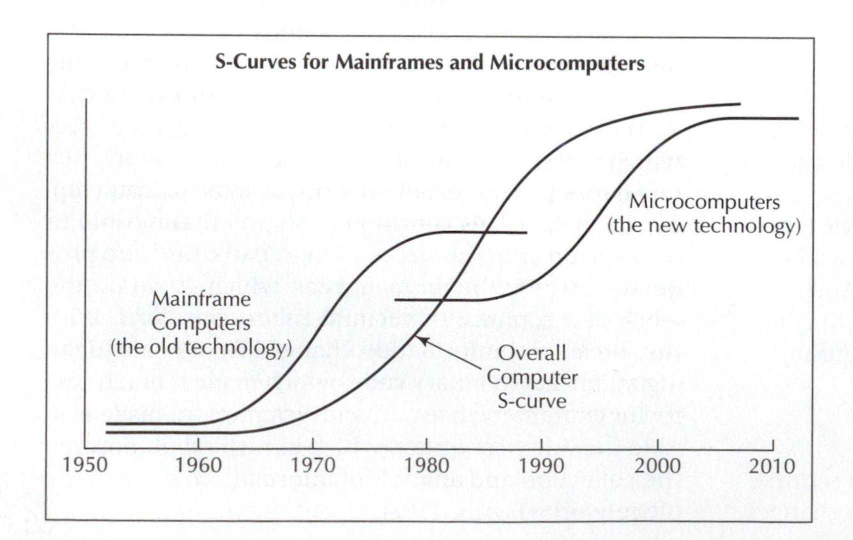

FIGURE 7.15
S-curves for mainframes and minicomputers reveal the transition of technologies within the computer industry. The microelectronics revolution had enormous effects, not only on industrial firms but also in agriculture and services, and generated much of the productivity increases of the United States in the late twentieth century.

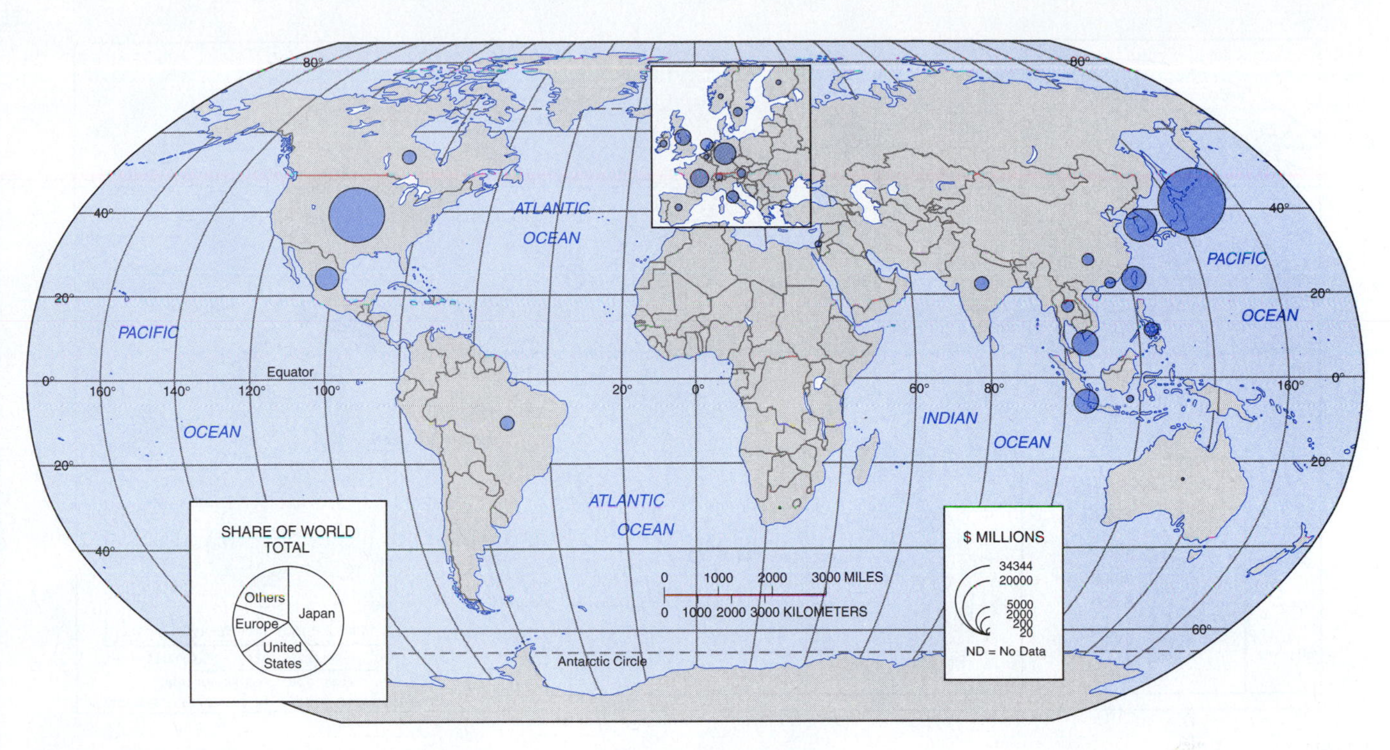

FIGURE 7.16
World production of electronic components, including semiconductors, integrated circuits, and microprocessors. In 2002, Japan produced 42 percent of the world's total electronic components, followed by the United States with 26 percent. Europe produced 12 percent of the world's total. Other leading producers included Germany, France, and the United Kingdom, with 31 percent, 19 percent, and 16 percent, respectively. Growth in the electronics industry has been greatest to the Pacific Rim with recent major centers of production developing in South Korea, Malaysia, Taiwan, and Thailand.

for 80% of the total world production, in the consumer electronics industry, these three regions account for much less. Developing countries, especially in East and Southeast Asia, are heavily involved in producing consumer electronics. Television production is an example: whereas the United States, Europe, and Japan combined produce only 30% of the world's televisions, China alone produces 26%, and South Korea produces another 14%. Singapore and Malaysia are also significant producers (Figure 7.17). As in the case of automobile and semiconductor manufacture, there has been a global shift from the developed nations to East Asia. Much of the television production that formerly occurred in the United States, Germany, and the United Kingdom now takes place in China and Malaysia. Outside Asia and North America, Brazil produces 87% of the televisions used in Latin America.

Biotechnology

Biotechnology may be defined as the application of molecular and cellular processes to solve problems, develop products and services, or modify living organisms to carry desired traits. Arising after the discovery in 1973 of recombinant DNA, biotechnology has been a rapidly growing industry worldwide, with extensive linkages to agriculture, health care, energy, and environmental sciences. In 2003, the U.S. biotech industry (excluding medical equipment firms) consisted of roughly 1,473 firms that employed 406,000 people and generated roughly $64 billion in output. There is a wide range in the size of firms in this industry, including single proprietorships and firms of more than 500 employees; the mean national annual salary in the industry is $62,500, which is well above the national average.

Pharmaceutical firms, which tend to be much larger than biotechnology companies, form the major market for biotechnology products. Large pharmaceutical firms are reliant upon biotech clusters for innovative drug solutions, and human therapeutics thus account for the vast bulk of the biotechnology industry's revenues. Other applications are found in agriculture, industry, and veterinary medicine. Many biotech firms enter into alliances with drug manufacturers, who may provide venture capital in return for marketing rights after the product is commercialized.

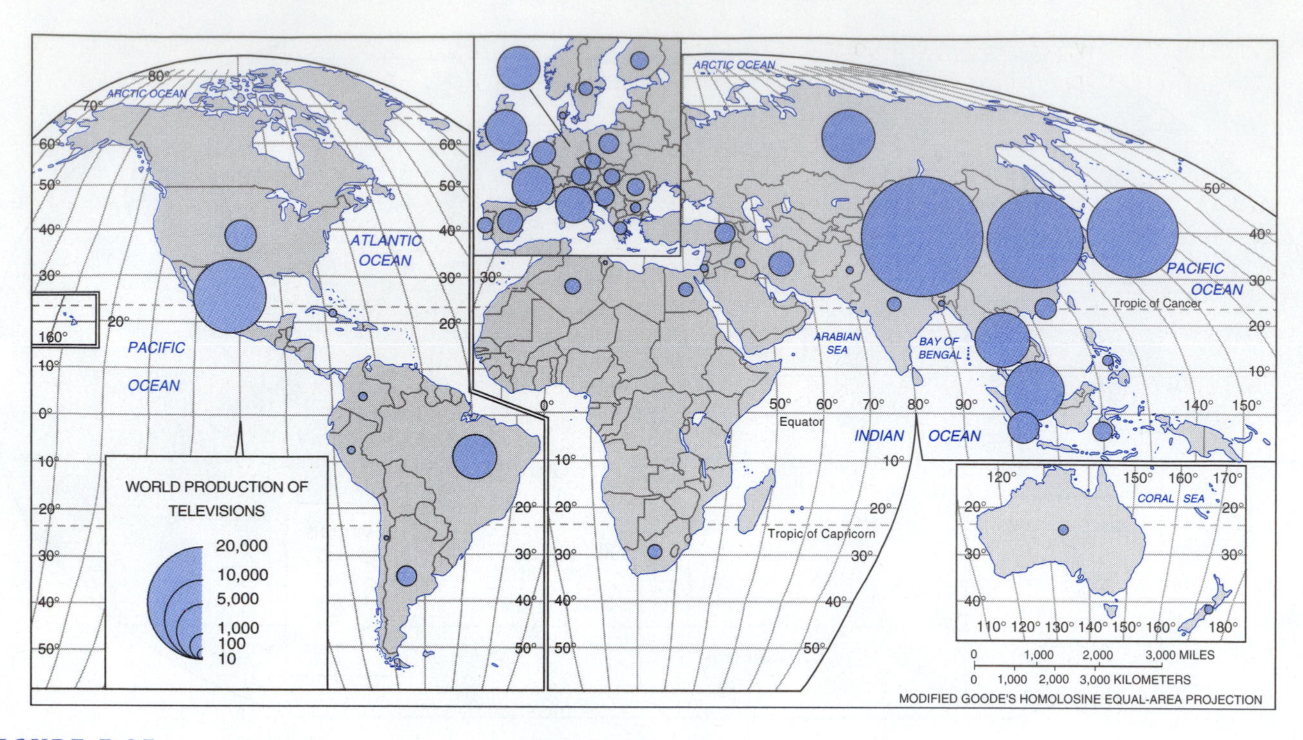

FIGURE 7.17
Worldwide television receiver production, 2000. East Asia has become the leading area in the manufacture of this commodity.

Venture capital is critical to making basic research in biotechnology commercial viable. Most small biotech firms lose money, given the high costs and enormous amounts of research necessary to generate their output and the long lag between R&D and commercial deployment (generally on the order of 12–15 years of preclinical development). Only one in 1,000 patented biotechnology innovations leads to a successful commercial product, and it may take 15 years. Venture capitalists may invest in many biotech firms, and one biotech firm may receive funding from several venture capitalists. Above all, venture capitalists look for an experienced management team when deciding in which companies they are willing to invest. Venture capitalists often provide advice and professional contacts, and serve on the boards of directors of young biotech firms. As a biotech firm survives and prospers, its relations with investors often become spatially attenuated, that is, venture capitalists gradually withdraw from day-to-day direct management.

There has been extensive state involvement in establishing biotechnology complexes since the industry began. Because of its rapid growth as well as demonstrated and potential technological advances, many national science policies target it as a national growth sector. The survival and success of biotechnology firms is heavily affected by federal research funds, primarily through institutions such as the National Science Foundation (NSF) and National Institute of Health (NIH). Other federal offices such as the Small Business Technology Transfer (STTR), Small Business Innovation Research (SBIR), Environmental Protection Agency, and the Food and Drug Administration are also significant. Federal policies regarding patents and intellectual property rights, subsidies for medical research, and national health care programs are all important. Roughly 83 percent of state and local economic development agencies have targeted the industry for industrial development. State level determinants are also critical, including regulatory policies, educational systems, taxation, and subsidies.

Biotechnology firms tend to cluster in distinct districts, and place-based characteristics are essential for the industry's success in innovation. Europe, for example, hosts the BioValley Network situated between France, Germany, and Switzerland. In Britain, Cambridge has assumed this role. Similarly, Denmark and Sweden formed Medicon Valley.

Geographically, the U.S. biotechnology industry is currently dominated by a small handful of cities, particularly Boston, San Diego, Los Angeles, San Francisco, New York, Philadelphia, Seattle, Raleigh-Durham, and

Washington, D.C., which together account for three-quarters of the nation's biotech firms and employment (Figure 7.18). All these cities have excellent universities with medical schools, state-of-the-art infrastructures (particularly fiber optics, airports), and offer an array of social and recreational environments.

Biotechnology firms tend to agglomerate for several reasons. In an industry so heavily research intensive, the knowledge base is complex and rapidly expanding, expertise is dispersed, and innovation is to be found in networks of learning rather than individual firms. This observation is at odds with the popular misconception that such firms are the products of heroic individual entrepreneurs. In a highly competitive environment in which the key to success is the rate of new product formation, and in which patent protections lead to a "winner take all" scenario, the success of biotechnology firms is closely related to their strategic alliances with universities and pharmaceutical firms. Although many biotechnology firms engage in long distance partnering, these tend to be complementary, not substitutes for, colocation in clusters where tacit knowledge is produced and circulated face to face, both on and off the job.

Because pools of specialized skills and a scientifically talented workforce are essential to the long process of research and development in biotechnology, an essential element defining the locational needs of biotech firms is the location of research universities and institutions and the associated supply of research scientists. Most founders of biotech companies are research scientists with university positions. Regional human capital may be measured by examining the prevalence of bachelor's degrees, which indicate a region's educational attainment, and the location and size of regional universities that grant Ph.D. degrees in biology and related fields. However, to a large extent local labor shortages can be mitigated through in-migration. Because knowledge is generated and shared most efficiently within close loops of contact, the creation of localized pools of technical knowledge is highly dependent on the detailed divisions of labor and constant interactions of colleagues in different and related firms. Successful biotechnology firms often revolve around the presence of highly accomplished academic or scientific "stars" with the requisite technical and scientific skills but also the vision and personality to market it. Often such individuals begin in academia and move into the private sector.

THE CHANGING GEOGRAPHY OF U.S. MANUFACTURING

The United States experienced massive industrial devolution in the 1970s and 1980s, a period during which its share of world manufacturing output decreased significantly. This trend points to a more rapid growth of manufacturing output in other countries, especially Asia. Thus, deindustrialization within the United States was

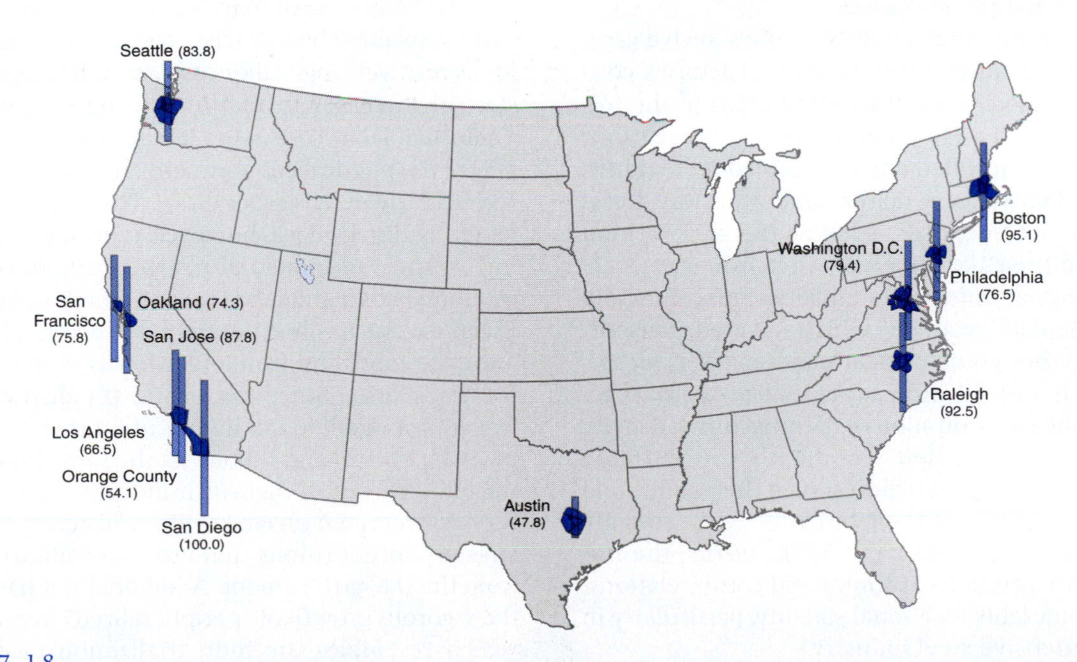

FIGURE 7.18
Major clusters of the U.S. biotechnology industry, 2004. The industry exhibits a pronounced tendency to cluster in major metropolitan areas, where access to universities and agglomeration economies is essential.

occurring at a time when American corporations were reacting to the prolonged economic crisis. Inside the United States, corporate profit rates were declining. As profit rates declined, corporations switched capital in space, going global in an effort to restore profitability. The effect of the globalization of manufacturing change has been most severe in the Manufacturing Belt, which lost millions of relatively well-paying jobs as factories closed, with devastating economic and social impacts on their communities.

North America has numerous manufacturing regions (see Figure 7.2). By far the largest is the Manufacturing Belt, which accounts for about 53 percent of the manufacturing capacity of the United States and Canada. The belt extends from the northeastern seaboard along the Great Lakes to Milwaukee, where it turns south to St. Louis, then extends eastward along the Ohio River valley to Washington, D.C. This great rectangle encompasses more than 10 districts, each with its own specialties that reflect the influences of markets, materials, labor, power, and historical forces.

The first major factories in the Belt—the textile mills of the 1830s and 1840s—clustered along the rivers of southern New England. When coal replaced water as a power source between 1850 and 1870, and when railroads integrated the Belt, factories were freed from the riverbanks. Between 1850 and 1870, many urban areas enjoyed rapidly expanding industrial production. Manufacturing employment in New York City, Philadelphia, and Chicago soared more than 200 percent between 1870 and 1900. The 10 largest industrial cities increased their share of national value added in manufacturing from less than 25 percent to almost 40 percent between 1860 and 1900.

Why did metropolitan complexes draw such a great proportion of manufacturing activity? Factories concentrated in large cities for a combination of the following reasons: (1) They could be near large labor pools, including unskilled and semiskilled immigrants; (2) they could secure easy railroad and waterway access to major resource deposits, such as the Appalachian coalfields and the Lake Superior—area iron mines; (3) they could be near industrial suppliers of machines and other intermediate products, which lowered transport costs; and (4) they could be near major markets for finished goods. In other words, *agglomeration economies* accounted for the concentration of manufacturing activity in the Manufacturing Belt. The highly concentrated pattern of industrial production served the nation (and much of the world) well for about 100 years—roughly the century between 1870 and 1970. *Inertia,* the immobility of the investment forces and social relations, ensured considerable locational stability, particularly in the capital-intensive steel industry.

However, cracks in this accumulation regime appeared as early as the late nineteenth century. Labor unrest intensified. After 1885, the number of strikes increased rapidly. Gradually, owners lost some power to labor, which allowed workers to negotiate higher wages, to organize high levels of unionization and extract better working conditions, and to command progressive welfare policies. By the early twentieth century, manufacturing started to move out of center cities to the suburbs, lured by the easy access afforded by the truck for freight transportation. But as transportation costs equalized across the nation, even the suburbs of the older manufacturing belt cities were unable to compete with more agreeable labor environments in the South and the West. The 1960s marked the start of the steady gain of manufacturing employment in the Sunbelt, including the South and parts of the Southwest. Since the early 1980s, however, there has been evidence that these areas where class conflict is low are being bypassed in favor of even cheaper labor regions in Mexico and East Asia.

The change in the location of manufacturing in the United States was particularly pronounced in the 1970s and 1980s. Virtually all states in the Manufacturing Belt experienced manufacturing job loss, and virtually all states in the South and the West registered manufacturing job gains (Figure 7.19). Labor costs in the South were often lower because it was less skilled and less unionized, in part because Southern states are "right to work" states, meaning that unions cannot force employees at an establishment to join. The migration of employment from areas of high labor costs to areas where the labor costs were less saved companies billions of dollars. From 1960 to 1980, roughly 1.7 million jobs shifted from states with high labor costs to states with low labor costs.

The expansion of manufacturing in the West is not easily explained by low labor costs. California is known for its relatively high labor costs, yet it has registered substantial increases in manufacturing employment. In California, labor costs were less important than the appeal of the physical environment and the role of the state (particularly defense spending). The West Coast manufacturing district does, however, represent an outstanding example of industrial restructuring in response to economic crisis and labor unrest. The Los Angeles–San Diego district has been extremely successful in negotiating the transition from older forms of manufacturing to newer ones. Since the 1960s, the district has shed much of its traditional, highly unionized heavy industry, such as steel and rubber. At the same time, it has attracted a cluster of high-tech industries and associated services, centered on electronics and aerospace and tied strongly to enormous defense and military contracts from the U.S. government. Additionally, it has witnessed the vigorous growth of "peripheralized" manufacturing, which resembles the industrialization of developing countries in that it depends on a highly controllable supply of cheap, typically immigrant or female labor from Mexico and East Asia.

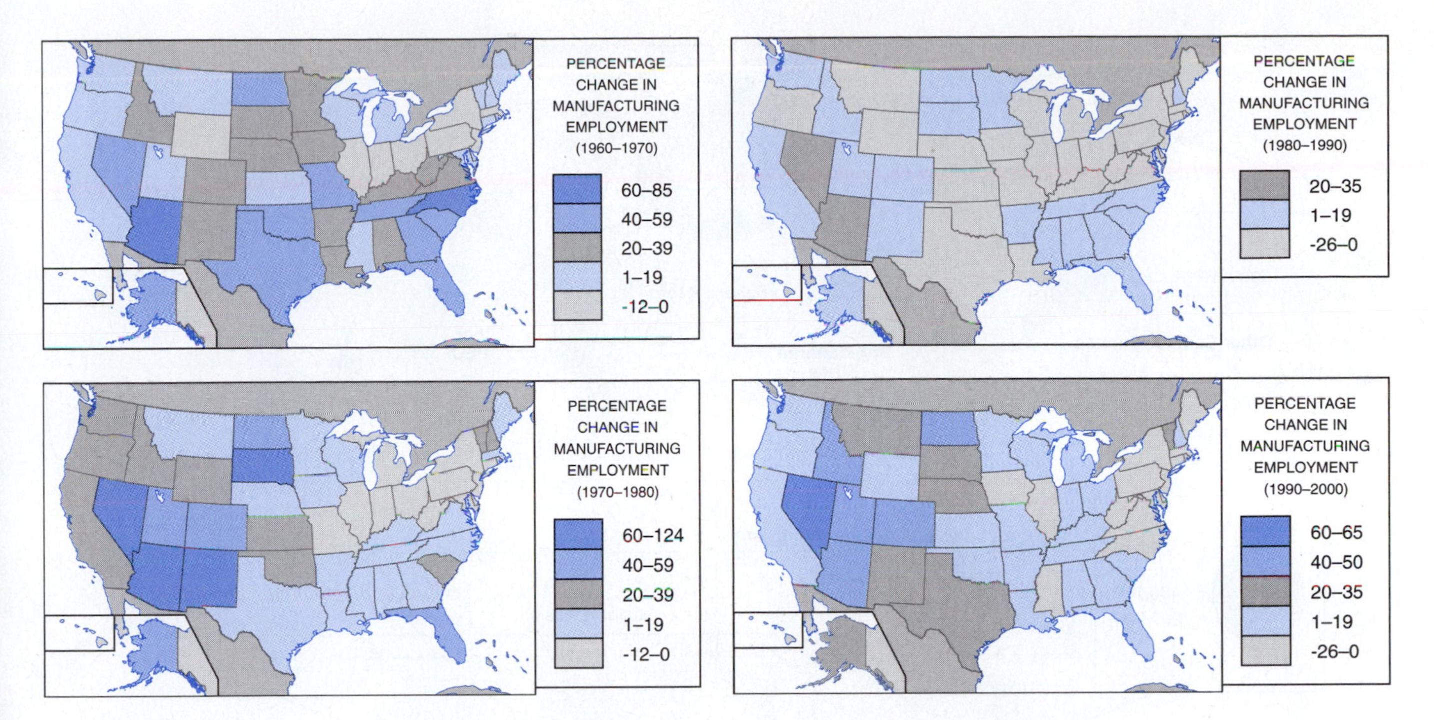

FIGURE 7.19
Changes in U.S. manufacturing employment by decade, 1960–2000. The steady losses in the northeastern Manufacturing Belt and rise of Sunbelt producers reflected the sustained impacts of globalization, technological change, and the shift into lower cost, less unionized parts of the country. The microelectronics revolution and information technology allow firms much greater locational flexibility than in the past and contribute to the low-density, decentralized landscapes of industry as well as urban areas.

Meanwhile, industrial restructuring continues to be a painful process throughout much of the Manufacturing Belt. The region contends with problems of obsolescence and reduced productivity, especially in such leading industries as steel, rubber, automobile manufacturing, and shipbuilding. Many of its inner-city areas are littered with closed factories, bankrupt businesses, depressed real estate, and struggling blue-collar neighborhoods. The effects of disinvestment on workers and their communities have been devastating. Victims of plant closings sometimes lose not only their current incomes but also often their accumulated assets as well. When savings run out, people lose their ability to respond to life crises, and often suffer depression and marital problems. Although job losses occurred in many occupations, some groups are more vulnerable than others. Unskilled workers are particularly likely to bear the costs of globalization, including job displacement. African American workers, many of them unskilled or semiskilled, were particularly hard hit, and by driving up unemployment, the deindustrialization of the inner city was in no small part responsible for the creation of the impoverished ghetto communities there.

Although the widespread manufacturing decline has produced a lasting effect on people and communities in the Manufacturing Belt, all is not lost in the region. There have been numerous attempts to respond to the economic crisis. Some old industrial cities such as Pittsburgh successfully built new bases for employment in services, and others, such as Cleveland, indicate the potential for doing so. Southern New England, which suffered high unemployment rates throughout much of the post–World War II period, underwent a new round of industrial expansion based on electronics, one that took advantage of its pool of highly skilled workers.

The International Movement of American Manufacturing

The relocation of manufacturing within the United States is only one aspect of a wider dispersal of manufacturing capital. Foreign direct investment by American enterprises was established as early as the end of the nineteenth century. But only since World War II have American enterprises become major foreign investors. The 1940s saw heavy investment in Canada and Latin America; the 1950s in Western Europe; the 1960s and 1970s in Europe, Japan, and the Middle East; and the 1980s and 1990s in South

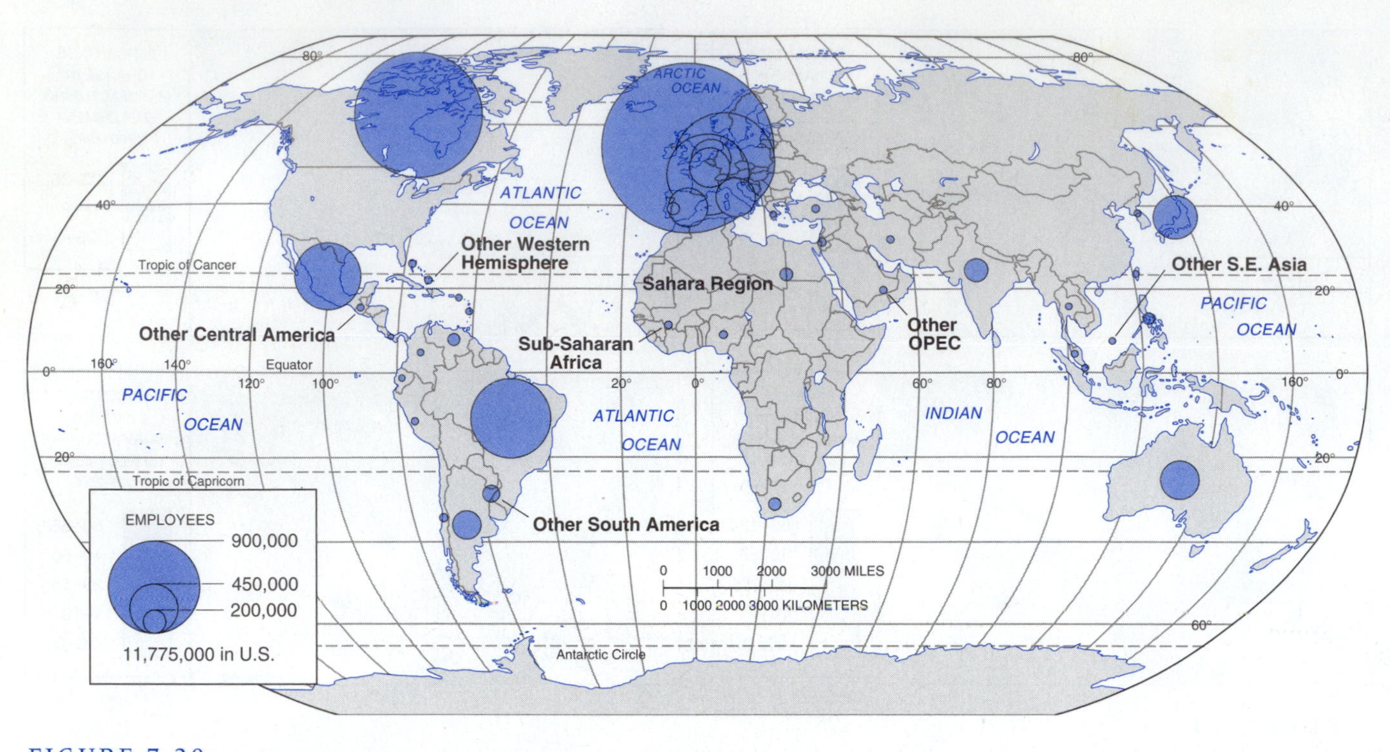

FIGURE 7.20
World employment in manufacturing of U.S.-based multinational corporations in 2000. Most MNCs invest in other developed countries.

and East Asia. Overall, most of U.S. manufacturing investments occurred in advanced industrialized countries rather than in developing countries (Figure 7.20).

Between 1945 and 1960, most U.S.-based companies were content to produce in the old industrial districts. But by 1960, western European countries and Japan had become competitors. Mounting international competition and falling profit rates at home coerced American companies to decentralize not only within the United States but also abroad. Thus domestic restructuring and internationalization can be seen as two sides of the same coin. By 1980, the 500 largest U.S.-based corporations employed an international labor force almost equivalent to the size of their labor force within the United States. Toy production, for example, has largely moved into East Asia, particularly China.

Steel offers a compelling example of the decline in industrial capacity. Between 1970 and 2005, the North American and Western European proportion of total global steel production declined from 67 percent to 42 percent, whereas developing countries' production levels increased from 10 percent to more than 30 percent. Many steel-manufacturing firms have gone out of business as the global steel-production capacity exceeds global demand. Because of government subsidies, steel mills in some countries, especially in Europe, have remained open in the face of dwindling quotas. The U.S. government, however, has been less willing to pay unemployment compensation to displaced workers and has allowed the U.S. steel industry to decline. Since the 1970s, U.S. production has decreased 33 percent, whereas employment in the steel industry has declined 66 percent.

The dispersal of manufacturing investment to foreign lands resulted in enormous savings for American firms and enormous losses for American workers, largely due to the differences in wage rates (Table 7.1). In terms of average wages, workers in Europe, particularly Scandinavia and Germany, fare the best, with wages often one-third higher than American workers (as well as longer paid vacation periods and universal health coverage). U.S. wages are comparable to Australia, slightly higher than Ireland or Japan, and well above the developing world.

Downsizing

Downsizing (sometimes euphemistically called "right sizing") is the most critical factor inhibiting wage inflation. Once seen mainly as a component of corporate reorganization aimed at achieving a healthier bottom line, the chilling effect of downsizing on workforce wage demands has effectively kept wage inflation in check too.

Table 7.1
Hourly Wage Costs for Factory Workers, U.S. Dollars, 2004

Norway	34.64
Denmark	33.75
Germany	32.53
Netherlands	30.76
Finland	30.67
Switzerland	30.26
Belgium	29.98
Sweden	28.42
Austria	28.29
Luxembourg	26.57
United Kingdom	24.71
France	23.89
United States	23.17
Australia	23.09
Ireland	21.94
Japan	21.90
Spain	17.10
Israel	12.18
South Korea	11.52
New Zealand	9.14
Singapore	7.45
Portugal	7.02
Taiwan	5.97
Mexico	2.50
Sri Lanka	0.51

Source: U.S. Bureau of Labor Statistics, 2004.

From 1989 through 2003 in the United States, more than 4.7 million job cuts were announced, and employees who remain are reluctant to push for pay increases. They know management has placed a ceiling on the amount allotted for wages in order to avoid the inflationary move of charging higher prices and the risk of becoming noncompetitive. The workforce is well aware that a request for a pay increase could be met with dismissal, as management turns the downsizing valve, eliminating employees deemed too costly to keep. Employees also know that any vacated position could quickly be filled by someone willing to accept the present wage. Uncertainty over when management will turn the downsizing valve has insecure employees scrambling for jobs in the most productive, high-priority areas of a company. No one wants to be targeted for dismissal simply because he or she stayed too long in an area perceived by management as too bureaucratic or obsolete. Even when employees manage to land a job in an area reasonably safe from the threat of downsizing, few are willing to ask for a raise.

Downsizing has erased the traditional bond of employer-employee loyalty, has depleted corporate memory, and in many cases has eroded the work ethic, productivity, and morale of employees. The number of executives and managers who have been laid off two or three times is growing. With each new, unexpected twist in their careers, they become more uncertain and insecure.

Downsizing has proven effective as a means of keeping inflation in check and is here to stay as a means of making companies leaner and more competitive. But how long can companies remain competitive with a disloyal, insecure workforce? Management now must ask itself if there are alternatives to turning the valve of downsizing in order to keep wages stable and inflation low. In the long run, this method of controlling inflation may not be worth the price.

Restructuring and downsizing means layoffs and hurting people at the basic survival level. Downsizing challenges people's ability to survive, to provide shelter, food, and the services needed by their families. Today white-collar as well as blue-collar jobs are being lost as globalization has extended competition from low-wage, low-skilled positions to increasingly high-skilled ones.

American Manufacturing Today

Since 1962, worldwide exports have increased from 12 percent to more than 33 percent of the world GNP, meaning roughly one-third of everything produced in the world is traded. Worldwide exports totaled $7 trillion in 2005. For manufactured goods, the proportion is much higher, ranging as high as 70 percent. Global corporations are developing a manufacturing network of decentralized plants based in large, sophisticated regional markets. Each plant will be smaller and more flexible than is typically found in today's manufacturing environment. The location of such plants will be based more on regional infrastructure, local skill levels, and government policies than on purely cost-based factors. Consider the following:

1. The development of large and sophisticated overseas markets dictates a global presence of leading manufacturers.
2. Increasing levels of nontariff barriers are forcing firms to localize production resources.
3. The evolution of a world trade system based on regional trading blocs creates incentives for firms to follow direct investment strategies that give

them a manufacturing presence in each region of significant demand.

4. Regionalization of trading economies is increasing the benefits to decentralized manufacturing organizations.
5. Exchange rates and other aspects of risks are forcing firms to be flexible in terms of capacity and location and to view their global networks in a holistic way.
6. The emergence of manufacturing technologies and methods, such as flexible manufacturing systems, just-in-time manufacturing, and total quality control, have reduced scale, increased the importance of worker education and skilled development, and placed demands on local infrastructure.
7. Large, centralized manufacturing facilities in low-cost countries with poorly skilled workers are generally not sustainable.

Traditional approaches to understanding the location of manufacturing are increasingly questionable. Large, centralized manufacturing facilities have given way to decentralized manufacturing structures, with smaller, lower scale plants serving demands and regional markets. Location depends increasingly on educational and institutional infrastructure.

To conclude, globalization, advances in technology, changes in management approach, and shifting market requirements are the new dynamics that shape manufacturing firms' organizational and locational decisions today. These trends suggest that global corporations of the future will move to networks of decentralized plants, based in large sophisticated regional markets, but linked with information technology to one another. Specific locations will be based more on local infrastructure, such as workforce capability, training programs, and government policies, than on traditional cost-base considerations. Plants will be smaller than current ones, more flexible, yet have significantly more ability to produce multiple products.

Flexible Manufacturing

In the aftermath of the turbulent 1970s—which brought on, among other things, the shift from fixed to floating exchange rates, the petrocrises, the rise of the NICs, massive deindustrialization in the West, and the microelectronics revolution—geographers and others began to recognize that capitalism in the late twentieth century was undertaking a new direction. The new world economy, characterized by mounting globalization and competition among nation-states, also included a profound shift in the nature of manufacturing, including changes in markets,

Table 7.2

Contrasts between Fordism and Post-Fordism

Fordism	*Post-Fordism*
Vertically integrated	Vertically disintegrated
Long-run contracts	Short-run contracts, JIT
Large firms	Small firms
Economies of scale	Specialized output
Competitive producers	Cooperative networks
Product price	Product quality
Mass consumption	Segmented markets
Unionized	Nonunionized
Unskilled labor	Skilled labor
Routinized work	Varied tasks
National linkages	International linkages

technologies, and location. Now, as we have seen, capitalism is a highly dynamic economy characterized by many such transformations in its history. In this sense, the creation of a new form of capitalism in the 1980s and 1990s was not particularly new. However, each age brings with it a new form of commodity production and consumption, and often a new terminology to describe these changes. A common set of terms used to sum up these epochal changes is the shift from "Fordism" to "post-Fordism," also variously known as "flexibilism" or "the flexible economy" (Table 7.2).

Fordism

Fordism, as the name indicates, is named after the American industrialist Henry Ford, who pioneered the mass production of automobiles in the early twentieth century using standardized job tasks, interchangeable parts (which date back to gun maker Eli Whitney), and the moving assembly line. Ford's methods, which were very successful, were widely imitated by other industries and soon became almost universal throughout the North American, European, and Japanese economies.

The precise moment when Fordism became the dominant form of production in the United States is open to debate. Some argue that it began as early as the late nineteenth century, during the wave of technological change in the 1880s and 1890s, when mass production first made its appearance, displacing the older, more labor-intensive (and less profitable) forms of artisanal production. For example, during this period, glass blowing, barrel making, and the production of rubber goods such as bicycle tires became steadily standardized, and the Bessemer process for fabricating steel was invented. Fordism, however, elevated this process to a whole new level, including highly refined divisions of labor within the factory, so that each worker engaged in highly repetitive tasks. Ford engaged the services of Frederick Taylor, the founder of industrial psychology, who applied time-and-motion studies to workers' jobs to organize them in the most efficient and cost-effective manner. By breaking down complex jobs into many small ones, Fordism made many tasks suitable for unskilled workers, including the waves of immigrants then arriving into the United States, and greatly increased productivity.

Others argue that Fordism was a particular kind of social contract between capital and labor, one that tolerated labor unions (such as the Congress of Industrial Organizations [CIO] that came into being in the 1930s), and so Fordism should be seen as beginning in the crisis years of the 1930s. Yet others make the case that Fordism was the backbone of the great economic boom in the three decades following World War II, when the United States emerged as the undisputed superpower in the West, and that it should only be dated back to the 1950s. Whenever its origins, Fordism is reflective of a historically specific form of capitalism that dominated most of the twentieth century.

Fordism came to stand for the mass production of homogeneous goods, in which capital-intensive companies relied heavily on economies of scale to keep production costs low and profits high. Thus, mass consumption and advertising would also come into being as the demand side of Fordism. Typically, firms working in this context were large and vertically integrated, controlling the chain of goods from raw material to final product. Ford's plants, for example, saw coal and iron ore enter one part of the factory and cars come out the other end. Well suited to large, capital-intensive production methods, this system of production and labor control was largely responsible for the great manufacturing complexes of the North American Manufacturing Belt, the British Midlands, the German Ruhr region, and the Inland Sea area of Japan.

While Fordism "worked" quite successfully for almost a century, ultimately it began to reach its social and technical limits. Productivity growth in the 1970s began to slow dramatically, and the petrocrises and rise of the NICs unleashed wave upon wave of plant closures in the United States. Because wages and salaries are often tied to the overall growth of productivity, these changes not only led to widespread layoffs, but also to declining earning power of American workers. Rates of profit in manufacturing began to drop in the 1970s and 1980s, and many firms faced the choice of either closing down, moving overseas, or reconstructing themselves with a new set of production techniques. It is in this context that Fordism began to implode, giving way to post-Fordist, flexible production techniques, which have become widespread throughout the economy today.

Post-Fordism/Flexible Production

Post-Fordism refers to a significantly different approach to the production of goods than that offered by Fordism. *Flexible manufacturing* allows goods to be manufactured cheaply, but in small volumes as well as large volumes. A flexible automation system can turn out a small batch, or even a single item, of a product as efficiently as a mass-assembled commodity. It appeared, not accidentally, at the particular historical moment when the microelectronics revolution began to revolutionize manufacturing; indeed, the changes associated with the computerization of production in some respects may be seen as capitalists' response to the crisis of profitability that accompanied the petrocrises. Post-Fordism also reflected the imperative of American firms to increase their productivity in the face of rapidly accelerating, intense international competition (Figure 7.21).

The most important aspect of this new, or lean, system is flexibility of the production process itself, including the organization and management within the factory, and the flexibility of relationships among customers, supplier firms, and the assembly plant. In contrast to the large, vertically integrated firms typical of the Fordist economy, under flexible production, firms tend to be relatively small, relying on highly computerized production techniques to generate small quantities of goods sold in relatively specialized markets. Microelectronics, in essence, circumvented the need for economies of scale.

The classic technologies and organizational forms of post-Fordism include robots and just-in-time inventory systems. The Japanese developed *just-in-time manufacturing* systems shortly after World War II to adapt U.S. practices to car manufacturing. The technique was pioneered by the Toyota Corporation (and hence sometimes called "Toyotaism"), which obviated the need for large, expensive warehouses of parts (the "just-in-case" inventory system), which saved rents in a country in which the price of land is high. Just-in-time refers to a method of organizing immediate manufacturing and supply relationships among companies to reduce inefficiency and delivery times. Stages of the manufacturing process are completed exactly when needed, according to the market, not before and not later, and parts required in the manufacturing process are supplied with little storage or warehousing time

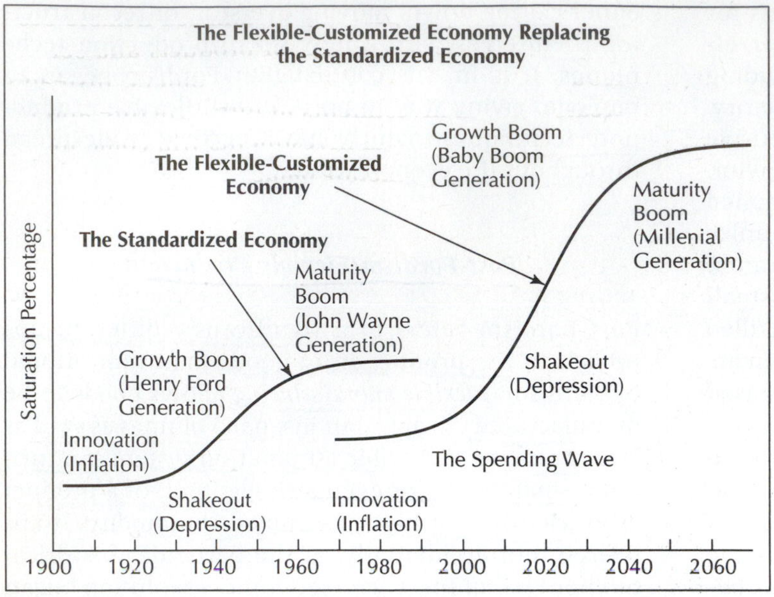

FIGURE 7.21
The flexible customized economy is replacing the standardized economy. Henry Ford's assembly-line approach led off the innovation phase of the standardized economy. The late twentieth century has been a comparable innovation period launched by the baby boomers. By 2000, microcomputer technology, comparable to the automobile of the early part of the century in innovation, is multiplied in its potential by powerful, flexible software. This software and hardware can revolutionize any industry, just as the assembly line revolutionized industries throughout the Fordist period of the last century. The growth boom and the spending wave create a time when innovations begin moving out of their niches into the mainstream, driven by the power of the individualist. New technologies put downward pressure on prices, but rising consumer demand from the spending wave exerts an upward pressure. A shakeout occurs, leading to survival of only the fittest companies.

and cost. This system reduces idle capital and allows minimal investment so that capital can be used elsewhere.

Occasionally machines are idle because they run only fast enough to meet output. If machines run more quickly than the market requires, they must be shut off and manufactured items warehoused. The manufacturing run proceeds only as far as the market demands, and no faster. Thus, suppliers and producers of raw materials must warehouse their inventories. Buffer stocks are very small and are only replenished to replace parts removed downstream. Workers at the end of the line are given output instructions on the basis of short-term order forecasts. They instruct workers immediately upstream to produce the part they will need just-in-time, and those workers in turn instruct workers upstream to produce just-in-time, and so on. In practice this means that buffers between workers are extremely small.

Thus, many firms in the late twentieth century engaged in significant downsizing, ridding themselves of whole divisions of their companies to focus on their core competencies. Many companies reversed their old principles of hierarchical, bureaucratic assembly-line (Fordist) processes as they switched to customized, flexible, consumer-focused processes that can deliver personal service through niche markets at lower costs and faster speeds.

In the process, the use of subcontracts accelerated rapidly. Firms always face a "make or buy" decision (i.e., a choice of whether to purchase inputs such as semifinished parts from another firm or to produce those goods themselves). Under the relatively stable system afforded by Fordism, most firms produced their own parts (i.e., decided to "make" rather than buy), justifying the cost with economies of scale, which lowered their long-run average cost curves. Large firms, for example, would have their own parts producers, trucks, or printing shops. Under post-Fordism, however, this strategy is no longer optimal: Given the uncertainty generated by the rapid technological and political changes of the late twentieth century, many firms opted to buy rather than make (i.e., to purchase inputs from specialized companies). This strategy reduces risk for the buyer by pushing it onto the subcontractor, who must invest in the capital and hire the necessary labor.

A key to production flexibility lies in the use of information technologies in machines and operations, which permit more sophisticated control over the process. With the increasing sophistication of automated processes and, especially, the new flexibility of the new electronically controlled technology, far-reaching changes in the process of production need not be associated with increase scale of production. Indeed, one of the major results of the new electronic computer-aided production technology is that it permits rapid switching from one process to another and allows the tailoring of production to the requirements of individual customers. Traditional automation is geared to high-volume standardized production; the newer flexible manufacturing systems are quite different.

As interfirm linkages grew rapidly in the 1980s, many firms found themselves compelled to enter into cooperative agreements with one another such as strategic alliances. Quality control (i.e., minimizing defect rates) became very important. Many firms succeed in this environment by entering into dense urban networks of interactions, including many face-to-face linkages, ties that the roles played by noneconomic factors such as tacit knowledge, learning, reflexivity, conventions, expectations, trust, uncertainty, and reputation in the interactions of economic actors. Post-Fordism thus highlighted the culturally embedded nature of economic linkages.

Post-Fordist approaches to production, which vary in their nature in different regions, came to dominate much of the electronics industry and automobile manufacturing, the minimills in the steel industry, and became closely associated with the new manufacturing spaces such as Silicon Valley, Italy's Emilia-Romagna region, Germany's Baden-Wurtenburg, the Danish Jutland, and the British electronics region centered around Cambridge.

The major impacts of the information and telecommunications revolution will extend well into the future. Because of the telecommunications revolution, products and services once thought to be confined to premium and niche markets will move rapidly into the mainstream. No technological innovation of the past, not automobiles, railroads, plastics, iron, or steel, ever came close to the power of the change being brought on by the information technologies. The power of 16 Cray supercomputers on a single chip is simply unprecedented in the history of the world, which will result in rapid changes in how computers increase business productivity, marking the enormous gains associated with the flexible revolution.

The flexible economy will provide an even greater innovation than the assembly line of Fordism, which in itself allowed an amazing array of standardized products to move into mass affordability and a 10-fold average wage increase for the American worker. In the coming decades, the computer will help to make customized, flexible products and services increasingly affordable. The microcomputer revolution and the innovative market demand by baby boomers will dictate the growth for decades to come and will be based on these principles: flexibility and customization to individual needs and wants, higher quality and higher value added, rapid response and delivery just-in-time, and improved personal service and follow-up.

Customization and flexibility are the watchwords of business in the immediate future. Companies that can adjust to the customized flexible economy are the ones that will prosper. The customized flexible economy means custom design of products and services around individual needs and customers. Instead of greater capital investments in infrastructure and machines, greater capital investments will come in software, allowing marketing databases to estimate needs of customers and identify niche markets. Such software will allow the production and service machinery to make "short runs" for individual and market niche needs, quickly changing markets without the setup costs and delays of the old standard assembly-line systems. The business trend of the customized flexible economy will mean that premium niche products and services of the past will move into mainstream affordability as the computer and software forces down the price and as bureaucracies become restructured and more efficient, causing a flexible labor force.

IT and Strategic/Competitive Advantage

The use of IT to increase the competitive advantage of organizations is an important issue faced by MNCs. Information technology (IT) contributes to strategic management in many ways. Consider these three:

1. IT creates applications that provide direct strategic advantage to organizations.
2. IT supports strategic changes such as reengineering. For example, IT allows efficient decentralization by providing speedy communication lines, and it streamlines and shortens product design time by using computer-aided engineering tools.
3. IT provides business intelligence by collecting and analyzing information about innovations, markets, competitors, and environmental changes. Such information provides strategic advantage because, if a company knows something important before its competitors, or if it can make the correct interpretation of the information before its competitors, then it can introduce changes and benefit from them.

Competitive strategy is the search for a competitive advantage in an industry, either by controlling the market or by enjoying larger than average profits. Such a strategy aims to establish a profitable and sustainable position against the forces that determine industry competition. The shift of corporate operations from a competitive to a strategic orientation (of which competition is only one aspect) is fundamental. IT has a significant impact on the profitability of an organization and even on its survival.

THE PRODUCT CYCLE IN MANUFACTURING

Product cycles and production systems help us to appreciate the importance of technological considerations in corporate spatial organization. The *product life cycle,* which begins with a product's development and ends when it is replaced with something better, is important geographically because products at different stages of production tend to be manufactured at different places within corporate systems. Moreover, at any given stage of the cycle, the various operations involved in the manufacture of a product such as a camera are not necessarily concentrated at a single factory. Production of a camera's complex components occurs at a different place from where the final product is assembled.

The famous economist Simon Kuznets developed the concept of the three-stage product cycle (Figure 7.22). In Stage 1, innovators discover, develop, and commercially launch a product. They also benefit from a temporary monopoly and all the special privileges—high profits—that result from it. In Stage 2, competitors buy or steal the new idea, which forces an emphasis on low-cost, standardized, mass-production technologies. Sales of the product increase for a while, but the initial high returns diminish. By Stage 3, the product begins to be superseded. Markets are lost to new products, and manufacturing capacity is reduced.

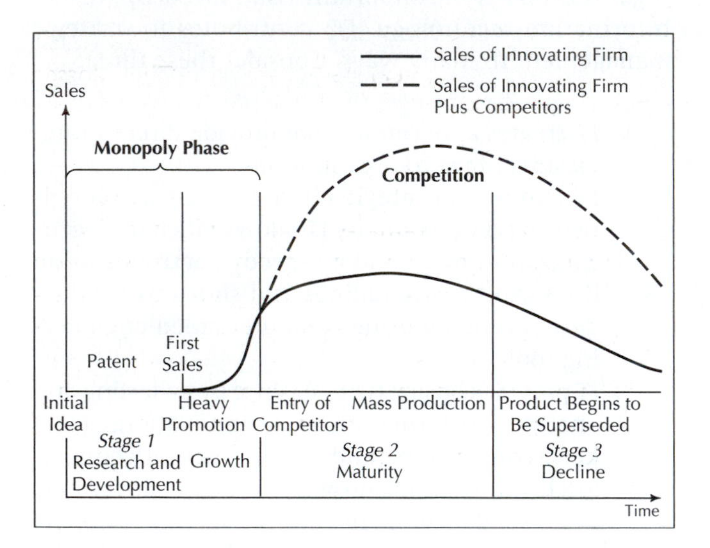

FIGURE 7.22
A typical product life cycle. Stage 1 is the monopolistic phase in which initial discovery and development are followed by the commercial launching of the product. Rapid sales ensue. The company may enjoy a monopoly during this period, at which time they attempt to improve the products. Stage 2 is characterized by the entry of competition. Emphasis is now on mass-produced, inexpensive items that are standardized and directed toward expansions of the market. Competition begins to erode a large share of the innovating firm's sales. In Stage 3, a large share of the market has been lost to new products and other companies. Overall sales of the product decline as alternative products and manufacturing processes are introduced.

Innovation begins in an advanced industrial country. These countries have the science, the technology, and the internal market to justify R&D. As a result, they also have an international advantage, and they export their product around the world. But as the technology becomes routinized, other producers appear on the scene, first in the other advanced countries, then on the periphery. Meanwhile, back in the rich country, investment in the newest generation of sophisticated technology is the cutting edge of the economy.

There is no doubt that developed countries are the innovators of the world economy and that LDCs increasingly specialize in the laborious task of transforming raw materials into commodities. But developed countries are also engaged in activities associated with the second and third stages of the product cycle. Indeed, Britain and Canada have expressed concern about their recipient status. This concern has also been voiced in the United States.

Not all manufacturing operations are fragmented. Corporate branch plants are often *clones,* supplying identical products to their market areas. For example, medium-sized firms in the clothing industry often have this structure, as do many multiplant companies manufacturing final consumer products. Part-process structures tend to be associated with certain industrial sectors, such as electronics and motor vehicles, characterized by complex finished products comprising many individual components.

Labor is an important variable in the location of facilities making components. Manufacturers seek locations where the level of worker organization, the degree of conflict, and the power of labor to affect the actions of capital are more limited than in long-established production centers such as Detroit, Coventry, and Turin. Starting in the 1970s, Fiat began to decentralize part of the company's production away from its traditional base in Turin to the south of Italy. Compared with the workers of Turin, who were relatively strong and well organized, the workers of the south were new to modern industry and had little experience of union organization. At the international level, Ford adopted a similar tactic when it invested in Spain and Portugal in the 1970s. Ford management perceived that it could operate trouble-free plants in a region of low labor costs. The labor factor is further emphasized by the practice of *dual sourcing*. To avoid total dependence on a single workforce that could disrupt an interdependent production system, companies such as Ford and Fiat are willing to sacrifice economies of scale for the security afforded by duplicate facilities in different locations.

Locational Adjustment

Corporate production systems undergo continuous locational adjustment. Shifts may be inspired by technical and organizational developments internal to an industry or by changes in the external environment in which they operate, such as the oil-price hikes of 1973. Particularly significant from a geographic viewpoint are adjustments in response to major shocks or stresses placed on an enterprise. For example, when faced with the challenge of competition from lower-cost regions and with a falling rate of expansion of global markets, an enterprise can adopt a number of strategies—rationalization, capital substitution, outright closure, reorganization of productive capacity associated with the closure of older plants—which all in one way or another result in losses of employment. The industrial decline of Britain provides many illustrations of painful corporate restructuring programs. The 10 largest manufacturing employers in the West Midlands reduced their British employment by 25 percent between 1978 and 1981 while increasing their overseas workforce by 9 percent. This shift in the productive base of these companies abroad undermined the economic well-being of this area. Such employment withdrawals are an aspect of the growing international integration of production and mobility of capital.

One of capital's crucial advantages over labor is geographic mobility; it can generally make use of distance and differentiation in a way that labor cannot. Corporations take advantage of such flexibility by shifting production to low-wage regions, setting up plants in areas with low levels of worker organization, or establishing plants in areas that offer incentive policies. Many LDCs offer tax relief and capital subsidies for new industries.

Summary

This chapter offered an overview of the changing empirical patterns of manufacturing, the reasons that underlay them, and emphasized that manufacturing geographies are always fluid, changing over time. Four major regions of the world account for approximately 80 percent of the world's industrial production—northeastern North America, northwestern Europe, western Russia, and Japan. The textiles, clothing manufacture, and consumer electronics industries have shifted globally from the developed world to the developing world. Automobile production and semiconductor manufacture have also experienced global shifts, from North America and Europe to Japan and East Asia.

We explored the social relations that lead to industrial change, described worldwide manufacturing trends, and examined the recent history of industrial devolution in the developed world and of industrial revolution in the developing world. The processes of *deindustrialization* and *industrialization* are not temporary tendencies within the global system; rather, they constitute a zero-sum global game played by multinational nomadic capital. Multinationals switched production from place to place because of varying relations between capital and labor and new technological innovations in transportation and communications. With improved air freight, containerization, and telecommunications, multinational corporations can dispatch products faster, cheaper, and with fewer losses.

In the United States, 500,000 manufacturing jobs per year were lost between 1978 and 2002. These losses were hidden to some extent by selective reindustrialization and the migration of manufacturing from the American Manufacturing Belt to the South and the West.

The latter part of this chapter explored the ways in which manufacturing landscapes change over time. Capitalism is constantly reinventing itself through the process of "creative destruction," and one of its prime strengths is the capacity to adjust to rapid change. In the late twentieth century, as the Fordist regime of production began to collapse, the petrocrises, deindustrialization, the rise of the NICs, and the microelectronics revolution spawned the emergence of post-Fordist, flexible production techniques and organizational forms. Mounting international competition called for a restructuring of the production process in order to restore the conditions of profitability, and many firms downsizing, subcontracted many functions, and adopted technologies such as just-in-time inventory systems. Concomitantly, the business world engaged in widespread reengineering. In many sectors, these changes may be seen in light of the product cycle, a metaphorical model that encapsulates the simultaneous economic, technological, organizational, and geographic changes that confront every firm in the course of its industry's evolution.

STUDY QUESTIONS

1. What are the forces of production and the social relations of production?
2. Why are capital-labor relations a necessary starting point for studying economic geography?
3. What are four major world regions of manufacturing?
4. Summarize the historical development of the U.S. Manufacturing Belt.
5. What is inertia in industrial location?
6. When and why did the Manufacturing Belt begin to lose industry? Where, specifically, did it go?
7. Where are most U.S. multinational investments concentrated? Why?
8. What are some major world industrial problems?
9. What is the flexible economy?
10. What is the product life-cycle model?

KEY TERMS

American Manufacturing Belt
assembly costs
assembly plants
backward integration
capital-labor relations
conglomerate merger
deindustrialization
dependent industrialization
diseconomies of scale
distribution costs
diversification
dual economy
externality
flexible labor
foreign sourcing
forward integration
franchising
horizontal integration
industrial inertia
industrialization
industrial park
industrial restructuring
industry concentration
industry life cycle
inertia
information technology
integrated circuit
international subcontracting
joint venture
just-in-time manufacturing
least-cost approach
licensing venture
locational interdependence
machinofacture
microprocessor
mode of production
multinational corporations (MNCs)
multiplant enterprises
nomadic capital
offshore assembly
oligopoly
outsourcing
postindustrial
postindustrial society
product life cycle
raw-material—oriented
real-time information systems
surplus value
Taylorism
transnational corporation
value added
vertical integration
zaibatsu

SUGGESTED READINGS

Bluestone, B. and B. Harrison. 1982. *The Deindustrialization of America: Plant Closings, Community Abandonment, and the Dismantling of Basic Industry.* New York: Basic Books.

Dicken, P. 2004. *Global Shift: Industrial Change in a Turbulent World,* 4th ed. London: Guilford.

Krugman, P. 1991. *Geography and Trade.* Cambridge, Mass.: University Press.

Massey, D. 1984. *Spatial Divisions of Labor: Social Structures and the Geography of Production.* New York: Methuen.

Knox, P., Agnew, J., and L. McCarthy. 2003. *The Geography of the World Economy,* 4th ed. London: Edward Arnold.

Linkon, S. and J. Russo. 2002. *Steel-Town U.S.A.: Work & Memory in Youngstown.* Lawrence: University Press of Kansas.

Child labor is the norm in most developing countries, where children are indispensable to the operation of farms. Children perform a variety of tasks, often differentiated by gender, including working on fields, tending to animals, carrying water and firewood, and assisting adults. Even in urban areas children are often needed to supplement low family incomes, often working in the informal economy. The prevalence of child labor is the single greatest factor underlying high fertility rates in the developing world.

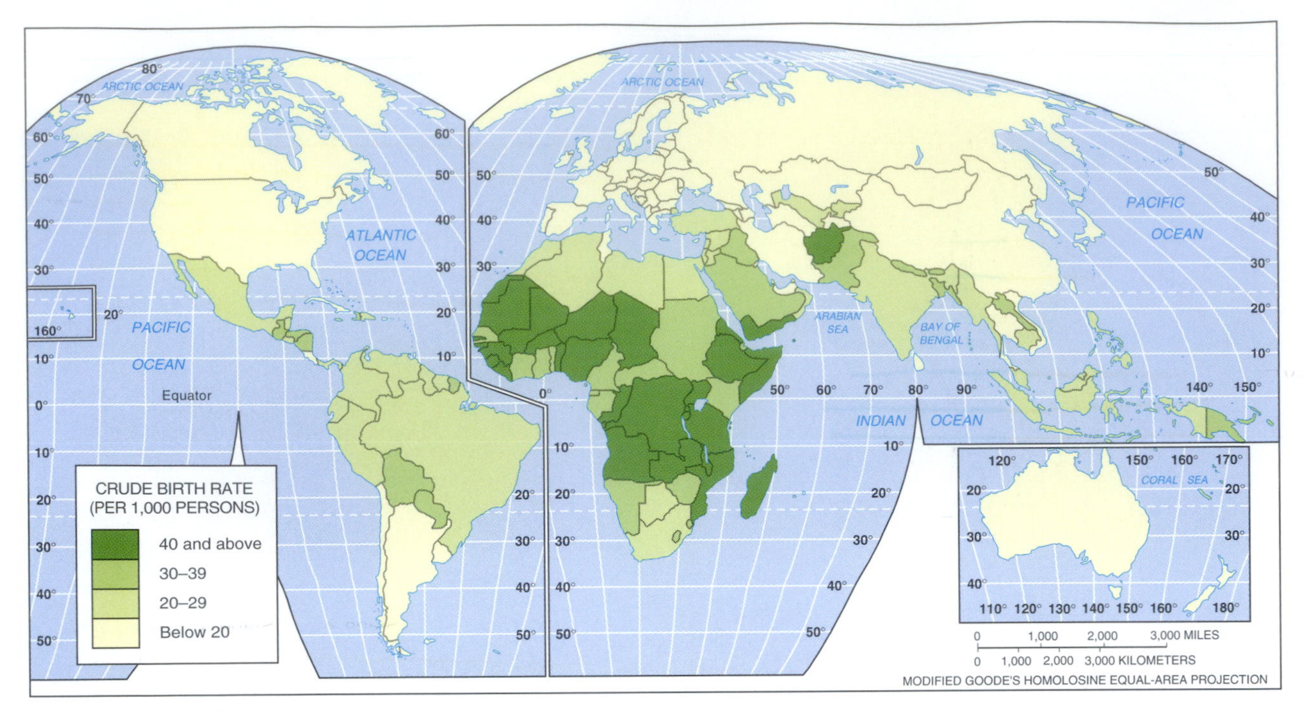

Crude birth rates measure the number of babies born per 1,000 people per year. Typically, the poorer a country is, the higher is its birth rate, as rural families need child labor, have many children in response to high infant mortality rates, and require offspring to care for them in old age. Thus, countries in SubSaharan Africa have the highest birth rates in the world. Economically developed countries, in contrast, in Europe, Japan, North America, and Australia and New Zealand, tend to have lower birth rates, a reflection of the costs and benefits of having children in urbanized contexts in which many women work outside the home.

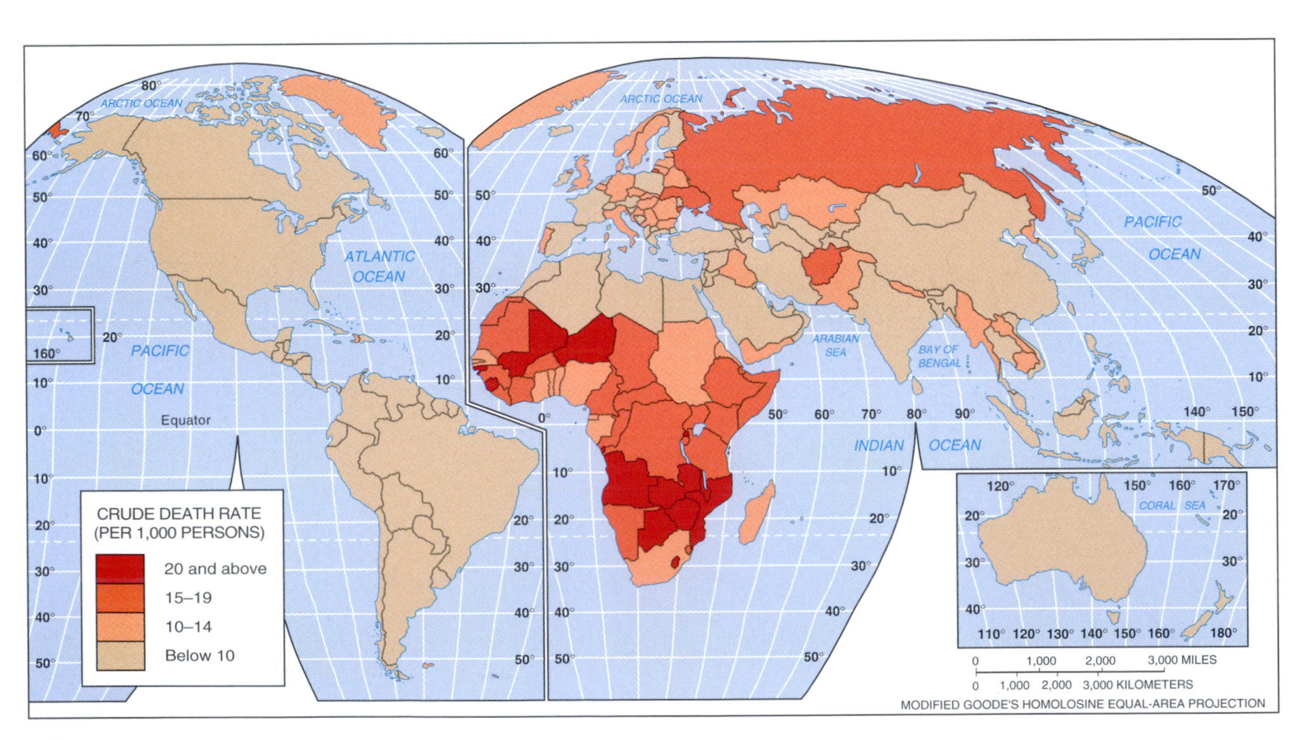

Crude death rates measure the number of deaths per 1,000 people per year. In economically underdeveloped countries, malnutrition, unclean drinking water, and infectious diseases are responsible for relatively high death rates and correspondingly low life expectancies. Thus, in Africa, death rates frequently exceed 20 per 1,000 annually. In economically developed countries, in contrast, death rates are lower and life expectancies are higher. Moreover, the causes of death shift as countries urbanize and develop; most deaths are "lifestyle" or behaviorally related ones associated with smoking, heart disease, obesity, and cancers of various sorts, followed by homicide, suicide, and accidents.

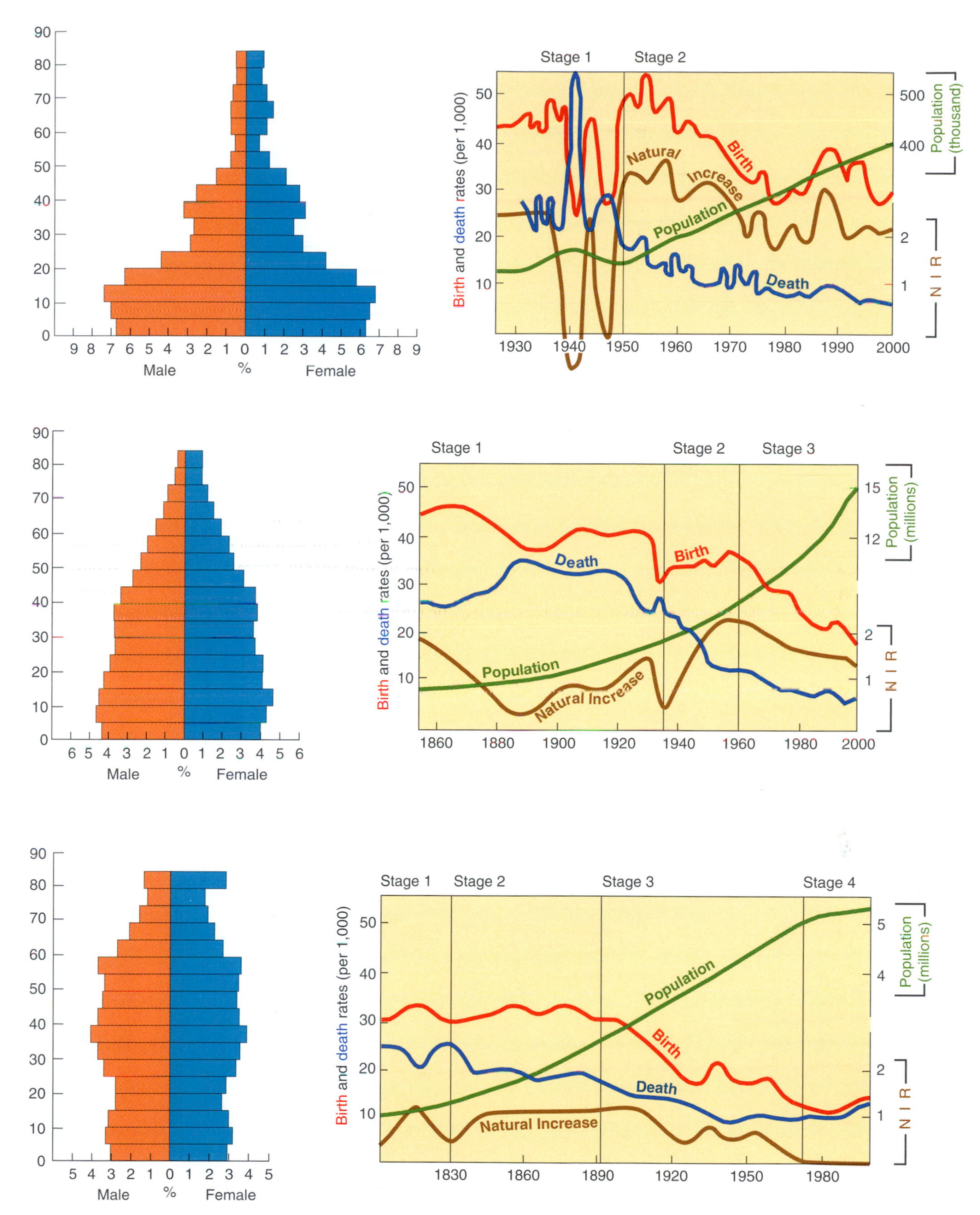

These three examples of the Demographic Transition illustrate the population composition (sex and age structure) and how it changes over time. Cape Verde exemplifies a country with relatively high birth rates and low death rates, and thus high natural growth. Chile illustrates a declining birth rate and hence declining rate of natural increase. Denmark is a case in a country where births and deaths are roughly equal, leading to zero natural growth.

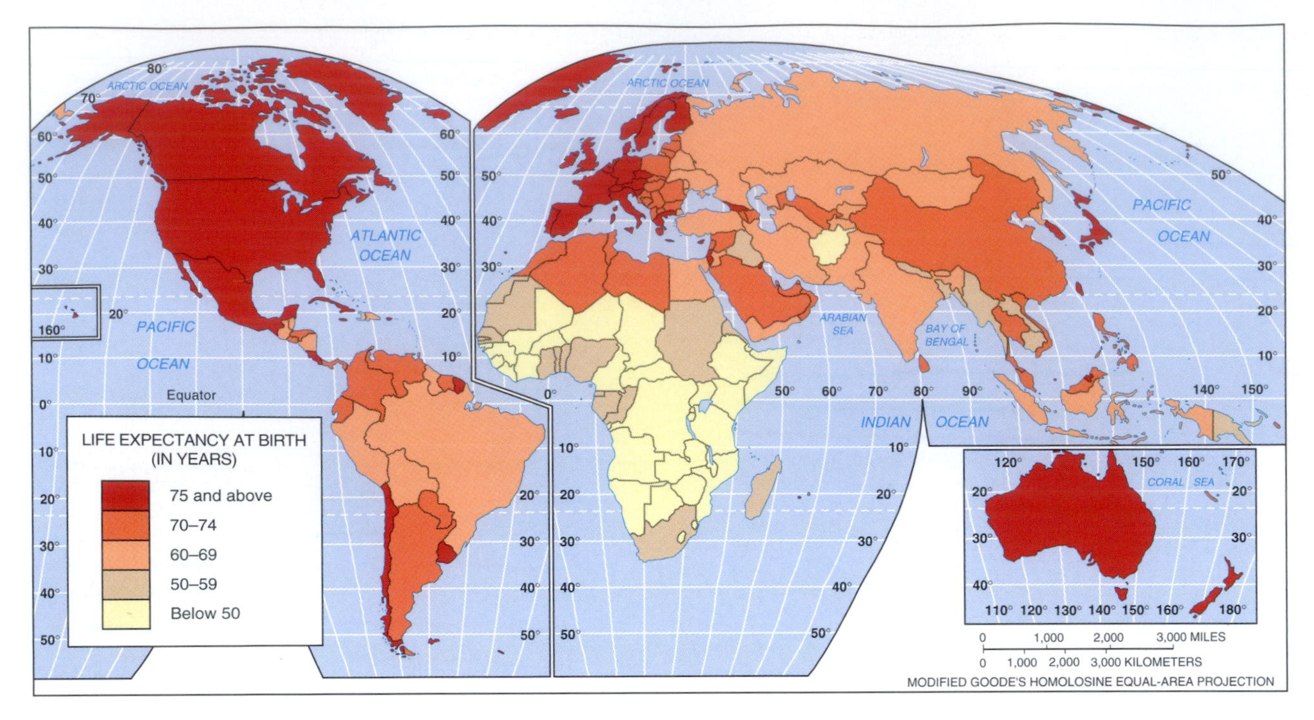

How long the average person can expect to live at birth is largely a function of a country's level of economic development, including access to an adequate diet, clean water, and health care. In the economically developed world, most people can expect to see their 75th birthday (although there are significant variations in terms of gender, class, and ethnicity). In the poorest countries in Africa, in contrast, most people will die before they reach the age of 50, a reflection of their poverty, malnutrition, the widespread wars on the continent, and the growing incidence of diseases, particularly AIDS.

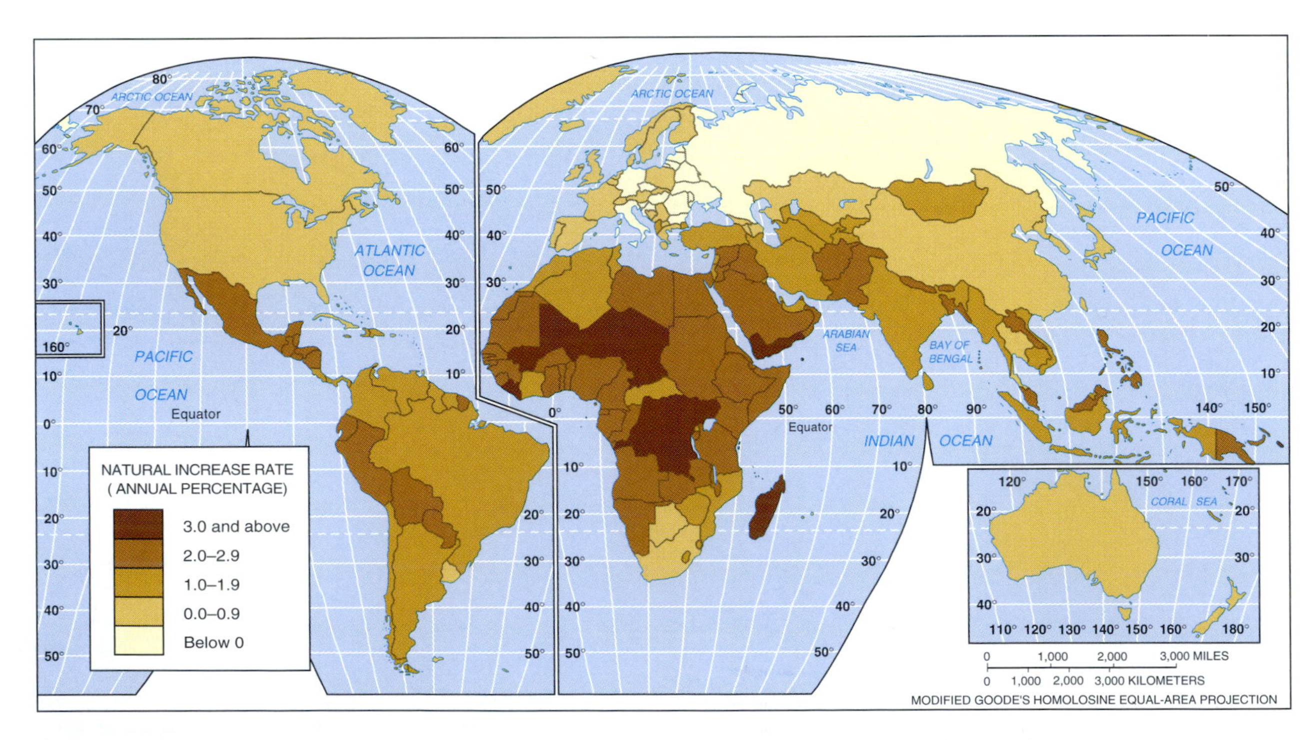

The rate of natural increase is the difference between birth and death rates. Almost always, the poorest countries grow the most rapidly, a reflection of the need to have many children and large families in agricultural contexts. Thus, Africa reveals the highest rates of natural growth, often 3.0 percent annually or higher, although the Arab world is also growing quickly. In contrast, economically wealthy countries in North America, Europe, Japan, and Australia and New Zealand, in which both birth and death rates are low, tend to hover around zero population growth. Russia and several countries in East Europe have death rates exceeding birth rates, indicating they are losing people and in a state of population decline.

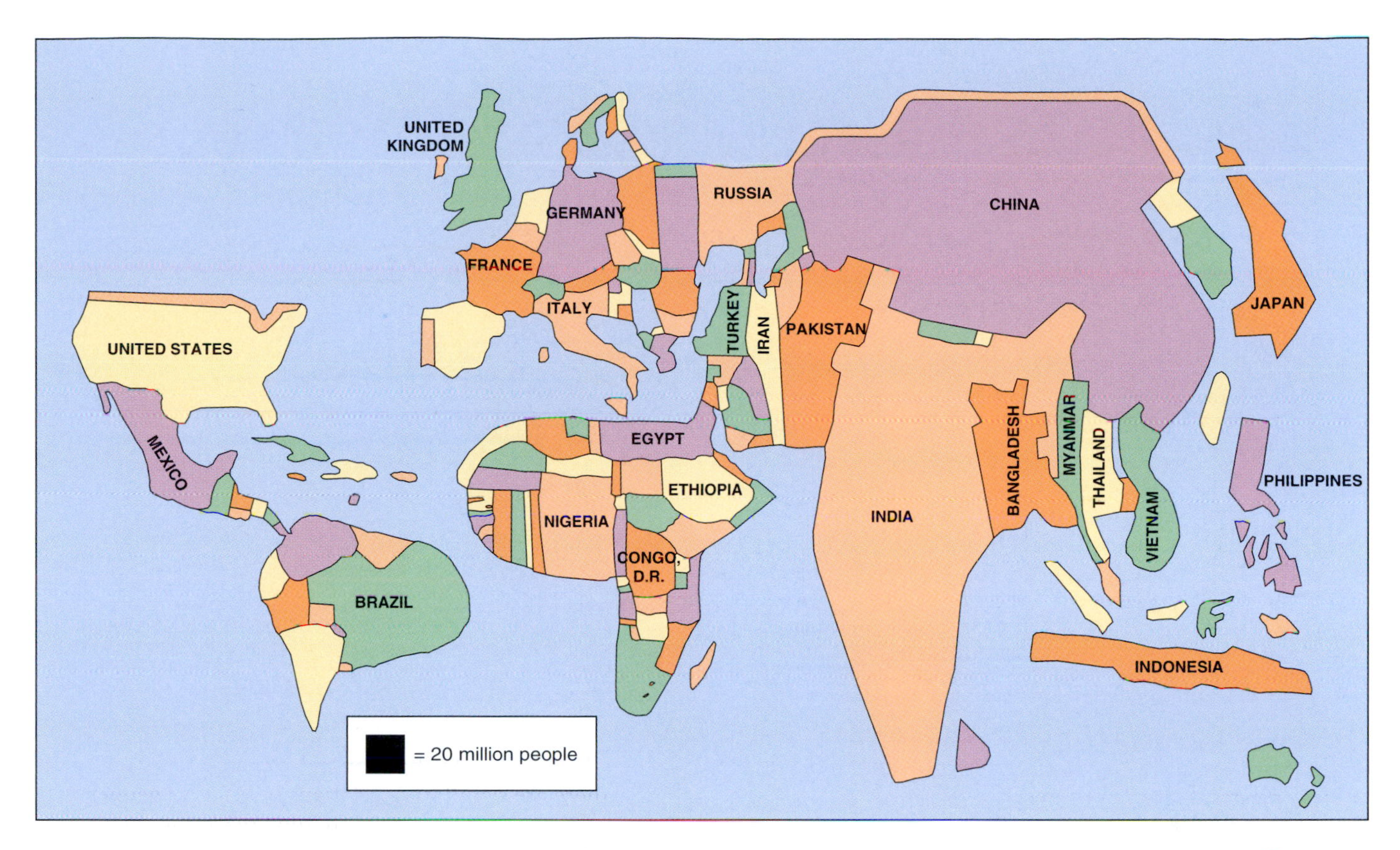

A cartogram is a map that deliberately distorts the size of units (in this case, countries) in proportion to a given variable (in this case, total population). Such a map reveals the vast numbers of people who live in East and South Asia, and the relatively small proportions in Africa and North and South America.

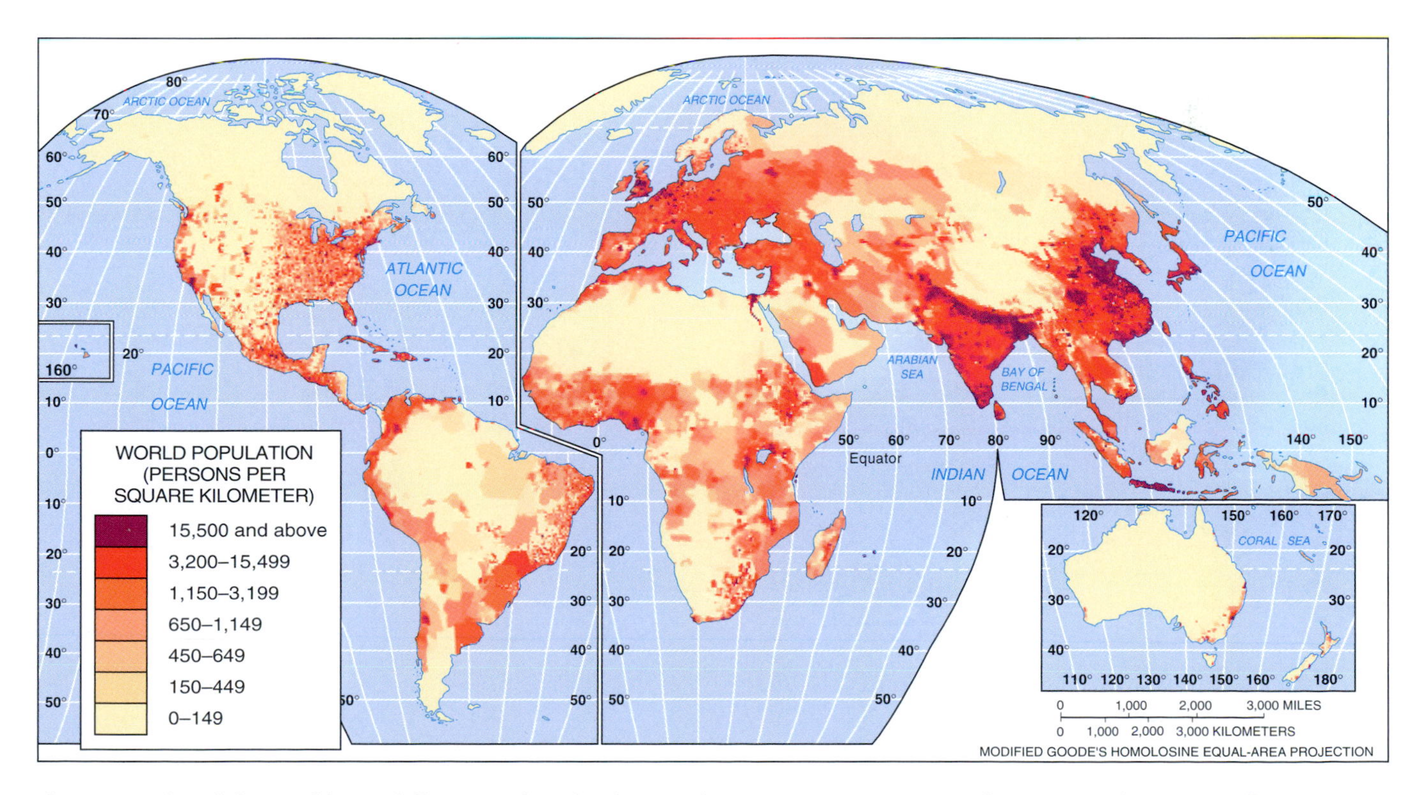

The geography of the world's 6.3 billion people is fundamental to its economic activity. Three major clusters stand out—East Asia, South Asia, and Europe. Vast areas in central Asia, Northern Africa, Australia, and the interior of South America are too hot, too cold, or too dry to sustain dense populations.

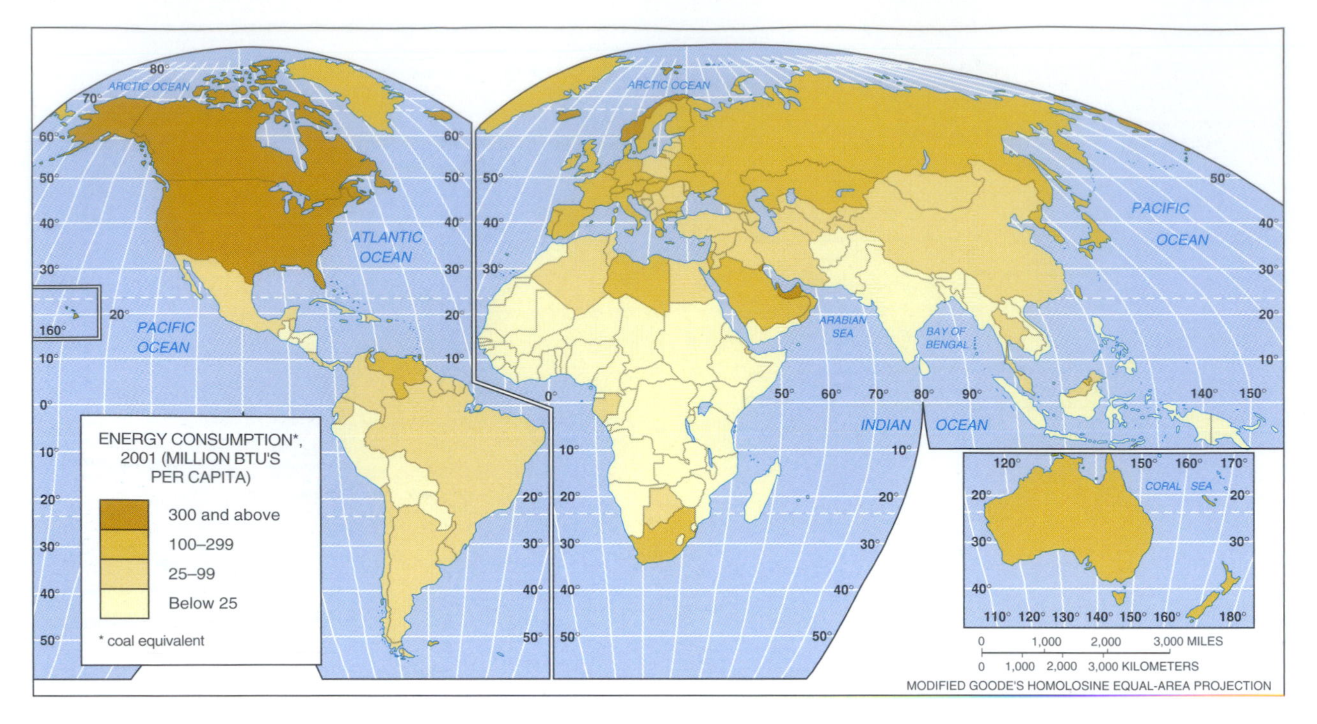

The average rate of energy consumption per person is a reliable measure of economic development. Urbanized, industrialized economies in North America, Europe, Japan, and Australia and New Zealand consume vast quantities for manufacturing, transportation, electricity generation, and heating. In contrast, in much of the developing world, energy consumption levels—and corresponding standard of living—are relatively low.

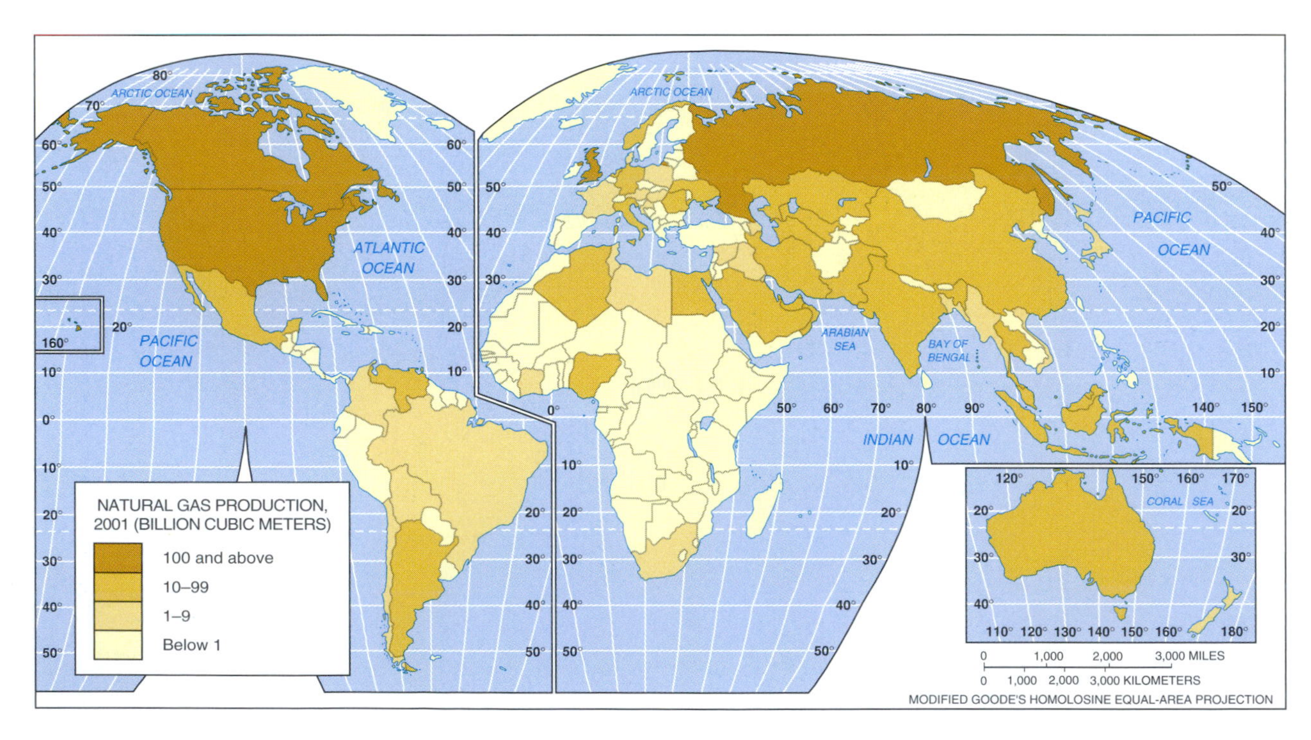

Along with petroleum and coal, natural gas is an important fossil fuel, and one that produces virtually no pollution. The world's major suppliers are in North America, Russia, Britain, and parts of Asia.

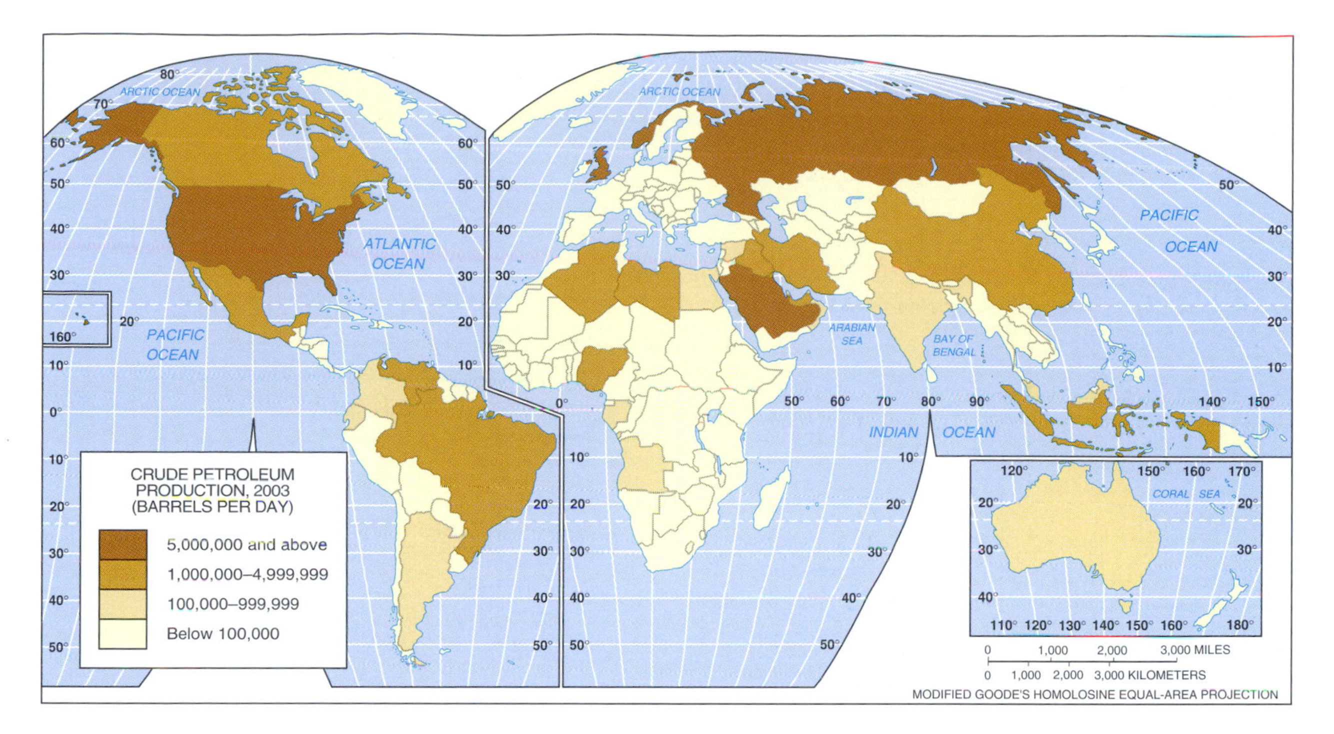

Petroleum—the lifeblood of the industrial world's economies—is produced in a select group of countries. Some of these, including those in the Middle East as well as Venezuela and Indonesia, belong to the Organization of Petroleum Exporting Countries (OPEC). Other producers, which do not belong to OPEC, include Russia, the United States, Mexico, Britain, and Norway.

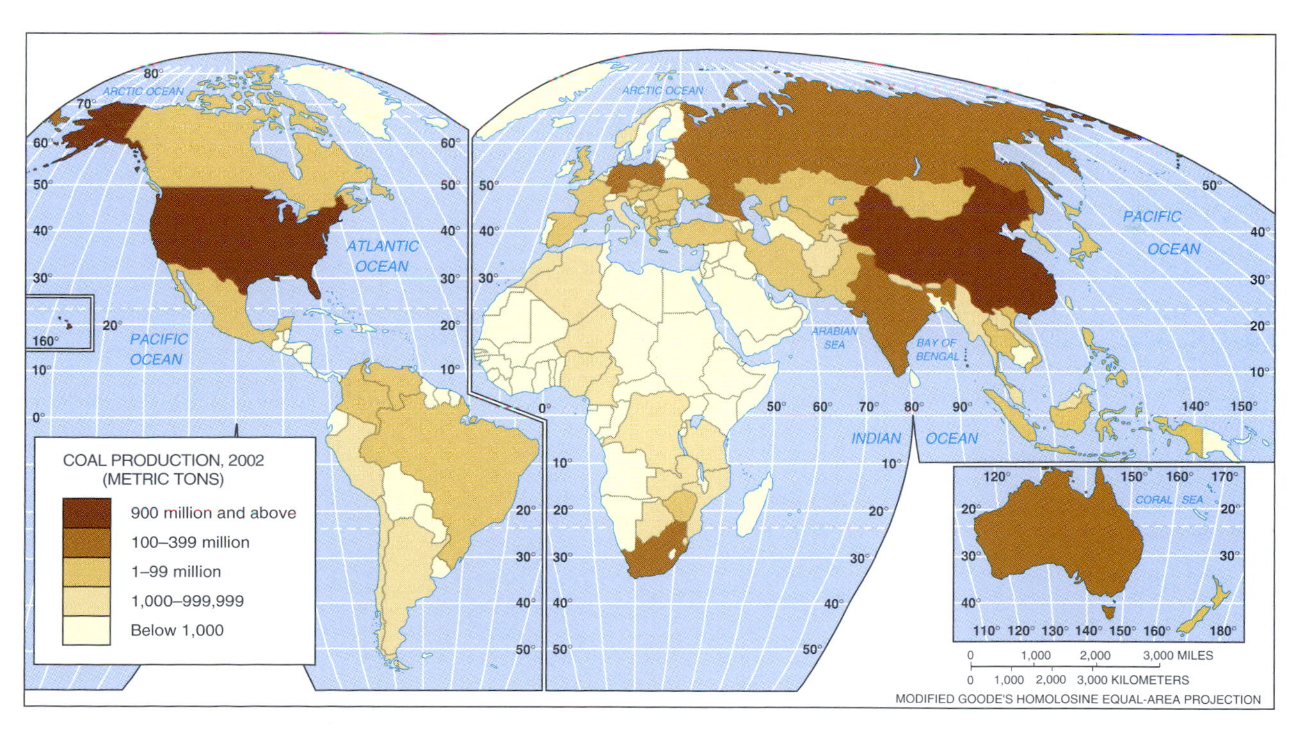

Coal, which made the Industrial Revolution possible, continues to be used in large quantities today. It is a major source of energy for the United States, and the largest source for China, the world's major producer and consumer. However, the burning of coal has serious environmental consequences, including urban air pollution, the release of vast quantities of carbon dioxide, which contributes to global warming, and acid rain, which destroys forests.

The U.S. and Canada are threatened by environmental problems that result from large industrial economies, including urban air pollution, acid rain, and the depletion of groundwater from aquifers.

In Latin America, poverty, poor land distribution, lack of rural opportunities, and exploitation of the environment by multinational corporations combine to threaten tropical rainforests, pollute water, and create low quality air in urban areas.

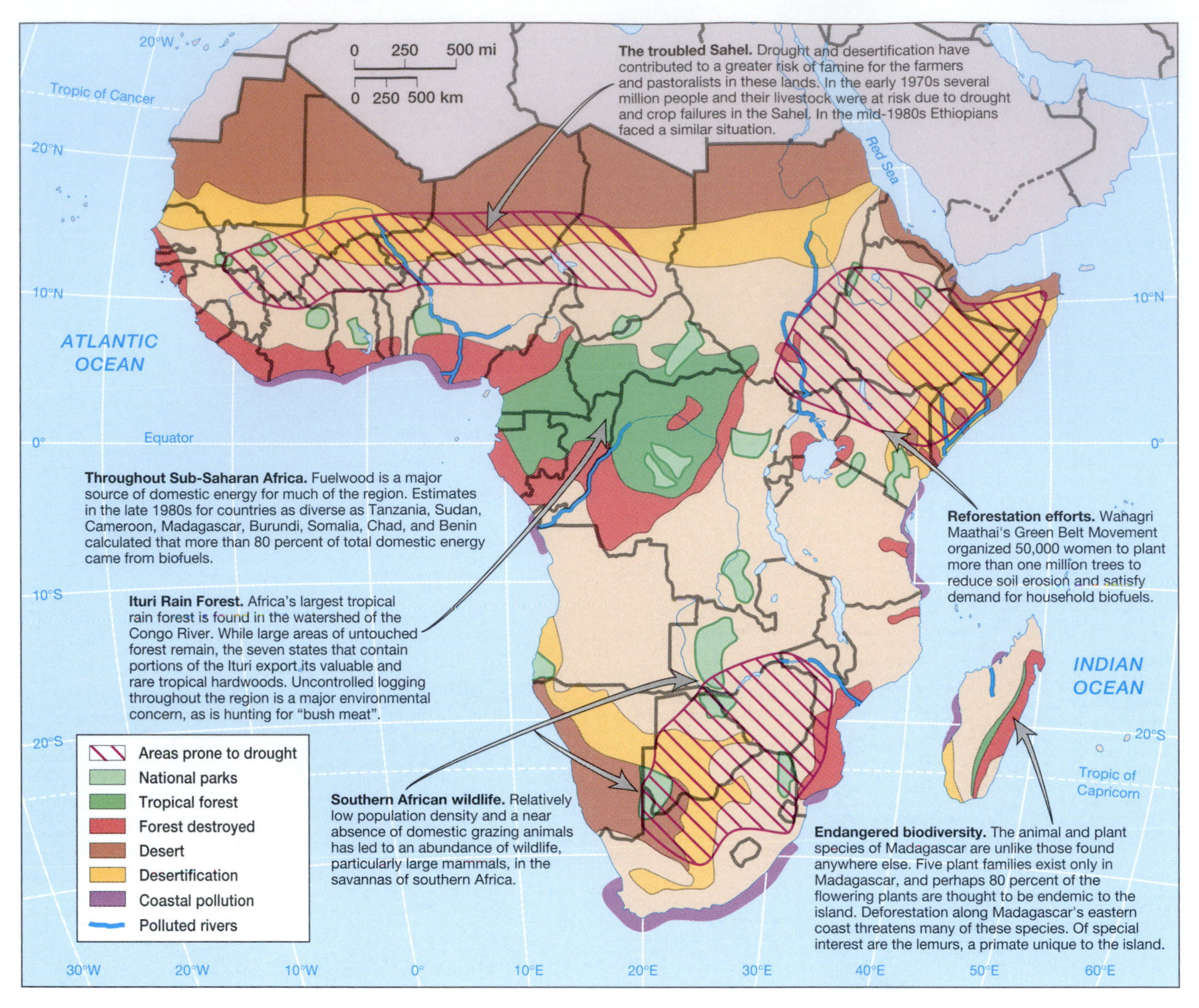

The poorest part of the world, Sub-Saharan Africa suffers from multiple environmental problems, including deforestation, desertification, overgrazing, soil erosion, and the extinction of wildlife.

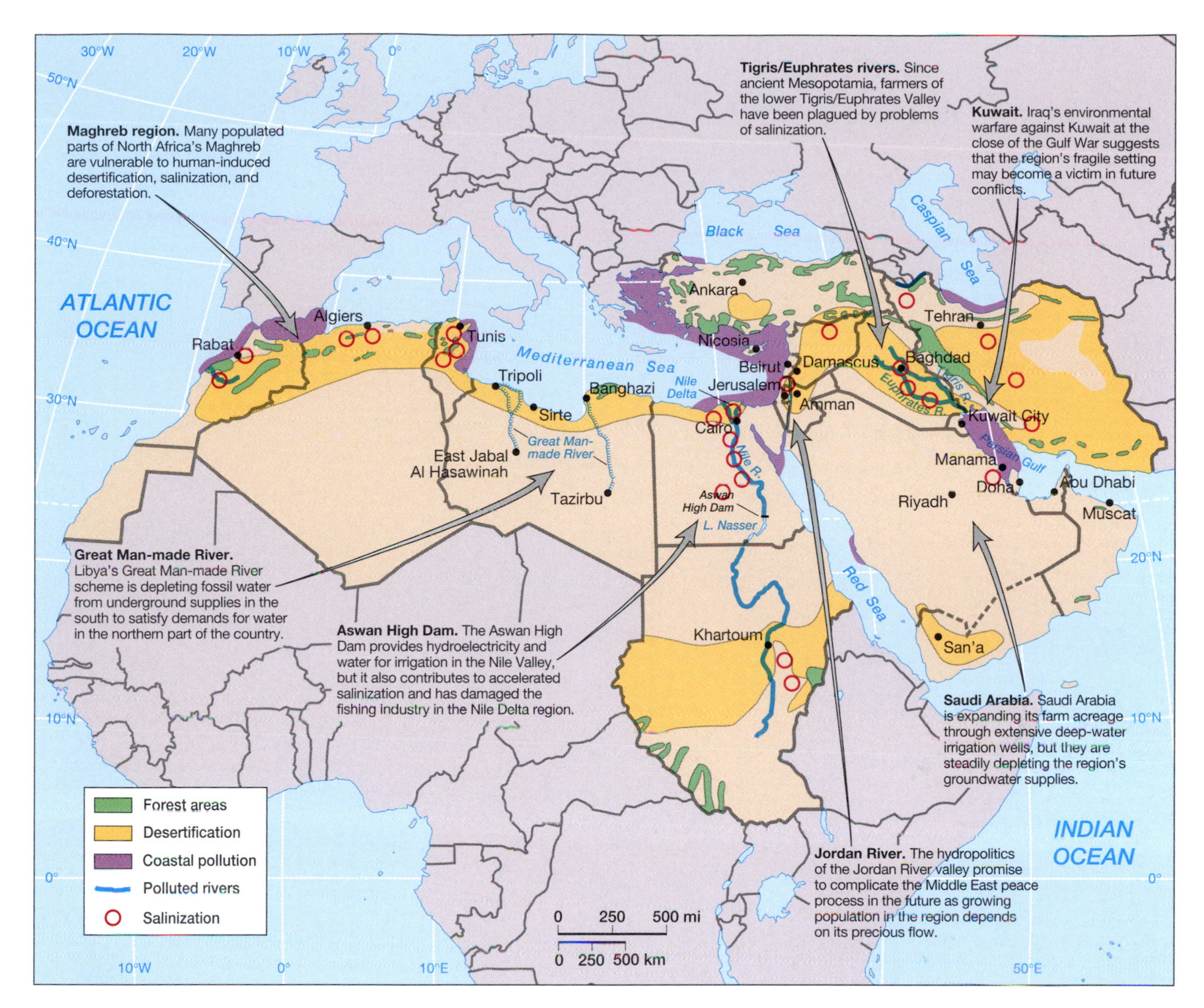

An arid part of the world with rapidly growing populations, North Africa and the Middle East suffers from inadequate water supplies, overgrazing, and irrigation-induced salinization.

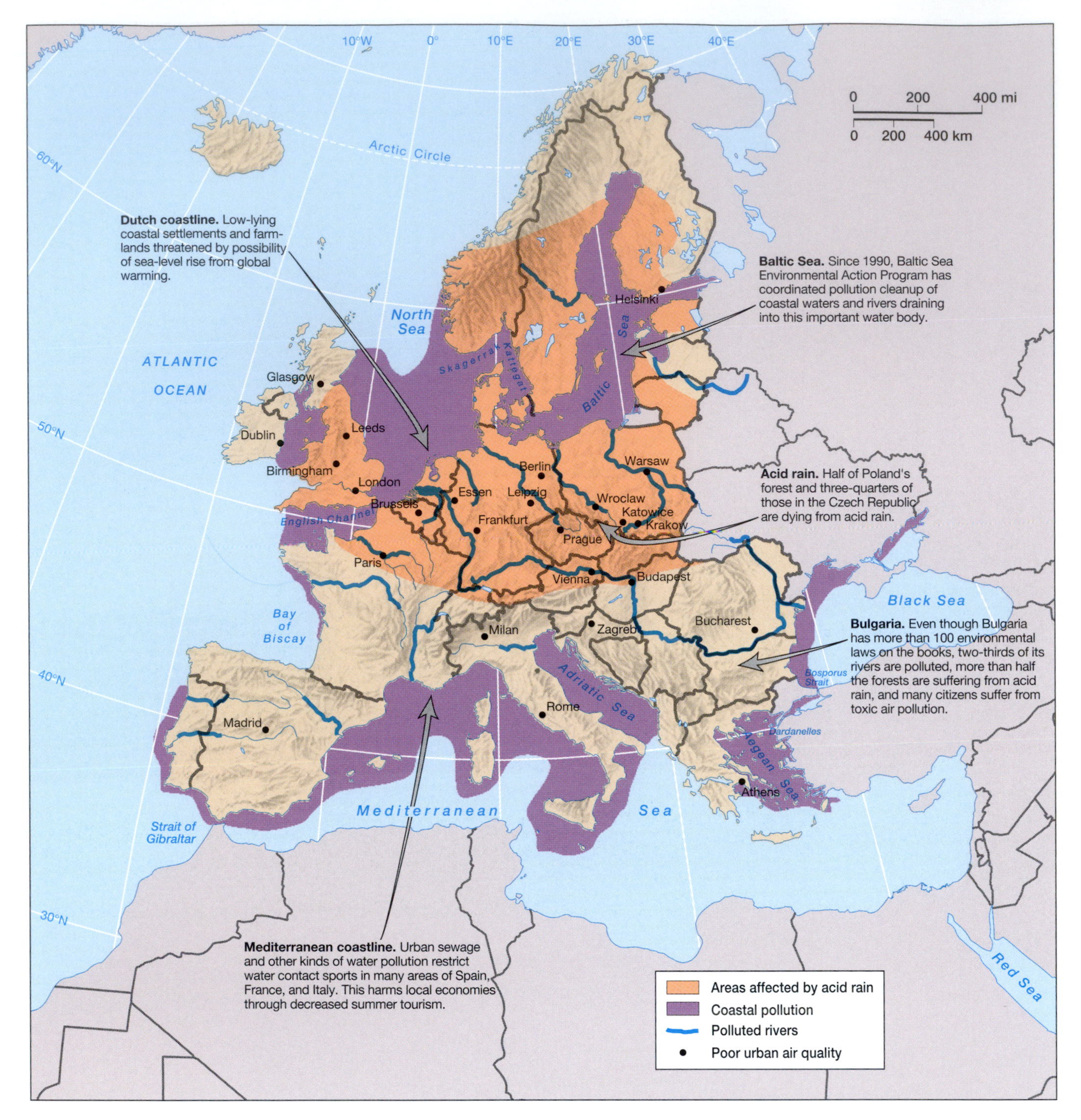

Europe suffers from numerous environmental problems, including pollution of its maritime environments. Eastern Europe suffers in particular due to decades of mismanagement by the communist governments there.

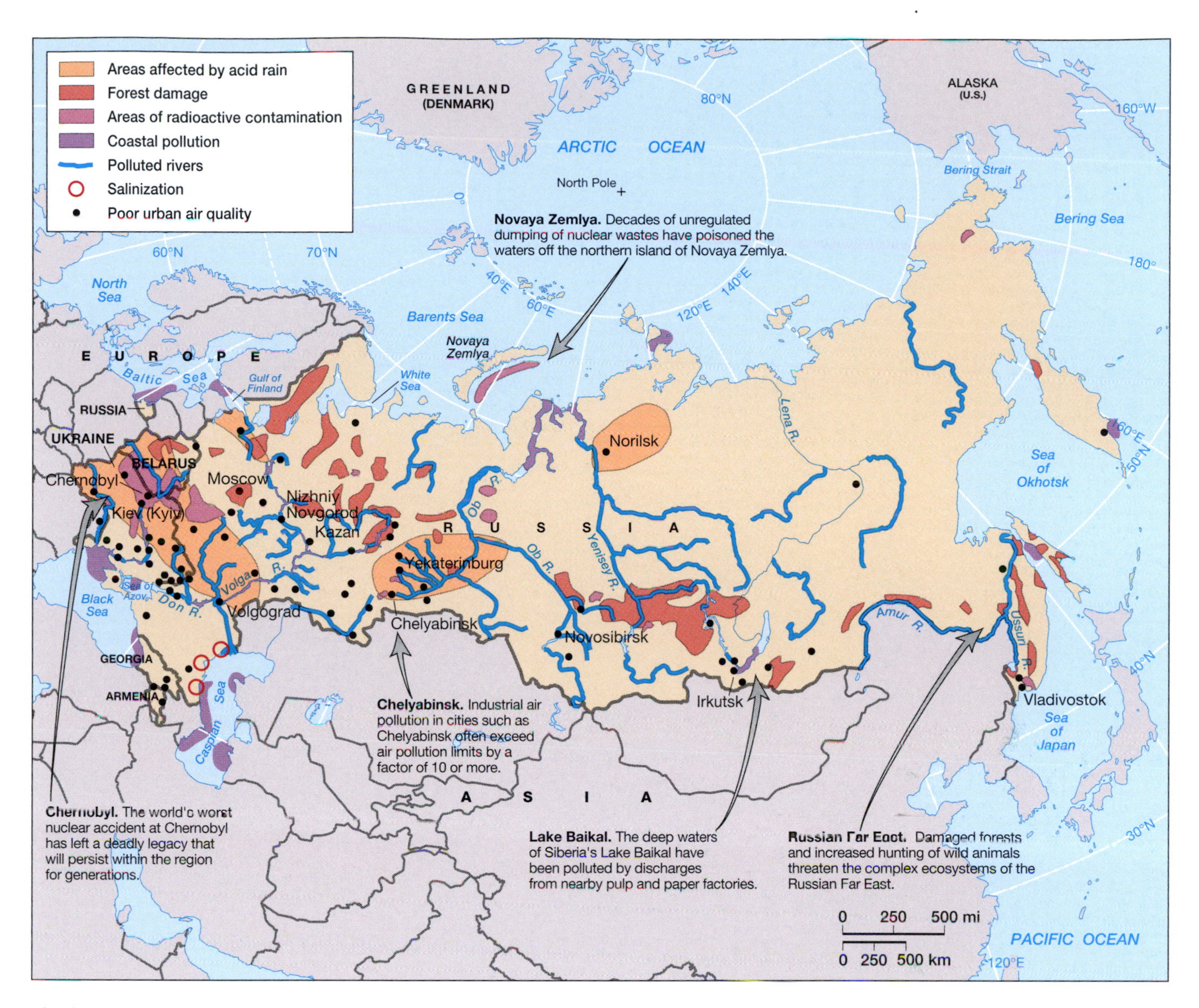

The largest country in the world in terms of area, Russia's decades of communist governments left landscapes often heavily polluted, including nuclear waste and polluted rivers.

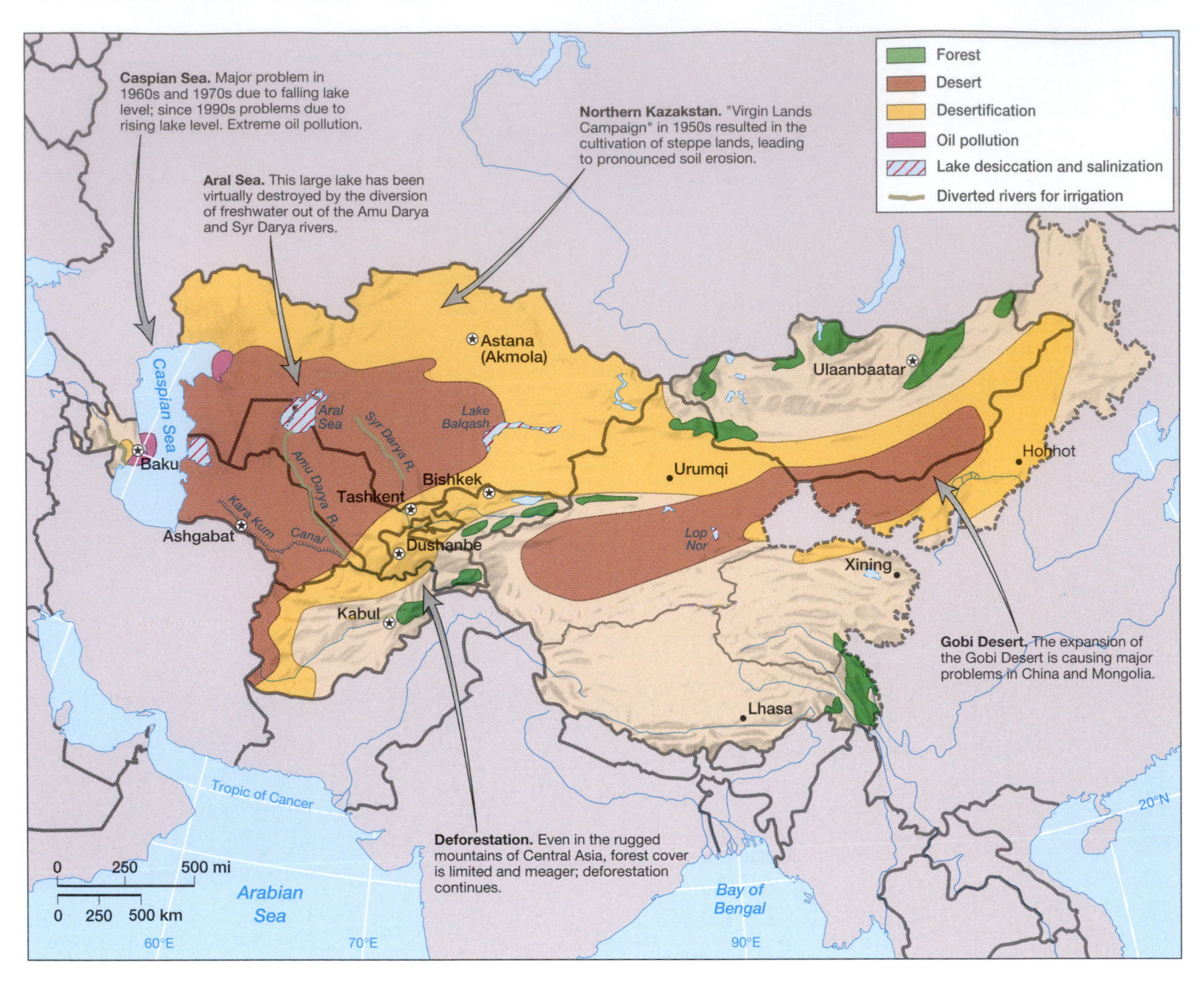

Central Asia, generally quite arid, suffers from mounting problems such as soil erosion, overgrazing, and the depletion of water in the Caspian Sea and Aral Sea.

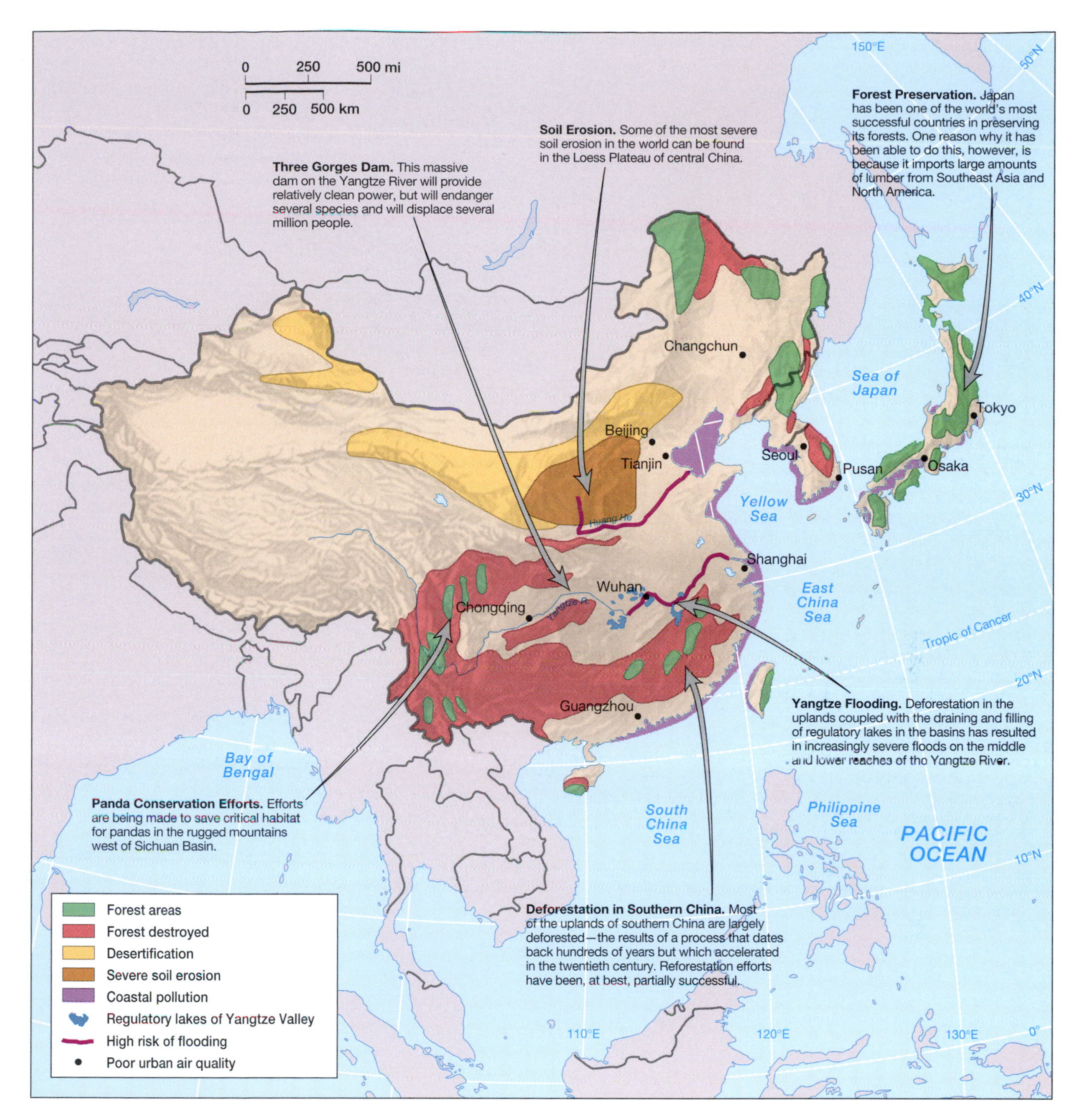

Home to one-third of the world, East Asia's natural environment has been extensively reworked over millennia. China in particular, with a rapidly growing economy, has witnessed extensive deforestation. Periodic floods along the Yangtze River valley have led the Chinese to construct the famous Three Gorges Dam, which, when completed, will be the world's largest.

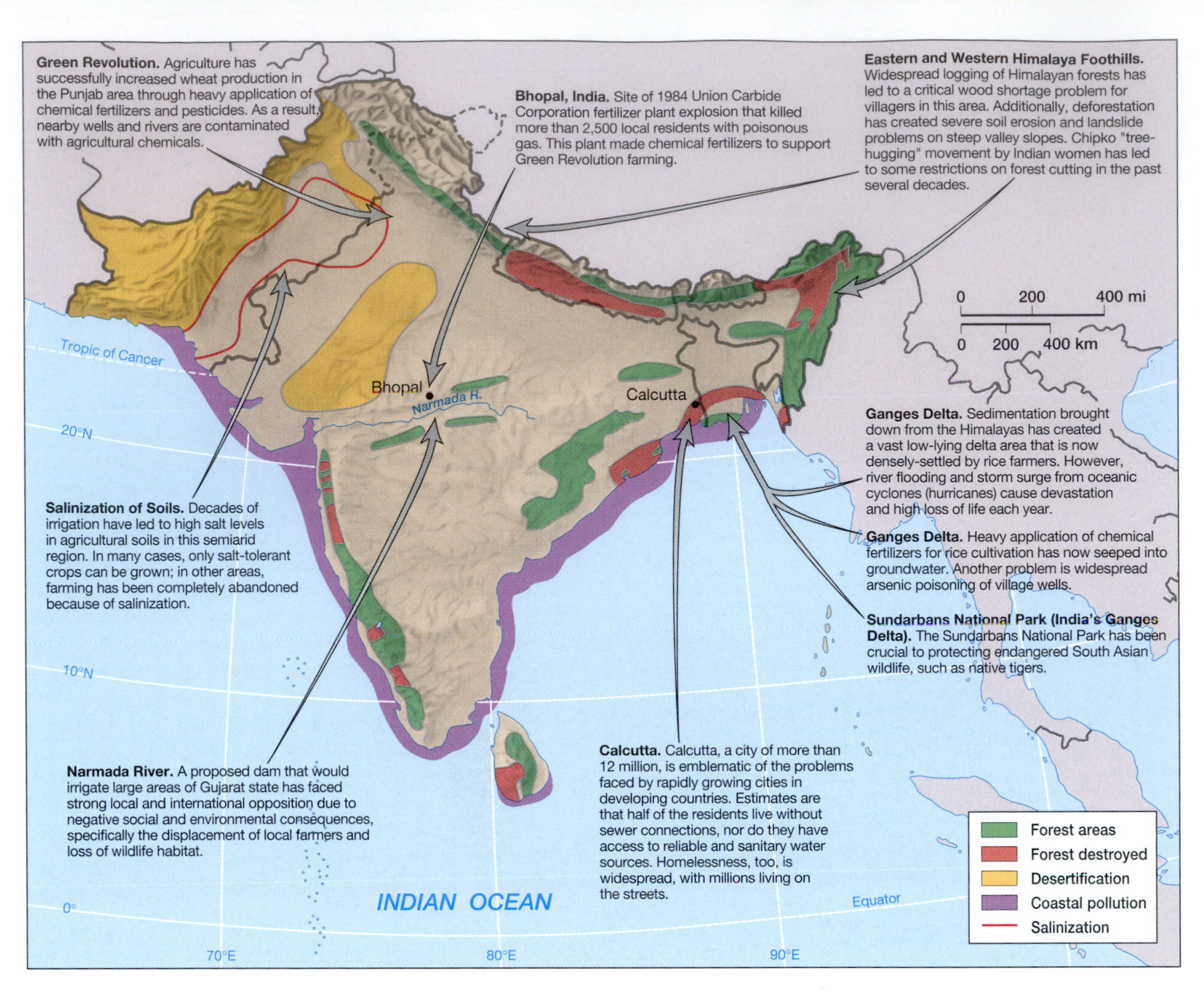

Home to 1.5 billion people, the Indian subcontinent wrestles with numerous environmental problems, including deforestation, flooding, and soil erosion.

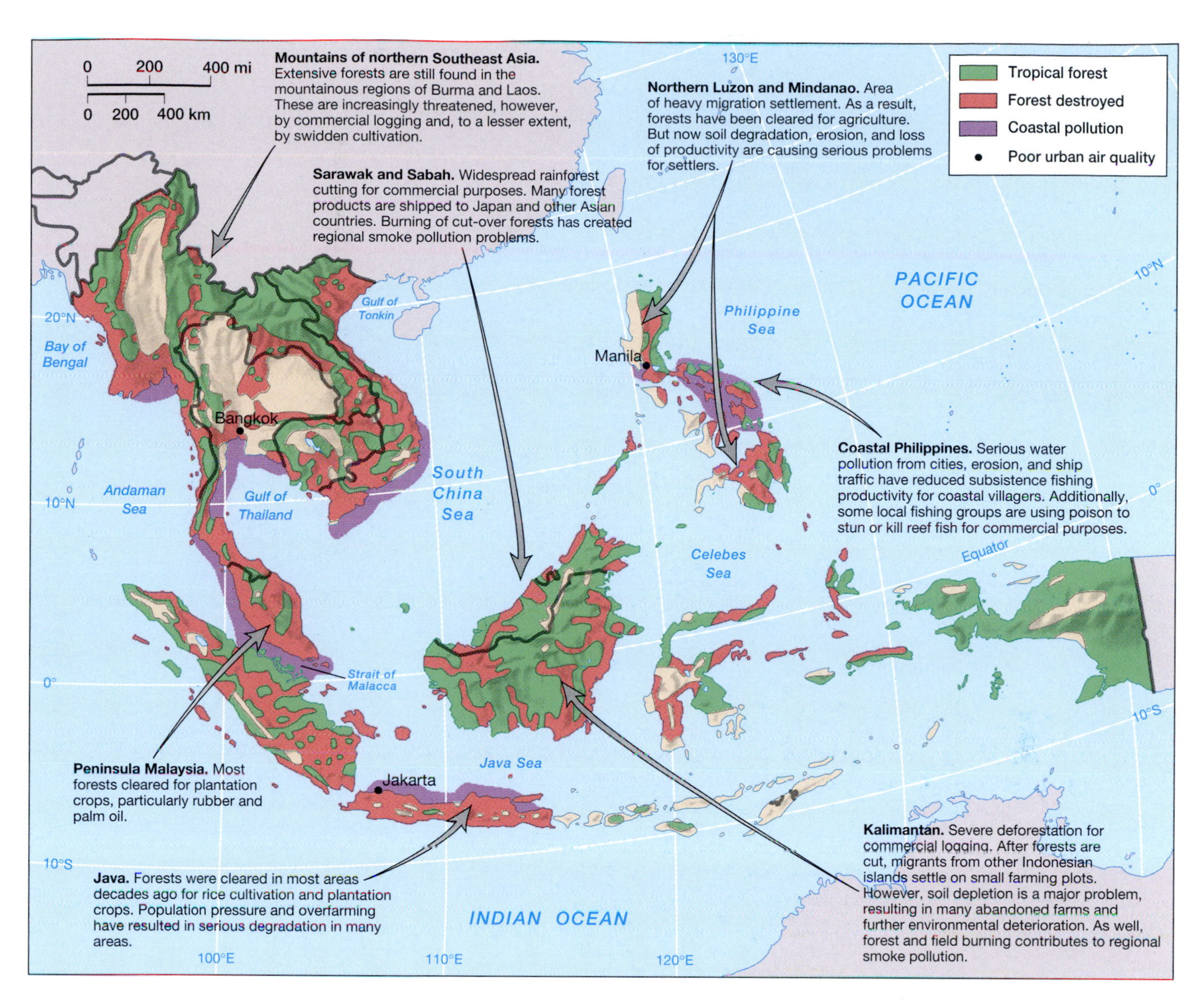

Logging and expanding farmlands are rapidly deforesting the lush rainforests of Southeast Asia, one of the world's three major areas for such ecosystems. Urban air pollution in the region's rapidly growing cities is also a problem.

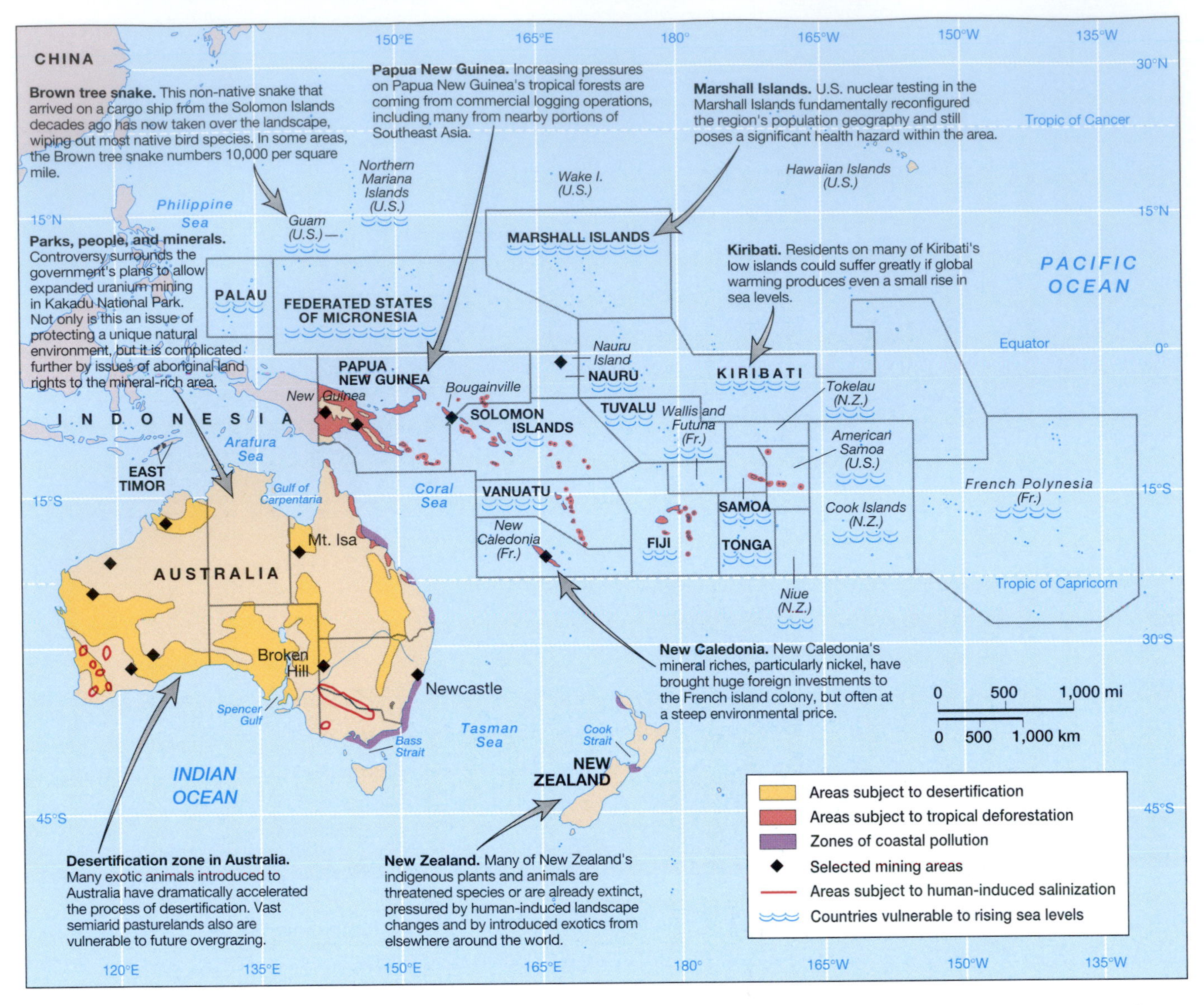

The natural ecosystems of Australia and Pacific Islands have been dramatically reconfigured by people. Whereas Australia suffers from the threat of desertification, other places, such as Papua New Guinea and many of the smaller islands face environmental challenges in the forms of deforestation and the legacy of open air nuclear testing in the region.

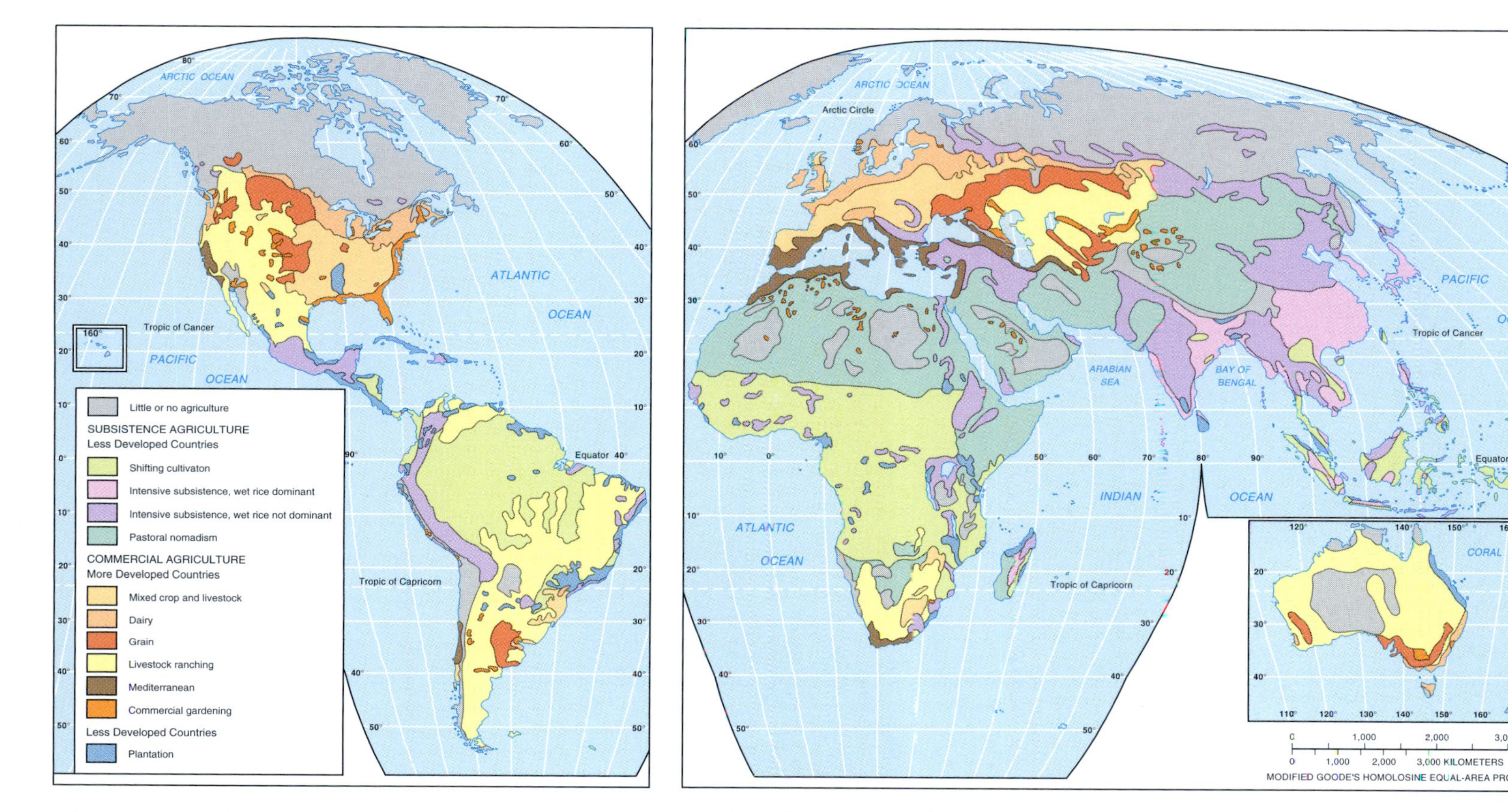

The geography of agriculture worldwide reflects an enormous number of forces, including, among others, climatic patterns and soil fertility, topography, the availability of water, as well as social forces such as the impacts of colonialism, the legal and cultural organization of property ownership, population growth, the market prices of crops, government subsidies, and the role of multinational corporations. Agriculture thus reflects the complex intersections of nature and society in such a way that they cannot be separated. There are a large variety of forms of agriculture, ranging from pre-industrial, subsistence production (such as shifting cultivation and intensive rice cropping) to highly industrialized, capital- and energy-intensive corporate agriculture such as is found in the United States.

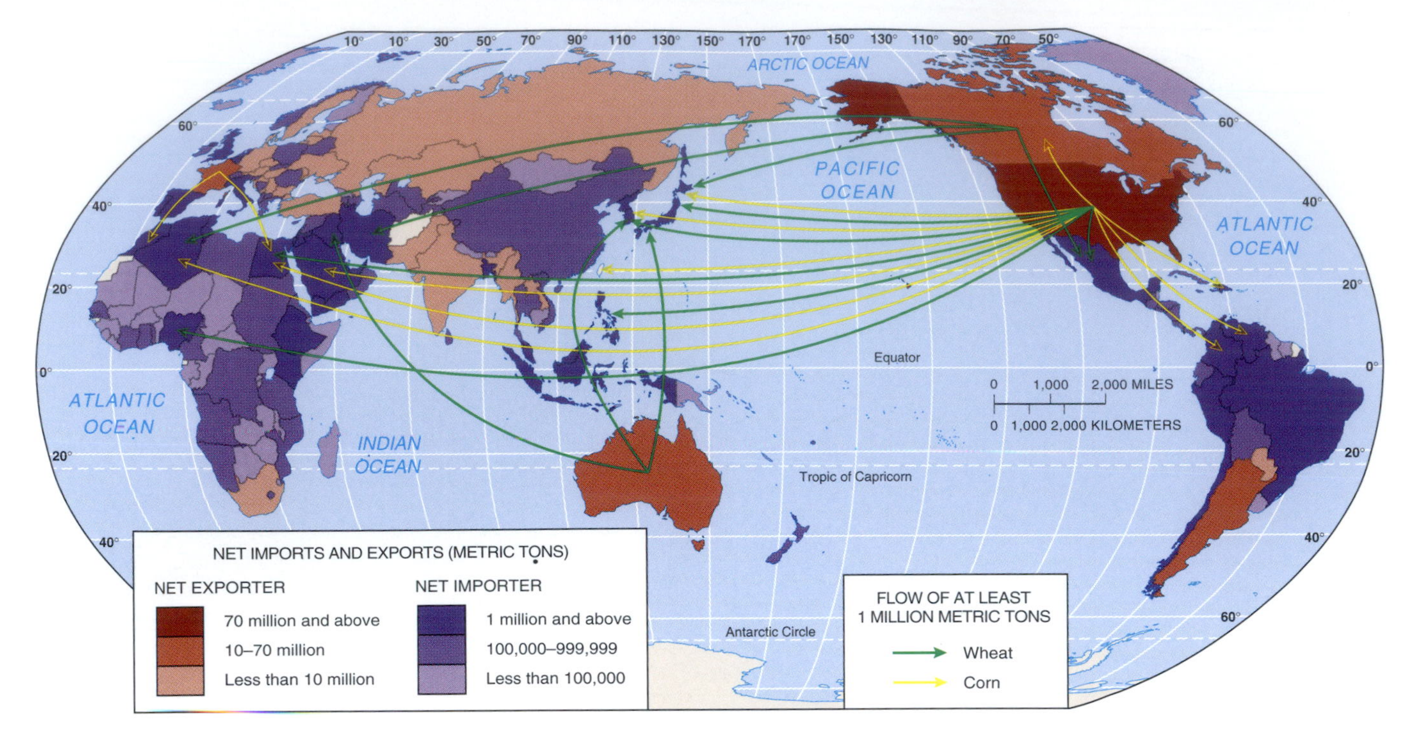

The geography of international trade in grains, including wheat, rice, and corn, reflects the highly uneven distribution of supply and demand over space. A small handful of countries, led by the United States but including Canada and Australia, are the major exporters, supplying food to vast numbers of people worldwide, especially in East Asia.

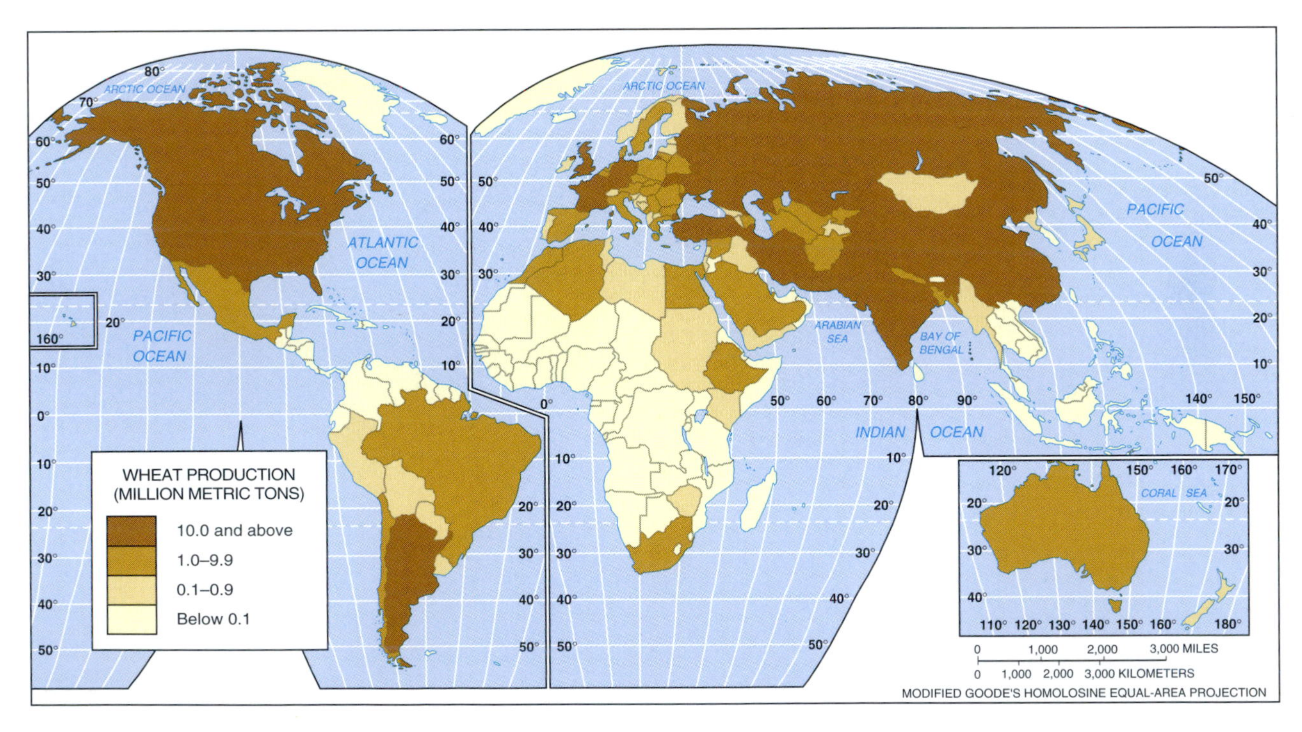

Wheat is one of the world's premier agricultural crops. While wheat is bought and sold extensively on the international market, much of it is consumed domestically. Wheat production levels tend to be highest in mid-latitude countries with vast open areas, suitable climate and rainfall, and rich soils, including those in North America, Argentina, Russia and Ukraine, as well as India and northern China.

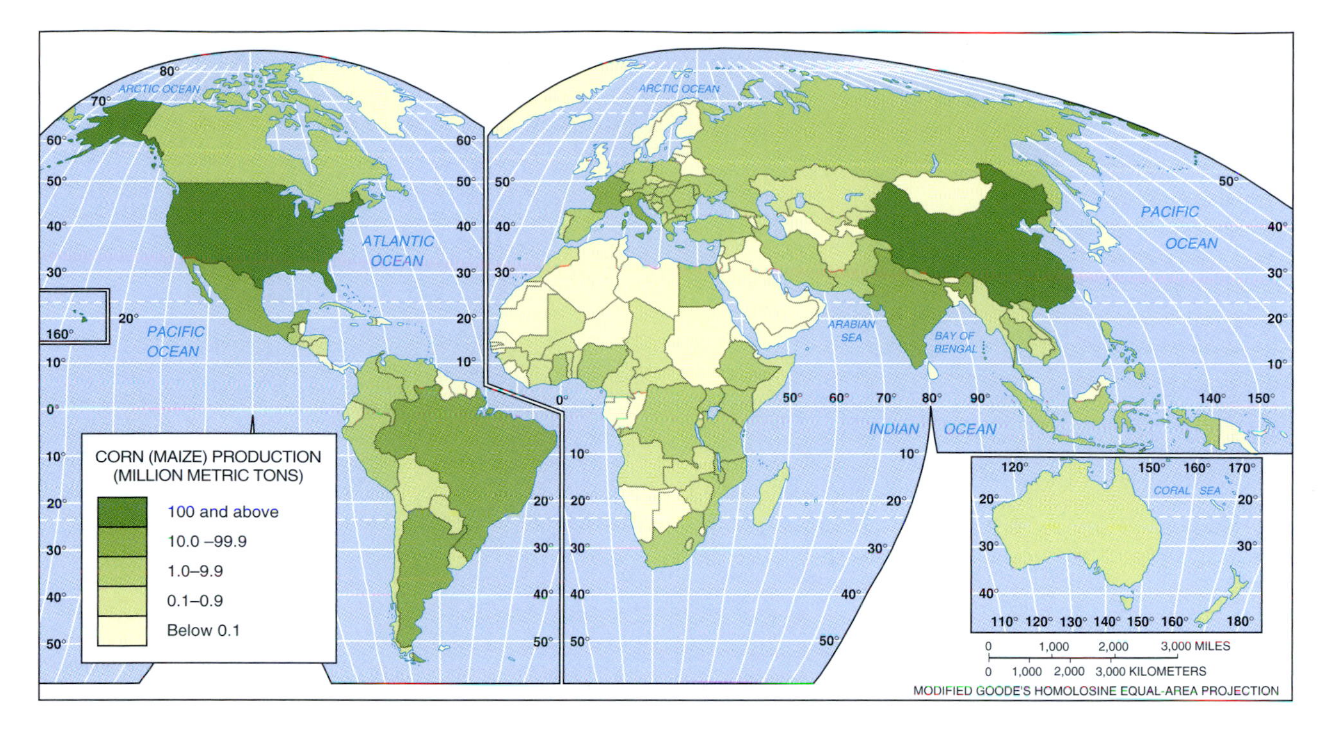

Corn is an important crop that is used to feed not only people but often animals as well. China and the U.S. dominate world corn production. In the U.S., the bulk of corn is used to fatten cattle, as well as to produce sweeteners.

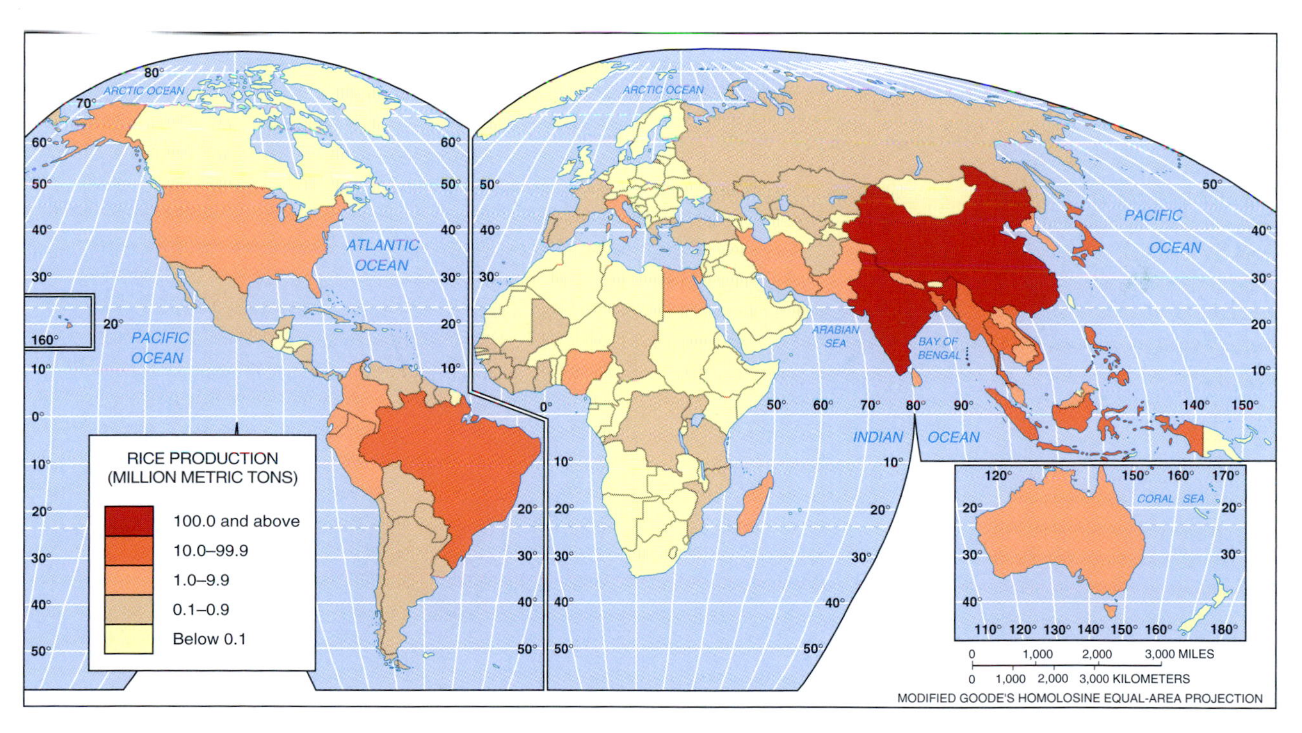

Rice is the most commonly grown crop in the world. In East Asia and eastern India, where rice is the staple food for one-third of humanity, it has long been grown in the labor-intensive form of irrigated rice paddies, often carved out of hillsides, where hundreds of millions of people engage in arduous tasks such as planting and harvesting. Because rice requires large amounts of water, this type of agriculture involves intricate irrigation systems.

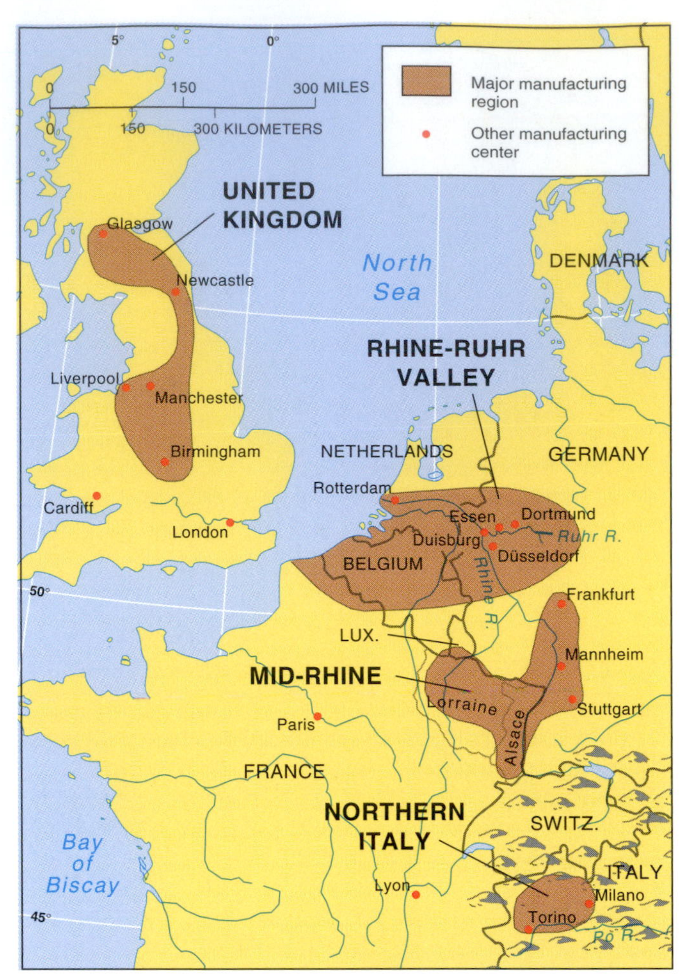

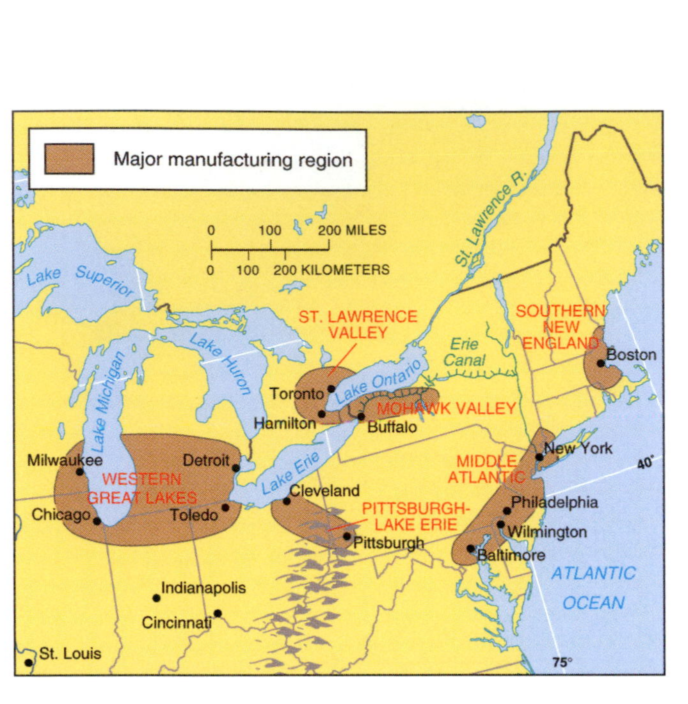

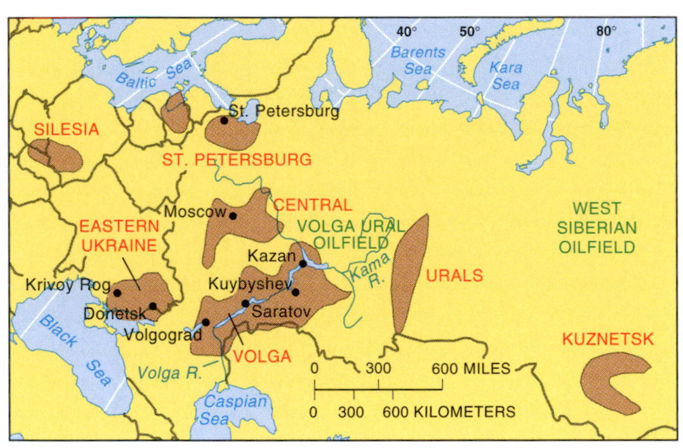

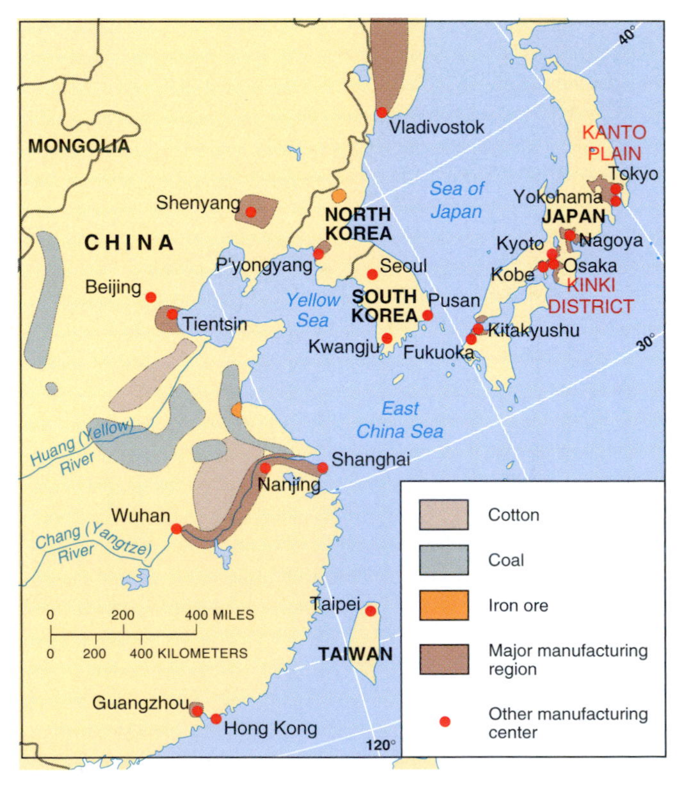

Manufacturing and industrial activity tends to be tightly concentrated in a few, select urban regions, where firms have access to agglomeration economies, specialized pools of labor, information, and infrastructure. In North America, several industrial districts form the Manufacturing Belt of the northeast and Midwest. In Western Europe, a broad north-south belt extends from central Britain to western Germany to northern Italy. In Eastern Europe and Russia, the complexes of Silesia, St. Petersburg, the Donetsk basin in Ukraine, and the Volga River basin stand out. In East Asia, manufacturing is clustered in central Japan, South Korea, and the rapidly growing districts of China, particularly along the southern Yangtze River basin.

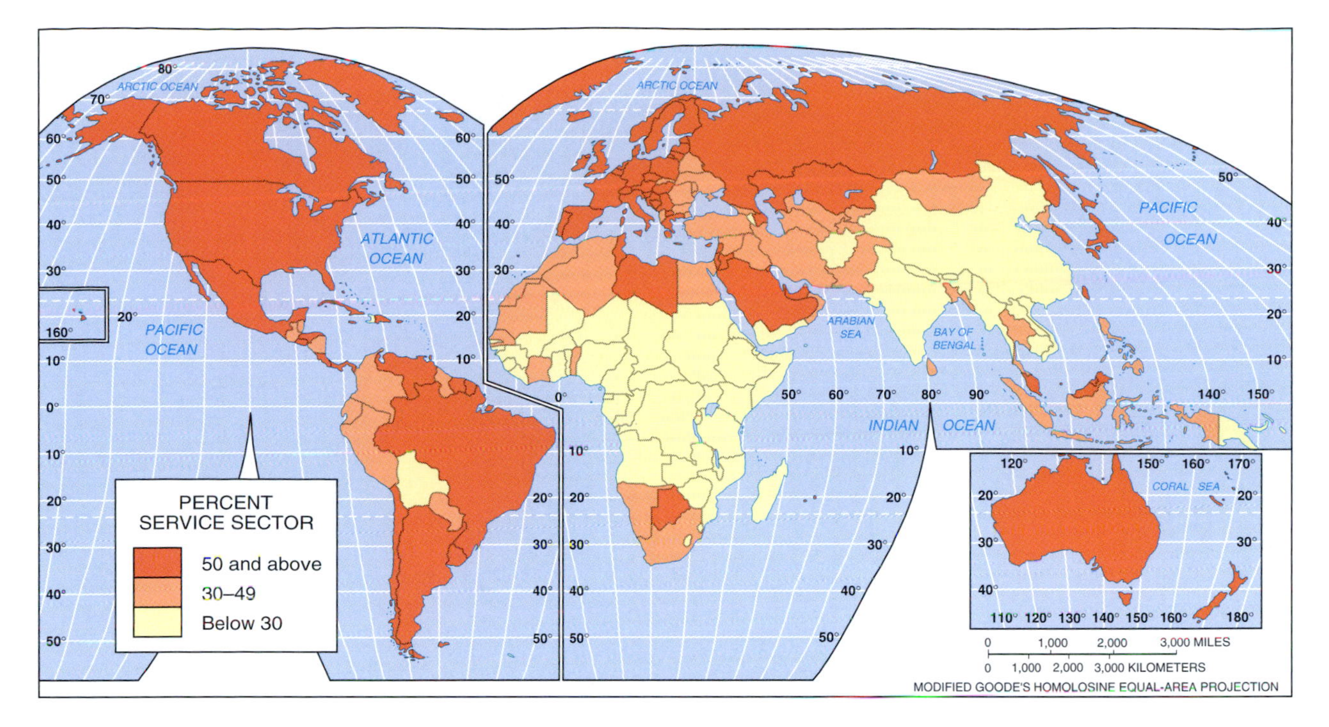

The proportion of workers in service occupations is closely associated with the degree of national development. Generally, in the First World (Europe, North America, Japan, Australia and New Zealand), the vast bulk of workers are involved in the production of intangibles, while in the developing world this proportion tends to be relatively low. The category "services," however, encompasses a vast array of different jobs ranging from prostitutes to professors.

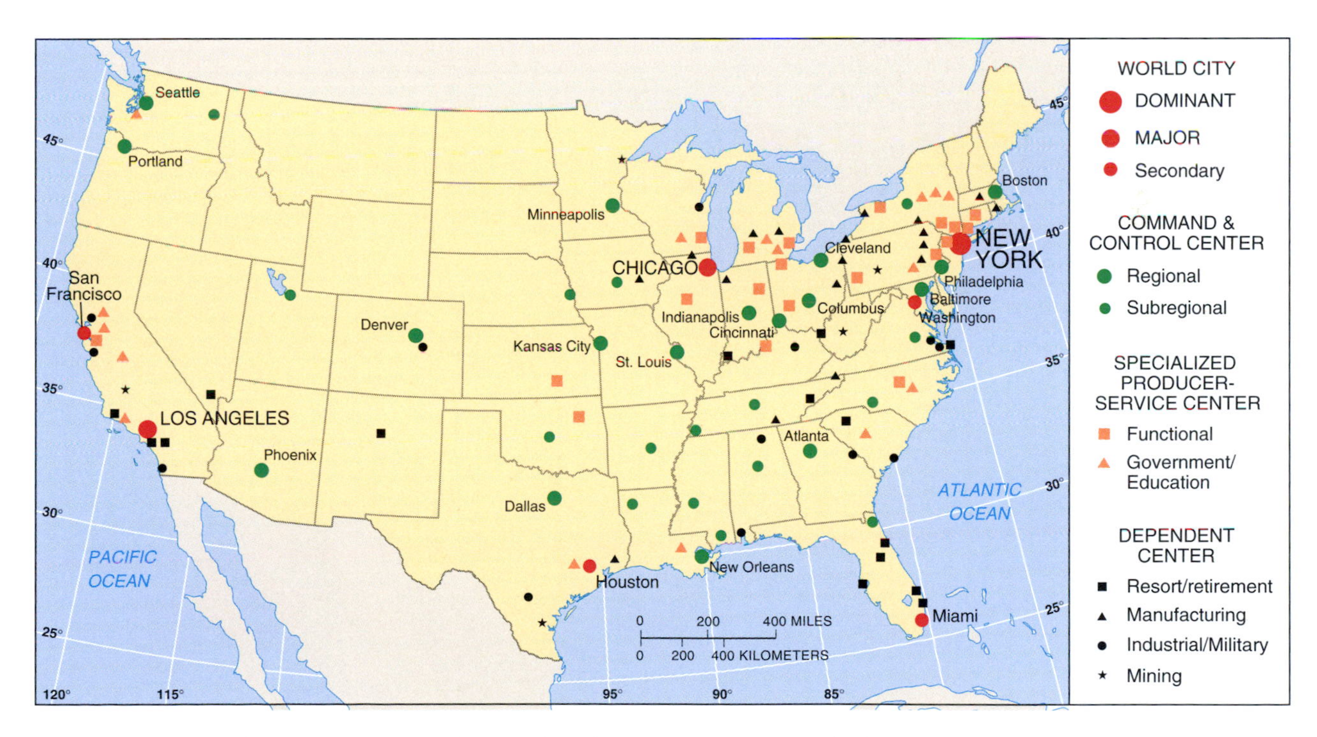

Business services, which cater to other firms rather than households, include firms that produce specialized expertise, such as legal services, advertising, accounting, engineering and architecture, marketing, and corporate research and development. These activities agglomerate in large cities, where they rely upon dense webs of connections to other firms, often in the form of face-to-face contact.

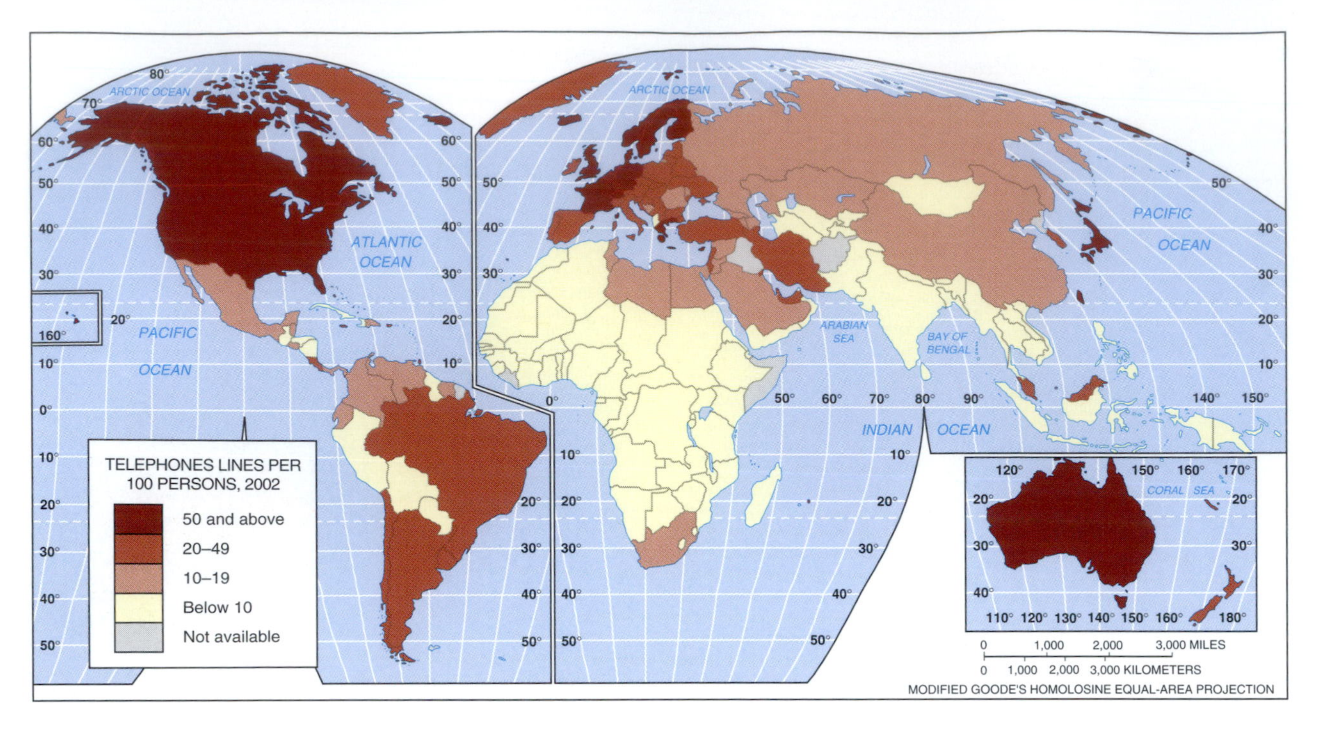

The workhorse of telecommunications is the telephone. While the ratio of people in the First World is relatively low (often as low as 2 per phone), in the developing world there may be 50 or more people per phone. One-half of the world has never made a phone call.

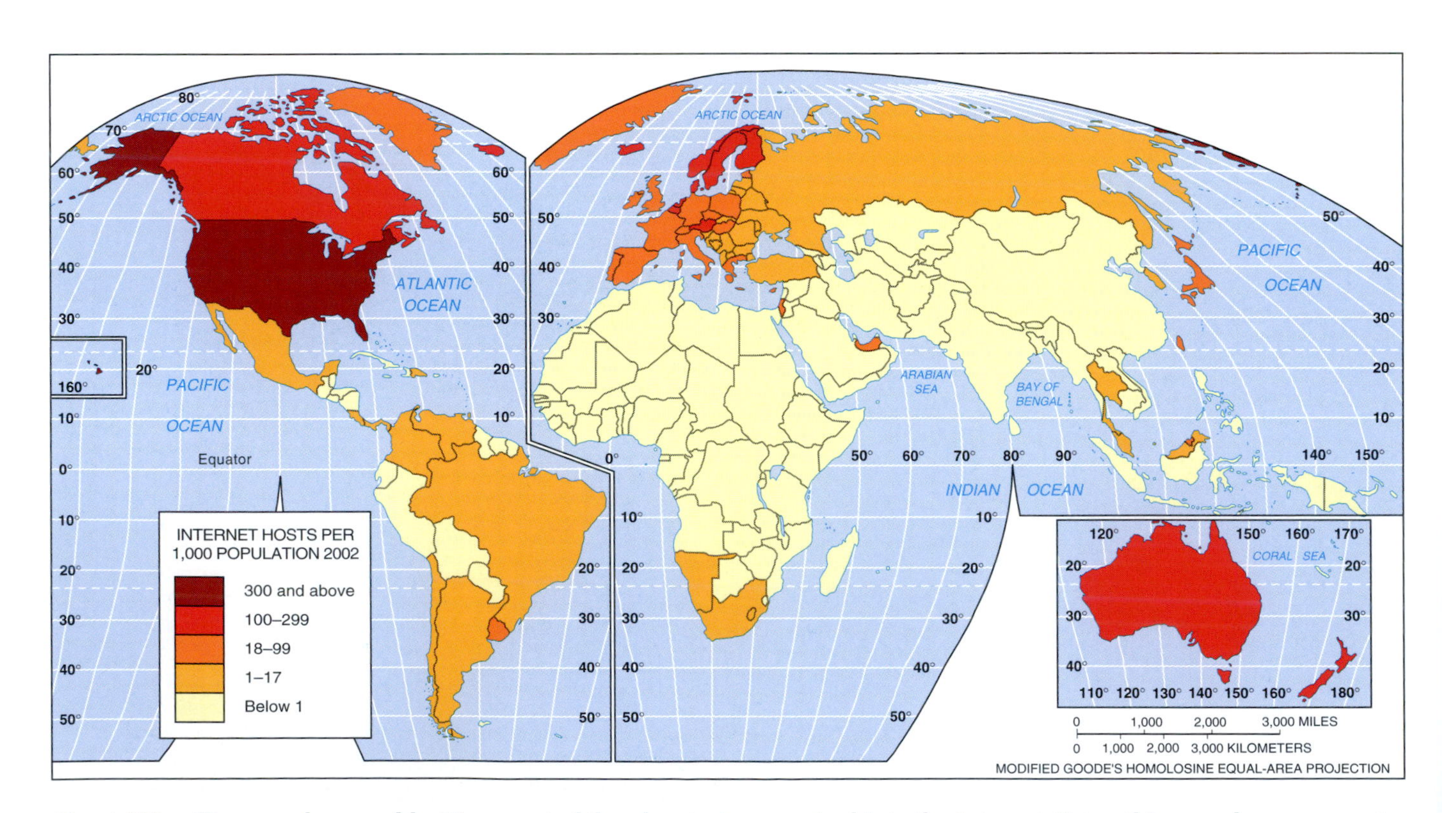

About 600 million people—roughly 10 percent of the planet—is now wired into the Internet. Yet as this map shows, access to the Internet is in many ways a map of the world's wealth and poverty. Internet penetration rate (% of people with access) range from as little as 0.9% in Africa to as high as 62% in North America. While citizens in the developed world suffer from information overabundance, vast numbers in impoverished countries will never get on-line.

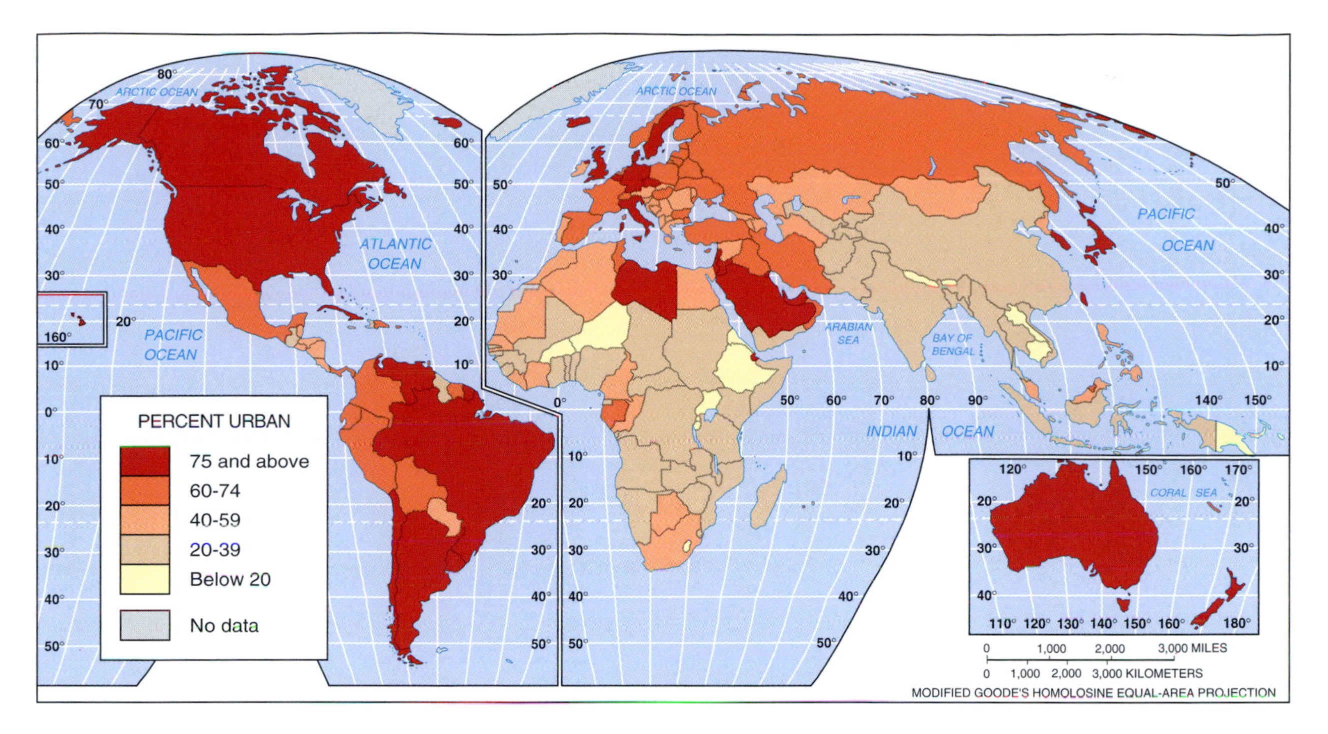

Today, one-half of the world's six billion people live in cities. Definitions of "urban" vary widely among countries. The proportion of urbanites is closely associated with countries' wealth, labor force and occupational structure, and degree of rural development, factors that in turn reflect their historical circumstances and position in the world economy. Generally, wealthy, industrialized countries are more heavily urbanized than poorer, pre-industrial ones.

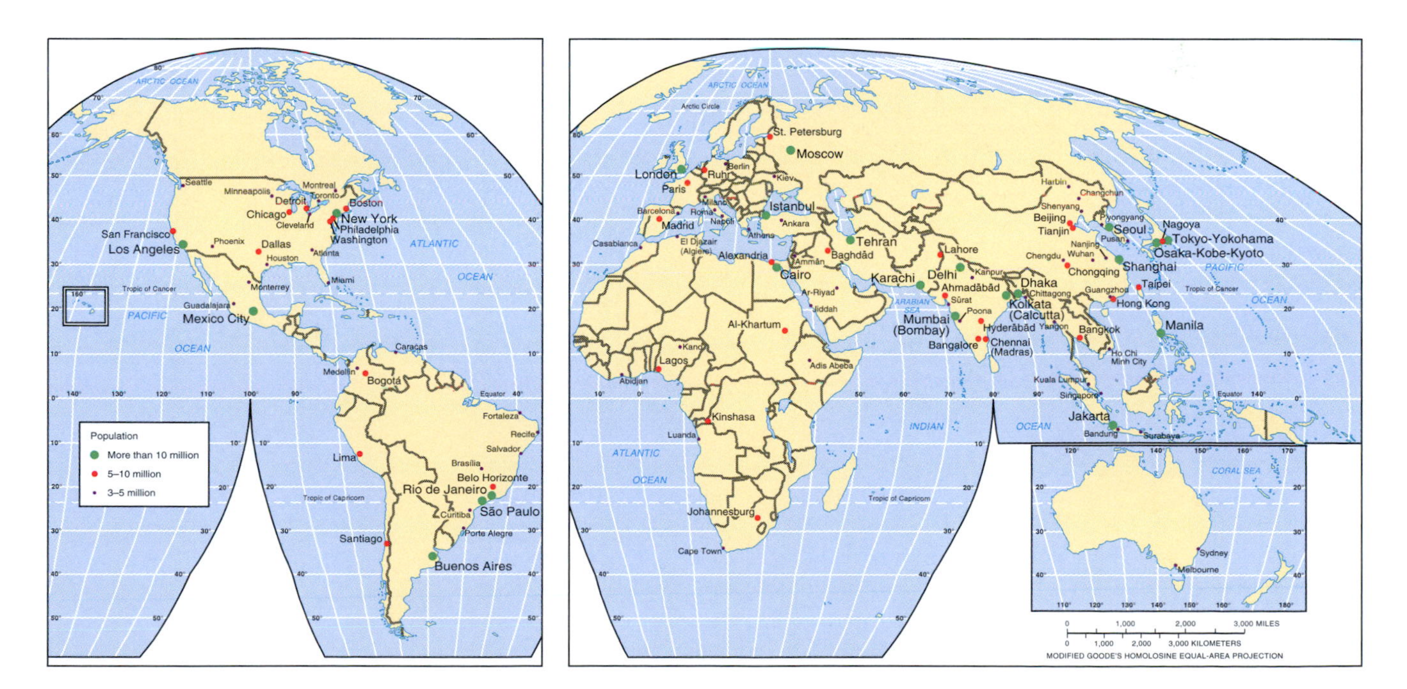

There are dozens of large cities around the world, with some reaching levels as high as 20 million people (e.g., Mexico City, Sao Paolo). Most urban growth, and most of the world's largest cities, are found in the developing world, where cities are swollen from relatively high natural rates of growth and rural-to-urban migration.

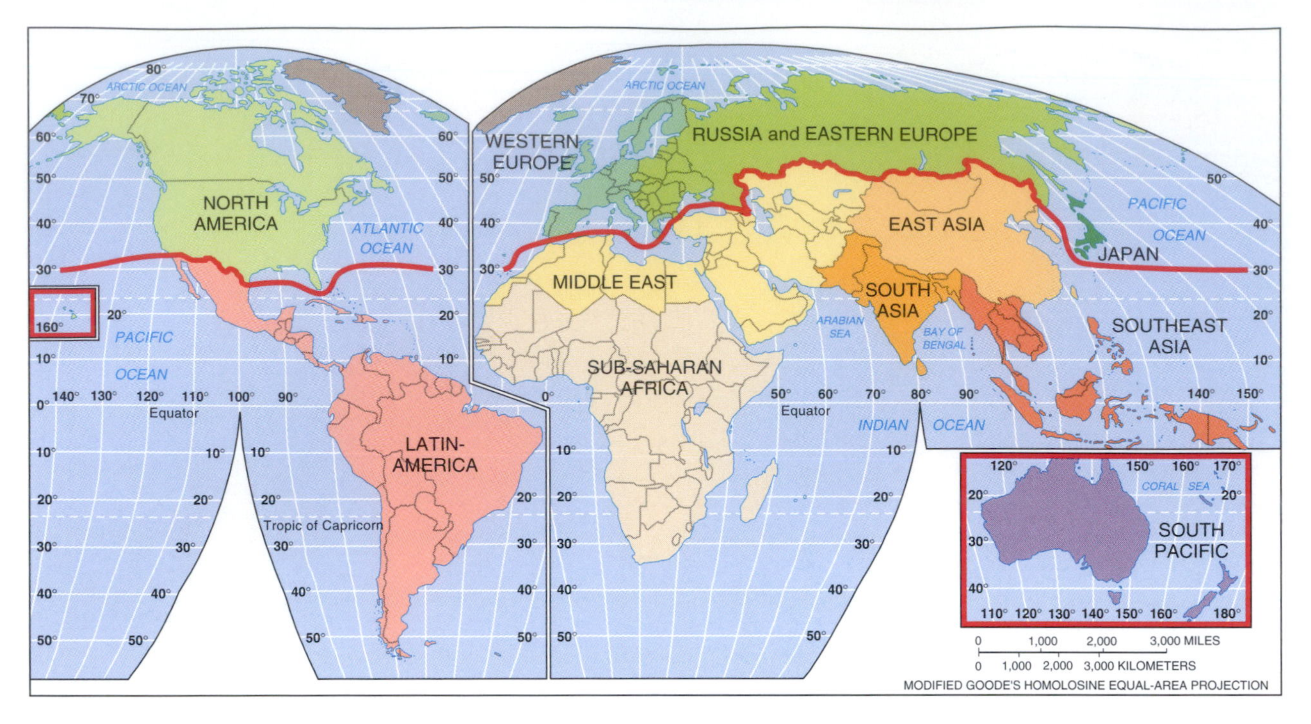

Perhaps the defining aspect of the geography of the global economy is the division of the world into two major parts, the economically developed and underdeveloped regions. During the cold war, the term First World was used to describe the economically advanced countries of western Europe, Japan, Australia and New Zealand, and the U.S. and Canada. The Third World consisted of the former European colonies in Latin America, Africa, and Asia (excluding Japan). The Second World consisted of the Soviet Union and its allies in eastern Europe; with the disintegration of the USSR in 1991, however, this world has disappeared, divided into the other two.

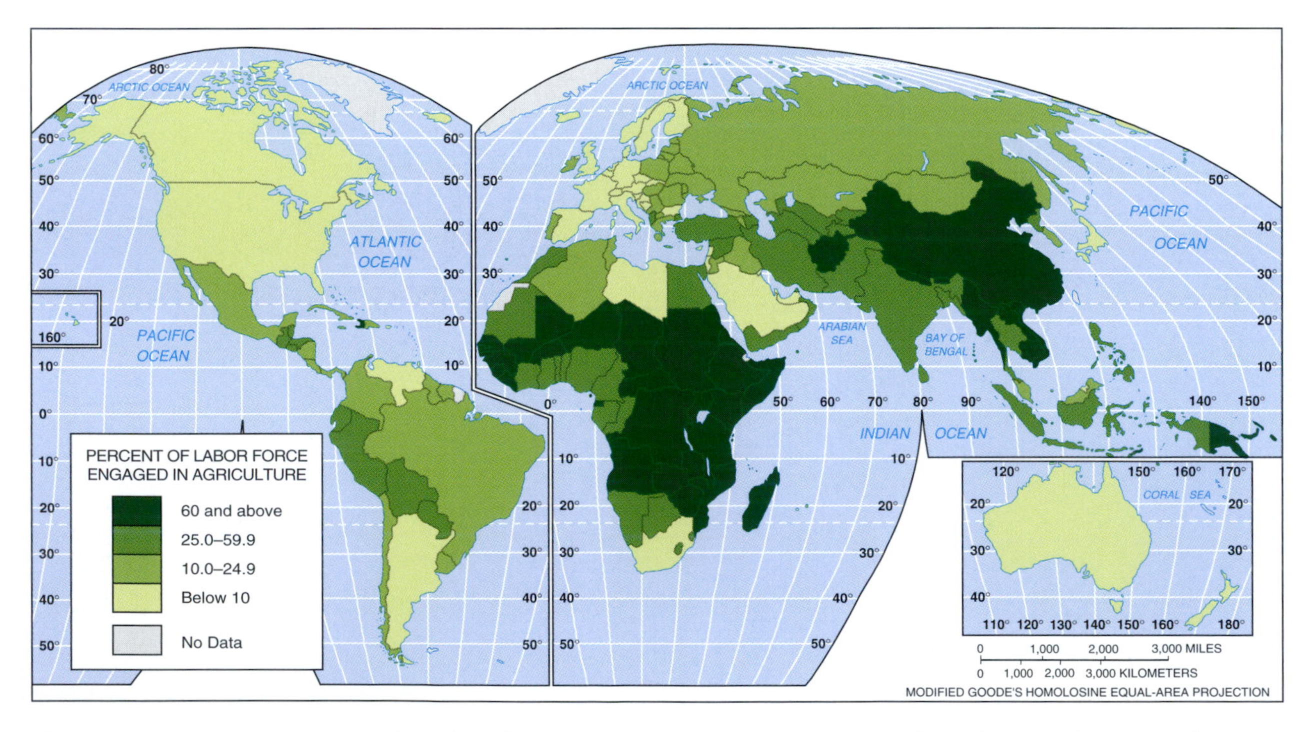

The primary economic sector consists of workers in extractive activities, i.e., agriculture, forestry, fishing, and mining. The relative size of this sector is a measure of the level of economic development of a country. In economically advanced countries, relatively few workers are employed in this way (in the U.S. it is roughly five percent). In developing countries, in contrast, the bulk of workers are often farmers, as in most of Africa and Asia.

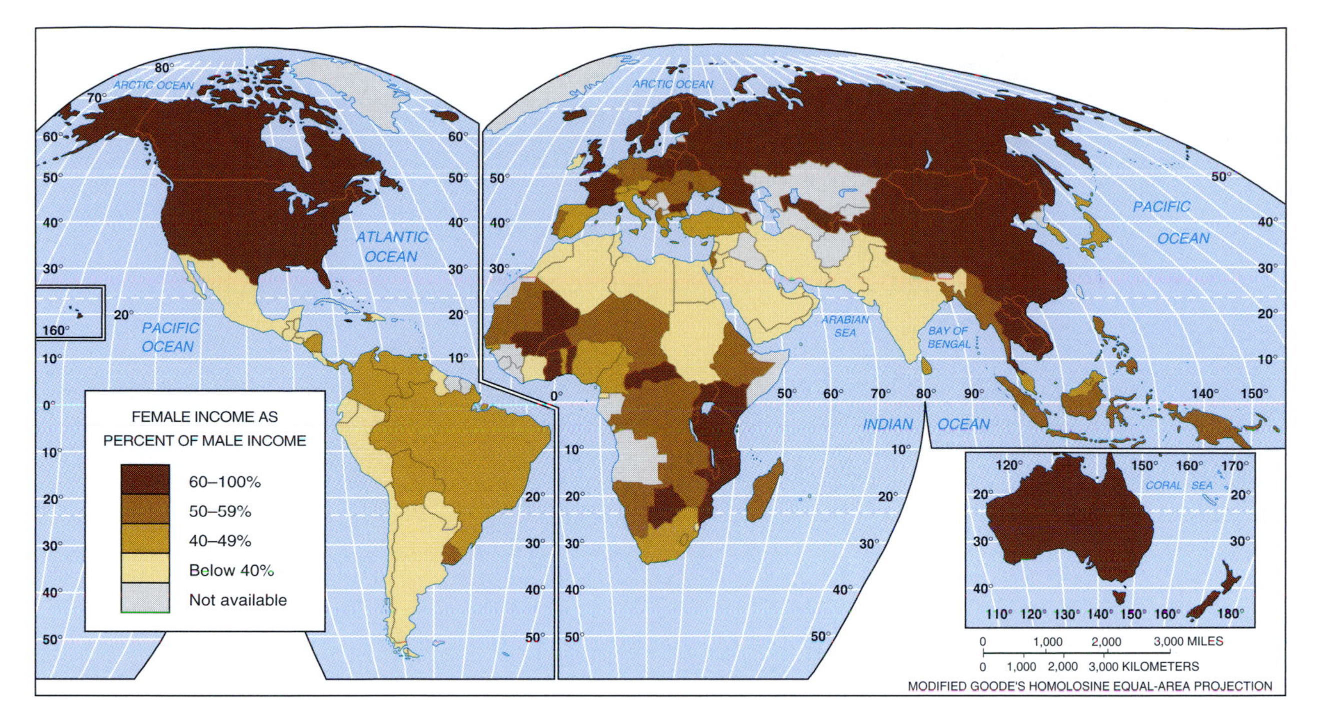

Gender is a major dimension of economic structure in time and space, and intersects with class and ethnicity in complex ways. In almost all societies, men enjoy a privileged social and economic status compared to women, with higher incomes and positions of responsibility. The gap between men's and women's incomes is thus a measure of the geography of patriarchy. Economically advanced countries tend, in general, to exhibit comparatively greater equality between men and women in this regard, although there are notable exceptions (e.g., Japan). In the developing world, men frequently enjoy much greater advantages than do women, although this relation is mediated by culture. In the Muslim world of North Africa and the Middle East, for example, the gap between men's and women's incomes is the world's greatest.

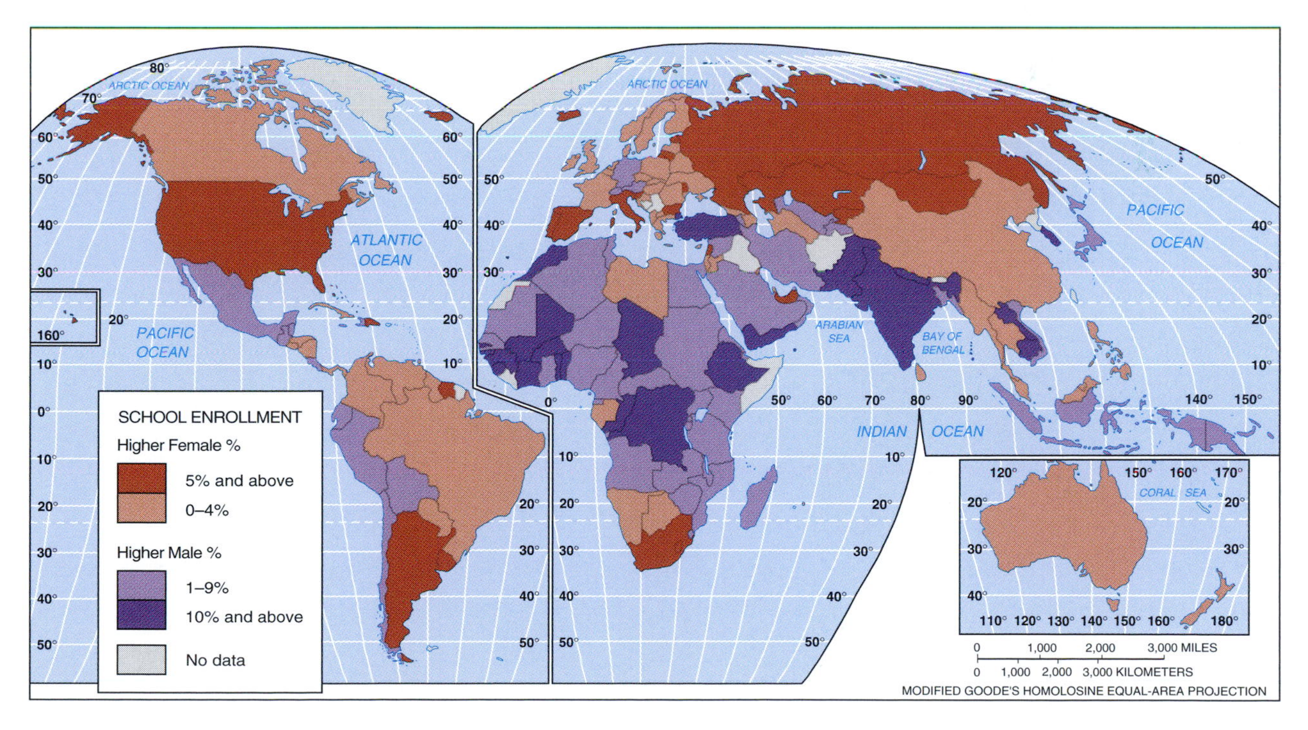

One of the resources that males typically enjoy more abundantly compared to women is education. In many developing countries, girls are much less likely to go to school than are boys, and female literacy rates are typically much lower. Low education levels handicap women's participation in the economy and public life, and raising women's literacy levels is a significant step in the process of economic development.

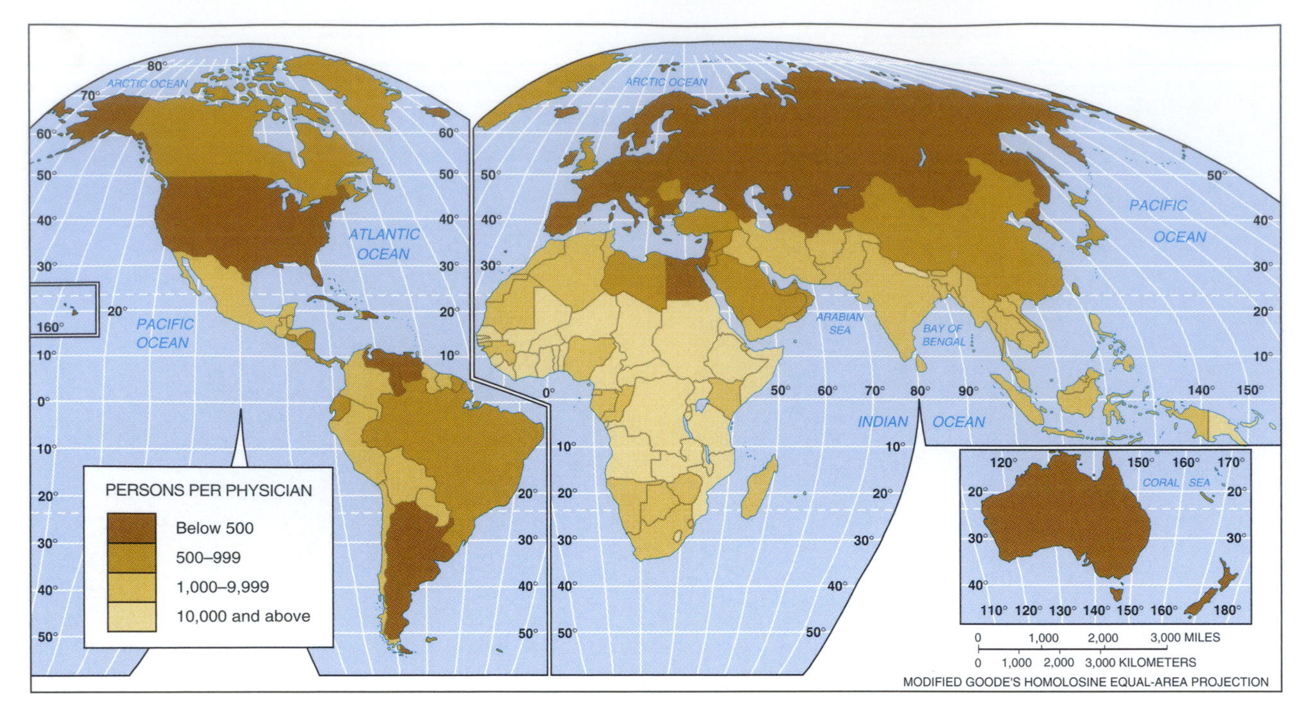

Access to health care, here measured in terms of the number of people per physician, is an important dimension of quality of life and economic development. In the economically developed world, this ratio is generally low, i.e., often 500 or less, meaning that most people can see a doctor when they need one (although accessibility varies widely within countries as well, depending on income and the form of national health care available). In most of the developing world, in contrast, population/physician ratios are high—in SubSaharan Africa they exceed 10,000—meaning that few people have access to health care, number that translates into high mortality rates (including infants), low life expectancy, and diminished quality of life.

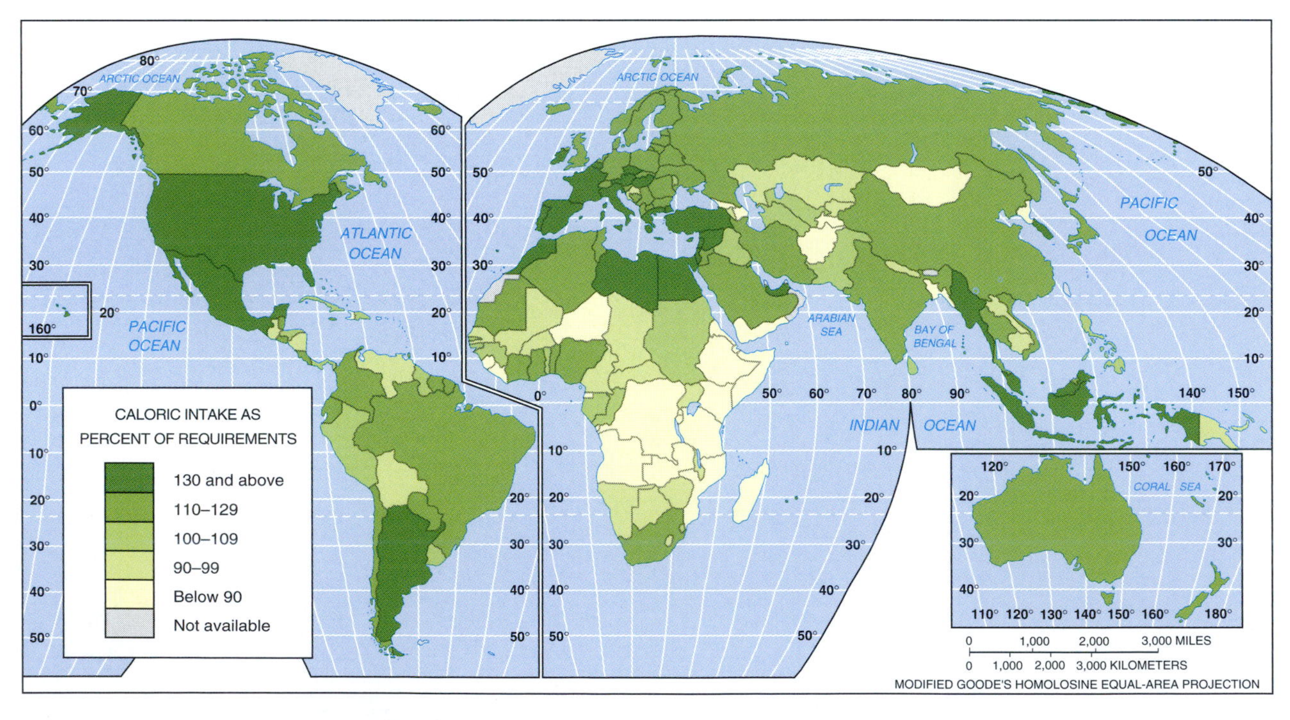

Caloric intake per capita reflects the quantity and quality of a nation's food supply as well as the demands placed upon it by its population. Generally, in wealthy, industrialized countries, relatively few people go hungry or are malnourished; indeed, in the United States, the more pressing problem is too much food and associated obesity rates. In the developing world, access to food varies widely, and many people suffer from less than 100% of their body's caloric requirements, leading to widespread malnutrition. It is worst in the impoverished nations of SubSaharan Africa, site of most of the world's famines.

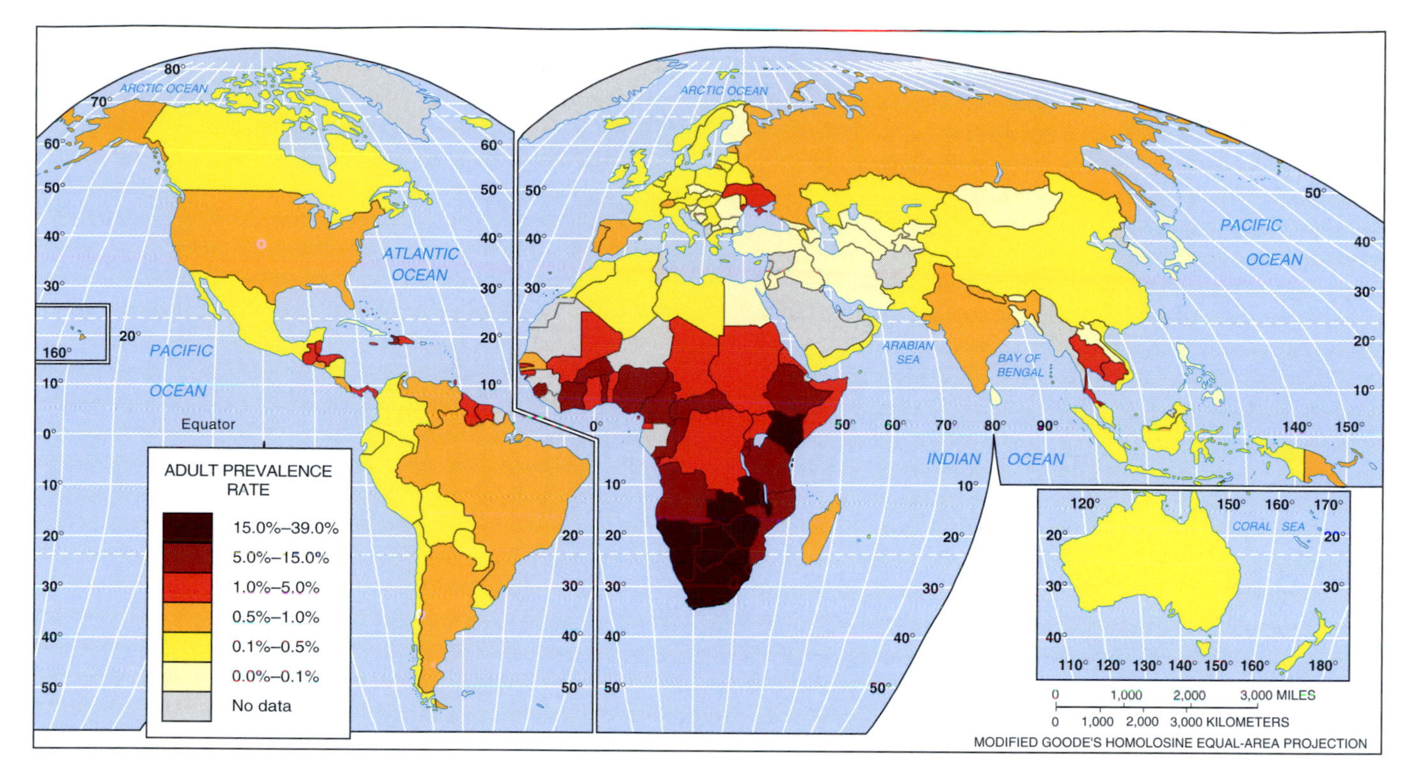

The Human Immunodeficiency Virus (HIV), which causes Aquired Immune Deficiency Syndrome (AIDS), has reached the levels of a global epidemic. More than 42 million people worldwide are infected with HIV. With a long lead time before AIDS symptoms appear, HIV is transmitted primarily through sexual contact. SubSaharan Africa is the epicenter of the world's AIDS pandemic, where in many countries HIV infection rates have reached 40 percent. Vast numbers of villages have lost adults in their prime working years and millions of children are orphaned as a result. The virus is just now beginning to penetrate the huge masses of people in Southern and Eastern Asia.

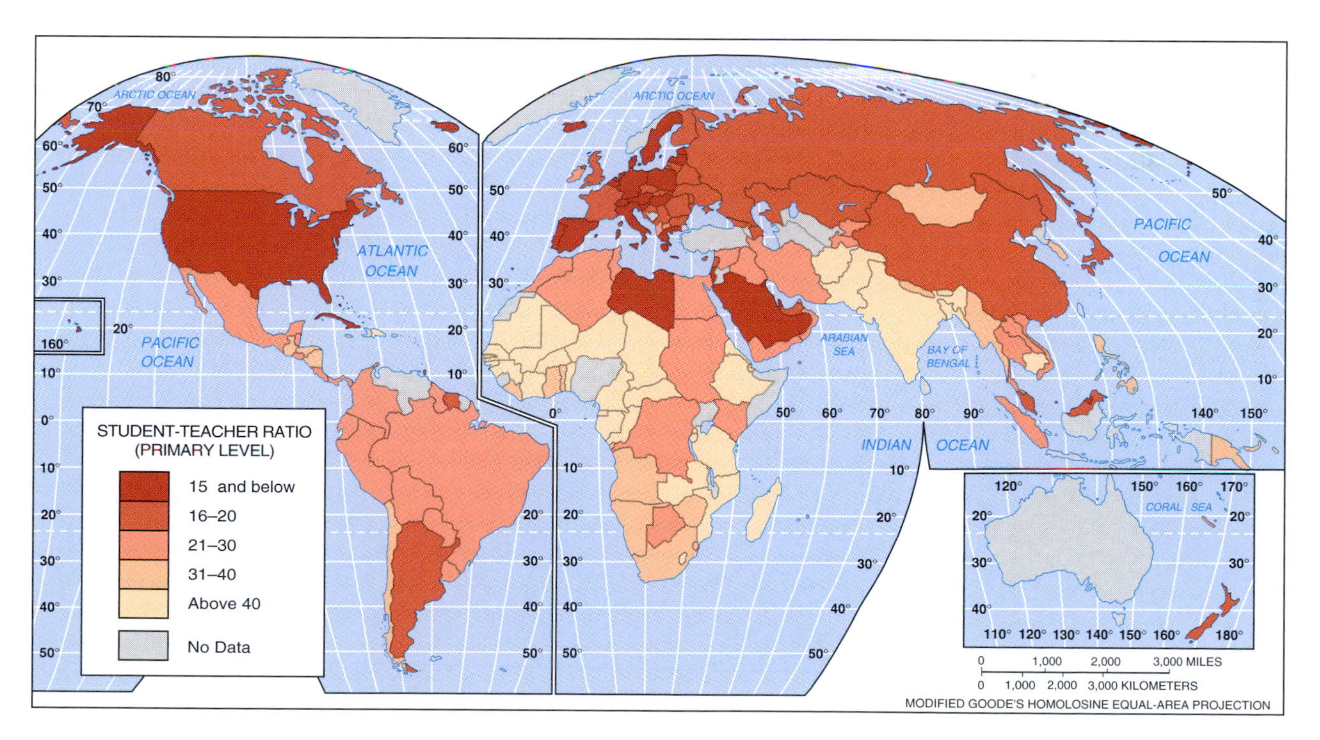

Access to education is often measured by the ratio of students to teachers. In the economically developed world, where education generally receives adequate (if just barely) funding, student/teacher ratios are relative small (i.e., below 15). In most developing countries, with government budgets strapped and priorities focused on the military, schools are badly underfunded, teachers underpaid, and student/teacher ratios high, a well known variable that inhibits effective education.

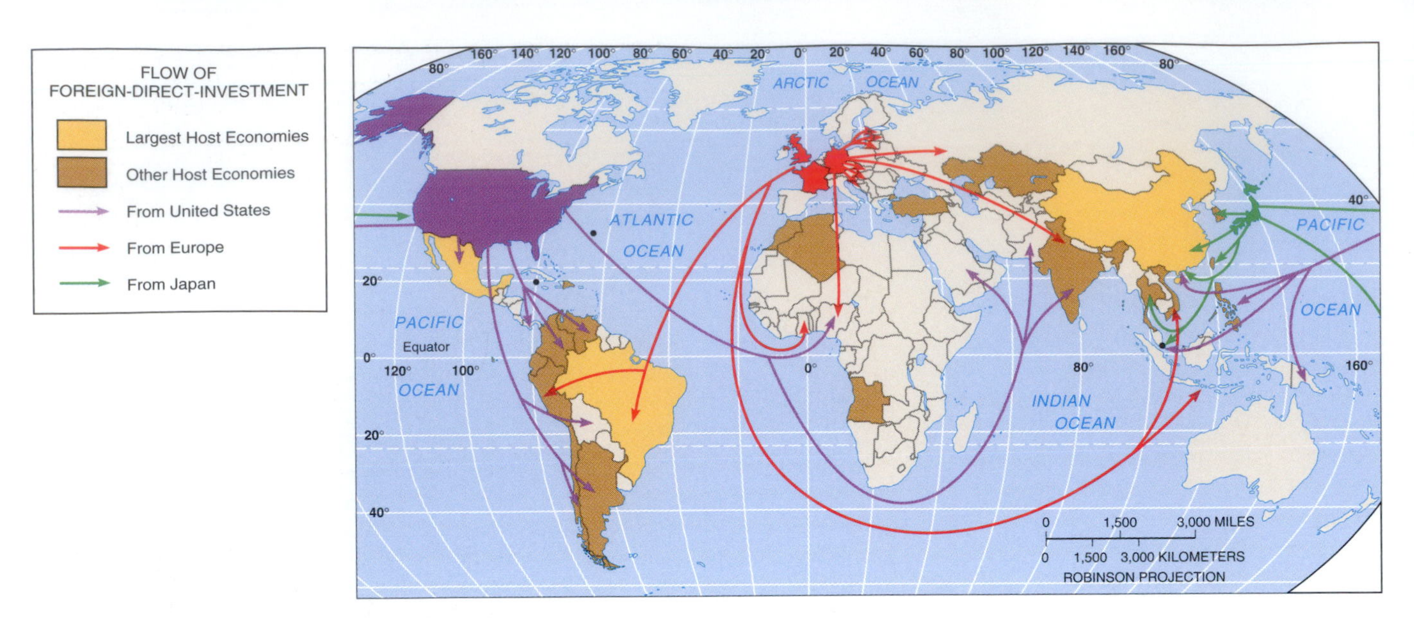

International flows of capital, or foreign investment, are critical to the economic health of countries, especially those that have insufficient reservoirs of domestic capital. Generally, flows of foreign investment take place through multinational corporations. They originate in countries where capital is abundant, i.e., the U.S., Japan, and western Europe. The destinations of foreign investment include other developed countries as well as parts of the developing world.

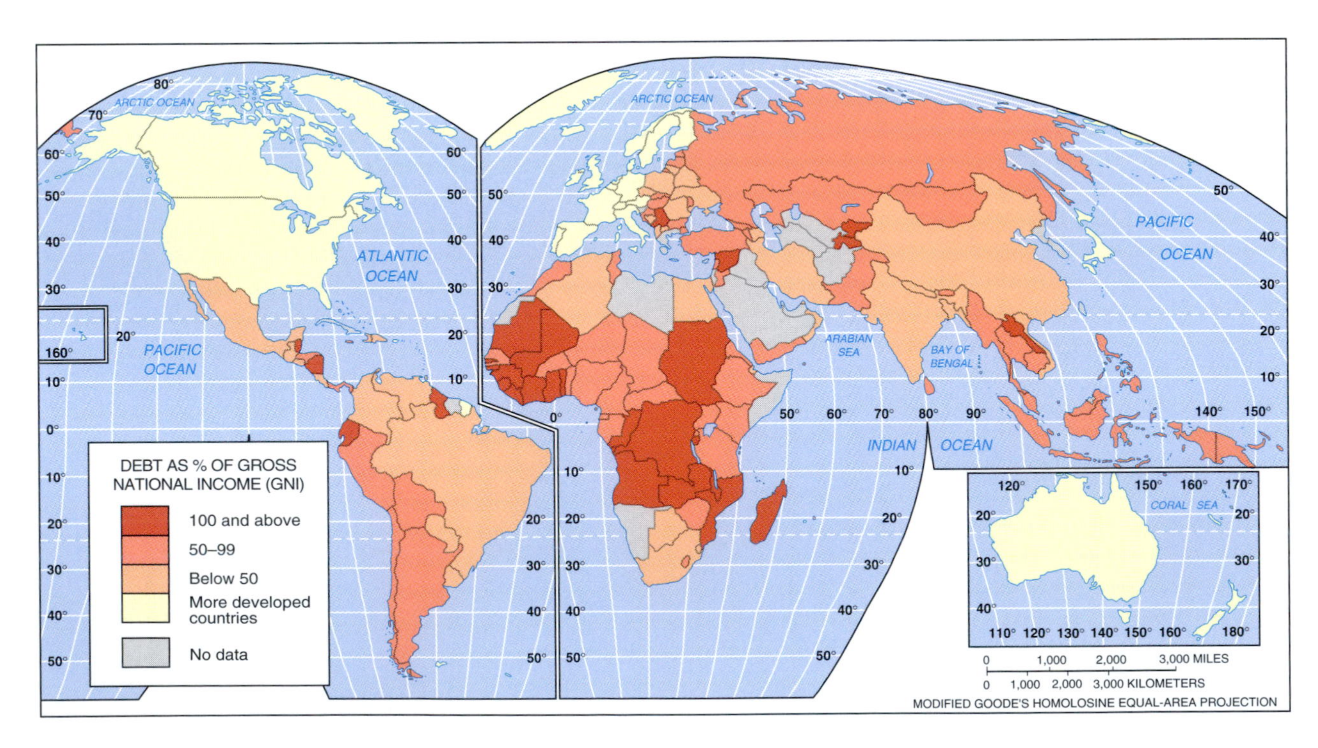

The global debt crisis originated with the recycling of petrodollars in the 1970s, as OPEC countries deposited billions in western banks, which in turn lent them to governments in the the developing world. Today, many countries struggle to pay back the interest and principal on their loans, which consumes a significant share of their export revenues. In absolute terms, the U.S. is the world's largest debtor. As a proportion of Gross Domestic Product, the most indebted countries are in SubSaharan Africa. Often the International Monetary Fund is heavily involved in determining the conditions under which debt is to be repaid.

With more than one-half of the world's 6.3 billion people living in cities, urban areas have become critical centers of social, political, and economic life. Most of the planet's largest cities are in the developing world, where they are often swollen through rural-to-urban migration and high rates of natural population growth. Large numbers of people live in intolerable conditions marked by substandard housing, poor infrastructure, inadequate employment opportunities, and little access to health and educational services. Many cities in the developing world are ringed by huge shantytown districts, sometimes with millions of inhabitants apiece. Improving the lives and hopes of people caught under these circumstances will require urban as well as rural development.

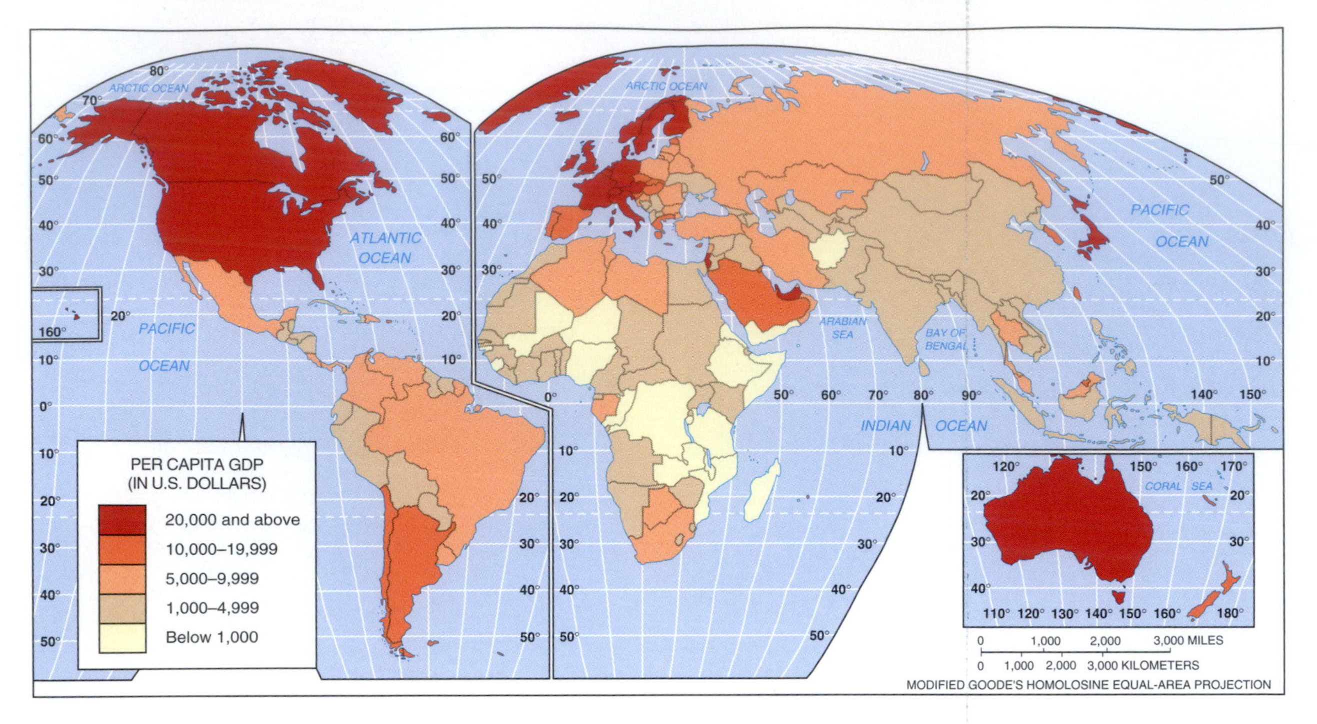

Annual Gross Domestic Product (GDP) per capita is the most widely used measure of economic development, or the lack thereof. It includes the sum total of a country's economic output (goods and services) divided by its population, and ranges from less than $5,000 in most of the developing world to more than $30,000 per person in North America, Japan, and parts of Europe. However, GDP per capita is an imperfect measure, failing to capture non-commodified forms of production (e.g., household labor, subsistence production) and being sensitive to exchange rate fluctuations.

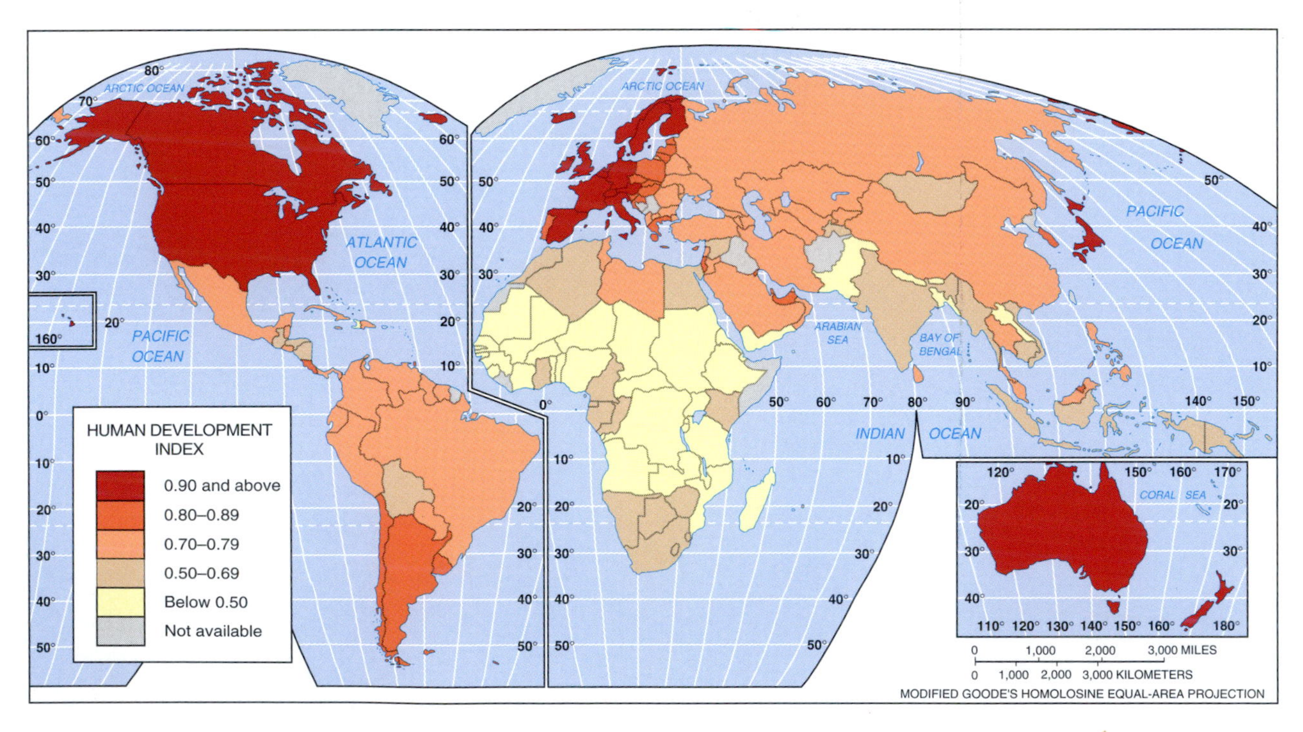

The Human Development Index (HDI), created by the United Nations, combines multiple measures of social and economic development, including Gross Domestic Product per capita, life expectancy at birth, and literacy rates. It thus encapsulates the historical forces that surround people's lives as they play out unevenly over space. The geography of the HDI reflects the world's sharp bifurcation into a small group of relatively well off countries (many of which contain internal pockets of poverty) and large numbers of poor countries in which the HDI is low.

WORLD WIDE WEB SITES

WORLD FACTBOOK ON COUNTRIES
http://www.odci.gov/cia/publications/pubs.html

WORLD TRADE ORGANIZATION
http://www.wto.org/
The principal agency of the world's multilateral trading system. Its home page includes access to documents discussing international conferences and agreements, reviewing its publications, and summarizing the current state of world trade.

WORLD BANK
http://www.worldbank.org/
A leading source for country studies, research, and statistics covering all aspects of economic development and world trade. Its home page provides access to the contents of its publications, to its research areas, and to related Web sites.

BUREAU OF LABOR STATISTICS
http://stats.bls.gov/
Contains economic data, including unemployment rates, worker productivity, employment surveys, and statistical summaries.

U.S. INTERNATIONAL TRADE IN GOODS AND SERVICES HIGHLIGHTS
http://www.census.gov/indicator/www/ustrade.html

U.S. DEPARTMENT OF COMMERCE
http://www.doc.gov/
Charged with promoting American business, manufacturing, and trade. Its home page connects with the web sites of its constituent agencies.

ECONOLINK
http://www.progress.org/econolink
Econolink has selective descriptions of economic issues; many sites themselves have web links.

WEBEC: WORLD WIDE WEB RESOURCES
http://netec.wustl.edu/WebEc/WebEc.html
An extensive set of site listings concerning the economy.

NEW
STOP

CHAPTER

8

SERVICES

OBJECTIVES

- To illustrate the difficulties in defining and measuring services
- To assess the diversity of services, including the range of industries and occupations
- To provide contrasting views on the reasons and consequences of services growth
- To outline the debate about productivity and services
- To describe the world of services work
- To provide case studies of finance and producer services sectors
- To examine the globalization of services
- To sketch the nature of consumer services and tourism

New York Stock Exchange: New York City is a major financial center in the global economy. Tokyo and Londan also play significant roles.

Broadly defined, the tertiary sector of the economy consists of all those sectors engaged in the provision of services of various sorts, intangibles that include retailing, banking, real estate, finance, law, education, and government. "Services" encompass an enormous diversity of occupations and industries, ranging from professors to plumbers to prostitutes. Indeed, so great is the variation among firms and occupations within services that the term threatens to lose any coherence whatsoever. For this reason, it is simplistic to use the term "*the* service sector," which masks the enormous diversity among and within different service industries.

The traditional perspective on services, derived from *postindustrial* perspectives in the 1950s and 1960s and now out of date, focused on services as information processing activities, including clerical activity, executive decision making, telecommunications, and the media. Such a view heralded information processing as a qualitatively new form of economic activity; thus services were held to represent a historically new form of capitalism. Unfortunately, this view is mistaken: While many service jobs do involve the collection, processing, and transmission of large quantities of data, clearly others do not. The trash collector, restaurant chef, security guard, and janitor all work in the service sector, but the degree to which these activities center around information processing is minimal.

More recent theorizations stress services as another form of capitalist commodity production, involving the same constraints to location, production, and consumption as other industries (i.e., manufacturing). In this light, services may be seen not as a qualitatively new phenomenon but as an extension of market relations into new domains of output and activity. Such a view does not deny that services may indeed possess an inner logic somewhat different from that exhibited by manufacturing; however, it stresses the embeddedness of services within the broader social contours and relations of capitalism generally.

Throughout most of the economically developed world, various forms of services have replaced manufacturing employment. In large part this trend reflects the persistent tendency of the international division of labor to shift manufacturing activities from economically developed countries to low-cost developing ones. The percent of national labor forces employed in various types of services varies considerably around the world (Figure 8.1); generally, the more economically advanced countries have the highest proportions of service workers; in the United States, more than 80 percent of people work in these activities, whereas in

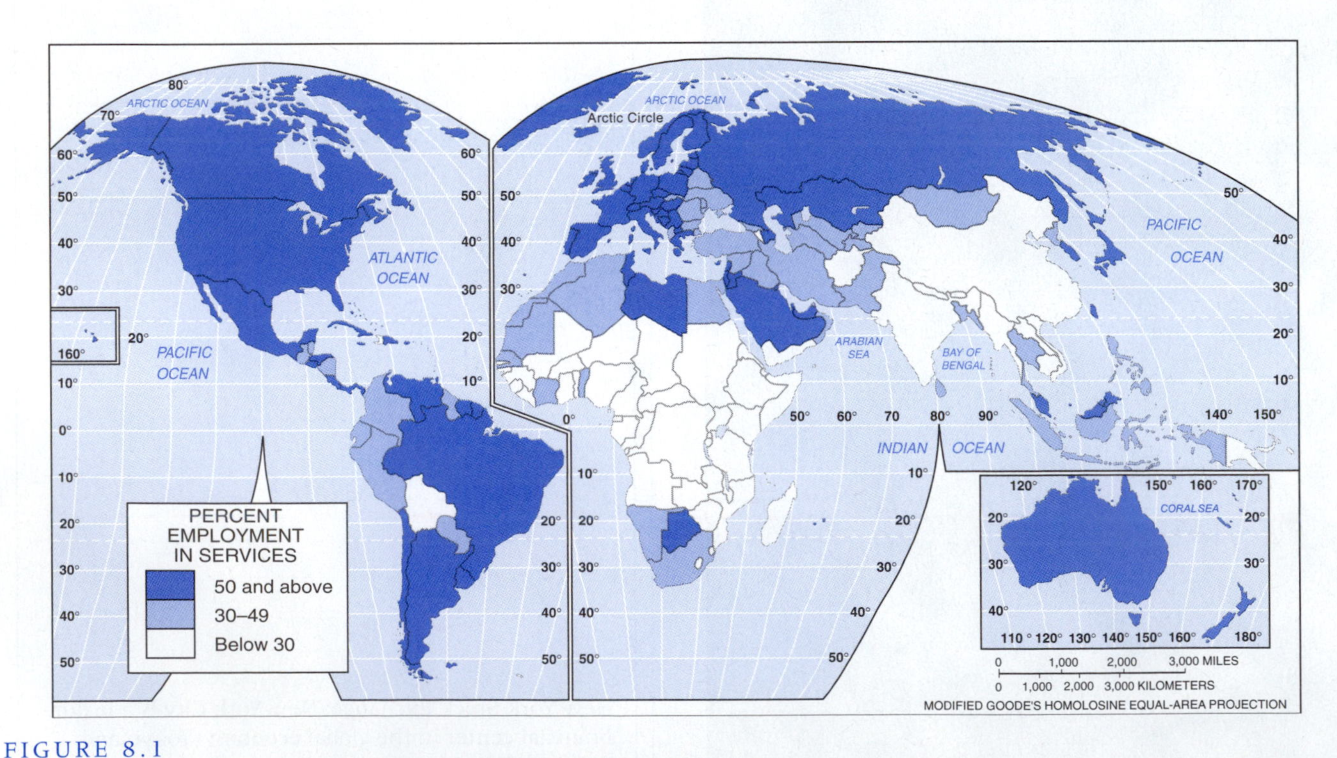

FIGURE 8.1
Proportion of workers in services by country. Note that developed countries have much higher proportions of service workers than in less developed countries. Less developed countries proportions are inflated because governments employ labor in primary and secondary government-owned ventures. (See Color Insert for more illustrative map.)

poorer countries, notably Africa, where most people work in agriculture, the share of service workers is relatively low.

Two related processes have been at work: structural change in the economy and alterations in the ways in which individuals experience work. The word *service* may be used in three distinct ways: *Service industries* consist of enterprises in which the final commodity (service or product) is intangible or immaterial; *service occupations* are forms of work that are not directly but may be indirectly involved in producing physical products; and *service functions* are the uses that consumers obtain from consuming the products of service industries.

This chapter has an ambitious agenda. It begins by pointing out problems in the definition of services and lays out some of the ways in which various kinds of services differ from one another. Next, it turns to the forces that underlay the growth of the service economy, emphasizing the increasingly complex division of labor in advanced economic systems. Third, it summarizes the reasons why many firms externalize services, or purchase them from subcontractors rather than produce services themselves. Fourth, it explores the productivity implications of services, and whether they are responsible for the overall decline in productivity growth over the last several decades. Fifth, the chapter offers case studies of some key sectors of services, including finance and *producer services* such as accounting, design, and legal services. Sixth, it addresses the nature of labor markets in services, contrasting them with manufacturing. Seventh, it explores the means by which services are traded among cities, regions, and countries, focusing on three major economic domains that telecommunications has transformed: electronic funds transfer, offshore banking, and the global back office.

DEFINING SERVICES

A broad consensus exists that services may be understood as the production and consumption of intangible inputs and outputs and thus stand in contrast to manufacturing, the product of which can be "dropped on one's foot." What, for example, is the output of a lawyer? A teacher? A doctor? It is impossible to measure these outputs accurately and quantitatively, yet they are real nonetheless. The U.S. federal government estimates services output using revenues as a proxy, yet revenues are determined by output and prices, so this measure is distorted by changes in the relative prices of service outputs. To complicate matters, many services generate both tangible and intangible outputs. Consider a fast-food franchise: The output is assuredly tangible, yet it is generally considered a service; the same is true for a computer software firm, in which the output is stored on disks.

The fact that it is difficult, if not impossible, to measure output in services has enormous implications for their analysis. For example, some critics of the service sector argue that the slowdown in U.S. productivity growth in the late twentieth century reflected the growth of services. Yet if output in services cannot be adequately measured, how can one argue that output per employee in services is high or low, rising or falling? If wages reflect productivity increases, what is the relation between services output and salaries?

If outputs do not differentiate services from manufacturing, then what about inputs? All forms of manufacturing as well as services involve inputs of *both* goods and services: An automobile producer must purchase legal services, advertising, banking, and public relations inputs, just as airlines or securities firms must purchase hardware and equipment to do what they do. Clearly the lines between services and manufacturing are blurry, and no simple definition will suffice.

Figure 8.2 indicates the major categories commonly used to depict the changing nature of the U.S. employment structure over time. A broad consensus exists as to the major components of the service sector, including the following:

1. *Producer services* are those primarily sold to and consumed by corporations rather than households. Many producer services also serve final demand, such as attorneys that cater to both commercial clients and individuals. Producer services are commonly divided into two major groups:

   a. The *finance, insurance, and real estate (FIRE)* sector includes commercial and investment banking, insurance of all types (e.g., property, medical, casualty), and the commercial and residential real estate industry.
   b. *Business services* subsume legal services, advertising, engineering and architecture, public relations, accounting, research and development, and consulting.

2. *Transportation and communications* include the electronic media, trucking, shipping, railroads, airlines, and local transportation (taxis, buses, etc.).
3. *Wholesale and retail trade* firms are the intermediaries between producers and consumers.
4. *Consumer services* such as eating and drinking establishments, personal services, and repair and maintenance services, all of which have locational

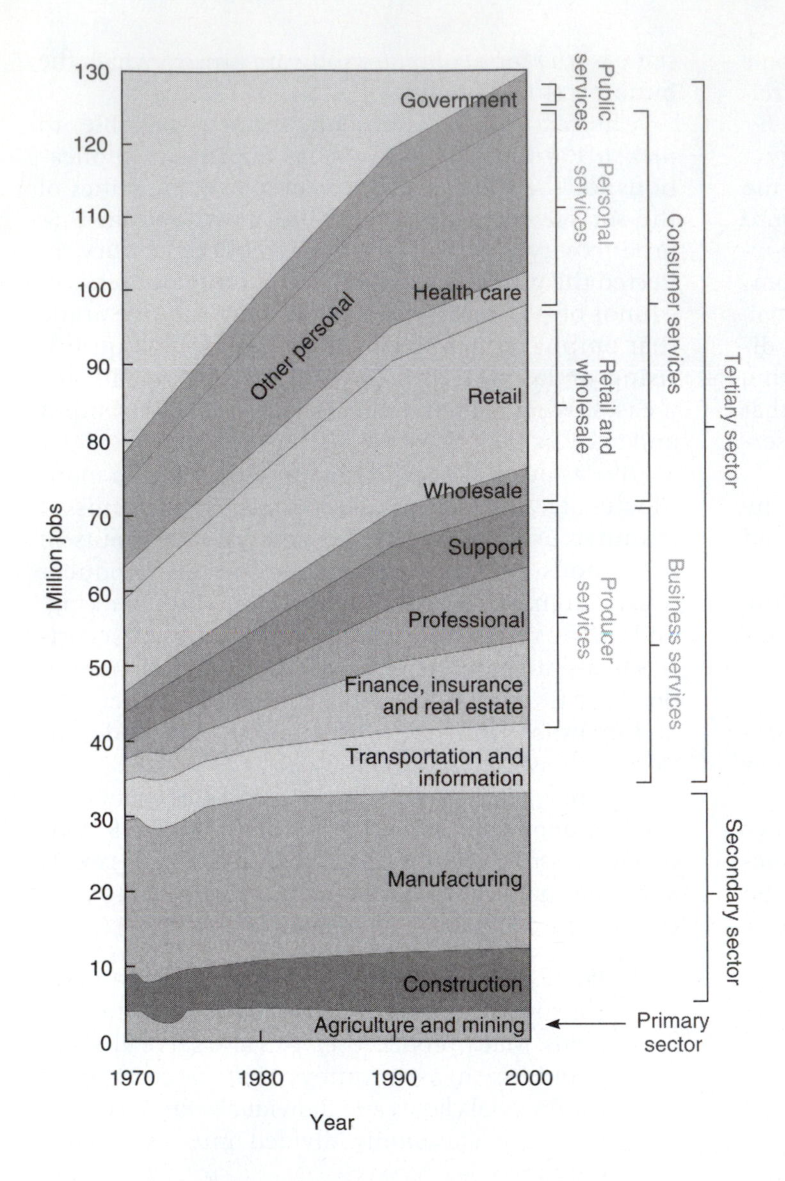

FIGURE 8.2
Employment change by economic sector in the United States 1960–2000. Since 1960 the greatest employment gains have come in professional services, personal services, and social services. Manufacturing has remained relatively constant in absolute terms, but declined as a proportion of the total, while agriculture and mining employment has declined in both relative and absolute terms.

requirements closely associated with local demographics and transport structures (Figure 8.3). *Entertainment,* hotels, and motels comprise elements of tourism, the world's largest industry in terms of employment.

5. *Government* at the national, state, and local levels includes public servants, the armed forces, and all those involved in the provision of public services (e.g., public education, health care, police, fire departments, etc.). While much has been written about the state, its forms and functions, the public sector has only recently been viewed as part of the service economy.
6. *Nonprofit* services include charities, churches, museums, and private, nonprofit health care agencies, many of which play influential roles in local economies.

Every definition of services is slippery, however. For example, does the term services refer to a set of industries or occupations? (The U.S. government sidesteps this problem by adopting industries.) Yet when measured on the basis of their daily activities, many workers in manufacturing are in fact service sector workers, including, for example, personnel in headquarters, administration, clerical functions, and research. Is the secretary who works for an automobile company part of manufacturing while the secretary who works for a bank part of services? The use of industrial versus occupational definitions is particularly critical given the growth of many *"nondirect" production* workers within many manufacturing firms, such as clerical, administrative, research, advertising, and maintenance functions. Clearly, different definitions of services have significantly varying implications for assessments of the size and composition of the service sector.

Using the standard definition of *intangible output,* services comprise the vast bulk of output and employment in most economically developed countries of the world.

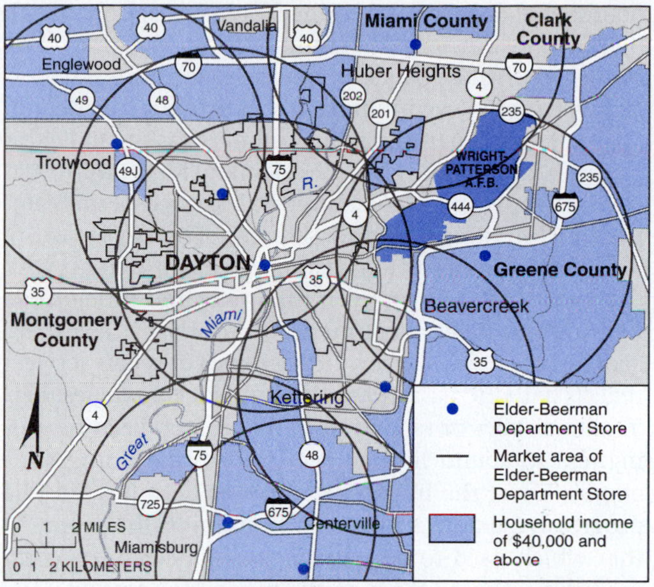

FIGURE 8.3
Locating a department store. Geographers apply location
ation systems to
es.

Indeed, more than 75 percent of the labor force of the United States is employed in services (Table 8.1); similar proportions hold in Europe, Canada, and Japan. Even as early as 1910, services exceeded manufacturing in the United States, indicating it was a "postindustrial" economy before becoming an industrial one! Further, services comprise the vast majority of all new jobs generated in these economies, indicating that they are not only predominantly service oriented but also are becoming increasingly more so. Even in much of the developing world services comprise a large share of the labor force, including much of the "informal" (untaxed, unregulated) economy; this fact belies earlier, simplistic assertions such as the Fischer-Clark thesis, which maintained that all economies inevitably are transformed in a series of rigid stages (i.e., agricultural to industrial to postindustrial).

FORCES DRIVING THE GROWTH OF SERVICES

Why have services grown so rapidly? In economically developed countries, services employment has increased steadily in the face of low rates of population growth,

Table 8.1

on of U.S. Labor Force (millions), 1950–2000

		1960	1970	1980	1990	2000
		4.8	3.9	3.6	3.5	3.3
		.7	.6	1.0	.7	.5
		2.9	3.6	4.3	5.1	6.6
		16.8	19.4	20.3	19.1	18.5
		4.0	4.5	5.2	5.8	7.0
		3.1	4.0	5.3	6.2	7.0
		8.2	11.0	15.1	19.6	23.3
		2.6	3.6	5.1	6.7	7.6
		7.3	11.4	17.5	27.3	39.9
		.8	.9	.8	1.1	1.2
		.6	1.4	2.5	5.1	9.7
		.1	.3	.8	2.0	
			.6	.9	1.2	
Health services		1.5	3.0	5.1	7.6	10.0
Legal services			.3	.5	.9	1.0
Education		.6	.9	1.1	1.6	2.3
Social services			.5	1.1	1.7	2.8
Engineering					2.5	3.3
Government	5.8	8.2	12.3	16.1	18.0	20.4
Total	48.5	59.2	75.0	94.7	124.3	145.6

[a]Transportation, Communications, and Public Utilities.
[b]Finance, Insurance, and Real Estate.
Source: U.S. Bureau of Labor Statistics.

slowly rising rates of productivity and income, and significant manufacturing job loss. While services and manufacturing are intimately intertwined, it is equally apparent that services exhibit growth and locational dynamics somewhat different from manufacturing, although both constitute commodity production in varying forms.

Fundamental to understanding the growth of services are the processes and pressures that are encouraging consumers (both final and intermediate demand, i.e., households and firms) to require more service products. There are different drivers at work in these sectors. Individuals are consuming more services for several reasons: (a) as physical commodities become more complicated (e.g., automobiles); (b) as work dominates a larger share of their lives, requiring increasing expenditure on types of personal services that are required to replace activities that they previously undertook for themselves; and (c) as new services develop that they are encouraged to consume (e.g., Internet-based entertainment). Conversely, enterprises are constantly reconfiguring the production process as well as the design of products, and these two processes have increased the need for services that are embedded in the production process. For example, many products will only function with software.

Six reasons for the increase in services employment throughout the world are suggested next.

Rising Incomes

First, gradually rising per capita incomes, particularly in the industrialized world, have contributed to rising services employment (Figure 8.4). The demand for many services is income elastic, that is, increases in real personal income tend to generate proportionately larger increases in the demand, in contrast to most manufactured goods (Figure 8.5). Services with particularly high *income elasticities* include entertainment and transportation. U.S. households spend slightly more on services (51% of disposable income) than they did on durable and nondurable goods combined (Figure 8.6). An important reason contributing to this growth is the increasing value of time that accompanies rising incomes (especially with two income earners per family). As the value of time climbs relative to other commodities, consumers generally will attempt to minimize the time inputs needed for the accomplishments of many ordinary tasks. While this phenomenon also explains the demand for washing machines, dishwashers, and automobiles, it is especially important for the growth of services. The explosion of fast-food restaurants, for example, has little to do with the quality of food (or even the price) and much to do with attempts by consumers to minimize time spent cooking at home. Similarly, the growth of repair services reflects both increasingly sophisticated technologies (e.g., in automobiles or televisions) and a generalized unwillingness to spend limited recreation time doing such chores. Thus, the increasing value of time has led to a progressive externalization of household functions, so that which used to be done in-house becomes a commodity purchased through the market for a profit.

Demand for Health Care and Education

Second, rising levels of demand for health and educational services comprise an important part of the broader growth of the service economy. Health services employment and output have increased steadily throughout Europe, North America, and Japan, often leading to political and economic conflicts about how to contain them, such as American debates about the rising costs of Medicaid and Medicare. In the United States, health care expenditures surpassed $1.6 trillion in 2002 (Figure 8.7), roughly 15 percent of gross domestic product and much higher than any other industrialized country (Figure 8.8). In many cities, health care is the largest and most important part of the local economy, including, for example, Pittsburgh, Pennsylvania, formerly steel capital of the world, and Birmingham, Alabama. The demand for health care workers has risen steadily, often proving to be recession proof.

The provision and consumption of health care has increased steadily in large part because of the changing demographic composition of industrialized countries.

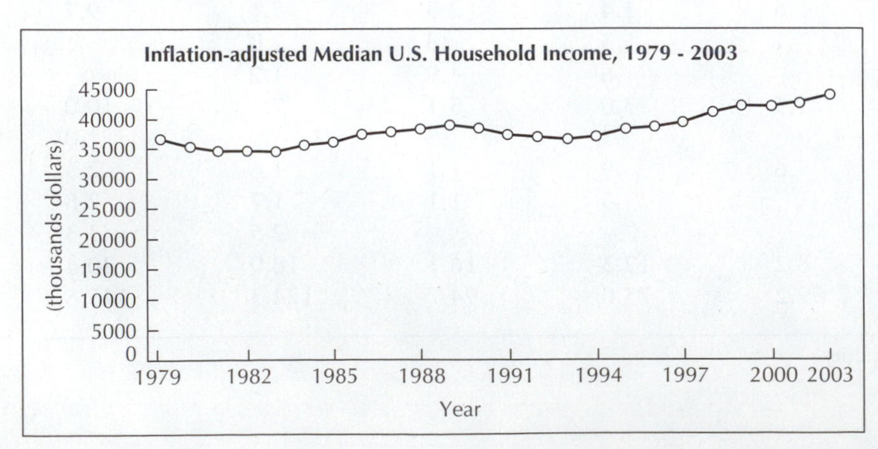

FIGURE 8.4
U.S. median household income, 1979–2003. Although productivity gains have keep prices low, the U.S. labor market has been besieged by deindustrialization and corporate restructuring; consequently, real income growth has been meager and mostly concentrated in the upper echelons.

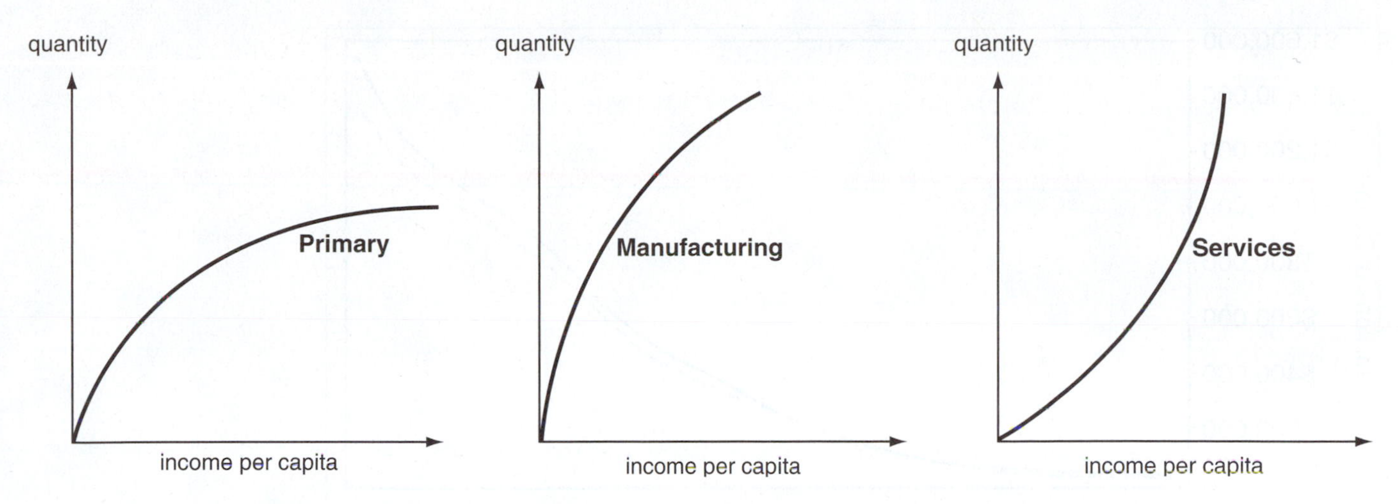

FIGURE 8.5
Income elasticities of demand for services are high compared with those for agricultural or manufactured goods.

Throughout the developed world, the most rapidly growing age groups today are the middle aged and the elderly, precisely those demographic segments that require relatively high per capita levels of medical care. As the baby boom enters its retirement years, the demand for health services will rise even higher. Higher life expectancies and soaring equipment, pharmaceutical, and research expenses have added to the costs of this sector.

Similarly, globalization, increasing technological sophistication, and increasing demand for more analytical skills (particularly numeracy and computer skills) at the workplace have driven the increasing demand for educational services at all levels, a process reflected in higher enrollments in universities, which have become prerequisites for an entrance to middle class jobs. Whereas the bulk of graduating high school students in the United States did not attend college in the 1960s, today almost 70 percent do so.

An Increasingly Complex Division of Labor

Third, the growth of services reflects the increasing complexity of the division of labor. Statistically, this process is manifested in a rising proportion of nondirect production workers, including firms in the manufacturing sector. All corporations today devote considerable resources to dealing with a complex marketplace and legal environment, including many specialized clients, complex tax codes, environmental and labor restrictions, international competition, sophisticated financial markets, and real estate purchases and sales. Deregulation—the lifting of state controls in many industries—increased the uncertainty faced by many firms and had significant impacts on the profitability, industrial organization, and spatial structure of numerous sectors. To negotiate this environment, firms require administrative bureaucracies to collect and

FIGURE 8.6
Relative consumption of goods and services for U.S. households, 1929–2004. The income elasticity of most services compared with goods have led them to constitute more than half of the average household's expenditures, particularly for education, transportation, health care, and entertainment.

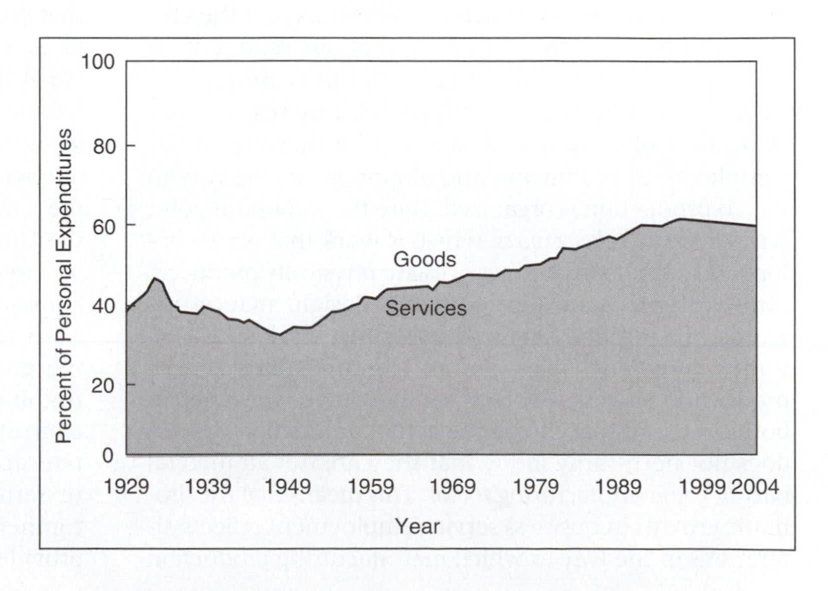

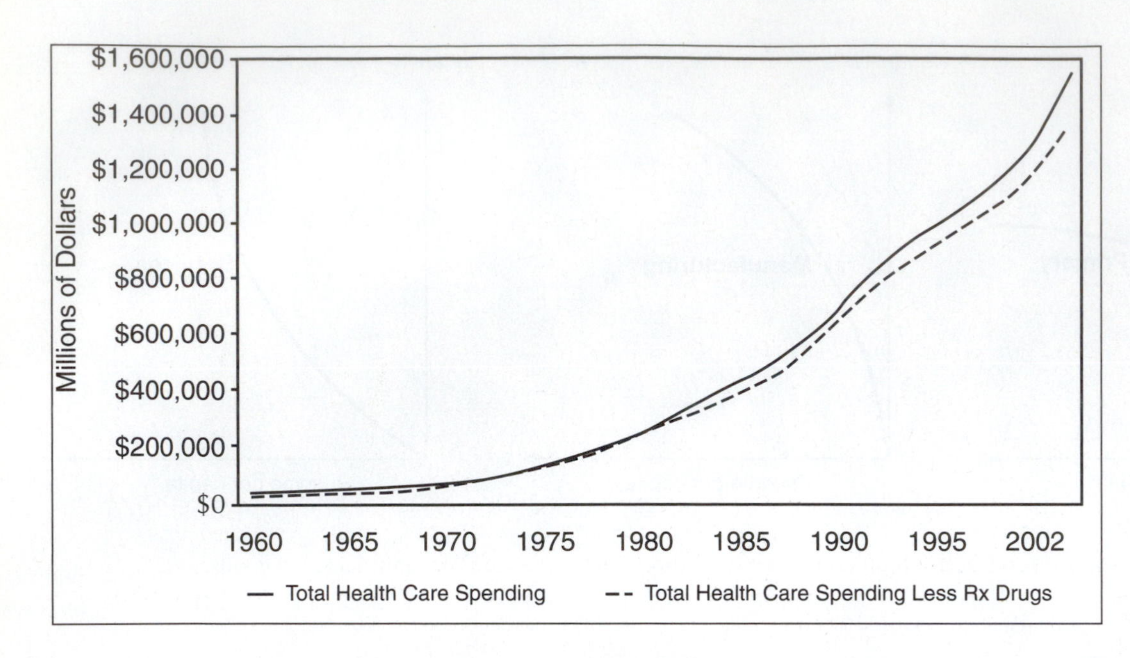

FIGURE 8.7
U.S. national health care expenditures ($ billions), 1990–2002.

process vast quantities of information and make strategic decisions, clerical workers to assist with mountains of paperwork, researchers to study market demand and create new products, advertisers and sales people to market their output, and legions of people engaged in public relations, accountants, lawyers, and financial experts to assist in an enormously complicated decision-making environment. Similarly, the introduction of sophisticated machinery requires maintenance and repair personnel, while offices or industrial plants require security and building maintenance staff—all nondirect production workers and all services. Unlike income-based or demographic arguments, this approach has the added appeal of explaining the growth in producer services, the most rapidly growing part of the economy of most developed countries.

In this line of thought, the growth of services reflects not the development of a new economy but rather the revolutionary force of capitalism as it has generated dramatic shifts in the division of labor in society. Thus, the outsourcing of service functions by clients and the creation of new types of service occupations represent an extension of the division of labor. An increasingly specialized division of labor reflects both increasing specialization of activity with a resultant increase in the complexity of production and alterations in the way in which production is organized. Here the important point is the *extended labor process*, which is work that occurs before and after goods and services are physically produced. Thus, research and development, design, market research, trial production, product testing, marketing, customer care, and sales are all essential parts of the production process. The fact that they can be separated in both time and space from the actual production process does not necessarily imply that they are not an integral part of the manufacturing sector. This means that the dramatic growth in business service employment reflects alterations in the way in which manufacturing production is organized rather than the development of a new type of service or knowledge economy.

The Public Sector: Growth and Complexity

A fourth reason underpinning the growth of services is the increasing size and role of the public sector and concomitant expansions in government employment. Despite the popular stereotype that capitalism consists only of "free markets" (a view challenged in Chapter 5), the state in fact acts as a major actor shaping markets, building infrastuctures, controlling money supplies, subsidizing firms, negotiating trade, and framing the broader legal and institutional context of economic activity. Governments contribute to the growth of services in two ways. First, government employment has increased steadily, especially since the 1930s, because the public demands the services that it provides. It is possible, as conservatives often insist, that governments are inefficient in providing these services, which occasionally may be more effectively provided through the private sector, which is the rationale behind the move toward privatization. Others argue that governments provide services that are socially necessary but not profitable, such as health care for the poor. Today, the federal government is the largest single employer in the United States, employing more than 2 million people, and it is dwarfed by local and municipal government employment (Figure 8.9).

A second way in which government contributes to the growth of services is indirectly, through a labyrinthine web of laws, rules, restrictions, and regulations, contributing to the growth of tax attorneys, accountants, consultants, and other specialists that assist firms (externally or internally) in negotiating the legal environment. The degree of regulation that the government provides generates changing levels of uncertainty in the

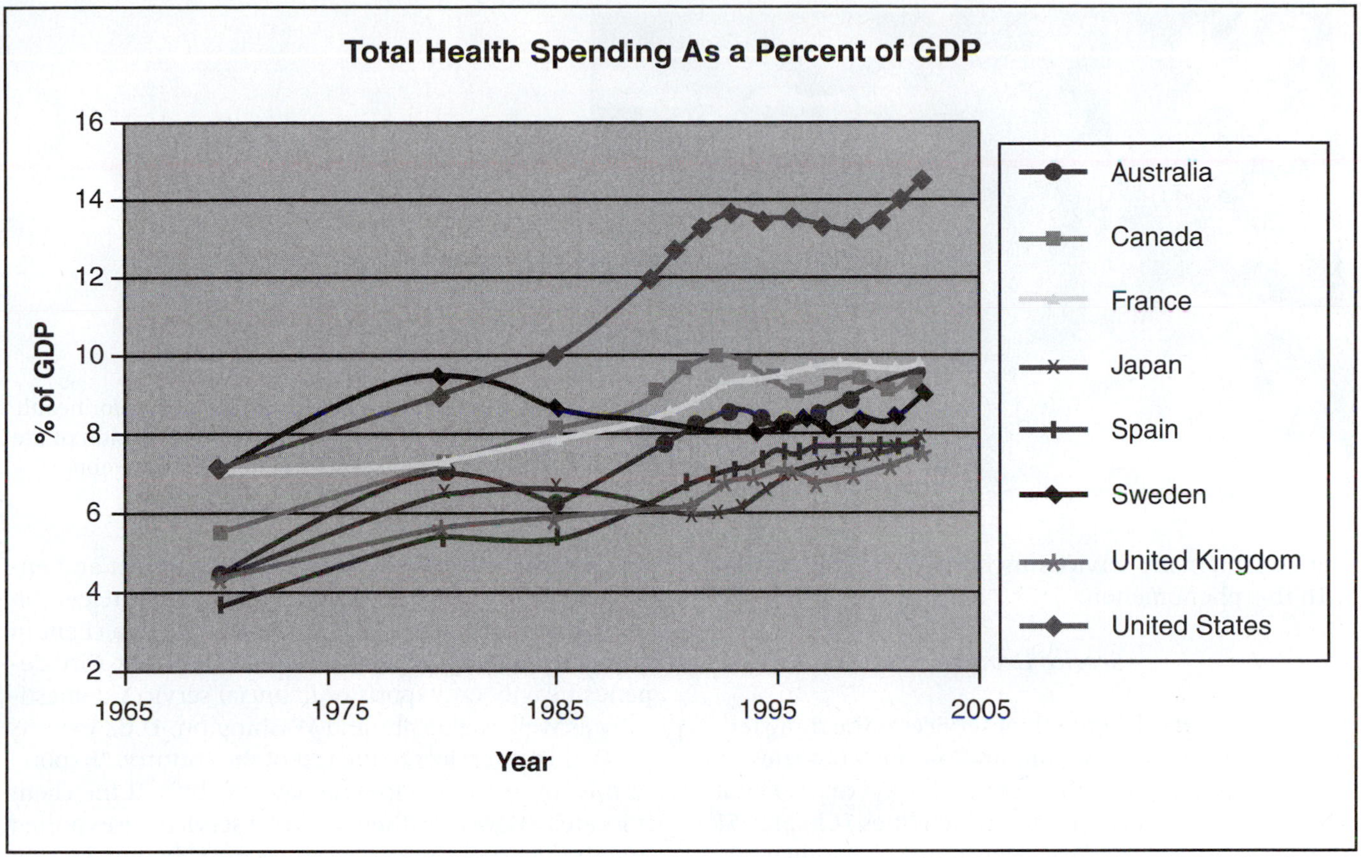

Total Health Spending As a Percent of GDP

% of GDP: 2, 4, 6, 8, 10, 12, 14, 16

Year: 1965, 1975, 1985, 1995, 2005

Australia
Canada
France
Japan
Spain
Sweden
United Kingdom
United States

FIGURE 8.8
Percent of GDP dedicated to health care expenditures, 1965–2005 for major industrial countries. As the population has aged and the cost of health services has risen, this sector consumes one-seventh of all dollars spent in the United States. Although the United States has the least equitable system of health care access in the industrialized world—one-quarter of adults have no health care insurance—its system is not the most efficient: European countries spend, on average, less on health care as a proportion of their total output.

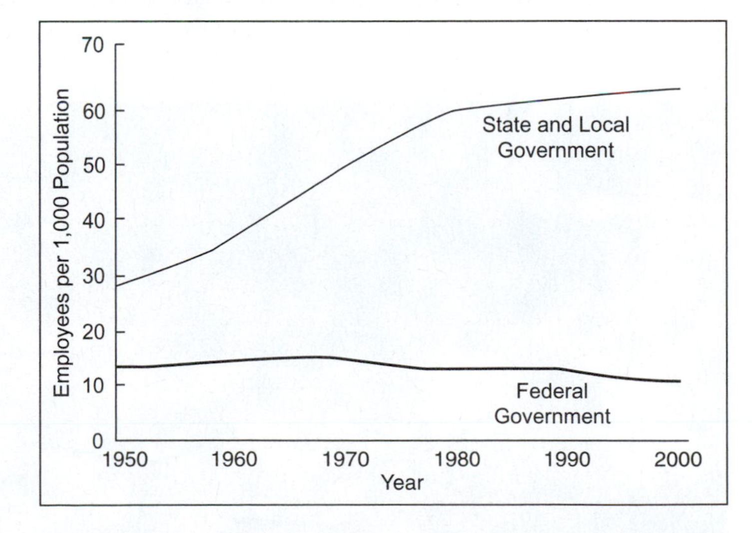

FIGURE 8.9
Employment in the federal and state and local governments in the United States, 1950–2000. Most public sector employment is at the local level.

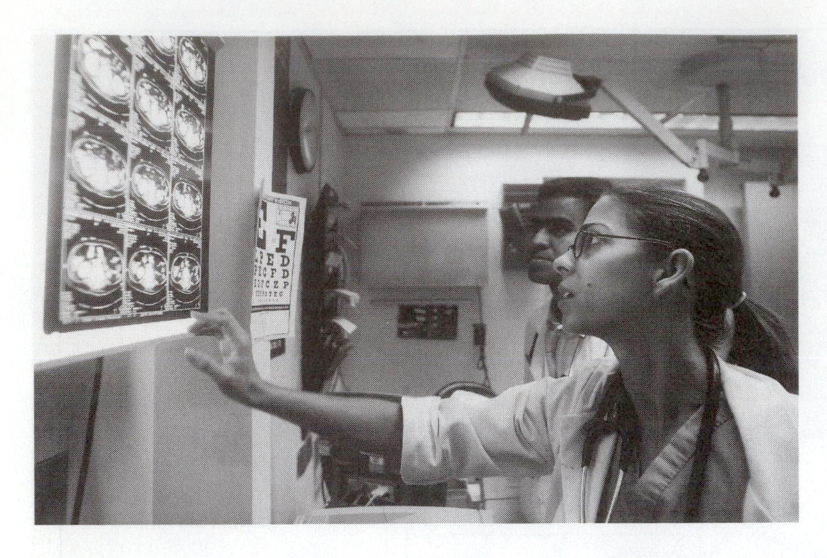

The steady and rapid growth in demand for health services is a significant part of the expansion of the service economy. An aging population, among other things, contributes to this process.

market, and many service functions allow firms to deal with this phenomenon.

Service Exports

A fifth reason for the growth of services is the rising levels of service exports within and among countries. A widespread myth exists that services always cater to local demand (i.e., they are nonbasic activities [Chapter 5] and are thus of secondary importance to manufacturing). This notion reflects the long-standing bias against services in economic thought noted earlier. In countries where the vast bulk of workers are engaged in the production of services, it would be astonishing if services were *not* traded among places. The economies of many cities, regions, and countries derive a substantial portion of their aggregate revenues from the sale of services to clients located elsewhere. Many urban areas export services to clients located to other parts of the same nation. Consider, for example, cities such as Las Vegas, which revolve exclusively around exports of tourism and entertainment services to visitors. Whenever a television company in one city sells advertising time to a client in another, that city exports services. New York City depends heavily on exports of financial services domestically as well as globally, and Washington, D.C., exports government services to the rest of the country. "Export" in this context refers to who pays the bills. If the client is located elsewhere, then sales of a service are exported and form part of the local basic sector (Chapter 5).

Services are also traded on a global basis, comprising roughly 20 percent of international trade. Internationally, the United States is a modest net exporter of services but runs major trade deficits in manufactured goods (Figure 8.10), which is one reason services employment has expanded domestically. Indeed, it could be said that as the United States has lost much of its comparative advantage in manufacturing, it has gained a new one in financial and business services. The data on global services trade are poor, but some estimates are that services

The microelectronics revolution revolutionized office work, increasing productivity and changing the demand for skills.

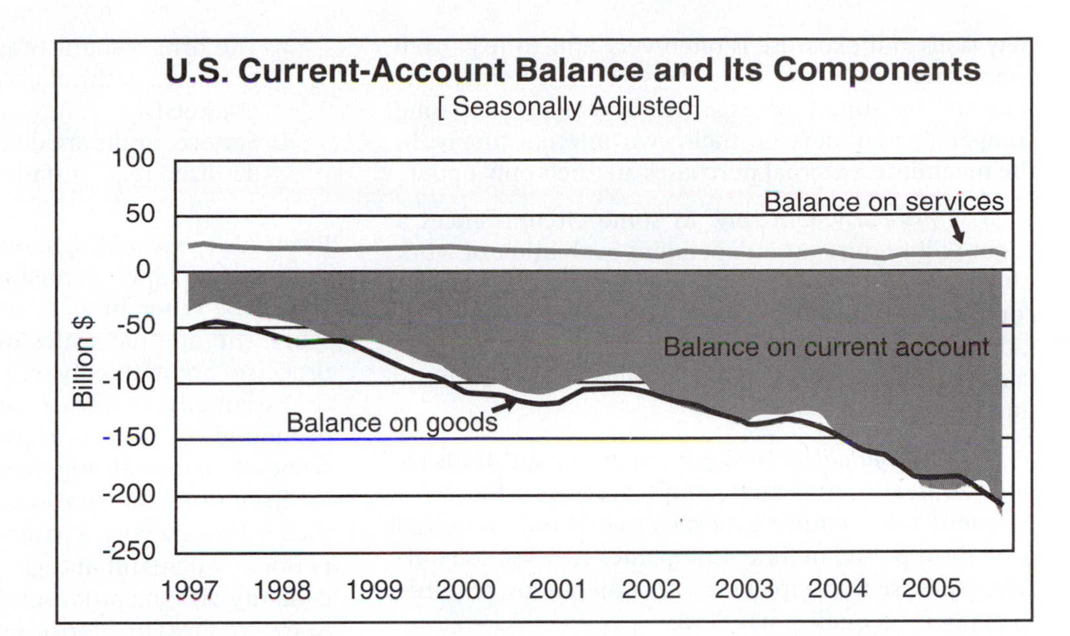

FIGURE 8.10
U.S. trade balance in goods and services, 1997–2005. Although the United States runs large trade deficits in manufactured goods, it has a small surplus in services.

comprise roughly one-third of total U.S. export revenues. These sales overseas take many forms, including tourism, fees and royalties, sales of business services, and profits from bank loans; unfortunately, the U.S. data in the Current Account include repatriated profits from manufacturing investments overseas as a "service," while in reality these comprise returns from capital investments. Foreign service exports do not generate jobs and revenues in all places equally: Cities such as New York, Los Angeles, and London, for example, which are critical to global capital markets, have benefited the most heavily.

The Externalization Debate

Yet another explanation for the growth of services concerns the externalization of many functions as large, previously vertically integrated firms downsized in the 1980s and 1990s in response to mounting international competition. As a result, the amount and degree of subcontracting grew explosively. One explanation of *externalization* holds that large firms use this strategy to minimize costs during economic downturns and periods of rapid restructuring. The alternative explanation is that changing business practices led to an increased need for external expertise. In the face of strategic downsizing, cost-cutting, and the desire to be "lean and mean," many large companies no longer had sufficient staff to internalize producer service functions. The result was the creation of large numbers of new firms.

There are several forces contributing to the external demand for producer services by corporations:

1. *Transaction costs.* Firms turn to the market for services in cases where it is less expensive to acquire them externally than internally (i.e., when it is more efficient to buy rather than to "make" them). Companies, for example, may subcontract with legal, repair, trucking, maintenance, or advertising firms, or management consultants and software engineers. The decision to use external providers of producer services may be made based on the lower costs of an outside supplier, or it may be based on a perception that external provision is of higher quality if subcontractors are highly specialized in a given area.

2. *Flexibility.* Firms employ external producer services where they face a requirement for types and quantities of expertise, information, and knowledge that are very different from their own. It makes limited sense to employ full-time staff if the demand is seasonal, unstable, or temporary in nature.

3. *Risk reduction.* By employing external providers of expertise, a firm is not exposed to the risks associated with full-time employees. It does not have to pay social security costs, provide training, and invest in buildings to house these functions. It also means that during downturns in the economy the firm can reduce its use of external expertise by not renewing contracts. Full-time workers are relatively difficult to remove from the firm; subcontracting spreads risks down the production chain.

4. *Concentration on core skills.* This is a well-known corporate strategy in which firms focus on what they do best rather than try to operate in-house departments designed to provide peripheral expertise. Thus, a company may emphasize activities in which it enjoys a comparative advantage rather than attempt to be too diverse. This strategy, and the outsourcing it produces, are often used during downturns in the business cycle.

5. *New types of services.* The growth of a service economy has been marked by the emergence of all sort sorts of niche markets, including those in which the supply of

new skills and expertise is often very limited (e.g., web page designers in the early days of the Internet). As soon as the supply of expertise develops, purchasing companies may develop their own internal supply. In the meantime, external purchases are their only option.

6. *Third-party objectivity.* In some circumstances a client will require an independent evaluation of work undertaken in-house or by another firm, such as a corporate audit. In this case, externalization lends legitimacy to an evaluation, and the decision to purchase the input reflects political constraints within the purchasing firm.

7. *New regulations.* New government regulations related to health and safety, employment, or the environment may require expertise that is only required for a short period of time. Companies will also use outside expertise to complete the documentation required to apply for a quality standard.

The externalization process witnessed the growth of many producer service firms that primarily offer expertise, reflecting the growing technological and administrative complexity of the workplace. As a result, many functions formerly performed in-house by manufacturing firms are subcontracted to small suppliers, some of which are in the service sector. Subcontracting and externalization are often studied in the context of manufacturing firms but have also become increasingly common among large service firms, which face similar competitive pressures to reduce costs.

The externalization of services occurs through several interrelated processes, including the following:

1. Product innovation: transformation of what is produced (e.g., shortening of product life cycles, technological change, emergence of stylized products)
2. Process innovations: transformation of how things are produced (e.g., process design and engineering);
3. An increasingly complex financial environment, including growth in activity centered on loans, corporate takeovers, leveraged buy-outs, and foreign exchange
4. Proliferation of internal management tasks such as strategic planning and coordination that impede the flow of information through the management hierarchy

Firms face a choice as to when to externalize, and how much. This choice reflects both the external environment of the firm and the internal role of its key decision makers. Externalization is most likely when

1. The firm faces severe in-house technical limitations (i.e., expertise).
2. The firm is an independently owned entity, not a branch plant; vertically integrated firms tend to provide more services in-house.
3. The firm is sophisticated relative to its competitors (e.g., technologically, or operates in foreign markets).
4. Service inputs are diverse, shifting, and nonstandardized (e.g., installation, repair, or consulting).

Firms also externalize services when they face rising uncertainty, rapid change in products or technology, when the labor process resists easy automation, or when the optimal scales of operation of production processes are markedly different.

Essentially, externalization allows for external economies of scale to replace internal economies of scope. By externalizing, firms substitute variable costs for fixed ones and spread the risks of production over their subcontractors, a particularly vital role during peak periods of demand and given mounting levels of uncertainty. Among producer services that firms are likely to externalize, insurance ranks as the most common, followed by accounting, advertising, research, management consulting, advertising and public relations, engineering/architecture, market research, headhunters or management/professional recruitment, computer installation/repair, commercial law, real estate, and temporary office help. Such services are ones that are not efficient or cost-effective to duplicate in-house and are only needed once or infrequently, on an unpredictable basis (e.g., troubleshooting). Typically such services augment or substitute for the internal resources of firms; when the demand is regularized and predictable, they are likely to be internalized within the firm. Further, the degree of externalization depends on whether the firm is a single establishment or a branch plant, as the latter has the lowest degree of local externalization.

Statistically, externalization has inflated the growth of services due to the limited languages employed to measure economic activity. For example, a steel company may lay off a janitor during a period of contraction but then subcontract with a janitorial service; similarly, a ball bearing company may lay off a secretary and contract out to a clerical services company. Nothing has substantively changed in the process—the same jobs are performed—but the manufacturing sector in which the janitor or secretary used to work has registered a decrease in employment while the service sector has registered an increase. Because statistics on employment rely on definitions based on industries and not occupations, this process results in a reduction of manufacturing employment and an increase in service employment. In terms of services employment, therefore, this process is something of a statistical mirage; were definitions of services based on occupations and not industries, this factor would cease to be an explanation of services growth.

Nonetheless, given the structure of national employment accounting systems, contracting out represents one reason why services employment has increased. This process serves as a reminder that the boundaries between services and manufacturing are often blurry. Others have argued that the vertical disintegration and externalization thesis has been exaggerated. One line of thought holds that the demand for such services is satisfied through the growth of many small firms that satisfy new needs, often consisting of a single individual filling a highly specialized niche with customized needs, low overheads, and few economies of scale.

THE PRODUCTIVITY DEBATE

One of the major issues concerning services is *productivity* (the ratio of economic outputs to inputs). It is generally accepted that improvements in the productivity of service industries have been slower and more difficult to measure than for other sectors of the economy. As to the size of the productivity gap—the difference between the average productivity of manufacturing and services—there remains a good deal of uncertainty and disagreement.

There are several ways of measuring the quantity of outputs from the service sector. Outputs from service industries can be measured by the value of the output, the physical quantities of the output, the wage bill, or the number of employees. Most interest is centered on the productivity of labor (i.e., the output per person or number of hours worked) and how it changes over time, varies among countries, or varies among different sectors within services. These issues also raise the problem of how to measure the productivity of public services, many of which do not have a market value. Moreover, there is the question of how the quality of a service should be measured and incorporated in any measure of productivity.

Any problems with measuring inputs into the production of services are matched by problems on the output side. This may be relatively easy in the case for services such as fast food, where the number of hamburgers served per employee per hour, for example, is clearly measurable. However, the majority of services, especially those that have been growing quickly and use the most highly skilled workers, produce outputs that are much more difficult to measure using conventional methods. This is because the producer and the consumer, or input and output, involve relationships that are rarely exactly the same for each transaction. In addition, while the production of a good can be readily separated from its consumption, this is not so for services. Take the example of a management consultant. How is the output of such a firm to be measured or how can its productivity be computed?

The success of an accountant advising a corporate client on how best to organize and present assets and liabilities in order to minimize liability to annual corporation tax depends on how the company interprets and implements the advice. Productivity in this case is not just the result of efficient representation of knowledge and expertise by the accountant but also on decisions and actions by the client and the institutions that act on the information that is sold to them. If measures of productivity are to be meaningful, it is therefore necessary to consider the system of producers and users as a whole or measuring the productivity of services with full reference to the production process *and* the social setting.

All this should not be taken to mean that it is not possible to see improvements in the productivity of service industries. Some services are amenable to the process of standardization. For many of the routine, back-office functions undertaken by banks or insurance companies, there are opportunities to substitute machines (computers and related equipment) to perform tasks previously undertaken by clerks or data processors, with dramatic improvements in productivity. But for most services, there are limits to achieving improvements in productivity growth. These can be summarized as follows:

1. The unique attributes of many services ensure that only limited rationalization of their production (and therefore improved productivity) is possible. The best examples are services that are heavily reliant on persons for their delivery, such as health, education, and care for the elderly.
2. Service producers and consumers often need to be co-present in time and space, which means that time is often incurred in unproductive traveling to meetings.
3. The fact that many services require personal proximity enables a degree of monopoly in supply that creates conditions not conducive to seeking improvements in productivity.
4. In some instances, service markets are opaque (i.e., the client does not know in advance what is actually being bought and how exactly it will affect the efficiency of his or her business).
5. The fact that many services are relational means that the outcomes from using them are as much dependent on the behavior of the client (interpretation of advice, implementation of specific actions) as they are on the producer.

Notwithstanding all the difficulties with defining and measuring the productivity of services, there is some evidence suggesting that the productivity of services in the United States has been improving. The explanation rests largely with investment by services in information technology (IT).

There has been a steady fall in the price of computer and communications hardware, as well is in the costs of national and international telecommunications services since the 1980s. Roughly 8 percent of the U.S. GNP in 2003 consisted of expenditures on IT, with 2 percent on hardware, a further 3.5 percent expended on telecommunications services and software, and 2.5 percent on telecommunications. Using computer equipment as a surrogate for IT capital, the five industries that were the largest purchasers (50% of all investment in computers) are all in the service-producing industries, including financial services, business services, wholesale trade, communications, and insurance. IT investment by financial services increased from $16 billion in 1990 to more than $75 billion in 2003. Wholesale services and retail trade services more than matched this trend and were investing almost $75 billion in 2003.

Since the 1960s, even following the relatively widespread availability of computers and data processing equipment, there seemed to be little impact on productivity gains by services. Part of the problem has been that conventional economic analysis has not been well suited to the difficult conceptual and empirical problems of constructing price indexes and real output measures and therefore measuring the productivity of service producers. The pace of IT development and the requirement for services such as communications or data processing to keep up with the latest technology and software make it difficult to assess their impacts on productivity. Retail services have been able to use IT to improve the efficiency of the ordering, invoicing, sorting, loading, transshipment, and unloading along the supply chain that includes raw material suppliers, components manufacturers and suppliers, assemblers, freight forwarders, and wholesalers—all this before the product reaches the retailer, who then uses IT to manage large inventories and to communicate orders for additional products or to issue invoices back up the supply chain. These are just a few of the possible examples of the ways in which IT has been integrated into the activities of service-producing firms in the United States.

However, recent evidence does suggest that there has been a productivity boost to the U.S. economy as a consequence of the link between innovations in IT and their widespread adoption by the service-producing sectors. Between 1995 and 2005, U.S. labor productivity improved by 1.6 percent annually, half of which resulted from faster productivity growth by service- and goods-producing firms outside the computer hardware industry. Between 1995 and 2005, value added per worker in finance, insurance, and real estate changed by 1.3 percent, in retail trade by 4.8 percent, and in wholesale trade by 6.4 percent.

LABOR MARKETS IN THE SERVICE ECONOMY

In the economically developed world, the vast majority—often more than 75 percent—of all jobs involve services of one form another. Further, new jobs—on the order of 90 percent—are overwhelmingly concentrated in services. Of course, the production of intangibles includes an enormously diverse array of industries and occupations. Nonetheless, despite these variations, the diverse labor markets in which most workers find employment demonstrate several common characteristics.

U.S. Employment in Services

Figure 8.11 shows employment change in the United States by economic sector from 1820 to 2005. The greatest increase by far occurred in services of various types. The greatest decrease occurred in the primary sector (farming, mining, etc.), which dropped substantially, and comprises only 6 percent of U.S. jobs today. Manufacturing and other secondary activities rose to its peak in the 1960s, only to decline to about 12 percent today. These structural shifts have enormous implications for labor markets, cities, and people's career opportunities.

Personal services increased dramatically during this period. With an aging population, large increases have been recorded in hospital staffs, clinics, nursing

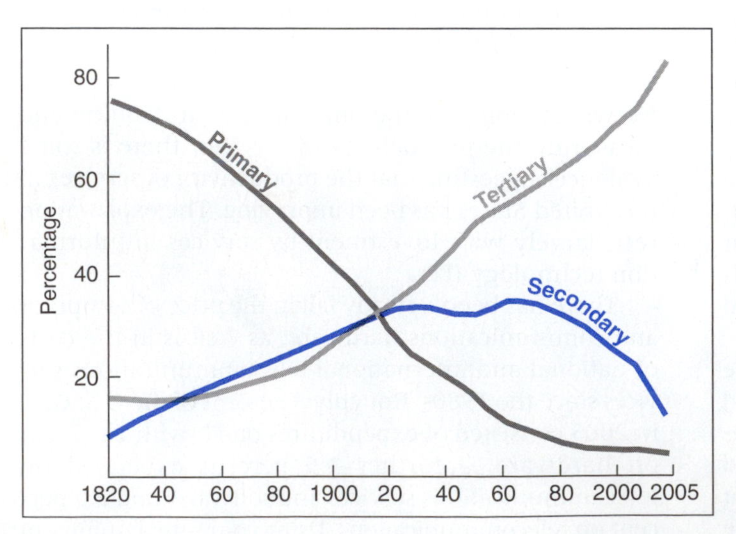

FIGURE 8.11
The changing composition of the U.S. labor force among the secondary and tertiary economic sectors, 1820–2005.

homes, retirement centers, and home health care services. Other large increases have occurred in recreation, maintenance, and social services. Retail services rose mainly due to the popularity of suburban shopping malls and a flurry of new restaurant construction. General merchandise and department stores have not increased as rapidly as specialty shops such as clothing, furniture, automotive dealerships, and electronics.

Transportation services saw only moderate increases because of new, more efficient movement technologies for transportation. Employment levels have increased in trucking and dramatically in air transportation services but have declined in railroads and shipping. While television-broadcasting employment increased during 1960–2005, new technological efficiencies in telephone and communications systems have lessened the demand for workers.

Federal government services remained almost constant during this period, contrary to the perception that the number of government workers has rapidly increased. State and local government has expanded much more rapidly than federal government employment to account for the decline in federal government services.

Characteristics of Services Labor Markets

Services exhibit a number of properties that simultaneously resemble and differ from those in manufacturing. Among these are their relatively labor-intensive nature, income distribution, gender composition, relative lack of unionization, and educational requirements.

Labor Intensity

In contrast to labor markets dominated by manufacturing firms, services tend to be relatively labor intensive, that is, they use relatively more labor per unit of output. Accordingly, the costs of wages and salaries for most services firms range from 70 percent to 90 percent of the total, compared with 5 percent to 40 percent in most manufacturing firms, depending on the degree of labor intensity or capital intensity each exhibits. Of course, some services, such as finance, can be quite capital intensive and generate huge outputs with minimal workers; others, however, such as education and medical care, require large numbers of employees. Whereas manufacturing output has been relatively easy to mechanize—witness the remarkable changes of the Industrial Revolution—many services are much more difficult or costly to replace workers with machines, particularly if they involve variations in tasks, judgment, or dexterity. Some services, obviously, have exhibited enormous technological change, including, for example, personal computers in the office or automatic scanners in retail stores. In poorly paying jobs, however, firms' incentives to replace workers with machines may be relatively low.

Income Distribution

The distribution of incomes in services occupations has been a major source of concern for many social observers. The standard argument holds that industrial economies generated a distribution of income that was relatively "normal" (in the statistical sense) (i.e., manufacturing created societies with a large middle class and relatively few rich or poor [Figure 8.12]). In contrast, services are frequently held to exhibit a bifurcated wage distribution polarized between well-paying, white-collar managerial/professional jobs, on the one hand, which require a university education, and unskilled, low-paying jobs on the other, which require little to no higher education (e.g., retail trade, many medical services, security guards, etc.). Indeed, in contrast to early, overly optimistic postindustrial expectations that a service-based economy would eliminate poverty, a large share of new service jobs pay poorly, offer few benefits, and are part time or temporary in duration, leading to widespread concerns about the

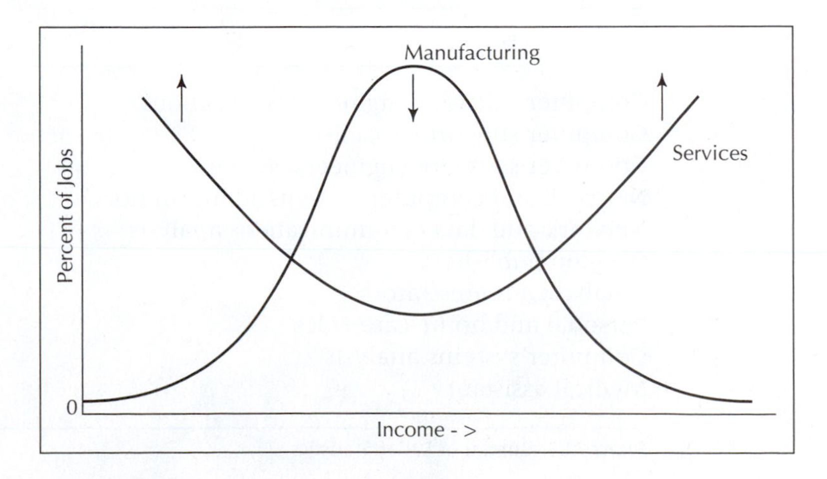

FIGURE 8.12
Depiction of the argument that manufacturing generates a middle-class set of incomes, whereas services incomes are polarized among the relatively rich and poor. In this line of thought, deindustrialization and the shift to a service economy are responsible for the mounting income inequality in the United States.

Table 8.2
Projected Employment Change in U.S. Labor Force by Occupation (millions)

	2000	2010	% Change
Management	15.5	17.6	13.5
Professional	26.8	33.7	25.7
Services	26.1	31.1	19.1
Retail sales	15.5	17.4	12.2
Office and administrative support	23.9	26.1	9.2
Farming, fishing, and forestry	1.4	1.5	0.1
Construction and extractive	7.5	8.4	12.0
Installation, maintenance, and repair	5.8	6.5	12.0
Production occupations	13.1	13.9	6.1
Transportation and material moving	10.1	11.7	15.8
Total	145.6	167.7	15.2

Source: U.S. Bureau of Labor Statistics.

"McDonaldization" or "K-Martization" of the economy. The most rapidly growing occupational groups in the United States, including over the next decade (Table 8.2), include professionals but also low-wage service workers and retail trade employees.

The occupations with the greatest relative projected job growth include almost anything having to do with computers, as well as health care related positions such as home personal care aids and medical assistants (Table 8.3). In absolute terms, the greatest number of new opportunities will be in low-wage, unskilled positions such as food preparation, customer service, retail sales, cashiers, clerks, and security guards (Table 8.4). Further, the United States has witnessed a steady growth of *"contingent" labor* (i.e., involuntary part-time jobs typically filled by women and minorities), a process that has heralded the birth of the "working poor" and helped to swell the ranks of the homeless caught between jobs that pay too little and housing that costs too much.

Concerns that the distribution of income in services is more bifurcated than that in manufacturing are augmented by the fact that deindustrialization has annihilated many jobs in manufacturing, while services continue to grow. In general, services have tended to pay poorly compared with manufacturing. The average clerical position in the United States generates only 60 percent of the annual income of a blue-collar industrial worker, and the average retail trade job only 50 percent as much. Bluntly, such fears are manifested in worries about the "declining middle class" and the polarization of postindustrial economies into distinct

Table 8.3
Fastest Growing Occupations (thousands)

	2000	2010	% Change
Computer software engineers, applications	380	760	100.0
Computer support specialists	506	996	96.8
Computer software engineers, systems	317	601	89.6
Network and computer systems administrators	229	416	81.6
Network and data communications analysts	119	211	77.3
Desktop publishers	38	63	65.8
Database administrators	106	176	66.0
Personal and home care aides	414	672	62.3
Computer systems analysts	431	689	59.8
Medical assistants	329	516	56.8

Source: U.S. Bureau of Labor Statistics.

Table 8.4
Occupations with Largest Job Growth (thousands)

	2000	*2010[*]*	*Absolute Change*
Food preparation & service	2206	2879	673
Customer service representatives	1946	2577	631
Registered nurses	2194	2755	561
Retail salespersons	4109	4619	510
Computer support specialists	506	996	490
Cashiers	3325	3799	474
Office clerks	2705	3135	430
Security guards	1106	1497	391
Computer software engineers	380	760	380
Waiters and waitresses	1983	2347	364

[*]Projected.
Source: U.S. Bureau of Labor Statistics.

groups of "haves" and "have nots." Statistically, however, there is little evidence to suggest that the distribution of incomes among services occupations differs significantly from that in manufacturing (Figure 8.13). Indeed, services generate the vast majority of employment that allows for millions of well-paid, middle-class suburbanites to live lives of relative ease. Others note that the statistical evidence pointing to the mounting inequality in income distributions, particularly in the United States (Figure 8.14), reflects the increasingly regressive nature of taxation in the United States (where the affluent escape their share), as well as the growth of unearned income (e.g., rents, royalties, stock dividends) from which the rich derive the vast bulk of their earnings.

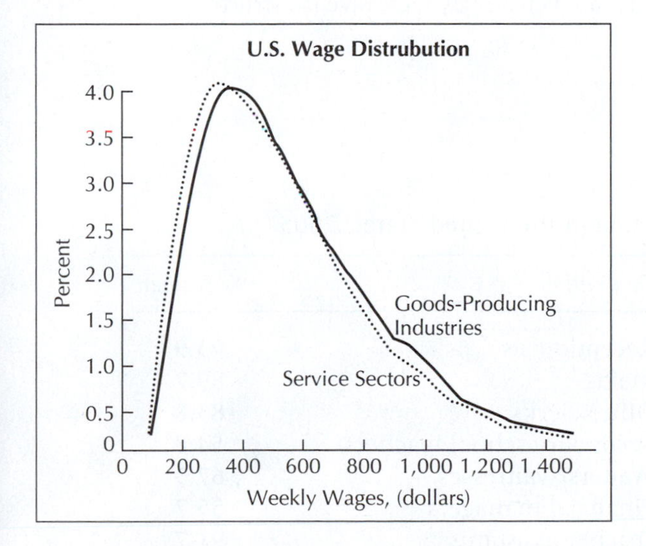

FIGURE 8.13
Distribution of wages in goods-producing and services industries in the United States, 2004. Contrary to the notion that services incomes are inherently polarized, in fact they closely resemble the distribution in manufacturing.

Gender Composition

A third widespread concern about the labor markets in services concerns the gender composition. Manufacturing-based economies predominantly employed males; while some women worked in factories (particularly textiles and garments), industrial economies characteristically saw relatively low rates of female labor force participation. Most women who worked outside the home did so either for brief periods prior to getting married or in a few specialized occupations such as teaching or nursing.

In contrast, the growth of services since World War II has been accompanied by the steady growth in women in the paid labor force throughout Europe, North America, and Japan. The most rapid rise in women's labor force participation rates has been among married women with children (Table 8.5; Figure 8.15). In the United States today, women comprise 45 percent of all full-time employees. However, women's entry into services jobs has been in large part limited to so-called pink-collar jobs, including clerical and secretarial work, retail trade, health care (other than doctors), eating and drinking establishments, teaching, and child care (Figure 8.16). Most of these jobs pay relatively poorly, leading to widespread concerns about the feminization of poverty. Indeed, until very recently, women's presence in well-paying occupations such as physicians, attorneys, or corporate management has been relatively low, leading to complaints about the "glass ceiling" many women face at the workplace.

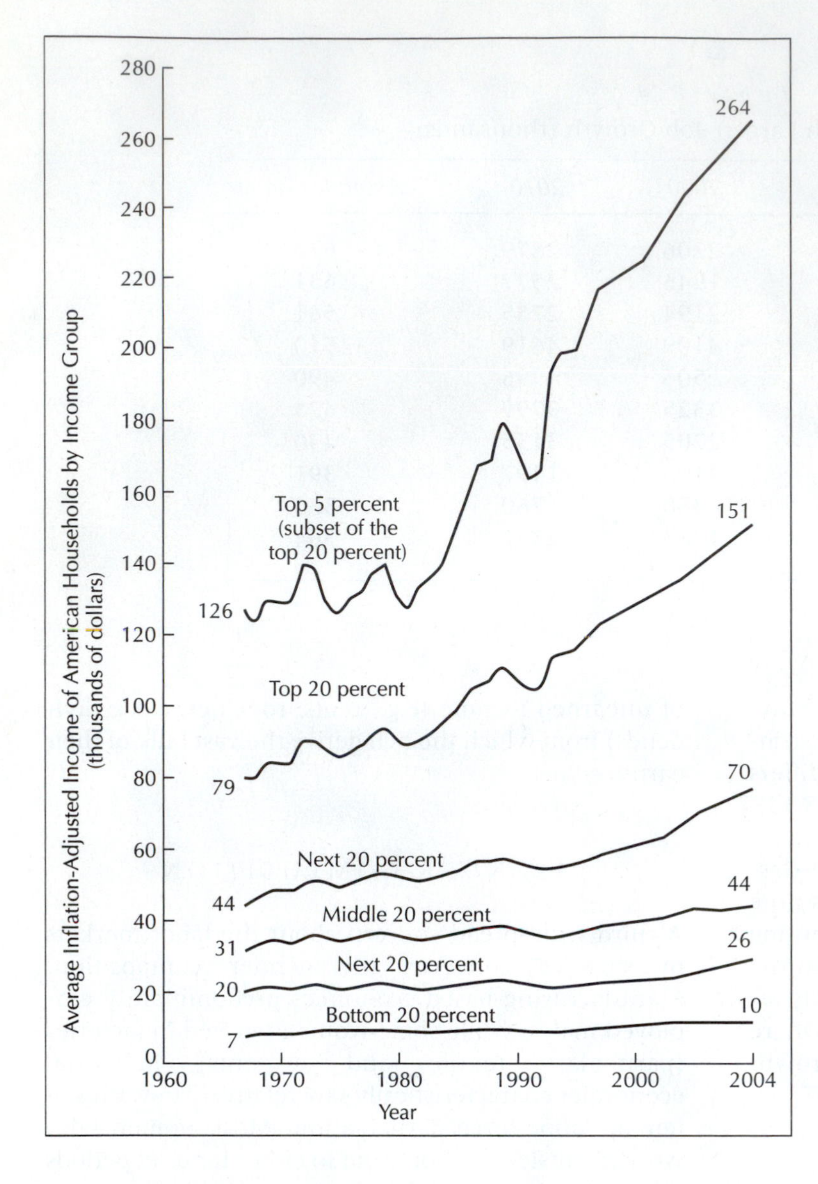

FIGURE 8.14
Income inequality in the United States, 1960–2004. Mounting income inequality reflects the impacts of globalization on the poor and the rapid rise in incomes of the top 5 percent who have benefited the most from the growth of unearned income and an increasingly regressive tax structure.

Table 8.5

20 Leading Occupations of Employed Women in the United States, 2005

Occupation	*% Women*	*Occupation*	*% Women*
Secretaries	96.7	Receptionists	93.9
Elementary school teachers	96.7	Maids	89.7
Nurses	91.8	Office clerks	83.8
Home health aides	88.3	Secondary school teachers	54.8
Cashiers	75.0	Waiters/waitresses	67.3
Administrative support	69.5	Financial managers	55.7
Retail sales	41.1	Teacher assistants	91.7
Customer service	70.1	Preschool teachers	97.7
Clerks	91.2	Social workers	76.1
Accountants	60.8		

Source: U.S. Department of Labor.

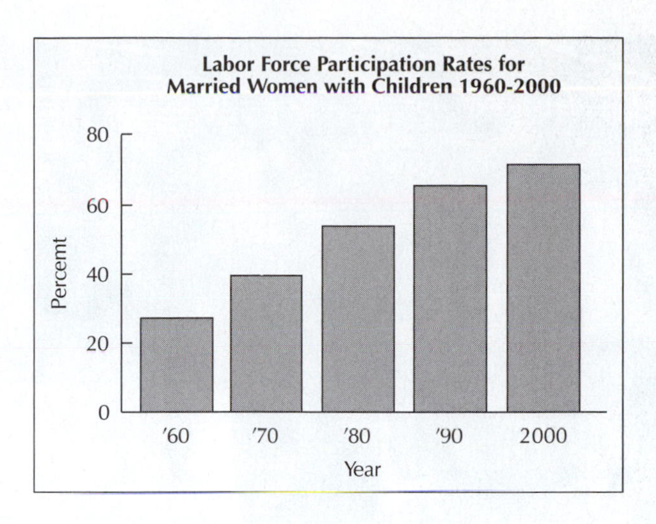

FIGURE 8.15
U.S. labor force participation rates for women with children with children under 18, 1960–2000.

Such a clustering in low-paying jobs is the major reason why women on average in the United States have incomes only 70 percent of that of men.

The rise of women's labor force participation in services carries several social implications. Some maintain that the increase in women's work serves as a strategic compensation to declining male incomes in the face of deindustrialization. The rise of the two-income family has significantly changed gender roles at home and led to (largely unsuccessful) pressures to redistribute housework to men. In countries without national day care systems, such as the United States, child care is an important constraint to women's job opportunities. For middle-class families capable of paying for private child care, this constraint can generally be overcome, although it represents a constant source of financial and emotional stress, complicating commuting patterns and changing the socialization patterns for the young. Most children today enjoy relatively little free time compared with earlier generations. For the poor, who typically cannot afford access to professional child care, adequate or otherwise, the rise in women's paid work has created a child care crisis. In many cases, unattended children, often called "latchkey kids" in the United States, spend long periods free from adult supervision, a phenomenon likely to lead to youths who engage in illicit or illegal activities.

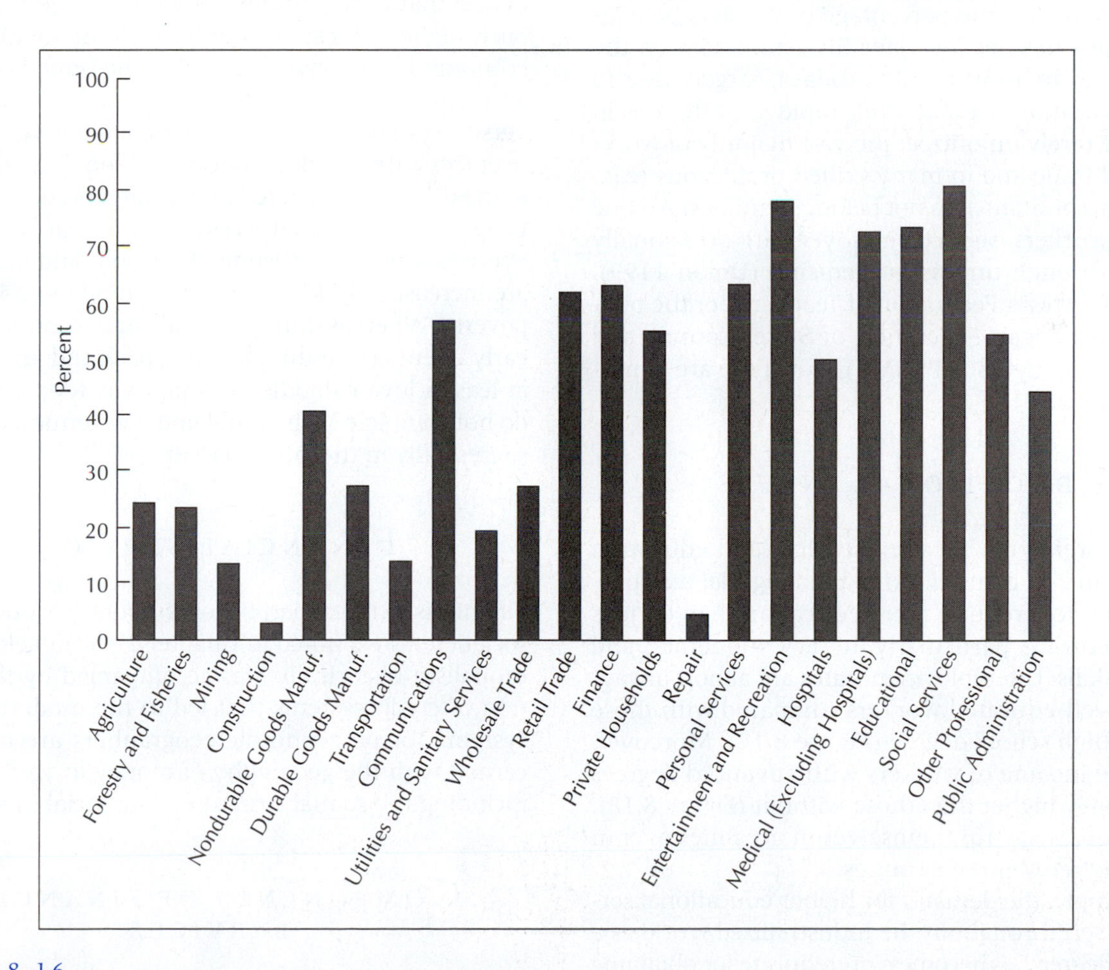

FIGURE 8.16
The gender composition of jobs varies widely among industries. Most women are concentrated in relatively low-paying pink-collar occupations.

Retail trade is a major source of employment in economically developed countries and one of the largest forms of unskilled, low-wage services work. Wal-Mart, with 1 million American employees, is the largest private employer in the United States.

Low degree of unionization

In general, most services are nonunionized. Indeed, since World War II, the percentage of workers belonging to unions has declined steadily (from 45% in the United States in 1950 to 12% today), largely due to deindustrialization. Despite their rapid growth, jobs in services are rarely unionized; the vast majority of workers in retail trade and in many skilled professions (e.g., attorneys, accountants) do not belong to unions. Among American workers, service employees are occasionally organized through unions in medicine (Union 1199), teaching (American Federation of Teachers), or the public sector (American Federation of State, County, and Municipal Employees, AFSCME), but these are generally exceptions.

Educational inputs

Finally, the relations between services and education are important. In general, skilled managerial and professional services require more education than do jobs in manufacturing, particularly literacy, numeracy, and computer skills. Unemployment rates are almost always lower for well-educated workers compared with those without a high school degree (Figure 8.17). Moreover, the average income of workers with advanced degrees is considerably higher than those without (Figure 8.18). College degrees pay for themselves many times over in terms of higher lifetime earnings.

Accordingly, the demand for higher educational services has risen throughout the industrialized world as a university degree has become a prerequisite for obtaining middle class jobs, leading to a surge in college enrollments. In the United States, for example, whereas only 20 percent of high school graduates went on to attend universities after graduating in the 1950s, today roughly 70 percent do so (although many do not graduate). Accordingly, many industrial countries have come to recognize that in the context of a knowledge-based economy, higher education is essential to national and local economic competitiveness. Within the United States, likewise, states that have spent to develop high-quality university systems have enjoyed a competitive advantage over those that neglect educational funding, particularly in the domain of high technology and producer services. People who are denied access to a university education—overwhelmingly consisting of the poor and minorities—are increasingly likely to be condemned to a lifetime of poverty. Whereas during the industrial economies of the early twentieth century a strong back and arms ensured at least a lower middle-class lifestyle, today those who do not complete high school find few venues to compete successfully in the labor market.

Financial services

Finance is a critical part of contemporary economies and societies. As we noted in Chapter 2, the development of capitalism historically was accompanied by the formation of credit systems that led to the modern banking system. Today, economic geographers are often concerned with the geography of money in various forms, including the spatial structure of financial industries.

Components of financial services

Finance is not a single industry but a set of closely related sectors that revolve around money in various forms. A central function of these institutions is to serve

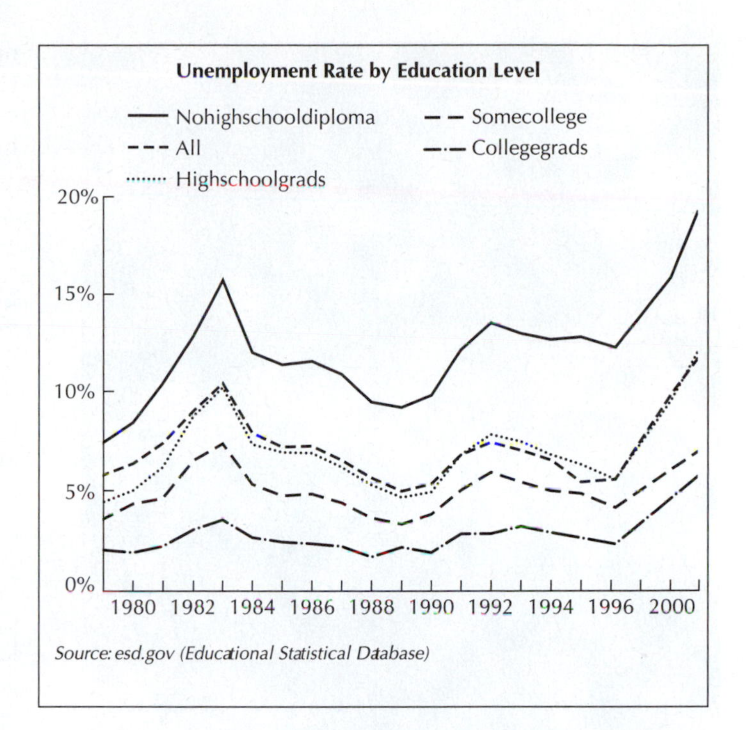

FIGURE 8.17
Unemployment rates by educational level in the United States, 1980–2000. The better educated workers are, the less likely they are to lose their jobs.

as intermediaries linking borrowers and savers together (i.e., individuals and institutions with excess capital to be lent and those who need capital for whatever purpose). Banking thus involves the exchange of money from savers to borrowers and assets (i.e., debt) from borrowers to savers. However, there are considerable differences in the markets involved, the size and behavior of firms, and the regulatory environment they face. The distinctions vary from country to country as they are often based on government-mandated regulations that prohibit some types of firms from operating in some markets. The following distinctions are based on the American financial system, and vary in other countries with different regulatory systems.

Commercial Banking

The largest component of financial services, and the one most people think of when they examine finance, is commercial banks. These are the institutions involved primarily in commercial loans (i.e., to firms, not individuals, including venture capital for start-ups and commercial real estate). Commercial banks provide most of the capital firms use for the building of office complexes, hotels, waterfront developments, and sports stadia. They also offer nonhousing personal loans, for example, to individuals who wish to purchase a boat or fund a college education, and provide retail banking services such as checking and saving accounts and

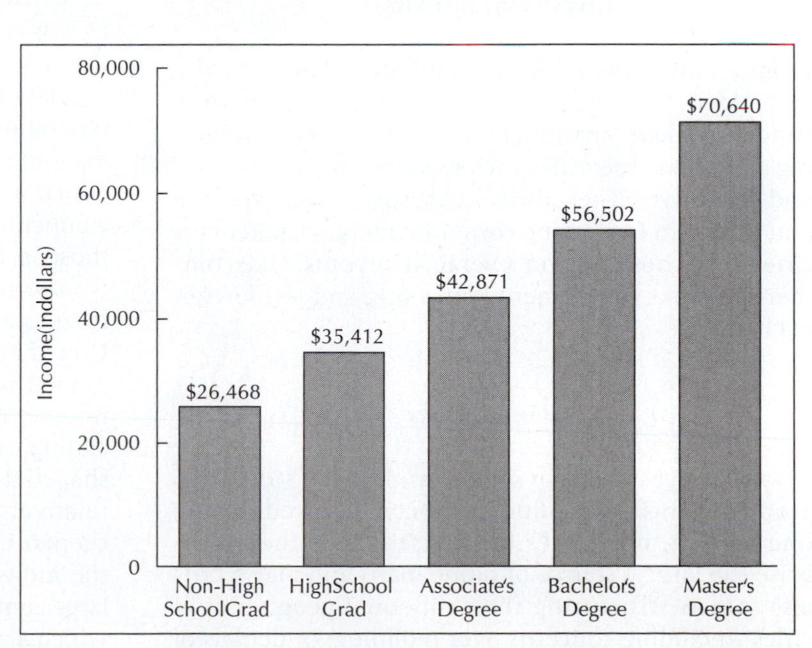

FIGURE 8.18
Median household income in the United States by educational level, 2003.

Because women have entered the labor force of services in large numbers and are also primarily responsible for raising children, they are often caught between conflicting obligations."

credit cards. However, retail banking is a relatively small part of their activities and often the segment that generates the least profit.

Investment Banking

Legally differentiated from commercial banks by the McFadden Act (1927) and Glass-Steagall Act of 1933 (since repealed), investment banks are involved in buying and selling securities such as stocks, bonds, futures, and derivatives. They also provide specialized expertise and capital in facilitating corporate mergers, takeovers (friendly or hostile), and leveraged buyouts. Like commercial banks, investment banks buy and sell foreign exchange.

Savings Banks

Also known as thrifts or savings and loans (S&Ls), this group of firms has traditionally been involved in only one market, mortgages for houses. Thus they comprise the largest source of home loans and play a critical role in structuring the residential geography of cities (including concerns over redlining, or denials of mortgages to low-income minority communities). However, with the deregulation of finance in the 1980s, savings banks began to diversify, penetrating the commercial real estate market.

Insurance

Insurance is a sector largely centered on the commodification of risk. Insurance policies, paid for by premiums paid by households and firms, provide protection from unexpected loss. There are many types of insurance, including property (e.g., for houses), casualty, life, health, automobile, and even nursing home and pet care!

THE REGULATION OF FINANCE

Compared with most industries, finance is highly regulated. The reasons for the comparatively high degree of government control lie in the centrality of finance to current economies. The behavior of financial firms profoundly affects the national money supply, which in turn shapes national interest, inflation, and exchange rates.

Virtually all countries have a national bank of some sort to regulate their financial sector. In Britain, it is the Bank of England; in Germany, the Deutschebank; in Japan, the Bank of Japan. In the United States, this role is played by the Federal Reserve, which was set up in 1913. The Federal Reserve, whose chair is appointed by the president, consists of 12 Federal Reserve Banks (Figure 8.19). It attempts to manage, among other things, interest rates and the money supply by regulating the federal funds rate, the reserve ratios of commercial banks, and by buying and selling government securities.

In the 1930s, as the banking system was devastated by a wave of bankruptcies created by the Depression, the federal government introduced a variety of new regulations to stabilize the system. These included the separation between commercial and investment banking and a series of laws prohibiting interstate banking, which were designed to protect small local banks from competition from larger out-of-state ones, particularly those in New York.

The result of this regulatory system was a highly fragmented, decentralized system of banking in the United States. There are today more than 8,000 banks in the United States, compared with 15 in Canada. The number of banks in the United States, however, is declining as a wave of mergers and acquisitions has reshaped the industry. Most banks are small in size, with relatively few employees and assets, and serve local clients (i.e., they are nonbasic). However, financially, the industry is dominated by "money center" banks, large commercial giants with international ties, most of which are headquartered in New York. Only 2 percent

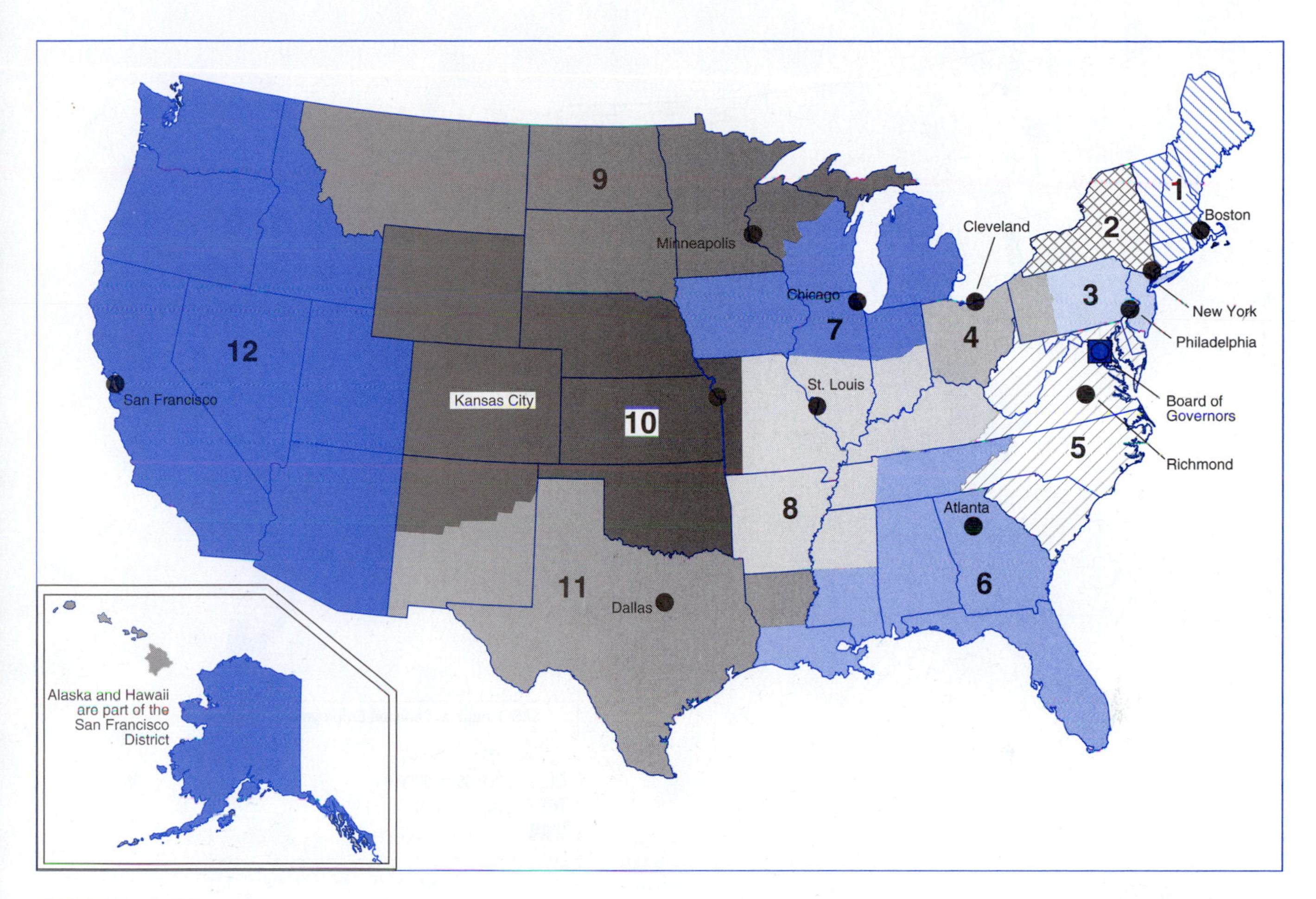

FIGURE 8.19
The 12 districts of the U.S. Federal Reserve, the agency responsible for controlling the nation's money supply. The head of the Federal Reserve is appointed by the president for a six-year term.

of all American banks, for example, own 50 percent of all bank assets. Thus, while financial employees are present in most towns throughout the United States, the largest concentrations are in large metropolitan areas near New York, Boston, Los Angeles, and San Francisco (Figure 8.20).

It should be noted that the regulated system of banking worked relatively well for roughly 50 years (i.e., the period between the Depression of the 1930s and the great changes that swept the world economy in the 1970s and 1980s). International banking was stabilized during this time by the Bretton-Woods Agreement of 1947, in which the United States and its allies set up the architecture for the global financial system after World War II (Chapter 12). The number of banking bankruptcies was very low.

In the 1970s and 1980s, however, much as manufacturing was shaken by the petrocrises and deindustrialization, the banking system was forced to confront a radically different environment. The collapse of the Bretton Woods agreement and shift to floating exchange rates opened up currency trading as a major source of revenues. Rising debt levels in the developing world were another factor that accentuated the globalization of banking as large banks extended loans to countries in Latin America and Asia. The microelectronics revolution introduced new ways of serving customers (e.g., automatic teller machines) and electronic funds transfer systems. As the relatively safe period following World War II drew to a close under mounting international competition and domestic deregulation, commercial banks found themselves in an increasingly hostile and competitive environment, to which many succumbed. As finance became increasingly globalized, the pressures at home to deregulate mounted accordingly.

THE DEREGULATION OF FINANCE

Beginning in the 1970s, the U.S. government undertook a series of actions that had far-reaching consequences in many industries, particularly real estate. In 1974, money market mutual funds were introduced, which sharply increased competition for core banking deposits. In 1979, the Federal Reserve changed from a policy of stabilizing interest rates to a policy of slowing money growth in order to combat inflation. In 1980,

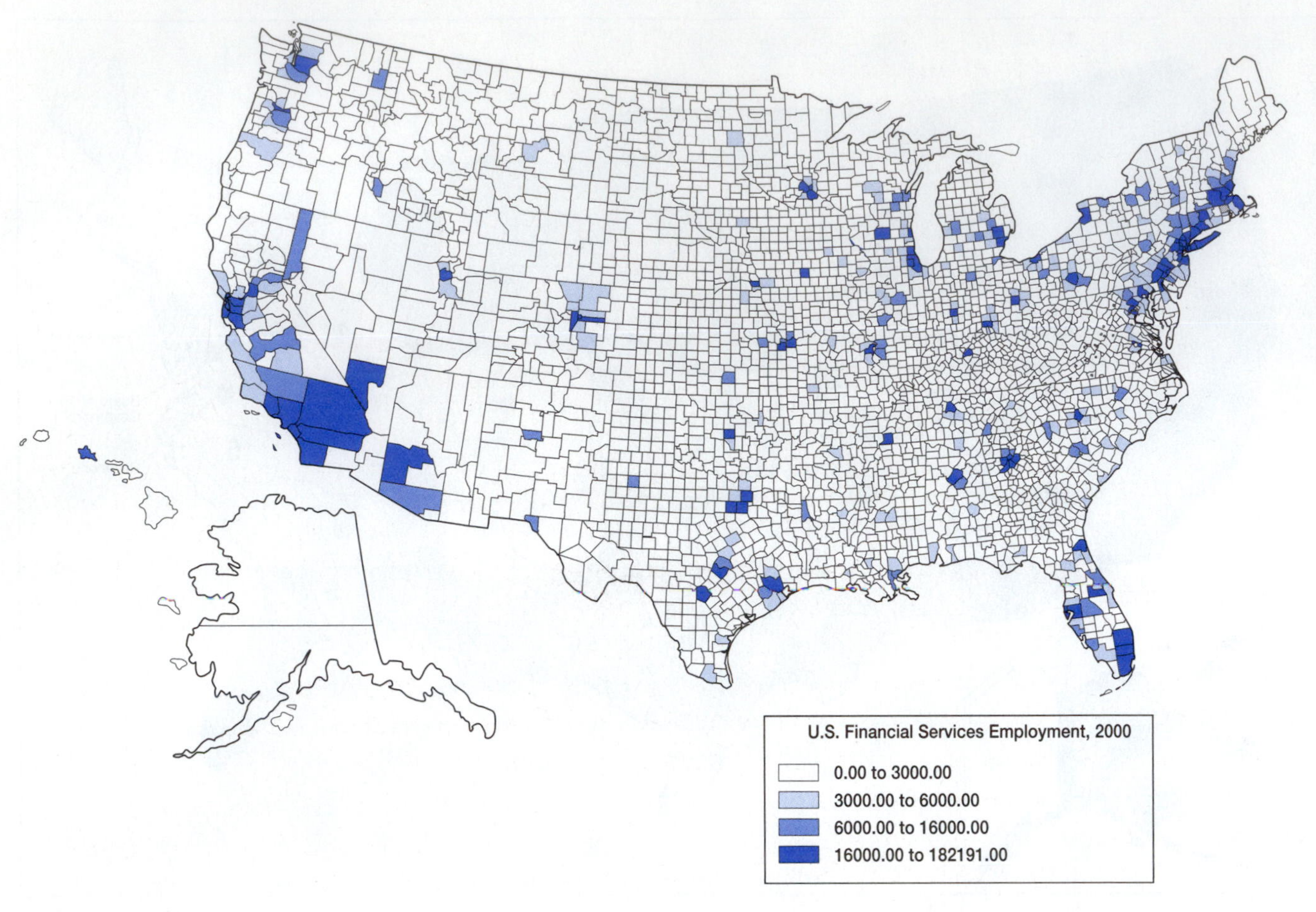

FIGURE 8.20
U.S. financial services employment, 2000.

Congress passed the Depository Institutions Deregulation and Monetary Control Act, and in 1982, the Garn-St. Germain Act, which permitted thrifts (savings and loans) to compete with commercial banks and eliminated geographic limitations on S&L lending. Deregulation included the removal of interstate banking regulations, which led to a national banking market dominated by a few giants. Finally, this process extended to the removal of restrictions governing pension and mutual fund portfolios, the abolition of fixed commissions on stock market transactions, the approval of foreign memberships on stock markets, and the repeal in 2001 of the Glass-Steagall Act, which had separated commercial from investment banking since the Great Depression.

These changes effectively liquidated the relatively stable division of markets and lending sources that had existed since the 1930s (commercial banks in commercial property lending, thrifts in residential lending), increased the liquidity and level of competitiveness of finance capital as new institutional players entered the commercial market, and markedly altered the role of the local state in attracting development. The two institutions most heavily affected by these developments were commercial banks and the savings and loans institutions.

An important consequence of deregulation was increased competition in banking, particularly as noncommercial banks invaded traditional commercial bank markets. S&Ls entered the commercial real estate market, and large retailers (e.g., Sears) began to offer their own credit cards. For many banks, this result was most unwelcome, reducing their profit margins. In response, many banks diverted funds in their investment and loan portfolios to higher risk/higher return opportunities, or by becoming more leveraged in commercial real estate. Many banks, desperate for new avenues of investment, began to raise funds from foreign banks or on the increasingly globalized securities markets, or moved away from traditional lending practices into higher value-added practices such as currency trading and cash.

Commercial real estate has always been the bread and butter of commercial banking; U.S. banks carry more than $800 billion in commercial real estate loans. Deregulation unleashed an enormous wave of investor-driven construction of commercial real estate in the 1980s and 1990s, as commercial banks and S&Ls, with funds augmented by junk bonds, joined mutual funds, insurance companies, and others in a massive surge of investments in lucrative commercial real estate projects, particularly office towers and shopping malls. Many banks have the vast majority of their net worth tied up in offices, shopping centers, and hotels. Consequently, in the 1990s, a serious nationwide office glut appeared, vacancy rates climbed, and commercial rents fell, all of which precipitated a wave of commercial bank and S&L failures and a hugely expensive federal government bailout.

SECTORAL STUDIES IN PRODUCER SERVICES

Services are a very broad group of industries with widely varying attributes. To appreciate this diversity, we now turn to several important service industries, including accounting, design, and legal services.

Accounting

Accounting is one of the most important professions operating in advanced capitalist economies. Modern financial auditing began to emerge in the middle of the nineteenth century, when the ownership and control of firms became separated. With this separation, increasingly detailed financial statements were required so that investors would have sufficient information to inform their decisions. For most companies, an annual audit undertaken by a professional accountant is a legal requirement, forcing them to enter into a short- to medium-term relationship with an accounting firm. This relation allows the accountant to sell related services such as tax advice, estate management, and consulting. The audit process provides an independent check for shareholders into the activities of publicly owned enterprises.

In 2002, a number of U.S. corporate scandals, including Enron, Johnson and Johnson, Tyco, and WorldCom, became examples of accounting failure in which accountants became too close to their clients. The collapse of the energy giant Enron highlighted the difficulties of regulating the relationships between companies listed on the stock exchange and their shareholders. In just 15 years, Enron transformed itself from a small company to America's seventh largest corporation. Enron's success, however, was founded on artificially inflated profits and highly dubious accounting practices. Arthur Andersen, one of the big-five global accountancy companies, had been Enron's external auditor since the 1980s but also acquired responsibility for the corporation's internal audits. The ties between Enron and Andersen were very close, with the in-house financial team dominated by former Andersen partners. When the firm began its rapid decline toward bankruptcy and exposure of the actual state of its balance sheet, Andersen employees began to shred documents. The end result was Enron's bankruptcy and effectively the closure of Arthur Andersen—the big five had become the big four accountancy firms. Andersen's major clients transferred their business to one of the other companies. Accountancy is based on trust and reputation, and Andersen had undermined its reputation. Andersen's global partnership collapsed, and its national partnerships were acquired by the other big accounting companies. Today, following waves of corporate mergers in the industry, the world's largest accounting firm is KPMG Peat Marwick, a joint U.S.-Dutch firm with offices worldwide (Figure 8.21).

Design and Innovation

Management consultants support the production process by improving management systems and by increasing the productivity and profitability of their clients. They also encourage client innovations, especially in

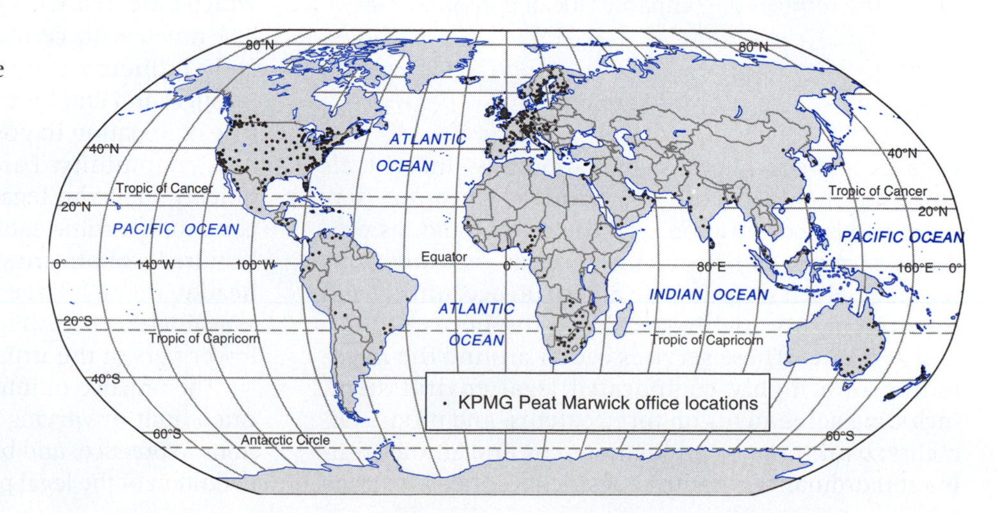

FIGURE 8.21
Offices of KPMG Peat Marwick, the world's largest accounting firm.

administrative practices. Consultants can act as catalysts for introducing or encouraging change. There are, however, producer services that play an important direct role in supporting, encouraging, and developing innovation. Traditionally, innovation has been thought to be vested in manufacturing rather than service firms, and the latter have only been considered as playing a supporting role. However, services now play an important role in innovation, including the role of education in providing a skilled workforce, universities in research and development, and research companies in product testing, product naming, brand development, and product or service development.

Some of the most important producer services directly involved in product innovation are design companies. The industrial designer is employed by a manufacturer to assist with changes that will increase the demand for a product. The task involves understanding public tastes and altering products to make them more attractive to the consumer. Design is an integral part of the production process but is separate from the actual manufacturing process; rather it informs and is informed by it. Small- to medium-sized firms frequently employ private design companies while large companies (e.g., automobile manufacturers) have in-house design departments but can still employ external designers.

Like most providers of producer services, informal or formal networks of small companies and sole practitioners develop. Networking with other companies allows information and experience to be shared but can also lead to collaborative partnerships. These allow small companies to provide a range of services that can compete with those provided by large companies. Small companies can be flexible as collaborations or temporary coalitions involving different service suppliers can be reformed on a client-by-client basis. This strategy enables small companies to provide a range of expertise but without having to employ full-time specialists. It also enables them to deliver services outside their local area by drawing on the services of companies located in other places.

Legal Services

Legal services may be consumed both by individuals and households, on the one hand, as consumer services and by corporations, on the other hand, as producer services. For firms, the services that law firms provide, which may include advice on accounting and investment, are highly specific to the needs of individual clients. These services center around the negotiation of a highly complicated legal environment, including agreements on torts, patents, and product liability; bankruptcy; employment law and antidiscrimination ordinances; antitrust restrictions; tax law; trade agreements; venture capital negotiations; joint venture agreements; technology transfer protocols; reinsurance; corporate mergers, buyouts, acquisitions, and takeovers; and intellectual property rights. In an increasingly uncertain production environment characterized by huge and unpredictable changes in markets, state restrictions, technology, and output, the demand for legal services reflects the broader corporate proliferation of management tasks such as strategic planning and coordination. Deregulation in many sectors has accentuated the uncertainty faced by many firms and increased the need for lawyers to assist in negotiating the resulting highly complex legal environment. Inputs of legal services also reflect the widespread wave of vertical disintegration of production that occurred in the late twentieth century, leading to a steady contracting out of many functions formerly performed "in-house." More recently, many large law firms have diversified into advising, consulting, accounting, debt restructuring, and deal brokering, competing directly against other large services providers based in other industries.

In the United States, the geography of legal services is characterized by high degrees of spatial and functional agglomeration, close interaction with clients and suppliers, and the prevalence of large cities (Figure 8.22). The prime motivation behind the agglomeration of such firms in metropolitan regions is the ready access they offer to clients, suppliers, and ancillary services, most of which is accomplished through face-to-face interaction. Because personal relationships in which trust and reputation are of paramount significance figure prominently in the practice of law, legal services are heavily embedded in local cultural relations and require frequent personal contacts and face-to-face meetings. For these reasons, and others, New York City remains the nation's undisputed capital of domestic corporate law while Washington, D.C., is the pinnacle of government-related legal practice.

The practice of law is dominated by large firms, which emerged with the birth of corporate law in the late nineteenth century and currently disproportionately influence the structure of incentives, rewards, prestige, and employment in the sector, shaping the nature of litigation in government, finance, and the business community. Large law firms have grown more quickly than the legal profession as a whole and are commonly multiestablishment firms with dozens, even hundreds, of attorneys. Such firms tend to have their headquarters in large cities, particularly New York and Washington, D.C. (Figure 8.23), and branch offices in lower tiers of the urban hierarchy.

The practice of international law involves a complicated quilt of varying national legal systems, qualifications to practice, and barriers to entry for foreigners. The definition of the legal profession varies around the world,

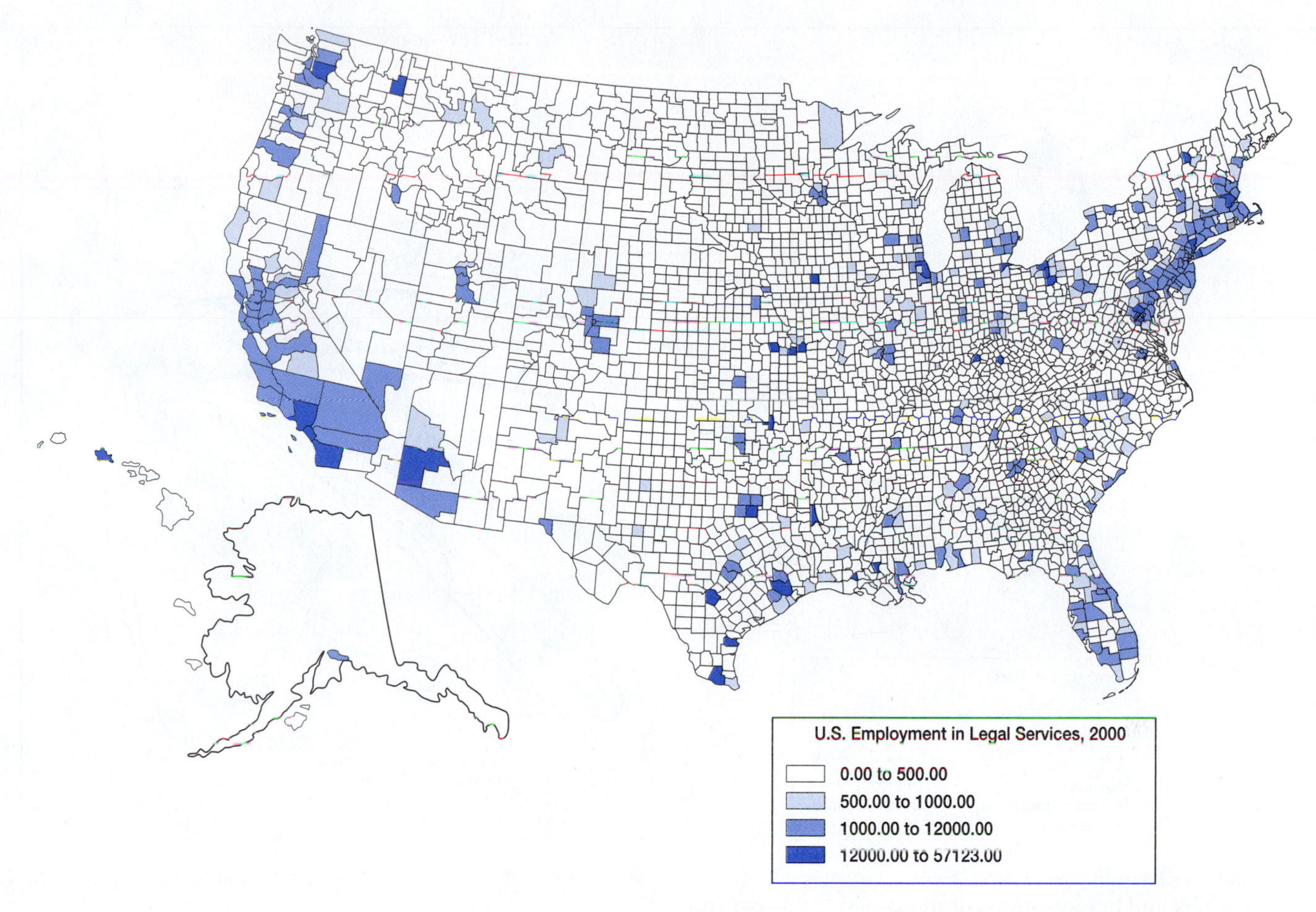

FIGURE 8.22
U.S. legal services employment, 2000.

with considerable diversity as to the requirements and obligations of lawyers, including education, licensing requirements, scope of practice, conduct standards, ethics, and forms of association. The motivations to seek foreign markets reflect a variety of circumstances at work in the contemporary global economy. Mounting international trade and investment and the increasing globalization of finance are the primary forces driving the demand for international legal services. After World War II, the outflow of U.S. capital across the world was accompanied by numerous service firms eager to service multinational clients. The globalization of services often followed the path set forth by manufacturing firms. While the foreign operations of international law firms were originally highly correlated with multinational corporate trade in goods, increasingly it is the global markets in finance and real estate that dictate the geography of demand. During the first wave of globalization, in the 1960s and 1970s, the bulk of firms handled foreign work exclusively through their home office, but throughout the 1980s and 1990s, offices located abroad came to be increasingly important. The principal markets tend to be highly developed countries with well-developed corporate economies, high degrees of international trade, and relatively high per capita income levels (as the demand for legal services is income-elastic). Table 8.6 indicates the number of U.S. attorneys employed overseas in 2002.

Significant national variations remain in the relative degree to which foreign attorneys may practice in different countries. Virtually all countries constrain foreign attorneys from practicing law within their borders to one extent or another. Law is a clubby work environment in which reputation is critical, and foreigners are often informally excluded, if not formally, as outsiders. The practice of law relies heavily on tacit local culture and networks, so international law firms must balance local knowledge with global expertise, reflecting the limited substitutability that characterizes this sector. Formal restrictions on foreign lawyers may take the form of nontariff barriers implemented at the behest of local firms fearful of foreign competition, such as Japanese attempts to exclude American firms from their domestic market, difficult licensing requirements, mandated cooperation with local firms, or outright prohibitions from certain kinds of legal practice. Perhaps for these reasons, the globalization of legal services has occurred more slowly than other services.

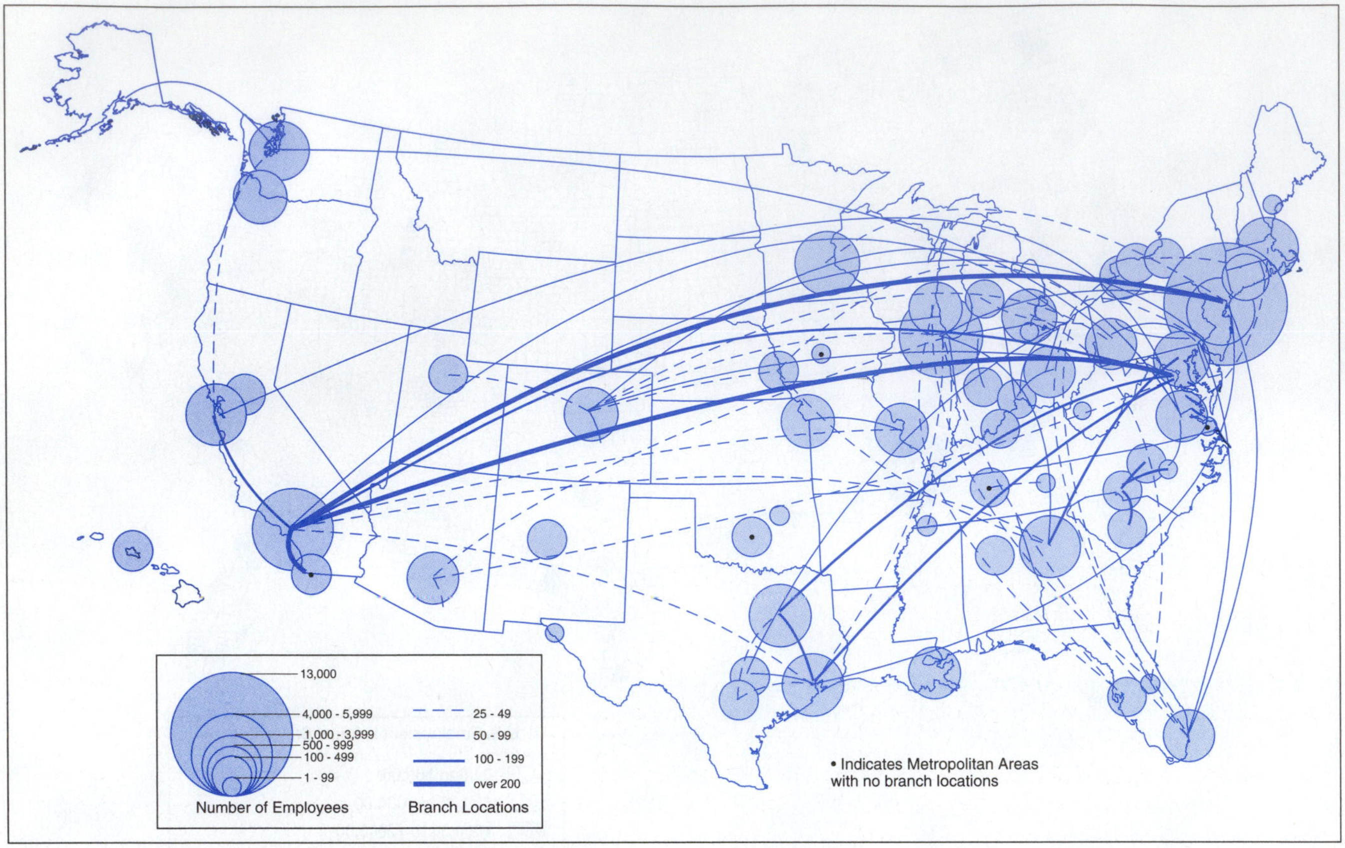

FIGURE 8.23
Headquarters and branch offices of the largest U.S. law firms.

Typically, it is large law firms with a long history of service catering to corporate clients that are best positioned to engage in international practice, which involve high start-up costs, short-term losses, and reliance on economies of scale. This fact favors a group of giants in a limited number of locations. Some of the largest law firms, such as Baker and McKenzie, are true multinationals in their own right. Size is a key asset in foreign markets because it offers "brand name" predictability and credibility (i.e., the necessary critical mass of expertise); reputation serves as an important proxy for reliability and quality of service.

The services that international law firms offer may be provided either directly from their headquarters office or via branch offices, subsidiaries, and affiliates located overseas, with varying degrees of autonomy in the control of their operations. The clients may be either foreign corporations operating abroad or local companies drawing on the resources of multinational law firms, which often have large pools of attorneys with highly specialized expertise. Often international legal services are provided through "cocounseling" agreements, strategic alliances between foreign and domestic firms, a common feature of post-Fordist capitalism. However, national variations in legal systems have made the formation of transnational law firms an exceedingly complicated matter.

The Location of Producer Services

Given their rapid growth, where producer services are located and the functions that they perform are increasingly important. Generally, highly skilled, high value-added functions such as financial and business services concentrate in metropolitan regions (Figure 8.24). Two out of every three business and professional services jobs in the United States are located in large urban areas. Agglomeration economies, such as the minimization of transaction costs between locally based suppliers and subcontractors, the lower costs of accessing a pool of labor with a wider range of skills than in nonurban areas, superior transport infrastructure and facilities, access to specialized information, and a range of office accommodations and telecommunications facilities play a large part in explaining this tendency.

Another theme to emerge from studies of the geography of producer services is the shift from metropolitan areas to nonmetropolitan areas. Large metropolitan areas in the United States experienced a fall in their share of services employment in the late twentieth century, while nonmetropolitan areas increased. This shift may take two forms: suburbanization around existing major service agglomerations, or

Table 8.6

U.S. Attorneys Employed Overseas, 2004

Europe	1,581
Former Soviet Union	97
North Africa/Middle East	44
Asia	710
Latin America	281
Africa	10
Australia/Pacific	149
Canada	5
Total	2877

Source: In Counsel 2002, v. 21.

higher relative growth in rural localities well away from the established service agglomerations. Improved life expectancy, early retirement, and/or an increased proportion of households with large retirement income or personal wealth have enabled individuals to continue providing high-value, high-knowledge services from localities offering a higher quality of life. In many cases the preference of service entrepreneurs may be for remote rural communities that, in reality, are as much a part of the national or international economy as any city because of information technology. The opportunities are similar for younger professionals and others who lose their jobs as a result of corporate downsizing or those who decide to invest in their specialist knowledge and skills to become self-employed.

Intraregional Trade in Producer Services

The long-held view among many planners, academics, and development officials is that services are nonbasic. They function in the local or regional economy essentially as support for the basic goods-related activities that are engaged in interregional trade and that generate the revenues that create the demand for services. This view holds that if one takes away the basic goods-producing activities, the demand for services will decline and they will contract accordingly. However, this view is erroneous. In fact, producer services also perform a basic role in regional development in their own right, although how they do so is less well understood.

Head offices are more likely to have an export orientation while its extent for independent, single-establishment firms varies according to their size. It is generally the case that the more specialized the service, the more likely it is to derive revenues from exports. Advertising, architecture and engineering, design, computing, research and development, real estate, management consulting, or transport services are more export oriented than legal, banking, or equipment rental services, which rely for a larger proportion of their revenues on sales to local clients. Overall, interregional service trade within individual countries is considerable, around 35 percent of total sales and rising.

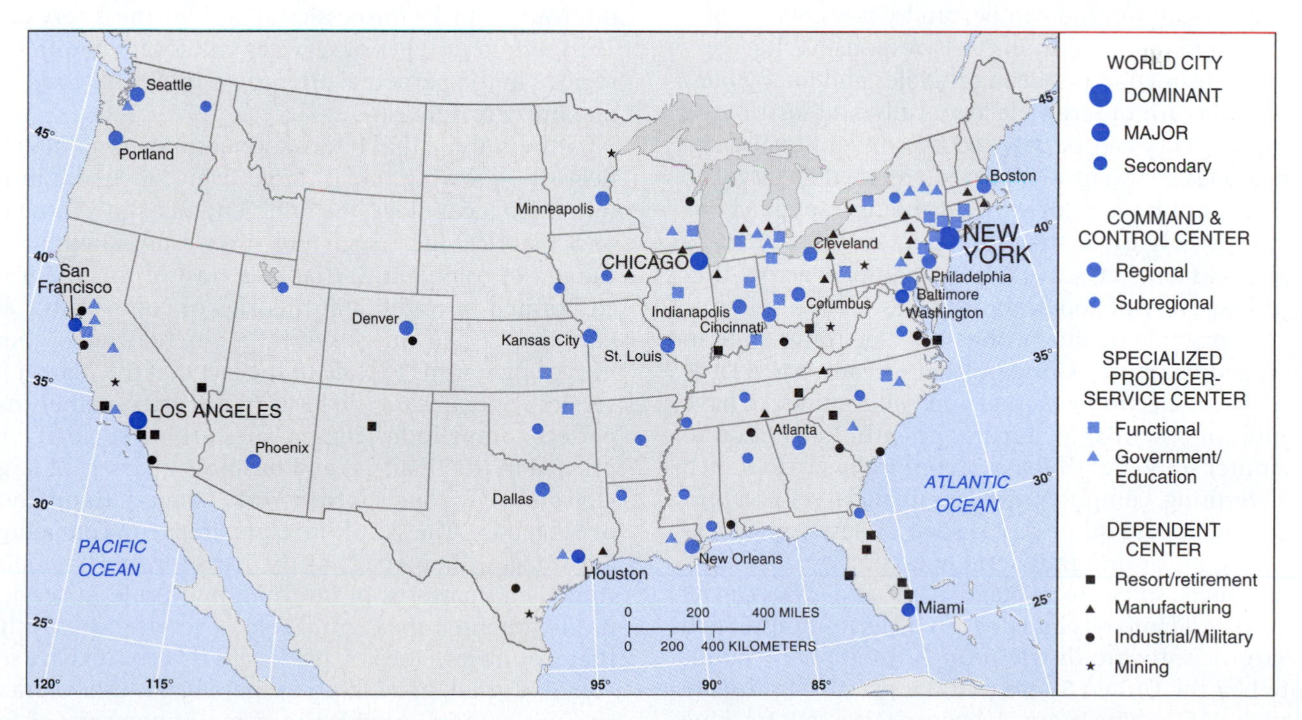

FIGURE 8.24

Business service cities in the United States.

INTERNATIONAL SERVICE TRANSACTIONS

Like agricultural and manufactured goods, services are traded internationally. Four modes through which services may be bought and sold internationally can be identified:

1. *Cross-border supply* occurs when suppliers of services in one country supply services directly to consumers in another country without either supplier or consumer moving into the territory of the other. This function is typically performed by mail, telephone, or the Internet, such as with computer software. This occurs where the service does not require direct face-to-face interaction.

2. A consumer resident in one country may move to another country to obtain a service, such as visits for difficult medical procedures.

3. Enterprises can supply services internationally through foreign affiliates, such as local advertising agencies.

4. Firms may send employees to the country of the consumer in order to provide a service on behalf of the employer, such as when law or accounting firms send representatives overseas.

The measurement of international services transactions has also been complicated by the developments in information technology. The Internet allows geography and location to be diluted as servers can be located in multiple locations and can be run by multiple companies. In addition, the Internet is responsible for blurring the distinction between tangibles and intangibles. Goods that are ordered electronically and distributed internationally can be tracked by conventional taxation and tariff systems without too much difficulty. The same cannot be said for services that are supplied online, especially those that are digital or can be rapidly digitized as an alternative to the traditional across-the-border delivery of goods and services.

An example of this problem is electronic trade in packaged computer software that already has a large share of the electronic market and will continue to have significant potential for further growth. In addition to computer software, this sector comprises services such as advertising, computer services, financial services, entertainment, and other media such as newspapers, periodicals, music and radio and television broadcasting. Newspapers, software packages, CDs, and DVDs can all be converted into bits and delivered electronically. Such electronic software distribution is traditionally dominated by the United States and the packaged software market is becoming more global each year, but tracking imports and exports is seriously hampered by measurement problems. It has been estimated that more than 50 percent of packaged software is directly downloadable from the Internet and that electronic sales represent some 5 percent to 6 percent of the world software market. Border valuations of the trade are based on the medium (CD-ROMs or floppy diskettes) rather than the content (the software), which will usually be significantly more valuable than the medium. Not only will the value of the software be underestimated, but also the value of copyrighted works sold in foreign markets will be overlooked. Computer software is often bundled with computer hardware (e.g., Microsoft software and Intel processors), and many other products, such as cameras, cars, and washing machines, also incorporate highly sophisticated electronic software.

How do we explain international trade in services? Does it resemble trade in goods, or does it have a qualitatively different logic? It is still an open question as to whether international trade theory, which has been derived primarily from studies of manufacturing, is a satisfactory basis for explaining patterns and trends in services trade. At the center of the debate is whether the theory of comparative advantage, originally devised by David Ricardo in 1817 (Chapter 12), can be applied to international trade in services. Different services require a combination of different factors of production, such as finance capital, knowledge capital, information technology and telecommunications, labor supply, educational qualifications in the labor force, and so on. Some services are labor intensive and dominated by simple tasks (call centers, data processing services) while others are dominated by knowledge-intensive, nonroutine tasks (professional and business services). Thus, different types of services will locate in different regions in the perpetual attempt to minimize costs and maximize revenues.

Others argue that international trade in services reflects the growing role of large firms, advances in information technology, the intervention of governments, and historical factors such as links established during the era of colonialism, that is, a host of issues not incorporated in traditional theories of comparative advantage. Trade in services is increasingly about increasing returns to scale or the fact that the market for services operates on the basis of imperfect rather than perfect competition. These issues are most clearly expressed in the financial and business quarters of large, globalized cities such as New York, London, Hong Kong, or Singapore. The shift from comparative to competitive advantage (Chapter 12) as the basis for understanding trends and patterns of international trade in services has highlighted the central role of service firms rather than countries. The key players in this context are service TNCs that, in addition to shaping patterns of trade, are also associated with the rising importance of foreign direct investment (FDI). Noneconomic determinants of services FDI and TNC growth include social and cultural differences between countries that also make it difficult to compete at arm's length (such as in law or advertising), thus encouraging direct representation in some form in order to match service supply with the nuances of local or regional demand characteristics.

Another way of approaching services TNCs is from the perspective of global value added chains, which include distribution, marketing, advertising, and similar functions. As business functions within TNCs have become more specialized, chains linking different stages in the production process have become increasingly widespread. In both services and manufacturing, high value-added functions tend to agglomerate in large cities; conversely, there has been a widespread geographic dispersal of support and clerical services. Clustering is now a global process whereby TNCs require colocation with information and knowledge providers, other service providers, and competitors.

Technological Changes in Services

Starting with the oil crises of the 1970s, global capitalism underwent an enormous round of technological changes that profoundly affected the production of services. Some of these changes, such as the collapse of the Bretton-Woods agreement and the shift to floating exchange rates, were driven by global geopolitics and remained primarily confined to the world of finance. Others, such as worldwide deregulation, had effects across many sectors. Yet other changes, in particular the microelectronics revolution, unleashed a storm of consequences through the rapid decline in the cost of and increase in the processing power of computers. A common expression of this process is Moore's law, which holds that the price for a comparable amount of computing power declines every 18 months. So enormous have the changes of the computer revolution been—including the elimination of some jobs (e.g., travel agents), the creation of new ones (e.g., computer repair), productivity increases, changes in necessary labor skills, and improvements in product quality, convenience, and safety—that they have been likened to the steam engines or automobiles of the information age.

The use of computers revolutionized the world economy. While the incentive to engage in this transformation lay in the need to reduce costs and enhance productivity, it was enabled by the technologies of the microelectronics revolution. The automation of the office beginning in the 1970s fundamentally reshaped the nature of office work and many nonoffice types of services, greatly increasing productivity. Mainframe computers and batch processing initiated a wave of capital intensification with the introduction of microcomputers into the workplace, making services far more capital intensive than they used to be. Today, computers are being adapted to automate the operations of a bureaucracy—repetitive, technical, clerical and professional work, including redundant calculations, allowing the labor force to organize, process, and make decisions concerning data that allows business to change and adapt to the new marketplace. *Networked computers* allow for tremendous quantities of information to be downloaded at each workstation in the firm. Huge databases linked through expert systems allow salespeople or service representatives to pursue higher-level tasks even stretched among different countries.

Most firms realize that conducting business in the information age is not simply making over old hierarchies or improving old methods little by little. It is about changing old methodologies and starting over. Computers have also helped to automate white-collar work. The computer revolution replaced many bureaucrats and middle managers in the 1990s instead of making them more efficient. This is what is propelling job restructuring.

In short, technological changes in services, closely linked to computers and telecommunications, thoroughly refashioned the world of work and the structure of capitalism worldwide. As computers and telecommunications merged in the late twentieth century, the demand for information processing capacity skyrocketed. By the end of the twentieth century, as it had done many times in the past, capitalism had thoroughly refashioned itself into post-Fordism (Chapter 7), becoming considerably more globalized, capital intensive, deregulated, flexible, and competitive.

Electronic Funds Transfer Systems

Financial markets worldwide were especially affected by the digital revolution, which essentially eliminated transactions and transmissions costs for the movement of capital much in the same way that deregulation and the abolition of capital controls decreased regulatory barriers. Banks, insurance companies, and securities firms, which are very information intensive in nature, have been at the forefront of the construction of an extensive network of leased and private telecommunications networks, particularly fiber-optics lines (see Chapter 9).

Electronic funds transfer systems, in particular, which form the nervous system of the international financial economy, allow banks to move capital around at a moment's notice, taking advantage of geographic differences in interest rates and exchange rates, and avoiding political unrest. Traveling at the speed of light, as nothing but formations of zeros and ones, global money dances through the world's telecommunications networks among the world's major financial centers 24

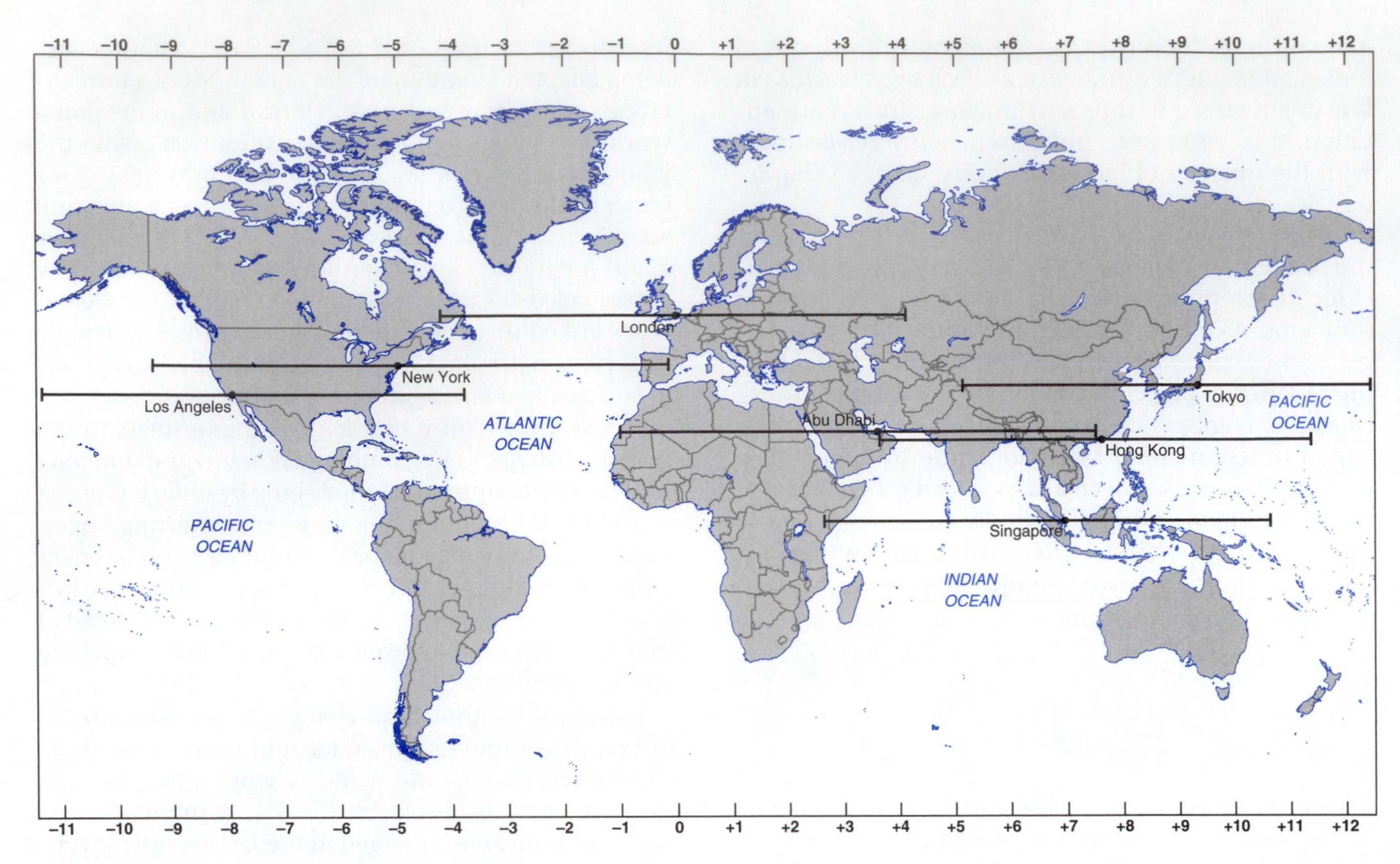

FIGURE 8.25
The world's major stock markets are linked by telecommunications networks.

hours per day (Figure 8.25). The total volume of electronic financial transactions worldwide exceeds $500 trillion annually. These networks include, in Europe, the Belgian-based SWIFT, the Society for Worldwide Interbank Financial Telecommunications, which now extends into 130 countries. The U.S. Federal Reserve Bank's Fedwire system transfers $250 trillion in the balances among private bank deposits annually, while the Clearing House Interbank Payments System (CHIPS), a consortium run by private firms to clear international transactions, processes $310 trillion annually.

Private firms have analogous systems. Citicorp, for example, erected its Global Telecommunications Network to allow it to trade $200 billion *daily* in foreign exchange markets around the world. Reuters, with 200,000 interconnected terminals worldwide linked through systems such as Instinet and Globex, alone accounts for 40 percent of the world's financial trades each day. Other systems include Securities Exchange Automated Quotation (SEAQ) in London; Soffex, the Swiss Options and Financial Futures Exchange; the Computer Assisted Order Routing and Execution System at the Tokyo Stock Exchange; and the Computer Assisted Trading System (CATS) in Toronto. Such networks provide the ability to move money around the globe instantaneously. Subject to the process of digitization, information and capital become two sides of the same coin.

The ascendancy of electronic money greatly altered the nature of financial investments. Foreign investments, for example, increasingly shifted from FDI to intangible portfolio investments such as stocks and bonds. Unlike FDI, which generates tangible levels of employment, facilitates technology transfer, and alters the material landscape over the long run, financial investments tend to create few jobs and are invisible to all but a few agents, often generating unpredictable consequences.

Globalization and electronic money had particularly important impacts on currency markets. Since the shift to floating exchange rates, trading in currencies has become a big business, driven by the need for foreign currency associated with rising levels of international trade, the abolition of exchange controls, and the growth of pension and mutual funds and institutional investors. The world's currency markets trade roughly $2 trillion every day, dwarfing the $25 billion that changes hands daily to cover global trade in goods and services. The vast bulk (72%) of foreign exchange transactions include only three currencies, the U.S. dollar, the Euro, and Japanese yen. London remains the premier world center for this practice, exchanging $500 billion in currencies annually, followed by New York ($300 billion), Tokyo ($200 billion), and Paris ($100 billion). The market opens each day in East Asia while it is evening in

North America; funds then travel east, bouncing from city to city over fiber-optic lines, typically from Tokyo to Hong Kong to Singapore to Bahrain to Frankfurt or Paris to London.

Electronic money can be exchanged an infinite number of times without leaving a trace, making it difficult for regulatory authorities to track down transactions both legal and illegal. The opportunities for money laundering are thus made all the more attractive. Tax evasion has become increasingly serious as electronic money has become the norm. Moreover, the jurisdictional question—who gets to tax what—is vastly more complicated. Digital counterfeiters can also take advantage of this situation, using the Internet to create currencies in any other place.

The neoclassical economic case for capital mobility holds that such fluidity allows countries with limited savings to attract financing for domestic investments, that it allows investors to diversify their portfolios, and that it spreads risk more broadly. Capital mobility implies that firms can smooth consumption by borrowing money from abroad when domestic resources are limited and dampen business cycles. Conversely, by investing abroad, firms can reduce their vulnerability to domestic disturbances and achieve higher risk-adjusted rates of return.

OFFSHORE BANKING

One of the most important effects of the telecommunications revolution on the world of finance concerns the emergence of offshore banking centers, loosely regulated or unregulated places. The emergence of these places reflects the growth of globalized "stateless money."

The origins of the contemporary, deregulated, globalized financial system lie in the Euromarket of the 1960s. The Euromarket arose in large part because of the large trade surpluses that western European countries enjoyed with the United States, which left them with large sums of excess dollars, known as Eurodollars. Thus, originally, the Euromarket comprised only trade in assets denominated in U.S. dollars but not located in the United States; today it has spread far beyond Europe and includes all trade in financial assets outside of the country of issue (e.g., Eurobonds, Eurocurrencies, etc.), not just currencies. One of the Euromarket's prime advantages was its lack of national regulations. Unfettered by national restrictions, it has been upheld by neoclassical economists as the model of market efficiency. U.S. banks invested in the Euromarket in part to escape domestic restrictions such as the Glass-Steagall Act, which prohibited commercial banks from buying and selling stocks. Further, the Euromarket lacked any reserve ratio requirements until 1987, when the world's central bankers met at the Bank for International Settlements in Basle, Switzerland, to agree on global reserve standards.

Beyond the Euromarket, globalization spawned the growth of offshore banking centers, often in very small countries (and not always offshore, i.e., on islands). Offshore banking centers offer regular commercial services (i.e., loans), foreign currency trades and speculation, access to electronic funds transfer systems, asset protection (insurance), investment consulting, international tax planning, and trade finance (e.g., letters of credit). There are five major world clusters of offshore finance, including the Caribbean (e.g., the Cayman Islands, Bahamas, Panama), Europe (e.g., Isle of Man, Jersey, Luxembourg, Liechtenstein, Andorra, San Marino), the Middle East (Cyprus, Lebanon, Bahrain), Southeast Asia (Hong Kong, Singapore), and the south Pacific (Nauru, Vanuatu) (Figure 8.26). The emergence of such places reflects the flows of electronic money across the global topography of financial regulation as large financial institutions shift funds to take advantage of lax regulations and low taxes found on the periphery of the global financial system. Many such places, although by no means all of them, have dubious reputations as centers of money laundering, that is, as places in which illicit funds garnered through international drug or weapons sales are converted into legitimate monies.

As the digital revolution allowed global capital to circulate more freely and rapidly, the technological barriers to moving money declined dramatically. Accordingly, spatial variations in the nature and degree of regulation rose in importance. Even small differences in regulations concerning taxes or repatriated profits may be sufficient to induce large quantities of capital to enter, or exit, particular places. Thus, many small states attempt to attract finance capital by deregulating as much as possible, lifting controls over currency exchanges, investment, repatriated profits, and eliminating taxes in the hope that global money will select their locale. By allowing funds to cross national borders at will, offshore banking reworks the notion of national sovereignty, inducing nation-states to retain political sovereignty but to sacrifice financial sovereignty.

BACK-OFFICE RELOCATIONS

Large service corporations generally divide their labor forces into two major segments: those associated with headquarters and those that carry out back-office functions. Headquarters functions typically involve skilled professionals, including executives and managers, as well as those who perform research, publicity, finance, legal work, accounting, engineering, marketing, and product development (Chapter 5). In contrast, back offices perform data entry of office records, telephone books, library catalogs, payroll and billing, bank checks, insurance claims, and magazine subscriptions. These tasks involve unskilled or semiskilled labor, primarily

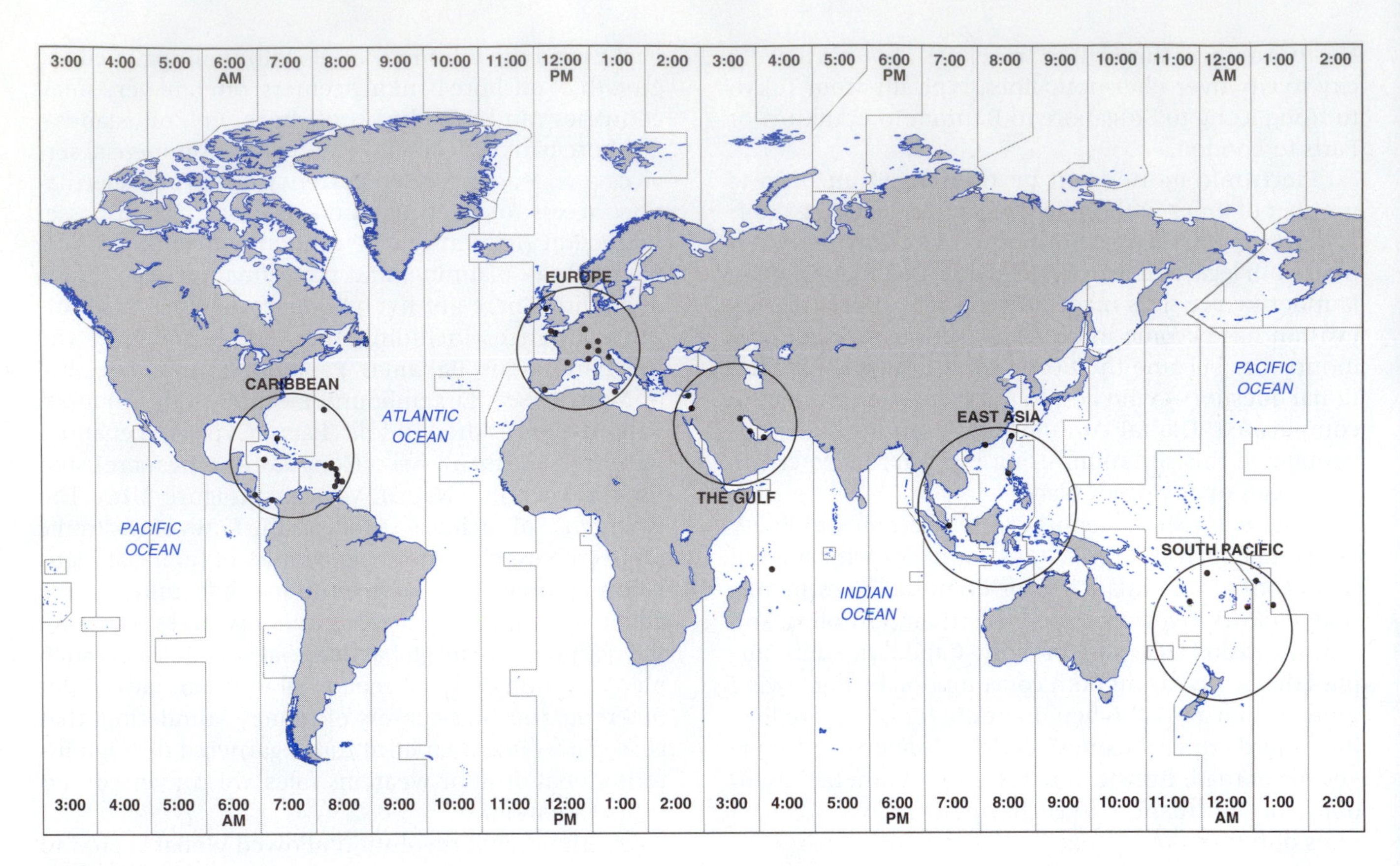

FIGURE 8.26
Five major world areas of offshore banking.

women, and frequently operate on a 24-hour-per-day basis. Unlike headquarters, back offices have few linkages to clients or suppliers. These facilities require extensive computer facilities, reliable sources of cheap electricity, and sophisticated telecommunications networks.

Historically, back offices were located next to headquarters activities (i.e., behind the "front office") in downtown areas to ensure close management supervision and rapid turnaround of information. However, as central city rents rose in the 1980s and 1990s and companies faced shortages of qualified (i.e., computer literate) labor, many firms began to uncouple their headquarters and back-office functions, moving the latter out of the downtown to cheaper locations on the urban periphery. This uncoupling was made possible by the introduction of digital telecommunications, including microwave transmission. Back offices that moved to the suburbs often found pools of female labor willing to work part time, generally consisting of married women with small children. Moreover, suburbs offered ample parking, lower crime rates, and cheaper taxes and electricity. A considerable part of the suburban commercial boom of the late twentieth century consisted of back offices.

More recently, given the increasing locational flexibility provided by satellites and fiber-optic lines, back offices have also begun to relocate on a much broader, continental scale. Many financial and insurance firms and airlines moved their back offices from New York, San Francisco, and Los Angeles to lower-wage communities in the Midwest and South. Phoenix, Atlanta, and Kansas City have been significant beneficiaries of this trend. Omaha, Nebraska, created thousands of telegenerated jobs because of its location at the crossroads of the national fiber-optic infrastructure. Similarly, with abundant cheap labor, San Antonio, Texas, and Wilmington, Delaware, have become well-known centers of telemarketing. Boulder, Colorado, Albuquerque, New Mexico, and Columbus, Ohio, have moved in much the same direction.

Internationally, back-office dispersal took the form of the offshore office, which remained insignificant until transoceanic fiber optic lines enabled their relocation on an international scale. The capital investments in such operations are minimal, and they possess great locational flexibility, maximizing their ability to choose among places based on minor variations in labor costs or profitability. Offshore back offices are established not to serve foreign markets but to generate cost savings for U.S. firms by tapping Third World labor pools where wages are low compared with those in the United States. Notably, many firms with offshore back offices are in industries facing competitive pressures to enhance productivity, particularly insurance, publishing, and airlines. Several New York–based life insurance companies, for example, relocated back-office facilities to Ireland (Figure 8.27), with the active assistance of the Irish Development Authority. Often situated near Shannon Airport, the principal international transportation hub, they ship in documents by Federal Express and export the digitized records back via satellite or one of the numerous fiber-optic lines that connect New York and London.

Likewise, the Caribbean, particularly Anglophone countries such as Jamaica and Barbados, has become a particularly important location for American back offices. American Airlines paved the way in the Caribbean when it moved its data processing center from Tulsa to Barbados in 1981; through its subsidiary Caribbean Data Services (CDS), it expanded operations to Montego Bay, Jamaica, and Santo Domingo, Dominican Republic, in 1987. Manila, in the Philippines, has emerged as a back-office center for British firms, with wages 20 percent of those in the United Kingdom. The microelectronics and telecommunications revolutions allowed for the dispersal not only of offshore banks and back offices but also of telephone operators, call centers, graphic designers, Internet service providers (ISPs), and readers of magnetic resonance imaging (MRI) records, all of which can be transmitted easily through the Internet. Such trends indicate that globalization and telecommunications may accelerate the offshoring of many low-wage, low-value added jobs from the United States, with dire consequences for unskilled workers.

Another form of low-wage, low-value-added services involves centers of telework, often labeled call centers, which are designed for the purpose of high-volume sales, marketing, customer service, telemarketing, and technical support. Call centers functions include telemarketing, customer assistance, and phone orders, often with designated 1-800 numbers. They range greatly in size, from as few as five to as many as several thousand employees. Like back offices, call centers are primarily screen based and do not require proximity to clients. The major cost consideration is labor, although the workforce consists primarily of low-skilled women, and high turnover rates are common. There are an estimated 80,000 to 100,000 call centers within the United States, which employ between 3 percent and 5 percent of the national labor force and the majority of which are located in urban or suburban locations. Cities that have recently established competitive niches in this domain include Omaha, Nebraska, San Antonio, Texas, Wilmington, Delaware, Albuquerque, New Mexico, and Columbus, Ohio.

As with so many industries, call centers have become increasingly globalized. India, for example, has attracted a significant number of customer service centers near its software capital of Bangalore, where workers are trained to speak with the U.S. dialect of English

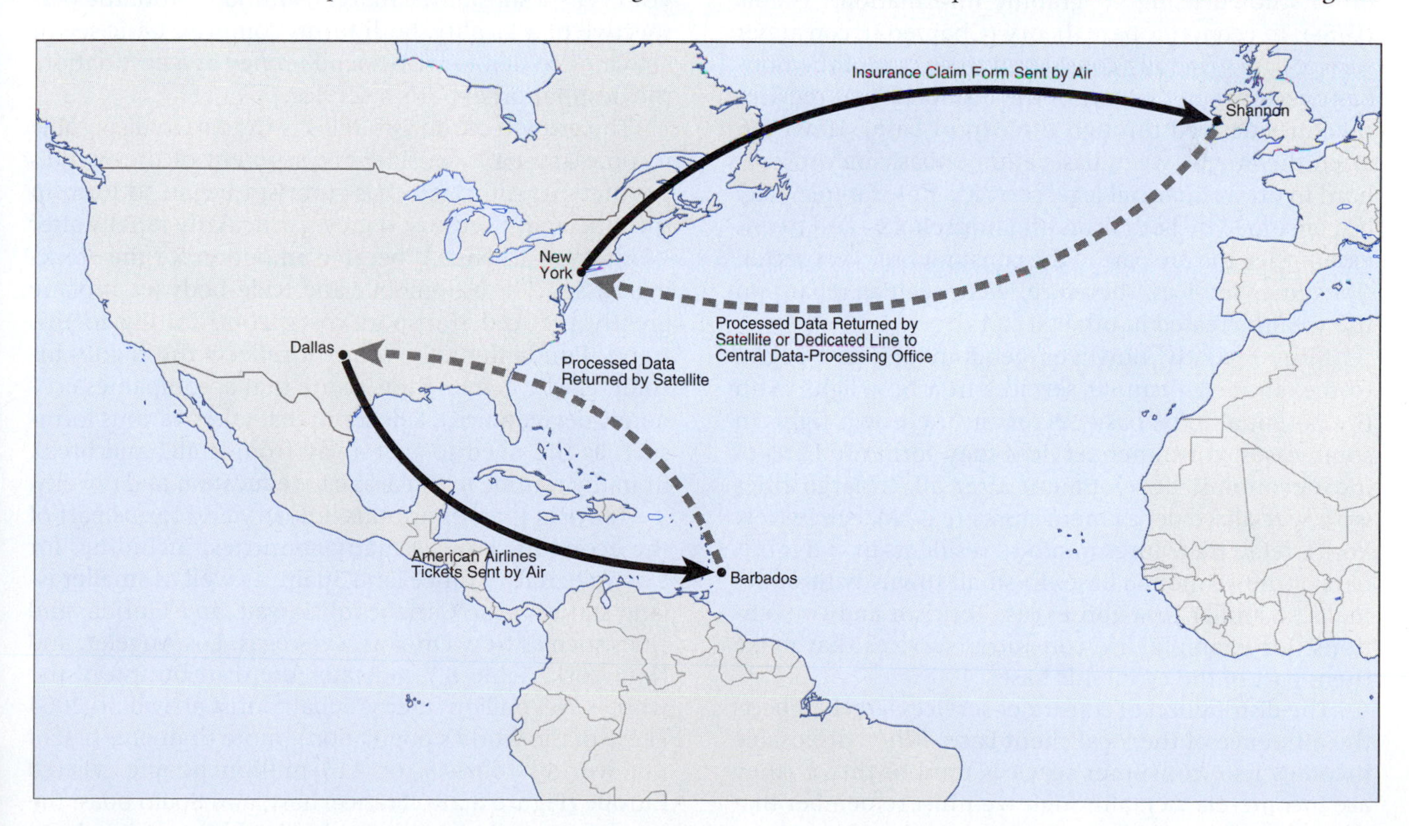

FIGURE 8.27
Mechanics of American back offices in Ireland and the Caribbean.

and are able to gossip with customers about pop culture. Wages there, which average $US 2000 per year, are higher than average Indian salaries but are only 10 percent of what equivalent jobs pay in the United States. The next time you speak with a company representative to change your cell phone, hard drive, or airline reservation, for example, you are likely to speak with a worker in India.

CONSUMER SERVICES

Most attention in economic geography concerning services has focused on producer services, those that are sold primarily to other firms. However, a large part of the service sector concerns consumer services—those sold to individuals and households. Consumer services include a wide variety of industries and occupations, including retail trade and personal services, eating and drinking establishments, tourism, and sports and entertainment facilities. To some extent, the health care and education fields comprise consumer services.

Traditional geographic interpretations of consumer services focused on the travel costs incurred in their consumption. Many large retail firms today deploy specialists in marketing to assess the market potential of stores, including the possible number of consumers, land costs, and access to transportation routes, an approach that is often studied using geographic information systems (GISs). In economic base theory (Chapter 2), consumer services are generally considered to be part of the nonbasic sector (i.e., the part of the economy that recycles revenues derived through the export base). However, often the line between basic and nonbasic industries is hard to draw; financial legal services, for example, may be consumed by both firms and households, and to this extent they too are part of the consumer services sector. Consumer services, therefore, were seen as reliant on the wealth created in other sectors.

More recently, however, geographers have begun to reexamine consumer services in a new light, with the potential to be basic sectors in their own right. In some cases, consumer services may form the basis of local economic development after all. In large cities with specialized department stores (e.g., Macy's in New York), retail trade lures nonlocal residents in and forms part of the economic base. In small towns with outlet malls, a similar function exists. Tourism and conventions, for example, are consumer services that make them part of the economic base.

The distribution of consumer services largely reflects the affluence of the local client base. When disposable incomes rise, consumer services tend to thrive (they are income-elastic), although we must remember that the expenditures on goods and services do not rise equally for all types. The study of consumer services is thus intimately linked to the analysis of local demographics, the associated tastes and preferences of people, their inclination to save or spend, the advertising and other sources of information that shape their preferences, and their age, gender, and ethnic dimensions that affect spending and buying habits (see Chapter 11 on Consumption). Other forces that shape consumer services include changing household work patterns, the growth of chains and franchises, regulatory frameworks, and the relative prices of imports and other goods, which themselves reflect a galaxy of variables in the global economy. The ever-changing geography of consumer services is also heavily influenced by the location of purchasing power of different types. For example, as suburbanization drew much of the middle class and its incomes to the metropolitan periphery, the urban geography of retail trade was reshaped, leading to an explosion of suburban malls of various sizes.

Tourism

Tourism is an important consumer service with enormous international and local implications. Tourism includes a large variety of visitors, including personal as well as business travelers and conventions, stretched over a wide range of durations of stays (the minimum is generally defined as one night away from home). Domestic tourism generally outnumbers international visitors by a substantial margin, although from the perspective of a locality, both forms comprise basic sector activities. When tourists spend money at a destination, the destination exports a service.

The growth of tourism reflects the rise in disposable income among a considerable segment of the world's population. Although mass tourism began to form in the nineteenth century, it grew particularly rapidly after World War II, when it became an option for the working classes. The automobile and wide-body jet airplane greatly reduced transport costs, contributing to this trend. Fundamentally, tourism reflects the highly income-elastic demand for leisure that accompanies economic development, a demand that takes various forms such as the need to "get away from it all" and break from the routine to the desire for education and novelty.

Tourism is an enormous industry and forms part of the economic base of many countries, including, for example, Italy, France, and Spain, as well as smaller island states in the Caribbean, Hawaii, and Florida, and cities such as New Orleans, Las Vegas, Los Angeles, and New York. Table 8.7 indicates the distribution of the world's 763 million international tourist arrivals in 2004 (12% of the world's population); more than one-half of the world's tourists, or 416 million people, visited Europe (Figure 8.28). France, Italy, and Spain boast the largest tourist industries in the world, attracting large numbers of tourists from colder northern climates. To

Table 8.7
International Tourist Arrivals, 2004

Region	*Visitors (Millions)*	*% of World*
Europe	416.4	54.5
Africa	33.2	4.3
North America	85.8	11.2
Caribbean	18.2	2.3
South & Central America	21.8	28.5
East Asia	87.6	11.4
Southeast Asia	47.3	6.2
South Asia	7.5	1.0
Oceania	10.2	1.3
Middle East	35.4	4.6
World	763.4	100.0

Source: World Tourism Association.

a lesser extent, East Asia and North America are also significant.

Economically, the tourism industry consists of a variety of sectors, including international and local transportation, hotels and motels, eating and drinking establishments, entertainment, and retail trade. To sustain these activities requires enormous investments in water, transportation, communications, and other infrastructures.

There is a vast diversity in the types of tourism and tourist destinations, ranging from low-impact ecotourism to highly urbanized urban cores, from simple backpacking to luxury cruises, from safaris to writers' camps to honeymoon packaged retreats, from individual explorers to package tours, from health resorts to Asian sex tourism, from dude ranches to gay night clubs, from museums and fine art institutes to tribal cultural events, from small ski resorts to tropical playgrounds organized by Club Med. The facilities that accommodate tourists are similarly varied, ranging from large hotel chains to quaint bed-and-breakfast establishments. Tourism is often highly seasonal, fluctuating greatly over the course of a year, with corresponding changes in prices for hotels and travel.

The volume of tourists to a given destination will reflect, among other things, the information available to potential clients, their disposable incomes and willingness to travel, currency exchange rates, transportation and lodging supply and costs, the relative cultural familiarity or degree of exotic appeal the destination may have, concerns over crime, and unpredictable events such as terrorist attacks. Political restrictions, for example, prohibit U.S. tourists from visiting Cuba, which attracts visitors from Europe and Canada.

The impacts of tourism are varied and often confined most heavily to localized sectors that deal with visitors such as hotels and retail trade. In cases where a large share of inputs is not purchased locally, tourism can generate relatively small positive impacts. "Tourist enclaves" are not uncommon, in which visitors have few contacts with local residents. Often the jobs involved are unskilled (such as janitors) and pay low wages, with low multiplier effects. Indeed, tourist-dependent economies may be trapped in the "low road" to economic development. At times, tourist developments may generate resentment among locals, who can witness wealthy foreigner frolicking in the sun but not enjoy the benefits. In some cases, particularly fragile ecosystems, tourism can have significant detrimental effects on local flora and fauna. Skiing in the Alps or backpacking in Nepal, for example, can damage local landscapes, generating garbage, and high-density

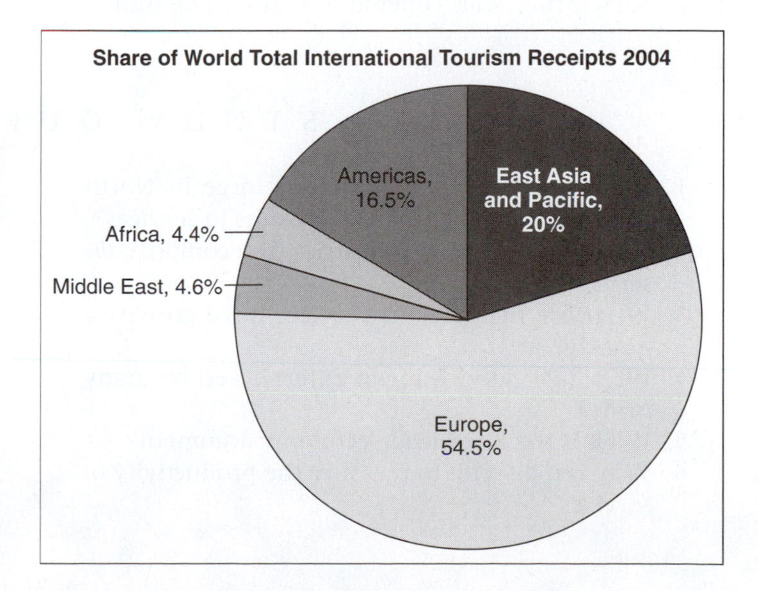

FIGURE 8.28
World tourism 2004, in billions of dollars. As a region, Europe is the largest international tourist destination because many small countries with affluent populations generate numerous world travelers. The United States is the country with the greatest number of arrivals and the greatest tourism expenditures.

tourist developments (e.g., Hawaii) can create large quantities of waste.

SUMMARY

Throughout the world, particularly in developed countries, the vast majority of people work in services such as shops, offices, restaurants, hospitals, airlines, banks, universities, and hotels. A service is any economic activity that provides a benefit to the consumer yet does not involve the transaction of a physical good or product (i.e., the production and consumption of intangibles).

This chapter explored the difficulties in defining services and emphasized the significant differences that exist among and within services industries. Services encompass both highly skilled, well-paying positions, such as international financier or brain surgeon, and low-paying, unskilled ones such as security guard, cashier, and janitor. The chapter explored the reasons for the massive structural shift from a manufacturing economy into services, noting the driving forces include changes in income and consumption patterns, demographics, international trade in services, and the increasingly complex production environment. It paid particular attention to the externalization of services as many companies, in an age of globalized post-Fordism, shed the functions they previously performed in-house (vertical disintegration).

The chapter delved into the debates about the productivity of services, which is notoriously difficult to measure. Services have been blamed for the slowdown in productivity growth in the world over the last few decades, but the extent to which this claim is true is questionable. Labor markets in services received scrutiny, including the empirical patterns of growth in the United States, as well as the features that differentiate services labor markets from those in manufacturing. It also offered several sectoral studies to demonstrate that the dynamics of services vary from sector to sector.

The geography of services is characterized by an ongoing tension between the pressures to agglomerate, typically in large metropolitan areas, and decentralizing forces associated with standardization, routinization, and the availability of information technology. While highly skilled, white-collar functions tend to remain in cities, where they rely on face-to-face contact and agglomeration economies, unskilled ones have increasingly dispersed into rural areas and less developed countries.

Contrary to popular, erroneous stereotypes that services are always consumed where they are produced, this chapter illustrated interregional and international trade in services. The ways by which companies sell services across national borders are varied and complex, including direct sales and subsidiaries. It focused on offshore banking and the globalization of back offices as examples of this process. The chapter addressed the multitude of technological changes in services, particularly those associated with the microelectronics revolution and the digitization of information. Finally, the chapter noted some of the economic implications of the global telecommunications infrastructure, including the role and significance of electronic funds transfer systems. The emergence of hypermobile, digitized money has been a defining feature of contemporary globalization and has brought with it new forms of global finance (offshore centers), as well as the internationalization of clerical work in the form of the offshore back office. These trends are likely to continue, if not accelerate, in the future.

Finally, the chapter noted the growing role of consumer services, not all of which are simply nonbasic in character. Tourism, arguably the world's largest industry in terms of employment, is very powerful in shaping national and local economies in a variety of ways.

STUDY QUESTIONS

1. What proportion of the labor force in North America, Japan, and Europe works in services?
2. What are the major industries that comprise *the* service sector?
3. What are five reasons services have grown so quickly?
4. Why have services been externalized by many firms?
5. What is the knowledge economy argument?
6. Why is it difficult to measure the productivity of services?
7. What are five ways in which labor markets in services differ from those in manufacturing?
8. What are the components of the FIRE sector?
9. Why do U.S. law firms do business overseas?
10. Where do producer services tend to locate, and why?
11. What forms does international trade in services take?
12. Describe some forms of technological change in services.
13. What are electronic funds transfer systems?

14. Where are the major areas of offshore banking?
15. How have many firms globalized their back offices?
16. Are all consumer services nonbasic?
17. Define tourism. What continent is the world's leader in this industry?

KEY TERMS

back offices
business services
call centers
consumer services
contingent labor
distance learning
electronic funds transfer
externalization
FIRE
income elasticity
knowledge economy
networked computers
nondirect production workers
offshore banking
outsourcing
postindustrial society
producer services
productivity
services

SUGGESTED READINGS

Beyers, W. and Lindahl, D. 1998. "Explaining the Demand for Producer Services: Is Cost Driven Externalisation the Major Factor?" *Papers of the Regional Science Association* 75:351–74.

Bryson, J., Daniels, P., and Warf, B. 2004. *Service Worlds: People, Organizations, Technologies.* London: Routledge.

Cohen, B. 1998. *The Geography of Money.* Ithaca, N.Y.: Cornell University Press.

Leyshon, A. and Thrift, N., eds. 1997. *Money/Space: Geographies of Monetary Transformation.* London and New York: Routledge.

Martin, R., ed. 1999. *Money and the Space Economy.* New York: Wiley.

Roberts, S. 1994. "Fictitious Capital, Fictitious Spaces: The Geography of Offshore Financial Flows." In *Money, Power and Space,* edited by S. Corbridge, R. Martin, and N. Thrift. eds Oxford: Blackwell.

Schiller, D. 1999. *Digital Capitalism: Networking the Global Market System.* Cambridge, Mass.: MIT Press.

Shaw, G. and Williams, A. 2004. *Tourism and Tourism Spaces.* London: Sage Publications.

Solomon, E. 1997. *Virtual Money: Understanding the Power and Risks of Money's High-Speed Journey into Electronic Space.* Oxford: Oxford University Press.

Solomon, R. 1999. *Money on the Move: The Revolution in International Finance since 1980.* Princeton, N.J.: Princeton University Press.

Strange, S. 1998. *Mad Money: When Markets Outgrow Governments.* Ann Arbor: University of Michigan Press.

Warf, B. 2001. "Global Dimensions of U.S. Legal Services." *Professional Geographer* 53:398–406.

Williams, C. 1997. *Consumer Services and Economic Development.* London: Routledge.

WORLD WIDE WEB SITES

RESER
Homepage of the European services research network
http://www.reser.net/

GUNTER KRUMME'S WEBPAGE
Webpage of University of Washington Geography professor Gunter Krumme with extensive readings and webpages on economic geography, including services.
http://faculty.washington.edu/krumme/readings/services.html

UNIVERSITY OF BIRMINGHAM RESEARCH GROUP ON SERVICES
http://w3.bham.ac.uk/geography/servicesector.htm

WORLD TOURISM ORGANIZATION
http://www.world-tourism.org

HOWRAH 55 B.GARDEN
HOWRAH

CHAPTER

9

TRANSPORTATION AND COMMUNICATIONS

OBJECTIVES

- To develop an understanding of modern transportation and communication systems
- To point out the historically specific nature of these systems
- To illustrate the nature of cost-space and time-space convergence
- To demonstrate the relationship between transport and economic development
- To emphasize the critical role of transportation policy.
- To consider recent innovations in transport development of U.S. metropolitan areas
- To examine communications innovations and online computer networks
- To summarize the social and economic impacts of the Internet
- To speculate realistically on likely future impacts of telecommunications technology

Traffic in Calcutta, India.

For most of human existence, people occupied narrowly circumscribed areas that were mostly isolated from other groups of people. Gradually, improvements in the efficiency of transportation systems changed patterns of human life. Control and exchange became possible over wider and wider areas and facilitated the development of more elaborate social structures such as far-flung empires. The course of human history, and people's relations to space and time, changed dramatically when capitalism conquered over the globe. From the sixteenth century onward, there were great revolutions in science and trade, great voyages of discovery and conquest, and a consequent increase in the amount of productive, commodity, and financial capital. During the Industrial Revolution of the late eighteenth and nineteenth centuries, the speed with which people, goods, and information crossed the globe accelerated exponentially when inanimate energy was applied to transportation. Capitalism required a world market for its goods; hence, it broke the isolation of preindustrial economies.

The engine that drove this economic expansion was capital accumulation, that is, production for the sake of profit. In an effort to increase the rate of accumulation, all forms of capital had to be moved as quickly and cheaply as possible between places of production and consumption. To annihilate space by time, some of the resulting profits of commerce were devoted to developing the means of transportation and communication. "Annihilation of space by time" does not simply imply that better transportation and communication systems diminish the importance of geographic space; instead, the concept poses the question of how and by what means space can be used, organized, created, and dominated to facilitate the circulation of capital. Time and space appear to us as "natural," that is, as somehow existing outside of society, but a historical perspective on how capitalism has changed our experience of them reveals time and space to be social constructs. Different societies experience and give meaning to time and space in different ways, and the changes unleashed by capitalism reconfigured these experiences worldwide.

The steady integration of production systems around the globe does not change their absolute location (*site*), but it does dramatically alter their relative location (*situation*). If we measure the distances between places in terms of the time or cost needed to overcome them (the friction of distance), then those distances have steadily shrunk over the last 500 years, particularly over the last 100. Transport improvements thus increase the importance of relative space. The progressive integration of absolute space into relative space means that economic development becomes less dependent on relations with nature (e.g., resources and environmental constraints) and more dependent on social relations across space.

Improvements in transportation promote spatial interaction; consequently, they spur specialization of location. The formation of local comparative advantages is facilitated by declines in transport costs (Chapter 12). By stimulating specialization, better transportation leads to increased land and labor productivity as well as to more efficient use of capital. As societies abandon self-sufficiency for dependency on trade, their wealth and incomes generally rise, although not equally for everyone.

In today's world, almost nothing is consumed where it is produced; therefore, without transport services, most goods would be worthless. Part of their value derives from transport to market. Transport costs, then, are not a constraint on productivity; rather, efficient transportation increases the productivity of an economy because it promotes specialization of location.

Transportation and communications are keys for understanding economic geography. How does the geographic pattern of transport routes affect development? How do changing transport networks shape and structure space? What is the impact of transport costs and transit time on the location of facilities? How is information technology (IT) changing the way we live, work, and conduct business? This chapter provides answers to these questions in discussions on transport costs and networks, transport development, transportation and communications innovation, and metropolitan concerns in transportation policy. It explores the historical development of modern transport systems, some general properties of transport costs, and the central role played by the state. Later, it turns to telecommunications and their impacts.

TRANSPORTATION NETWORKS IN HISTORICAL PERSPECTIVE

Prior to the development of railroads, overland transportation of heavy goods was slow and costly. Movement of heavy raw materials by water was much cheaper than by land. For this reason, most of the world's commerce was carried by water transportation, and the important cities were maritime or riverine cities.

To bring stretches of water into locations that needed them, canals were constructed in Europe beginning in the sixteenth century, with the height of technology represented by the pound lock developed in the Low Countries and northern Italy. Until the ninteenth century, canals were the most advanced form of transportation and were built wherever capital was available. Road building was the cheap alternative where canals were physically or financially impractical.

The most active period of canal building coincided with the early Industrial Revolution in the eighteenth century. The vast increase in manufacturing and trade fostered by the canals paved the way for the Industrial

Revolution. The canals were financed by central governments on the Continent and by business interests in England, where a complex network was built during the last 40 years of the eighteenth century and the first quarter of the nineteenth century. Somewhat later, artificial waterways were constructed in North America (e.g., the Erie Canal). They supplemented the rivers and Great Lakes, the principal arteries for moving the staples of timber, grain, preserved meat, tobacco, cotton, coal, and ores.

At sea, efforts before the Industrial Revolution concentrated on expanding the known seas and on improving ships (e.g., better hulls and sails) to allow for practical transport over increasing distances. By 1800, or a few decades later in the case of the technology of sail, the traditional technology of transport reached its ultimate refinement. Subsequently, the rapid expansion of commerce and industry overtaxed existing facilities. The canals were crowded and ran short of water in dry periods, and the roads were clogged when traffic in wet periods. These problems contributed to a general crisis of profitability by the late eighteenth century. The result was an effort to utilize mechanical energy as the motivating power.

The invention of the steam engine by James Watt in 1769 paved the way for technical advances in transportation. Its application to water in 1807 and to land in 1829, through the development of the steamship and locomotive, respectively, heralded the era of cheap transportation. In Europe, an expanding network of railways helped to create markets and provided urban populations with an excellent system of freight and passenger transportation (Figure 9.1). In the United States, the railroad was an instrument of national development; it preceded virtually all settlement west of the Mississippi, helped to establish centers such as Kansas City and Atlanta, and integrated regional markets. Today the Amtrak system (Figure 9.2) forms the passenger rail system in the United States. However, relatively low demand and public subsidies have rendered the U.S. rail network very poorly developed compared with the high-speed trains of Europe. In developing countries, railroads linked export centers to the economies of Europe and North America.

Until the 1880s, cities were mainly pedestrian centers requiring business establishments to agglomerate in close proximity to one another. This usually meant about a 30-minute walk from the center of town to any given urban point; hence, cities were extremely compact. The transformation of the compact city into the modern metropolis depended on the invention of the electric traction motor by Frank Sprague. The first electrified trolley system opened in Richmond, Virginia, in 1888. The innovation, which increased the average speed of intraurban transport from 5 to more than 15 miles per hour, diffused rapidly to other North American and European cities, as well as to Australia, Latin America, and Asia. Electric trolleys were the primary form of urban commuting until the widespread adoption of the automobile in the 1920s.

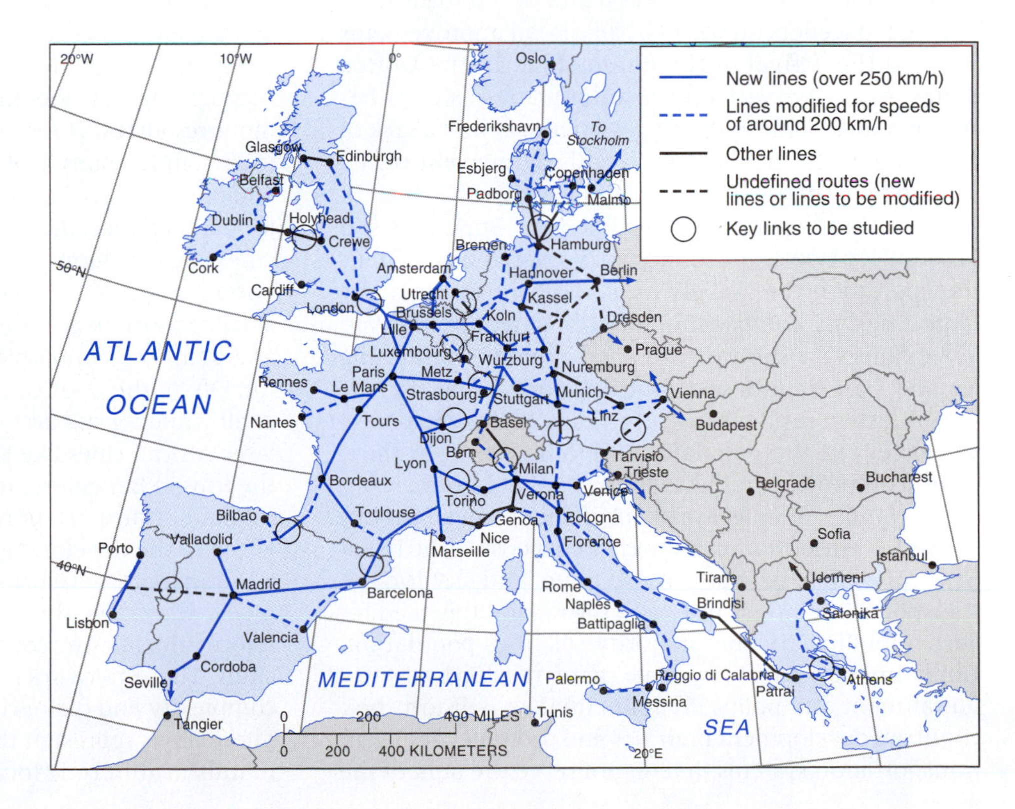

FIGURE 9.1
Europe's planned new rail network. As part of a worldwide mission to encourage more travelers onto rail from other modes, railway companies are developing improved passenger and freight information services. Real-time information at stations and on trains is now a reality in some quarters, with more services to come. High-speed passenger transport applications are driving technology innovation in the rail industry. The focus is now on extensions into eastern Europe.

FIGURE 9.2
Amtrack's network.

The nineteenth century was a time when roads were reduced to feeders for the railroads. Road improvements awaited the arrival of the automobile. In the United States, heavy reliance on the automobile is a cross between a love affair with the passenger car and a lack of alternatives. In most cities, roughly 90 percent of the working population travels to and from work by car; in the less auto-dependent cities like New York, cars still account for two-thirds of all work-related trips. Public transportation is relatively rare, and confined to a few large cities. By comparison, in Europe, where cities are less extensively suburbanized and average commuting distances are half those of North America, only 40 percent of urban residents use their cars. In Tokyo, a mere 15 percent of the population drives to work. In these cases, commuting trains and buses are the norm.

In the developing world, insufficient capital investments in transportation have created a crisis—the result of a mismatch between inadequate budgets for the transportation infrastructure, services, and the need for vast mobility of the majority of the population. Governments that favor private car ownership by a small but affluent and politically influential elite distort their country's development priorities and promote inefficient transportation systems that do not serve the bulk of the population. Importing fuels, car components, or already assembled cars consumes foreign revenues and negatively impacts trade balances. Similarly, building and maintaining an elaborate highway system devours enormous resources. The twentieth century saw a road-building boom in many LDCs, to the detriment of railroads and other forms of transport. With insufficient resources for maintenance, many roads in LDCs are in disrepair. In cities, bus systems and other means of public transportation are often also in a poor state, meeting only a small proportion of transportation needs; crowded buses and trains are symptoms of underinvestment in this sector. Often, the poor cannot afford public transportation at all. Walking still accounts for two-thirds of all trips in large African cities like Kinshasa and for almost one-half the trips in Bangalore, India. Pedestrians and traditional modes of transportation are increasingly being marginalized in the developing countries.

Transport networks are constructed to facilitate *spatial interaction,* the movement of goods, people, and information among countries and cities as well as within cities. These networks range greatly in their degree of complexity and connectivity among nodes (Figure 9.3). These flows represent the exchange of supplies and demands at different locations. The term *distance decay*

Five Simplified Networks

A E D B C

(a) BRANCHING NETWORK: shortest total interpoint connections, lowest construction costs; poor connections for nearby points (e.g., developing country road or rail system).

A E D B C

(b) CIRCUIT NETWORK: shortest connection between points; lowest user costs (e.g., developed country road or air network).

A E D B C

(c) HIERARCHY NETWORK: the shortest set of connections between a central point and all other points; the hub-and-spoke system of airlines (e.g., connection to primate city).

A E D B C

(d) PAUL REVERE'S RIDE: the shortest path between a beginning point and all other points; a minimum distance solution to pipeline placement or energy transmission lines.

A E D B C

(e) TRAVELING SALESMAN NETWORK: the shortest route around a set of points; the most efficient shopping trip pattern (e.g., truck delivery pattern).

FIGURE 9.3
Five simplified transport networks.

An Amtrak conventional passenger train at Harper's Ferry, West Virginia. New magnetic levitated trains will shuttle passengers between American cities at over 300 mph. Using far less energy and time than automobile and air travel, one will go by train from Los Angeles to San Francisco in an hour and a half, or between Washington, D.C., and Boston in less than an hour.

describes the attenuation or reduction in the flow or movement among places with increasing distance between them. Most food shipments, passenger trips, natural resource flows, and commodity movements occur within regions and within countries, rather than between them. The underlying principle of distance decay is the *friction of distance.* There are time and cost factors associated with extra increments of distance for all types of flow or movement. Figures 9.4 and 9.5 illustrate the distance decay for the movements of goods and people, respectively, in the United States, indicating that most trips occur over a relatively short distance. For individuals, the out-of-pocket costs of operating a vehicle or truck are combined with the cost of a person's time. With longer distance, it is more expensive to ship

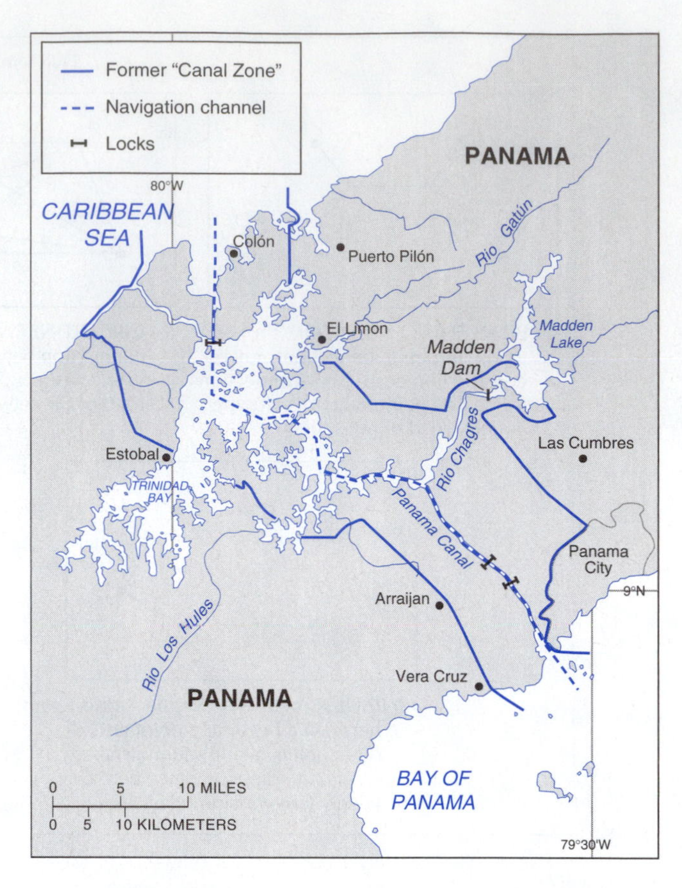

FIGURE 9.5
The Panama Canal, with 14,000 transits in 2004, has enormous strategic and commercial importance. The government of Panama now controls the canal after 85 years of operation by the U.S. government.

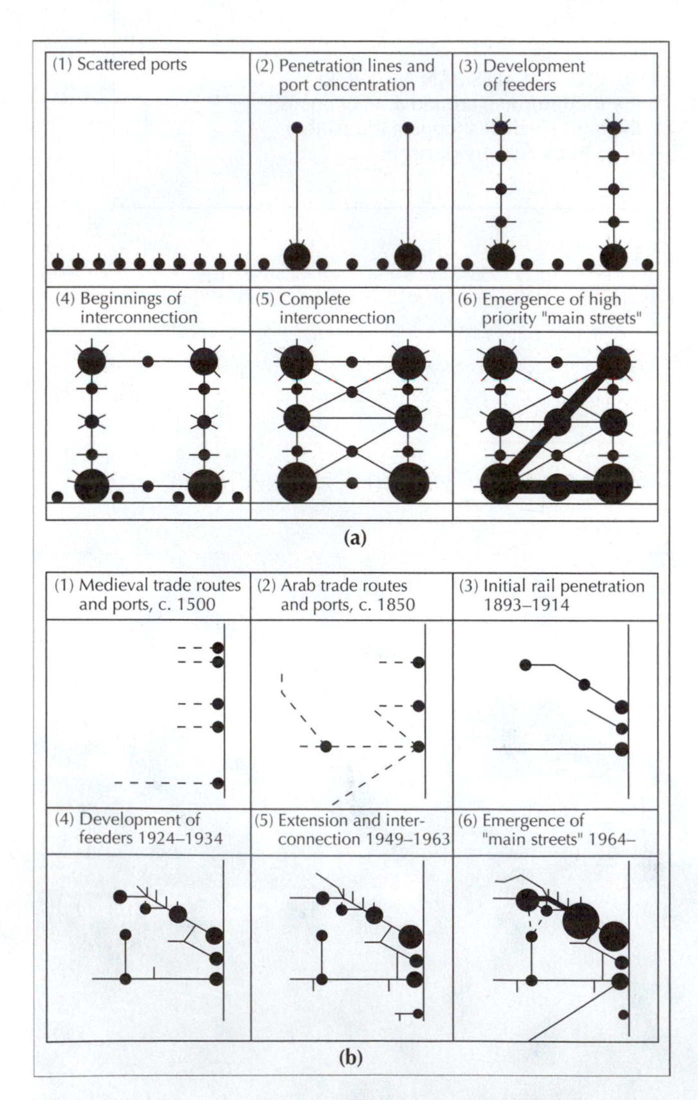

FIGURE 9.4
An idealized sequence of transportation development: (a) the Taaffe, Morrill, and Gould model, 1963; and (b) an adaptation to East Africa by Hoyle, 1973.

commodities because of labor rates of drivers and operators, as well as over-the-road costs of vehicle operation. Telephone calls and parcel deliveries, likewise, are generally more expensive for more distant locations.

Historically, the development of transport networks has reflected and induced settlement, industrialization, and urbanization. The impact of transport networks on regional economic development is demonstrated in a famous stage model of network change in underdeveloped countries created by Taaffe, Morrill, and Gould (1963). This model illustrates how the interplay between the evolution of a transport network and urban growth is self-reinforcing (see Figure 9.4). The ideal-typical sequence begins the first stage when early colonial conquest creates a system of settlements and berthing points along the seacoast. Gradually, a second stage evolves with the construction of penetration routes that link the best-located ports to the inland mining, agricultural, and population centers. Export-based development stimulates growth in the interior, and a number of intermediate centers spring up along the principal access routes. This process results in the third stage of transport evolution—the growth of feeder

Commuter traffic in Bejing. For half of the world's population, the bicycle is a principal means of transportation. In China, less than one person in 1000 owns an automobile. Public buses are cheap and moderately efficient, but human power is still very important for intraurban movement by foot and bicycle. Pedestrian movement dominates urban areas in terms of the number of trips. If the price of gasoline or the price of parking doubled, would you be willing to travel by bicycle to get to work or school?

routes and links from the inland centers. By the fourth stage, lateral route development enhances the competitive position of the major ports and inland centers. A few nodes along the original lines of inland penetration (i.e., N_1 and N_2) become focal points for feeder networks of their own, and they begin to capture the hinterlands of smaller centers on each side. The fifth stage evolves when a transport network interconnects all the major centers. In the sixth and last stage, the development of high-priority linkages reinforces the advantages of urban centers that have come to dominate the economy.

The idealized model of network change describes one typical sequence of development. It shows that a transport network has the short-run purpose of facilitating movement but that its fundamental effect is to influence the subsequent development and structure of the space economy through the operation of cumulative causation. *Cumulative causation* refers to the process by which economic activity tends to concentrate in an area with an initial advantage. The stage model, therefore, illustrates how a space economy roots itself ever more firmly as initial locational decisions that shaped the system are subsequently reinforced by other decisions. The result is a concentrated and polarized pattern of development.

Transport changes in the last 175 years have not been confined to railroads and roads. At sea, ships equipped first with steam turbines and then diesel engines facilitated the rapid expansion of international trade. In addition, the opening of the Suez Canal in 1869 and the Panama Canal in 1914 dramatically reduced the distance of many routes (see Figure 9.5), reconfiguring trade networks and changing the geography of port cities. In ocean shipping, containerships, which use standardized containers that can be efficiently stacked and easily switched between ships and trucks, have become the basic transoceanic carrier. Planes have ousted passenger liners and trains as the standard travel mode for long-distance passengers. The shipment of cargo by air, however, is still in its infancy; while heavy, bulky, and low-valued goods are always shipped by water (e.g., oil, flour), only perishable, high-value, or urgently needed shipments are sent by air freight (e.g., pharmaceuticals).

COST-SPACE AND TIME-SPACE CONVERGENCE

If geography is the study of how human beings are stretched over the earth's surface, a vital part of that process is how we know and feel about space and time. Although space and time appear as "natural" and outside of society, they are in fact social constructions; every society develops different ways of dealing with and perceiving them. In this reading, time and space are socially created, plastic, mutable institutions that profoundly shape individual perceptions and social relations. Transport improvements have resulted in what geographers call *cost-space* and *time-space convergence* or *time-space compression*—that is, the progressive reduction in cost of travel and travel time among places. Similarly, if we measure transport costs in terms of the cost of overcoming the friction of distance, ever-cheaper movement of people and goods leads to *cost-space convergence.*

Transport improvements have brought significant cost reductions to shippers, creating a cost-space compression that altered the geographies of centrality and peripherality of different places. For example, the opening of the Erie Canal in 1825 reduced the cost of transport between Buffalo and Albany from $100 to $10 and, ultimately, to $3 per ton. Railroad freight rates in the United States dropped 41 percent between 1882 and 1900. Between the 1870s and 1950s, improvements in the efficiency of ships reduced the real cost of ocean transport by about 60 percent.

Cheaper, more efficient modes of transport widened the range over which goods could be shipped economically and contributed to the growth of cities. They enabled cities to obtain food products from distant places and facilitated urban concentration by stimulating large-scale production and geographic division of labor. Furthermore, transportation improvements changed patterns of urban accessibility. North American cities have grown from compact walking- and horse-car cities (pre-1800–1890), to electric streetcar cities (1890–1920), and, finally, to dispersed automobile cities in the recreational automobile era (1920–1945), the freeway era (1945–1970), the edge city era (1970–1990), and the exurban era (1990–present) (Figure 9.6).

Developments in transportation have also cut travel times extensively. For example, the travel time between Edinburgh and London, a distance of 640 kilometers, decreased from 20,000 minutes by stagecoach in 1658 to less than 60 minutes by airplane today (Figure 9.7). Time-space convergence was marked during the period of rapid transport development; for example, in the 1840s, travel time between Edinburgh and London was longer than 2,000 minutes by stagecoach, but by the 1850s, with the arrival of the steam locomotive, the travel time had been reduced by two-thirds, to 800 minutes. By 1988, the rail journey between Edinburgh and London took 275 minutes. When the line was electrified in 1995, travel time was reduced to less than 180 minutes. In communications, the steadily declining cost of long-distance telephone calls increased the interactions among cities such as New York and San Francisco (Figure 9.8).

Air transportation provides spectacular examples of time-space convergence. In the late 1930s, it took a DC-3 between 15 and 17 hours to fly the United States from coast to coast. Modern jets now cross the continent in about 5 hours. In 1934, planes took 12 days to fly between London and Brisbane. Today the Boeing 747 SP is capable of flying any commercially practicable route nonstop. The result is that any place on earth is within less than 24 hours of any other place, using the most direct route.

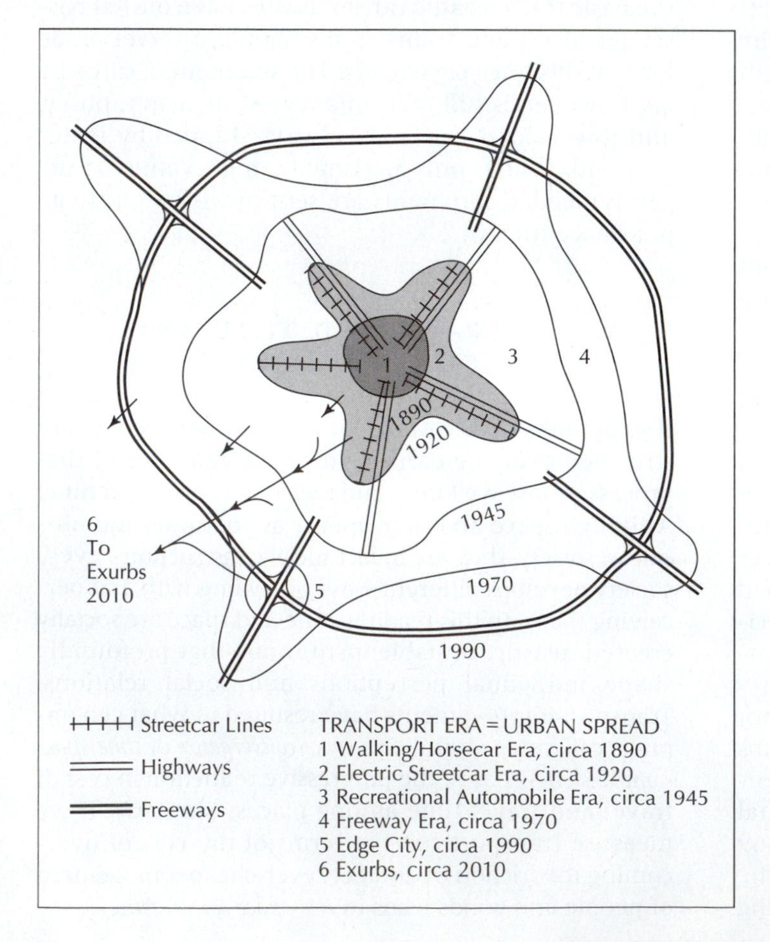

FIGURE 9.6
Stages of metropolitan growth and transport development in a North American city.

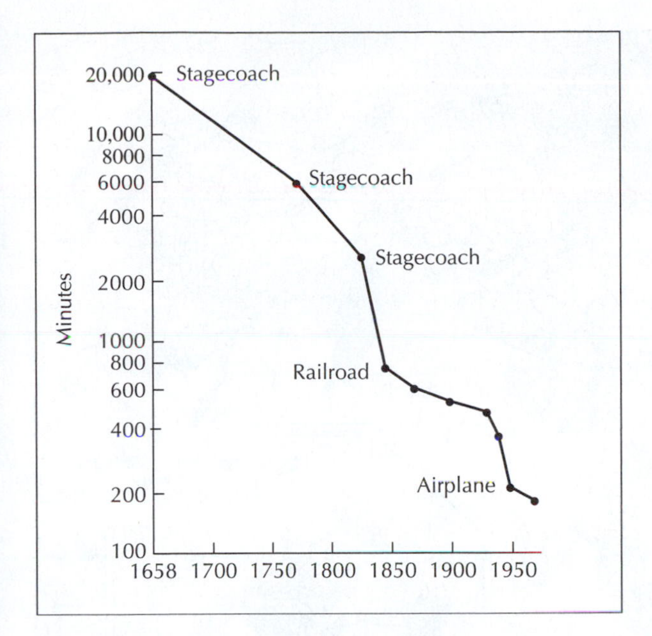

FIGURE 9.7
Time-space convergence between London and Edinburgh.

TRANSPORTATION INFRASTRUCTURE

Transportation and communication infrastructures allow countries to specialize in production and trade. This regional division of labor is comparable to the task division of labor among its workers. The transportation and communications infrastructure of a country influences its internal geography. In some countries, transportation and communications are slow and difficult. Some regions are totally inaccessible (Figure 9.9). Much of the developing world has poorly developed infrastructures.

Fast and efficient transportation systems release capital for productive investment and allow the development of natural resources, regional specialization of production, and internal trade among regions. Even though India appears to be well connected, it is a country whose economic growth is harnessed by an inadequate transportation and communications infrastructure. Passenger and commodity traffic on India's roads has increased 30-fold since independence in 1948. Roads are overcrowded, and 80 percent of villages lack all-weather roads. Improved accessibility would allow regional specialization of production and cash crop farming, especially of high-value crops such as fruits and vegetables that spoil quickly.

Conversely, the well-developed infrastructures of North America and Europe bespeak their level of economic development. However, most countries of the world have simple transport networks penetrating the interior from ports along the ocean. These railroads and highways are called tap routes. Tap routes are the legacy of colonialism or the product of neocolonialism. Such routes facilitate getting into and out of a country, but they do not allow for internal circulation or circulation between countries in the same region.

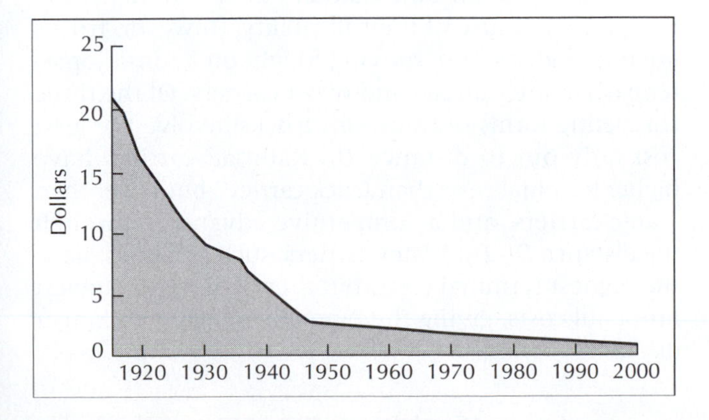

FIGURE 9.8
Cost-space convergence in telephone calls between New York and San Francisco, 1915–2000.

GENERAL PROPERTIES OF TRANSPORT COSTS

Transportation costs appear deceptively simple but are far more complex upon further scrutiny. They can be categorized as either *terminal costs* or *line-haul costs* (Figure 9.10). Terminal costs must be paid regardless of the distance involved. They include the cost of loading and unloading, capital investment, and line maintenance. Line-haul costs, in contrast, are strictly a function of distance. For example, fuel costs are proportional to the distance a load must be moved.

The most recent development has been the provision of container-handling facilities and roll-on/roll-off terminals. The world's first containerized service tied trucks and ships together in 1956. By the early 1970s, numerous carriers entered into the containership business. At first, the greatest appeal of the containership was its speed and economy in port. Moreover, it facilitated the multimodal transport of goods. For example, commodities from Japan and other Pacific Rim countries could be transported economically to Europe via North America. Later, container operations sped up the ocean voyage as well—top usable speeds increased from 15 knots in the 1950s to 33 knots in the 1970s.

With the emergence of a new international division of labor, ports continue to modernize their methods of handling cargo as they compete with each other for shares of global commodity traffic. In developed countries served by many ports, competition has decreased the relevance of the traditional concept of the port hinterland (i.e., the area served by the port). On the West Coast of the United States, for example, ports in

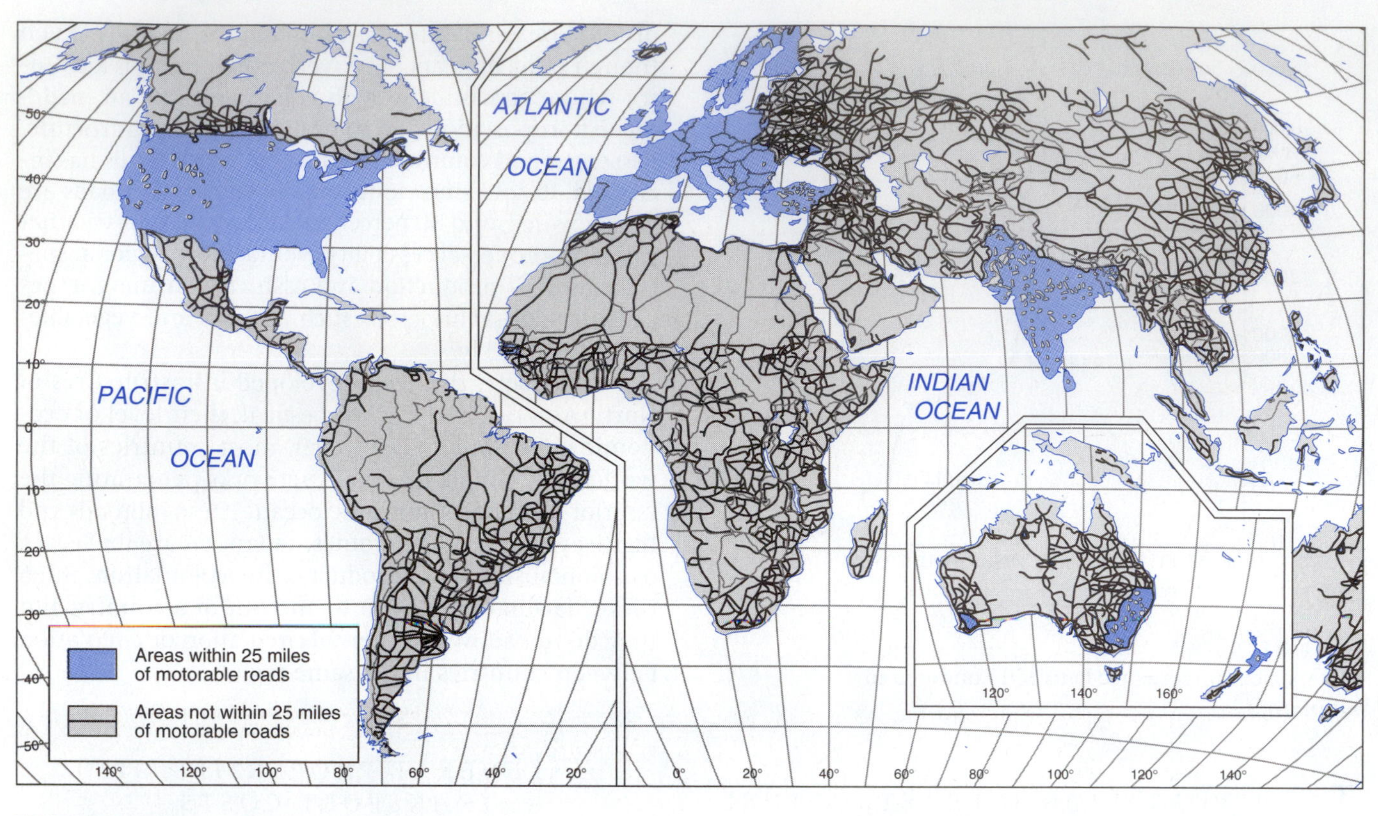

FIGURE 9.9
Major world roads and highways. Note areas within the United States, Europe, India, Southeastern Australia, and New Zealand are virtually all within 25 miles of roads and highways. Only a few exceptions exist in the Western mountain regions of the United States.

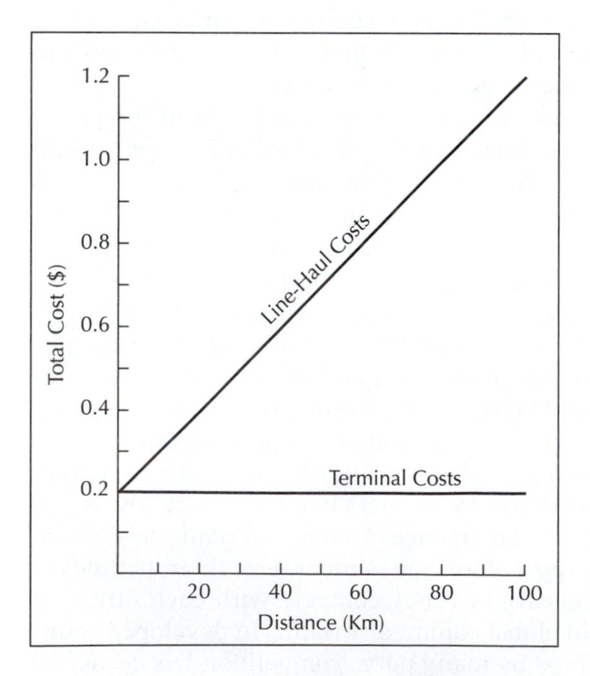

FIGURE 9.10
Terminal and line-haul costs. Terminal costs are also fixed costs. Line-haul costs incurred "over the road" are also variable costs.

California, Oregon, and Washington compete fiercely for the mounting trade with East Asia.

Carrier Competition

Competitive differences in terminal and line-haul costs among various transport modes lead firms to use different forms over different distances (Figure 9.11). Trucks have low terminal costs partly because they do not have to provide and maintain their own highways and partly because of their flexibility. However, trucks are not as efficient in moving freight on a ton-kilometer basis as are railroad and water carriers. Of the three competing forms of transport, trucks involve the least cost only out to distance D_1. Railroad carriers have higher terminal costs than truck carriers, but lower than water carriers, and a competitive advantage through the distance D_1–D_2. Water carriers, such as barges, have the highest terminal costs, but they achieve the lowest line-haul costs, giving them an advantage over longer distances.

Elasticity of Demand

The *elasticity of demand* for transportation is the degree of responsiveness of a good or service to changes in its

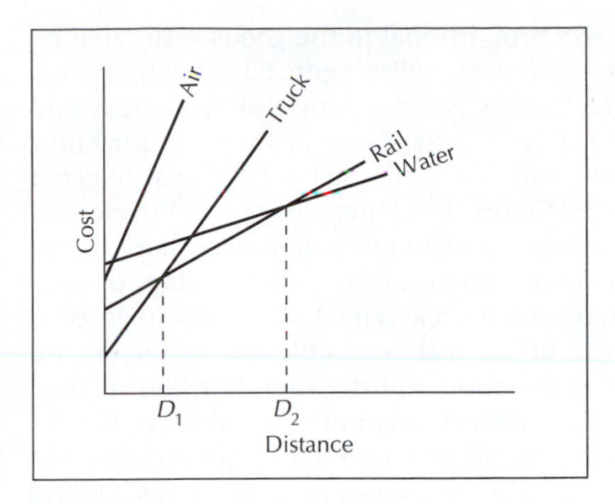

FIGURE 9.11
Variations in terminal and line-haul costs for air, truck, rail, water, and pipeline.

price (i.e., the percentage change in demand that a percentage change in price causes). Carriers generally charge what the market will bear. Goods with a very high value per unit of weight, such as televisions, are able to bear a higher transportation rate than goods with a very low value per unit of weight, such as coal (Figure 9.12). The left-hand graph illustrates transportation price inelasticity for television sets. An increase in the rate from P_1 to P_2 produces only a slight change in the quantity of shipments. Coal, however (in the right-hand graph), exhibits a great change in the quantity of shipments, with only slight change in the transportation rate (P_1 to P_2).

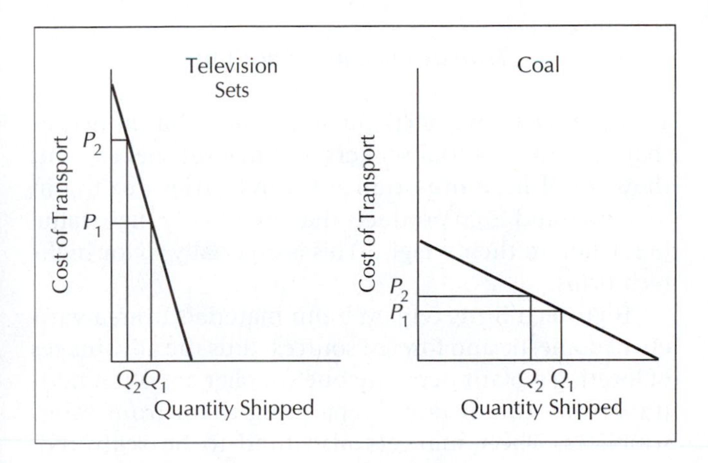

FIGURE 9.12
Demand elasticity for transportation. Higher-valued television sets are more valuable and less elastic than quantities of coal. A large price increase to ship television sets results in a small reduction in the quantity shipped. A small price increase to ship coal results in a large reduction in the quantity shipped.

Container cargo handling at the Maersk Line Terminal, Port Newark, New Jersey. Containerization has greatly improved the operation, management, and logistics of conventional oceangoing freight. The impact of the container evolution has gone far beyond shipping and international trade alone. Newly designed cellular vessels have much faster ship turnaround times in ports as well as improved cargo-handling productivity at ports. An expanded interface between water and land transportation has occurred. Container trains have also enhanced the economy and scale of rail transportation.

Freight Rate Variations and Traffic Characteristics

An absence of competition between transport modes means a carrier can set rates between points to cover costs, and in the absence of government intervention, a carrier may set unjustifiably high rates. Intermodal competition or government regulation reduces the likelihood of such practices. Competition among carriers reduces rate differences among them. For example, the opening of the St. Lawrence Seaway in 1959 resulted in lower rail freight rates on commodities affected by low water-transport rates.

Many carriers face heavy demand only in a specific direction. Consider the large volume of produce shipped from Florida to New York. Trucks must often return empty for the next load. The cost of the total trip, however, is used to determine the transportation rate. Because carriers must make return trips anyway, they are willing to charge very low rates on the backhaul. Any revenue on backhaul is preferred to returning empty. Rates are higher where there is little or no possibility of backhauling; most such runs occur in the transportation of raw materials from resource points to production points. An example is the railroad that carries iron ore pellets from Labrador to the port of Sept Iles, Quebec. This railroad may be likened to a huge conveyor belt that operates in one direction only. By contrast, the distribution of finished products generally involves traffic among many cities, creating a reciprocal flow and lower rates.

Regimes for International Transportation

In the international arena, transport rates and costs are affected by the nature of the regime governing the transport mode. To illustrate, consider the contrasting regimes of civil aviation and shipping. The international regime for aviation is dominated by nation-states; in contrast, the international regime for shipping has been shaped by large shipping corporations. The regime for shipping evolved over more than 500 years and has been more concerned with facilitating commerce than with national security. The regime for civil aviation developed in the early twentieth century and primarily reflects a concern for national security.

The fundamental principle governing international aviation is that states have sovereign control over their own air space. From this principle, rules and procedures have developed that permit countries to regulate their routes, fares, and schedules. As a result, many countries, developed and developing, have secured a market share that is more or less proportional to their share of world airline traffic. Developing countries have been able to compete with companies based in the industrialized world on an equal footing; for example, Air India, Avianca, and Korean Air Lines can challenge Delta, Air France, and British Airways.

The international regime for shipping has left many developing countries in a weak position with regard to establishing and nurturing their own merchant fleets. In a world of markets, few underdeveloped countries have much influence when it comes to setting commodity rate structures. Lack of control over international shipping is an important area of concern in the Third World's quest for development.

Although the regime for shipping is dominated by firms, the market is inherently unfair—it favors developed countries over developing countries. Hence, LDCs are faced with rate structures that work against them, inadequate transport services, a perpetuation of center-periphery trade routes, and a lack of access to decision-making bodies. Those LDCs generating cargoes such as petroleum, iron ore, phosphates, bauxite/alumina, and grains cannot penetrate the bulk-shipping market, which is dominated by the vertically integrated MNCs based in developed countries. Cartels of ship owners called liner conferences set the rates and schedules for liners (freighters that ply regularly scheduled routes).

Developing countries have attempted to change the international rules of shipping. They want to generate fleets of a size proportional to the goods generated by their ports. Their accomplishments have been limited, however. The United Nations Commission on Trade and Development (UNCTAD) Code of Conduct for Liner Conferences, which was adopted in 1974, was rejected by the United States. The Liner Code gives developing country carriers a presumptive right to a share of the market; however, proposals to eliminate flags of convenience have not been accepted. Flags of convenience assume little or no real economic link between the country of registration and the ship that flies its flag. They inhibit the development of national fleets, but for ship owners they offer a number of advantages, including low taxation and lower operating costs. Liberia and Panama are the most important open-registry, or flags-of-convenience, countries. Flags of convenience are used mainly by oil tankers and bulk-ore carriers controlled by MNCs.

The global pattern of container ports reflects the geography of production and trade. Traditionally, the largest ports were located in Europe and North America. For many years, Rotterdam, at the mouth of the Rhine River in the Netherlands, was Europe's primary port and the largest in the world. Up until the 1980s, the largest ports in the United States were located on the Atlantic coast (e.g., New York), as most American trade was with Europe. The rapid economic growth of East Asia, however, changed these patterns. Today, the world's largest ports are located in Hong Kong and Singapore (Figure 9.13). Because most U.S. trade is across the Pacific Ocean, West Coast ports such as Los Angeles, Oakland, and Seattle have surpassed East Coast ports in trade volume. These changing patterns reflect the ways in which globalization is changing the domestic economic geography of the United States.

Transit Time and Location

Transport costs are of crucial importance for industries that are raw material seekers and market seekers, but they are of little importance for industries dealing in materials and final products that are of very high value in relation to their weight. This is especially so for high-tech firms.

High-tech firms rely on input materials from a variety of domestic and foreign sources; thus the advantages of locating a plant near any one supplier are often neutralized by the distance separating them from other suppliers. Their markets also tend to be scattered. Transport is a factor of some locational significance for these firms, but transit time is more crucial than cost. High-technology firms require access to high-level rapid-transport facilities to move components and final products, as well as specialized and skilled personnel. For this reason, they are often attracted to sites near

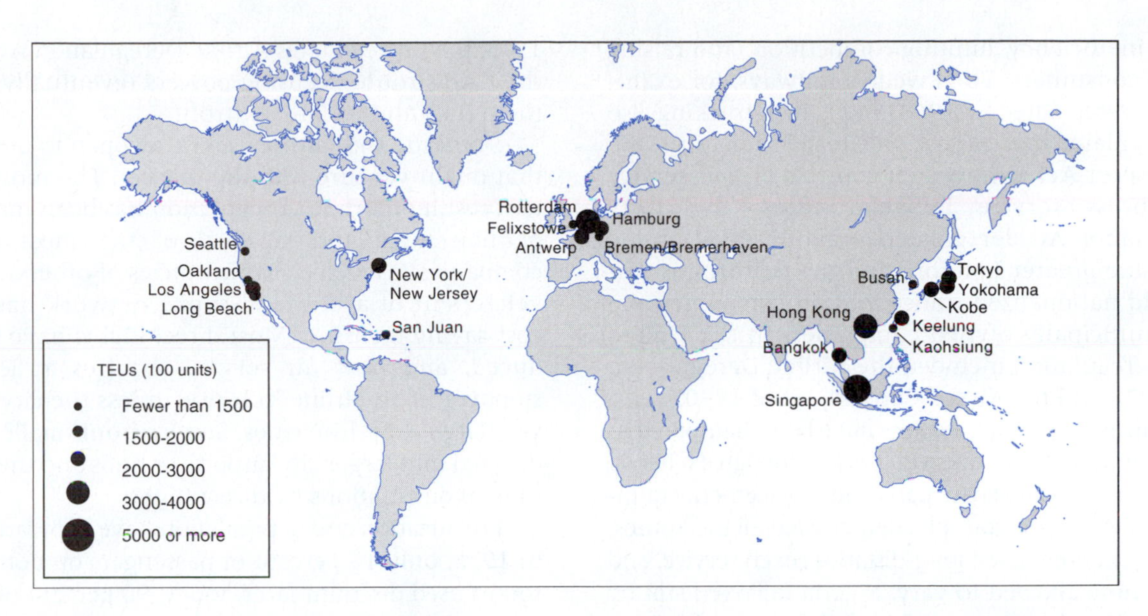

FIGURE 9.13
Most of the world's container ports are located in East and Southeast Asia and in Europe. New York, Los Angeles, and Vancouver represent major container ports in North America. Such ports are "hub ports" and act as major centers where container traffic splits into feeder flows to and from centers within the hub's respective hinterland.

major airports with good national and international passenger and air-cargo facilities. Concentrations of high-tech firms and research and development facilities are located in Silicon Valley near San Francisco, along the M4 motorway in Britain to the west of London, and in Tsukuba Science City, situated northeast of Tokyo.

Transport Improvements and Location

Transport innovations have reduced circulation costs and fostered the new international division of labor. They have encouraged the decentralization of manufacturing processes in industrialized countries, both from major cities toward suburbs and smaller towns, and from central regions to those more peripheral. They have also encouraged the decentralization of manufacturing processes to those LDCs with an abundance of weakly unionized, low-wage labor.

The "container revolution" and bulk-air cargo carriers enabled MNCs based in the United States, Japan, and Europe to locate low-value-added manufacturing and high-pollution manufacturing processes "offshore" in more than 80 Third World free-trade zones. Almost one-half of these zones are in Asia, including Hong Kong, Taiwan, Malaysia, and South Korea. Free-trade zones are areas where goods may be imported free of duties for packaging, assembling, or manufacturing and then exported. These global workshops are geared to export markets, often with few links to the national economy or the needs of local consumers. They tend to be located near ports (e.g., La Romana, Dominican Republic), near international airports (e.g., San Bartola, El Salvador), and in areas virtually integrated into global centers of business (e.g., Mexico's northern border or maquilla zone) in Tijuana–San Diego.

Transportation Policy

Well-established national transportation policies and regulation were the norm until the 1970s. The purpose of regulation of airlines and rail carriers was to ensure quality control, protect companies and customers, and establish quality and safety control standards throughout the industry. During this period, providers not only provided basic transportation services but also met a social obligation, such as providing service to low-income, unprofitable areas. For example, Britain's Road Traffic Act of 1930 introduced a system of licenses and rates that effectively regulated the sporadic and unsafe market for bus services in that country.

Deregulation and Privatization

By the late 1970s, with the growth of conservative neoliberalism, international trade regulators required that there be free entry of new transportation operators into the market to ensure efficiency and welfare maximization. The move toward *privatization* had begun. Regulation was criticized by advocates of markets as creating inefficiency, limiting competition, and raising prices to consumers. The Swedish railways, for example, were deregulated by 1968, and British trucking also was deregulated in that year. In Great Britain, in 1980, the Transport Act removed all controls on bus service and express service between cities; the 1985 Transportation Act deregulated local bus service inside and outside greater London; and the British government sold nationalized transportation companies and many municipally owned companies. In the United States, deregulation included the Airline Deregulation Act of 1978 and the Motor Carriers Act of 1980.

Privatization and *deregulation* have been hampered in developing countries because of a lack of foreign exchange to purchase necessary spare parts and replacement equipment. Sri Lanka, for example, deregulated all bus routes, while China deregulated long-distance coach service, and fares are now allowed to vary. Nigeria followed suit by privatizing Nigeria Airways and its National Shipping Line, while Singapore privatized Singapore Airlines and started to privatize its mass transit corporation.

Deregulation of the U.S. Airlines

The Civil Aeronautics Board (CAB) of the United States regulated the U.S. airline industry from 1938 until recently. During most of this period, the CAB's goal was to preserve the 16 trunk line airlines that existed in 1938 and to provide good service at fair prices with a high level of quality control. More recently, the 16 companies were reduced to 11 companies by mergers.

Air passenger traffic increased 1000% between 1950 and 1970. Airfares remained almost constant because of the lower cost of operating more efficient planes. However, the oil embargos of 1973–1974 and 1978 increased operating costs, leading to a crisis of profitability and leading airlines to pressure the federal government for deregulation of domestic air services. In 1978, the United States Airline Deregulation Act limited the CAB's route licensing powers (eventually phasing them out) and its fare controls.

Domestic U.S. airlines are now open to any carrier that might venture into the market. The most important result of airline deregulation has been more competitive fares and survival of the most efficient companies (as well as bankruptcies of others). The development of a hub-and-spoke network has been a cost-saving measure. Most direct flights have been reduced, and now air service requires at least one stopover in an airline hub city, unless the city pair are very large American cities. Service from smaller cities is directed into larger city airports or hubs and then linked to final destinations by direct flights.

Privatization and deregulation have kept fares down. In 1976, only 15 percent of passengers on domestic air routes used discount fares; today, 90 percent of passengers use discounted tickets. However, as average fares have fallen on long-haul routes, fares on short routes have risen. Load factors have increased substantially with a hub-and-spoke system, and the number of flights has declined, leading to lower overall costs.

Hub-and-Spoke Networks

In order to remain competitive, the airlines that survived the shake-out following deregulation restructured their networks so that they could reduce direct flights between most city pairs. They made their operations more efficient and cost-effective by using a hub-and-spoke network model. Hubs serve central locations that collect and redistribute passengers between sets of original cities. Extremely large passenger volumes are funneled through hubs, and this allows the airlines to fly larger and more efficient aircraft and to offer more frequent flights between major hubs, increasing load factors (Figure 9.14).

Delta Airlines' network has hubs in Atlanta and Cincinnati.

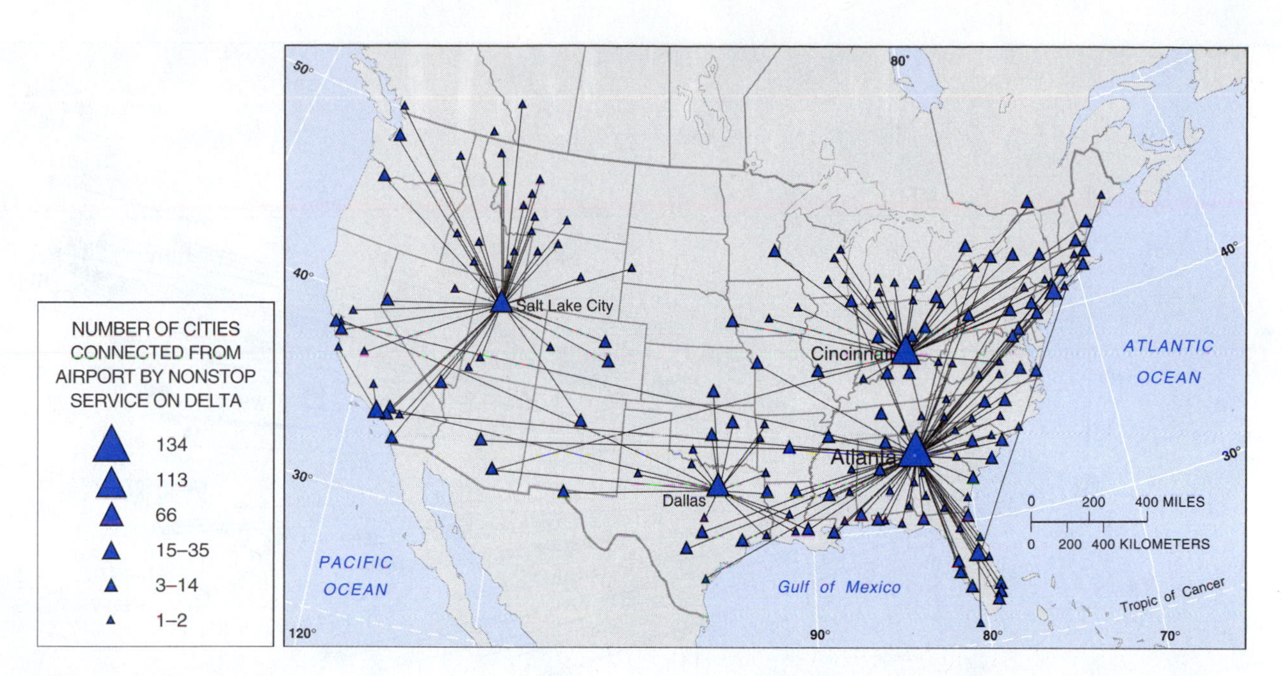

FIGURE 9.14
Hub-and-spoke networks for two major U.S. air carriers. (These figures were first published in the *Journal of Transport Geography* Vol. 1, no. 1 [March 1993]: pp 51–54, and are reproduced here with the permission of Butterworth-Heinemann, Oxford, U.K.)

However, *hub-and-spoke networks* can provide disadvantages, especially to the travelers who find the number of links in their trips increased, frequently with a change of planes, and fewer direct flights available. Also, congestion is created at the main hub cities, and this affects efficiency both in the air and on the ground. It is important for airlines to make careful decisions as to the location and exact number of hubs so that their operation is competitive with other airlines. There are a large number of optimum hub location studies in the literature. These mathematical optimization approaches attempt to capture the real-world realities of air passenger networks and the design problems that face most airlines.

Not all cities have fared equally well. Some airports showed a precipitous decline in traffic after deregulation began. Resulting from this mad scramble to reduce fares and elevate efficiency for megahubs, Atlanta, Chicago, Dallas, and Denver surfaced as major hubs for two or more airlines; Salt Lake City, Minneapolis–St. Paul, Memphis, and Detroit have also emerged. Different airlines use different cities in their respective hub-and-spoke networks (Table 9.1).

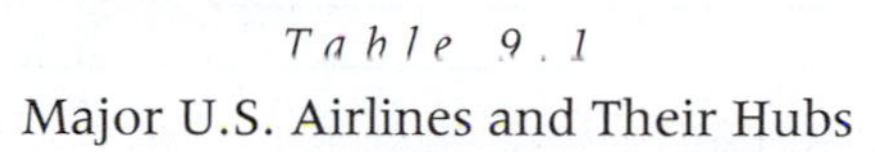

Table 9.1
Major U.S. Airlines and Their Hubs

American	Dallas–Fort Worth, Chicago
Continental	Houston, New York, Denver, Cleveland
Delta	Atlanta, Cincinnati, Dallas
Northwest	Detroit, Memphis, Minneapolis –St. Paul
United	Chicago
U.S. Air	Pittsburgh, Charlotte

Transportation of Nuclear Wastes

Nuclear wastes produced during the fission process include reactor metals, such as fuel rods and assemblies, coolant fluids, and gases found in the reactor. Fuel rods are the most highly radioactive waste found on earth today, and their dangerous level of radioactivity requires that they be transported to a special site for storage. The storage site must be located so that the radioactive material will not contaminate the groundwater or the biosphere in any way. Geologic stability is a must. Most experts support deep underground geologic disposal in salt domes or other rock formations. Presently, the United States has 100 sites where radioactive waste has been temporarily stored. These include Hanford, Washington; Livermore, California; Beatty and Las Vegas, Nevada; Idaho Falls, Idaho; Los Alamos and Albuquerque, New Mexico; Amarillo, Texas; Weldon Springs, Missouri; Sheffield, Illinois; Paducah, Kentucky; Oakridge, Tennessee; Aiken and Barnwell, South Carolina; and Niagara Falls and West Valley, New York. Recently, however, the federal government

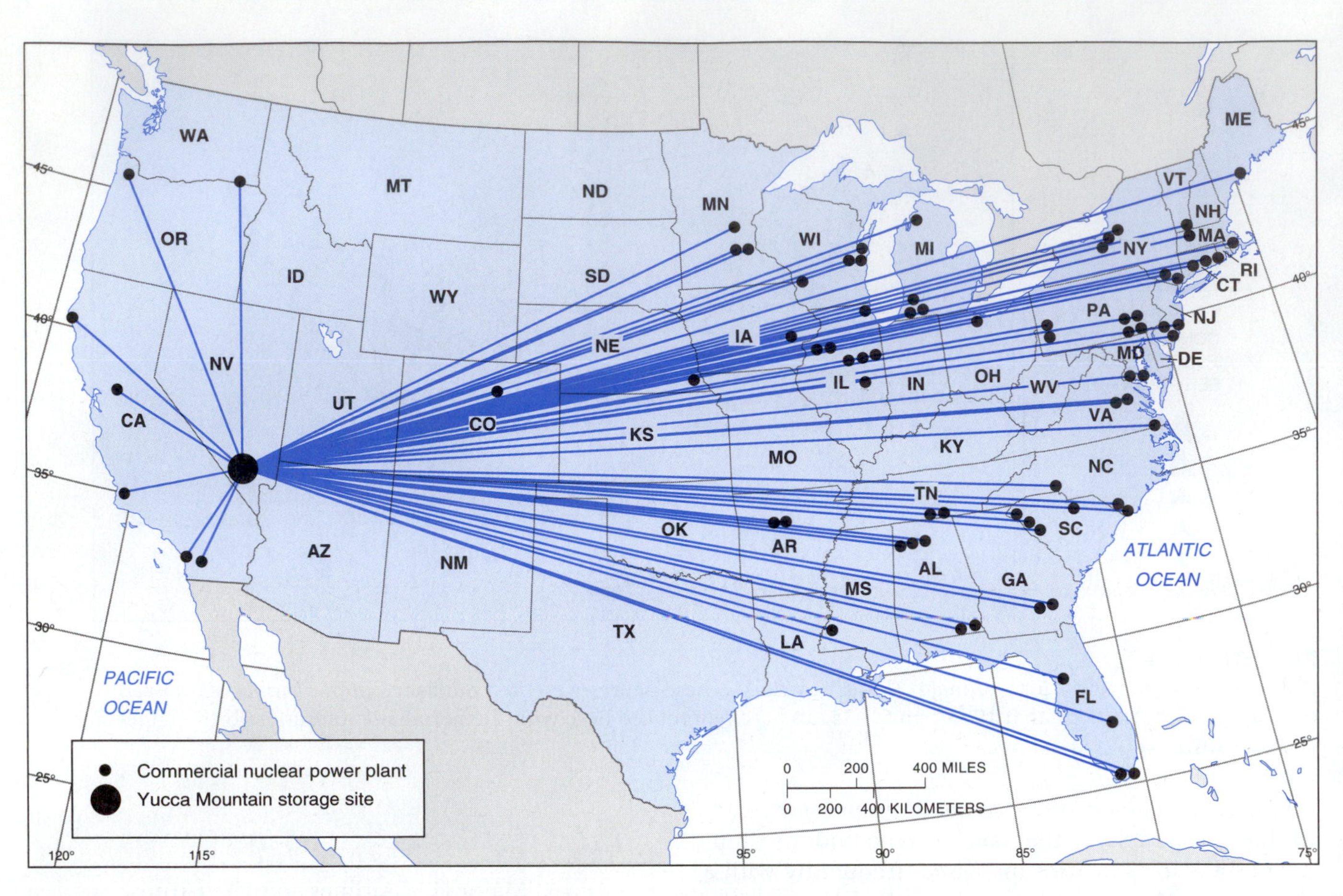

FIGURE 9.15
High-level nuclear waste from commercial power plants will be transported to Yucca Mountain, Nevada.

selected Yucca Mountain, Nevada, as a permanent facility for the storage of high-level nuclear wastes from commercially operated power plants in America (Figure 9.15).

Special containers on rail and truck will be used, and the routes will avoid high population areas. One of the factors responsible for the selection of Yucca Mountain is the so-called *NIMBY* ("not in my backyard") effect. The problem with nuclear waste deposition is that, because no official or engineer can assure that each site will be completely safe, local residents want no possibility of an accident. Politicians are quite sensitive to their constituencies' pleas and concerns for safety and usually vote to remove nuclear wastes and power plants from their districts.

Personal Mobility in the United States

One important dimension of urban transportation concerns changes in the personal mobility in the United States. Personal mobility in the United States is at its highest point in history, with individuals making more and longer trips and owning more vehicles. Three factors account for this greater level of mobility in general. The first is the overall increased performance of the national economy. When more people have more money to spend on transportation, greater automobile ownership and greater travel distances result. The second factor is the increasing growth of cities and their spread over the surrounding countryside through low-density exurban expansion. On the average, distances between home and job have increased, leading to longer commutes. A third major reason for increased mobility is the changing role of women in the workforce. Many more women own their own vehicles, have entered the workforce full time, and have increased their travel demands during the last 40 years.

Because of increased mobility, individuals have benefited in the social and economic sense, but society as a whole suffers the negative consequences. New concerns about rising levels of air pollution, congestion on the freeways, and the movement of goods are being posed. The new levels of mobility have created a set of problems that are very difficult to address. Two techniques to address the issue of greater congestion and slower average speeds are being forwarded. One is to increase volumes on present roadways through *intelligent vehicle highway systems* (IVHS), and the other is to reduce travel demand by planning a land-use mix in localities so that trip origins and destinations are less

Rush-hour traffic fills the northbound lanes of the San Diego freeway in California. Rapid growth of automobile ownership in Western cities in the second half of the twentieth century has been met by unparalleled congestion. Greater levels of wealth in Western countries, more drivers, more auto ownership, and more women entering the labor force have contributed to high levels of car ownership. In California, auto ownership approaches one car for every person, whereas in cities worldwide the figure rarely exceeds 10 per 100, and the figures are much lower figures than that for rural areas.

separated. This approach is called *transit-oriented development* (TOD). However, before we discuss these two measures of reducing congestion and increasing the greater volumes of flow, we first examine trends occurring in personal mobility in the United States.

In the late twentieth century, the number of households, drivers, vehicles, vehicle trips, vehicle miles traveled (VMT), person trips, and the person miles of travel all increased at a much faster rate than did the population. In addition, the number and percentage of households that did not own a vehicle dropped. However, the number of households with three or more vehicles increased dramatically.

The *journey-to-work trip*, both in terms of total miles of travel and in terms of number of trips, continued to account for the largest proportion of travel by U.S. households. Both annual vehicle miles traveled and annual number of vehicle trips per household increased from 1970 to 2004 (Figure 9.16). The average home-to-work trip was 12 miles, while social and recreation trips averaged just over 11 miles per trip. Other family or personal business trip lengths averaged 7 miles, and shopping trips averaged 5.6 miles. The declines in vehicle occupancy are explained partially by the increased number of vehicles per household and the decrease in average household size during this period.

Because of the escalation in average vehicle price, Americans retain their vehicles for a longer period. For example, in 1969, 42% of household automobiles were two years old or less; by 2005, only 14 percent were two years old or less. The number of automobiles that were 10 or more years in age increased from 6 percent in 1969 to 37 percent by 2005. The usage of older vehicles, which have been shown to burn energy less efficiently and cleanly than newer vehicles, contributes to energy and air pollution problems.

Intelligent Vehicle Highway Systems (IVHS)

During the next 30 years, traffic volume in the United States is expected to double. Yet each year, some 135 million drivers spend about 2 billion hours stuck in highway traffic. An estimated $46 billion is lost by American drivers trapped in traffic delays, by detours, and by wrong turns. However, IVHS could greatly ameliorate this situation. "Smart" cars are equipped with microcomputers, video screens, and other technologies that reduce the frustration of driving. Through the use of in-vehicle computers and navigation systems, drivers are guided step-by-step to their destinations. Fast, accurate

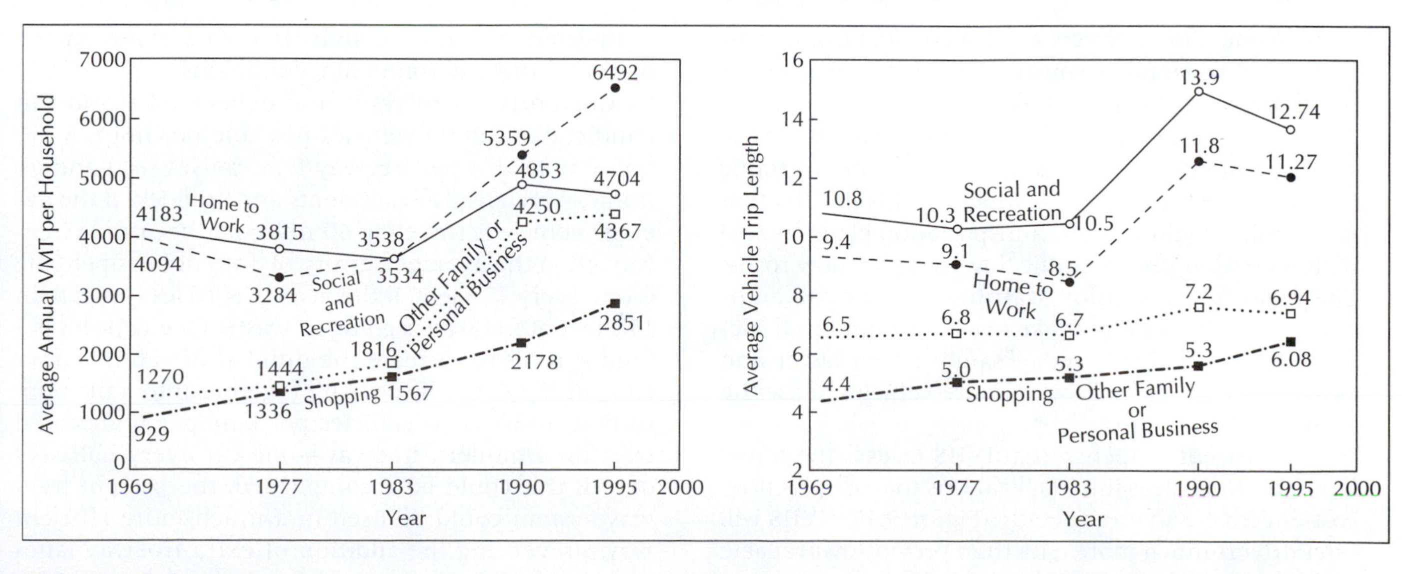

FIGURE 9.16
Vehicle miles traveled (VMT) and average vehicle length in the United States, 1970–1995.

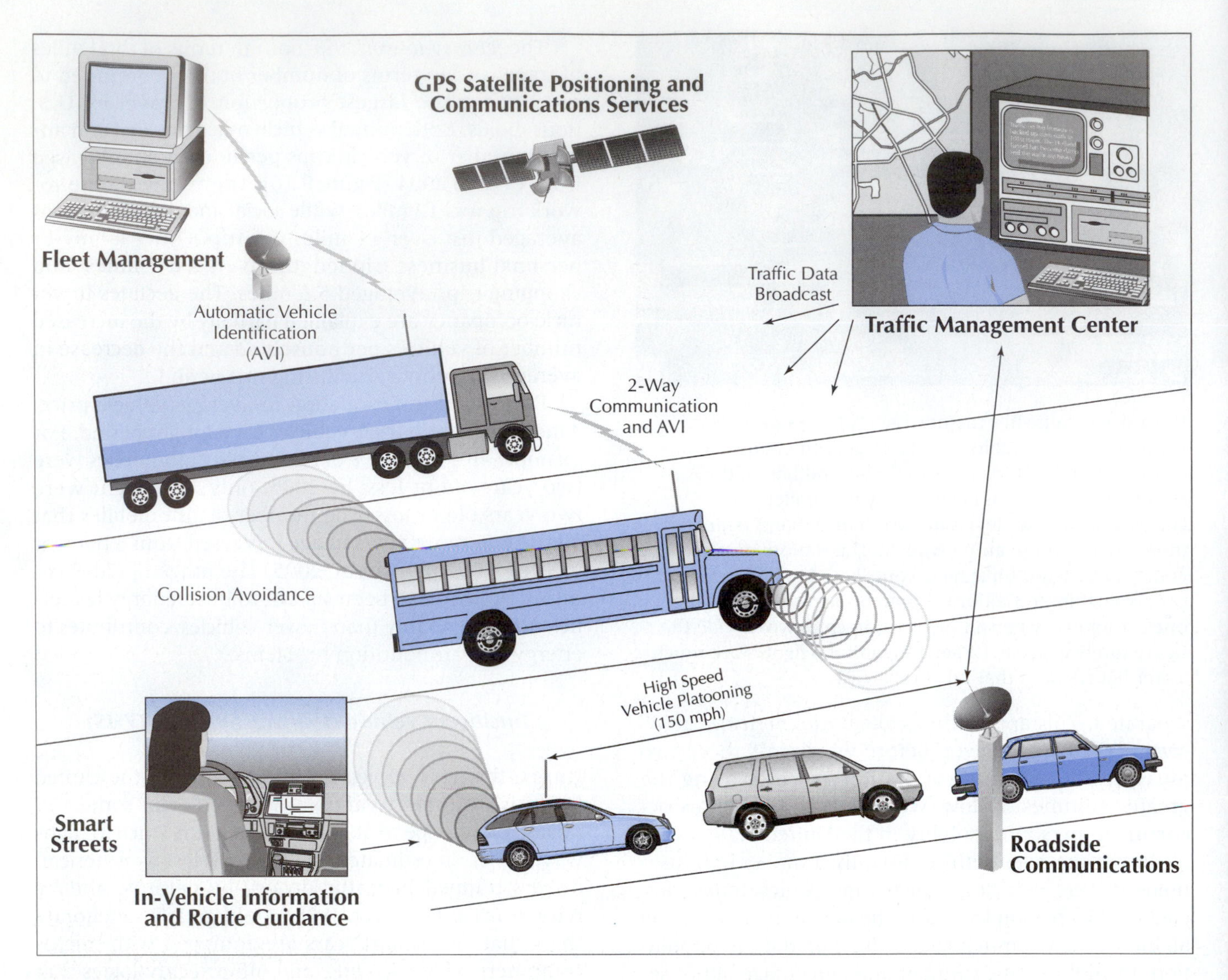

FIGURE 9.17
Elements of IVHS.

information allows drivers to avoid accidents and congestion while simultaneously offering information on restaurants, hotels, attractions, and emergency services.

"Smart" highways are created by installing vehicle sensing systems. These sensing systems monitor traffic volume, speed, and vehicle weights. This information helps traffic engineers and transportation planners regulate signals to control traffic flow and plan new roads. The radar-based collision warning systems will automatically signal a car to brake to avoid collisions. If successful, cars could someday safely travel faster and closer together, thus allowing more vehicles to use the road at the same time.

The current challenge that IVHS faces is the translation of these ideas into applications that are practical, cost-effective, and user friendly (Figure 9.17). IVHS will offer drivers much more data than previously available with paper maps and atlases, including data that can be updated on a continuous basis, such as status reports on traffic and environmental conditions.

Currently, freeways in the developed world can handle about 2000 vehicles per lane per hour. More traffic than this per freeway lane causes stop-and-go traffic, which leads to accidents and gridlock. If the average vehicle is traveling 60 miles per hour on a freeway, then the average density of automobiles per lane is one every 135 feet. If the vehicle is 16 feet long, then 118 feet of freeway is going to waste. One concept behind IVHS is to increase the number of vehicles traveling at high speeds, packing together into a platoon, so that up to 7000 vehicles per lane per hour could exist on a modern freeway—one car every half-second. If this could be accomplished, the present freeway system could be used in a much more efficient way, preventing the addition of extra freeway lanes and the double-deckering of freeways through the

The dashboard in a smart car.

and German governments have spent approximately $1 billion each on magnetic levitation research. In the future, *maglevs* may transfer passengers between U.S. cities separated by up to 300 miles, at over 300 miles per hour, using far less energy and time than automobiles, Amtrak, or even air carriers. One could shuttle between Boston and Washington, or Chicago and Minneapolis, or Los Angeles and San Francisco in less than an hour. Maglevs presently are twice as fuel efficient as automobiles and four times as efficient as airliners, producing little or no air pollution. In the future, maglevs may be built alongside highways and will occupy far less room than airports (the Dallas–Fort Worth airport consumes as much land as a 65-foot-wide right-of-way coast to coast).

In the meantime, high-speed conventional rail systems have been improved to include *tilting train technology* (TTT). An example of this technology is found in the Swedish X2000 train that can travel up to 150 miles per hour and give service in the northeast corridor of the United States. The passenger car carriage tilts inward on curves, allowing increased speeds on existing track curvatures. Amtrak, the national railroad passenger corporation train service, is presently making heavy investments in tilting train technology. ISTEA, the Intermodal Surface Transportation Efficiency Act of 1991, has identified five existing rail corridors selected for development of high-speed trains. These

most congested urbanized areas. Double-deckering in a variety of U.S. cities, including San Francisco, Seattle, and New York, has always been met with strong environmental opposition.

IVHS technologies range from real-time routing and congestion information being broadcast to the auto driver via radio to allowing the car to drive by itself on an automated roadway. New electronics associated with IVHS provide real-time information on accident, congestion, and roadway incidents. Traffic controllers, which have information, beam it to motorists, who can select new routing strategies or use roadside services. Collision avoidance systems using radar, lane tracking technologies that platoon or stack vehicles at high density, and readout terminals on the dashboard that display a map of the city, as well as locations of accidents and the shortest route between two points based on real-time traffic flow information, will be given.

High-Speed Trains and Magnetic Levitation

Magnetic levitation technology eliminates mechanical contact between a vehicle and the roadbed, thus eliminating wear, noise, and alignment problems. The vehicle floats on a cushion of air one-half foot above the guideway supported by magnetic forces. The Japanese

TGV Express Train in France. In 1983, the Trans Grande Vitesse was introduced by the French Railway with service between Paris and Lyon at speeds of up to 200 mph on an entirely new track. It soon captured millions of new passengers from the highways and from domestic airlines. Proposed high-speed trains from Naples to Milan, from Lisbon to Marseille, from Bordeaux to Glasgow via the Channel Tunnel, and from Geneva to Amsterdam are to be opened by the year 2005. Despite these advances, interurban rail will continue to occupy a subordinate role in the United States. But in China, Russia, and India, where private car ownership is low, the railway still carries the bulk of interurban traffic.

include SanDiego–LosAngeles–San Francisco, Dallas–Houston–San Antonio, Miami–Orlando–Jacksonville, Pittsburgh–Chicago–Minneapolis, and Washington–New York–Boston.

TELECOMMUNICATIONS

The transmission of information is every bit as important as the movement of people and goods. An abundance of information availability facilitates and accompanies economic development and political liberty. Modernization of transportation has integrated the economic world, but equally important are technical developments in communication. Communications is the invisible layer of transport supplementing the physical transport links among and within cities, regions, and countries. Because the circulation of information is critical to the operation and success of large, complex economies, the history of capitalism has been accompanied by wave after wave of innovations in communications (Figure 9.18).

Telecommunications are not a new phenomenon. Starting in 1844, the first form of telecommunications began with Samuel Morse's invention of the telegraph, which allowed communications to become detached from transportation. The telegraph made possible the worldwide transmission of information concerning commodity needs, supplies, prices, and shipments—information that was essential if international commerce was to be conducted on an efficient basis. Telegraphy grew rapidly in the United States, from 40 miles of cable in 1844 to 23,000 miles by 1852. Starting with the first transcontinental telegraph wire in 1861, the telegraph

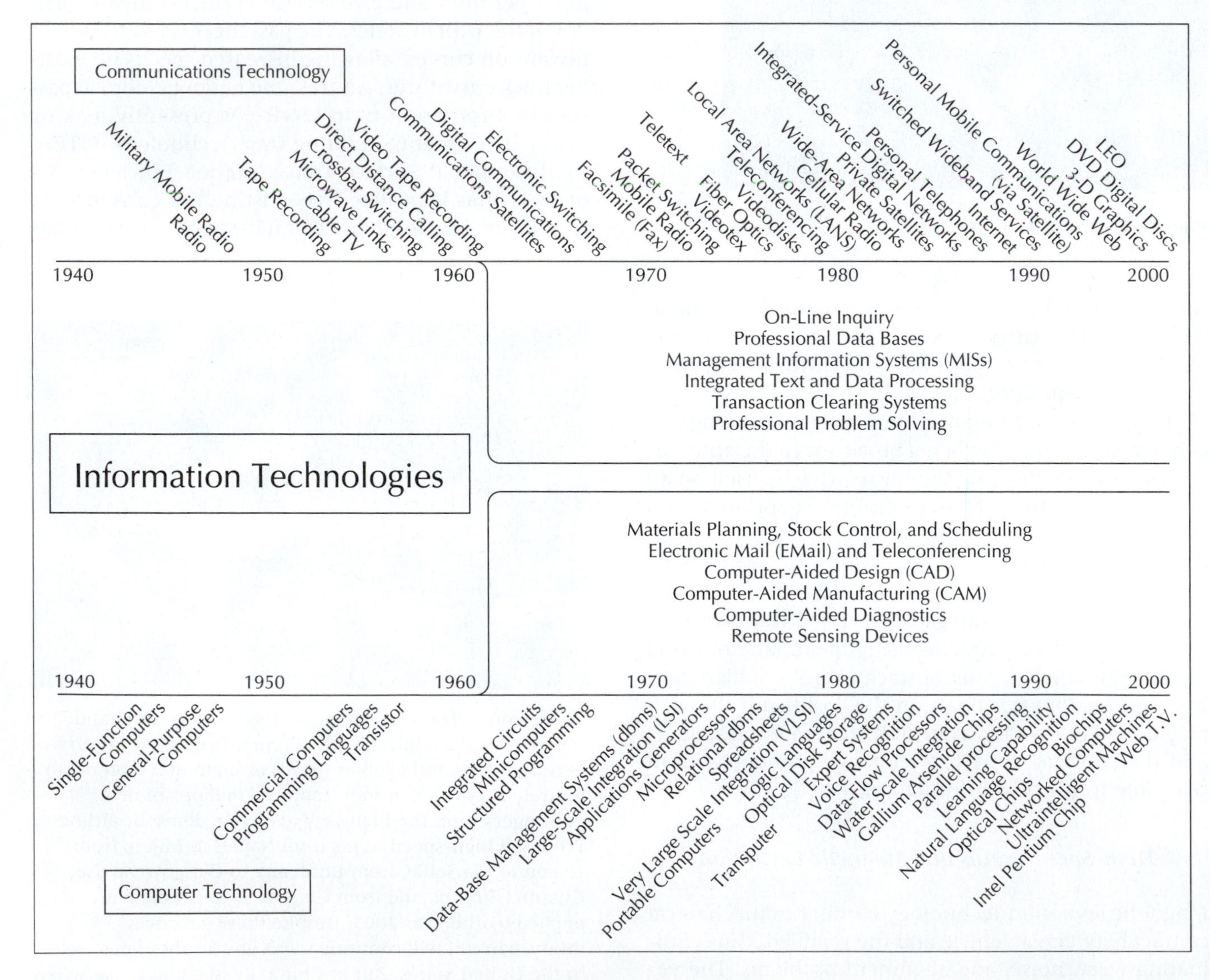

FIGURE 9.18
Innovations in the history of computer and communication technologies.

was important in the American colonization of the West, where it displaced the Pony Express, and helped to form a national market by allowing long-distance circulation of news, prices, stocks, and other information. In 1868, the first successful trans-Atlantic telegraph line was laid, part of the round of international time-space compression that accompanied the Industrial Revolution.

For decades after the invention of the telephone in 1876, telecommunications was synonymous with simple telephone service. Just as the telegraph was instrumental to the colonization of the American West, in the late nineteenth century the telephone became critical to the growth of the American city-system, allowing firms to centralize their headquarters functions while they spun off branch plants to smaller towns. Growing multiestablishment corporations utilized the telephone to coordinate production and shipments. In the 1920s, the telephone, like the automobile and the single-family home, became a staple of the growing middle class, with significant social effects on friendship networks, dating, and other ties. In the 1950s, direct dialing eliminated the need for shared party lines, and the first international phone line was laid across the Atlantic Ocean. Even today, despite the proliferation of several new forms of telecommunications, the telephone remains by far the most commonly used form of telecommunications for businesses and households.

From 1933 to 1984, the American Telegraph and Telephone Company (AT&T) enjoyed a monopoly over the U.S. telephone industry. Congress exempted AT&T from antitrust laws in return for its commitment to guarantee universal access among the population, eventually resulting in a 98 percent penetration rate among U.S. households. The widespread deregulation of industry extended to telecommunications, and in 1984, AT&T was broken up into one long-distance and several local service providers ("Baby Bells"). New firms such as MCI and Sprint entered the field. Faced with mounting competition, telephone companies have steadily upgraded their copper cable systems to include fiber-optic lines, which allow large quantities of data to be transmitted rapidly, securely, and virtually error free.

The provision of telephones is a common measure of a nation's communications infrastructure (Figure 9.19). Telephone penetration rates—the availability per 1000 people—are highest in the economically developed world. Africa has less than 1 percent of the world's telephones and comprises 15 percent of the world's population. One-half of the world's population has never made a telephone call. Landlines, however, are rapidly being supplemented or replaced by wireless technologies; indeed, there are more cell phones today than landlines. For developing countries, wireless technologies offer the possibility of "leapfrogging," that is moving directly into newer, lower cost forms of technology.

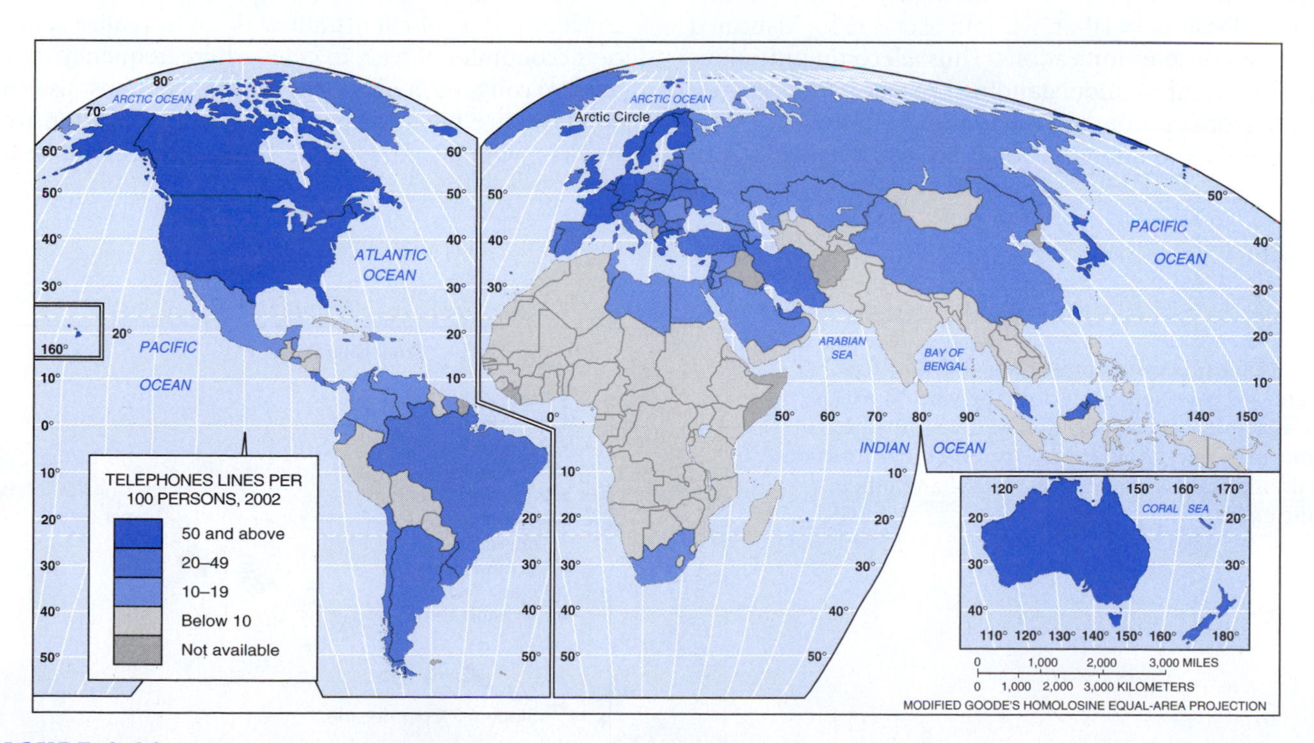

FIGURE 9.19
Telephones per 1000 people worldwide, 2002. (See Color Insert for more illustrative map.)

A shopkeeper in Ivory Coast has steady customers for mobile phones.

The microelectronics revolution of the late twentieth century was particularly important in the telecommunications industry, which is arguably the world's most rapidly expanding and dynamic sector today. Innovations in processing power have led to exponential increases in the ability of computerized systems to analyze and transmit data (Figure 9.20). The ability to transmit vast quantities of information in real time over the planetary surface is crucial to "digital capitalism." No large corporation can operate in multiple national markets simultaneously, coordinating the activities of thousands of employees within highly specialized corporate divisions of labor, without access to sophisticated channels of communications. Thus telecommunications are important to understanding broader issues pertaining to globalization and the world economy, including the complex relations between firms and nation-states.

Fiber-optic and Satellite Systems

Today, two technologies—satellites and fiber-optic lines—form the primary technologies deployed by the global telecommunications industry. The transmission capacities of both of these grew rapidly in the late twentieth century as the microelectronics revolution began to unfold. Multinational corporations, banks, and media conglomerates typically employ both technologies, often simultaneously, either in the form of privately owned facilities or leased circuits from shared corporate networks. Roughly 1000 fiber optic and two dozen public and private satellite firms provide international telecommunications services. The network of fiber lines linking the world constitutes the nervous system of the global financial and service economy, linking cities, markets, suppliers, and clients around the world (Figure 9.21).

Although they overlap to a great extent, satellite and fiber-optic carriers exhibit market segmentation. Fiber is heavily favored by large corporations for data transmission and by financial institutions for electronic funds transfer systems. Satellites tend to be used more often by international television carriers. Telephone and Internet traffic use both. These two types of carriers are differentiated geographically as well: Because their transmission costs are unrelated to distance, satellites are optimal for low-density areas (e.g., rural regions and remote islands), where the relatively high marginal costs of fiber lines are not competitive. Fiber-optic carriers prefer large metropolitan regions, where dense concentrations of clients allow them to realize significant economies of scale in cities where frequency transmission congestion often plagues satellite transmissions. Satellites are ideal for point-to-area distribution networks, whereas fiber-optic lines are preferable for

FIGURE 9.20
Computer processing power per unit cost has increased exponentially (note the vertical axis is logarithmic). The application of new software that links manufacturing, inventory, and other functions is dramatically increasing industrial and office efficiency.

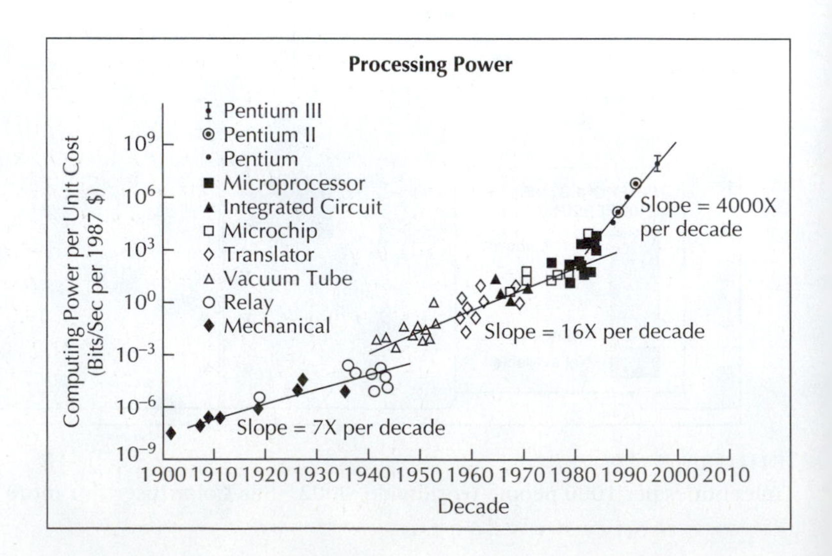

FIGURE 9.21
The global network of fiber optic lines. Starting in the 1980s across the Atlantic, many telecommunications giants have laid an expanding grid of fiber optics with exponentially increasing power to transmit enormous quantities of data.

point-to-point communications, especially when security is of great concern.

Historically, the primacy of each technology has varied over time. From 1959 to 1980 (i.e., before the invention of fiber optics), satellites enjoyed limited competition from transocean copper cable lines with low-capacity rates. From the 1970s onward, the microelectronics revolution allowed fiber-optic lines to erode the market share of traffic held by satellites (Figure 9.22). New techniques of data transmission, such as the so-called frame delay format, raise speeds of transmission nearly 30-fold over the 1990s technology.

However, more recently the growth of wireless and cellular phone traffic has led to a resurgence of low-orbiting satellites. Large scale, *low earth orbit* (LEO) satellite systems provide telephone communications to and from anyplace on earth and are ideal for the rapidly expanding cell phone market. These private, global satellite constellations transmit television, radio, fax, computer, and voice images.

One of the satellites in the tracking and data relay satellite system in orbit around the earth is the TDRSS satellite. The system provides advanced tracking and telemetry services for a number of other satellites, as well as commercial telecommunications services.

Telecommunications and Geography

There exists considerable popular confusion about the real and potential impacts of telecommunications on

An important improvement in global communications has been the development of satellite technology. The first communication satellites date from 1965, with the launching of Early Bird, able to carry 250 telephone conversations or two television channels simultaneously. Since then, more advanced communication satellites carry 100,000 circuits of simultaneous telephone communications or television channeling.

spatial relations, in part due to the long history of exaggerated claims made in the past. We keep read, for example, that telecommunications means "the end of geography." Often such views hinge on a simplistic, utopian technological determinism that ignores the complex relations between telecommunications and local economic, social, and political circumstances. For example, repeated predictions that telecommunications would allow everyone to work at home via telecommuting, dispersing all functions and spelling the obsolescence of cities, have fallen flat in the face of the persistent growth in densely inhabited urbanized places and global cities. In fact, telecommunications are usually a poor substitute for face-to-face meetings, the medium through which most sensitive corporate interactions occurs, particularly when the information involved is irregular, proprietary, and unstandardized in nature. Most managers spend the bulk of their working time engaged in face-to-face contact, and no electronic technology can yet allow for the subtlety and nuances critical to such encounters. It is true that networks such as the Internet allow some professionals to move into rural areas, where they can conduct most of their business online, gradually permitting them to escape from their longtime reliance on large cities where they needed face-to-face contact. Yet the full extent to which these systems facilitate decentralization is often countered by other forces that promote the centralization of activity.

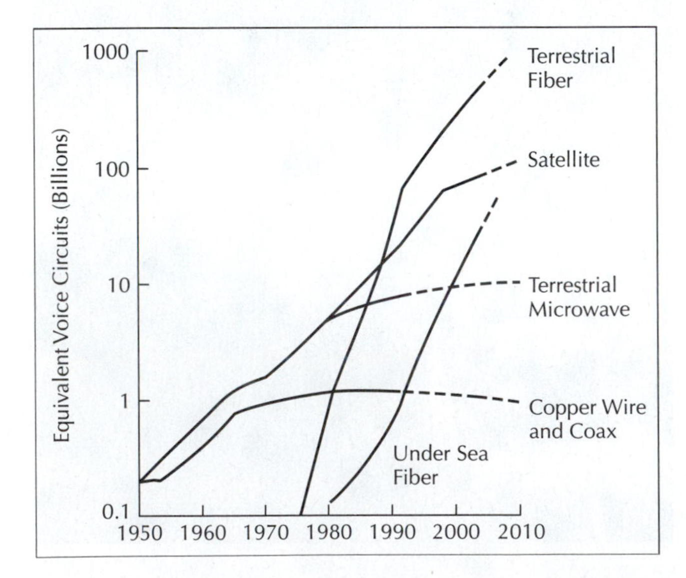

FIGURE 9.22
Global telecommunications capacity. Fiber-optic and satellite transmission have surpassed copper wire and the coaxial cable. The increase has been an exponential one regarding voice circuitry.

For this reason, a century of telecommunications, from the telephone to fiber optics, has left most high-wage, white-collar, administrative command and control functions clustered in downtown areas. In contrast, telecommunications are ideally suited for the transmission of routinized, standardized forms of data, facilitating the dispersal of functions involved with their processing to low-wage regions. In short, there is no particular reason to believe that telecommunications inevitably lead to the dispersal or deconcentration of functions; by allowing the decentralization of routinized ones, information technology actually enhances the comparative advantage of inner cities for nonroutinized, high-value-added functions that are performed face to face. Thus telecommunications facilitate the simultaneous concentration and deconcentration of economic activities.

Thus, popular notions that "telecommunications will render geography meaningless" are simply naïve. While the costs of communications have decreased, as they did with transportation, other factors have risen in importance, including local regulations, the cost and skills of the local labor force, government policies, and infrastructural investments. Economic space, in short, will not evaporate because of the telecommunications revolution. Exactly how telecommunications are deployed is a contingent matter of local circumstances, public policy, and local niche within the national and world economy.

Within cities, digital networks have contributed to an ongoing reconstruction of urban space. Telecommunications networks tend to be largely invisible to policy makers and planners and receive little attention. While many city governments are willing to invest in new roads or water control projects, urban planners and economic development officials have often overlooked or ignored altogether the role that telecommunications can play in stimulating economic growth. For example, only 5 percent of U.S. municipalities have explicit plans for telecommunications. One reason is that there is no statistical correlation between local investment in telecommunications and economic growth; the idea that "if you build it, they will come" is not necessarily true. However, while the telecommunications industry per se is relatively small and capital intensive, generating few jobs, and while telecommunications do not guarantee economic development, such systems have become necessities for many firms. In short, telecommunications are necessary but not sufficient to induce economic growth.

Teleworking is often touted as the answer to reduced transportation costs, easing demands on energy and reducing environmental impacts. There is a growing trend toward wireless terminals because the wiring of computers and peripherals to networks is costly. Wireless terminals include computers and other devices that can communicate with machines with infrared or electromagnetic signals. This allows computers to function within company or international networks. The first generation of wireless terminals is now popular in the form of desktop and laptop PCs. Palmtop units can send large data files or e-mail using satellite communications technology, and will eventually replace pagers and cellular phones. The trend toward wireless terminals is significant because it allows more portability and eliminates the need to be connected and disconnected from local area networks.

Although large cities typically have much better developed telecommunications infrastructures, the technology has rapidly diffused through the urban hierarchy into smaller towns and is becoming increasingly equalized among regions. In the future, therefore, the marginal returns from investments in this infrastructure are bound to diminish, minimizing competitive advantages based on information systems infrastructure alone and forcing competition among localities to occur on other bases, such as the cost and quality of labor, taxes, and regulatory framework. Regions with an advantage in telecommunications generally succeed because they have attracted firms for other reasons.

A growing set of impacts of information systems on urban form concerns transportation informatics, including a variety of improvements in surface transportation such as smart metering, electronic road pricing, synchronized traffic lights, automated toll payments and turnpikes, automated road maps, information for trip planning and navigation, travel advisory systems, electronic tourist guides, remote traffic monitoring and displays, and computerized traffic management and control systems, all of which are designed to minimize congestion and optimize traffic flow (particularly at peak hours), enhancing the efficiency, reliability, and attractiveness of travel. Wireless technologies such as cellular phones allow more productive use of time otherwise lost to congest. Such systems do not so much comprise new technologies as the enhancement of existing ones.

Among U.S. cities, telecommunications have accelerated the spatial reorganization of financial services. By relying on economies of scale, large firms can combine services in a few centralized database management systems. American Express, for example, shifted its credit card processing to three facilities, and Aetna Insurance consolidated 55 claims adjustment centers to 22 metropolitan regions. Allstate Insurance consolidated 28 processing centers into three (Charlotte,

Teleworking has allowed busy parents to juggle professional and domestic responsibilities by working at home either full time or part time.

Dallas, and Columbus); CNA centralized theirs in Reading, Pennsylvania; and Travelers Insurance established two in Knoxville, Tennessee, and Albany, New York. Other insurers are developing online marketing via the Internet. Among telecommunications carriers, AT&T has six megacenters, and Sprint opted for low-cost places such as Jacksonville, Florida, Dallas, Texas, Kansas City, Missouri, Phoenix, Arizona, and Winona, Minnesota. Meanwhile, local sales offices in small towns have experienced a steady decline. This phenomenon exemplifies the manner in which telecommunications can simultaneously centralize as well as decentralize different economic activities.

With the digitization of information, telecommunications steadily merged with computers to form integrated networks (Figure 9.23). New technologies such as fiber optics have complemented and at times substituted for telephone lines. Fax services and 800 number free toll calls are now standard for virtually all companies, and even newer technologies such as Electronic Data Interchange and wireless services are becoming increasingly popular. Like the railroad system of the nineteenth century and the interstate highway system of the twentieth century, the information highway of fiber-optic cables, satellites, and wireless grids links billions of computers, telephones, faxes, and other electronic products all over the world.

Geographies of the Internet

Among the various networks that comprise the world's telecommunications infrastructure, the largest and most famous is the Internet, a vast web of electronic networks nicknamed the *information superhighway.* This system delivers large amounts of services to homes, offices, and factories, including e-mail, telephone calls, TV programs and other video images, text, and music. The system enables students at rural schools to use computers to tap resources at distant universities or researchers in small colleges to use supercomputers at located far away. The superhighway allows doctors to check patients from their homes, and it permits doctors in several remote cities to collaborate on a patient's care by immediately sharing multimedia computer screens.

Origins and Growth of the Internet

The Internet originated in the 1960s under the U.S. Defense Department's Defense Agency Research Projects Administration (DARPA), which designed it to withstand a nuclear attack. Much of the durability of the current system is due to the enormous amounts of federal dollars dedicated toward research in this area. In the 1980s, control of the Internet was transferred to the

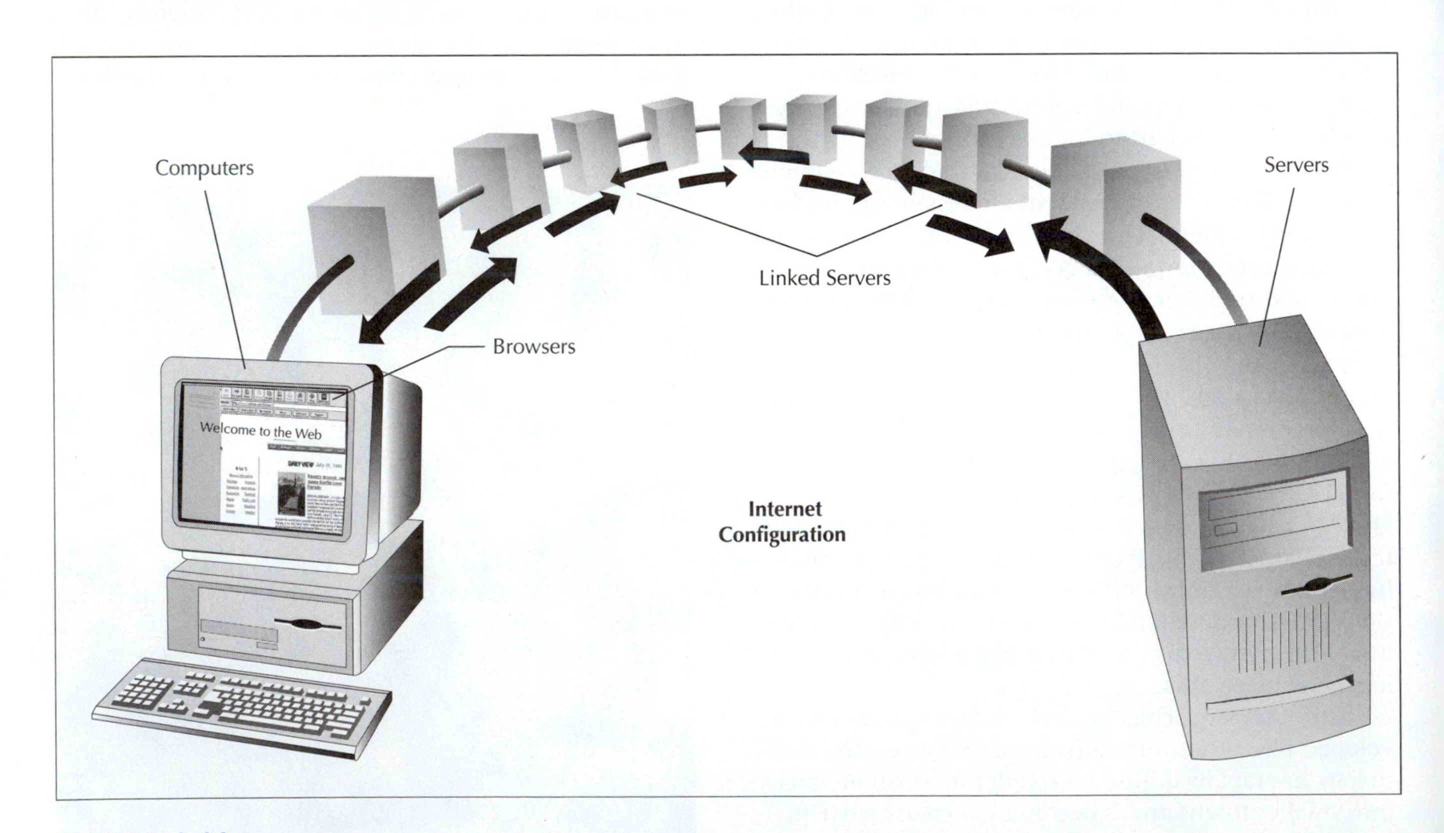

FIGURE 9.23
The convergence of computer technology and communications technology yields information technologies (IT). Two initially distinct technologies have now combined to impact the world economy. Communications technology is concerned with the transmission of information, while computer technology is concerned with the processing of information.

National Science Foundation, and in the 1990s, control was privatized via a consortium of telecommunications corporations. The Internet emerged on a global scale through the integration of existing telephone, fiber-optic, and satellite systems, which was made possible by the technological innovation of packet switching, TCP/IP (Transmission Control Protocol/Internet Protocol), and *Integrated Services Digital Network* (ISDN), in which individual messages may be decomposed, the constituent parts transmitted by various channels, and then reassembled, virtually instantaneously, at the destination. In the 1990s, graphical interfaces developed in Europe greatly simplified the use of the Internet, leading to the creation of the World Wide Web.

The growth of the Internet has been phenomenal (Figure 9.24); indeed, it is arguably the most rapidly diffusing technology in world history. By the end of 2005, penetration was 50 percent of U.S. households and 63 percent of the population. Like the automobile 80 years ago—which inspired the development of the moving assembly line, a fundamental and far-reaching innovation in manufacturing practice—the Internet has significantly changed the ways we work, consume, and live. The Internet is well on its way to becoming an indispensable mass communications technology for the average American.

Technological changes will further increase the utility and popularity of the Internet in the future. Mobile phones make it possible for consumers to access the Internet from any location, not just in the home or at the office. Broadband connectivity is becoming increasingly mainstream, allowing for innovations such as on-demand television. This next phase of Internet expansion will produce the real information revolution for everyday consumers. TV-quality video and voice-activated commands will enjoy the Internet as a practical home appliance, useful for entertainment, communication, and shopping.

In 2005, an estimated 1018 million people, or roughly 15 percent of the world's population (including 185 million in the United States) in more than 200 countries, were connected to the Internet (Figure 9.25).

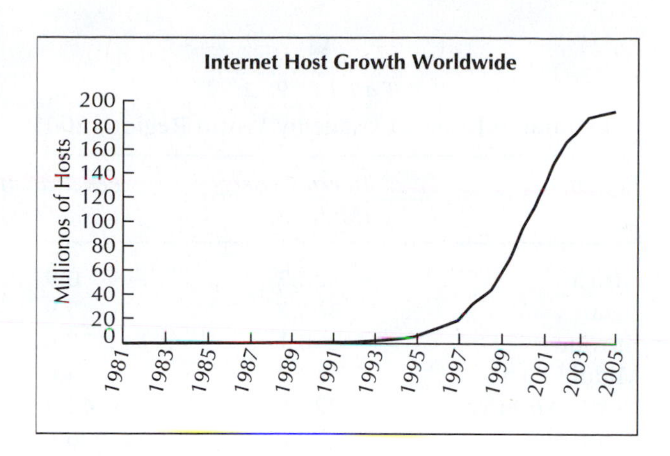

FIGURE 9.24
The growth of the Internet, 1981–2005. Roughly 1 billion people, or 15 percent of the planet, use the Internet today.

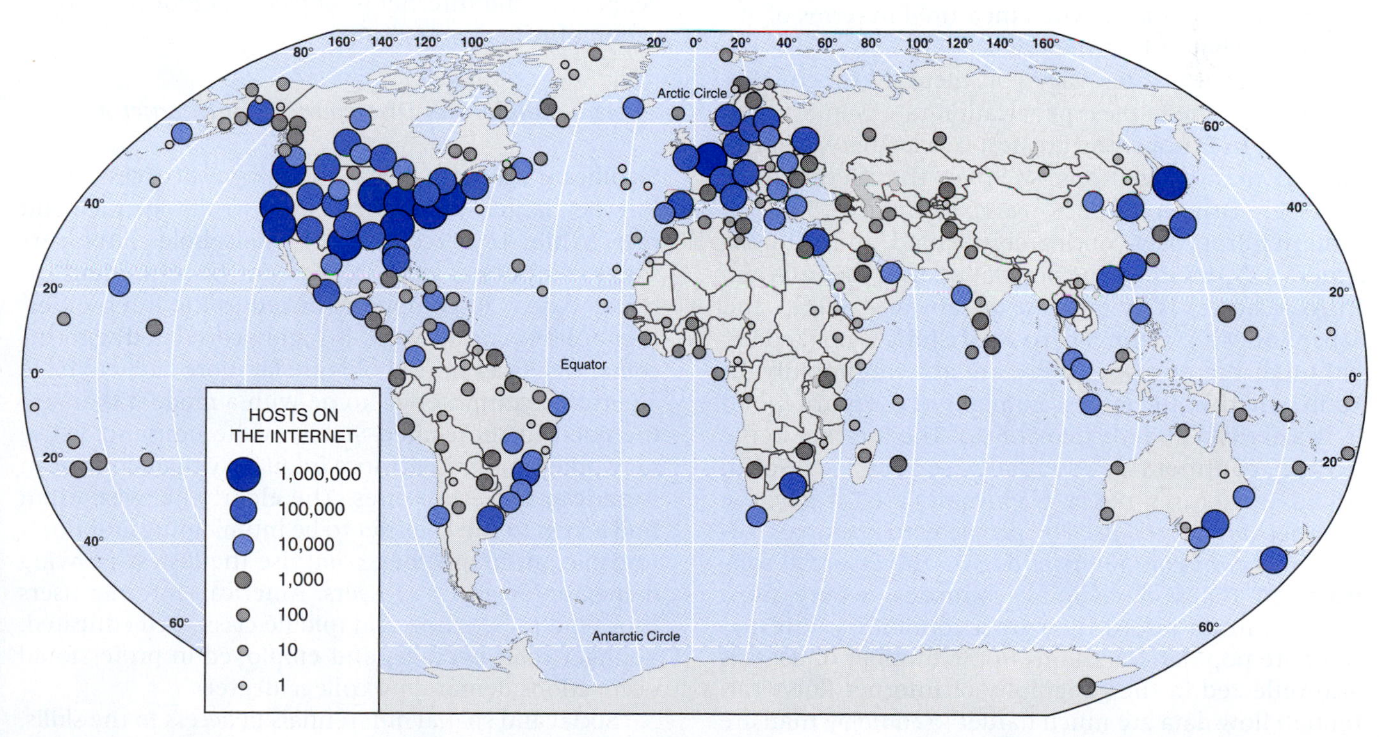

FIGURE 9.25
The geography of Internet hosts, 2003.

Table 9.2

Estimated Internet Usage by World Region, 2005

Region	Internet Users (Millions)	Penetration Rate (%)
Africa	22.7	0.9
Asia	364.2	5.9
Europe	290.1	27.6
Middle East	18.2	4.6
North America	225.8	63.2
Latin America	79.0	6.6
Oceania	17.7	47.9
World	1,018.0	15.7

Source: Internet World Statistics, www.internetworldstats.com.

However, there are large variations in the Internet penetration rate (percentage of people with access) among the world's major regions (Table 9.2), ranging from as little as 0.9 percent in Africa to as high as 63 percent in North America. Inequalities in access to the Internet internationally reflect the long-standing bifurcation between the First and Third Worlds. While virtually no country is utterly without Internet access (although portions of Africa come close), the variations among and within nations in accessibility are huge. Given its large size, the United States—with more than 185 million users—dominates when measured in terms of absolute number of Internet hosts.

Internet use rates vary considerably by country (Table 9.3). The highest penetration rate is in Sweden (75.8%). In Europe, the greatest connectivity is in relatively wealthy nations such as the Netherlands (63.7%), Britain (58.2%), and Germany (53.9%); Eastern Europe lags considerably behind, and in Russia a mere 4.2 percent of the population uses the Internet. In Asia, access is by greatest in Taiwan (49.1%) and Japan (46.4%); about 5 percent of China is hooked up, although the numbers there are growing rapidly. In Latin America, the largest numbers of users are found in Brazil (8%) and Mexico (4.6%). The Internet in the African continent is essentially confined to South Africa. In all cases, per capita incomes are the key; the Internet can only be used by people with resources sufficient to own computers and learn the essential software. In many developing countries, where most people cannot afford their own computers, Internet cafés are popular. Variations in the number of users is also reflected in the geography of Internet flows (although flow data are much harder to come by than are place-specific attribute data): 80 percent of all international traffic on the Internet is either to or from the

Table 9.3

Largest 20 Internet Users, 2005

	Users	Penetration Rate
United States	184.5	63.2
China	68.0	5.2
Japan	59.2	46.4
Germany	44.1	53.9
United Kingdom	34.4	58.2
South Korea	26.3	56.1
France	22.0	37.2
Italy	19.2	34.2
Canada	16.8	53.1
India	16.6	1.6
Brazil	14.3	8.0
Spain	14.0	33.7
Australia	12.8	64.2
Taiwan	11.6	49.1
Netherlands	10.3	63.7
Malaysia	24.0	32.5
Sweden	8.9	75.8
Russia	6.0	4.2
Turkey	4.9	6.7
Thailand	4.8	7.6

Source: Internet World Statistics, www.internetworldstats.com.

United States (Figure 9.26), fueling fears among some people that the Internet is largely a tool for the propagation of American culture.

Social and Spatial Discrepancies in Internet Access

Significant discrepancies exist in terms of access to the Internet, largely along the lines of wealth, gender, and race. While 40 percent of U.S. households have personal computers, only 20 percent have modems at home. Access to computers linked to the Internet, either at home or at work, is highly correlated with income; wealthier households are far more likely to have a personal computer at home with a modem than are the poor. In the United States, white households use networked computers more frequently than do African American or Latino ones. The elderly likewise often find access to the Internet to be intimidating and unaffordable, although they comprise the fastest-growing demographic group of users. American Internet users thus tend to be white and middle class, well educated, younger than average, and employed in professional occupations demanding college degrees.

Social and spatial differentials in access to the skills, equipment, and software necessary to get onto the electronic highway threaten to create a large, predominantly

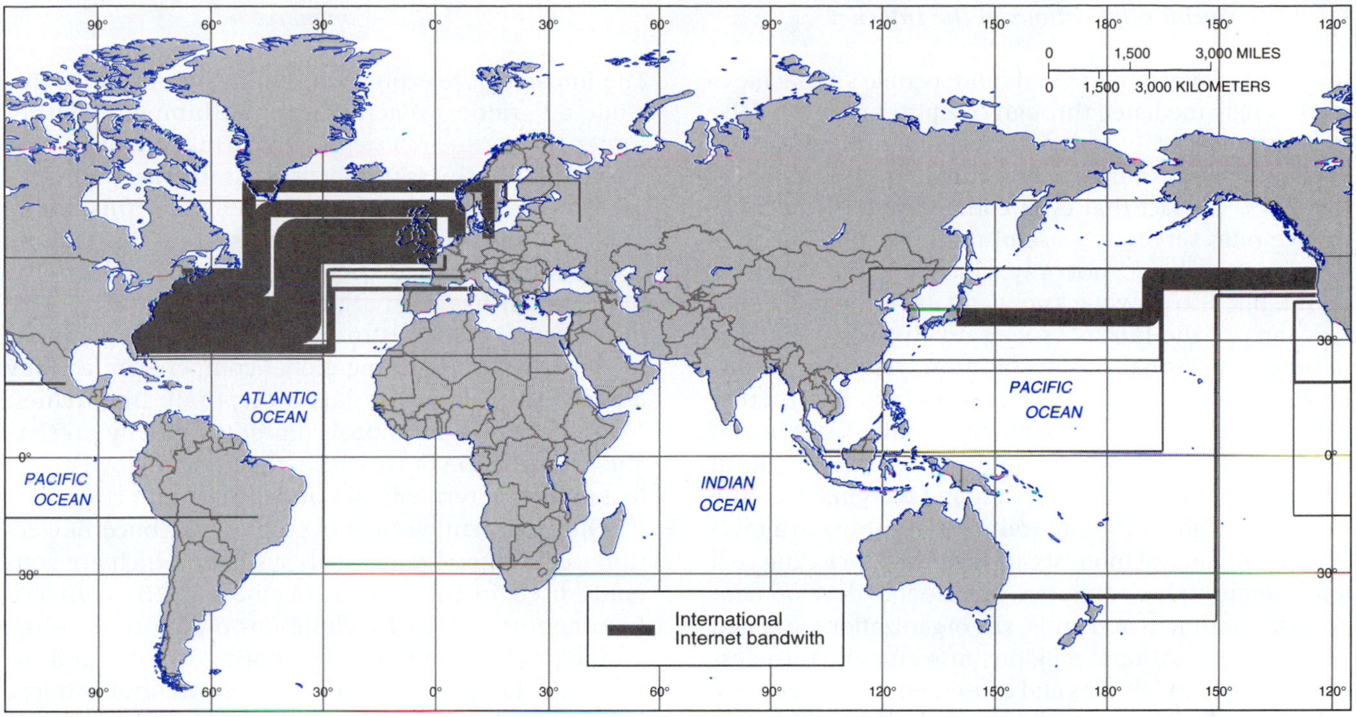

FIGURE 9.26
International information flows on the Internet. Eighty percent of all traffic is to or from the United States.

minority underclass deprived of the benefits of cyberspace. This phenomenon must be viewed in the broader context of the growing inequalities throughout industrialized nations generated by labor market polarization (i.e., deindustrialization and growth of low-income, contingent service jobs). Modern economies are increasingly divided between those who are comfortable and proficient with digital technology and those who neither understand nor trust it, disenfranchising the latter group from the possibility of citizenship in cyberspace. Despite the falling prices for hardware and software, basic entry-level machines for Internet access cost roughly $1000, an exorbitant sum for low-income households. Internet access at work is also difficult for many. For employees in poorly paying service jobs (the most rapidly growing category of employment) that do not offer access to the Internet at their place of employment, the obstacles to access are formidable.

The public educational system cannot offer an easy remedy to the problem of the digital divide. Even in the United States, the wide discrepancies in funding and the quality of education among school districts, particularly between suburban and central city schools, may reproduce this inequality rather than reduce it. Some public libraries offer free access to the Internet; mounting financial constraints in many municipalities, moreover, have curtailed the growth of these systems. Even within the most digitized of cities there remain large pockets of "offline" poverty. Those who need the Internet the least, already living in information-rich environments with access through many non-Internet channels (e.g., newspapers and cable TV), may have the most access to it, while those who may benefit the most (e.g., through electronic job banks) may have the smallest chance to log on.

Internationally, access to the Internet is deeply conditioned by the density, reliability, and affordability of national telephone systems. Most Internet communications occurs along lines leased from telephone companies, some of which are state regulated (in contrast to the largely unregulated state of the Internet itself), although the global wave of privatization is ending government ownership in this sector. Prices for access vary by length of the phone call, distance, and the degree of monopoly. In nations with telecommunications monopolies, prices are higher than in those with deregulated systems, and hence usage rates are lower. The global move toward deregulation in telecommunications will lead to more use-based pricing (the so-called pay-per revolution), in which users must bear the full costs of their calls, and fewer cross-subsidies among different groups of users (e.g., between commercial and residential ones), a trend that will likely make access to cyberspace even less affordable to low-income users.

Social Implications of the Internet

In an age in which more and more people's social life is increasingly mediated through computer networks, the reconstruction of interpersonal relations around the digitized spaces of cyberspace is of the utmost significance. However, the fact that cybercontacts differ from face-to-face ones serves as a useful reminder that telecommunications change not only what we know about the world, but also how we know and experience it.

Many of the Internet's uses revolve around entertainment, personal communication, research, downloading files, and online games. However, the Internet can also be used to challenge established systems of domination and legitimate and publicize the political claims of the relatively powerless and marginalized. The Internet has given voice to countless groups with a multiplicity of political interests and agendas, including civil and human rights advocates, sustainable development activists, antiracist and antisexist organizations, gay and lesbian rights groups, religious movements, those espousing ethnic identities and causes, youth movements, peace and disarmament parties, nonviolent action and pacifists, animal rights groups, and gays living in homophobic local environments. By facilitating the expression of political positions that otherwise may be difficult or impossible, the Internet allows for a dramatic expansion in the range of voices heard about many issues. In this sense, it permits the local to become global. Within the Internet itself one finds all the diversity and contradictions of human experience. Cyberpolitics mirrors those of its nonelectronic counterparts, although the boundaries between the two realms are increasingly fuzzy. Indeed, in a sociopsychological sense, cyberspace may allow for the reconstruction of "communities without propinquity," groups of users who share common interests but not physical proximity.

Finally, there is also what may be called the "dark side" of the Internet, in which it is deployed for illegal or immoral purposes. Hackers, for example, have often wreaked havoc with computer security systems. Such individuals are typically young men playing pranks, although others may unleash dangerous computer viruses and worms. Most hacks—by some estimates as much as 95 percent—go unreported, but their presence has driven up the cost of computer firewalls. The dark side also includes unsavory activities such as identity theft of counterfeit drivers' licenses, passports, Social Security cards, identities, securities swindles, and adoption scams. Credit card fraud is a mounting problem; 0.25 percent of Internet credit card transactions are fraudulent, compared with 0.08 percent for non-Internet transactions. Some Internet sites even offer credit card "marketplaces," where people who hack into merchant accounts may steal large numbers and sell them wholesale.

E-Commerce

The impacts of telecommunications on businesses include a varieties of activities often lumped together under the term *e-commerce,* which may include both business-to-business transactions, as well as those linking firms to their customers. In general terms, information technology lowers the transaction costs among corporations, which helps to spur productivity. Moreover, it has been argued that such systems were instrumental in the restructuring of many corporations in response to mounting global competition, as they downsized in favor of flatter corporate hierarchies. Many firms sought improved productivity by accelerating information flows within the firm and lower costs by reducing intermediaries and distribution costs.

One important version of ecommerce concerns electronic data interchange (EDI) systems, which are generally used in business-to-business (B2B) contacts. Common uses of EDI include up-to-date advertising, online product catalogs, the sharing of sales and inventory data, submissions of purchase orders, contracts, invoices, payments, delivery schedules, product updates, and labor recruitment. E-commerce reduces delays and marketing and delivery costs and has led to a greater emphasis on connectivity, ideas, creativity, speed, and customer service. In the same vein, "e-tailing" or electronic retailing, reveals the growing commercialization of the Internet. In 1993, 2 percent of all Web sites were commercial (i.e., "dot com") sites; by 2005, 60 percent were so categorized. Shopping by the Internet requires only access (e.g., a modem), a credit card, and a parcel delivery service, and allows effortless comparison shopping. The most successful example perhaps is Amazon.com, started by Seattle entrepreneur Jeff Bezos, which now is responsible for 60 percent of all books sold online. Other examples include online auctions (e.g., eBay), Internet-based telephony (e.g., Skype), and Internet music (e.g., downloading of MP3 music files), which has provoked a firestorm of opposition from music companies complaining about infringement of their intellectual property rights and declining over-the-counter music sales.

Internet sales have also provoked worries about tax evasion and sales of illegal goods (e.g., pharmaceuticals from abroad). Despite predictions that "click and order" shopping would eliminate "brick and mortar" stores, e-tailing has been slow to catch on, however, comprising only 3 percent of total U.S. retail sales, perhaps because it lacks the emotional content of shopping. Shoppers using this mode tend to be above average in income and relatively well educated. Web-based banking has experienced slow growth, even though it is considerably cheaper for banks than automatic teller machines, as have Internet-based bill payments, mortgages, and insurance. Internet-based sales of stocks (e.g., E. Schwab, E* Trade) now comprise 15 percent of all trades. One particularly successful application has been in the travel reservation and ticketing business, where Web-based purchases of hotel rooms and airline seats (e.g., through services such as Travelocity and Priceline.com) have caused a steady decline in the number of travel agents. Electronic publishing, including more than 700 newspapers worldwide, has been extended to e-books and e-magazines, which, unlike printed text, can be complemented with sound and graphics. Other services offer Internet searches of databases and classified ads. Webcasting, or broadcasts over the Internet (typically of sports or entertainment events), demands high-bandwidth capacity but comprises a significant share of Internet traffic today. Web-logs, or "blogs," have become increasingly important sources of personal, social, and political commentary, alternatives to the mainstream media and a voice for independent views.

Internet advertising has proven to be difficult, in part because the Internet reaches numerous specialized markets rather than mass audiences. Cyberspace does allow specialized companies to reach global niche markets. E-advertising comprises only 1 percent of total revenues in the United States and is overwhelmingly focused on computer and software firms. Indeed, many users are now wary of "spam" e-mail (unwanted commercial messages), which constitute an ever-larger, and increasingly annoying, share of e-mail traffic (by some estimates as high as 75%).

Another version of e-commerce concerns universities, many of which have invested heavily in Web-based distance learning courses. Although such programs are designed to attract nonlocal and nontraditional students, many of whom may not be able to take lecture-based courses in the traditional manner, they also reflect the mounting financial constraints and declining public subsidies that many institutions face, which may see distance learning as a means of attracting additional students, and tuition, at relatively low marginal costs. The largest example of Web-based teaching is the University of Phoenix, based in Arizona but with students located around the world; with more than 100,000 students, it is now the largest university in the world. Distance learning has provoked fears that it opens the door to the corporatization of academia and the domination of the profit motive, while others have questioned whether the chat rooms that form an important part of its delivery system are an effective substitute for the face-to-face teaching and learning that classrooms offer. It remains unclear whether Web-based learning is an effective complement or substitute for traditional forms of instruction. Others suggest that distance-learning programs may be better suited to professional programs in business or engineering than in the liberal arts.

More morally ambiguous is the growing role of Internet-based gambling systems, which include a variety of betting services, especially concerning sports events, and even online slot machines in which gamblers may use their credit cards. (Some complain that online gambling doesn't adequately substitute for the heady experience of a gaudy casino in Las Vegas, Nevada, or Atlantic City, New Jersey). Because the geography of legal gambling is highly uneven, the existence of such systems challenges the laws of communities in which gambling is illegal. Offshore gambling centers have grown quickly, particularly in the Caribbean, which started when Antigua licensed its first Internet casino in 1994. By 2004, an estimated 700 online casinos, mostly in the Caribbean, attracted roughly 8 million users. In the United States, gambling is permitted in some states (e.g., Nevada and Mississippi) but not others; oddly, running an online casino in the United States is legal but using one is not.

Electronic Data Interchange

EDI can be defined as the electronic movement of standard business documents between and within firms. EDI uses a structured machine-retrievable data format that permits data to be transferred between networked computers without rekeying. Like e-mail, EDI enables the sending and receiving of messages between computers connected by a communication link such as a telephone line.

However, EDI has some special characteristics:

1. *Business transaction messages.* EDI is used primarily to transfer repetitive business transactions. These include purchase orders, invoices, approvals of credit, shipping notices, and confirmations. In contrast, e-mail is used mainly for nonstandard correspondence.
2. *Data formatting standards.* Because EDI messages are repetitive, it is sensible to use some formatting standards. In contrast, there are no data formatting standards for e-mail because it is usually not formatted.
3. *EDI translators.* The conversation of data sent into standard formats is done by special EDI translators.
4. *EDI uses Value Added Networks (VAN).* In contrast to e-mail, which uses regular telephone lines, EDI uses value-added networks, specialized companies with expertise in maintenance that help customers develop the necessary interfaces.

The major advantages of EDI are as follows:

1. EDI enables a company to send and receive large amounts of information around the world in real time.
2. Companies have the ability to access partners' computers to retrieve and store standard transactions.

3. EDI fosters a true partnership because it involves a commitment to a long-term investment and refinement of the system over time.
4. EDI creates a paperless environment, saving money and increasing efficiency.
5. The time for collecting payments can be shortened by several weeks.

FUTURE IMPACTS OF INFORMATION TECHNOLOGIES

For all that telecommunications technology has provided, the biggest changes are still ahead. The coming digital revolution will redefine lives, work, education, commerce, and leisure. Whole industries will vanish, and new ones will spring up overnight. The restraints of time and space, long the adhesive that has held many communities together, will be reconstructed dramatically.

Today, the "virtual transportation network" of computers and cables promises to remake cities and nation-states yet again. Communities unprepared for these changes risk being consigned to geopolitical obsolescence. In the process, they are likely to suffer a fate similar to that of many of the great industrial cities of the past, becoming "electronic ghost towns" on the virtual frontier, abandoned by corporate and human citizens seeking a more electronically hospitable environment.

Smart Cities

Linking parts and functions of a city through networked computers may be one answer to urban decline. Such computer-networked cities are known as *smart cities.* In the past, cities prospered as geopolitical entities because of their importance as transportation crossroads or as centers of industrial production. But telecommunications developments like telephones, fax machines, and electronically linked computers weakened the once inextricable connections between transportation systems and mobility.

The new telecommunications network made it possible for businesses to produce, consumers to purchase, and workers to interact with one another without the need for a common physical location, allowing corporate and human citizens to choose where to reside based on a wide array of factors beyond mere physical convenience. Worse, communities suddenly found themselves competing for these residents not just with neighboring municipalities, but with cities across the country—or around the globe.

At the same time, telecommunications advances are transforming a world of discrete local and national economic spheres into a single global economy, placing local businesses and workers alike into direct competition with their counterparts on distant shores. Government and business leaders watched helplessly as they began to lose control over their communities' economic destinies even as shifting business and population patterns eroded their economic base and their place in the urban hierarchy. As a result, economic and social institutions ranging from governments to economic development agencies to local schools, whose increasingly time-worn practices already were proving insufficient to the rapid technological changes and complex social and economic challenges of the late twentieth century, were left with neither adequate resources nor adequate responses for solving the very difficult problems within their midst, further loosening their hold on the people and businesses who remain.

Government

Placing government and social service information (office locations, departmental telephone numbers, city council minutes, government documents, and so on) online is a valuable first step toward building an Internet-savvy community—but only a first step. Increasing people's access to such information serves the important function of creating a more well-informed citizenry.

Information technology's greatest potential in the government arena may lie in transforming the very nature of local government, making it possible to reconfigure traditionally monolithic, downtown City Halls into a network of small, neighborhood-based "branches" linked electronically to a slimmed-down city "headquarters." Under this scenario, almost all government and social services would be dispensed in the neighborhood—either from kiosks or in small, multifunction neighborhood service centers. Such structural reengineering could further reduce government staffing, operational, and office costs; minimize traffic congestion and pollution; and increase people's access to government officials and services while creating a government more institutionally sensitive to neighborhood concerns.

E-government takes a variety of forms, ranging from simple broadcasting of information to integration (i.e., allowing user input) to integration, in which network integration minimizes duplication of efforts. E-government allows, for example, for the digital collection of taxes, voting, and provision of some public services, particularly the provision of information. Such steps boost the efficiency and effectiveness of public services, allowing, for example, online registration of companies and automobiles; electronic banking; utility bill payments; applications for government programs, universities, and licenses; access to census data; and reducing the waiting time as paperwork filters through government bureaucracies.

Business

Hundreds of thousands of companies, from small start-ups to the *Fortune* 500, use the Internet to promote their businesses. Web sites are a passive form of advertising, and potential customers often do not encounter a particular company's site unless they happen to be looking for it—turning the World Wide Web into something of a high-tech Yellow Pages. In an effort to reach a broader Web audience, therefore, many companies have taken a cue from the television advertising model and have begun to sponsor high-profile, Web-based news, information, and entertainment sites where they can gain exposure to customers that they otherwise might never reach.

For all their promotional potential, however, information networks are more than effective advertising and marketing vehicles. They can, in fact, change the very nature of work. In much the same way that government telecommunications networks may allow government agencies to distribute their workforces among communities, large businesses can establish dedicated neighborhood *telework centers* that give them access to potential employees, such as rural residents, homemakers, people with disabilities, or individuals without transportation, whom they otherwise might not be able to attract. By moving operations out of congested and high-priced central cities, telework centers simultaneously can increase workers' productivity while reducing companies' rent, transportation, and labor costs.

Telecommuting, of course, is an idea that has been around for decades, and the practice has not yet taken off in the way many of its advocates forecast. This has been due in part to business's lack of interest in telecommuting and to its reliance on traditional, fixed-site management practices. But the growing number of home-office, temporary, and contract workers makes telework not only feasible but increasingly necessary. With workers demanding more flexible schedules, shorter commutes, and relief from traffic congestion, companies may find that telecenters are a powerful tool for retaining or attracting high-quality workers. What's more, with videoconferencing, e-mail, and high-speed computer networks, telecommuting finally has become technologically practical in a wide range of job categories and work situations.

Education

Information technology's most obvious potential in education lies in giving students access to computers and the Internet. That is, in fact, the substance of most of the smart community educational initiatives now taking place at local and federal levels. Bringing computers into the classroom will increase young people's access to information, help to equalize educational resources among poorer and wealthier school districts, and better prepare today's students for tomorrow's workforce.

Indeed, technology is forcing educational planners to reevaluate the entire concept of mass-produced, discipline-based education, in which students are herded around schools to assigned classrooms according to fixed schedules and one-size-fits-all curricula. In the smart communities, students—adults as well as young people—will learn when they want to, how they want to, and at their own pace. Already, predominantly rural states like Montana and Iowa are using telecommunications to link schools in sparsely populated areas into a statewide educational network, allowing instructors and students to interact from around the state within the confines of a single "virtual classroom." Similarly, library systems are remaking themselves into comprehensive information and research centers, while others are being planned from the start as "smart libraries," relying on multimedia, virtual reality, and global networks.

Health Care

Escalating health care costs have severely strained the nation's health care system. Much of this cost increase is due to an enormous volume of recordkeeping and data transfer—a problem well suited to a technological solution. In the United States alone, billions of dollars per year could be saved by using advanced communications technology for the routine transfer of laboratory tests and more orderly collection, storage, and retrieval of patient information. Thus telemedicine, in which physicians at one location use remote viewing techniques to diagnose and provide advice to patients located somewhere else, has grown in popularity, particularly in rural areas with inadequate access to health care. In some cases, such technologies facilitate the training of physicians engaged in virtual surgeries. The Internet has also become an important source of medical information, dramatically changing the traditional doctor-patient relationship: two-thirds of everyone online have searched for health-related information there, although much of it is inaccurate or misleading.

Summary

This chapter examined two major systems of circulation—transportation of people and goods and communication of information—that are critical to the ever-changing structure of global capitalism. We considered some of the factors other than distance that play a role in determining transport costs—the nature of commodities, carrier and route variations, and the regimes governing transportation. Transport costs remain critical for material-oriented and market-oriented firms, but they are of less importance for firms that produce items for which transport costs are but a small proportion of total costs. For these firms, transit time is more crucial than cost. Modernized means of transport and reduced costs of shipping commodities have also made it possible for economic activities to decentralize. Multinationals have taken full advantage of transport developments to establish "offshore" branch-plant operations.

Movements of goods and people take place over and through transport networks. We focused on the historical development of transportation and explained how improvements over the centuries have resulted in time-space and cost-space convergence. Improved transport and communications systems integrated isolated points of production into a national or a world economy. Although the friction of distance has diminished over time, transport remains an important locational factor. Only if transportation were instantaneous and free would economic activities respond solely to aspatial forces such as economies of scale. The chapter also considered the role of the state in transportation policy, which has varied around the world and among different sectors (rail, airlines, trucking, etc.).

Innovation in urban transportation systems is necessary because of the tremendous increase in travel demand in large cities of the developed world. For example, in the United States, vehicle miles traveled, automobile ownership, and total vehicle trips are increasing rapidly.

Communications and information technology (IT) are transforming the world economy at rates never before thought possible. Profound implications, even many that the world cannot yet measure, accompany this IT explosion. At the center of this information explosion are the microprocessor, networked computers, and the Internet, which connected more than 1 billion people worldwide in 2005, about 15 percent of the planet. The chapter explored the origins, growth, and size of the Internet; noted the uneven access to it that people around the world and within the United States have; and touched on some of its many consequences, including the growth of cybercommunities. It pointed out that the social divisions that exist offline are replicated online. It also explained the nature and impacts of e-commerce.

The telecommunications revolution may only be starting to have its real impacts. The chapter noted a number of ways the IT revolution may carry forward into the future, including employment, health care, and government services.

STUDY QUESTIONS

1. What are curvilinear and stepped freight costs?
2. What are terminal and line-haul costs?
3. How do transport costs enter into location theory?
4. What network accessibility and connectivity?
5. What is the Taaffe, Morrill, and Gould model of transportation and urban development?
6. What is the gravity model?
7. What are cost-space and time-space convergence?
8. How did deregulation affect the structure of airline networks?
9. When did telecommunications begin?
10. How did the microelectronics revolution affect telecommunications?
11. What are fiber optics and what role to they play in international telecommunications?
12. Do telecommunications mean the end of geography?
13. How did the internet begin?
14. How does access to the Internet reflect social divisions?
15. What are some social impacts of the Internet?
16. What is e-commerce and why do many firms like it?
17. What is electronic data interchange?
18. What are some likely future impacts of information technology?

KEY TERMS

accessibility index
artificial intelligence
backhaul
break-of-bulk point
connectivity
cost-space convergence
deregulation
distance learning
distance-decay effect
elasticity of demand for transportation
electronic data interchange
expert systems
friction of distance
global office
gravity model
hub-and-spoke networks
information warehouse
intelligent vehicle highway systems (IVHS)
Internet
ISDN
journey to work
line-haul costs
Maglev
multimedia
networked computers
outsourcing
privatization
smart cars
smart highways
spatial interaction
stepped freight rates
telepresence
terminal costs
time-space compression or convergence
transport costs

SUGGESTED READINGS

Cairncross, F. *The Death of Distance.* 1997. Boston, Mass. Harvard Business School Press.

Crang, M., Crang, P., and May, J. 1999. *Virtual Geographies: Bodies, Space and Relations.* London: Routledge.

Dodge, M. and Kitchin, R. 2001. *Mapping Cyberspace.* London: Routledge.

Graham, S. and Marvin, S. 2001. *Splintering Urbanism: Networked Infrastructures, Technological Mobilities and the Urban Condition.* London: Routledge.

Kitchin, R. 1998. *Cyberspace: The World in the Wires.* New York: John Wiley.

Standage, T. *The Victorian Internet.* 1998. New York: Walker and Company.

Warf, B. 2006. "International Competition between Satellite and Fiber Optic Carriers: A Geographic Perspective." *The Professional Geographer* 58:1–11.

WORLD WIDE WEB SITES

ATLAS OF CYBERSPACE

http://www.cybergeography.org/atlas/

Most comprehensive collection of maps of the Internet around.

INFOSPACE

http://www.infospace.com

Search for addresses, phone numbers, e-mail addresses. Also a "My Town" profile that, using phone book data, gives you a personal profile of any town in the United States. Want to know the names of Chinese restaurants in Boise? Click. Want to see them on a map? Click. Want written directions on how to get there? Click.

MAPQUEST

http://www.mapquest.com

Type in an address and get a clickable, zoomable map of that location.

TRIPQUEST

http://www.tripquest.com

Type in a starting location and an ending location. The program provides door-to-door or city-to-city directions.

ETRADE

http://www.etrade.com

Flat rate broker allows buying or selling of stock for a flat $14.95 fee for 5000 shares or less, a penny a share more above that.

INTERNET WORLD STATS

http://www.internetworldstats.com/stats2.htm

Up-to-the-minute data on Internet users internationally.

EARTHCAM

http://www.earthcam.com

The locations of nearly every live camera on the Internet. Spy on a classroom, watch people work, check out a live skyline.

RAND MCNALLY

http://www.randmcnally.com/home/

Good travel site enhanced by a searchable index of major road construction projects all over the United States.

AMERICAN TELECOMMUTING ASSOCIATION

http://www.knowledgetree.com/ata.html

TELECOMMUTING JOBS WEB PAGE

HTTP://WWW.TJOBS.COM/

ing's Cross
OXFORD CIRCUS STATION

CHAPTER

10

CITIES AND URBAN ECONOMIES

OBJECTIVES

- To explore the relationship between urban growth and capitalist development
- To analyze how cities are linked together through their economic bases and export sectors
- To describe how the supply and demand for housing is related to space and population in cities
- To summarize the causes and consequences of suburbanization and urban sprawl
- To address the reasons, costs, and benefits of gentrification
- To illustrate the reasons for inner city poverty and the multiple problems of the ghetto
- To discuss global cities in light of the current round of globalization

Oxford Street is a major shopping district in London's West End. London is not only the seat of government, but also it houses the headquarters of 80 percent of major British transnational corporations. It is the center of banking, insurance, publishing, fashions, advertising, and the legal system for Britain. London's influence stretches far beyond the city to all of the British Isles and even northwest Europe.

Cities lie at the heart of economic geography. For thousands of years, and particularly since the emergence of industrial capitalism, cities have played a uniquely important economic, political, and social role. A city is many things: It is a built environment—a tangible expression of religious, political, economic, and social forces that houses a host of activities in proximity to one another. Cities consist of dense webs of social relationships that are fundamental to the play of economic ties and the reproduction of labor. Cities are also depositories of cultural meaning, where the symbolic systems that people use to negotiate the world are produced and consumed. Cities, the foundation of modern life, represent humanity's largest and most durable artifact. They are living systems—made, transformed, and experienced by people.

This chapter provides a summary of some major issues pertaining to urbanization and economic geography. It opens with a historical overview of the role of cities in capitalist development. Next, it dwells on the issue of intraurban social and spatial organization, including the residential location decision, the filtering process of housing, how population densities are distributed, and the influential concentric ring model. Third, it explores the dynamics and consequences of suburbanization and urban sprawl, noting the reasons that underlie it and the impacts for metropolitan cores and peripheries. Then it turns to gentrification and the resurgence of certain inner-city cores. The chapter also addresses the persistent problems of the inner city, including the crisis of the African American ghetto. Finally, it concludes with some observations about global cities and the international urban hierarchy.

Cities in Historical Perspective

The vast bulk of human existence was spent hunting and gathering, as small nomadic groups without settled communities. In this type of society, without agriculture, the efficiency of caloric production was insufficient to sustain the dense populations found in cities. Cities in any recognizable form have only existed for the last 5 percent (or less) of the time people have been around. Thus, cities are in that light relatively new. It is useful to situate urban areas in their historic context by examining cities at previous eras in time.

The First Cities

The first cities emerged in the Mesopotamian area of the Middle East about 7000 years ago during the Neolithic Revolution, which saw the invention or discovery of agriculture, metalworking, writing, the state, and the first class societies (Figure 10.1). Thus cities were part of a new social and spatial division of labor, largely based on slavery, in which the production of an agricultural surplus enabled a relatively small minority to live in urban areas to become bureaucrats, warriors, scribes, priests, and craft workers. Cities also developed early in the Nile Valley (about 3000 B.C.); in the Indus Valley in southern Asia (by 2500 B.C.); in the Huang He River Valley of China (by 2000 B.C.); in eastern, western, and southern Africa at various times in the first millennium A.D.; and in Central America and Peru (by A.D. 500). The raison d'être of cities from the start was to exchange goods and services with surrounding communities, although many were also important religious centers. As urbanization spread from its ancient hearths, it was incorporated into the cultures of various regions.

In Europe, urban life began more than 2000 years ago. Apart from their own city-states, the ancient Greeks were responsible for the foundation of other Mediterranean cities such as Naples and Marseilles. The Roman Empire generated an integrated division of labor across the Mediterranean and western Europe. At its height in the second century A.D., Rome had more than a million people. The Romans also founded other cities such as Naples, Florence, Paris, Lyon, Bordeaux, Lisbon, Toledo, London, York, Vienna, Budapest, Belgrade, and Lubljana.

The collapse of the Roman Empire in the fifth century A.D. ushered in a period of urban decline throughout Europe. From the fall of the Roman Empire to the early modern period, cities in Europe grew slowly or not at all. They ceased to be important during the period loosely referred to as the Dark Ages, a time when long-distance trade and rural-urban interaction drastically declined. Feudalism, as we saw in Chapter 2, was primarily an agricultural-based society centered on local lords, their manors, and networks of serfs and peasants. City size was correspondingly small. Not until the first signs of capitalism in the fourteenth and fifteenth centuries, and the trade systems that it ushered in, did city size begin to revive.

Before A.D. 1500, Europe was a mere waystation in a world system that included major interlocking subsystems of central places stretching from the Mediterranean region to China. These subsystems were dominated by cities such as Constantinople, Baghdad, Samarkand, and Tashkent on the Silk Road routes connecting the Middle East and Asia. Delhi in India and Hangchow in China were much larger than any city in Europe and played a more central role in the world economy than their European counterparts did.

Prior to the Industrial Revolution, market exchange was an appendage to the redistributive economy of the feudal society, in which the state, ruled by the aristocracy, enforced the extraction of surplus value from the peasantry. Under feudalism, European cities were usually extensions of the aristocracies that governed medieval

FIGURE 10.1
Location of the world's urban hearth areas. Cities began in several places around the world in the aftermath of the Neolithic Revolution. The invention of agriculture generated a social surplus and more refined division of labor that sustained a small class of people who could live off of the labor of others.

societies. For example, Venice, Italy, was the city of the leaders known as *doges*. Located at the seaward margins of the marshy Po Delta, Venice became one of medieval Europe's most important centers of manufacturing and long-distance trade, as well as an incipient center of capitalism. The dominant economic institutions of Venice, Florence, and other European towns were the *guilds*—craft, professional, and trade associations.

The Rise of the Modern City

Cities revived from the sixteenth century onward under the emerging capitalist economy. A network of new towns spread across the continent, fueled by population growth and rising productivity. A new merchant class developed, and the revolution in the countryside reduced the peasant class; in cities, capitalist labor markets (i.e., the commodification of labor time) established the working class. The accumulation of capital, the growth of new social classes, new trade networks, the use of inexpensive labor and raw materials in the colonies, and scientific and technological breakthroughs destroyed the feudal barriers to production. The capitalist city reflected the lower transportation and communication costs for firms that needed to interact with one another; hence, most commercial and industrial enterprises concentrated in and around the most accessible part of the city.

A canal in the merchant city of Bruges, Belgium. This Hanseatic town was the north European counterpart to Venice during the mercantile period.

With the Industrial Revolution came a steady penetration of market relations throughout European societies. The industrialization of agriculture displaced thousands of rural workers, resulting in waves of rural-to-urban migration. In Britain, home of the Industrial Revolution, industrialization saw the rapid growth of the manufacturing centers in the Midlands as well as London, Glasgow, and Belfast in Northern Ireland (Figure 10.2).

The rise of capitalism was accompanied by colonialism and the European conquest of much of the world. In the process, European colonization threatened the urban civilizations of Asia, Africa, and the Americas. Often, land-based cities in Asia, Africa, and the Americas were undermined by the growth of colonial ports (e.g., Lima, Peru; Rangoon, Burma; Jakarta, Indonesia). Centuries of European penetration and occupation resulted in the growth of many cities that owe their origins to colonial foundations or to trading requirements, such as Calcutta, Hong Kong, and Singapore. From the sixteenth to the nineteenth centuries, *colonial cities* dominated the urban patterns of Africa, Asia, and Latin America. Political independence and the development of the new international division of labor saw in colonized countries a transformation of the urban process as profound as that in nineteenth-century Europe and North America. Thus, the construction of a global, maritime-based world economy, mercantilism, and the growth of port cities in Latin America, Asia, and Africa can be seen as different facets of one process (Figure 10.3).

The nineteenth century saw waves of urban growth sweep through Europe and North America. In the United States, the emerging Industrial Revolution saw

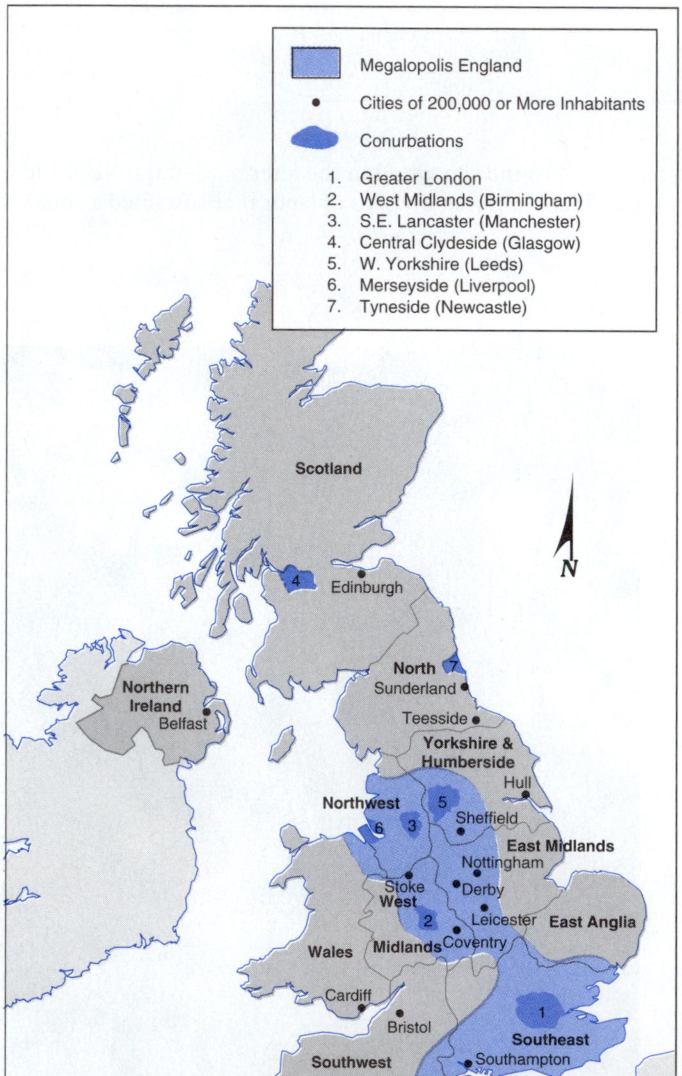

FIGURE 10.2
Urban Britain, home of the Industrial Revolution. Britain's early lead in industrialization created a network of cities that made it the most urbanized country in the world in the nineteenth century.

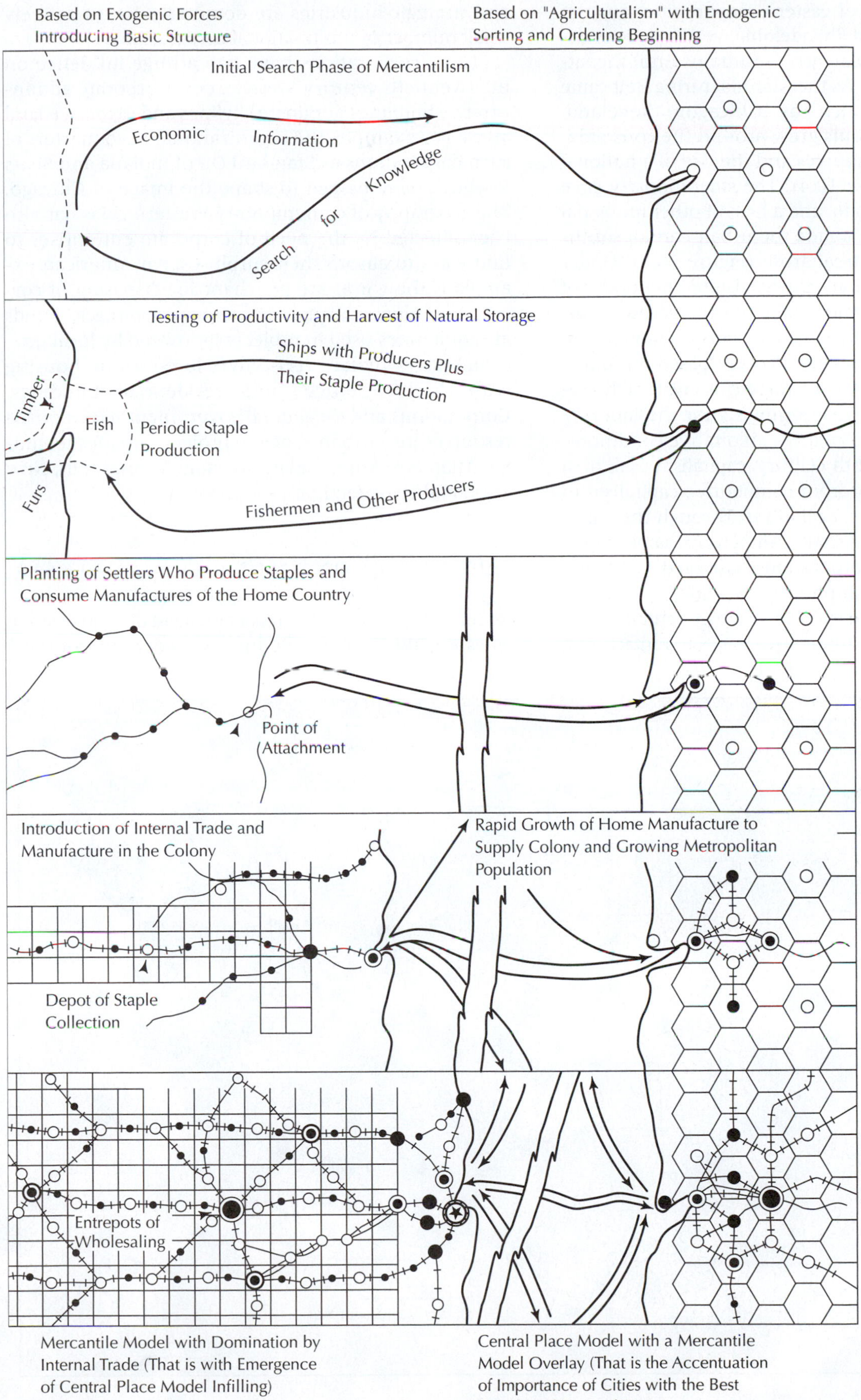

Based on Exogenic Forces Introducing Basic Structure

Based on "Agriculturalism" with Endogenic Sorting and Ordering Beginning

Initial Search Phase of Mercantilism

Economic Information Search for Knowledge

Testing of Productivity and Harvest of Natural Storage

Ships with Producers Plus Their Staple Production

Timber

Fish

Periodic Staple Production

Furs

Fishermen and Other Producers

Planting of Settlers Who Produce Staples and Consume Manufactures of the Home Country

Point of Attachment

Introduction of Internal Trade and Manufacture in the Colony

Rapid Growth of Home Manufacture to Supply Colony and Growing Metropolitan Population

Depot of Staple Collection

Entrepots of Wholesaling

Mercantile Model with Domination by Internal Trade (That is with Emergence of Central Place Model Infilling)

Central Place Model with a Mercantile Model Overlay (That is the Accentuation of Importance of Cities with the Best Developed External Ties)

FIGURE 10.3 Urban evolution under colonial mercantilism.

the explosive growth of eastern seaport cities such as New York, Boston, and Philadelphia; very few cities existed west of the Appalachian mountains until the latter half of the century. As the Manufacturing Belt came into being, cities such as Chicago, Cleveland, Milwaukee, and St. Louis grew around the consolidation of agricultural resources and the rise of a national railroad system (Figure 10.4). The steel industry gave rise to Pittsburgh, Buffalo, and a host of other industrial centers. Similarly, in Canada the urban core of southern Ontario and Montreal arose (Figure 10.5). Under this regime of production, cities registered high rates of industrial innovation and prodigious increases in productive power, fueled by large numbers of immigrants. Their standards of achievement were based on industry and technology. However, these cities were ugly creations and horrifying environments for the laboring poor, who toiled under highly exploitative conditions.

In the late nineteenth century, capitalism took on a different form, shifting from competitive capitalism to monopoly, or corporate, capitalism. Through the elimination or absorption of small competitors, large industrial and financial corporations emerged, rescaling production and consumption on a national scale. In today's economically developed countries (and in many underdeveloped ones), important areas of manufacturing and strategic industries are dominated by a relatively small number of multinational corporations.

Large corporations have had a huge influence on the twentieth-century Western city. Corporate administrative buildings dominate skylines and extensive land areas. For example, the organizational headquarters of such corporations as Standard Oil of Indiana and Sears Roebuck have helped to shape the image of Chicago. The geography of contemporary Western cities has also been affected by the need of corporate enterprises to find ways to absorb their surpluses. An American example is the corporate penchant for disposing of corporate surpluses through urban renewal projects. Funds are sometimes used in projects sponsored by local governments to replace run-down, low-income housing with luxury office and residential buildings. Corporations and the federal government have poured resources into urban renewal projects in cities including Atlanta, Boston, Dallas, Houston, Minneapolis, New York, and Philadelphia.

URBAN ECONOMIC BASE ANALYSIS

Because they are part of a broader capitalist division of labor, cities are always integrated into a wider environment

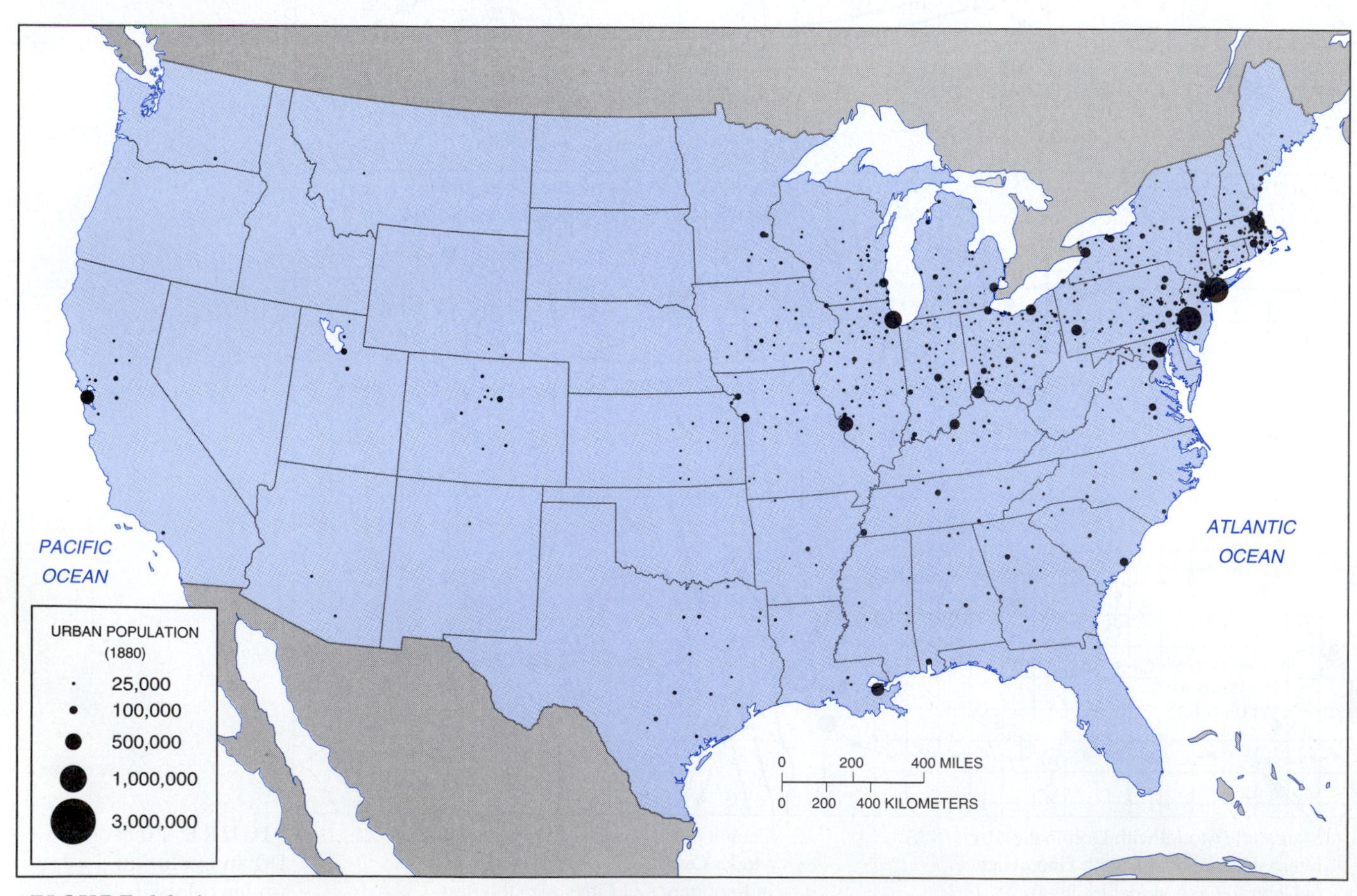

FIGURE 10.4
Urban population of the United States, 1880, just as the Industrial Revolution was creating the Manufacturing Belt.

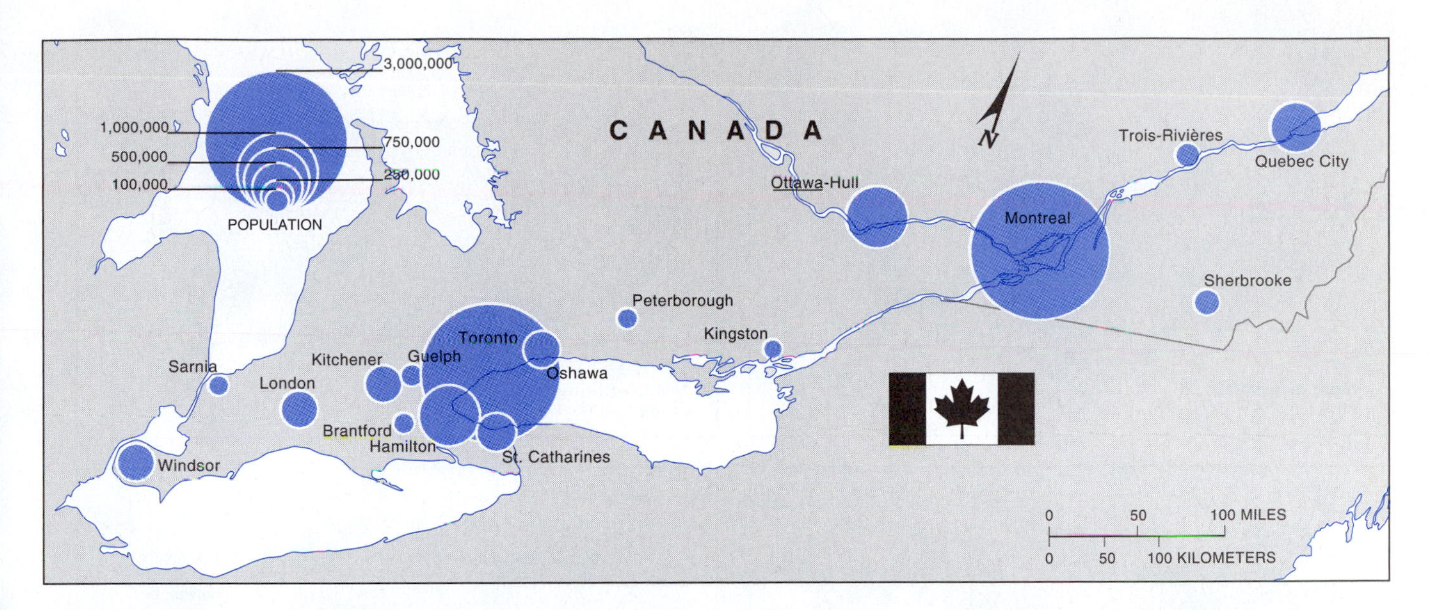

FIGURE 10.5
Canada's urban core.

that both reflects and shapes their structure, location, and behavior in time and space. Linkages among cities are largely linkages among firms, which buy and sell parts, goods, and services to one another. Firms thus simultaneously constitute and create geographies, that is, they are shaped by and in turn shape their local, regional, and global surroundings. Economic base analysis (sometimes also called export base theory) offers a means of understanding how cities are integrated with one another.

An important part of the geography of capitalism is the tendency for cities and regions to specialize in the production of some outputs and not others (i.e., develop a comparative advantage; see Chapter 12). U.S. history is replete with numerous examples of urban specialization in the past and present. For example, New York has long been known as a garment-producing center and the capital of finance; Detroit arose around automobile production; Akron, Ohio, was once the epicenter of rubber manufacturing; Pittsburgh, Buffalo, and Youngstown were renowned as centers of the steel industry; Minneapolis originated as a flour-milling center; Corning, New York, has been an important glass-producing city; Cincinnati was known as "porkopolis"; Memphis was the cotton seed oil capital of the Mississippi; and Seattle has long been closely tied to aerospace through the Boeing Corporation. Often urban specialization takes the form of small towns that arise in connection with agricultural processing, timber and lumber, or mining centers. Some contemporary examples of specialization are depicted in Figure 10.6.

The *economic base* (or export base) of a city is the part of its economy that links it to markets in other regions and countries. Thus, economic base analysis has important ties to the theory of comparative advantage (Chapter 12). The economic base exports a city's or region's products to a wider market, selling its output to clients at competitive prices and deriving revenues in return. It is important to note that *export* in this context does not necessarily mean foreign exports. Cities that sell their output to clients in other parts of the same country are, from the perspective of the producer, effectively exporting their output. For example, Dalton, Georgia, exports carpets to every state in the nation, and New York exports financial and advertising service to clients across the United States. This approach to the urban economy views the economic base as the engine that drives the remainder of local economic activity. The economic base is thus vital to the health of a city's or region's economy.

A common myth about cities is that they only export agricultural and manufactured goods. Often, planners and academics view services as only produced and consumed locally. Yet this conception is mistaken. As we saw in Chapter 8, services are actively bought and sold on an interregional and international basis. Given that the vast bulk of the labor force is involved in services, it would be astonishing if they were *not* traded among cities. For example, New York sells financial services; Pittsburgh, Birmingham, Alabama, and Gainesville, Florida, export health care services to clients in other cities and states; and Los Angeles exports television and film services. University towns export educational services to students who move there from other places to attend school. Whenever a radio station in one city sells advertising time to a client located in another, it exports a service. When tourists descend on Disneyworld, New Orleans, Las Vegas, or San Francisco, those places export services. Even Washington, D.C. (as well as state capitals) exports government services in the sense that the costs of producing these (as well as the benefits) are

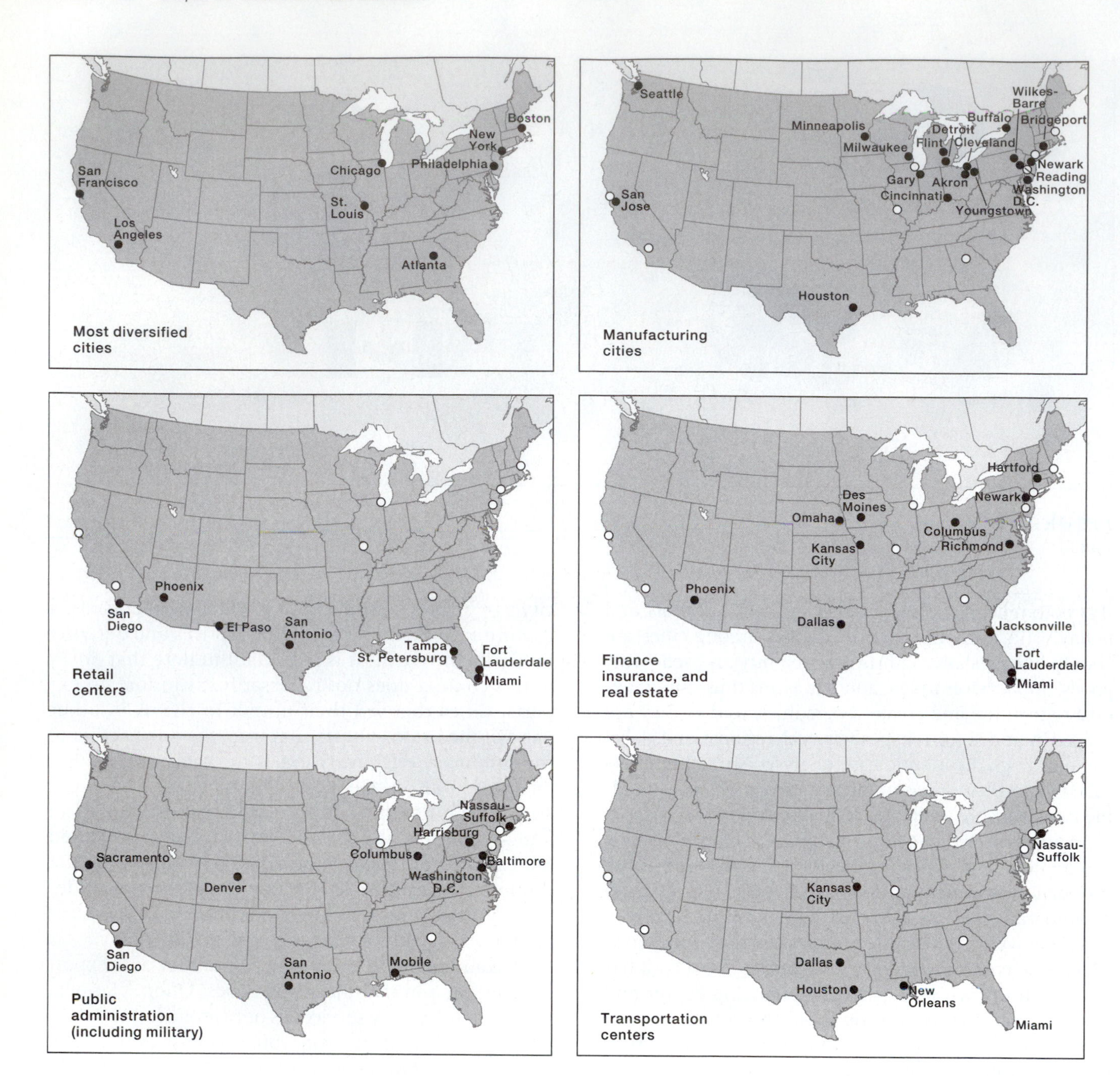

FIGURE 10.6
Specialization within the U.S. urban hierarchy.

paid for (and enjoyed) by people and firms located elsewhere. Thus, cities certainly can and do have service industries as an export base.

It is useful, therefore, to classify the urban economy into two broad segments, the economic base, or basic sector (B), and the nonbasic sector (NB). If the basic sector is export oriented (selling agricultural, manufactured goods, or services), the nonbasic sector recycles corporate and personal incomes in ways that meet the demands of firms and households locally. Nonbasic economic sectors include retail trade, eating and drinking establishments, real estate, and personal services. Essentially, these functions rely on local earnings and cater to local demand. From the standpoint of economic base analysis, therefore, the nonbasic sector occupies a less important role in shaping the urban economy than does the economic base. Of course, sometimes the division between these two sectors is not so simple. Some cities, such as New York or outlet malls, for example, may export retail services, and some sectors, such as legal services, may be both basic and nonbasic (e.g., a law firm that handles corporate mergers in another city and local divorces).

The relative sizes and relations between the economic base and the nonbasic sector are important to understand. We can understand these relations using a simple economic

Stamford Town Center, Connecticut. The shopping malls and retail stores here exemplify the urban nonbasic sector, which recycles incomes in the local economy to meet the needs of households and firms.

base model, in which total employment (T) equals the sum of the basic (B) and nonbasic (NB) sectors, or

$$T = B + NB.$$

Now let us introduce the concept of a *multiplier, m.* The multiplier in its simplest form reflects the degree to which the basic sector "drives" the total economy. There are many types of multipliers and ways of estimating them, but for the sake of simplicity, let us define the multiplier (m) as the ratio of total to basic sector employment,

$$m = T/B.$$

Substituting ($B + NB$) for T in the preceding equation yields

$$m = (B + NB)/B = 1 + NB/B.$$

(Note this is the simplest possible definition of an average multiplier; more sophisticated approaches use marginal multipliers, which are calculated slightly differently.) This approach is useful because it allows us to estimate the degree to which changes in the basic sector (ΔB) cause changes in the total economy (ΔT), that is,

$$\Delta T = m \Delta B.$$

Using employment, let us consider a simple example. Say that a small lumber town has 1000 people in the labor force. (Total population will be higher, depending on the labor force participation rate and demographic structure of the community.) Of that 1000, 400 work for a local paper mill. If the paper mill hires an additional 100 people, what will be the impact on the community? We can answer this question simply enough by defining $B = 400$; therefore, $NB = 600$. Thus, the paper mill's multiplier

$$m = 1 + 600/400 = 2.5.$$

The change in total employment that an additional 100 in the basic sector generates is thus

$$\Delta T = m \Delta B = 2.5 \cdot 100 = 250.$$

The additional 100 jobs in the paper mill created 250 jobs in the community (including the 100 new jobs in the basic sector). Note that since the multiplier is always greater than zero, changes in the basic sector always create somewhat larger changes in the rest of the local economy; that is, since $m > 0$, $\Delta T > \Delta B$.

How did this process operate? The economic base model holds that when the basic sector expands (or contracts), those initial changes reverberate through the local economy in a series of interfirm linkages, as well as changes in consumer spending. The total changes can be decomposed into three constituent parts. First, changes in total employment include changes in the basic sector itself, or direct effects. Second, changes in total employment include changes in firms that sell goods and services to the export base through subcontracts, or indirect effects. For example, when the paper mill in the preceding example expands, it may increase its purchases of equipment, repair and maintenance services, office machinery, advertising services, legal services, and other inputs. Thus, the size of a multiplier—and the impacts of the basic sector on the rest of the urban economy—is closely tied to the linkages among firms. Every firm purchases inputs from other firms, that is, every purchase is a sale, depending on whether you take the perspective of the buyer or the seller. Firms that supply the paper mill with inputs will enjoy increases in sales as the paper mill buys more inputs from them. These are the indirect effects.

The size of the multiplier in many ways reflects the spatial distribution of backward linkages from the basic sector, that is, the geography of subcontracts. To the degree that the basic sector firm subcontracts with suppliers located far away, many of the benefits of its growth will be exported as well as "leakages." Conversely, the more the firms in the basic sector subcontract locally, the higher the local multiplier effect will be. Sometimes public subcontracts (e.g., for construction) mandate the use of local employers for this reason.

Finally, the third part of total changes resulting from multiplier effects are changes in consumer spending resulting from changes in the basic sector. When workers' incomes (in the exporting firm and its subcontractors) go up, they have more to spend. These expenditures, or induced effects, will occur primarily in the nonbasic part of the economy. Thus, higher wages will lead to greater savings rates (depending on workers' propensity to save or spend), as well as higher expenditures on real estate, food, entertainment, transportation, and other consumer goods and services. The size of induced effects will depend on the number of workers affected, their average salaries, and their spending habits and preferences.

It is important to note that multipliers are double-edged razors, that is, they may work against as well as for a region. Changes in the basic sector can be negative as well as positive, such as when a steel plant or auto assembly plant shuts down. Thus, if $\Delta B < 0$, then $m \Delta B$ will be < 0. For example, if a pulp mill closes down, or if a tool and die factory reduces its labor force (perhaps as part of a strategy of industrial reorganization and relocation overseas), then the multiplier effects will take hold, amplifying the initial losses. Subcontracts to supplying firms dry up, reducing the indirect effects. Lower workers' incomes in turn lead to negative induced effects, causing local businesses to suffer and real estate prices to decline. For this reason, the downtowns of many deindustrialized cities often have storefronts that have been closed and housing that is relatively cheap.

Finally, the diversity of the local economy is important. Typically, the basic sector declines, and the nonbasic sector increases as cities rise in population, that is, from the lower to the upper tiers of the urban hierarchy (Figure 10.7). This pattern reflects the fact that large cities usually have a diversified economic base. Los Angeles, for example, has an economic base in the film industry, as well as finance, the port, and aerospace. Conversely, small towns often have very narrow economic bases centered on one industry, and often only one firm, such as agricultural processing centers, mining towns, and lumber-dependent cities low in the urban hierarchy.

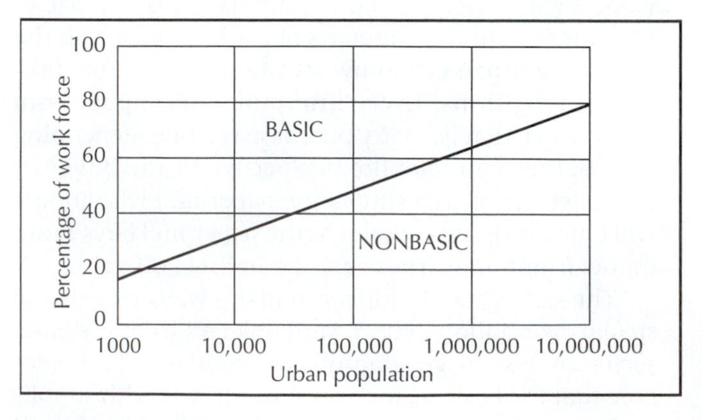

FIGURE 10.7
The relative sizes of the basic and nonbasic sectors vary greatly across the urban hierarchy. Generally, the economies of small towns are more specialized around the export of a single good, including those that rely on mining, agriculture, and lumbering; hence their basic sectors are relatively large. In large metropolitan areas, in contrast, diverse economies and large internal markets reduce the relative proportion of basic sectors.

INTRAURBAN SPATIAL ORGANIZATION

In addition to the hierarchy among cities and their interrelations, there exists a definite spatial ordering of economic and social activities within cities. This ordering is reflected in the nonrandom nature of urban land use and the activities of the people who inhabit cities, including their residential locations and employment. Four dimensions of intraurban spatial structure are addressed here: the residential location decision, the filtering process of housing, the population density gradient, and the concentric ring model.

The Residential Location Decision

Typically, the most important criterion when people select a home is accessibility to where they work in the city. The choice of a residential location depends in part on how much money a family can afford to spend. For purposes of our discussion, we assume a single-centered city; that is, a city with only one center of employment—the CBD (central business district).

First, consider patterns of residential land use and the cost of commuting to work. A family's budget must account for living costs, housing costs, and transportation costs. Poor families, who have relatively little money to spend on commuting after living and housing expenses are deducted from their income, have sharply negative bid-rent curves. They attach much importance to living close to where they work. The only way the inner-city poor can afford to live on high-rent land is to consume less space. On the other hand, relatively wealthier families have plenty of money to spend on transportation; therefore, the proximity of residential sites to their places of employment is of little consequence to them. They can trade access for larger housing lots farther away from the center of the city. This segregation process generates the familiar, ironic pattern in which the relatively prosperous live in the cheaper land in the suburbs and the poor are concentrated in the inner city.

Although cost, time, and mode of travel to work have important implications for residential location, many other dimensions to accessibility, such as nearness to services, and general supply and demand factors must be considered. These and other factors influence bid prices. In every situation, however, the rich can outbid the poor.

The results of the competitive-bidding process in market societies are always relatively advantageous to the rich and relatively disadvantageous to the poor.

The Filtering Model of Housing

The geography of housing markets is fundamental to the ordering of residential space in urban areas. For most people, housing is acquired through units that have been previously owned or occupied by someone else. The *filtering model* describes the process by which houses pass from one household to another. The model is based on the principle that there is a decrease in the housing quality through time, through physical deterioration, deferred maintenance, technological obsolescence of fixtures, facilities and design, and changes in housing construction fashion. The filtering model suggests that houses are occupied by families with progressively lower incomes as the house quality declines over time. The dwelling becomes affordable to households that require progressively lower quantities of housing services because of their more limited incomes (Figure 10.8). The filtering model describes why low-income households occupy used rather than new housing. This model also shows that poor households benefit from zoning and land use policies that encourage the building of new high-end housing throughout the city but especially in the suburbs. The low-income households benefit from new construction policies because medium-quality housing declines in value, allowing some of the lower-income households to move up. Housing filters down from the medium-quality submarket to the low-quality submarket, decreasing the price of low-quality housing. Naturally, the escalation of housing costs can reduce the positive benefits of filtering for medium- and low-income families.

Key demand factors in the housing market are the number of people and household size; the price of housing, including interest rates; inflation-adjusted, or "real," income per household, which reflects the number of wage earners and the dynamics of the labor market; and, in the case of owner-occupied housing, the home buyers' expectations of the future change in home prices (*speculation*). Important supply factors are the availability of items needed in building new housing units (land, labor, and construction equipment and materials) and construction standards or building and zoning codes. Availability of financing may be viewed both as a demand factor (long-term mortgages) or as a supply factor (construction loans). In both supply and demand, the price of housing is critical; given that most people must take out mortgages to buy their homes, interest rates are a key variable to the behavior of housing markets.

As an example of the effect of a demand factor, consider an increase in the number of households in an area as a result of in-migration. If vacant housing of the desired kind is limited, the price of existing housing will be bid up. This will induce new construction and create new demand for land, labor, and raw materials. If these items are abundant, their prices should remain constant. However, if supplies are limited, their prices will be bid up. This increase in builder cost will lead to even further increases in housing prices, with the amount depending on how willing home buyers are to pay the higher prices. Thus, in places with rapidly growing populations (e.g., Florida), housing prices rise quickly. Booming urban economies, typically those that

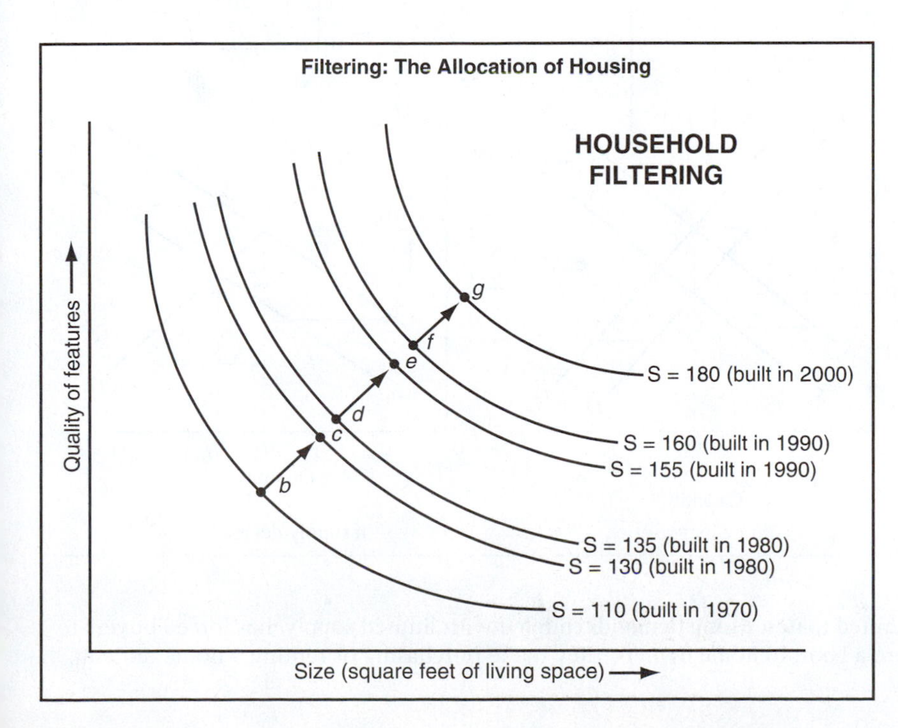

FIGURE 10.8
The filtering model of housing. As incomes rise, either over time or across socioeconomic groups, the ability to purchase more space at a higher quality also rises. Lower-income groups tend to rely on housing that has been previously occupied. Thus, the population filters up through the housing stock, or conversely, the housing supply filters downward through the population.

are globally oriented with a concentration in producer services, will likewise generate explosive increases in rents and housing prices. Increases in income and a heightened expectation of future increases in home prices will have similar effects.

On the supply side, consider a case of vacant land where the local authorities suddenly change the zoning from residential to nonresidential. The reduction in available supply will raise land prices and, hence, builder costs. The ability of builders to pass on the higher costs depends on the willingness of buyers to pay them. The more willing the buyers are to pay, the higher the demand pressures remain on land, labor, and so forth, and the more limited these items are relative to the demand, the greater the upward pressure on home prices. Similar responses occur for reduced vacancies; higher labor, equipment, and material prices; and stricter building codes.

Better access to amenities—the CBD, jobs, shopping, schools, and high-rent districts—adds to the willingness of buyers to pay higher prices and to overall demand. The three graphs at the bottom of Figure 10.9 describe how supply and demand curves shift upward, thus raising the price of housing. At the bottom left, the demand curve shifts to the right, raising the price of housing, whereas the supply curve remains fixed. The graph at the lower right describes a situation in which the supply curve shifts upward to the left, raising the price of housing, whereas the demand curve stays fixed. When these processes occur simultaneously, the center graph describes an even larger increase in price. Figure 10.10 shows average U.S. home prices in 2002.

Population Density Gradients

The labor and housing markets in cities serve to concentrate people to different degrees in different parts of the city. As distance increases away from the CBD, *population density gradients* of the city decrease (Figure 10.11). Decline in density with distance from the center is understandable because both land value and access decline outward from

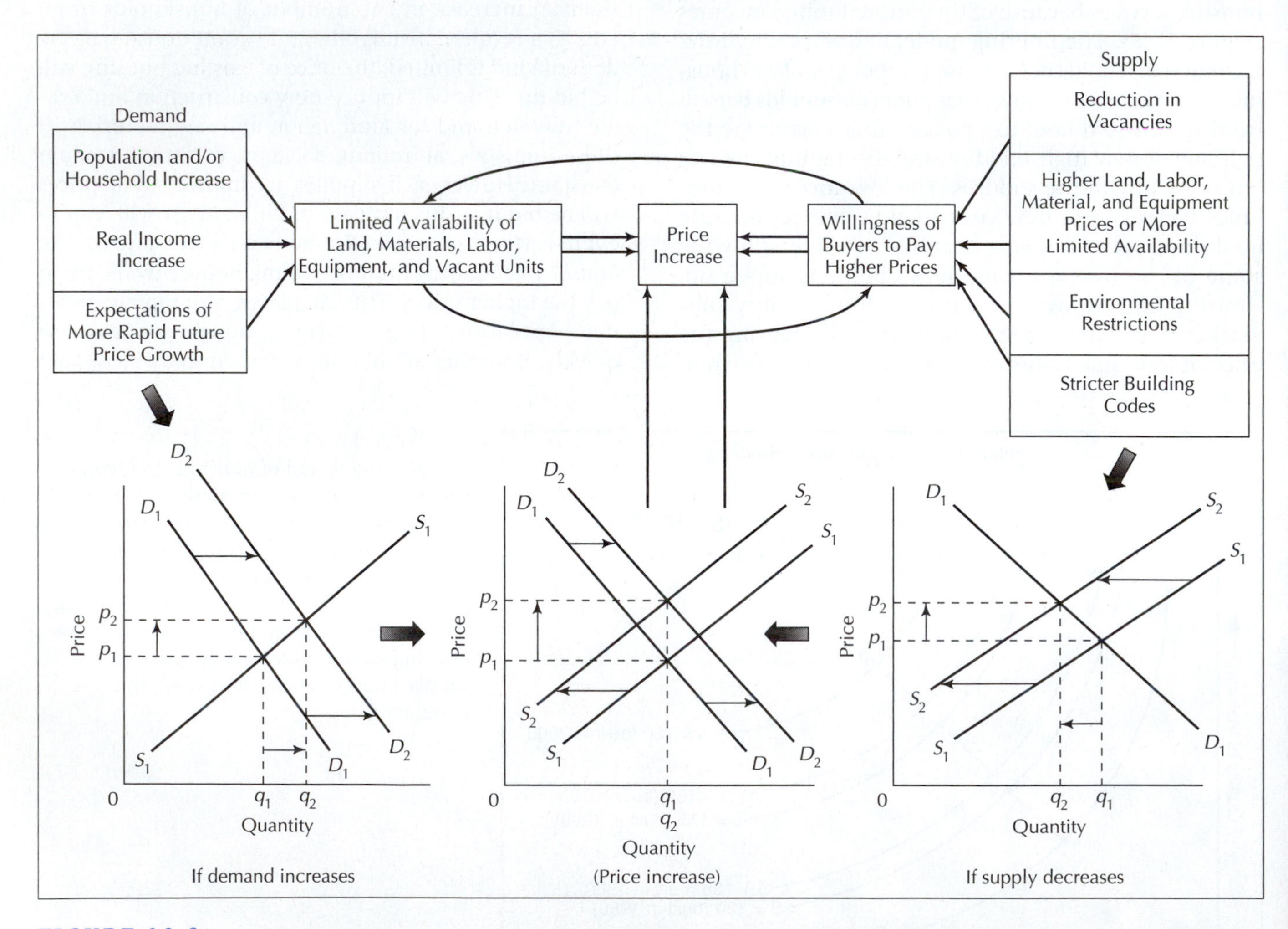

FIGURE 10.9

The dynamics of rising housing prices in the United States. Rising demand, coupled with limited supply, has forced buyers to pay ever higher prices. While such increases are a boon to home owners, they make purchasing or renting a home difficult, particularly for low-income residents.

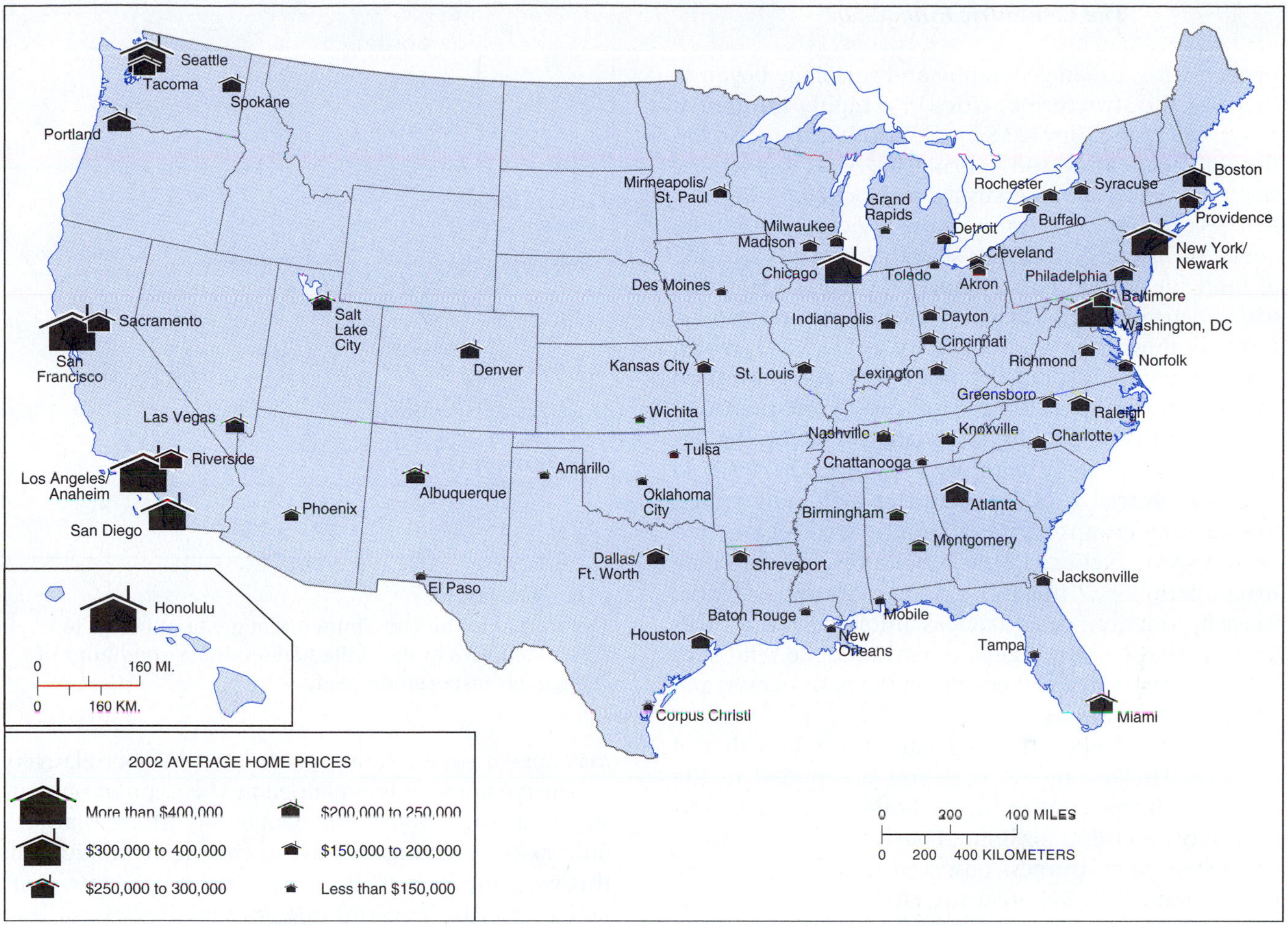

FIGURE 10.10
Housing affordability in the United States, 2002. Cities of the northeastern megalopolis and the Pacific Coasts in California and Hawaii show the highest prices.

the city whereas the amount of space increases. The exception to this role is the tendency for a hollow crater to be formed at the center of the city as land values at this location reflect commercial activities rather than housing. Many urbanites trade off higher commuting costs for lower-priced land on the edge of the city. Through time and with improved transportation access via freeways, population densities flatten as the move to the suburbs intensifies. Subsidiary centers compete with downtown for workers, customers, and business. Densities in the downtown area decline while agricultural and open land on the city's fringe converts to housing.

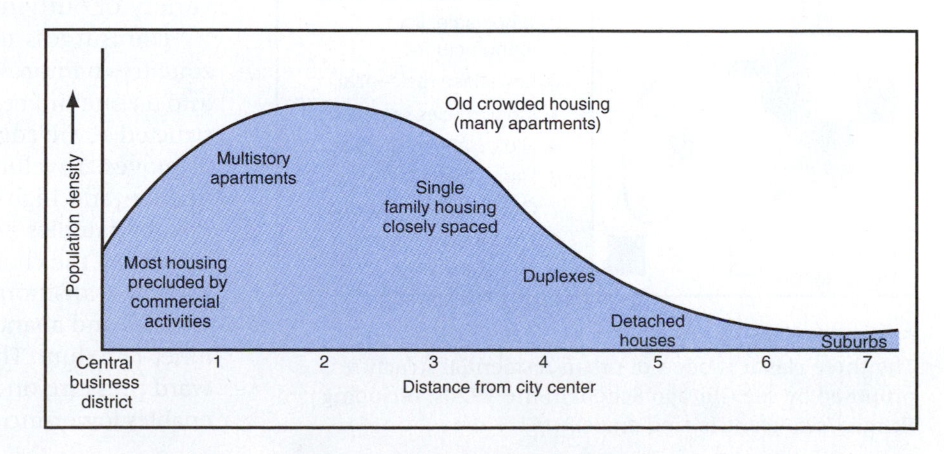

FIGURE 10.11
Population density gradients for typical U.S. city. Urban cores are typically densest, offering maximum accessibility of people to one another. Because the supply of land increases rapidly with distance away from the CBD, the population density declines in negative exponential fashion.

The Concentric Ring Model

In the 1920s, urban geography and sociology began to theorize the structure of cities in a highly influential group called the Chicago School. Within this constellation of academics and journalists, who essentially started modern urban analysis, the sociologist Earnest Burgess developed a famous and widely influential *concentric-ring model* of urban structure and growth, one of three highly influential models of urban social structure (Figure 10.12). This view emphasizes centripetal forces that focus economic activity on the CBD, which was the dominant center of urban spatial organization in the industrial city. Burgess suggested a sequence of zones from center to periphery that reflected the decreasing cost of space with distance from the CBD, so that commercial uses occupied the high-rent center, low-income groups clustered around that region, and progressively higher income populations inhabited the urban periphery. Thus Burgess charted the residential space of American cities that was unfolding in the 1920s as industrial labor markets, the automobile, and suburbanization segregated people on the basis of class and ability to afford housing.

In some respects this approach resembles that of the von Thünen model (Chapter 6) applied to the urban landscape (Figure 10.13); both approaches focus on the competition for limited parcels of land of varying accessibility. Burgess observed that lower-income groups tended to live near the city center; while this may appear to be a paradox—the poor live on relatively expensive land—it is explained by the population density gradient. Low-income people can afford relatively little space, meaning they live in crowded conditions; in this way, the land market maximizes aggregate rents and profits by constraining low-income users into densely packed communities. (The implications for human welfare and health are important as well.) Thus, the dynamics of land markets—operating through the price of housing—explain why in American cities low-income immigrant and minority neighborhoods tend to be near the city center. These observations must be put in the context of urban labor markets, transport systems, and government policy. This approach can be extended to include multiple urban nodes (Figure 10.14), in which the urban rent gradient does not simply decline with distance from the central business districts but rises and falls around a variety of outlying centers.

The Burgess model suggests that residents of one zone try to improve their situation by moving outward into a zone of better housing units. New housing constructed at the edge of the city triggers a complex chain of moves. Dwellings vacated by the out-migration of middle- and high-income families are filled by lower-income families moving from the next inner zone. At the end of the chain, the working poor move out of the zone of transition, leaving behind the least fortunate families and abandoned housing units. The result is an inner-city slum. This *filtering process,* which exerts downward pressure on rents and prices of existing housing, enables lower-income families to obtain better housing.

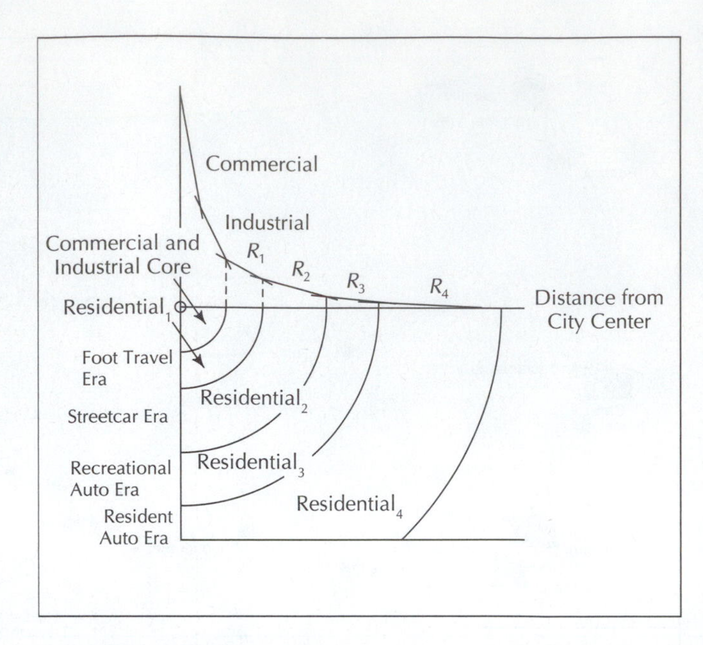

FIGURE 10.13
Application of the von Thünen rent gradient model to urban land uses in the United States under conditions of changing transportation regimes.

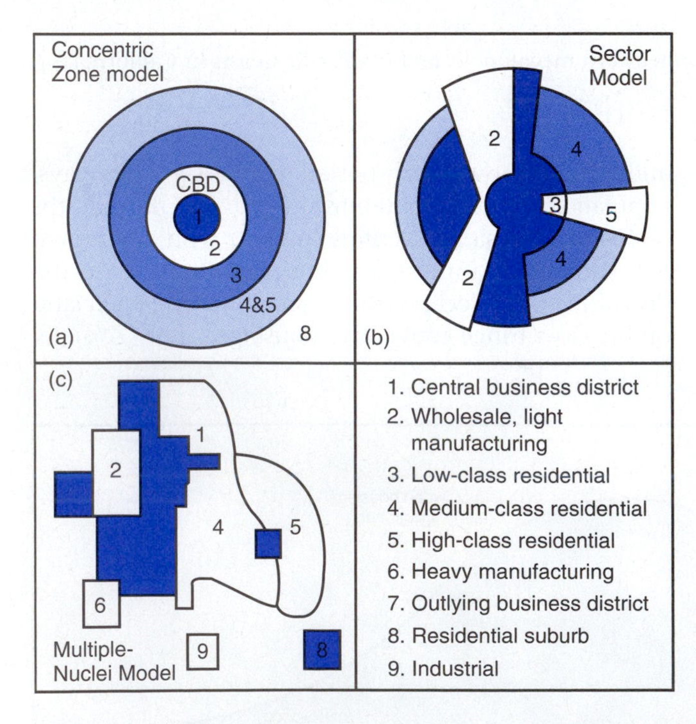

FIGURE 10.12
The three classic models of urban residential structure proposed by the Chicago School in the 1920s, including Burgess's concentric-ring model.

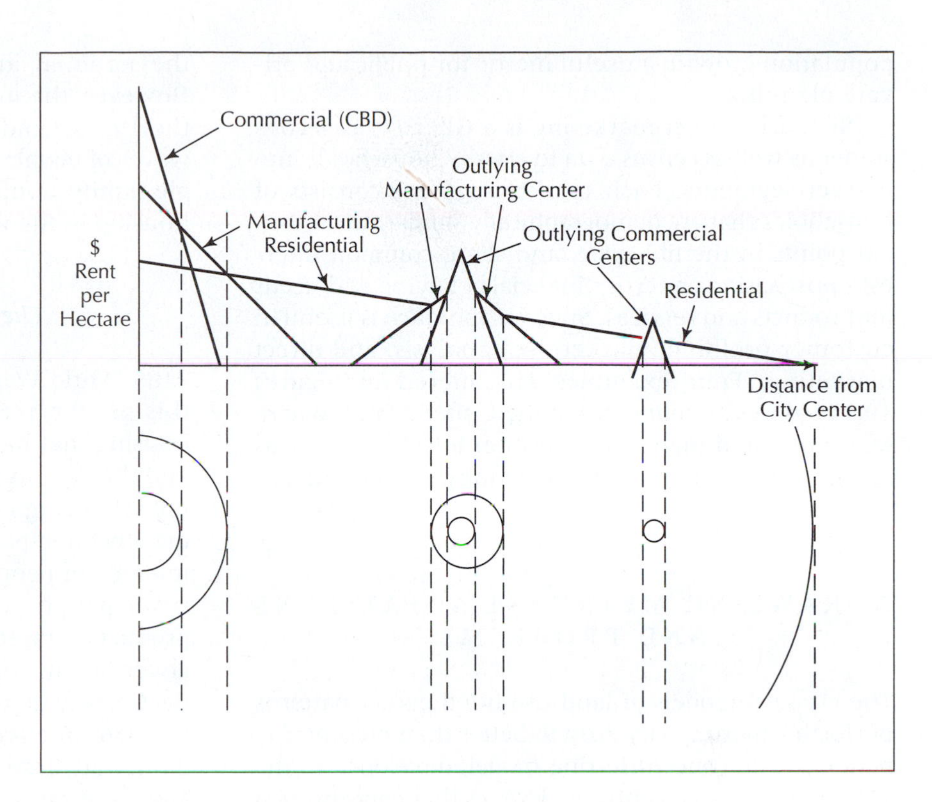

FIGURE 10.14
Land-use patterns resulting from multiple urban nodes, either distinct centers with overlapping hinterlands or regional nodes within a large metropolis.

The major reason the filter-down process occurs is that the poor, with the strongest latent demand for housing, are the least able to afford new housing. In contrast, the rich can most easily afford to move into new housing and leave their old homes to others. The demand for high-income housing is generally elastic—a new demand generates a quick response from the private housing industry.

The Sector and Multiple-Nuclei Models

Two other models attempted to make urban land use models more realistic. The *sector model* takes into account differences in accessibility and, therefore, in land values along transport lines radiating outward from the city center. According to Hoyt's model, a city grows largely in wedges that radiate from the central business district. One wedge may contain high-rent residential; another, low-rent residential; and still another, industrial. Hoyt believed the contrasts in activity along various sectors usually became apparent early in the city's history and continued to be marked as the city grew.

The concentric and sector models describe single-centered cities. However, most modern cities have *multiple nuclei:* a downtown with satellite centers on the periphery. In 1945, Harris and Ullman described a model city that develops zones of land use around discrete centers. Their model of urban structure encompasses five areas: (1) the central business district, (2) a wholesaling and light-manufacturing area near interurban transport facilities, (3) a heavy industrial district near the present or past edge of the city, (4) residential districts, and (5) outlying dormitory suburbs.

Harris and Ullman recognized that the number and location of differentiated districts depend on the size of the city and its overall structure and peculiarities of historical evolution. They also gave reasons for the development of separate land-use cells. The pattern of multiple cells might result from specialized requirements of particular activities, repulsion of some activities by others, differential rent-paying ability of activities, and the tendency of certain activities to group together to increase profit from cohesion.

Applied Urban Geography

Urban geographers use theoretical models of urban structure, change, and land use in a variety of applied career settings. Geographers work for zoning boards, utility companies, real estate developers, large retail chains, and marketing firms to assess current and future patterns of urban land use. Using geographic information systems (GIS), geographers can construct and implement sophisticated models of urban land use, including population structure and change, transport routes, water and energy usage, zoning codes, labor markets, public services, housing, and consumer behavior. Such models attempt to replicate the forces that shape cities over time, including economic and demographic changes that are reflected in the supply and demand for housing, schools, and transportation. Careful calibration of such models allows for reliable projections and simulations of future population growth, a useful metric for public and private planning.

Similarly, target marketing is a GIS that uses consumer as well as census data to classify households into market segments. Each market segment consists of households that are demographically similar, are at similar points in the life cycle, and share common interests, purchasing patterns, financial behavior, and needs for products and services. Such an approach is useful in customer profiling, market area analysis, and direct marketing. Four examples are offered in Figures 10.15a–d in different cities of the United States, where sales areas and market penetration for different firms are illustrated. Thus, urban geography has a useful applied dimension.

SPRAWLING METROPOLIS: PATTERNS AND PROBLEMS

The classical models of land use fitted earlier patterns of North American city growth better than present-day patterns. The concentric-ring model, developed in the 1920s, emphasized centripetal forces that concentrated economic activity in the downtown of the inner city. In the second half of the twentieth century, however, centrifugal forces gained the ascendancy in the form of massive suburbanization. As a result of automobile-based dispersal and the decentralization of jobs, the North American city has evolved into a *multicentered metropolis.* Classical models do not realistically accommodate this new urban reality.

The roots of suburbanization actually extend deep into American history. As early as the late eighteenth century, the urban elite constructed country homes on the outskirts of cities. Brooklyn arose in much this fashion outside of Manhattan. Throughout the nineteenth century, those who could escape the filth, crowding, and squalor of the inner city often did, situating themselves along horse-drawn omnibus lines. By the 1880s and 1890s, with the introduction of the electric streetcar, the growing middle class increasingly relocated to the periphery. Precisely during this time, many cities were being filled by waves of impoverished immigrants. Thus, early suburbanization created a spatial bifurcation between American-born and foreign-born populations. In the culture of mass consumption that became hegemonic during this period, movement outward from the inner city became deeply associated with movement upward socially (i.e., suburbs were linked to aspirations of escape from working-class life). It is important therefore to note that suburbanization is not simply produced by changes in transportation technology, important as those may be, but also involved transformations in the urban division of labor, class structure, housing markets, and cities' positions in the national and international urban hierarchies. However, the mass introduction of the automobile in the 1920s rapidly accelerated this process, leading to waves of people settling on the urban fringe as the single family home became the most common type of housing in the nation.

The Postwar Suburban Boom

After World War II, suburbs exploded in size. In part, this growth reflected years of pent-up demand for housing that had accumulated during the Depression and the war, as well as the baby boom that was coming into being in the 1950s. With an economy enjoying low unemployment and rising real incomes, tens of millions of people made the choice to relocate to the urban periphery. Yet suburbanization is not simply the product of a free market but owes much to the role of government, particularly the radial and circumferential freeway network started by the federal government in 1956. In effect, the freeway system eroded the regional advantage of the CBD, making most places along the expressway network just as accessible to the metropolis as the downtown used to be (Figure 10.16). Other factors that reinforced suburban sprawl are (1) low mortgage interest rates, (2) loan guarantees provided under federal housing and veterans' benefit programs, (3) property-tax reductions for owner-occupied homes, (4) cheap transportation, (5) massive highway subsidies, and most of all (6) cheap land. The freeway-dominated automobile era has removed virtually all restrictions on intraurban population mobility so that residential land use is feasible almost anywhere in a metropolitan area, especially with automobile companies buying and closing down intraurban streetcar lines across the country. In Los Angeles, for example, a well-functioning public streetcar system was dismantled in the 1930s, forcing the public to adopt the automobile. As a result of industrial and demographic shifts to the periphery, downtown areas gave way to an ever-widening belt of suburban cities, including new neighborhoods, new business centers and office complexes, and new shopping malls. Far from simply being residential centers, suburbs have become commercial, financial, and retail centers in their own right. More Americans now live, work, play, shop, and dine within the confines of suburbs than in city centers or rural areas (Figure 10.17).

Suburbanization of retailing was a response to the residential flight to the suburbs, new merchandising techniques, and technical obsolescence of older retailing areas. The automobile provided customers with a convenient mode of transportation to shopping places, but downtown parking facilities were scarce and expensive. A need to improve the parking situation and

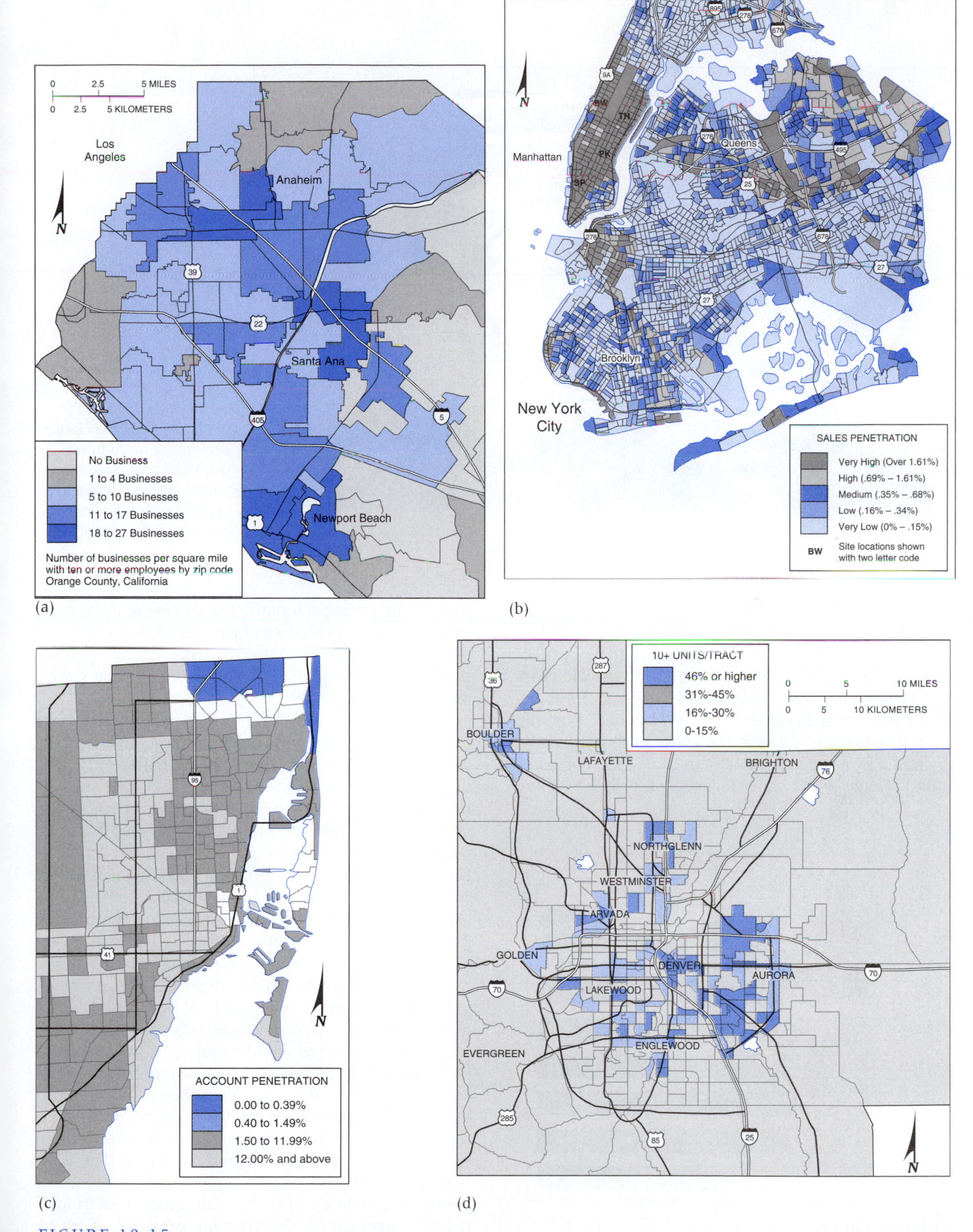

FIGURE 10.15

Applications of applied urban geography: target marketing in four cities: (a) Xerox potential sales in Orange County, California; (b) True Value Hardware sales penetration in New York City; (c) Atlantic Federal Savings in Miami; and (d) Duds and Suds in Denver and Boulder, Colorado.

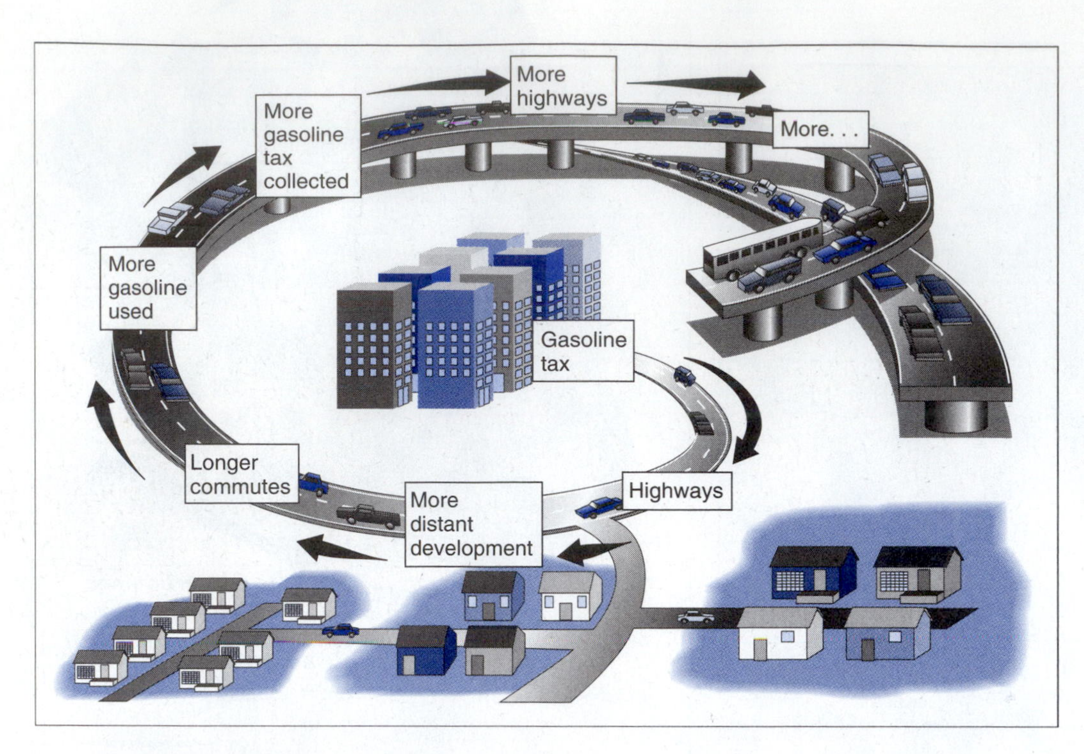

FIGURE 10.16
The development cycle ushered in by the Highway Trust Fund. Because federal funds are used so heavily to subsidize highways, the demand grows ever larger, generating more funds through taxes on gasoline, which perpetuate the cycle of suburban growth and inner-city decline.

to increase profits impelled retailers to the suburbs. The decentralization trend began in the 1920s as stores began spreading out from the downtown along main thoroughfares. Yet it was not until the postwar years that retailers moved to the suburbs in large numbers. First came the strip center, or neighborhood shopping center, consisting of a string of 10 to 30 shops, usually anchored by a supermarket. Then came the larger community center with a small department store or a variety store as the principal tenant. The success of these early centers, which catered to a limited trade area, depended on a main-road location, free parking, and the persuasiveness of "drive-in" everything, self-service stores, and discount outlets.

Neighborhood and, later, larger community shopping centers became vulnerable to more attractive regional shopping centers that appeared after 1955. The newest and biggest of these centers in distant suburbs have several floors, three or more department stores, and scores of specialty shops. Surrounded by huge parking lots, these shopping complexes are usually enclosed so that customers can shop in climate-controlled comfort. Unlike early suburban shopping centers, the giant regional shopping centers are catalysts attracting a variety of activities to the area.

Decentralization of manufacturing began before the turn of the century. Technical advances, such as the development of continuous-material flow systems, induced many manufacturers, especially those engaged in large-scale production of industrial goods, to spread out along suburban railway corridors where land was relatively cheap and abundant. Nonetheless, most manufacturers, despite truck transportation, decided to remain in or near the central city until the 1960s, when two technological breakthroughs occurred. These innovations involved (1) the completion of the urban expressway system and (2) the scale economies in local trucking operations. The freeway network—built by the state, not the market—helped to neutralize the transportation cost differential between inner city and

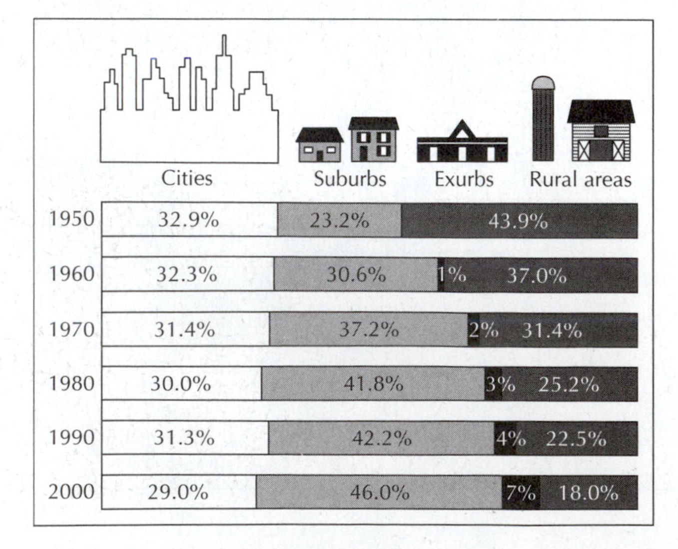

FIGURE 10.17
Proportions of Americans living in central cities, suburbs, exurbs, and rural areas, 1950–2000.

suburb. As the locational pull of central city water and rail terminals declined, most of the remaining urbanization economies of downtown were nullified.

Expansion of offices into the suburbs began in the early post–World War II years, when large corporations began looking for new headquarters on the urban periphery. For example, General Foods, IBM, Reader's Digest, Union Carbide, and ESSO-Standard Oil left New York for the suburbs, where rents are lower and access to transportation much easier than in congested downtowns. The major factor in office-site selection typically is accessibility to an expressway. This trend of the large corporations prompted an avalanche of similar moves by a host of smaller firms precisely when the growth of business services was generating considerable demand for new office space. Suburban office parks became a mainstay of employment growth on the metropolitan fringe.

Without doubt, the defining feature of the *edge city* is the huge regional shopping center. Super shopping malls are catalysts for other commercial, industrial, recreational, and cultural facilities. The result is the emergence of miniature downtowns called edge cities. In many metropolitan areas, edge cities are unplanned, loosely organized, multifunctional nodes, and they are strongly shaping the geography of suburbia. Local and regional trends combine with the general shift of the U.S. population and economic activity from the Northeastern and Midwestern Snowbelt to the Southern and Western Sunbelt, where U.S. metropolitan areas have been growing particularly rapidly (Figure 10.18).

Suburbanization had huge consequences for people and landscapes on both the urban periphery and the center. At the edges of metropolitan areas, the steady conversion of agricultural land to residential and commercial property has eroded one of the nation's prime resources, farmland. Because many cities were established on river floodplains as agricultural centers, urban sprawl eats up the choicest farmland with the best soils. Building a road and water infrastructure in relatively low-density areas such as suburbia is also expensive. Many suburbs are faced with

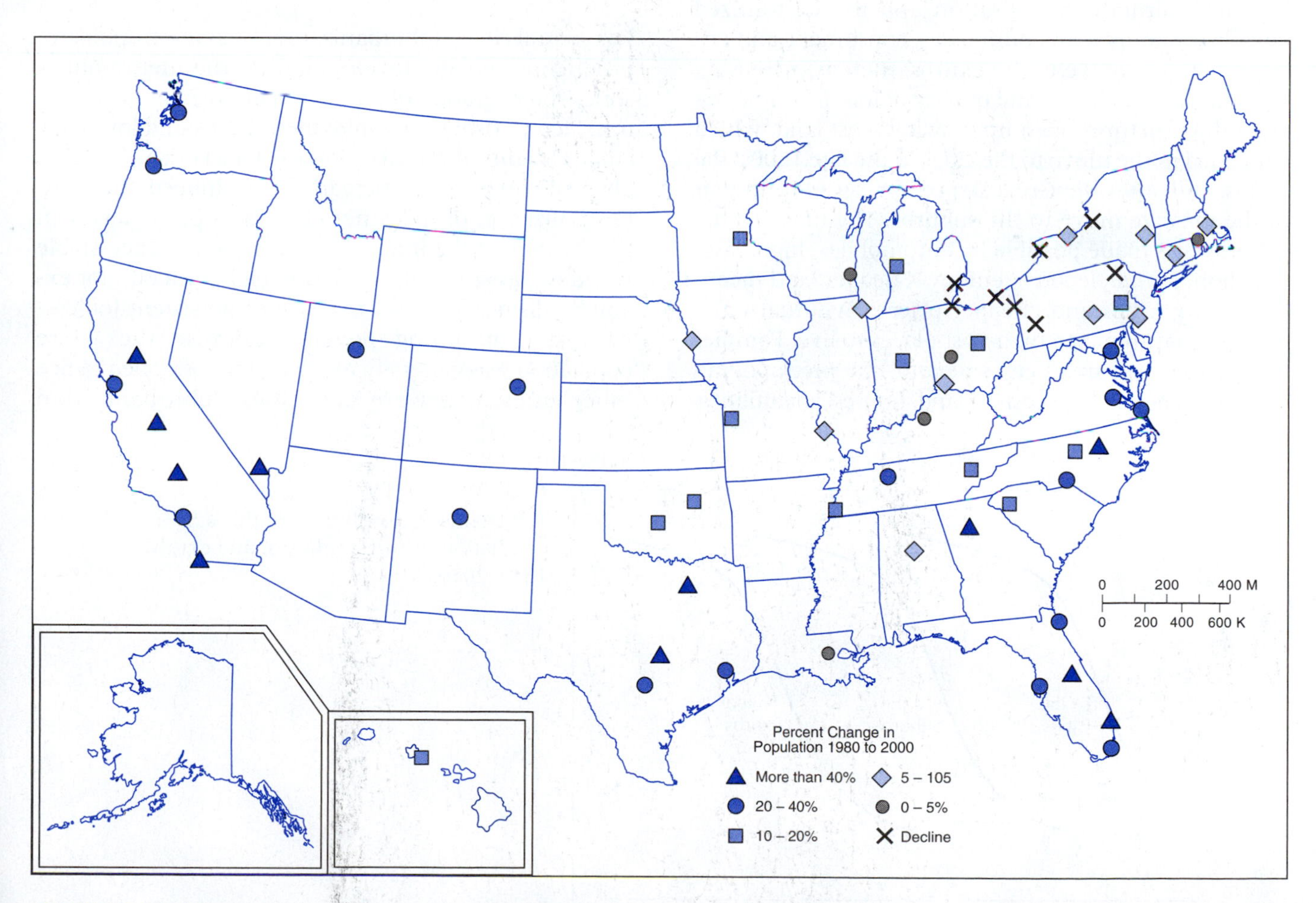

FIGURE 10.18
Metropolitan growth and decline in the United States, 1980–2000.

persistent traffic problems, access to water, and housing costs increasingly out of the range of their residents.

Out in the Exurbs

Beyond the suburbs, Americans have been pulled in increasing numbers to the outermost edges of metropolitan areas, the exurbs (Figure 10.19). Affordable housing for most workers is not being provided in adequate quantities near growing suburban employment centers of edge cities, commuting trips are becoming longer, and pollution levels from automobile emissions continue to increase. Rapidly growing communities, *exurbs,* lack urban amenities and are "boomtowns" in remote areas for only one reason—access to affordable housing. The trade-off for people living in the exurbs is the stress of the long commute and the strain it places on family relations. This problem has now spread to most metropolitan areas of the country. Long work commutes have fueled the clamor for growth management in which the jobs/housing balance and traffic congestion play important roles.

The information revolution and the customized flexible economy allow businesses, consumers, and entrepreneurs to move to the exurbs, thereby increasing their standard of living and quality of life. Technologies and infrastructures open up new low-cost land, which happened in the move to the cities in the late 1800s, the 1920s, and, after the Great Depression and World War II, the massive move to the suburbs. The move to the suburbs was made possible by automobiles, highways, telephones, and electrical energy. Cheaper land meant larger properties and cheaper prices. Eventually the suburbs matured into high-cost places to live. Families were looking to small cities beyond the metropolitan area to escape the high costs of land, houses, and utilities, and high taxes, congestion, and crime. Thus, exurban growth is the latest chapter in a long series of moves from the urban center to the periphery.

Typically, exurbs are 50 to 150 miles beyond the metropolitan fringe. They include small resort towns, recreationally oriented towns that have lower costs and high appeal, perhaps with a university or with potential for business infrastructure. Businesses will relocate to the exurbs or small towns to reduce their costs and to become more competitive, provided that they can still reach their markets. Such businesses that move to smaller towns will receive much higher average growth and less competition at the outset. Retirees, of course, as well as baby boomers, also flock to the exurbs. The result is an urban geography of the United States dominated by sprawling conurbations that encompass thousands of square miles and millions of people (Figure 10.20). The effect of this shift on central cities and suburbs can be disastrous as it depletes the tax base and purchasing power of the metropolitan core.

Suburbanization and Inner-City Decline

For central cities, suburbanization was catastrophic, reproducing uneven development at the metropolitan scale. The migration of jobs, particularly manufacturing ones, led to rising unemployment levels and lower incomes in the inner city (Figure 10.21). This process affected minorities particularly badly. Indeed, the deindustrialization of the inner city was a prime factor in the creation of the inner city ghetto. Many once-stable working class African American neighborhoods, for example, disintegrated in the face of persistent joblessness and mounting poverty. Because the white population was generally much better positioned to flee inner cities than were minorities, suburbanization

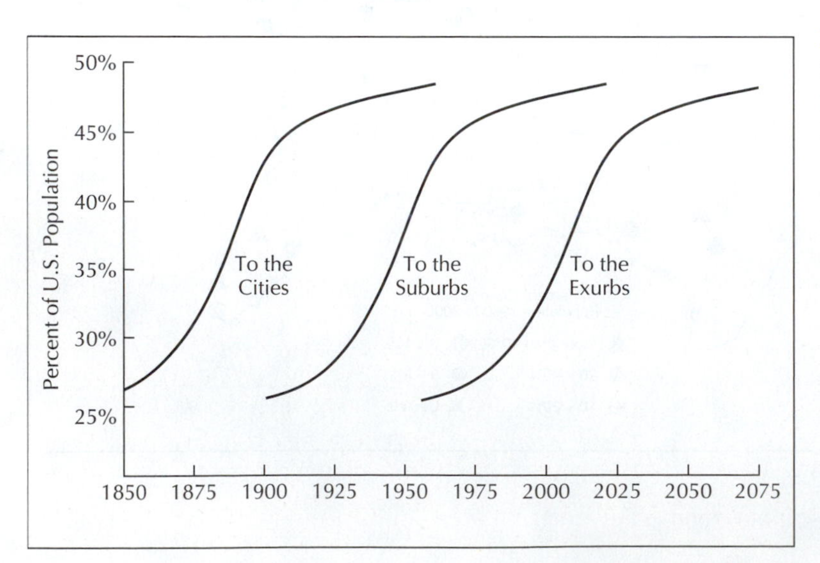

FIGURE 10.19
The changing distribution of the U.S. population in cities, suburbs, and exurbs, 1850–2050.

Cheap transportation, particularly through the automobile, has fostered waves of decentralization of jobs and people to the urban periphery. American urban form is thus locked into access to low-cost gasoline. Urban sprawl contributes to long commuting times and rapid rural-to-urban land conversion as local farmland is converted into housing developments and shopping centers.

contributed to a marked increase in racial segregation within metropolitan areas. Declining property, income, and sales tax revenues made it increasingly difficult for central city governments to offer adequate public services (particularly education) precisely when the demand for them rose steadily. Suburbs typically are incorporated as independent municipalities to avoid paying property taxes to central cities, a process that creates a metropolitan political geography in which large inner cites are surrounded by archipelagoes of small suburbs on the fringes. Declining populations also erode the political representation of inner-city governments in state and federal circles and their capacity to acquire funds for housing or transportation. In short, suburban expansion and inner-city decline may be seen as two sides of the same coin.

Gentrification

During the 1980s and 1990s, just as the collapse of the inner city was widely perceived as irreversible, a resurgence of growth began in many downtown areas, a process known as *gentrification*. Explanations of gentrification point to changes in the growth of producer services that accompanied the late twentieth century round of globalization. Producer services often rely heavily on agglomeration economies and cluster in centralized locations, that is, downtowns. Just as suburbanization had its basis in the changing urban division of labor after World War II (if not before), so too did gentrification reflect the reworking of cities in the face of a globalized, service-based economy. The explosive growth of producer services—finance, legal services, accounting, advertising, engineering, sales, and so forth—created large numbers of well-paid positions demanding university educations. These jobs were typically filled by young urban professionals—"yuppies"—who brought with them considerable disposable income and purchasing power. Often the initial wave of gentrification, consisting of gays and "urban homesteaders" in relatively inexpensive lofts, was succeeded by couples employed in professional services and living in luxury condominiums or brownstones. Finally, commercial gentrification followed in the form of publicly subsidized office and hotel complexes, waterfront developments, convention centers, and sports stadia. Gentrification is thus not just a market-driven phenomenon.

Other explanations center on the changing demographic structure of families, particularly the growth of two-income families, and residential preferences, such as the love of the diversity and historic architecture of inner-city areas. The rise of the baby boom involved smaller families, two-income families (including "DINKS," or "double income, no kids" couples), many of whom had a strong preference for the recreational opportunities of the inner city. Some gentrification involves a "back to the city" movement from the suburbs; the number of families involved was relatively small, however. For every family that moved back to the inner city, eight moved to the suburbs. Residential revitalization in and around the CBD clearly did not spawn a return-to-the-city movement by suburbanites. In fact, most reinvestment was undertaken by those already living in the central city.

From the perspective of many longtime residents of inner-city communities, who often consist of the working class or relatively low-income minorities, gentrification represents a tidal wave of change. Rents

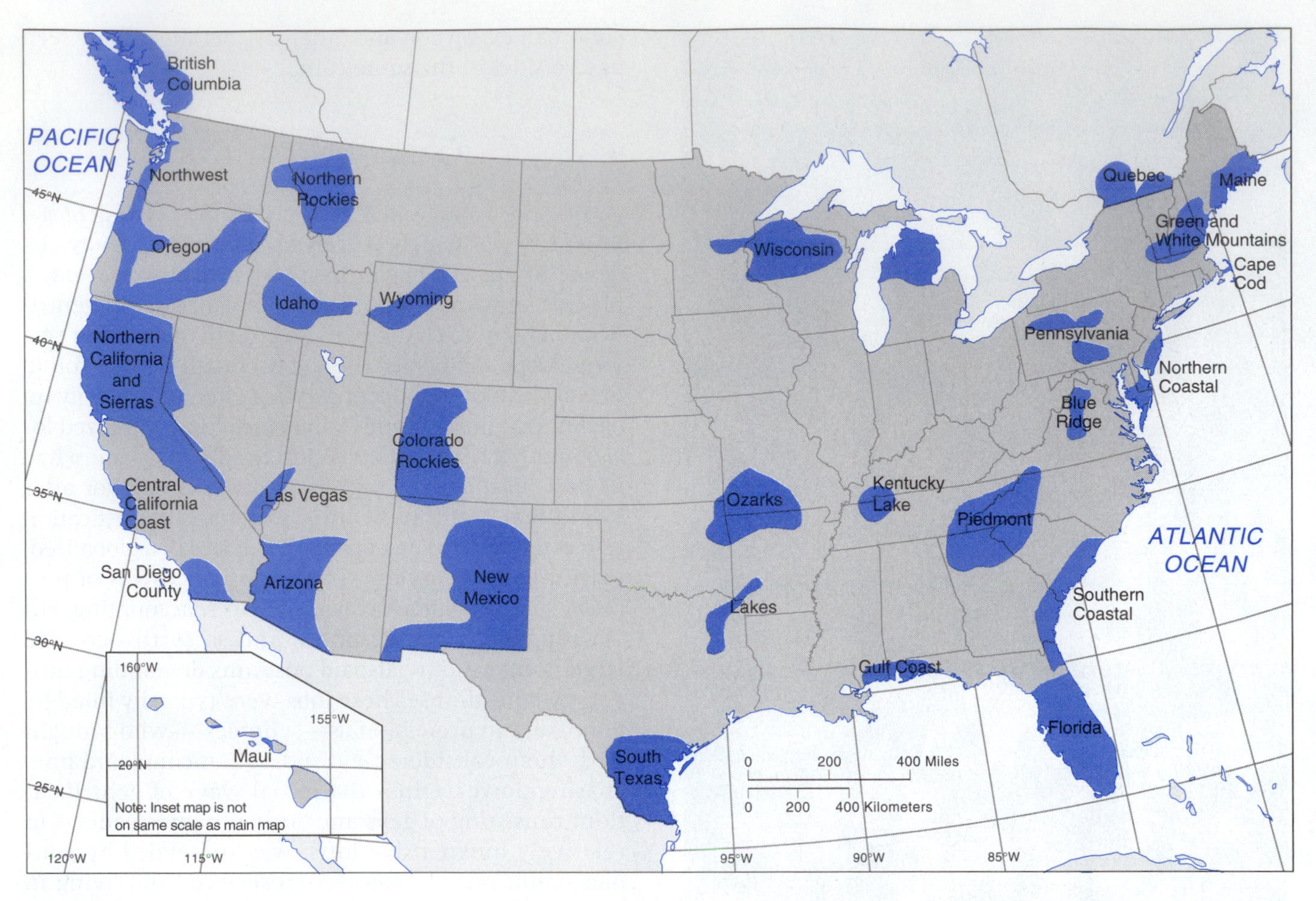

FIGURE 10.20
Exurban sprawl in the United States is producing vast conurbations with millions of people apiece.

and housing prices go up, displacing those who cannot afford them and contributing to urban homelessness. Local community-oriented stores may disappear, unable to compete with expensive trendy boutiques. Often this trend involves the displacement of ethnic minorities by higher income whites. For these reasons, gentrification is often despised and resisted by those who have little to gain and much to lose.

Poverty Rates by Race and Location for 2000

Category	Poverty rate (percent)
All persons	13.5
White	10.7
Black	31.9
Hispanic	28.1
In metropolitan areas	12.7
In central cities	19
In suburban rings	8.7
Outside metro areas	16.3

FIGURE 10.21
Poverty rates in the United States, 2000, by race and location. Black and Hispanic poverty rates are three times that of whites, while central city rates are twice that of suburban areas.

Propelled by the surge in producer services over the last three decades, gentrification occurs in many forms, including residential construction of luxury co-ops and commercial construction of hotels, sports arenas, and waterfront developments. Corporate America reclaimed the inner cities of the United States in the 1990s, generating both new levels of wealth and inequality simultaneously.

Gentrification, then, reflects the reworking of the inner city as the types of capital associated with globalized services reconquered once-dilapidated areas of the inner city. Because it involves the in-migration of people who are very different from local residents in terms of their class and ethnicity, gentrification is simultaneously an economic, political, cultural, and geographic phenomenon.

Problems of the U.S. City

The socially and geographically uneven development of the contemporary metropolis has produced a pattern of suburban sprawl and inner-city decline. From a rare social entity at the beginning of the twentieth century, suburbs have evolved into major growth centers for industrial and commercial investment, and a suburban way of life has been adopted by millions of people. The decentralized metropolis fostered large-scale consumption and prosperity in the past, but it is causing real problems now. In some instances, the urban fringe has pushed out farther than workers are willing or able to commute. Urban sprawl has generated externalities such as uneven development, traffic congestion, pollution, and the irrational use of space, which increasingly impinge on the life of urban residents. Furthermore, recurrent fiscal crises threaten to bankrupt central cities.

In the United States, the development of metropolitan-wide governance systems to tackle metropolitan-wide problems (e.g., congestion, air pollution) is blocked by the political independence of the suburbs. The suburbs have resulted from the differential ability of some groups (i.e., the predominantly white middle class) to organize and protect their advantages. They are not willing to abdicate clear-cut, short-run benefits for less certain long-run gains. Thus, metropolis-wide planning in the face of a bewildering multitude of rigid and outdated municipal boundaries—1200 of them in the New York metropolitan area alone—is extremely difficult, if not impossible, to implement. Yet without planning, without redrawing areas of municipal authority, the continued profitability and stability of the metropolis and capitalist society are threatened.

Most governments of Western Europe and North America have concentrated, often for decades, on relieving congestion and welfare pressures by demolishing block upon block of old housing, factories, and other buildings. A major problem for local governments has been where to rehouse the people affected by clearance. One solution has been to replace crowded terraced streets with blocks of tall apartments.

Urban Decay

Affluent residents moving to the suburbs begin the process of *exurb migration* and *urban blight.* Local governments' major source of revenue to pay for municipal services—public schools, road maintenance, police and fire protection, garbage collection and disposal, fresh water and sewers, welfare services, libraries and local parks—comes from local property taxes. The local property tax is the yearly proportional assessment according to real estate market value. If property values increase or decrease, property taxes are adjusted accordingly. Most central city governments are separate entities from the governments of the surrounding suburbs.

Therefore, exurban migration causes, through the laws of supply and demand, escalating suburban property values because of the influx of affluent people. Suburban governments enjoy increasing tax revenue with which they can expand local services. However, property values in the central city decline because of declining demand, leading to an eroding tax base. Those who are unable to move to the suburbs are usually lower income and disadvantaged people who are non–property owners requiring public assistance; thus, cities must carry a disproportionate burden of welfare obligations, as well as increasing crime and deterioration of infrastructure due to declining tax revenues (Figure 10.22).

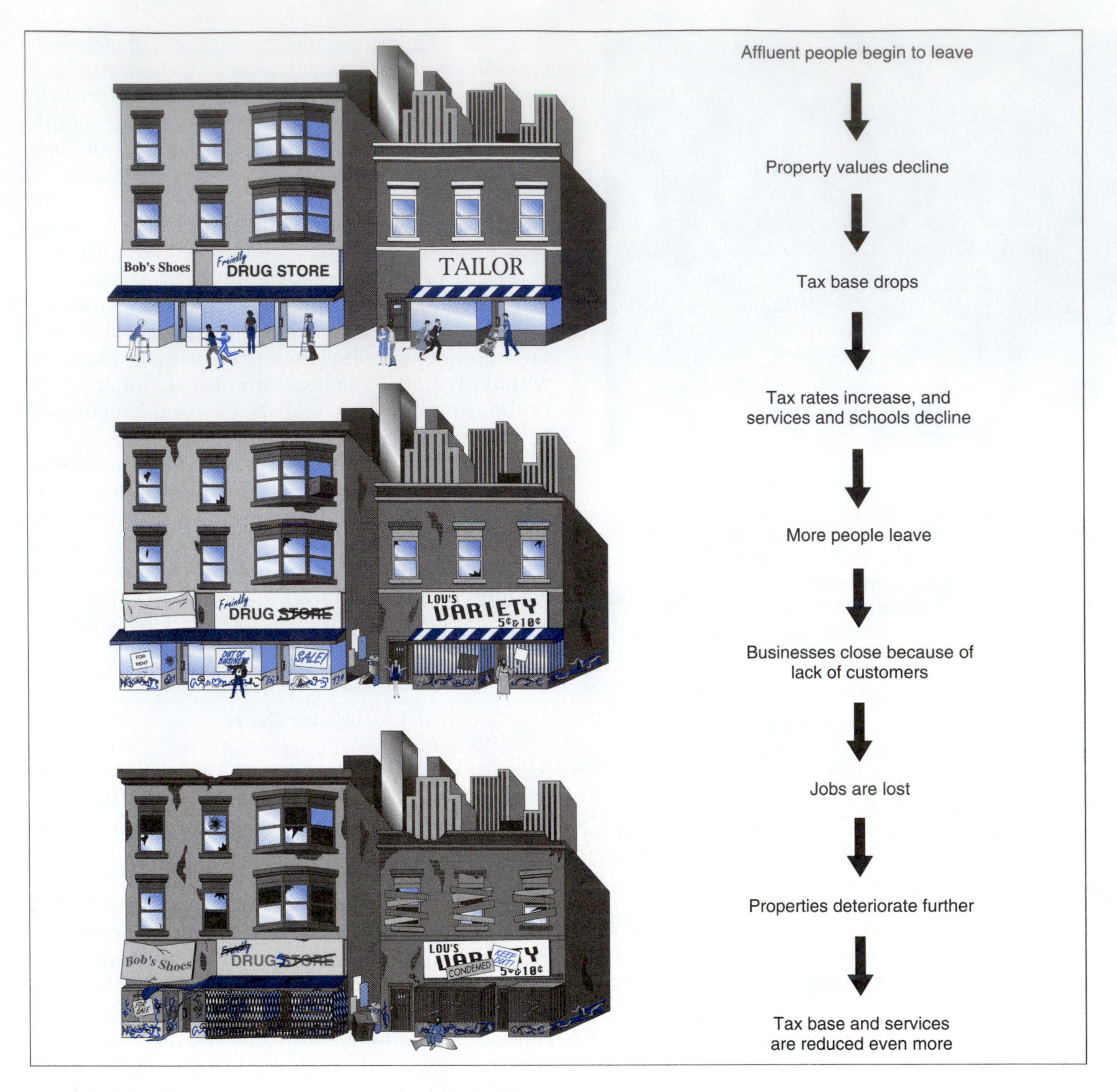

FIGURE 10.22
The cycle of suburban migration and urban decay. People moving to the suburbs from the city create the cycle of urban blight, which continues in most U.S. cities to this day. Most cities are responsible for provision of public schools, maintenance of local roads and facilities, police and fire protection, waste collection, public water and sewer, welfare services, libraries, and local parks. The major source of revenue to pay for these services is property taxes.

The eroding tax base forces local governments to cut local services and raise the property tax rate. Therefore, property taxes on a home in the central city are often several times as high, proportional to assessed valuation, as a home in the suburbs. This difference could be $5000 to $8000 per year in many cities based on an average size home. At the same time, parks, schools, streets, libraries, and garbage collection services deteriorate from neglect, leading to the vicious cycle of exurban migration and urban decay (Table 10.1). This downward spiral of conditions is referred to as *urban decay.*

The Crisis of the Inner-City Ghetto

While the cores of American metropolitan areas have been centers of poverty for a long time, including many immigrant neighborhoods in the late nineteenth century,

Table 10.1

Decline in Population of American Central Cities, 1950–2000

					Population (Thousands)	Percentage change
City	1950	1960	1970	1980	2000	1950–2000
Baltimore, MD	950	939	905	787	651	−32
Boston, MA	801	697	641	563	589	−26
Buffalo, NY	580	533	463	358	292	−50
Cleveland, OH	915	876	751	574	478	−48
Detroit, MI	1850	1670	1514	1203	951	−49
Louisville, KY	369	391	362	298	256	−31
Minneapolis, MN	522	483	434	371	382	−27
Oakland, CA	385	365	362	339	399	03
Philadelphia, PA	2072	2003	1949	1688	1517	−27
St. Louis, MO	857	705	662	453	348	−59
Washington, DC	802	764	757	638	572	−29

the formation of the African American ghetto can be traced back to the 1920s, when several circumstances combined to send millions of workers north. In the South, the mechanization of agriculture reduced job opportunities there; in the North, a steady growth in manufacturing jobs and congressional quotas on immigration restricted the supply of labor, creating relatively high wages. As a result, large numbers of African Americans moved north into cities such as Boston, New York, Baltimore, Philadelphia, Pittsburgh, Cleveland, Gary (Indiana), Chicago, and Milwaukee, taking jobs in railroads, steel mills, meat packing, shipyards, stockyards, machine tools, auto assembly, glass and rubber plants, and other industrial enterprises (Figure 10.23). These jobs, while still paying less than those filled by whites, nonetheless created relatively stable working-class communities with homeowners and low crime rates.

This window of opportunity, however, was very limited in time. Following World War II, inner cities were besieged by waves of economic dislocation. The flight of manufacturing to the suburbs, accompanied by the exodus of the white middle class, created a sustained downturn in the economic fortunes of inner cities. Trapped in unskilled or semiskilled jobs, many minority inhabitants were unable to flee to the suburbs, constrained by their lack of skills in a deteriorating labor market. Neither the low-paying jobs of consumer services (e.g., retail trade) or the well-paying jobs of producer services, which required advanced educations, offered alternatives. Low-wage service jobs often pay the minimum wage and carry few or no health benefits, pensions, or paid holidays; nonetheless, each advertisement often draws large numbers of desperate applicants.

Unemployment rates rose steadily in inner cities from the 1960s onward, accompanied by deepening poverty and a degeneration of the social fabric. Working-class neighborhoods became transformed into impoverished ghettos. Thus, inner-city poverty rates, especially among minorities, are often twice or three times the levels found among whites. African American incomes today are, on average, only two-thirds of those of white families. While the overall U.S. poverty rate in 2000 was 12 percent, it was 7 percent for whites and 26 percent for African Americans and Latinos/Hispanics, revealing how class and racial inequality are deeply intertwined.

One result of the exodus of whites was that minorities became an increasing share of the population of inner-city areas. In 2005, for example, African Americans comprised 78 percent of the population of Detroit, 38 percent of Chicago, and 24 percent of New York. Including Latinos and Asian Americans, minorities frequently comprise over one-half of the populations of large cities. One result has been the growth of minority politicians, including numerous African American mayors.

While many observers point to deindustrialization as the primary cause of the inner-city ghetto, conservative

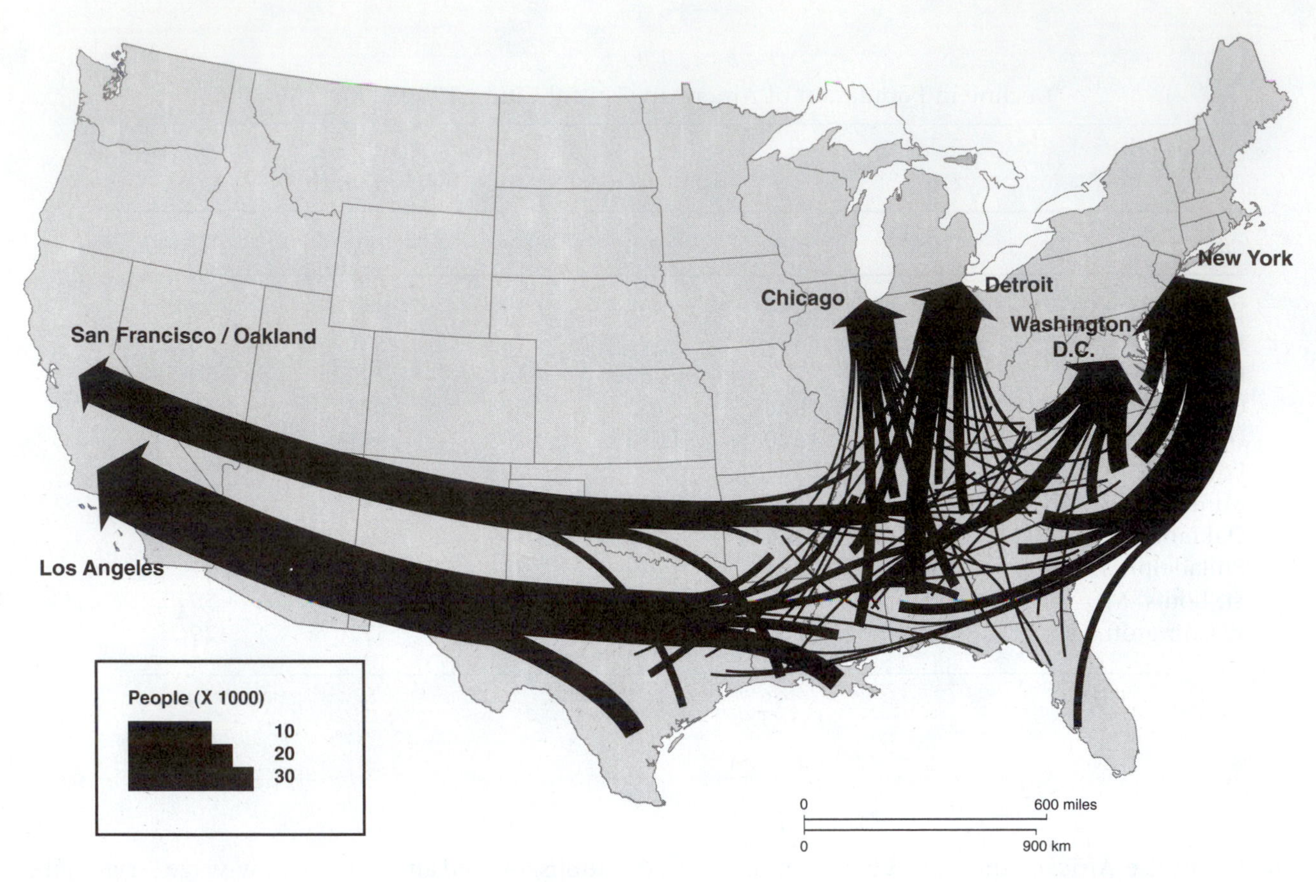

FIGURE 10.23
African American migration streams to the North, 1920s–1940s. Lured by relatively well paying industrial jobs and flight from discrimination in the South, black migrants in the early twentieth century formed the nuclei of many contemporary African American communities in large U.S. cities today.

Abandoned by capital, with high unemployment rates, the inner cores of American metropolises are increasingly divided between a high-wage sector centered around globalization and producer services and a low-wage sector that keeps the unskilled trapped in poverty. These economic conditions are mirrored in the landscapes and lives of the people who inhabit them.

explanations often focus on the role of welfare and public assistance. In this perspective, transfer payments to low-income families (e.g., Aid to Dependent Children, formerly AFDC) allegedly create a disincentive to work. This view then tends to blame the victim, attributing poverty to the behavior of poor people. Such a perspective ignores the macroeconomic context of urban poverty, the decline in available jobs, and the considerable drop in earning power among low-income workers brought on by globalization and technological change. It overlooks the fact that public assistance has declined substantially in postinflationary dollars over the last four decades.

Others note that in the face of high, and rising, unemployment in the late twentieth century, the African American family underwent a severe contraction in its scope and viability. Long a strong institution during the centuries of slavery, in which women played a crucial role, the African American family underwent a profound transformation in the face of urban deindustrialization. Essentially, the supply of employed males declined steadily, leaving relatively few marriage partners for women of the same age cohort. As a result, teenage pregnancy and out-of-wedlock births increased steadily; the poorest, most economically vulnerable families are often headed by women. While many mothers struggle heroically to raise their children in the face of low wages and persistent poverty, children in such an environment are often deprived of the role models of stable, two-income families that have become the norm among white Americans.

As a result of the changes in inner-city labor markets, including poverty, unemployment, low wages, hopelessness and despair, many (but by no means all) residents in low-income inner-city communities may be seen to belong to an urban "underclass." The term itself is the subject of criticism for dehumanizing its members, but the concept nonetheless retains validity, pertaining to unskilled, poorly educated, unemployed, and often unemployable people without access to schooling, skills, or adequate jobs. The perpetuation of the underclass remains one of urban America's greatest challenges today.

In addition to a crisis in employment, inner cities came to suffer other, related problems that lowered the quality of life for their inhabitants. Major problems include the following:

A homeless person on the streets of New York. Rising housing costs, deindustrialization, unemployment, globalization, and reductions in government funding of low-income housing all contribute to the growth of people on the streets. Roughly 1 percent of the population of the United States is homeless today, and the fastest-growing group is women with children.

- A crisis in *housing:* In the 1950s, urban renewal, a federal program to clear space in inner cities for freeways and corporate capital, destroyed millions of housing units in inner cities, contributing to an ongoing housing shortage. Many private landlords disinvested in their communities, investing the profits from their rents in other parts of the city, and on occasion even burning units for the insurance funds. As a result, the physical landscape in many inner-city areas became increasingly decrepit. Inner-city housing markets are unable to provide sufficient housing for the poor. In addition to private disinvestments and lack of new construction (although units for the wealthy continued to be produced), these areas suffered through cutbacks in federal and state housing assistance. Gentrification also exacerbated the problem. The most vulnerable members include the 1 to 2 million Americans without homes, including those caught between jobs that pay too little and housing that costs too much, the long-term unemployed, children recently out of the foster care system, those with drug problems, military veterans, and rising numbers of homeless women.

- The *inadequate funding of inner-city public schools,* which are often plagued by large classes, violence, underpaid and overworked teachers, and buildings that need serious repair, leads many students to drop out, perpetuating the skills mismatch.

- *Health care* for low-income residents in inner cities is typically very poor. Many residents do not have jobs that carry medical insurance as a benefit. Forty million Americans lack health insurance, a population disproportionately concentrated in inner cities. Often, residents must use overcrowded hospitals as their primary source of care, or go without, a factor that contributes to relatively high infant mortality rates and low life expectancy among African Americans. Illegal drug use has compounded this problem, as has the epidemic of AIDS.

- *Crime rates* in inner-city areas are substantially higher than in suburban areas and disproportionately involve young minority males. However, white fears of crime tend to be exaggerated, and the victims are overwhelmingly also minorities. Crime rates are largely determined by the status of the labor market. When jobs are plentiful, crime goes down. Nonetheless, harsh incarceration policies have led to the imprisonment of more than 2 million Americans, and 10 percent of all African American males are either in prison or on probation.

Employment Mismatch

Private and public discrimination (e.g., redlining by lending institutions, discriminatory actions of realtors, screening devices adopted by subdivision developers) and exclusionary zoning practices give the poor, especially minorities, little recourse but to locate in inner-city areas. Meanwhile, most new employment opportunities matched to the work skills of these people are created in the suburbs. This disjunction is known as the *spatial mismatch*. The poor are faced with the problem of either finding work in stagnating industrial areas of the inner city or commuting longer distances to keep up with the dispersing job market. Although *reverse commuting* has increased, barriers abound. These barriers include transportation constraints—such as increased time and cost of the daily journey to work and inadequate public transportation for those without cars—and communication constraints—such as difficulty in obtaining timely information concerning new job opportunities. Other serious obstacles to suburban employment faced by the inner-city poor include few and substandard work skills and biased hiring practices. In the face of these problems, many otherwise employable persons give up job hunting altogether and contribute instead to the growing number of unemployed in inner-city neighborhoods.

The spatial mismatch is an important factor in the high unemployment levels of black youths. Overall, unemployment levels for both black and white youths increase with greater distances to job opportunities. The principal reason for higher black unemployment rates is that the commuting time is higher for blacks than for whites. However, the mismatch hypothesis accounts for only approximately half the unemployment, leaving another 50 percent to be explained by labor-market discrimination, differences in education, and so forth.

Global Cities

To this point, we have considered cities at the national and regional levels. Now let us shift our attention to the international level, where cities function as centers of international business. Around the world, the proportion of people living in urban areas (Table 10.2, Figure 10.24) reflects countries' economic wealth and power; industrial and postindustrial societies tend to be highly urbanized, whereas most developing countries have a large share of the labor force in agriculture and rural areas. This trend is not universally true, as evidenced by the rapid growth of cities in the developing world (Chapter 13).

The international division of labor is forging the cities of the world into an integrated, composite system. Figure 10.25 depicts the location of the largest cities in the world, many of which are located in developing countries. However, while cities in Asia, Africa, and Latin America are important as centers of

Table 10.2
Proportion of Population in Urban Areas

Region	*1950*	*1970*	*1980*	*1990*	*2000*
World	29	37	41	45	50
Developed countries	54	67	72	75	76
Underdeveloped countries	17	25	31	35	39
Africa	16	23	30	31	33
Latin America	41	57	69	71	73
East Asia	17	27	29	36	38
South Asia	16	21	28	30	32
North America	64	74	74	75	75
Soviet Union/Russia	39	57	66	68	70
Europe	56	67	72	74	75
Oceania	61	71	71	71	72

Source: Population Reference Bureau, 2001.

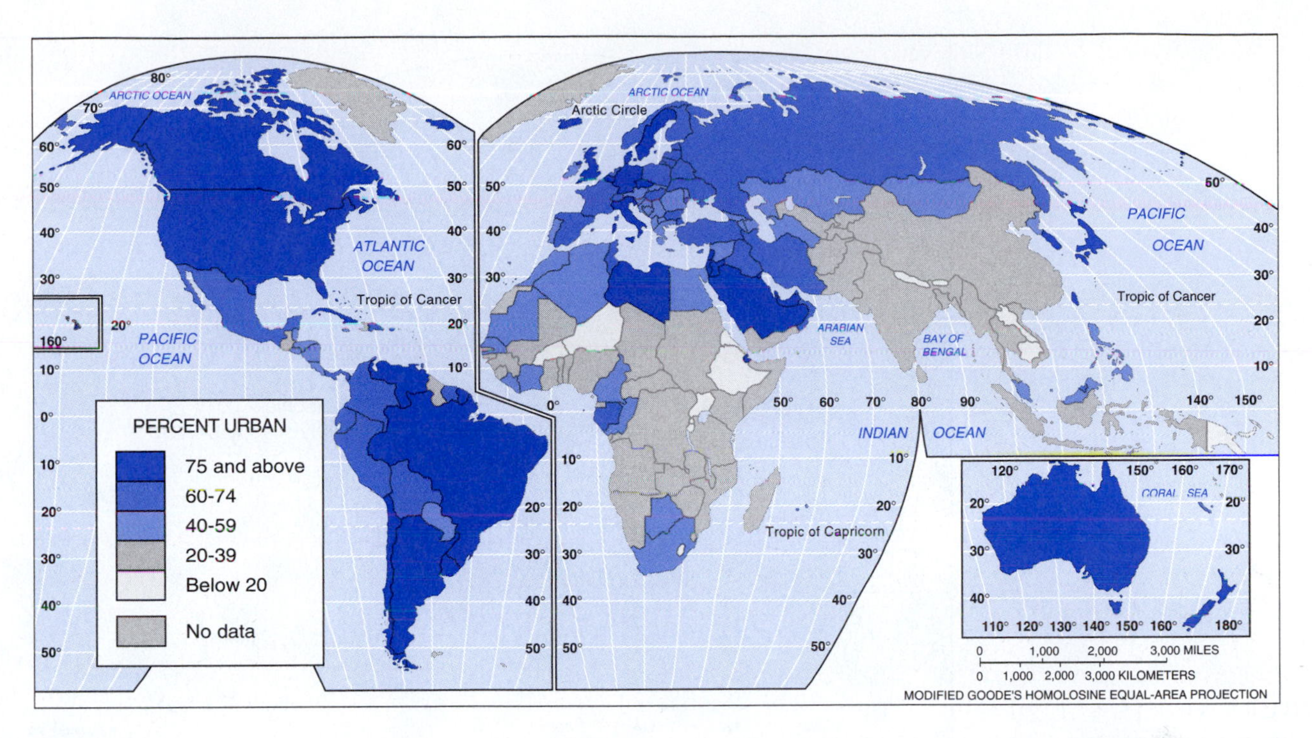

FIGURE 10.24
Urbanization rates around the world. Usually, more economically advanced countries have a higher proportion of people living in cities. (See Color Insert for more illustrative map.)

people and shape the everyday lives of billions, by and large they are not the major centers of control in the global economy, which lie in the economically developed world. Position in the worldwide urban hierarchy, therefore, depends less on population size than it does on a city's role in the global division of labor, corporate hierarchies, and access to the political and financial wealth that controls much of the world economy.

Restructuring and the Urban Hierarchy

The wave of intense *industrial restructuring* that started in the 1970s is changing the global urban hierarchy. The process of restructuring involves the movement of industrial plants from developed to developing areas within or between countries; the closure of plants in older, industrialized centers, as in the American Rust Belt; and the technological improvement of industry to increase productivity. Forces behind restructuring include the need for multinationals to develop strategies to locate new markets and to organize world-scale production more profitably, the national policies of developed countries to improve their future international competitive position, and the national policies of developing countries to attract subsidiaries of multinationals. These multinational strategies and governmental policies have contributed to major shifts in employment and trade. The greatest effects have been felt in the urban centers of developed countries and in the larger cities of underdeveloped countries.

One of the most significant repercussions of the internationalization of economic activity has been the growth of "global cities," particularly London, New York, and Tokyo (Figure 10.26). Global cities act as the "command and control" centers of the world system, serving as the home to massive complexes of financial firms, producer services, and corporate headquarters of multinational corporations. In this capacity, they operate as arenas of interaction, allowing face-to-face contact, political connections, artistic and cultural activities, and elites to rub shoulders easily. While other cities (e.g., Paris, Toronto, Los Angeles, Osaka, Hong Kong, and Singapore) certainly can lay claim to being national cities in a global economy, the trio of New York, London, and Tokyo has played a disproportionate role in the production and transformation of international economic relations in the late twentieth century.

At the top of the international urban hierarchy, global cities are simultaneously (a) centers of creative innovation, news, fashion, and culture industries; (b) metropoles for raising and managing investment capital; (c) centers of specialized expertise in advertising and marketing, legal services, accounting, computer services, and so on; and (d) the

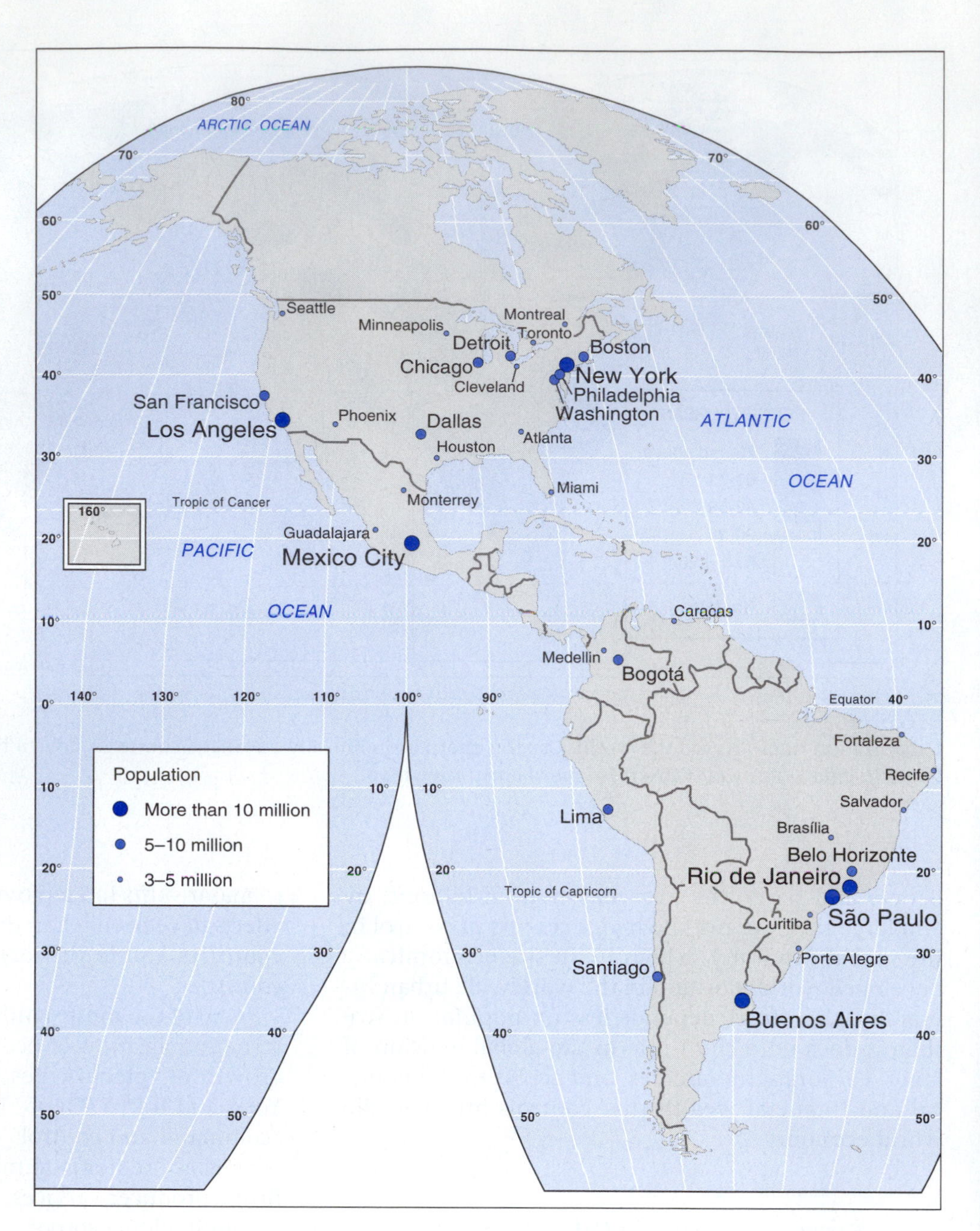

FIGURE 10.25
Major cities of the world. Cities with populations of more than 2 million are shown on this map. While urbanization of a country usually increases with per capita income, more than 10 million cities are located in the developing world, especially in East and South Asia. The growth of large cities in developing countries suggests the overall rapid growth of populations and the migration from rural areas to the city as more jobs are offered in the manufacturing sector and fewer labors are required in the agricultural sector.

management, planning, and control centers for corporations and nongovernmental organizations (NGOs) that operate with increasing ease over the entire planet. At their core, global cities allow the generation of specialized expertise on which so much of the current global economy depends. Each city is tied through vast tentacles of investment, trade, migration, and telecommunications to clients and markets, suppliers and competitors, scattered around the world. All three metropolises are endowed with enormous telecommunications infrastructures that allow corporate headquarters to stay in touch with global networks of branch plants, back offices, customers, subcontractors, subsidiaries, and competitors. This phenomenon again illustrates how geographic centralization can be facilitated by telecommunications.

Singular and economically preeminent world cities are basing points for global capital flows. These *world cities* are usually the largest and most important cities in the international system that measures capital, population, employment, and output. These cities are control points of the world economy where the critical mass of capital and articulation of production and marketing create dramatic world economic development. High levels of economic development have been primarily in developed countries; the globalization of economic activity has little touched developing countries. Therefore, the world cities—dominant centers of transnational corporations,

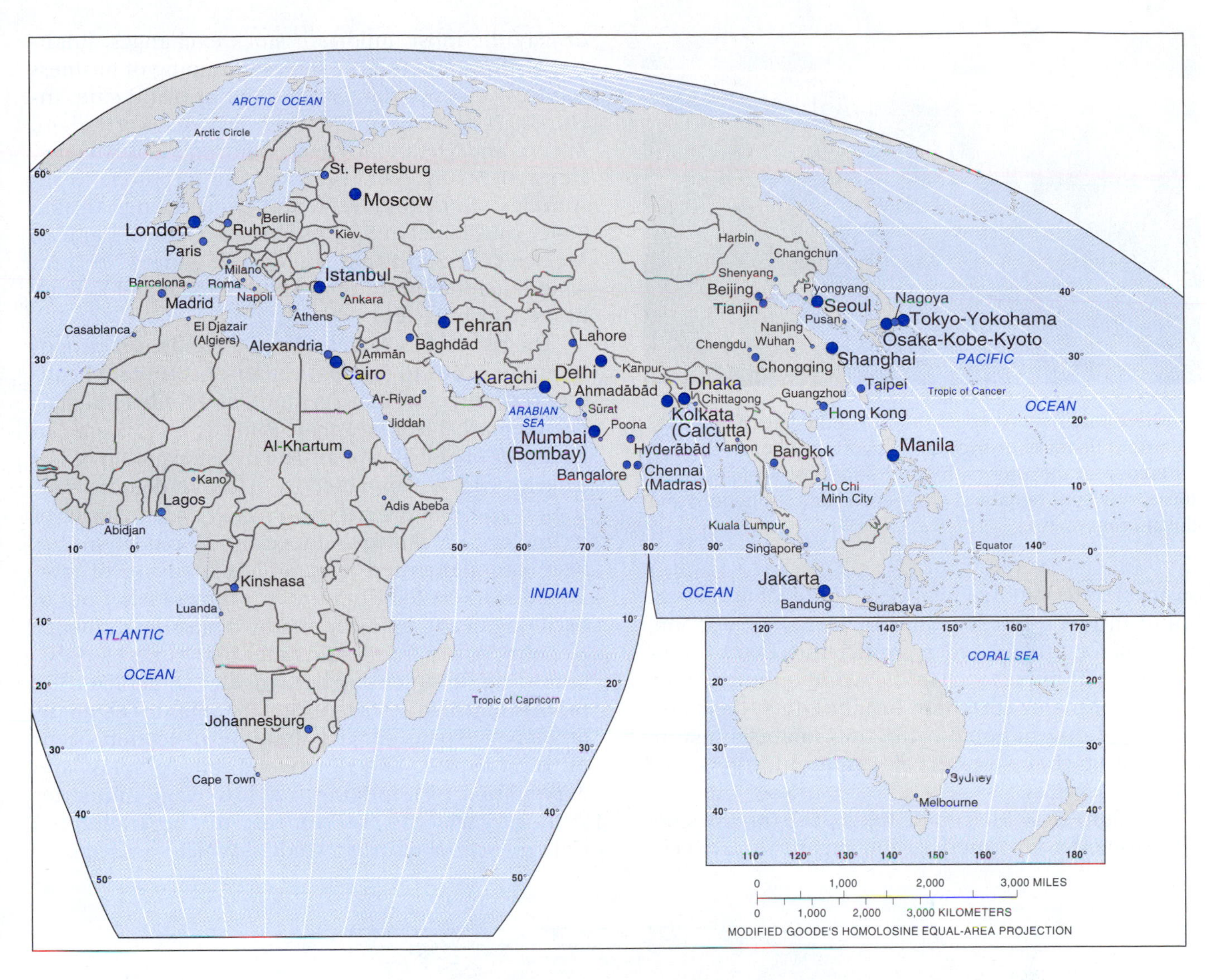

FIGURE 10.25

Downtown Minneapolis-St. Paul reflects not only the agglomeration of producer services in the Central Business District but also the city's role as the premier center of services in the northern Midwest.

Southern Manhattan represents one of the largest and most important financial districts in the world. Although the city suffered from the aftermath of the attacks of September 11, 2001, New York remains a vital center of the American and global economies.

business, international finance, and capital flows—are found in just a few locations. A hierarchy of dominant, major, and secondary world cities exists on the basis of their importance in the world economy, their hosting of major corporate headquarters, their possession of international banks and financial institutions, and their wealth of communications and business services.

London, New York, and Tokyo, the largest cities of Europe, North America, and Japan, respectively, display the most important stock exchanges, financial institutions, and the largest grouping of business services. An important second tier of world cities includes Paris, Berlin, Madrid, Milan, Rotterdam, Zurich, and Vienna in Western Europe; Los Angeles, Houston, Miami, San Francisco, and Toronto in North America; Bangkok, Bombay, Hong Kong, Osaka, Seoul, and Taipei in East Asia; Buenos Aires, Caracas, Mexico City, Sao Paolo, and Rio de Janeiro in Latin America; Johannesburg in South Africa; and Sydney in the South Pacific.

The rise of Los Angeles from a regional metropolis in the 1960s to a global center of corporate headquarters, financial management, and trade today has been remarkable. The Pacific Rim city has become an epicenter of global capital. The transformation of Los Angeles was accompanied by selective deindustrialization and reindustrialization. A growing cluster of technologically skilled and specialized occupations has been complemented by a rapid expansion of low-skilled workers fed from the recycling of labor out of declining heavy industry and by a massive influence of Third World immigrants and part-time workers. Sprawling, low-density Los Angeles symbolizes the process of urban restructuring: It combines elements of Sunbelt expansion, globalization, an economy centered on finance and Hollywood film complex, and exploitation of immigrants in low-wage industries such as garments. In contrast to the traditional Chicago School of urban studies, which focused on a

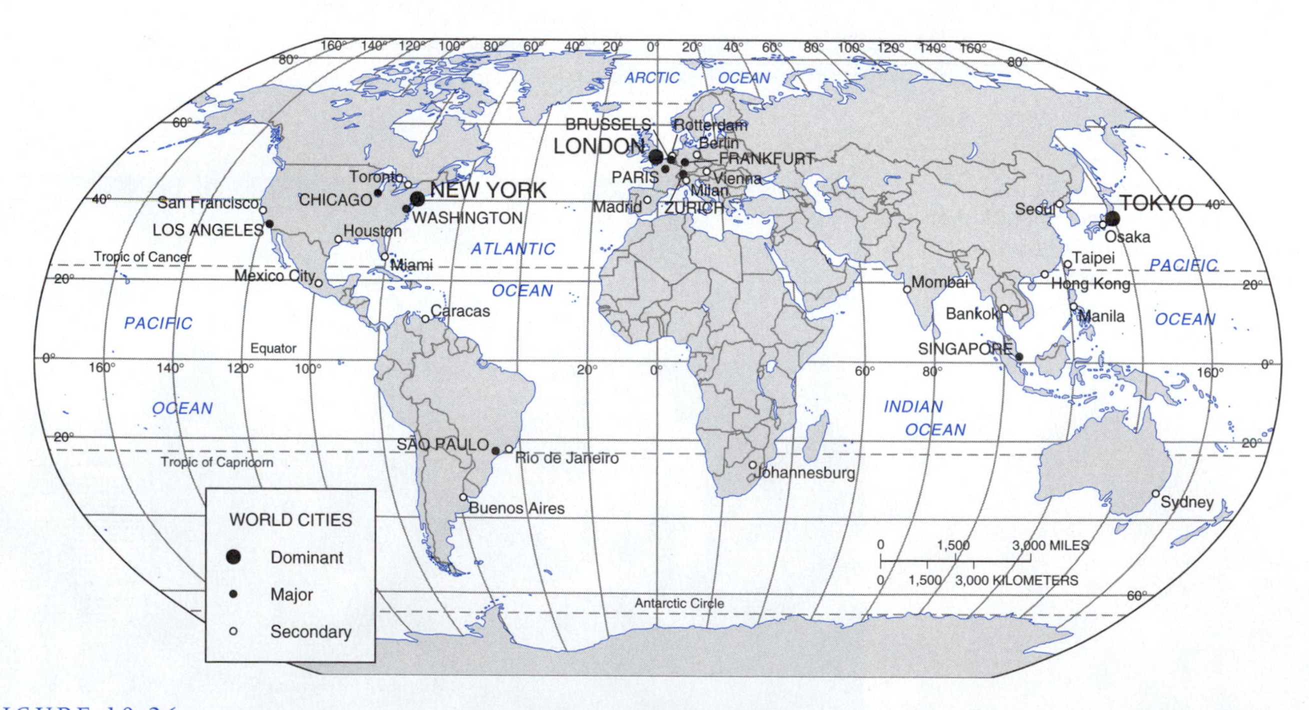

FIGURE 10.26
Global cities such as London, New York, and Tokyo dominate the world system of cities. While these are not the largest in terms of population, they are the primary command and control centers of the world economy.

The Canadian National Tower looms over Toronto at sunset, with the Toronto sky dome beside it. Toronto is Canada's largest city and leading financial center.

manufacturing-oriented city embedded in the national division of labor, many scholars point to the incipient "Los Angeles" model as suggestive of the shape and form of contemporary globalized, service-oriented urbanization.

Specialized producer service centers provide a more specialized variety of services to specific industries. Research and development (R&D) services cluster around motor vehicle agglomerations in Detroit; steel agglomerations in Chicago, Pittsburgh, and Baltimore; office equipment in Rochester, Atlanta, and Los Angeles; and semiconductors in San Jose, Phoenix, and Boston. Centers of government and education services include college towns and most state capitals.

Summary

Cities exist in societies that create the conditions necessary for the appropriation of the *surplus product.* In the nineteenth century, the stratified societies of Europe and North America experienced a massive urban transformation during the Industrial Revolution. During this period, cities, especially large manufacturing ones, were ugly creations and horrifying environments for the poor. Denied access to the fruits of rapid economic growth, the worker bore the social costs of urban industrialization. The early nineteenth-century industrial city was characterized by many small, relatively powerless enterprises. Toward the end of the nineteenth century, however, the market mode of economic integration took on a different appearance. There was a drift from individual to monopoly capitalism. Nonetheless, the mechanization of agriculture and lure of factories concentrated the majority of the populations of Europe, Japan, and North America in urban areas by the 1920s.

Cities are organized into urban hierarchies, webs of interaction that stratify them based on their economic significance. At the intraurban scale (i.e., within the metropolitan area) spatial ordering exists because firms and people sort themselves in response to the differential distribution of opportunities and constraints. We examined the residential selection process, how the filtering of housing reflects changes in the supply and demand, and how land and labor markets generate dense concentrations of people in urban cores and relatively few on the periphery, and we studied the well-known concentric-ring model of urban structure.

Suburbanization and urban sprawl reflect the pronounced, prolonged tendency toward decentralization characteristic of most cities in the industrial world. This trend reflects many combined forces, including changes in incomes and residential preferences; the search by manufacturing and office firms for cheaper land, taxes, and labor on the urban periphery; and the role of the state. The flip side of urban sprawl, although much smaller in scope, is gentrification, which in part reflects the rapid growth of producer services agglomerated in inner cities.

Urban areas, especially inner cities, are beset by a multitude of ills and problems. We traced the massive impacts that suburbanization and deindustrialization had on metropolitan cores, which were depleted of many well-paying jobs and suffered rising unemployment and poverty. Thus, suburban growth and inner-city decline are two sides of one coin. These issues have particularly affected minority dominated areas, including African American and Latino neighborhoods, which, in addition to a low number of jobs, also suffer from a crisis in low-income housing, inadequate schools and health care, and high crime rates.

Finally, we considered the reconfiguration of the world's urban hierarchy in light of the huge wave of contemporary globalization. Global cities, the command and control centers of the world economy, are sites where corporate headquarters are concentrated, along with specialized producer services. These cities, dominated by New York, London, and Tokyo, are places whose decisions affect most people on the planet. A secondary tier of important local command centers complements these major nodes.

STUDY QUESTIONS

1. How did the Industrial Revolution change cities?
2. What causes housing prices to fluctuate in cities?
3. How does the house filtering model work? Give examples from your community.
4. Why are urban land uses arranged the way they are?
5. How do cities in developing countries differ from those in the developed world?
6. What are some causes of suburbanization and urban sprawl?
7. What are some major urban problems today and what causes them?
8. What is an urban hinterland?
9. What are some problems of the inner city?
10. What is gentrification and what causes it?
11. What are world cities? What and where are the largest ones?

KEY TERMS

basic sector
global city
hierarchy
industrial restructuring
multiplier
nonbasic sector
central business district
concentric ring model
edge city
employment mismatch
exurbs
filtering
gentrification
megalopolis
population density gradient
residential location decision
urban sprawl

SUGGESTED READINGS

Garreau, J. 1991. *Edge City: Life on the New Frontier.* New York: Doubleday.

Knox, P. and McCarthy, L. 2005. *Urbanization: An Introduction to Urban Geography.* Englewood Cliffs, N.J.: Prentice Hall.

Knox, P. and Pinch, S. 2005. *Urban Social Geography: An Introduction,* (5th ed.). Englewood Cliffs, N.J.: Prentice Hall.

Lauria, M. 1997. *Reconstructing Urban Regime Theory: Regulating Urban Politics in a Global Economy.* Thousand Oaks, Calif.: Sage Publications.

Pacione, M. 2001. *Urban Geography: A Global Perspective.* London: Routledge.

Sassen, S. 1991. *The Global City: New York, London, Tokyo.* Princeton, N.J.: Princeton University Press.

Taylor, P. 2000. "World Cities and Territorial States under Conditions of Contemporary Globalization." *Political Geography* 19:5–32.

WORLD WIDE WEB SITES

CENTER FOR URBAN POLICY RESEARCH
http://www.policy.rutgers.edu/cupr/

USA CITYLINK PROJECT
http://usacitylink.com//
Comprehensive listing of WWW pages featuring American cities.

URBAN GEOGRAPHY ON THE WEB
http://www.geog.buffalo.edu/ugsg/
The official home page of the Urban Geography Specialty Group of the Association of American Geographers. It has facts on the subdiscipline as well as links to Web sites of interest.

THE CALIFORNIA GEOGRAPHICAL SURVEY
http://130.166.124.2/CApage1.html
A digital atlas of California that contains more than 300 maps at the moment and many more to come.

TOYOTA
QUALIS
Defines Today's Style and Luxury
TOYOTA
COROLLA
The Global Sensation
MGF TOYOTA
9810273500 GURGAON
9810244224 FARIDABAD
9810322464 MGF TOYOTA SERVICE
rush
as yet?
CITIGOLD
citibank
Pizza Hut
provogue
Light
Artlite

CHAPTER

11

Consumption

OBJECTIVES

- To offer an historical overview of consumption and consumerism
- To summarize sociological, neoclassical, and marxist views of consumption
- To analyze the geographies of consumption at multiple spatial scales
- To note the environmental impacts of mass consumption

Consumption is simultaneously a spatial as well as economic, cultural, and psychological process.

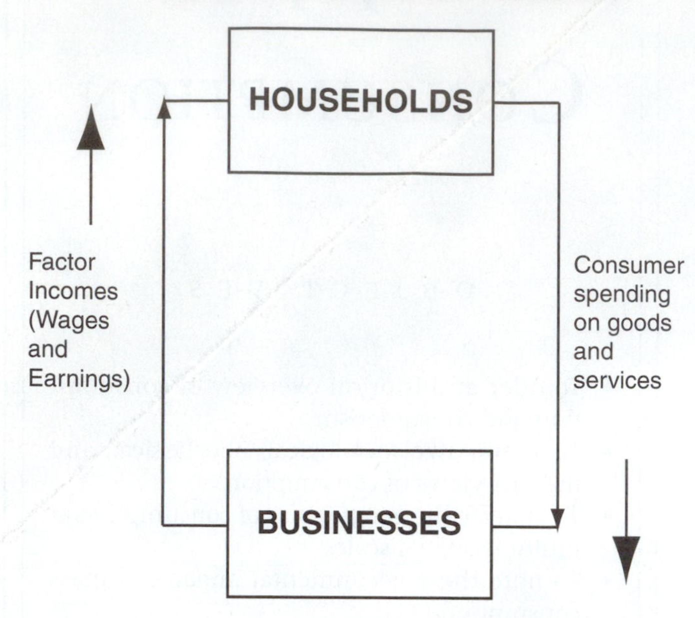

FIGURE 11.1
Production and consumption constitute two intertwined dimensions of capitalism. Without demand, there will not be a supply. Because both production and consumption produce and are shaped by complex geographies, their interrelations across space and time are complex, contingent, and ever changing.

In contrast to production, which has been studied in exhaustive detail in economic geography, consumption has long been ignored. While consumption played a role in traditional models of space and location theory, only recently have economic geographers come to view the issue in social terms. Yet consumption is critically important to understanding how economies are stretched over the earth's surface, their impacts, and how they change. Consumption and production cannot be neatly separated and are closely intertwined: Most people work in order to consume, and consume in order to live. Thus, we can think of consumption and production as two facets of one circular process (Figure 11.1).

The term "consumption" means a variety of things: As long as there have been people, there has been consumption. The Latin word *consumere* meant "to use up," and for centuries consumption referred to the disease tuberculosis. Today, consumption is typically taken to mean the use of goods and services to satisfy individual and collective wants and needs. Consumption is a complex topic because it lays at the intersection of various spheres of social life, such as production, class, gender, the relations between state and market, and the individual and society. Far more than simply an "economic" phenomenon, consumption is also a cultural and psychological construction that reflects and affects people at both the social and personal levels. Consumption is a major interface between the individual and society.

Economically, consumption is enormously important, constituting the bulk of the economic activity of most countries. For example, roughly 55 percent of the U.S. GDP consists of consumer or household purchases of different types (Figure 11.2); in comparison, business spending (investment) is only 15 percent, foreign trade is 18 percent, and government is 12 percent.

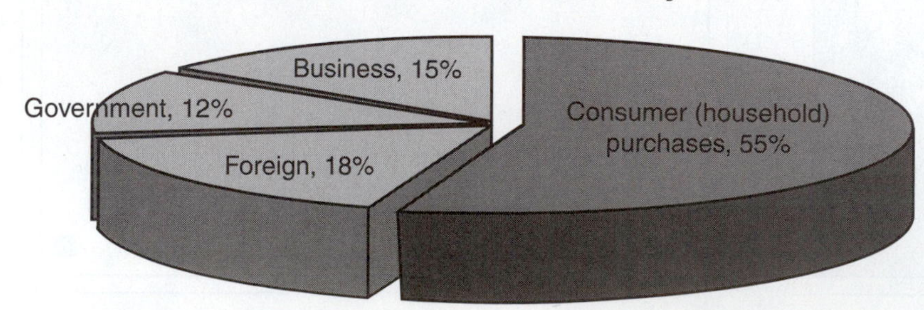

FIGURE 11.2
More than half of the U.S. economy consists of consumption by households and individuals. Small changes in consumer spending have enormous repercussions for the economy as a whole. For this reason, consumer confidence levels, shopping habits, changing preferences, and inclinations to spend versus save are closely monitored by social scientists and policy-making officials.

The Historical Context of Consumption

Although it may seem "natural," consumption as a social process has a history that reflects the extent and ways in which people have consumed over time.Most of what we know of premodern consumption focuses on elites, who had the greatest purchasing power. In such societies, in which there was relatively little trade and most goods were produced and consumed locally, trade was largely confined to luxury goods that were affordable only to the wealthy. High levels of consumption were hobbled by low average spending power. In medieval Europe, there was no single word for consumption; only with the rise of taxation in the sixteenth and seventeenth centuries did diverse activities like clothes and eating come to be seen as having something in common. We may thus posit that consumption as a social category is essentially an invention of capitalism. The rise of the colonial world economy in the sixteenth century (Chapter 2) simultaneously created a class of people with significantly large incomes to spend and the ability to make available products from a wide array of places around the globe, such as sugar from the Caribbean, cod from New England, and spices from Asia.

The rise of the modern form of consumption—consumerism—occurred largely through the emergence of the Industrial Revolution in the late eighteenth century. Industrialization, inanimate energy sources, and mechanized production generated a sudden, dramatic breakthrough that significantly lowered the prices of many goods and lifted millions of people above subsistence levels. Food and clothing, for example, became affordable to the growing middle class. Simultaneously, gradually increasing incomes began to transform workers into consumers, leading to the birth of the consumer society, with enormous implications for class consciousness. In the 1830s, the first department stores appeared, in Paris, which were the retail trade version of the factory. Historically, the growth of mass production in the nineteenth century was accompanied by mass consumption. Because capitalism thrives on newness, advertising was an integral part of this process, forever generating new needs and desires and converting luxuries into necessities. In the late twentieth century, changes in the world economy, including deindustrialization and the explosive growth of producer services, induced concomitant changes in consumption, including increasingly specialized niche markets and sophisticated consumers.

It should be noted that this process was far from smooth: Consumerism threatened much older ways of looking at the world. In particularly, ideas of thrift, embodied in slogans such as "waste not, want not," gradually gave way to morals that reflected the role of money as the measure of social status and consumption as the means to obtain it. Objections to consumerism had religious overtones, often couched in a scorn for the material world and an emphasis on ascetism. Ultimately, however, the ethic of the rising bourgeois class triumphed, a process that occurred during the growth of capitalism, the Industrial Revolution, the Enlightenment, and the growing dominance of individualism. Mass consumption entailed the commodification of consciousness, in which the self is defined through want, not social obligation. Consumerism was thus an integral part of the ascendancy of modernity; if tradition is the regulation of desire, modernity tends to unleash it.

Consumerism involved not simply the purchase of goods but the entire experience of shopping, including early catalogs and mail orders made possible by the growth of postal system, as well as printed advertisements (which were useful in societies with low literacy rates). As workers fought for higher wages and lower working hours, including the eight-hour work day and the five-day work week, consumption increasingly came to be equated with leisure. In the process, everyday life changed dramatically. Child rearing, for example, increasingly included the purchase of goods to comfort children. Clean clothes became the norm expected of an increasingly larger fraction of society. By the early twentieth century, the association of romance with consumption led to the invention of dating among couples. Buying and purchasing increasingly came to be naturalized as the keys to happiness, success, relief from pain and boredom, and a positive self-image.

Consumerism started primarily in Britain and the United States, although in the nineteenth century it had engulfed Western Europe. After World War II, consumerism has spread to large parts of the world, where it displaced traditional value systems. In many ways the spread of this phenomenon reflects the enormous influence of the United States in the world economy and the diffusion of capitalism. Multinational corporations, increased international trade, rising literacy levels, and television played key roles in this process. The triumph of consumerism was uneven geographically, and often resisted, yet almost always unsuccessfully. Typically the growth of consumerism has been slowest in societies in which entrenched religious objections were difficult to overcome. As ever larger shares of the world's population embraced this lifestyle, consumerism came to be the model for many kinds of social relations, such as education, religion, and family life. As commodity relations dominate more societies, everyone, it seems, has become essentially a buyer or seller, for better or worse.

After World War II, the expansion of consumerism was closely linked to the growth of credit. In consumerist societies, credit cards can enhance many people's quality of life, and are useful for emergencies or the purchase of important luxuries. Psychologically, credit cards are often portrayed as synonymous with

Consumer using credit card.

freedom, a rite of passage for adolescents. In the United States, the size of the "credit card nation" is staggering, including 158 million Americans and 1.5 billion cards in 2001 or 9.5 per card holder. The average card has $6,648 in debt. The rising tide of credit card debt in the 1990s, including small businesses, has generated a wave of personal bankruptcies; indeed, more people go bankrupt each year than go to college. Credit card companies such as American Express make a profit by charging merchants fees and by charging consumers relatively high interest rates.

The explosion of consumer debt undermined the traditional view of credit, which was grounded in the Puritan work ethic. This perspective emphasized frugality, delayed gratification, and saving, and debt was often regarded as sinful. Consumption was morally supposed to be held in line with one's income.

This worldview was annihilated by the onslaught initiated by banks and credit card companies in the 1970s and 1980s, with marketing campaigns that centered on the commodification of fun. Credit cards were sold as a marker of social sophistication and a carefree lifestyle. In the context of the deregulation of banking, which produced a more competitive environment, credit cards became the golden goose for many banks, earning higher rates of return than other avenues. Banks aggressively targeted many social groups that they had previously ignored, such as college students and the elderly. Moreover, consumers faced growing problems that arose with the end of the postwar boom, including deindustrialization, globalization, and stagnant incomes (Chapter 7). As spending continued to rise but incomes did not (Figure 11.3), U.S. savings rates declined steadily (in 2004, it was zero) (Figure 11.4) and many households found themselves swimming in debt they could not afford.

THEORETICAL PERSPECTIVES ON CONSUMPTION

Consumption may be viewed from several conceptual angles. Three perspectives are discussed here, the sociological, neoclassical economic, and marxist notions. There are merits as well as weaknesses to each of these.

Sociological Views

Sociologists have long examined the role of consumption in relation to individuals' and households' standard of living, class, and status. In the early twentieth century, for example, Max Weber argued that delayed gratification lay at the heart of the Protestant ethic; consumerism mutated this ethic into the unmitigated pursuit of pleasure. American sociologist Thorsten Veblen, studying the profligate consumption of the rich during the late nineteenth century era of the robber barons, coined the famous term "conspicuous consumption" to denote consumption as a social statement to others, that is, as a display of wealth and status. This line of thought led to a long series of works concerned with the relations between consumption and identity, emphasizing shopping as something much more than simply the purchase of goods but as a set of signals used to define who we are and what we mean. It is ironic, for example, that in societies such as the United States many people find their individual identities through the purchase of mass-produced commodities.

Sociologists have noted how consumption varies by social category, a phenomenon of vital importance to marketers. For example, consumption is highly gendered: men and women buy very different items, and shop in different ways. Because women are typically responsible for buying necessities for the household, they comprise the bulk of consumers, and many advertisers appeal to them. Likewise, shopping preferences vary predictably by age across the lifestyle. For example, as people age, they tend to spend more on medical care (Chapter 8). Race and ethnicity are also

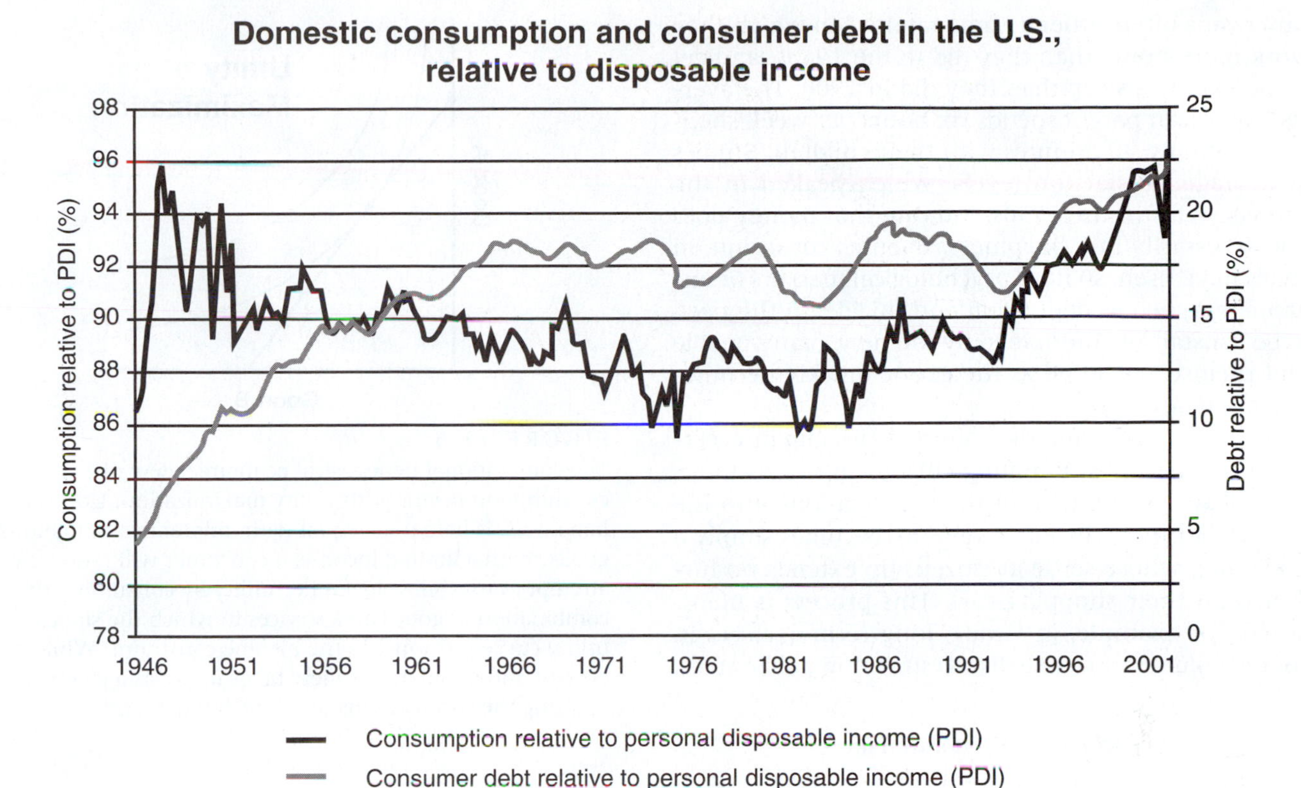

FIGURE 11.3
Faced with stagnant average real (post-inflation) incomes, rising costs, and ferocious advertising, and rising consumer debt, the U.S. savings rate has steadily declined over time. Savings forms the pool of capital necessary for investment.

critical sociological categories that are intertwined with the class and gender dimensions of consumption, as different ethnic groups have different preferences for food, clothes, automobiles, and other goods.

Critical to the sociological understanding of consumption is the role of advertising. Advertising is essentially a mechanism that funnels consciousness to the commodity, generating new wants and desires. Consumerist societies are saturated by advertising, which plays a major role in the construction of roles and role models. For example, two-thirds of newspapers typically consist of advertising. The average American child has seen 1 million television advertisements by age 20. Advertising encourages us to meet nonmaterial needs through material ends, generating a faith that commodities will impart a desired status, sexuality, physical prowess, lifestyle, and so forth.

So extensive has consumerism become that it has generated an epidemic of "affluenza," which may be variously described as the dogged pursuit of more, an obsessive quest for pleasure, materialism run amuck. Many people find themselves on an endless treadmill of permanent discontent. Ironically, never has so much meant so little to so many. To pay for this lifestyle,

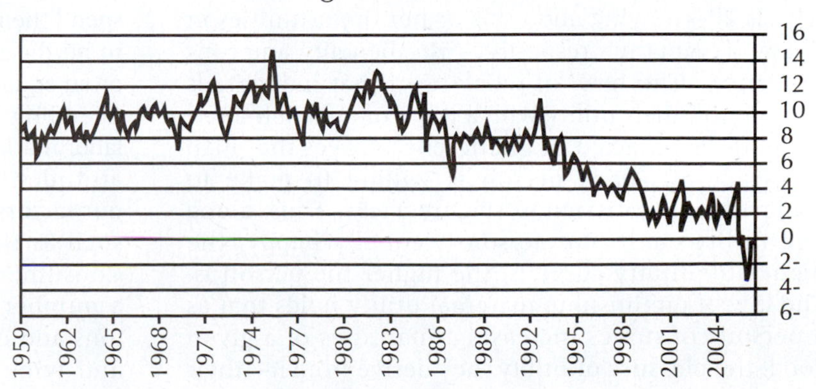

FIGURE 11.4
In the face of mounting international competition, deindustrialization, and low income growth but sustained consumption, the U.S. personal savings rate has declined steadily.

Americans often suffer a "time famine" in which they work more hours than they did in the 1960s, and get 20 percent less sleep than they did in 1900. The average American parent spends six hours per week shopping and only 40 minutes with their children. Studies of average satisfaction levels, which peaked in the United States in the 1950s, indicate that money does not necessarily buy happiness. Even as consumption levels have risen, so have psychological markers of unhappiness such as depression and suicide. In the need to be constantly entertained by the new, many people find themselves bored as the exotic quickly becomes commonplace.

Moreover, consumerism often carries into interpersonal relationships, mutating citizens into consumers and collapsing the field of social obligations into the self. Rather than citizens, everyone becomes simply a consumer, whose sense of citizenship extends no further than their shopping cart. This process is manifested, for example, in steady, long-term declines in voting, volunteering, and disinvestment in public areas.

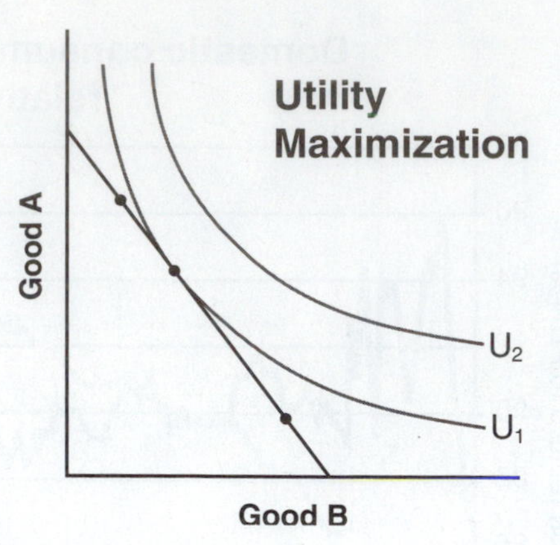

FIGURE 11.5
The conventional neoclassical economic view of consumption begins with utility maximization. Utility, or happiness, is held to be equal with different combinations of goods. With a limited income, a consumer will rationally attempt to maximize his or her utility by consuming the combination of goods and services in which the slope of the utility curve just touches the income constraint. While elegant and popular, this view takes utility as a given, ignoring the social origins of desire and demand.

Neoclassical Economic Views

The historically dominant view of consumption came from neoclassical economics, which analytically privileges demand over production. Neoclassical economics, which emerged in the 1870s, detached economics from political economy, and drew upon a long tradition of utilitarianism initiated by the philosopher Jeremy Bentham (1748–1832). This school of thought, which is dominant in the United States Canada, and much of Europe, begins with the individual. Bentham offered what he called a "hedonic calculus" based on mythical units called "utils" that he suggested standardized our understanding of happiness. Arguing that people are inherently self-interested, he suggested the principle of greatest happiness, that is, people will maximize pleasure and minimize pain. In the nineteenth century, neoclassical economists began to model this process in highly simplified but powerful ways, often invoking complex mathematics.

In this perspective, individual consumers are modeled using a model of human beings, *Homo economicus,* who is all-knowing about his or her opportunities in the world and their respective costs, benefits, and consequences. The level of satisfaction that individuals experience, their utility with a particular combination of goods, is reflected in indifference curves that map the trade-offs each person is willing to make to achieve equal satisfaction (Figure 11.5). Thus, along one utility curve the person is equally happy; the higher the utility curve is, the higher the person is. The law of diminishing marginal utility holds that as a person consumes increasing quantities of a given good, the pleasure or utility they derive from it—their utility—increases, but at a decreasing rate. Because consumption is always limited by income, each person must maximize his or her total utility or happiness by allocating their income among different goods. Since we cannot spend more than we earn (at least in the long run), the model does not allow the consumer to rise from Utility line 1 to Utility line 2. The slope of the income constraint reflects the relative prices of Good A and Good B, that is, the amount one must give up of one in order to purchase a unit of another. The optimal amount consumed (point z) occurs when the consumer maximizes utility by equating the marginal amount derived from the consumption of each good with the marginal cost as reflected in the income line. Essentially, utility curves reflect what consumers would *like* to do, income constraints depict what they *can* do, and the point of tangency explains what they *actually* do.

Aggregate consumption patterns in this model reflect the multitude of individual choices about how to spend their incomes. Because consumer demand is held to be the motor that drives the economy, this view is often equated with the notion of consumer sovereignty. Spending levels, in turn, reflect the dynamics of the labor market, unemployment rates, wage levels, prices, attitudes toward saving and spending, and the various factors that determine the shape of utility curves, such as the demographic composition of a society. Consumer tastes and preferences will be molded by a number of social issues, including fads and fashions and concerns over health and safety (e.g., beer and wine are displacing hard liquor, chicken has surpassed beef consumption in the United States). Consumption patterns will reflect how income-and price-elastic the demand for different commodities is. We noted in Chapter 8 that increased consumption of services is a driving force in the growth of retail trade and personal services.

The neoclassic view of consumption is that markets are always optimally efficient economically (and hence morally optimal as well). While the neoclassical view is internally consistent within its own terms of reference, it suffers from several problems, including the lack of any historical or social context and its inability to do justice to the rich semiotics and social dimensions of consumption. In part, this failure arises because neoclassical economics does not represent the consumer, or consumption, as a *social* act, that is, one embedded within broader relations of class, gender, ethnicity, and power, but as a purely individual one. For example, neoclassical economics offers no account of the origins of utility curves or why they assume their particular form; they are simply taken as given. "Individual" choices are always part of web of public policy choices; for example, the simple act of buying butter reflects federal subsidies to dairy farmers, health laws, the use of artificial chemicals and hormones, and so on. Social categories in neoclassical economics, if they arise at all, are defined largely by their relations to consumption: Class in conventional social analysis, for example, refers to income and socioeconomic status. Adam Smith's famous hidden hand assumes individual's choices are independent, not socially produced, and that they don't affect each other.

Finally, even when individuals act rationally, that is, maximize utility, they can lead to collective irrationalities or market failures. These occur, for example, when the market price does not reflect all of the social costs of a good. If each person at a concert stands up for a better view, for example, everyone is forced to do so and is therefore worse off. Each person's decision to drive is individually rational, but collectively these decisions create irrational traffic jams and gridlock.

Marxist Views

A third interpretation of consumption comes from marxism, which argues that social science must penetrate the veneer of outer appearances to reveal the social relations that lie beneath them. The expansion of capitalism historically has been predicated upon a widening and deepening of commodity relations, that is, the development of new markets and the transformation of goods and services that were formerly outside of the market into commodities (e.g., housing, child care, transportation, education). While marxism has generally privileged production and labor over consumption, Karl Marx did argue that unless products are consumed in the market, firms and employers cannot realize the profits generated at the workplace.

In this vein, Marx argued that commodities are not simply *things*, but embodiments of social relations. To view commodities separately from their social origins is commit the error of commodity fetishism, the opaqueness by which market relations obscure relations among producers is functional for capitalism. Marx argues that the social character of labor appears as objective, given the nature of products:

> The relations connecting the labour of one individual with that of the rest appear, not as direct social relations between individuals at work, but as what they really are, material relations between persons and social relations between things. . . . To [producers], their own social action takes the form of the action of objects, which rule the producers instead of being ruled by them. (1976, pp. 73, 75)

When seen in this way, commodities are not simply items on the shelf or advertised on television, they become complex combinations of labor, nature, and ideology.

Marxism drew upon classical economics to differentiate the use value of commodities—the qualitative, subjective dimensions—from their exchange value, the quantitative price they command on the market. For example, the use value of an apple is its taste and relief from hunger it offers, its exchange value is what it sells for. Critically, for marxists, labor too is a commodity whose use value to employers is less than its exchange value in wages. Class is thus defined by relations to production, not consumption.

Marxism suggests that the extraction of surplus value by employers inevitably leads to underconsumption by the working class and the tendency toward crisis. Employers, in this view, cannot by definition pay their workers the value of their output, or no profit would be generated. Thus, capitalism is perpetually faced with the problem of producing too much, which drives down prices, and ultimately profits. For marxists, this line of thought is the most severe of the internal contradictions that capitalism faces.

Geographies of Consumption

Drawing upon the work of sociologists, historians, philosophers, and anthropologists, geographers have engaged in numerous lines of thought that suture commodities to their social and spatial origins. This body of work has tended to fall into three major categories.

First, drawing upon the tradition of humanistic geography, some geographers have examined the relations between consumption, the body, and individual experience. The body is the most intimate of geographies, the

Fast food preparation exemplifies standardization of production that accompanied the stanardization of consumption. The 'McDonaldization' of many forms of social activity is widespread.

site of intentionality, where mind resides and a basis of identity. Bodies appear natural but are social constructions, inscribed with social meanings. The geography of the body locates it within social relations, as a place within a network of places. A considerable literature, for example, has looked at food, its origins and cultural meanings in different geographic contexts, and its role in the unfolding of daily life. Eating is the most intimate relation between body and environment, and food consumption plays a major role in shaping bodies, such as the epidemic of obesity plaguing the United States today, in part due to the widespread consumption of fast food.

Similarly, geographers have examined the shopping mall not simply as an economic phenomenon, but as a cultural site pregnant with meanings. For many Americans, time spent in shopping centers is third only to time at home or at work or school; many prefer shopping to sex. Shopping is the nation's dominant form of public life, and the shopping mall the only remaining pedestrian space in which to congregate. Thus, the mall is not just a place to consume, but a metaphor for public life in general. For example, the West Edmonton mall in Canada, the first and largest megamall in the world, has 600 stores, 18,000 employees, and generates 1 percent of all of Canada's retail trade. Inside it contains a golf course, skating rink, fantasy-land hotel, and four submarines. Similarly, the Mall of America in Bloomington, Minnesota, has 520 stores, chapels, a roller coaster, aquarium, and rain forest. In this environment, fantasy, fun, and the commodity are merged into a seamless whole. Malls are carefully engineered to maximize throughput of people and turnover of goods, including the locations of anchor stores and minimally visible exits. Symbolically, this environment is designed to transport shoppers to a mythologized looking glass world where the only thing that matters is commodity. All "backstage" functions involving the production, transportation, and storage of goods are carefully hidden.

Third, geographers have focused on consumption in the context of the global economy, particularly the manner in which commodities are produced, distributed, and consumed through the use of commodity chains (also called value chains). A commodity chain is a network of labor and production processes that gives rise to a commodity; it extends from the raw material to various stages in processing, delivery, and ends in consumption. For example, the coffee commodity chain begins with

West Edmonton Mall in Canada exemplifies the ways in which spaces of consumption lie at the boundary of reality and fantasy, enticing consumers into a world in which status and happiness are allegedly guaranteed by purchasing commodities.

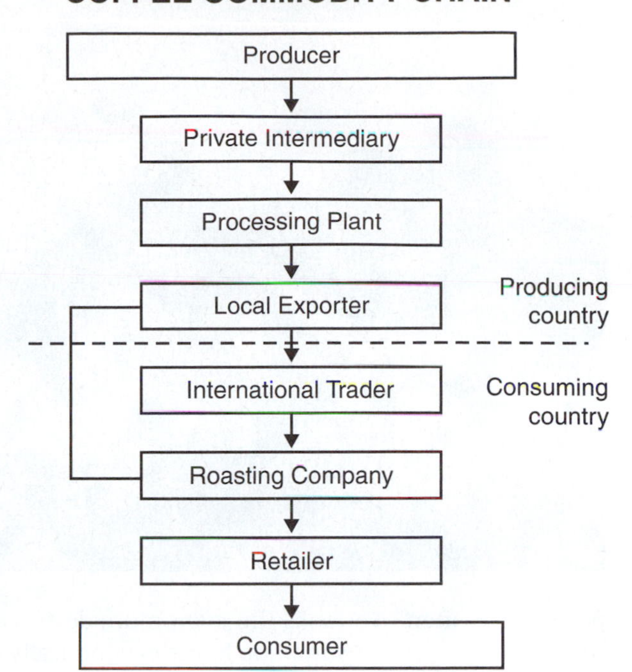

FIGURE 11.6
The coffee commodity chain represents the multiple steps involved in growing, harvesting, and roasting the bean, the various intermediaries involved in transporting and marketing it, and finally, the consumer. Different goods have different commodity chains associated with them, which stretch unevenly over time and space, suturing together producers, consumers, and those in between in complex networks of causality.

the grower, typically an impoverished farmer, and extends through the processing plant, exporters, traders, roasting companies, retailers, and, finally, the consumer (Figure 11.6). At each stage, the commodity is transformed in some way, and value is added. The same company may control one or more stages in a commodity chain depending on how vertically integrated or disintegrated the production process is. Because different nodes where these activities are carried out are spatially separated, commodity chains are geographic as well as economic and cultural phenomena.

Commodity chains are thus a means of depicting the ways economic activity reverberates through the production process, the linkages among different economic sectors, flows of value over time and space, and overcome the artificial separation between consumption and production. They allow us to see the commodity as more than just a thing, but as an embodiment of processes at different spatial scales. Essentially, commodity chains are mechanisms that allow us to trace the impacts of consumption decisions backward through the production and distribution process, broadening our scale of analysis from the local to the global. They trace the commodity through complex, contingent lines of causality linking sellers and buyers across multiple spatial scales.

Over time, with the expansion of capitalism globally, commodity chains have become longer and longer. This device allows us to understand, for example, the ways in which globalization has unleashed a tidal wave of cheap imports that has propelled the high rates of consumer spending in societies such as the United States. By uniting consumption with production, they point to the sacrifices made by low-wage labor trapped in sweatshops in the developing world in order to provide American consumers with cheap goods. Such a perspective reveals consumption as being simultaneously an economic, cultural, psychological, and environmental act that simultaneously reproduces both the world's most abstract space, the global economy, and the most intimate, the individual subject and body.

Environmental Dimensions of Consumption

In addition to its status as an economic and cultural phenomenon, consumption is also a deeply environmental one. Because consumption is linked to other domains such as transportation and production, every act of consumption imposes changes on the environment. These impacts occur through direct consumption by households and individuals and indirect consumption of resources in the production process. Often, because commodity chains are long, the environmental impacts may be felt thousands of miles away or on the other side of the world.

Traditional views of environmental destruction typically focused on the poor. With origins in Malthusianism (Chapter 3) and the high birthrates found in many developing countries, the large numbers of the world's impoverished have often been blamed for clearing forests for farms, soil erosion, and other environmental predicaments. While there is no question that rapid population growth contributes to environmental destructive, this

Large trash dump.
If consumption is the purchase of inputs to the household, this large trash dump illustrated the flip side, the enormous quantities of waste that industrialized societies generate, with significant social and environmental consequences.

perspective often overlooks the even more destructive role played by overconsumption in the economically developed world, which generates the vast bulk of the planet's environmental problems.

Inevitably, as standards of living rise, and thus consumption, a society puts increasing stress on the environment, using more resources and producing more waste. For example, in 2000, the United States consumed 14 times more raw materials than it did in 1900, but its population only tripled. In many respects, mass consumption is disastrous for the world's ecologies, with enormous repercussions. Whole ecosystems are sacrificed to support our lifestyle, including the deforestation of tropical regions to produce plywood and newspaper. Entire landscapes are ravaged by the tremendous hunger of the developed world, and, increasingly, the developing world, for minerals, water, and energy to produce houses, hamburgers, and clothes. Our demand for seafood is emptying the world's oceans: 75 percent fish stocks are overfished, and the size of the catch is not limited by our fishing technology but by reduced stocks in the seas. Pollution, deforestation, and habitat destruction are generating the greatest loss of biodiversity in the planet's history. For economically developed countries, the effects of these processes are often not experienced directly because globalization allows us to export problems.

Of the many things we consume, private transportation via the automobile is perhaps the most wasteful. Urban sprawl and long commutes have locked us into an energy-intensive lifestyle. In the United States, for example, there are more cars than drivers. Transportation consumes 40 percent of the country's total energy budget, largely due to fuel-inefficient cars. With 5 percent of the world's people, the United States consumes 25 percent of its energy. The consumption of petroleum is a primary generator of greenhouse gases such as carbon dioxide that are a major force in the creation of the greenhouse effect and global warming (Chapter 4).

A fundamental feature of the world economy is the uneven distribution of purchasing power and resource consumption, a pattern that reflects the legacy of colonialism and the uneven development generated by capitalism. Today, roughly 20 percent of the world consumes about 80 percent of its vital materials, including 40 percent of the world's meat, 60 percent of its energy, and 80 percent of its paper and vehicles (Figure 11.7). In return, the First World generates

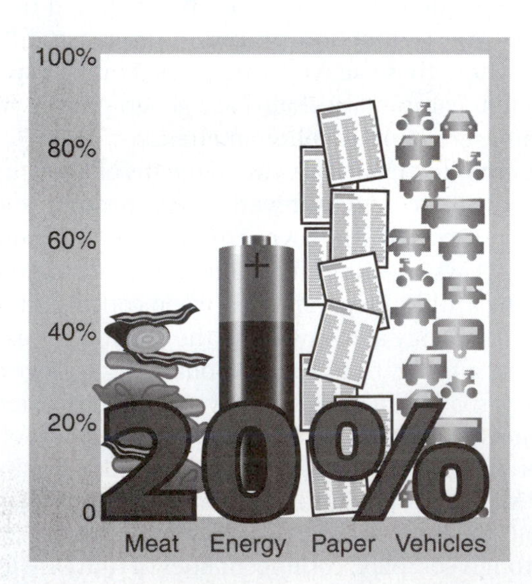

FIGURE 11.7
The economically developed world, which represents only about one-fifth of humanity, nonetheless consumes the bulk of many of the world's resources, testimony to its higher incomes and privileged position internationally. This unevenness reflects the historical dynamics of capitalism, which simultaneously produced wealthy and impoverished societies worldwide.

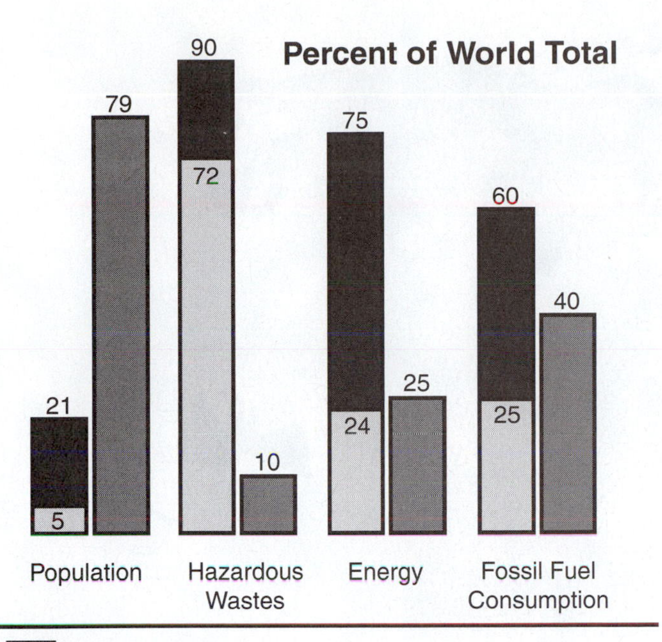

FIGURE 11.8
The flip side of consumption is the production of waste. By consuming a vast share of the world's resources, the economically advanced world also generates most of its pollution.

disproportionate amounts of the world's hazardous wastes (Figure 11.8). As the consumerist lifestyle has spread to much of the developing world, their demand for energy, meat, and other materials has risen accordingly. The growing middle class in China and India, for example, is putting huge stress on their ecosystems. Many ecologists are highly doubtful that the world could generate the resources necessary to support the entire planet's population at any level approaching that of the United States today.

One way to analyze this issue is through the notion of an ecological footprint, which estimates the amount of resources necessary to support a person's lifestyle. For example, during his or her lifetime, one U.S. citizen has 200 times the environmental impact that child in Mozambique has. (Go to ecofoot.org to estimate your own footprint size, which is a function of the size of your house, your diet, mode of transportation, etc.) The economically developed world has much larger footprints than do societies in Asia, Africa, or Latin America (Figure 11.9). The global distribution of total footprints generated by each country thus reflects total population and the per capita environmental impact (Figure 11.10).

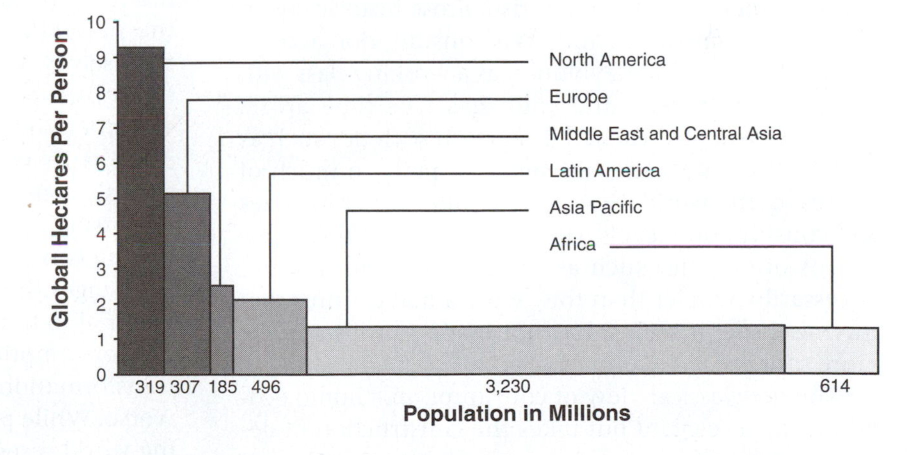

FIGURE 11.9
Ecological footprints reflect the use of the world's resources and their biophysical impacts on the planet. Their size is directly proportionate to the levels of income and consumption that occur in each society. Consumption is thus simultaneously an economic, social, psychological, geographical, and ecological phenomenon.

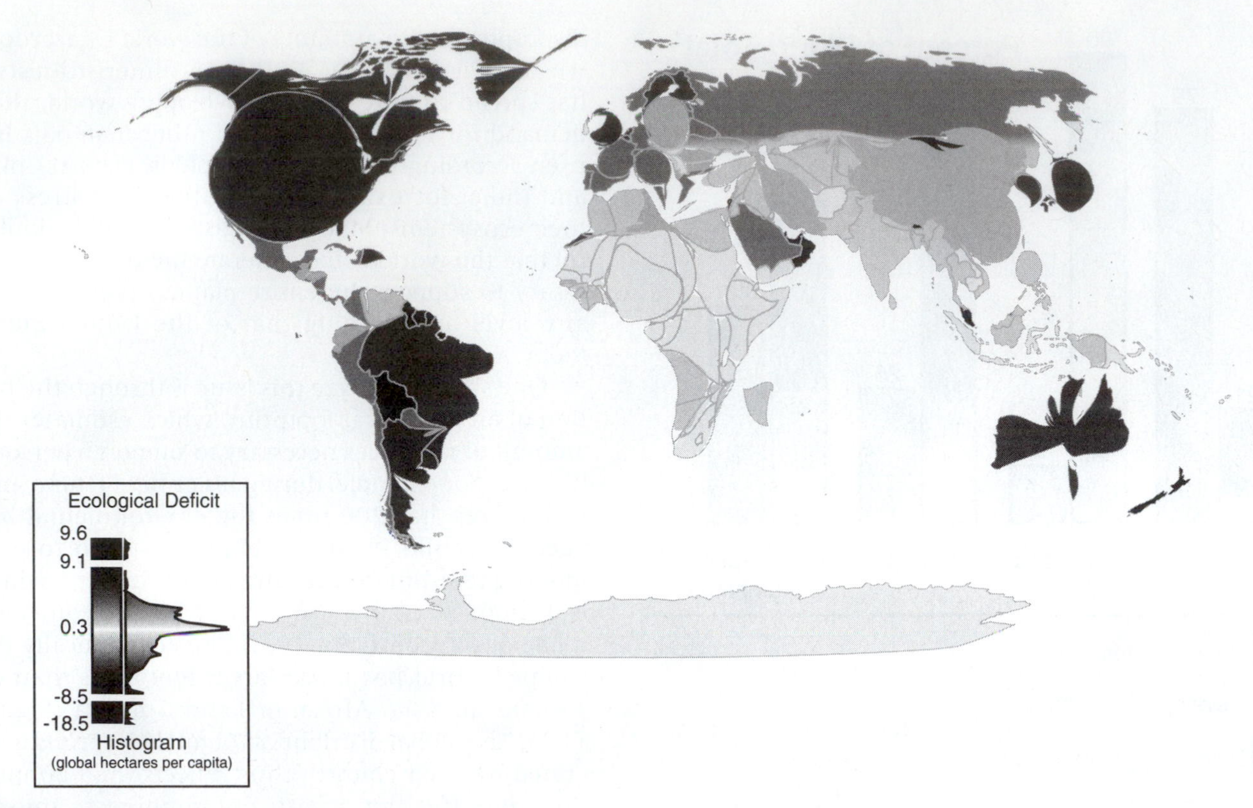

FIGURE 11.10
Because the First World consumes so much, it has much greater ecological footprints on the world's ecosystems than does the developing world.

SUMMARY

Consumption is a complex process with psychological, economic, cultural, political, geographic, and environmental dimensions. Consumerism arose historically on the heels of capitalism, and mass consumption accompanied the Industrial Revolution as a working class with a disposable income and sufficient free time arose. Consumerism has reached its apex in societies such as the United States, and has spread rapidly to much of the rest of the world. However, despite higher incomes and consumption levels, surveys reveal that the populations of societies such as the United States are not necessarily happier than they were a half century ago. Buying more, it seems, has not made as substantially more satisfied.

The neoclassical view of consumption, "homo economicuo," is elegant but takes the construction of demand as given. Sociological views of consumption portray it as a social act, in which individual choices are produced and constrained by their relative social status. In this view, demand does not simply appear in the individual consumer, but is generated by pressures such as advertising, which is exceptionally adept at turning luxuries into necessities. In neoclassical economics, demand is modeled using utility curves and the desolate figure of *homo economicus*. In marxism, the labor theory of value leads to the conclusion that employers always extract more value from workers than they can consume, leading to a chronic problem of overproduction and underconsumption.

Geographically, consumption can be understood at multiple spatial scales. The body, for example, reflects the networks of consumption in which people find themselves. While bodies appear natural, they are social constructions. Shoppers engage in consumption in spatial environments such as the mall that are saturated with webs of symbolic meanings, often consciously constructed to entice people to spend as much as possible. Commodity chains are a useful means of linking consumption with production, noting the various stages through which commodities pass from raw material to final product.

Consumption inevitably entails some degree of transformation of the environment, generally for the worse. While popular opinion often holds the developing world responsible for most of the world's environmental ills, statistically it is the mass consumption lifestyle of the First World that generates the bulk of environmental problems such as global warming. A relatively small proportion of the world—around 1.7 billion people or 20 percent—consume up to 80 percent of the world's resources. As consumerism has spread to much of the world, the impacts on ecosystems have multiplied accordingly. Ecological footprints are one means of analyzing the environmental costs of consumption.

STUDY QUESTIONS

1. When and where did consumerism originate historically?
2. Why don't rising levels of consumption make everyone happier?
3. What are the major theoretical perspectives on consumption?
4. How is the body a geographic locus of consumption?
5. Are most of the world's environmental problems generated in the developing world?
6. What is an ecological footprint?

KEY TERMS

affluenza
commodification
commodity chains
conspicuous consumption
consumerism
ecological footprint
exchange value
homo economicus
underconsumption
use value
utility maximization

SUGGESTED READINGS

Bell, D. and Valentine, G. 1997. *Consuming Geographies: We Are Where We Eat.* London: Routledge.

De Graaf, J., Wann, D., and Naylor, T. 2001. *Affluenza: The All-Consuming Epidemic.* San Francisco, Calif: Berrett-Koehler Publication.

Goss, J. 1993. "'The Magic of the Mall': An Analysis of Form, Function, and Meaning in the Contemporary Retail Built Environment." *Annals of the Association of American Geographers* 83:18–47.

Goss, J. 1999. "Once Upon a Time in the Commodity World: An Unofficial Guide to the Mall of America." *Annals of the Association of American Geographers* 98:45–75.

Hartwick, E. 1998. "Geographies of Consumption: A Commodity-Chain Approach." *Environment and Planning D: Society and Space* 16:423–37.

Hartwick, E. 2000. "Towards a Geographical Politics of Consumption." *Environment and Planning A* 32:1177–92.

Stearns, P. 2001. *Consumerism in World History: The Global Transformation of Desire.* London: Routledge.

WORLD WIDE WEB SITES

EARTHDAY NETWORK:
ECOFOOT.ORG

Calculate your ecological footprint or that for any part of the world.

VHS
Panasonic
Colour
COCA-COLA
Enjoy
Coca-Cola
Coke
Carlsberg Probably the best in the world. Carlsberg

WIMPY

WIMPY
Hamburgers

CHAPTER

12

INTERNATIONAL TRADE AND INVESTMENT

OBJECTIVES

- To explain the theoretical bases of international trade and factor flows, including comparative and competitive advantage
- To examine the effects of trade barriers such as tariffs and barriers
- To examine the dynamics of foreign direct investment
- To understand the financing of international trade, including the impacts of exchange rates
- To know the role of trade organizations such as cartels, the World Trade Organization (WTO), and regional trade agreements

Multinational corporate advertising in Piccadilly Circus, London, England.

Since World War II, most national economies around the world have become more integrated than ever. *International integration* refers to the specialization of different countries within a global division of labor. Contributing to this process have been technological breakthroughs in transportation and communications, the decline in protectionism, and massive transformations in corporate behavior. These innovations and developments have greatly enhanced the role of international businesses, which are any form of business activity that crosses a national border. International business includes the international transmission of goods, services, information, and capital. International trade is expanding, and its composition and patterns are changing. But in many respects, it is now less significant in the global business structure than the international movement of capital and services. Increasingly large numbers of companies are investing in foreign countries to acquire raw materials, to penetrate markets, and to exploit cheap labor. The expansion of production overseas has been matched by a parallel, symbiotic expansion of service enterprises, which now account for an increasing share of foreign direct investment (FDI).

The late twentieth century marked a watershed as the world economy entered a prolonged period of massive change. First, industrialized countries experienced a slowdown in their economic growth rates, in part due to the petroshocks and ensuing deindustrialization of the 1970s. Increases in oil prices reduced real income in the advanced countries and dealt a particularly harsh blow to the oil-importing Third World countries. Although oil prices declined in the 1980s, the shocks left a permanent imprint on the structure of global finance, trade, and investment. Second, competitive rivalry among industrialized countries increased significantly. This competition was stimulated by the slower and more unstable growth rates and, in turn, contributed to them. The rivalry gave rise to an increase in restrictive policies as each country sought to overcome its national crisis at the expense of others. Third, global financial markets underwent a profound series of alterations (Chapter 8). In 1973, the collapse of the Bretton-Woods monetary arrangement, involving the replacement of fixed exchange rates and the convertibility of the dollar into gold with a system of fluctuating exchange rates, permitted the United States to devalue its currency in an effort to retrieve lost competitiveness with its trading rivals. A global network of fiber optics allowed vast sums of money to be traded internationally, creating an almost seamless financial market around the planet. The World Trade Organization became a permanent body to regulate barriers to trade. The fourth structural change saw massive geopolitical realignments. Japan and other East Asian newly industrialized countries (NICs) enjoy rapid industrial growth. China reentered the global economy after half a century of isolation. The Soviet bloc collapsed, sending waves of turmoil throughout central Asia and eastern Europe. And Europe pursued a relentless strategy of economic unity through the European Union, a path followed to some extent in North America under the North American Free Trade Agreement.

The increasingly complex international business environment warrants the attention of geographers. Knowledge of the international sphere of business helps us to understand what is occurring in the world, as well as within our own countries. This chapter examines the concepts and patterns that underlie the expanding world of international business. It seeks answers to two questions: What theories shed light on the processes of international interaction? What are the dynamics of world trade and investment? First it reviews the major theories of trade, including classical comparative advantage, the Heckscher-Ohlin modification of this theme, and competitive advantage. Second, it turns to international capital markets, including exchange rates and FDI. It focuses on the role played by multinational corporations. Then the chapter considers major obstacles to trade, particularly tariffs but also quotas and nontariff barriers. Finally, it examines ways in which the world economy has sought to diminish protectionism, including the roles of trade organizations such as the General Agreement on Trade and Tariffs (now the World Trade Organization) and regional trade agreements such as the European Union and the North American Free Trade Agreement.

International Trade

Trade among countries has long been a central part of capitalism and a major factor linking various parts of the world together. World trade jumped from $2 trillion annually in 1980 to over $5 trillion today. Why are so many countries, large and small, rich and poor, deeply involved in international trade? (Figure 12.1). One answer lies in the unequal distribution of productive resources among countries. Trade offsets disparities with regard to the availability of productive resources. However, whether a country can export successfully depends not only on its resources but also on the conditions of the economic environment; the opportunity, ability, and effort of producers to trade; and the capacity of local producers to compete abroad.

Production factors—labor, capital, technology, entrepreneurship, and land containing raw materials—vary from country to country. Some countries have populations large enough to support industrial complexes; others do not. One country is home to an enormous pool of workers adept at running modern machinery; another abounds with scientists and engineers specializing in research-laden products. In some countries, entrepreneurs are more able and knowledgeable than

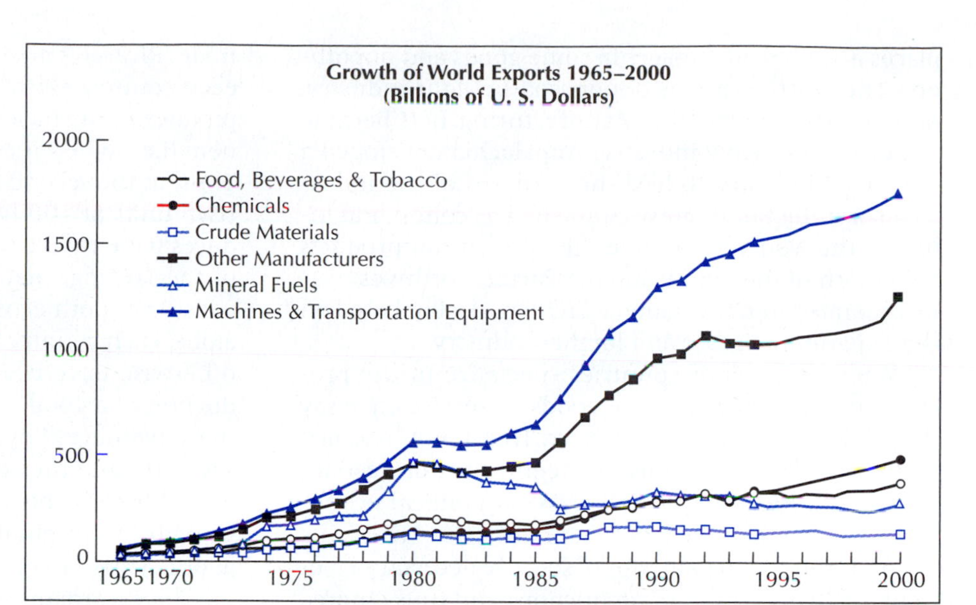

FIGURE 12.1
World growth of exports by value, 1965–2000. Despite the concerns of some in the labor movement, the reality is that most U.S. manufacturing employees work in plants that export. These jobs pay, on average, 10 percent to 15 percent more than nonexport jobs. Export firms expand employment nearly 20 percent faster than other firms. Moreover, productivity in export firms is one-fifth higher than in nonexport firms. Companies that export are 10 percent less likely to fail than others.

in others. The imbalance in natural and human-made resources accounts for much of the international interchange of production factors and the products and services that the factors can be used to produce.

A country well endowed in natural and human resources has an advantage over countries that lack these assets. But assets in and of themselves are insufficient to guarantee economic success. American producers, for example, tend to be less successful exporters than their Japanese or European counterparts. Numerous factors may reduce the ability of countries to best use their productive advantages, including inflation, exchange rates, labor conditions, governmental attitudes, and laws. Other countries are hobbled by the legacy of colonialism, drought, or political violence.

Inflation, which is a rise in the general level of the prices of goods and services, can be detrimental to a country's ability to compete domestically or internationally. Exchange rates, the prices of currencies in foreign-exchange markets, can also influence competitiveness. For example, if a currency is overvalued in relation to other currencies, local producers may find it difficult not only to compete with foreign imports but also to export successfully. In addition, recurring labor disputes that interrupt production can create serious obstacles for exporters, and governments can encourage or discourage their export sectors. Finally, the competitiveness of exporters is affected by labor laws, tax laws, and patent laws.

Trade by Barter and Money

At one time, trade was conducted on a barter basis. *Barter,* or *countertrade,* is the direct exchange of goods or services for other goods or services. It still occurs in some traditional markets in underdeveloped countries and is of increasing importance in the modern world economy. Roughly 30 percent of world commerce is now countertrade. Russia and eastern European countries use barter to trade among themselves and with underdeveloped countries. Major oil-exporting countries such as Iran and Nigeria barter oil and gas for manufactured goods.

Despite its widespread use, particularly by governments that have turned toward national economic protectionism, barter is a cumbersome way of conducting international exchange. Even within a country, consumers would find it difficult to barter the goods or services that their families produce for goods and services to satisfy their daily needs. In fact, more time would be devoted to exchange than to production. Money simplifies exchange and trade between countries. Money does present some problems, such as those associated with exchange rates, but introducing it as an exchange medium does not alter the theoretical bases for international trade.

COMPARATIVE ADVANTAGE

One of the hallmarks of capitalism is its tendency to generate uneven economic landscapes, that is, great differences in the types of economic activity from place to place, as well as the standards of living and life chances that those activities create. Different regions have long specialized in the production of different types of goods and services. In Europe during the Industrial Revolution, for example, Britain became a major producer of textiles, ships, and iron; France produced silks and wine; Spain, Portugal, and Greece generated citrus, wine, and olive oil; Germany, by the end of the nineteenth century, was a major exporter of heavy manufactured goods and chemicals; Czechs were selling glass and linens; Scandinavia sold furs and timber; and Iceland exported cod to the growing middle classes. Within the United States, similarly, different places acquired advantages in some goods and not others: The Northeast was dominated by light industry, particularly textiles; the Manufacturing Belt became the center of heavy industry; Appalachia developed a large coal industry to feed the furnaces of the industrial core; the South grew crops such as cotton and tobacco; the Midwest became the agricultural products behemoth of the world; and the Pacific Northwest was incorporated into the national division of labor based on the expanding timber and lumber industry.

When regions or countries specialize in the production and export of some goods or services, we say they enjoy a *comparative advantage.* This notion was first introduced by the famous nineteenth-century economist David Ricardo. Like all classical political economists, he assumed the labor theory of value (the value of goods reflects the amount of socially necessary labor time that goes into their production) and thus ignored demand. Ricardo concluded that nations will specialize in the production of a commodity that they can produce using the least labor compared with other nations.

Ricardo's classic example of this process is demonstrated in Table 12.1, which illustrates the allocation of labor time in England and Portugal, two long-time allies and trading partners, before and after they specialized. In the first table, which depicts the labor hours per unit of wine or cloth that England and Portugal must each dedicate to the production of one unit of each good, it is evident that Portugal has an absolute advantage in both goods (i.e., it can produce both of them with fewer labor hours than can England). If Portugal is more efficient, does it make sense for Portugal to trade? The answer is clearly yes, implying that even the most efficient producer benefits from trade. Ricardo's analysis examined what happens when each country allocates its resources to the good it can produce most efficiently compared with its trading partners (i.e., when it acquires a comparative advantage). Thus, in the second table, England only produces cloth (two units at 100 hours each) and Portugal only produces wine (two units at 80 hours each). In the process of specializing, that is, of producing for a market that consists of both economies together rather than either alone, each country frees up some resources that would otherwise have been dedicated to the inefficient production of a good in which it did not have a comparative advantage. England saves 20 labor hours, Portugal saves 10, and the combined trading system thus saves 30, which can be reallocated toward investment (although the original model is static and says nothing about change over time).

The Ricardian model—the simplest of many, more complex notions of comparative advantage—has important implications for economic geography. First, it shows how powerfully trade and exchange shape local production systems. It demonstrates that trade allocates resources to the most efficient (i.e., profitable) ends at minimum cost. The only costs of free trade are borne by inefficient producers, in this case, English wine growers

Table 12.1

Ricardian Example of Comparative Advantage

Before specialization (labor hours/unit):

	Wine	*Cloth*	*Total*
England	120	100	220
Portugal	80	90	170
Units	2	2	390

After specialization (labor hours/unit):

	Wine	*Cloth*	*Total*	*Savings*
England	0	200	200	20
Portugal	160	0	160	10
Units	2	2	360	30

Ricardo was one of the most famous of the founders of political economy, and hugely influential for his work on comparative advantage, among other ideas.

and Portuguese textile producers. Second, Ricardian notions of comparative advantage reveal that specialization reduces the total costs of production; thus, trade improves efficiency even without reallocating resources. For this reason, the vast majority of economists favor free trade as beneficial to all parties concerned. Third, this approach points out that large markets allow more specialization than do small ones. Adam Smith, the great political economist of the eighteenth century, noted the same thing when he stated that the "division of labor is governed by size of the market." In this case, when the market expanded from one country to two, it allowed firms to specialize and become more efficient in the process.

Transport Costs and Comparative Advantage

Clearly, just as there is no specialization without trade, likewise there can be no trade without transportation. Goods must be moved across space from producer to consumer, and these transport costs must ultimately be borne by those who consume them. To the degree that transport costs affect the delivered price of commodities, they also influence consumers' willingness to buy them, and thus the competitiveness of the regions that export them. If transport costs are low, their impacts on the division of labor will be low.

However, sometimes, particularly for heavy and bulky goods, transportation costs may increase the market prices of exports/imports prohibitively, as demonstrated in Figure 12.2. In the two regions here, A and B, the demand for the same good is identical, but the production costs in A are lower, and thus, so is the market price (MP). Producers in A can sell their good in region B for a high price, and thus earn a higher rate of profit. Unfortunately, the market price plus transport costs (MP + TC) of the imported goods in region B exceed the domestic production price in region A; in other words, the transport costs make the exports too expensive to ship across regions. Throughout the history of capitalism, declines in transport costs have made it progressively easier for regions to realize their comparative advantage; thus, lower transport costs have contributed to lower production costs and higher standards of living. For example, New Zealand became a major producer of lamb following the introduction of refrigerated shipping in the late nineteenth century. Similarly, the Pacific Northwest began to export vast quantities of wood and paper to the cities of the Midwest and East Coast following the completion of the transcontinental rail lines in the 1890s.

Heckscher-Ohlin Trade Theory

David Ricardo's two-country, two-product theory of comparative advantage can be expanded by taking into account several countries and commodities and by allowing several production factors. The multifactor approach to trade theory derives from work by two Swedish economists, Eli Heckscher and Bertil Ohlin. The *Heckscher-Ohlin theory* takes the view that a country should specialize in producing those goods that demand the least from its scarce production factors and that it should export its specialties in order to obtain the goods that it is ill equipped to make. Unlike the original Ricardian model, it includes demand and allows for the

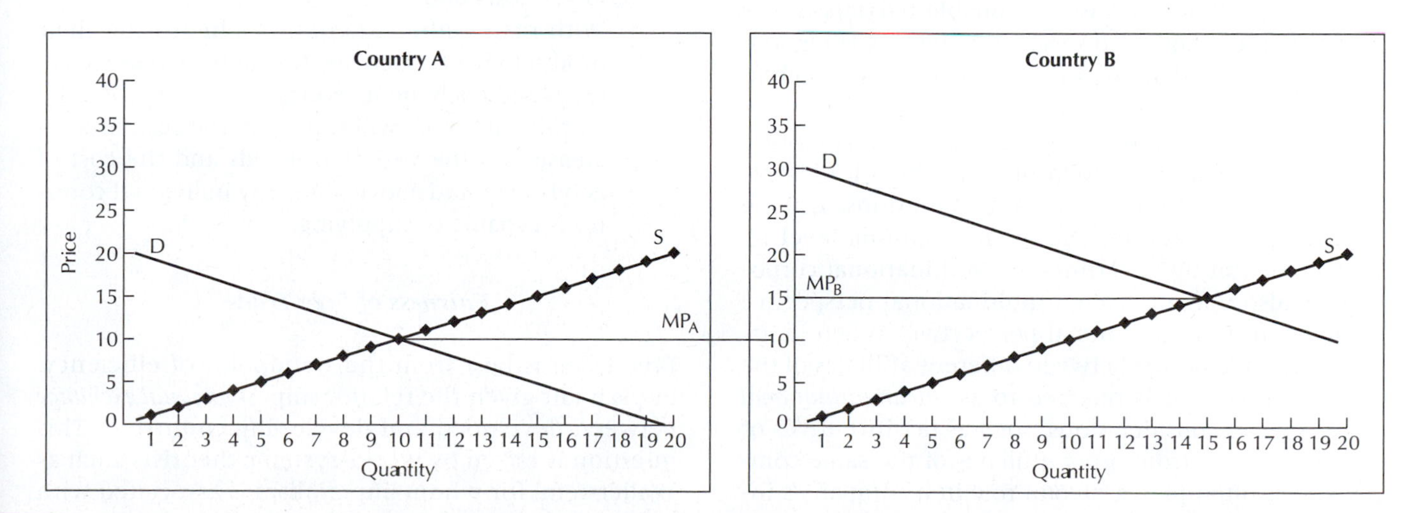

FIGURE 12.2
The influence of transport costs on comparative advantage. When the costs of shipping exports becomes greater than the differential in market prices between two countries, trade is inhibited. However, declining transport costs have steadily allowed for comparative advantages to flourish worldwide.

production of more than one good. Typically, in this formulation, specialization of production will be incomplete, that is, countries may continue to produce some of a good even if they do not enjoy complete superiority in the costs of production.

The Heckscher-Ohlin theory argues not only that trade results in gains but also that wage rates will tend to equalize as the trade pattern develops. The reasoning behind this *factor-price equalization,* as it came to be called, is as follows: If a country specializes in a labor-intensive good, its abundance of labor diminishes, the marginal productivity of labor rises, and wages increase. Conversely, a different country specializes in capital-intensive goods, labor becomes less scarce, the marginal productivity of labor falls, and wages also fall.

Inadequacies of Trade Theories

Trade theories are based on restrictive assumptions that limit their validity. They generally ignore considerations such as scale economies and transportation costs. Scale economies improve the ability of a country to compete even in the face of higher factor costs. Trade theories assume perfect knowledge of international trading opportunities, an active interest in trading, and a rapid response by managers when opportunities arise. However, corporate executives are often ignorant of their trading opportunities. Even if they are aware, they may fear the complexities of international trade.

Other inadequacies of trade theories include the assumptions of homogeneous products, perfect competition, the immobility of production factors, and freedom from governmental interference. Products are not homogeneous. Oligopolies exist in many industries. Production factors such as capital, technology, management, and labor are mobile. Governments interfere with trade; they can raise formidable barriers to the movement of goods and services, as well as labor, and affect interest, inflation, and exchange rates, which in turn affect the prices of imports and exports.

The most important shortcoming of trade theories, however, is their failure to incorporate the role of firms, especially that of multinational corporations. Trading decisions are made on the microeconomic level by managers, not by governments. Multinational corporations also operate from a multinational perspective rather than from a national perspective. When international trade occurs between different affiliates of the same company, it is referred to as *intramultinational trade.* Special considerations, such as tax incentives or no competition from other affiliates of the same company, can often play a pivotal role in a company's international decisions.

Despite their limitations, traditional trade theories provide an essential basis for our understanding of international business. They still underlie the thinking of many scholars, managers, labor leaders, and government officials. They offer a background for understanding the barriers to international business. They also frequently explain commodity trade, such as the international movement of wheat.

An Alternative Theory of International Trade and Transactions

Dissatisfaction with the explanatory powers of traditional theory generated alternative approaches to the study of trade. A new theory of foreign direct investment set about to explain both the overseas production of multinationals and the international trade conducted by these enterprises. The new trade theory allows trade between countries even when they have identical factor endowments. Because of imperfect competition, there is a possibility of increasing returns to scale (scale economies) that offer gains from trade over and above those attainable from conventional theory alone. These gains are summarized as follows:

1. Countries will benefit from trade if they are able to generate *increasing returns to scale* (agglomeration or scale economies)—then they are able to increase their output, thereby reducing the unit cost and reducing prices, therefore increasing trade.
2. Increasing returns (scale economy) industries will result and arise in productivity efficiencies in the host country, thereby enlarging global output and reducing prices worldwide.
3. Increased competition will reduce profits, which will thin out competing companies from trade. With increasing returns, this effect will result in greater overall productivity and sufficiency and lower costs and prices.
4. With many different countries increasing their productivity efficiencies through increasing returns and exchanging goods with each other, the world as a whole will gain from the resulting increase and the variety of goods and the cost of goods over and above what any individual country is capable of supplying.

Fairness of Free Trade

Free trade is best from the standpoint of efficiency, but is it fair given the relationship of *unequal exchange* between developed and developing countries? This question is raised by world-systems theorists such as Wallerstein, for whom imperialism is associated with relatively free trade (Chapter 14). Their argument is that an artificial division of labor has made earning a good income from free trade difficult for most LDCs.

The British were instrumental in creating an unfair global division of labor. Implicit in the early nineteenth-century argument for free trade was the notion that what was good for Great Britain was good for the world. But free trade was established within a framework of inequality among countries. Britain found free trade and competition agreeable only after becoming established as the world's most technically advanced industrial nation. Having gained an initial advantage over other countries, Britain then threw open its markets to the rest of the world in 1849. Other countries were instructed or lured to do the same. The pattern of specialization that resulted was obvious. Britain concentrated on producing manufactured goods, such as vehicles, engines, machine tools, paper, and textile yarns and fabrics, and exporting them in exchange for a variety of primary products. Imports included specialized cargoes such as Persian carpets, furs, wines, silks, and bulk imports such as timber, grains, fruit, and meat. Although many countries gained from the application of this artificial division of labor, none gained more than Britain.

The only way other countries could break out of this artificial division of labor was by interfering with free trade. The United States and Germany did so in the 1870s by adopting protectionist policies. France and a few other European countries with embryonic industries did the same. Dependent countries, however, failed to escape, either because of colonialism or because it was not in the interest of their ruling groups to do so.

The original division of labor changed little until after World War II, when a new structure began to evolve. Some underdeveloped countries were given a limited license to industrialize. The basic trend was export-led industrialization, concentrated in a few countries. For the best-off poor countries, industrial growth is geared to the needs of the old imperial powers. Thus, the growth of manufacturing in the Third World, under multinational corporate auspices, is not a portent of its emancipation from an artificial division of labor.

Worsening Terms of Trade

A deterioration in the *terms of trade*—the prices received for exports relative to the prices paid for imports—exemplifies the problem of unequal exchange for LDCs. By and large, LDCs export raw and semiprocessed primary goods—agricultural commodities and minerals. Primary commodities account for about 70 percent and 47 percent of the total exports of low- and middle-income countries (excluding China and India), respectively. The proceeds from these exports are needed to pay for imports of manufactured goods, which are vital for continuing industrialization and technological progress. Shifts in the relative prices of commodities and goods can therefore change the purchasing power of the exports of LDCs dramatically. The situation is exacerbated because many of these low- and middle-income countries are vulnerable, single-commodity dependent countries (Table 12.2).

In the late twentieth century, LDCs experienced a worsening in their terms of trade. This was caused by a decline in the prices of primary commodities and an increase in the prices of manufactured goods. In some years, the adverse shift was offset by an increased volume of LDC exports. For the period as a whole, however, import volume growth exceeded export volume growth, and, given the overall deterioration in the terms of trade, the result was large current account deficits. Maintenance of these deficits was possible only because the LDCs had access to external finance sources.

The economies of many LDCs are characterized by structural rigidity. They cannot alter the composition of exports rapidly in response to changing relative prices. Thus, if their commodity export prices decrease, they have no alternative but to accept declines in their terms of trade (Figure 12.3).

Another factor that may lead to worsening terms of trade is technological advances in developed countries. Advanced technology enables industrial economies to (1) reduce the primary content of final products; (2) produce high-quality finished products from less valuable or lower-quality primary products; and (3) produce substitutes for existing primary products (e.g., synthetic rubber for naturally grown rubber). These developments are irreversible. The demand for many primary products may be inelastic for price decreases, but in the long run it may be very elastic for price increases. A rise in the price of a raw material provides an incentive for industrial research geared to economizing on the commodity, or substituting something else for it, or producing it in the importing country.

COMPETITIVE ADVANTAGES OF NATIONS

The traditional theory of comparative advantage is very simplistic and unrealistic. Ricardo never gave an adequate account of why regions specialize in some goods and not others, instead offering a picture that is static with respect to time, overemphasizes labor and climate, ignores consumption as well as the role of economies of scale and agglomeration, says nothing about the nature of competition, and is silent concerning the impacts of public policy. Some of these issues have been addressed by neoclassical models in successively more sophisticated and complex models of comparative advantage.

A different route, advocated by Michael Porter (1990), is called the theory of *competitive advantage*. Unlike the Ricardian model, which was useful for understanding the simpler economies of the early Industrial Revolution, this approach focuses on the social creation

Table 12.2

Single-Commodity Dependent Countries

Product's Percentage of Total Export Earnings

40 to 59	*60 to 80*	*More than 80*
Agriculture and Fishing		
Benin (cotton)	Bhutan (spices)	Dominica (bananas)
Burkina Faso (cotton)	Burundi (coffee)	Rwanda (coffee)
Burma (lumber, opium[a])	Ethiopia (coffee)	Uganda (coffee)
Chad (cotton)	French Guiana (seafood)	
Cocos (Keeling) Islands (copra)	Guadeloupe (bananas)	
Comoros (spices)	Malawi (tobacco)	
El Salvador (coffee)	Maldive (seafood)	
Equatorial Guinea (cocoa, lumber)	Martinique (bananas)	
Finland (wood products)	St. Lucia (bananas)	
Ghana (cocoa)	Seychelles (seafood)	
Grenada (spices)		
Honduras (bananas)		
Iceland (seafood)		
Kiribati (copra, seafood)		
Mali (cotton)		
Mauritania (seafood)		
Nicaragua (seafood)		
Sudan (cotton)		
Crude Oil and Petroleum Products		
Congo	Bahrain	Algeria
Ecuador	Gabon	Angola
Syria	Libya	Brunei
Yemen	Venezuela	Iran
		Iraq, Kuwait
		Nigeria
		Oman
		Qatar
		Saudi Arabia
		Trinidad & Tobago
		United Arab Emirates
Metals and Minerals		
Central African Republic (diamond)	Botswana (diamonds)	Nauru (phosphates)
Chile (copper)	Guinea (aluminum)	Zambia (copper)
Jamaica (aluminum)	Niger (uranium)	
Liberia (iron ore)	Papua New Guinea (copper)	
Mauritania (iron ore)	Suriname (aluminum)	
Togo (phosphates)		

[a]Although impossible to quantify, Myanmar's opium exports may exceed 40 percent of total exports.

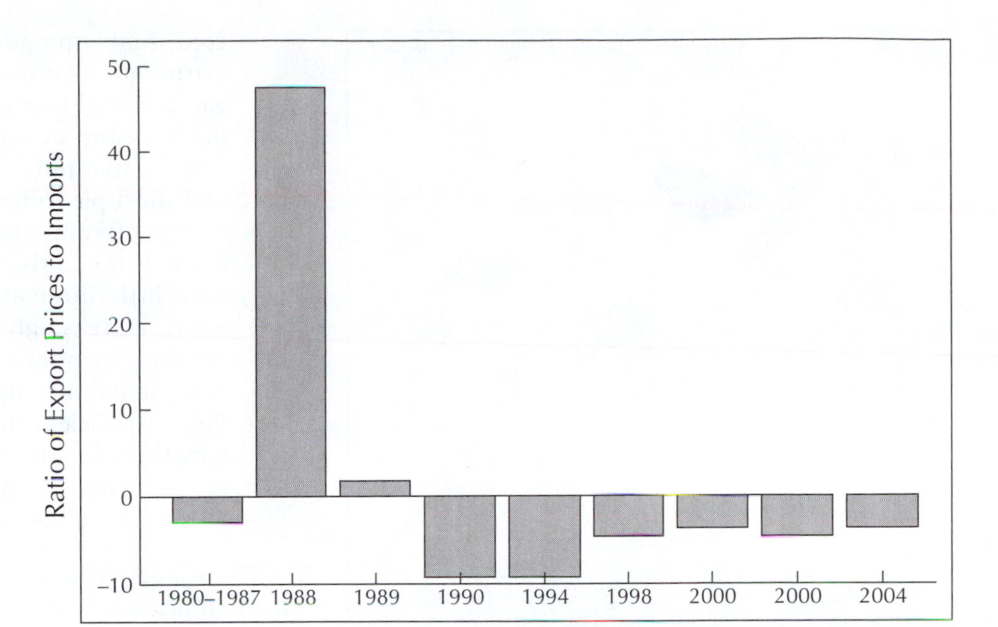

FIGURE 12.3
The worsening terms of trade for ores, minerals, and nonferrous metals, 1980–2000. Many underdeveloped countries cannot alter the composition of exports rapidly in response to changing relative prices.

of innovation in a knowledge-based economy. Porter begins by dismissing two commonly held myths about the source of national competitiveness, cheap labor, and abundant natural resources. Is cheap labor central to economic success? Cheap labor is, on a worldwide scale, virtually ubiquitous. Countries that have succeeded, such as Germany or Japan, have done so with labor costs well above those of their competitors. Why? Because their labor is productive, well educated, and because their economies endow workers with sufficient capital. In contrast, countries with the cheapest labor, say most of Africa, have done poorly in the global economy. Neither are the abundant natural resources necessary for economic success. Japan, for example, has done well despite having virtually no resources, and many developing countries with resources are trapped in low-wage economies that export raw materials. In a global economy, flows of oil, minerals, and foodstuffs are available anywhere.

What, then, does determine economic success? The key, in Porter's formulation, is productivity growth: Over the long run, rising productivity creates wealth for everyone, if not equally. Productivity growth in turn reflects many factors, including the education and skills of the labor force, available capital and technology, government policies and infrastructure, and the presence of scale economies (as discussed earlier). In the context of global markets, all firms can maximize scale economies.

Porter emphasizes that competitive advantage, unlike the Ricardian view, is dynamic and changes over time. The goal of national development strategies is to move into high-value-added, high-profit, high-wage industries as rapidly as possible. Such goods have high multiplier effects and do the most to trigger rounds of growth. To accomplish this goal, firms and countries should seek to sell high-quality goods at premium prices in differentiated markets. Quality is a key variable here; countries often acquire reputations for producing high- or low-quality goods, earning (or not earning) brand loyalty as a result. Finland is well known for its production of cell phones, for example, just as Japan is well regarded for its automobiles. By moving into high-value-added goods, nations should seek to automate low-wage, low-skill functions and retain knowledge-intensive ones.

Porter argues that although the global economy is increasingly seamless, competitive advantage is created in highly localized contexts (i.e., within individual metropolitan areas). Globalization does not eliminate the importance of a home base. Thus, countries that succeed internationally do so because a few regions within them move into cutting-edge products and processes. Within the current Kondratiev wave, for example, the propulsive regions of the U.S. economy include Silicon Valley, Boston's Route 128, and New York's position in finance and producer services; in Europe, they are Italy's Emilia-Romagna (Figure 12.4), that continent's largest high technology region, as well as Germany's Baden-Wurtenburg, Denmark's Jutland peninsula, and the Cambridge region of the United Kingdom; in Japan, the government has actively constructed a series of technopolises toward this end.

The overall determinants of competitive advantage include the following:

1. Skilled labor, good educational systems, and adequate technical training.
2. Agglomeration economies, including pools of expertise, webs of formal and informal interactions, trust, linkages, strategic alliances, trade associations,

FIGURE 12.4
Italy's Emilia-Romagna area, Europe's largest conglomeration of high-technology firms, produces a variety of high-quality items ranging from shoes to electronics. It exemplifies the regional basis of competitive advantage and the critical role of agglomeration economies.

integrated networks of suppliers and ancillary services.

3. A culture that rewards innovation: adaptation, experimentation, risk tolerance, and entrepreneurship; this includes heavy levels of corporate and public research and development and the continual upgrading of capital and skills. Corporations must engage in ongoing and organizational learning, anticipating changes in markets and demand; rigid corporate bureaucracies, like public ones, lead to complacency and short planning horizons.
4. Competitive markets at home; uncompetitive markets (i.e., private or public monopolies) exhibit little innovation. In the world economy today, increasingly sophisticated buyers spur a constant upgrading in the quality of output.
5. Adequate financing and venture capital.
6. Public policies that encourage productivity growth, including subsidized research, export promotion, educational systems, and an up-to-date infrastructure (i.e., airports, telecommunications).

The theory of competitive advantage concludes that four attributes of a nation combine to increase or decrease its global competitive advantage and world trade: (1) factor conditions, (2) demand conditions, (3) supporting industries, and (4) firm strategy, structure, and competition.

Factor Conditions

Factor conditions, or *production factors*, include land, labor, capital, technology, and entrepreneurial skill:

1. **Human resources:** The quantity of labor, the skill, educational level, productivity, and cost of labor.
2. **Physical resources:** Raw materials and their costs, location, access, and transport costs.
3. **Capital resources:** All aspects of money supply and availability to finance the industry and trade

Manufacturing electronics in Shinjuku, near Tokyo, Japan. Shinjuku is one of the busiest manufacturing locales in Japan. Japan enjoys a $60 billion export surplus. Its largest trading partner is the United States. Electronics and automobiles have been its most important export products. Because Japan is in a severe recession, having lost 50 percent of its equity on the Japanese stock exchange, Japan refuses to lower trade barriers to U.S. products. It has had a long history of such seemingly unfair trade practices.

from a particular country, including the amount of investment capital available; the savings rate; the health of money markets and banking in the host country; government policies that affect interest and exchange rates and the money supply; levels of indebtedness; trade deficits; public and international debt; and so forth.

4. **Knowledge-based resources:** Research, development, the scientific and technical community within the country, its achievements and levels of understanding, and the likelihood for future technological support and innovation.
5. **Infrastructure:** All public services available to develop the conditions necessary for producing the goods and services that provide a country with a competitive advantage. Included are transportation systems, communications and information systems, housing, cultural and social institutions, education, welfare, retirement, pensions, and national policies on health care and child care.

These five factors are identified in current international and economic circles as the keys to the competitive advantage of a nation in the foreseeable future.

Demand Conditions

Demand conditions are the market conditions in a country that aid the production processes in achieving better products, cheaper products, scale economies, and higher standards in terms of quality, service, and durability. Demand conditions cause firms to become innovative and therefore produce products that will sell not only in the domestic market but also in the world market.

Supporting Industries

To be competitive internationally, firms require access to networks of other firms specialized in different tasks in the economy. For example, large financial institutions require law firms, marketers, and advertisers. Often large companies use management consultants or similar business services, subcontracting tasks that require heavy investments in human capital. Access to these industries, which generally provide expertise, is often done through face-to-face contact.

Firm Strategy, Structure, and Competition

Firm strategy, structure, and competition relates to the conditions under which firms originate, grow, and mature. For example, because stockholders demand U.S. companies to show short-term profits, U.S. corporate performance may be less successful it would be if it were judged over a much longer time period, as is Japanese and German corporate performance.

State support of corporate strategy and performance is important. For example, a country can regulate taxes and incentives so that investment by firms is high or low. In addition, competition within a country can impose demands on company performance; new business formations often pressure existing firms to improve products and lower prices and thus increase competitiveness.

International Money and Capital Markets

In addition to trade, *capital markets,* or long-term financial markets, form another component of the international financial system. Stock exchanges, futures exchanges, and tax havens have proliferated. American, European, and Asian multinational corporations take advantage of *tax-haven countries,* countries where taxes on foreign-source income or capital gains are low or nonexistent (see Chapter 8).

The global expansion of the financial system has three components: the internationalization of (1) currencies, (2) banking, and (3) capital markets. *International currency markets* developed with the establishment of floating exchange rates in 1973 and with the growth in private international liquidity.

Capital movements take two major forms. The first type involves lending and borrowing money. Lenders and borrowers may be in either the private or the public sector. The public sector includes governments or international institutions such as the World Bank and agencies of the United Nations. The second type of capital movement involves investments in the equity of companies. If a long-term investment does not involve managerial control of a foreign company, it is called *portfolio investment.* If the investment is sufficient to obtain managerial control, it is called *direct investment.* Multinational corporations are the epitome of direct investors.

Monetary capital is the result of historical development. Unlike a natural resource like iron ore, it must be accumulated with time as a result of the willingness of a society to defer consumption. Low-income countries have low capacities to generate investment capital; all the capital that they do generate is usually employed domestically. Developed countries have much greater capacities for generating investment capital. They provide most of the world's private-sector capital, although a few fast-growing countries such as the NICs are also capital exporters.

Optimally, financial markets should produce an efficient distribution of money and capital throughout the world. However, there are many barriers to optimal distribution. Personal preferences of investors, practices of investment banking houses, and governmental intervention and controls confine money and capital movements to well-worn paths. They flow to some areas and not to others, even though the need in neglected areas may be greater.

International Banking

Paralleling the internationalization of domestic currency is the *internationalization of banking* (Chapter 8). International banks have existed for centuries; for example, banking houses such as the Medici and the Rothschilds helped to finance companies, governments, voyages of discovery, and colonial operations. The banks of the great colonial powers—Britain and France—have long been established overseas. American, Japanese, and other European banks went international much later. Major American banks moved into international banking in the 1960s, and the Japanese banks and their European counterparts in the 1970s.

Banks were enticed into international banking because of the explosion of foreign investment by industrial corporations in the 1950s and 1960s. The banks of different countries "followed the flag" of their domestic customers abroad. Once established overseas, many found international banking highly profitable. From their original focus on serving their domestic customers' international activities, banks evolved to service foreign customers as well, including foreign governments.

Euromarkets

Eurocurrencies are bank deposits that are not subject to domestic banking legislation. With relatively few exceptions, they are held in outside countries, "offshore" from the country in which they serve as legal tender. They have accommodated a large part of the growth of world trade since the late 1960s. The Eurocurrency market is attractive because it provides funds to borrowers with few conditions; it also offers investors higher interest rates than can be found in comparable domestic markets.

At first, Euromarkets involved U.S. dollars deposited in Europe; hence, they were called *Eurodollar* markets. Although the dollar still represents about 80 percent of all Eurocurrencies, other currencies, such as the deutsche mark and yen, are also vehicles of international transactions. Therefore, *Eurocurrencies* is preferred to the less accurate term *Eurodollar*. However, even *Eurocurrencies* is a misnomer. Only 50 percent of the market is in Europe, the major center of which is London. Other Eurocenters have developed in the Bahamas, Panama, Singapore, and Bahrain.

FINANCING INTERNATIONAL TRADE

In international trade, the buyer country must swap its currency, in proportion to the value of the product, for the currency of the exporting country. If a retail chain in the United States wants to buy televisions, video camcorders, or DVD players from Japanese firms, the buyer must convert U.S. dollars to Japanese yen in order to satisfy the terms of the purchase. As in the case with most international transactions—exports and imports—the seller receives payment in the currency of his or her own country, not in the currency of the purchasing country.

The value of the U.S. dollar, compared with a foreign currency, is called the dollar's *exchange rate*. An exchange rate is the number of dollars required to purchase one unit of foreign money. The U.S. dollar strengthened in Europe during the 1990s as a result of economic stagnation abroad and currency devaluation by European governments. However, the value of the dollar decreased in Asia and the Pacific during the same period. For American travelers, this meant taking a cruise to the Greek Islands or examining the ruins of Rome but postponing the hiking trip to Mount Fuji. More recently, the dollar has declined against the Euro as well, making travel to Europe more expensive for Americans.

Determining Exchange Rates

If the value of a currency fluctuates according to changes in supply and demand for the currency on the international market, a *floating exchange rate* is in effect. Figure 12.5 shows the changing relationship between the Mexican peso and the U.S. dollar as the peso was devalued. The demand curve, *D*, shows dollars sloping downward to the right. U.S. citizens will demand more Mexican pesos if they can be purchased with fewer dollars. Point p_0 suggests that fewer pesos can be purchased with a dollar, whereas Point p_1 suggests that many more pesos are available per dollar.

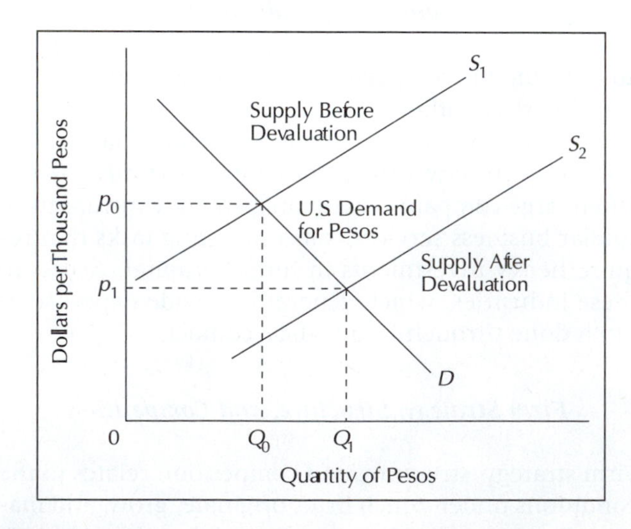

FIGURE 12.5
Determining the exchange rate of the U.S. dollar and the Mexican peso.

The city of London is a major center of Eurocurrencies, international banking, and capital markets. Originally called Eurodollars, offshore deposits, wherever they may originate, are now Eurocurrencies. They were gladly received by most international banks because they were not regulated by national banking controls. London became the center of Eurocurrencies. Because banks were able to make loans on the basis of their Eurodollar accumulations, offshore banking business expanded rapidly during the 1970s and 1980s, leading to the internationalization of world finance. Eurocurrency deposits increased much more rapidly than official country reserves, allowing international money markets and international banks to take over international finance from their own governments.

The demand for Mexican pesos in the United States is based on the amount of goods and services that a U.S. citizen wants to purchase in Mexico. A lower exchange rate for the peso makes Mexican goods less expensive to Americans.

In Figure 12.5, the supply of pesos is upward sloping to the right, which suggests that, as the number of dollars increases per 10,000 pesos, more pesos are offered in the marketplace. Mexican residents desire more goods from the United States when the dollar exchange rate (price) for the peso is high. The more dollars per 10,000 pesos, the relatively cheaper American products are for Mexicans. Therefore, Mexican residents will demand more dollars with which to purchase American goods and will consequently supply more pesos to foreign exchange markets when the exchange rate for the peso increases. The equilibrium position is reached when supply and demand conditions for foreign exchange is, therefore, based on supply and demand of international goods produced in Mexico demanded by Americans and American goods demanded by Mexicans.

Line S_2 represents devaluation of the peso because of economic restructuring in Mexico. This restructuring effectively reduced the number of dollars per 10,000 pesos on the international market. The result was that fewer American products could be purchased for the same amount of pesos because American goods and services became relatively more expensive. Cross-border purchases by Mexican border residents decreased dramatically, as did the international flow of goods and services from America to Mexico. At the same time, the quantity of pesos available to Americans increased from q_0 to q_1. American purchasers poured into border communities and increased their travel to the main tourist destinations within Mexico. The international flow of goods from Mexico to the United States increased because Mexican goods and services were relatively less expensive.

When the dollar appreciates in foreign-exchange markets relative to other currencies, it can buy more foreign currency and, therefore, more goods and services from other countries (Figure 12.6). As a result, American retailers import more goods to the United States. At the same time, the appreciated U.S. dollar means more costly American goods and therefore less demand for them. Exports from America decline under these circumstances. Thus, a strong dollar, which means that the dollar can buy more units of foreign currency, is not always desirable for U.S. trade. Conversely, when the dollar depreciates and is a "weak" dollar, it can buy fewer units of foreign currency and therefore fewer goods and services. Imports usually decline under these circumstances.

Why Exchange Rates Fluctuate

Exchange rates fluctuate for five reasons. First, as a country becomes wealthier and increases its real output and efficiency compared with that of other countries, it imports more goods from abroad. The result is that the increased demand for foreign currency raises

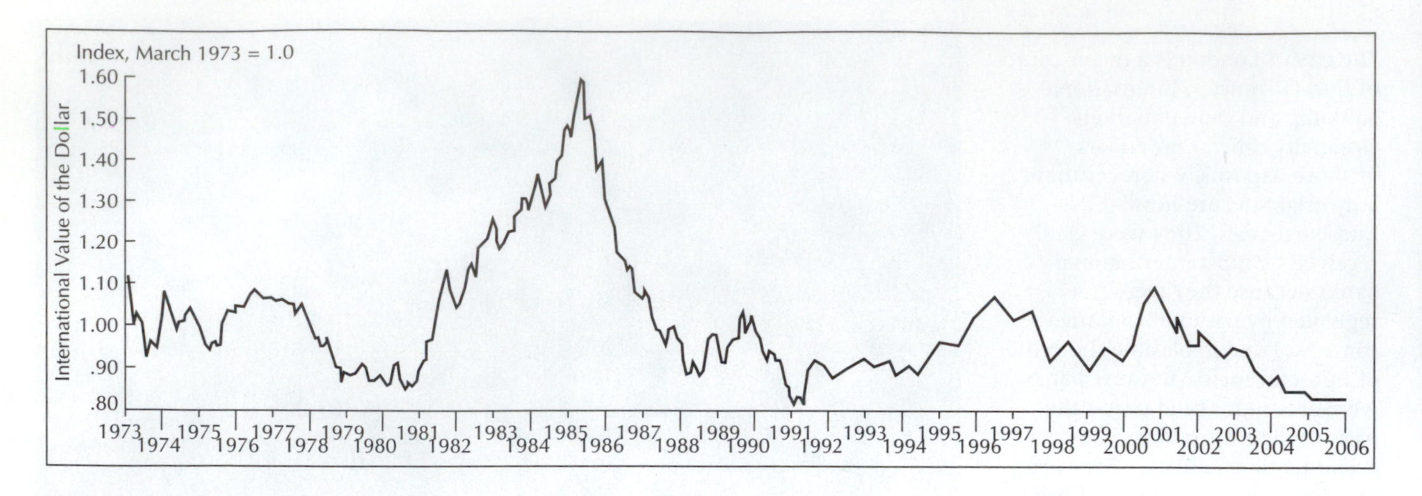

FIGURE 12.6
International value of the dollar, 1973–2006. From 1981 to 1986, the value of the dollar was relatively high compared with other major currencies. Since foreign goods were cheap, the U.S. imports increased, and their exports decreased. The dollar fell abruptly from 1986 to 1988 as a result of the Plaza Accord. In the 1990s, however, the dollar rose against all the other major currencies, with the exception of the Japanese Yen.

the exchange rate of the currency and decreases the value of the dollar internationally.

A second factor is the inflation rate of a nation. If the inflation rate of one nation increases faster than that of its trading partners, the currency of the nation with high inflation will depreciate compared with the currency of its trading-partner nations. Consequently, the products of the trading-partner nations will be more attractive to consumers in the country with high inflation. For example, when the U.S. inflation rate increases, the demand for Canadian dollars increases as investors seek a currency that will not lose its value so quickly. This demand raises the U.S. dollar price of both the Canadian dollar and Canada's exports to the United States. At the same time, Canadians demand fewer of the comparatively higher priced U.S. goods.

Third, domestic demand is a factor in determining exchange rates. Real income growth and the relative price levels between countries affect domestic demand. However, domestic demand also depends on consumer tastes and preferences. Americans will pay higher prices for specialty items and technologically advanced foreign goods than for comparable products at home. Examples of these foreign goods are VCRs, DVD players, and electronic consumer products from Asia, French wines and perfumes, German automobiles, and Italian shoes. A shift in the direction of foreign goods decreases domestic demand and causes the dollar to depreciate.

Fourth, the dollar may appreciate on foreign exchange markets if interest rates in America increase and, therefore, provide a higher yield to foreign investors who are interested in U.S. assets. Foreign investors increase their demand for dollars in order to purchase American companies and thus supply more of their foreign currency in exchange for U.S. dollars in the world foreign-exchange markets. When U.S. interest rates are low, foreign investments there yield lower rates of return than in other nations. Foreign investors, although still drawn to companies, land, and fixed assets, are not drawn to U.S. government securities. Conversely, as interest rates in the U.S. rise, the dollar becomes more attractive to investors elsewhere.

Fifth, currency speculation helps determine exchange rates worldwide. Real events are important in determining exchange rates, just as they are in determining home prices in any U.S. city. However, the expectation of future events is almost as important as actual events. The expectation that economic events will cause the dollar to appreciate or depreciate promotes currency speculation that may, in the not-too-distant future, be a self-fulfilling prophecy. If a major event such as a sudden war or the assassination of a major political figure occurs in a foreign country, it can trigger fear, which encourages individuals to sell the currency of that country to buy dollars. If all people reacted in the same manner to such events, the market would be driven down for that currency against the dollar, and the anticipated depreciation would actually occur.

U.S. TRADE DEFICITS

The United States enjoyed a trade surplus throughout most of its history. Starting in 1976, however, the volume of imports began to exceed the volume of exports (Table 12.3). The merchandise trade deficit was $25 billion in 1980, but by 2005, it had jumped to $782 billion. Three causes of the U.S. trade deficit are generally cited:

Table 12.3
Trade in the U.S. Economy, 1960–2000

	1960		1975		2005	
	Amount†	Percent of GDP	Amount†	Percent of GDP	Amount†	Percent of GDP
Exports of goods and services	$23.3	4.9	$136.3	8.6	$1,275	10.2
Imports of goods and services	22.3	4.4	122.7	7.7	1,991	15.9
Net exports	4.4	0.5	13.6	0.9	–716	–5.7

Data are on a national income accounts basis.
†In billions of dollars.
Source: Statistical Abstracts of the United States, 2001.

the increase in the value of the dollar, the rapid growth of the American economy, and the decrease in the volume of goods exported to less developed countries.

As shown in Figure 12.6, the international value of the dollar has changed widely over time. Increases in the value of the dollar mean that U.S. currency and products are relatively expensive to foreign nations, whereas foreign goods are less expensive to Americans. The U.S. dollar peaked in value internationally in 1985, which increased the amount of foreign imports to the United States and decreased the amount of its exports. After 1985, the value of the dollar began to fall, partly as a result of the Plaza Accord, an international agreement broached by the United States in the Plaza Hotel, New York City.

While the U.S. demand for imported goods remains strong, its exports have not kept pace, leading to severe trade deficits (Figure 12.7). The modest trade surplus the U.S. enjoys in services trade could hardly compensate for this deficit, leading to a deterioration of the U.S. current account (Figure 12.8).

There are several reasons for the explosion of the U.S. trade deficit. Fundamentally, Americans import a larger number of products from other countries than they export to them. Many households and corporations finance these purchases through debt, as does the federal government. The U.S. trade deficit also grew due to the reduction in exports to less developed countries. As the international debt of less developed countries increased, many countries, under pressure from the International Monetary Fund, restricted the amount of foreign imports and restructured their debts, agreeing to lessen their trade deficits. These austerity programs reduced the demand for U.S. exports. With devalued currencies, these countries could not afford U.S. goods. The combined effect was that the United States could export fewer goods to developing countries while importing a large amount of goods from them.

Results of the U.S. Trade Deficit

The U.S. trade deficit has three impacts: First, it reduces aggregate demand in the United States. An increase in the volume of imports pushes prices of domestic goods downward as domestic goods competed with imports for consumer demand. Second, the trade deficit hurts industries that are highly dependent on international trade. For example, automobile manufacturers, steel manufacturers, and the American farmer struggle the most because of the trade deficit.

The United States is now a net debtor instead of a creditor nation, owing foreign governments more than they owe it. The U.S. foreign debt was $5 trillion in 2002, making the United States by far the largest debtor nation in the world. In other words, American consumers have been subsidized because more goods and services flowed into the country than flowed out of the country. The reverse is true for its lenders, including particularly Japan and China. The United States has been living above its means, and its consumers have received an economic boost. This situation, however, is not likely to be tenable in the long run. Large federal budget deficits and huge balance of trade deficits have led to the so-called selling of America to foreign investors. Foreign investors now own 26 percent of America's total domestic assets.

Capital Flows and Foreign Direct Investment

In the nineteenth and early twentieth centuries, foreign portfolio investment overshadowed FDI. Theorists concentrated therefore on foreign portfolio investment, which was directed toward raw-material extraction, agricultural plantations, and trade facilities. Given the massive scale of FDI today, however, an understanding

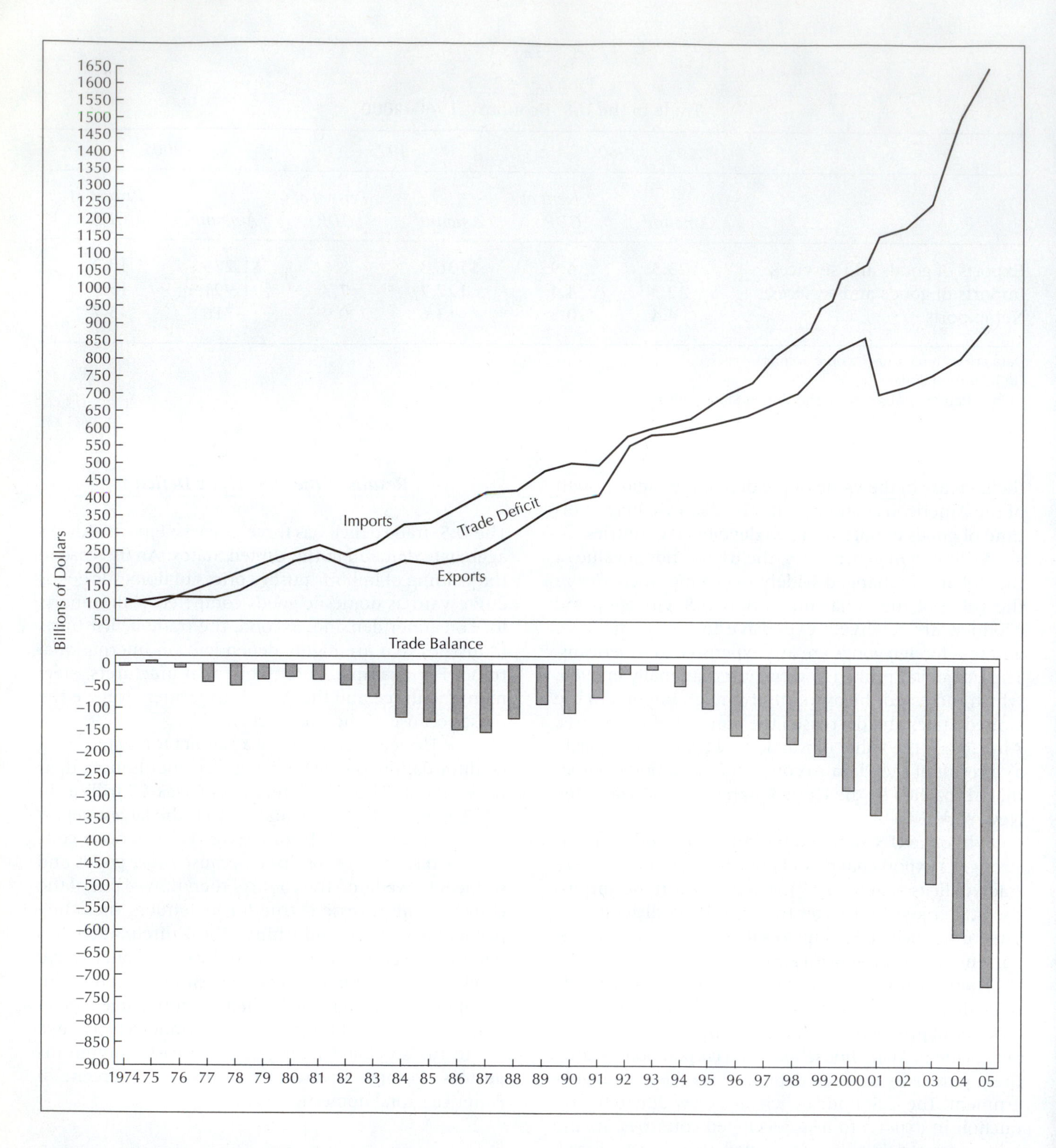

FIGURE 12.7
The U.S. merchandise trade deficit, 1974–2005. Because imports have exceeded exports, the United States regularly runs a deficit of almost $800 billion annually.

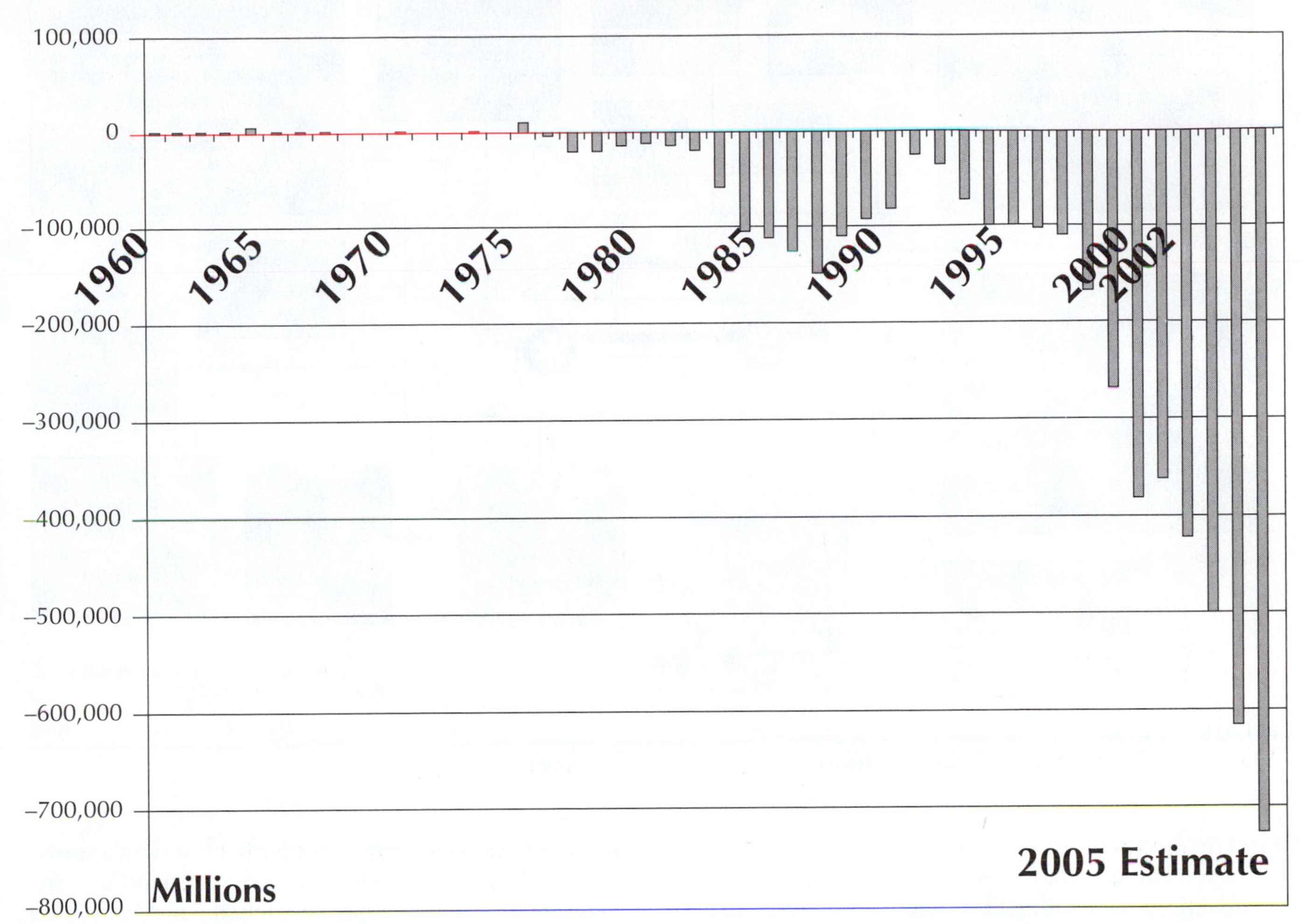

FIGURE 12.8
U.S. international transaction balances, 1960–2000.

of the rationale for such investment is important. The only help that classical capital theory gives us is that capital will flow from where it is abundant to where it is in short supply or, in other words, from where the rates of return are low to where they are high.

As with classical trade theory, classical capital theory is macroeconomic. However, in reality international money and capital flows are dependent on managerial decisions. Therefore, it is necessary to examine international capital movements from a microeconomic perspective. Even a brief examination reveals that although foreign investment is a simple process conceptually, a complex of motivations is involved.

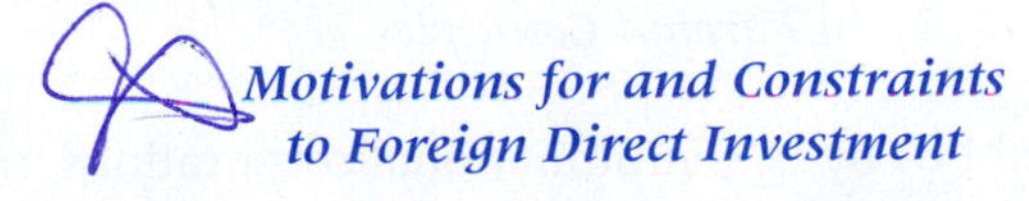

Motivations for and Constraints to Foreign Direct Investment

The primary reason for a firm to go international is profit. Three strategic profit motives drive a firm's decision to operate abroad. One motive for many direct investments is to obtain natural or human resources. Resource seekers look for raw materials or low-cost labor that is also sufficiently productive. A second motive is to penetrate markets. The third goal is to increase operating efficiency. These three motives are not mutually exclusive. Some segments of a corporation's operation may be aimed at obtaining raw materials, whereas other segments may be aimed at penetrating markets for the products made from the raw materials. Automobile producers such as Ford, for example, rely on a global network of parts suppliers (Figure 12.9). These operations may also result in some productive and market efficiencies.

There may be strong motivations for a firm to internationalize, but there are also compelling constraints. Prominent among these are the uncertainties of investing or operating in a foreign environment. Consumers' incomes, tastes, and preferences vary from country to country. Japanese consumers, for example, are wary of

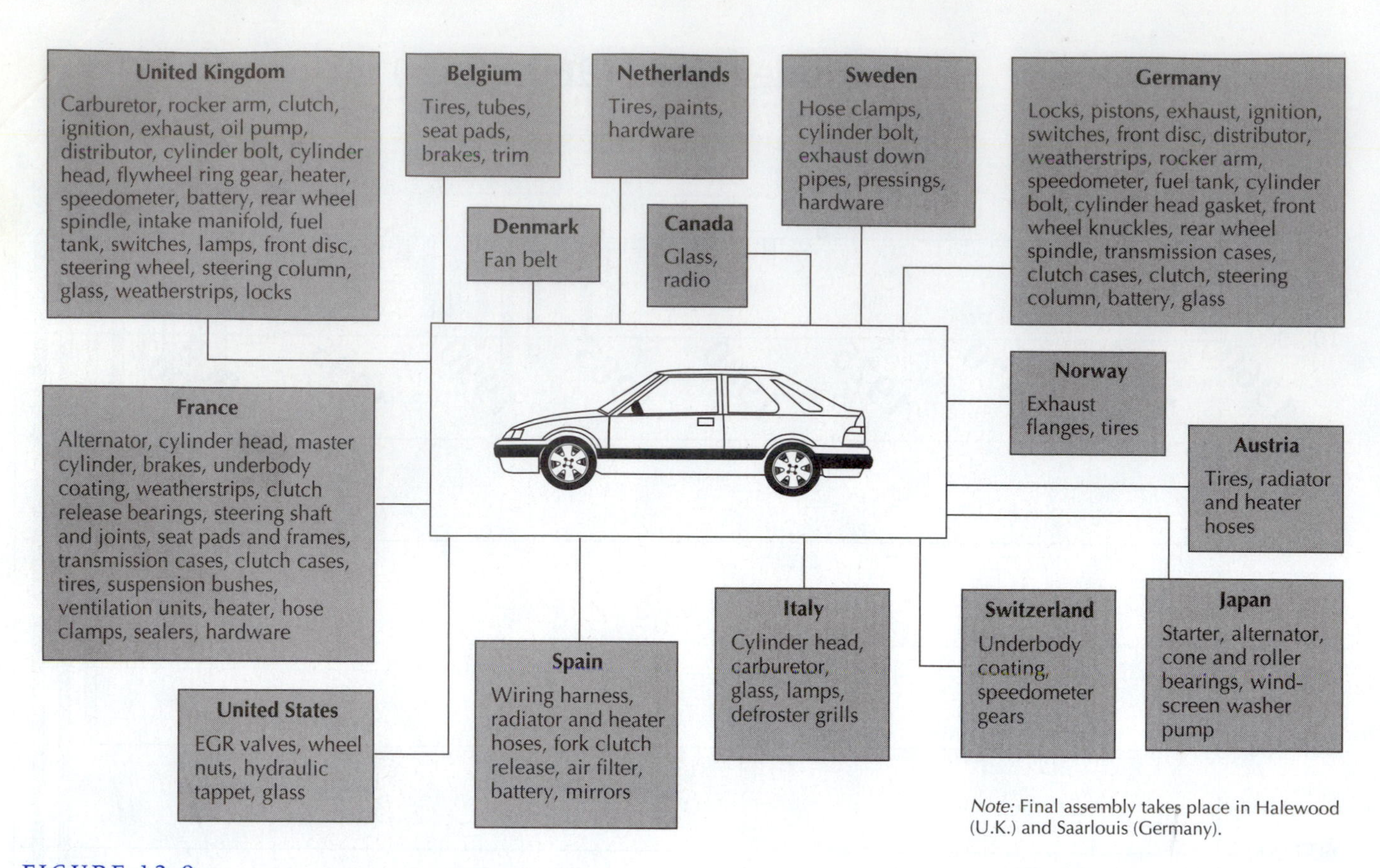

FIGURE 12.9
The international car: the component network for the Ford Escort (Europe).

foreign products, at least those that are not name brands. Cultural differences in business ethics and protocol, attitudes regarding time, and even body language in interpersonal relationships complicate the task of conducting business in two or more languages. Added to these barriers are problems relating to currencies, laws, taxation, and governmental restrictions.

World Investment by Multinationals

American firms lead the world in FDI, but their share of the total is slipping. Until the early 1970s, U.S.-based multinationals accounted for nearly two-thirds of the world's corporate investment abroad. This figure decreased throughout the late twentieth century as corporations headquartered in other countries stepped up their rates of FDI. The rate of increase has been most rapid for companies domiciled in Western Europe and Japan; however, some underdeveloped countries have also increased their outflow of FDI (Figure 12.10). Major bases are Hong Kong, Brazil, Singapore, South Korea, Taiwan, Argentina, Mexico, and Venezuela.

In 2003, the share of stocks held abroad by U.S. companies fell to approximately one-third of the world total. The United Kingdom ranks in second place, with approximately 15 percent of world FDI, followed by Japan at 13 percent. Germany is not far behind with approximately 10 percent, followed by Switzerland and the Netherlands, each at approximately 6.5 percent. Other industrialized countries make up 15 percent, whereas less developed countries make up only 2 percent.

Significant changes in the destination of FDI are the increased flow to the United States and to the LDCs (Figures 12.11 and 12.12). In 1975, the United States accounted for only a small proportion of the stocks held by foreign companies; 10 years later, the United States emerged as a major host country. Investment in the developing world has focused mainly on eight countries—China, Brazil, Mexico, Singapore, Indonesia, Malaysia, Argentina, and Venezuela—which accounted for more than one-half the stock for foreign investment in underdeveloped countries. Availability of natural resources, recent strong growth, and political and economic stability were among the factors that attracted foreign investment to these LDCs.

Investment by U.S. Multinationals in Foreign Countries

The pace of FDI by U.S. multinational corporations increased dramatically since the 1960s. In 1960, almost two-thirds of foreign investment by U.S. companies was

Foreign direct investment. Ford Motor Company headquarters in Britain. The Ford Motor Company is a true multinational corporation with 330,000 employees scattered throughout the world in 30 countries. Although accused of exploiting local labor in developing countries, Ford and other multinationals helped to stimulate the economy and, therefore, provide foreign sources of revenue for a struggling economy.

U.S. Ford plant in Europe. General Motors and the Japanese manufacturers of automobiles produce two-thirds to nine-tenths of their products in their home countries. The Ford Motor Company produces two-thirds of its more than 5 million manufactured vehicles abroad. Ford Motor Company has been one of the most prolific of the American multinational corporations. The 1960s saw the most rapid expansion, with over 300 new subsidiaries being set up annually by American companies in foreign countries. American firms saw the advantage of multinational expansion because postwar markets were developing rapidly for their goods in Europe, Japan, and throughout the world. By establishing manufacturing plants in those foreign countries, companies could reduce expensive transportation costs of finished products as well as import duties. A major factor is sometimes overlooked, and that is transfer pricing. Japan and European countries have higher rates of corporate taxation than does America. Shrewd multinationals can invoice their subsidiaries in these foreign countries in such a way to show low profits in high tax countries and therefore shelter their income.

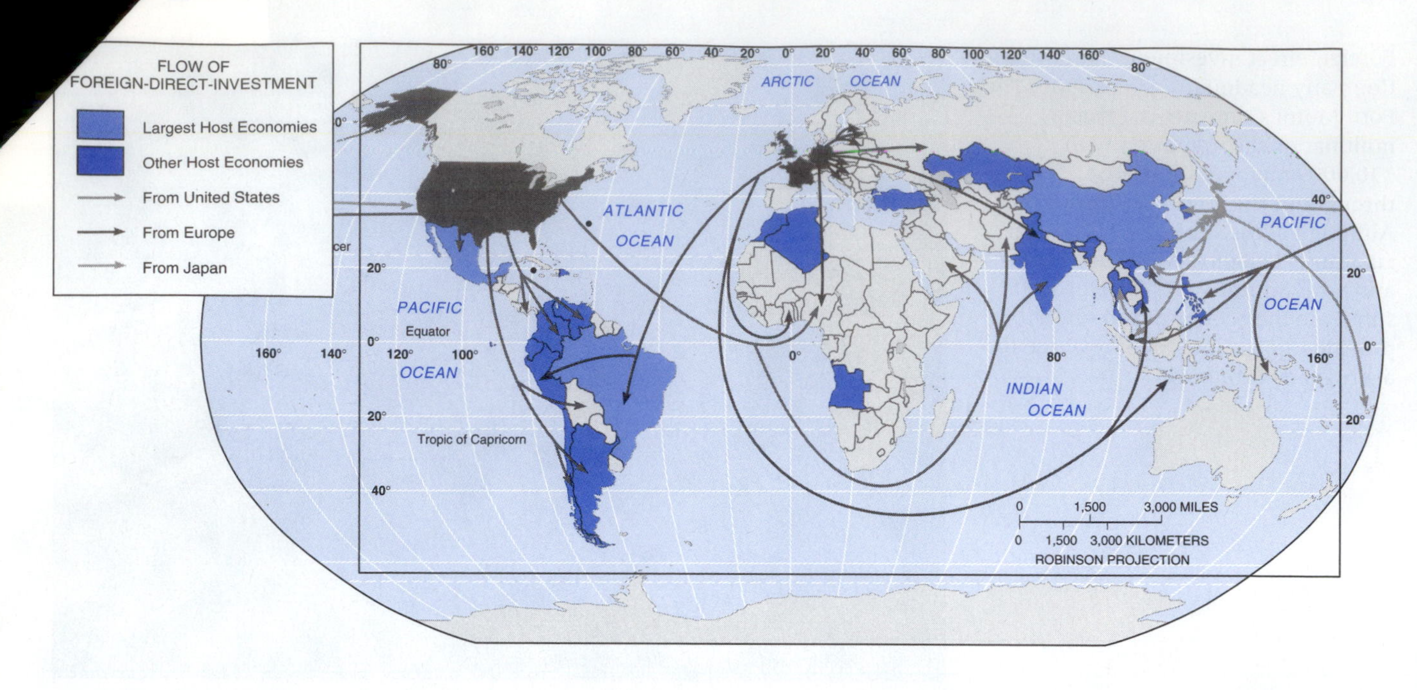

FIGURE 12.10
Flows of foreign direct investment. Multinationals' foreign direct investment originates, for the most part, in the United States, Japan, the United Kingdom, Germany, and France. These transnational corporations have invested most of their resources in other developed countries. In addition, U.S. multinational corporations are more likely than the Japanese or Europeans to invest in Latin America. European multinational corporations are more likely to invest in eastern Europe and the Middle East, while Japanese transnationals are more likely to invest resources in East Asia.

in nearby Canadian ventures. Other foreign investment was located in Europe and in Latin America. East Asia saw little investment by U.S. multinationals, partly because of distance, cultural, and political barriers. By 1980, U.S. FDI was positioned chiefly in Europe in principally manufacturing operations.

Figure 12.13 shows that in 2000 Europe was still the preeminent area for U.S. FDI. During this period investment in Europe increased by 284 percent, whereas U.S. investment in Canada increased only 85 percent. The greatest proportional increase, however, occurred in East Asia, primarily in the Four Tigers of

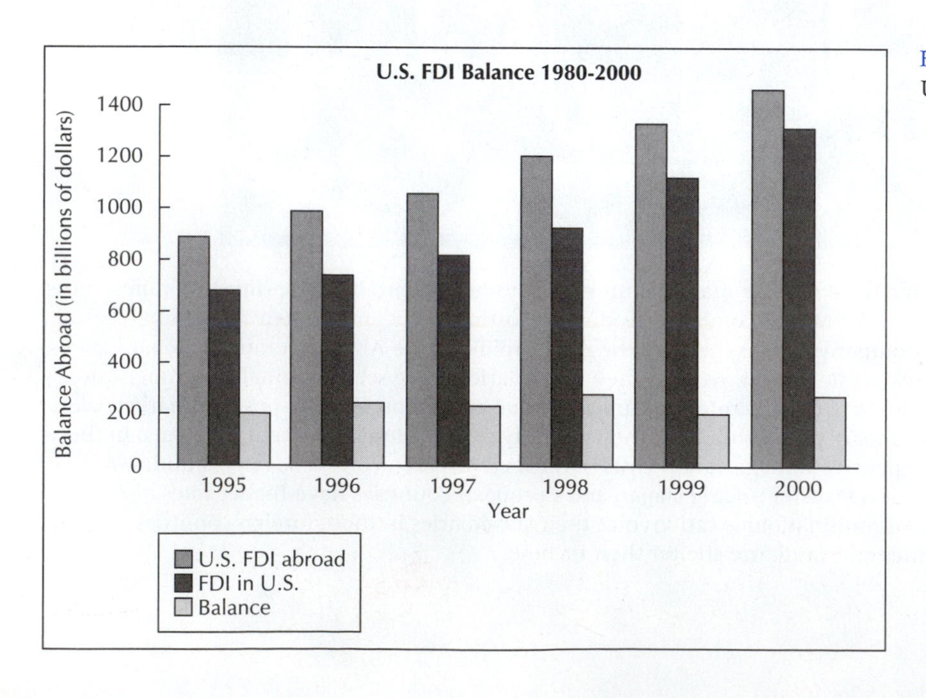

FIGURE 12.11
U.S. FDI balance, 1980–2000.

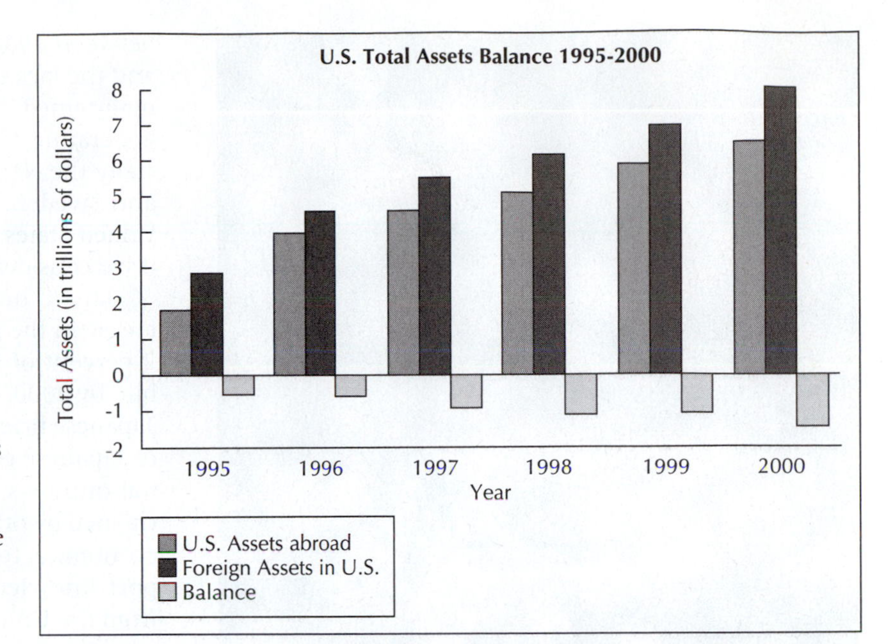

FIGURE 12.12
U.S. total assets balance, 1995–2000. Foreign held assets in the United States have accumulated as a result of the mounting U.S. indebtedness to the rest of the world. This is shown by a comparison of total U.S. overseas assets, both governmental and private, with total assets held in the United States by foreign interests. After many years of positive asset balances with the world, the United States became a net debtor nation in the 1980s.

South Korea, Taiwan, Singapore, and Hong Kong, but also in Malaysia and the Philippines. Part of the reason for the increase in investment in East Asia is that the labor-intensive semiconductor assembly plants are located in these cheap-labor countries. Because of strong Japanese barriers to foreign investment, Japan entertains a relatively small proportion of total FDI by U.S. firms but numerous overseas affiliates. Another region, Africa, received only marginal increases in U.S. FDI, further hurting its opportunities for development. Until the 1980s, FDI by U.S. firms centered on manufacturing and mining activity; more recently, investment has targeted real estate service activities.

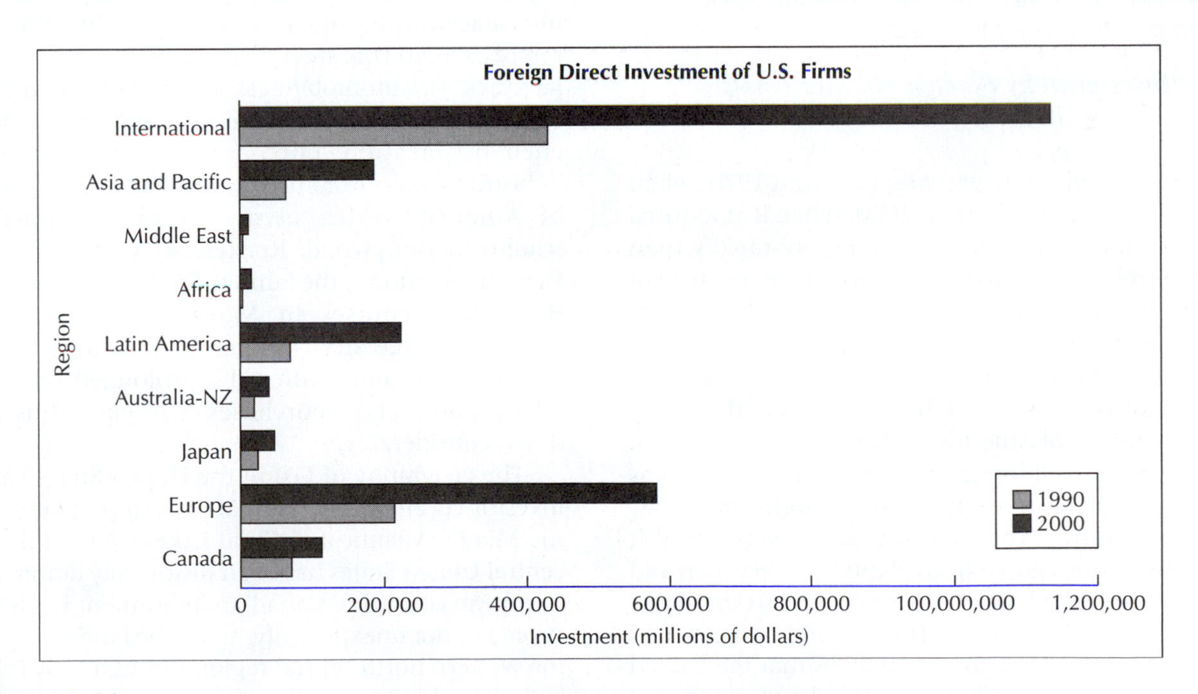

FIGURE 12.13
Foreign direct investment of U.S. firms by region and percent change, 1980–2000. Europe remained the favorite destination for U.S.-based multinational investments. Latin America remains in second place, Australia is the only contender for third place, and Canada remains a steady fourth. The largest increase proportionally for foreign direct investment of U.S. companies occurred in Australia and the Pacific Rim countries, especially the Four Tigers, where the pace of U.S. investment accelerated sharply in the 1990s.

Foreign direct investment by U.S. firms in the Pacific includes the ubiquitous McDonald's.

Investment by Foreign Multinationals in the United States

FDI in the United States grew rapidly from 1970, when it was a skimpy $13 billion, to 2000, when it amounted to $800 billion. In fact, it increased more rapidly than did U.S. foreign investment overseas. The popularity of investing in the United States was a result of the power and stability of the country economically and politically and the relatively inexpensive dollar in world markets.

The trade deficit declined but is still large. The deficit led to the outflow of American dollars into foreign hands. This money allowed foreign governments and corporations to buy American real estate and factories. Ironically, many foreign firms that sell to U.S. markets found it cheaper to produce goods from plants that they own and operate *in* the United States. For instance, foreign multinational corporations such as Honda and Mazda opened plants in America to build automobiles that the United States formerly imported from Japan. Land, labor, and capital were cheaper in the United States; therefore, the product was less expensive to produce there.

As a result, foreign investment in the United States increased sharply. The average increase was 33 percent between 1980 and 2000. Because of cultural affinities and the lack of language barriers, the United Kingdom maintained its lead as America's chief foreign investor, generating 19 percent. Other European nations, especially the Netherlands, Germany, Switzerland, France, and Sweden, in that order, are strong investors in the United States. The Netherlands is especially noteworthy if we consider its relatively small size.

Japan, however, had the largest proportional increase in the last 20 years. In 1970, Japan owned only 2 percent of foreign investment in the United States, but, by 2000, its share had risen to 18.2 percent. FDI by Japanese firms is a relatively recent occurrence because of Japanese governmental policy, which regulates capital outflows in an effort to restrict foreign exchange claimed by other countries. The Japanese have found it economical to invest in U.S. companies and thereby export knowledge-intensive activities and technology from the United States to Japan. Many of these knowledge-intensive industries are now well placed in Japan, but because of import restrictions and the high price of Japanese currency compared with the dollar, Japanese firms have found it even more profitable to sell to their chief buyer, the United States, from Japanese-owned manufacturing plants in America. The most notable example is the accumulation of Japanese-owned automobile-assembly plants and autoparts industries in the U.S. Midwest.

FDI in the United States is centered on manufacturing, chemicals, electrical machinery, electronics, pharmaceuticals, and services. Japan has controlling interests in 60 U.S. steel operations, 25 rubber and tire factories, 10 automobile-assembly plants, and 300 autoparts distributors. U.S. public attention was heightened by Japanese purchases of retail and service industries such as Columbia Records; the Music Company of America, which owns several motion-picture studios in Hollywood; Rockefeller Center; Federated Department Stores; the famous Spyglass Hill and Pebble Beach golf courses in Monterey, California; and Yosemite Lodge and National Park Company. As the Japanese economy suffered a prolonged crisis in the 1990s, however, its purchases of foreign firms slowed down considerably.

The geography of FDI in the United States varies by investor country. The original pattern of investment in the Middle Atlantic and Great Lakes states of the north central United States has been historically dominated by European countries. Canadian investment has been the strongest, not unexpectedly, along the border states from the western north central region through New England, including the South Atlantic states. But the Pacific region of the United States leads overall in FDI, primarily because of recent heavy investment by the Japanese and other East Asian countries (Figure 12.14). Overall, California was the FDI leader with approximately $92

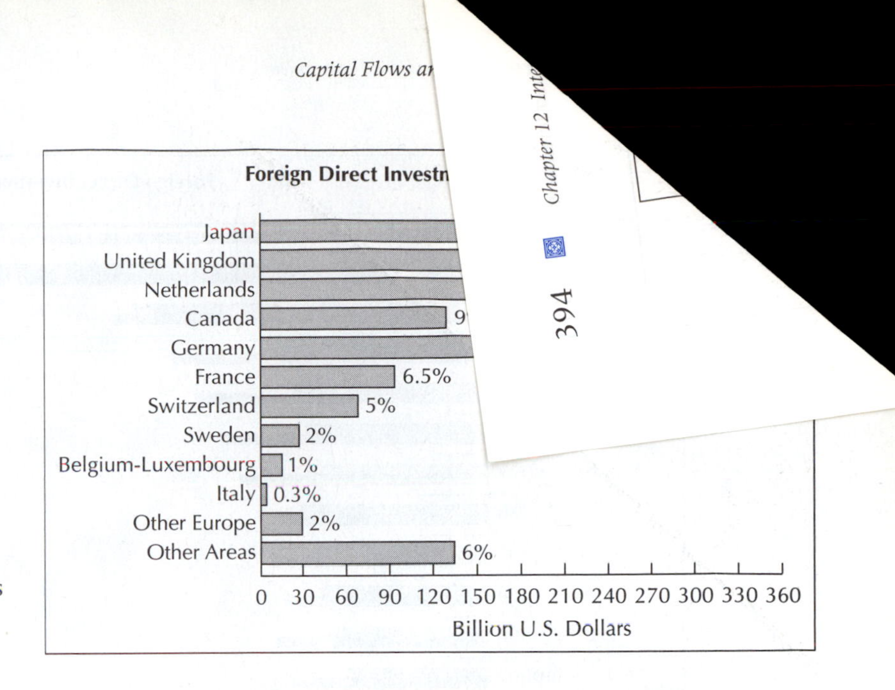

FIGURE 12.14
Foreign direct investment (FDI) in the United States by source area, 2000. FDI in the United States is defined by the U.S. Bureau of the Census as all U.S. companies in which foreign interest or ownership is 10 percent or more.

billion in 2000 (Figure 12.15). Texas was the second largest state for FDI, with $78 billion, followed by New York as a distant third, with $54 billion. In the Pacific region, Alaska was second with $26 billion.

Effects of Foreign Direct Investment

Are the effects of widespread FDI desirable? Should the operations of multinationals be controlled? There is no unanimity of opinion, particularly when LDC development is the issue. There are two polar attitudes regarding the presence of multinationals in LDCs. Those on the right of the political spectrum argue that the multinational firm has a high potential to aid the economic development process. In this view, the transnational corporation (TNC) is an efficient social, economic, and political institution that accomplishes the following tasks for the less-developed nations:

1. Raising, investing, and reallocating capital
2. Creating and managing organizations
3. Innovating, adopting, perfecting, and transferring technology
4. Distributing, performing maintenance, marketing, and sales
5. Furnishing local elites with suitable career choices
6. Educating and upgrading both blue-collar and white-collar labor
7. Serving as a source of local savings and taxes and in furnishing skilled graduates to the local economy
8. Facilitating the creation of vertical organizations or vertical arrangements that allow for the progression of goods from one stage of production to another
9. Finally, providing both a market and a mechanism for satellite services and industries that can stimulate local development

This view is in marked contrast to that of scholars who argue that the multinational corporation is counterproductive to development. In the view of *dependency theorists,* modern capital-intensive industry does not result in rapidly increasing employment (Chapter 14). Foreign firms can bankrupt local producers, who may be small or undercapitalized, and establish local monopolies. MNCs that use extensive networks of foreign suppliers have few local linkages and low employment multipliers. They may not improve the host country's balance of payments because of heavy profit repatriations. Indeed, some countries attempt to restrict repatriation of profits to keep the benefits of investment at home. Although the balance-of-payments problem could be avoided in part if multinational firms reinvested more of their profits in the host country, it is uncertain that the national interest would be served. Reinvestment causes growing foreign control of the economy and the denationalization of local industry. Some argue that the multinational firm is an assault on political sovereignty, including demands for government subsidies, training programs, and tax breaks. This latter issue is particularly important given the long history of MNC involvement in the politics of developing countries, ranging from bribes to helping military officers overthrow democratically elected governments. Moreover, the transnational system internationalizes the tendency to unequal development and to unequal income.

To be sure, multinationals are imperfect organs of development in the developing countries, and their potential for the exploitation of poor countries is tremendous. There is, therefore, an inherent tension between the multinational's desire to integrate its activities on a global basis and the host country's desire to integrate an affiliate with its national economy. Maximizing corporate profits does not necessarily maximize national economic objectives.

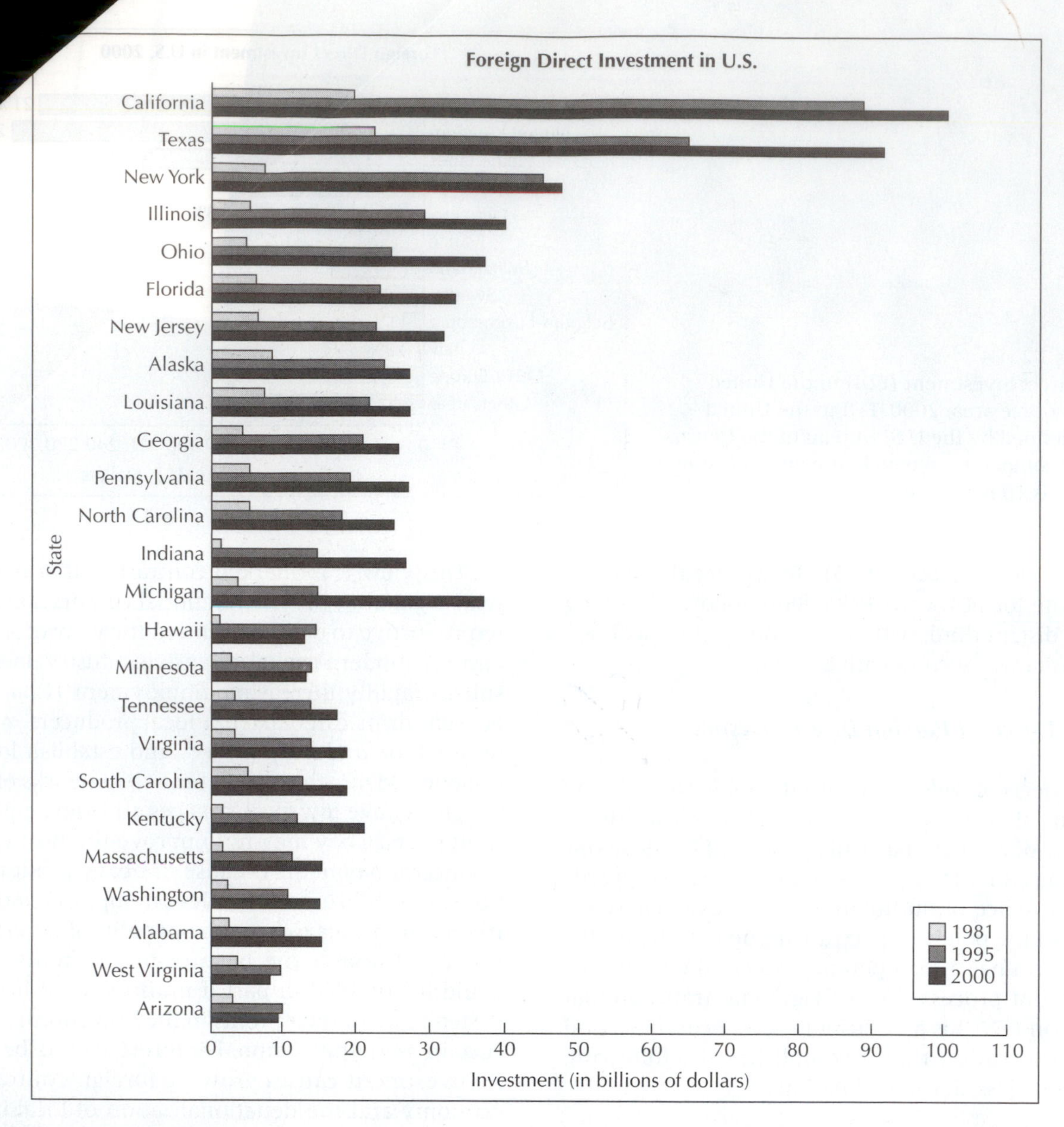

FIGURE 12.15

Foreign direct investment in the United States, 2000. The value of foreign direct investment (FDI) during this period has quadrupled. The top four states accounted for more than a third of the total FDI, and California, Texas, and New York led all states. A large influx of foreign direct investment has brought prosperity to the Rust Belt states of Michigan, Illinois, Indiana, and Ohio, with extensions into Kentucky and Tennessee.

The impacts of MNCs depend on both the characteristics and bargaining power of the corporation and the host country. Both larger and wealthier LDCs and MNCs have more bargaining leverage. A consumer-products manufacturing corporation will accept more controls to gain access to a country with a large market. Similarly, the degree of host-country control varies across industries and states. Manufacturing industries with advanced and dynamic technologies are more difficult to control than firms in the raw-materials area. Host governments' policies range from open hostility to open encouragement. Corporations prefer to invest in countries that follow an outward-oriented, export-led development strategy, impose few controls, offer incentives, and appreciate the employment, skills, exports, and import substitutes that foreign investment can bring, an outlook exemplified by the newly industrializing countries (NICs).

Barriers to International Trade and Investment

Just as trade can, international flows of the production factors can help to reduce imbalances in the distribution of natural resources. Whereas trade *offsets* differences in factor endowments, factor movements *reduce* these differences. International trade and factor flow would occur more commonly if barriers did not exist. The main barriers relate to management, distance, and government.

Management Barriers

A number of managerial characteristics reduce trade and investment expansion. These characteristics include limited ambition, unawareness of opportunity, lack of skills, fear, and inertia.

Firms may have the potential to expand but fail to do so because they are *satisficers*—settle for less than the optimal. Until the economic crisis of the 1970s, many U.S. firms paid little attention to foreign markets. They were satisfied with the large domestic market. Firms may have the will to go international but may lack knowledge of potential markets. The burden of recognizing export opportunities rarely falls on the managers of individual companies, however. Most national and local governments are actively involved in increasing international awareness and in promoting exports.

Firms may have the potential and will to go international *and* an awareness of the opportunities, but they may be thwarted by the complexity of international business and ignorance of foreign cultures. Governments and universities can aid companies by providing education. Knowledge of intermodal rate structures, freight forwarders, shipping conferences, and customshouse brokers (firms that contract to bring other companies' imported goods through local customs) is vital for the conduct of international business. Just as necessary is a knowledge of foreign cultures. In the past, for example, U.S. firms could not penetrate the Japanese market, partly because of a failure to appreciate Japanese culture.

Government Barriers to Trade

No country permits a free flow of trade across its borders. Governments have erected barriers to achieve objectives regarding trade relationships and indigenous economic development. Trade barriers include *tariffs*—schedules of taxes or duties levied on products as they cross national borders—and *nontariff barriers*—quotas, subsidies, licenses, and other restrictions on imports and exports. These kinds of obstacles (apart from political bloc prohibitions) are the most pervasive barriers to trade.

Free market enthusiasts advocate free trade because it promotes increased economic efficiency and productivity as a result of international specialization. They argue that trade, a substitute for factor movements, benefits each participating nation and that deviation from free trade will inhibit production. It follows, then, that *protectionism* adversely affects the welfare of the majority.

What are some of the major arguments in favor of protectionism? One of the most common is the cheap foreign-labor argument, which suggests that a country such as the United States with its high union wages must protect itself against a country such as Taiwan with its low-paid workers. This argument contradicts the principle of comparative advantage. Other more compelling arguments appeal to national gain. One argument asserts that a country with market power can improve its terms of trade with a tax that forces down the price at which other countries sell to it. Another argument is that tariffs can be used to divert demand from foreign to domestic goods so as to shift a country's employment problem onto foreign nations. Still another argument is that tariffs can be used to protect an infant industry that is less efficient than a well-established industry in another country. The *infant-industry argument* was invoked to justify protectionist policies in nineteenth- and twentieth-century America and nineteenth-century Germany. It was also used to justify the protection that developed in the LDCs in the 1960s. Although these arguments have some merit, free marketeers recommend other approaches to attain desired goals. For example, they suggest that if grounds exist for protecting an infant industry until it has grown large enough to take advantage of scale economies, protection could be given through a subsidy rather than through a protective tariff.

Despite long-standing arguments for free trade and low tariffs, throughout the history of the United States tariff rates have been unquestionably high (Figure 12.16). The Compromise Tariff of 1833 and the Smoot-Hawley Tariff of 1930 put rates at almost 70 percent of dutiable imports; that is, 70 percent of imports were subject to the tariff. The reasons for tariffs are clear. Special-interest groups who stand to gain economically from tariffs and quotas press the government for protection through the use of high-powered, politically savvy lobbyists. The public, which must then absorb these tariffs and quotas as a surcharge on all imported products, are politically uninformed and not well represented in Washington, D.C.

In 1947, however, the United States and 25 other nations signed the *General Agreement on Tariffs and Trade* (GATT). GATT established multinational reductions of tariffs and import quotas and now has more than 110 signatories (World Trade Organization [WTO]), which is discussed in more detail later. The *Uruguay Round*, which began in 1986, once again reduced tariffs, so U.S. tariff rates currently average approximately 5 percent of dutiable imports.

Tariffs, Quotas, and Nontariff Barriers

Tariffs are the most visible of all trade barriers, and they can be levied on a product when it is exported, imported, or in transit through a country. The tariff structure

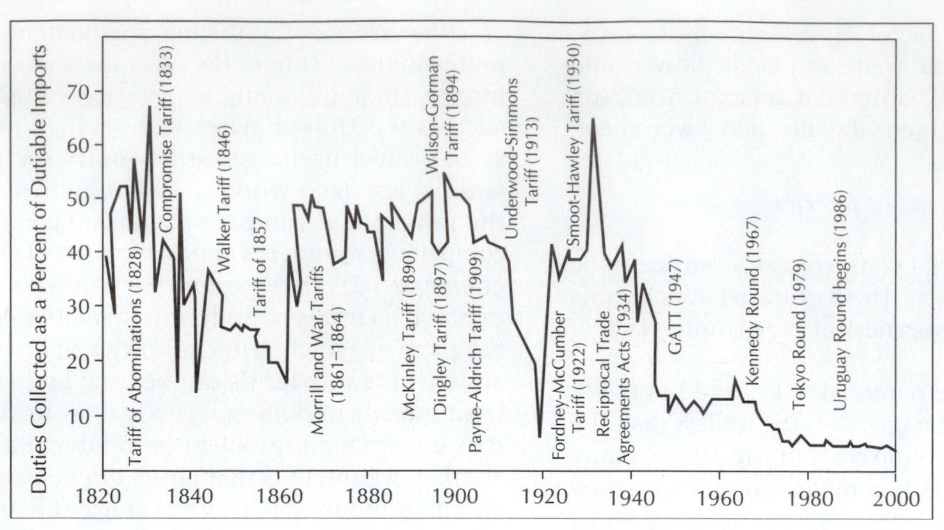

FIGURE 12.16
U.S. average tariff rates, 1820–2000. Compared with the inception of the General Agreement on Tariffs and Trades (GATT) in 1947, U.S. tariff rates have been historically high. They reached a peak during this century in the mid-1930s, with the Smoot-Hawley tariff, designed to protect U.S. markets from foreign goods, precipitating a trade war with Europe. Today, the United States has the lowest barriers to trade in the world, although on occasion American administrations will impose tariffs for political gain, as the Bush administration did in 2001 with steel imports, which prompted the World Trade Organization to rule them illegal and forced them to be rescinded in 2003.

established by the developed countries in the post–World War II period works to the detriment of underdeveloped countries. The underdeveloped countries encounter low tariffs on traditional primary commodities, higher tariffs on semimanufactured products, and still higher tariffs on manufactures. These higher rates are, of course, intended to encourage firms in industrial countries to import raw materials and process them at home. They also discourage the development of processing industries in the developing world.

In recent years, the relative importance of tariff barriers has decreased, whereas nontariff barriers have gained significance. The simplest form of nontariff barrier is the *quota*—a quantitative limit in the volume of trade permitted. A prominent example of a product group subject to import quotas in developed countries is textiles and clothing. Since the early 1970s, textiles and clothing have been subject to quotas under successive Multifibre Arrangements (MFAs). These arrangements have created a worldwide system of managed trade in textiles and clothing in which the quotas severely curtail underdeveloped-country exports. Another common nontariff barrier is the *export-restraint agreement.* Governments increasingly coerce other governments to accept "voluntary" export-restraint agreements, through which the government of an exporting country is induced to limit the volume or value of exports to the importing country. The United States has employed this special type of quota—*an export quota*—extensively.

Other nontariff barriers include discretionary licensing standards; labeling and certificate-of-origin regulations; health and safety regulations, especially on foodstuffs; calendriers, which allow foodstuffs to be imported only during certain seasons to avoid competing with the peak production of the importers; and packaging requirements. Increasingly, loose, or break-bulk, cargo is unacceptable to developed-country mechanized transportation handlers. Dockers and longshoremen often demand bonuses for handling such items as unpacked skins and hides. Consumers too demand agricultural products in packaging that requires more investment on the part of the exporting country. These examples represent only a few of the hundreds of nontariff barriers devised by governments. The evidence indicates that these barriers in developed countries are higher for exports from developing countries than they are for exports from rich, developed countries.

Effects of Tariffs and Quotas

The economic effect of tariffs and quotas in the host country is the development and expansion of inefficient industries that do not have comparative advantages. At the world level, tariffs and quotas penalize industries that are relatively efficient and that do have comparative advantages. The result is less international trade and penalized consumers.

Figure 12.17 shows the economic effects of a protective tariff and an import quota. Let's first deal with the case of a protective tariff. Line D_d represents domestic demand in, for example, the United States for cassette players, whereas line S_d is the domestic supply. (Disregard the S_d + Quota line for now.) The domestic equilibrium position is at price p_3 and quantity q_3. Now

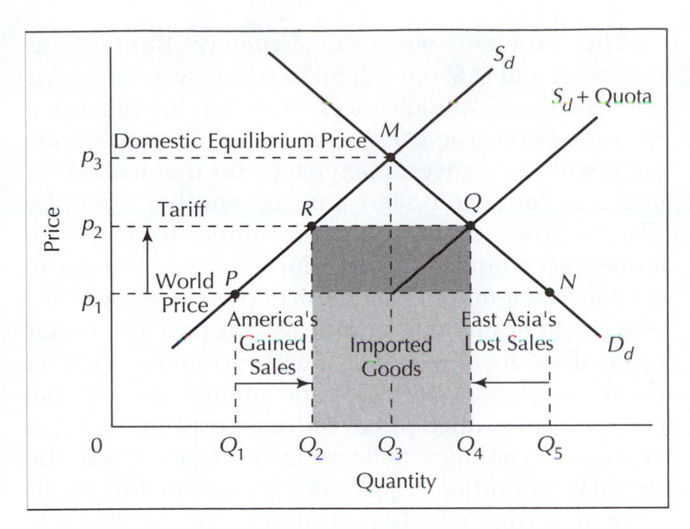

FIGURE 12.17
The economic impacts of tariffs and quotas.

assume that the United States market for CD players is open to world trade. The world price is lower than the domestic price because, compared with Japan, Malaysia, or Taiwan, the United States has the comparative disadvantage of high labor costs. The world equilibrium price is p_1. At this price, Americans will consume the quantity q_5 at point *N*. With the low price, the domestic supply is only q_1 at point *P*, with the quantity $q_1 - q_5$ supplied by foreign imports. Next, let's say that the United States imposes a tariff on the import of CD players. This tariff raises the price from p_1 to p_2. The equilibrium price and quantity are now p_2, q_4, respectively, at point *Q*.

The first reaction will be a decline in quantity demanded by American consumers, from q_5 to q_4, as they lock up their demand curve toward the higher price. American consumers are hurt by the tariff because they can buy fewer goods at a higher price. While the consumers move back up the demand curve to point *Q*, domestic producers, with a higher price opportunity, increase their production and move up their supply curve from point *P* to point *R*. Domestic production has increased from q_1 to q_2. Consequently, we can understand why domestic producers send lobbyists to Washington to invoke tariffs that give the producers a relative advantage in the market.

The increased tariff reduces the number of East Asian imports from $q_1 - q_5$ to $q_2 - q_4$. The U.S. government, not the East Asian supplier, receives the tariff monies, $p_1 - p_2$. At the same time, the market decreases because of reduced demand, and domestic supply increases. The shaded area represents the amount of tariff revenue paid to the U.S. government. This revenue is an economic transfer from the consumers of the country to the government.

The result of levying a tariff is reduced world trade and reduced efficiency of the international economic system, which hurts foreign suppliers, aids domestic producers, and costs the consumer. The indirect effect is that the supplying countries have a smaller market in America and thus earn fewer dollars with which to exchange or invest in American resources. As FDI decreases, trade deficits may increase.

Next, let us consider the effects of levying of an import quota. The difference between a tariff and an import quota is that a tariff yields extra revenue to the host government, whereas a quota produces revenue for the foreign suppliers. Imagine that the United States subjects a foreign nation to an import quota, rather than imposing a tariff (Figure 12.17). The import quota in this case is $q_2 - q_4$. The quantity $q_2 - q_4$ is the number of CD players that foreign producers are allowed to supply. Note that for easy comparison this example limits the quota to the exact amount of imported goods in our tariff example. The quota establishes a new supply curve, S_d + Quota, with an equilibrium position at point *Q*. The new supply of CD players is the result of domestic supply, plus a constant amount, $q_2 - q_4$, which is supplied by importers.

The chief economic outcomes are the same as with the tariff example. The price is p_2 rather than p_1, and domestic consumption is reduced from q_5 to q_4. The American manufacturers enjoy a higher price for their goods, p_2 rather than p_1, and increased sales, q_2 rather than q_1. But the main difference is that the shaded box, paid by the domestic consumer on imports of $q_2 - q_4$, is not paid to the U.S. government. Instead, the extra revenue in the shaded box is paid to the foreign supplier. That is, no tariff exists, and consequently, the foreign supplier keeps all the revenue in the box $q_2 - q_4 - Q - R$.

The result is that for local consumers, and their government, a tariff produces a better revenue situation than a quota does. Tariff money can be used to lower the overall tax rate and provide social services and infrastructure for the population as a whole. However, either case is detrimental to international trade and economic efficiency and takes away from the comparative advantages of supplying nations.

Government Stimulants to Trade

Not only do governments attempt to control trade, but they also attempt to stimulate trade. Examples of governmental assistance to promote exports include market research, provision of information about export opportunities to exporters, international trade shows, trade-promotion offices in foreign countries, and free-trade zones, areas where imported goods can be processed for reexport without payment of duties. Advocates of free trade believe that government intervention to promote trade is yet another obstacle to free trade and a subsidy to politically influential corporations. In their view, gains from trade should result from economic efficiencies, not from government support.

Reductions of trade barriers

Several efforts have been made to eliminate some of the trade barriers that were erected in the past. GATT and the Uruguay Round were two of these efforts. In addition, the United States continues to try to penetrate Japan's closed markets.

General Agreement on Tariffs and Trade

The most notable effort to reduce trade barriers was a multilateral effort known as GATT. GATT was put into operation in 1947. When 23 countries signed the agreement, they thought that they were putting in place one part of a future WTO. The organization would have wide powers to police its trading charter and regulate international competition in such areas as restrictive business practices, investments, commodities, and employment. It was to be the third in the triad of Bretton Woods institutions charged with overseeing the postwar economic order—along with the International Monetary Fund and the World Bank. But the draft charter of the WTO was never ratified by the U.S. Congress and GATT remained a treaty without an organization.

GATT was administered on behalf of more than 100 member countries, which make decisions by a process of negotiation and consensus. This process resulted in a substantial reduction of tariffs. However, GATT's rules proved inadequate to cope with new forms of nontariff barriers such as export-restraint agreements. Areas such as services, which now account for about 30 percent of world trade, were not covered by GATT at all. GATT was also of little help to developing countries with limited trading power. Since the 1970s, the liberal trading order that GATT helped to uphold has been steadily challenged by renewed threats of protectionism, especially in the guise of nontariff barriers. Between 1981 and 1990, the proportion of imports to North America and the European Union (EU) that were affected by nontariff barriers increased by more than 20 percent. Trade between developed and underdeveloped countries is increasingly affected by nontariff barriers; roughly 20 percent of LDC exports were covered by such measures in 2000. In the coming years, pressure on governments in developed countries to protect domestic jobs through trade barriers is likely to mount.

In 1986, the eighth round of GATT opened in Montevideo, Uruguay. This Uruguay Round of GATT negotiations centered on (1) removing barriers to international trade and services, which now account for 21 percent of all international trade; (2) ending limits on foreign economic investments; (3) establishing and policing patent, copyright, and trademark rights (intellectual property rights) on an international basis; and (4) reducing agricultural trade barriers and domestic subsidies.

The problem with the Uruguay Round was European and U.S. opposition to reducing and phasing out agricultural subsidies, a problem that has reoccured with the WTO. The EU wanted to maintain *export subsidies,* which are government payments that reduce the prices of goods to buyers abroad. Another type, the *domestic farm subsidy,* constitutes direct payments to farmers according to their production levels in order to subsidize their output. The result of these subsidies is increased domestic food output, which provides unfair competition for U.S. and LDC agricultural products on the world market. Both types of subsidies are artificial barriers that reduce prices on the world market and provide advantages to local farmers. In 1993, the Uruguay Round of negotiations reconvened in an attempt to further resolve trade disagreements. The U.S. effort was an attempt to pry open foreign markets, especially those considered to be unfair traders. Because most of the industrialized world was in the midst of an economic slowdown or an outright recession in 1992, there was much interest in the international trade measures of GATT at Uruguay.

Finally, in Geneva in 1993, 117 nations agreed to reduce worldwide tariffs, lower subsidies, and eliminate other barriers to trade. In reducing worldwide protectionism, the GATT nations have chosen to improve resource allocation, which will eventually increase trade and employment and raise wages and standards of living. The United States obtained most of what it sought: the opening of agricultural markets, cuts in industrial tariffs, intellectual property rights, and the opening of markets in the world's service industries. There were both losers and gainers. America's heavy equipment manufacturers, toy makers, and beer brewers were joyful, but the pharmaceutical industry, Hollywood filmmakers, and textile manufacturers were not pleased.

World Trade Organization

In 1995, the last round of GATT established a World Trade Organization (WTO), to which most of the world's countries now belong, including China (Figure 12.18). Until the WTO was enacted, countries with conflict over trade had to resolve their own problems. WTO describes arbitration by a third party to settle conflicts between a pair of nations over trade disagreements. The judgments will be enforced through retaliatory trade sanctions by all members of the WTO, meaning that a part of a country's sovereignty is given up to this multilateral world organization. With the enactment of the WTO, a country's power to control flows across its borders is to some extent lost. For some countries not abiding by international trade agreements, this is a scary proposition.

Under the WTO, quantitative limits on imports become illegal. For example, with this provision, Japanese and Korean import bans on rice from the United States,

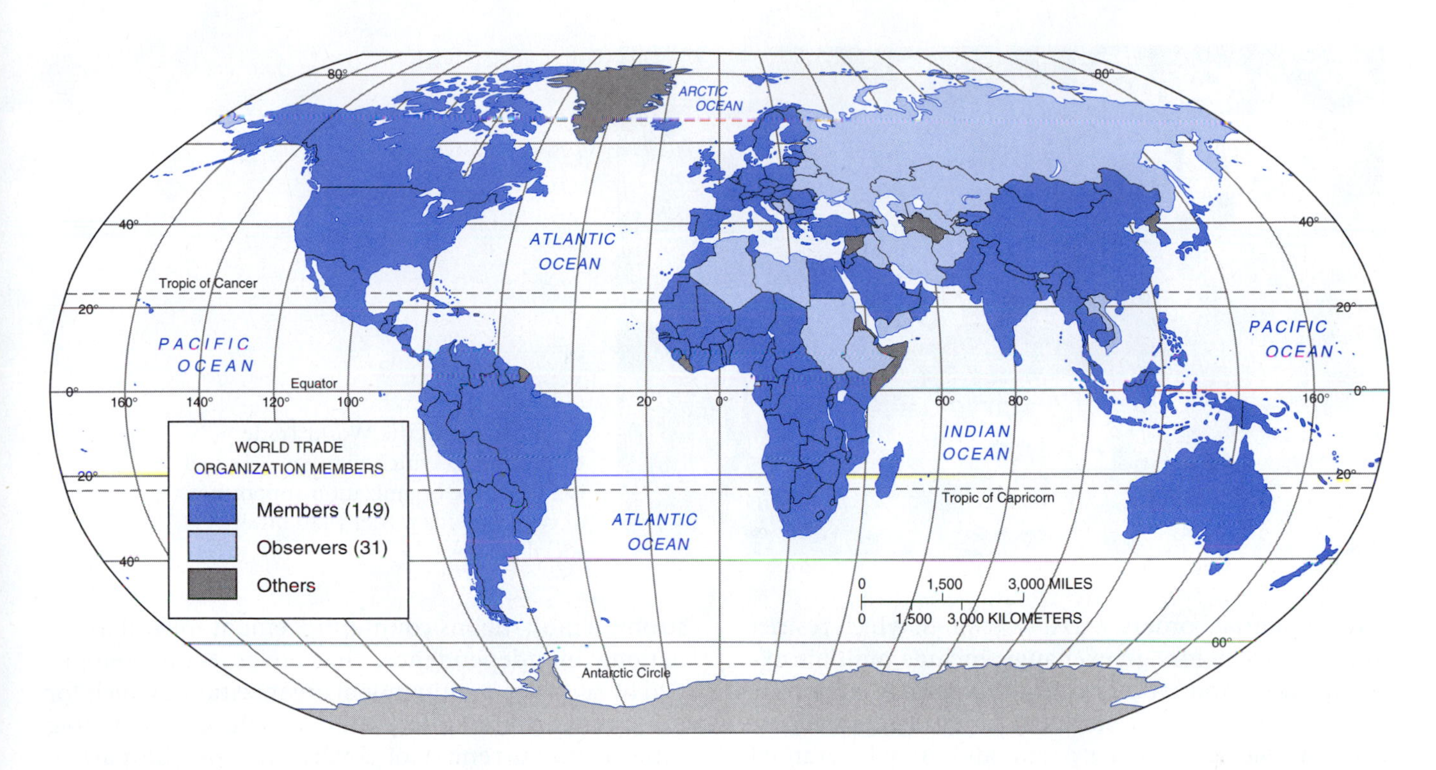

FIGURE 12.18
Member states of the World Trade Organization, 2005.

peanuts, dairy products, and sugar are now banned. Intellectual property rights is perhaps the most important aspect of the WTO to the United States, which has a strong competitive advantage in this arena. All signatories of the WTO are required to protect patents, copyrights, trade secrets, and trademarks. This measure is designed to end the wholesale pirating of computer programs, video cassettes, musical recordings, books, and prescription drugs widely practiced in some developing countries (e.g., China). The WTO calls for free trade and financial services, shipping, and audiovisual product—movies, television, programs, and musical recordings. The WTO also prohibits members from requiring a certain proportion of content in products manufactured within their borders. This practice was widely employed as a device to limit the use of imported parts and components and, thus, to bolster local employment.

Criticism of the WTO has centered on the lack of environmental production standards. A good can be entered into international trade without meeting production standards of another country. Many signatories have argued that environmental standards should be a consideration for a good on the international market. Others argue that globalization, dominated by large corporations, is secretive and undemocratic. Labor unions are concerned about the exploitative conditions in sweatshops and the downward pressure on wages that globalization has created. Others protest the austerity programs of the IMF. For these reasons, and others, globalization is often protested, including the notable "Battle in Seattle" that occurred in 1999 and subsequent demonstrations at meetings of ministers of the Group of Eight (G-8).

Closed Japanese Markets

Japan, despite being a major world exporter, has long practiced mercantilist trade policies designed to keep imports out. The accusation of closed Japanese markets is a contentious issue that is difficult to prove. A number of econometric studies have failed to support unequivocally the conclusion that Japan's imports of American products are less than can be expected given its level of income and resources. In the recent past, Japan increased its imports and was the third largest importer of world merchandise as of 1990. It ranked first in imports of commercial services. However, part of Japan's reluctance to open its markets may be because it is experiencing a lingering recession and continuing political turmoil and scandal, and it is not as economically robust as it was earlier.

Government Barriers to Production-Factor Flows

Although not as complex as trade barriers, obstacles to the free flow of capital, labor, and technology constrain international managerial freedom. Exchange controls and capital controls are the main types of controls that interfere with the movement of money and capital

Protests against the World Trade Organization which many people feel is a secretive organization supportive of corporations rather than other social groups.

across national borders. *Exchange controls,* which restrict free dealings in foreign exchange, include multiple exchange rates and rationing. In multiple-exchange-rate systems, rates vary for different kinds of transactions. For example, a particular commodity may be granted an unfavorable rate. Foreign exchange may also be rationed on a priority basis or on a first-come, first-served basis. Thus, exchange rates are political tools bearing little relationship to economic reality. *Capital controls* are restrictions on the movement of money or capital across national borders. They are typically designed to discourage the outflow of funds.

All countries regulate immigration, but the movement of workers from poorer to richer countries was the dominant pattern during the long postwar boom. When the boom ended, jobs moved to the workers. One reason for this change was tighter immigration laws in the advanced industrial countries. These tighter laws strengthened the position of labor and resulted in a growth in managed trade and a decline in managed migration.

Technology, which is highly mobile, can be transferred in many ways: export of equipment, provision of scientific and managerial training, provision of books and journals, personal visits, and the licensing of patents. Political and military controls regulate the export of technology. Although these controls are not yet terribly onerous, demands for more stringent controls are on the increase. One source of demand for control is labor unions in advanced industrial countries. These unions attribute domestic job loss to the export of high technology.

Multinational Economic Organizations

As nations turn inward to concentrate on problems of economic growth and stability, we are witnessing a resurgence of protectionism. But also in evidence is a strong, simultaneous countermovement toward international interdependence. This movement is exemplified by scores of multinational organizations, which for the most part are loosely connected leagues entailing little or no surrender of sovereignty on the part of member nations.

Some of these international organizations are global in scale. The most inclusive is the United Nations (UN), with 191 member nations that account for more than 99 percent of humankind. Much of the UN's work is accomplished through approximately two dozen specialized agencies such as the World Health Organization (WHO) and the International Labor Organization (ILO). Other international organizations have a regional character; for example, the Association of South-East Asian Nations (ASEAN) and the Asian Development Bank (ADB). Many international organizations are relatively narrow in focus—mostly military, such as the North Atlantic Treaty Organization (NATO), or economic, such as the Organization of Petroleum Exporting Countries (OPEC). Some international organizations are discussion forums with little authority to operate either independently or on behalf of member states; for example, GATT and the Organization of Economic Cooperation and Development (OECD).

Others, such as the IMF and the World Bank, have independent, multinational authority and power, performing functions that individual states cannot or will not perform on their own. Some international organizations integrate a portion of the economic or political activities of member countries—as, for example, the EU. International organizations to promote regional integration are the most ambitious of all. Some observers believe that regional federations are necessary to the process of weakening nationalism and developing wider communities of interest. However, if a rigid, inward-looking regionalism is substituted for nationalism, the ultimate form of international integration—world federation—will be difficult to achieve.

This section examines international economic organizations that affect the environment in which firms operate and thus influence the development of developing countries. We look at international financial institutions, groups that foster regional economic integration, and groups such as commodity cartels, which deliberately manipulate international commodity markets.

International Financial Institutions

International financial institutions are largely a phenomenon of the post–World War II period. The IMF and the International Bank for Reconstruction and Development (IBRD), or World Bank, were established in 1945. Regional development banks—the Inter-American Development Bank (IADB), Asian Development Bank (ADB), and African Development Bank (AFDB)—were established in the 1960s. Various other multilateral facilities, of which the United Nations Development Program (UNDP) is the most important, were also established after 1960. These institutions are significant sources of multilateral capital, especially aid for developing countries. Multilateral capital is particularly important for the poorer developing countries that do not have access to private capital markets.

The IMF is an international central bank that provides short- to medium-term loans to member countries, and the IBRD is an international development bank that provides longer-term loans for particular projects. Both institutions are clusters of governments, each government paying a subscription or quota determined by the size of its economy. Because quotas determine a member's voting power, the banks are dominated by the most powerful economies—particularly, by the United States.

The IMF and the World Bank were originally established to prevent a recurrence of the crisis of the 1930s. At first, they embodied Keynesian principles, which offered a rationale for state intervention in the market. Starting in the 1980s, however, the IMF and the World Bank became firmly wedded to neoliberal, as opposed to dependency, interpretations of development. Under pressure from the United States, these institutions adopted adamantly market-oriented positions and imposed them on the governments of LDCs that needed their assistance, often at the cost of enormous human suffering. Loans from the IMF and the World Bank, therefore, tend to uphold the basis of U.S. economic and foreign policy. The neoliberal "Washington Consensus" that these international organizations uphold revolves around requiring LDCs to engage in tight monetary policy to combat inflation, which serves investors and bondholders well but raises interest rates for consumers, as well as liberalization of financial markets, including a variety of deregulatory programs. In the name of budget balancing, governments are often forced to reduce subsidies for the poor, including public transportation, kerosene, or cooking oil. Privatization policies encourage or require governments to sell off public assets to private investors, often at reduced prices; all over the world, formerly state-owned or operated power plants, hydroelectric facilities, bus routes, airlines, telecommunications firms, and other assets are rapidly being sold to the private sector on the assumption that it is more efficient than the public sector. Yet public services often exist precisely to overcome market failures, particularly the inability of markets to provide adequate services to the poor. Similarly, IMF policies require trade liberalization, including the end of tariffs, quotas, nontariff barriers, and subsidies. Critics note that such liberalization is often just a smokescreen for increased penetration of LDC markets by firms from the United States, so the IMF is accused of doing the dirty work of American capital. Even as LDCs are forced to give up subsidies, the U.S. government lavishes subsidies on its farmers, giving them an unfair advantage in selling low-priced crops to foreign markets. Thus, the IMF imposes requirements on LDCs that governments of developed countries would never accept.

Critics of these policies deride it as neoliberal "market fundamentalism" (see Stiglitz, 2002). By the IMF's

The atrium of the International Monetary Fund (IMF) headquarters in Washington, D.C. The IMF attempts to maintain foreign exchange balances and to promote economic modernization and growth in the Third World. Adjustment programs generally include measures to manage demand, improve the incentive system, increase market efficiency, and promote investment.

Headquarters of the World Bank, Washington, D.C.

own admission, its policies have often exacerbated the problems of LDCs, such as during the Asian financial crisis of the late 1990s. Tight monetary policies can generate recessions and lead to high unemployment. The macroeconomic models employed to buttress these policies are often highly oversimplified and underestimate the complexity of local countries political and social contexts. Many LDCs lack a proper institutional environment for privatization to work, including bankruptcy procedures, protection of property rights, and debt repayment programs. Moreover, the Washington Consensus increases inequality in LDCs by protecting and rewarding investors at the expense of the poor. In pursuing policies that dramatically emphasize economic stabilization over job creation, the IMF is eager to bail out bankers but never the impoverished masses. Finally, market fundamentalism tends to be overly optimistic about the private sector and overly pessimistic about the public sector.

Some developing countries have obtained more resources through two subsidiary World Bank organizations: the International Finance Corporation (IFC), founded in 1956, and the International Development Association (IDA), founded in 1960. These organizations provide loans with stipulations less stringent than those of the IBRD. For example, the IDA may provide loans with no interest charges, 10-year grace periods (no repayment of principal for the first 10 years), or 50-year repayment schedules for poorer developing countries. Because the IDA is much less creditworthy than the IBRD, all its resources must come from member-government contributions. These banks reflect the desire of developing countries to enhance their control. Of the three banks, the ADB is most under the control of developed countries, in particular Japan. The AFDB has been the most independent, but it is also the smallest.

International financial institutions are important for international business. The IMF, the World Bank, and regional development banks annually finance billions of dollars of the import portion of development projects. This can be valuable business for foreign companies involved in the projects, either occasionally as part owners or, more commonly, as contractors or suppliers.

Although international financial institutions facilitate international business, their project aid may not promote development. Conservative and progressive critics agree that development must be primarily an indigenous process. Foreign aid has served many underdeveloped countries as an easy substitute for devising means to generate domestic development. It is also apparent that aid from institutions heavily influenced by developed countries is a palliative designed to ensure the continuity of unequal exchange in the world economy.

Regional Economic Integration

Regional integration is the international grouping of sovereign nations to form a single economic region. It is a form of selective discrimination in which both free-trade and protectionist policies are operative: free trade among members and restrictions on trade with nonmembers. Four levels of economic integration are possible. At progressively higher levels, members must make more concessions and surrender more sovereignty (Figure 12.19). The lowest level of economic integration is the *free-trade area,* in which members agree to remove trade barriers among themselves but continue to retain their own trade practices with nonmembers. A *customs union* is the next higher level of integration. Members agree not only to eliminate trade barriers among themselves but also to impose a common set of trade barriers on nonmembers. The third type is the *common market,* which, like the customs union, eliminates internal trade barriers and imposes common external trade barriers; this regional grouping, however, permits free production factor mobility. At a still higher level, an *economic union* has the common-market characteristics plus a common currency

	Removal of trade restrictions between member states	Common external trade policy towards non-members	Free movement of factors of production between member states	Harmonisation of economic policies under supra-national control
Free Trade Area	*			
Customs Union	*	*		
Common Market	*	*	*	
Economic Union	*	*	*	*

Types of regional economic integration

FIGURE 12.19
Types of regional economic integration.

and a common international economic policy. The highest form of regional grouping is *full economic integration,* which requires the surrender of most of the economic sovereignty of its members.

There are a variety of trade organizations in effect throughout the world (Figure 12.20). These groups range from loosely integrated free-trade areas such as the Latin American Free Trade Association (LAFTA) to common markets such as the EU. There are North-South ties between the EU and LAFTA, and South-South ties between LAFTA and ASEAN. Most of these links are bilateral; that is, agreements between nations within different regions. Fully fledged interregional integration has yet to be achieved. Indeed, regional groups are more concerned with closer economic integration *within* regions than *among* regions.

Barriers to successful regional integration are stronger in developing countries than in developed countries. The most significant barriers are political—an unwillingness to make concessions. Without concessions to the weaker

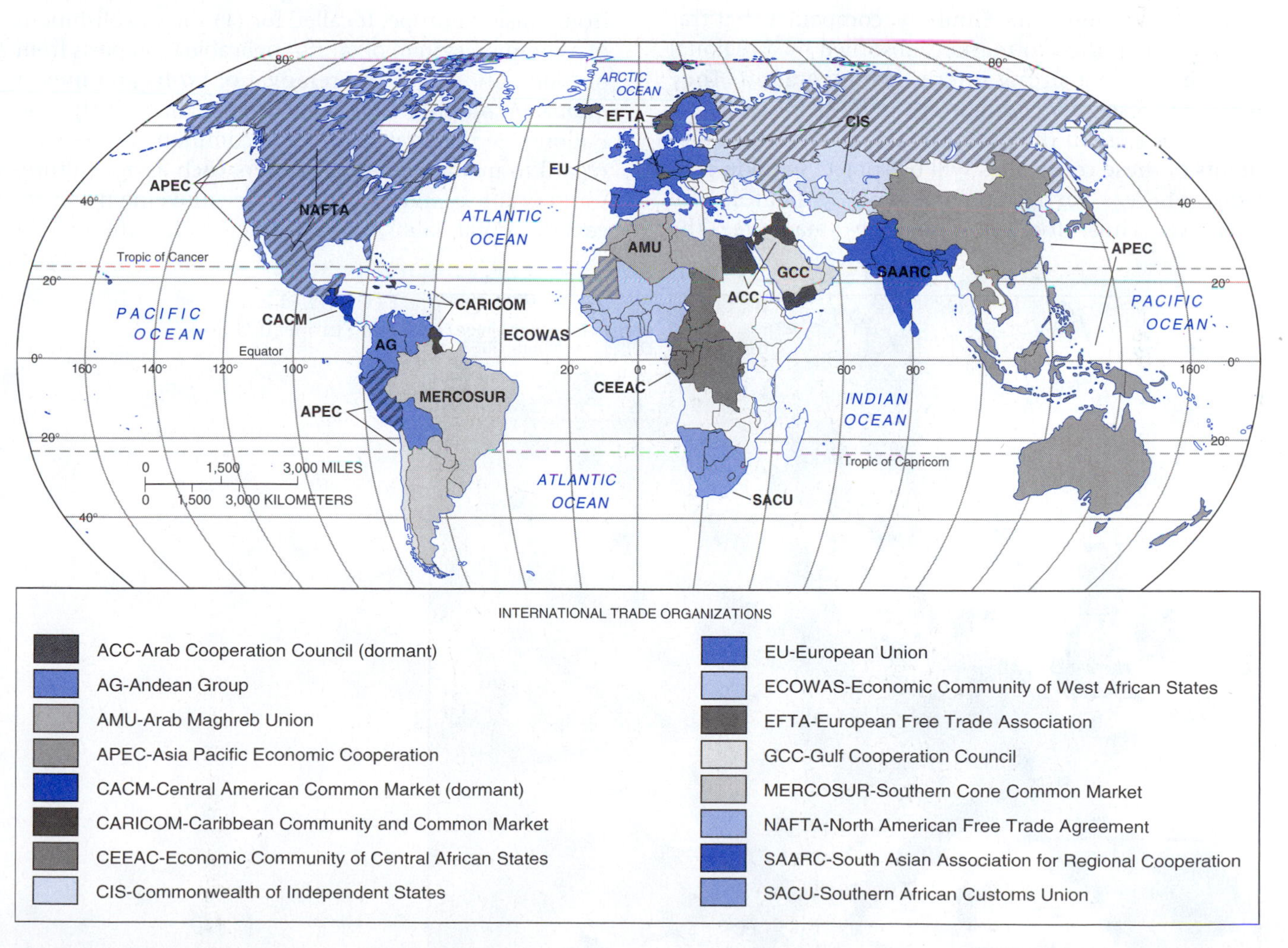

FIGURE 12.20
International Trade Organizations around the world.

partners of a regional group, the benefits from cooperation pile up in the economically more prosperous and powerful countries. Another difficulty is that developing countries have not historically traded extensively among themselves. Still another obstacle to integration is the lack of sufficient transportation and electrical networks. Nonetheless, much potential for integration exists in developing countries, particularly because many are too small and too poor to grow rapidly as individual units.

Many developing countries turned to regional integration schemes in the 1960s and 1970s. Reasons for integration included a need to gain access to larger markets, to obtain more bargaining power with the developed countries than they could if they adopted a "go it alone" policy, to create an identity for themselves, to strengthen their base for controlling multinational corporations, and to promote cohesive solidarity. The effect of regional groupings differs from one company to another. Companies that enjoy a secure and highly profitable position behind national tariff walls are unlikely to favor removal of these barriers. Conversely, companies that see the removal of trade barriers as an opportunity to expand their markets see integration as a favorable development. Similarly, companies that traditionally exported to markets absorbed by a regional grouping have a strong interest in integration. They perceive these enlarged markets to be more attractive than they were in the past. But as outsiders, the shipments of these companies will be subject to trade controls, whereas barriers for internal competitors will decrease. Thus, foreign companies may face the prospect of losing their traditional markets because they are outside the integrated group of countries. As a result, there is an incentive to invest inside the regional grouping. This is why many U.S. firms invested directly in the EEC (now EU) countries.

The European Union

The most successful example of economic integration is the EU. It began in 1957 with six nations: France, West Germany, Italy, Belgium, the Netherlands, and Luxembourg, and since then has added United Kingdom, Ireland, Denmark, Greece, Sweden, Finland, Spain, and Portugal. In addition, most eastern European nations have applied for membership in the EU, and several have been accepted, so that by 2006 it had 25 members and a total population of more than 500 million (Figure 12.21). Romania and Bulgaria will join in 2007. The EU today is the largest single trade bloc in the world and, along with the EFTA, accounts for 40 percent of international trade, which is three times its world share of population.

The intent of the EU was to give its members freer trade advantages while limiting the importation of goods from outside Europe. It called for (1) the establishment of a common system of tariffs applicable to imports from outside nations; (2) the removal of tariffs and import quotas on all products traded among the participating nations; (3) the establishment of common policies with regard to major economic matters such as agriculture, transportation, and so forth; (4) free movement and access of capital, labor, and currency within the market

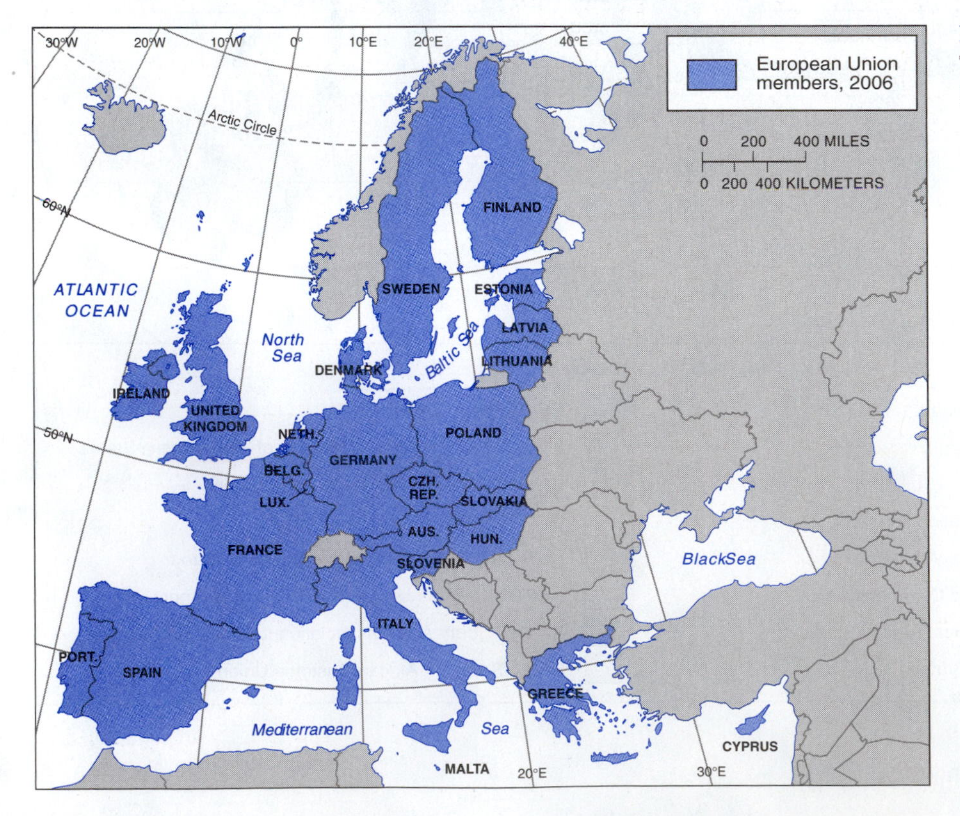

FIGURE 12.21 The European Union, 2006.

countries; (5) transportation of goods and commodities across borders with no inspection or passport examination; and (6) a common currency.

The EU has made tremendous progress toward its stated goals. The member nations have been afforded more efficient, large-scale production because of potentially larger markets within the EU, permitting them to achieve scale economies and lower costs per manufactured unit, something that pre-EU economic conditions had denied them. However, the current stumbling block seems to be the lack of a common economic unit of currency. In addition, some northern countries scoffed at opening borders that would allow southern Europeans to immigrate and thereby take advantage of social welfare programs of economic assistance from cradle to grave.

Americans are often concerned about the economic power of the EU. Present tariffs have been reduced to zero among EU nations, whereas tariffs for American-made products have been maintained. Consequently, importing goods to EU nations is relatively more difficult. At the same time, increased prosperity among EU nations allows these nations to become potential customers for more American exports.

The EU's Single Currency

On New Year's Day 1999, many members of the EU adopted a unified currency, the Euro. Some countries, such as Britain, Denmark, and Sweden, opted to retain their own currencies. The exchange rate of participating countries' currencies is locked in against one another and against the Euro. However, the Euro floats against non-EU currencies such as the U.S. dollar (Figure 12.22). When members of the European Union adopted a single currency, they were aware that the sacrifice would be great. Each country effectively surrendered the right to independently balance its own budget and manage its own debt. Each country relinquished its individual monetary identity.

There is no question that when the economic efforts of 25 countries and 500 million people are combined, European goods and services will be better represented in the world economy by a currency capable of maintaining price stability in member countries.

U.S.-Canadian Free Trade Agreement

Even before signing the U.S.-Canadian Free Trade Agreement in 1989, Canada and the United States were each other's largest trading partners. According to the agreement, tariffs, quotas, and nontariff barriers were almost totally eliminated by 1999. The agreement helps producers on both sides of the border. For example, for Canadian producers, their potential markets increase by a factor of 10. The population of Canada is about 30 million, but that of the United States is 300 million. At the same time, the Canadian market became a large group of consumers for U.S. producers. The reduced Canadian tariffs help American producers gain access to the Canadian consumer.

North American Free Trade Agreement

The economic pressure placed on the United States by the EU led the United States to promote freer trade through GATT and the U.S.-Canadian Free Trade Agreement. In 1992, the North American Free Trade Agreement (NAFTA) was signed by the U.S. and Mexican presidents and the Canadian prime minister, and in 1994 the U.S. Congress passed it.

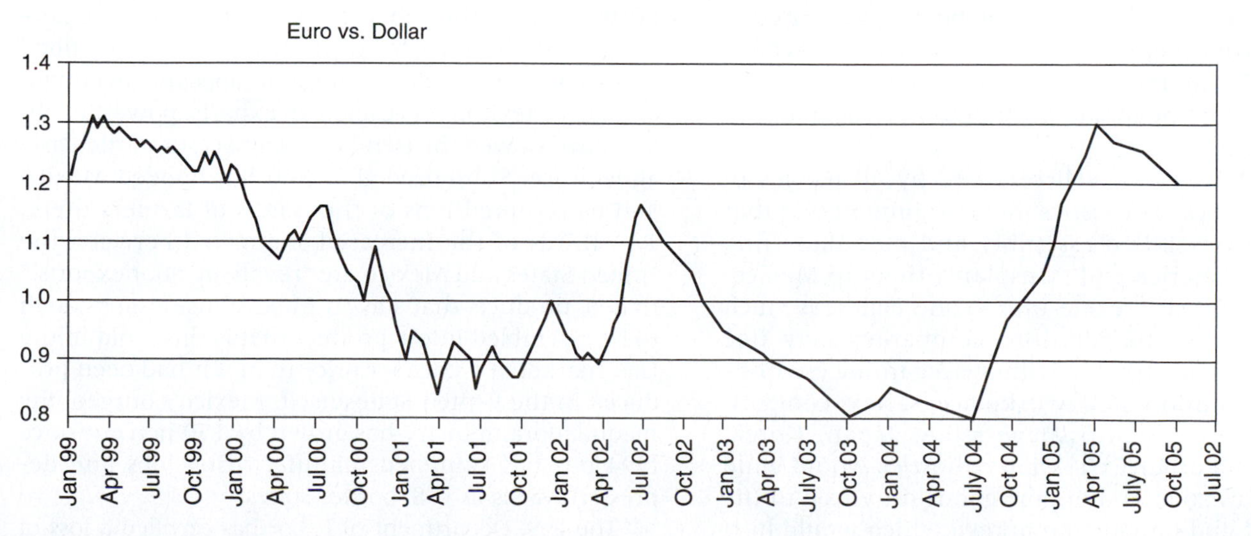

FIGURE 12.22
Value of the Euro in terms of the U.S. dollar, 1998–2002.

The Euro. A single currency has replaced national currencies in most, but not all, of the member states of the European Union.

NAFTA encompasses 300 million Americans, 30 million Canadians, and 100 million Mexicans. Unlike the EU, NAFTA includes a developing country. Access to the North American market is coveted by EU countries and Japan. Proponents of NAFTA argued that EU nations would negotiate a free-trade agreement between the two blocs.

NAFTA was not well received by all parties in North America. The critics' main argument was that it robs lower-skilled assembly and manufacturing jobs from America and transplants them to Mexico, where labor costs are one-fifth to one-eighth as much as in America. In addition, companies may flee America's more stringent climate regarding environmental pollution and workplace safety controls. Critics of NAFTA also suggest that Japan, Korea, Taiwan, and other East Asian countries would build plants in Mexico and import goods duty free to the American and Canadian markets, which would hurt U.S. firms and workers.

The principal argument in favor of NAFTA was that free trade would enhance U.S., Canadian, and Mexican comparative advantages: Raise per capita income in Mexico, and increase Mexican demand for goods from the United States and Canada. Another argument suggests that higher living standards in Mexico would help control the flow of undocumented aliens crossing the U.S. border, which is now estimated at 1 million per year. With free trade, wages would rise in Mexico; therefore, undocumented aliens could stay home and work in their native country.

NAFTA frees up trade with its elimination of performance standards on foreign producers in Mexico, who are often required to export a certain proportion of their Mexican output. NAFTA also frees up financial and investment trade by reducing restrictions on establishment of United States and Canadian financial service subsidiaries. Thus, NAFTA deregulates Canadian and U.S. banking, securities brokering, and insurance operations in Mexico. Truck transport service allows free access to the Mexican market opened to Canadian and American trucking companies, as well as free access to markets north for Mexican truckers. Before NAFTA, the United States' tariff rates on imports from Mexico averaged approximately 5 percent. Therefore, NAFTA did not substantially raise the incentive for Canadians and United States firms to invest in Mexico. Before NAFTA, a free trade zone, 100 miles south of the border, called *"the maquiladora zone,"* had been in operation, and NAFTA has in effect extended the maquiladora zone to all of Mexico, allowing imports from manufacturing plants to America with duties imposed only on the value added by manufacture in the maquiladora plants.

After 10 years, NAFTA has expanded trade and investment with Mexico through lower tariffs, but American imports have grown more quickly than exports, turning what was once a U.S. trade surplus with Mexico into a deficit. Mexico saw its U.S.-bound exports jump in the ten years since 1994 by 241 percent, while imports from the United States climbed by 170 percent. While U.S. tariffs disappeared overnight on many Mexican and Canadian exports, powerful lobbies have slowed the trend in some sectors, particularly agriculture. Subsidized U.S. coin has flooded Mexico and bankrupted tens of thousands of farmers there. Two-thirds of the increased shipments between the United States and Mexico are "revolving door exports," that is, products that stay in Mexico just long enough to be assembled into a product that is then sold in the U.S. market at a cheaper price than if it had been produced in the United States itself. Mexico's burgeoning maquiladora industry has grown by 150 percent since 1994 but has skimmed manufacturing jobs and depressed wages in U.S. border states.

The U.S. Department of Labor has certified a loss of more that 600,000 U.S. jobs to Mexico as a result of

NAFTA, many of them in manufacturing—a small fraction of the 139 million jobs that make up the U.S. labor force. About 15 million of those are in manufacturing. About an equal number of jobs have been created on the U.S. side of the border but are concentrated in lower-paying service sector occupations. Meanwhile, the United States continues to lure Mexican workers, many of whom were uprooted from rural communities when Mexico opened its markets to subsidized U.S. agricultural goods

NAFTA was grossly oversold on both sides of the border as an instrument for reducing unwanted immigration. Illegal border crossings have increased steadily since l994, mainly because of Mexico's disappointing economic performance and a large and growing wage disparity between the two countries. A very low minimum wage, 20 percent unemployment, and the lure of higher wages in the United States continue to draw millions of Mexican across the border. Full of undocumented immigrants, many communities along the U.S. side of the border remain as mired in poverty as they were a decade ago. The net impacts of these immigrants, who are largely unskilled and often illiterate, has been hotly debated. Many note that they take jobs typically unwanted by U.S. citizens, particularly picking fruits and vegetables, the garment industry, construction, and retail trade. Many concede that they increase the supply of unskilled labor and force wages down in those labor markets. Others note that they pay income and Social Security taxes but are often ineligible to receive benefits such as Social Security. Many live in desperate destitution, vulnerable to employers who take advantage of their precarious situation.

Canada and NAFTA

Canada has done a better job than Mexico in increasing its auto exports. Yet for a time, NAFTA was alleged to create a "giant sucking sound" of industries moving to Mexico to reap high profits and low wages. However, the largest exporter of cars and trucks to United States is no longer Japan, but Canada. A decade after the NAFTA agreement was signed, which many feared would send U.S. industry to Mexico, free trade is having the opposite effect of what was expected. The United States actually gained jobs in the auto industry, though not higher paying ones. Meanwhile, Canada has been gaining jobs. Canada has doubled the number of cars and trucks it exports to the United States since 1989. In 1986, it sold $11.1 billion more in vehicles to the United States than it bought. Canada employs almost 60,000 workers assembling mostly United States parts into motor vehicles—about 85 percent of which were exported to United States.

By 1995, Mexico had lost 13 percent of its 137,000 auto manufacturing jobs when foreign competition wiped out Mexican parts suppliers. One year after NAFTA began, the peso collapsed, and auto sales plummeted 70 percent. But due to the peso devaluation and reduced labor costs, Mexican production by Nissan, Volkswagen, GM, Ford, and Chrysler hit 1.4 million vehicles in 2003, approximately 85 percent of which was exported to the United States. Yet new equipment and more efficient production has meant fewer workers than were expected to produce cars in Mexico.

NAFTA has been beneficial for the auto industry's competitiveness because the trade agreement allowed U.S. automakers to win back a share of the domestic market, a loss to cheaper Asian and European cars. In order to do this, they had to send some jobs to neighboring countries to keep costs down, but large numbers of U.S. workers have been laid off. The pressure to keep reducing the cost of labor to produce a car is continuing.

Currently, negotiations have begun on extending NAFTA to include 34 nations across the western hemisphere in the form of the Free Trade Area of the Americas (FTAA). Given the long U.S. hostility toward Cuba, that country would be excluded. If the FTAA is successful, it would create the largest free trade zone in the world, exceeding even the EU.

OPEC

A *cartel* is an agreement among producers that seeks to artificially increase prices by arbitrarily raising them, by reducing supplies, or by allocating markets. The most successful commodity cartel is the Organization of Petroleum Exporting Countries (OPEC). Founded in 1960, OPEC consists of 13 countries—Saudi Arabia, Iran, Venezuela, Kuwait, Libya, Nigeria, Iraq, Indonesia, Algeria, Gabon, Qatar, Ecuador, and the United Arab Emirates (Figure 12.23). The success of OPEC at raising oil prices encouraged other underdeveloped countries to create new non-oil cartels.

The first oil shock occurred in 1973, followed by another in 1979. The price for a barrel of crude petroleum oil peaked in 1981 at $36. Major new exploration by American oil companies commenced, and billions of dollars of infrastructure were erected in territories of North America that were rich in oil, low-grade oil, shales, and tar sands, notably Colorado, Wyoming, Alberta, and Montana. With the decline in world oil prices, these new oil operations and oil explorations likewise declined, which sent shock waves through the oil industry and depreciated home and business values throughout Houston, Texas, and other oil-dependent cities.

In the late 1980s, the price of a barrel of crude petroleum on the world market declined to $15, but it rose again gradually in the 1990s to $21 because of the First Persian Gulf War and the closing of wells in Iraq and Kuwait. Output has not returned to prewar production levels because of the devastation in Iraq and

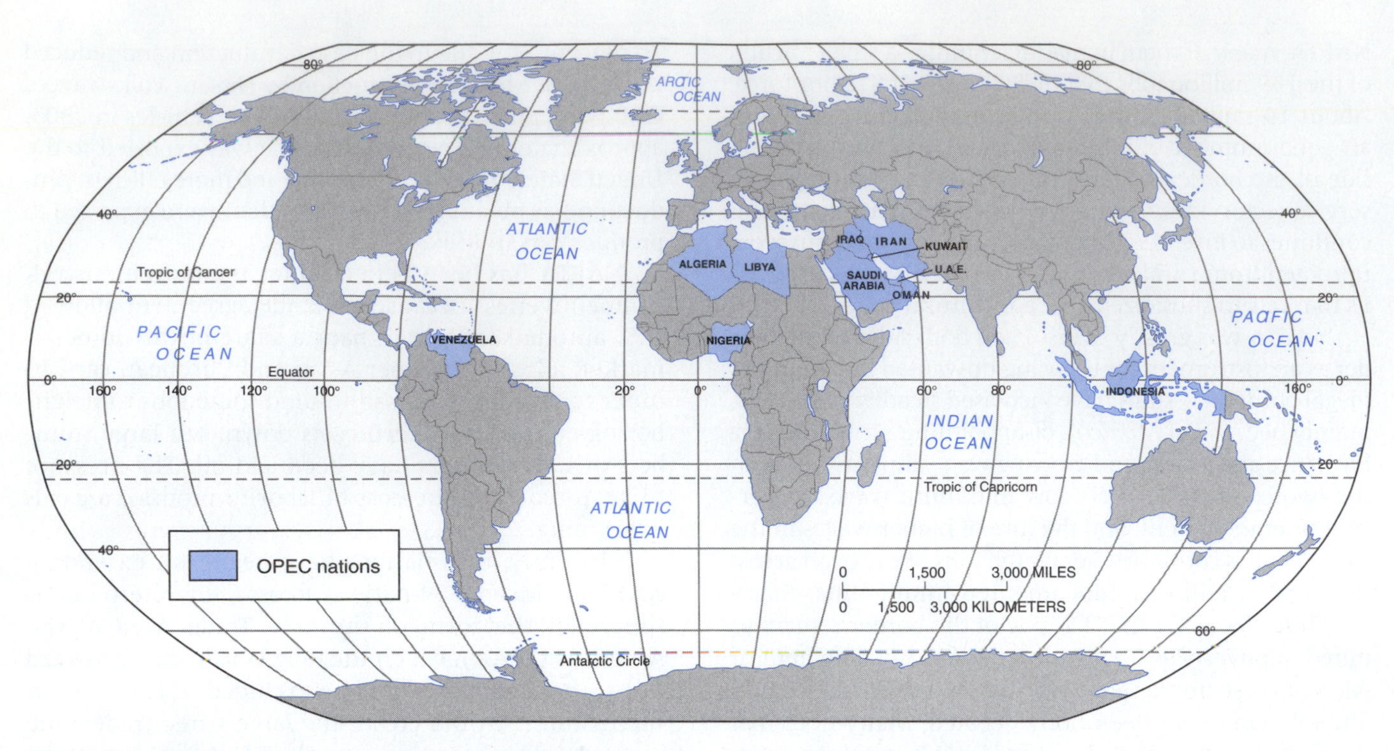

FIGURE 12.23
The Organization of Petroleum Exporting Countries, OPEC. Although dominated by Arab and Middle Eastern states, OPEC includes Venezuela, Nigeria, and Indonesia, its only Asian member.

Kuwait. With the lower output especially by Iraq after the war there in 2003, higher levels of production or refining capacity are required by other OPEC nations. By 2004, prices exceeded $40 per barrel.

World oil prices depend on the resumption of economic growth in not only the developed nations, but also the less developed nations, and the demand for crude petroleum. The resumption of production and refining in Russia and Central Asia will also affect world petroleum prices.

The future demand for OPEC oil at the beginning of the twenty-first century is likely to be substantial and increasing because of the expected recovery of world economic output and OPEC's large reserves. At present, OPEC produces less than 50 percent of the world's output but controls 80 percent of the proven reserves. Despite the tremendous oil reserves of several trillion barrels, generating enough money to maintain its production and increased capacity will be a problem for OPEC nations because, surprisingly, most have serious financial debt problems. Much of their petrodollar accumulation is spent on military equipment to keep the wealthy sheiks in power.

Globalization and Business Cycles

Globalization may be less vulnerable to the normal fluctuation of business cycles than most national economies. Companies today, large and medium size, are emerging as global players procuring parts and materials worldwide, as well as selling their products overseas. The result may be an economy that responds in new ways to swings in supply and demand.

The companies are managing their inventories much more efficiently than in the past because they have adopted just-in-time techniques and new computerized systems to better track the delivery of products. The ability of producers to expand abroad, at the same time that non–North American countries establish a presence in North America, obviates the decade old notions of how the regional economy operates. More than a quarter of the economy in North America now depends in some way on exports and imports.

Globalization has also changed the relationship between American producers and consumers. Before foreign competitors, such as the Japanese, saturated the North American market, North American companies had more power to make products of less quality and to charge a price that they wanted. The poor quality of many American cars prior to the Japanese onslaught is testimony to this fact.

Because of globalized competition, booms and busts are more difficult. If Dow or Dupont tries to raise prices, customers will now reflexively seek price quotes from Germany's companies or other suppliers. If a foreign competitor underbids a North American supplier, components can be air-freighted for delivery within 24 to 48 hours. In short, raising prices has never been tougher.

This trend suggests that the economic cycles may not carry the inflationary risk they once did.

The decline in the proportion of American workers represented by unions—from 34 percent to 15 percent since 1980—goes a long way toward explaining why wages have not increased in real terms recently. Some economists blame unions for pushing up wages in good times and for not being flexible enough when bad times hit, thus causing wilder fluctuations in the business cycle.

SUMMARY

This chapter examined some aspects of the international sphere of economic geography. *International business* is any form of business activity that crosses a national border. It includes the international movement of almost any type of economic resource—merchandise, capital, or services.

Our discussion of traditional trade theory introduced *comparative advantage* as the underlying explanation for international trade. The original Ricardian conception argued that specialization within a larger international division of labor led to higher standards of living for all parties involved. Beyond predicting that everyone gains something from trade, including the most and least efficient producers, classical trade theory neglects too many issues to be realistic, including the roles of innovation, government, and economies of scale. Free trade was established in the nineteenth century within a colonial framework of inequality among countries. Developing countries as primary goods producers became dependent on foreign demand and, therefore, vulnerable to the business cycle of expansion and contraction in developed countries. Thus, some critics argue that it is a patina that disguises the ability of rich countries to take advantage of less developed ones. One country's free trade may appear like another's exploitation.

A major extension of the original theory of comparative advantage, Porter's notion of competitive advantage includes the roles of labor skills, knowledge and innovation, agglomeration economies, financing, and government policy. Unlike comparative advantage, which is static, the theory of competitive advantage is dynamic, noting that regional and national patterns will always change over time.

The discussion of international trade and investment was extended in a consideration of the basis of *production-factor flows*. Production factors that are most readily movable among countries are capital, technology, and labor. We focused primarily on *capital movements*, enhancing understanding of FDI by using a managerial perspective. In many respects, the international movement of capital, technology, and managerial know-how is now more important than international trade. Theories of international trade and production-factor movements emphasize the benefits of a relatively deregulated, market-oriented business environment. However, a number of obstacles significantly impede flows of merchandise, capital, technology, and labor. These obstacles include distance, managerial behavior, and governmental barriers. Much progress was made in reducing governmental barriers—*tariffs, quotas,* and *nontariff barriers*—during the long postwar boom.

International trade—the movements of outputs—is only one facet of globalization. Another is investment. There are many types of investment, including, for example, purchases of government debt, but most attention has focused on foreign direct investment (FDI). Most FDI is organized through multinational corporations (MNCs) and originates in the economically developed countries. However, most MNCs invest in developed, not developing, countries. For example, various west European, Canadian, and Japanese firms invest heavily in the United States, where they generate jobs, wages, output, and profits.

Money must be exchanged on international markets for goods and services, and so exchange rates play a critical role in influencing the prices of imports and exports among countries. Determination of exchange rates allows world trade to function. Explaining why exchange rates fluctuate is no easy matter but is related to levels of real output, inflation rates, demand factors, and currency speculation in trading-partner countries.

Despite pressures for protectionism, countries continue to participate in myriad multinational operations. Major organizations that can be important to international business are *international financial institutions* and groups that promote *regional integration*. A common way of shaping global output and trade in a given sector is *commodity cartels*. Leaders in developed countries view commodity cartels as an unfortunate departure from market-oriented principles. In contrast, most LDC leaders view commodity cartels as a means to reduce their vulnerability in a world of unequal exchange. The most famous cartel, OPEC, has played a fundamental role in affecting the price of petroleum worldwide.

The General Agreement on Tariffs and Trade (GATT) reduced trade barriers worldwide. The Uruguay Round of GATT made progress in a number of difficult issues—farm policy, intellectual property rights, and trade barriers related to a growing international provision of services. The United States' current, most troublesome world trade problems, however, are the sluggish world economy and the closed Japanese markets. In the 1990s, GATT transitioned into the World Trade Organization, a permanent institution designed to minimize trade barriers and arbitrate disputes among countries. A current issue that lies at the center of WTO policies are the huge subsidies paid by the governments of countries such as the United States and western European states to their agribusinesses, resulting in low prices that cripple farmers in LDCs.

The most widely acclaimed regional integration to date is the European Union, which involves 25 states. The EU involves the most complete economic integration among countries the world has ever seen, including the absence of trade and investment barriers among its member states, as well as flows of labor. The adoption of a common currency, the Euro, has been accepted by most but not all of its members. The EU includes most eastern and western European nations in a powerful trade bloc with more than 500 million people and that manages almost 50 percent of worldwide trade. Starting in 1994, the U.S.-Canadian free trade agreement was extended to include Mexico in the form of the North American Free Trade Agreement (NAFTA).

STUDY QUESTIONS

1. Why does international trade occur?
2. What are the inadequacies of existing trade theory?
3. What is meant by the terms of trade? Why have they decline for many Third World nations?
4. Where are world largest debtor nations located?
5. What forces have driven the internationalization of banking?
6. How do exchange rates affect trade?
7. How has the U.S. dollar changed globally over the last two decades?
8. What caused the huge U.S. trade deficits?
9. What is foreign direct investment? What are its impacts?
10. How has the U.S. FDI balance changed over time? Why?
11. What are the principal barriers to international business?
12. What are tariffs, quotas, and nontariff barriers?
13. What was the GATT and what is the WTO?
14. Why were the IMF and World Bank established?
15. What is regional economic integration and why does it occur?
16. Compare and contrast the EU and the NAFTA.

KEY TERMS

absolute advantage
brain drain
capital controls
cartel
common market
comparative advantage
countertrade
development bank
direct investment
domestic farm subsidy
economic integration
economic union
Eurocurrency
European Union (EU)
exchange controls
exchange rate
export quota
export restraint agreement
export subsidies
factor of production
floating exchange rate
foreign direct investment (FDI)
free-trade area
General Agreement on Tariffs and Trade (GATT)
import quota
infant industry
intellectual property rights
international currency markets
International Monetary Fund (IMF)
intramultinational trade
market seekers
multinational corporations (MNCs)
NAFTA
nontariff barriers
Organization of Petroleum Exporting Countries (OPEC)
petrodollars
portfolio investment
production factors
protectionism
regional integration
tariff
tax-haven country
terms of trade
trade deficit
transfer costs
transfer pricing
unequal exchange
UNCTAD
World Bank
World Trade Organization (WTO)

SUGGESTED READINGS

Dicken, P. 2004. *Global Shift: Industrial Change in a Turbulent World,* 4th ed. London: Harper & Row.

Friedman, T. 2005. *The World Is Flat: A Brief History of the Twenty-first Century.* New York: Farrar, Straus and Giroux

Hugill, P. 1993. *World Trade since 1431: Geography, Technology, and Capitalism.* Baltimore, Md.: Johns Hopkins University Press.

Peet, R. 2003. *Unholy Trinity: The IMF, World Bank, and WTO.* New York: Zed Books.

Porter, M. 1990. *The Competitive Advantage of Nations.* New York: Free Press.

Stiglitz, J. 2002. *Globalization and Its Discontents.* New York: W. W. Norton.

WORLD WIDE WEB SITES

THE WORLD TRADE ORGANIZATION
http://www.wto.org/
The principal agency of the world's multilateral trading system. Its home page includes access to documents discussing international conferences and agreements, reviewing its publications, and summarizing the current state of world trade.

THE WORLD BANK
http://www.worldbank.org/
A leading source for country studies, research, and statistics covering all aspects of economic development and world trade. Its home page provides access to the contents of its publications, to its research areas, and to related Web sites.

U.S. DEPARTMENT OF COMMERCE
http://www.doc.gov/
Charged with promoting American business, manufacturing, and trade. Its home page connects with the Web sites of its constituent agencies.

BUREAU OF LABOR STATISTICS WEB SITE
http://stats.bls.gov/
Contains economic data, including unemployment rates, worker productivity, employment surveys, and statistical summaries.

OOCL KOREA
8
7

CHAPTER 13

INTERNATIONAL TRADE PATTERNS

OBJECTIVES

- To describe the evolving pattern of international commerce
- To document the emerging markets for global exports
- To examine global trade flows of six major commodities groups

Containerships capable of carrying thousands of tons of cargo exemplify the importance and growth of international trade, linking far-flung economies around the world. This one, of Chinese origin, reflects the growing importance of East Asia in the world economy.

Table 13.1
Distribution of World Output ($ Billion)

Region	*1980*	*1990*	*2000*	*Percentage of Total*	*Percentage Change 1980–2000*
World Total	**11,982.7**	**25,442.3**	**28,854.0**	**100.00**	**240.8**
North America	3,234.1	6,495.4	8,809.4	30.5	272.4
Other Western Hemisphere	842.7	1,505.3	1,994.4	6.9	236.7
Middle East	471.4	876.2	964.7	3.3	204.6
Asia and Oceania	2,497.3	6,839.5	8,626.8	29.9	345.4
Western Europe	2,999.7	5,939.1	8,739.9	30.3	291.4
East Europe and Russia	1,399.2	2,759.4	1,784.9	6.2	127.6
Africa	538.4	1,027.3	1,089.1	3.8	202.3

Note: Figures for each region's share of world output are purchasing power parity estimates based on the UN International Comparison Project.
Source: International Monetary Fund.

As we have seen repeatedly in this book, capitalism is an economic and political system forever in flux. Incessant change is the norm in market-based societies, and this pattern continued in the late twentieth century. The focus of this chapter is largely empirical and is designed to add substance to the theoretical discussions offered in Chapter 12.

Total world output, roughly $28.8 trillion in 2000, was highly unevenly distributed among the planet's major regions (Table 13.1). The turbulent decades of the late twentieth century saw major changes in the volume and composition of international trade. World growth has averaged about 3.7 percent annually for the last three decades (Figure 13.1). World exports jumped from $2 trillion in 1980 to over $8.8 trillion in 2003, or more than 36 percent of gross world product (GWP), a clear sign of the increasing integration of national economies (Figure 13.2). Of these exports, agricultural goods comprised 7.6 percent, mining ores, fuels, and minerals another 10.8 percent, all manufactured goods 61.3 percent, and all services 20.2 percent. The growth in exports was unevenly distributed around the world. East Asian NICs led the way, with China surging to the forefront. Manufacturing exports, with the exception of a dip in the mid-1970s, continued their rapid growth and now account for about 60 percent of world exports by value.

The changing structure of trade has affected various world regions differently, leading to fluid geographies of exports (Figure 13.3). Changes in international supply and demand, prices, production and transportation technology, production techniques, and government policies all play out differently in unique local contexts. For example, OPEC countries recorded a meteoric rise in the value of their exports in the 1970s and a precipitous decline in the 1980s and 1990s; by the 2000s, oil prices had climbed again as economies around the world, including China in particular, increased their demand for energy. North America, Europe, and East Asia experienced a drop in their export earnings after the oil crisis, but as a group they recovered and now account for 80 percent of the value of world trade. With the exception of the major oil exporters, LDCs that depend

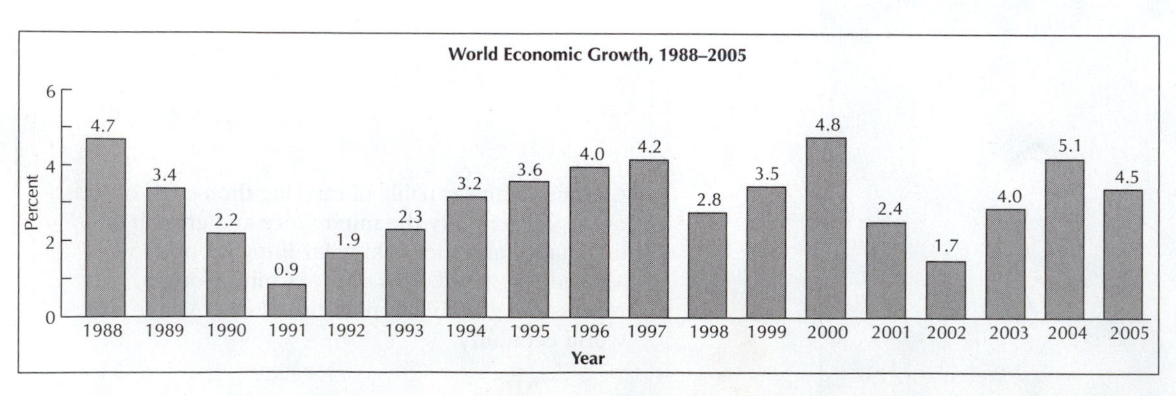

FIGURE 13.1
World economic growth, 1988–2005.

FIGURE 13.2
Postwar growth in world output and exports.

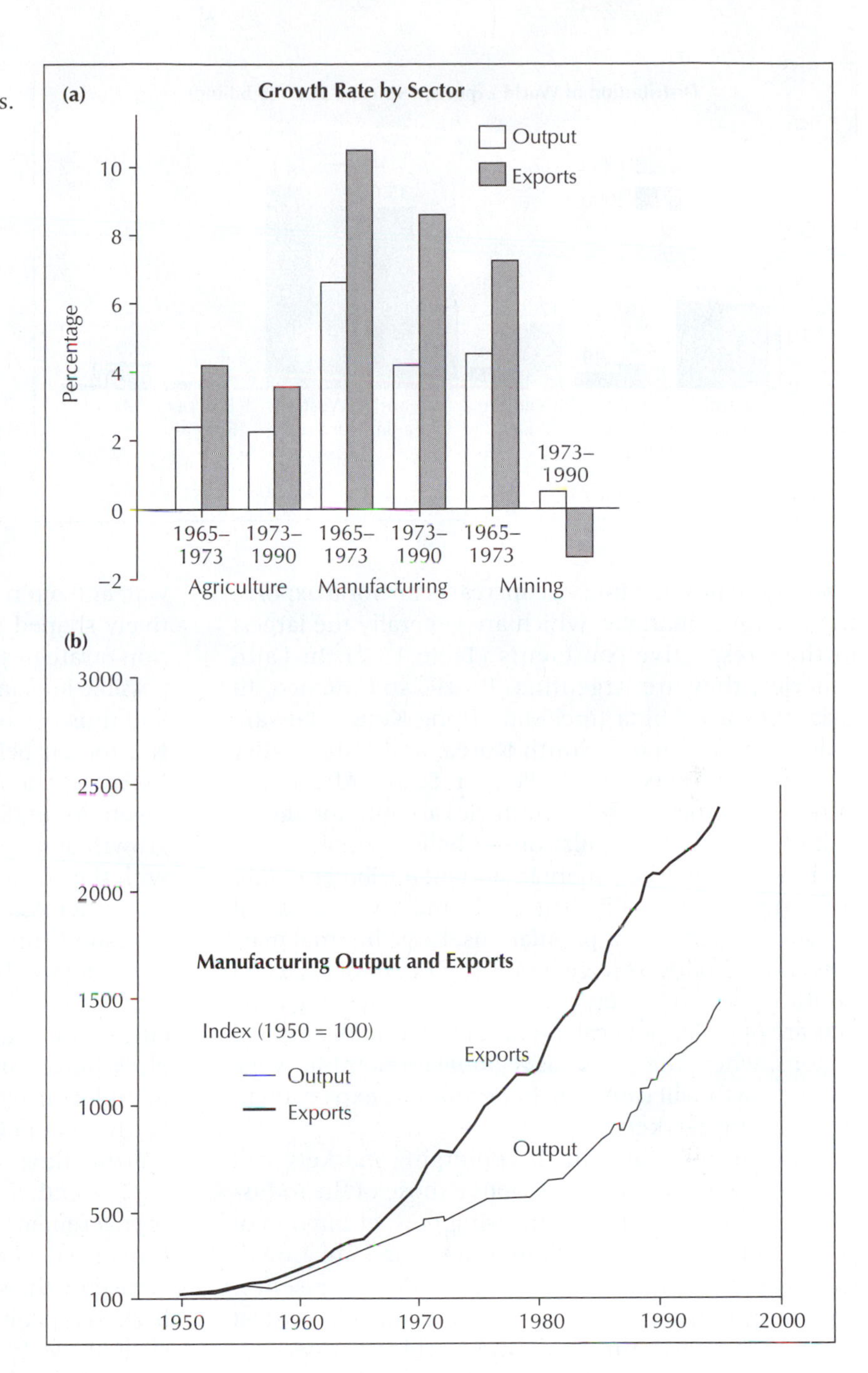

heavily on the export of a few primary commodities have fared badly. For many of them, the growth in primary commodity exports has been negative since 1980.

In addition to the expanding volume of trade, increased diversification of trade ties represents one of the most significant developments in the contemporary world economy. Advanced industrial countries still trade primarily among themselves, but the proportion has declined from more than 75 percent in 1970 to around 66 percent today. They have increased their share of exports to LDCs, and their imports from LDCs have increased still more. Another major development has been the growth of manufacturing exports from LDCs to developed countries and, to a lesser extent, to other LDCs. Manufacturing exports now account for about 40 percent of total nonfuel exports of these countries, compared with 20 percent in 1963, and LDCs now supply 13 percent of the imports of manufactures by developed countries, compared with only 7 percent in 1973. Yet only a handful of Asian and Western Hemisphere countries are involved in this development.

EMERGING MARKETS

Exports of the industrial countries have been concentrated among the member countries of Europe, Japan, and North America. Another category of countries

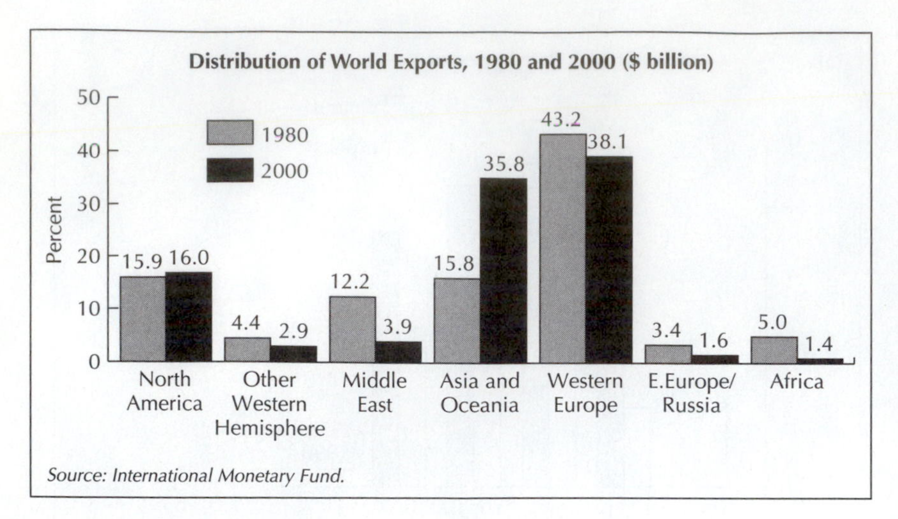

FIGURE 13.3
Distribution of world exports, 1980 and 2000. Western Europe still leads the world in the distribution of world exports. Many small countries supply neighbors with manufactured goods, taking advantage of scale economies and satisfying the need for aerial specialization.

holds more promise for large increases in world exports, the emerging markets, which are generally the largest in their respective continents (Table 13.2). In Latin America, they are Argentina, Brazil, and Mexico. In Asia, they are China (including Hong Kong), Taiwan, Indonesia, Singapore, South Korea, and India. Other emerging markets include Poland, South Africa, and Turkey. Together, these 12 countries account for almost half of the world's population—3 billion people.

Emerging markets share important attributes in that they are all physically large, offering a vast array of products to serve their populations. Large internal markets allow producers to generate economies of scale and produce goods relatively cheaply. All the emerging markets are of major political importance within their world regions, where they serve as regional economic drivers. Their growth will cause further economic expansion in neighboring markets.

GDP growth rates in the emerging markets will probably enjoy growth rates above those of the industrialized countries. Thus, emerging market imports of goods and services will be important sources of growth for exporters in the developed world. In essence, emerging markets will probably become the fastest growing markets throughout the world in the twenty-first century. Such developing countries make up about 40 percent of the market for U.S. exports today.

Of the world's major regions, Asia has been growing the most quickly, experiencing 8 percent to 10 percent annual GDP growth. Since World War II, these countries, originally led by Japan and now by China, have attracted huge sums of foreign investment and industrialized at rates that compare with those of the United States in the nineteenth century. Never before in the history of humanity have so many people been raised from poverty in such a short period of time. The epitome of the newly industrializing countries in Asia are the Four Tigers (Singapore, South Korea, Taiwan, and Hong Kong). Heavily subsidized by the United States as part of its geopolitical strategy during the cold war, and often with authoritarian governments that actively shaped their growth (e.g., with export promotion strategies), the Tigers demonstrated that it was possible for some poor countries to industrialize quickly and thus rise on the ladder of economic development. Not too far behind are the achievements of Thailand, Malaysia, and Indonesia, which saw growth despite the severe Asian financial crisis of 1997. The high economic growth of these Asian countries is contrasted markedly with the economic stagnation of Pakistan, Bangladesh, and other Asian countries that pursued policies of import substitution and isolationism, producing excessive bureaucratic barriers and tariffs and fostering inefficiency and low growth. While India is changing its policies in this regard, and experiencing rapid growth, elsewhere, especially in Africa, inefficient state-run manufacturing and agricultural enterprises continue to be the dominant mode of economic activity, thus penalizing those countries' citizens.

The end of the twentieth century saw the reduction of government influence throughout the world. This conservative policy, based on a faith that only markets generate growth, is called neoliberalism, a term that harkens back to capitalism prior to the emergence of the welfare state in the 1930s. It was initiated in the United States under the Reagan administration in the 1980s, in Britain under Margaret Thatcher, and has challenged welfare states in the remainder of Europe. Globally it has been fostered by the United States, World Bank, International Monetary Fund, and the World Trade Organization. Throughout the world, neoliberal programs, implemented at the behest of the United States and the IMF, have led governments to sell off many publicly owned assets (e.g., telecommunications companies) and deregulate their economies. While such moves may generate gains in efficiency, they may also create inequality and typically penalize those who lack the resources to compete effectively in markets. Thus, economic growth is often accompanied by rising inequality and can be deleterious for the poor and unskilled.

Table 13.2

Economic Growth in the Emerging Markets, 1976–2000

Item	Annual Percentage Change		Average Growth in Real GDP
	1976–1985	1987–1993	2000
World	**3.4**	**2.7**	**2.0**
Industrial countries	2.8	2.2	1.8
Developing countries	4.5	5.1	2.5
Emerging markets	6.0	6.4	2.3
China	7.8	9.3	7.5
Taiwan	8.6	8.3	4.8
India	4.6	4.1	6.0
Indonesia	5.7	6.5	−2.5
South Korea	8.0	8.1	4.5
Turkey	4.0	4.5	3.5
South Africa	2.1	0.9	1.3
Argentina	−0.5	4.4	1.5
Brazil	4.0	3.4	−1.5
Mexico	4.3	2.3	3.0
Poland	1.8	1.5	3.5
Other developing countries	0.3	1.8	3.0
East Europe and Russia	3.9	1.3	0.3

Sources: International Monetary Fund, Department of Commerce; World Economic and Social Survey 2001.

Economic growth in Latin America has varied widely among countries. Chile, which adopted neoliberal reforms under the Pinochet dictatorship, has enjoyed relatively rapid growth, including exports of wine and textiles to the United States. Argentina, in contrast, suffered an economic collapse in 2002. The low value of the peso has made Mexican goods more competitive on world markets, reversing the trend toward an increasing balance of payments and deficits for Mexico. Mexico was the only country to double its manufacturing exports in the 1990s, largely due to NAFTA.

Central Europe experienced positive but uneven growth in the last decade. Some, such as Poland, made a transition to a market-based economy relatively smoothly and haved joined the European Union. These "economies in transition" remain vulnerable to external economic forces. Several, such as the Czech Republic, Hungary, and Poland, are showing modest, positive growth. Russia and the Ukraine, two leading former Soviet republics, however, suffered severe economic decline after the collapse of the Soviet Union in 1991. Caught between an old, discredited system that at least fed people and a new, corrupt, market-based one in which the government has been incapacitated by corruption and its assets sold at bargain basement prices to investors, Russia has taken much longer than expected to recover from the transition from communism to a market economy. Despite its vast resources and skilled labor force, it will be years before Russia joins the world as a fully industrialized, competitive economic power.

The economic growth in Asia over the past three decades, first in Japan, followed by the Four Tigers (newly industrialized economies of South Korea, Taiwan, Hong Kong, and Singapore), and now more recently by China, Indonesia, Malaysia, and Thailand, has brought with it a boost of international trade. It is impossible to separate increasing trade and economic growth. Economic growth leads to expanded levels of domestic consumption and investment, which result in higher levels of imports of consumer goods, raw materials, and capital goods.

World Patterns of Trade

Trade simultaneously reflects economic and spatial differences in production and consumption among nations and in turn helps to generate those differences. As we saw in Chapter 12, international trade can be understood on the basis of the theory of comparative and competitive advantage and has grown more rapidly than the output of individual countries, reflecting the ways in which they are tied together by globalization. However, the volume, growth, and composition of trade vary widely among the world's major trading countries and regions.

Table 13.3

Total World Exports of Goods and Services, 2003

	Value $ Billions	%	Annual Percentage Change 1995–2000	2001	2002	2003
Agricultural products	674	7.6	−1	0	6	15
Food	543	6.1	−1	3	6	16
Raw materials	130	1.5	−3	−9	4	15
Mining products	960	10.8	10	−8	−1	21
Ores and other minerals	79	0.9	1	−4	3	24
Fuels	754	8.5	12	−8	0	23
Nonferrous metals	127	1.4	3	−9	−3	13
Manufactured goods	5,437	61.3	5	−4	5	14
Iron and steel	181	2.0	−2	−7	9	26
Chemicals	794	9.0	4	3	11	19
Other semimanufactures	529	6.0	3	−3	6	14
Machinery and transport equipment	2,894	32.6	6	−6	3	13
Textiles	169	1.9	0	−5	4	11
Clothing	226	2.5	5	−2	4	12
Other consumer goods	644	7.3	5	−2	5	15
All commercial services	1,795	20.2	4	0	7	13
Transportation	405	4.6	3	−1	5	13
Travel	525	5.9	3	−2	4	10
Other commercial services	865	9.8	6	3	10	15
Total Goods and Services	8,866	100.0				

Source: World Trade Organization.

The United States

The United States is the world's largest trading nation, accounting for more than $2 trillion worth of exports and imports in 2003. During the 1950s, the United States accounted for 25 percent of total world trade but now accounts for only 19 percent, a reflection of the growing competitiveness of other countries, particularly Europe and East Asia. From 1960 to 1970, the United States enjoyed a net trade surplus as a result of its strength in manufacturing, low oil prices, and the weak value of the dollar. However, following the petroshocks of the 1970s and deindustrialization, this surplus turned into growing trade deficits.

U.S. MERCHANDISE TRADE

Figure 13.4 shows the composition of U.S. merchandise trade with the world in 1964 and 2000. Machinery and transportation equipment accounted for the largest single proportion of exports (45%). Chemicals and other manufactures added another 31 percent, whereas agricultural products amounted to 8 percent. More than 70 percent of U.S. exports are manufactured items. Despite decades of deindustrialization, the United States remains the world's largest manufacturer today. In part this status reflects the rounds of investment that accompanied the microelectronics revolution, in which computerization dramatically raised productivity levels in steel, automobiles, and other sectors. The continued strong performance by the U.S. economy and the continuation of controlled inflation with low unemployment rates, coupled with a favorable dollar exchange rate, will ensure that U.S. goods and services remain highly competitive in world markets. U.S. firms continued to restructure and downsize to reduce costs and are continuing to reengineer with information technologies at a rapid pace.

Manufacturing goods also account for approximately 75 percent of the traded goods flowing to the United States from foreign companies. America has essentially farmed out much of its labor-intensive manufacturing to developing countries, shipping semifinished goods to countries such as Mexico for manufacturing and reimport. At the same time, it acquired an expensive taste for foreign-made items such as automobiles from Germany and Japan; shoes from Italy and Brazil;

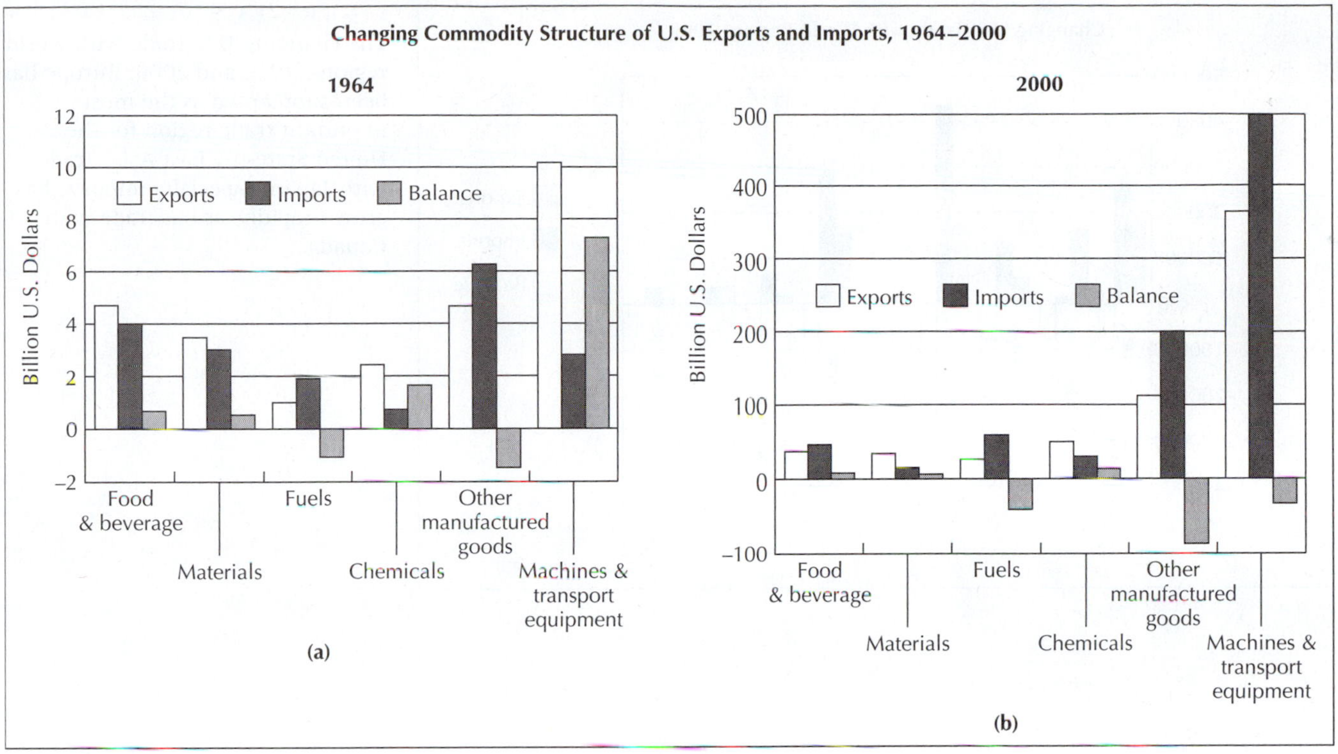

FIGURE 13.4
Changing structure of U.S. commodity exports and imports, 1964 and 2000. In the 1960s, the United States enjoyed a very large trade surplus in machinery and transport equipment. By 2000, however, the trade deficit was at least $230 billion because imports of manufactured goods, especially automobiles, exceeded exports.

electronic items, apparel, and toys from the Far East; and perfume and wine from France. Other imports include fossil fuels, for which the United States is not self-sufficient (Chapter 4).

The geography of U.S. merchandise trade with the world is shown in Figure 13.5. Canada, with 30 million people, represents the single largest trading partner, including 23 percent of its exports and 19 percent of its imports. U.S. trade with Canada is dominated by transportation equipment. The European Union accounts for 24 percent of U.S. exports and 21 percent of U.S. imports. The United States has a competitive disadvantage and trade deficit with Japan and China, however. During the last 30 years, the pattern of trade linkages between the United States and the world has shifted from Western Europe toward the Asian and Pacific regions. Ties with Latin America remain strong, and Mexico ranks third behind Canada and Japan as the leading trade partners of the United States, followed by China, Germany, the United Kingdom, Taiwan, France, South Korea, Singapore, Italy, Malaysia, the Netherlands, Brazil, and Belgium (Table 13.4).

The U.S. trade deficit is increasing and will remain large. In 2005, the trade deficit was $780 billion compared with $75 billion in 1993. Trade in services includes financial and business services, as well as royalties/license fees, which include entertainment, videos, cable TV, compact disks, and recordings. Trade in services generates a surplus of revenues, but not enough to keep the trade deficit from ballooning. In 2003, the United States had a surplus of $86 billion in services trade, up from a mere $7.5 billion in 1987.

The United States has gradually undergone an important shift in the destination of its exports—away from the traditional European markets and toward Asia, Mexico, and Canada. The reasons behind this change include the rapidly growing economies of Asia, so that trans-Pacific trade now exceeds that across the Atlantic. Another reason is the progress made under the World Trade Organization in the process of liberalization of tariffs in many countries. The North American Free Trade Agreement (NAFTA) of 1994 was such a liberalization agreement and has increased trade among Mexico, the Unites States, and Canada. In addition, there have been increasing amounts of U.S. and foreign investment in the Asian economies, and this has produced rapid increases of trade in capital and intermediate goods. Investment has also come from Chinese and Japanese multinational corporations.

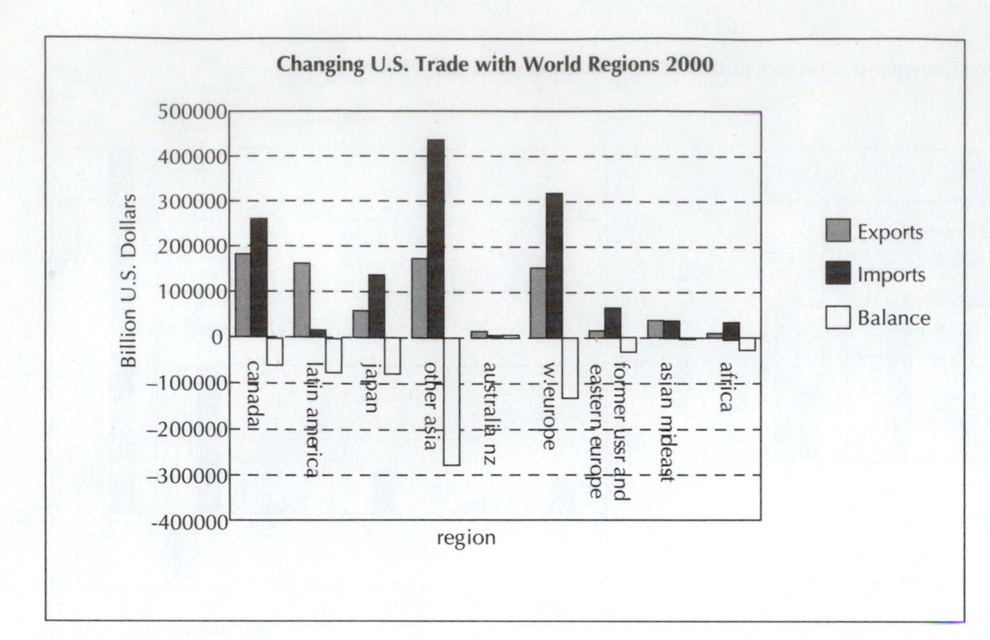

FIGURE 13.5
The changing U.S. trade with world regions, 1970 and 2000. Europe has been supplanted as the most important trade region for the United States by East Asia. Trade with Japan, especially imports, has grown rapidly, as has trade with Canada.

Table 13.4
Economic Growth in Top 20 Markets for Exports of U.S. Manufactured Goods, 1980–2000

Country	*Rank 2000*	*Exports ($ Billion)*	*Share of U.S. Exports*	*Growth in GDP 1980–2000*
Top 20 markets		556.9	81.5	3.8
Canada	1	156.3	22.9	2.8
Mexico	2	79.0	11.6	1.6
Japan	3	57.9	8.5	2.7
United Kingdom	4	39.1	5.7	2.7
Germany	5	26.6	3.9	1.9
Netherlands	6	19.0	2.8	2.5
Taiwan	7	18.2	2.7	N/A
France	8	17.7	2.6	1.9
South Korea	9	16.5	2.4	7.8
Singapore	10	15.7	2.3	7.3
Brazil	11	15.2	2.2	3.0
China	12	14.3	2.1	10.7
Belgium	13	13.9	2.0	1.8
Australia	14	11.9	1.7	3.5
Saudi Arabia	15	10.5	1.5	0.8
Italy	16	9.0	1.3	1.8
Malaysia	17	9.0	5.8	6.5
Switzerland	18	7.3	2.3	1.2
Israel	19	7.0	N/A	5.4
United States			**2.5**	**2.7**
World		683.0	3.0	2.8

Sources: International Monetary Fund, World Bank, Bank for International Settlements, United Nations, Organization for Economic Cooperation and Development, Department of Commerce.

Another shift in trade for the United States has been an expansion of the Mexican market for U.S. goods. Since 1990, Mexico has been the second largest market for U.S. manufactured goods after Canada, while Japan dropped to third in rank (but continues as the second largest partner when exports and imports are considered together). The value of U.S. manufactured goods exported to Mexico increased fourfold between 1983 and 2003. During the same period, U.S. manufactured exports to developing economies of Asia increased by 18 percent, U.S. exports bound for Europe dropped slightly, and the share exported to Japan and Canada held constant. Despite this global shift, the most important trading partners for U.S. exports remained almost unchanged from 1983 to 2003: Canada, Mexico, Japan, the United Kingdom, and Germany.

There are several sectors within manufacturing in which the United States is competitive internationally, including medical equipment, transportation equipment, and computer hardware. The medical equipment sector is one of the most competitive sectors of the U.S. economy, with export growth averaging 15 percent per year over the last five years. In 2003, exports were $17 billion, far exceeding the $6 billion of imports. Medical equipment markets are shifting, however, and industrialized countries, particularly in Europe, have traditionally been the best markets.

Two major exports of the United States are aircraft and motor vehicles. In 2003, the auto sector had a U.S. trade deficit of $120 billion, while the aircraft sector had a $20 billion surplus (Figure 13.6). Half of the motor vehicle exports are shipped outside of North America, including the three large markets of Japan, Taiwan, and Saudi Arabia. The European markets are essentially saturated. World sales of motor vehicles grew just 1.2 percent annually over the last 10 years in the developed markets because of the prolific manufacturers.

The U.S. computer equipment industry commands more than 75 percent of the world's computer sales through global operations, but the United States has had a trade deficit in this sector since 1992 as semiconductor production and assembly of consumer electronics has moved to East Asia and Mexico. In 2003, the U.S. trade in computer equipment was a deficit of $18 billion. Foreign sales now account for more than half of the total revenues of many leading U.S. computer equipment suppliers.

U.S. SERVICES TRADE

The United States is a world leader in computer software, supplying 49 percent of the $153 billion world market for packaged software per year. The world market for software grew 35 percent annually between 1995 and 2003, benefiting world suppliers, and reached almost $180 billion in market demand by the year 2003. Asia and Latin America are expected to be the fastest growing markets for computer software. The computer software industry is moving toward *multimedia*, which combine video, animation, voice, music, and text. Implementation of *intellectual property rights* through the World Trade Organization will reduce piracy, which has been a major problem for the industry.

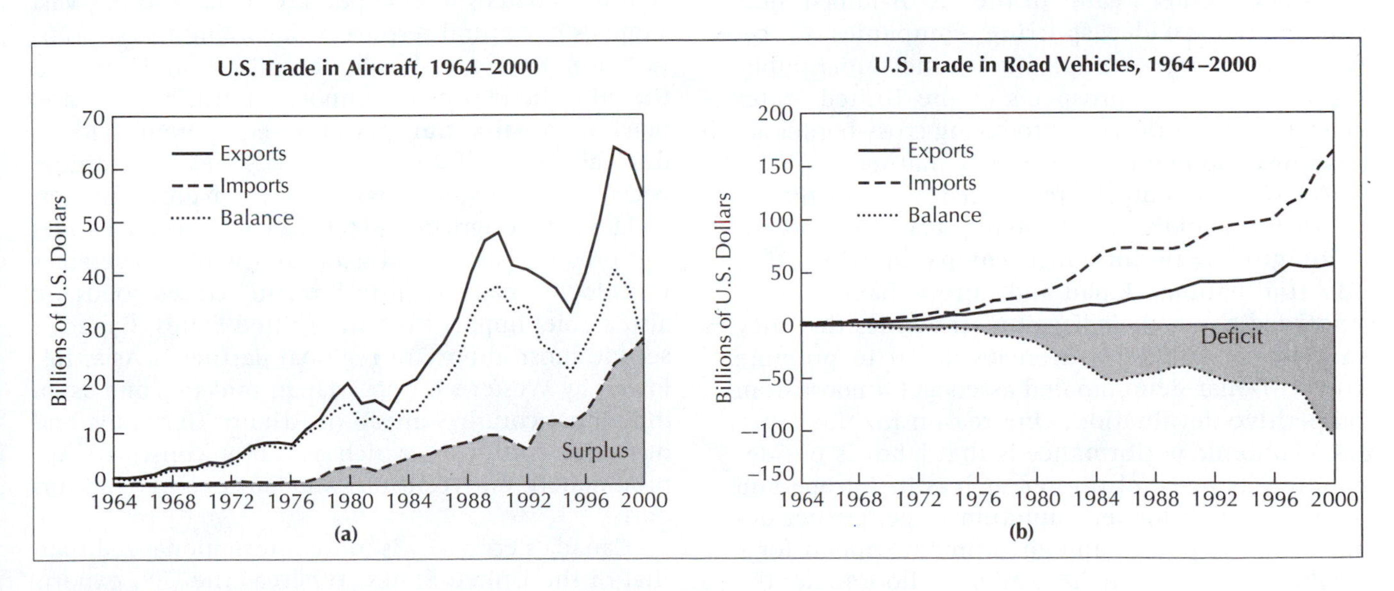

FIGURE 13.6
U.S. balance of trade in aircraft and motor vehicles, 1964–2000. The United States is the world leader in aircraft manufacturing and has maintained an important surplus since 1992. At the end of the cold war, both exports and imports dipped slightly. Road vehicles have been a bright spot in U.S. trade up until 1970. From that point onward, a trade deficit developed. Foreign auto companies sell more cars in the United States than U.S. companies sell abroad.

The United States is also the world leader in the production, use, and export of information services and commanded 50 percent of the world market by 2003. U.S. firms produce and use the most advanced software, and exports have grown rapidly. The information industry is especially affected by government policy regarding market access, intellectual property rights, privacy protection, data security, and telecommunications services. The largest markets for U.S. information services will be the industrialized countries, especially the United Kingdom, Japan, and Canada. The most important emerging markets will be China and Korea, followed by Mexico, Brazil, and Argentina.

Services are not just food and travel, but information software, telecommunications, advertising, and entertainment. Services in North America account for about a quarter of gross domestic product and 30 percent of total exports. Canada and the United States have 150 of the world's top 500 service corporations. On the cutting edge of industry—software, information technology, and entertainment, which account for one-third of the world services sales—the United States' and Canada's position is unchallenged. For example, U.S. service exports dwarf auto exports, $266 billion to $43 billion in 2003. *Soft power* is services power, the power of Microsoft to write programs, of Hollywood to make movies, and of U.S. cultural ideals, products, and practices to become known around the world through the information revolution. Soft power gives North America the edge over every other region of the world and may replace heavy industry as the motor of the world economy of the future.

Finance is another important area of services. The U.S. stock market's gains in the 1990s rallied most equities worldwide, spurring companies to set records for mergers, acquisitions, and initial public offerings. Brighter prospects in the United States boosted trade worldwide, producing cross-border acquisitions and helping other stock markets.

The U.S. economy has reemerged as a growth magnet for the world's surplus savings because of its superior growth performance compared with other industrial nations. Japan and Europe have been so concerned about their flagging economies that they want to knock their currencies down to promote growth. Dollar-denominated assets get a boost from competitive devaluation. One reason for the better U.S. economic performance is that labor is not demanding its piece of the pie as it is in Europe and Japan. Europe's higher minimum wage, greater degree of union power, and government support for a social market economy have all put a floor under the wages of the lowest paid. To be sure, this means that income polarity is far more pronounced in the United States than elsewhere. In the United States, male wage earners in the top 10 percent of incomes make 4.4 times what those in the bottom 10 percent make, compared with 2.5 in Europe and Japan. Between 1979 and 2003, the poorest one-fifth of American families saw their income drop 11 percent, while the wealthiest one-fifth enjoyed a 28 percent increase. In the United States, capital clearly dominates labor. Overseas, that's not so, at least to the same degree. Profit ratios of publicly held U.S. companies sprinted to record levels in the 1990s, while average family incomes remained stagnant. Similarly, while U.S. public sector entitlement problems are severe, they pale in contrast to those abroad.

Canada

The United States is Canada's most significant trading partner, accounting for 82 percent of Canada's exports and 69% of its imports (Figures 13.7 and 13.8). Canada and the United States have the largest bilateral (between two countries) trade relationship in the world. Canadian exports have grown rapidly in recent years and held about $170 billion in 2000. The overall trade surplus amounts to over $10 billion annually. Trade with the United States amounted to over $290 billion in goods and services in 2003. For a country roughly one-tenth the size of the United States, Canada accounts for 21 percent of U.S. exports and 20 percent of U.S. imports. The United States runs a large merchandise deficit with Canada; in 2003, it was $20 billion.

Canada exports automobiles and transportation equipment, industrial supplies, and industrial plant and machinery parts, which combined account for 60 percent of Canada's total exports. Canada also has vast supplies of natural resources, including forest products, iron ores, metals, oil, natural gas, and coal. On the other hand, Canada imports industrial plant and machinery parts, transportation equipment, and industrial supplies from the United States. In addition, because of a longer growing season, balmy climates, and temperate agricultural territories, the United States can produce subtropical fruits and winter vegetables for colder Canada. High-tech manufactured goods are also a chief import from the United States. Canada's second most important regional partner is Asia, followed by Western Europe. Japan ranks second as an individual country, ahead of Britain, Germany, and other EU countries, which as a bloc constitute approximately 7 percent of Canadian exports and imports.

Canada's economy is more internationalized than that of the United States. Whereas the U.S. exports approximately 11 percent of its output, Canada exports approximately 20 percent of its GDP. Because of its relatively small population (30 million), Canada cannot attain the economies of scale (Chapter 5)

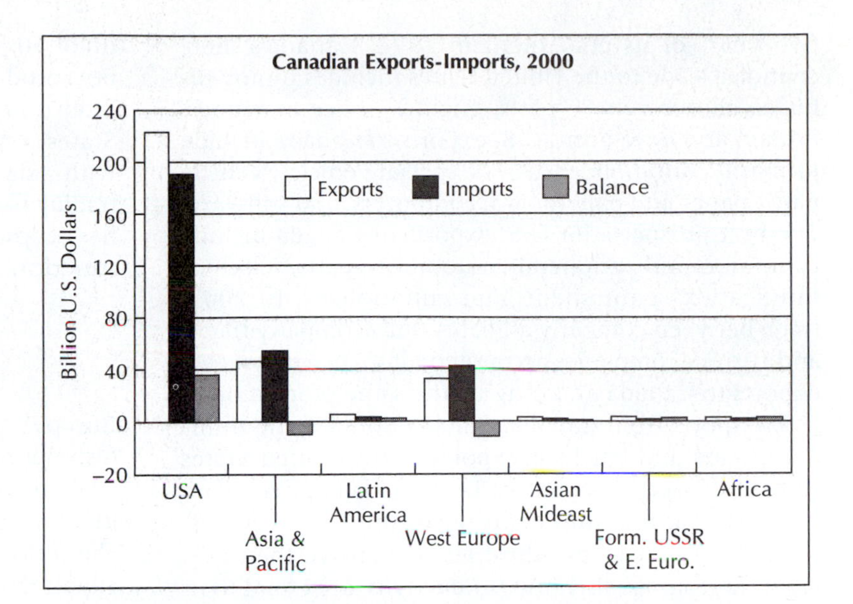

FIGURE 13.7
Canadian exports and imports by world region, 2000. Most of the foreign trade produced by the Canadian economy is with the United States. Canada enjoyed a $30 billion trade surplus with the world. The United States took 82 percent of Canadian exports and supplied 68 percent of its imports. The second most important trade partner for Canada is Asia and the Pacific, followed by Western Europe.

necessary for highly efficient plants and, therefore, must import many of its goods. As with many small countries, especially in Europe, the smaller the country, the more dependent it is on foreign markets for imports and exports.

Canada exports energy resources because of its vast amounts of hydroelectric power and its ability to manufacture hydraulic turbines and electric generators. It also produces high-tech communications equipment, including fiber-optic cables. Transportation and telecommunications equipment are required because of the vast territories that must be overcome to interconnect with the second largest country in the world. For Canada, automobiles and automobile parts represent the largest category of exports to the United States. This is a result of the U.S.-Canadian Free Trade Agreement (FTA), which favored the export of automotive industrial goods from Canada to the United States. The FTA, enacted in 1989, lowered trade barriers between the United States and Canada as part of the 10-year phasing out process and became the nucleus for NAFTA. Today, Canadian exports and imports to the United States outpace every other world region. Free trade has come at a cost, however: Canada's competitiveness has been improved by industrial restructuring, although that process laid off many manufacturing workers. Unemployment, while declining, still remains high, at approximately 8 percent in 2003, making job creation the government's chief objective.

NAFTA sparked U.S. and Canadian merchandise trade, which saw a 14 percent increase over the first

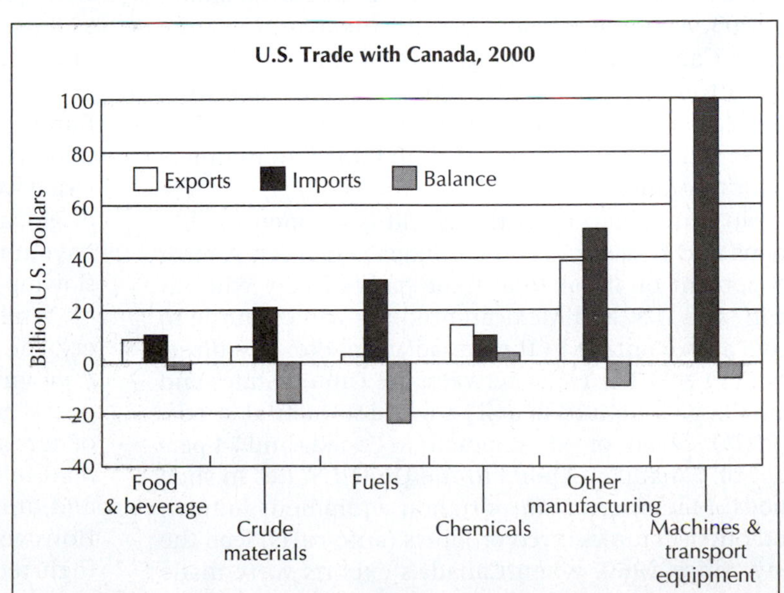

FIGURE 13.8
Canadian-U.S. exports and imports, 1970–2000. Asia and the Pacific became the leading export destination for both Canadian and U.S. trade by 1980, following a long period of dominance by Europe and the Atlantic countries. Countries most responsible for this shift are Japan, Hong Kong, Singapore, Taiwan, and South Korea. To this list we now must add Australia, Indonesia, Malaysia, Thailand, and the Philippines.

two years of its enactment in 1994. Canada's merchandise trade to the United States includes motor vehicles, motor vehicle parts, engines, office machines, timber, and newsprint. U.S. exports to Canada include primarily automobiles, trucks, special vehicles, vehicle parts, paper and paperboard, computers, and software. The best prospects for U.S. exports to Canada include computers and peripherals, automotive parts, telecommunications equipment, and automobiles. In 2003, trade between company affiliates (intracompany trade) and current companies accounted for 45 percent of U.S. exports to Canada and roughly the same proportion of U.S. imports from Canada. Canada doubled the number of cars and trucks it exports to the United States since 1989 as a result of NAFTA. Canada has overtaken Japan as the top auto exporter to the United States. Canadian labor is one-third less expensive than U.S. labor, and the quality and productivity are equal to or better than that of the United States. Automakers save $300 for each car made in Canada because of its national health care financing system, which offers for free most of the health insurance that some U.S. companies buy for their employees. (In contrast, 40 million Americans lack health care insurance.) GM spends $5,000 per year to provide health care to each current and retired employee and less than $1000 per current Canadian worker. Canada's cheap dollar ($.70 U.S.) allows the dollar of the U.S. manufacturers to go further.

A fast-growing sector of U.S.-Canadian trade is U.S. service exports, which jumped from $10 billion in 1988 to $17 billion in 2000. Canada's financial service market continues to expand as a result of the 1987 accord between the Toronto Securities Commission and the U.S. Department of Finance. This agreement allows the deregulation and integration of the financial securities industry, removing the distinctions among banks, trusts, insurance companies, and brokerages. U.S. direct foreign investment in Canada was more than $110 billion in 2003. About one-half of it was is the manufacturing sector. Canadian direct foreign investment in the United States totaled about $79 billion in 2003. Therefore, Canada's investment income balance with the United States is the single largest deficit in Canadian nonmerchandise trade.

No other market in the world is as open to U.S. goods and services as is the Canadian market. Almost 98 percent of all bilateral trade passes freely without tariff, and U.S. and Mexican products will continue to have an advantage in the Canadian market as a direct result of NAFTA. Trade between the United States and Canada is weighted heavily toward industrial goods, with 90 percent of U.S. shipment to Canada and 74 percent of Canadian exports to the United States in such goods. Machinery, transportation equipment (autos), and other manufactured products (auto parts) lead the way. Since 1960, when Canada's exports were made up of 50 percent primary products—forest, mine, and field products—Canada has shifted its emphasis toward industrial merchandise (Figure 13.9). Like the United States, Canada has found more recent trade growth with Asia and the Pacific Rim than with its traditional trading partner, Western Europe. Since 1980, the Pacific has eclipsed the Atlantic as the leading arena of North American commerce.

The European Union

Europe's trade, as a proportion of total world trade, is disproportionately large compared with its population, one-third of a billion. The EU is the largest trading block in the world, with exports totaling $2.6 trillion in 2003, about the same as imports. The EU ranks second as an export market to the United States, after Asia. Although it possesses only one-fifteenth of the world population, it accounts for 50 percent of world trade because of (1) the strength of the EU, (2) short distances and well-developed transport systems among member countries, and (3) complementary trade flows among its smaller states. Some European countries are comparable in population and size to individual U.S. states. The proximities and complementarities of Europe, with many relatively small countries close to one another, make intraregional trade ideal. This type of trade has increased from 55 percent of all European exports to 75 percent in 2003, that is, 75 percent of all exports from European nations go to other European nations. Italy, France, and the United Kingdom have populations of 60 million each, while Germany has roughly 80 million (Figure 13.10a, b, c, d). Other countries are much smaller. Some have food resources, such as Denmark and France; some have energy resources, such as Norway, the Netherlands, and the United Kingdom; some produce iron, steel, and heavy equipment, such as Italy, Germany, France, and Spain; and others produce high-value consumer goods.

As the EU grows to include new members in Eastern Europe and the Mediterranean, European intraregional trade will continue to increase. The countries that are faring worst economically are those that resist the forces of globalization. For example, France and Germany have not allowed reform of their rigid labor markets, slowing growth of their economies.

While the EU is in the midst of its economic recovery, the structural problems, such as high labor costs, economic rigidity that stalls the growth of smaller companies, industrial obsolescence, difficulty in making use of new technologies, and costly social welfare programs, tend to restrain growth below its economic potential and are keeping unemployment above 10 percent. However, the EU is the fastest growing market for U.S. high-technology exports and remains the principal

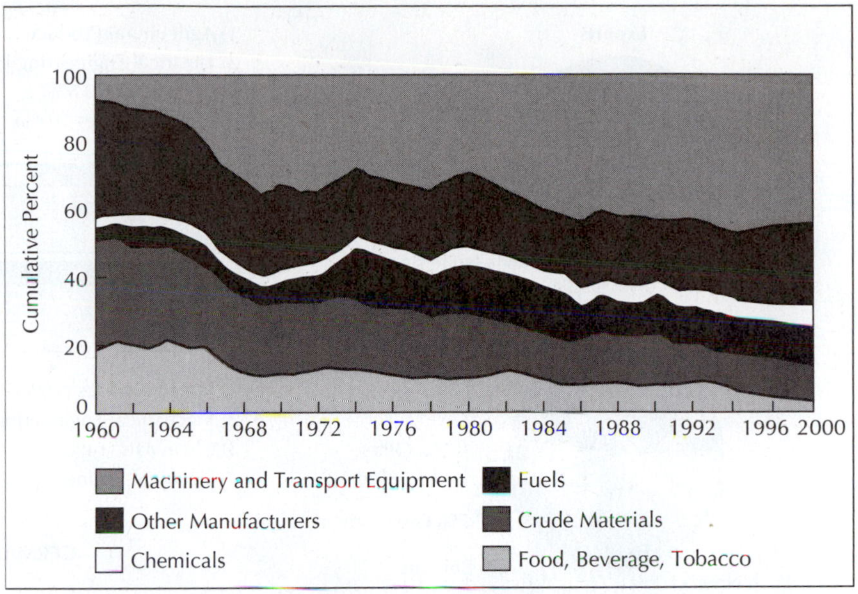

Cumulative Percent

100 80 60 40 20 0

1960 1964 1968 1972 1976 1980 1984 1988 1992 1996 2000

Machinery and Transport Equipment
Fuels
Other Manufacturers
Crude Materials
Chemicals
Food, Beverage, Tobacco

FIGURE 13.9
Canada's changing export composition, 1960–2000. In 1960, 50 percent of Canada's exports were made up of primary products: from the mine, forest, and field. Since that time, Canada has shifted its exports to industrial and manufactured products.

destination for U.S. FDI. The EU is as well the largest source area of FDI in the United States. The leading U.S. exports to the EU in 2003 were aircraft; data processing and office equipment; engines and motors; measuring, checking, and analyzing equipment; and other electronic equipment. The most promising sectors for U.S. exports to the EU include telecommunications equipment, computer peripherals, software, electronic items, pollution control, machinery, medical equipment supplies, and aircraft.

The leading economies of the EU include Germany, a major exporter as well as importer of automobiles. Britain, which has not accepted the Euro, is a significant exporter and importer of manufactured goods. The economy of France, which is more agricultural and less productive than Germany, is nonetheless a significant producer and consumer of industrial and consumer goods. Italy, the smallest of the four, exhibits a growing strength in the exports of engineering products.

Most international trade takes the form of *intraindustry trade*—investment in foreign affiliates that produce abroad, rather than shipments of U.S.-produced goods to target export markets. These sales accounted for more than half of U.S. affiliate sales worldwide and were four times as large as U.S. affiliate sales in Canada or Asia. Consequently, Europe's overall importance for U.S. com-

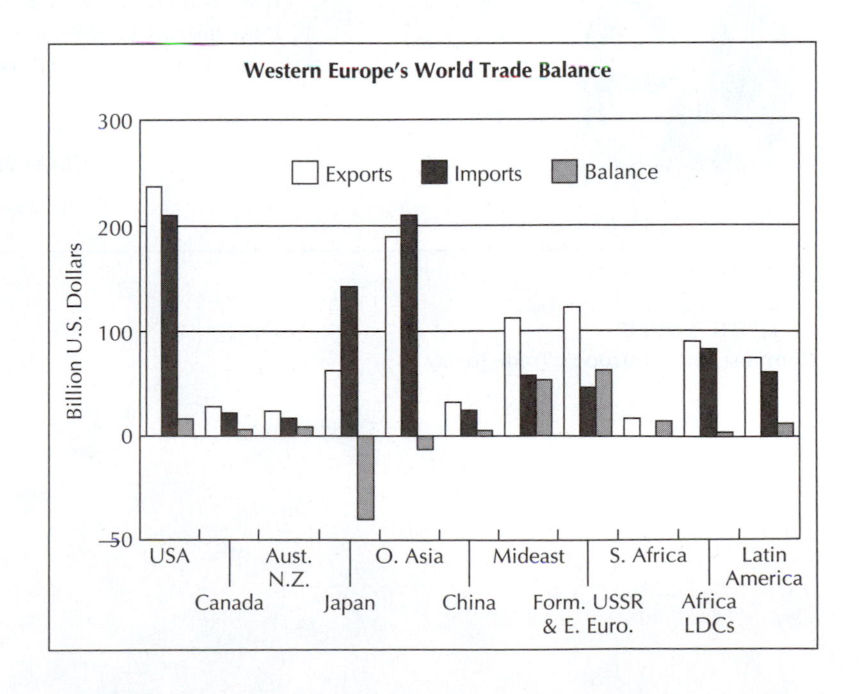

FIGURE 13.10
Composition of Europe's Track.

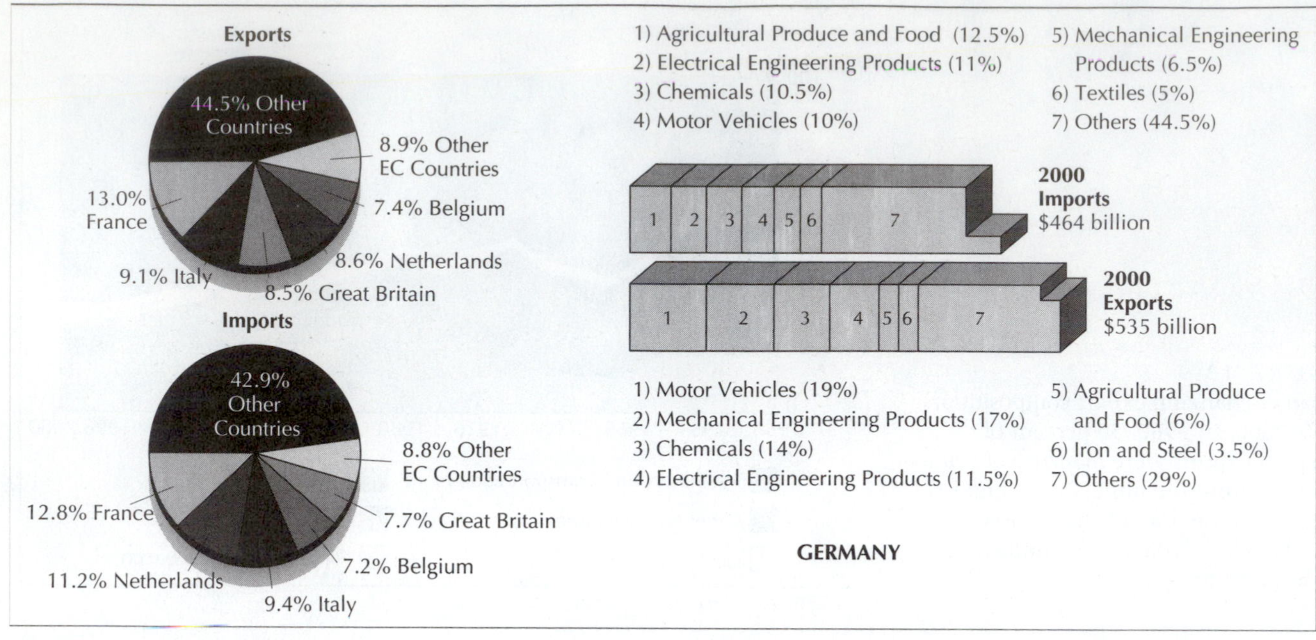

(a)

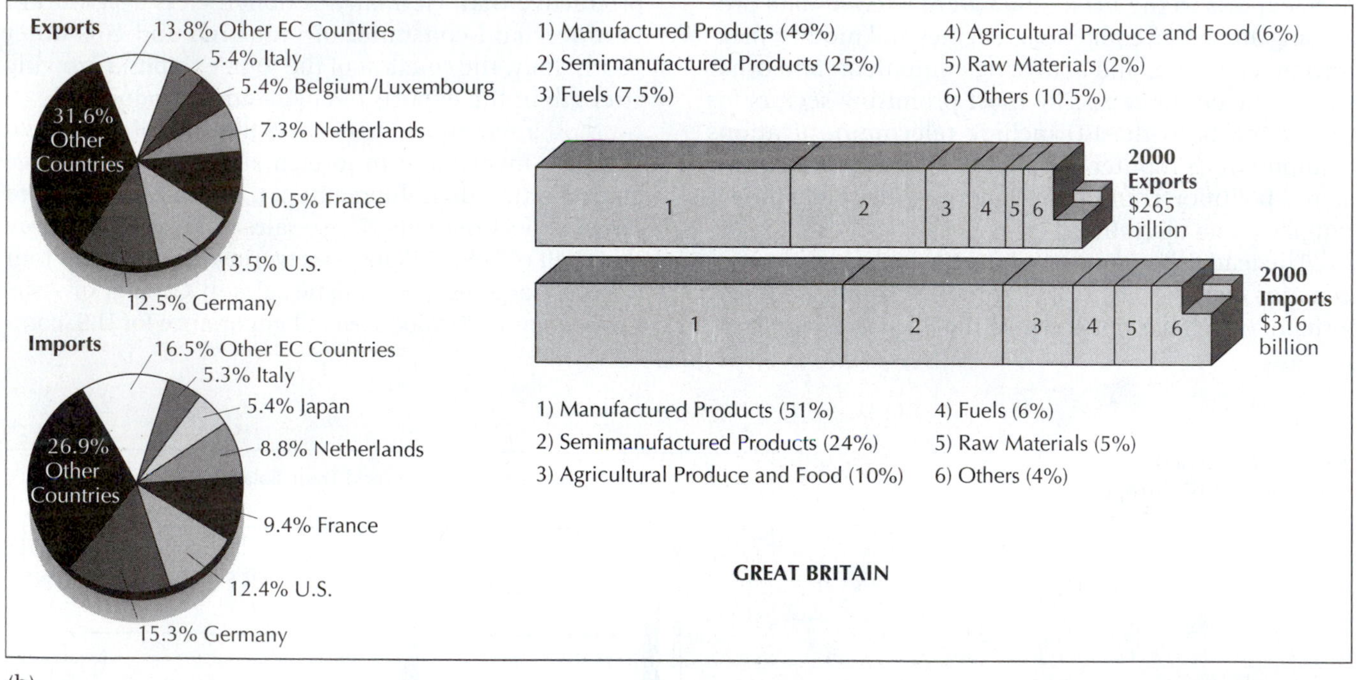

(b)

FIGURE 13.10
Compostion of Europe's Trade *(cont.)*

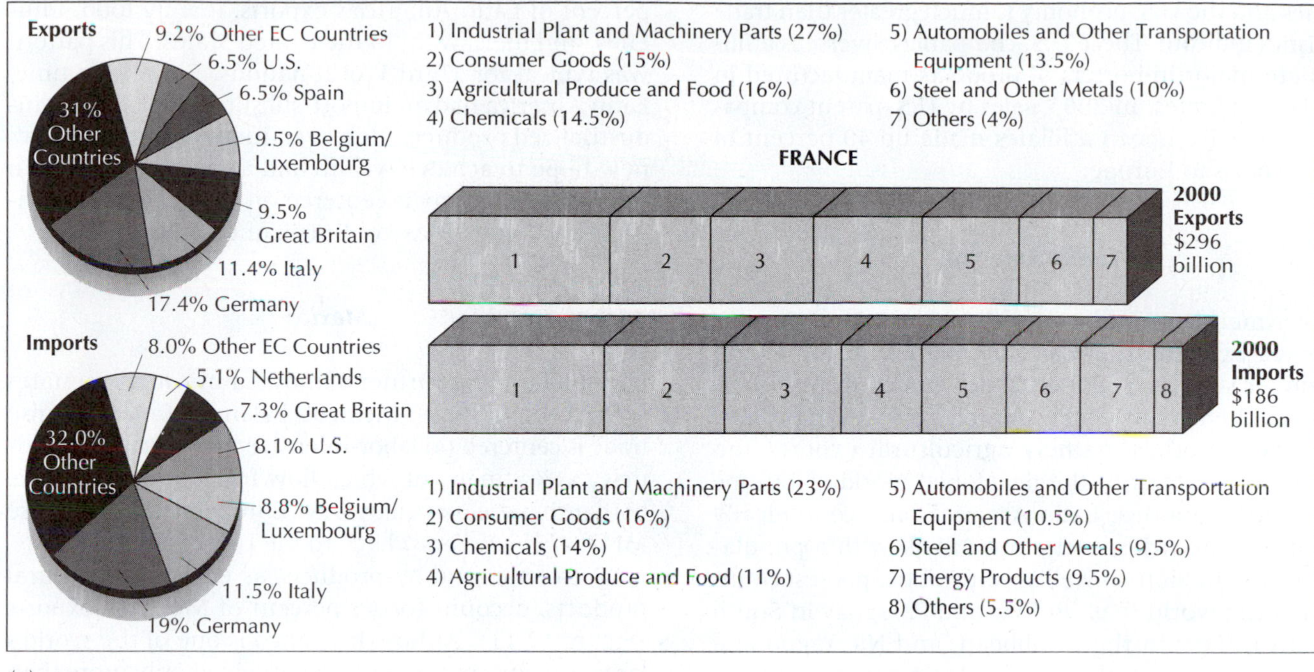

(c)

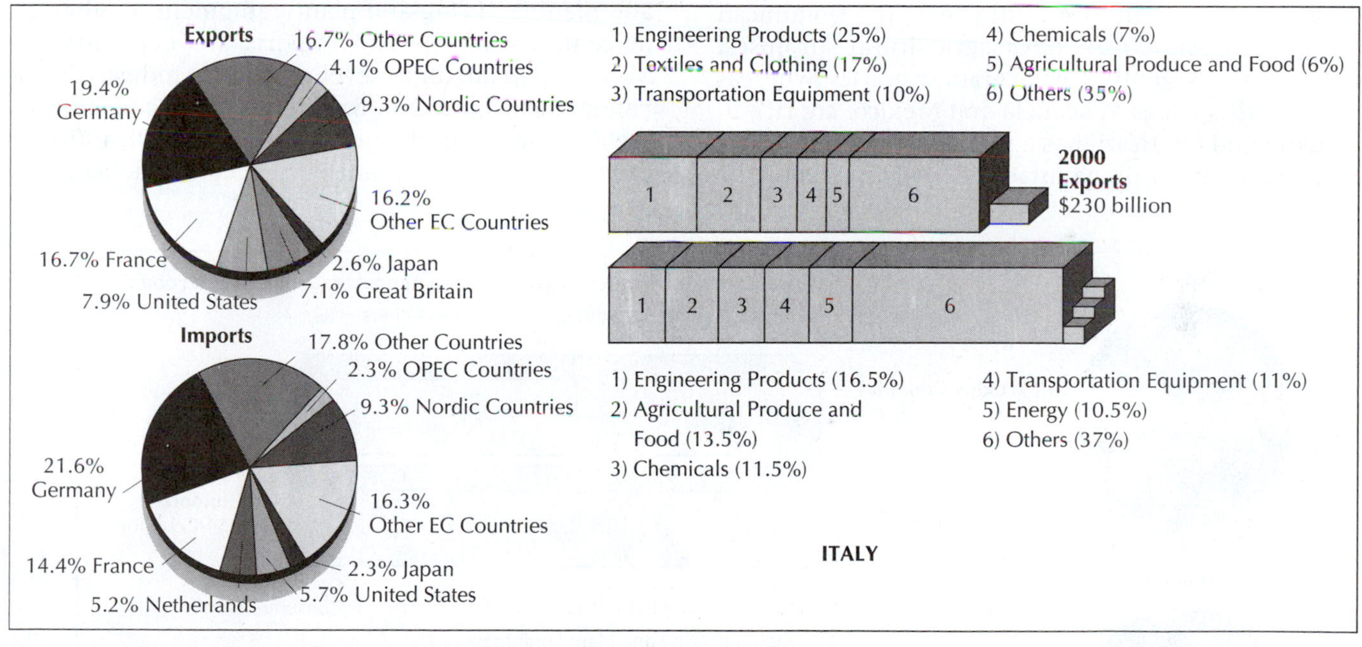

(d)

FIGURE 13.10
Composition & Europes's Trade *(cont.)*

panies and the U.S. economy is much greater than trade statistics indicate. These U.S. companies' overseas affiliates are major importers of products manufactured in the United States. In 2003, sales by U.S. parent companies to the European affiliates made up 40 percent of U.S. exports to Europe.

Latin America

Latin America comprises a series of developing countries with different levels of population, income, and economic structures. For centuries, under Spanish colonialism, their traditional economic role was to provide primary materials, namely agricultural exports and mineral resources, to the developed world of Europe and North America. Latin America has economically advanced countries, including Brazil, with a population of 161 million, as well as some of the poorest countries in the world (e.g., Bolivia and Paraguay in South America, Haiti in the Caribbean, and Nicaragua and Guatemala in Central America). Latin American countries are diverse not only with regard to population and size but also with respect to development and natural resources. Some countries, such as those in the Caribbean and Central America (e.g., the Dominican Republic, Costa Rica), have agricultural surpluses. Others, as in Argentina, have grain surpluses, whereas still others, such as Venezuela and Mexico, are rich in iron ore and oil. Brazil has a wealth of minerals and is a strong producer of manufactured goods. In 2003, 62 percent of Latin America's exports, mainly food, minerals, and fuels, went to the United States. This pattern was typical for Third World nations. For a long time, Latin America had an import-substitution policy for industrialized products. However, today, Latin America's new hope to achieve wealth and a prominent place in the world economy is centered on export-led industrialization, led by Mexico, Brazil, and Chile.

Mexico

The chief trading partner of most Latin American states is the United States. Mexico's balance of merchandise trade is centered on labor-intensive manufactured products (52%), many of which flow from plants along the border that are owned by U.S., European, and Japanese MNCs (*maquiladoras*) back to the United States.

Petroleum and by-products, as well as agricultural products, account for 45 percent of Mexico's exports (Figure 13.11). Although Mexico is one of the world's largest exporters of energy, oil provided only more than 35 percent of its export revenue in 2003. Semifinished industrial supplies that act as input materials for final production compose 60 percent of Mexican imports, and manufacturing and plant equipment another 23 percent. These types of imports are necessary for Mexico to maintain its level as a newly industrializing nation. Mexico's economic woes were punctuated with the devaluation of the peso in 1995. With insufficient capital inflows to finance its account deficit, the

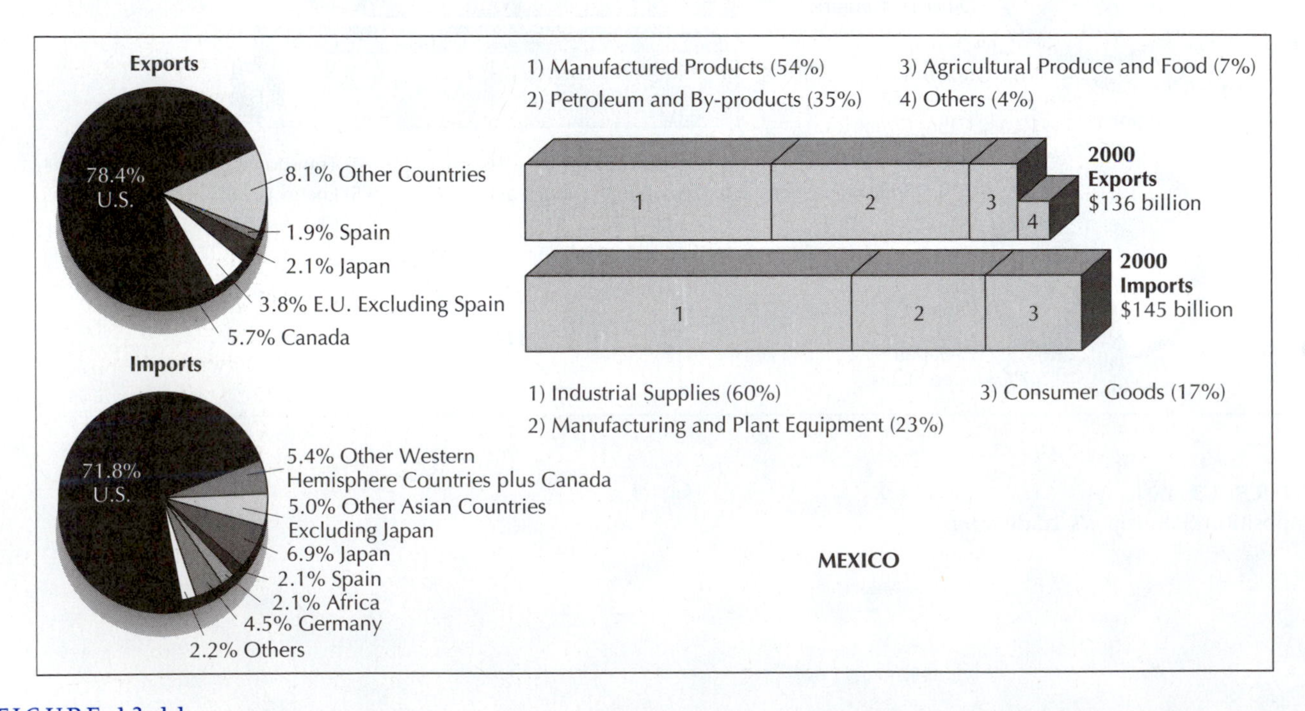

FIGURE 13.11
Composition of Mexico's world merchandise trade. The U.S. import share of the Mexican market also increased 10-fold. The United States currently dominates the Mexican market, but the Europeans and Japanese are looking at Mexico as a possible springboard to the NAFTA and South American markets.

Mexican government let the peso devalue in the face of rapidly diminishing foreign reserves. Major government programs included reduced government spending, tax and price increases, and strict control of credit. The new government programs included accelerated privatization and liberalization of key industries.

Mexican markets continued to open to foreign competition, following the enactment of the North American Free Trade Agreement (NAFTA) on January 1, 1994. NAFTA turned the Mexican economy around. Inflation dropped to about 7 percent, down from its peak at more than 150 percent in 1987. Mexico is experiencing a modernizing economy that is no longer protected from foreign competition and an improved investment environment. In 1997, rather than default on a $30 billion foreign debt, Mexico announced the return of its creditworthiness by the prepayment of the remaining $4 billion owed to the United States from the $13 billion emergency aid package negotiated in 1995. Mexico is among the fastest growing export markets for U.S. products. In 1986, Mexico became the third largest market for U.S. exports after Canada and Japan. Foreign investment opportunities are particularly strong in Mexico in infrastructure development, where the government invested $38 billion between 1995 and 2003, including airline privatization, highway construction, railroad services, and water and energy projects.

Under NAFTA, preferential duty treatment of U.S. origin goods gives them an edge over products from European or Japanese firms, which are often the principal competitors. U.S. and Canadian manufacturers must pay duty only on the value added by manufacturers in Mexico, not the products or parts shipped as semifinished or raw materials to plants in Mexico from North America. Under NAFTA, half of U.S. exports to Mexico have been eligible for no Mexican tariffs—semiconductors, computers, machine tools, aerospace equipment, telecommunication and electronic equipment, and medical supplies. Another important feature of NAFTA was the gradual phaseout of the Mexican Auto Decree, which helped the U.S. triple exports of passenger cars to Mexico since 1994. Since 1994, autos, auto parts, semiconductors, machine tools, and certain fruits and vegetables, including apples, realized export increases of between 100 percent and 10,000 percent when Mexican barriers were sharply reduced or eliminated under NAFTA.

In short, NAFTA has neither lived up to the fervent expectations of its proponents nor fulfilled the dire warnings of its opponents. It has enhanced Mexico's ability to supply U.S. manufacturing firms with low-cost parts but has not made Mexico economically independent. To some extent, Mexico's growth has been limited by American barriers to Mexican exports, including tomatoes, avocadoes, and truck drivers. Finally, it must be remembered that compared with the EU, NAFTA is a much more modest, failing to lift regulations on movements of labor, for example.

South America

South America represents a large and diverse picture of economic growth, change, and stagnation. The southern countries of Argentina, Uruguay, Chile, and Bolivia have had strong ties to Western Europe. The East Asian NICs are currently strengthening their economic ties with Latin America. Unfortunately, Latin American trade within the region is not nearly so strong as that within North America, Western Europe, or the Pacific Basin. Each country is more tied economically to Europe, the United States, and East Asia than to one another.

As a result of high indebtedness, high interest rates, and low foreign revenues, some Latin American countries can barely keep up with debt service on their international loans. Mexico, Brazil, and Argentina each owe over $100 billion to the developed world, and several other Latin American countries are close behind. Most of the loan money was put into urban infrastructures, but high world interest rates, oil prices, tariffs, and agricultural subsidies in the developed world and international recessions have minimized exports and thus foreign revenues, which are necessary to repay the debt. Brazil is a case in point. Exports are a little more than one-tenth those of the United States, but the population is approximately two-thirds that of the United States. In general, the 1990s were a disadvantageous time for Latin America because Latin American governments were forced by the International Monetary Fund (IMF) to devalue their currencies and to invoke austerity programs by restructuring their economies, raising taxes, decreasing public expenditures, and selling government-owned business enterprises, such as state banks, power companies, metal refineries, and transportation and airline companies.

Argentina, which after World War II enjoyed a standard of living comparable to Europe, suffered a steady decline in the late twentieth century. To find favor with working class voters, General Juan Peron engaged in import substitution (Chapter 11), outlawed outsourcing, developed legislation against employers dismissing employees, and strengthened unions' ability to strike, a pattern that operated in South America during the 1940s through the 1960s. The Peronist populist actions, put forward as humanistic measures to reduce the gaping inequalities in Argentine incomes, did not achieve their stated purposes: reduction of poverty and real income disparities. Instead, they resulted in hyperinflation lasting for decades. Since 1990, the government moved to correct a floundering economy hit by hyperinflation, high budget deficits, and an overcapitalized public sector. Brazilian exports to Argentina are more competitive as a result of the Mercosur Trade Agreement, which eliminated tariffs on most goods traded among the four member countries (Argentina, Brazil, Paraguay, Uruguay). The average tariff of Argentina is now 9 percent. The key sectors of import are computers, communications, and capital goods. The new export opportunities for the Argentina market include telecommunications equipment, electric power generation, transmission equipment, and medical equipment supplies and services.

In Brazil, after a surge in economic growth that averaged 5 percent during the early 1990s, the economy slowed down in the aftermath of the Asian financial crisis. Brazil is still not on sound economic footing because of erratic domestic policies and high inflation. To counterbalance high inflation, Brazil introduced a new national currency, the *real*, in 1995. With it came strict monetary controls, and consumer price increases have averaged 3 percent per month since then. Part of Brazil's policy continues to be pressured by the desire for high wages and the need to maintain high interest rates to finance domestic government debt and borrowing and to prevent capital outflows from the country.

Through all of this turmoil, Brazil maintained its trade competitiveness; imports rose almost 50 percent to $54 billion in 1996, and the EU was the most important regional market for Brazilian exports receiving 25 percent of the total. Brazil continues to export primary products, such as soybeans, iron ore, and coffee, but also exports transportation equipment and metallurgical products. Imports include industrial plant and machinery parts, fuel and lubricants, chemicals, and iron and steel. The most rapidly growing sectors in Brazil's economy include computers and peripherals (Figure 13.12). The largest single country market for Brazilian exports was the United States, accounting for $9 billion, or 20 percent, of total exports in 2000. Brazil remains the United States' largest market in South America as well, and the third largest market in the Western Hemisphere after Canada and Mexico.

Japan

The fastest growing world trade region is East Asia and the Pacific. After Europe, this region has the largest amount of internal world trade. Exports and imports in 2000 amounted to $2.9 trillion, or 23 percent of the world's total. Japan traditionally took the lead role in the development of East Asia and the Pacific. The growth of its economy after World War II is nothing short of an economic miracle. Since the 1960s, Japan has been joined by the Four Tigers of Taiwan, South Korea, Singapore, and Hong Kong. In the 1990s, however, new emerging dragons followed suit with rapidly growing economies: Thailand, Malaysia, Indonesia, and China.

While the rest of the world reeled from two major oil-price hikes in the 1970s, East Asia and the Pacific forged ahead with unprecedented growth (Chapter 12). Several factors contributed to this success, including U.S. economic and military subsidies, the Confucian culture dedicated to learning, and governments that actively promoted a shift into export promotion. In addition, Japan and other countries protected home markets with high import duties. Unlike America, where short-term profits were important to satisfy stockholders, banks, and financial institutions, Japan encouraged reinvestment and long-term growth cycles. These long cycles allowed firms time to develop products and to reinvest in the highest-quality production systems before the owners or employees could reap any of the profits. Further, many of the Asian/Pacific countries acted as resource supply centers for the United States from 1965 to 1975, during the Vietnam War, which allowed them to collect a heavy inflow of U.S. dollars.

All these factors combined allowed Japan to develop the world's second largest economy, after the United States. Japan is also tremendously more prosperous than China, which has nearly nine times the population, although China's economy is growing more quickly. Between 1960 and 2000, the combined domestic product of East Asia and the Pacific increased 25-fold, which changed their economies from developing nations to NICs. The governments initiated a switch from import substitution to export promotion, with a new emphasis on electronics, automobiles, steel, textiles, and consumer goods, whereas other developing countries in Latin America, Africa, and South Asia did not have such policies. From 1970 to 2000, foreign investment in the region, especially in Japan and the Four Tigers, grew 10-fold. This investment was led not only by United States, British, German, Canadian, and Australian firms, but also by the Japanese.

The United States plays an important role in East Asian trade. Until 1970, Western Europe was North America's chief trading partner. From 1970 to 1980, however, Asia and the Pacific had caught up with Western Europe in terms of total trade with North America, and since 1979, North American trade with Asia and the Pacific increased more rapidly than trade with Western Europe. The trade gap continued to increase, and in 2003, North American trade with Asia and the Pacific outpaced trade with Western Europe by 30 percent, nearly hitting the $450 billion mark.

Japan is the world's second leading international trading nation. Japanese exports manufactured goods and imports raw materials, food, and industrial components. Japan's economy essentially involves exchanging raw materials into high-value-added products, with high inputs of technology and labor. Unlike Western Europe and North America, Japan shows a huge export trade surplus. Although the

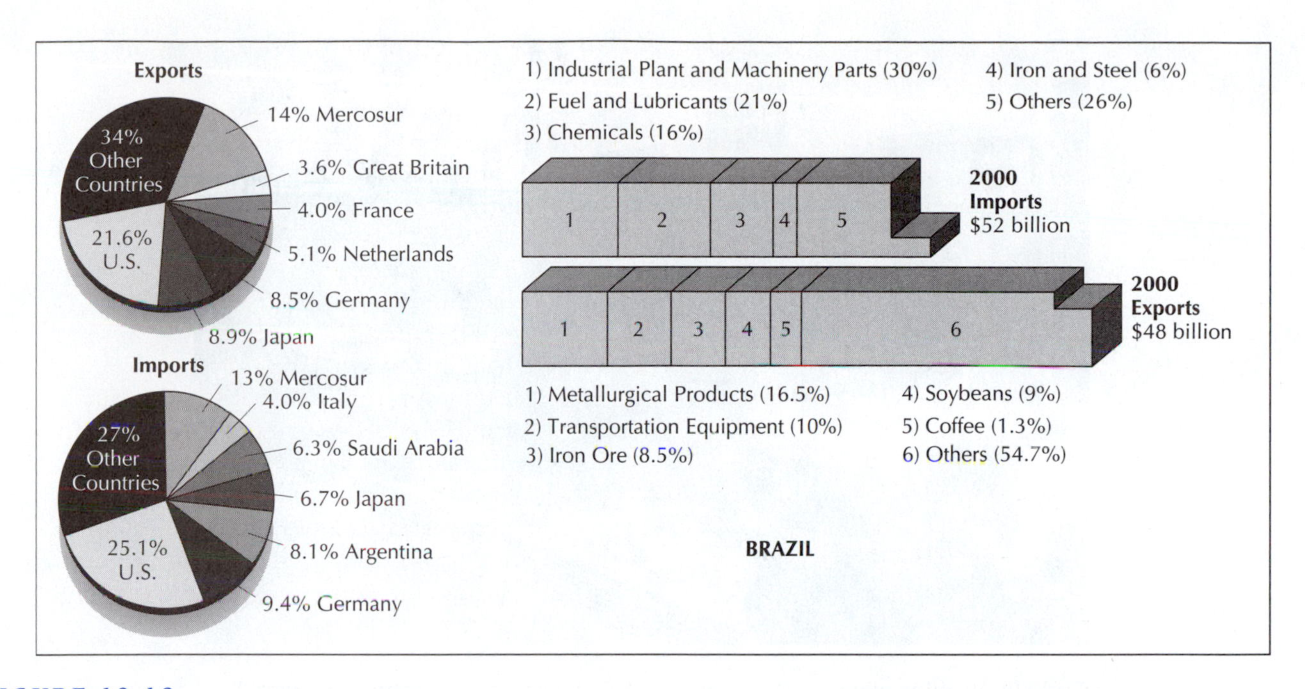

FIGURE 13.12
Composition of Brazil's merchandise trade. Computers and peripherals are the fastest-growing sectors of the Brazilian economy, followed by plastic materials and resins.

United States is its principal trading partner, both for exports and for imports, Japan has a tremendously diversified trading base, with almost 50 percent of exports and imports going to countries each composing less than 4 percent individually (Figure 13.13). The United States is by far Japan's largest trading partner, and in 2003, 30 percent of Japanese exports went to the United States and Canada, while 25 percent of imports came from Canada and the United States. After Canada, Japan has been the second largest market for U.S. exports for many years, but the U.S.-Japan trade deficit widened to $100 billion. This constituted more than one-third of the U.S. merchandise trade deficit (the other largest U.S. trade deficit was with China). The United States sells fewer exports to Japan than it imports, creating large U.S. trade deficits. The United States ships primary sector goods such as grains, feed, fruit, lumber, and non-oil commodities to Japan as a way of accounting for its imports of high-value-added manufactured items—microelectronics and automobiles, primarily. The United States does sell to Japan some high-value-added goods such as Boeing aircraft. U.S. products still account for the largest single proportion of goods imported by Japan. Indonesia, Australia, China, South Korea, and Germany follow. The United States and Indonesia fill a large need for energy that Japan cannot meet domestically. In addition, because of Japan's mountainous terrain, agricultural and food products compose 15 percent of total imports. Chemicals, textiles, and metals are also imported.

Japan's trade surplus with Southeast Asian countries exceeded its surplus with that of the United States. As a group, Southeast Asian countries and the new industrializing countries (NICs) are now a more important export market for Japan than is the United States. To counterbalance the impact of the high-priced yen, Japanese companies have been moving labor-intensive assembly manufacturing operations to Southeast Asian countries—mostly to Thailand and Malaysia, but more recently to China.

The world dominance of Japan in the manufacture of motor vehicles is phenomenal (Figure 13.14). Fully 22 percent of its exports are motor vehicles, followed by high-tech office machinery, chemicals, electronic tubes, iron and steel products, and scientific and optical equipment. Diversification is a key word used to describe the breadth of Japanese exports. Similar to America's exportation of manufacturing jobs to Mexico and East Asia, Japan has done the same with automobile-assembly plants and autoparts firms in the United States, which now number near 400.

By 1980, Japan had become the world's leading creditor nation and a dominant player in the world financial scene. In 2003, Japanese banks accounted for 2 of the 10 largest in the world (a drop from the height of the "bubble economy," when they formed 6 out 10), although Tokyo is still one of the world's premier financial centers,

Forklifts awaiting export to America from Yokohama, Japan. Japan is the world's most prolific producer of motorized vehicles, automobiles, and commercial vehicles. More spectacular than the sheer volume of automobiles produced is the rate of industrial growth in this area. From 1960 to 2000, Japanese auto production increased by more than 5000%.

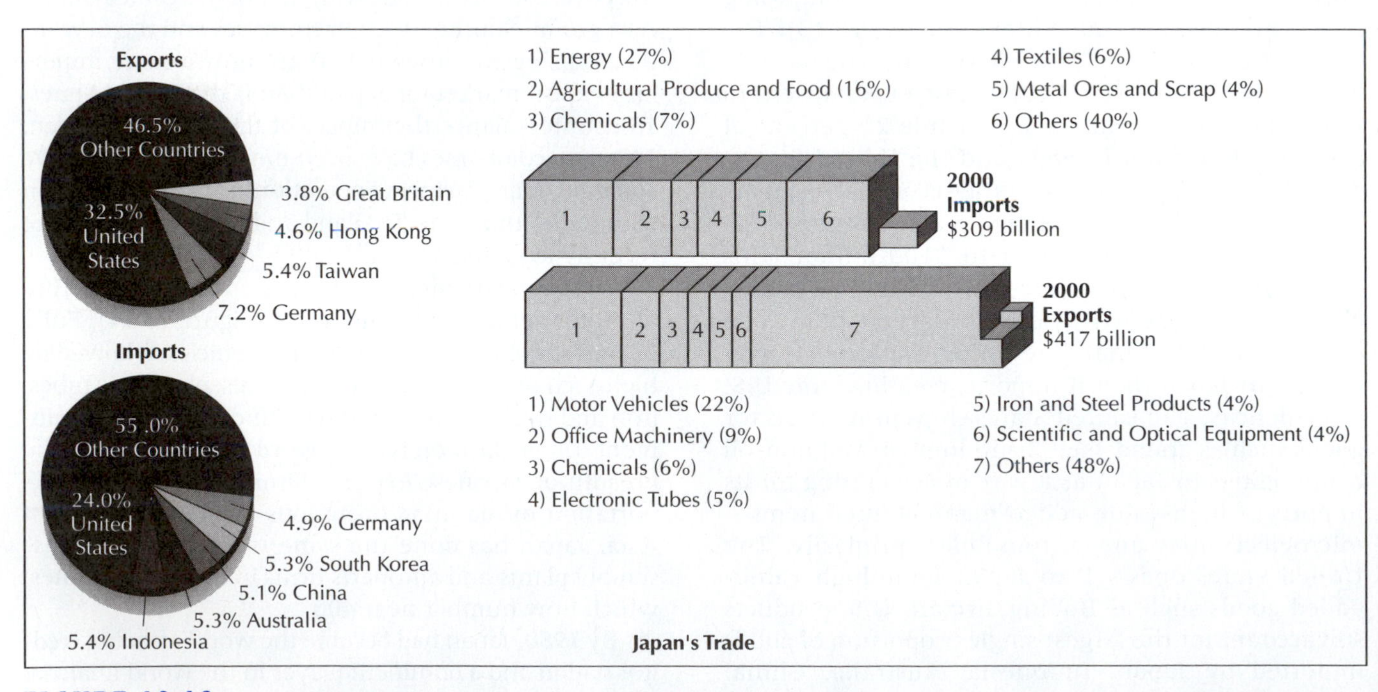

FIGURE 13.13
Composition of Japan's world merchandise trade, 2000. Japan now stands as the world's second largest economy, after that of the United States. Most important, Japan has a huge trade surplus. Japan's leading trade partner for both exports and imports is the United States.

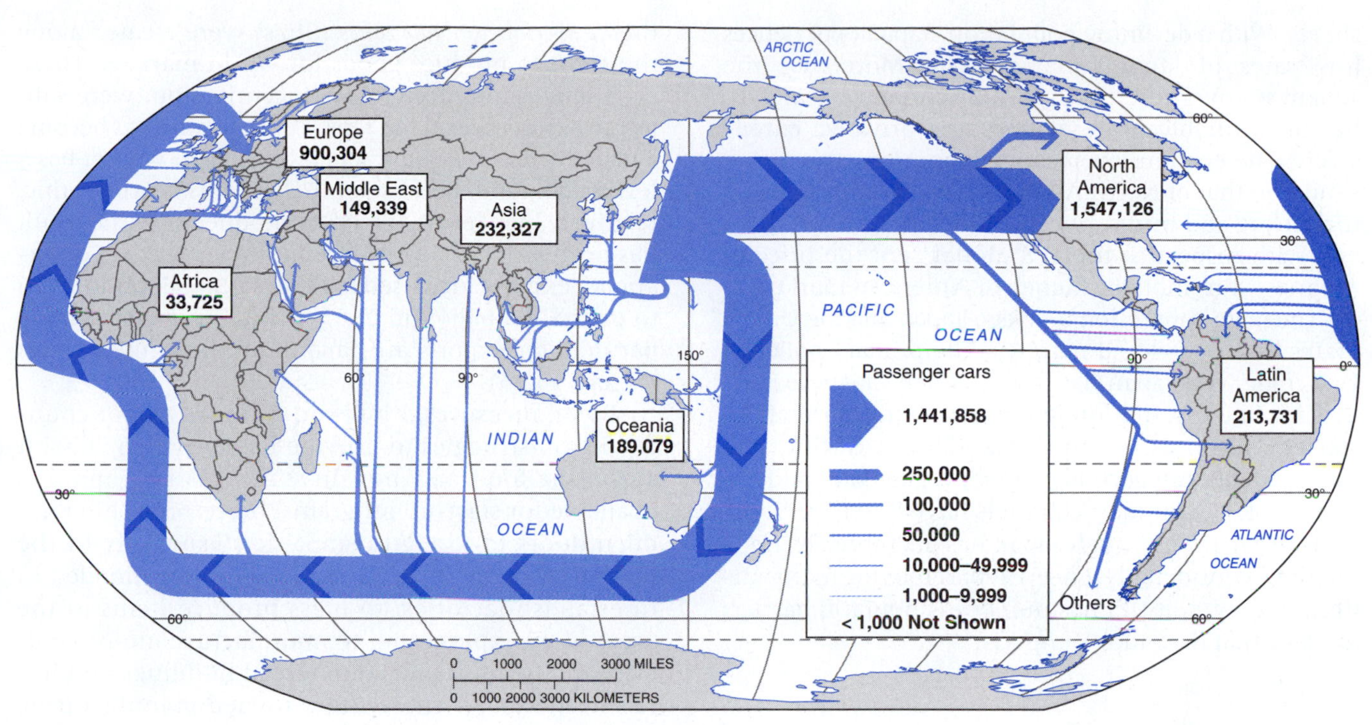

FIGURE 13.14
Geography of Japanese car exports. As the world's largest producer of automobiles, Japan controls a substantial share of the market in North America, Australia, and Europe.

alongside London and New York. Japan has managed all this with almost a total lack of food, mineral, and energy resources. It has shown the way for other East Asian and Pacific countries to follow suit. Singapore and Hong Kong are also major players in the world banking scene and have important money markets.

From 1980 to 2003, Japan attained a trade surplus of $750 billion, by far the largest in the world. It enjoyed a $100 billion bilateral trade surplus with the United States, $35 billion with Western Europe, and $25 billion with Asia and the Pacific. Both the EU, which received 15 percent of Japanese exports, and the United States, which received 28 percent, contributed to this trade surplus. As a result, protectionist voices in Europe and in America can be heard periodically accusing Japan of establishing new markets for itself by temporarily undercutting prices of foreign competitors. This monopolistic approach to competition is illegal in North America. The Japanese have also been accused of selling goods by dumping them on local markets to weaken rivals and force them to sell their market shares to Japanese firms. At the same time, the Japanese have restricted foreign firms from selling in Japan by a huge amalgam of import duties, tariffs, and regulations.

In 1985, the United States, Japan, Britain, France, and Germany began selling dollars on the world market to try to drive down the currency's value. The result was a big success, and the dollar is now quite weak, making American products competitive on world markets. The desirability of American goods has not eliminated the U.S. trade deficit yet, but movement is in that direction. At the same time, the yen soared, which caused the prices of Japanese goods in Europe and America to skyrocket. Japanese sales dropped, and the deficit narrowed even more. However, although the American and Japanese deficits narrowed from 1988 to 2003, they still represent a huge surplus for the Japanese.

The Japanese response to the decreased demand for their goods was to establish manufacturing plants in America and in Europe to reduce the prices of their marketed items. They also established manufacturing locations in Southeast Asia, where costs were lower. Japanese multinational corporations such as Nippon erected plants in Europe, North America, and Southeast Asia. This strategy—in North America and in Southeast Asia—allowed Japan to continue trading competitively with cheaper products. Thus, Japan has relegated Southeast Asia and the United States to Third World status by using its vast amounts of inexpensive labor to produce its manufactured products.

In the 1990s, the Japanese economy endured a prolonged, serious recession prompted by the collapse of the so-called bubble economy. Equity values plunged to 50 percent of their paper value; heavy investment in North American properties tapered off to almost zero. The declining asset values left Japanese banks, which numbered seven of the largest 10 in the world, with nonperforming loans and precarious balance sheets. With a declining population, Japan experiences low rates of labor force growth. Moreover, the Japanese government, plagued by corporate scandals, has been unable to deregulate and privatize extensively. The economy is facing economic restructuring similar to that of North America (i.e., away from manufacturing and into services). The appreciation of the yen with respect to the U.S. dollar sent the price of Japanese automobiles higher in American markets.

The main contention between Japan and the United States has been closed markets. The primary policy of U.S. trade with Japan has been to open Japanese markets for imports and foreign investment. Informal obstacles such as testing standards, certification requirements, intellectual property regulations, and impenetrable distribution channels have limited U.S. exports to Japan. Although Japan has liberalized its trade relations considerably since 1990, removing many tariffs and quotas, it still has numerous nontariff barriers in place that discourage imports.

China

Following a long period of isolation in the 1950s and 1960s, China opened up to international trade and investment after the death of Mao Zedong in 1976. During the 1980s, under the policies of Deng Xiaoping, China allowed foreign companies to set up joint ventures there. *Special economic zones* (SEZs) were created along the coast to produce goods for world markets. These economic zones received tax incentives but were subject to a host of legal red tape. Today, China has become a major actor in world trade. It has a large worker base, low wages, and, because of the East Asian work ethic, relatively high levels of worker productivity. The result has been a dramatic increase in foreign trade. For example, exports increased from only $5 billion in 1976 to over $200 billion in 2003. China's primary trading partners for exports are Japan and the United States (Figure 13.15).

By all measures, however, China is a poor country. It still struggles to provide its many people with sufficient food and housing. Almost no capital is available for start-up programs; therefore, China has open doors to foreign companies, especially in the industrial sector. Foreign investment has flooded in to establish factories, to mass produce items in the areas of oil exploration, to manufacture motor vehicles, and to construct commercial buildings and hotels in the major cities. State-owned manufacturing plants account for approximately 50 percent of manufacturing exports. The other 50 percent, and an increasing proportion, are small-scale industrial plants that are owned by rural townships but leased to private individuals for profit. Textiles, clothing, and industrial products accounted for 70 percent of exports in 2003.

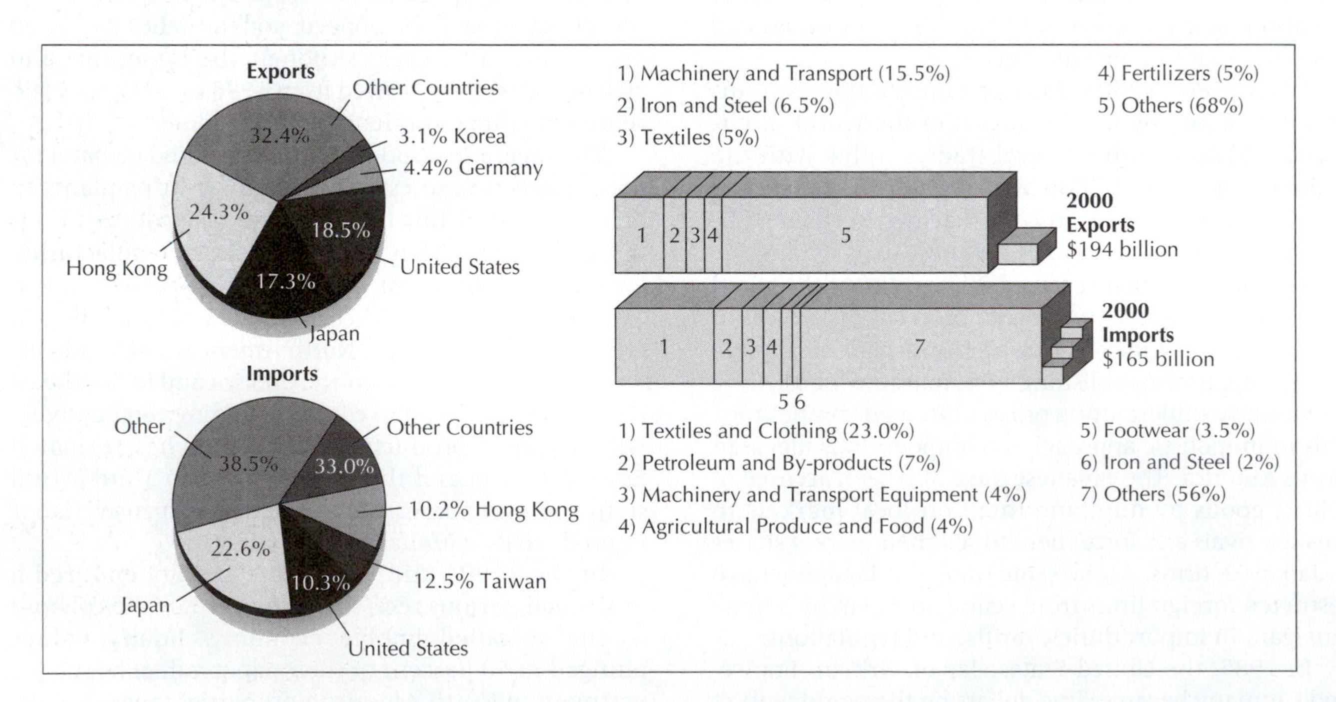

FIGURE 13.15
Composition of China's world merchandise trade. World trade is tiny compared with its population size of more than 1.2 billion people, yet it enjoys a trade surplus, exporting more than it imports, including textiles, clothing, petroleum, and by-products.

Over the last two decades, China recorded a GDP growth of 8 percent to 14 percent, continuing the surge in investment-led growth that began in 1992. Industrial output soared due in part to collectives and firms that are partially foreign owned. China's principal economic plan has been the continuation of high growth to generate jobs and to spread the benefits of economic reform from the cities to the countryside, in addition to controlling inflation.

Chinese manufactured goods lead export growth, whereas agricultural and primary products lead import growth. SEZs in southern and eastern coastal areas continue to outpace the rest of the nation in economic growth and trade. Japan continues to be China's top trading partner and source of imports. Hong Kong is the main port for exports and imports of China. Imports from the United States have been growing rapidly, but the United States ranks as the fourth largest source of Chinese imports after Japan, the EU, and Taiwan. China's major imports from the United States include cotton, fats and oils, manufactured fibers, fertilizer, aircraft, wood pulp, and leather. The most rapidly growing sectors of the Chinese economy include electrical power systems, telecommunications equipment, and automobiles (Figure 13.16). The U.S. trade deficit with China has increased steadily and stood at about $100 billion in 2003. Imports to the United States from China have grown 30 percent per year and exports 15 percent. Part of this imbalance is because China maintains an intricate system of import controls. Most products are subject to quotas, licensing requirements, or other restrictive measures.

Three out of four toys sold in America are foreign made, and 60 percent of those imports come from China. Sixty percent of all shoes sold in America come from China. In the near future, China is expected to dislodge Japan as the country giving America its biggest trade deficit. Other major goods imported into the United States from China include clothing, telephone and other telecommunication equipment, household appliances, televisions and computer chips, computers, and office equipment. China's higher trade barriers protect inefficient state-run companies that still employ two-thirds of the urban workers. China's trade patterns will change following its recent entry into the World Trade Organization (WTO).

China's transition to economic superpower status continues unabated. The economy has grown at an annual clip of about 9 percent. In 2003, China began to institute private property laws allowing home ownership of land and buildings. Real estate tycoons abound in China today, especially in coastal cities like Shanghai. China is becoming the world's factory, with growing trade surpluses to prove it. Ford and General Motors plants churn out autos for sale in East Asia. Most Chinese cannot yet afford autos, but televisions, washers, and stoves are selling well. The web of interconnections between foreign companies headed by overseas Chinese investors and a nominally communist China is growing.

Taiwan

Taiwan, with a population of 21 million and a GDP of $200 billion, has sustained a 6 percent to 7 percent economic growth rate since 1987. Long-term growth and prosperity have led to increasing land and labor costs in Taiwan and a gradual restructuring of the economy as low-wage sectors have fled. Manufacturers of

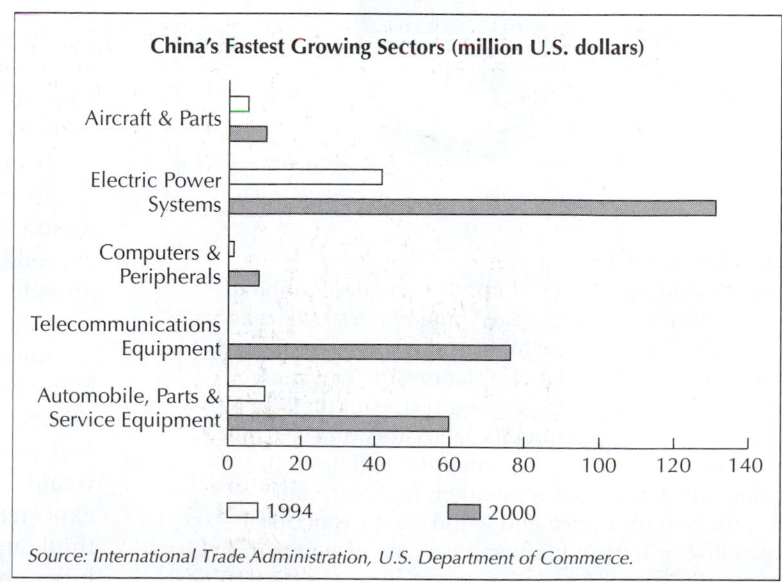

FIGURE 13.16
China's fastest growing sectors of the economy. China's fastest growing sectors include those associated with the development of power systems to produce electricity, followed by telecommunications equipment and automobile parts and service equipment. What used to be a Chinese unified national market has now separated into a number of regional markets, each pursuing its own development plan and each competing with the others to attract foreign investment and technology.

labor-intensive products such as toys, apparel and shoes, and circuit boards have moved offshore, mainly to Southeast Asia and to China. Large state-run enterprises still account for one-third of Taiwan's GDP, but major privatization efforts are underway to release power generation.

Exports account for over 35 percent of GDP in Taiwan's export-oriented economy—one of the four newly developing NICs, or Four Tigers. Manufacturing growth is now concentrated in technology-intensive industries, including petrochemicals, computers, and electronic components (Figure 13.17). In the 1990s, Taiwan's imports grew at a 14 percent annual increase. Historical trade patterns in the Pacific Rim are likely to keep Taiwan dependent on capital goods from imports from Japan. Over the last 10 years, Taiwan has made great progress in lowering trade barriers and improving market access of foreign goods.

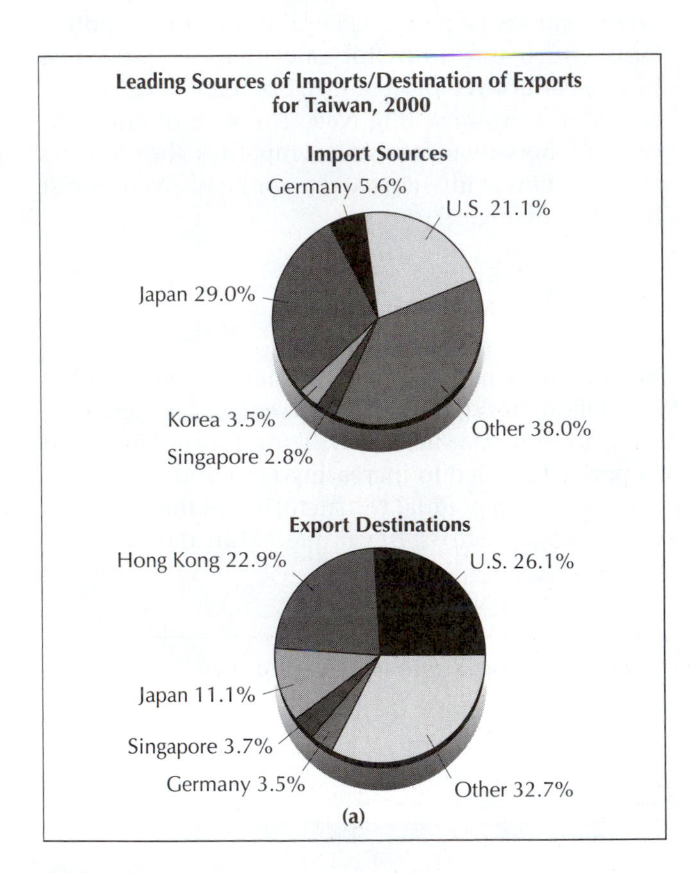

FIGURE 13.17
Taiwan's leading sources of imports and destination of exports. While Japan and the United States rank number one and two for Taiwan imports, the United States and Hong Kong are the chief destinations for Taiwanese exports. Taiwan's fastest growing sectors of the economy include computer peripherals, laboratory and scientific equipment, telecommunications, and pollution equipment. The consumer market, from cars and computers to insurance and world travel, is not only expanding in Taiwan in overall size, but also new niches continually open up as Taiwan's consumers become more discerning and more sophisticated in the world economy.

South Korea

Another miracle of the Pacific Rim is South Korea, a country of 47 million, with a GDP of $340 billion in 2003. Its GDP growth averaged over 8 percent over the last 10 years, and its GDP per capita is $6800 (which compares favorably with the $300 GDP per capita in North Korea), putting South Korea's standard of living on a par with parts of southern Europe (e.g., Portugal). South Korea's rapid advancement has turned this nation into one of the most economically powerful in the world.

The fastest growing sectors in South Korea include transportation services and computers and peripherals (Figures 13.18 and 13.19). In 1995, South Korea opened power production to the private sector. These areas are limited to the production of electricity by coal, LNG (liquid natural gas), and water power, benefiting Canadian and U.S. companies. South Korea is implementing an ambitious transportation infrastructure development program that includes major high-speed rail and transit programs, airport development, and highway construction. Once the world's largest producer of tennis shoes, South Korea's success has driven up labor costs to the point where companies such as Nike have fled to lower-cost countries such as China, Vietnam, and Indonesia.

Australia

Australia's main trading partners are the United States, the EU, and Japan (Figure 13.20). Exports go primarily to Japan, which accounts for 20 percent. The next largest share, 10 percent, goes to South Korea, followed by New Zealand, the United States, China, and Singapore. The EU accounts for approximately another 11 percent. Because of its small population, 19 million, industrial supplies, automobiles, and industrial equipment account for more than 60 percent of imports. Japan has made its greatest market penetration into Australia and accounts for 50 percent of all vehicles purchased. The United States leads the list of importers, providing 23 percent, while EU countries supply 35 percent. Japan follows with 13 percent.

Australia is one of the leading raw-material suppliers in the world. It exports *primary products*—mainly ores and minerals, coal and coke, gold, wool, and cereals ("rocks and crops"). Almost all of its exports are from the vast wealth of land and resources that it enjoys. It is the largest exporter of iron ore and aluminum and the second or third largest exporter of nickel, coal, zinc, lead, gold, tin, tungsten, and uranium. Consequently, Australia's current problem is to withstand the declining world prices of raw

Samsung VCR assembly line in Seoul, South Korea. Foreign national corporations, like Samsung, have raised foreign exchange. Recent downturns in the global economy have slowed South Korea's progress. New problems now exist with North Korea's nuclear capability and sullen response to trade.

Pusan, South Korea is a major shipbuilding port that facilitates that country's exports of steel.

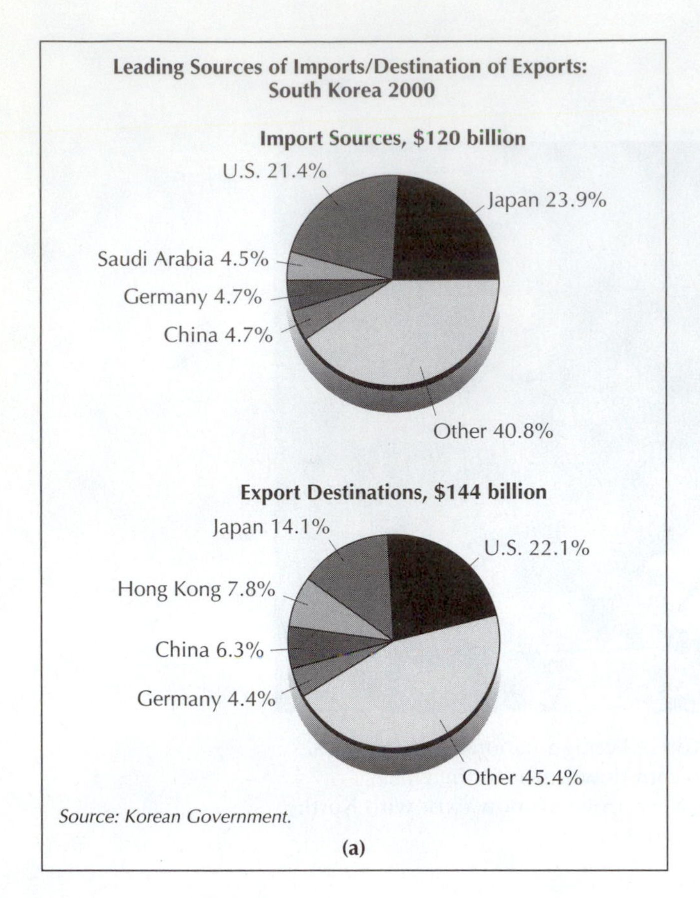

FIGURE 13.18
South Korea's leading sources of imports and destination of exports. Leading sources of imports included Japan and the United States, accounting for roughly 43 percent of imports. The leading export destinations included the United States and Japan, accounting for 30 percent of destinations, but a wide variety of other countries contributed to both import and export destinations for South Korea.

materials. To cushion against fluctuations in these prices, Australia needs to industrialize so that it can transform its raw materials into finished products and become an exporter of higher-value items. It has, however, become a significant exporter of wine. However, doing so is nearly impossible with a small industrial base that demands consumer products before industrial products.

India

In South Asia, India, with 1 billion people, has the world's second largest population but a relatively small economy, a reflection of the huge pools of poverty found there. Its world trade is minuscule but growing. As a result of the Green Revolution, India is self-sufficient in food production. Today, it is an exporter of primary products, gems and jewelry, textiles, clothing, and engineering goods (Figures 13.21 and 13.22). In order for its factories to operate, it must import industrial equipment and machinery, and crude oil and byproducts, as well as chemicals, iron, and steel. Twelve percent of its imports include uncut gems for its expanding jewelry trade.

In international trade, India is no longer dominated by its former colonial overseer, Britain. Its leading countries of export include the United States and Japan. Britain and other EU countries account for 21 percent of total exports and 27 percent of total imports. Because India represents such a large pool of demand, most manufactured goods and consumer goods are consumed locally, not exported. Since 1990, India has also become an exporter of cereals and grains, and textiles and clothing are now the chief exports.

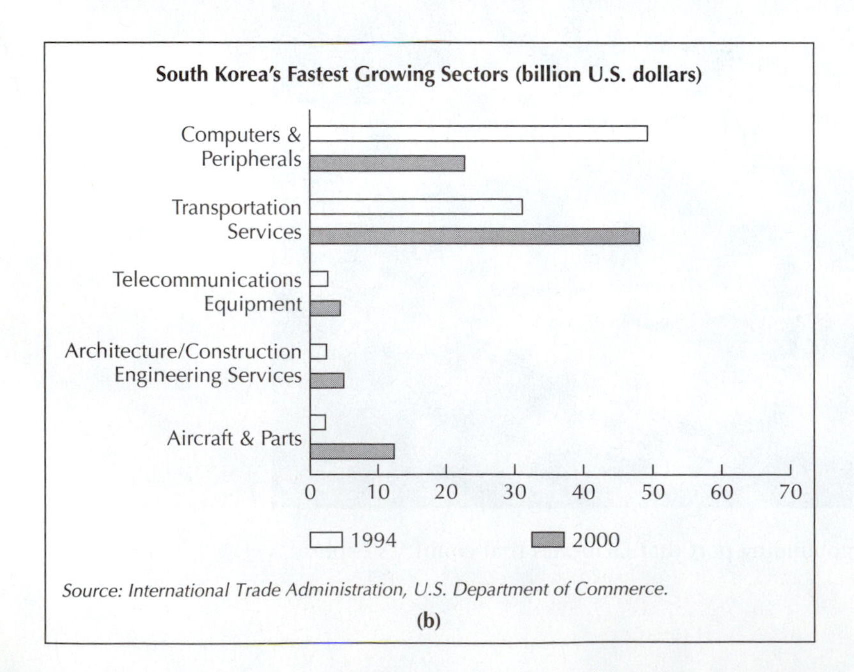

FIGURE 13.19
The fastest growing sectors in South Korea include transportation services and computers and peripherals. South Korea is implementing an ambitious transportation infrastructure development program that includes major high-speed rail and transit programs, airport construction, and highway development. While South Korea's production technologies have reached nearly the level of advanced countries in key heavy and high-tech industries, they still lag somewhat behind.

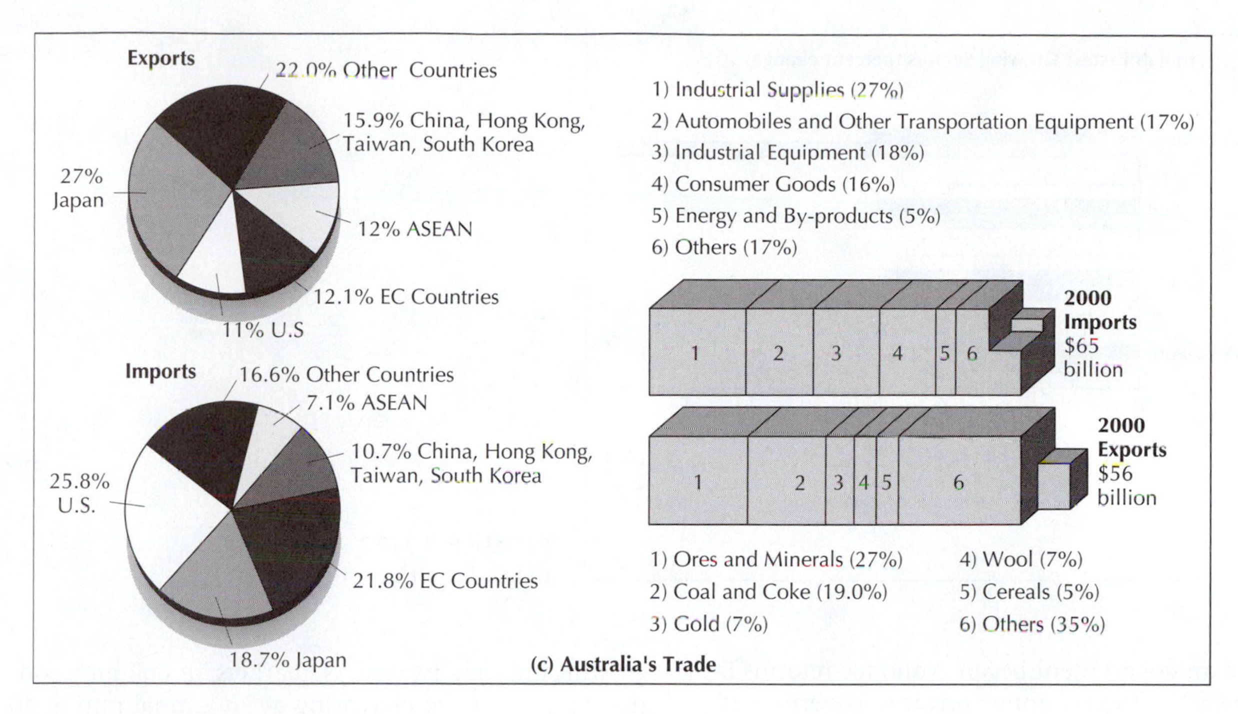

FIGURE 13.20
Composition of Australia's merchandise trade.

India's GDP has continued at a 5 percent to 6 percent increase in the 1990s. Record levels of foreign capital (over $6 billion in portfolio and direct investment) stimulated a capital market over the last decade. But Indian public investment and infrastructure continue to be insufficient for a country with developmental goals.

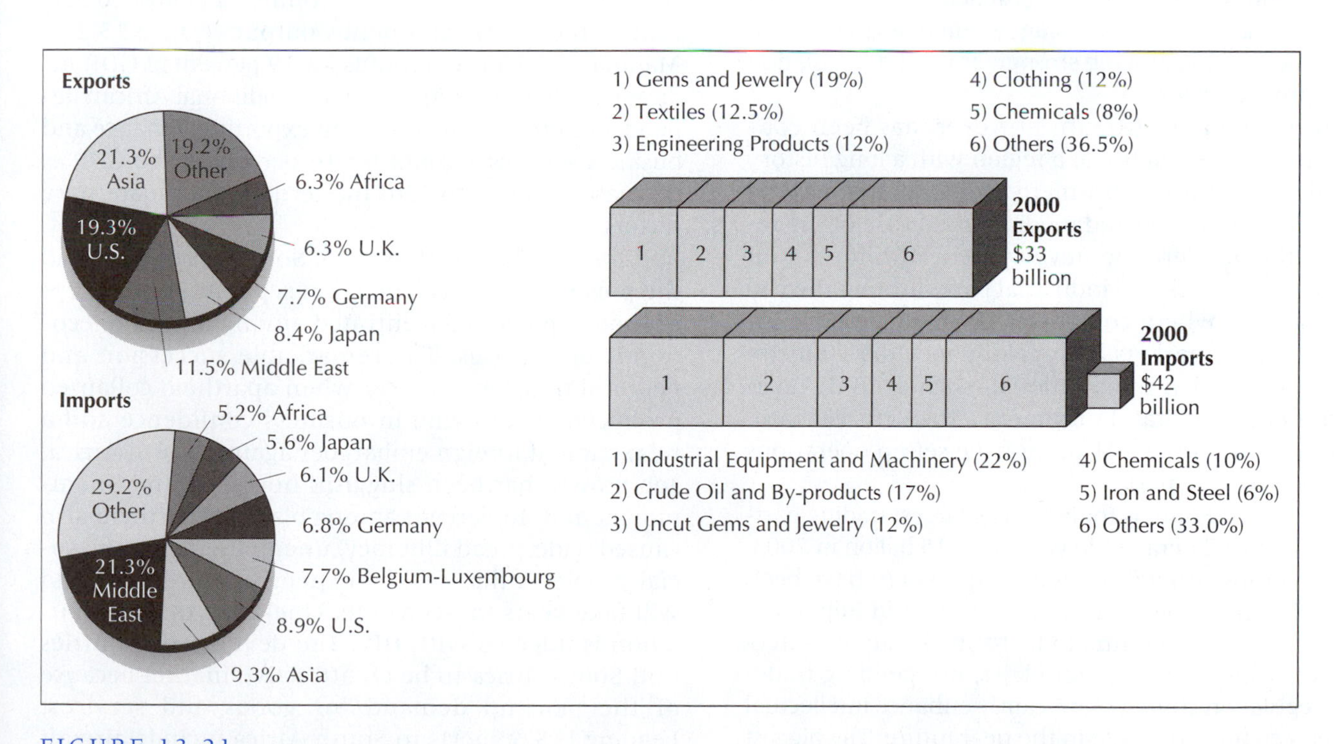

FIGURE 13.21
Composition of India's world merchandise trade, 2002. India exports gems, jewelry, and textiles. The country imports badly needed industrial equipment and machinery, as well as crude oil for fuel. Its total trade is tiny compared with its 1 billion population.

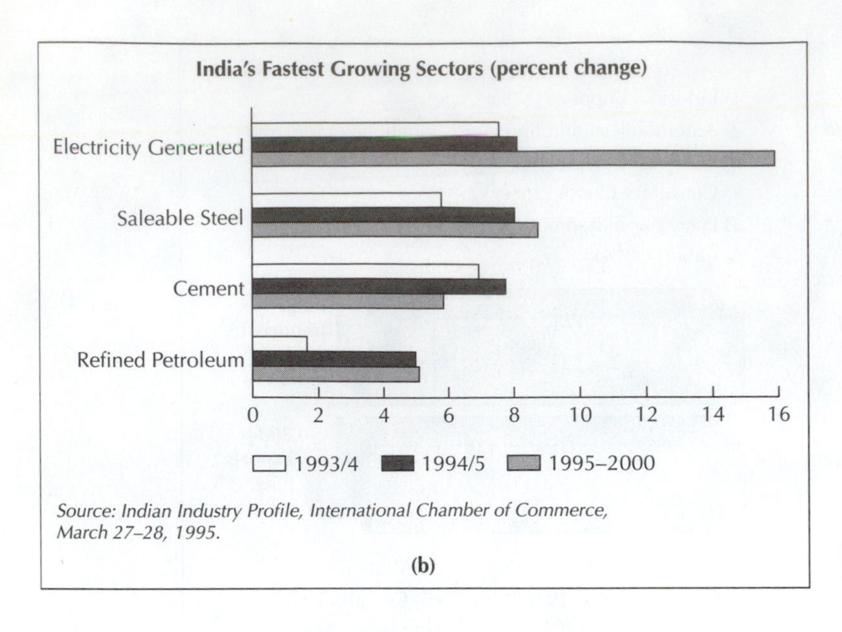

FIGURE 13.22
India's fastest growing sectors.

The Indian government began economic reforms in 1991 to liberalize the economy, privatize government-owned industry, and open India to international competition. As part of the global ascendancy of neoliberalism, the International Monetary Fund (IMF) persuaded India to turn its back on a policy of trade protection and import substitution that had been in place since the country became independent in 1949. Since 1991, India has used macroeconomic tools to improve its government deficits, inflation, and balance of payments. Import tariffs were slashed, and the government loosened its hold on business by dismantling the licensing system that governed all economic activity, moving strongly toward deregulation of the private sector.

Most economic growth, however, has been concentrated in western India, a region with a long history of trade ties. Mumbai (formerly Bombay) has become a significant financial and media center. One of the results is the accelerating development of information technology centers and industrial parks around the city of Bangalore, where companies produce software for international markets. Bangalore is India's Silicon Valley, and foreign multinationals such as IBM, Texas Instruments, Digital Equipment, Hewlett Packard, Motorola, 3M, and Qualcomm have set up operations in a huge science park.

The United States is India's single largest trading partner, and total bilateral trade was over $15 billion in 2003. Tariffs on those capital goods and equipment have been reduced from 35 percent to 25 percent, and imports of consumer goods continue to be banned. Patent protection is lacking new corporate laws, but pending trademark legislation should significantly enhance intellectual property rights protection in the near future. The biggest export opportunities to India from developed regions of the world include large electronic components. Because of increasingly stringent environmental regulations and growing industry awareness, markets for pollution control equipment are increasing at an annual rate of 40 percent. Food processing and packaging equipment is in demand as India's agricultural sector employs 70 percent of the country's workforce.

South Africa

The Republic of South Africa, with a population of 41 million and a GDP of $133 billion, is the most productive economy in all of Africa and accounts for almost 50 percent of the entire continent's output (Figure 13.23). Manufacturing now accounts for 19 percent of GDP, indicating a diversification from the traditional African dependence on gold and diamond exporting. Finance and business services account for 16 percent of the GDP as the nation moves toward the tertiary and quaternary sectors.

From World War II to 1995, South Africa practiced the policy of *apartheid,* or racial separation, that kept economic power concentrated among few large economic enterprises. The remarkable succession and peaceful transfer of power when apartheid collapsed produced an upswing in business confidence and a relaxation of foreign embargoes against South Africa. Job growth has been sluggish, however, and unemployment is 40 percent among blacks. Apartheid also caused widespread illiteracy, unemployment, and social problems that will be expensive to remedy and will take years to overcome. One-third of the population is infected with HIV. The developed countries find South Africa to be an attractive market because of the pent-up demand for goods and services. Leading U.S. exports to South Africa include aircraft and parts, industrial chemicals, computer software, pharmaceuticals, medical equipment, telecommunications equipment, and building and housing products.

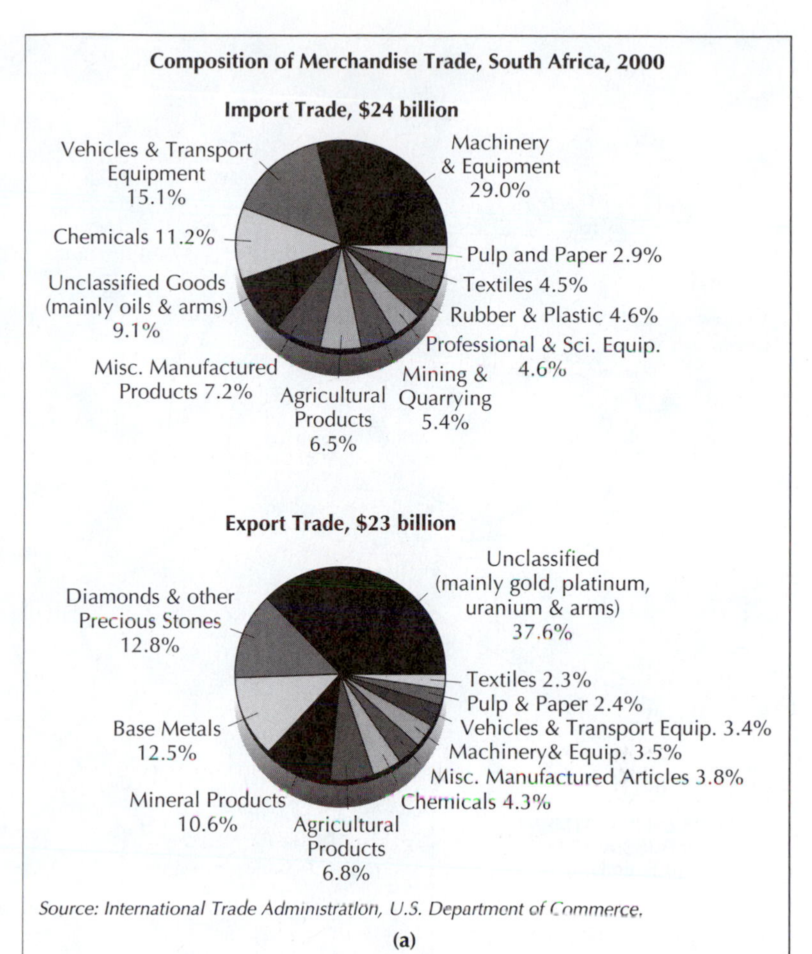

FIGURE 13.23
South Africa's composition of merchandise trade, imports and exports. U.S. and Canadian firms involved in South Africa's redevelopment will contribute to the country's revitalization process while gaining a strong commercial foothold in the new South Africa and the Africa beyond.

The United States is South Africa's largest trading partner. South Africa recently has paid attention to reorienting its own economy away from import substitution toward international competition.

Russia

The disintegration of the Soviet Union in 1991 was one of the most momentous occurrences in recent history. Russia and its former client states in Eastern Europe began a long, slow, and painful transition to market economies. While Eastern Europe, to one degree or another, has managed this transition relatively well, in Russia, transition led to economic collapse. The old central-planning systems have broken up, but the new systems that will replace them are not yet fully developed.

In 2003, the national income was 60 percent less in Russia than it was in 1990, and investment fell to pre-1970 levels. For the majority of the population, this resulted in rising unemployment, crime, and poverty. Although rich in resources such as gold, natural gas, petroleum, and timber, Russia's government and economy have been so disorganized and corrupt that the benefits of the sale of these assets have been concentrated in the hands of a tiny, wealthy oligarchy. With the collapse of trade under the Council for Mutual Economic Assistance (COMECON), orders from nearby countries decreased by more than 50 percent, seriously hurting internal production and trade. There is a slowdown in the production of necessary raw materials, equipment, and replacement parts and a massive shortage of food, pharmaceuticals, textiles, garments, footwear, and machinery. Russia and its neighboring republics have been wrecked by inflation.

Although Russia was the basket case of the 1990s, globalization started to transform the corrupt society of postcommunism. One sign is the integration of Russia's oil industry into the world economy. In 2003, British Petroleum bought a $10 billion stake in the Russian oil giant TNK, and ExxonMobil invested $25 billion in oil development in Siberia. Russia's new oligarchies have decided that they can make more money selling shares of their fortune in global financial markets than by larceny. Russia's economic output has now reached and surpassed the level of 1991.

What does the future hold for this region with regard to international trade? There is a certain complementarity in Euro-Asia. Western Europe needs the minerals, oil, natural gas, and other raw materials that

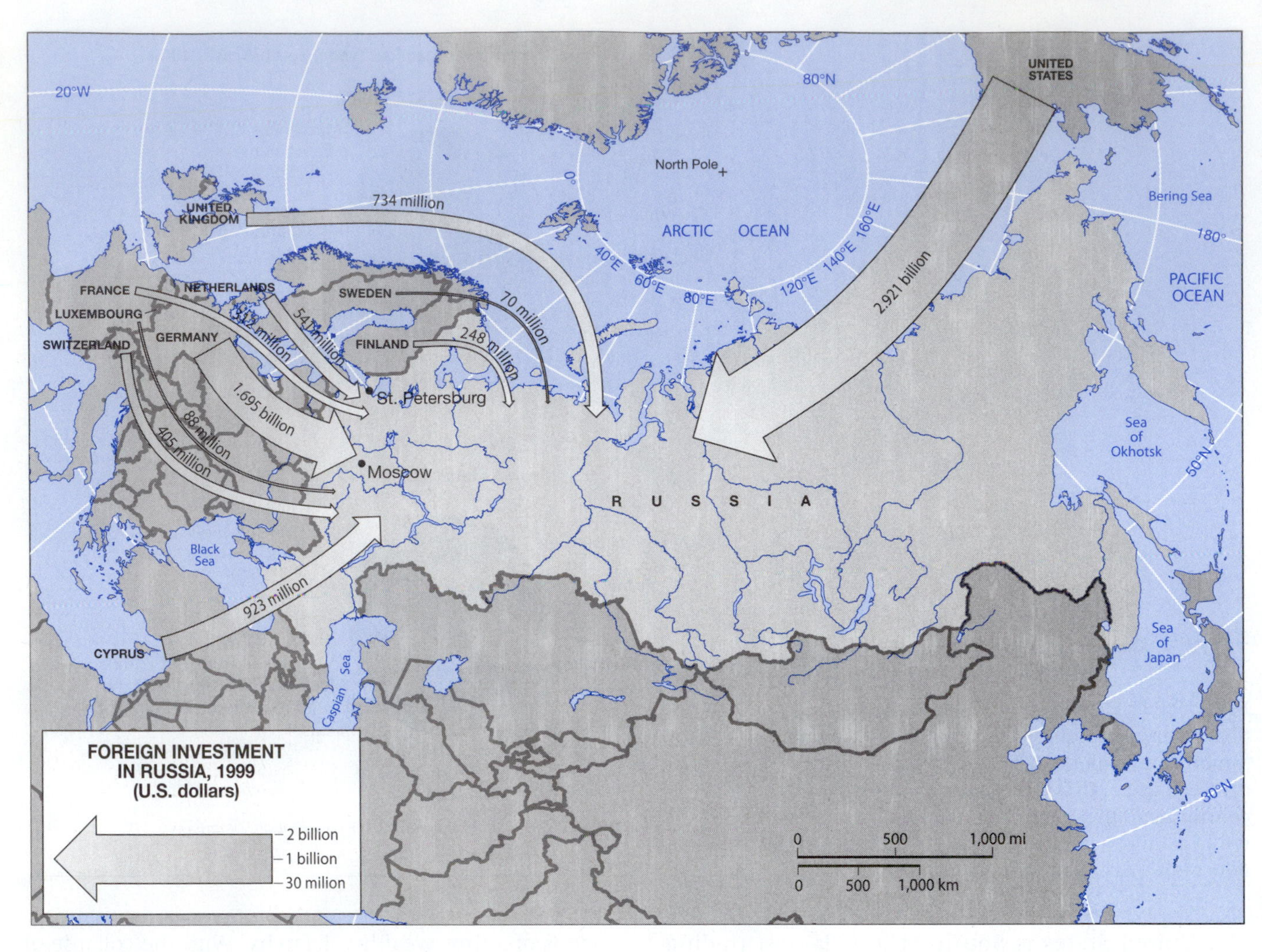

FIGURE 13.24
Foreign investment flows into Russia, 1999.

are in vast supply in Russia and the republics of the former Soviet Union. At the same time, the eastern bloc nations need foodstuffs and industrial equipment and machinery to resume their powerhouse of economic production.

Multinationals from every OECD country are investigating the potential for investment in the former Soviet Union (Figure 13.24). Automobile manufacturers from Western Europe, Japan, and the United States are also investigating their opportunities, as are consumer electronics producers. However, uncertainties still exist, and these multinationals must be cautious because of the gigantic economic uncertainties in the political and economic transformation.

Although much business and property have been privatized, agriculture remains largely in state hands as well as a third of all manufacturing plants. In 2003, Russia's population was 145 million (and falling), with a GDP of roughly $270 billion. Unemployment has leveled off, but difficult inflationary battles continue and the infrastructure is decaying. Russian privatization has attracted investment targets, including Aeroflot, the National Electric Utility, and Gazpron (Russia's gas company).

Russia is an emerging market and constitutes one of the 10 fastest growing world markets for manufactured goods. Ordinarily, manufacturing exports to Russia continue to be infrastructure related: engineering equipment, automatic data processing, and telecommunications equipment. In 2000, the largest single export to Russia was aircraft and related equipment. Total U.S. exports to Russia have been increasing steadily and were more than $3.5 billion in 2003. Hindrances to increased world exports to Russia are substantial value added taxes, high import duties, and high excise levees. The most promising prospects for developed country exports to Russia are telecommunications equipment, computers and computer peripherals, pollution control equipment, oil and gas field machinery, construction equipment, medical equipment, electrical power systems equipment, automotive parts and services, building products, and food processing and packaging equipment.

The Middle East

The Middle East contains approximately 64 percent of the world's oil reserves, with Saudi Arabia containing 264 billion barrels, more than one-third of the world's total. Other oil producers and exporters are Bahrain, Kuwait, the United Arab Emirates, Oman, and Qatar. Four other countries—Iraq, Iran, Egypt, and Syria—have very small amounts of oil. Countries in the region without oil supplies include Egypt, Israel, Jordan, Lebanon, Turkey, Yemen, Morocco, and Tunisia.

Inexpensive oil from the Middle East fueled the world for a long time. In fact, the United States and the western European nations have enjoyed a large supply. At $4 to $5 per barrel, Middle Eastern oil helped rebuild Europe after World War II. However, in 1973 and 1979, oil supplies were interrupted by Arab boycotts of the West for its support of Israel in the 1973 Yom Kippur War, by the overthrow of the shah of Iran in 1979, and by the Iraq-Iran War, and prices climbed to $30 per barrel. OPEC's revenues reached $300 billion, and a worldwide recession was triggered. However, because of squabbling among OPEC members and because U.S.-backed Saudi Arabia decided to undercut the market to provide Western stability, oil prices decreased to less than $10 per barrel in 1986. Revenues plummeted. In the 1990s, prices declined, and OPEC nations had lost much of their stranglehold on the world market, dropping to approximately equal to OPEC's pre-1973 levels. By 2004, however, oil climbed to $40 per barrel.

While all the fluctuations in oil prices were occurring, other sources of oil, synthetic fuels, and solar, geothermal, and nuclear power sources were being explored. OPEC's largest market was Western Europe, which traditionally had poor supplies of fossil fuels. Even before the oil crises, the Middle Eastern nations fulfilled 45 percent of Western Europe's energy needs. By the 1990s, however, that proportion dropped to 20 percent as a result of not only the exploitation of the North Sea oil fields but also increased coal production in central Europe.

The region's second most important activity after oil is agriculture, but water is scarce and the few sources that do exist are heavily tapped. The Tigris and Euphrates rivers, for example, as well as the Jordan River, are argued over by countries such as Turkey, Lebanon, Syria, Israel, and Iraq. In any case, wheat, barley, vegetables, cotton, and citrus fruits can be grown and supplied to Europe, which is a short distance away. Agriculture is also the basis of Turkey's export economy. Turkey is a large exporter of wheat and mineral resources, including iron, copper, and zinc.

Egypt is now one of the world's leading exporters of cotton, yarns, textiles, and denim. Egypt has been criticized because much of its agricultural base is devoted to cotton at the expense of food crops needed to feed its people.

Israel exports cut and polished diamonds, machinery, computer software, telecommunications equipment, and is the only high-tech economy of the Middle East. Throughout the region, tourism has been impor-

Iranian oil field and peasant herder. Iran remains an important source of world oil supplies, but squabbles among OPEC members have kept world prices low.

The Suez Canal remains a vital link for distribution of oil from the Persian Gulf to Europe. Despite the fact that the Middle East was one of the early hearths of civilization and city development, it did not share in the capitalist expansion and prosperity of the last 300 years that was centered on Europe and North America. It was only with the discovery of vast oil deposits by U.S. companies in the Persian Gulf area in the second half of the twentieth century that the focus of world attention returned to the Middle East. Today, the Middle East contains the greatest extremes of wealth, versus poverty, to be found anywhere in the world, all based on whether or not a country has oil supplies.

tant; however, the continued conflict between Israel and the Palestinians has depressed the tourist industry there.

Major Global Trade Flows

Six major commodity groups merit further attention as fundamental to understanding international patterns of trade: microelectronics, automobiles, steel, textiles and clothing, grains and feed, and non-oil commodities.

Microelectronics

Microelectronics includes semiconductors, integrated circuits and parts for integrated circuits, and electronic components and parts. Japan and the East Asian countries, especially the Four Tigers of South Korea, Taiwan, Hong Kong, and Singapore, together account for the predominant flow of microelectronics in the world (Figure 13.25). The single largest flow from this group is from the developing countries of the Western Hemisphere other than the United States and Canada, most notably Mexico, and the countries in South Asia and East Asia, which send more than $24 billion worth of microelectronics to the United States. Japan sends another $20 billion worth of microelectronics to developing countries and more than $7 billion worth to the United States. The single largest flow of microelectronics in the world is from the United States to developing countries and is worth $21 billion.

Although the United States no longer leads the world in the manufacture of semiconductors, it is still a major player in the global trade flow of microelectronics. Canada and the western European nations of the EU and the European Free Trade Area (EFTA) account for a much smaller proportion of overseas trade in this category. However, intra-European trade in microelectronics accounts for $15 billion.

Automobiles

Global trade within the EU accounts for the largest single flow of automobiles. More than $100 billion worth of automobiles were shipped among EU countries in 2003, and more than another $10 billion worth were sent to other European countries (Figure 13.26). The United States imported $11 billion worth as well, including Mercedes, Audis, Porsches, Volkswagens, Peugeots, Fiats, and Renaults. Japan made major inroads into the European automobile market, shipping $8 billion worth to EU. The single largest volume of flow is also accounted for by Japan and its shipment to the United States, $20 billion worth.

Japanese automobile manufacturers made major penetrations in the world automobile market between 1960 and 2000 (Figure 13.27). During the same time, the big three automakers in America scaled down operations substantially. General Motors and Ford remained

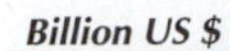

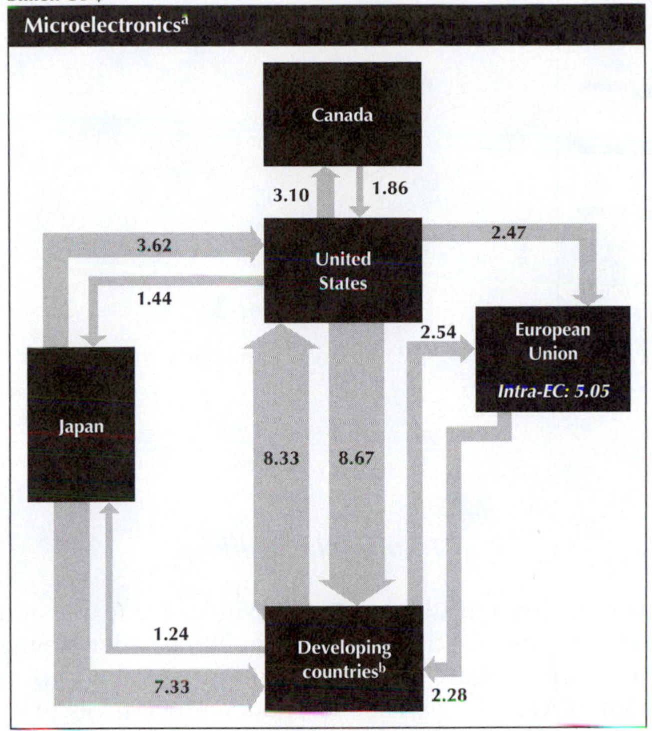

[a]Including integrated circuits and IC parts, semiconductors, and electronic component parts.
[b]Including Western hemisphere other than the United States and Canada; Africa and the Middle East; South Asia; and East Asia other than Japan, Mongolia, North Korea, Laos, and Cambodia.
Note: Width of arrows scaled to dollar volume. Trade flows less than $1 billion not shown.

FIGURE 13.25
Global trade flows of microelectronics, in billions of U.S. dollars.

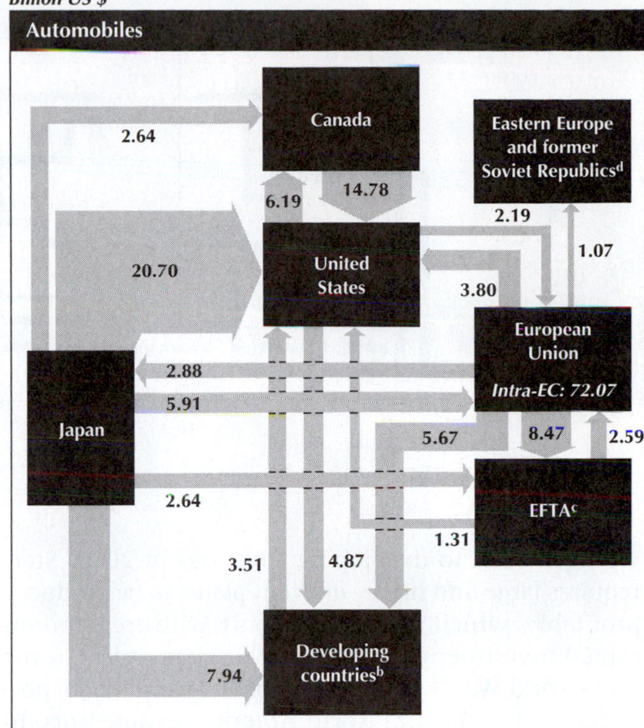

[c]Including Austria, Switzerland, Sweden, Norway, Finland, Iceland, and Liechtenstein.
[d]Including Poland, Hungary, Czech Republic, Slovakia, Albania, Romania, Bulgaria, and former Soviet Republics.
Note: Width of arrows scaled to dollar volume. Trade flows less than $1 billion not shown.

FIGURE 13.26
Global trade flow of automobiles, in billions of U.S. dollars.

the world's largest automobile manufacturers in 2000, even though the Japanese captured almost 26 percent of the American automobile and light truck market. American car builders, after years of struggle, are finally turning out higher-quality competitive products. Porter (1990) described demand conditions and taste as an important factor in global comparative advantage. Given the low cost of oil, American consumers are largely bypassing passenger cars for light trucks, minivans, four-wheel drives, and sporty utility vehicles.

Japanese MNCs such as Toyota, Nissan, Mitsubishi, Mazda, and Honda have invested heavily in factories in the United States, where they escape import quotas and have access to skilled, compliant labor. Starting with the Honda plant in 1982 in Marysville, Ohio, Japanese-made vehicles that are assembled and built in America account for nearly half the Japanese car sales in the United States. Locally made Japanese cars are cheaper than those imported from Japan because they escape the high-priced parts and labor associated with the yen. Nissan's newest plants are in low-wage Smyrna, Tennessee, and produce Sentras and Ultimas. These American-made Japanese cars are called *Japanese transplants*.

Exports of automobiles from Europe have been heavy. Germany is Europe's largest producer of cars, followed by France, Spain, and the United Kingdom. U.S. companies have diversified, so both Ford and General Motors now have substantial international automobile production (Figure 13.28).

Steel

Whereas America has lost as much as two-thirds of its steel employment in the last 20 years and now is a net importer of steel, Western Europe continues to lead the world in steel production and trade. The EU accounted for more than $44 billion worth of steel traded internally in 2000, and trade with the other European countries accounted for another $4 billion (Figure 13.29).

The single largest flow of steel was from Japan to developing countries. In addition, the EU sent $10 billion

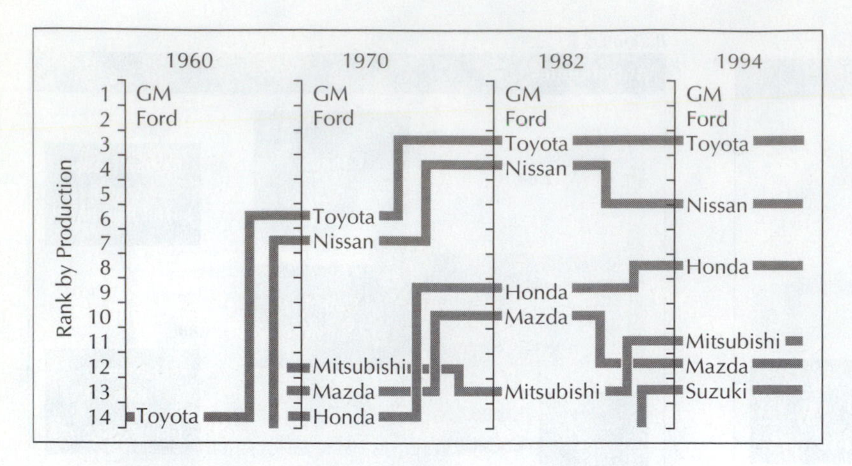

FIGURE 13.27
The rise of Japanese automobile manufacturers, 1960–1994.

worth of steel to developing countries in 2000. Steel requires large and highly efficient plants to be produced profitably, which are possible only with tremendous capital investments and large-scale economies. In the post–World War II period, steel made by traditional producers in Europe and North America became uncompetitive as new production centers began to emerge in Brazil, South Korea, Taiwan, and Japan. The migration of steel production to the Third World reflected the growing importance of labor costs, government subsidies, and taxes to the delivered cost of steel. In Chapter 7, we discussed the problems of the British and U.S. steel industries: insufficient reinvestment, reluctant unions, narrow-minded management, and lack of government support of an ailing industry. However, under the impetus of the microelectronics revolution, minimills have restored some of the efficiency of U.S. steel producers. Tariffs on European steel imports by the George W. Bush administration in 2002 threatened to provoke a trade war until the WTO forced the United States to remove them in 2003.

Textiles and Clothing

As discussed in Chapter 7, labor-intensive textile and clothing manufacture has largely shifted to developing countries (Figure 13.30), including Central America, South Asia, and parts of East and Southeast Asia (e.g., China, Thailand, Indonesia), where labor costs are much lower (Figure 13.31). Correspondingly, textile production in the last 40 years declined in the United States and Western Europe, which were the dominant producers as late as 1950. International trade in textiles reflects these shifts in production. Developing countries accounted for more than $85 billion worth of exports in 2003.

Surprisingly, Germany and Italy lead the world in textile exports, even though the main manufacturing centers of older industrialized nations have given ground to the developing world, particularly East Asia. Western Europe accounts for more than $71 billion worth of textile and clothing trade among nations within Western Europe.

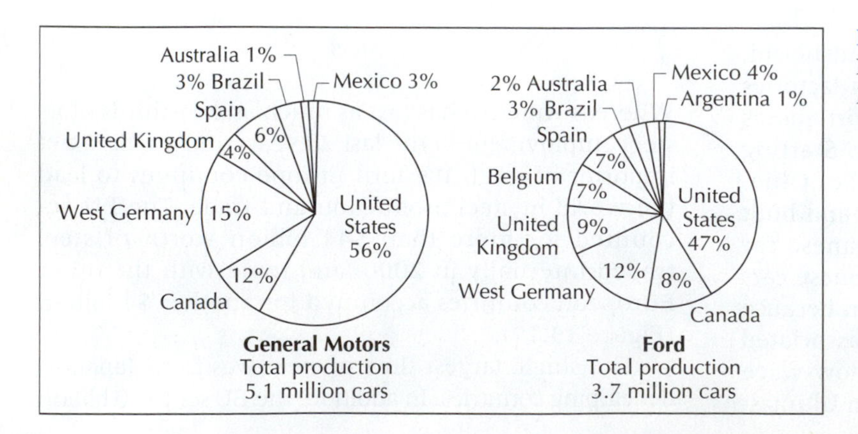

FIGURE 13.28
International automobile production by General Motors and Ford, 2000.

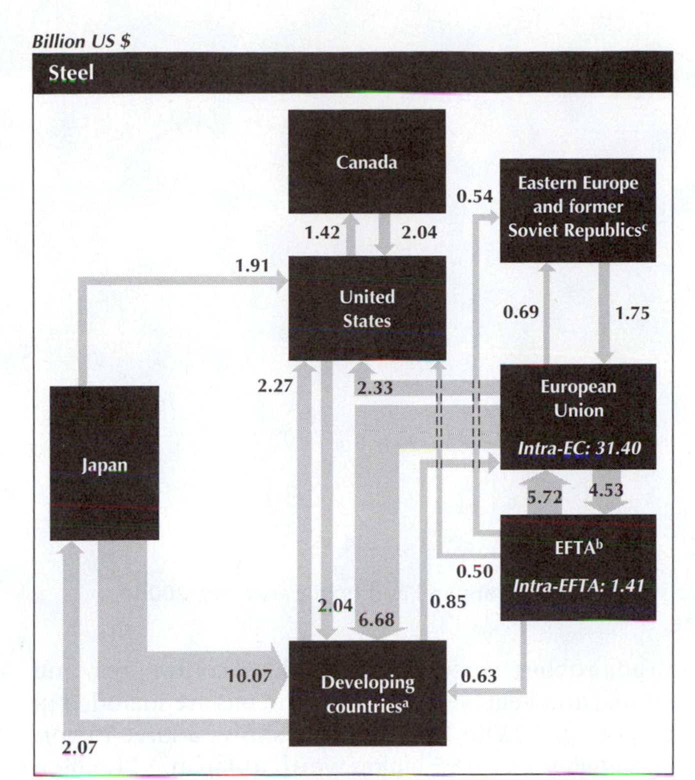

[a]Including Western Hemisphere other than the United States and Canada; Africa and the Middle East; South Asia; and East Asia other than Japan, Mongolia, North Korea, Laos, and Cambodia.
[b]Including Austria, Switzerland, Sweden, Norway, Finland, Iceland, and Liechtenstein.
[c]Including Poland, Hungary, Albania, Romania, Bulgaria, Czech Republic, Slovakia, and former Soviet Republics.
Note: Width of arrows scaled to dollar volume. Trade flows less than $500 million not shown.

FIGURE 13.29
Global trade flow of steel, in billions of U.S. dollars.

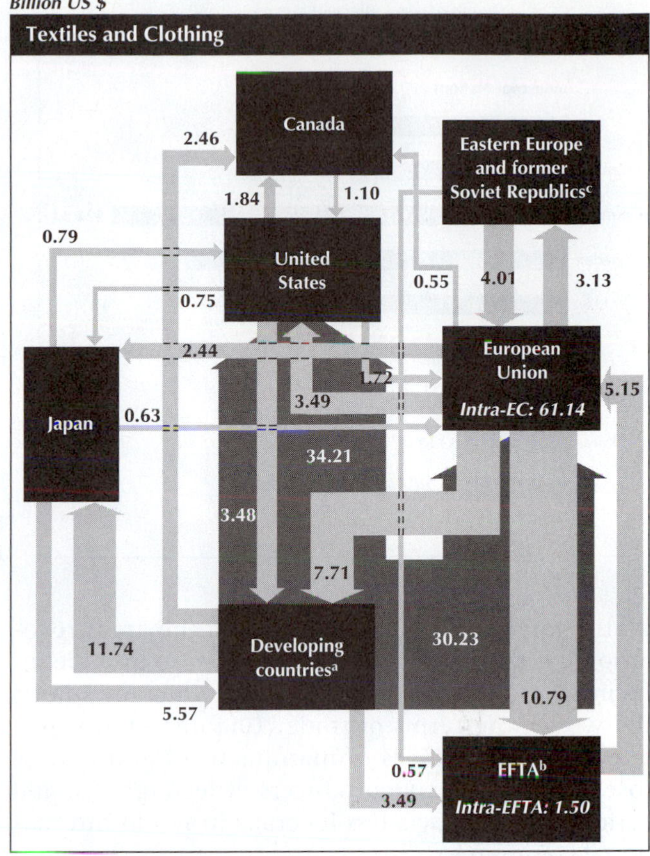

[a]Including Western Hemisphere other than the United States and Canada; Africa and the Middle East; South Asia; and East Asia other than Japan, Mongolia, North Korea, Laos, and Cambodia.
[b]Including Austria, Switzerland, Sweden, Norway, Finland, Iceland, and Liechtenstein.
[c]Including Poland, Hungary, Albania, Romania, Bulgaria, Czech Republic, Slovakia, and former Soviet Republics.
Note: Width of arrows scaled to dollar volume. Trade flows less than $500 million not shown.

FIGURE 13.30
Global trade flow of textiles and clothing, in billions of U.S. dollars. Globalizers claim that cheaper imported goods will more than compensate for the disruption and downward pressure on wages caused by increasing international trade and investment. This trade includes, of course, what is produced overseas by U.S. and Canadian multinational corporations like Nike, who pay $1.60 a day to their workers in Vietnam. But it hasn't worked out that way. A declining real wage means exactly that: Whatever benefit we have gotten from cheaper imported goods has been marginalized by other forces, including runaway factories and increased global competition.

Major gainers during the last 30 years include the East Asian countries of China, Hong Kong, South Korea, and Taiwan. By 2003, European nations accounted for 47 percent of the world trade flow, whereas East Asian countries accounted for another 43 percent. However, a much larger proportion of textiles and clothing flowed from developing countries to the United States than they did from Western Europe to the United States. Eastern Europe, Russia, Japan, and Canada are relatively small players in the world textile and clothing trade.

Grains and Feed

The primary products of wheat, corn, rice, other cereals, feed grains, and soybeans are included in the category of grains and feed. The United States exports more than $28 billion worth of feed and grains and thus is a world leader in this category (Figure 13.32). Canada is also a major exporter.

Japan, with its small base of agriculture and arable land, is a net importer, as is Eastern Europe and Russia. Trade within the EU is large ($17 billion in 2003). Developing countries such as India, Egypt, and Argentina are some of the largest net exporting developing countries in this category.

World trade in grains, feeds, and food products had been as high as 30 percent in 1965, but these commodities slipped to less than 15 percent by 2003. Some of this reduction is because Western seeds, grains, and fertilizers are now commonplace in Third

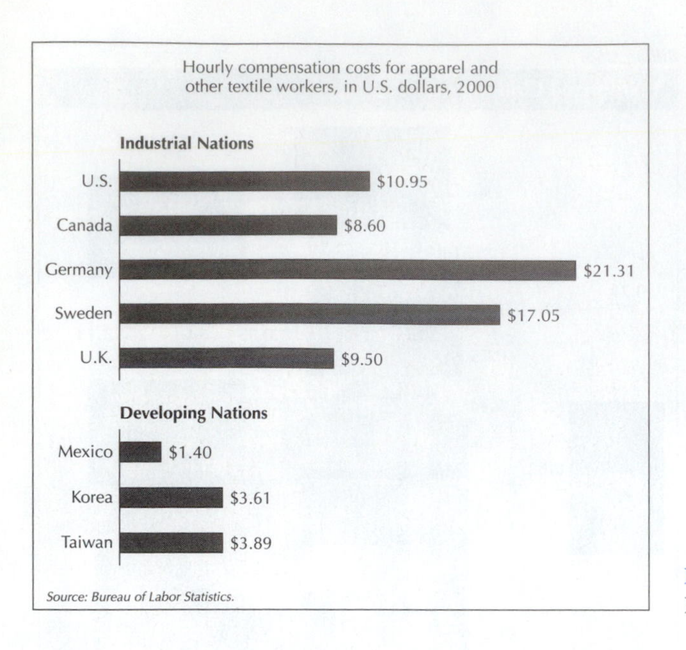

FIGURE 13.31
Hourly labor costs for apparel and textile workers, 2000.

World nations, and technology from the grain revolution has taught developing countries to provide for themselves. Another portion of the reduction reflects the worsening terms of trade (Chapter 11) for primary goods, as prices of manufactures and energy rose rapidly and gave producers of feed, grains, and agricultural products less leverage in world international commerce.

Non-Oil Commodities

Non-oil commodities include copper, aluminum, nickel, zinc, tin, iron ore, pig iron, uranium ore, and alloys. Crude rubber, wood and pulp, hides, cotton fiber, and animal and vegetable minerals and oils are included in this category. The United States shows a large export potential, sending $5 billion worth to Japan, $11 billion worth to developing countries, and another $5 billion worth to the EU.

However, the developing countries of the world lead in the export of raw materials (Figure 13.33). The largest single flow of raw materials outside the more than $43 billion worth exchanged within the EU is from developing countries to Japan ($21 billion) and from developing countries to Western Europe ($24 billion). Since the early days of Europe, the international

Combined grain is off loaded on a farm near Wataga, Illinois. The United States is a world supplier of primary products, including grains, timber, and other agricultural products. Most every country in the world grows grain, and approximately a quarter of the 600 million tons of wheat produced each year enters world trade. By 1960, the Third World began to outstrip its own food production, and Europe had long been a grain importer by then. The United States and Canada became the only large grain exporters, providing 80 percent of the world's exports.

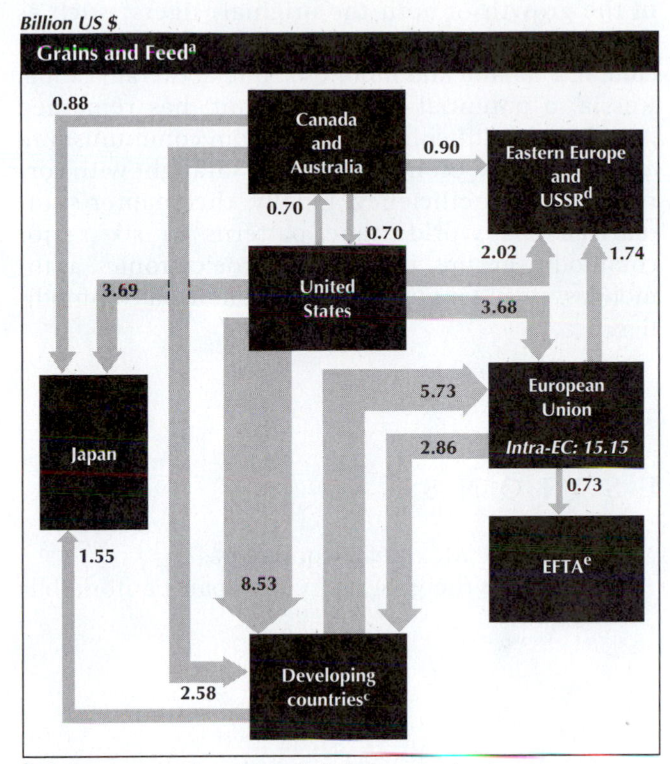

[a]Including wheat, corn, rice, other cereals, feedgrains, and soybeans.
[b]Including unwrought copper, aluminum, nickel, zinc, and tin, and ores thereof; iron ore and pig iron; uranium ore and alloys; other nonferrous ores, unwrought metals, and crude minerals; crude rubber, wood and pulp; hides; cotton fiber and other textile fibers; crude animal and vegetable materials.
[c]Including Western Hemisphere other than the United States and Canada; Africa and the Middle East; South Asia; and East Asia other than Japan, Mongolia, North Korea, Laos, and Cambodia.
[d]Including Poland, Hungary, Czech Republic, Slovakia, Albania, Romania, Bulgaria, and former Soviet Republics.
[e]Including Austria, Switzerland, Sweden, Norway, Finalnd, Iceland, and Liechtenstein.
Note: Width of arrows scaled to dollar volume. Trade flows less than $500 million not shown for grains/feed; trade flows less than $2 billion not shown for nonoil commodities.

FIGURE 13.32
Global trade flow of grains and feed, in billions of U.S. dollars.

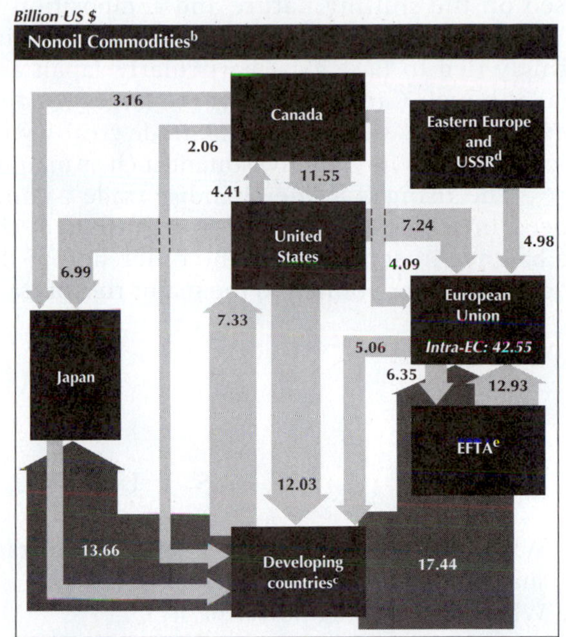

[a]Including wheat, corn, rice, other cereals, feedgrains, and soybeans.
[b]Including unwrought copper, aluminum, nickel, zinc, and tin, and ores thereof; iron ore and pig iron; uranium ore and alloys; other nonferrous ores, unwrought metals, and crude minerals; crude rubber, wood and pulp; hides; cotton fiber and other textile fibers; crude animal and vegetable materials.
[c]Including Western Hemisphere other than the United States and Canada; Africa and the Middle East; South Asia; and East Asia other than Japan, Mongolia, North Korea, Laos, and Cambodia.
[d]Including Poland, Hungary, Czech Republic, Slovakia, Albania, Romania, Bulgaria, and former Soviet Republics.
[e]Including Austria, Switzerland, Sweden, Norway, Finalnd, Iceland, and Liechtenstein.
Note: Width of arrows scaled to dollar volume. Trade flows less than $500 million not shown for grains/feed; trade flows less than $2 billion not shown for nonoil commodities.

FIGURE 13.33
Global trade flow of non-oil commodities, in billions of U.S. dollars.

division of labor was based on international trade. Under this unfair program, less developed countries traded their non-oil commodities—grains, feed, food stocks, and energy sources—for industrialized goods from primarily Europe and other developed nations.

Summary

International trade has grown rapidly over the last 30 years and now comprises roughly one-quarter of the world's total output. As we saw in Chapter 11, trade patterns simultaneously reflect and shape the specialization of production among and within different countries. Thus, trade must be viewed within the context of national changes in the level and composition of GDP. The growth of trade has occurred more quickly than the growth in national output, meaning that countries have become increasingly interconnected. However, important shifts in production internationally have contributed to changes in trade patterns, particularly the offshoring of many low-wage, low-value-added sectors in manufacturing from developed countries to LDCs. More generally, recent changes in global trade must be seen in light of the broader dynamics of globalization, including trade deregulation under the WTO, the rise of the NICs, and changes in global finance, including currency exchange rate fluctuations.

This chapter charted several major dimensions of international trade. It noted the growth of emerging markets in some of the largest developing countries, such as China, Brazil, South Korea, and India, all of which have exhibited steady, if uneven, growth. It focused on the shifting nature and composition of U.S. exports and imports, which have become increasingly tied to East Asia, particularly Japan and China, although Canada remains the largest trading partner. NAFTA has expanded U.S. trade greatly with Mexico, as well as foreign investment such as maquiladores. The rising U.S. merchandise trade balance has been offset to some extent by a surplus in the trade balance in services. On the other side of the Pacific, the chapter pointed to the major role of trade in the growth of both the original "tigers" such as South Korea and Taiwan as well as new tigers, including Thailand and Indonesia, but, above all, China. Russia, a potential emerging giant, has remained mired in the difficult transition from communism to a market-based economy, a process fraught with corruption and inefficiency. Finally, the chapter summarized the world trade patterns in six major commodity groups, including microelectronics, automobiles, steel, textiles, grains, and non-oil commodities.

STUDY QUESTIONS

1. What has happened to the level of world trade since World War II? Why?
2. What is purchasing power parity?
3. Does the United States trade more with Europe or East Asia? Why?
4. What are Mexican maquiladoras?
5. Describe the geography of Japan's automobile exports.

KEY TERMS

capital resources
demand conditions
factor conditions
factor-driven stage
Five Little Dragons
Four Tigers
General Agreement on Tariffs and Trade (GATT)
human resources
import substitution
infrastructure
innovation drive stage
investment-driven stage
Japanese transplants
knowledge-based resources
Little Dragons
microelectronics
modernization theory
Organization of Economic Cooperation and Development (OECD)
physical resources
purchasing power parity
wealth-driven stage

SUGGESTED READINGS

Dicken, P. 2004. *Global Shift: The Internationalization of Economic Activity*, 4th ed. New York: Guilford Press.

Grant, R. 2000. "The Economic Geography of Global Trade." In *A Companion to Economic Geography*. Oxford: Blackwell.

Hiscox, M. 2001. *International Trade and Political Conflict: Commerce, Coalitions, and Mobility*. Princeton, N.J.: Princeton University Press.

Krugman, P. 1997. *The Age of Diminished Expectations*. Cambridge, Mass.: MIT Press.

Porter, M. E. 1990. *The Competitive Advantage of Nations*. New York: Free Press.

Van Marrewijk, C. 2002. *International Trade and the World Economy*. Oxford: Oxford University Press.

WORLD WIDE WEB SITES

OFFICE OF TRADE AND INDUSTRY INFORMATION
http://www.ita.doc.gov/td/industry/otea/

WORLD BANK
http://www.worldbank.org/

TRADE STATISTICS
http://www.census.gov/indicator/www/ustrade.html

CHAPTER

14

DEVELOPMENT AND UNDERDEVELOPMENT IN THE DEVELOPING WORLD

OBJECTIVES

- To outline the multiple definitions of development
- To acquaint you with the major economic problems inhibiting development in this part of the world
- To introduce major theories and perspectives on international development
- To examine the causes of poverty in the world today
- To explore the role of women in the world economy and gender roles in the workplace
- To shed light on development strategies that work and do not work

Women collecting water from a well in Ethiopia.

The modern world has its origin in the European societies of the late fifteenth and early sixteenth centuries, when capitalism began in earnest and eventually displaced feudalism throughout the continent (Chapter 2). True "modernity," however, arrived largely on the heels of the Industrial Revolution, the Enlightenment, and the massive political, social, economic, cultural, and technological changes of the nineteenth and twentieth centuries.

One of the most enduring and striking characteristics of the modern world is the division between rich and poor countries. By the nineteenth century, this division was achieved through an international system in which the wealthy minority of countries industrialized, using primary products produced by the impoverished majority of their colonies. More recently, this original global division of labor gave way to a new one: The wealthy minority is increasingly engaged in office work while parts of the developing world find hands-on manufacturing jobs on the global assembly line as well as in agriculture and raw-material production. The creation of today's world, with a rich core and a poor periphery, was not the result of conspiracy among developed countries but was the outcome of a systemic process—that process by which the world's political economy functions and, in so doing, reproduces uneven spatial development. In short, the birth of the modern world system and the schism between rich and poor countries are two sides of the same process that lies at the heart of global political economy.

This chapter deals with how this world of unequal development came about and how present structures are the result of the past. We begin by problematizing the word and idea of *development*, noting how it embodies many different concepts and measures. Next we survey the major regions of the developing world. The chapter then turns to a discussion of the characteristics of less developed countries (LDCs) and some of the barriers to their development. It compares and contrasts three major schools of thought on this issue, modernization theory, dependency theory, and world systems theory. The last part turns toward development strategies, including the central role of trade, and concludes with an examination of the potentials and pitfalls of Third World industrialization.

WHAT'S IN A WORD? "DEVELOPING"

If Europe, the United States, Australia, and Japan, all of which enjoy relatively high standards of living and material consumption, are described as developed countries, then what adjective should we use to describe the poor countries of the world? Certainly, there are many from which to choose. In the past half-century, each of the following terms has flourished in succession: *primitive, backward, undeveloped, underdeveloped, less developed, emerging,* and *developing*. Today, many people and policy makers use the word *developing* and, increasingly, the phrase *less developed countries,* but social scientists in the marxist tradition favor the term *underdeveloped*.

Underdeveloped was formerly used to describe situations in which resources were not yet developed. People and resources were seen as existing, respectively, in a traditional and "natural" state of poverty. The problem with this view is that it takes poverty to be natural and inevitable, rather than a social product, and masks the origins and historical contexts and processes that make people poor. Many scholars argue that poverty is produced, much like a building or a shirt or a TV show. Those working in the marxist tradition use the adjective *underdeveloped* to describe not an initial state but rather a condition arrived at through the agency of imperialism, which set up the inequality of political and economic dependence of poor countries on rich countries. Thus, instead of viewing underdevelopment as an initial or *passive state,* marxists view it as an *active process*.

While economic development may seem like a straightforward concept, in fact it involves several complex, even contradictory goals and concepts. Broadly speaking, development entails the growth of per capita income and the reduction of poverty. However, some countries that have experienced rapid growth of per capita income also see increases in poverty, unemployment, and inequality. Beyond income, other measures of development include income equality, nutrition, health, infant mortality, access to education, and civil liberties. In some cases, development may actually accelerate income inequality. For example, in India, the much heralded Green Revolution, which depends on fertilizer and water inputs, benefited mainly the farmers in the Punjab who were already wealthy and who owned large tracts of land. Yet other measures of development include capital inflows, the capacity to produce capital goods, trade balances, and trade reliance on one major trading partner.

Food, health, adequate shelter, and protection are essential to human well-being. When they are sufficient to meet human needs, a state of development exists; when they are insufficient, a degree of underdevelopment prevails. Even more intangible measures include life sustenance, esteem, and freedom. Mass poverty prevents people and societies from receiving due recognition or esteem. These people may even reject development. For example, if people are humiliated or disillusioned through their contacts with the "progress" introduced by foreigners, they may advocate a return to their traditional ways of life. Suspicions of western institutions that disrupt traditional lifestyles and balances of power help to fuel opposition to globalization worldwide.

More down-to-earth development goals include the following:

1. A balanced, healthful diet
2. Adequate health care

3. Environmental sanitation and disease control
4. Labor opportunities commensurate with individual talents
5. Sufficient educational opportunities
6. Individual freedom of conscience and freedom from fear
7. Decent housing
8. Economic activities in harmony with the natural environment
9. Social and political milieus promoting equality

In conventional usage, *development* is a synonym for economic growth. But growth is not development, except insofar as it enables a country to achieve the nine goals. If these goals are not the objectives of development, if modernization is merely a process of technological diffusion, and if the spatial integration of world power and world economy is devoid of human referents, that is, insensitive to people's psychological and political needs, then development should be redefined.

How Economic Development Is Measured

Geographers and other social scientists measure economic development through a number of social, economic, and demographic indexes. The principal ones are GDP per capita, the distribution of the labor force by economic sector, ability to produce consumer goods, educational and literacy levels, the status of health of a population, and its level of urbanization.

GDP per Capita

By far the most common measure of wealth and poverty internationally is gross domestic product (GDP) per capita, that is, the sum total of the value of goods and services produced by a national economy divided by its population. As shown in Figure 14.1, GDP per capita is more than $15,000 in most highly developed nations. At the same time, the United Nations estimates that 1 billion people live on less than $2 a day, or $750 per capita per year. Japan, North America, Western Europe, Australia, and New Zealand have the highest per capita incomes in the world. The Middle East, Latin America, South Asia, East Asia, Southeast Asia, and sub-Saharan Africa have the lowest. However, GDP is a flawed measure in several respects: It does not capture nonmarket, noncommodified economic activity (e.g., barter and subsistence production or household domestic labor) and is vulnerable to fluctuations in exchange rates and the costs of living.

Per capita purchasing power is a more meaningful measure of actual income per person (Figure 14.2). The relative purchasing power in developed nations is more than $10,000 per capita per year, whereas in Africa it is much less than $1000 per capita per year. Per capita

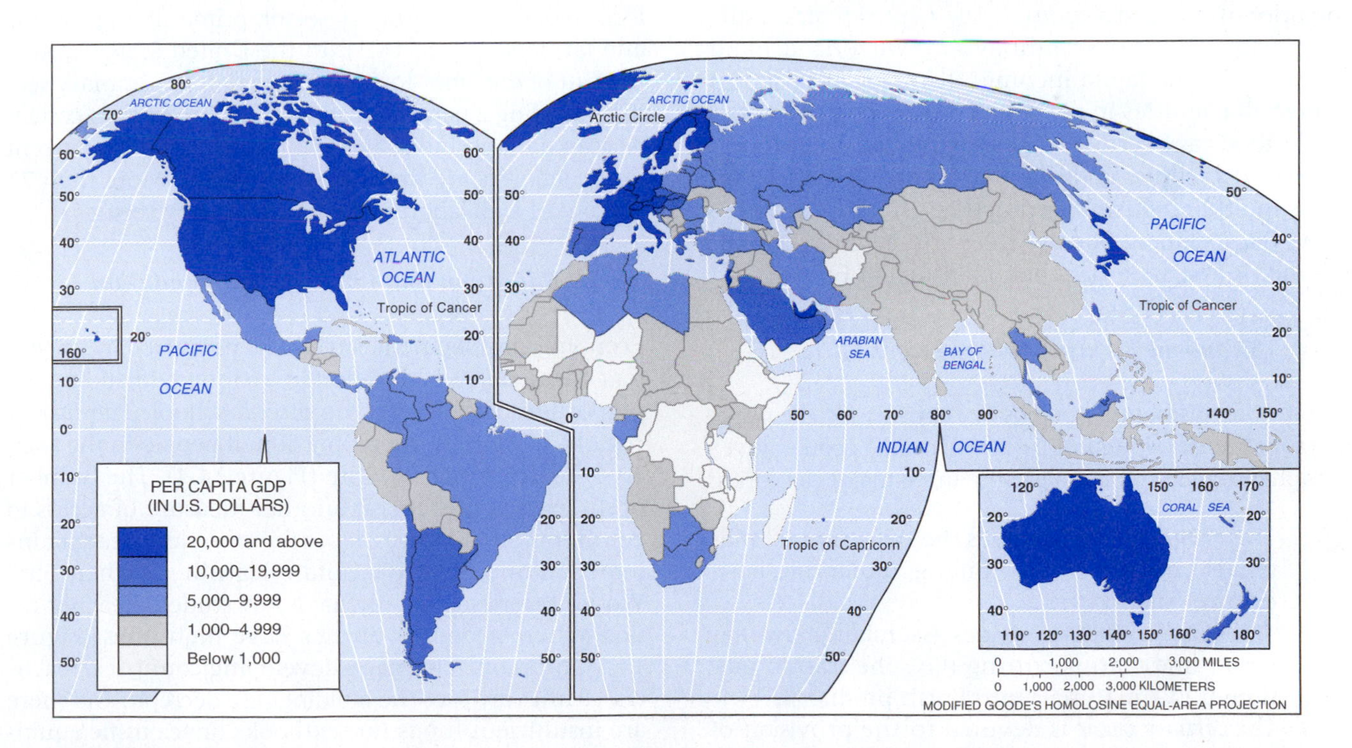

FIGURE 14.1
Annual gross domestic product (GDP) per capita. GDP per capita—the sum of a country's output divided by its population—is the most common, if flawed, measure of standards of living. GDP per capita varies from less than $250 per year in the most impoverished nations to more than $20,000 per person annually in much of Europe, Japan, and North America.

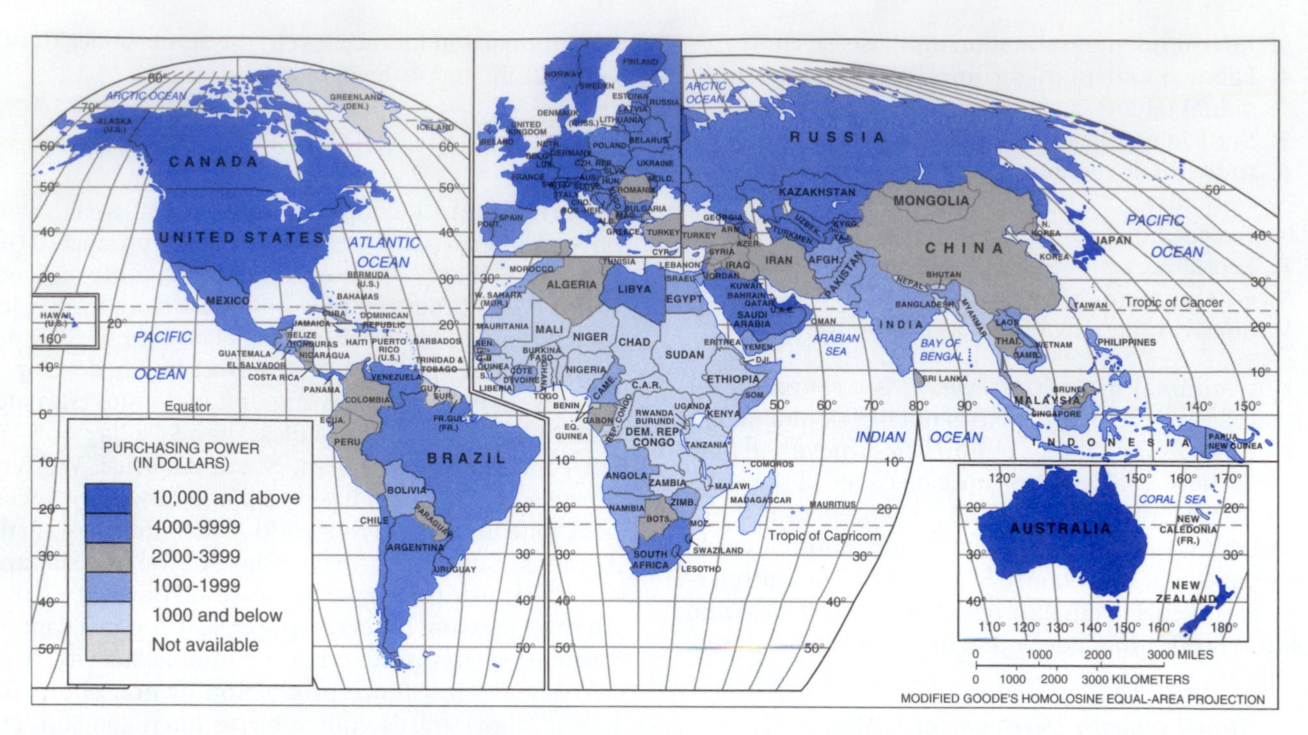

FIGURE 14.2
Per capita purchasing power. Per capita purchasing power is a better measure of a country's relative wealth than is GDP per capita because it includes the relative prices of products. For example, Switzerland, Sweden, and Japan have higher per capita GDPs than the United States. However, the United States has the world's highest per capita purchasing power because of relatively low prices for food, housing, fuel, and services.

purchasing power includes not only income, but also the price of goods in a country. The United States is surpassed by Japan, Scandinavia, Switzerland, and Germany in per capita income. However, it surpasses almost all countries in per capita purchasing power because goods and services, particularly housing, are relatively inexpensive in the United States compared with those in other industrialized nations. In other respects, however, including poverty rates and income inequality, the United States lags behind much of Europe.

Economic Structure of the Labor Force

The sectoral distribution of jobs of a country also bespeaks its economic development. Economists and economic geographers divide employment into three major categories:

1. The *primary sector* involves the extraction of materials from the earth—mining, lumbering, agriculture, and fishing.
2. The *secondary sector* includes assembling raw materials and manufacturing (i.e., the transformation of raw materials into finish products).
3. The *tertiary sector* is devoted to the provision of services—producer services (finance and business services) wholesaling and retailing, personal services, health care and entertainment, transportation, and communications.

In less developed countries, a large share of the labor force works in the primary sector, primarily as peasants and farmers (Figure 14.3). In the United States, only 5 percent of the labor force is engaged in the primary sector, including 2 percent in agriculture, whereas in certain African nations, India, and China, more than 70 percent of the laborers are in the primary sector. More than 75 percent of U.S. laborers are in the tertiary sector.

Education and Literacy of a Population

Economic development can also be measured by the extent and quality of education in a country, including the proportion of children who attend school. The *literacy rate* of a country is the proportion of people in the society who can read and write (Figure 14.4). The number of students per teacher is another measure of access to education; small classes allow more student-teacher interaction and greatly facilitate learning. Richer, First World societies generally have low student-teacher ratios, whereas poor countries have high ones (Figure 14.5). Moreover, in many developing countries, teachers' salaries are low, the buildings are decrepit, and there are insufficient funds for textbooks or scientific equipment. Notably, despite these underinvestments, many countries have ample funds for their militaries, indicating that insufficient investment in human capital is a policy choice, not a "natural" limitation.

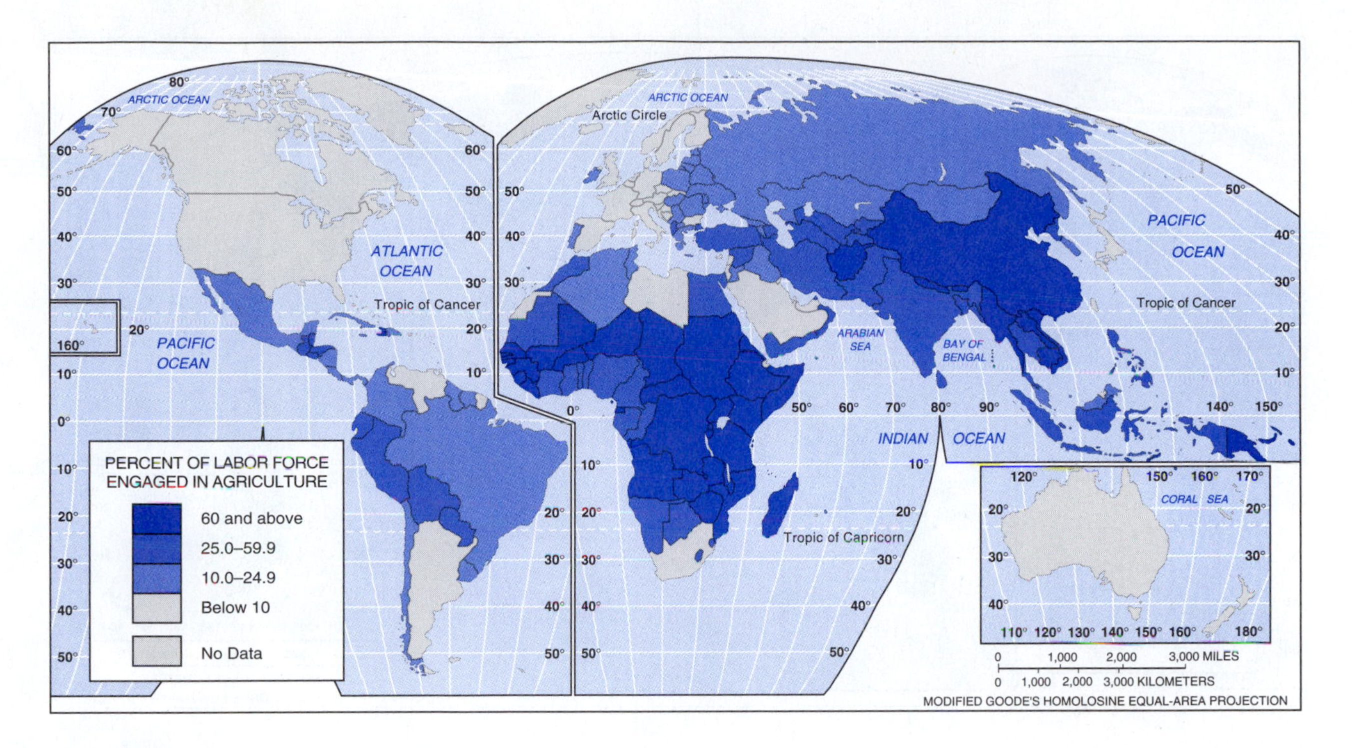

FIGURE 14.3
Percent of labor force in agriculture. Typically, poor countries have large shares of their workers in agriculture, often working in preindustrial conditions, whereas in economically advanced countries a small fraction of the labor force is so employed.

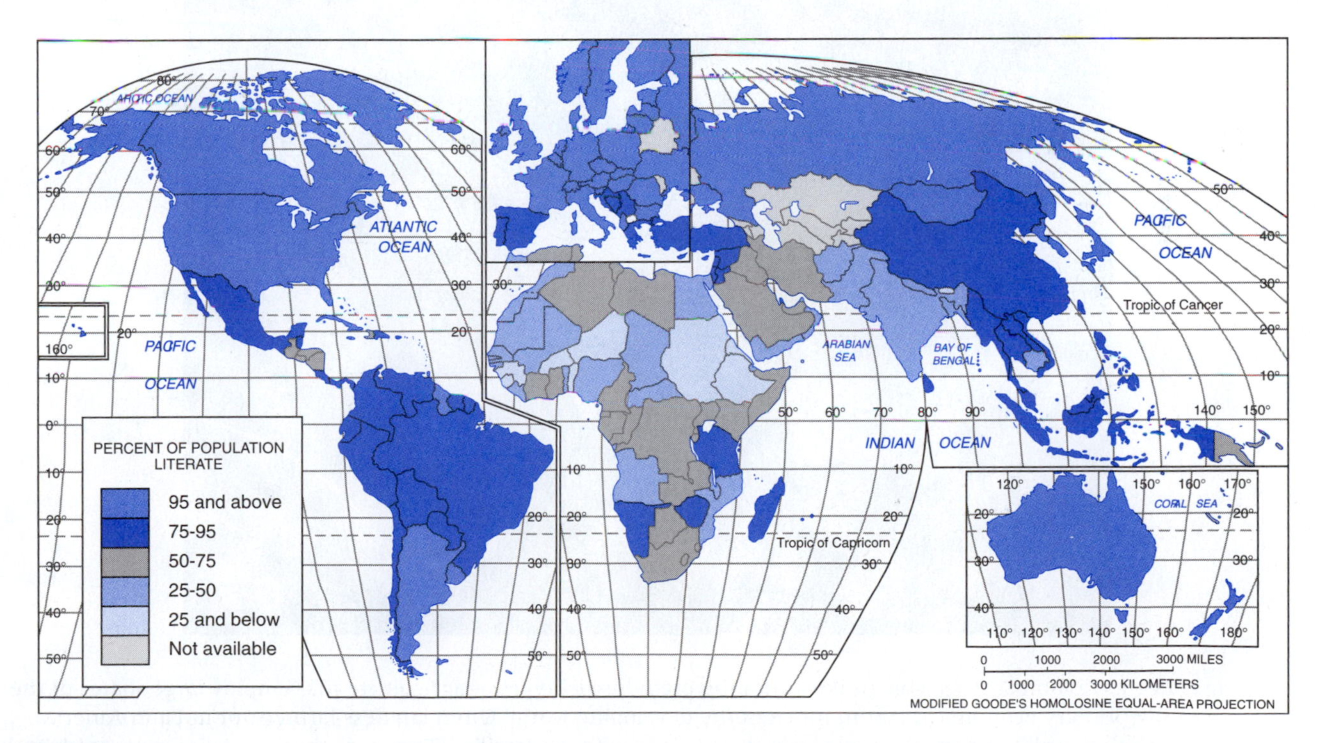

FIGURE 14.4
Literacy rates. A great disparity in literacy rates exists between inhabitants of developed countries and less developed countries. In the United States and other highly developed countries, the literacy rate is more than 98 percent. Notice the large number of countries in Africa and South Asia where the literacy rate is less than 50 percent.

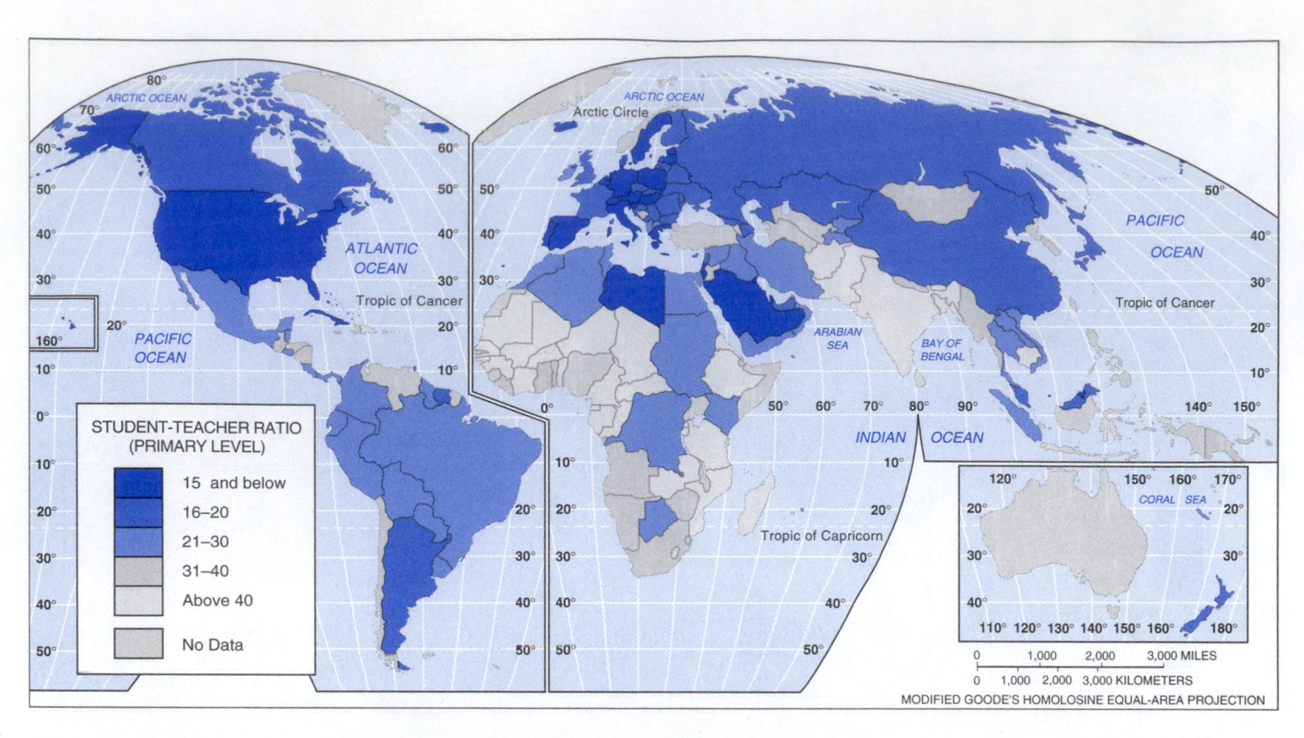

FIGURE 14.5
Students per teacher in primary school. This measure of average class size reflects the quality of each country's educational system. Small class sizes facilitate more student-teacher interaction and are one of the best measures of the resource base of a school system.

Harvesting coffee in Colombia. This woman illustrates the preindustrial types of agriculture that employ large shares of the labor force in the primary economic sector in much of the developing world. Often landless farmers, or in Latin America, *campesinos,* must sell their labor power at low wages in order to feed their families. These are the populations most vulnerable to globalization, low export prices, and exploitative local landowners. Their poverty reflects the ways in which global, national, and local forces are telescoped into places to generate both wealth and suffering simultaneously.

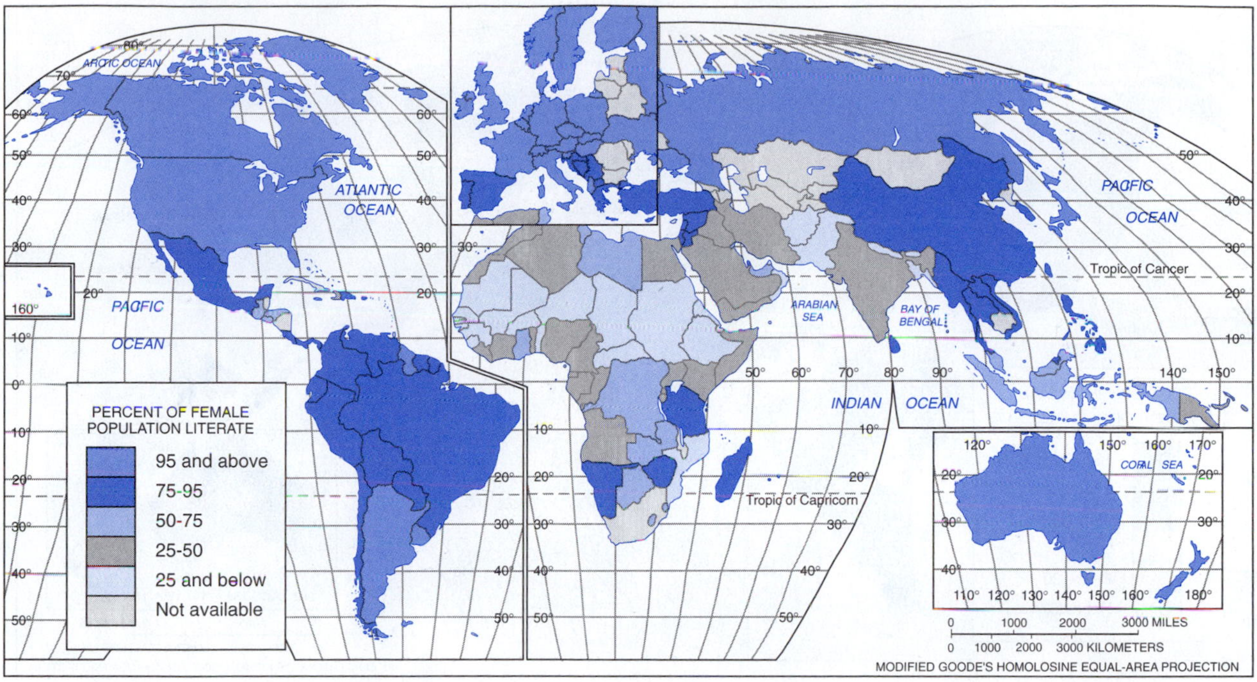

FIGURE 14.6
Literacy rate of women. The gender gap for literacy is most pronounced in the developing world, especially Africa, South Asia, and the Middle East. In most developing countries, the female literacy rate is much below the rate for men. Raising women's literacy rates is one of the most efficient means of stimulating economic and social development.

There are typically vast gender differences in literacy within developing countries. In many impoverished societies, desperately poor rural families can often only afford to send one child to school (the others are working the fields), and that child is very likely to be male. Unfortunately, worldwide, only 75 girls attend school for every 100 boys. In areas particularly disadvantageous to women, their literacy rates tend to be much lower than men's (Figure 14.6). In many nations, the literacy rate of women is less than 25 percent, whereas the literacy rate of men is between 25 percent and 75 percent. The Middle East and South Asia, where the role of women is clearly subservient to men, show the greatest disparities. However, in the highly developed world, the literacy rates of men and women are almost identical. The regions that have low percentages of women attending secondary school also generally have poor social and economic conditions for women. Indeed, raising women's literacy rates has been shown to lower fertility rates and to empower women economically and politically. In addition, because more people can read and write, a proliferation of newspapers, magazines, and scholarly journals improve and foster communication and exchange, which leads to further development by informing them of opportunities, circulating best practices, and so forth.

Health of a Population

Measures of health and welfare, in general, are much higher in developed nations than in LDCs. One measure of health and welfare is diet, typically measured as caloric consumption per capita (Figure 14.7). Most people in Africa do not receive the UN daily recommended allowance. However, in developed nations, the population consumes approximately one-third more than the minimum daily requirement and is therefore able to maintain a higher level of health. In some areas of each country, calories and food supplies are insufficient, even in the United States, where significant pockets of hunger and malnutrition exist. Conversely, an overabundance of cheap food and inadequate exercise have generated an obesity epidemic in this country, where lack of exercise and high-fat diets have caused waistlines to balloon. Obesity has become an epidemic worldwide; there are today, worldwide, more obese people than malnourished ones.

People in developed nations also have better access to doctors, hospitals, and health care providers. Figure 14.8 shows worldwide access to physicians as measured by persons per physician. For relatively developed nations, there is one doctor per 1000 people, but in developing countries, each person shares a doctor with many thousands of

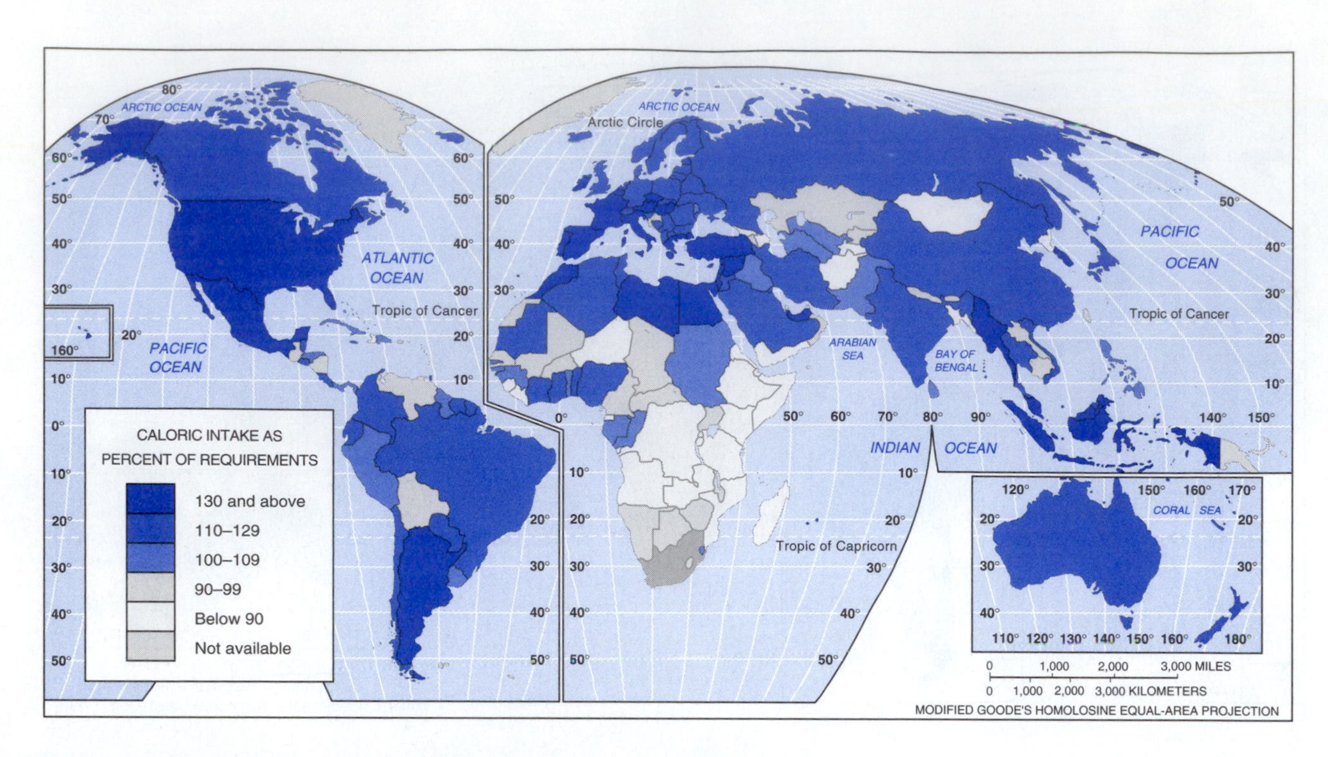

FIGURE 14.7
Daily caloric consumption per capita as a percentage of minimum nutritional requirements. Access to adequate nutrition is another measure of economic development. In many impoverished nations, malnutrition may be chronic. Conversely, many economically advanced countries, particularly the United States, suffer from an epidemic of obesity.

others. Africa by far has the worst access to health care (in some countries there are more than 15,000 people per physician, effectively meaning that most people *never* see one), followed by Southeast Asia and East Asia. Portions of the Middle East are also lacking in medical care. Everywhere, wealthier societies have better access to health care, although there are huge discrepancies within them as well, such as in the United States.

Infants and children are the most vulnerable members of any society, in part because their immunological systems are not as well developed as adults and partly because they lack effective political power to access health care resources. In developed nations, on the average, fewer than 10 babies in 1000 die within the first 100 days; in many less developed nations, more than 100 babies die per 1000 live births, the result of poor prenatal care, malnutrition, and infectious diseases. When there are economic downturns, droughts, or disruptions of the food supply brought on by war, they are generally the first to die. The geography of infant mortality (Figure 14.9)—the

An all-girls school in Kenya reflects that country's investment in human capital. The education of girls is an important means of raising standards of living and reducing birthrates, as well as providing economic opportunities for women.

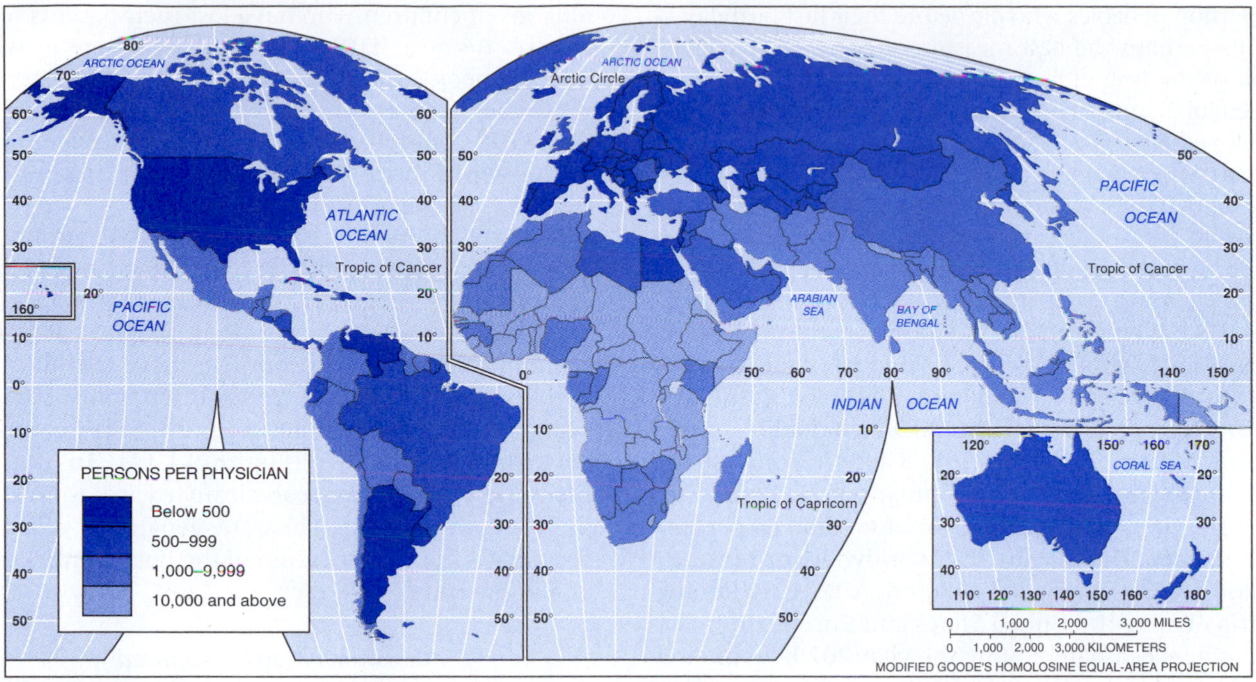

FIGURE 14.8
Persons per physician. An important measure of economic development is the number of persons per physician in a country. This measure is a surrogate for health care access, which includes hospital beds, medicine, and nurses and doctors. Most of Africa exhibits 10,000 people per physician, or more, while Europe and the United States average one doctor for every 200 people.

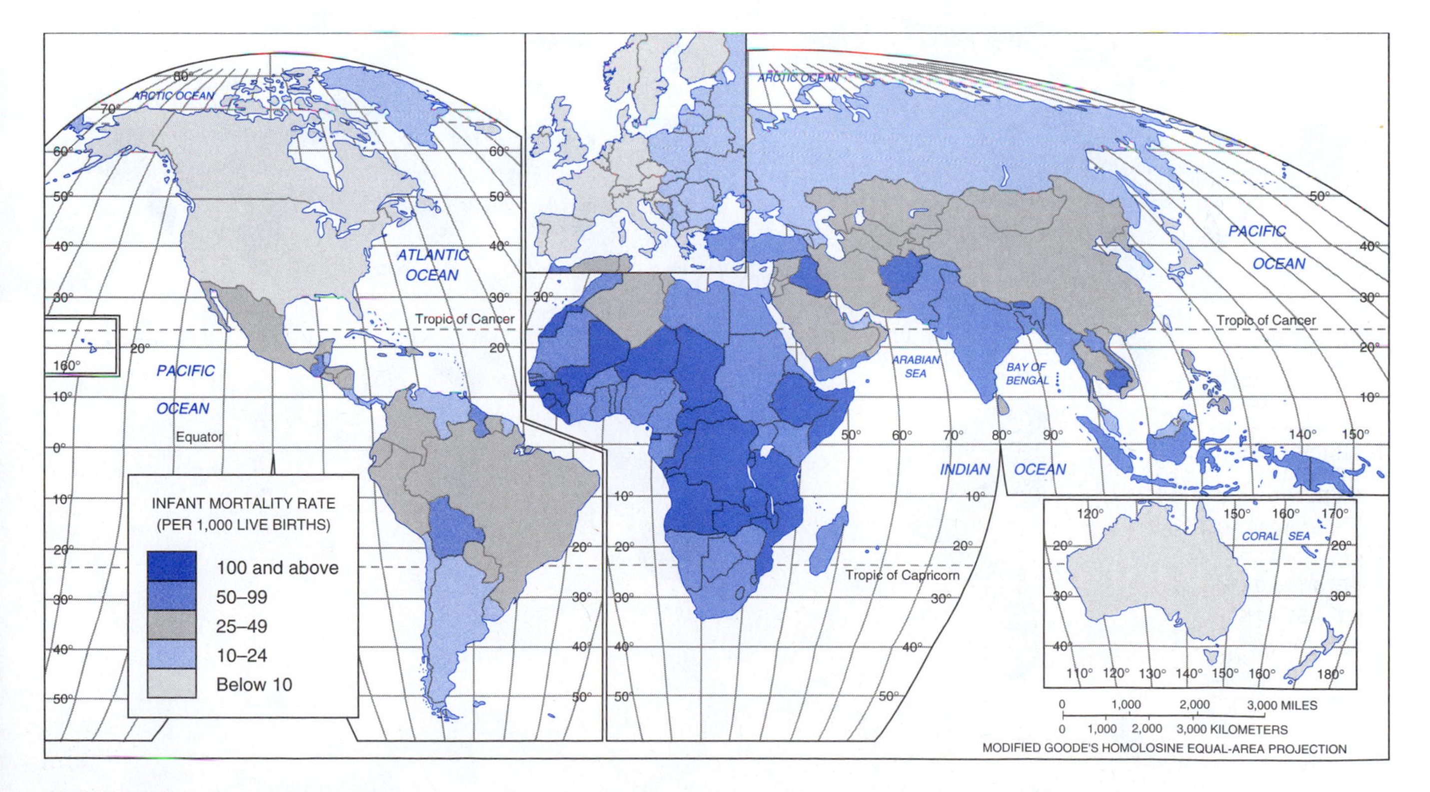

FIGURE 14.9
Infant mortality rates. Babies are the most vulnerable members of any society, and the percent who die before their first birthday is another measure of economic development and well-being. In the poorest nations, particularly Africa, more than 10 percent of babies die before their first birthday.

proportion of babies who die before their first birthday—is thus perhaps the best measure of economic development or the lack of it. In much of Africa, more than 10 percent of infants do not live through their first year. In the developed First World, by contrast, infant mortality rates are very low. Notably, Cuba's infant mortality rate—6.0—is lower than that of the United States, which is 7.0, a discrepancy that reflects the former's investment in the health care and the latter's widespread unavailability of health insurance amid its poor.

AIDS has emerged as a significant threat worldwide. More than 23 million people have died of this disease, and an additional 45 million are infected with the HIV virus. The epicenter of the AIDS epidemic is sub-Saharan Africa (Figure 14.10), where in some countries 40 percent of the adult population is infected. The sub-Saharan region accounts for more than 60 percent of the people living with HIV worldwide, or some 25 million men, women, and children. AIDS is tightening its grip outside the United States and Europe. In India, researchers estimate that by the year 2020, 50 million people could be HIV positive. Half the prostitutes in Bombay are already infected, and doctors report that the disease is spreading along major truck routes and into rural areas, as migrant workers bring the virus home. In China, AIDS is spreading rapidly. The social consequences of this die-off are catastrophic, including millions of children who have lost their parents to the disease. Because AIDS has a long lead time in which infected people do not show symptoms, and because sexual behavior is very difficult to change, many fear that AIDS could lead to a depopulation of large parts of the world in the future comparable to the Black Death of the fourteenth century.

The reliability and quality of the food supply, access to clean drinking water, public health measures, ability to control infectious diseases, and access to health services all shape how long one lives. People in economically advanced countries have the luxury of living a relatively long time—often more than 75 years, on average—while those in poorer countries live considerably less (Figure 14.11). In parts of Africa, few adults can realistically expect to live past their fiftieth birthday. Thus, the geography of life expectancy is another measure of the health and welfare of a population.

Consumer Goods Produced

The quantity and quality of consumer goods purchased and distributed in a society is another measure of the level of economic development in that society. Easy availability of consumer goods means that a country's economic resources have fulfilled the basic human

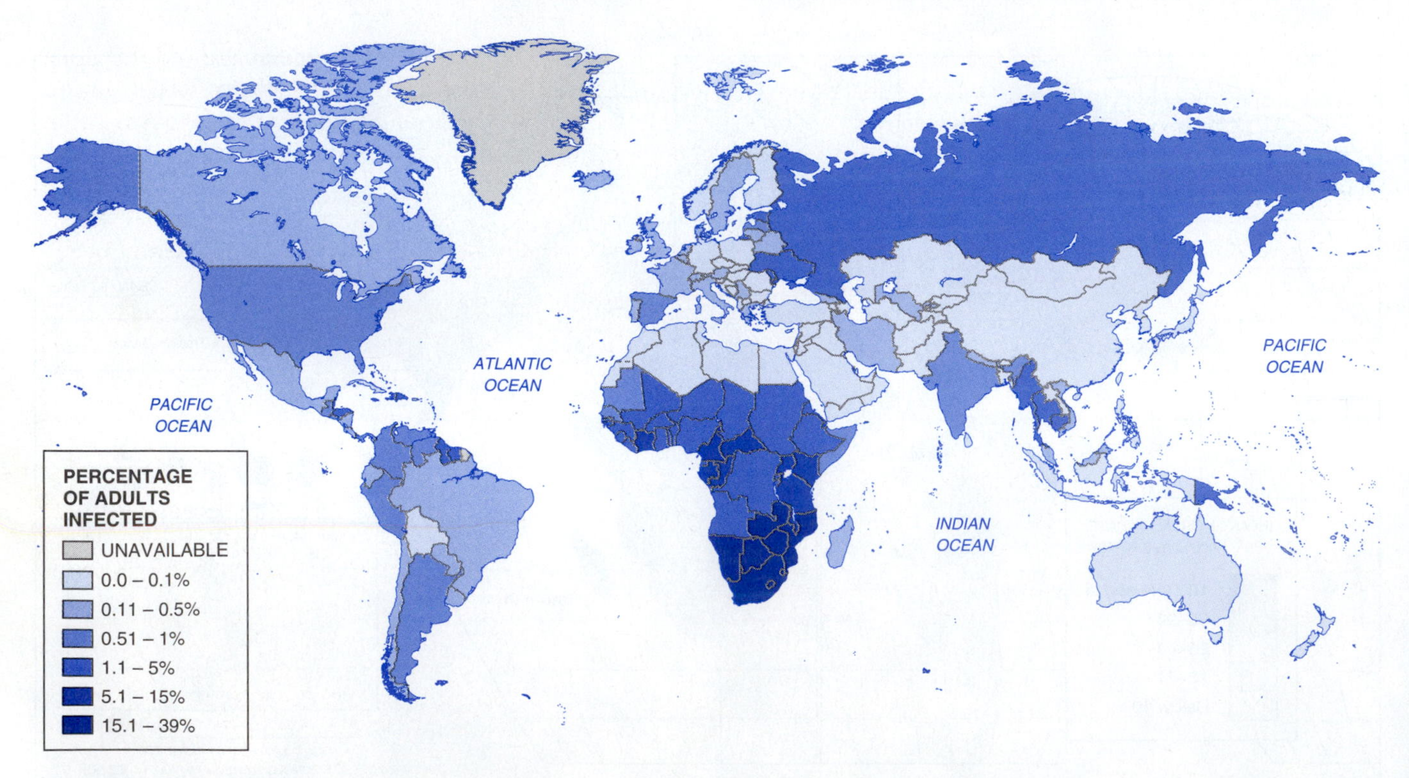

FIGURE 14.10
The geography of AIDS, 2005. Although AIDS does not kill as many people as malnutrition or heart disease, it ranks among the world's leading killers today, and the most rapidly growing. More than 23 million people have died of AIDS, and another 45 million are infected. The epicenter of the epidemic is Africa, where in some countries 40 percent of adults are HIV infected. The disease is only now making inroads into the huge populations of India and China and has the potential for creating a catastrophic depopulation in the future.

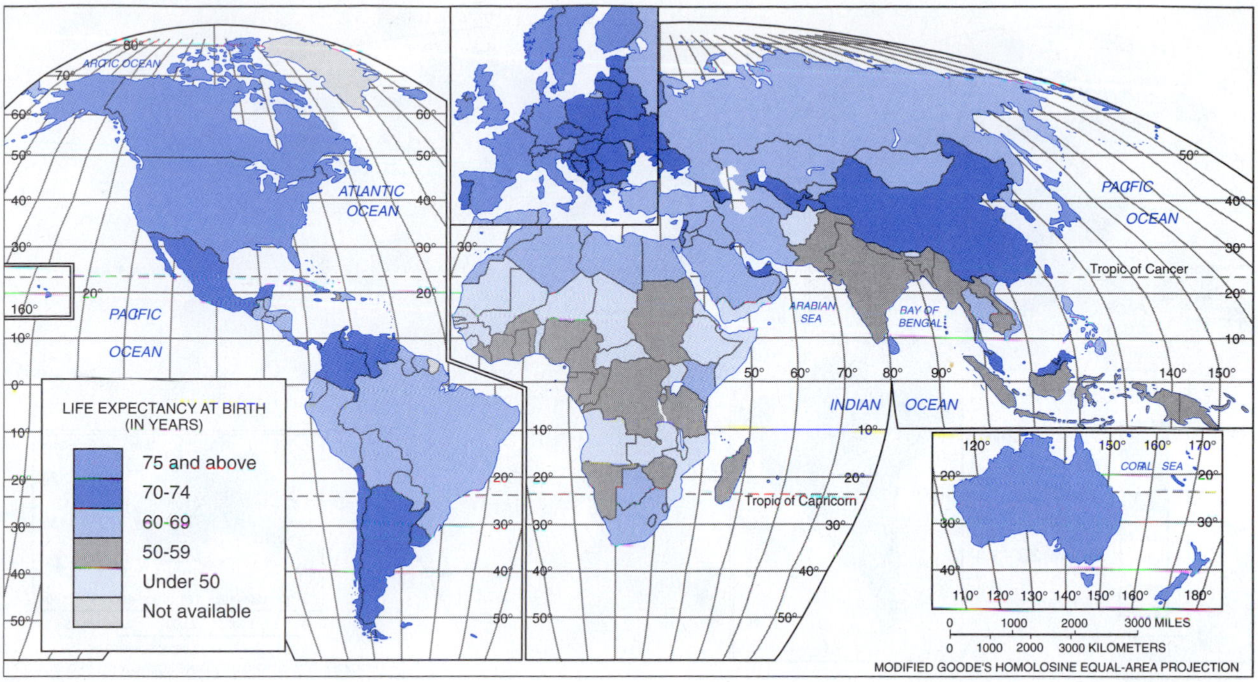

FIGURE 14.11
Life expectancy at birth. How long we can hope to live is another measure of economic development or the lack of it. In societies where people enjoy sufficient access to food, public health measures, and medical care, life expectancies are often over 75. In the poorest countries, in contrast, most people cannot expect to live beyond 50.

needs of shelter, clothing, and food, and more resources are left over to provide nonessential household goods and services. Automobiles, textiles, home electronics, jewelry, watches, refrigerators, and washing machines are some of the major consumer goods produced worldwide on varying scales. In industrialized countries, more than one television, telephone, or automobile exists for every two people. In developing nations, only a few of these products exist for a thousand people. For instance, the ratio of persons to television sets in developing countries is 150 to 1, and population to automobiles is 400 to 1. In California, the ratio for these consumer items is almost 1 to 1. The number of consumer goods such as telephones and televisions per capita is a good indicator of a country's level of economic development.

Urbanization in Developing Countries

In the industrialized West, urbanization occurred on the heels of the Industrial Revolution, that is, it was synonymous with industrialization (Chapter 2). While parts of the developing world are urbanized, in general LDCs lag behind the economically advanced countries in this regard. However, cities throughout the developing world are growing quickly, much more so than those in Europe, North America, Japan, or Australia. Today, about 56 percent of the world's people live in cities (Figure 14.12), an all-time high. However, the proportion of each country's people that lives in cities varies widely around the globe (Figure 14.13). In parts of Latin America, urbanization rates resemble those of North America or Australia, where 75 percent or more of the people live in cities. In Asia and Africa, however, the proportions are much lower. Only 20 percent of China's people live in cities, and in wide swaths of Africa less than 20 percent do so.

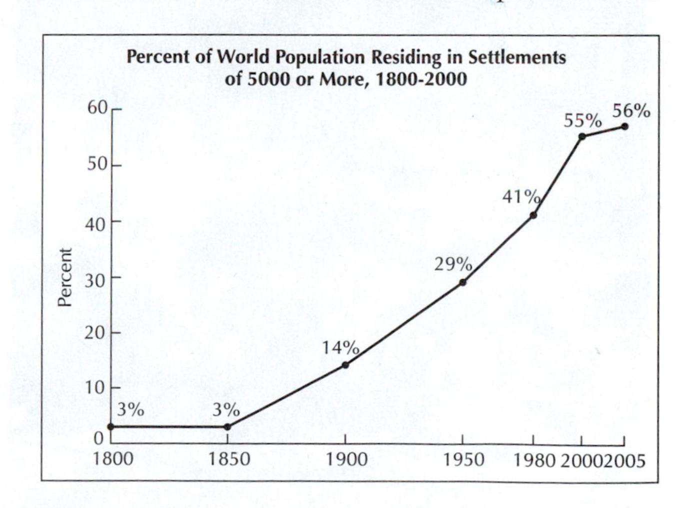

FIGURE 14.12
Percent of the world living in cities, 1800–2005. Today more than half of the planet's population lives in urban areas. Most urban growth occurs in the developing world, where large cities are fueled by waves of rural to urban migrants.

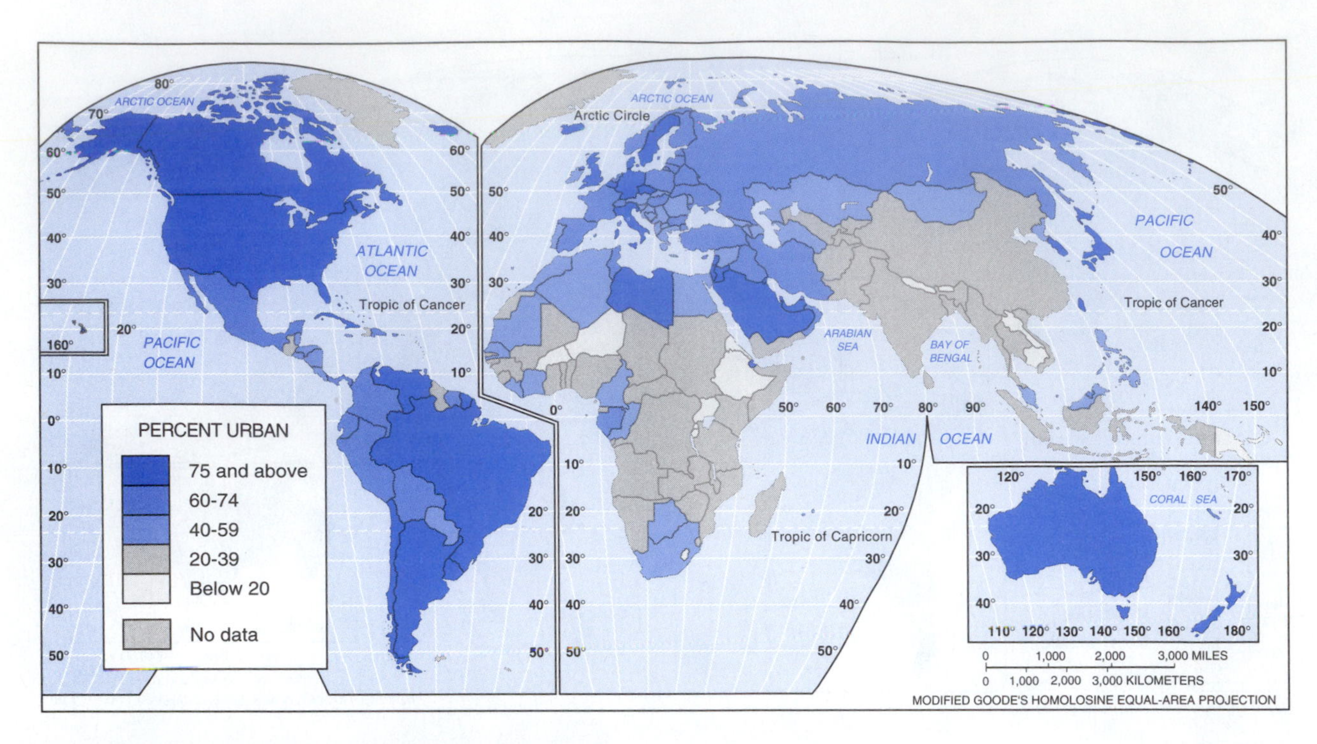

FIGURE 14.13
Urbanization rates. The proportion of people living in cities is another measure of economic development, although not a very accurate one. Generally, poor countries are less urbanized than wealthy ones. In China, for example, only 20 percent of the population lives in cities. However, in Latin America even relatively poor countries such as Brazil have high urbanization rates, where much of the poverty is clustered in large urban conurbations such as Sao Paolo and Rio de Janiero.

African children watching television. Although literacy levels increased between 1970 and 2000 in all but one country, literacy rates for women are still below 50 percent in 45 of the countries for which data are available. Literacy rates for men are below 50 percent in only 17 countries.

Urbanization in the developing world differs significantly from that in the West. Because the historical contexts of LDCs differ from that of the West, particularly through the impacts of colonialism, and because their mode of incorporation into the global division of labor was very different, the patterns of urban growth are also qualitatively different. Some countries in the developing world had well-established traditions of urbanization before the Europeans came, such as the Arab world and China. In others, powers such as Britain played a major role, constructing cities such as Calcutta in India, Rangoon in Myanmar (Burma), or Singapore; similarly, the Dutch started Batavia, Indonesia, which later became Jakarta.

Today, the vast bulk of the world's urban growth is in the developing world. The world's largest cities, for example, are found in Latin America and Asia, not Europe or North America (Table 14.1). In 2005, Mexico City and Sao Paolo, Brazil, vied for the rank of the world's largest metropolitan area, with roughly 25 million inhabitants each. Many of the others, with populations over 5 million, are located in China and India (Figure 14.14).

Moreover, cities in the developing world are growing much more rapidly than their counterparts in the developed countries. While natural growth rates in cities in the developing world is a little higher than that in the West, urbanization in LDCs is primarily due to the

Table 14.1

World's Largest Urban Areas, 1950, 1980, and 2000 (Millions)

1950		*1980*		*2000*	
New York	12.3	Tokyo	19.0	**Mexico City**	25.8
Shanghai	10.3	**Mexico City**	16.7	**Sao Paolo**	24.0
London	10.2	New York	15.6	Tokyo	22.2
Tokyo	6.7	**Sao Paulo**	15.5	**Calcutta**	16.5
Beijing	6.6	**Shanghai**	12.1	**Bombay**	16.0
Paris	5.4	**Buenos Aires**	10.8	New York	15.8
Tianjin	5.4	London	10.5	**Shanghai**	14.3
Buenos Aires	5.1	**Calcutta**	10.3	**Seoul**	13.8
Chicago	4.9	**Rio de Janiero**	10.1	**Tehran**	13.6
Moscow	4.8	**Seoul**	10.1	**Rio de Janiero**	13.3
Calcutta	4.4	Los Angeles	10.0	**Jakarta**	13.2
Los Angeles	4.0	Osaka	9.6	**Delhi**	13.2
Osaka	3.8	**Bombay**	9.5	**Buenos Aires**	13.2
Milan	3.6	**Beijing**	9.3	**Karachi**	12.0
Rio de Janiero	3.4	Moscow	8.9	**Beijing**	11.2
Philadelphia	2.9	Paris	8.8	**Dacca**	11.2
Bombay	2.9	**Tianjin**	8.0	**Cairo**	11.2
Mexico City	2.9	**Cairo**	7.9	**Manila**	11.1
Detroit	2.8	**Jakarta**	7.8	Los Angeles	11.0
Sao Paolo	2.7	Milan	7.5	**Bangkok**	10.7
Naples	2.7	**Tehran**	7.2	London	10.5
Leningrad	2.6	**Manila**	7.1	Osaka	10.5
Birmingham, U.K.	2.5	**Delhi**	7.0	Moscow	10.4
Cairo	2.4	Chicago	6.9	**Tianjin**	9.7
Boston	2.2	**Karachi**	6.2	**Lima**	9.1

Note: Cities in less developed countries are shown in boldface.
Source: National Research Council, 2003.

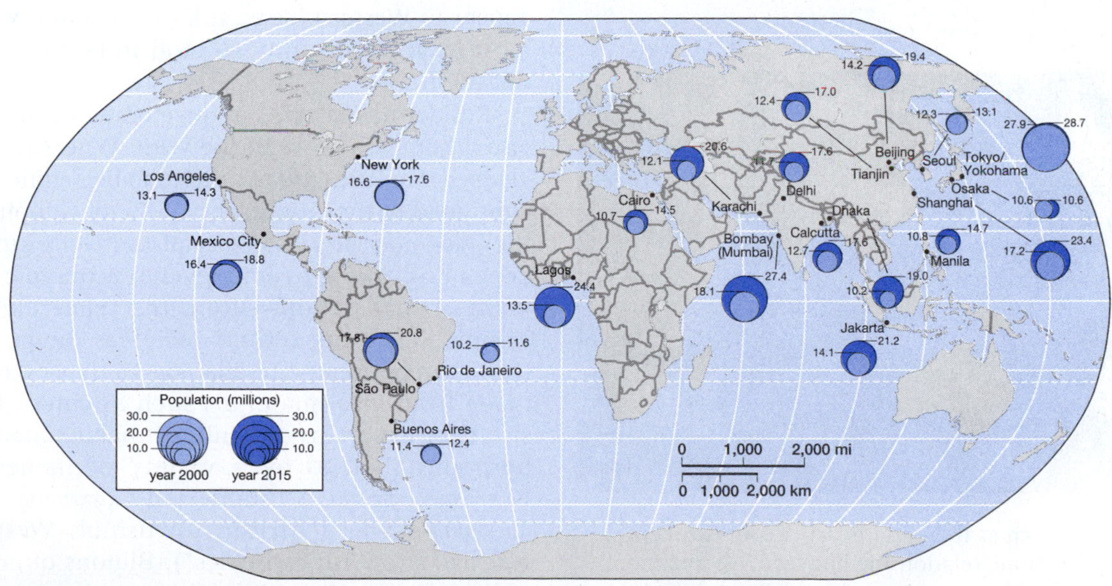

FIGURE 14.14
The location of the world's largest cities. The majority of the world's biggest cities are located not in the developed but in the developing world, particularly in China and India.

The informal economy—unregulated, untaxed, and generally consisting of low-paying service jobs—comprises a variety of occupations that employ large numbers of the urban poor in both the developing and industrialized countries. In LDCs, opportunities in the formal sector, including export-oriented multinational firms and the government, are often relatively scarce.

massive influx of rural-to-urban migrants, many of whom are displaced by agricultural mechanization, unequal land distribution, low crop prices, war, and high population growth in rural areas. To some extent, this pattern was true of cities in the developing West as well. However, because few developing countries have generated the industrial job growth that the West did during the nineteenth and early twentieth centuries, the labor markets and employment conditions in the developing world are fundamentally different. Thus, we must see LDC urban growth and rural crisis as two sides of one coin. Because many migrants move to the cities on the basis of their perceptions that there are greater opportunities to be found there (perceptions that may be erroneous due to imperfect information), they often find themselves plunged into desperate circumstances.

Assembly plants such as these in Mexico are an excellent example of a symbiotic relationship between less and more developed countries. The latter get cheap labor, while the former benefit through a boost in the local economy.

LDC urban labor markets generally do not generate sufficient employment opportunities, leading to high unemployment rates, or underemployment, in which migrants do not utilize their skills. Many find work in low-paying service jobs in the informal economy, including mining trash piles for recyclables. In Manila, the Philippines, for example, 30,000 people earn their living this way on the Payatas garbage dump, as adults and children work together, stepping over rotting debris to scavenge bits of plastic or tin cans. Others sell trinkets and food on the streets. Others find marginal incomes in prostitution, selling illegal drugs, the black market, illicit currency exchange, or as casual day laborers doing construction. Export-oriented jobs that tie cities to the global economy, such as the garment industry or electronic assembly plants—which, however exploitative they appear to Western ideas, still offer higher wages than most local opportunities—tend to be the exception, not the norm.

Residential patterns in developing countries' cities also differ from those in the West. Whereas in the developed countries the poor tend to be a minority, often consigned to the city center, in the developing world, the poor may be a majority of the city's inhabitants, depending on the overall level of economic development of that nation. Often, the relatively wealthy command the city centers, whereas the poor live in the urban periphery. In dilapidated houses, often self-made from local materials such as cinder blocks or sheets of tin, many urbanites inhabit squalid neighborhoods that go by a variety of names: slums, Brazilian *favelas,* Indonesian *kampong,* Turkish *gacekondu,* South African townships, West African *bidonvilles* (or "tin can cities"). Billions of people live today in such conditions, without adequate housing, electricity, roads, transportation systems, schools,

Most of the urban poor in the developing world live in decrepit, makeshift shelters, a reflection of their low incomes, inadequate employment opportunities, the lack of profit to lure builders of commodified housing, and their generally marginalized position within the world economy.

Favelas, or slum districts, on the outskirts of Rio de Janiero. This photo exemplifies the tendency of urban labor and housing markets in the developing world to cluster the wealthy in the center and the poor along the periphery, creating an inverse of the social geography of the American city.

clean water, or sewers. Often densities in such places are far higher than in any Western city. When migrants seize buildings or land that belongs to wealthy landowners, the government may deem such "squatter settlements" to be illegal and tear them down with bulldozers. Such communities can become more than just cesspools of misery, they can turn into breeding grounds of resentment and political activism against corrupt and uncaring governments.

Human Development Index

One measure of social and economic development that combines several types of information is the United Nation's Human Development Index (HDI), which includes life expectancy at birth, GDP per capita, and indices of schooling and literacy. Not surprisingly, the HDI varies widely among countries of the world (Figure 14.15; Table 14.2), reflecting standards of living and access to both private and public resources, which in turn are produced through various places' position in the world economy. The highest levels are found in western Europe, Japan, North America, Australia, and New Zealand, that is, the First World. sub-Saharan African countries dominate the lowest rungs of the HDI, indicative of the widespread poverty and misery found on that continent. In between lie most of Latin America, the Arab world, the former Soviet bloc, China, and South Asia. The HDI reflects the geography of human welfare and suffering, the ways in which the global economy, the legacy of colonialism, international investments (or lack thereof), and state policies shape the everyday lives of the planet's people.

Most of the world's people live in economically underdeveloped countries, and the world's largest cities are usually found there. Often the poor comprise the bulk of the population, and driven from agricultural regions by unequal land distribution and low commodity prices, flock to urban areas in chains of rural to urban migration. Within inadequate employment opportunities, low wages, and insufficient access to capital and public services, billions of people live in decrepit shantytowns without adequate water or sanitation.

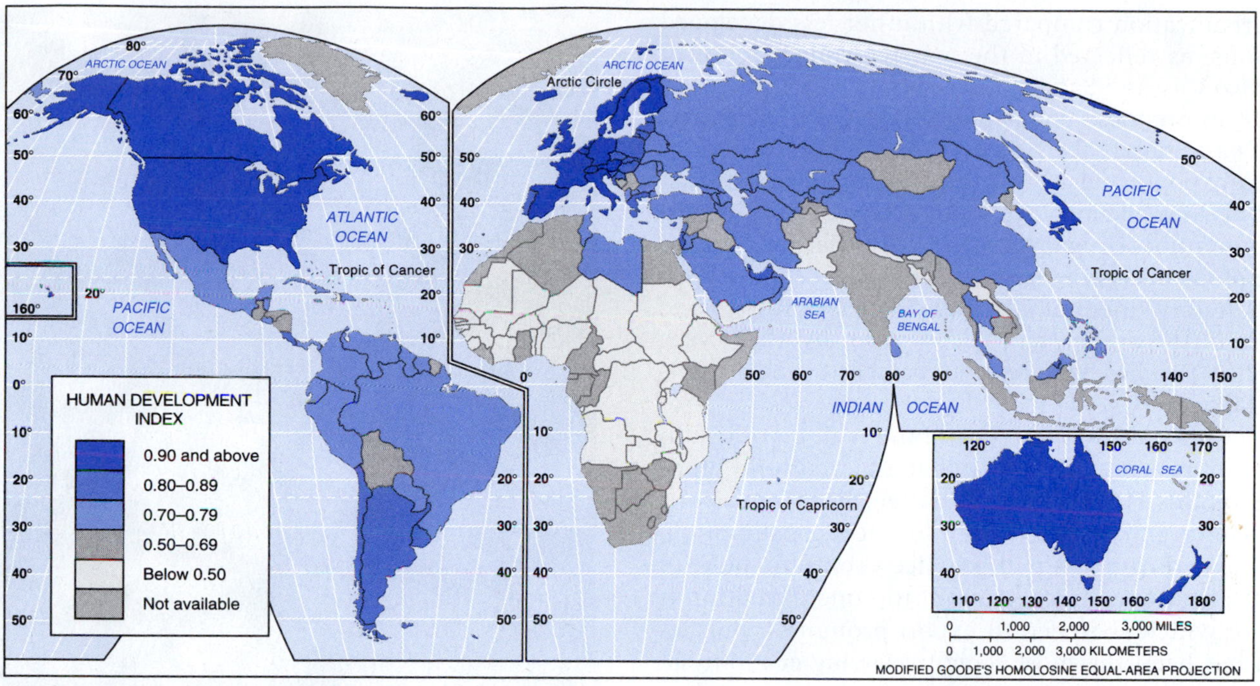

FIGURE 14.15

The Human Development Index is perhaps the best overall measure of economic development. The United Nations constructs a single index measuring life expectancy at birth, per capita purchasing power, years of schooling, and literacy rates. Note the massive difference between the First World and the Third. Western Europe is clearly distinguished from Eastern Europe and the former Soviet block. Central America and Bolivia have the lowest scores for Latin America. Sub-Saharan Africa, South Asia, and Southeast Asia have the lowest scores overall.

THE LOCATION OF UNDERDEVELOPMENT

The developed and less developed countries of the world are clearly separated on a map of the planet. A line drawn at 30° north latitude would put most of the developed countries to the north and the underdeveloped to the south, a division often known as the *North-South split.* Note that this dichotomy is not synonymous with the northern and southern hemispheres of the globe; most of the world's landmass is in the northern hemisphere, including much of the developing world (e.g., India and China). In this categorization, Australia and New Zealand, which are in the southern hemisphere, nonetheless belong economically to the North. *North* and *South* are thus shorthand terms to describe the First and Third Worlds, respectively, that is, they are social products, not referents to the world's physical geography.

The LDCs of the world—which contain the vast majority of the planet's population—include diverse societies in Latin America, East Asia (except Japan), Southeast Asia, South Asia, the predominantly Muslim world of southwest Asia and North Africa, and sub-Saharan Africa.

Table 14.2
Human Development Index, 2001

Less Developed Regions		*Industrially Advanced Regions*	
Latin America	.71	Japan	.98
East Asia	.61	North America	.98
Southeast Asia	.52	Australia/ New Zealand	.97
Middle East	.51	Eastern Europe	.87
South Asia	.29	Western Europe	.95
Sub-Saharan Africa	.23		

Latin America

Most inhabitants of this region are descendants of either Spanish or Portuguese colonists, slaves brought from Africa, and various indigenous peoples. Many people are *mestizos,* or mixed race. Even though many different native languages are still spoken by the native American peoples, Spanish and Portuguese are the dominant languages and Catholicism is the most widespread form of religion. This region shows a higher level of urbanization compared with other less developed nations, as reflected in the massive conurbations of Mexico City and Sao Paulo.

Latin America's population is largely clustered along the coast, mainly the Atlantic Ocean, with the interior scarcely populated. Most of the region's economic activity is located along the south Atlantic coast, including Argentina, Uruguay, Brazil, Venezuela, and Mexico. One of the most striking characteristics of Latin America is its highly unequal distribution of wealth. A large share of the arable land is controlled by a few very wealthy families, who rent out parcels to landless tenant farmers.

Latin America is also one of the major world suppliers of raw materials. The major agricultural products of this region are tea, coffee, beef, and fruits grown for export rather than for local consumption. Many governments still practice economic policies that fail to encourage growth; the one exception is Chile, which switched to export promotion policies and has been the most rapidly growing economy in the region.

Southeast Asia

This region is a potpourri of nationalities and islands spread over a large area. Culturally, it is a mix of Buddhism and Islam, with several major languages. Indonesia is the most populous country in the region and comprises over 13,000 islands and 230 million people.

Indochina was long plagued by horrific warfare, which shattered the economies of Vietnam and Cambodia. Despite modest growth, they remain poor. Myanmar/Burma remains an isolated country trapped in poverty. Rice and plantation crops are the most common agricultural products.

The area is abundant in tin and contains substantial petroleum reserves, including Indonesia, the only Asian member of OPEC, and the tiny oil-rich sultanate of Brunei. The region's cheap labor has made it a leading manufacturer of textiles and clothing, but there are widespread discrepancies among states in terms of their economic development. Malaysia and Singapore are now major producers of electronic goods, and Singapore is a center of finance and telecommunications of worldwide significance. Thailand and Indonesia have recently joined the ranks of the newly industrializing countries (NICs), attracting foreign capital and developing a more diverse export capacity. Cities such as Bangkok have become congested with automobiles and expensive. The Philippines, in contrast, remains mired in severe poverty, a reflection of the Spanish land grant system and endemic government corruption.

Markets in Asia. Frequently, markets in the developing world are poorly developed, reflecting the partially commodified nature of social life there. Effective markets require enforced property rights and an infrastructure to work well.

East Asia (Excluding Japan)

East Asia is a vast, heavily populated region that has enjoyed the most rapid rate of economic growth in the world since World War II. Culturally, it is dominated by Buddhism.

China is the region's giant, with 1.3 billion people; despite its economic success, it still ranks among the world's poorest countries. The bulk of Chinese—80 percent—live in rural areas, including 700,000 agricultural villages. From the ascendancy of communism in 1949 to the death of Mao Zedong in 1976, China was an isolated country ruled by a Communist Party pursuing unadulterated socialism. Major events of this period include the Great Leap Forward, in which 30 million starved to death, and the Cultural Revolution of 1966–1976, a period of massive social disruption.

Since 1979, China has undergone an enormous economic and social transformation. Most communist regulations have been loosened, and market imperatives dominate in an ostensibly socialist country; farmers can now own land, control their own production, and sell their products in the marketplace. The government still has a firm grasp on the everyday lives of the citizens, but a majority of Chinese feel the country is better off today than it was several decades ago. China is the largest recipient of foreign direct investment (FDI) in the developing world, and its coastal regions have experienced rapid economic growth, including the miraculous transformation of Guangdong province in the south, which has close ties to Hong Kong.

Outside of China, Taiwan and South Korea have enjoyed rapid rates of growth since the 1960s, making them textbook examples of NICs. They benefited from being front-line states during the cold war, with massive U.S. military and economic assistance. With little arable land, they relied heavily on skilled labor and export promotion policies to advance into high-value-added goods, including steel, ships, and electronic goods, and have sizeable middle classes as a result. South Korea has become one of the most successful countries of the developing world, with a large middle class. North Korea, in contrast, suffers deepening poverty and famine.

South Asia

This area boasts the world's second largest population and some of the world's poorest people. Formerly a British colony, it split into India and Pakistan in 1947; Bangladesh seceded from Pakistan in 1971. India and Pakistan have fought several wars in the past, and continue to dispute possession of the region of Kashmir; both now possess nuclear weapons.

India is dominated by Hinduism, whereas Pakistan and Bangladesh, with roughly 160 million and 144 million people, respectively, are overwhelmingly Muslim. Population densities are high, as it is the natural rate of population increase, and the region has huge pools of poor people. India, the largest country of the region with over 1 billion people, overshadows its neighbors demographically and economically. India is the leading producer of certain world crops such as cotton. Its government long practiced policies of import substitution and protectionism that greatly slowed economic growth. Recently, as the government turned toward more export-oriented policies, western India has enjoyed modest economic growth, including the production of automobiles, movies, and software. In contrast, eastern India lags behind, as has Bangladesh.

The majority of South Asia's people live in villages and subsist directly off of the land, growing rice under quasi-feudal social relations. Many cities in the region, such as Calcutta, Madras, Delhi, Bombay, Dacca, and Karachi, contain large numbers of urban poor. Economic development is progressing slowly as most of the achievements have been erased by the rapidly growing population. Women tend to suffer harsh social and economic circumstances, and their opportunities for advancement are limited.

Middle East and North Africa

Dominated physically by deserts, this region lacks a substantial agricultural base; due to chronic shortages of water, many food products must be imported. Islam is the overriding religion, with the exception of Israel. The region's major asset is its immense petroleum reserves, the revenues from which are used to finance economic development. Culturally, this domain is dominated by Arab states, which exhibit considerable variations in economies and standards of living. Not all countries in the Middle East have large oil reserves, which are concentrated mainly in the Persian Gulf region. In general, Arab countries with large populations (e.g., Egypt) lack oil reserves, while the major petroleum exporters (e.g., Saudi Arabia, Kuwait, the United Arab Emirates) have relatively small populations. Population growth in the Arab states continues at very high levels, creating societies with large numbers of unemployed young people.

The reasons for the region's low level of development despite the presence of petroleum reserves include repressive governments that do not favor growth. Most Arab governments are repressive and at best only quasi-democratic. Some argue that the revenues from petroleum inhibit motivations to diversify their economies; in this sense, resources can be a curse rather than a blessing. Islamic traditionalism also challenges economic development in some areas; for example, women's roles in business and public life are sharply restricted, and financial markets are hindered by the Koranic prohibition against interest. The region has been severely affected by political instability. Fundamentalist Shiite Muslims took over Iran in 1979 and have promoted revolutions to cleanse the land of Westernized values and institutions elsewhere. Several wars hampered economic growth in the region, including the 1979–1989 war between Iraq and Iran, in which 1 million people died; the two American wars against Iraq (1991 and 2003), the latter of which is ongoing; civil wars in Lebanon, which decimated its banking industry; several wars between Israel and its Arab neighbors, (1948, 1956, 1967, and 1973); and constant strife between Israel and the Palestinians.

Israel is the region's only country not dominated by Muslims and its sole democracy. Israel is also the Middle East's most economically advanced state, in part because of large American subsidies, and has become a significant exporter of high-technology products and computer software. However, Israel has long been mired in conflict with the Palestinians, which has depressed the political stability of the region, including tourism and foreign investment.

Sub-Saharan Africa

Sub-Saharan Africa is by far the poorest region in the world. Despite the fact that it has a surprisingly low population density and an abundance of natural resources, most people live in poverty and suffer from poor health and a lack of education. The legacy of the long European colonial era still lives on in African countries, whose states are relatively artificial constructions wracked by tribal conflicts. Africa's artificial boundaries are the source of numerous wars, often tribal in nature. The governments of most of Africa have proved to be too corrupt and indifferent to the needs of their populations, spending most of their limited funds on the military rather than their civilian populations. Over the last 30 years, wars in Angola, Mozambique, Rwanda, Congo/Zaire, Liberia, Sierra Leone, Somalia, Ethiopia, Eritrea, and Sudan have killed over 30 million people. Famine claims the lives of another 15 million annually. Africa's economies are largely centered on the production of raw materials, including copper, uranium, diamonds, and gold. Low prices on the world resources and commodities market have contributed to the continent's economic malaise.

Africa also suffers from high rates of population growth, as its fertility rates are the highest in the world. The demand for farmland and overgrazing have stripped many agricultural areas of their potential to grow crops. Deforestation has decimated the continent's rich ecosystems and contributed to the worldwide crisis in biodiversity loss. AIDS has claimed the lives of millions, and in many countries more than one-third of all adults are infected with HIV. The epidemic has produced millions of orphans and shows little sign of slowing down. Malaria and other diseases are endemic in much of the continent. As a result of numerous wars, corrupt governments, rapid population growth, economic mismanagement, disease, drought, and lack of foreign investment, Africa is the only region in the world that has become poorer since World War II. By any measure, Africans are the poorest people in the world and the ones most in need of economic growth.

Characteristic Problems of Less Developed Countries

The developing world encompasses a vast array of societies with enormous cultural and economic differences. Nonetheless, several characteristics tend to be found throughout these nations to one degree or another. Obviously, the more economically developed a country is, the less likely it is to exhibit these qualities.

Rapid Population Growth

Can we ascribe the problems of LDCs to rapid population growth? We must be careful to avoid the simplistic Malthusian error of ascribing all of the world's problems to population growth. Nonetheless, in many countries, the rate of growth does exceed productivity gains in agriculture and other sectors, diminishing the average standard of living. Many scholars argue that rapid population growth must be controlled if development is to succeed.

In many LDCs, particularly Africa, rapid population growth rates tax the available food supplies. Some LDCs (particularly in Africa) approach levels of subsistence and starvation. Rapid population growth also reduces the ability of households to save; therefore, the economy cannot accumulate investment capital (Figure 14.16). In addition, with rapid population growth, more investment is required to maintain a level of real capital growth per person. If public or private investment fails to keep pace with the population growth, each worker will be less productive, having fewer tools and equipment with which to produce goods. This declining productivity results in reduced per capita incomes and economic stagnation. Rapid population growth in agriculturally dependent countries, such as China and India, means that the land must be further subdivided and used more heavily than ever. Smaller plots from subdivision inevitably lead to overgrazing, overplanting, and the pressing need to increase food production for a growing population from a limited amount of space. Many LDCs are rapidly urbanizing. Rapid population growth generally entails large flows of rural farmers to urban areas and more urban problems. Housing, congestion, pollution, crime, and lack of medical attention are all seriously worsened by the rapid urban population growth.

Assuredly, a rapid increase in population—especially the number and proportion of young dependents—creates serious problems in terms of food supply, public education, and health and social services; it also intensifies the employment problem. However, a high rate of population growth was once a characteristic of present-day developed countries, and it did not prevent their development. This observation makes it difficult to argue that population growth necessarily leads to underdevelopment or that population control necessarily aids development. Thus it is erroneous and simplistic to blame all the LDCs' problems on population growth, which is one of several factors that contribute to poverty. Focusing on rapid population growth (i.e., the Malthusian argument laid out in Chapter 3) detracts

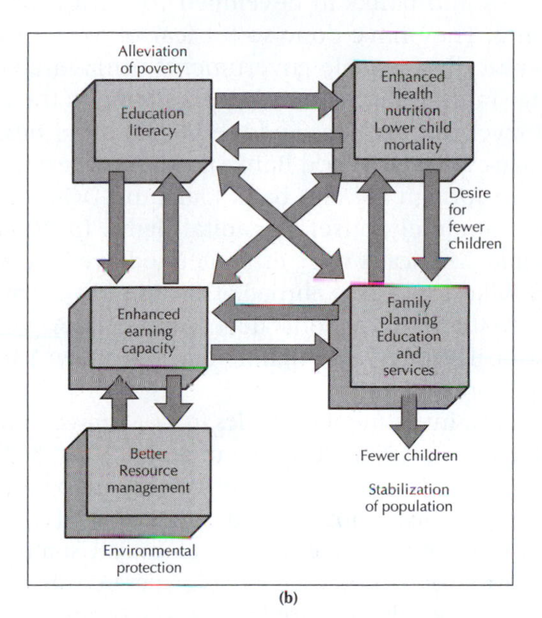

FIGURE 14.16
The cycle of poverty. Economic development means intervention in a host of intertwined variables that conspire to keep people poor. These include the historical context of a developing country and its linkages to the world system, as well as internal factors such as population growth, investment levels, and government policies.

from other, more important but politically controversial factors such as the legacy of colonialism and governments' indifference to investing in their populations and infrastructure.

Unemployment and Underemployment

Unemployment and underemployment are major problems in LDCs. Put bluntly, their economies rarely generate enough jobs for all, for a variety of reasons. *Unemployment* is a condition in which people who want to work cannot find jobs. *Underemployment* means that working people are not able to work as many hours as they would like, usually much less than 8 hours per day, and that their skills and talents are underutilized. Reliable statistics on unemployment and underemployment in LDCs are difficult to obtain, but unemployment in many developing countries often exceeds 20 percent.

Many cities in LDCs have recently experienced rapid flows of migrants from rural areas as a result of low agricultural prices and lack of land reform. Large number of migrants are attracted to cities by the expectation of jobs and higher salaries, expectations that may not be met in reality. Thus, these flows exemplify the influence of imperfect information upon spatial decision making. Once in the cities, many migrants cannot find work and contribute to the crisis of unemployment. Other migrants find limited employment as shop clerks, handicraft peddlers, or street vendors, often forming part of the unregulated, low-wage, informal economy.

Unemployment and underemployment are not the sole reasons for the problems of LDCs. Jobs are generated only when there is sufficient investment, which is lacking in most LDCs. Thus unemployment is much more complex than simply the willingness of people to work; it involves a vast, complex institutional framework in which jobs are generated, occupations and skills created and maintained, and investment capital is channeled.

Low Labor Productivity

Are the problems of LDCs a result of low labor productivity? It is true that a day's toil in a typical developing country produces very little commercial value compared with a day's work in a typical developed country. This is particularly evident in agriculture. Human productivity in a developing country may be as little as one-fiftieth of that in a developed country. Why is this?

One reason for low levels of productivity is the small scale of farming operations often found in LDCs, which generate few *economies of scale* (Chapter 5). Another is that capital investment rates are often very low, interest rates are relatively high, and most capital is generated by foreign-owned firms, whose major incentive is exports and profits, not job generation. Most developing countries lack the machines, engines, power networks, and factories that enable people and resources to produce at maximum levels of efficiency. In addition, LDCs are less able to invest in *human capital.* Investments in human capital—such as education, health, and other social services—prepare a population to be productive workers. Schools are often inadequate and literacy rates may be comparatively low.

Low labor productivity in LDCs is exacerbated by a lack of organizational skills and the absence of adequate management, both of which are necessary for increased productivity. Many of the most skilled workers immigrate to economically advanced countries, where they are more highly paid for their labor. The United States and Europe, for example, are replete with some of the best-trained labor from LDCs, including doctors, engineers, mathematicians, and scientists who have come looking for better salaries from companies or the government. This immigration has contributed to a so-called *brain drain,* whereby LDCs lose talented people to the developed world.

Although it must be acknowledged that low labor productivity is a universal attribute of LDCs, it is not a *causative* factor. The important question to consider is this: What prevents labor productivity from improving in developing countries?

Lack of Capital and Investment

Most LDCs suffer from a lack of capital accumulation in the form of machinery, equipment, factories, public utilities, and infrastructure in general. The more capital, the more tools available for each worker; thus there is a close relationship between output per worker and per capita investment and, in turn, between investment and income. If a nation hopes to increase its output, it must find ways to increase per capita capital investment.

In most cases, increasing the amount of arable land for an LDC is no longer a possibility. Most cultivable land is already in use (Chapter 6). Therefore, capital accumulation for an LDC must come from savings and investment. If an LDC can save, rather than spend, all its income and invest some of its earnings, resources will be released from the production of consumer goods and be available for the production of capital goods. But barriers to saving and investing are often high in LDCs, including high interest rates, political unrest, corruption, and inefficient regulations. (The United States also had notable problems with low savings and investing rates.) An LDC has even less margin for savings and investing, particularly when domestic output is so low that all of it must be used to support the many needs of the country. Ethiopia, Bangladesh, Uganda, Haiti, and Madagascar save between 2 percent and 3 percent of their domestic outputs. In 2003, India and China managed to save an average of 21 percent of their domestic outputs, compared with 20 percent for Japan, 15 percent for Germany, and 5 percent for the United States.

Unpaved road near Kumasi, Ghana, exemplifies a transportation infrastructure suffering from insufficient investment, an obstacle to development.

Many LDCs suffer from *capital flight,* which means that wealthy individuals and firms in these countries have invested and deposited their monies in overseas ventures and banks in developed countries for safekeeping. They have done so for fear of expropriation by politically unstable governments, unfavorable exchange rates brought on by hyperinflation in the LDCs, high levels of taxation, and the possibility of business and bank failures. World Bank statistics suggest that inflows of foreign aid and bank loans to LDCs were almost completely offset by capital flight. In 2000, for example, Mexicans were estimated to have held about $100 billion in assets abroad. This amount is roughly equal to their international debt. Venezuela, Argentina, and Brazil also have foreign holdings between $30 and $60 billion each.

Finally, investment obstacles in LDCs have impeded capital accumulation. The two main problems with investment in LDCs are (1) lack of investment opportunities comparable to those available in developed countries and (2) lack of incentives to invest locally. Usually LDCs have low levels of domestic spending per person, so their markets are weak compared with those of advanced nations. Without domestic industries, consumers must often turn to imports to satisfy their needs, especially for high-value-added products. Factors that keep the markets weak are a lack of effective demand backed by purchasing power, a lack of trained personnel to manage and sell products at the local level, and a lack of government support to ensure stability. There is also a lack of infrastructure to provide transportation, management, energy production, and community services—housing, education, public health—which are needed to improve the environment for investment activity.

Inadequate and Insufficient Technology

Typically, LDCs lack the technologies necessary to increase productivity and accumulate wealth. Some LDCs acquire new production techniques through technology transfer that may accompany investment by multinational corporations, as happened in many of the NICs of East Asia. OPEC nations benefited from foreign technology in oil exploration, drilling, and refining. Unfortunately, for LDCs to put this available technology

The trains are crowded in Bangladesh. Typically, resources for feeding and moving the poor are inadequate. Such systems generate low rates of profit and do not attract large investors. Thus, transport systems in impoverished countries generally are provided by financially strained governments, many of which are reeling under low export prices and neoliberal programs imposed by the IMF. Thus crowding is not simply a result of population growth but reflects the political economy of public services in developing countries.

to use, they must have a certain level of capital goods (machinery, factories, etc.), which they by and large do not possess. The need is to channel the flow of technologically superior capital goods that have high levels of reliability to the developing nations so that they can improve their output.

In developed countries, technology has been developed primarily to save labor and to increase output, resulting in capital-intensive production techniques. However, in LDCs, capital intensification tends to displace workers, eliminating critically needed jobs. Thus LDCs need capital-saving technology that is labor intensive. In agriculture, much of the midlatitude technology of the developed countries (e.g., wheat harvesting combines) is unsuited for low-latitude agricultural systems with tropical or subtropical climates and soils.

Unequal Land Distribution

In most developing countries, in which a large part of the population lives in rural areas, land is a critical resource essential to survival. Unfortunately, land is often highly unequally distributed, and a small minority of wealthy landowners controls the vast bulk of arable (farmable) land. In Brazil, for example, 2 percent of the population owns 60 percent of the arable land; in Columbia, 4 percent owns 56 percent; in El Salvador, 1 percent owns 41 percent; in Guatemala, 1 percent owns 36 percent of the land; and in Paraguay, 1 percent owns an astounding 80 percent of all land suitable for farming. These numbers reflect the historical legacy of the Spanish land grant system imposed over centuries of colonialism. Wealthy rural oligarchs often control vast estates and plantations designed primarily for export crops, not domestic consumption. The majority of the rural population, consequently, is landless and must sell its labor, often at very low (below subsistence) wages, and live in serflike conditions. Shortages of land are accentuated by agricultural mechanization and high population growth. The result is frequent political turmoil, including demands for land redistribution; in much of Latin America, peasant social movements to regain control of the land are very active, and violence over land possession is frequent. Further, rural areas with high natural growth rates and widespread poverty become the source of waves of rural-to-urban migration, which often swamps cities with desperately poor people seeking jobs.

Throughout much of the developing world, land reform, which is vitally important to increased agricultural output, is lacking because the government is too inept or corrupt to redistribute the land owned by a few wealthy families. For some nations, such as the Philippines, land reform is the single most pressing problem deterring them from economic development.

In contrast, strong action taken by the South Korean government after the Korean War allowed for increased productivity and the development of an industrial and commercial middle class that made South Korea one of the success stories of the Pacific Rim.

Poor Terms of Trade

A country's terms of trade refer to the relative values of its exports and imports (Chapter 12). When the terms of trade are good, a country sells, on average, relatively high-valued commodities (e.g., manufactured goods) and imports relatively low-valued ones (e.g., agricultural products). This situation generates foreign revenues that allow the LDC to pay off debt or to reinvest in infrastructure or new technologies. Unfortunately, many LDCs, with economies distorted by centuries of colonialism, in which they became suppliers of minerals and foodstuffs, export low-valued goods, and must import expensive high-valued, manufactured ones. Many LDCs rely heavily on exports of goods such as petroleum, copper, tin and iron ore, timber, coffee, and fruits and must import items such as automobiles, pharmaceuticals, office equipment, and machine tools. The worldwide glut in raw materials, including petroleum, wheat, and many mineral ores, has depressed the prices for many LDC exports. Without indigenous manufacturing, many are forced to sell low-valued goods such as bananas for high-valued ones such as computers. This situation makes it difficult to generate foreign revenues and exacerbates a country's debt problems. Further, poor terms of trade perpetuate a country's cycle of poverty: Low foreign revenues yield little to reinvest, which helps to create a shortage of capital, resulting in low rates of productivity.

Foreign Debt

Much of the developing world is deeply in debt to foreign governments and banks. In 2005, total world debt amounted to more than $2.6 trillion (Figure 14.17). Some nations, notably Argentina and Nigeria, have debts in excess of 100% of their annual GDPs (Figure 14.18). Having borrowed hundreds of billions of dollars, they find themselves unable to repay either the interest or the principal. Debt repayments often consume a large share of their export revenues, which might otherwise be used for development.

Debt restructuring policies imposed by the International Monetary Fund typically tie debt relief to "structural adjustment," which includes reductions in government subsidies to the poor and devaluations of currencies (which drive up the costs of imports), policies that lower the quality of life for hundreds of millions, if not billions, of people. For example, a poor mother in an LDC with a sick infant relies on imported pharmaceuticals when her country lacks a domestic industry; IMF-imposed currency devaluations will drive

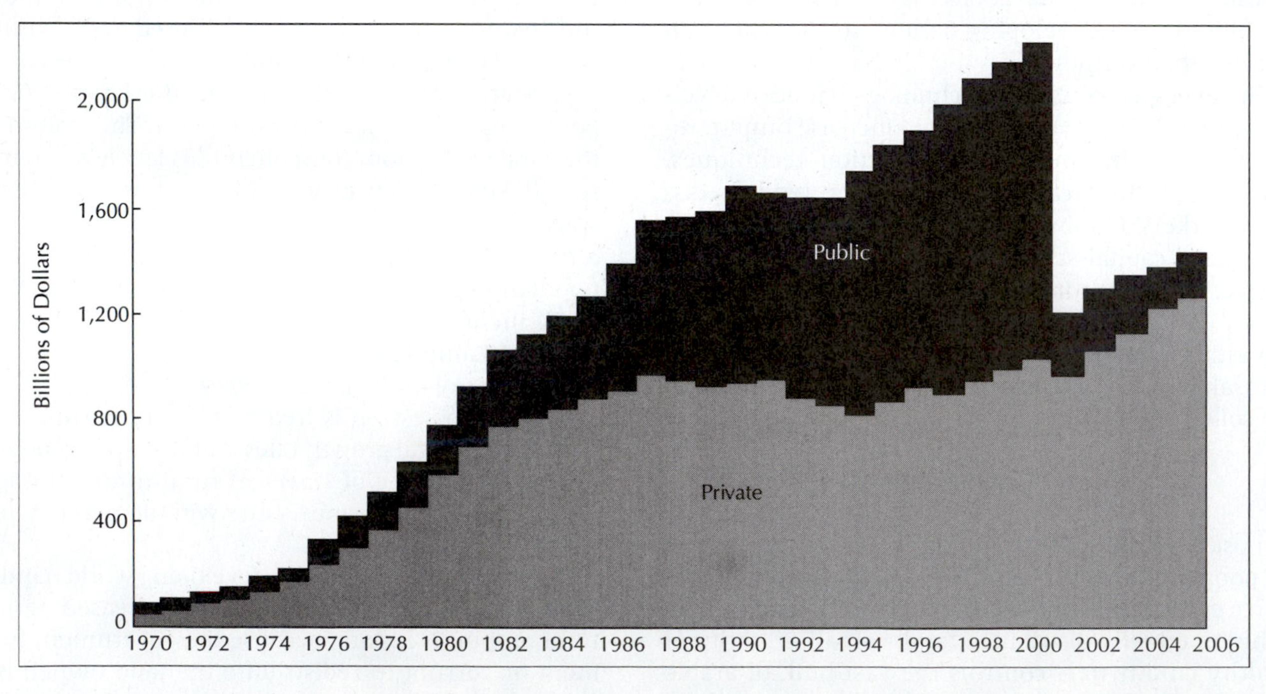

FIGURE 14.17
The external debt of all developing countries grew substantially between 1970 and 2005. It consists of two parts: the public debt, which is owed to foreign governments, and the private debt, which is owed to private banks.

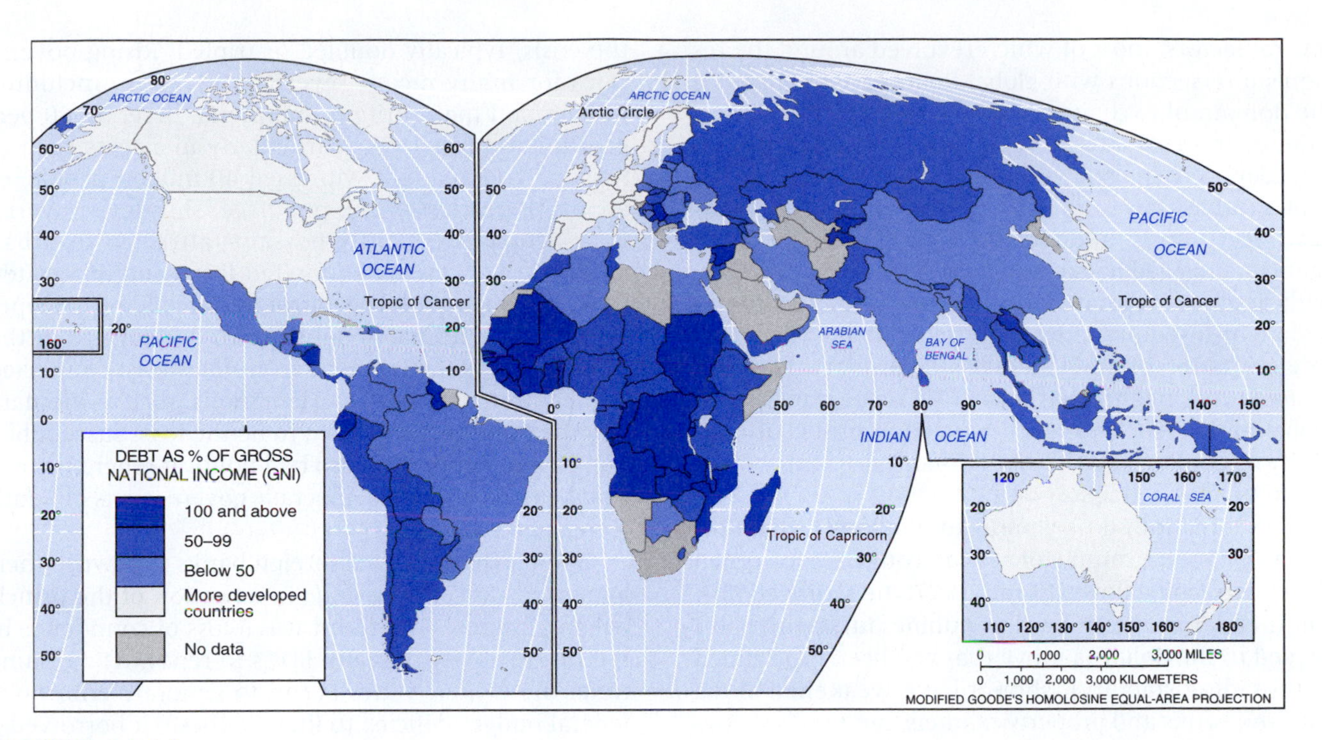

FIGURE 14.18
Debt as a percentage of gross national income, 2004. Many developing countries became indebted to finance future development, only to find themselves confronted with low prices for their exports and repatriated profits from foreign direct investment. A large share of government revenues and earnings from exports thus must pay the principal and interest of their debt. Although the poor often did not benefit from these loans, it is their taxes that are often used to pay them back.

up the costs of these necessities, often making them out of reach for the poor, who need them the most. For these reasons, scholars and politicians involved in international development often call for a debt moratorium to allow LDCs a respite from what are often crushing debt burdens.

The origins of the debt crisis lay in the 1970s and 1980s, when many LDCs took out large loans with the expectation of paying them off when their economies improved in the future. Western banks had a huge influx of petrodollars and were eager for borrowers; developing countries were happy to take advantage of this unaccustomed access to cheap loans with few strings attached. The borrowing enabled them to maintain domestic growth.

But a series of major economic changes in the world scene meant that many of these loans could not be repaid. The OPEC oil embargoes of 1973 and 1979 sent the price of crude oil skyrocketing, creating a cash-flow crisis for oil-importing LDCs. Furthermore, monies spent for oil could have been used for economic development, such as for increased infrastructure, improved education, and needed agricultural reforms. The debt of oil-importing LDCs grew from $150 billion in 1973 to $800 billion by 1985. Even oil-exporting nations such as Mexico, Libya, and Nigeria overborrowed based on their expectations of rapidly inflated oil prices. When oil prices fell significantly in the 1980s, these nations found themselves saddled with debts that they could not afford.

The mounting debt caused concerns about the stability of the international monetary system. The cause of this instability was the overexpansion of credit, particularly through the Eurocurrency market in the 1960s and 1970s, which led to a crisis that had its roots in the overaccumulation of capital and the declining rate of profit. In 1982, Mexico ran into difficulties meeting interest and capital payments on its debts. Brazil and Argentina also appeared ready to default. A collapse of the financial system was forestalled by a series of emergency measures designed to prevent large debtor countries from defaulting on their loans. These measures involved banks, the IMF, the Bank for International Settlements, and the governments of lending countries in massive bail-out exercises that accompanied debt reschedulings.

The debt overhang persisted because debt-service ratios—annual interest and amortization payments as a percentage of total exports—remained at dangerously high levels. This factor caused a second financial crisis of global proportions in 1997 that began in Southeast Asia. The origins of the Asian crisis have been attributed to several factors, most of which revolved around the region's intersections with global finance, particularly in the domain of exchange rates. Asian financial markets suffered from insufficient regulation, leading to political considerations in credit allocation based heavily on corrupt, cronyist ties to ruling families. Often badly managed and poorly audited, banks and other lending agencies engaged in excessively risky lending practices, with assets exaggerated by inflation rates, leading to high debt-to-equity ratios. Poor recordkeeping masked a deterioration of fundamentals, the primary measures of economic health that attract or repel capital on a global basis. When exposed to sudden onsets of illiquidity or bankruptcy, the Asian financial system turned out to be highly fragile.

The crisis found its genesis in Thailand, which offers a textbook example of a small country confronted with large capital flows. In July 1997, the saturated Thai commercial property market bubble burst; the baht, pegged to the dollar (a move that had been proclaimed as providing stability), should have weakened along with the banks and property markets, but the Thai government, determined to avoid the embarrassment of devaluation, stubbornly supported the baht. Thai money market managers, knowing their currency was overvalued, sought to make a quick killing by selling baht. When Thailand's investors began buying dollars, they attracted the attention of foreign speculators, particularly foreign hedge fund speculators, who, sensing the nation's difficulty, bet heavily against the baht. Fresh from the Mexican currency crisis of 1995, American speculators in particular sought greater returns in the economically greener pastures of the Asian NICs. Determined to preserve its currency at all costs, the Bank of Thailand risked everything to protect the nation's currency against the foreign onslaught, until it was finally forced to let go, sending the baht into a free fall of devaluation.

The Thai "blood baht" soon spiraled into a vortex of financial instability that included all of East Asia except China, with secondary effects on Australia, North America, and Europe. The "Asian flu" leapfrogged in 1997–1998 from Thailand to Malaysia to Indonesia to South Korea to Taiwan to Hong Kong. A massive reversal of capital flows, which had swept into the region for decades, saw tens of billions of dollars leave for the United States and Europe; in 1997 alone, net capital flows out of Thailand were equivalent to 10.7 percent of the GDP. Interest rates rose throughout the region, stimulating a banking crisis. Decreased earnings led to deflationary shock. Corporate bankruptcies soared, especially among firms with the most debt exposure and highest debt-to-equity ratios. Stock exchanges throughout Asia suffered a repeated series of drubbings, falling as much as 75 percent by the end of 1998. Unemployment rates, generally below 5 percent before the crisis, typically doubled or tripled. Rising poverty rates for many meant severe deprivation, including hunger and malnutrition; Indonesia, beset by 60 percent inflation in 1998 (particularly in staples such as rice and cooking oil), witnessed 40 million people, or about 20 percent of the populace, sink into poverty. Poor people, frequently peasants attracted to urban areas by the presence of foreign firms and jobs, often took the brunt of the burden, forced into deeper poverty or a return to subsistence agriculture in the hinterlands. Notably, those economies least connected to the international financial markets, such as Vietnam and the Philippines, proved to be the least susceptible. The economic turmoil also had political repercussions, leading to new, more democratic governments in South Korea and Indonesia.

Some LDCs required foreign banks to rewrite their loans and cancel or *write down* a portion of the principal and interest. The result was a loss of confidence in the future ability of many LDCs to repay. At the same time, the United States began to generate enormous federal budget deficits; to finance these, it borrowed a large portion of available investment revenue, driving up interest rates worldwide. Higher oil prices, declining prices for raw material exports, higher world interest rates, an increase in the value of the U.S. dollar, and a decline in public and private lending to LDCs because of loss of confidence all contributed to the debt crisis that prevails today.

Restrictive Gender Roles

Deeply entrenched social stratification systems work against people, especially women, in many developing countries. Gender refers to the socially reproduced differences—including both privileges and obligations—between women and men, and permeates every society's allocation of resources and people's life chances. Typically, gender roles work to the advantage of men and the disadvantage of women.

The economic, political, and social status of women around the world is spatially variable. Worldwide, nowhere can women claim the same rights and opportunities as men. Women account for most of the world's 1.3 billion people living in extreme poverty. In most countries, women do most of the field work while still being responsible for household chores such as cooking, carrying water, and raising children. In some countries with more mechanized forms of agriculture, such as most of Latin America, men assume the job of plowing, with female participation strongly diminished in the field. In these cases, women work more in the market.

Women still spend more time working than men in all regions except Anglo America and Australia, and

In Mali, women often must travel several miles a day from their villages to gather firewood. This task is usually a woman's chore in most countries and exemplifies the fact that women almost everywhere work harder and longer than do men.

their wages are lower everywhere. In Muslim areas, women are not very economically active outside of the home because of religious prohibitions. In Latin America, labor force participation of women in the economy is increasing but mostly outside of the agricultural sector. Sub-Saharan Africa, India, and Southeast Asia are highly dependent on female farm and market income as well as commodified labor in the waged labor market. At a world scale, women generally garner only 60 percent to 70 percent of what men earn, often for the same work. This ratio is remarkably widespread, including the United States, in which women comprise the bulk of the poor and where women occupy less than 3 percent of the highest levels of management and ownership.

Corrupt and Inefficient Governments

Often, LDC governments are controlled by bureaucrats whose primary interest is catering to the wealthy elite, creating governments that are at best indifferent and often outright hostile to the needs of their own populations. Many LDC public policies are ineffectual or counterproductive, frequently ignoring the rural areas in favor of cities. Government jobs are frequently allocated through patronage, not a merit system. For example, in Africa, government jobs often go disproportionately to members of the same tribe as the president. The military is often the most well-funded and well-organized institution and may be a source of political instability, as during military takeovers of the government in coups d'état. Corruption often is endemic and may become a way of life, generating inefficiency and inequality. During the cold war, both superpowers backed oppressive regimes throughout the developing world with subsidies and arms. In poor countries, dictatorial governments often curtail their citizens' civil liberties, censoring the media and imprisoning or executing dissidents. Attacks on journalists, labor union leaders, student movements, religious groups, and ethnic minorities are often common under such situations. Thus, poor nations often have oppressive governments; widespread and well-protected civil liberties are more common with economic development. Understandably, many people feel alienated from their own governments under these conditions and may sympathize with various resistance movements, contributing to frequent political instability.

LDC governments, often with insufficient resources, frequently have great difficulty building or maintaining their nation's infrastructures. Roads, bridges, tunnels, and highways may fall into disrepair, driving up transportation costs. Dams and airports likewise may be neglected. Electrical power stations may not be maintained correctly, leading to power outages. Unsanitary water supplies become major carriers of infectious diseases such as cholera. Because the infrastructure is essential to economic development, governments that do not reinvest in their nation's transportation, water, and communication lines do not facilitate the process of economic growth.

Similarly, public services in many LDCs are often underfunded and inadequate. This means that public schools—the major avenue of upward mobility for many—may fail to educate the nation's young, leading to high illiteracy rates, especially for girls. Salaries for teachers are often too low for them to support themselves. Inadequate health care, including severe shortages of physicians and nurses, depresses the health of the labor force and lowers the productivity of labor. Underfunded transportation systems make circulation within and among cities difficult and expensive in terms of forgone time. Thus, images of crowded buses and trains in countries such as India testify to government priorities as much as population growth. Only the military tends to be a well-funded public service in much of the world.

Trends and Solutions

Worldwide, the gap between the rich and the poor is widening. The World Bank estimates that in 2004 the global poverty rate was 24 percent or about 1.5 billion people. In relative terms, the two regions of the world with the greatest proportions of people living in poverty are Latin America and sub-Saharan Africa. However, in absolute numbers, Asia is home to the greatest number of the world's poor, including vast numbers on the Indian subcontinent. There are indicators that the numbers of the poor in Asia may be declining, while the percentage of sub-Saharan Africa living in poverty is increasing quickly.

There is clearly no one-size-fits-all approach to ending poverty in poor countries. Policy analysts generally argue that poverty-reducing strategies must include the following: investing in health care and education, protecting land tenure rights for the rural poor, incorporating informal sectors of the economy into the formal sector, establishing political stability, democratizing decision making through representative government, debt relief, ending developed countries' trade restrictions against imports from less developed countries, and increasing women's decision-making power in the household. Obviously, these steps are difficult to implement, often because powerful, entrenched interests benefit from the status quo. Perhaps the best hope lies in a smoothly functioning world economy that can pull hundreds of millions of people out of poverty, as has happened in much of East Asia.

MAJOR PERSPECTIVES ON DEVELOPMENT

Theories of development have existed for many years. The earliest of them can be traced to the classical economists of the eighteenth century such as Adam Smith. But discussion of the term *development* in the social sciences is fairly recent and flourished after World War II. Three perspectives on development in economic geography hold widely varying assumptions and conclusions, including modernization theory, dependency theory, and world-systems theory. All of these grapple with the complex question of how the global economy shapes patterns of opportunities for development in individual countries.

Modernization Theory

The first and most widely accepted theory of development is *modernization theory.* This perspective starts with the central question: Does the development of the less-developed world follow the same historical trajectory as the West? The intellectual origins of this line of thought lie with the famous sociologist Max Weber, who attributed the success of northern Europe during the Industrial Revolution to the Protestant ethic (Chapter 2). Weber's work established a precedent that viewed economic development and social change as a function of people's ideas, culture, and beliefs rather than their social relations and historical context.

Weber's ideas were enormously popular in the United States following World War II; they were translated and elaborated by Talcott Parsons. The American version tended to equate the "modern" with Western, denying the possibility of modern, non-Western cultures. Writing during a period of intense competition between the United States and the Soviet Union for influence in the developing world during the cold war, Parsonian modernization theory explicitly advocated capitalism as the best possible path any country could choose to follow (i.e., the only one that leads to modernity, wealth, and democracy). This theory argued that if LDCs adopted Western values, market relations, and government institutions, they would become affluent, democratic societies. Hence, the path to progress from traditional to modern is unidirectional. In this view, rich industrial countries, without rival in social, economic, and political development, are modern, whereas poor countries must undergo the modernization process to acquire these traits.

By changing their way of looking at the world, in particular giving up the traditions that kept them trapped in an irrational past, this theory argued that LDCs would eventually achieve the same prosperity that Europe and North America enjoyed. Modernization theory advocated stability and gradual change, not revolutionary leaps. It is worth noting that whereas Weber held that capitalism was a unique institution to Europe, modernization theorists maintained that the triumph of capitalism is inevitable worldwide.

Modernization theory posited that economic development occurred through a series of stages. At the broadest level, these include the widespread view that all economies pass through stages in which the labor force is employed first in agriculture, then manufacturing (i.e., the Industrial Revolution), then services (a postindustrial world). However, while this line of thought has some accuracy in the Western world, many developing countries witness a leapfrogging effect in which many agricultural workers plunge directly into services. This discrepancy points to the dangers of simplistically generalizing from the experience of the West to the developing world, which has a very different historical context and trajectory.

The diffusion of progress was also a major line of thought in modernization theory: New ideas, technologies, and institutions were held to flow from the core to the periphery at both the international and national scales. Internationally, diffusion occurred from the world's core, that is, the developed countries, to the global periphery as multinational corporations invested in LDCs. Investment was held to realize a country's comparative advantage and allow it to carve a niche for itself in the world economy; thus modernization theory advocated free trade. Within countries, diffusion occurred from the urban core, that is, cities, which were held to be the foci of modernity, to the rural periphery, that is, the countryside. Thus modernization theory held that urbanization was inherently a good thing, that rural areas were trapped in cycles of tradition and stagnation.

Poverty, in this conception, is held to be the result of the incomplete formation and diffusion of markets. Markets, in this view, promote an equalization of standards of living in poor and rich regions through the free flow of capital and labor. Thus modernization theory lies in direct contrast to marxist claims that capitalism inevitably generates uneven development. Economic development in this conception is likened to a race, in which the rural areas must catch up with the cities much as LDCs must catch up with the developed world.

Modernization theory also advocated major social, cultural, and political changes on the road to capitalism. Population growth, for example, must be brought under control or else the demand for resources would exceed the productivity gains of markets. Thus modernization theorists advocated family planning programs, a line of thought similar to the demographic transition (Chapter 3). During the cold war, population control policies were advocated as a means to raise standards of living and reduce the appeal of the communist bloc's alternative to capitalism. Culturally, societies (or parts of them) were divided between the traditional and the modern: Tradition was held to be synonymous with irrationality and fear of change, whereas modernity implied the belief that change is good and necessary for the sake of progress. Thus cities were alleged to be repositories of the modern, and rural areas identified with the traditional, to be transformed and brought into the modern age. The culture of modernity was to be diffused through mass education, the media, and growing literacy rates.

Politically, modernization theory equated capitalism with democracy, arguing that only market-based societies could protect civil liberties. In contrast to the centralized power of many precapitalist societies, markets tend to create multiple centers of power. The rise of a middle class was often held to be a central event in protecting civil liberties. Despotic societies tend to be poor, with a small dictatorial elite in control. By generating an educated, informed citizenry, markets were held to be antithetical to dictatorship. Modernization theorists point to countries such as South Korea as evidence of this thesis.

A major advocate of this approach was W. Rostow, an influential economist and presidential policy advisor in the 1960s who proposed a famous five-stage model of development (Figure 14.19). Rostow's model likened economic growth to an airplane taking off and proposed five stages in this process:

1. **Traditional society**—This term defines a country that has not yet started the process of development and is mired in poverty. It includes traditional societies with a very high percentage of people engaged in agriculture and a high percentage of national wealth allocated to what Rostow called "nonproductive" activities, such as the military and religion. Contemporary examples include Mali or Bhutan.

2. **Preconditions for take-off**—Under the international trade model, the process of development begins when an elite group initiates innovative economic activities. Under the influence of these well-educated leaders, the country starts to invest in new technology and infrastructure, such as water supplies and transportation systems. These projects will ultimately stimulate an increase in productivity. A culture of growth begins to take root. Examples include Thailand, Mexico, and Indonesia.

3. **The take-off**—Rapid growth is generated in a limited number of economic activities, such as textiles or food products. These few take-off industries achieve technical breakthroughs and become efficiently productive, while other sectors of the economy remain dominated by traditional practices. Examples today include China and Chile.

4. **The drive to maturity**—Modern technology, previously confined to a few take-off industries, diffuses to a wide variety of industries, which then experience rapid growth comparable to the take-off industries. Workers become more skilled and specialized. Examples include Singapore, Taiwan, and South Korea.

5. **The age of mass consumption**—The economy shifts from production of heavy industry, such as steel and energy, to consumer goods, like motor vehicles and refrigerators, and advanced services. Examples include the United States and Germany.

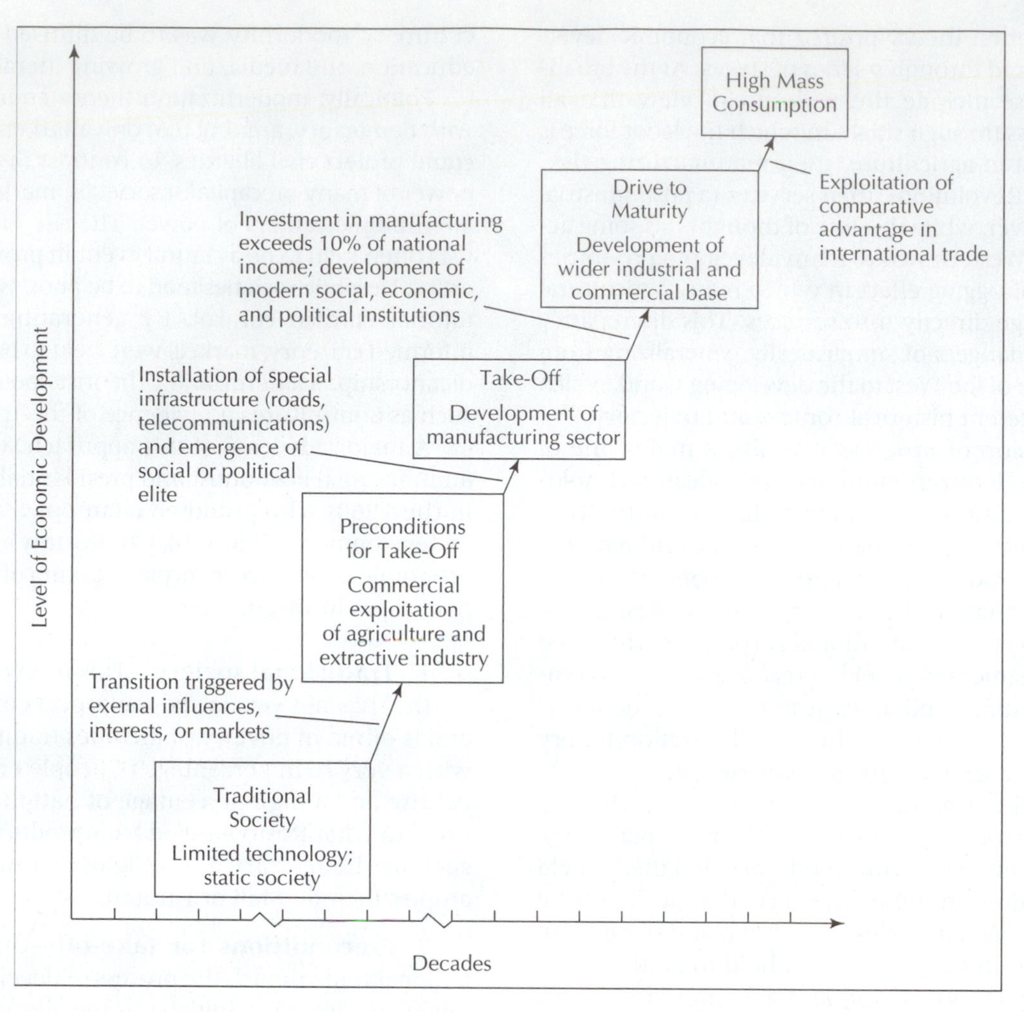

FIGURE 14.19
Rostow's five stages of the modernization process. The conventional approach to economic development viewed the process as akin to an airplane taking off, in which economies progress through a series of stages culminating in mass consumption. Others have criticized this approach for being ethnocentric (holding up the West as the only model of development) and ignoring the global and historical context of development.

Modernization theory lumped every country in the world into one of these stages. LDCs reside in the primary few stages while developed countries have passed through the preliminary stages and are now in stage 5 or beyond. It is assumed that LDCs will use this model to follow the steps of the developed countries. The international trade approach keeps countries competitive by forcing the industries to adapt and develop along world standards.

The policy implications of modernization theory are straightforward. Overall it advocates the formation of unfettered markets. Thus barriers to trade and investment should be removed, and foreign capital in the form of multinational corporations should be welcomed. Urban development should be stressed at the expense of rural areas. Internationally, this view was central in the switch from import substitution to export promotion that occurred in much of East Asia in the 1970s and 1980s (Chapter 12). More recently, this theory is used to justify neoliberal structural adjustment policies of the International Monetary Fund, including currency devaluations and reductions in government subsidies.

Modernization theory has been soundly criticized on a number of grounds. Critics note that it is ethnocentric: It posits the history of the West as an ideal to be imitated, and everyone else's culture as inferior. It offers a simplistic, unidimensional view of history that culminates only in the experience of Europe and North America. LDCs are seen as little more than backward versions of the West, not as unique entities with their own distinct cultures and histories. Critics note that hundreds of years of colonialism produced a world economy that greatly disadvantaged the developing world; development is hardly a fair race when not all countries enjoyed the same starting point (i.e., the West has enjoyed great, deeply entrenched advantages over LDCs). Finally, by uncritically celebrating markets as mechanisms that only produce wealth and not poverty, modernization theory only celebrates the benefits but ignores the costs of capitalist development. Markets look fine to the winners, who preach free trade, but appear less rosy to those who have not benefited from them. Finally, modernization theory focuses only on the internal dynamics within countries (i.e., their cultures) and ignores the external context, the countries' colonial legacies, and the countries' position in the global division of labor. In this way it ends up blaming the victim (i.e., attributing poverty to poor people), which is politically easy but not accurate.

Dependency Theory

In the 1960s, numerous scholars from the developing world, particularly Latin America, began to question the promises and assumptions of modernization theory. What appeared as a comparative advantage and interdependence to Western scholars, for example, appeared to many in the LDCs as exploitation. Drawing from the heritage of Marxism, *dependency theorists* claimed that the development of the core countries was intrinsically dependent on the underdevelopment of the periphery countries. The theory suggests that unequal development of the world economy stems directly from the historical experience of colonialism. The development of Europe and North America, especially during the nineteenth and early twentieth centuries, relied on the systematic exploitation of underdeveloped areas by means of unequal terms of trade, abuse of low-skilled and low-paid labor, and profit extraction.

In this reading, poverty didn't just happen, but is a historic product (i.e., like a shirt or pair of tennis shoes, poverty is actively made by the dynamics of capitalism, not just something that happens to appear). This process is described as the "development of underdevelopment," which emphasizes that underdevelopment is not a static state but an active process. Dependency theorists thus argue that the LDCs were *made to be poor* by the West. Exploitation of the periphery occurs through the process of uneven exchange, in which LDCs produce low-valued goods in the primary sector and purchase high-valued goods from the core, a market mechanism that appears like an exchange of opposites but conceals the appropriation of surplus value that low-income workers in developing countries produce.

Unlike modernization theory, therefore, dependency theory focuses on external, not internal, causes of poverty. Because they occupy very different roles in the capitalist world economy, developed and underdeveloped countries are very different animals. Thus, unlike its colonies, the West was undeveloped, but not underdeveloped; LDCs do not temporarily "lag" behind West but are mired in poverty produced by the West, a situation naturalized as inevitable by the economics of comparative advantage. Dependency theorists thus argue that development and underdevelopment are two sides of global capital accumulation. The wealth of developed countries is derived from the labor and resources of the LDCs. Whereas modernization theory holds that markets eradicate uneven development, dependency theorists maintain it perpetuates it. Development and underdevelopment are therefore two sides of the same coin, a zero-sum game: Development somewhere (i.e., the West) requires underdevelopment somewhere else (i.e., its colonies). The political and social structures created under this vision of the world economy imply that independent development is impossible.

The policy implications of dependency theory, directly opposite of those espoused by modernization theorists, emphasize self-reliance; countries should exclude transnational corporations and practice import substitution to promote domestic production. Some even advocate defaulting on foreign debt.

Like modernization theory, this view is also subject to criticisms. Dependency theory tends to view the global periphery as passive and incapable of taking action; it lumps all LDCs together as if they were victimized by capitalism to the same degree. It ignores the internal causes of poverty, such as rural aristocracies that inhibit development, and its explanation for the core's wealth is simplistic. For example, dependency theory has not offered an adequate account of technological change and productivity growth. The claim that global capitalism always generates poverty in LDCs was unsustainable. Empirically, in the 1970s and 1980s, the rapid growth of the East Asian NICs in particular showed that development on the global periphery is indeed possible and that capitalism does not automatically reduce all LDCs to impoverished states.

World-Systems Theory

It is important to keep in mind that the forces driving the world economy are in a continual state of flux. The rise of the United States from a periphery to a core country, the fall of the Soviet Union, the appearance of the NICs in Asia, and the increasing importance of transnational firms exemplify the notion of permanent spatial disequilibrium. A third body of theory, *world-systems theory*, started by the sociologist Immanuel Wallerstein, takes into account this disequilibrium in explaining the changing structure of the world economy.

Unlike dependency theory, world-systems theory allows for some mobility within the capitalist world economy. Its focus is on the entire world rather than individual nation-states. Fundamentally, this view maintains that one can't study the internal dynamics of countries without also examining their external ones. Thus the boundary between foreign and domestic effectively disappears.

World-systems theory distinguishes between large-scale, precapitalist world empires, such as the Romans, Mongols, or Ottomans, which appropriate surplus from their peripheries through the state, and the capitalist world system, which arose in the "long sixteenth century" (1450–1650). Under global capitalism, there is no single political entity to rule the world (i.e., there is a single market but multiple political centers, meaning there is no effective control over the market). The political geography of capitalism is thus not the nation-state but the interstate system. Occasional attempts to reassert a world empire included the Hapsburgs, Napoleon, and Germany in World Wars I and II.

World-systems theorists maintain that capitalism takes many forms and uses labor in different ways in different regions. In the core, labor tends to be waged (i.e., organized through labor markets), while in developing countries there is considerable use of unfree labor, ranging from slavery to indentured workers to landless peasants working on plantations. The world economy structures places in such a way that high-valued goods are produced in the core and low-valued ones in the periphery. It is the search for profits through low-cost labor that drives the world system forward to expand into uncharted territories.

Unlike the bifurcation between developed and less developed countries that both modernization and dependency theories advocate, in world-systems theory there is a tripartite division among core, periphery, and semiperiphery (Figure 14.20). The core and periphery are the developed and undeveloped countries, respectively: One is wealthy, urbanized, industrialized, and democratic, the other rural, impoverished, agriculturally based, and dominated by authoritarian governments. The semiperiphery has characteristics of both core and periphery and includes states at the upper tier of the LDCs, such as the NICs, Brazil, Mexico, Thailand, and Saudi Arabia. Core processes high wages, high levels of urbanization, industrialism and postindustrialism, the quaternary sector of the economy, advanced technology, and a diversified product mix. The world periphery processes are low wages; low levels of urbanization, preindustrial, and industrial technology;

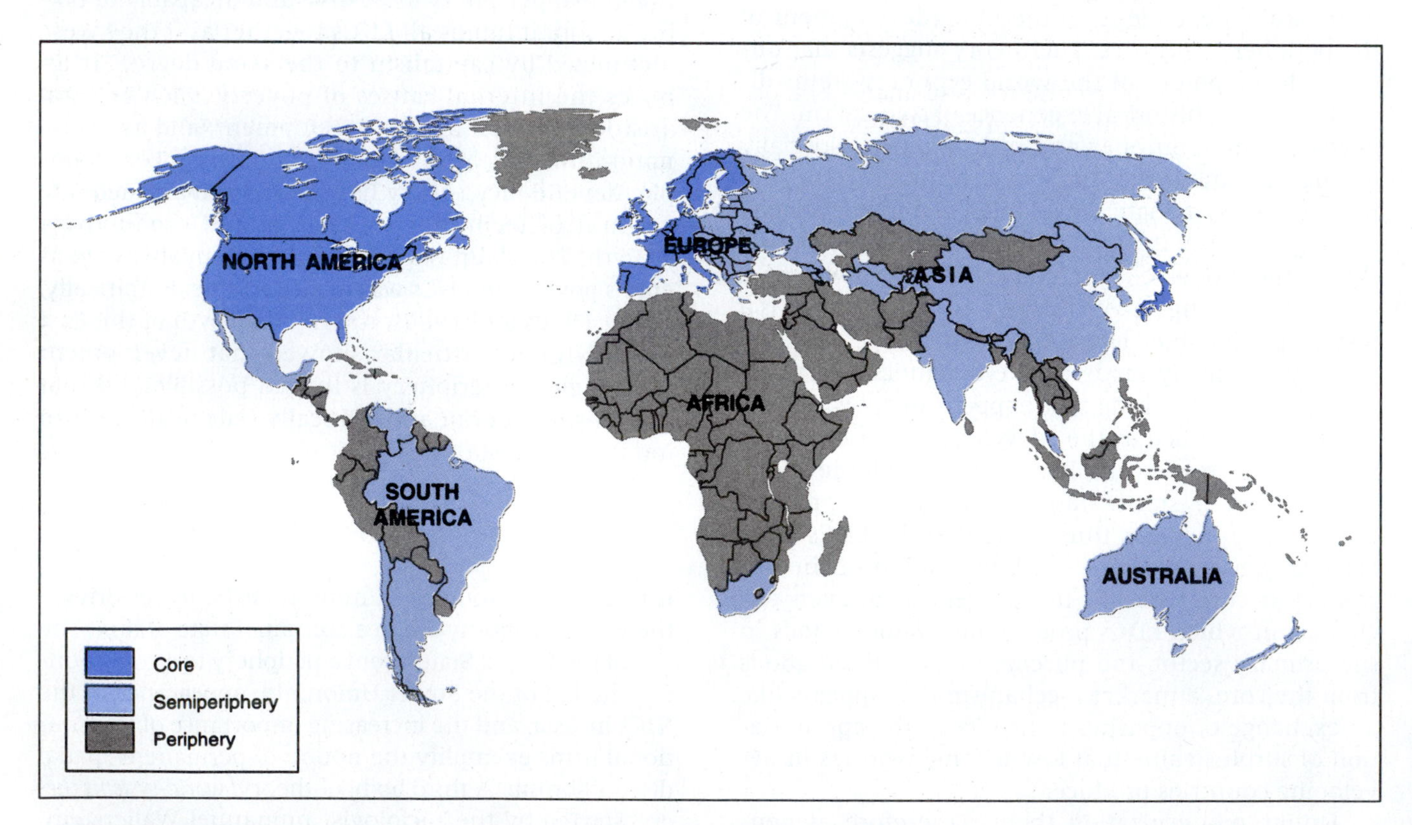

FIGURE 14.20
The geography of the world system. In world-systems theory, the global network of states and markets is dominated by a hegemonic power and a core of powerful, wealthy countries in Europe, North America, and Japan. The rest of the world is divided into an impoverished periphery, which produces raw materials, and a semiperiphery, including the newly industrialized countries, which has aspects of both the core and periphery. This approach recognizes the movement of states up and down the world system as global capitalism creates new layers of uneven development worldwide.

and a simple production mix. Consumption is low. In between are states that are part of the semiperiphery where both sets of processes coexist to a greater or lesser degree. The theory suggests that the semiperiphery countries are exploited by the core countries with regard to raw material and product flow while at the same time exploiting periphery countries.

World-systems theory pays particular attention to the role of a single hegemon that dominates the global political and economic system. The hegemon "sets the rules of the game," so to speak. During the period of Spanish dominance in the sixteenth and seventeenth centuries, for example, mercantilism was the dominant ideology. Under the Pax Britannica of the nineteenth century, free trade was the norm. And since the rise to dominance of the United States, especially since World War II, neocolonialism has been the typical pattern (although during the cold war there were two superpowers and a bifurcated world system). Hegemonic powers may overextend themselves militarily, leading to an erosion of their economic base. When powers in the core have conflicts among themselves, they open opportunities both for new hegemons and for countries on the periphery. The Napoleonic Wars of the early nineteenth century and World War II in the twentieth were thus openings for nationalist anticolonial movements worldwide, first in Latin America and later in Asia and Africa.

Hegemony exists when one core power enjoys supremacy in production, commerce, and finance and occupies a position of political leadership. The hegemonic power owns and controls the largest share of the world's production apparatus. It is the leading trading and investment country, its currency is the universal medium of exchange, and its city of primacy is the financial center of the world. Because of political and military superiority, the dominant core country maintains order in the world system and imposes solutions to international conflicts that serve its self-interests. Britain played this role in the nineteenth century, and the United States has done so since World War II. Consequently, hegemonic situations are characterized by periods of relative peace (e.g, the nineteenth century). During a power's decline from hegemony, rival core states, which can focus on capital accumulation without the burden of maintaining the political and military apparatus of supremacy, catch up and challenge the hegemonic power. Thus, in the early twentieth century, Germany challenged Britain for global leadership, with catastrophically violent results.

Regional Disparities within Developing Countries

In addition to the bifurcation between the developed and underdeveloped worlds, which forms the primary axis of the global economy, there are also profound spatial differences within developing countries. The geographies of colonialism, including profound rural-urban differences, are one major dimension of this predicament. Major cities of developing countries operate largely as export platforms linking the rich industrial countries and their sources of raw materials. Modern large-scale enterprises tend to concentrate in capital and port cities. Injections of capital into these urban economics attract new migrants from rural areas and provincial towns to principal cities. Migrants, absorbed by the system, are maintained at minimal levels. There is little incentive to decentralize urban economic activities. The markedly hierarchical, authoritarian nature of political and social organization retards the diffusion of ideas throughout the urban hierarchy and into rural areas.

Because economic landscapes are produced by social relations, regional inequalities within developing countries are reflections of social inequalities. The class relations in much of the less developed world are often typified by a small, powerful elite and large numbers of poor peasants, with a small middle class. Countries with highly unequal distributions of income, such as Brazil, tend to have highly unequal standards of living among their various regions. Those countries that have achieved more equality economically, such as the NICs of East Asia, tend to have fewer disparities spatially. In this way, economic landscapes mirror and contribute to social bifurcations, revealing how geographies are socially produced and socially producing.

The center-periphery concept echoes the marxist argument that the center appropriates to itself the surplus of the periphery for its own development. The center-periphery phenomenon may be regarded as a multiple system of nested centers and peripheries. At the world level, the global center (rich industrial countries) drains the global periphery (most of the underdeveloped countries). But within any part of the international system, within any national unit, other centers and peripheries exist. Centers at this level, although considerably less powerful, still have sufficient strength to appropriate to themselves a smaller, yet sizable, fraction of remaining surplus value. A center may be a single urban area or a region encompassing several towns that stand in an advantageous relation to the hinterland. Even in remote peripheral areas local, regional imbalances are likely to exist, with some areas growing and others stagnating or declining.

There are reverse flows from the various centers to the peripheries—to peripheral nations, to peripheral rural areas. Yet these flows, themselves, may further accentuate center-periphery differences. For example, World Bank, U.S. Aid for International Development (USAID), and International Development Association (IDA) loans support major infrastructural projects such as roads and power stations, which are proven money earners and reinforce the centrality and drawing power of cities and the export sectors. USAID strongly supports projects dealing with agriculture, health and family planning, school construction, and road building; industrialization projects are seldom financed.

Development Strategies

The economically developed countries must come to the aid of LDCs today. How can this occur peacefully? Three methods are generally cited for developed countries to help LDCs: (1) expand trade with LDCs, (2) invest private capital in LDCs, and (3) provide foreign aid to LDCs.

Expansion of Trade with Less Developed Countries

Economists suggest that expanding trade with LDCs is one way to help them. It is true that reducing tariffs and trade barriers with LDCs will improve their situation somewhat. Free trade, therefore, can have its costs. Eliminating protectionism levied against developing country producers gives them access to the large, wealthy markets in the United States, Canada, Europe, and Japan, increasing their export volumes and revenues. With the North American Free Trade Agreement (NAFTA), the United States removed all trade barriers with Mexico, for example. Mexican manufacturing flourished as a result. However, trade liberalization also opened the doors for U.S. imports, particularly heavily subsidized agricultural products, which wreaked havoc with Mexican farmers.

Private Capital Flows to Less Developed Countries

LDCs are also a destination for investments from MNCs, private banks, and large corporations. Some of these are foreign direct investment. For example, major U.S. automakers have now built numerous plants in Brazil and Mexico. In Tijuana alone, 500 U.S. labor-intensive manufacturing plants now take advantage of the low hourly wage rate. Other types of capital flows are purely financial: Citicorp and Chase Manhattan Bank have made loans to numerous LDC governments. Since the debt crisis of the 1980s and 1990s, however, investments and private capital flows have decreased substantially because of concerns about returns on investment and fears of debt default.

An international trade climate must also be supported by financial and marketing systems, a favorable tax rate, an adequately maintained infrastructure, and a reliable flow of sufficiently skilled labor. Often, however, LDCs cannot guarantee that a politically stable environment will prevail, which is difficult in many countries torn by tribal conflicts, ethnic strife, religious struggles, and civil wars. These obstacles often thwart the major capital flows that might otherwise exist. African states in particular have not been able to tap private capital flows from large corporations and commercial banks because of problems with these conditions.

Foreign Aid from Economically Developed Nations

In order to reverse the vicious cycle of poverty, foreign aid is needed in the form of direct grants, gifts, and public loans to LDCs. Capital is necessary to finance companies, build the infrastructure, generate jobs, to increase productivity, and to retrain the labor pool.

In absolute terms, the United States has been a major world player in foreign-aid programs. U.S. foreign aid averages $15 billion per year, for example, less than 1 percent of GDP (well behind the shares of all other developed countries). The majority of this aid was administered through the State Department's USAID. Additional direct aid included food programs to needy countries under the U.S. government's Food for Peace program. Other nations have also rallied. Developed countries as a group contribute a total of $50 billion annually. In addition, OPEC nations contribute $2.5 billion. For many countries in Africa, aid can form as much as 15 percent of their national output (Figure 14.21).

Unfortunately, most U.S. aid has stipulations such as purchase requirements that make the LDCs patronize American products and services. Almost 75 percent of U.S. foreign aid is military in nature, and the vast bulk of it flows to close political allies such as Israel, Egypt, and Turkey on the basis of political ties as opposed to economic need. For example, Israel, Egypt, and Turkey each receive nearly $3 billion in aid per year. These nations are neither the most populous nor the most needy, but they occupy strategic areas of the Middle East where vast oil deposits exist and where the United States struggles against Islamic fundamentalism. In addition, the United States has guaranteed its support of Israel in its hostile environment.

Unfortunately, developed country contributions amount to only 0.5 percent of their collective GDPs. This amount is much too small to make a meaningful difference to the fortunes of the developing world. To make matters worse, Russia and eastern European nations are now making strong pleas for increased aid from the West. Many LDCs fear that aid that would normally be channeled to them will now go toward supporting privatization in former communist areas. The developed countries' fear is that the cost of failure of democratization in Russia will be far greater than the cost of foreign aid. If Europe and America agree, the larger portion of their foreign aid (through grants, loans, and direct aid) will be siphoned from the LDCs.

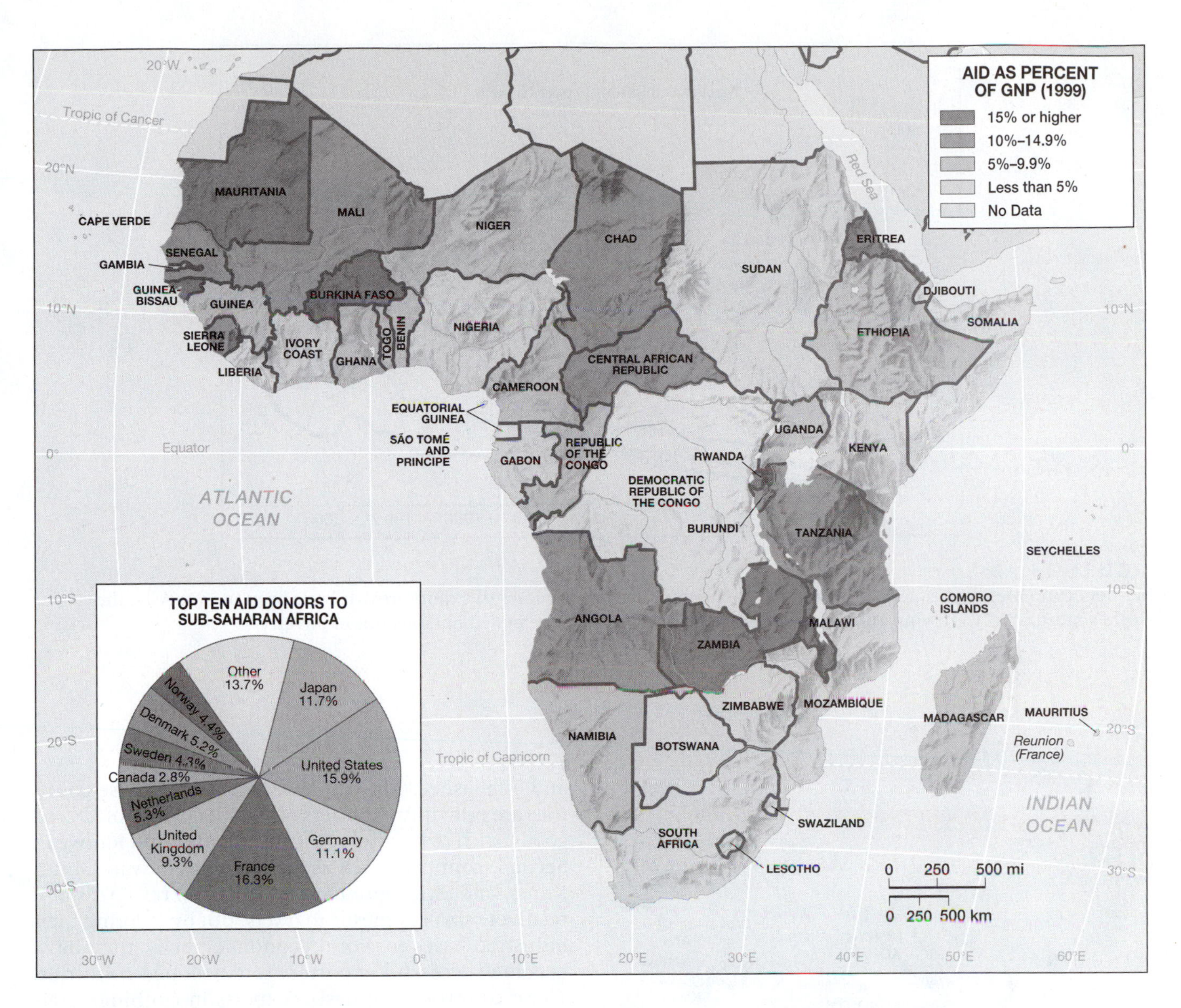

FIGURE 14.21
Aid dependency in Africa. In parts of sub-Saharan Africa, which lack domestic industries and export low-value primary sector goods, foreign aid comprises more than 15 percent of total output. The largest donors are France and the United States. Less than 1 percent of U.S. GDP goes to foreign aid, and most of that is military assistance to allies such as Israel.

Industrialization in the Developing World

Deindustrialization in the economically developed world did not induce widespread industrialization in the developing world. In 2004, 40 countries accounted for 70 percent of manufacturing exports from developing countries; the top 15 alone accounted for about 60 percent. Even more striking is that about one-third of all exports from the LDCs came from four Southeast Asian countries—Hong Kong, South Korea, Singapore, and Taiwan, the original "tigers" of the East Asian miracle (Figure 14.22). As the most rapidly growing economies in the world since World War II, East Asian countries have formed a growing belt of manufacturing that may soon become the largest in the world (Figure 14.23). Most undeveloped countries saw zero or very little manufacturing growth. Industrialization occurred, therefore, only in selected parts of the developing world.

Manufacturing was slowest to take hold in the poorest countries of the periphery, most of which are in Africa. It grew fastest in the newly industrialized countries (NICs) that made a transition from an industrial strategy based on import substitution to one based on exports. The exporters can be divided into two groups. First, countries such as Mexico, Brazil, Argentina, and India have a relatively large domestic industrial base

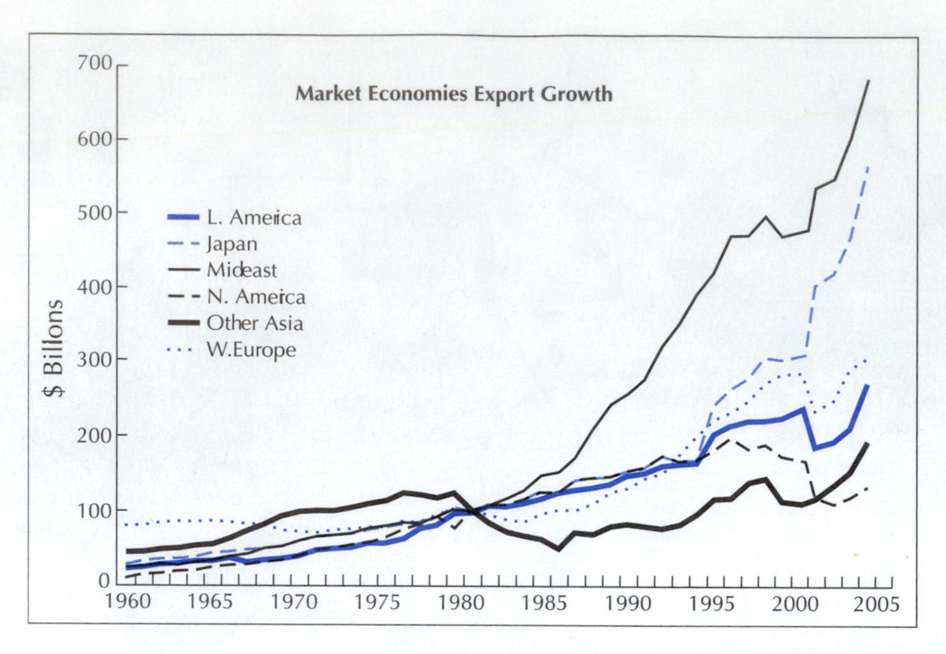

FIGURE 14.22
Market economies export growth, 1960–2004. The most spectacular rates of export growth have been in East Asia, the world's most rapidly growing area economically, where the NICs have pulled millions out of poverty.

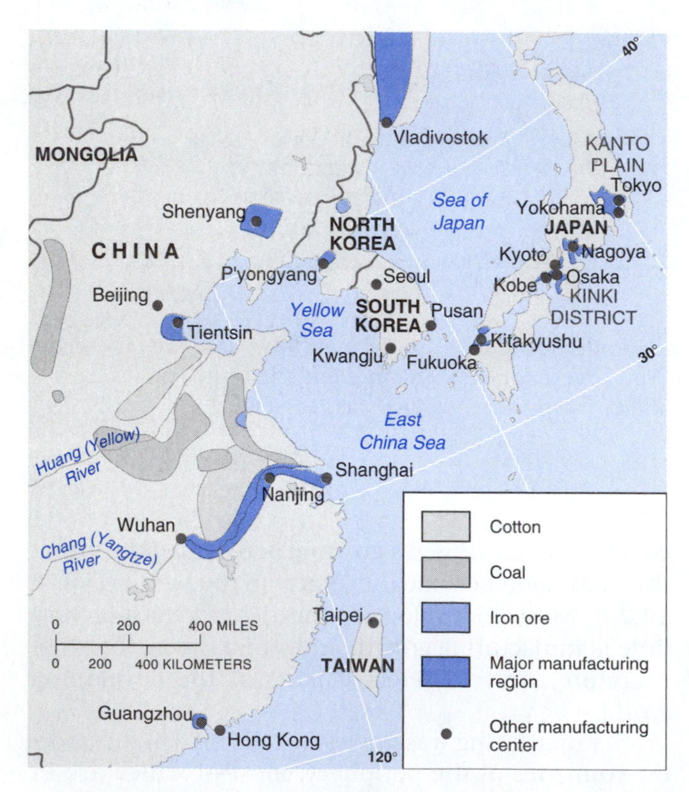

FIGURE 14.23
Manufacturing centers in East Asia. The industrial districts of China, Korea, and Japan are becoming the largest aggregate complex of manufacturing in the world.

and established infrastructure. All four of these countries are primarily exporters of traditional manufactured goods, such as furniture, textiles, leather, and footwear. Second, countries such as Hong Kong, Taiwan, South Korea, and Singapore have few natural resources and relatively small domestic markets. But by tailoring their industrial bases to world economic needs, they have become successful exporters to developed countries. These countries emphasize exports in clothing, engineering, metal products, and light manufactures. Their success encouraged other LDCs to adopt a similar program of export-led industrialization.

Import-Substitution Industrialization

In the post–World War II period, newly independent developing countries sought to break out of their domination by, and dependence on, developed countries. Their goal was to initiate self-expanding capitalist development through a strategy of *import-substitution industrialization*. This development strategy involved the production of domestic manufactured goods to replace imports. Only the middle classes could support a domestic market; thus, industrialization focused on luxuries and consumer durables. The small plants concentrated in existing cities, which increased regional inequalities. These "infant industries" developed behind tariff walls in order to reduce imports from developed countries, but local entrepreneurs had neither the capital nor the technology to begin their domestic industrialization. Foreign multinational corporations came to the rescue. Although projects were often joint ventures involving local capital, "independent" development soon became *dependent industrialization* under the control of foreign capital. Many countries experienced an initial burst in manufacturing growth and a reduction in imports. But after a while, the need to purchase raw materials and capital goods and the heavy repatriation of profits to the home countries of the multinationals dissipated foreign-exchange savings.

Export-Led Industrialization

By the 1960s, it was apparent to many leaders of LDCs that the import-substitution strategy had failed to generate economic growth. Only countries that had made an early transition to *export-led industrialization,* such as the Asian NICs, were able to sustain their rates of industrial growth. Once again, LDC development became strongly linked to the external market. In the past, export-oriented development had involved the export of primary commodities to developed countries. Now, export-oriented development was to be based on the production and export of manufactured goods.

The growth of export-led industrialization coincided with the international economic crisis of the 1970s and 1980s. It took place at a time when the demand for imports in the advanced industrial countries was growing despite the onset of a decline in their industrial bases. It was, in short, a response to the new international division of labor. Export-oriented industrialization tends to concentrate in *export-processing zones,* where four conditions are usually met:

1. Import provisions are made for goods used in the production of items for duty-free export, and export duties are waived. There is no foreign exchange control, and there is freedom to repatriate profits.
2. Infrastructure, utilities, factory space, and warehousing are usually provided at subsidized rates.
3. Tax holidays, usually of five years, are offered.
4. Abundant, disciplined labor is provided at low wage rates.

The first export-processing zone was not established in the developing world but in 1958 in Shannon, Ireland, with the local international airport at its core. In the late 1960s, a number of countries in East Asia began to develop export-processing zones, the first being Taiwan's Kaohsiung Export-Processing Zone, set up in 1965, a strategy soon imitated across East Asia (Figure 14.24). By 1975, 31 zones existed in 18 countries. By 2001, at least 96 zones were established in developing countries. Most of them are in the Caribbean, Central and Latin America, and East Asia.

Central to the growth of LDC manufacturing exports to the developed countries are multinational corporations, which establish operating systems between locally owned companies and foreign-owned companies. The arrangement is known as *international subcontracting* or offshore assembly and *outsourcing.* Although numerous legal relationships exist between the multinational and the subcontractor, from wholly owned subsidiary to independent producer, the key point is that developing country exports to developed countries are part of a unified production process controlled by firms in the advanced industrial countries. For example, Sears Roebuck Company might contract with an independent firm in Hong Kong or Taiwan to produce shirts, yet Sears retains control over design specifications, advertising, and marketing.

Consequences of Export-Led Industrialization for Women

Export-led industrialization moves work to the workers instead of workers to the work, which was the case during the long postwar boom. In some countries this form of industrialization has generated substantial employment. For example, since their establishment, export-processing zones have accounted for at least 60 percent of manufacturing employment expansion in Malaysia and Singapore. However, in general, the number of workers in the export-processing zones' labor forces is modest. It is unlikely that these zones employ more than 1 million workers worldwide.

Much of the employment in export-processing zones is in electronics and electrical assembly or in textiles. Young, unmarried women make up the largest part of the workforce in these industries—nearly 90 percent of zone employment in Sri Lanka, 85 percent in Malaysia and Taiwan, and 75 percent in the Philippines. Explanations for this dominance of women in the workforce vary; it is often attributed to sexual stereotyping, in which the docility, patience, manual dexterity, and visual acuity of female labor are presupposed. Of more significance is the fact that women are often paid much less than men are for the same job. Thus the cheap labor so essential to the labor-intensive industries of the global assembly line also reproduces the patriarchal relations that keep many women in particularly low-paying jobs.

Export-led industrialization may lead to an imposed economic system at odds with the cultural and political institutions of the people that it exploits. Often people produce things that are of no use to them. How they produce has no relation to how they formerly produced.

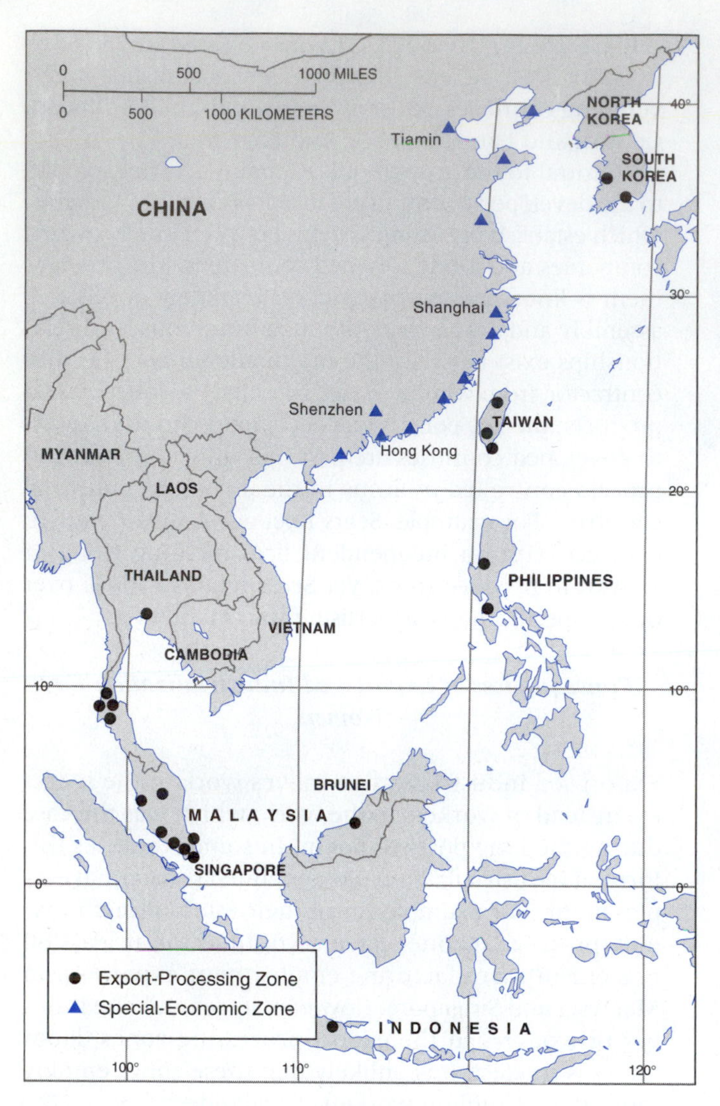

FIGURE 14.24
Export processing zones in East and Southeast Asia. These platforms, constructed by governments to attract foreign capital and facilitate exports, have become widely adopted throughout the developing world as part of the broader shift from import substitution to export-led industrialization.

Workers are often flung into an alien labor process that violates their traditional customs and codes. For example, female factory workers often pay a high price for their escape from family and home production, especially in Asia, where women's family roles have been traditionally emphasized. Because of their relative independence, Westernized dress, and changed lifestyles, women may be rejected by their clan and find it hard to reassimilate when they can no longer find employment on the assembly line.

Although export-oriented industrialization leads to growth in production and employment, as well as to increases in foreign exchange, it will not lead to the creation of an indigenous, self-expanding capitalist economy. The economic linkages between export platforms and local economies are minimal. Some scholars cite South Korea as an example of a country that has completed a successful transition to industrial capitalism. But so far Korean industrial expansion has not taken place because of domestic demand. Rather, it has occurred because the Koreans have sought to increase exports and international competitiveness. This expansion is changing, partly because of the general global tendency to stagnate.

Economic stagnation in developed countries is a major concern of developing countries that have enjoyed success with the export-led industrialization strategy. For developed countries, where production and investment are moving out, purchasing power will be lost. The resultant spiraling down of general economic activity will choke off dependent industrialization and increase poverty and suffering for workers and peasants in LDCs.

Sweatshops

Some Guess clothing is made by suppliers that use underpaid Latino immigrants in Los Angeles, sometimes working in their own homes. Mattel makes tens of millions of toys each year in China, where young female Chinese workers who have migrated hundreds of miles from home are alleged to earn less than the minimum wage of $1.99 a day. Nike is criticized for manufacturing many of its shoes in tough labor conditions in Indonesia, and some of Disney's hottest seasonal products are being made by suppliers in Sri Lanka and Haiti—countries with unsavory reputations for labor and human rights. Soccer balls are sewn together by child laborers in Pakistan, who work up to 12 hours per day. In an era of the global economy, it is impossible for consumers to avoid products made under less than ideal labor conditions. Moreover, what may appear to be horrific working environments to most citizens in the world's richest nation are not just acceptable but actually attractive to others who live overseas or even in Third World pockets of the United States. Anyone even casually familiar with how some Americans recompense their (usually immigrant) housekeepers understands their desire to work and support their families.

One icon of American culture whose manufacturing practices seem out of sync with its brand name is Disney. Disney maintains almost 4000 contracts with other companies that assume the right to manufacture Disney paraphernalia, some of which are then sold in Disney stores. These licensees go to some of the world's lowest-cost labor countries, including Sri Lanka and Indonesia, to produce stuffed animals and clothes. Disney itself rarely takes a direct hand in manufacturing. Sears, which carries 200,000 products from manufacturers operating in virtually every country, is tightening up on buying goods from suppliers with dubious records. The Gap, after enduring criticism, also has become a model for manufacturing and sourcing products abroad. In contrast to Disney, Mattel does most of its own manufacturing. It makes a staggering 100 million Barbie dolls a year in four factories, two in China and one each in Malaysia and Indonesia. The Barbie craze produced $1.4 billion in annual revenues for the firm.

Does a global economy mean consumers face no choice but to buy products made under conditions Americans don't want to think about? A number of U.S. companies say that intense global competition is no excuse for keeping working standards at the lowest possible level. Levi Strauss, for example, imposes its own "terms of engagement" on manufacturers who make its jeans products in 50 countries.

In many ways, what Americans buy is their most direct and intimate connection with the global economy. In a post–cold war era in which governments seem to be losing their power to shape the lives of people, U.S. consumer spending can be an important tool in extending American values. The silver lining is that if Americans respond to even some of these concerns, they could enjoy their shopping and improve the conditions that millions of people around the world encounter in their daily lives.

The East Asian Economic Miracle

What does it take to turn an LDC into a developed nation? Who is marching successfully forward to development? The most successful examples have been the trading states of East Asia. South Korea, Taiwan, Singapore, and Hong Kong followed the path pioneered by nearby Japan, which began to modernize in the nineteenth century. In the 1980s and 1990s, Malaysia, Thailand, Indonesia, and the Pacific coast of China headed down the same road. After the devastation of World War II, Japan's economy was in ruins. Today, Japan's economy is the second largest in the world, two-thirds the size of that of the U.S. economy (Japan's population is only half as large). In the mid-1960s, South Korea was a land of traditional rice farmers who made up over 70 percent of the country's workforce. Its GDP per capita—$230—was the same as Ghana; today, South Korea's GDP per capita was 20 times larger, and over 70 percent of its people lived and worked in urban areas rather than on farms. South Korea's economy is now the eleventh most powerful in the world, ahead of such countries as Sweden, the Netherlands, and Australia. South Korea has become the world's largest shipbuilding nation and the world's fifth largest auto manufacturer. Its iron and steel and chemical industries are thriving, and with the largest number of PhDs per capita in the world, South Korea has become a formidable competitor in research and development of semiconductors, information processing, telecommunications, and civilian nuclear energy. Few other countries have achieved as much economically in so short a time.

Making a Jackie Kennedy doll in China. The use of low-wage labor in developing countries may generate jobs for people, cheap goods for consumers in the developed world, and profits for multinational corporations, but it also frequently involves the blatant exploitation of human beings working long hours under horrendous conditions.

How has East Asia emerged as the hub of the increasingly prosperous Pacific Rim? There are several characteristics that these societies share, which, taken together, help to explain their sustained economic growth. In addition, unique circumstances in the world political economy helped to facilitate their growth.

First, East Asian countries have exhibited an enormous commitment to education. Some of this phenomenon may be derived from the Confucian respect for learning and scholarship. East Asian educational mores encourage a docile, well-trained workforces, often consisting of easily exploited, low-wage female labor, means that they have relatively few days lost to strikes and labor unrest.

Second is a high level of national savings. East Asian governments have encouraged personal savings by restricting the movement of capital abroad, maintaining low tax rates while keeping interest rates above the rate of inflation, and limiting the importation of foreign luxury goods. The result has been the accumulation of large amounts of low-interest capital that allowed Asian countries to finance education, infrastructure, manufacturing, and commerce. Many countries in East Asia save one-third or more of their GDP (in America, in contrast, the domestic savings rate is between zero and 3 percent of the GDP).

Only recently, after economic take-off was well underway, have East Asian governments loosened financial policies to allow increased consumption and capital investment in consumer durables like new homes. Such purchases, rather than savings, may finance Asian prosperity. As Asian economies mature and their populations age, it is possible that their saving rates decline and imports may increase. Will East Asians tend more and more to consume rather than create wealth, as Americans tend to do?

Third, East Asia has enjoyed a strong political framework within which economic growth is fostered. Industries targeted for growth were given a variety of supports—export subsidies, training grants, and tariff protection from foreign competitors. Low taxes and energy subsidies assisted the business sector. Trade unions were restricted and democracy was constrained. In Japan, powerful government bureaucrats largely beyond public control promoted industrial expansion with little regard for the opinions or needs of Japanese consumers. Military governments in South Korea and Taiwan dealt harshly with industrial unrest and political dissidents. Authoritarian regimes long ruled Singapore. In no way did Asian governments follow a laissez-faire model; it is a common myth that the Asian NICs demonstrate a "free market" in operation. It is only lately that multiparty politics have been permitted outside Japan.

Fourth, the NICs engaged in widespread land reform in the 1950s. In part, this was made possible by the turmoil of the Japanese occupation and the Korean war, which dislodged the rural aristocracies that owned much of the arable land. As a result, rural land ownership in these nations is relatively egalitarian, with high rates of productivity, unlike Latin America, which still suffers from the legacy of the Spanish land grant system. Besides a thriving agricultural economy, which has dwindled in the face of rapid industrialization, the NICs have benefited from comparatively low rates of rural-to-urban migration, which helps to prevent the urban areas from being flooded by desperate peasants seeking to escape poverty.

Fifth, East Asian NIC countries exhibited a sustained commitment to exports (export-led industrialization) in contrast to the import-substitution policies that characterized India, Africa, and Latin America until very recently. Instead of encouraging industrialists with low labor costs to target foreign markets and compete there, governments in India, Latin America, and Africa decided to protect their economies rather than open them to international competition. They shielded firms from foreign competition by protective tariffs, government subsidies, and tax breaks. As a result, their products became less competitive abroad. While it was relatively easy to create a basic iron and steel industry, it proved harder to establish high-tech industries such as computers, aerospace, machine tools, and pharmaceuticals.

Most import-substituting states depend on imported manufactured goods, whereas exports still chiefly consist of raw materials such as oil, coffee, and soybeans, a condition that creates poor *terms of trade*. A country that relies on the export of raw materials—mineral and agricultural commodities, "rocks and crops"—with little or no value added to the products through finishing or manufacturing, and then has to import expensive (because of the immense value added) high-tech products, is not headed toward development unless the country commits its earnings to investment in quality exports and competitive high-tech manufacturing. Such countries need to get the fundamentals right: Keep inflation low and fiscal policies prudent, maintain high savings and investment rates, improve the educational level of the population, trade with the outside world, and encourage foreign investment.

Throughout Latin America and Africa that sort of economic strategy was often missing. Governments poured money into state-owned enterprises, large bureaucracies, and oversized armed forces, paying for them by printing money and raising loans from Western banks and international agencies. Public spending soared, price inflation accelerated, domestic capital took flight to safe deposits in American and European banks, and indebtedness skyrocketed. By the 1990s, payments on loans consumed about half of Africa's export earnings. By 1989, Argentina owed developed country banks and governments a staggering $1800 for each man, woman, and child in the country. Zambia's debt rose to 334 percent of the GDP.

Just when Latin America and Africa needed capital for economic growth, countries there found themselves overwhelmed by debt, starved of foreign funds for investment, with currencies made worthless by hyperinflation. By the 1990s, poverty in developing countries outside East Asia had increased dramatically. Lands as well as people paid the price. Forests have been recklessly logged, mineral deposits carelessly mined, fragile lands put to the plow, and fisheries overexploited in a desperate effort to escape the poverty trap.

In addition to their domestic structures, their position in the world economy also played a role in the growth of the Asian NICs. Thus, not all of the success of the NICs is due to internal, domestic factors. For example, despite the brutality of war and the untold suffering it caused, many NICs benefited from the legacy of their occupation by Japan between 1895 and 1945. Japan initiated the steel, chemicals, and textiles industries in Korea and Taiwan, as well as that in Manchuria. The Japanese also built much of the industrial infrastructure in the NICs, including roads, bridges, tunnels, ports, and airports, which, while designed to maximize the extraction of raw materials such as coal, nonetheless became important after the war. Japan also centralized the state bureaucracies of these countries, which displaced the reactionary rural aristocracies that controlled much of the economies and hindered growth. Subsequent to the war, many NICs, particularly South Korea, established imitations of the Japanese corporate zaibatsu (e.g., the Korean chaebol), many of which were closely linked to banks and obtained easy credit. Finally, Japanese foreign investment in East Asia, which exceeds that of the United States, also has accelerated the industrialization of these countries.

In addition to Japan, the United States also is responsible for much of the growth of the NICs. Throughout the cold war (1945–1991), the United States provided generous economic and military aid, freeing resources that could be harnessed for economic development. The Containment Doctrine, also known as the Truman Doctrine, positioned Japan, South Korea, Taiwan, and, to a lesser extent, other countries, as frontline states in the war against communism. The United States gave copious grants and subsidies to the NICs and awarded them preferential trade status, such as exemptions from tariffs, which allowed them easy access to the American market and significant export earnings. The roles of the United States and Japan are important in a theoretical sense as well. To some extent, the geopolitical conditions that enabled the take-off of the NICs were a unique constellation of circumstances that could not be duplicated elsewhere.

In the 1990s, the original Four Tigers of South Korea, Taiwan, Hong Kong, and Singapore were joined by a new group, including Thailand, Malaysia, and Indonesia. Each of them replicated the experience of the original NICs to one extent or another. Thailand, which long enjoyed close economic and military ties to the United States, saw a wave of investment in textiles, automobiles, toys, and tourism, propelling Bangkok into a prosperous but crowded urban center. Malaysia, under an authoritarian government, launched its "Plan 2020" to become a fully industrialized country by the year 2020. Already this strategy has succeeded, at least in the western half of the country. The country is the world's largest exporter of refrigerators and semiconductors and has close ties to banks and firms in Singapore as well. Kuala Lumpur has become a thoroughly modernized city. Indonesia, struggling to escape decades of poverty, has enjoyed some of the growth experienced in the Singapore–Kuala Lumpur corridor, including investments in textiles, shoes, electronics, and even aerospace.

Finally, no discussion of the NICs could be complete without mentioning China, the "800-pound gorilla" looming behind all of the NICs. Following decades of isolation under the communists, China began to open itself to the world economy in the late 1970s under the leadership of Deng Xiaoping. The policies implemented in the 1980s revolutionized the economy and society of the most populous nation in the world, encouraging the growth of private property and markets. Whereas the communists shunned contacts with the West, the Chinese in the last 20 years have welcomed it, making the country the single largest recipient of foreign direct investment in the developing world. The government targeted coastal areas in particular, designating them special economic zones in which investors were showered with subsidies and tax breaks. These regions include Guangdong province, located near and benefiting greatly from its ties to Hong Kong (e.g., cities such as Shenzen); Fujian province across the straits from Taiwan; and Pudong, the financial center near Shanghai (Figure 14.25). In such regions, industries such as electronics, toys, garments, and other types of light industry have exploded. Indeed, China's economy has grown an average of 8 percent annually for the last two decades, one of the highest rates in the world, and China enjoys large trade surpluses with most of the developed world. This growth has been geographically

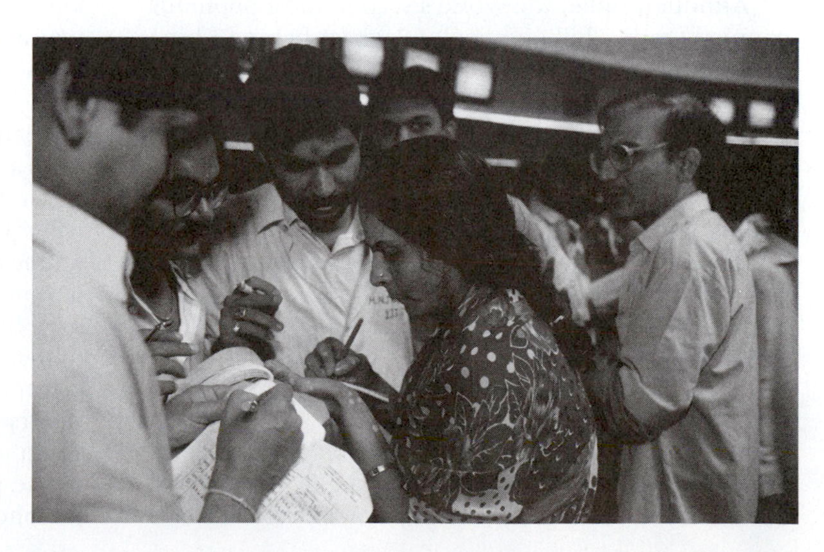

Mumbai (Bombay), India, stock exchange provides evidence for this city's central role in the rapidly globalizing South Asian economy.

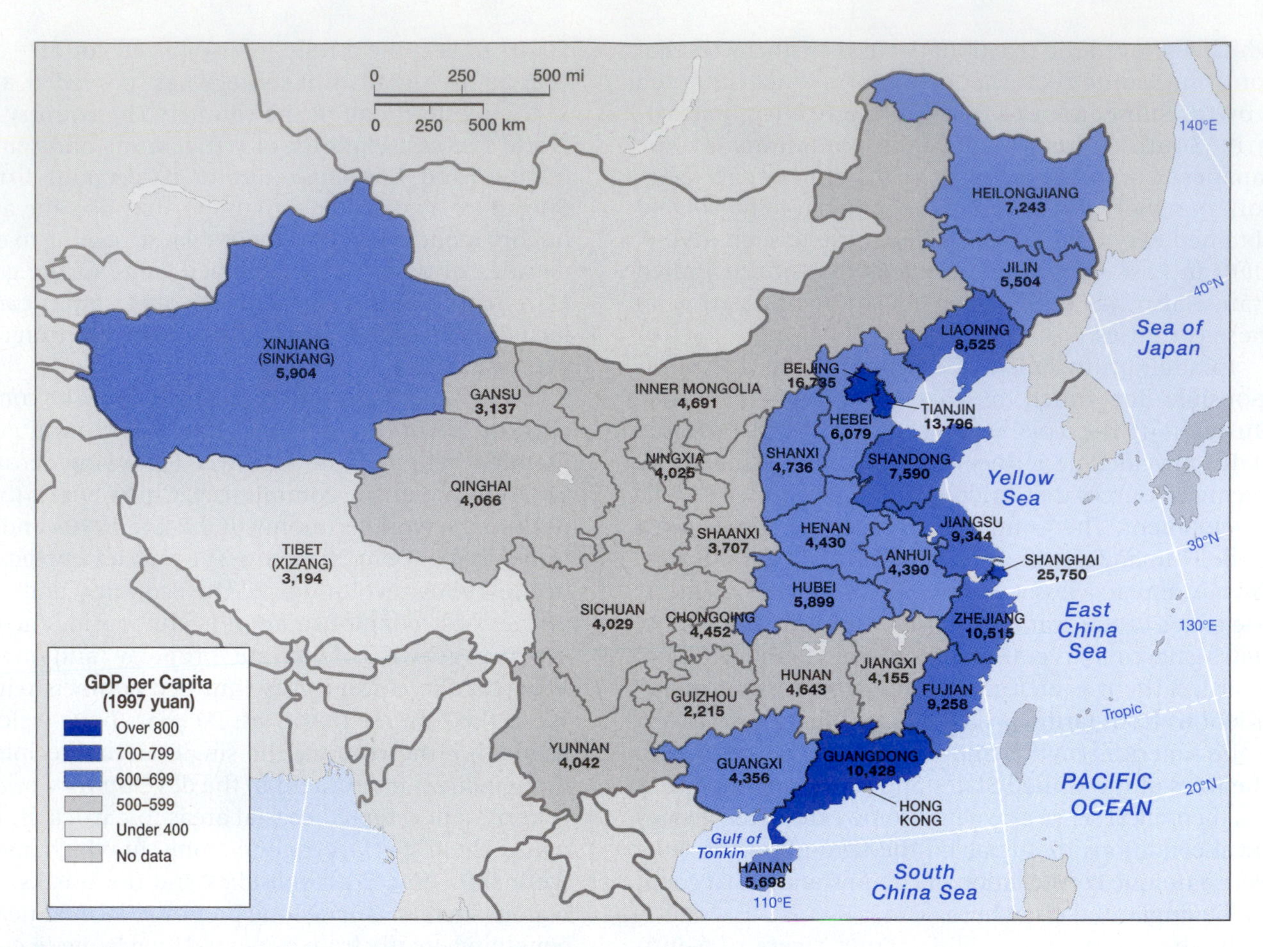

FIGURE 14.25
Unequal development in China as measured by GDP per capita. As China reentered the global economy in the late twentieth century and rapidly adopted market based economic systems, it experienced a surge in social and spatial inequality. The most rapid growth has been along the coasts, most famously in the southern province of Guangdong, while much of the country's interior remains mired in poverty. Not surprisingly, the labor force has responded to these geographic differentials by moving in large numbers to the cities, where recent migrants are often vulnerable to exploitation.

uneven (i.e., as a form of uneven spatial development), and in response it has attracted waves of peasants moving from less densely inhabited, poorer regions in the interior toward the prosperous coasts (Figure 14.25).

Although India, the world's second most populous country (with 1 billion people), lags far behind China in its level of economic development, India too has seen its pace of economic growth recently increase. Much of this growth is located on the western half of the Indian peninsula (Figure 14.26), including financial centers such as Mumbai (formerly Bombay), one of the world's largest cities, and the famous software complex in Bangalore, India's answer to Silicon Valley. Indeed, India today is the world's largest producer of software as well as films.

SUMMARY

In this chapter, we considered the issue of Third World development. We began by noting how slippery the term *development* is and that there are a variety of ways to measure it. We discussed goals for development by listing objectives that are by and large universally endorsed. We then surveyed the locations of underdevelopment in various regions of the developing world. Next we explored typical characteristics and development problems of LDCs—overpopulation, lack of resources, capital shortages, insufficient foreign revenues and poor terms of trade, unequal land distribution, inadequate infrastructures and public services, and corrupt governments.

In discussing major perspectives on development, we saw how modernization theories, which stress economic growth and Westernization, have obscured important aspects of underdevelopment, particularly the long history and effects of colonialism and neocolonialism. Two major views derived from political economy—dependency theory and world-systems theory—explain why the Third World does not develop, attributing this lack of development to the role of the planet's core in dominating and exploiting its periphery.

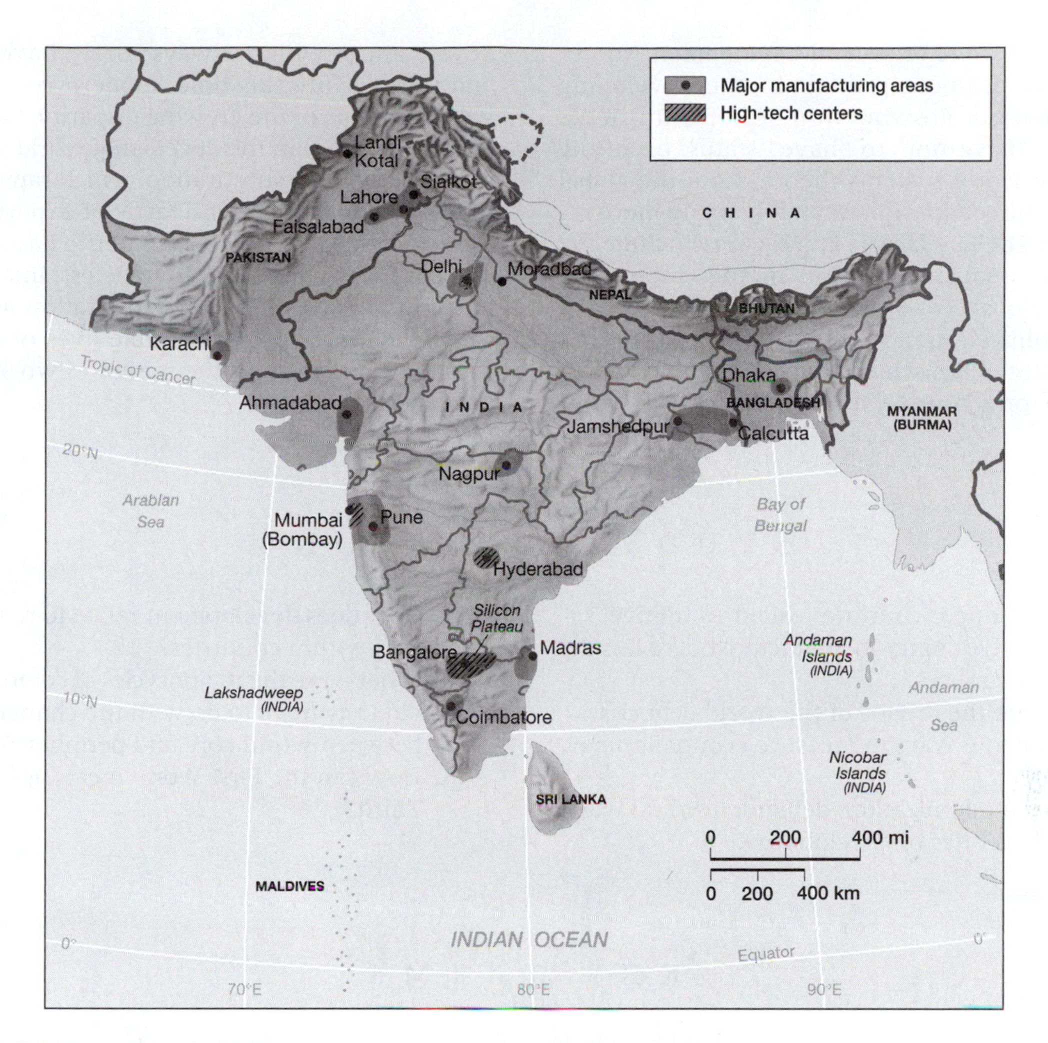

FIGURE 14.26
Uneven development in India. As India deregulated its economy and lured foreign investment, it too experienced uneven growth. The western parts of the country (including the financial and media center of Mumbai formerly Bombay) and the software district of Bangalore have done much better than the eastern half.

Bangalore, India's Silicon Plateau. With a growing population of well-educated workers, India has become the largest software producer in the world. Many of California's Silicon Valley firms use engineers and programmers in India.

Finally, we turned to several development strategies that may promote growth in the developing world. A small but growing number of countries is moving from "have-not" to "have" status, or, in the framework of world-systems theory, from the global periphery to the semiperiphery, while many more remain behind. The key factors at play here include the history of colonialism, position in the commodity chains of the global economy, cultural attitudes toward education, political stability, and capacity to carry out long-term plans, all of which shape economic performance from one country to another. The race to development will as always surely have its winners and losers. Only this time, modern communications will remind us of the growing disparity. We noted how industrialization in the developing world started under policies of import substitution, which rapidly gave way to the far more successful tactic of export promotion. This shift was most successful in the East Asian newly industrialized countries. Nonetheless, industrialization has brought with it brutal exploitation as in the case of sweatshops. The comfortable lives of many in the West often depend on the misery of workers in many poor countries.

STUDY QUESTIONS

1. Why aren't poor countries called "primitive"?
2. What are five common characteristics of less developed countries?
3. What were the origins of the world debt crises?
4. What are four ways to measure economic development?
5. Contrast modernization, dependency, and world systems theories of development.
6. How does development relate to regional disparities within countries?
7. What were the major cycles of colonialism? What did they have to do with the change in relations between world core and periphery?
8. How can the First World meaningfully assist the Third?

KEY TERMS

backwash effect
basic needs
capital flight
circular and cumulative causation
colonial division of labor
colonial organization of space
core-periphery
dependency theory
development
direct foreign aid
dual economy
export-led industrialization
import substitution
less developed countries (LDCs)
modernization theory
neocolonialism
newly industrializing countries
primary sector of the economy
privatization
secondary sector of the economy
squatter settlements
tertiary economic activity
trickle-down effects
underdevelopment
underemployment
world-systems theory

SUGGESTED READINGS

Barber, B. 1996. *Jihad v. McWorld.* New York: Ballantine.

Bhagwati, J. 2004. *In Defense of Globalization.* Oxford: Oxford University Press.

Chase-Dunn, C. 1998. *Global Formation: Structures of the World-Economy,* 2d ed. Lanham, Md.: Rowman and Littlefield.

Friedman, T. 2005. *The World Is Flat: A Brief History of the Twenty-first Century.* New York: Farrar, Straus and Giroux?

National Research Council, 2003. *Cities Transformed: Demographic Change and Its Implications in the Developing World.* Washington, D.C.: National Academies Press.

Porter, P., and Sheppard, E. 1998. *A World of Difference Society, Nature, Development.* New York: Guilford Press.

So, A. 1990. *Social Change and Development: Modernization, Dependency, and World-System Theories.* Newbury Park, Calif.: Sage.

Stiglitz, J. 2002. *Globalization and Its Discontents.* New York: W. W. Norton.

WORLD WIDE WEB SITES

TERRA: BRAZIL'S LANDLESS MOVEMENT—NEW YORK TIMES [REALPLAYER]

http://www.nytimes.com/specials/salgado/home/

The *New York Times* has recently opened this Web special by documenting the plight of Brazil's landless in both words and pictures.

ATLAPEDIA

http://www.atlapedia.com/index.html

Atlapedia Online contains key information on every country of the world. Each country profile provides facts and data on geography, climate, people, religion, language, history, and economy, making it ideal for students of all ages.

UNITED NATIONS HUMAN DEVELOPMENT

http://www.undp.org

Report includes excellent information for developing countries.

CIA FACT BOOK

http://www.odci.gov/cia/publications/95fact/

This site includes data from every country in the world with a map.

LIBRARY OF CONGRESS

HTTP://WWW.LOC/.GOV

Astounding variety of resources and exhibits.

GLOSSARY

absolute advantage The ability of one country to produce a product at a lower cost than another country.

abstract space A geographic space, homogeneous in all respects. Movement over this space is equally easy in all directions.

accessibility index A measure of the shortest path between one vertex and all others.

accessibility A measure of aggregate nearness. It refers to the nearness of a given point to other points.

achieved characteristics Sociodemographic characteristics, such as education, occupation, income, marital status, and labor force participation, over which we have some degree of control.

acid rain Acid rain, snow, or fog derives from the combustion of coal, releasing sulfur and nitrogen oxides that react with water in the earth's atmosphere.

agglomeration economies The benefits gained by firms by clustering near other firms, including reduction of transport costs of inputs and outputs, access to specialized labor and ancillary services, access to specialized information, and the ability to access a particular type of infrastructure.

agribusiness Food production by commercial farms, input industries, and marketing and processing firms that contribute to the total food sector.

American Manufacturing Belt The manufacturing core region of the United States extending from Boston westward through upstate New York, southern Michigan, and southeastern Wisconsin. At Milwaukee it turns south to St. Louis, then extends eastward along the Ohio River valley to Washington, D.C.

artificial intelligence The application of new decision-making technologies with the aid of computers.

assembly costs The costs of bringing raw materials together at a factory.

assembly plants Plants where components are assembled together.

baby boom The dramatic rise in the U.S. birthrate following World War II, between the years 1946 and 1964.

baby bust The years immediately following the baby boom in which U.S. fertility rates fell dramatically, or any period of low population growth.

backhaul A carrier's return trip.

backward integration The process of purchasing productive capacity "upstream" in the production process, that is, in the creation of inputs, through establishing a unit in-house or purchasing an existing company.

backward integration A firm takes over operations that were previously the responsibility of its suppliers.

backwash effect The phenomenon of perpetuating growth in expanding regions and retarding growth elsewhere.

balance-oriented life-style A mind-set that insists that because resources are finite, they must be recycled, and input rates slowed down to prevent ecological overload.

basic sector The part of an urban or regional economy engaged in the export of goods or services to clients located elsewhere.

biomass The total dry weight of all living organisms within a unit area; plant and animal matter that can in any way be used as a source of energy.

birthrate (crude) The number of live births per 1000 population per year.

brain drain Less developed countries lose talented people to industrially advanced nations through immigration politics.

break-of-bulk point The stage at which a shipment is divided into parts. This typically occurs at a port where the shipment is transferred from water transport to land transport.

Bubonic plague The massive epidemic that swept through Europe from 1347 to 1351, killing roughly 25 million people and forever changing the continent.

burghers The emerging merchant class of early capitalist Europe, which grew wealthy and powerful from the expanding markets and trade routes.

business process reengineering (BPR) BPR includes management realignments, mergers, consolidations, operation integration, downsizing, and reorienting distribution practices to become more efficient and more profitable in the world economy.

capital A factor of production, including tools, buildings, and machines used by labor to fashion goods from raw material.

capital controls Restrictions on the movement of money or capital across national borders.

capital flight Local individuals and whole countries have invested their monies in overseas ventures and in foreign banks for safekeeping.

capital intensive A term that applies to an industry in which a high proportion of capital is used relative to the amount of labor employed.

capitalism The social system in which markets and production for profit are the primary (but not only) means of organizing resources. Ownership of private property is a key institutional characteristic. The importance of markets varies over time and space, giving rise to many different forms of capitalism, and the state always plays a role.

capital-labor relations The relations between the labor force and the capitalists that influence locational decision making.

capital resources Money supply and availability to finance an industry.

carrying capacity The maximum population an ecosystem can support.

cartel An organization of buyers and sellers, capable of manipulating price and/or supply.

circular and cumulative causation Myrdal's theory that continuing changes move the system in the same direction as initial changes, but much further, resulting in industrial and urban agglomeration and innovation.

circular flowing market system The flow of labor and resources from households to firms and the flow of products and wages from firms to households.

colonial division of labor The form of labor specialization imposed on underdeveloped countries by colonial powers.

colonial organization of space The European organization and zoning of land at all scales—urban, regional, national—to serve Europe's own interests during the colonial period.

command economy A society in which a central authority or government establishes the rules of economic behavior and decision making and usually owns the means of production.

commercial agriculture Agricultural goods produced for sale in the city or on the international market.

commodities Goods and services that are produced for sale on a market (i.e., to make a profit). Most goods under capitalism are commodities, but not all are (e.g., public sector goods or those that have not been commodified such as air).

common market A form of regional economic integration among member countries that disallows internal trade barriers, provides for common external trade barriers, and permits free factor mobility.

Commonwealth of Independent States (CIS) Twelve of the former Soviet Union's fifteen socialist republics (all except the three Baltic states), the largest of which is Russia.

comparative advantage The theory that stresses relative advantage, rather than absolute advantage, as the true basis for trade. Comparative advantage is gained when countries focus on exporting the goods they can produce at the lowest relative cost.

connectivity A measure of the interrelation between places in a network.

conservation Careful management and use of resources to assure continuing availability in the longer run.

consumer services Services oriented toward households and individuals rather than firms, such as retail trade and personal services (e.g., haircutters).

contingent labor Workers who are employed part time or temporarily rather than full time.

core-periphery An economic and spatial relationship between regions and countries where those on the outside export raw materials to industrialized regions at the center. Core regions are self-sustaining whereas peripheral areas are dependent on the core.

cost-space convergence The reduction of travel costs between places as a result of transport improvements.

countertrade The direct exchange of goods or services for other goods and services.

daily urban system Trade areas around cities or employment centers that have a large area of dominant influence.

death rate (crude) Annual number of deaths per thousand population.

deforestation The clearing and destruction of forests (especially tropical rain forests) to make way for expanding settlement frontiers and the exploitation of new economic opportunities.

deindustrialization The economic transformation of manufacturing industry to service occupations.

demand The quantity of a good that buyers would like to purchase during a given period at a given price in a competitive market economy.

demand conditions The market conditions in a host country that aid the production process.

demographic transition The historical shift of birth and death rates from high to low levels in a population.

dependence A conditioning situation in which the economies of one group of countries are underdeveloped by the development and expansion of other groups.

dependent industrialization Economic development, mainly in less developed countries, stimulated by investment from foreign multinational corporations.

depletion curves Graphs used to project lifetimes of resources.

deregulation The reduction of government controls on economic activity within a country.

desertification Overuse, overpopulation, and drought cause the expansion of desert lands and noncultivatable regions.

development A historical process that encompasses the entire economic and social life of a nation, resulting in change for the better. Development is related to, but not synonymous with, economic growth.

development bank An investment and/or loan fund that aids the development of underdeveloped countries.

direct foreign aid Cash flows from rich governments to poor governments to improve social economic welfare.

direct investment The purchase of enough of the equity shares of a company to gain some degree of managerial control.

diseconomics of scale The disadvantages from trying to produce at a scale larger than the optimum point on the long-run average cost curve, including overcrowded facilities, production pressure that leads to equipment breakdowns, and rising costs of inputs.

distance decay effect The decline of the frequency of interaction with increasing distrance from the starting point.

distance-decay effect With increase in distance, the decline in the level of interaction between two places.

distance learning Virtual instruction, class attendance, or participation via telecommunications allowing instruction source and student to be separated from one another.

distribution costs The costs of transporting finished products to the market.

diversification A strategy by which a firm enters a different product market from the one in which it has traditionally been engaged.

division of labor The specialization of production within or among firms, regions, and countries.

doubling time The time in years required for the population of a country to double.

dual economy In the study of industrial location, a term used to refer to two types of business enterprises, fundamentally different from one another: large, organically complex center firms and small, simply structured periphery firms. In the study of development, the term is used to refer to two types of social and economic systems existing simultaneously within the same territory: one system modern, the other traditional.

economic geography The subdiscipline concerned with the spatial organization of economic activity, including production, consumption, and transportation of goods and services, raw materials, people, and information.

economic integration The ultimate form of regional integration. It involves removing all barriers to interbloc movement of merchandise and factors of production, and unifying the social and economic policies of member nations. All members are subject to the binding decisions of a supranational authority consisting of executive, judicial, and legislative branches.

economic union A form of regional economic integration having all the features of a common market, as well as a common central bank, unified monetary and tax systems, and a common foreign economic policy.

ecopolitical economy Economic theories that concentrate on the ecological and cultural consequences of economic growth.

elasticity of demand The responsiveness of demand for a good to changes in price.

electronic data interchange The electronic movement of standard business documents between and within firms.

energy Usable power.

entrepreneurial skill The human know-how and skill that combines other resources to create a product or service using innovations and decisions.

environmental determinism The notion that human behavior is environmentally prescribed.

environmental perception The ways in which people form images of other places, and how these images influence decision making.

European Union (EU) Supranational organization constituted by fifteen European countries to further their common economic interests.

exchange controls Restriction of free dealings in foreign exchange, including multiple exchange rates and rationing.

exchange rate The value of the U.S. dollar compared with foreign currency.

expert systems Rule-based instruction in the form of computer software enabling computers and machines to help make decisions. Also, software systems that can control operations or manufacturing using programmed knowledge from experts.

export quota The government of an exporting country limits the volume or value of exports to an importing country voluntarily.

export restraint agreement A nontariff barrier whereby governments coerce other governments to accept voluntary trade export restraint agreements.

export subsidies Export payments made by governments or forms that lower the final cost of goods and services.

extensive margin of cultivation The general boundaries for rational zones of agricultural production established from the knowledge of total, average, and marginal productivity.

externality An effect on society, economy, environment, or region surrounding a firm or business, impacting it in a positive or negative way. This occurs when costs or benefits of an action are not captured in the market price.

externalization Purchase of a service by a firm rather than production in-house, usually through subcontracting.

factor conditions Land, labor, capital, enterprise, and other factors of production.

factor-driven stage Processing and exporting natural resources and primary products.

factors of production The economic inputs—land, labor, capital, entrepreneurship, technology—essential to production.

fertility rate The number of live births per 1000 people in a country per year.

feudalism The social system in Europe, as well as some other places such as Japan, that preceded capitalism. The class relations, culture, and geography of feudalism differed markedly from those of capitalism. The system was organized primarily through the power of the state, not the market.

Fifth wave A common nickname for the Kondratiev cycle that began in the late twentieth century fueled by information technology.

FIRE Finance, insurance, and real estate.

fixed capital Capital investments that are sunk in one place and difficult to change, including buildings and equipment. Fixed capital can remain constant over small expansions in output, as opposed to variable capital, which rises as a function of output.

flexible labor One of the outcomes of business process reengineering (BPR). See *business process reengineering.*

flexible production Also known as post-Fordism, this term refers to the form of capitalism that took shape starting in the 1970s and is characterized by vertical disintegration, computer technologies (e.g., just in time inventory systems), and lack of reliance on economies of scale.

floating exchange rate The value of a currency fluctuates according to changes in supply and demand for the currency on the international market.

footloose industries Firms that possess considerable locational mobility, that is, relatively little inertia; such companies are generally labor intensive with few barriers to entry and exit.

forces of production A marxist term that refers to the ensemble of productive forces, including technology, around which the production process is organized at different historical junctures.

foreign direct investment (FDI) Investing in companies in a foreign county, with the purpose of managerial and production control.

foreign sourcing An arrangement whereby firms based in advanced industrial countries provide design specifications to producers in underdeveloped countries, purchase the finished products, and then sell them at home or abroad.

forward integration The process of purchasing productive capacity "downstream" in the production process, that is, in the creation of outputs, through establishing a unit in-house or purchasing an existing company.

fossil fuels Fuels, including oil, coal, and natural gas, that are formed from plant and animal remains.

Four Tigers South Korea, Taiwan, Hong Kong, and Singapore.

four-field rotation system Rotating three crops among four fields over a period of years, while allowing a fourth rotated field to remain fallow, thus resting the soil for that year.

franchising A license venture allowing a licensee to manufacture and sell under the original company's name.

free-trade area A form of regional economic integration in which member countries agree to eliminate trade barriers among themselves but continue to pursue their independent trade policies with respect to nonmember countries.

friction of distance There are time and cost factors associated with movement across space, which exerts a friction to movement and flow.

General Agreement on Tariffs and Trade (GATT) An international agency, headquartered in Geneva, Switzerland, supportive of efforts to reduce barriers to international trade; replaced by World Trade Organization in 1995.

geographic information system (GIS) A technique for storing and transforming geographic data on a wide variety of variables to produce digital maps with hardware and associated computer software.

geothermal energy Energy produced by heat from deep inside the earth as water interfaces with heated rocks from the earth's core, producing steam.

global city A city that occupies a preeminent international position for business decision making and corporate services.

global office Workplaces interlinked by satellite communications offering telecommunications and computer-based methods of information transfer.

graphs Idealized transportation networks that are comprised of vertices, which represent points or intersections, and edges, which represent interconnecting routes.

gravity model The product of the masses or the populations of two cities, divided by the distance squared, is proportional to the traffic flow between them.

greenhouse effect The warming of the atmosphere due to increased amounts of carbon dioxide, nitrous oxides, methane, and chlorofluorocarbons.

Green Revolution A popular term for the greatly increased yield per hectare that followed the introduction of new, scientifically bred, and selected varieties of such food crops as wheat, maize, and rice.

gross domestic product (GDP) The monetary value of final output produced by businesses, individuals, and government located inside a country in a given year.

gross national product (GNP) The market value of all goods and services produced by businesses, individuals, and government owned by citizens of a country, whether located inside and outside, in a given year.

growth-oriented lifestyle A mind-set that insists on maximum production and consumption. It assumes an environment of unlimited waste and pollution reservoirs and indestructible ecosystems.

gross raw materials Inputs that lose weight in the production process, such as petroleum.

growth stage theories A developmental sequence of the national economic improvement of underdeveloped countries based on sufficient surplus to generate and sustain economic growth.

guilds The medieval system of handicraft production. Guilds were organized around different types of goods (e.g., paper, leather, iron, etc.). Master artisans or craftsmen ran workshops in which apprentices worked and learned the trade.

hegemonic power In international political economy, the most powerful country in the world, which "sets the rules" that others follow. Examples include Britain in the nineteenth century and the United States in the twentieth.

hierarchy The arrangement of urban settlements in a series of discrete classes, the rank of each determined by the level of specialization of functions.

highest and best use The notion that land is allocated to the use that earns the highest location rent.

high value added Commodities that, due to the manufacturing process, are much more valuable than the raw materials used to produce them.

hinterland Service area around a city.

horizontal integration A business strategy to increase a firm's scale by buying, building, or merging with another firm at the same stage of production of a product.

hub-and-spoke networks Hubs are major cities that collect passengers from small cities, in the local vicinity, via spoke lines. Hubs redistribute passengers between sets of original major cities.

human capital The sum of skills, education, and experience that makes labor productive.

human resources The skill and cost of labor, the cultural factors of the term, and the motivation for the work ethic.

human suffering index A useful descriptive measure of the differences in living conditions among countries, including ten measures of human well-being.

import quota Quantity restrictions imposed on imports.

import substitution A trade strategy, now largely discredited, that puts high tariffs on imports as a way to stimulate domestic production of goods. The opposite of export promotion.

income elasticity The percentage change in demand that accompanies a change in income.

industrial-complex economies Conglomerations of interrelated firms held together with intricate linkages of inputs and outputs, often engaged in the production of similar goods (e.g., electronics, textiles).

industrial inertia The resistance of some types of firms to change their spatial location, often due to heavy fixed capital investments.

industrialization Movement from an agricultural economy to a manufacturing-based, export-oriented economy.

industrial park A planned or organized district with a comprehensive plan which is designed to ensure compatibility between the industrial operation therein and the existing activities and character of the community in which the part is located.

industrial restructuring A term used to refer to the alternating phases of growth and decline in industrial activity. It emphasizes changes in employment between regions and links these with change in the world economy.

industry concentration Clusters of firms in locations that lower costs or increase returns. Reasons for concentrations include highly skilled human resources, high-quality service industries and infrastructure, advanced transportation and communications facilities, industrial concentration, and developed external economies.

industry life cycle The typical sequence of developmental stages in the evolution of an industry.

inelastic supply A supply schedule that does not show large increases due to a price increase.

inertia The immobility of the investment forces and social relations.

infant industry A young industry that, it is argued, requires tariff protection until it matures to the point where it is efficient enough to compete successfully with imports.

infant mortality rate Number of deaths during the first year of life per 1000 live births.

information technology Microelectronics technologies, including microprocessors, computers, robotics, satellites, and fiber-optic cables.

information warehouse A decision support tool collecting information from multiple sources and making that information available to users in a consolidated, consistent manner.

infrastructure The transportation and communication system and other public goods (e.g., dams, sewers) necessary for an economy to function.

innovation drive stage Innovation in new product design derived from high levels of technology and skill.

integrated circuit Transistors connected on a single, small silicon chip acting as a semiconductor of electrical current.

integration The process of expanding either vertically (upstream or downstream) in the production process or horizontally (within the same product market).

intellectual property rights Establishing and policing patent, copyright, and trademark rights on an international basis.

intensive margin of cultivation The stage related to agricultural production where additional units of labor still result in an increase in total production. That is, where marginal-product values are still positive.

intensive subsidence agriculture A high-intensity type of primitive agriculture practiced by densely populated areas of the developing world.

interest Payments made to owners of capital as a compensation for its contribution to production.

intermediate technology Low-cost, small-scale technologies intermediate between primitive stick-farming methods and complex Western agri-industrial technical packages.

international currency markets The internationalization of domestic currency, banking, and capital markets.

international economic order The placement and position of countries within the world economy based on capital, trade, and production.

international economic systems The system of the world economy based on flow across international boundaries.

internationalization of capital Capital is transcending international boundaries and is available for use in the global marketplace.

International Monetary Fund (IMF) An international financial agency that attempts to promote international monetary cooperation, facilitate international trade, make loans to help countries adjust to temporary international payment problems, and lessen the severity of international payments disequilibrium, often by imposing Structural Adjustment Policies.

international subcontracting The arrangement by multinational corporations (MNCs) to use Third World firms to produce entire products, components, or services in order to cover markets in an advanced industrial country.

Internet A computing and data network that connects computers through fiber optic, telephone, and satellite to allow users to share information.

intramultinational trade Cross-shipment of goods from the same industries between countries—automotive products for example.

investment-driven stage Using foreign technology and scale economies to produce standardized products with mass labor inputs provided by the local population.

involution The ability of the peasantry or the protoproletariat in the Third World to absorb an unusual number of people. The process of involution is characterized by a tenacity of basic pattern, internal elaboration and ornateness, technical hairsplitting, and unending virtuosity.

ISDN Integrated Services Digital Network, the technical format that allows data to be exchanged on the Internet.

isotropic surface A plane that is homogeneous in all respects, with equal ease of movement in all directions from every point.

joint venture An enterprise undertaken by two or more parties. It may be a jointly owned subsidiary, a consortium, or a syndicate.

journey to work Travel by individuals to work, yielding the largest proportion of travel by American households.

just-in-time inventory Quick delivery and response of parts and inventory delivery from component plants to final assembly operations.

knowledge-based resources Research, development, and scientific and technical skill within the country.

knowledge economy A theory of services that maintains they are qualitatively different from other types of com modity production.

Kondratiev cycles Named after the Russian economist who discovered them, these refer to long-term (roughly fifty to seventy-five years) oscillations in the capitalist economy linked to major waves of technological change, as measured by fluctuations in prices, output, profits, productivity, and employment.

labor A factor of production that includes human physical exertion performed in the creation of a good or service.

labor force Those in society who work, both employed and unemployed.

labor intensive A term that applies to an industry in which a high proportion of labor is used relative to the amount of capital employed.

labor migration theory Theories to explain the process of changing residences from one geographic locale to another, due to economic factors.

labor theory of value Originally conceived by Adam Smith but popularized by Marx, this view holds that value is produced through the amount of socially necessary labor time embodied in the production of goods (including capital goods), not supply and demand.

land A factor of production that includes not only a geographic portion of the earth's surface but also the raw materials from this region.

law of demand An empirical regularity of consumer behavior that presumes the quantity of a good demanded and the price of the good are inversely related.

law of diminishing returns The law according to which, when factors of production (land, labor, and capital) are doubled, output doubles; but if one factor of production or only some factors are doubled, output increases, but fails to double. The law assumes given levels of technological knowledge.

law of supply An empirical regularity of producer behavior that presumes the quantity of a good produced and the price of that good are directly related.

least-cost approach Firms, especially in competitive markets, choosing a location that minimize production costs.

less developed countries (LDCs) The Third and Fourth Worlds, encompassing Latin America, Africa, and most of Asia, characterized by relatively high rates of population growth and low per capita income.

licensing venture The rental of patents, trademarks, or technology by a company in exchange for royalty payments.

limits to growth The optimum population size for the world, provided by the Club of Rome, shows that growth must be limited. A gloomy forecast by Paul Ehrlich suggests worldwide famine and war as the inevitable results of continued increases in world population.

line-haul costs Costs involved in moving commodities along a route.

liquid or variable capital Capital that can be converted to other forms easily, that is, corporate savings, bonds, stocks, and loans.

localized raw materials Inputs that are found in only a few select locations and must be transported to the point of production.

locational costs Costs involved in bringing inputs to the point of production and sending outputs to the markets.

locational factors Major elements that shape the decision of the firm to locate in some places and not others, including cost and productivity of labor, land, and other inputs (see factors of production).

locational interdependence A concept that implies that competition from rival producers can lower potential revenues at a given point and space.

location rent The advantage of one parcel of land over another because of its location; the concept of declining rents with an increase in distance from the market.

location theory A compilation of ideas and methods dealing with questions of accessibility.

logical positivism The philosophy of science centered on the formation of hypotheses, data collection, and reproducible results.

machinofacture The phase of the developing division of labor where mechanization and division of labor within production occurs.

Maglev A magnetically levitated train that operates with a linear induction engine and cruises on a cushion of air at high speeds on a detached right-of-way and heralded as the state of the future in ground transportation systems.

marginal product The addition to output resulting from an increase of one unit of increased input.

margin of cultivation The location at a certain distance from the market at which marginal products are produced and, therefore, the edge of profitability is reached.

marine fisheries Harvesting of fish and shellfish from the ocean.

market Institutions composed of buyers and sellers of commodities. Just as there is a large array of different types of producers and consumers, there are many kinds of markets for different goods and services, ranging dramatically in size and sophistication.

marketing linkages A form of agglomeration economy in which a cluster of firms in an industry attract specialized services such as advertising or distribution that would be difficult or expensive to obtain alone.

material index The ratio of the weight of the localized raw material to that of the finished product; goods with high indices tend to be weight losing and firms producing them locate near the source of raw materials, while goods with low indices are weight gaining and their producers locate near the final market.

matrix structure The organizational structure of a firm, that is, its internal divisions based on the types of products produced, types of customers services, types of regions served, or some combination of these.

maximum sustainable yield Maximum production consistent with maintaining future productivity of a renewable resource.

Mediterranean cropping Agriculture producing specialty crops because of mild climates, including citrus, grapes, nuts, avocados, tomatoes, and flowers.

microelectronics Semiconductors, integrated circuits, and electronic components and parts.

microprocessor A computer the size of one's fingernail used for applications, including calculators, electric typewriters, computers, industrial robots, and aircraft guiding systems.

migration A change in residence intended to be permanent frequently across international boundaries.

minerals Natural inorganic substances that have a definite chemical composition and characteristic crystal structure, hardness, and density.

mine tailings Leftover ore wastes from which minerals have been extracted.

mixed crop and livestock farming The raising of beef cattle and hogs as the primary revenue source, with the production of crops fed to the cattle.

mixed economic systems Economic systems that are a hybrid form of capitalists, command economy, or a traditional system, usually where both government and private decisions determine how resources are allocated.

mode of production A marxist term that refers to the basic forces and social relations of production. Slavery comprised one mode, feudalism another, capitalism yet a third. Each is typified by an ensemble of class relations, culture, technologies, and geographic landscape.

modernization A replacement of traditional approaches to production with new technologies and techniques.

modernization theory The approach to development that maintains countries should embrace global capitalism, reduce trade barriers, invite foreign investment, and diffuse markets to stimulate growth.

multimedia Computers with TVs, VCRs, CD players, and other entertainment devices in a human-machine communications media with combined applications.

multinational corporations (MNCs) Companies based in one country that do business in one or more other countries; also known as transnational corporations.

multiplant enterprises Companies with factories and offices in widely scattered locations, sometimes in other countries.

multiplier The effect on total employment (or output, wages, and profits) generated by changes in an industry, including interindustry linkages and expenditures resulting from changes in personal income (wages and salaries).

NAFTA North America Free Trade Agreement between Mexico, Canada, and the United States.
natural increase Population growth measured as the excess of live births over deaths per 1000 individuals per year. Natural increase of a population does not reflect emigrant or immigrant movements.
negative population growth A falling level of population where out-migration and death exceed in-migration and births.
neo-Malthusian Someone who accepts the Malthusian principle but hopes to avoid famine with the intervention of government controls.
networked computers The linking of computers, servers, and databases from around the country and around the world, allowing each computer work station more power and information than it contains in a stand-alone mode.
NIMBY "Not in my backyard" is the cry of many local movements resisting unwanted land uses, such as trash incinerators.
nomadic capital The switching of production from place to place because of innovation in transportation and communication savings.
nonbasic sector The part of an urban or regional economy that caters to local demand (i.e., it is not export oriented), including retail sales, real estate, and consumer services.
nondirect production workers Workers in a manufacturing firm who are not directly involved in the production process (e.g., in management, administration, research, marketing, and sales).
nonrenewable resources Resources that are fixed in amount—that cannot regenerate—such as fossil fuels and metals.
nontariff barriers Restrictions that limit entry into an industry by competitive firms or countries.
normal lapse rate The rate at which the atmosphere cools with increasing elevation (3.6 degrees per 1000 feet).
offshore assembly An arrangement whereby firms based in advanced industrial countries provide design specifications to producers in underdeveloped countries, purchase the finished products, and then sell them at home or abroad.
oligopoly The control of a market by a small number of firms or producers that can affect the market price.
opportunity cost The cost of foregone alternatives given up in order to produce other activities or goods.
optimum population size The theoretical number of people that would provide the best balance of population and resources for a desirable standard of living.
Organization of Economic Cooperation and Development (OECD) The international organization of most developed countries.
Organization of Petroleum Exporting Countries (OPEC) The international cartel of oil-producing countries.
outsourcing Subcontracting and the shifting of work to other locations and firms outside the principal corporation.
overpopulation A level of population in excess of the "optimum" level relative to the food supply or rate of consumption of energy and resources.
parity price Equality between the prices at which farmers could sell their products, and the prices they could spend on goods and services to run the farm.
pastoral nomadism Animal herds used for subsidence, moved from one forage area to another, in a cyclical pattern of migration.
peasant agriculture Subsidence agriculture, using little mechanical equipment and producing meager, labor-intensive crops.
petrodollars OPEC oil surpluses poured into the major Euro banks.
phenomenology The philosophy that human life can be best understood through the world of subjective individual experiences and meanings.
physiologic density The number of people per square mile of arable or farmable land.
polarization effects The negative influences prosperous regions exert on less prosperous regions.
political economy The approach to studying society that views social relations as a unified whole organized along lines of class, gender, ethnicity, and other lines of power.
population composition Age, gender, and reproductive demographics of the people.
population pyramid A special type of bar chart indicating the distribution of a population by age and sex.
portfolio investment Capital investment in the equity of a company, not involving managerial control of a foreign country.
postindustrial society A theory that modern society is dominated by services that maintains they form a historically new type of economic and social system, and landscape, in which a class of knowledge professionals will end traditional scarcity through rising productivity.
price ceiling A legally mandated price level below the typical market price.
price floor A guaranteed price above the market price.
primary economic activity An economic pursuit mainly involving natural or culturally improved resources, such as agriculture, livestock raising, forestry, fishing, and mining.
primate city A country's leading city economically, culturally, historically, and politically, and much larger than competing cities in population, wealth, and power.
privatization Process by which government-owned companies are transferred to private ownership and management.
producer services Services that sell their output (primarily expertise and specialized information) to other firms rather than to households, including financial services and business services (e.g., legal services, advertising, accounting, public relations, etc.).
production factors Labor, capital, technology, entrepreneurship, and land containing raw materials.
production linkages Purchases and sales of tangible inputs and outputs by firms, as opposed to other types of interfirm linkages such as specialized services and information.
production possibilities analysis A table or curve that shows various combinations of goods or services that can be produced given employment, resources, and technology are held constant.
productivity The ratio of outputs to inputs; productivity growth is the change in productivity over time.
product life cycle The typical sequence through which a product passes, from its introduction into the market to when it is replaced by a new product.
product market The market where households buy and firms sell the products and services they have produced.
profit The difference between gross revenues and production costs.
protectionism An effort to protect domestic producers by means of controls on imports such as tariffs and quotas.

purchasing power parity Income per capita compared to costs of living.

pure raw materials Inputs in the production process that lose no weight when used, such as parts.

push-and-pull factors The conditions in the source area, which tend to drive people away, and the perceived attractiveness of the destination that simultaneously stimulate migration flows.

quaternary sector of the economy Information processing, finance, insurance, real estate, education, and computer and telecommunication fields.

rank-size rule An empirical rule describing the distribution of city sizes in an area. It states that the population of any given city tends to be equal to the population of the largest city in the set divided by the rank of the given city. For example, if the population of the largest city numbers 10,000, the population of the fifth largest city will be 2000—that is, 10,000 divided by 5.

rate of natural increase The excess of births over deaths, or the difference between the crude birthrate and the crude death rate.

raw materials A substance in the physical environment considered to have value of usefulness in the production process.

raw-material-oriented Firms located and concentrated next to the sources of their key factors of production.

real-time information systems Systems that allow data retrieval and data information access in an instantaneous format, allowing processing of information as the information is gathered and recorded: instantaneous communication.

recycling The reuse of disposed materials after they have passed through some form of treatment (e.g., melting down glass bottles to produce new bottles).

regional growth forest Long-range urban forecasting of population housing and the economic activity for the region and small geographic areas within it.

regional integration The international grouping of sovereign nations to form a single economic region.

relative location Position with respect to other locations.

renewable resources Resources capable of yielding output indefinitely if used wisely, such as water and biomass.

rent Payments made to land owners as a productive factor for their contribution to the production process and the operation of the economy.

reserve A known and identified deposit of earth materials that can be tapped profitably with existing technology under prevailing economic and legal conditions.

resource A naturally occurring substance of potential profit that can be extracted under prevailing conditions.

rule of seventy Dividing the average annual rate of growth by 70 (technically 100 times the natural log of 2, or 69.3147), yields the doubling time of population for a country.

scale As opposed to spatial scale, scale in production refers to varying levels of output over which firms can spread their fixed costs and achieve economies of scale.

secondary economic activity The processing of materials to render them more directly useful to people; manufacturing.

second law of thermodynamics The law according to which any process has as a consequence a net increase in disorder or entropy. It can also be expressed as the degradation of energy into a less useful form, such as low-grade heat.

serfs The basis of the rural labor force under feudalism. Unlike slaves, serfs were not owned by a master but were bound to the land by law and custom. Unlike wage workers, they did not receive payment for their work; rather, they paid their lord rent and kept any surplus that may remain.

service linkages Purchases and sales of intangible inputs and outputs by firms such as information and expertise, as opposed to production linkages in tangible goods.

services Economic activity associated with the buying and selling of intangibles, including expertise and information.

shifting cultivation Temporary use of rain forest land for agriculture by cutting and burning the overgrowth.

slash-and-burn agriculture Temporary use of rain forest land for agriculture by cutting and burning the overgrowth.

social relations of production A marxist term that refers to ownership and control of the means of production, that is, the pattern of class in a given mode of production.

solar energy Radiation from the sun, which is transformed into heat primarily at the earth's surface and secondary in the atmosphere.

spatial integration Linkages between a city or firm and the members of its economic environment.

spatial interaction The movement, contact, and linkage between points in space; for example, the movement of people, goods, traffic, information, and capital between one place and another.

spatial margins of profitability The region over which a firm may expect to generate a profit as defined by the market price and spatial variations in the costs of production.

spatial organization A theme in geography emphasizing how space is organized by individuals and societies to suit their own designs. It provides a framework for analyzing and interpreting location decisions and spatial structures in a mobile, interconnected world.

spread effects The beneficial influences prosperous regions exert on less prosperous regions.

squatter settlements Residential areas that are home to the urban poor in underdeveloped countries. The various terms used to identify squatter settlements include the following: callampas, tugurios, favelas, mocambos, ranchos, and barriadas in Latin America; bidonvilles and gourbivilles in North Africa; bustees, jhoupris, jhuggis, kampongs, and barung barong in south and southeast Asia.

stages of production According to the law of diminishing returns, the three stages that total product passes through as successive units of variable input are applied to a fixed input. In Stage 1, the average product curve rises to its peak; in Stage 2, it declines; and in Stage 3, the total product curve declines.

stationary state The dynamic state of a system in which input and output are balanced at a point below the maximum limits of the system and its surroundings.

stepped freight rates A zonal-rate system with rates increasing with distance.

strategic minerals Those minerals deemed critical to the economic and military well-being of the nation.

strategy A firm's plan or vision for future growth, including the means by which it anticipates dealing with changes in market opportunities, the prices of inputs and outputs, technology, competition, regulation, and other variables, risk, information, and so forth.

structuralism The philosophy that holds the best way to understand society and space is through social relations, historical context, and political economy.

subsistence agriculture Peasant agriculture, using little mechanical equipment, producing meager and labor-intensive crops.

surplus value A marxist term for the value of output a worker generates but for which he or she is not compensated. Based on the labor theory of value, this view holds that workers ultimately create all wealth, but are not compensated for their efforts by the market price of their labor. All employment relations are thus held to be inherently exploitative.

target pricing The government pays a farmer directly the difference between the market selling price and the price that the government has set artificially.

tariff A schedule of duties placed on products. A tariff may be levied on an ad valorum basis (i.e., as a percentage of value) or on a specific basis (i.e., as an amount per unit). Tariffs are used to serve many functions—to make imports expensive relative to domestic substitutes; to retaliate against restrictive trade policies of other countries; to protect infant industries; and to protect strategic industries, such as agriculture, in times of war.

tax-haven country Tax-related inducement given by local governments to guide banking or industrial location.

Taylorism The application of scientific management principles to production.

technique The method a firm uses to produce a given output, including its choice of factor inputs (land, labor, capital, information).

telepresence Virtual face-to-face communication using networked computers.

terminal costs Costs incurred in loading, packing, and unloading shipments, and preparing shipping documents.

terms of trade The relative price levels of exports to imports for a country.

tertiary economic activity An economic pursuit in which a service is performed, such as retailing, wholesaling, servicing, teaching, government, medicine, and recreation.

Third World debt crisis The dangerous economic position of certain Third World nations that carry an enormous debt to overseas banks, private banks, and governments, the interest payments of which rob the host country of needed investment.

time-space compression The reduction in travel time between places that results from transport improvements.

trade area The area dominated by a central place; sometimes called a hinterland or tributary area.

trade deficit The excess of imports over exports for a country for any specific year.

traditional market An economic system in which culture, tradition, and folkways determine how scarce resources will be used by the economy.

tragedy of the commons Public resources are frequently ruined by the cumulative isolated actions of individuals in that overuse is practiced, rather than conservation, thus ruining resources held in common.

transfer costs Terminal costs and other fixed costs, plus line-haul costs or over-the-road costs equal transfer costs.

transfer pricing The transfer of taxable profits to lower tax assessing countries through price setting, thus minimizing corporate tax penalties and maximizing profit.

transit oriented development Development guidelines to help communities reduce dependency on automobiles.

transit time The amount of time needed to move something from a place to another.

transmaterialization Technology and innovation has reduced the demand for minerals as inputs to mature industries that undergo replacement by higher-quality, smaller, and technologically more advanced materials producing high-tech, less expensive, and more durable products, such as fiber optics, smart metals, transistors, and computers.

transnational corporation Companies that operate factories or service centers in countries other than the country of origin; also known as multinational corporation.

transportation-cost surface Costs that arise as a function of the distance from the least-cost location.

transport costs The alternative output given up when inputs are committed to the movement of people, goods, information, and ideas over geographic space.

triage The partitioning into three groups of nations, those so seriously deficient in food reserves that they cannot survive, those that can survive without food aid, and finally, those that can be saved by immediate food relief measures.

tributary areas The area dominated by a central place; the area to which a central place services and from which it draws raw materials and labor supply.

trickle-down effects The beneficial impact of prosperous regions on less prosperous regions.

ubiquitous raw materials Inputs that are found everywhere, such as water, and thus have no transport costs in the production process.

underdeveloped Countries characterized by high rates of population growth and low per capita income.

underemployment Shortage of job opportunities forcing people to accept less than full-time employment or being employed well beneath one's training and ability.

undernutrition A state of poor health in which an individual does not obtain enough calories.

unemployment Share of the labor force that is actively seeking but cannot find employment.

unequal exchange The argument is that an artificial division of labor has made earning a good income from free trade difficult for most LDCs.

urban peasants The poorest class of the city population.

urbanization The process of a society changing from rural to urban and economic activities concentrating in cities.

urbanization economies The benefits firms accrue from locating in large cities, with access to other firms, information, labor, and infrastructure.

use value The usefulness of a commodity to the person who possesses it.

value added The difference between the revenue of a firm obtained from a given volume of output and the cost of the input (the materials, components, services) used in producing that output.

vertical integration Expansion of a firm's in-house productive capacity "upstream" (i.e., of inputs) or "downstream" (i.e., of outputs).

Von Thünen model A famous nineteenth-century land use model that revealed how land markets reflect the interaction of agricultural prices, production costs, and transportation as users seek to maximize their income or rent.

wealth-driven stage Stage of economic development described by a population that has a high overall standard of living and high levels of mass consumption and technological virtuosity.

wheat belts Areas, such as those in North America, in which wheat predominates as an agricultural product.

wind farm Capturing wind energy with wind turbines and converting it to electricity.

World Bank A group of international financial agencies, including the International Bank for Reconstruction and Development, the International Finance Corporation, and the International Development Association.

world cultural realms Giant world regions that possess broad similarities of culture, economy, and historical development.

world economy A multistate economic system created in the late fifteenth and early sixteenth centuries by European capitalism and, later, its overseas progeny.

world political economy theories The structure of political and economic relationships between dominant and dominated countries.

world-systems theory Countries and regions practice economic activities and succeed based on their ability to produce needed goods and services for the world economy.

World Trade Organization (WTO) The world trade union which came into existence following the Uruguay Round of the GATT Treaty. WTO enforces trade rules and assesses penalties against violators.

write-down A cancellation or reduction of a portion of the principal and/or interest by a creditor bank for a less developed country's loans because of the possibility of default.

zaibatsu A large Japanese financial enterprise, similar to a conglomerate in the West.

zero population growth (ZPG) As a result of the combination of births, deaths, and migration, the population of a country is level, not rising or falling from year to year.

REFERENCES

Abu-Lughod, J. 1989. *Before European Hegemony: The World System A.D. 1250–1350.* New York: Oxford University Press.

Barber, B. 1996. *Jihad v. McWorld.* New York: Ballantine.

Barnes, T., and D. Gregory. 1997. *Reading Human Geography.* New York: John Wiley.

Barros, R., F. Louise, and R. Mendonca. 1997. Female-Headed Households, Poverty, and the Welfare of Children in Urban Brazil. *Economic Development And Cultural Change* 45:231–58.

Bender, W. 1997. How Much Food Will We Need in the 21st Century? *Environment* 39:6–11.

Beyers, W., and D. Lindahl. 1998. Explaining the Demand for Producer Services: Is Cost Driven Externalisation the Major Factor? *Papers of the Regional Science Association* 75:351–374.

Bhagwati, J. 1998. The Capital Myth: The Difference Between Trade in Widgets and Dollars. *Foreign Affairs.* May/June.

Bingham, R., and E. Hill. eds. 1997. *Global Perspectives on Economic Development.* New Brunswick, N.J.: Center for Urban Policy Research.

Brett, E. A. 1985. *The World Economy Since the War.* London: Macmillan.

Britton, J., ed. 1996. *Canada and the Global Economy.* Montreal: McGill-Queens University Press.

Brohman, J. 1996. Postwar Development in the Asian NICs: Does the Neoliberal Model Fit Reality? *Economic Geography* 72:107–130.

Bryson, J. 1997. Business Service Firms, Service Space and the Management of Change. *Entrepreneurship and Regional Development* 9:93–111.

Bryson, J., and P. Daniels. 1998. *Service Industries in the Global Economy: Service Theories and Service Employment.* Cheltenham: Edward Elgar.

Bryson, J., P. Daniels, and B. Warf. 2004. *Service Worlds: People, Organizations, Technologies.* London: Routledge.

Cairncross, F. *The Death of Distance.* 1997. Boston: Harvard Business School Press.

Camagni, R. 2002. On the Concept of Territorial Competitiveness: Sound or Misleading? *Urban Studies* 39: 2395–2411.

Chase-Dunn, C. 1998. *Global Formation: Structures of the World-Economy.* (second edition). Lanham, Md.: Rowman and Littlefield.

Chirot, D. 1985. The Rise of the West. *American Sociological Review* 50:181–195.

Clark, D. 1996. *Urban World/Global City.* London: Routledge.

Clawson/Fisher. 1998. *World Regional Geography: A Development Approach, Sixth Edition.* Upper Saddle River, N.J.: Prentice Hall.

Coffey, W., and R. Shearmur. 1997. The Growth and Location of High Order Services in the Canadian Urban System, 1971–1991. *The Professional Geographer* 49:404–417.

Cohen, B. 1998. *The Geography of Money.* Ithaca, N.Y.: Cornell University Press.

Conway, D., and J. Cohen. 1998. Consequences of Migration and Remittances for Mexican Transnational Communities. *Economic Geography* 74:26–44.

Cox, K., ed. 1997. *Spaces of Globalization: Reasserting the Power of the Local.* New York: Guilford.

Crang, M., P. Crang, and J. May. 1999. *Virtual Geographies: Bodies, Space and Relations.* London: Routledge.

Cruz, W., M. Munasighe, and J. Warford. 1996. Greening Development: Environmental Implications of Economic Policies. *Environment* 38:163–82.

Daniels, P. W. 1996. *The Global Economy in Transition.* White Plains, N.Y.: Longman.

Diamond, J. 1999. *Guns, Germs, and Steel.* New York: W.W. Norton.

Dicken, P. 2004. *Global Shift: The Internationalization of Economic Activity.* 4th edition. New York: Guilford Press.

Dodge, M., and R. Kitchin. 2001. *Mapping Cyberspace.* London: Routledge.

Dutt, A., et al., eds. 1994. *The Asian City: Processes of Development, Characteristic and Planning.* Dordrecht/Boston/London: Kluwer Academic Publishers.

Edwards, S. 1995. *Crisis and Reform in Latin America: From Despair to Hope.* New York: Oxford University Press.

Ethier, W., and J. Markusen. 1996. Multinational Firms, Technology Diffusion and Trade. *Journal of International Economics* 41:1–28.

Frank, R. 1999. *Luxury Fever.* Princeton: Princeton University Press.

Friedman, T. 1999. *The Lexus and the Olive Tree.* New York: Farrar Straus Giroux.

Friedman T. 2005. *The World is Flat.* NY: Farrar, Straus, Giroux.

Garreau, J. 1991. *Edge City: Life on the New Frontier.* New York: Doubleday.

Glasmeier, A., and M. Howland. 1995. *From Combines to Computers: Rural Services and Development in the Age of Information Technology.* Albany: State University of New York Press.

Graham, S., and S. Marvin. 1996. *Telecommunications and the City: Electronic Spaces, Urban Places*. London: Routledge.

Grigg, David. 1995. *An Introduction to Agricultural Geography*. New York: Routledge.

Hall, D., ed. 1993. *Transport and Economic Development in the New Central and Eastern Europe*. London: Belhaven.

Hanson, S., ed. 1995. *The Geography of Urban Transportation*. 2nd ed. New York: Guilford.

Harrington, J., and B. Warf. 1995. *Industrial Location*. New York: Routledge.

Herod, A., G. Tuathail, and S. Roberts, eds. 1997. *An Unruly World? Globalization, Governance, and Geography*. London: Routledge.

Hindess, B., and P. Hirst. 1975. *Pre-Capitalist Modes of Production*. London: Routledge and Kegan Paul.

Hirsh, M., and E. Henry. 1997. The Unraveling of Japan Inc. *Foreign Affairs* 76:11–32.

HOST Network, P. Bye, and A. Mounier. 1996. Growth Patterns and the History of Industrialization. *International Social Science Journal* 48:537–49.

Hsu, S. 1997. The Agroindustry: A Neglected Aspect of the Location Theory of Manufacturing. *Journal of Regional Science* 37:259–74.

Hugill, P. 1993. *World Trade since 1431: Geography, Technology, and Capitalism*. Baltimore: Johns Hopkins University Press.

Kirk, D. 1996. Demographic Transition Theory. *Population Studies* 50:361–88.

Kitchin, R. 1998. *Cyberspace: The World in the Wires*. New York: John Wiley.

Knox, P. 1994. *Urbanization: An Introduction to Urban Geography*. Englewood Cliffs, N.J.: Prentice Hall.

Knox, P. 1995. "World Cities and the Organization of Global Space." in *Geographies of Global Change*. R. J. Johnston, P. Taylor, and M. Watts, eds. Oxford: Blackwell.

Knox, P., J. Agnew, and L. McCarthy. 2003. *The Geography of the World Economy*. ed. London: Edward Arnold.

Kramer, R., and D. Mercer. 1997. Valuing a Global Environmental Good: U.S. Residents Willingness to Pay to Protect Tropical Rain Forests. *Land Economics* 73:196–210.

Krugman, P. 1991. *Geography and Trade*. Cambridge, Mass.: University Press.

Krugman, P. 1997. *The Age of Diminished Expectations*. Cambridge, Mass.: MIT Press.

Landes, D. 1969. *The Unbound Prometheus: Technological Change and Industrial Development in Western Europe from 1750 to the Present*. Cambridge: Cambridge University Press.

Lauria, M. 1997. *Reconstructing Urban Regime Theory: Regulating Urban Politics in a Global Economy*. Thousand Oaks, Calif.: Sage Publications.

Leichenko, R., and R. Erickson. 1997. Foreign Direct Investment and State Export Performance. *Journal of Regional Science* 37:307–29.

Leyshon, A., and N. Thrift, eds. 1997. *Money/Space: Geographies of Monetary Transformation*. London and New York: Routledge.

Leyshorn, A., and N. Thrift. 1997. Spatial Financial Flows and the Growth of the Modern City. *International Social Science Journal* 151:41–54.

MacPherson, A. 1997. The Role of Producer Services Outsourcing in the Innovation Performance of New York State Manufacturing Firms. *Annals of the Association of American Geographers* 87:52–71.

Mann, M. 1986. *The Sources of Social Power*. Cambridge: Cambridge University Press.

Manning, R. 2000. *Credit Card Nation*. New York: Basic Books.

Martin, R., and P. Sunley. 1998. Slow Convergence? The New Endogenous Growth Theory and Regional Development. *Economic Geography* 74:201–227.

Martin, R., ed. 1999. *Money and the Space Economy*. New York: Wiley.

Massey, D. 1984. *Spatial Divisions of Labor: Social Structures and the Geography of Production*. New York: Methuen.

Massey, D., and P. Jess. 1996. *A Place in the World: Places, Cultures, and Globalization*. New York: Oxford University Press.

Meyer, S., and M. Green. 1996. Foreign Direct Investment from Canada: An Overview. *The Canadian Geographer* 40:219–37.

National Research Council. 2003. *Cities Transformed: Demographic Change and ItsImplications in the Developing World*. Washington, D.C: National Academies Press.

Nebel, B., and R. Wright. 1996. *Environmental Science, the Way the World Wood Works*. Upper Saddle River, N.J.: Prentice Hall.

Newbold, K. 2002. *Six Billion Plus: Population Issues in the Twenty-First Century*. Boulder, Colo.: Rowman and Littlefield.

O'Uallachainm, B. 1997. Restructuring the American Semiconductor Industry: Vertical Integration of Design Houses and Wafer Fabricators. *Annals of the Association of American Geographers* 87:217–37.

O'Brien, R. 1992. *Global Financial Integration: The End of Geography*. Washington, D.C.: Council on Foreign Relations.

Office of Technology Assessment. 1995. *The Technological Reshaping of Metropolitan America*. Washington, D.C.; U.S. Government Printing Office.

Pandit, K. 1997. Cohort and Period Effects in U.S. Migration: How Demographic and Economic Cycles Influence the Migration Schedule. *Annals of the Association of American Geographers* 87:439–50.

Peet, R. 2003. *Unholity Trinity: The IMF, World Bank & WTO*. NY: Zed Books.

Porter, M. 1990. *The Competitive Advantage of Nations*. New York: Free Press.

Porter, M. 2003. The Economic Performance of Regions. *Regional Studies* 37:549–578.

Porter, P., and E. Sheppard. 1998. *A World of Difference Society, Nature, Development*. New York: Guilford Press.

Ritzer, G. 2000. *The McDonaldization of Society*. Thousand Oaks, CA: Pine Forge Press.

Roberts, S. 1994. Fictitious Capital, Fictitious Spaces: the Geography of Offshore Financial Flows. in *Money, Power and Space*. S. Corbridge, R. Martin, and N. Thrift, eds. Oxford: Blackwell.

Rosenfeld, S. 1992. *Competitive Manufacturing*. New Brunswick, N.J.: Center for Urban Policy Research.

Rubenstein, J. 2001. *The Cultural Landscape: An Introduction to Human Geography*, 7th ed. New York: Macmillan.

Sassen, S. 1991. *The Global City: New York, London, Tokyo*. Princeton, N.J.: Princeton University Press.

Sayer, A., and R. Walker. 1992. *The New Social Economy*. Cambridge, Mass.: Blackwell.

Schiller, D. 1999. *Digital Capitalism: Networking the Global Market System*. Cambridge, Mass.: MIT Press.

Schlosser, E. 2001. *Fast Food Nation*. New York: Perennial.

Sidaway, J., and J. Bryson. 2002. Constructing Knowledges of Emerging Markets: UK Based Fund Managers and their Overseas Connections. *Environment and Planning*, 34: 401–416

Singh, R. 1996. Female Agricultural Workers Wages, Male-Female Wage Differentials, and Agricultural Growth in a Developing County, India. *Economic Development and Cultural Change* 45:89–123.

So, A. 1990. *Social Change and Development: Modernization, Dependency, and World-System Theories*. Newbury Park, Calif.: Sage.

Soja, E. 2000. *Postmetropolis*. Oxford: Blackwell.

Solomon, E. 1997. *Virtual Money: Understanding the Power and Risks of Money's High-Speed Journey into Electronic Space*. Oxford: Oxford University Press.

Solomon, R. 1999. *Money on the Move: The Revolution in International Finance since 1980*. Princeton, N.J.: Princeton University Press.

Standage, T. *The Victorian Internet*. 1998. New York: Walker and Company.

Stearns, P. 2001. *Consumerison in World History*. London: Routledge.

Stiglitz, J. 2002. *Globalization and Its Discontents*. New York: W.W. Norton.

Storper, M. 1997. *The Regional World: Territorial Development in a Global Economy*. New York: Guilford.

Strange, S. 1998. *Mad Money: When Markets Outgrow Governments*. Ann Arbor, Mich.: University of Michigan Press.

Taylor, P. 2000. World Cities and Territorial States under Conditions of Contemporary Globalization. *Political Geography* 19:5–32.

Telegeography. 2002. *Global Telecommunications Traffic Statistics and Commentary*. Washiington, D.C.: Telegeography, Inc.

Thompson, E. P. 1966. *The Making of the English Working Class*. New York: Vintage Press.

Thrift, N., and K. Olds. 1996. Reconfiguring the Economic in Economic Geography. *Progress in Human Geography* 20:311–337.

Timberg, T. 1997. Big and Free Is Beautiful: China and India, the Past 40 Years And the Next. *Economic Development and Cultural Change* 45:435–42.

Turner, B. II, and S. Brush, eds. 1987. *Comparative Farming Systems*. New York: Guilford.

Van Marrewijk, C., J. Stibora, and J. Viaene. 1997. Producer Services, Comparative Advantage, and International Trade Patterns. *Journal of International Economics* 42:195–220.

Walker, R. 1985. Is There a Service Economy? The Changing Capitalist Division of Labor. *Science and Society*, 42–83.

Wallerstein, I. 1979. *The Capitalist World-Economy*. Cambridge: Cambridge University Press.

Warf, B. 1995. Telecommunications and the Changing Geographies of Knowledge Transmission in the Late 20th Century. *Urban Studies* 32:361–378.

Warf, B. 1999. The Hypermobility of Capital and the Collapse of the Keynesian State. In *Money and the Space Economy*. R. Martin, ed., pp. 227–240. London: Wiley.

Warf, B. 2001. Global Dimensions of U.S. Legal Services. *Professional Geographer* 53:398–406.

Watts, M. 1996. Development III: The Global Agrofood System and Late Twentieth-Century Development. *Progress in Human Geography* 20:230–45.

Whyte, M. 1996. The Chinese Family and Economic Development: Obstacle Or Enigma. *Economic Development and Cultural Change* 45 (1996):1–30.

Williams, C. 1997. *Consumer Services and Economic Development*. London: Routledge.

World Resources Institute. 2004. *World Resources*. New York: Oxford University Press.

Wright, M. 1997. Crossing the Factory Frontier: Gender, Place and Power in the Mexican Maquiladora. *Antipode* 29:278–302.

Yapa, L. 1996. What Causes Poverty?: A Postmodern View. *Annals of the Association of American Geographers* 86:707–28.

CREDITS

Chapter 1

Figure 1.1 Countries & Regions, World Bank, 1999.
Figures 1.2, 1.3, 1.12 Rubenstein, *The Cultural Landscape: An Introduction to Human Geography, 8/e,* New Jersey: Pearson Education.
Figure 1.5 CIA World Fact Book, 2002.
Figure 1.7 Statistical Abstract of the United States, 1999.
Photo 1.0 © Charles O'Rear / Corbis. All rights reserved.
Photo 1.1 PhotoEdit Inc.
Photo 1.2 United Nations

Chapter 2

Figure 2.1 The Image Works.
Figure 2.2 Knox & Marsen, World Regions in Global Context: Peoples, Places , and Environments, 2/e, p. 49. New Jersey: Pearson Education.
Figure 2.6 Rijksmuseum.
Figure 2.8 North Wind Picture Archives.
Figure 2.13 Dave King © Dorling Kindersley. Courtesy of The Science Museum London.
Figures 2.16, 2.24 Rubenstein, *The Cultural Landscape: An Introduction to Human Geography, 8/e,* New Jersey: Pearson Education.
Figure 2.32 Michael Holford Photographs.
Figure 2.20 George Eastman House
Photo 2.0 © Bettmann/CORBIS
Photo 2.1 CORBIS- NY
Photo 2.2 The British Library / Topham-HIP / The Image Works
Photo 2.3 Adam Woolfitt, Corbis
Photo 2.4 Rijksmuseum
Photo 2.5 North Wind Picture Archives
Photo 2.6 Dave King © Dorling Kindersley
Photo 2.7 Lewis W. Hine, George Eastman House
Photo 2.8 Getty Images Inc.–Hulton Archive
Photo 2.9 Michael Holford, Michael Holford Photographs

Chapter 3

Figures 3.2, 3.3, 3.14, 3.15, 3.16, 3.17, 3.19, 3.24, 3.25, 3.31, 3.36, 3.37, 3.38, 3.40, 3.42 Rubenstein, *The Cultural Landscape: An Introduction to Human Geography, 8/e,* New Jersey: Pearson Education.
Figure 3.7 Corbis/Bettman
Figure 3.26 Weeks, 2002, p. 35.
Photo 3.0 Michael J. Okoniewski, The Image Works
Photo 3.1 Corbis/Bettmann
Photo 3.2 World Bank Photo Library
Photo 3.3a J. L. Bulcao, Joao Luiz Mendonca Bulcao
Photo 3.3b Tyler Hicks, New York Times Agency
Photo 3.3c Masterfile Corporation
Photo 3.4 Paolo Koch/Rapho, Photo Researchers, Inc.
Photo 3.5 J. P. Laffont, Gamma Press USA, Inc.

Chapter 4

Figures 4.3, 4.5, 4.11, 4.14, 4.15, 4.18, 4.20, 4.23, 4.25 Rubenstein, *The Cultural Landscape: An Introduction to Human Geography, 8/e,* New Jersey: Pearson Education.
Figure 4.8 World Resources Institute, 1993, p. 179.
Figure 4.9 Based on data from U.S. Bureau of Mines, 1986.
Figure 4.10 Adapted from L. Waddell and W. Labus, *Transmaterialization: Technology and Materials' Demand Cycles.* Morgantown: West Virginia University, 1988.
Figure 4.13 Data from Statistical Abstracts, U.S. Department of Commerce, 1996.
Figure 4.17 BP Amoco, 1999.
Figure 4.21 Fisher, 1992, p. 121.
Figure 4.27 Mesoule 1991, *Conservation: Tactics for a Constant Crisis,* Science, 253, p. 744.
Figures 4.28, 4.29, 4.30, 4.31, 4.32, 4.33, 4.34, 4.35, 4.36 Rowntree, *Diversity Amid Globalization: World Regions, Environments, Development, 2/e,* New Jersey: Pearson Education.
Photo 4.0 Trip, Alamy Images
Photo 4.1 Andrew Holbrooke, Corbis/Stock Market
Photo 4.2 World Bank Photo Library
Photo 4.3 World Bank Photo Library
Photo 4.4 UN/DPI PHOTO/Ruth Massey
Photo 4.5 Photo Researchers, Inc.
Photo 4.6 Shell Oil Company
Photo 4.7 U.S. Department of Energy

Chapter 5

Figure 5.6 Rubenstein, *The Cultural Landscape: An Introduction to Human Geography, 8/e,* New Jersey: Pearson Education.
Figure 5.14 Expanded from Dicken, 1998.
Photo 5.0 ArtPix, Alamy Images
Photo 5.1 Allen Birnbach, Masterfile Corporation
Photo 5.2 Ric Francis, AP Wide World Photos
Photo 5.3 Owen Franken, Stock Boston
Photo 5.4 Ford Motor Company
Photo 5.5 David Parker, Photo Researchers, Inc.

Chapter 6

Figure 6.2 Bergman, *Human Geography,* 1995, p. 196.
Figures 6.3, 6.6, 6.7, 6.9, 6.10, 6.13 Rubenstein, *The Cultural Landscape: An Introduction to Human Geography, 8/e,* New Jersey: Pearson Education.
Figure 6.8 U.S. Bureau of the Census, 1994, p. 241.
Figure 6.11 Food and Agricultural Organization, United Nations, 1999.
Figure 6.12 Rowntree, *Diversity Amid Globalization: World Regions, Environments, Development, 2/e,* New Jersey: Pearson Education.
Figure 6.18 Industrial & Commodities Statistical Yearbook, United Nations, 1998, p. 176.
Figure 6.21 Based on Wheeler and Muller, 1981, p. 314.
Photo 6.0 Keren Su, Corbis/Bettmann
Photo 6.1 Bridgestone/Firestone, Tire Sales Co.

Photo 6.2 Bridgestone/Firestone, Tire Sales Co.
Photo 6.3 U.S. Department of Agriculture
Photo 6.4 United Nations
Photo 6.5 World Bank Photo Library
Photo 6.6 Deborah Harse, The Image Works
Photo 6.7 Michael Lawton, U.S. Department of Agriculture
Photo 6.8 U.S. Department of Agriculture
Photo 6.9 Joe Munroe, Photo Researchers, Inc.
Photo 6.10 U.S. Department of Agriculture
Photo 6.11 Joe Soehm, The Stock Connection

Chapter 7

Figures 7.1, 7.8, 7.10, 7.12 Rubenstein, *The Cultural Landscape: An Introduction to Human Geography, 8/e*, New Jersey: Pearson Education.
Figure 7.2 Lional Delevingne, Stock Boston.
Figures 7.5, 7.14 Rowntree, *Diversity Amid Globalization: World Regions, Environments, Development, 2/e*, New Jersey: Pearson Education.
Figure 7.7 Williams, 1989a, p. 331.
Figure 7.13 World Motor Vehicle Data, American Automobile Manufacturing Association, Detroit, Michigan, 1997.
Figure 7.21 Dent and Smith, 1993.
Photo 7.0 Susan Meiselas, Magnum Photos, Inc.
Photo 7.1 Lionel Delevingne, Stock Boston
Photo 7.2 San Jose Convention & Visitors Bureau
Photo 7.3 World Bank Photo Library
Photo 7.4 Susan Meiselas, Magnum Photos, Inc.
Photo 7.5 AP Wide World Photos
Photo 7.6 Cary S. Wolinsky, Stock Boston
Photo 7.7 Tom Wagner, Corbis/SABA Press Photos, Inc.
Photo 7.8 Spencer Grant, Photo Researchers, Inc.

Color Insert

Photo CI.1 AP Wide World Photos
Photo CI.2 Masterfile Corporation
Photo CI.3 Tyler Hicks, New York Times Agency
Photo CI.4 J. L. Bulcao, Joao Luiz Mendonca Bulcao
Photo CI.5 Catherine Karnow, Corbis-NY
Photo CI.6 Paul Conklin, PhotoEdit Inc.
Photo CI.7 David G. Houser, Corbis-NY
Photo CI.8 Howard Davies, Corbis-NY

Chapter 8

Figures 8.1, 8.2, 8.3, 8.11, 8.25 Rubenstein, *The Cultural Landscape: An Introduction to Human Geography, 8/e*, New Jersey: Pearson Education.
Figure 8.14 U.S. Bureau of Labor Statistics, 2000.
Figure 8.21 U.S. Bureau of Labor Statistics.
Figure 8.23 U.S. Bureau of Labor Statistics.
Figure 8.24 Warf, B., and C. Wije. 1991. "The Spatial Structure of Large U.S. Law Firms." *Growth and Change* 22: 157-174.
Figure 8.26 Redrawn from Warf, B. 1989. "Telecommunications and the Globalization of Financial Services" *Professional Geographer*, 31:257-271.
Photo 8.0 New York Convention & Visitors Bureau
Photo 8.1 Mario Tama, Getty Images, Inc.
Photo 8.2 IBM Corporation
Photo 8.3a William Thomas Cain, Getty Images, Inc.
Photo 8.3b Chris O'Meara, AP Wide World Photos
Photo 8.4 Michael Krasowitz, Getty Images, Inc. - Taxi

Chapter 9

Figures 9.1, 9.2 Bergman: Human Geography, 1995, p. 470.
Figure 9.3 Adapted from Fellman, Getis, and Getis, 1999, p. 68.
Figure 9.5 Bergman: Human Geography, 1995, p. 491.
Figure 9.7 Adapted from Janelle, 1968, p. 6.
Figure 9.8 Modified from Fellman, Getis, and Getis, 1999, P. 67.
Figure 9.9 Bergman: Human Geography, 1995, p. 443.
Figure 9.13 Fleming and Hayuth, 1994, p. 10.
Figures 9.14, 9.19, 9.25 Rubenstein, *The Cultural Landscape: An Introduction to Human Geography, 8/e*, New Jersey: Pearson Education.
Figure 9.21 *Atlas of Cyberspace*, http://www.cybergeography.org/atlas.cables.html.
Figure 9.23 Expanded from Freeman, C., 1987, *The Challenge of New Technologies NOECD, Interdependence and Cooperation in Tomorrow's World*, OECS, Paris, pp. 123-156.
Figure 9.26 Telegeography 2001.
Photo 9.0 World Bank Photo Library
Photo 9.1 Jerome Wexler, Photo Researchers, Inc.
Photo 9.2 Paolo Koch, Photo Researchers, Inc.
Photo 9.3 Port Authority of New York & New Jersey
Photo 9.5 Jeff Greenberg, PhotoEdit Inc.
Photo 9.6 Spencer Grant, Photo Researchers, Inc.
Photo 9.7 Trimble Navigation Limited
Photo 9.8 Rapho-Beaune, Photo Researchers, Inc.
Photo 9.9 Andre Ramasore, Galbe.com
Photo 9.10 NASA Headquarters
Photo 9.11 ML Sinibaldi, Getty Images
Photo 9.12 Ron Chapple, Getty Images, Inc. - Taxi

Chapter 10

Figure 10.2 Based on Herbert, 182.
Figure 10.3 Vance, 1970, p. 151.
Figure 10.5 Yeates, 1984, p. 242.
Figure 10.9 Stutz, 1990.
Figure 10.10 Updated from Stutz and Kartman, 1982.
Figure 10.24, 10.25 Rubenstein, *The Cultural Landscape: An Introduction to Human Geography, 8/e*, New Jersey: Pearson Education.
Photo 10.0 Alan Becker, Getty Images Inc.–Image Bank
Photo 10.1 Marvullo, Belgian National Tourist Office
Photo 10.2 The Taubman Company
Photo 10.3 Bonnie Kamin, PhotoEdit Inc.
Photo 10.4 David Forbert, SuperStock, Inc
Photo 10.5 SuperStock, Inc
Photo 10.6 Robert Harbison, Robert Harbison
Photo 10.7 Greater Minneapolis Convention & Visitors Association

Photo 10.8 Mark Lennihan, AP Wide World Photos
Photo 10.9 John Mitchell, Photo Researchers, Inc.

Chapter 11

Photo 11.0 Robert Nickelsberg/Getty Images, Inc.
Photo 11.1 Jupiter Images - PhotoArts Corporation/Brand X Photos Royalty Free
Photo 11.2 Nancy Siesel, New York Times Agency
Photo 11.3 Walter Bibikow, Danita Delimont Photography
Photo 11.4 Andrew Holbrooke, CORBIS- NY

Chapter 12

Figure 12.1 United Nations *Yearbook of International Trade Statistics*, 2000.
Figure 12.3 1980-1990 data: *Handbook of International Trade Statistics*, 1993, CIA, p. 170; 1994-1998 data: UN International Statistics Commodities, 1998.
Figure 12.6 Federal Reserve Bank of Minneapolis, 2000.
Figure 12.7 Statistical Abstracts of the United States, 2003, p. 798.
Figure 12.9 Dicken, 1992, p. 301.
Figure 21.10 Rubenstein, *The Cultural Landscape: An Introduction to Human Geography, 8/e*, New Jersey: Pearson Education.
Figure 12.11 Statistical Abstracts of the United States, 2000.
Figure 12.13 Statistical Abstracts of the United States, 2001.
Figure 12.15 Statistical Abstracts of the United States, 2000.
Figure 12.19 Dicken, 1998.
Photo 12.0 Michael Coyne, Michael Coyne Photography
Photo 12.1 Getty Images Inc.–Hulton Agency
Photo 12.2 Kuninori Takahashi
Photo 12.3 Matthew Weinreb, Getty Images Inc. - Image Bank
Photo 12.4 Ford Motor Company
Photo 12.5 Ford Motor Company
Photo 12.6 Greg Davis, Japan National Tourist Organization
Photo 12.7 Steven Rubin, The Image Works
Photo 12.8 International Monetary Fund
Photo 12.9 Alex Wong, Getty Images
Photo 12.10 Corbis Digital Stock

Chapter 13

Figure 13.2 World Bank, 2000, p. 134.
Figure 13.4 U.S. Statistical Abstract, 2001.
Figure 13.5 U.S. Statistical Abstract, 2001.
Figure 13.7 United Nations, 1998.
Figure 13.11 Stutz, 1994; San Diego Economic Development Corporation, 1993.
Figure 13.12 U.S. Department of Commerce, 2002.
Figure 13.13 U.S. Department of Commerce, 2002.
Figure 13.17 U.S. Department of Commerce, 1997.
Figure 13.18 U.S. Department of Commerce, 1997.
Figure 13.25 Rubenstein, *The Cultural Landscape: An Introduction to Human Geography, 8/e*, New Jersey: Pearson Education.
Figure 13.26 Data from United Nations, 1998.
Figure 13.27 Data from United Nations, 1996.
Figure 13.28 World Motor Vehicle Data, 2001.
Figure 13.30 United Nations, 1998.
Figure 13.31 United Nations, 1998.
Figure 13.33 United Nations, 1998.
Figure 13.34 United Nations, 1998.
Photo 13.0 Elaine Thompson, AP Wide World Photos
Photo 13.1 Pam Hasegawa, Pam Hasegawa, Photographer
Photo 13.2 T. Matsumoto, Corbis/Sygma
Photo 13.3 World Bank Photo Library
Photo 13.4 Paolo Koch, Photo Researchers, Inc.
Photo 13.5 Frederic Neema
Photo 13.6 Martha Tabor, Carlotta Joyner

Chapter 14

Figure 14.1, 14.3, 14.5, 14.7, 14.8, 14.14, 14.15, 14.18, 14.23 Rubenstein, *The Cultural Landscape: An Introduction to Human Geography, 8/e*, New Jersey: Pearson Education.
Figure 14.13, 14.21, 14.25, 14.26 Rowntree, *Diversity Amid Globalization: World Regions, Environments, Development, 2/e*, New Jersey: Pearson Education.
Figure 14.17 World Bank, 2000.
Photo 14.0 World Bank Photo Library
Photo 14.1 Diego Goldberg, Corbis/Sygma
Photo 14.2 Betty Press, Woodfin Camp & Associates
Photo 14.3 Raccah, United Nations
Photo 14.4a AP Wide World Photos
Photo 14.4b AP Wide World Photos
Photo 14.5 Getty Images
Photo 14.6 World Bank Photo Library
Photo 14.7 Alain Keler/Sygma, Corbis/Sygma
Photo 14.8a Catherine Karnow, CORBIS- NY
Photo 14.8b Paul Conklin, CORBIS- NY
Photo 14.8c Dave G. Houser, CORBIS- NY
Photo 14.8d Michael Brennan, CORBIS- NY
Photo 14.8e Howard Davies, CORBIS- NY
Photo 14.8f CORBIS- NY
Photo 14.9 Susan McCartney, Photo Researchers, Inc.
Photo 14.10 James Strachan, Robert Harding World Imagery
Photo 14.11 World Bank Photo Library
Photo 14.12 Ian Steele, United Nations
Photo 14.13 Paul Fusco, Magnum Photos, Inc.
Photo 14.14 Viviane Moss, CORBIS- NY
Photo 14.15 Brian Lee, CORBIS- NY

INDEX

U

V

W

Y

Z